Condition Perception and Performance Evaluation of Engineering Structures

Condition Perception and Performance Evaluation of Engineering Structures

Editors

Jingzhou Xin
Hong Zhang
Yan Jiang
Simon X. Yang

MDPI

Basel • Beijing • Wuhan • Barcelona • Belgrade • Novi Sad • Cluj • Manchester

Editors

Jingzhou Xin
Chongqing Jiaotong University
Chongqing
China

Hong Zhang
Chongqing Jiaotong University
Chongqing
China

Yan Jiang
Southwest University
Chongqing
China

Simon X. Yang
University of Guelph
Guelph, ON
Canada

Editorial Office
MDPI AG
Grosspeteranlage 5
4052 Basel, Switzerland

This is a reprint of articles from the Topic published online in the open access journals *Atmosphere* (ISSN 2073-4433), *Buildings* (ISSN 2075-5309), *Materials* (ISSN 1996-1944), *Remote Sensing* (ISSN 2072-4292), and *Sensors* (ISSN 1424-8220) (available at: https://www.mdpi.com/topics/Smart_System).

For citation purposes, cite each article independently as indicated on the article page online and as indicated below:

Lastname, A.A.; Lastname, B.B. Article Title. *Journal Name* **Year**, *Volume Number*, Page Range.

ISBN 978-3-7258-2637-7 (Hbk)
ISBN 978-3-7258-2638-4 (PDF)
doi.org/10.3390/books978-3-7258-2638-4

Cover image courtesy of Jingzhou Xin

Contents

Preface

Engineering structures encompass various engineering facilities constructed above ground, underground, and in water, such as bridges, roads, buildings, tunnels, dams, and other similar structures. These infrastructures play a vital role in daily life and social activities. However, due to limitations at the technological development level and management level, and the influence of adverse environmental factors, problems regarding the security and usability of engineering structures have become increasingly prominent, resulting in serious casualties, property losses, and adverse social impacts. Therefore, there is an urgent need to establish an effective system for evaluating the condition and performance of engineering structures to monitor and evaluate their health status. Fortunately, with the rapid development of the Internet of Things, big data, artificial intelligence, and remote sensing, more advanced technologies have been applied to the evaluation of the condition and performance of engineering structures. This Special Issue contains thirty-three papers covering new research and development activities related to the condition perception and performance evaluation of engineering structures.

All published papers primarily concern engineering structures such as bridges, roads, and buildings. Twenty-three papers focus on the two key phases of the life cycle of bridges, namely construction and maintenance. Among them, six papers focus on the construction phase of bridges. Based on differences in research methods, these papers can be broadly classified into two categories. The first regards numerical simulation-based methods, which take finite element models as their main analytical tools to simulate and evaluate the safety of bridges during construction. For example, Qian et al. (the 1st article) used a finite element model to analyze stress changes during the welded construction of steel bridge slabs, which revealed the residual stress distribution after the re-welding of fatigue cracks in welds of orthotropic anisotropic steel bridges. Chen et al. (the 2nd article) used finite elements to simulate the sudden failure of steel truss bridge members during construction, and proposed a structural modification method to improve the robustness of the structure. Zhang et al. (the 3rd article) discussed the vibration response and control countermeasures under three different construction conditions of cable-stayed bridges by simulating the three-dimensional wind field. Jiang et al. (the 4th article) developed a multi-objective cable optimization strategy for arch bridges during cable-stayed buckle construction, which improved the construction safety. The second category of research uses the model testing as a research tool to simulate the behavior of structures during the actual construction. For example, Dong et al. (the 5th article) employed a 1:10 scale model of a dual-coupled steel pipe arch bridge to analyze the deflection of the buckle tower during the installation of the arch rib segments, and established a method to control the buckle tower deflection of the arch bridge during the construction process. Fan et al. (the 6th article) designed a 1:10 scale model test using the 600 m Tian'e Longtan Bridge as the prototype to investigate the evolution of the concrete interface stresses during the construction process and their influence on the ultimate bearing capacity of the main arch ring. There are fourteen papers on bridge operation and maintenance, which can be further subdivided into overall performance evaluation and local condition perception. For example, Yue et al.(the 7th article) investigated the quasi-isolated behavior of bridges using the pushover and incremental dynamic analysis methods, showing the changes in the behavior of the bridge system under different shear key strengths. Yang et al. (the 8th article) analyzed the rule for the change in the overall aerodynamic performance of the bridge when the fairing parameters were changed by combining a wind tunnel test and numerical simulation. Wang et al. (the 9th article) compared the dynamic amplification factors of the deflection and bending moment of a continuous girder bridge by a mode-superposition method, and analyzed the effects of different-order modes on the dynamic amplification factors. Zhou et al. (the 10th article) proposed a high-precision bridge deformation monitoring method based on terrestrial laser scanning technology, which provided an accurate measurement of the overall morphology of the bridges studied. Jiang et al (the 11th article) collected five hundred and thirty-nine sets of bridge inspection data and used the whale optimization algorithm and extreme learning machine to establish a nonlinear mapping relationship between each influencing factor and bridge condition indicators. He et al.(the 12th article) developed a bridge finite element model updating method by combining a wavelet neural network and wind-driven optimization algorithm, which can reflect the nonlinear relationship between structural responses and parameters. He et al. (the 13th article) proposed a new method to calculate the equivalent damping ratio by using the dissipated energy in the maximum-displacement hysteresis loop and the cut-line stiffness, which provided a new perspective for evaluating the seismic performance of bridge piers. The perception of the local state is conducted by establishing the correlation between different physical signals and the state of local components of bridges. For example, Liu et al. (the 14th article) introduced the metal magnetic memory technique into the stress detection of steel strands, and the correlation between the self-magnetic flux leakage signals and stress levels was obtained through the graded tensile test of steel strands. Li et al. (the 15th article) proposed a method based on microwave radar to intelligently identify the cable force of inclined cable, and deduced the formula for calculating the cable force of inclined cable with a damper, which solved the problem that the traditional method failed to accurately measure the cable force of inclined cable with a damper. Tian et al. (the 16th article) proposed an absolute stress measurement method for steel structures based on a laser ultrasonic technique, and obtained the linear relationship between ultrasonic relative wave velocity and absolute stress. Chen et al. (the 17th article) used the time-varying and frequency-varying characteristics of acoustic emission signals to visualize the damage evolution process of concrete, and found that the heterogeneity of concrete should be considered in the exploration of concrete failure precursors and damage assessment methods and evaluated using acoustic emission technology. Zhang et al. (the 18th article) carried out an experimental analysis of corroded steel strand using the self-magnetic flux leakage detection technique and proposed a characterization index suitable for the corrosion detection of steel strands under vibration. Chen et al. (the 19th article) established a finite element model of a bridge with cracks based on fracture mechanics theory and analyzed the extension stability of cracks under the effects of temperature, vehicle loading, and prestress loss. Lu et al. (the 20th article) used a stochastic configuration network approach to effectively identify the localized damage in bridges by collecting the acceleration and crack-width data. In addition, Yang et al. (the 21st article) provided an overview of damage monitoring for the localized issues and damage identification methods for the overall performance of arch bridges, which helped researchers to understand the existing damage detection methods for arch bridges. Wu et al. (the 22nd article) analyzed the ultimate span of a concrete-filled steel tube arch bridge under different combinations of design parameters, providing theoretical support for bridge design optimization. Xin et al. (the 23rd article) proposed a method to assess the operational status of bridge monitoring systems, which can better reflect their construction and operational status.

There are two papers on road engineering which mainly focus on road safety. For example, He et al. (the 24th article) used an integrated multisource and multiscale monitoring technique for assessing the health status of high-speed railway subgrades. Wang et al. (the 25th article) investigated the mechanism of improved performance of asphalt binders by shaped phase-change materials, which can help improve the safety of asphalt pavements.

There are five papers on the investigation of building structures. Wang et al. (the 26th article) investigated wind-induced responses of the Shanghai World Financial Centre during the passage of Super Typhoon Lekima and analyzed the effect of wind loads on the modal characteristics of the structure. Li et al. (the 27th article) constructed a nonlinear finite element model of a tension cable-supported power transmission structure to investigate the wind-induced fatigue damage of structures. Qi et al. (the 28th article) proposed a beam–column connection joint that can be rapidly repaired under earthquake incidence, and the seismic performance of connection joints was studied through experiments. Hu et al. (the 29th article) used a neural model combining convolutional enhancement and a graph feature fusion network for frame structure damage detection. Yao et al. (the 30th article) utilized a numerical simulation method to optimize the construction sequence for reducing the stresses in long-span spatial steel structures and ensuring the safety and stability of the structure during construction.

In addition to the above structural response studies for specific structural objects, external environmental influences also have an important impact on the safety and durability of structures, and three articles are focused on this topic. For example, Wang et al. (the 31st article) investigated the laws influencing different elements of a typhoon wind field on typhoon profile characteristics based on field measurements in coastal areas. Wang et al. (the 32nd article) conducted a forecast analysis of daily surface air temperature and precipitation in the Tibetan Plateau region and estimated the trend in variation in the future. Liu et al. (the 33rd article) proposed an improved ensemble-strategy-assisted method for the prediction of high-precision online wind speeds.

This book covers topics on the perception of conditions and evaluation of performance in engineering structures, featuring 33 papers. These studies provide some novel methods, models, and technological applications for bridges, roads, and buildings, which provide innovative solutions and technical approaches to the key problems of these engineering structures.

Jingzhou Xin, Hong Zhang, Yan Jiang, and Simon X. Yang
Editors

buildings

MDPI

Article

Rewelding Residual Stress of Fatigue Crack at U-Rib-to-Deck of an Orthotropic Steel Deck

Ji Qian [1,2], Shi-Jie Song [1], Lin-Qiang Zhou [1], Pei-Yun Zhang [1] and Zhen-Bo Xu [3,*]

1 School of Civil Engineering, Chongqing Jiaotong University, Chongqing 400074, China; jiqian@cqjtu.edu.cn (J.Q.); songshijie411@163.com (S.-J.S.); 611220080025@mails.cqjtu.edu.cn (L.-Q.Z.); 18784418447@163.com (P.-Y.Z.)
2 State Key Laboratory of Mountain Bridge and Tunnel Engineering, Chongqing Jiaotong University, Chongqing 400074, China
3 College of Civil and Transportation Engineering, Hohai University, Nanjing 211100, China
* Correspondence: 622210970013@mails.cqjtu.edu.cn

Abstract: The orthotropic steel deck is a prevalent stiffening girder structure utilized in long-span cable-stayed bridges and suspension bridges. Nonetheless, the issue of fatigue cracking has persisted in in-service orthotropic steel decks, significantly impacting the longevity of bridges. This study examines the analysis of the distribution of residual stress during the rewelding process of a fatigue crack at the U-rib-to-deck-plate joint of an orthotropic steel bridge deck. Additionally, the impact of the initial welding and the stiffness of the boundary constraint on the residual stress field during rewelding is discussed. The findings indicate that the removal of the fatigue crack prior to rewelding alleviates the transverse residual stress caused by the initial welding. After undergoing the rewelding procedure, both the transverse residual stress and the longitudinal residual stress exhibited a significant stress peak. More precisely, the transverse tensile stress underwent a rise from 21 MPa to 385 MPa, while the longitudinal tensile stress experienced an increase from 345 MPa to 525 MPa. Furthermore, the range of tensile stress within the longitudinal residual distribution expanded by 88%. Moreover, the stress redistribution during the rewelding of the local fatigue crack varied depending on the constraints imposed on the steel bridge deck. Notably, the transverse residual stress increases by 40.6% when compared to the absence of constraints. The findings of this research offer valuable insights for the implementation of rewelding repair techniques on steel bridge decks, emphasizing the significance of considering the effects of residual stresses induced during the rewelding process.

Keywords: orthotropic steel bridge deck; residual stress; fatigue crack; rewelding repair; boundary constraint stiffness

Citation: Qian, J.; Song, S.-J.; Zhou, L.-Q.; Zhang, P.-Y.; Xu, Z.-B. Rewelding Residual Stress of Fatigue Crack at U-Rib-to-Deck of an Orthotropic Steel Deck. *Buildings* **2023**, *13*, 2515. https://doi.org/10.3390/buildings13102515

Academic Editor: Mohammad Adil Dar

Received: 2 August 2023
Revised: 13 September 2023
Accepted: 2 October 2023
Published: 4 October 2023

1. Introduction

The utilization of orthotropic steel bridge decks with closed U-shaped stiffening ribs (U-ribs) is prevalent in the construction of wide, flat steel box girder bridges. During the welding procedure of a steel bridge deck, non-uniform heating of its components leads to the development of thermal stresses. When the thermal stress exceeds the yield strength of the material, plastic deformation occurs in the welding area. During the subsequent cooling process, the constrained deformation of the welding zone by the surrounding area leads to the formation of residual stress within the component. This phenomenon, known as residual stress, has been extensively studied [1,2]. The presence of residual stress has the potential to diminish the ultimate capacity and fatigue strength of the structure [3,4]. Research has indicated that fatigue cracks commonly emerge at the weld connection between the deck plate and the U-rib under wheel load. Once a fatigue crack initiates, it propagates along the thickness direction of the deck plate and gradually extends through the pavement layer [5–7].

When fatigue cracks manifest in steel bridge decks, it becomes imperative to employ an appropriate treatment approach to mitigate crack propagation and prolong the lifespan of the structure. Presently, prevalent crack treatment methodologies encompass borehole crack arrest, adhesive bonding reinforcement, and rewelding repair [8–10]. Recent decades have witnessed extensive research on the repair parameters of borehole crack arrest and adhesive bonding reinforcement technology, aiming to delay crack growth by either eliminating the crack tip or enhancing the local capacity. Additionally, the evaluation of the fatigue performance of repaired structures has been conducted using these parameters [11–14]. Nevertheless, the existing body of research on rewelding repair technology remains inadequate, and the impact of rewelding residual stress on the fatigue behavior of a bridge deck remains ambiguous.

Many studies have investigated the residual stress and fatigue properties of fillet welds, particularly the initial fillet weld in this study, at the U-rib-to-deck-plate joint of an orthotropic steel bridge deck during its construction phase. Puymbroeck et al. [15], Chen et al. [16], and Zhao et al. [17] conducted an experimental study to measure the residual welding stress of U-rib stiffened deck plates. It should be noted that limitations such as measurement errors, variations in model size, and limited test data hinder the ability to establish a universal law. In light of the advancements in computer technology, numerical simulation of the welding process has gained significant popularity. Kainuma [18], Puymbroeck et al. [15], Xiong et al. [19], and Gu et al. [20] conducted a comprehensive investigation on the distribution of residual welding stress in orthotropic steel bridge decks. Their research has successfully validated the accuracy of numerical simulations through experimental analysis. Furthermore, these studies have examined the residual stress and fatigue behavior resulting from initial fillet weld welding while also assessing the influence of welding residual stress on the fatigue life of the bridge decks. This research focuses on the initial fillet weld residual stress and incorporates the distribution of rewelding residual stress during the fatigue crack rewelding process. At present, there exists a scarcity of research pertaining to the residual stress field that remains at the site of fatigue cracking, as well as the newly formed residual stress field resulting from the implementation of secondary welding techniques for fatigue crack repair.

Investigating the residual stress induced by rewelding in the repair of fatigue cracks presents a more intricate challenge, as the rewelding process is conducted under the influence of an initial stress field and subjected to a complex boundary constraint. The findings of previous studies on butt welds in stainless steel tubes indicate that the rewelding procedure has a notable impact on the distribution of axial residual stress in neighboring heat-affected zones [21]. A distinct inference can be drawn from these findings, which suggests that the alteration in transverse stress is more pronounced when compared to the welding of new components with less stringent constraints [22,23]. Moreover, previous studies have indicated a positive correlation between the length of the weld and the magnitude of the increase in transverse residual stress [24]. Edwards et al. [25] and Leggatt [26] carried out relevant studies on repair welding of high-strength steels using various research methods, with the aim of obtaining the optimal repair welding principles. Bouchard et al. [27] performed fatigue tests again on the repaired specimens (specimens of rib-to-deck welds and diaphragm-to-rib welds in steel bridge decks) and discussed the position and propagation of the new cracks to evaluate the efficacy of the repairs. The results indicated that new cracks occurred in the rewelded zone, and the fatigue life after rewelding was determined by the new weld. Consequently, further investigation into the distribution of residual stress in the rewelded zone and the fatigue life after rewelding was determined by the new weld.

This study focuses on investigating the prevalent fatigue crack that occurs at the U-rib-to-deck-plate joint of an orthotropic steel bridge deck. Initially, the residual stress of the initial fillet weld is examined through both finite element simulation and experimental methods. Subsequently, the process of crack removal and multi-layer rewelding is analyzed, and the impact of the initial stress field and the stiffness of the boundary constraint on the resulting residual stress field is discussed.

2. Rewelding Repair Method

As a result of the impact of fatigue cracking, the deck plate thickness of recently constructed steel bridges exceeds 16 mm. Conversely, existing long-span cable-stayed bridges and suspension bridges commonly employ deck plates 12 mm thick. A comprehensive investigation has revealed significant issues with fatigue cracking in these bridge decks, with the most severe occurrences observed at the intersection of the longitudinal rib and deck plate (Figure 1). The occurrence of fatigue cracks at the weld in the orthotropic bridge deck is primarily attributed to the presence of stress concentration and elevated residual stress levels induced during the welding process. The fatigue damage on the outer side of the welded joint between the longitudinal rib and the bridge deck is large. At this time, fatigue cracks easily occur on the weld between the roof and the longitudinal rib and extend to the bridge deck, while the roof measured inside the longitudinal rib does not have obvious cracks. Addressing the issue of fatigue cracking in steel bridge decks that are currently in use poses a significant challenge for the management and maintenance of expensive bridges. Among the various repair methods available, crack removal and rewelding offer prompt and cost-effective solutions. The rewelding repair process adheres to specific guidelines outlined in the German steel bridge deck repair procedure DVS 1709. This procedure encompasses several essential steps, including the elimination of existing cracks, thorough cleaning of the weld root, and subsequent rewelding, as illustrated in Figure 2. However, the process of repairing cracks through rewelding will inevitably result in the introduction of additional residual stress. This newly introduced residual stress interacts with the pre-existing residual stress, thereby forming an unknown residual stress field that impacts the fatigue performance of a repaired steel bridge deck. Consequently, it becomes imperative to conduct research on residual stress, taking into account both the initial welding residual stress and the residual stress resulting from rewelding, under various constraint conditions.

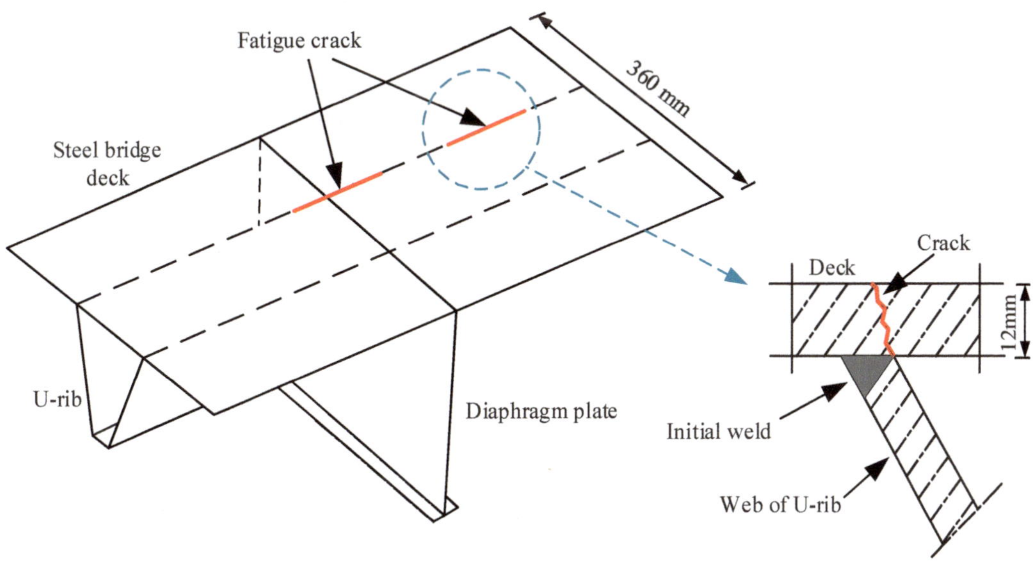

Figure 1. Fatigue crack at the U-rib-to-deck-plate joint of an orthotropic steel bridge deck.

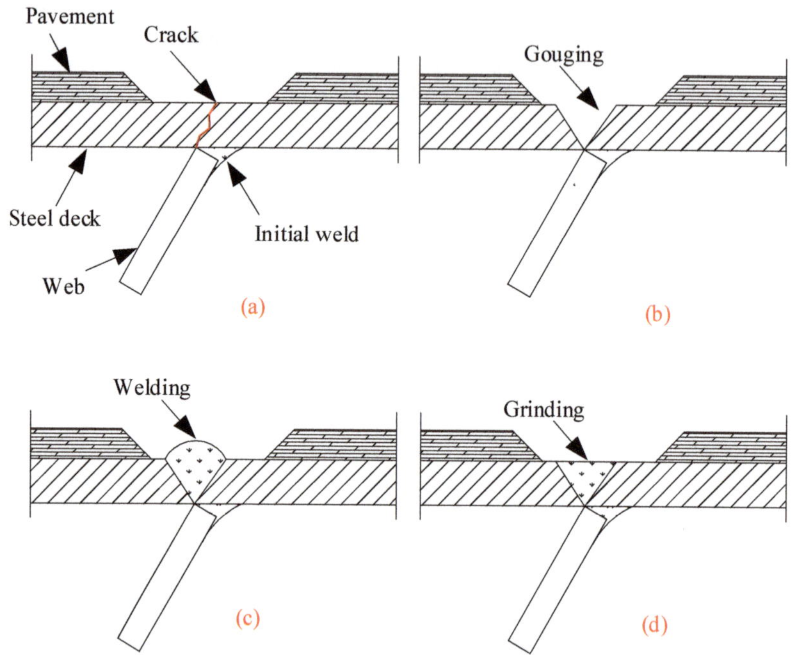

Figure 2. Rewelding repair procedure of cracks.

3. Finite Element Analysis of Rewelding Residual Stress

3.1. Geometric Model and Material Parameters

In this research, we analyze the structural form and the section geometry (as shown in Figure 3) of an orthotropic steel bridge deck, specifically focusing on the Sutong Yangtze Bridge, which holds the distinction of being the world's second-longest cable-stayed bridge. The analytical model utilized in this research has a length of 300 mm, and the material chosen for the bridge deck is Q345D low-alloy steel. For the sake of simplifying the analysis, it is assumed that the weld material is identical to the base metal. The material's parameters can be found in Table 1 [28].

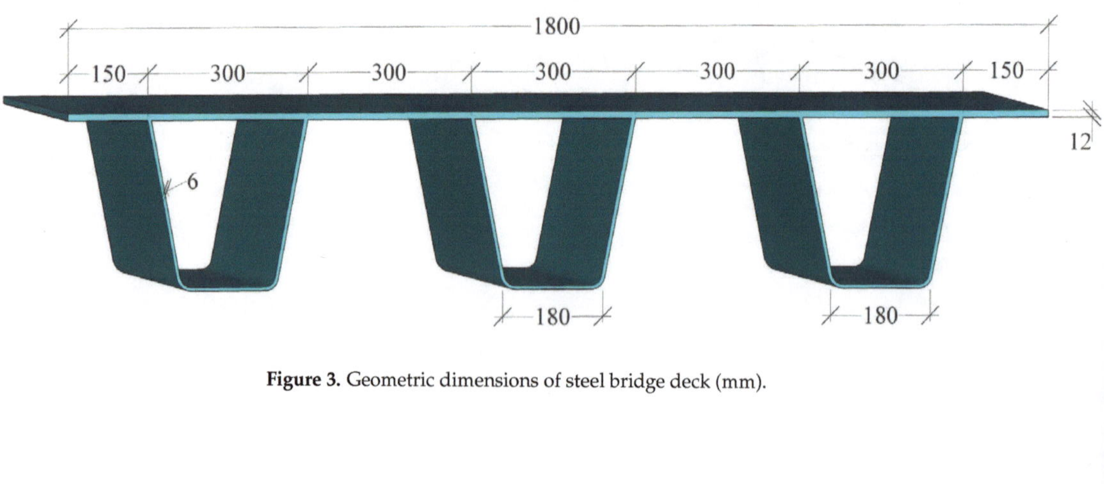

Figure 3. Geometric dimensions of steel bridge deck (mm).

Table 1. Mechanical properties of Q345D steel at different temperatures.

T/°C	Mechanical Properties of Q345D Steel			
	Yield Stress *f*/**MPa**	**Elastic Modulus** *E*/**GPa**	**Poisson's Ratio** v	**Coefficient of Thermal Expansion** α
0	345.0	202	0.28	11.0
400	224.3	202	0.30	13.38
500	182.9	258	0.31	--
600	103.5	95	0.32	13.39
700	44.9	46	0.35	--
800	24.2	22	0.36	--
900	17.3	23	0.37	--
1000	10.4	8	0.38	--
1100	6.9	4	0.39	--
1200	0.0	0	0.40	20.05

In the course of welding, intricate physical and chemical reactions transpire within the weld pool, wherein each phenomenon interrelates, rendering comprehensive simulation of the welding process arduous. The objective of this study is to examine the temperature field engendered by welding and the consequent stress field. Consequently, the finite element simulation omits the consideration of influential factors like fluid dynamics and surface tension in the fluid state during welding, and instead relies on the following assumptions: (1) The chemical reaction, stirring, and flow phenomena occurring within the molten pool are disregarded. (2) A constant welding speed is employed. (3) Only unidirectional thermo-mechanical coupling is taken into account. Although the actual welding process involves thermal-stress coupling, the plastic deformation heat and latent heat generated by the specimen during welding are lower than the input welding heat. Consequently, it is assumed that the stress field does not impact the temperature field. In other words, only the unidirectional influence of the temperature field on the stress field is considered.

3.2. Finite Element Heat Source Model

The configuration of the welding heat source plays a crucial role in determining the energy distribution within the weld pool, thereby impacting the spatial variation of the welding temperature gradient. This, in turn, affects both the quasi-steady temperature field and the cooling temperature field. In order to achieve more precise results in finite element simulations of welding, it is recommended to employ a double ellipsoid heat source model. The Goldak double ellipsoid model [29] has been found to effectively depict the density distribution of the heat source during its forward progression. The double ellipsoid heat source model, which is frequently employed in gas-shielded welding and submerged arc welding processes, is utilized to describe the welding repair of the steel bridge deck. The specific double ellipsoid heat source model employed in this study is depicted in Figure 4.

The heat flux densities of the front and rear ellipsoids are determined through the following calculations [30]:

$$q_i(x,y,z) = \frac{6\sqrt{3}f_i\eta UI}{abc_i}\exp(-\frac{3x^2}{a^2} - \frac{3y^2}{b^2} - \frac{3z^2}{c_i{}^2})\ i = 1,2 \tag{1}$$

where f_1, f_2 is the energy fraction of the front and rear hemispheres, and $f_1 + f_2 = 2$. η is heat input efficiency, U is voltage, I is current. c_1 and c_2 are the half lengths of the front and rear ellipsoids; a and b correspond to the width and depth of the heat source. q_1 and q_2 represent the heat input density before and after the molten pool (W/mm^3).

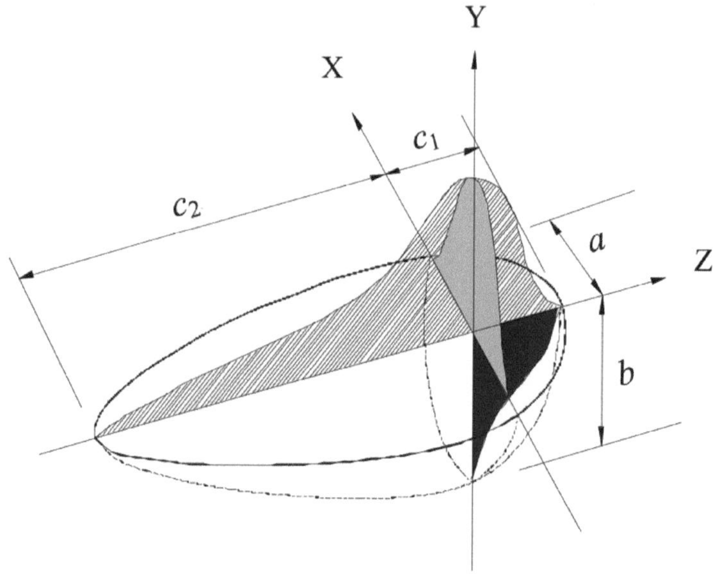

Figure 4. Double ellipsoid heat source model.

3.3. Initial Welding Stress Field

During crack rewelding, the presence of a residual stress field resulting from the initial welding around the crack is observed in contrast to the typical sequential welding process employed in construction. Shahani et al. [31] have demonstrated that the residual stress surrounding the crack is alleviated during the initiation and propagation of fatigue cracks. Generally, the initiation and propagation of cracks lead to the release of certain residual stresses. The removal of a crack and the subsequent formation of a V-shaped groove prior to rewelding facilitate the release of a portion of the residual stresses in the vicinity of the crack region. At present, a comprehensive elucidation regarding the persistence of stress after partial stress release and its impact on the rewelding procedure remains elusive. Consequently, this study endeavors to examine the effect of the residual stress field from an initial weld through the simulation of ANSYS finite element software.

To streamline the model, the U-rib's symmetry is exploited, typically involving the analysis of half of a U-rib, as depicted in Figure 5. The displacement boundary conditions of the finite element model are established to ensure that the free deformation of the welding process is not impeded. These conditions are based on the actual welding constraints and are designed to prevent rigid body displacement during the finite element calculation process. In the coordinate system of the finite element model, the X-direction represents the vertical weld, the Y-direction represents the deck thickness, and the Z-direction represents the weld length. The symmetric constraint is implemented along the symmetric centerline of the U-rib, whereby the displacement in the Y direction on both sides of the bottom surface of the deck plate is constrained to emulate the support platform. Conversely, the displacement in the Z direction is restrained on the opposite side of the deck plate.

The thermo-mechanical sequential coupling analysis method is employed to sequentially conduct thermal analysis and structural analysis. In the thermal analysis, the generation of the finite element model involves the utilization of the eight-node hexahedral thermal element (SOLID70), wherein only one degree of freedom is assigned for the temperature at each node. Conversely, in the Ansys analysis, due to the identical geometric shape shared by the thermal element (SOLID70) and the structural element (SOLID185), the thermal element (SOLID70) can be substituted directly with the corresponding structural element (SOLID185). The latter is an eight-node three-dimensional solid element that possesses three displacement degrees of freedom per node, thereby facilitating the creation of the structural finite element model.

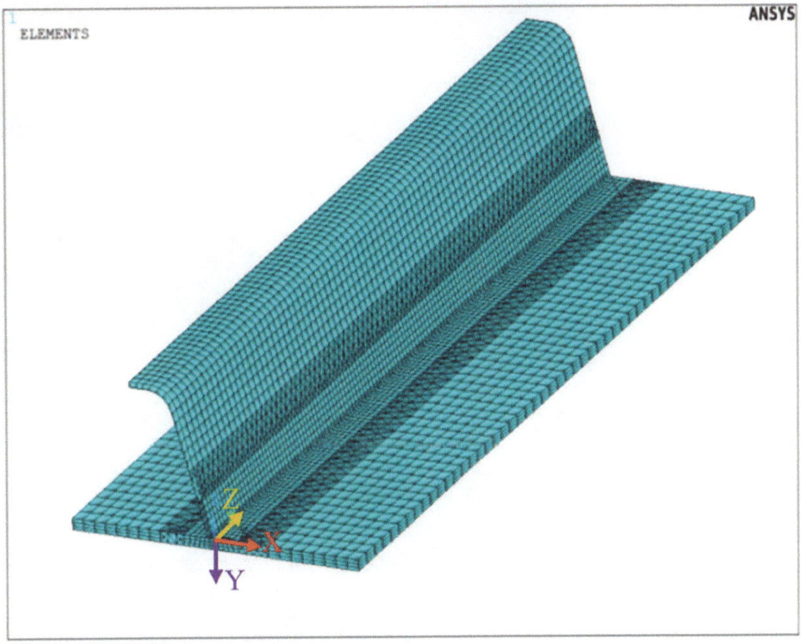

Figure 5. Model of finite element calculation.

The simulation of the fillet weld at the U-rib-to-deck-plate joint of an orthotropic steel bridge deck employs the double ellipsoid welding heat source as the input for welding heat. The parameters for the heat source, namely, a, b, c_1, and c_2, are determined as 0.008, 0.012, 0.012, and 0.048, respectively, while the initial temperature is set at 20 °C. Numerous research findings have been published on the residual stresses of a steel bridge deck calculated using the standard welding process [32,33]; thus, this paper does not delve into a detailed discussion of this matter. The resulting Mises equivalent residual stress after the initial weld cooling is shown in Figure 6.

As shown in Figure 7, the blind hole method was used to measure the residual stresses along the weld direction of the bridge deck after the initial weld cooling. The detailed test process and the data are shown in Reference [18]. Because the measuring points of each section were discontinuous due to the limitations of feasibility and cost, five measuring points were arranged along the measuring path. Hence, drawing upon the available literature and experimental findings, the analysis incorporated a line connecting each measuring point to account for the self-balancing nature of compressive/tensile residual stress and to visually represent the distribution of residual stress more effectively. Thus, the distribution can be reflected distinctly. A comparison between the measured data and the finite element data is shown in Figure 8.

The comparison between the yield strength of the deck plate and the maximum discrepancy of 5.2% observed in the finite element analysis results of the longitudinal residual tensile stress, as depicted in Figure 8, indicates that the measured longitudinal residual tensile stress aligned well with the numerical simulation outcomes. This finding underscores the accuracy of the employed finite element simulation approach in terms of predicting the welding residual stress.

Further, the removal of groove elements, as depicted in Figure 2, results in the creation of an empty welding channel to be rewelded based on the initial welding residual stress.

Subsequently, during the rewelding procedure, a portion of the stress is alleviated, leading to the formation of a residual stress field after redistribution. This field serves as the initial state for analyzing the stress evolution during the rewelding process.

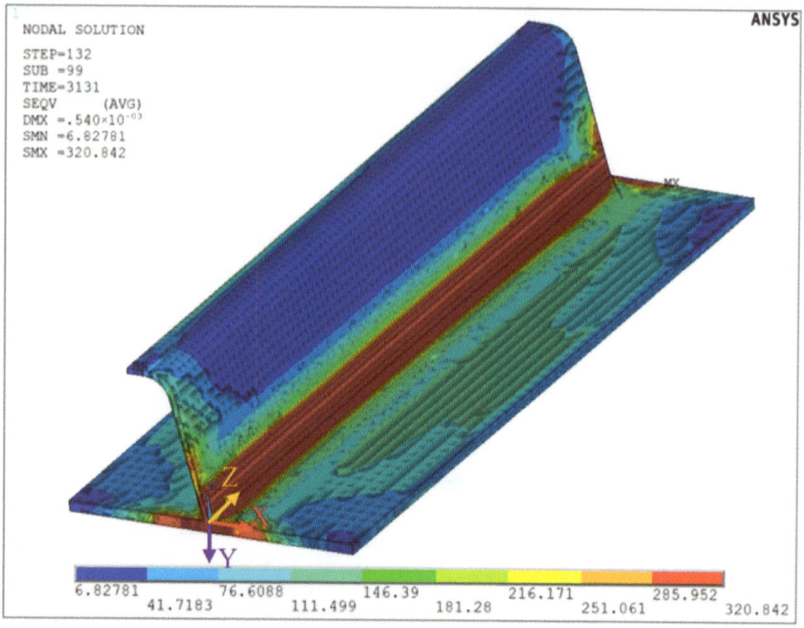

Figure 6. Mises equivalent stress field.

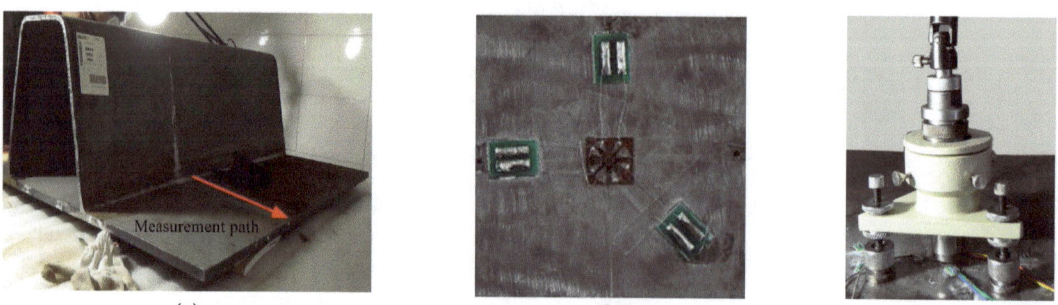

(a) (b) (c)

Figure 7. Test of the residual stress of steel bridge plate with the blind hole method. (**a**) Schematic diagram showing the measurement path; (**b**) strain rosette; (**c**) drilling.

3.4. Boundary Conditions of Rewelding

The formation of the reweld bead occurs after the removal of the pre-existing fatigue crack. Consequently, the boundary conditions for the rewelding process differ from those of the initial welding, necessitating consideration of the impact of the constraint stiffness of the deck surrounding the weld. Due to the greater bending stiffness of the longitudinal ribs in comparison to the deck plate, the stress experienced by the deck plate between the ribs is treated as that of a one-way plate. This is illustrated in Figure 9, where a segment of the deck with a unit width is selected in the direction of the longitudinal rib. Due to the requirement of the #2 rib exposing the groove at the crack location, rendering it no longer a closed section, its stiffness is disregarded. Given the substantial stiffness of the closed section, the #1 and #3 closed rib sections are established as fixed constraint boundary conditions. Consequently, the simplified constraint stiffness of the constraint conditions during the local welding of the #2 rib is computed. The elastic constraint boundary stiffness of the finite element model corresponds to the stiffness of the free end when the #1 and #3 longitudinal ribs are consolidated, as shown in Figure 9.

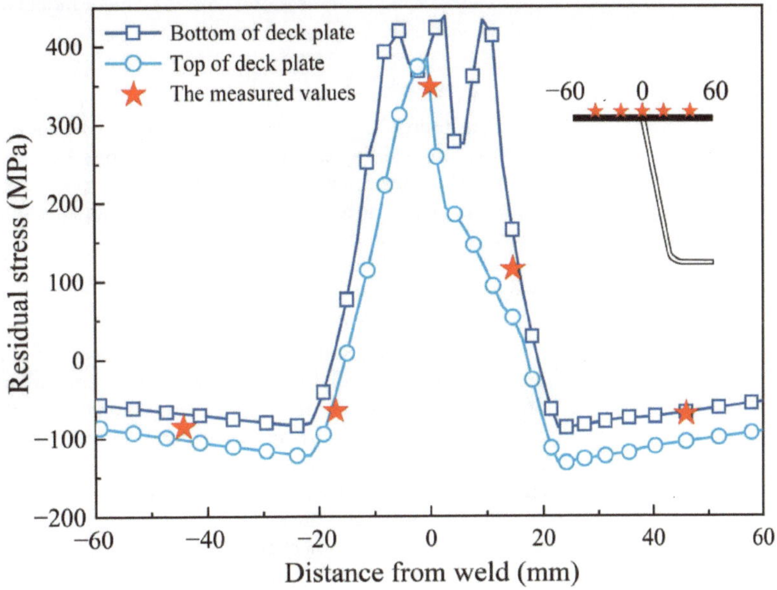

Figure 8. Comparison of the longitudinal residual stress of finite element and measured data.

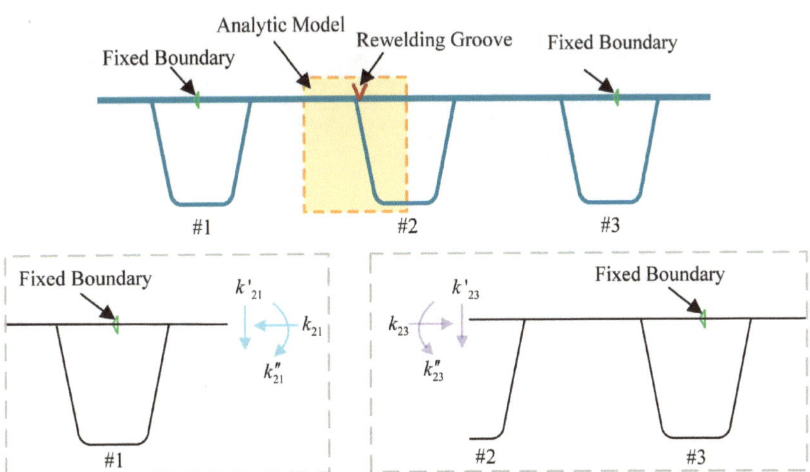

Figure 9. Diagram of weld elastic boundary calculation (mm).

The axial stiffnesses of the finite element model boundary restrained by the #1 rib and the #2 rib are:

$$k_{21} = EA/l_1,\ k_{23} = EA/l_3 \quad (2)$$

The shear stiffnesses of the finite element model boundary restrained by the #1 rib and the #2 rib are:

$$k'_{21} = GA,\ k'_{23} = GA \quad (3)$$

The bending stiffnesses of the finite element model boundary restrained by the #1 rib and the #2 rib are:

$$k''_{21} = EI/l_1,\ k''_{23} = EI/l_3 \quad (4)$$

where k_{21}, k'_{21}, and k''_{21} are the axial stiffness, shear stiffness, and bending stiffness of the model restrained by the #1 rib, respectively. k_{23}, k'_{23}, and k''_{23} are the axial stiffness, shear stiffness, and bending stiffness restrained by the #3 rib, respectively. E is Young's modulus, G is the shear modulus, I is the moment of inertia of the section, A is the cross-sectional area, l_1 is the distance from the #1 rib to the model boundary, and l_3 is the distance from the #3 rib to the model boundary.

4. Results and Discussion

4.1. Rewelding Temperature Field

The modeling approach for rewelding is analogous to that of the initial welding process, and the model was subjected to an equivalently simplified constraint condition. Initially, the birth–death element as eliminated, redirecting its heat flux to the node connecting the remaining model section and the removed area, thereby simulating the dissipation of residual stress in the groove region of the weld. Subsequently, the double ellipsoid heat source was incrementally shifted layer by layer, corresponding to the three layers of the weld, and the residual stress resulting from the rewelding process was computed through thermo-mechanical coupling. The dimensions of the weld zone mesh were 1 mm × 1 mm × 2 mm, comprising a total of 73,397 solid185 solid elements (Figure 10). The rewelding parameters remained consistent with those outlined in the preceding initial welding process.

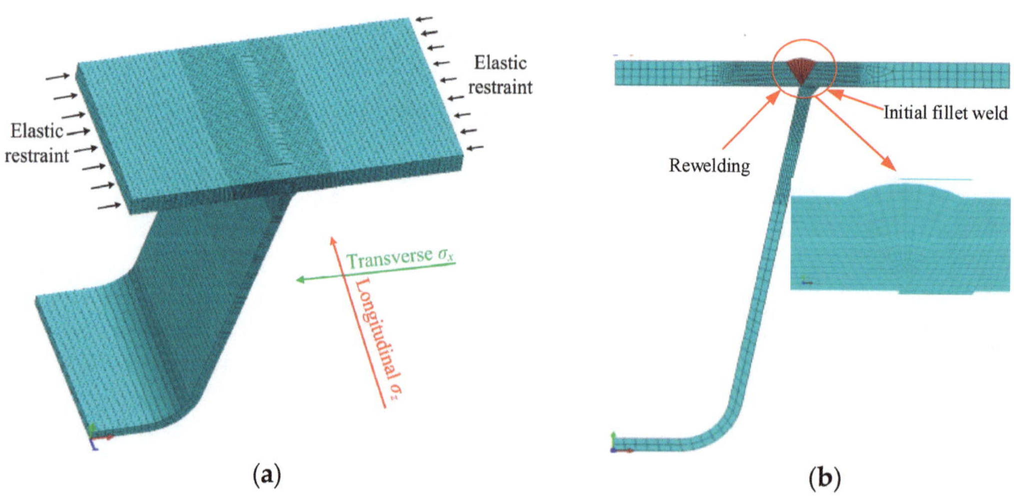

Figure 10. Meshing and constraint setting of finite element model. (**a**) Boundary constraint condition; (**b**) Position of welding.

Figure 11 depicts the temperature contours of the three-layer weld bead during the rewelding process. It can be observed that the central temperature of the weld pool exceeded 1550 °C, which aligns with the distribution characteristics. Specifically, the temperature field gradient of the double ellipsoid heat source intensified in the direction of its movement, while it noticeably diminished in the opposite direction. Moreover, the temperature field exhibited a symmetrical pattern along the weld direction and facilitated heat transfer towards the periphery.

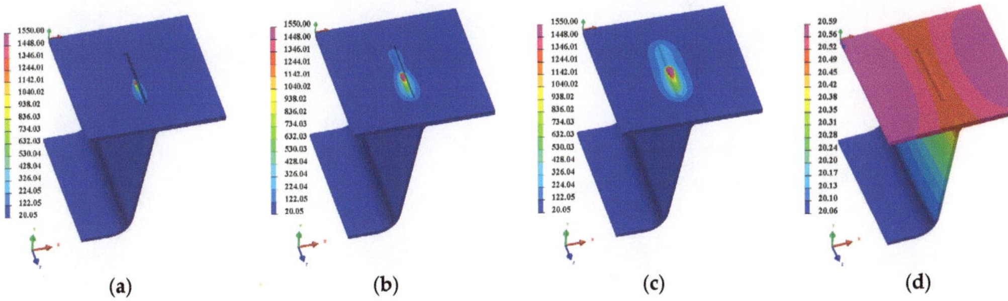

Figure 11. Transient temperature field of rewelding (°C). (**a**) The heat source movement in the first bead; (**b**) the heat source movement in the second bead; (**c**) the heat source movement in the third bead; (**d**) temperature of U-rib at cooling.

Within an 8 cm vicinity surrounding the weld seam, three layers were chosen, with each layer consisting of five selected nodes. Consequently, a total of fifteen nodes were designated for the purpose of analyzing temporal fluctuations in temperature. The specific locations of these selected nodes are visually depicted in Figure 12. Subsequently, the temperature variations over time for each individual node were computed and are presented in Figure 13. The temperature field between the weld bead and the temperature changes of each node with time exhibited variations due to the welding time sequence and the filling of the weld bead. Nodes located on the same bead, such as A1–A5, experienced a delay in the arrival time of the peak temperature as the distance from the bead increased, accompanied by a rapid decrease in the peak temperature. Furthermore, notable disparities were found in the temperatures observed across different weld beads. The third weld bead attained a maximum temperature of 600 °C at the A bead nodes, whereas the first weld bead registered a maximum temperature below 200 °C at the C bead nodes. The heat conduction effect of weld filling resulted in a larger heat-affected zone in the cap weld.

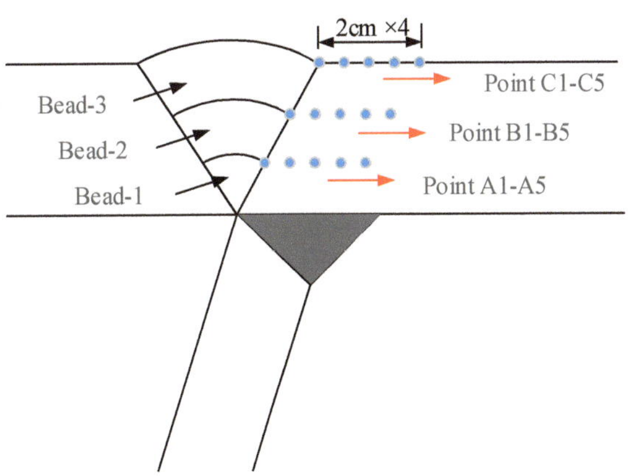

Figure 12. The position of the temperature analysis node (blue circles).

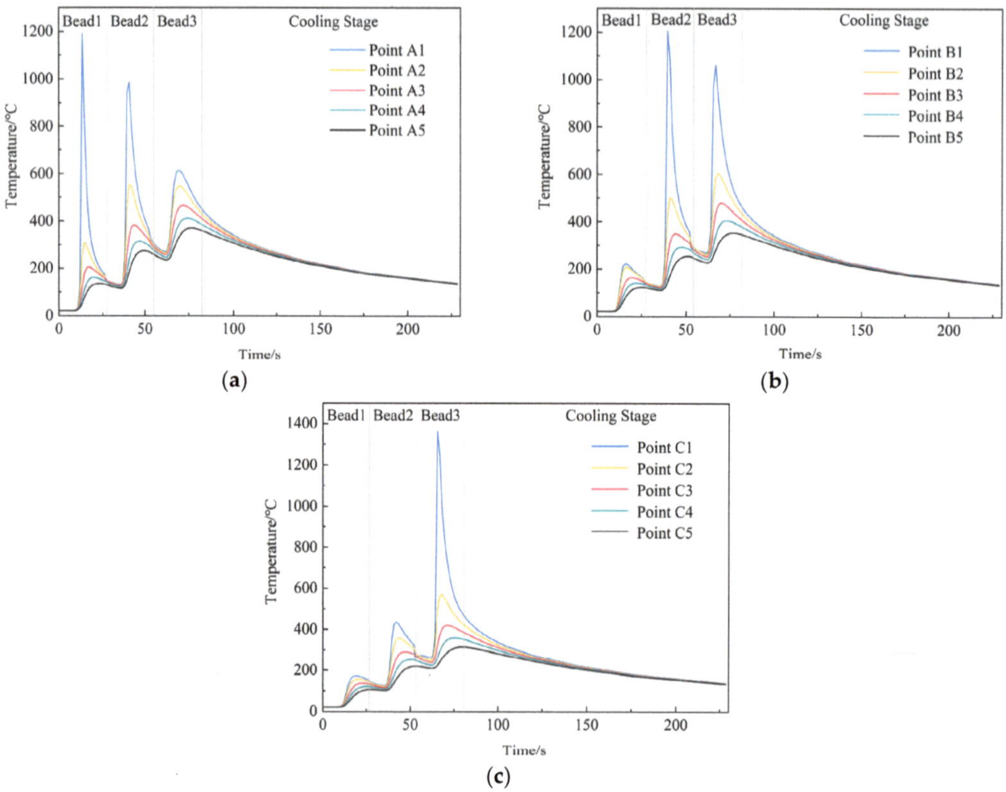

Figure 13. Diagrams of the temperature changes at different times. (**a**) Temperature variation curve of nodes A1–A5; (**b**) temperature variation curve of nodes B1–B5; (**c**) temperature variation curve of nodes C1–C5.

The temperature field was symmetrical in the center of the weld. During the welding procedure, the majority of the top and web components remained at ambient temperature. The application of high-density energy resulted in a significant disparity in temperature between the weld and the base metal region. Consequently, the material within the weld area experienced varying degrees of constraint from the base metal during the expansion and contraction phases of heating and cooling. This phenomenon plays a pivotal role in determining residual distribution.

4.2. Rewelding Residual Stress

Rewelding repairs involve conducting welding repairs after the creation of a groove on the top plate. The welding seam undergoes inherent constraints from the adjacent material during the heating and cooling procedure. As a result, the welding residual stress that arises upon cooling to ambient temperature will differ from the stress distribution observed during the initial fillet weld processing. The temperature field obtained from the thermo-mechanical analysis of rewelding through sequential coupling is presented in Figure 11, whereas the residual stress field is depicted in Figure 14.

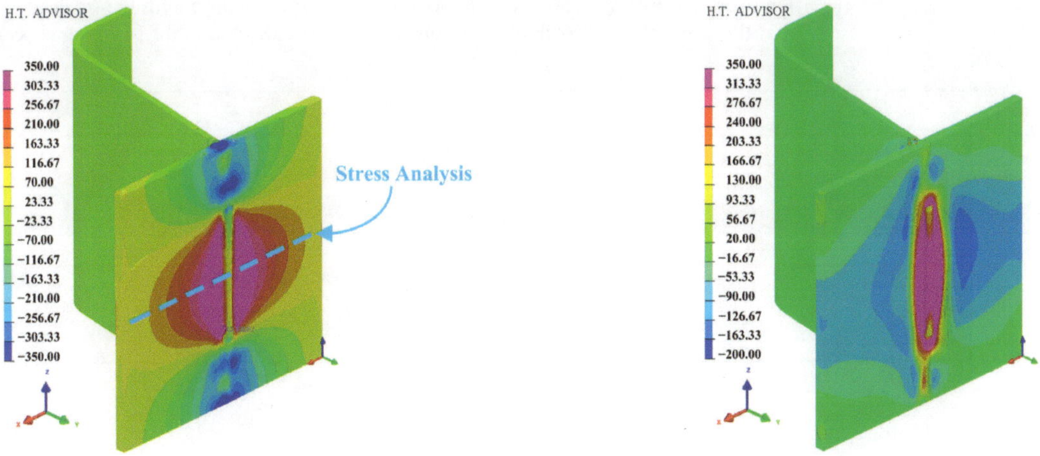

(a) (b)

Figure 14. Distribution of residual stress field (MPa). (**a**) Transverse residual stress; (**b**) longitudinal residual stress.

Figure 14 depicts the residual stress field of the weld after cooling, taking into account the initial stress and elastic boundary constraints. The welding process imposed a robust constraint around the deck, resulting in the persistence of longitudinal tensile stress of approximately 80 MPa at the deck's edge. This disparity in residual stress distribution between the rewelded deck and the free boundary deck under initial construction conditions was substantial.

In order to assess the variations in residual stress throughout the entirety of the rewelding process, the stress analysis was partitioned into three distinct stages based on the aforementioned rewelding repair procedure. The initial stage involved the creation of a primary weld, the residual stress of which is examined in Section 3.3. Subsequently, in the second stage, the fatigue crack was eliminated and a rewelding groove was established, resulting in the partial release of the residual stress accumulated in the preceding stage. The residual stress subsequent to crack removal is visually depicted in Figure 15. The residual stress analysis in this study involved the examination of transverse and longitudinal residual stresses. The orange line in the graph represents the transverse residual stress, whereas the brown line represents the longitudinal residual stress during the initial stage. The second and third stages of the analysis encompassed initial residual stress and boundary constraints. Figure 15 illustrates the residual stress that arose from the rewelding process. The transverse residual stress of beveling is represented by the pink line, while the blue line represents the longitudinal residual stress of beveling. Furthermore, the red line signifies the transverse residual stress of reweld, and the purple line represents the longitudinal residual stress of the reweld. The positions of the stress analysis node in Figure 14 can be described in the following manner. A 300 mm long straight line was selected along the vertical weld direction, specifically in the middle of the weld length direction. The nodes located on this line were designated as stress analysis nodes, as depicted in Figure 14a.

First, the influence of the beveling on the initial weld stress field was analyzed. It should be emphasized that the stress field that formed after beveling is the initial residual stress in the rewelding calculation. Following the process of beveling, a significant alteration in the transverse stress (the orange line) adjacent to the weld region was observed, in contrast to the initial transverse stress (the pink line). This discrepancy suggests a substantial release of transverse stress predominantly in the vertical direction of the groove. The longitudinal stress (yellow line) did not change much compared with Figure 8, suggesting that the stress along the direction of the groove is less affected in the beveling process.

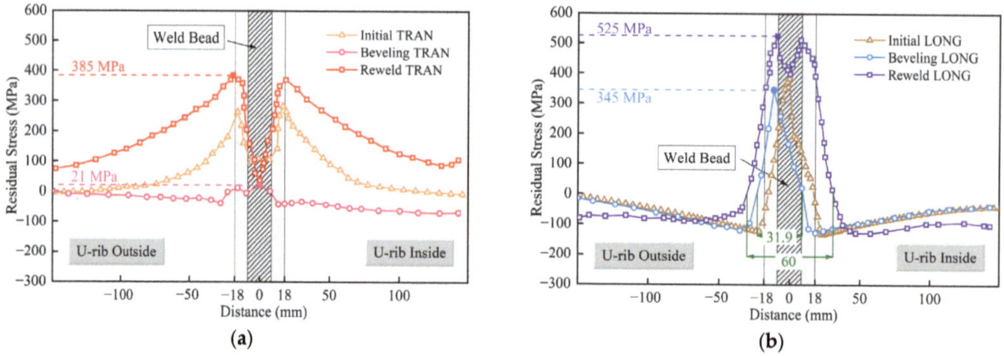

Figure 15. Residual stress distribution before and after rewelding on the upper surface of the deck. (**a**) Transverse residual stress; (**b**) longitudinal residual stress.

The residual stress field after rewelding is further analyzed. After rewelding, there is significant residual stress in both the longitudinal and transverse directions. However, the longitudinal residual stress (purple line) is higher than the transverse residual stress (red line) because the longitudinal residual stress was not fully released before rewelding, while the transverse residual stress was released more. However, the transverse residual stress increased more during rewelding. Compared with the residual stress generated by the beveling during construction, the residual longitudinal tensile stress in rewelding was higher than the yield strength in the range of about 18 mm from the centerline of the weld. The peak value of the tensile stress increased from 345 MPa to 525 MPa, and the distribution width of the tensile stress increased from 31.9 mm to 60 mm, with an increase of 88%. Meanwhile, the peak value of the transverse tensile stress in rewelding increased from 21 MPa to 385 MPa. In general, the transverse residual stress changes more obviously in rewelding, while the peak value and the influence range of longitudinal residual stress increased further. Therefore, the rewelding process should consider effective residual stress elimination measures.

During the rewelding cooling process, martensitic transformation occurred when the weld bead pool temperature was lower than the martensitic temperature. The volume expansion caused by the transformation was much greater than the thermal strain caused by austenite cooling. The solid phase transformation produced compressive stress in the hot melt pool area, thus greatly reducing the tensile stress in the weld pool position, as shown in Figure 15.

4.3. Influence of Boundary Constraints

In the determination of the initial weld residual stress for a steel bridge deck, it is customary to employ a calculation model that is statically determined in its entirety, disregarding the impact of boundary conditions. Nevertheless, the rewelding procedure entails partial welding across the entire bridge deck, and neglecting the boundary constraints will yield imprecise calculation outcomes. To examine the effect of boundary conditions on the residual stress resulting from rewelding, four distinct boundary constraints are taken into account. The four types of boundaries considered in this study include the free boundary, the boundary with solely axial constraints, the boundary with both axial and tangential constraints, and the fully constrained boundary. In the preceding calculation, the axial and tangential elastic constraints were employed. The longitudinal and transverse residual stress distributions for each of the four boundary conditions are presented in Figure 16.

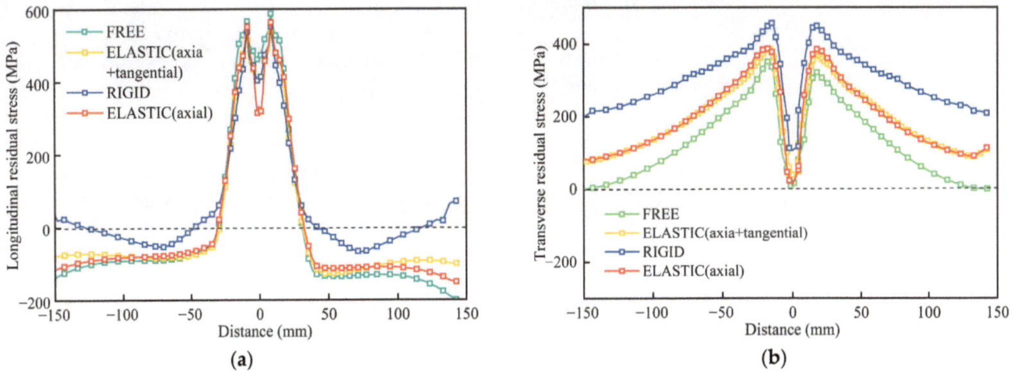

Figure 16. Comparison of rewelding residual stress for different constraints. (**a**) Longitudinal residual stress with four boundary constraints; (**b**) transverse residual stress with four boundary constraints.

The longitudinal residual stress near the heat-affected zone was minimally influenced by the boundary constraint, as depicted in Figure 16a, while the compressive stress at the model's edge was affected. When a fully constrained boundary was imposed, tensile stress was observed at the model's edge. Comparatively, the transverse residual stresses were more influenced by the boundary conditions. Figure 16b illustrates a similar distribution of transverse residual stress under different boundary conditions. The maximum tensile stress experienced a 40.6% increase, rising from 320 MPa to 450 MPa, as the stiffness of the boundary constraint increased. Additionally, the tensile stress at the edge position increased from 0 MPa to 200 MPa. It is important to acknowledge that the elastic constraint boundary utilized in this study yielded simplified calculation outcomes, with the constraint stiffness falling between that of a free boundary and a fully constrained boundary. The calculation results depicted in Figure 16 demonstrate that the residual stress lies within the range of the free boundary and the fully constrained boundary, thereby indicating the reasonableness of the obtained results.

Typically, the transverse residual stress arising from the elastic constraint lies within the range of the free boundary and the fully constrained boundary. Consequently, when assessing fatigue strength after rewelding, it is imperative to account for the impact of boundary constraints in order to attain more precise outcomes and prevent the underestimation of the influence of residual stress by employing free boundary conditions.

5. Conclusions

The rewelding repair technology offers a convenient and efficient solution for addressing longitudinal fatigue cracks occurring at the U-rib-to-deck-plate joint of an orthotropic steel bridge deck. It should be noted that this repair process inevitably generates new residual stresses, which exhibit distinct characteristics compared to the original weld. During the removal of the fatigue crack prior to rewelding, a partial release of the residual stress from the original weld occurred. The findings of the analysis indicate that the transverse residual stress in the vicinity of the beveling zone was effectively alleviated, whereas the longitudinal residual stress remained relatively stable. Upon completion of the rewelding process, significant transverse residual stress was anticipated, with a peak value approaching the initial transverse residual stress. In comparison to the residual stress induced by the initial welding process during construction, the residual longitudinal tensile stress in rewelding surpassed the yield strength within a span of approximately 18 mm from the weld's centerline. The peak value of the residual tensile stress escalated from 345 MPa to 525 MPa, while the distribution width of the residual tensile stress expanded from 31.9 mm to 60 mm, signifying an 88% increase. And the peak value of the transverse tensile stress in rewelding increased from 21 MPa to 385 MPa.

There was minimal disparity observed in the longitudinal tensile stress within the heat-affected zone across various boundary conditions. It should be noted that distinct variation is evident in the longitudinal compressive stress at the periphery of the model When a fully constrained boundary was employed, tensile stresses were even observed at the model's edge. In relative terms, the transverse stress was more susceptible to alterations in the boundary conditions. In comparison to the free boundary condition, the tensile stress escalated from 320 MPa to 450 MPa when a fully constrained boundary was implemented The present study employed an elastic constraint boundary condition lying between the free boundary condition and the rigid boundary condition. It is imperative to ensure that all actual conditions fall within this range. Consequently, when evaluating the fatigue strength following rewelding, it is crucial to account for the impact of boundary constraints and the initial welding stress field. This consideration is necessary in order to attain more precise outcomes and prevent the underestimation of the influence of residual welding stress, which may occur due to the simplification of assuming no initial stress field and free boundary conditions.

Author Contributions: Conceptualization, J.Q. and Z.-B.X.; methodology, S.-J.S.; software, L.-Q.Z., validation, P.-Y.Z., S.-J.S. and L.-Q.Z.; formal analysis, S.-J.S.; investigation, L.-Q.Z.; resources, P.-Y.Z., data curation, Z.-B.X.; writing—original draft preparation, Z.-B.X.; writing—review and editing, S.-J.S.; visualization, L.-Q.Z.; supervision, J.Q.; project administration, J.Q.; funding acquisition, J.Q All authors have read and agreed to the published version of the manuscript.

Funding: This work was supported by the National Natural Science Foundation of China (Grant No. 52378283); China Postdoctoral Science Foundation (Grant No. 2021M702782); Natural Science Foundation of Chongqing, China (Grant No. CSTB2023NSCQ-MSX0633); Team Building Project for Graduate Tutors in Chongqing (Grant No. JDDSTD2022003).

Data Availability Statement: The data supporting the article's findings are available from the corresponding author upon reasonable request.

Conflicts of Interest: The authors declared no potential conflict of interest with respect to the research, authorship, and/or publication of this article.

References

1. Wang, Q.D.; Ji, B.H.; Fu, Z.Q.; Ye, Z. Effect of crack-closure treatment on fatigue durability of cracked rib-to-deck welded joints in steel bridge decks. *J. Cent. South Univ.* **2019**, *26*, 2554–2568. [CrossRef]
2. Liu, L.; Sun, J.; Chen, W.; Sun, P. Modified Layer-Removal Method for Measurement of Residual Stress in Pre-stretched Aluminium Alloy Plate. *J. Harbin Inst. Technol.* **2015**, *22*, 34–40.
3. Krebs, J.; Kassner, M. Influence of welding residual stresses on fatigue design of welded joints and components. *Weld. World* **2007**, *51*, 54–68. [CrossRef]
4. Zhang, Z.N.; Zhang, Y.C. Effects of welding residual stresses on the stability capacity of the monosymmetric I-beams. *J. Harbin Inst. Technol.* **2007**, *39*, 1864–1868.
5. Wang, Q.; Ji, B.; Li, C.; Zhongqiu, F. Fatigue evaluation of rib-deck welds: Crack-propagation-life predictive model and parametric analysis. *J. Constr. Steel Res.* **2020**, *173*, 106248. [CrossRef]
6. Yong, Z.; Yu, Q.; Anbang, G. Large sectional fatigue experiment of the orthotropic steel deck from a box-girder suspension bridge *J. Harbin Inst. Technol.* **2016**, *48*, 95–100.
7. Aljabar, N.J.; Zhao, X.L.; Al-Mahaidi, R.; Ghafoori, E.; Motavalli, M.; Powers, N. Effect of crack orientation on fatigue behavior of CFRP-strengthened steel plates. *Compos. Struct.* **2016**, *152*, 295–305. [CrossRef]
8. Zhong-Qiu, F.U.; Bo-Hai, J.I.; Shu-Hui, X.; Liu, T.-J. Crack stop holes in steel bridge decks: Drilling method and effects. *J. Cent. South Univ.* **2017**, *24*, 2372–2381.
9. Yuan, Y.; Wu, C.; Jiang, X. Experimental study on the fatigue behavior of the orthotropic steel deck rehabilitated by UHPC overlay *J. Constr. Steel Res.* **2019**, *157*, 1–9. [CrossRef]
10. Kim, Y.J.; Harries, K.A. Fatigue behavior of damaged steel beams repaired with CFRP strips. *Eng. Struct.* **2011**, *33*, 1491–1502. [CrossRef]
11. Ishikawa, T.; Shimizu, M.; Tomo, H.; Kawano, H.; Yamada, K. Effect of compression overload on fatigue strength improvedby ICR treatment. *Int. J. Steel Struct.* **2013**, *13*, 175–181. [CrossRef]
12. Kuhlmann, U.; Bove, S.; Hubmann, M. Instandsetzung und Verstärkung von Stahlbrücken mit Kategorie-2-Schäden. *Stahlbau* **2017**, *86*, 574–586. [CrossRef]

13. Stranghöner, N.; Lorenz, C.; Friedrich, H. Instandsetzung und Verstärkung von Stahlbrücken mit Kategorie-1-Schäden. *Stahlbau* **2017**, *86*, 562–573. [CrossRef]
14. Freitas, S.T.; Kolstein, H.; Bijlaard, F. Fatigue behavior of bonded and sandwich systems for strengthening orthotropic bridge decks. *Compos. Struct.* **2013**, *97*, 117–128. [CrossRef]
15. Puymbroeck, E.V.; Staen, G.V.; Iqbal, N.; De Backer, H. Residual weld stresses in stiffener-to-deck plate weld of an orthotropic steel deck. *J. Constr. Steel Res.* **2019**, *159*, 534–547. [CrossRef]
16. Chen, S.J.; Yang, K.C. Inelastic behavior of orthotropic steel deck stiffened by U-shaped stiffeners. *Thin-Walled Struct.* **2002**, *40*, 537–553. [CrossRef]
17. Zhao, Q.; Guo, Z.; Shen, X.; Briseghella, B. Test study on residual stress distribution of hybrid steel u-rib stiffened plates. *J. Constr. Steel Res.* **2016**, *121*, 261–267. [CrossRef]
18. Kainuma, S.; Jeong, Y.S.; Yang, M.; Inokuchi, S. Welding residual stress in roots between deck plate and U-rib in orthotropic steel decks. *Measurement* **2016**, *92*, 475–482. [CrossRef]
19. Xiong, Y.; Li, C.; Chen, Z.Y.; He, J.; Xin, H. The Evolution of Residual Stress in Rib-Diaphragm Joints of Orthotropic Steel Decks Subjected to Thermal Cutting and Welding. *Materials* **2020**, *13*, 3804. [CrossRef]
20. Gu, Y.; Li, Y.; Zhou, Z.; Ren, S.; Kong, C. Numerical Simulation and Measurement of Welding Residual Stresses in Orthotropic Steel Decks Stiffened with U-Shaped Ribs. *Int. J. Steel Struct.* **2020**, *20*, 856–869. [CrossRef]
21. Dong, P.; Hong, J.K.; Bouchard, P.J. Analysis of residual stresses at weld repairs. *Int. J. Press. Vessel. Pip.* **2005**, *82*, 258–269. [CrossRef]
22. Chen, Z.; Duan, Y.; Wang, P.; Qian, H. Residual Stress Redistribution Analysis in the Repair Welding of AA6082-T6 Aluminum Alloy Joints: Experiment and Simulation. *Materials* **2022**, *15*, 6399. [CrossRef] [PubMed]
23. Jiang, W.; Luo, Y.; Zeng, Q.; Wang, J.; Tu, S.T. Residual stresses evolution during strip clad welding, post welding heat treatment and repair welding for a large pressure vessel. *Int. J. Press. Vessel. Pip.* **2021**, *189*, 104259. [CrossRef]
24. Fu, Z.; Wang, Q.; Ji, B.; Yuanzhou, Z. Rewelding repair effects on fatigue cracks in steel bridge deck welds. *J. Perform. Constr. Facil.* **2017**, *31*, 04017094. [CrossRef]
25. Edwards, L.; Bouchard, P.J.; Dutta, M.; Wang, D.Q.; Santisteban, J.R.; Hiller, S.; Fitzpatrick, M.E. Direct measurement of the residual stresses near a 'boat-shaped' repair in a 20 mm thick stainless steel tube butt weld. *Int. J. Press. Vessel. Pip.* **2005**, *82*, 288–298. [CrossRef]
26. Leggatt, R.H. Residual stresses in welded structures. *Int. J. Press. Vessel. Pip.* **2008**, *85*, 144–151. [CrossRef]
27. Bouchard, P.J.; George, D.; Santisteban, J.R.; Bruno, G.; Dutta, M.; Edwards, L.; Kingston, E.; Smith, D.J. Measurement of the residual stresses in a stainless steel pipe girth weld containing long and short repairs. *Int. J. Press. Vessel. Pip.* **2005**, *82*, 299–310. [CrossRef]
28. *CEN, DAFT ENV 1993*; Eurocode 3: Design of Steel Structure. European Committee for Standardization: Brussels, Belgium, 1995.
29. Goldak, J.; Chakravarti, A.; Bibby, M. A new finite element model for welding heat sources. *Metall. Trans. B* **1984**, *15*, 299–305. [CrossRef]
30. Glickstein, S.S.; Friedman, E. Characterization and modeling of the heat source. In *ASM Handbook*; ASM International: Almere, The Netherlands, 1994; pp. 6710–6736.
31. Shahani, A.R.; Shakeri, I.; Rans, C.D. Effect of residual stress redistribution and weld reinforcement geometry on fatigue crack growth of butt welded joints. *Int. J. Fatigue* **2020**, *139*, 105780. [CrossRef]
32. Wang, F.; Lyu, Z.D.; Zhao, Z.; Chen, Q.-K.; Mei, H.-L. Experimental and numerical study on welding residual stress of U-rib stiffened plates. *J. Constr. Steel Res.* **2020**, *175*, 106362. [CrossRef]
33. Zhang, Q.H.; Ma, Y.; Cui, C.; Chai, X.Y.; Han, S.H. Experimental investigation and numerical simulation on welding residual stress of innovative double-side welded rib-to-deck joints of orthotropic steel decks. *J. Constr. Steel Res.* **2021**, *179*, 106544. [CrossRef]

buildings

MDPI

Article

Robustness of a Steel Truss Bridge Subjected to Sudden Member Breakage during the Continuous-to-Simply-Supported Transition

Yanming Chen [1], Yong Liao [2], Liming Zhu [1,*], Lingkun Chen [3] and Yilian Chen [1]

[1] College of Transportation Engineering, Nanjing Tech University, Nanjing 214000, China; ymchen@njtech.edu.cn (Yanming Chen)
[2] China Railway No. 2 Engineering Group Co., Ltd., Chengdu 610031, China
[3] College of Architecture Science and Engineering, Yangzhou University, Yangzhou 225000, China
* Correspondence: zhulm@njtech.edu.cn

Abstract: Steel truss bridges are especially vulnerable in the event of a sudden loss of a load-carrying element, which can trigger a chain of failures. This paper describes a unique case study of a steel truss bridge under construction subjected to sudden member breakages with an extensive monitoring system. The failures occurred during the dismantlement of temporary members that had been used to transform a three-span simply supported steel truss bridge into a three-span continuous structure during incremental launching. These temporary members needed to be removed once the bridge reached its final position. The robustness of the bridge was assessed using computer simulations of various failure scenarios to evaluate its capacity to effectively activate alternative load paths (ALPs). The results demonstrated the structural redundancy of the steel truss bridge. However, the dynamic response resulting from the failure of the temporary upper chord, due to the initially high tension in the rods, should not be overlooked. To mitigate this issue, a structural retrofitting method was proposed, involving jacking the truss girder above the side pier to reduce the tension in the temporary upper chord above the middle pier. The effectiveness of this method was demonstrated through both simulated and formal experimental tests.

Keywords: failure; robustness; robustness-enhanced method; steel truss bridge; sudden member breakage

Citation: Chen, Y.; Liao, Y.; Zhu, L.; Chen, L.; Chen, Y. Robustness of a Steel Truss Bridge Subjected to Sudden Member Breakage during the Continuous-to-Simply-Supported Transition. *Buildings* **2024**, *14*, 3035. https://doi.org/10.3390/buildings14103035

Academic Editor: Oldrich Sucharda

Received: 8 August 2024
Revised: 10 September 2024
Accepted: 14 September 2024
Published: 24 September 2024

1. Introduction

Steel truss bridges became widely used structures due to their simplicity and load-carrying efficiency. However, steel truss bridges are especially vulnerable in the event of a sudden loss of a load-carrying element, in which the bridge immediately experiences an alteration of its load paths, developing progressive collapse [1–3]. For example, the Skagit River bridge (2013) collapsed due to a truck impact and drag of one sway frame member [4]; the Chauras bridge (2012) collapsed after buckling of top chord compression members [5]; the I-35 W Mississippi truss bridge (2007) collapsed due to the fracture of undersized gusset plates; the Seongsu bridge (1994) collapsed due to an undetected crack propagation and the subsequent failure of a key connection; and the Quebec bridge (1907) collapsed due to the failure of one of its chords [6]. A study conducted in the US [7] identified more than 500 steel truss-type bridge failures in a period of 11 years (1989–2000). Several causes for these failures exist, including fatigue, overloading, corrosion, and collisions [8,9].

An initial local failure in a structural component of a steel truss bridge can commonly lead to a rapid chain reaction of subsequent failures within a short period [10,11]. The effects of initial local failure have raised the interest of many researchers, and collapse databases have been created, providing valuable insights for preventing such failures [12–14]. Russell reviewed robustness considerations in Eurocodes and other international codes [15]. Adam conducted a comprehensive review of the major advancements made in the 21st century with regards to progressive collapse [16]. Santiago López collected information from 25 case studies to build a novel database of detailed information on the initial damage and its propagation, as well as the consequences of the collapse. The results allowed the identification of the most frequent initial constituted damage states or failures leading to progressive collapse [3]. Patricia Vanova assessed the dynamic responses of a bridge under random excitation, in which damage scenarios with various crack locations and severities were considered for a member [17]. Several scaled models were built to study structural robustness and more realistic structural responses [18]. Padil et al. built a scaled model of a steel truss bridge in a laboratory, where they simulated three damage scenarios by cutting specific members using several cuts [19]. Teng et al. studied both a scaled laboratory model and a real steel truss bridge by testing the acceleration signals of the physical bridge considering five structural scenarios [20]. Buitrago and Bertolesi examined a full-scale 21-metre bridge span, which was tested under laboratory conditions to provide practical recommendations for the early detection of local failures that could potentially lead to a progressive collapse [21,22]. Caredda examined the structure's capacity to activate alternative load paths, and then noted that continuous structural monitoring is also recommended with the optimal arrangement of sensors to be able to predict in time the occurrence of local failures and prevent the complete failure of the entire structure [23].

This paper describes the research team's unique opportunity to study a full-scale steel truss bridge with the aims of (1) simulating and experimentally analyzing its robustness, (2) evaluating the performance of an unconventional continuous-to-simply-supported incremental launching method, and (3) testing the effectiveness of the proposed structural retrofitting method for robustness enhancement. This study is novel and permitted an advance in two areas, namely, (i) an innovative incremental launching method for a steel truss bridge, and (ii) the analysis and experimental testing of the robustness in a steel truss bridge retrofitted with a robustness-enhanced method subjected to sudden member breakages.

To comply with these aims, after the Introduction, Section 2 presents a description of the real case used to carry out the continuous-to-simply-supported incremental launching method. Section 3 describes the computational simulation together with a description of the finite element (FE) model, the loads considered, and the verification process. Section 4 contains the main results of the robustness tests of the steel truss bridge subjected to sudden temporary member breakages. Section 5 contains a discussion of the structural retrofitting method for robustness enhancement obtained with simulated and formal experimental testing. Finally, Section 6 provides a summary of the conclusions.

2. Unconventional Incremental Launching Method

The incremental launching method has the characteristics of large span capacity, high safety, and strong applicability, and it is increasingly used in long-span bridges, especially steel bridges [24]. There are many application cases in long-span simply supported steel truss bridges [25,26]. The Yuyao River Bridge begins at the south of Zhuangqiao Station and the north of Lijiang West Road, crosses the Yuyao River, and ends on the south bank. The main bridge features a three-span simply supported steel truss girder with spans of 80 m + 128 m + 80 m across the Yuyao River. The panel length of the 128 m span is 16.0 m, and the truss height is 16.0 m. The panel length of the 80 m spans is 10.0 m, and the truss height is 11.6 m. The center-to-center distance between the two main trusses is 11.8 m. The main girder structure is constructed from Q370q and Q345q steel. Due to the limitations of the construction site conditions, including navigation and flood safety constraints, temporary piers cannot be added between the main piers. Consequently, conventional construction methods are inadequate. This project adopts an unconventional continuous-to-simply-supported incremental launching method for the first time. Temporary members are used to connect the three-span simply supported steel truss bridge, transforming it into a three-span continuous structure during incremental launching construction. These temporary members include temporary upper chords, horizontal connections, and lower chords. Temporary members are installed between the 128 m and 80 m spans on both the north and south sides. Temporary upper chords and horizontal connections are assembled with adjacent node plates, while the temporary lower chords consist of two additional webs and one top plate added to the interior of the node plate at the end of the 128 m span, which are butt-welded to the 80 m span node plate on-site. After the bridge has been pushed to its design position, the temporary members are dismantled. The schematic diagram of the steel truss bridge is shown in Figure 1.

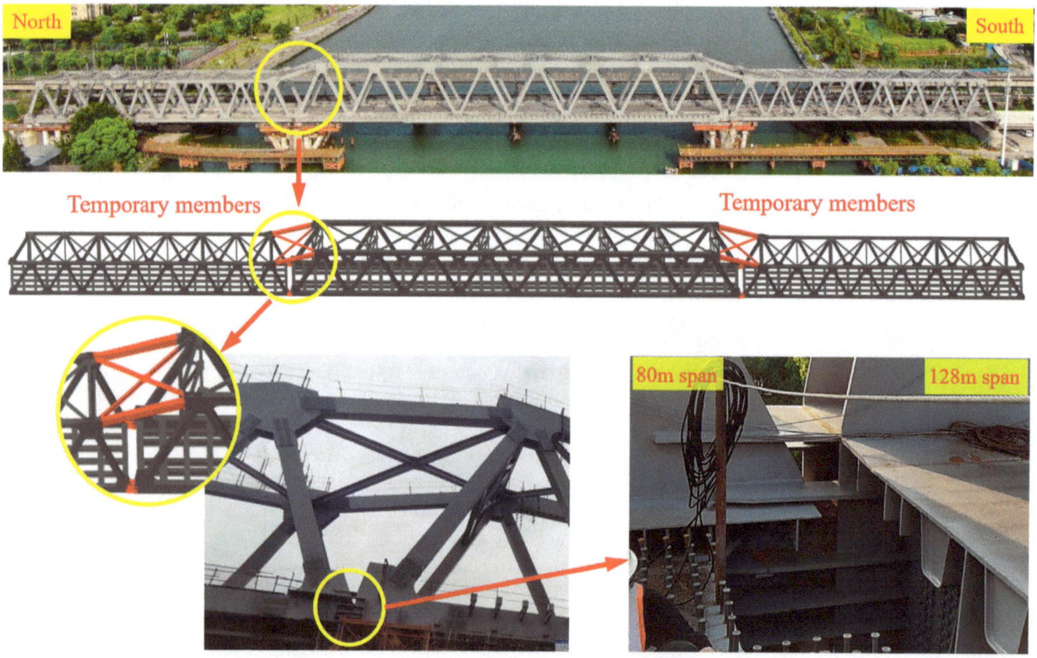

Figure 1. Schematic diagram of the steel truss bridge.

The incremental launching construction phase involves several key steps: assembling the steel truss girders in sections between temporary piers 10# and 12#, and using a cantilever assembly method for the entire process. The three-span simply supported steel truss girders are initially assembled into a three-span continuous configuration with the aid of temporary members. Subsequently, the steel truss girders are pushed forward over a significant distance. Incremental launching equipment is installed on the slipway girders of piers 11#, 12#, 13#, 14#, and 15#. Once the bridge reaches its design position, the temporary members are removed, converting the structure from a three-span continuous steel truss bridge to a three-span simply supported steel truss bridge. The detailed construction phases are outlined in Table 1.

Table 1. Continuous-to-simply-supported incremental launching construction phase.

Construction phase 1: assembling 80 m span steel truss girder

Table 1. *Cont.*

Construction phase 2: achieve pier 13# 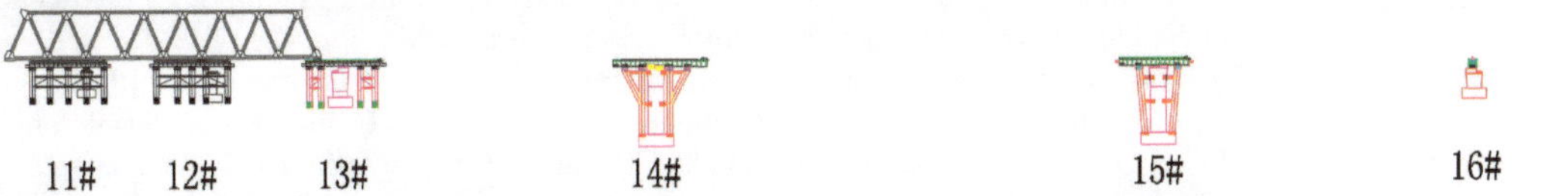
Construction phase 3: push forward 80 m; the 4.5 span of 128 m steel truss girder assembly was completed 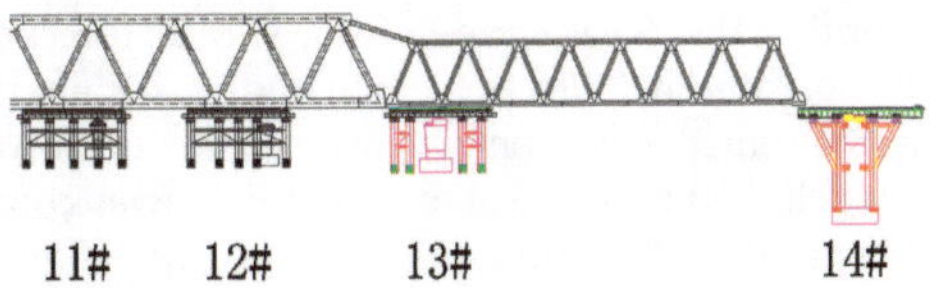
Construction phase 4: achieve pier 14# 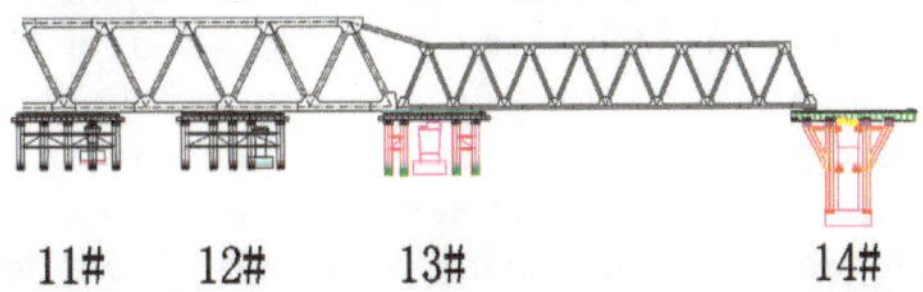
Construction phase 5: push forward 128 m; the whole structure of steel truss bridge assembly was completed 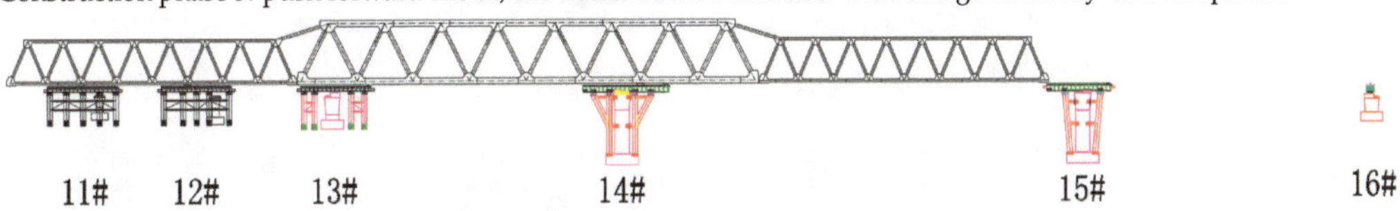
Construction phase 6: achieve pier 15# 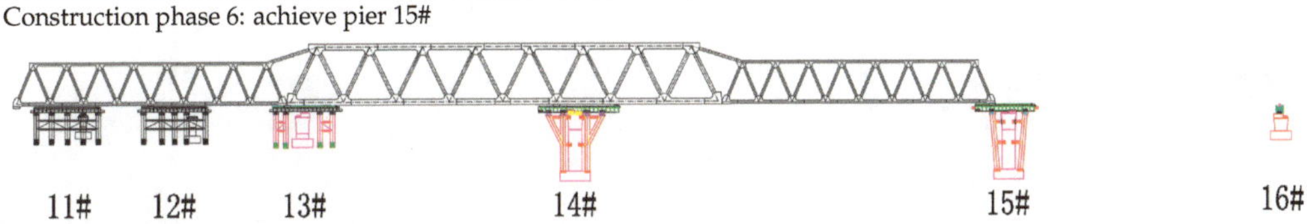
Construction phase 7: push forward 80 m, approach pier 16# 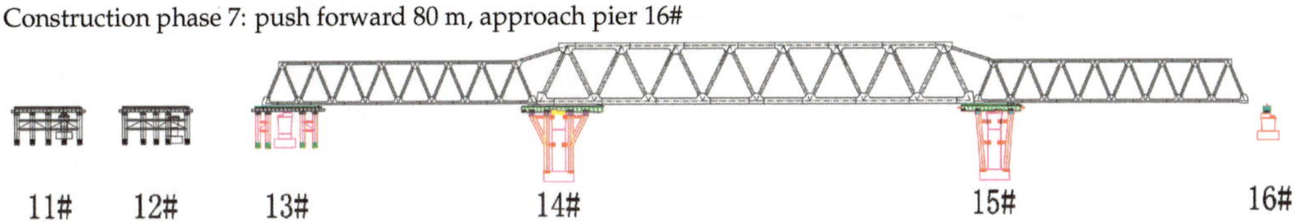
Construction phase 8: achieve pier 16#, release the vertical jacks, dismantle temporary members, and the system transformation is completed 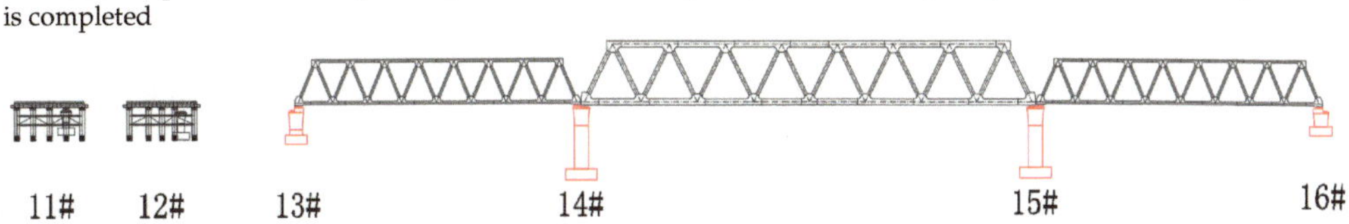

3. Model and Verification

3.1. FEA Model

To analyze the structural response during the sudden failure of temporary members, a detailed finite element model of the entire bridge was created using ABAQUS 2022 software. This model employs shell elements (S4R) to represent all members and gusset plates, with a grid size ranging from 100 mm to 200 mm. The steel truss beam has a material density of 7.85 g/cm^3, an elastic modulus of 206 GPa, and a Poisson ratio of 0.3. Contact settings were defined as binding.

The finite element model utilizes a rectangular coordinate system where the X-axis represents the horizontal direction of the bridge, the Y-axis represents the vertical direction of the bridge, and the Z-axis represents the longitudinal direction of the bridge. Within this model, a reference point is established at the center of the contact area between the lower chord and the slider. This reference point is linked to the local region of the lower chord via "coupling" constraints. The reference point restricts translation in the Y-direction to simulate the slider support. Additionally, another reference point is positioned accordingly to constrain the displacement of the steel truss girder in both the transverse and longitudinal directions; this reference point restricts translation in the X- and Z-directions and rotation in the Y- and Z-directions. The construction stage loads primarily include the self-weight of the steel truss girders and the loads imposed by construction machinery. Since work on the girders is prohibited during the incremental launching construction process, live loads are not considered. The overall model utilizing refined shell elements is illustrated in Figure 2.

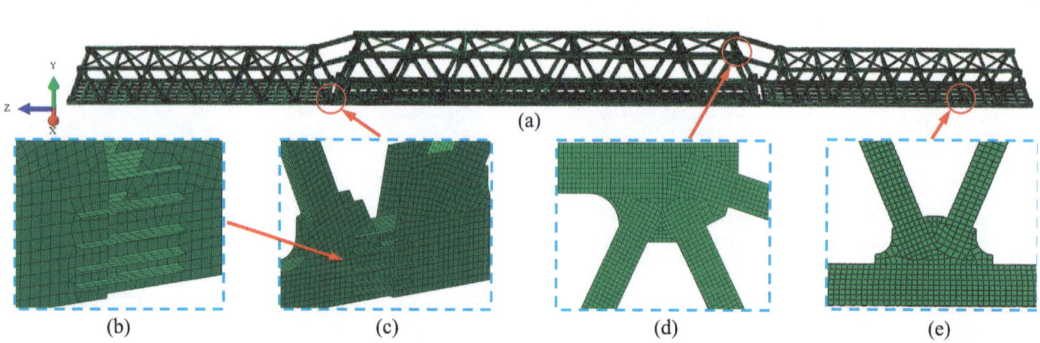

Figure 2. FEA model schematic diagram of steel truss bridge: (**a**) overall; (**b**) enlarged view of temporary lower chord; (**c**) temporary lower chord; (**d**) upper chord; (**e**) lower chord.

The quasi-dynamic simulation method, full-dynamic simulation method, and semi-dynamic simulation method are commonly employed to model the fracture of bars [27]. The accuracy of results across these three methods is generally comparable. In comparison to the quasi-dynamic and full-dynamic simulation methods, the semi-dynamic simulation method offers advantages such as reduced computational time and simpler implementation. Consequently, this paper employs the semi-dynamic simulation method to analyze the dynamic response of the entire structure after the temporary members have been severed. ABAQUS features the "life-and-death unit" technology, which allows for the simulation of member failure within a predefined time frame. The semi-dynamic simulation analysis is conducted as follows: First, the static internal forces of the members prior to fracture are determined through static analysis of the model. Subsequently, the members are severed using the "life-and-death unit", and a transient dynamic analysis is performed to assess the dynamic response of the structure to the sudden fracture of the members.

3.2. Measurement Arrangements

To ensure the safety of the steel truss bridge during incremental launching and system transformation construction, two types of measurements are collected during testing:

strains and displacements at key points previously identified from theoretical calculations. Strains are measured during the launching phase, converted to stresses, and compared with results obtained from finite element analysis.

A total of 19 measurement sections are selected for the upper chord, lower chord, web, and temporary members of the steel truss bridge. Measurement points are arranged for all sections on the east and west sides of the truss, with one section of each chord having measurement points on both the upper and lower edges. In total, 76 stress gauges are deployed, as shown in Figure 3. Vertical displacement at the front end of the main beam during the incremental launching process is also a crucial control index. The cantilever end of the bridge experiences vertical deformation due to the self-weight of the structure; therefore, a deformation monitoring point, specifically a prism, is established at the front end of the main girder and observed using a total station. During assembly, the elevation of the main girder before pushing is recorded as the initial value, with the measurement reference point set at the contact point between the front fulcrum pad and the bottom of the girder. The actual vertical deflection of the main girder is calculated as the difference between the bottom elevation of the beam at the front fulcrum and the cantilever end.

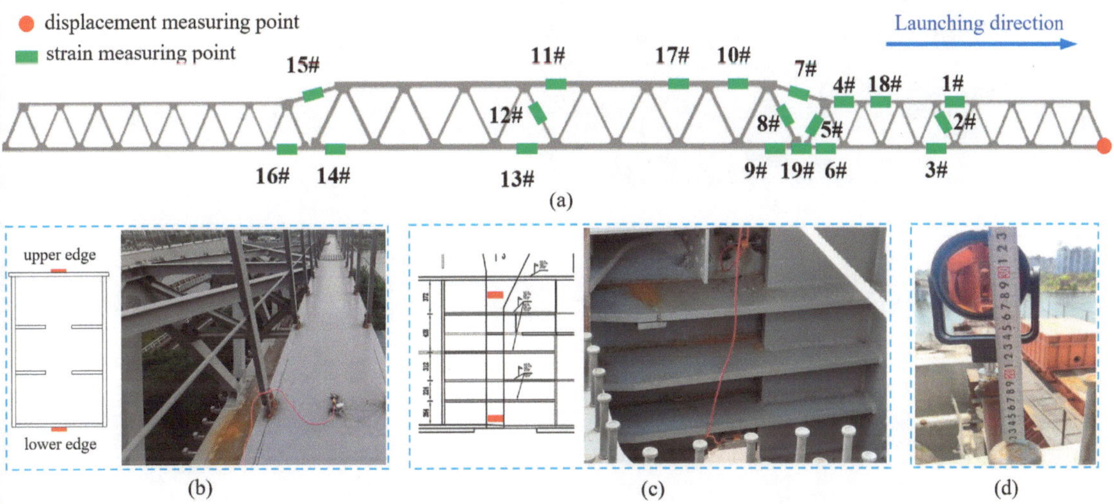

Figure 3. Layout of measuring points: (**a**) overall; (**b**) upper chord; (**c**) temporary lower chord; (**d**) displacement.

3.3. Verification

The comparison between the measured and calculated stress of the temporary upper chord (section 7#) is shown in Figure 4b, while the comparison between the measured and calculated vertical displacement at the front end is presented in Figure 4c. In these figures, positive values indicate uplift, and negative values represent downward deflection. Throughout the entire launching construction, the trends of the measured and calculated values are generally consistent. Just before the 15# pier, the main girder is in its maximum cantilever state, with the deflection of the front end reaching 922.70 mm, while the corresponding calculated value is 795.22 mm, resulting in a difference of 13.8%. The measured stresses at the upper and lower edges of the temporary upper chord are 101.4 MPa and 95.1 MPa, respectively, compared to calculated values of 96.5 MPa and 85.7 MPa, leading to differences of 6.2% and 11.2%. The differences between the measured and calculated stress values are mainly due to variations in the height of the bearing pads placed on the piers during the launching construction. Specifically, when the bridge front end reaches the pier, multiple steel bearing pads are stacked on top of the pier as temporary supports for the steel truss bridge. However, during the actual construction process, due to factors such as uneven contact surfaces between the bearing pads and minor height differences in each pad, the total height of the steel bearing pads placed on each pier is not exactly uniform. In contrast, in the model calculations, the bearing pad height was considered to be consistent leading to the observed differences between the measured and calculated stress values. Temporary supporting devices of the same specification can be used in subsequent similar cases, which can reduce the influence of this factor. Overall, the measured results verify the effectiveness of the ABAQUS model, which can be used as a reliable basis for the next research work.

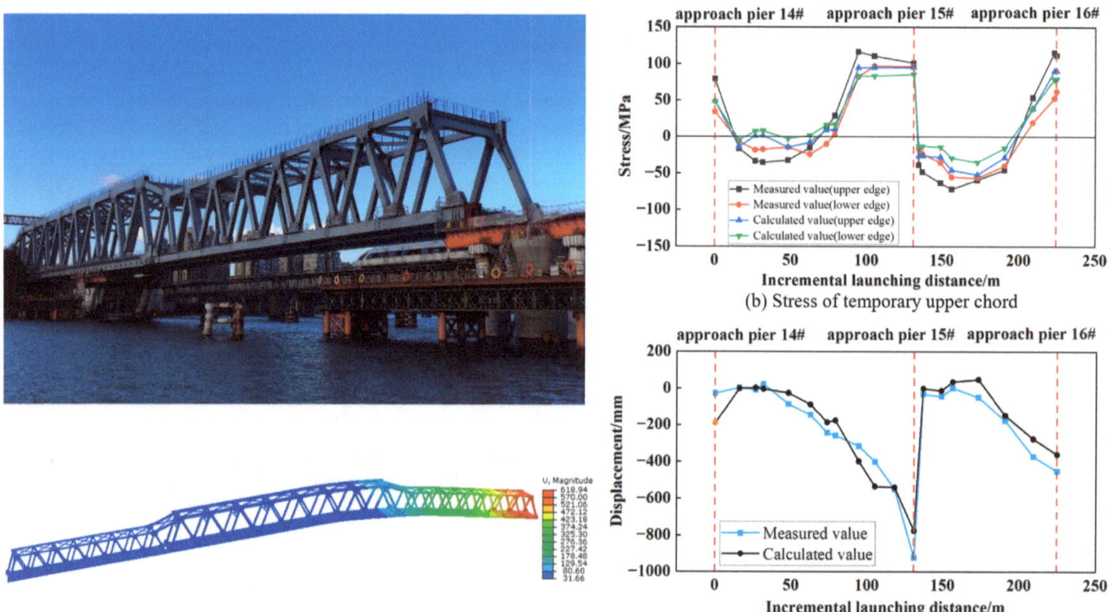

(a) Deformation schematic diagram of FEM

(c) Deformation of the main girder front end

Figure 4. Calculated and measured results during incremental launching construction.

4. Robustness

This section examines failure scenarios of temporary members based on field engineering conditions. The robustness of the structure was evaluated through computer simulations of various damage scenarios to assess the bridge's ability to effectively activate alternative load paths. Additionally, dynamic response analyses were conducted for the specific failure scenarios identified.

4.1. Failure Scenarios

Unlike some studies where a designed sequence of structural damage is necessary to analyze behavior with the evolution of deliberately induced damage [21], the case described in this paper is constrained by construction methods and engineering principles. The breakage sequence must ensure the clarity of the bridge's structural system. Specifically, all temporary members on either the south or north side must be dismantled first, converting the three-span continuous bridge into a two-span continuous bridge and a single 80 m span simply supported bridge. Subsequently, all temporary members on the remaining side are removed to transform it into a three-span simply supported bridge. Figure 5 illustrates the failure scenarios associated with different dismantling steps. On one side, the temporary members fail in the sequence of horizontal connection, one upper chord, another upper chord, and finally the lower chord, followed by the same sequence of failure on the other side.

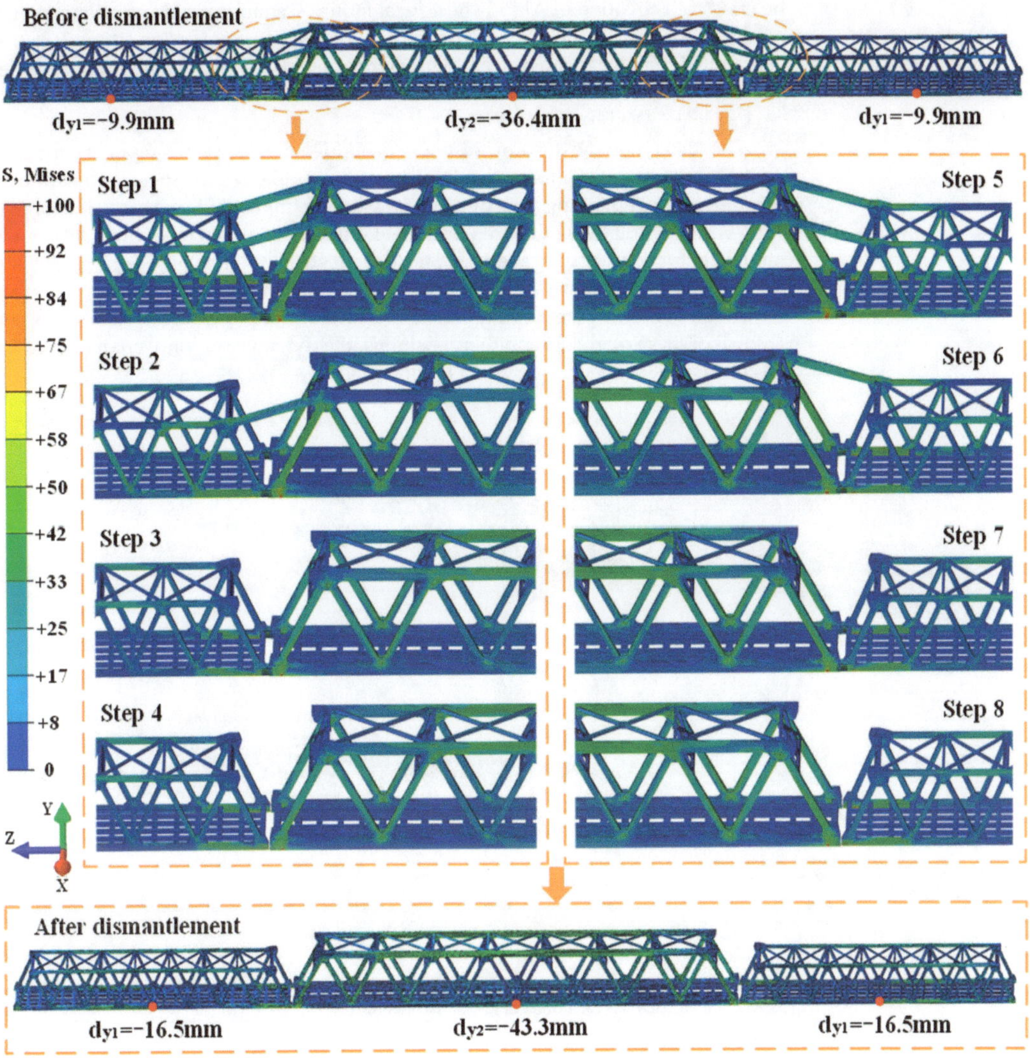

Figure 5. Von Mises stress for different failure scenarios. Units: Mpa.

4.2. Alternative Load Paths

The impact of dismantling temporary members on the overall structure is complex, as evidenced by the sudden and uncontrollable changes in stress redistribution [28,29]. Figure 5 illustrates the stress contour maps from FE analysis under various failure scenarios. The failure of temporary horizontal connections and lower chords has a minimal effect. In contrast, failures of temporary upper chords significantly affect the nearby upper chords and members in the midspan. Specifically, stresses in the upper chords near the failure site increased by up to 270%, rising from 5.61 MPa to 20.73 MPa. Stresses in the upper chords at the midspan increased by up to 43%, changing from 27.08 MPa to 38.65 MPa. Despite these stress increases, the structural stress remains within a safe range during the construction stage, as there are no vehicle loads. The diagram also indicates that the maximum deflection increased by up to 19.0% in the middle span and 66.7% in the side span.

Given that the final failure scenario, where the bridge becomes a three-span simply supported structure, was pre-designed and calculated, the structure proves reliable due to the effective activation of ALPs. The critical failure scenarios requiring attention are those involving the dismantling of upper chords under significant tension (steps 2, 3, 4, and 6), where the dynamic response to sudden member breakage must be carefully considered.

4.3. Dynamic Response

The focus is on the dynamic response of sudden breakages of temporary upper chords. The failure of these members within a short time frame results in a significant dynamic response in the remaining structure. To analyze the impact of temporary members on the entire structure following their removal, the failure times are set to 1 s (model 1#) and 0.01 s (model 2#) for comparative analysis. Subsequent structural response analysis steps are set to 5 s. The four temporary upper chords are designated as ①, ②, ③, and ④, corresponding to the order of their dismantling. Based on the static stress changes before and after the removal of these members and their positions, the dynamic response analysis locations for the chords are selected as shown in Figure 6.

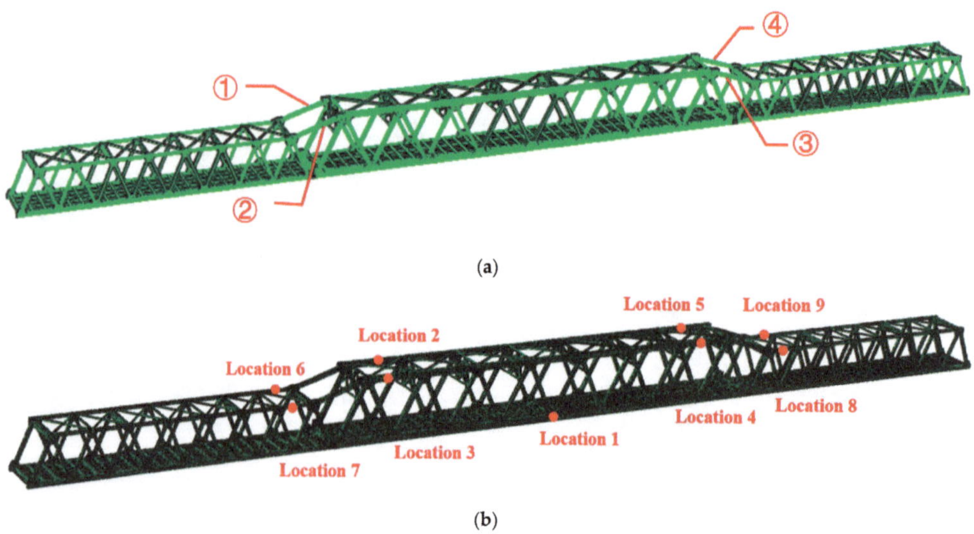

Figure 6. Schematic diagram of dynamic response analysis. (**a**) Diagram of dismantlement sequence. (**b**) Node location diagram of steel truss bridge chord.

Figure 7 presents the dynamic responses at position 1 for the two failure times, which are compared and analyzed. The initial stress value at the joint located at the mid-span position of the lower chord is 28.3 MPa for both failure times (0.01 s and 1 s). After the dismantling of rod ①, the peak stress response in Model 1# is 38.3 MPa, while in Model 2# it is 32.1 MPa, eventually stabilizing at 31.1 MPa. Following the removal of rod ②, the peak stress response in Model 1# is 41.6 MPa, compared to 37.2 MPa in Model 2#, with a stable stress value of 35.2 MPa. After the dismantling of rod ③, the peak stress response reaches 47.9 MPa in Model 1# and 43.3 MPa in Model 2#, eventually stabilizing at 41.8 MPa. After rod ④ is removed, the peak stress response is 55.1 MPa in Model 1# and 47.8 MPa in Model 2#, stabilizing at 44.7 MPa. The calculation results indicate that the stress peak at the joint is reached within 5 s of dismantling a temporary member, with stress fluctuations gradually decreasing and stabilizing, while the stable stress value increases progressively. For rods ①, ②, and ④, the nodes reached peak stress within 0.5 s, whereas for rod ③, peak stress appeared more slowly, approximately 3 s after dismantling. The model with a failure time of 0.01 s exhibits smaller differences in peak stress fluctuations compared to the 1 s failure time, with the smallest difference (10.2%) occurring after dismantling rod ③ and the largest (19.3%) after dismantling rod ①. This suggests that shorter failure times of rods lead to more significant dynamic responses in the structure. To further assess the adverse effects of sudden member breakages on the structure, the results from the dynamic model with a 0.01 s failure time are analyzed below.

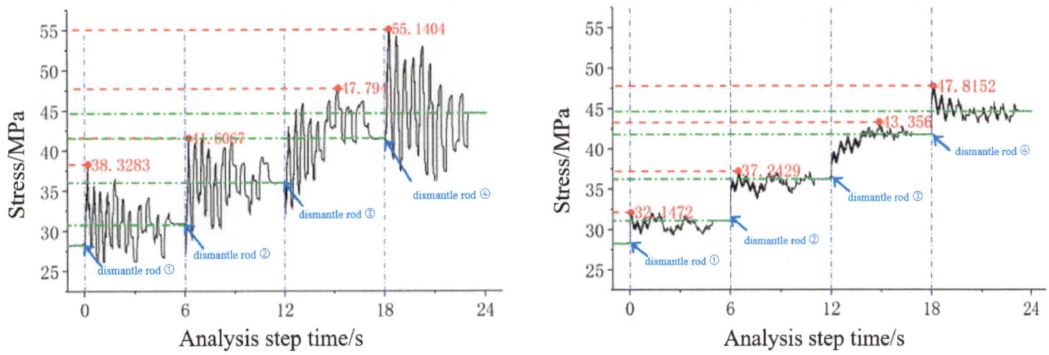

(**a**) failure time 0.01 s (**b**) failure time 1 s

Figure 7. Comparison of stress time-history at location 1.

Figure 8 illustrates the stress time-history response of nodes positioned along the chord of the main truss during the temporary member dismantlement process. For the 128 m span, nodes at positions 2, 3, 4, and 5 of the upper chords adjacent to the removed member were selected for analysis. The figure reveals that the stress at these nodes increases from approximately 6.5 MPa to around 15 MPa following the dismantlement of the temporary member. This increase in stress is primarily observed after the failure of the adjacent temporary member. The dynamic response of the upper chord nodes to the failure of the temporary member is significant. The timing and location of the stress peak are influenced by the distance of the dismantled rod. When the node and the dismantled rod are on the same side of the truss, the stress peak occurs within 0.25 s after the rod failure. Conversely, if the dismantled rod is on the opposite side of the truss, the stress peak appears significantly later compared to the same side. For instance, when rod ② is dismantled, the stress at node position 4 peaks after 3 s. The maximum stress peak observed throughout the process occurs at position 4 following the dismantlement of rod ③. Additionally, it is evident from the figure that, after the removal of rods ① and ③, the stress at positions 3 and 5, which are not adjacent to the upper chord on the same side of the truss, exhibits considerable fluctuation without a stable trend over a short period.

For the 80 m span, nodes at positions 6, 7, 8, and 9 of the upper chords adjacent to the dismantled rod were selected for analysis. The figure indicates that the stress at these nodes decreased from approximately 50 MPa to about 20 MPa following the temporary member's failure, with the stress reduction primarily occurring due to the failure of the adjacent temporary member. The dynamic response of the chord nodes in the 80 m span to the dismantlement of the temporary member is notably pronounced, and the stress fluctuation amplitude for some nodes is greater than that observed in the 128 m span. After the dismantlement of rod ①, position 7 exhibits the largest stress fluctuation, with an amplitude reaching 60 MPa. Following the removal of rods ① and ②, the stress at positions 6 and 7 is completely released, and further dismantlement of additional rods has minimal impact. The stress response at positions 8 and 9 is more complex, with prolonged periods of significant fluctuation. Throughout the process, the maximum stress peak occurs at position 9 after the removal of rod ③. When the node and the dismantled rod are on the same side of the truss, the stress peak is reached within 0.3 s; otherwise, the peak appears noticeably later. The pattern of peak occurrence is generally consistent with that observed in the 128 m span chord.

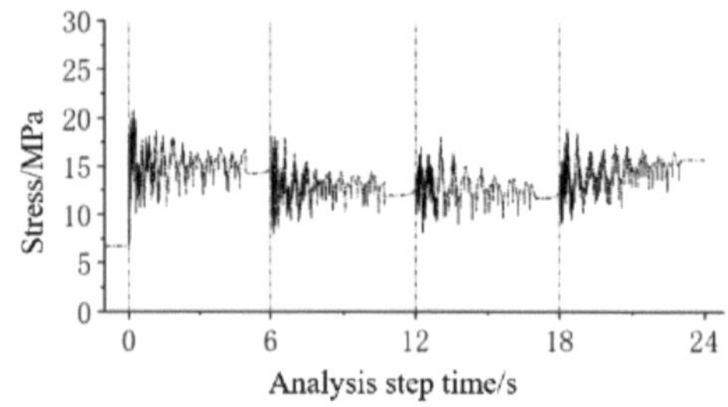

(**a**) stress at position 2

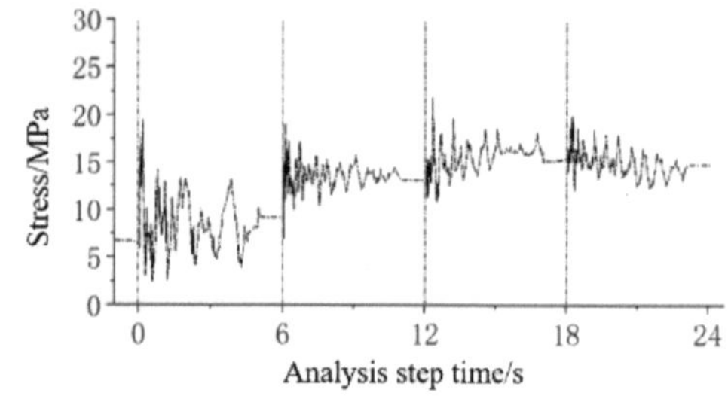

(**b**) stress at position 3

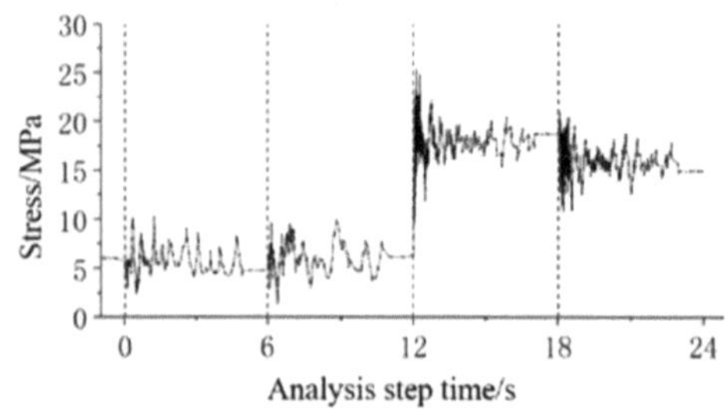

(**c**) stress at position 4

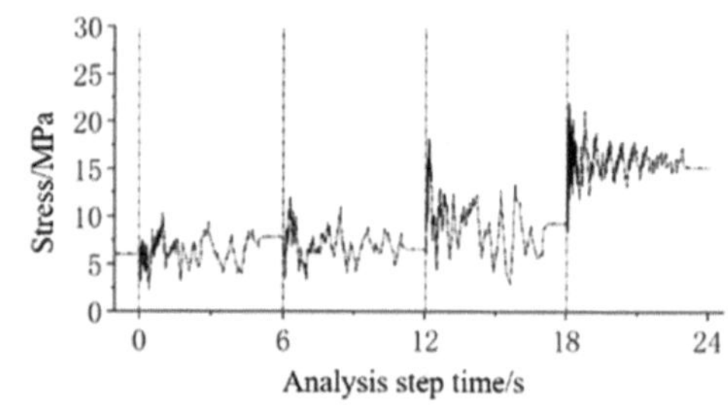

(**d**) stress at position 5

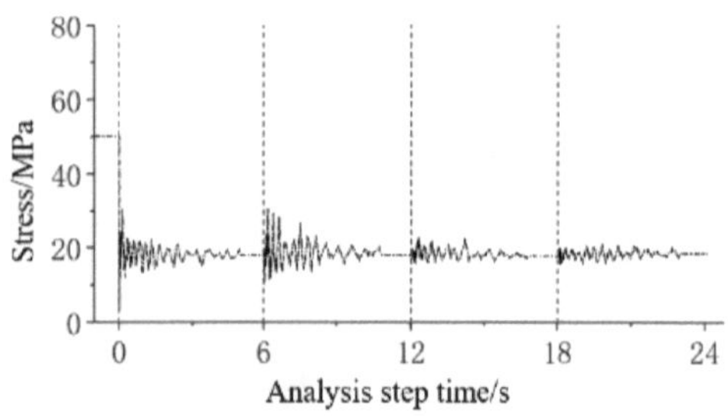

(**e**) stress at position 6

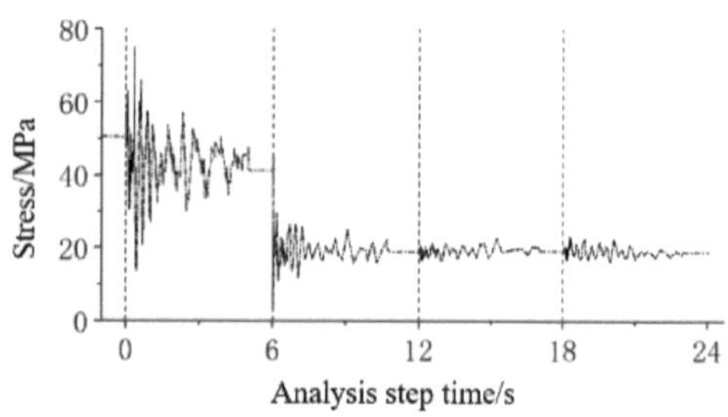

(**f**) stress at position 7

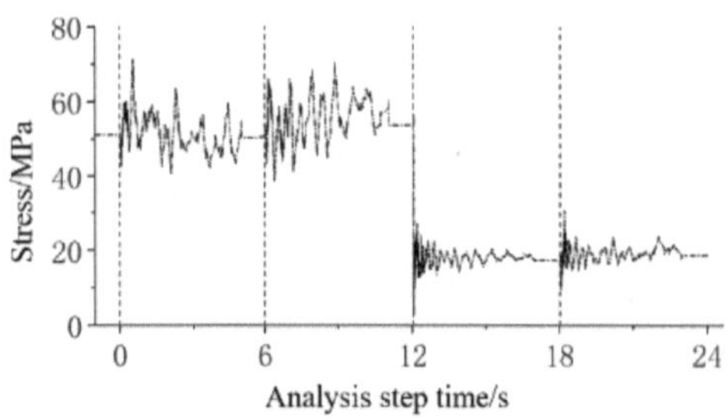

(**g**) stress at position 8

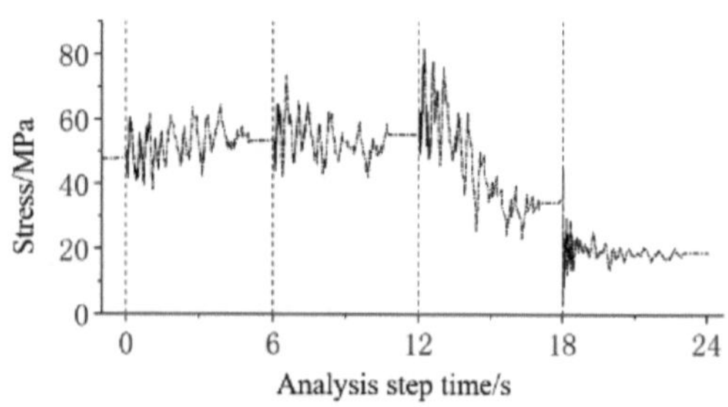

(**h**) stress at position 9

Figure 8. Stress time-history response of the upper chord on the steel truss bridge.

Figure 9 presents the stress time-history response of nodes at various positions within the main truss web during temporary member failure. The dynamic responses of four web position nodes for the 128 m span, adjacent to the temporary upper chord, are compared. Prior to the removal of the temporary members, the stress at these web position nodes is 42 MPa. This stress is alleviated by the failure of the adjacent temporary members, reducing to 29 MPa after all temporary members have been dismantled. The peak stress at the web position adjacent to the dismantled rod is observed to develop more rapidly, with the maximum stress peak occurring at position 4 after the dismantling of rod ③, measuring 59.65 MPa.

Similarly, the dynamic responses of four web position nodes adjacent to the temporary upper chord of the 80 m span are analyzed. Initially, the stress at these web position nodes is 15 MPa before the failure of the temporary members and increases to 21.5 MPa afterward. The dynamic response of the 80 m web is more pronounced compared to the 128 m web, with the greatest impact effect observed at the web adjacent to the dismantled rod, exhibiting a maximum fluctuation amplitude of approximately 30 MPa. Additionally, the peak time of the web located on the side opposite to the dismantled rod occurs significantly later than that of the web on the same side. The stress fluctuation amplitude for the web on the opposite side is within 10 MPa, indicating less pronounced stress changes.

(**1**) Node location diagram of steel truss bridge web rod

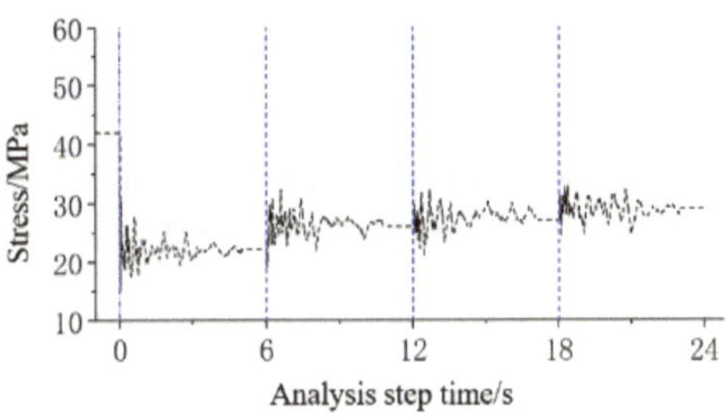

(**a**) stress at position 1

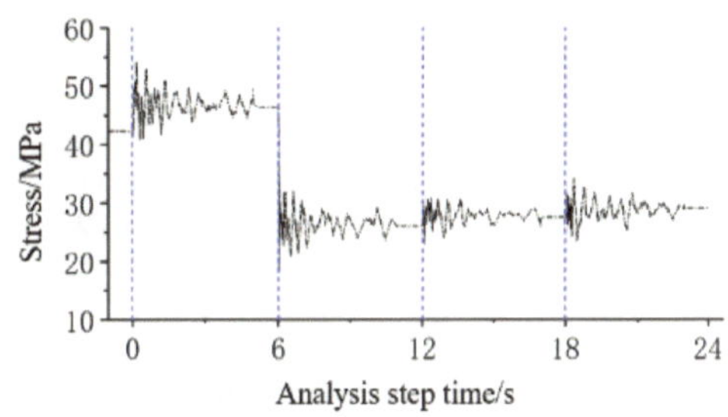

(**b**) stress at position 2

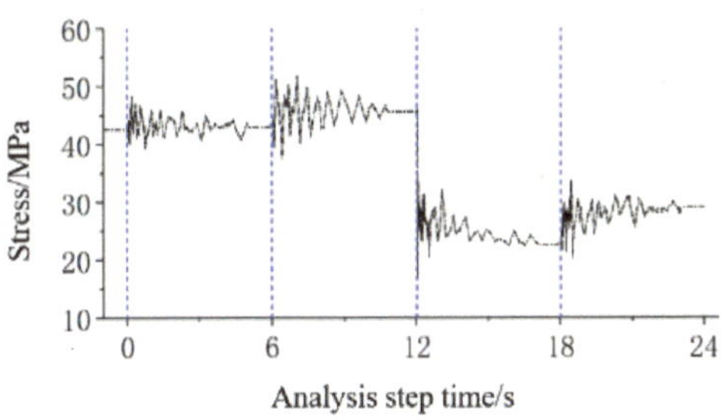

(**c**) stress at position 3

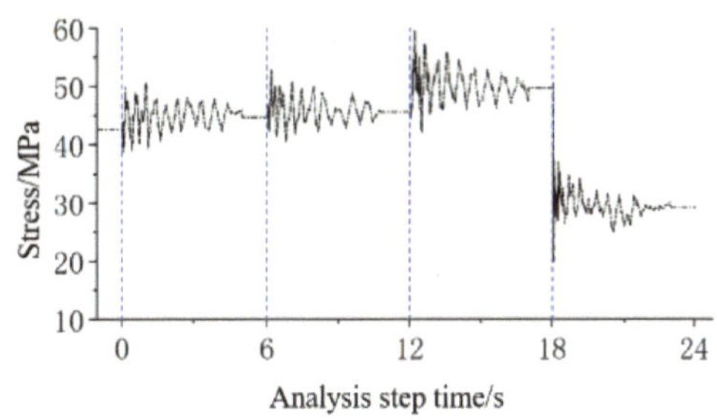

(**d**) stress at position 4

Figure 9. *Cont.*

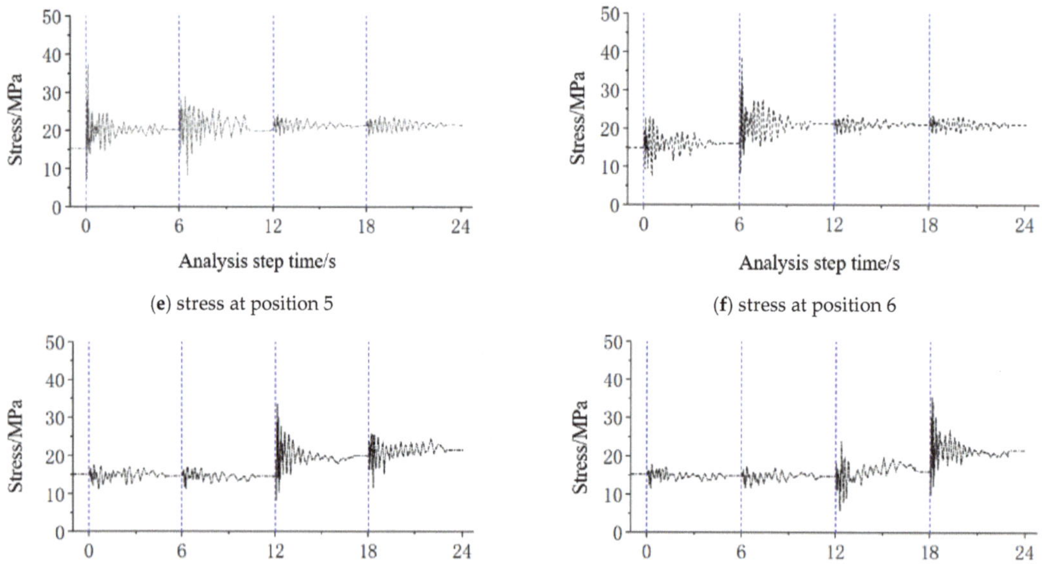

(**e**) stress at position 5

(**f**) stress at position 6

(**g**) stress at position 7

(**h**) stress at position 8

Figure 9. Stress time-history response of the web rod on the steel truss bridge.

The main structural components are made of Q370 steel, which has an allowable stress value of 220 MPa. During the construction process, this allowable stress can be increased by 1.2 times under dead load and construction load conditions, resulting in a maximum allowable stress of 264 MPa. Figure 10 displays the maximum stress values observed in the structure following the dismantling of rods ①, ②, ③, and ④. According to the calculation results, the maximum stress values after dismantling each rod are as follows: 232 MPa for rod ①, 217.6 MPa for rod ②, 237.1 MPa for rod ③, and 200.1 MPa for rod ④. These values, while all below the allowable stress limit of 264 MPa, are close to this threshold. The maximum stress peaks occur within 1 s after the failure of the temporary members. Notably, the dynamic stress generated by the removal of rods ① and ③ is greater than that resulting from the removal of rods ② and ④.

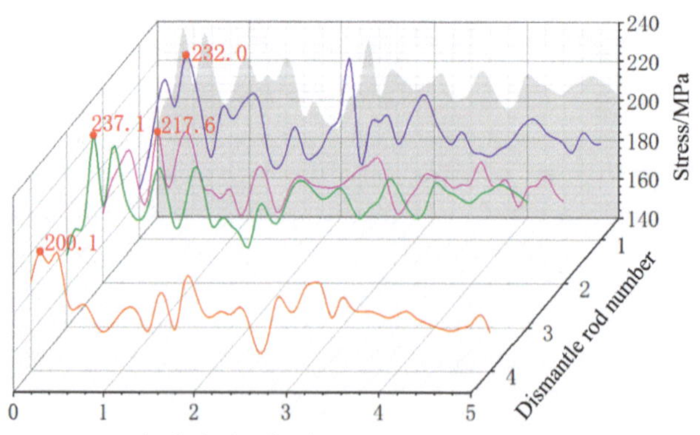

Figure 10. Maximum stress time-history diagram of the steel truss bridge structure.

5. Robustness-Enhanced Method and Tests

Given that the maximum stress response of the structure is close to the allowable value, this section presents a structural retrofitting method for robustness enhancement. A relevant simulation test was conducted to evaluate the effectiveness of this robustness-enhancement method, which was subsequently applied in failure scenarios equipped with a comprehensive monitoring system.

5.1. Structural Retrofitting Method for Robustness Enhancement

The study of the structural retrofitting method for robustness enhancement can be a challenging task due to the particularity of different configuration of the structure, nature of the triggering event and size of the initial failure [10]. The strengthening and retrofitting techniques can be categorized on the three criteria definitions: (i) preventing initial failure, (ii) controlling collapse propagation, and (iii) controlling the final collapse status [30]. The majority of strengthening techniques focus on redistribution-type collapses, in which alternate load paths (ALPs) have a determining role. Therefore, measures that try to increase the structural robustness by adding new load paths (or strengthening the existing ones), namely, improving catenary behavior in beams or cable strengthening of beams, adding trusses, or enhancing connections [31], are usually useful mainly in the case of redistribution-type collapses [32]. For other scenarios, i.e., impact-type collapses, the effectiveness of these techniques is under question and other techniques, namely, energy absorber devices [33], should be utilized instead [30,34].

In the failure scenario of this steel truss bridge, the members' fracture location, i.e., the initial failure, is determined, so preventing the initial failure is more economical and effective. Considering that the dynamic response mainly comes from the initial tension of the broken members, the fluctuation amplitude of which is up to 60 MPa, based on the idea of reducing the initial tension to prevent the initial failure, a structural retrofitting method of jacking a truss girder above the side pier to reduce the tension of the temporary upper chord above the middle pier was proposed. The application of bridge jacking technology in bridge retrofitting has gradually developed into a promising mature practical technology [35]. The internal force distribution of the steel truss can be changed through jacking technology [36], but the specific applied method and effect need to be verified [37–39].

5.2. Simulated Jacking Experimental Test

The relationship between the jacking height and the internal force of the temporary upper chord requires verification. For this purpose, a simulated jacking experiment was conducted under the working condition of the steel truss girder reaching pier 14#. The boundary conditions of the 80 m span in this experiment were maintained consistent with those prior to the system transformation. During the simulated jacking experiment, the front end of the truss girder above pier 14# was jacked, and both the jacking height and the stress in the temporary members above the adjacent pier 13# were monitored. The goal of this simulated jacking experiment was to gather relevant data to support the subsequent dismantlement of temporary members and to assess the feasibility of the robustness-enhancement method.

In the simulated jacking experiment, five working conditions were examined: before lifting, lifting by 46 mm, lifting by 96 mm, lifting by 122.8 mm, and restoring to the original height. The calculation results indicate that the upper chord transitions from tension to compression under the condition of lifting by 122.8 mm. Measurement points were defined as follows: point 7-1 represents the upper edge of the temporary upper chord, point 7-2 represents the lower edge of the temporary upper chord, point 19-1 represents the upper edge of the temporary lower chord, and point 19-2 represents the lower edge of the temporary lower chord.

Figure 11 illustrates the relationship between the jacking height and the stress in the temporary members. The test results indicate that as the jacking height increases, the stress in the temporary upper chord decreases, while the stress in the temporary lower chord gradually increases. Specifically, when the jacking height reaches 122.8 mm, the stress at the upper edge of the upper chord shifts from 9.6 MPa to −32.2 MPa, and the stress at the lower edge of the upper chord changes from 29.2 MPa to 10.3 MPa. After completing the jacking operation and restoring the front end of the steel truss girder to its original height, the internal forces in the members return to levels comparable to the initial state. The monitoring results from the simulated jacking experiment are consistent with the calculation results, confirming that the structural retrofitting method involving jacking effectively reduces the internal force in the upper chord.

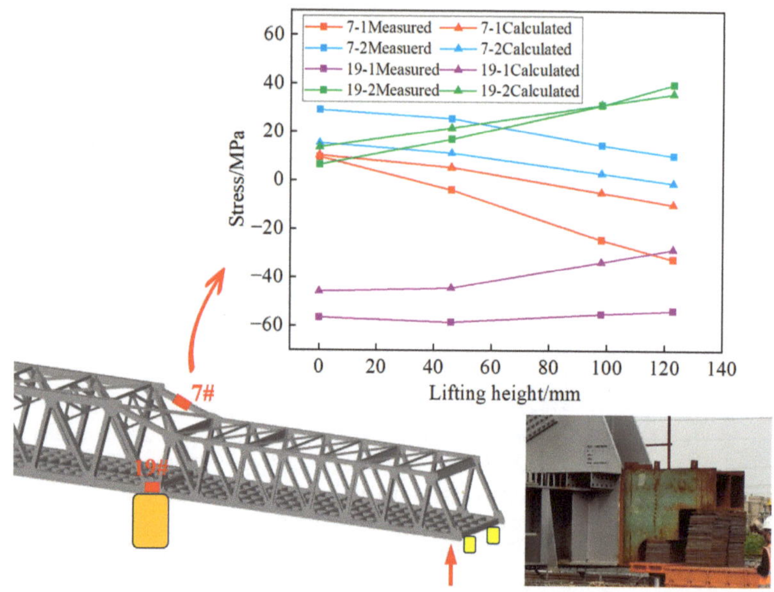

Figure 11. Relationship between jacking height and stress of temporary members.

5.3. Formal Jacking Experimental Test

The effectiveness of the robustness-enhanced method was verified by simulated jacking experiments. The sequence for dismantling the temporary members was determined as follows: First, the truss girder above pier 16# was lifted, and the temporary members above pier 15# (north side) were dismantled. Next, the truss girder above pier 13# was lifted, and the temporary members above pier 14# (south side) were dismantled. The specific steps for dismantling temporary members on one side are outlined below, as shown in Figure 12:

Step 1: Monitor the stress of the temporary members, and then use the 80 m span end lifting jack and pads to adjust the elevation, aiming to bring the temporary members close to a stress-free state;

Step 2: Cut temporary short flat joint welded joints, and then remove high-strength bolts connected with long flat joints;

Step 3: Weld and remove the lifting lugs on the temporary rods, cut the chords in sections avoiding the transverse partition, and remove the temporary upper chords on the west side first by using the crawler crane;

Step 4: Remove the temporary lower chord on the east side;

Step 5: Use flame cutting to remove the temporary connection of the lower chord, and grind the area to restore the original shape of the steel truss beam.

Step 6: Complete the dismantlement work and return the structure to the original height.

Figure 12. Dismantlement process.

In the formal test, the jacking height of the truss girder above pier 16# was set to 46 cm. Stress measurements of the temporary members on both the east and west sides of the truss above pier 15# were recorded. The monitoring results are presented in Table 2. The initial combined stress reflects the actual stress resulting from the combined effects of bending moments, axial forces, and other factors; the axial stress calculation is based on the average stress values at both the upper and lower edges of the member. Following the jacking operation, the internal forces in the temporary upper chords on both the east and west sides shifted from tension to compression. While the stress in the temporary lower chords increased, they remained in a compressed state.

Table 2. Monitoring results of temporary members in the final test. Units: MPa.

Position	Section	Initial Combined Stress	Initial Axial Stress	Combined Stress after Jacking	Axial Stress after Jacking
East	7-1	127.3	88.9	−43.9	−29.8
	7-2	50.5		−15.8	
	19-1	17.2	−53.8	−59.3	−27.6
	19-2	−124.8		4.1	
West	7-1	164.1	101.8	−20.3	−25.5
	7-2	39.4		−30.8	
	19-1	60.4	−48.1	−17.3	−3.2
	19-2	−156.6		10.9	

In the final dismantlement state, the internal forces of temporary upper and lower chords are pressure forces, so there will be no extra dynamic response during dismantlement. Changes in the internal forces of adjacent rods during the dismantlement process should be monitored [36]. Initially, the temporary horizontal connections had been removed. The dismantlement process began with the temporary upper chord on the west side, followed by the temporary upper chord on the east side. Monitoring was conducted on the upper chord (4#), webs (5#), and lower chord (6#) on both the east and west sides adjacent to the dismantled members.

Figure 13 displays the stress monitoring results for the adjacent members during the dismantlement process. Generally, the stress in most adjacent members decreased. Due to the differing dismantlement sequences of the temporary upper chords on the east and west sides, stress changes in adjacent members varied between the two sides at different stages of dismantlement. However, once both temporary upper chords were fully dismantled, the stress change values for the adjacent upper and lower chords on both sides were similar. Specifically, the stress in the upper chords decreased by 36.9 MPa on the east side and 32.9 MPa on the west side, while the stress in the lower chords decreased by 25.9 MPa on the east side and 18.3 MPa on the west side.

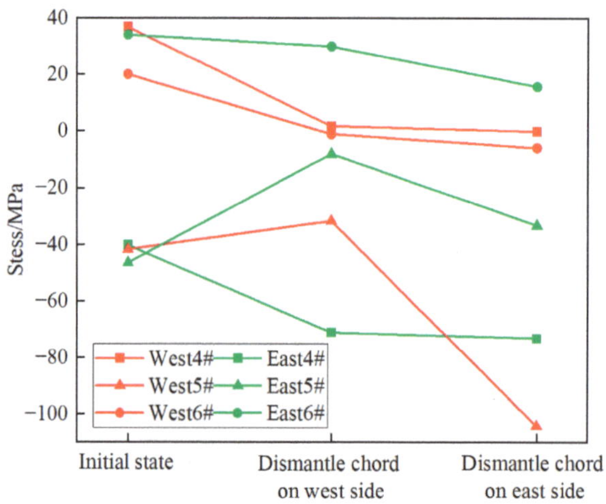

Figure 13. Stress monitoring results of adjacent members during temporary members' dismantlement.

6. Conclusions

This paper presents a unique case study of a steel truss bridge under construction that experienced sudden member failures. These failures occurred due to dismantlement during an unconventional continuous-to-simply-supported incremental launching process, and were monitored extensively using a comprehensive monitoring system. The robustness of the structure was evaluated through computer simulations of various failure scenarios to analyze the bridge's capacity to effectively activate alternative load paths. Additionally, both simulated and formal experimental tests were conducted to verify the effectiveness of the proposed robustness-enhancement method. The study's conclusions are as follows:

1. During the analysis of robustness in temporary member failure scenarios determined by the dismantlement sequence, it was found that
   - The stresses on the upper chords adjacent to the dismantled rod experienced the highest increase of up to 270%.
   - The maximum deflection increased by up to 19.0% at the middle span and 66.7% at the side span.
   - The structural bridge had structural redundancy based on the effective activation of ALPs.
2. The dynamic response primarily stemmed from the initial tension in the temporary upper chords, with fluctuation amplitudes reaching up to 60 MPa. The shorter the failure time of the rods, the more pronounced the dynamic response of the structure.
3. A structural retrofitting method involving jacking the truss girder above the side pier to reduce the tension in the temporary upper chord above the middle pier was

proposed. The effectiveness of this method was demonstrated through both simulated experimental tests and formal experimental tests.

4. The proposed unconventional continuous-to-simply-supported incremental launching method for a steel truss bridge has been proven to be safe and feasible. However, dismantling members with significant initial tension requires careful attention during the system transformation process. The proposed structural retrofitting method, which aims to enhance robustness, effectively reduces the initial tension in members, thereby preventing initial failure. Based on the findings of this study, future research could explore potential improvements and new approaches in the study of steel truss bridge robustness, particularly in preventing initial failure.

Author Contributions: Y.C. (Yanming Chen): Conceptualization, Methodology, Writing—review & editing. Y.L.: Project administration, Resources. L.Z.: Supervision, Resources, Writing—review & editing. L.C.: Supervision, Resources, Validation. Y.C. (Yilian Chen): Data curation, Writing—review & editing. All authors have read and agreed to the published version of the manuscript.

Funding: This research was funded by the Natural Science Foundation of Jiangsu Province of China under Grant No BK20200705 and the Key Technologies Research and Development Program (CN) under Grant No. 2021YFB2600600.

Data Availability Statement: The original contributions presented in the study are included in the article, further inquiries can be directed to the corresponding authors.

Conflicts of Interest: Author Yong Liao was employed by the company China Railway No. 2 Engineering Group Co., Ltd. The remaining authors declare that the research was conducted in the absence of any commercial or financial relationships that could be construed as a potential conflict of interest.

References

1. Chen, X.; Li, H.; Agrawal, A.K.; Ettouney, M.; Wang, H. Alternate load paths redundancy analysis of steel truss bridges. *J. Bridge Eng.* **2022**, *27*, 04022106. [CrossRef]
2. Okur, E.K.; Okur, F.Y.; Altunişik, A.C.; Günaydin, M.; Adanur, S. Progressive collapse assessment of Osmangazi Suspension Bridge due to sudden hanger breakage under different loading conditions. *Eng. Fail. Anal.* **2023**, *149*, 107269. [CrossRef]
3. López, S.; Makoond, N.; Sánchez-Rodríguez, A.; Adam, J.M.; Riveiro, B. Learning from failure propagation in steel truss bridges. *Eng. Fail. Anal.* **2023**, *152*, 107488. [CrossRef]
4. Stark, T.D.; Benekohal, R.; Fahnestock, L.A.; LaFave, J.M.; He, J.; Wittenkeller, C. I-5 Skagit river bridge collapse review. *J. Perform. Constr. Facil.* **2016**, *30*, 04016061. [CrossRef]
5. Birajdar, H.S.; Maiti, P.R.; Singh, P.K. Failure of Chauras bridge. *Eng. Fail. Anal.* **2014**, *45*, 339–346. [CrossRef]
6. Deng, L.; Wang, W.; Yu, Y. State-of-the-art review on the causes and mechanisms of bridge collapse. *J. Perform. Constr. Facil.* **2016**, *30*, 04015005. [CrossRef]
7. Wardhana, K.; Hadipriono, F.C. Analysis of recent bridge failures in the United States. *J. Perform. Constr. Facil.* **2003**, *17*, 144–150. [CrossRef]
8. Simoncelli, M.; Aloisio, A.; Zucca, M.; Venturi, G.; Alaggio, R. Intensity and location of corrosion on the reliability of a steel bridge. *J. Constr. Steel Res.* **2023**, *206*, 107937. [CrossRef]
9. Zhang, G.; Liu, Y.; Liu, J.; Lan, S.; Yang, J. Causes and statistical characteristics of bridge failures: A review. *J. Traffic Transp. Eng. (Engl. Ed.)* **2022**, *9*, 388–406. [CrossRef]
10. Kiakojouri, F.; Sheidaii, M.R.; De Biagi, V.; Chiaia, B. Progressive collapse of structures: A discussion on annotated nomenclature. *Structures* **2021**, *29*, 1417–1423. [CrossRef]
11. Torres, B.; Poveda, P.; Ivorra, S.; Estevan, L. Long-term static and dynamic monitoring to failure scenarios assessment in steel truss railway bridges: A case study. *Eng. Fail. Anal.* **2023**, *152*, 107435. [CrossRef]
12. Derseh, S.A.; Mohammed, T.A. Bridge structures under progressive collapse: A comprehensive state-of-the art-review. *Results Eng.* **2023**, *18*, 101090. [CrossRef]
13. Caredda, G.; Makoond, N.; Buitrago, M.; Sagaseta, J.; Chryssanthopoulos, M.; Adam, J.M. Learning from the progressive collapse of buildings. *Dev. Built Environ.* **2023**, *15*, 100194. [CrossRef]
14. Makoond, N.; Shahnazi, G.; Buitrago, M.; Adam, J.M. Corner-column failure scenarios in building structures: Current knowledge and future prospects. *Structures* **2023**, *49*, 958–982. [CrossRef]
15. Russell, J.M.; Sagaseta, J.; Cormie, D.; Jones, A. Historical review of prescriptive design rules for robustness after the collapse of Ronan Point. *Structures* **2019**, *20*, 365–373. [CrossRef]

16. Adam, J.M.; Parisi, F.; Sagaseta, J.; Lu, X. Research and practice on progressive collapse and robustness of building structures in the 21st century. *Eng. Struct.* **2018**, *173*, 122–149. [CrossRef]
17. Vanova, P.; Sun, Z.; Odinson, O.; Jiang, Z. Dynamic response analysis of a model truss bridge considering damage scenarios. *Eng Fail. Anal.* **2023**, *151*, 107389. [CrossRef]
18. Buitrago, M.; Makoond, N.; Moragues, J.J.; Sagaseta, J.; Adam, J.M. Robustness of a full-scale precast building structure subjected to corner-column failure. *Structures* **2023**, *52*, 824–841. [CrossRef]
19. Padil, K.H.; Bakhary, N.; Abdulkareem, M.; Li, J.; Hao, H. Non-probabilistic method to consider uncertainties in frequency response function for vibration-based damage detection using Artificial Neural Network. *J. Sound Vib.* **2020**, *467*, 115069 [CrossRef]
20. Teng, S.; Chen, X.; Chen, G.; Cheng, L. Structural damage detection based on transfer learning strategy using digital twins of bridges. *Mech. Syst. Signal Process.* **2023**, *191*, 110160. [CrossRef]
21. Buitrago, M.; Bertolesi, E.; Calderón, P.A.; Adam, J.M. Robustness of steel truss bridges: Laboratory testing of a full-scale 21-metre bridge span. *Structures* **2021**, *29*, 691–700. [CrossRef]
22. Bertolesi, E.; Buitrago, M.; Adam, J.M.; Calderón, P.A. Fatigue assessment of steel riveted railway bridges: Full-scale tests and analytical approach. *J. Constr. Steel Res.* **2021**, *182*, 106664. [CrossRef]
23. Caredda, G.; Porcu, M.C.; Buitrago, M.; Bertolesi, E.; Adam, J.M. Analysing local failure scenarios to assess the robustness of steel truss-type bridges. *Eng. Struct.* **2022**, *262*, 114341. [CrossRef]
24. Sun, Y.; Tai, X.; Liu, K.; Zhu, A.; Zhu, H. Determining launching construction parameters of superlong superwide bridges: A multiobjective optimization method using machine learning techniques. *Structures* **2022**, *41*, 15–28. [CrossRef]
25. Yu, X.; Deng, Y.; Yan, B. Case study of the 156 m simply supported steel truss railway bridge. *Struct. Eng. Int.* **2017**, *27*, 563–568 [CrossRef]
26. Wu, X.; Wu, T.; Chen, W. Analysis of height difference between three trusses of a steel truss bridge during incremental launching *Stahlbau* **2018**, *87*, 910–922. [CrossRef]
27. Wu, G.; Qiu, W.; Wu, T. Nonlinear dynamic analysis of the self-anchored suspension bridge subjected to sudden breakage of a hanger. *Eng. Fail. Anal.* **2019**, *97*, 701–717. [CrossRef]
28. Soto, N.; Cid, C.; Baldomir, A.; Hernández, S. Fail-safe optimum cable system under cable breakage in cable-stayed bridges. Application to the Queensferry Crossing Bridge. *Eng. Struct.* **2023**, *279*, 115557. [CrossRef]
29. Ruiz-Teran, A.; Aparicio, A. Response of under-deck cable-stayed bridges to the accidental breakage of stay cables. *Eng. Struct.* **2009**, *31*, 1425–1434. [CrossRef]
30. Kiakojouri, F.; De Biagi, V.; Chiaia, B.; Sheidaii, M.R. Strengthening and retrofitting techniques to mitigate progressive collapse: A critical review and future research agenda. *Eng. Struct.* **2022**, 262. [CrossRef]
31. Wang, J.; Ke, K.; Yam, M.C.; Teng, M.; Wang, W. Improving structural robustness of steel frame buildings by enhancing floor deck connections. *J. Constr. Steel Res.* **2023**, *204*, 107842. [CrossRef]
32. Anitori, G.; Casas, J.R.; Ghosn, M. Redundancy and robustness in the design and evaluation of bridges: European and North American perspectives. *J. Bridge Eng.* **2013**, *18*, 1241–1251. [CrossRef]
33. Li, Y.; Li, S.; Tan, P. A novel tuned mass damper inerter: Optimal design, effectiveness comparison, and robustness investigation. *Structures* **2023**, *55*, 1262–1276. [CrossRef]
34. Kiakojouri, F.; De Biagi, V.; Chiaia, B.; Sheidaii, M.R. Progressive collapse of framed building structures: Current knowledge and future prospects. *Eng. Struct.* **2020**, *206*, 110061. [CrossRef]
35. Gao, H.; Li, B.; Jian, J.; Yu, T.; Liu, H. Integral jacking of concrete continuous box beam bridge. *Structures* **2023**, *54*, 1026–1045. [CrossRef]
36. Wang, Y.; Thrall, A.P.; Zoli, T.P. Behavior of the Delaware River Bridge during repair. *J. Constr. Steel Res.* **2021**, *177*, 106448. [CrossRef]
37. Xiao, F.; Hulsey, J.L.; Chen, G.S.; Xiang, Y. Optimal static strain sensor placement for truss bridges. *Int. J. Distrib. Sens. Netw.* **2017**, *13*, 1550147717707929. [CrossRef]
38. Xu, C.; Guo, C.; Xu, Q.; Yang, Z. The global collapse resistance capacity of a seismic-damaged SRC frame strengthened with an enveloped steel jacket. *Structures* **2021**, *33*, 3433–3442. [CrossRef]
39. Xiao, F.; Hulsey, J.L.; Balasubramanianm, R. Fiber optic health monitoring and temperature behavior of bridge in cold region. *Struct. Control. Health Monit.* **2017**, *24*, e2020. [CrossRef]

buildings

MDPI

Article

Buffeting Characteristics of a Long-Span Cable-Stayed Bridge Crossing a Deep Canyon during Erection: Response Evaluation and Vibration Control

Yichi Zhang [1], Tianyi Zhang [1,*] and Yi Su [2]

[1] Sichuan Road & Bridge Group Co., Ltd., Chengdu 610041, China; shwayspistachio@gmail.com
[2] Key Laboratory of New Technology for Construction of Cities in Mountain Area, School of Civil Engineering, Chongqing University, Chongqing 400045, China; windsu888@cqu.edu.cn
* Correspondence: zty_swjtu@126.com

Abstract: The stiffness of a long-span cable-stayed bridge under construction may be much lower than that observed in service, making it more susceptible to wind effects, especially for a bridge designed using high piers crossing a deep canyon. To study the buffeting characteristics of such cable-stayed bridges under construction, a long-span cable-stayed bridge (the main span is 575 m) is taken as the engineering background. In this study, the buffeting responses and vibration countermeasures at three different construction states were systematically studied using time-domain analysis. It was found that the buffeting response enlarges with an increase in the wind attack angle. The RMS values of the vertical buffeting of the bridge deck end are relatively greater at the maximum double cantilever construction state and maximum single cantilever state. At maximum double cantilever construction state, the traditional wind-resistant cable connecting the bridge deck end to the bridge pile cap significantly reduces the vertical buffeting response, while the suppression effect on lateral and torsional buffeting is limited. When the bridge deck nears completion, wind-resistant cables installed at both cantilever ending in the 'soft connection' method would effectively suppress the vertical, lateral, and torsional buffeting. The suppression effect of cross-arranged wind-resistant cables is superior to that of the parallel arrangement. It is recommended that a reasonable wind-resistant cable layout scheme according to different construction conditions is selected.

Keywords: long-span cable-stayed bridge; construction state; buffeting response evaluation; time-domain analysis; vibration control; wind-resistant cable

Citation: Zhang, Y.; Zhang, T.; Su, Y. Buffeting Characteristics of a Long-Span Cable-Stayed Bridge Crossing a Deep Canyon during Erection: Response Evaluation and Vibration Control. *Buildings* **2024**, *14*, 305. https://doi.org/10.3390/buildings14020305

Academic Editor: Mizan Ahmed

Received: 5 November 2023
Revised: 2 December 2023
Accepted: 3 December 2023
Published: 23 January 2024

1. Introduction

In the atmospheric boundary layer, the oncoming turbulent wind induces flexible structures to undergo random forced vibrations called buffeting. In general, it is impossible to directly destroy the structure. However, for long-span bridges, the prolonged vibration under frequently encountered wind speeds would risk the security of the vehicles and cause discomfort for pedestrians. More seriously, it may cause fatigue damage to the components of the bridge [1–4]. Nowadays, balanced cantilever methods have been widely adopted to construct long-span cable-stayed bridges. Noting that the stiffness of a cable-stayed bridge during erection may be much lower than that in service. The cable-stayed bridge under construction is more susceptible to wind effects [5–7]. Moreover, long-span bridges usually employ the design of high piers to cross deep canyons, resulting in larger buffeting responses.

The theoretical analysis of buffeting responses can adopt the frequency-domain method or time-domain method. Of the two, the frequency-domain method, which depends on the probability statistical concept and pertains to linear structures, was established by Davenport in the 1960s [8–10]. On this basis, the effect of structural motion was considered by introducing the concept of flutter derivative proposed by Scanlan [11–15]. The flutter derivatives can be used to express the relationship between the structural buffeting response and the aerodynamic self-excited force. Thereafter, many scholars further developed frequency-domain analysis methods. The conventional frequency-domain method is on the basis of the linear hypothesis. Noting that for long-span bridges, the nonlinearity caused by geometric or aeroelastic effects needs to be considered in the buffeting calculation. Therefore, the frequency-domain method is not so suitable for long-span bridges [16]. As an alternative method, the time-domain method would take into account the nonlinearities caused by aerodynamic forces and large deformations. It is critical to transform the aerodynamic force from a spectral form to a time history and to simultaneously calculate the transient aerodynamic force caused by the motion. Baluarte Bridge is a cable-stayed bridge with a mid-span length of 520 m in Mexico. The study on the buffeting response of this bridge showed that the calculation results using the frequency-domain analysis are less than 30% of the full aeroelastic model wind tunnel test results, while the results of time-domain analysis are close to the experimental results [17–19]. Nowadays, the time-domain analysis method has been widely used with the improvement of numerical calculation ability [20–25].

In addition to accurately predicting the buffeting response of the bridge, controlling excessive buffeting responses is a crucial aspect of constructing a cable-stayed bridge to ensure the safety of both personnel and structures [26]. Many researchers have studied countermeasures to suppress the buffeting of long-span cable-stayed bridges. For cable-stayed bridges crossing sea or river with shallow water depth, the installation of temporary piers or steel buttresses under the bridge deck cantilever is widely used to improve the vertical stiffness of the bridge, thereby improving the stability of the bridge [27,28]. It is worth noting that cable-stayed bridges crossing deep canyons are characterized by high piers, and the construction of temporary piers or steel buttresses is not suitable due to high construction costs and poor safety. Moreover, the installation of large mechanical control devices on the bridge deck, such as the tuned mass dampers (TMDs), would decrease the wind-induced vibrations by increasing the damping ratio [29,30]. However, this method has the problems of short installation time, poor economy, and occuping the working space of the bridge deck. Currently, various forms of wind-resistant cables have been adopted to control the excessive buffeting responses of cable-stayed bridges during erection. Among these methods, diagonal wind-resistant cables anchored to the top surface of the pile cap are typically used to mitigate wind-induced responses [31–33]. In addition, the vertical wind-resistant cables anchored to the foundation or ground were also used in the Busan-Geoje Fixed Link Project and the Chishi Bridge during erection [34]. The wind-resistant cables are not sensitive to the topography of the bridge site and are therefore suitable for use in mountain areas. However, when the cantilever length of the bridge deck continues to increase, the length of the wind-resistant cable also increases, resulting in inconvenience during construction.

It is worth noting that the previous studies on the wind-induced vibration of cable-stayed bridges usually focused on plains and coastal areas and often only studied a single construction state. At present, the construction of expressways is gradually expanding to mountain and plateau areas, especially in the mountain areas of southwest China. There are more challenges for the construction of long-span cable-stayed bridges with high piers in complex terrains. However, the research on buffeting response analysis and targeted countermeasures is still insufficient. This paper focuses on the cable-stayed bridge with a high pier design crossing a deep canyon and studies the buffeting characteristics under the representative state throughout the whole construction process. The purpose is to propose a targeted buffeting suppression scheme suitable for the different construction states of cable-stayed bridges in mountain areas to ensure the safety and convenience of bridge construction. To this aim, a long-span cable-stayed bridge with a main span of 575 m crossing a "V" shaped deep canyon is taken as the engineering background. The time-domain FEM method is used to calculate the buffeting response under the different construction states, which include maximum double-cantilever construction state, single-cantilever mid-term construction state, and maximum single-cantilever construction state. The suppression effects of various forms of wind-resistant cables on the buffeting response are also systematically investigated.

2. Time-Domain Method for the Buffeting Response Evaluation of Long-Span Bridges

The present study adopts the time-domain method to analyze the buffeting response of the long-span cable-stayed bridge, which takes into account the nonlinear factors that the frequency-domain method cannot include. The calculation procedures of this method are as follows. First, the discrete three-dimensional turbulent flow field around the bridge is simulated according to the meteorological data and terrain at the bridge site. Second, the aerodynamic coefficients of the cross-sections of the bridge deck and tower are identified to calculate the time histories of the aerodynamic forces acting on the structure. Finally, the finite element method (FEM) is used to analyze buffeting responses through the step-by-step time series analysis. It is assumed that the oncoming flow is orthogonal to the bridge deck in the analysis. The theoretical method of bridge buffeting analysis is briefly introduced as follows.

2.1. Three-Dimensional Turbulent Wind Field Simulation

The turbulent wind can be decomposed in three different directions, including longitudinal direction (u), lateral direction (v), and vertical direction (w). Furthermore, three-dimensional turbulent wind field can be simplified into three orthogonal one-dimensional turbulent flow fields if the correlation between the wind fields is not considered. Thus, the three-dimensional wind fields can be expressed in the Cartesian coordinate system as:

$$\begin{cases} U = \overline{U}(z) + u(x,z,t) \\ v = v(x,z,t) \\ w = w(x,z,t) \end{cases} \tag{1}$$

where U is the longitudinal velocity of the oncoming wind; $\overline{U}(z)$ is the mean longitudinal velocity; x and z are the longitudinal and lateral coordinates, respectively; u, v, and w are longitudinal, lateral, and vertical fluctuating velocities; and t is the time parameter.

To simulate the wind field around a bridge, the field can be discretized using a three-dimensional wind field with discrete finite points in a certain area. Then, the time history of the velocity of each discrete point can be simulated by superimposing the mean and fluctuating velocities of the oncoming wind. In the atmospheric boundary layer, the mean longitudinal velocity varies with the height from the ground, which can be expressed by exponential law or logarithmic law. Moreover, the fluctuating wind speeds in three directions would be calculated by the harmony superposition method. For this method, it is critical to generate random process samples through the superposition of random trigonometric functions according to the given wind speed spectrum and correlation function.

2.2. Wind Loads Simulation

- Aerostatic force

The aerostatic force acting on a structure is expressed using Formula (2),

$$\begin{aligned} F_D &= \tfrac{1}{2}\rho U^2 C_D(\alpha_0) D \\ F_L &= \tfrac{1}{2}\rho U^2 C_L(\alpha_0) B \\ F_M &= \tfrac{1}{2}\rho U^2 C_M(\alpha_0) B^2 \end{aligned} \tag{2}$$

where ρ is the air density; α_0 is the average wind attack angle of oncoming flow; C_D, C_L, and C_M are drag, lift, and lift moment coefficients; and D and B are height and width of bridge deck.

- Buffeting force

Based on the quasi-steady theory, the aerodynamic force under fluctuating wind can be expressed as

$$\begin{aligned} L_b(t) &= \tfrac{1}{2}\rho U^2 B[C_L(\alpha_0)\chi_{Lu}\tfrac{u(t)}{U} + (C_L'(\alpha_0) + C_D(\alpha_0))\chi_{Lw}\tfrac{w(t)}{U}] \\ D_b(t) &= \tfrac{1}{2}\rho U^2 B[2C_D(\alpha_0)\chi_{Du}\tfrac{u(t)}{U} + C_D'(\alpha_0)\chi_{Dw}\tfrac{w(t)}{U}] \\ M_b(t) &= \tfrac{1}{2}\rho U^2 B^2[2C_M(\alpha_0)\chi_{Mu}\tfrac{u(t)}{U} + C_M'(\alpha_0)\chi_{Mw}\tfrac{w(t)}{U}] \end{aligned} \quad (3)$$

where C_D', C_L', and C_M' are the slopes of the curves of the aerostatic coefficients, which are identified using sectional model wind tunnel tests or numerical simulation, and χ_{Du}, χ_{Dw}, χ_{Lu}, χ_{Lw}, χ_{Mu}, and χ_{Mw} are aerodynamic admittances of drag, lift, and lift moment.

- Aerodynamic self-excited force

According to the theoretical expression proposed by Scanlan, the aerodynamic lift force L_{ae}, drag force D_{ae}, and lift moment M_{ae} per unit length can be written as

$$\begin{aligned} L_{ae} &= \tfrac{1}{2}\rho U^2(2B)\left[KH_1^*\tfrac{\dot{h}}{U} + KH_2^*\tfrac{B\dot{\alpha}}{U} + K^2H_3^*\alpha + K^2H_4^*\tfrac{h}{B} + KH_5^*\tfrac{\dot{p}}{U} + K^2H_6^*\tfrac{p}{B}\right] \\ D_{ae} &= \tfrac{1}{2}\rho U^2(2B)\left[KP_1^*\tfrac{\dot{p}}{U} + KP_2^*\tfrac{B\dot{\alpha}}{U} + K^2P_3^*\alpha + K^2P_4^*\tfrac{p}{B} + KP_5^*\tfrac{\dot{h}}{U} + K^2P_6^*\tfrac{h}{B}\right] \\ M_{ae} &= \tfrac{1}{2}\rho U^2(2B)\left[KA_1^*\tfrac{p\dot{h}}{U} + KA_2^*\tfrac{B\dot{\alpha}}{U} + K^2A_3^*\alpha + K^2A_4^*\tfrac{h}{B} + KA_5^*\tfrac{\dot{p}}{U} + K^2A_6^*\tfrac{p}{B}\right] \end{aligned} \quad (4)$$

where h represents vertical position; α represents torsional position; p is horizontal displacement; and H_i^*, P_i^*, and A_i^* $(i = 1,2,3\cdots,6)$ are aerodynamic derivatives, which only related to the cross-section shape. In this paper, the aerodynamic self-excited force is calculated using the method proposed by Ding [35].

3. Research Background and Numerical Setup

3.1. Research Background

A cable-stayed bridge crossing a typical deep 'V'-shaped canyon was taken as a research engineering background. Figure 1 illustrates the bridge's general arrangement. This bridge is located in the mountainous area of Southwest China. The main span of the bridge is 575 m, and its total length is 1089 m. As shown in Figure 2, the bridge deck is an opening steel-concrete composite cross-section with two 'I'-shaped longitudinal edge girders. It is 38.0 m in width and 3.0 m in height. Two 'L'-shaped deflectors are installed at both edges of the cross-section to suppress vortex-induced vibration. The two concrete towers have heights of 261.0 m and 217.0 m, respectively. The relative height difference between the bridge deck and the ground is about 330 m. The bridge is constructed using the traditional cantilever assembly method. Figure 3 shows three typical states during erection, including maximum double cantilever construction state, mid-term single cantilever construction state, and maximum single cantilever construction state. The maximum double cantilever state is the state when the cantilever length of the main beam is the largest before the connection between the main beam and the side pier. When the bridge deck is installed except for the closure section, it is in the maximum single cantilever construction state, and the bridge deck has the maximum cantilever length. The mid-term single cantilever construction state is the intermediate state of the transition from the maximum double cantilever state to the maximum single cantilever state. In the present study, these three typical construction states were considered.

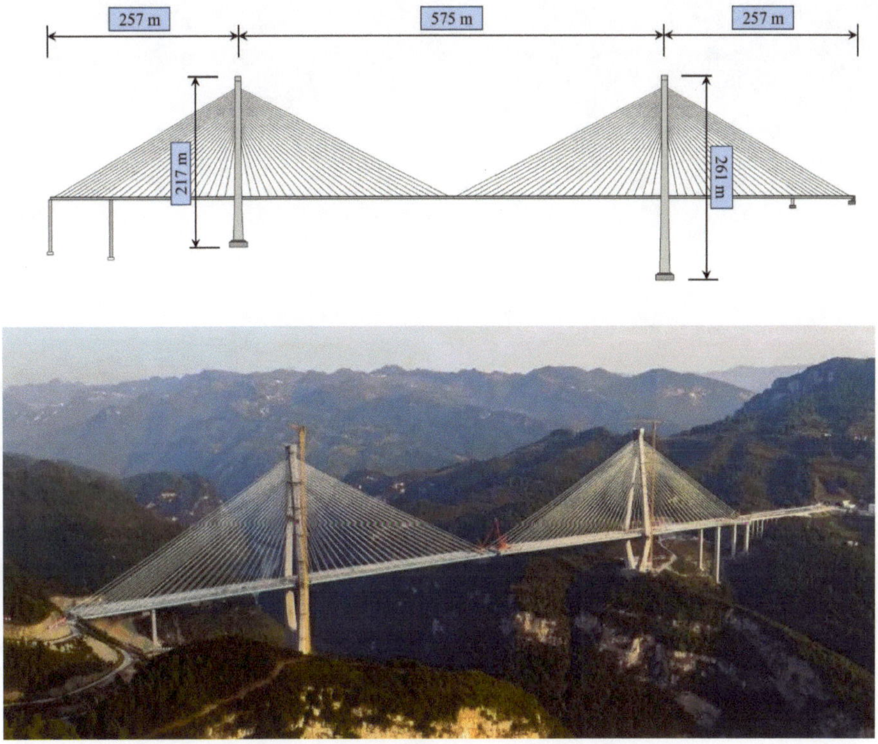

Figure 1. The general arrangement and photo of the bridge.

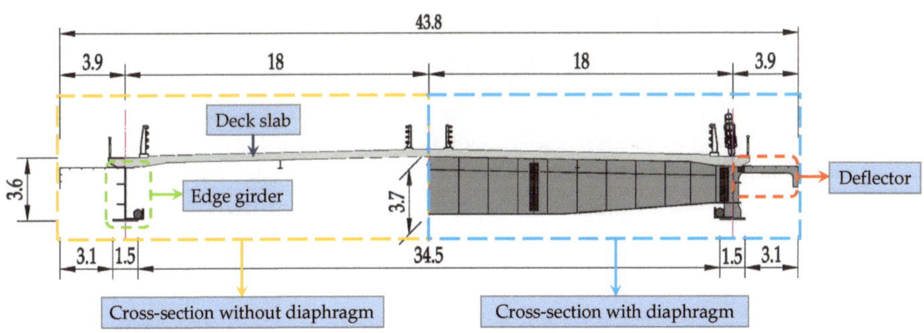

Figure 2. The cross-sections of the bridge main beam (unit: m).

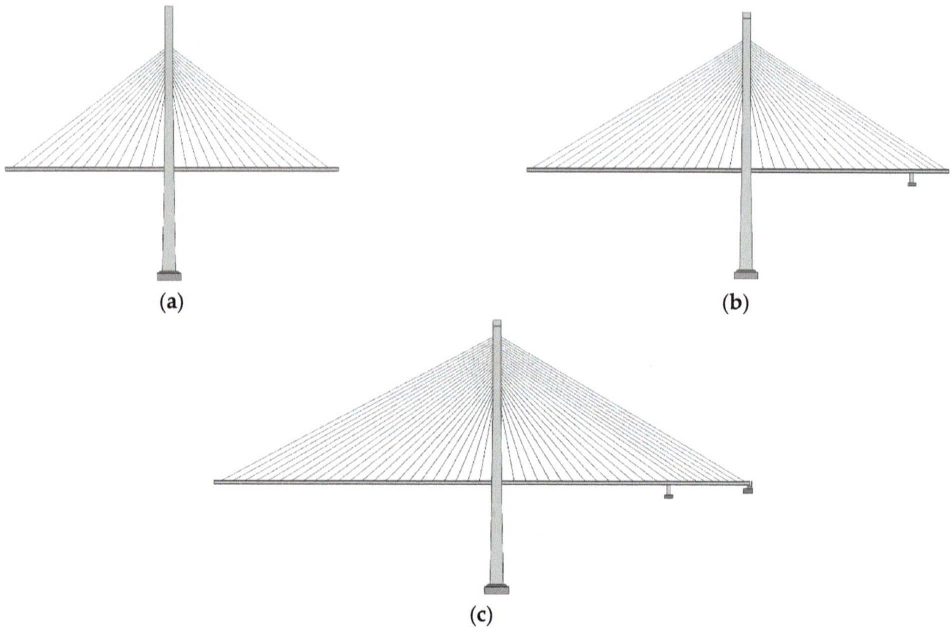

Figure 3. Schematic diagrams of the bridge at three construction states. (**a**) maximum double cantilever construction state; (**b**) mid-term single cantilever construction state; (**c**) maximum single cantilever construction state.

3.2. *Dynamic Characteristics*

The FEM software ANSYS 15.0 was used to construct the three-dimensional model and solve the dynamic characteristics results. In the model, the sliding bearing of the cable-stayed bridge was fixed during erection. The frequencies and mode shapes of typical vibration modes of the bridge deck in three different states are listed in Table 1.

Table 1. Dynamic characteristics of the bridge deck at different construction states.

Construction State	Order	Vibration Mode	Frequency (Hz)	Mode Shape
maximum double cantilever state	1	Vertical bending–symmetric	0.1297	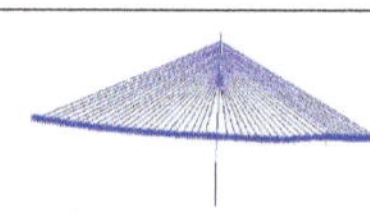
	2	Lateral bending–symmetric	0.2322	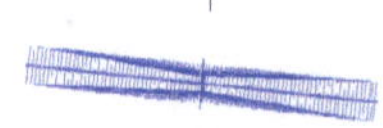
	10	Torsion–symmetric	1.2387	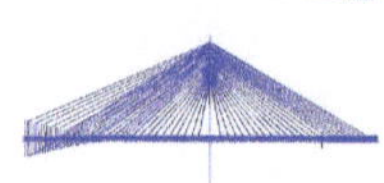

Table 1. *Cont.*

Construction State	Order	Vibration Mode	Frequency (Hz)	Mode Shape
single cantilever medium-term state	1	Vertical bending–symmetric	0.2491	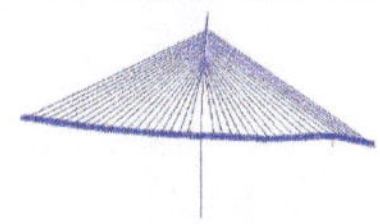
	2	Lateral bending–symmetric	0.2858	
	9	Torsion–symmetric	1.0193	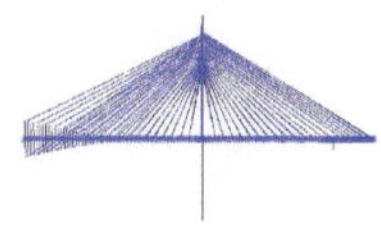
maximum single cantilever state	1	Lateral bending of tower	0.1297	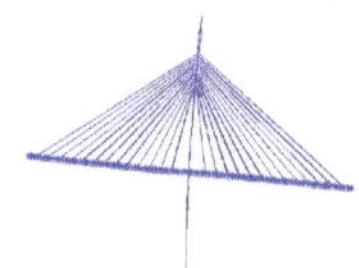
	2	Lateral bending–symmetric	0.2322	
	4	Vertical bending–symmetric	0.4535	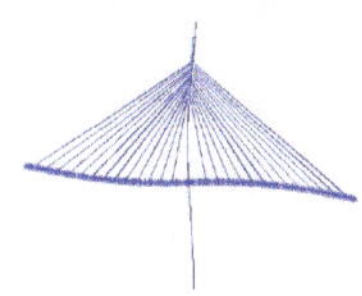
	10	Torsion–symmetric	1.2387	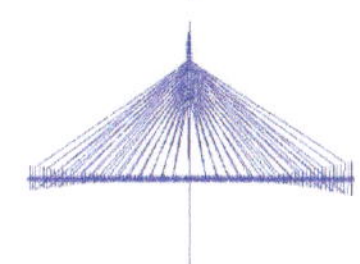

3.3. Setup of the Buffeting Response Calculation

- Wind field simulation results

Based on the topographical features, the design mean velocity of the bridge deck height was calculated to be 31.4 m/s during erection using the Chinese wind-resistant design specifications for highway bridges [36]. The surface roughness coefficient α_0 and surface roughness height z_0 were selected to be 0.22 and 0.3 m. The three-dimensional wind field simulation used the Kaimal wind spectrum recommended by Chinese specifications. The power spectrum of the longitudinal fluctuating velocity at the height z is expressed as

$$\frac{nS_u(n)}{(u^*)^2} = \frac{200f}{(1+50f)^{5/3}} \tag{5}$$

The wind power spectrum of vertical fluctuating velocity is written as

$$\frac{nS_\omega(n)}{(u^*)^2} = \frac{6f}{(1+4f)^2} \tag{6}$$

where $S_u(n)$ is the power spectral density function of longitudinal fluctuating velocity; $S_\omega(n)$ is the power spectral density function of vertical fluctuating velocity; n is the frequency of wind; and $f = \frac{nz}{U(z)}$ is the Moning coordinate.

A total of 42 discrete wind simulation points were set along the height direction of the bridge tower. A total of 28, 38, and 48 wind simulation points were set along the bridge deck at the maximal double cantilever construction state, mid-term single cantilever construction state, and maximal single cantilever construction state, respectively. The fluctuating wind time interval was set to 0.0625 s during the simulation. Figure 4 shows the longitudinal and vertical fluctuating velocity time history results of the simulated point of the bridge deck end at the maximum single cantilever state. Figure 5 shows the target and simulated wind spectrums of longitudinal and vertical velocities. The results indicate that the fluctuating velocity of the oncoming wind agrees well with the target spectrum, indicating that the three-dimensional wind field is accurately simulated.

- Determination of aerodynamic coefficients

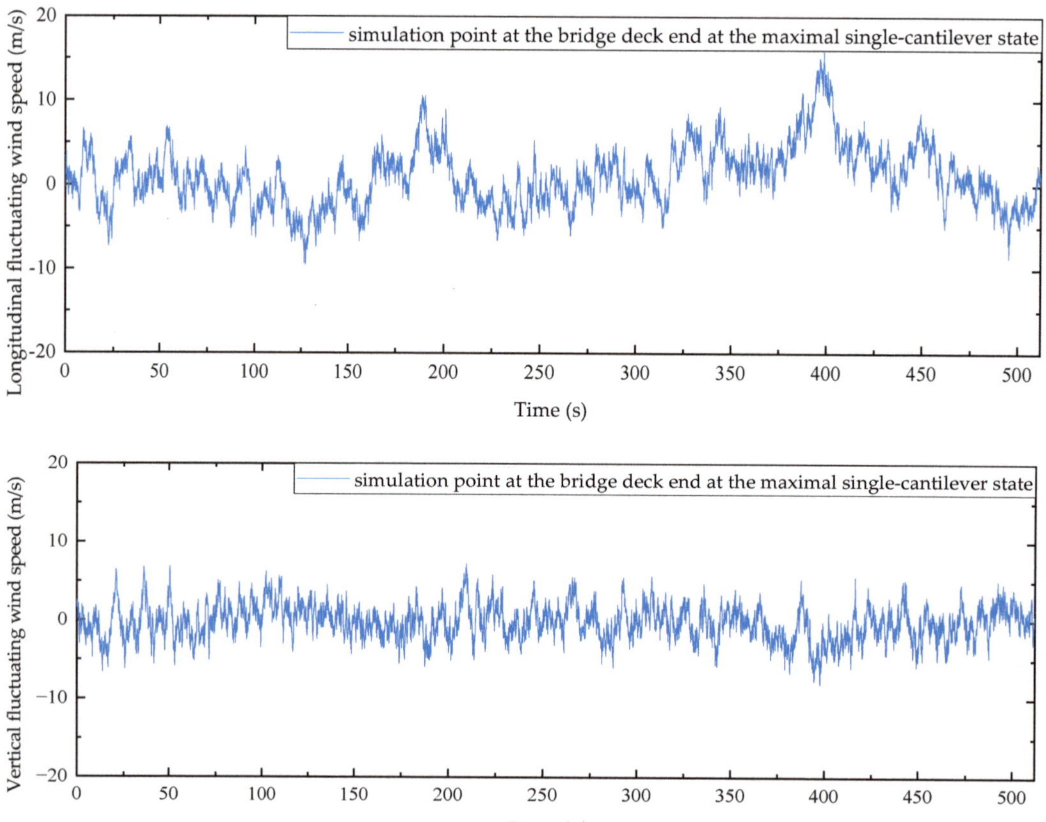

Figure 4. Velocity time history of simulation point at bridge deck end at the maximum single cantilever state.

In this study, the aerodynamic coefficients during erection were obtained through sectional model wind tunnel tests carried out at Tongji University. The identified aerodynamic coefficients are shown in Table 2. The aerodynamic coefficients of the tower were calculated by the CFD method. The aerodynamic admittances were set to 1 to simplify the calculation.

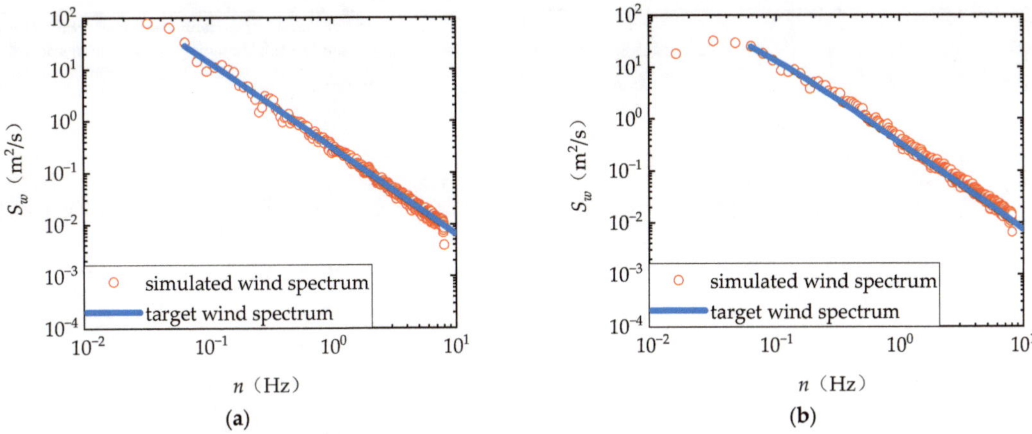

Figure 5. Wind spectrum of simulation point at the bridge deck end at the maximum single cantilever state. (**a**) Longitudinal direction; (**b**) vertical direction.

- Parameters in the time-domain analysis

Table 2. The aerodynamic coefficients of the bridge deck during erection.

Attack Angle	Drag Coefficient C_H	$dC_H/d\alpha$	Lift Coefficient C_V	$dC_V/d\alpha$	Lift Moment Coefficient C_M	$dC_M/d\alpha$
−3°	1.540	−5.133	−0.319	3.644	0.019	0.467
0°	1.414	0.898	−0.062	6.122	0.044	0.408
+3°	1.684	10.340	0.327	8.536	0.050	−0.213

The time-domain analysis in the present study was based on the APDL module in the ANSYS 15.0 FEM software. The calculation time step was 0.125 s in the simulation. In order to ensure the accuracy of the further spectrum analysis results, the number of sampling floating points in the spectrum calculation should be an integer exponential multiple of 2. In addition, through the trial calculation, it can be seen that the root mean square (RMS) value of the structural buffeting response tends to be stable after the calculation time is about 400 s. Therefore, the total simulation time was set to 512 s. Rayleigh damping was adopted to simulate the damping ratio, and the damping ratio was set to 1.0%, according to the Chinese specifications. The windward area of the stay cable is much smaller than that of the bridge deck and tower, the aerodynamic force has limited influence on the calculation results. Thus, the effect of vortex shedding on the stay cable was not considered in the calculation. To fully consider the nonlinearity characteristics of the structure, the gravity and mean aerodynamic forces were initially applied to the model and considered as the first state in the dynamic simulation.

4. Buffeting Responses in Three Different Construction States

The buffeting responses under three wind attack angles (−3°, 0°, and +3°) at different construction states, including maximum double cantilever state, mid-term single cantilever construction state, and maximum single cantilever state, were systematically calculated. Table 3 lists the mean displacement and the RMS induced by the wind effect at the bridge material positions. The displacement time histories at 0° wind attack angle are illustrated in Figure 6.

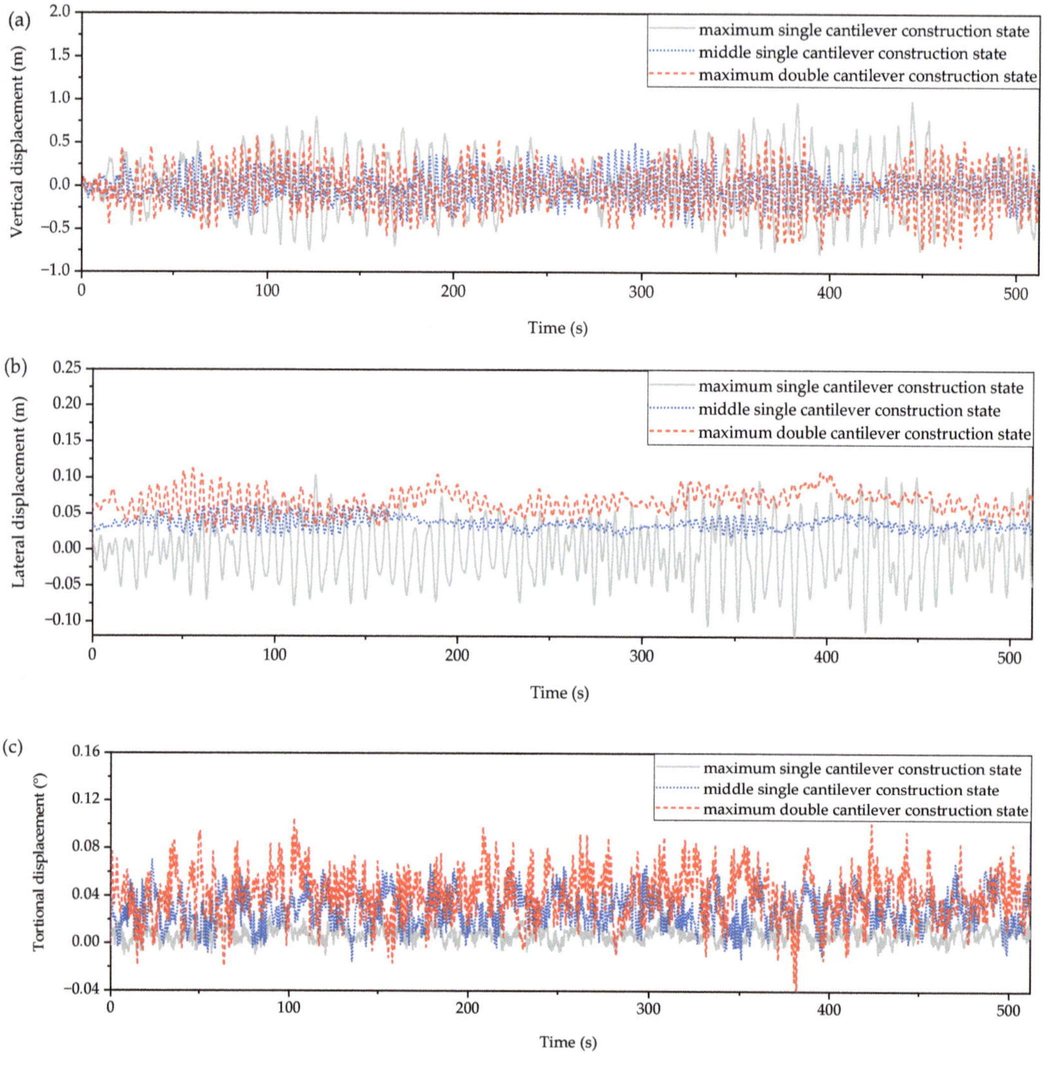

Figure 6. Displacement time histories of the end of the bridge deck cantilever at three construction states at 0° wind attack angle.

It can be seen that the buffeting response at the wind attack angle of +3° is significantly larger than that at the wind attack angles of −3° and 0°, which contributed to the larger slopes of the curves of the aerostatic coefficients. In the following section, the buffeting performance of the bridge at three construction states is investigated based on the simulation at 0° wind attack angle. It is shown that the lateral mean displacements of the bridge deck are significantly greater than the vertical mean displacement at two single cantilever construction states. The vertical RMS value of buffeting responses of the bridge deck cantilever end is largest at the maximum double cantilever state, reaching 0.3390 m. And it is close to the RMS value at the top of the bridge tower. This may indicate that the bridge deck vertical buffeting at the maximum double cantilever construction state mainly relies on the lateral bending of the bridge tower. While the bridge deck vertical displacements at the other two construction states may be participated by bridge deck vibration modes. The RMS values of the lateral and vertical buffeting responses are the largest at the maximum double cantilever state, reaching 0.0402 m and 0.3390 m, respectively. Moreover, the torsional buffeting response is the largest at the maximum single cantilever state, and the RMS value at the end of the bridge deck cantilever is 0.0205°.

Table 3. The mean displacements and RMS at main positions at three different states.

State	Attack Angle	Position	Longitudinal (m)		Vertical (m)		Lateral (m)		Torsional (°)	
			Mean	RMS	Mean	RMS	Mean	RMS	Mean	RMS
Maximum double cantilever state	−3°	Deck end	/	/	0.0136	0.2538	−0.0014	0.0291	0.0028	0.0069
		Tower top	0.0280	0.2665	/	/	0.0033	0.0041	−0.0019	0.0077
	0°	Deck end	/	/	0.0075	0.3390	−0.0001	0.0402	0.0082	0.0058
		Tower top	0.0121	0.3554	/	/	0.0029	0.0027	−0.0012	0.0104
	+3°	Deck end	/	/	−0.0021	0.4324	0.0016	0.0638	0.0098	0.0052
		Tower top	−0.0127	0.4525	/	/	0.0043	0.0063	−0.0009	0.0182
Medium single cantilever state	−3°	Deck end	/	/	−0.0488	0.1152	0.0398	0.0142	0.0132	0.0170
		Tower top	−0.0140	0.0417	/	/	0.0123	0.0046	0.0027	0.0022
	0°	Deck end	/	/	−0.0113	0.1674	0.0372	0.0087	0.0282	0.0143
		Tower top	−0.0030	0.0610	/	/	0.0119	0.0030	0.0016	0.0013
	+3°	Deck end	/	/	0.0458	0.2293	0.0434	0.0232	0.0331	0.0123
		Tower top	0.0137	0.0839	/	/	0.0135	0.0085	0.0019	0.0023
Maximum single cantilever state	−3°	Deck end	/	/	−0.0863	0.1837	0.0732	0.0325	0.0179	0.0244
		Tower top	−0.0232	0.0588	/	/	0.0151	0.0064	0.0036	0.0030
	0°	Deck end	/	/	−0.0290	0.2561	0.0675	0.0156	0.0409	0.0205
		Tower top	−0.0077	0.0824	/	/	0.01438	0.0032	0.00191	0.00161
	+3°	Deck end	/	/	0.0592	0.3386	0.0773	0.0401	0.0513	0.0182
		Tower top	0.0161	0.1093	/	/	0.0163	0.0086	0.0020	0.0029

To further study the buffeting characteristics at three construction states, the spectrums of the buffeting displacement response of the bridge deck cantilever end at 0° wind attack angle were calculated using FFT, as illustrated in Figure 7. For lateral and torsional buffeting responses, the peak point of the displacement response spectrum is at the natural frequency point, which verifies the correctness of the bridge buffeting response analysis program. It can be seen that, due to the high pier design for the bridge crossing the 'V'-shaped canyon in the mountainous area, the vibration induced by the lateral bending and torsion of the tower dominates the buffeting of the bridge deck at the double cantilever construction state. The vibration caused by the bridge tower is mainly located in a low-frequency region. At single cantilever states, the vibration frequency of the bridge deck increased, in which the vertical buffeting response has multi-order higher frequency participation, especially at the maximum single cantilever state. Meanwhile, the lateral and torsional buffeting responses are dominated by the first-order mode.

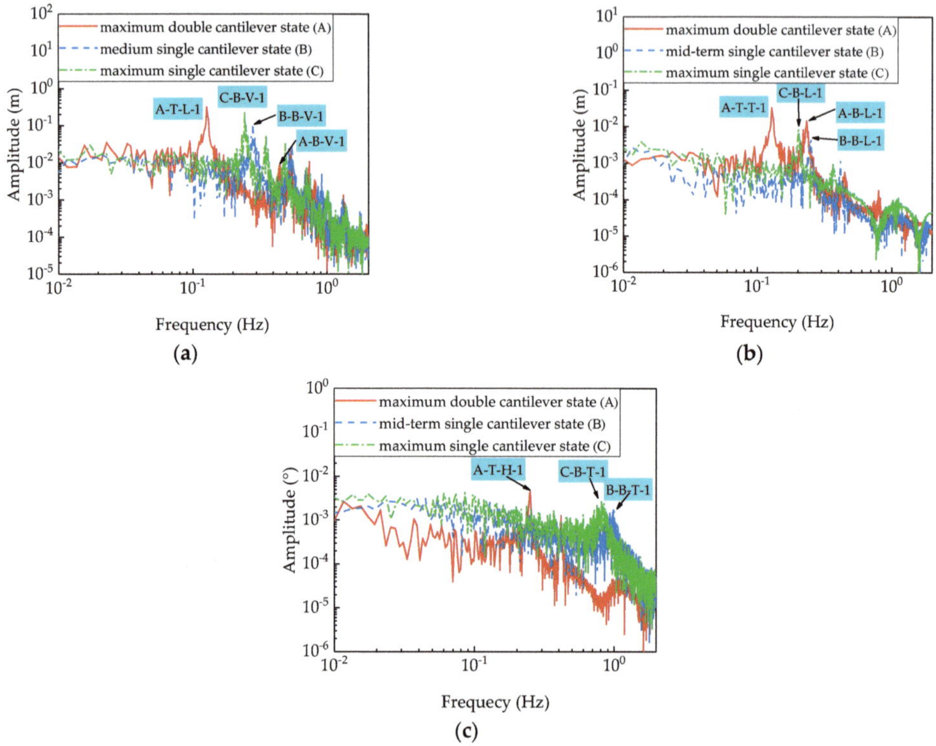

Figure 7. Spectrum of buffeting response of bridge deck end at three construction states, in which A, B, and C represent construction state, respectively; T and B represent tower and bridge, respectively; V, L, and T represent vertical, lateral, and tortional mode, respectively. (**a**) Spectrum of vertical buffeting response at bridge deck end; (**b**) spectrum of lateral buffeting response at bridge deck end; (**c**) spectrum of torsional buffeting response at bridge deck end.

5. Suppression Effect of Wind-Resistant Cables on Buffeting Response

To suppress the effect of buffeting response on cable-stayed bridges during erection, many bridges have been set up with vibration countermeasures. Among them, temporary support piers, TMDs, or various forms of wind-resistant cables are commonly used. Since long-span cable-stayed bridges in mountainous areas usually cross deep canyons, the bridge deck is usually hundreds of meters away from the bottom of the valley, making it difficult to install temporary support piers. In addition, the TMD device is expensive and cannot be recycled. In this investigation, according to the characteristics of long-span cable-stayed bridges crossing deep canyons located in mountainous areas, a variety of buffeting countermeasures based on wind-resistant cables were designed.

As shown in Figure 8a,b, four wind-resistant cables are installed symmetrically at both bridge deck ends at maximum double cantilever state. Two ends of the cable are fixed to the anchorage point of the bridge deck and the top surface of the pile cap, respectively. Considering the reduction in the demolition and installation work of the wind-resistant cables, the arrangement of wind-resistant cable at the single cantilever mid-term state is the same as that at the maximum double cantilever state in this study. Figure 8e,f show that the connection positions between the wind-resistant cables and the bridge deck are located at three-quarters of the bridge deck cantilever length at the maximum single cantilever state. Due to the connection between the bridge deck of the side span and the auxiliary pier, the wind-resistant cables are only set on the mid-span side at the two single cantilever construction states. In addition, considering the short duration of the maximum single cantilever stage, the 'soft connection' method proposed by Hu, Yang, Li, and Wang [37], as illustrated in Figure 8g,h, is introduced to suppress the buffeting response, consequently avoiding the installation of long wind-resistant cables. The edge girder is a vital force component in the composite bridge deck cross-section, and it needs to bear the large tension of the stay cable. Thus, it usually has great strength in the structural design. Similarly to the stay cable, the anchorage points of the wind-resistant cable are set on the longitudinal edge girder. A total of 120 Φ7 mm hot-dip galvanized steel wires which have the characteristics of good toughness and strong reliability are selected for the wind-resistant cables, and the area of a single cable is 4618.14 mm^2. The pre-tension of each wind-resistant cable is set to be 3 kN. The suppression effects of both parallel and crossed wind-resistant cables are investigated at each construction state.

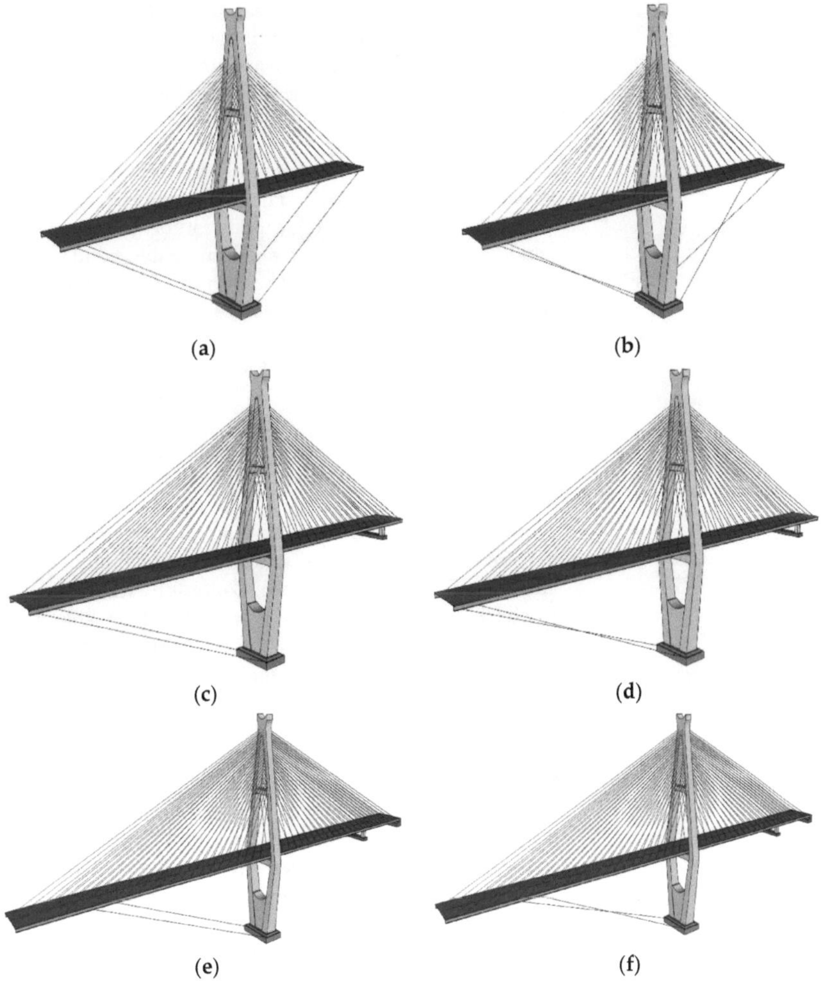

Figure 8. *Cont.*

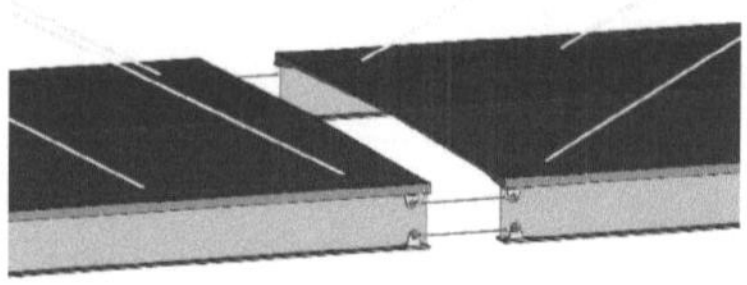

(g)

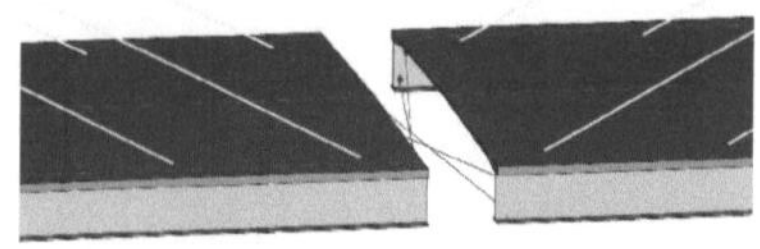

(h)

Figure 8. Design scheme of wind-resistant cables during erection. (**a**) Parallel arrangement at maximum double cantilever state; (**b**) cross arrangement at maximum double cantilever state; (**c**) parallel arrangement at mid-term single cantilever state; (**d**) cross arrangement at mid-term single cantilever state; (**e**) parallel arrangement at maximum single cantilever state; (**f**) cross arrangement at maximum single cantilever state; (**g**) parallel arrangement in 'soft connection' method; (**h**) cross arrangement in 'soft connection' method.

The frequencies of typical bridge deck vibration modes with and without wind-resistant cables are shown in Figure 9. It can be found that the frequencies of the first-order torsion and lateral bending modes can be improved with the installation of traditional wind-resistant cables at the maximum double cantilever state. Compared with the original construction state, the lateral bending and torsion frequencies are increased by 9.5% and 3.9% with the wind-resistant cables, respectively. The installation of the cross-arranged wind-resistant cables in the 'soft connection' method at the maximum single cantilever state can significantly increase the lateral bending frequency and slightly raise the vertical bending and torsion frequencies. The wind-resistant cables have a limited increase in the vibration frequency at other construction conditions, which indicates that installing the wind-resistant cables cannot significantly increase the structural stiffness.

The buffeting time-domain method was adopted to study the suppression effects of buffeting countermeasures. The calculation results in Figure 10 show that, with the symmetrical installation of the four wind-resistant cables at both bridge deck ends at the maximum double cantilever state, the vertical buffeting response of the bridge deck and the lateral bending of the tower are significantly reduced and the vibration reduction rate can reach 55.3%, indicating that this wind-resistant cable arrangement can effectively control the tower lateral bending and the bridge deck vertical buffeting during this construction state. Moreover, with the installation of this countermeasure, the decreased rate of lateral buffeting response is about 20%, and the suppression effect of torsional buffeting response is only within 10%. The bridge deck transforms to the mid-term single cantilever construction state when the bridge deck is connected to the auxiliary pier. The overall stiffness of the bridge is significantly improved, and the buffeting response is remarkably reduced. The application of the traditional wind-resistant cable only has a certain effect on the bridge deck vertical buffeting response and the tower lateral bending buffeting response in cases (c) and (d), and the suppression effect in other directions is not obvious. As the bridge deck cantilever continues to extend, the buffeting response obviously improves. Similarly to the mid-term single cantilever state, the traditional wind-resistant cables have a limited effect on the buffeting at the maximum single cantilever construction state, while the wind-resistant cables in the 'soft connection' method can significantly reduce the buffeting response. Moreover, the suppression effect of cross-connected wind-resistant cables is greater than that of parallel-connected cables. Using cross-connected wind-resistant cables, the reduction rates of vertical, lateral, and torsional buffeting responses of the bridge deck end are 35.6%, 39.3%, and 8.5%, respectively, and the lateral response of the tower top reduces by 27.0%.

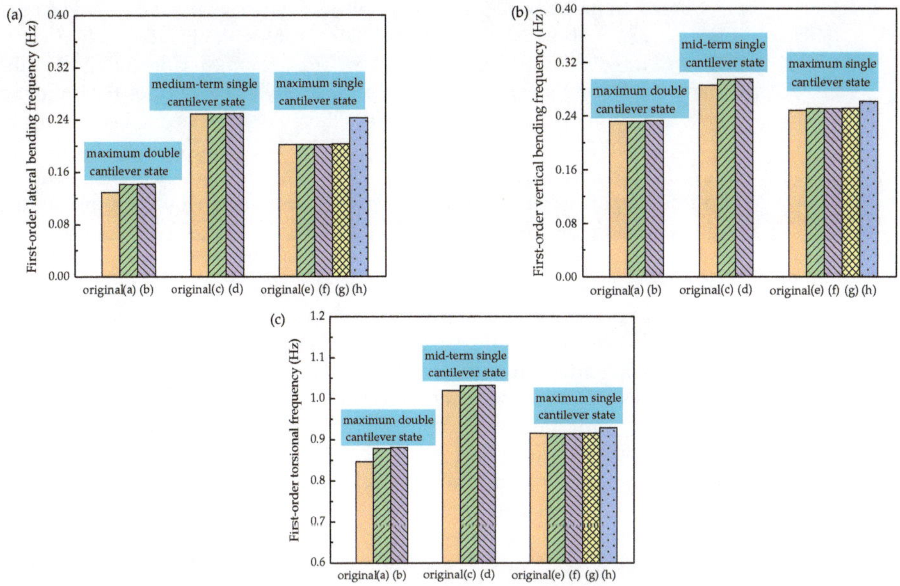

Figure 9. Frequencies of typical vibration modes of bridge deck with and without wind-resistant cables. The lowercase letters in the horizontal axis represent the wind-resistant cable arrangement shown in Figure 8; (**a**) First-order lateral bending frequency (Hz); (**b**) First-order vertical bending frequency (Hz); (**c**) First-order torsional frequency (Hz).

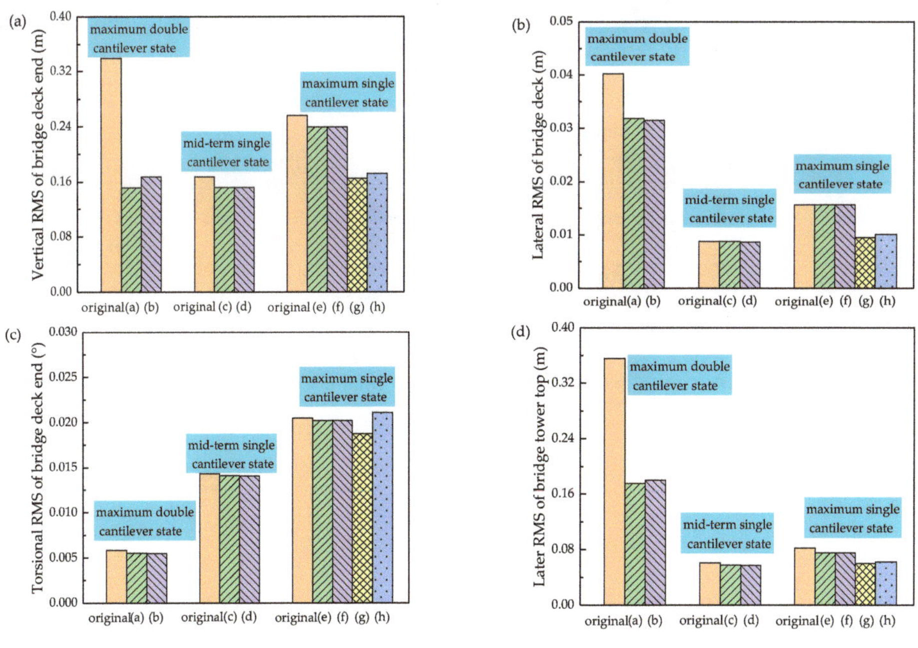

Figure 10. Buffeting responses of the material positions with the wind-resistant cables. The lowercase letters in the horizontal axis represent the wind-resistant cable arrangement shown in Figure 8; (**a**) Vertical RMS of bridge deck end (m); (**b**) Laterla RMS of bridge deck (m); (**c**) Torsional RMS of bridge deck end (°); (**d**) Later RMS of bridge tower top (m).

Based on the above calculation results, for the long-span cable-stayed bridges crossing canyons, reasonably installing wind-resistant cables would effectively decrease the buffeting response during bridge construction. The symmetrical installation of the traditional wind-resistant cables connected to the pile cap effectively reduces the buffeting response at the double cantilever construction state. When the side span connects to the auxiliary pier, the increase in the overall stiffness of the bridge will greatly reduce the buffeting response. The buffeting of the bridge deck end is most significant at maximum single cantilever state, and it is effectively suppressed by installing cross-arranged wind-resistant cables in the 'soft connection' method.

6. Conclusions

A typical long-span cable-stayed bridge was taken into account as the research background to systematically investigate the buffeting characteristics and propose economical vibration countermeasures for long-span cable-stayed bridges crossing deep canyons during erection. The time-domain analysis method was adopted to calculate the buffeting responses at three representative construction states. The main conclusions are as follows.

The results showed that the buffeting response at the wind attack angle of +3° is significantly larger than that at the wind attack angle of −3° or 0° due to the larger slopes of the curves of aerostatic coefficients. Driven by the lateral bending of the bridge tower, the bridge deck vertical buffeting response is the largest at the maximum double cantilever construction state. Under the design mean velocity (31.4 m/s), the vertical buffeting displacement RMS value at 0° attack angle reaches 0.3390 m. The overall stiffness of the bridge would be increased by connecting the bridge deck to the side pier, significantly reducing buffeting responses. At the maximum single cantilever construction state, the bridge deck cantilever is the longest, and the vertical buffeting displacement RMS value at 0° attack angle is 53.0% higher than that at the mid-term single cantilever state.

The traditional wind-resistant cables would greatly reduce the bridge deck vertical buffeting response at the maximum double cantilever construction state. The RMS value of the vertical displacement is reduced by 55.3%. The suppression effects on the lateral and torsional buffetings are limited, and the vibration reduction rate is less than 20%. Moreover, the traditional wind-resistant cable has no obvious suppression effect at the mid-term single cantilever construction state and the maximum single cantilever construction state, and the decreased rates are all not greater than 12%. When the bridge deck is nearing completion, the wind-resistant cables installed using the 'soft connection' method can significantly reduce buffeting responses. Using cross-connected wind-resistant cables, the decreasing rates of vertical, lateral, and torsional buffeting responses of the bridge deck end are 35.6%, 39.3%, and 8.5%, respectively.

It is recommended that a reasonable wind-resistant cable layout scheme according to different construction conditions is selected. In the future, the accuracy of buffeting response calculation will be further verified using wind tunnel tests or field measurements. The method of suppressing the buffeting response by changing the aerodynamic shape of the bridge needs to be investigated.

Author Contributions: Conceptualization, Y.Z. and T.Z.; software, Y.Z.; validation, Y.Z., T.Z. and Y.S.; formal analysis, T.Z.; investigation, Y.Z., T.Z. and Y.S.; writing—original draft preparation, Y.Z.; writing—review and editing, T.Z. and Y.S.; funding acquisition, Y.S. All authors have read and agreed to the published version of the manuscript.

Funding: The authors gratefully acknowledge the National Natural Science Foundation of China under grant Nos. 52208462 and U22A20230, the China Postdoctoral Science Foundation under grant No. 2022M720578.

Data Availability Statement: The data presented in this study are available on request from the corresponding author. The data are not publicly available due to reasons of confidentiality.

Conflicts of Interest: Authors Yichi Zhang and Tianyi Zhang were employed by the company Sichuan Road & Bridge Group Co., Ltd. The remaining author declares that the research was conducted in the absence of any commercial or financial relationships that could be construed as a potential conflict of interest.

References

1. Tao, T.; Wang, H.; Yao, C.; He, X. Parametric Sensitivity Analysis on the Buffeting Control of a Long-Span Triple-Tower Suspension Bridge with MTMD. *Appl. Sci.* **2017**, *7*, 395. [CrossRef]
2. Lystad, T.M.; Fenerci, A.; Øiseth, O. Buffeting response of long-span bridges considering uncertain turbulence parameters using the environmental contour method. *Eng. Struct.* **2020**, *213*, 110575. [CrossRef]
3. Xin, J.; Zhou, C.; Jiang, Y.; Tang, Q.; Yang, X.; Zhou, J. A signal recovery method for bridge monitoring system using TVFEMD and encoder-decoder aided LSTM. *Measurement* **2023**, *214*, 112797. [CrossRef]
4. Zhu, S.; Li, Y.; Yang, Y.; Ju, N. Stochastic Buffeting Analysis of Uncertain Long-Span Bridge Deck with an Optimized Method. *Buildings* **2022**, *12*, 632. [CrossRef]
5. Ma, C.; Duan, Q.; Li, Q.; Liao, H.; Tao, Q. Aerodynamic characteristics of a long-span cable-stayed bridge under construction. *Eng. Struct.* **2019**, *184*, 232–246. [CrossRef]
6. Yan, L.; Ren, L.; He, X.; Lu, S.; Guo, H.; Wu, T. Strong Wind Characteristics and Buffeting Response of a Cable-Stayed Bridge under Construction. *Sensors* **2020**, *20*, 1228. [CrossRef] [PubMed]
7. Zan, S.J.; Tanaka, H.; Yamada, H.; Wardlaw, R.L. Parametric investigation of wind-induced cable-stayed bridge motion using an aeroelastic model. *J. Wind Eng. Ind. Aerodyn.* **1989**, *32*, 161–169. [CrossRef]
8. Davenport, A.G. The spectrum of horizontal gustiness near the ground in high winds. *Q. J. R. Meteorol. Soc.* **1961**, *87*, 194–211. [CrossRef]
9. Davenport, A.G. The response of slender, line-like structures to a gusty wind. *Proc. Inst. Civ. Eng.* **1962**, *23*, 389–408. [CrossRef]
10. Davenport, A.G. The application of statistical concepts to the wind loading of structures. *Proc. Inst. Civ. Eng.* **1961**, *19*, 449–472. [CrossRef]
11. Scanlan, R.H. The action of flexible bridges under wind, II: Buffeting theory. *J. Sound Vib.* **1978**, *60*, 201–211. [CrossRef]
12. Scanlan, R.H. Interpreting Aeroelastic Models of Cable—Stayed Bridges. *Eng. Mech.* **1987**, *113*, 555–575. [CrossRef]
13. Scanlan, R.H.; Jones, N. Aeroelastic Analysis of Cable—Stayed Bridges. *J. Struct. Eng.* **1990**, *116*, 279–297. [CrossRef]
14. Yang, Y.; Li, M.; Ma, C.; Li, S. Experimental investigation on the unsteady lift of an airfoil in a sinusoidal streamwise gust. *Phys. Fluids* **2017**, *29*, 051703. [CrossRef]
15. Li, M.; Yang, Y.; Li, M.; Liao, H. Direct measurement of the Sears function in turbulent flow. *J. Fluid Mech.* **2018**, *847*, 768–785. [CrossRef]
16. Su, C.; Fan, X.; He, T. Wind-induced vibration analysis of a cable-stayed bridge during erection by a modified time-domain method. *J. Sound Vib.* **2007**, *303*, 330–342. [CrossRef]
17. King, J.P.C.; Gomez, R.; Pozos-Estrada, A.; Sánchez-García, R. Experimental and analytical evaluation of the aeroelastic behavior of the Baluarte bridge. In Proceedings of the 13th International Conference on Wind Engineering, Amsterdam, The Netherlands, 10–15 July 2011.
18. Pozos-Estrada, A.; Flores, R.; Gómez, R. Parametric Study of Stay Cables of a Bridge Under Simulated Spatially Correlated Turbulent Wind. *Lat. Am. J. Solids Struct.* **2016**, *13*, 1450–1463. [CrossRef]
19. Landero, H.H. Simulation of non-homogeneous wind processes. In *Comparison between Numerical and Experimental Model of a Bridge*; The University of Western Ontario: London, ON, Canada, 2022.
20. Wu, T.; Kareem, A. Modeling hysteretic nonlinear behavior of bridge aerodynamics via cellular automata nested neural network. *J. Wind Eng. Ind. Aerodyn.* **2011**, *99*, 378–388. [CrossRef]
21. Chen, X.; Kareem, A. Nonlinear response analysis of long-span bridges under turbulent winds. *J. Wind Eng. Ind. Aerodyn.* **2001**, *89*, 1335–1350. [CrossRef]
22. Li, Y.; Kareem, A. Simulation of Multivariate Random Processes: Hybrid DFT and Digital Filtering Approach. *J. Eng. Mech.* **1993**, *119*, 1078–1098. [CrossRef]
23. Diana, G.; Rocchi, D.; Argentini, T. An experimental validation of a band superposition model of the aerodynamic forces acting on multi-box deck sections. *J. Wind Eng. Ind. Aerodyn.* **2013**, *113*, 40–58. [CrossRef]
24. Wang, H.; Li, A.; Hu, R. Comparison of Ambient Vibration Response of the Runyang Suspension Bridge under Skew Winds with Time-Domain Numerical Predictions. *J. Bridge Eng.* **2011**, *16*, 513–526. [CrossRef]
25. Diana, G.; Resta, F.; Rocchi, D. A new numerical approach to reproduce bridge aerodynamic non-linearities in time domain. *J. Wind Eng. Ind. Aerodyn.* **2008**, *96*, 1871–1884. [CrossRef]
26. Yan, L.; Ren, L.; He, X.; Li, Y.; Du, B.; Zhong, R. Experimental Study of Buffeting Control of Pingtang Bridge during Construction. *J. Bridge Eng.* **2020**, *25*, 05020004. [CrossRef]
27. Ge, Y.; Yang, Y.; Zhao, L.; Cao, F. Study of Wind Resistant Stability of Main Navigable Span Bridge of Shanghai Changjiang River Bridge. *World Bridges* **2020**, *S1*, 62–72.
28. Su, Z.; Peng, J. Buffeting-Restraining Measures for Long Span Cable-Stayed Bridges at the Longest Double Cantilever Construction Stage. *Highw. Eng.* **2018**, *43*, 146–151.
29. Shen, Z.; Hu, Z.; Li, J.; Xue, X.; Gao, G. Controlling the Modal Coupled Buffeting Response of the Long Span Dense-frequency Cable-stayed Bridge by MDTMD. *China J. Highw. Transp.* **2019**, 32, 135–149.

30. Zapateiro, M.; Pozo, F. *Advances on Analysis and Control of Vibrations: Theory and Applications*; IntechOpen: London, UK, 2012.
31. Moir, G.; Edmonds, C.; Walser, P.; Romberg, M. Construction Engineering of Phu My Cable-Stayed Bridge, Vietnam. *Struct. Eng. Int.* **2010**, *20*, 331–337. [CrossRef]
32. Kim, B.-J.; Lee, S.-H.; Kim, H.-K. Mokpo Bridge: New Landmark in Mokpo City. *Struct. Eng. Int.* **2012**, *22*, 29–31. [CrossRef]
33. Svensson, H.S. *Cable-Stayed Bridges: 40 Years of Experience Worldwide*; Ernst & Sohn: Berlin, Germany, 2012.
34. Kim, Y.M.; Jeong, S.K.; Joo, S.J. Wind-resistant measures of the Busan~Geoje Fixed Link Bridges during construction. In Proceedings of the 18th Congress of IABSE, Seoul, Republic of Korea, 19–21 September 2012; IABSE: Seoul, Republic of Korea, 2012; pp. 660–667.
35. Ding, Q. *Refinement of Coupled Flutter and Buffeting Analysis for Long-Span Bridges*; Tongji University: Shanghai, China, 2001.
36. Ministry of Transport of the People's Republic of China. *Wind-Resistant Design Specification for Highway Bridges*; China Communications Press: Beijing, China, 2018.
37. Hu, X.; Yang, S.; Li, Y.; Wang, Q. Study of Buffeting Response and Suppression Measures for Composite Girder Cable Stayed Bridge in Longest Cantilever State During Construction. *Bridge Constr.* **2021**, *51*, 21–27.

buildings

MDPI

Article

Rib Alignment Control of Long-Span Arch Bridge in Cable-Stayed Buckle by Multi-Objective Optimization

Mengsheng Yu [1,2], Xinyu Yao [2,*], Longlin Wang [2], Tianzhi Hao [3] and Nianchun Deng [4]

1 Guangxi Beitou Gulf Investment Group Co., Ltd., Nanning 530029, China; xiaoluo19890316@163.com
2 Guangxi Transportation Science and Technology Group Co., Ltd., Nanning 530007, China; gxjkqlywll@163.com
3 Guangxi Beitou Traffic Maintenance Technology Group Co., Ltd., Nanning 530029, China; htz0537@163.com
4 College of Civil Engineering and Architecture, Guangxi University, Nanning 530004, China; dengnch@gxu.edu.cn
* Correspondence: xinyu1052007892@163.com; Tel.: +86-15678822242

Abstract: The construction duration of long-span arch bridges is excessively prolonged due to insufficient closing precision and the non-convergence of traditional cable adjustment calculation methods. This study investigates cable force management in long-span concrete-filled steel tubular (CFST) arch bridges during cable-stayed buckle construction, aiming to improve construction safety and precision in arch rib alignment. Using the Pingnan Third Bridge and Tian'e Longtan Bridge as practical examples, the research develops a multi-objective optimization method for cable forces that integrates influence matrices, constrained minimization, and a forward iterative approach. This method offers a robust strategy for tensioning and cable-stayed buckling, enabling real-time monitoring, calculation, and adjustment during the construction of large-span CFST arch bridges. The results reveal that the iterative approach notably enhances calculation efficiency compared to conventional methods. For instance, field measurements at the Pingnan Third Bridge show a minimal arch closure error of only 3 mm. Additionally, the study addresses concerns about excessive stress in exposed steel tubes during concrete casting. By optimizing the sequence of main arch closure and concrete casting, stress in the exposed steel tube is reduced from 373 MPa to 316 MPa, thus meeting specification requirements. In summary, the multi-objective cable force optimization method demonstrates superior efficiency in determining cable tension and controlling rib alignment during cable-stayed buckle construction of long-span CFST arch bridges.

Keywords: long-span CFST arch bridge; influence matrix; constrained minimization; construction monitoring

Citation: Yu, M.; Yao, X.; Wang, L.; Hao, T.; Deng, N. Rib Alignment Control of Long-Span Arch Bridge in Cable-Stayed Buckle by Multi-Objective Optimization. *Buildings* **2024**, *14*, 3281. https://doi.org/10.3390/buildings14103281

Academic Editors: Grzegorz Ludwik Golewski and Mizan Ahmed

Received: 20 August 2024
Revised: 24 September 2024
Accepted: 14 October 2024
Published: 17 October 2024

1. Introduction

The utilization of cable-stayed buckles has demonstrated effectiveness as a widely employed technique for constructing long-span arch bridges. Presently, eight arch bridges with spans exceeding 400 m are operational on a global scale. Notably, the Pingnan Third Bridge and the Tian'e Longtan Bridge feature impressive spans of 560 m (comprising 44 segments) and 600 m (comprising 48 segments), respectively. In CFST arch bridges, the span increases proportionally with the number of arch rib segments, thereby rendering the determination of reasonable cable force and the alignment control of arch rib an intricate process. As such, the construction monitoring of large-span arch bridges poses a significant challenge and calls for precise cable force determination and arch rib alignment control [1–7].

The present study investigates the optimization and implementation of cable force in the process of cable-stayed buckle construction of a CFST (Concrete Filled Steel Tubular) arch bridge, with primary reference to cable-stayed bridges [8,9]. Various methods have been proposed, including the influence matrix, forward iteration, and backward iteration [10]. However, these methods have limitations such as low computational efficiency, ignoring cable force uniformity, and arch alignment accuracy. To address these limitations, Xu et al. [11] proposed an improved iteration method but did not consider cable force uniformity and arch alignment accuracy. In contrast, Zhang et al. [12] argued that the backward iteration method is incapable of obtaining an accurate solution in complicated constructions due to unknown parameters and nonlinear effects. Therefore, this study proposes an approach to calculate the displacement influence matrix under the arch self-weight and cable forces based on the influence matrix [13–19]. The objective optimization function is the deviation between the control point and target alignment during construction.

In contrast, the deviation between the arch and its target alignment after arch closure serves as the constrained function. While this method effectively manages the uniformity of cable forces, it does not sufficiently address the accuracy of arch closure or the alignment of upstream and downstream segments. Another approach, known as the influence matrix method, involves adjusting and optimizing cable forces based on the influence degree matrix at each stage. The cable forces for the hoisting segments are calculated according to the construction sequence. However, without appropriate constraints on the influence matrix, the resulting cable forces can be non-uniform, leading to misalignment of the arch and failure to meet the desired objectives.

This study develops a rapid cable regulation method using the influence matrix to achieve high-precision adjustments in the linear shape of cable-structure bridges, such as arch bridges, during construction. This method ensures that the bridge achieves its optimal shape upon completion, aligning with engineering needs while offering notable social and economic benefits. Real-time monitoring throughout construction allows for the identification of deviations from the ideal state, enabling precise regulation and control. This paper presents a study that uses the Pingnan Third Bridge and Tian'e Longtan Bridge as case studies to develop a multi-objective cable force optimization method. The method integrates the influence matrix, constrained minimization, and forward iterative techniques to determine the optimal cable forces during the hoisting of arch rib segments in long-span CFST (Concrete-Filled Steel Tubular) arch bridges. This approach is especially useful for monitoring and optimizing the construction of arch bridges that require one-time tensioning without subsequent cable adjustments. By comparing the in situ measured data, it is found that the measured errors in cable force are within specification limits, and the alignment of the arch rib during installation and after arch closure is satisfactory. Therefore, this multi-objective cable force optimization method provides an effective means of controlling cable forces in long-span CFST arch bridges and offers a valuable reference for similar bridge construction projects.

2. Method and Theory of Cable-Stayed Buckle

2.1. General Solution of Initial Cable Tension and Hoisting Cable Force

The cable-stayed buckle arch bridge's force calculation is documented in the literature pertaining to cable-stayed bridges [20–23]. Presently, the optimization and determination of cable force during construction predominantly employ the influence matrix method. The control objective $\{\Delta R\}$ considers the influence of all structural dead load and cable force. The main purpose of the objective function is to reduce the difference between the target value and the model prediction value. An overarching equation to solve for the initial tension of the buckle cable in a CFST arch bridge is established as follows:

$$[A]\{T_0\} + \{\Delta S\} = \{\Delta R\} \tag{1}$$

where $[A]$ represents the unit influence matrix for cable force adjustments relative to control targets; T_0 represents the initial cable tension that needs to be determined; $\{\Delta S\}$ represents the displacement caused by the dead load during the construction of the arch cable-stayed buckle; $\{\Delta R\}$ represents the target displacement for the arch bridge design alignment.; m and n are the numbers of elements of $\{\Delta R\}$ and $\{\Delta T\}$, respectively, and m = n.

For the nonlinearity of a long-span arch bridge, the sag effect of the cable is mainly considered at the stage of cable-stayed suspension, and the sag effect is simulated by the equivalent elastic modulus method. In the context of cable-stayed buckle arch bridges, cable tension is determined via iterative forward calculations that utilize the initial tension as a starting point. Notably, the installation of new components occurs tangentially to the previous components. Consequently, the impact of tangential displacement resulting from this installation on the finite element calculation must be accounted for to ensure the calculation results conform to actual engineering requirements. The nonlinearity of the superposition principle is thus introduced, which leads to discrepancies between the calculated cable forces obtained from the influence matrix and the actual tension experienced during installation. Therefore, further optimization and adjustment are required to minimize errors in the tensioning of cables for cable-stayed buckle arch bridge construction that considers the influence of tangential installation displacement.

2.2. Arch Rib Alignment Control Method Based on Multi-Objective Optimization

The innovative multi-objective cable force optimization method proposed herein is based on the iterative analysis of rigid frame structures. This approach simultaneously controls the cable forces in the completed bridge, its geometry, and the displacement of the towers, achieving a one-time tensioning effect without subsequent adjustments to cable forces. It eliminates the need for a sequential adjustment process for orderly tensioning, fundamentally addressing the technical challenges associated with traditional empirical methods that require strict adherence to a predetermined tensioning sequence. If an inappropriate order is chosen for cable tensioning, it necessitates repeated adjustments to achieve design specifications, resulting in complex construction procedures and difficulties in ensuring precision.

When constructing a long-span CFST arch bridge, several factors must be considered to ensure the safety and uniformity of cable forces. These factors include the alignment of the arch rib, the uniformity of cable forces, and safety considerations. The alignment of the arch rib is primarily evaluated based on the cable closure and target alignment, the displacement error of the north and south segments following cable closure, the displacement error of the upstream and downstream sections of the cross brace, and the displacement error of the maximum cantilever end before cable closure. To achieve cable force uniformity and ensure safety during construction, cable forces are optimized by adjusting alignment. The least square method is used to establish the square error function Q, which optimizes cable forces:

$$\begin{gathered} Q(X) = \|[A_s]\{\Delta X\} + \{dis_{01} - dest\}\|^2 \\ Q = \sum_{i=1}^{m} [\{dis_{01} - u_{dest}\} - \sum_{j=1}^{n} \Delta\alpha_{ij} \cdot \Delta X_j]^2 \end{gathered} \tag{2}$$

where $\Delta\alpha_{ij}$ represents the change in the jth control target caused by a unit adjustment in the ith cable force.

According to the extreme value theorem, when the objective function $Q(\{\Delta X\})$ takes the minimum value, ΔX_k needs to satisfy the following equation,

$$\frac{\partial Q(\{\Delta X\})}{\partial \Delta X_k} = 0(k = 1, 2, \cdots, n) \tag{3}$$

$$2\sum_{i=1}^{m} \{[\{dis_{01} - u_{dest}\} - \sum_{j=1}^{n} \Delta\alpha_{ij} \cdot \Delta X_j] \cdot (-\alpha_{ik})\} = 0 \tag{4}$$

Simplified Equation (4) can be obtained:

$$\sum_{j=1}^{n} \sum_{i=1}^{m} \Delta\alpha_{ij} \cdot \Delta X_j = \sum_{i=1}^{m} \alpha_{ik} \{dis_{01} - u_{dest}\}_i \tag{5}$$

Equation (2) can be written in matrix form:

$$[A]_1{}^T[A]_1\{\Delta X\} = [A]_1{}^T\{dis_{01} - u_{dest}\} \tag{6}$$

Equation (4) is a system of linear equations with n unknowns and n equations and the final optimized one-time tension initial cable force $\{T\}$ can be obtained by solving the $\{\Delta X\}$ according to Equation (6):

$$\{T\} = \{T_0\} + \{\Delta X\} \tag{7}$$

where $\{T_0\}$ represents the initial cable tension of the control target before adjustment.

The constrained optimization problems are usually transformed into simpler subproblems that can be solved and used as the basis of the iterative process. Based on the solution of the Kuhn–Tucker (K–T) equation, the constrained optimization problem is transformed into an unconstrained optimization problem, and the constraint variables and functions are:

$$\begin{cases} |[A_0]\Delta X + dis_0 - u_0| \le \Delta_0 \\ |[C_1]\Delta X + u_1| \le \Delta_1 \\ |[C_2]\Delta X + u_2| \le \Delta_2 \\ dis_1 - \Delta_x \le [A_0]\Delta X \le \Delta_s - dis_1 \\ k = \frac{f_i^0}{f_{lim}} \end{cases}$$

$$A_0 = \begin{bmatrix} \delta_{11}^0 & \delta_{12}^0 & \cdots & \delta_{1n}^0 \\ \delta_{21}^0 & \delta_{22}^0 & \cdots & \delta_{2n}^0 \\ \vdots & \vdots & \ddots & \vdots \\ \delta_{n1}^0 & \delta_{n2}^0 & \cdots & \delta_{nn}^0 \end{bmatrix}, A_1 = \begin{bmatrix} \delta_{11}^1 & \delta_{12}^1 & \cdots & \delta_{1n}^1 \\ \delta_{21}^1 & \delta_{22}^1 & \cdots & \delta_{2n}^1 \\ \vdots & \vdots & \ddots & \vdots \\ \delta_{n1}^1 & \delta_{n2}^1 & \cdots & \delta_{nn}^1 \end{bmatrix}, A_2 = \begin{bmatrix} \delta_{11}^2 & \delta_{12}^2 & \cdots & \delta_{1n}^2 \\ \delta_{21}^2 & \delta_{22}^2 & \cdots & \delta_{2n}^2 \\ \vdots & \vdots & \ddots & \vdots \\ \delta_{n1}^2 & \delta_{n2}^2 & \cdots & \delta_{nn}^2 \end{bmatrix} \tag{8}$$

$$C_1 = \begin{bmatrix} \delta_{11}^1 & \delta_{12}^1 & \cdots & \delta_{1n}^1 \\ \delta_{21}^1 & \delta_{22}^1 & \cdots & \delta_{2n}^1 \\ \vdots & \vdots & \ddots & \vdots \\ \delta_{n1}^1 & \delta_{n2}^1 & \cdots & \delta_{nn}^1 \end{bmatrix}, C_2 = \begin{bmatrix} \delta_{11}^1 & \delta_{12}^1 & \cdots & \delta_{1n}^1 \\ \delta_{21}^1 & \delta_{22}^1 & \cdots & \delta_{2n}^1 \\ \vdots & \vdots & \ddots & \vdots \\ \delta_{n1}^1 & \delta_{n2}^1 & \cdots & \delta_{nn}^1 \end{bmatrix}, C_3 = \begin{bmatrix} \delta_{11}^1 & \delta_{12}^1 & \cdots & \delta_{1n}^1 \\ \delta_{21}^1 & \delta_{22}^1 & \cdots & \delta_{2n}^1 \\ \vdots & \vdots & \ddots & \vdots \\ \delta_{n1}^1 & \delta_{n2}^1 & \cdots & \delta_{nn}^1 \end{bmatrix}$$

where dis_0 represents the actual displacement after cable closure under the initial force t0; dis_1 represents the actual displacement of the current tension segment under the initial forcee t0; dis_2 represents the actual displacement of the current transverse brace segment under the initial force; ΔX represents the optimized cable force incremental vector; Δ_0 represents the allowable error in target displacement between the control point after cable closure and the bare arch; Δ_1 represents the upstream and downstream displacement error of the transverse brace segment; Δ_2 represents the displacement error of the largest cantilever before cable closure; C_1 represents the upstream and downstream displacement difference control matrix for the current transverse brace segment; C_2 represents the south and north displacement difference control matrix after cable closure; C_3 represents the displacement difference control matrix of the closure segment; u_0 represents the difference between the total displacement of the initial model and the actual displacement after cable closure under the initial force t0; u_1 represents the displacement difference vector of the south and north arch rib after cable closure under the initial force t0; u_2 represents the displacement difference vector of the upstream and downstream arch rib under the initial force t0; u_3 represents the displacement difference vector of the closure segment; Δ_x represents the lower limit of the current constraint tensile segment; Δ_s represents the upper limit of the current constraint tensile segment and denotes the cable tension force; f_{lim} represents the cable-breaking force.

With the aim of achieving alignment, uniformity, and safety in cable tension while maintaining the established status of the bridge, adjustments to the linear configuration of long-span cable-stayed bridges are made by optimizing cable forces during construction. A rapid solution method for determining appropriate installation cable tensions is developed to ensure that the number of tensioning operations is minimized. This approach guarantees that cable forces remain uniform and within permissible safety limits, ultimately leading to an optimal shape for the bridge.

3. Calculation Steps of Arch Rib Alignment Control Based on Multi-Objective Optimization

For long-span CFST arch bridges with numerous segments and intricate construction processes, cable force adjustments must be made repeatedly to satisfy multiple objectives concurrently. This necessitates the involvement of individuals with extensive experience in cable adjustment calculations. However, the tediousness of manual cable adjustment can be eliminated by utilizing MATLAB R2016a. The mathematical formulation implemented in MATLAB utilizes the cyclic loading method to tackle the problem [24–27]. It is important to ensure that the objective and constraint functions are continuous to prevent the occurrence of a locally optimal solution. In addition, the objective and constraint functions must be real numbers when the fmincon function is used for maximum constraint minimization. The steps involved in solving the alignment control problem of long-span bridge arch ribs using cable-stayed construction and multi-objective optimization are shown in Figure 1.

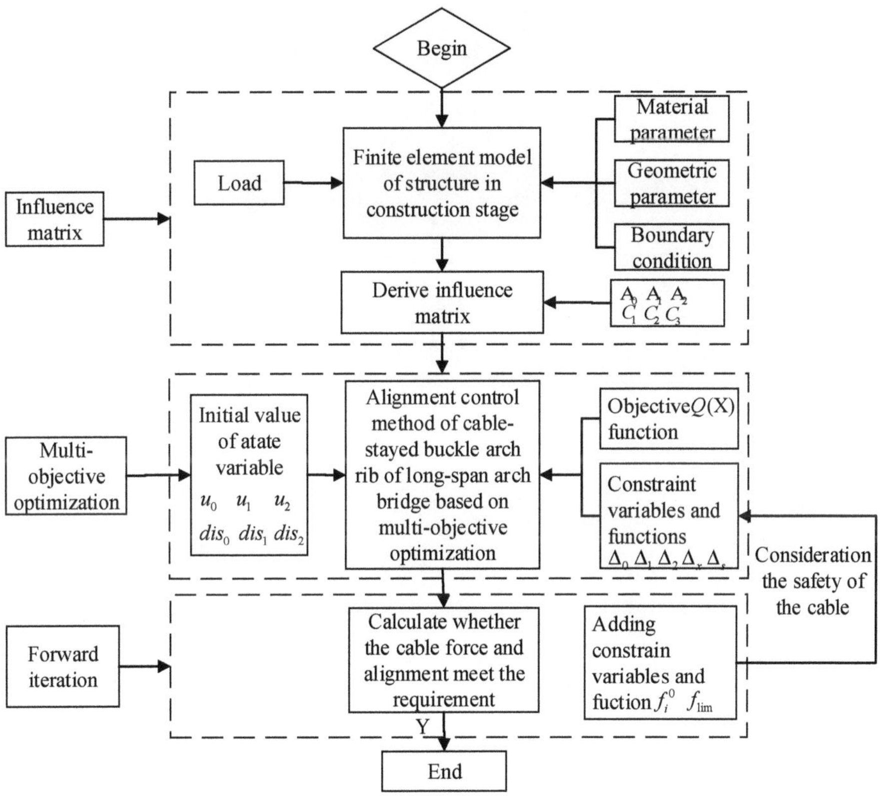

Figure 1. Iterative calculation steps of cable force adjustment.

4. Engineering Applications

4.1. Engineering Overview and Calculation Model Acquisition

The Pingnan Third Bridge's cable-stayed buckle system comprises two components: the buckle and the horizontal force adjustment cable. The cable buckle supports the cantilever assembly of the arch rib segment. At the same time, the ground stretching form is incorporated into the design for the safety and convenience of the cable buckle. The horizontal force adjustment cable consists of active and passive adjustment cables. In the event that the buckle tower is affected by unbalanced horizontal forces, the horizontal force adjustment cable regulates the tower's deviation within a small range, mitigating the influence of the hoisting segment on the buckle segment elevation. The buckle cable was constructed from high-strength, low-relaxation 1860 MPa Φ15.24 mm steel strands, with a single strand's area and elastic modulus (E) being 140 mm^2 and 1.95 × 105 MPa, respectively. The anchor end of the buckle cable is secured with a low retraction anchorage situated close to the hoisting arch rib segment's end at the buckle point. The overall layout of the cable-stayed buckle method is depicted in Figure 2. The Pingnan Third Bridge's finite element model comprises 12,321 nodes and 24,251 elements, with material properties, geometric characteristics, boundary conditions, and external load information consistent with the design drawing data. Due to the bridge's symmetry, the cable buckles are labeled based on the construction stage's sequence. The modeling assumptions employed include steel's ideal elastic material property and the plane section assumption for cross-section deformation. The finite element model is presented in Figure 3.

Figure 2. The elevation view of the Pingnan Third Bridge.

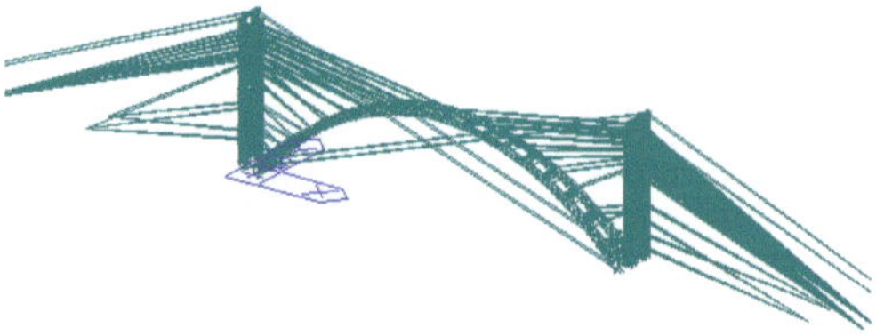

Figure 3. The schematic diagram of the displacement control point.

The Tian'e Longtan Bridge, which spans 624 m, is a rib-skeleton concrete arch bridge with a cable buckle design featuring divided cables, whereby the pulling end cable force is equal to the anchoring end. A single-bundle cable force was used during construction, whereas each arch rib segment consisted of two-bundle cables. The buckle cable was constructed of high-strength, low-relaxation steel strands with 1860 MPa Φ15.24 mm. A single steel strand has an area of 140 mm^2 and an elastic modulus of 1.95 × 105 MPa. The anchor end comprises a low-retraction anchorage located at the buckle point close to the end of the hoisting arch rib segment. The control point is the intersection point of the arch rib upper axis and the top vertical web axis horizontally translated outward from the main chord during construction. The finite element model of the Tian'e Longtan Bridge has 7614 nodes and 15,545 elements, and its material properties, geometric characteristics, boundary conditions, and external load information are consistent with the design drawing data. The buckles are numbered based on the bridge's construction stage sequence, owing to its symmetry. The modeling method is similar to that of Pingnan Third Bridge, the solid model is shown in Figure 4 below, and the finite element model is shown in Figure 5.

Figure 4. Elevation of Tian'e Longtan Grand Bridge.

Figure 5. The schematic diagram of the cable-stayed buckle construction.

4.2. Results of Arch Rib Alignment Based on Multi-Objective Optimization

The cable-stayed buckle method finite element model of the Pingnan Third Bridge and Tian'e Longtan Bridge was utilized to extract the displacement change influence matrix of the different construction segment control points, namely, the current tension segment, the current transverse brace segment, and the cable closure segment. This was achieved by considering the cable force of the current arch rib installing segment. In addition, the displacement influence matrix of the bare arch resulting from the weight change of the arch rib and the transverse brace of each segment was also extracted. The alignment operation was employed to solve the influence matrices C_1, C_2, C_3, which were solved by A_0, A_1, A_2.

In this study, the arch rib installation stage was simulated based on the final construction sequence, which involved using the tower cable-stayed buckle cantilever assemblage method. The Pingnan Third Bridge and Tian'e Longtan Bridge hoisting stages were divided into 44 and 48 sections, respectively. The cable force was used to balance the arch rib segment self-weight load during the cantilever assemblage. The construction approach involved symmetrical installation on both sides, which started with installing the upstream and downstream arch ribs, followed by transverse connection between the segments, and finally, installing the next segment until cable closure. The objective function aimed to simultaneously satisfy three criteria, including minimizing the sum of squares of the difference between the mean value of the actual displacement of the cantilever end in the arch rib segments installation and the control point displacement of the cantilever end in the arch transverse brace installation caused by the bare arch weight. Furthermore, it sought to minimize the variance between the mean value and target displacement and reduce the influence of the tension buckle cable of the uninstalled segment of the arch rib on the displacement control point of the installed segment. A comparative analysis reveals that for the Pingnan Third Bridge, the inflection point of the optimization function dispersion occurred at *delta* = 30 mm, with the dispersion increasing exponentially as the delta decreased. When the *delta* > 30 mm, the dispersion delta tended to smooth; therefore, it is recommended that the difference between the actual and the target displacement of the control point should not be large, and the delta of the Pingnan Third Bridge should be selected between 30 mm to 35 mm. Similarly, the *delta* of Tian'e Longtan Bridge should be 40 mm~50 mm. This is illustrated in Figures 6 and 7, which show the dispersion and delta variation curves for the Pingnan Third Bridge and Tian'e Longtan Bridge, respectively.

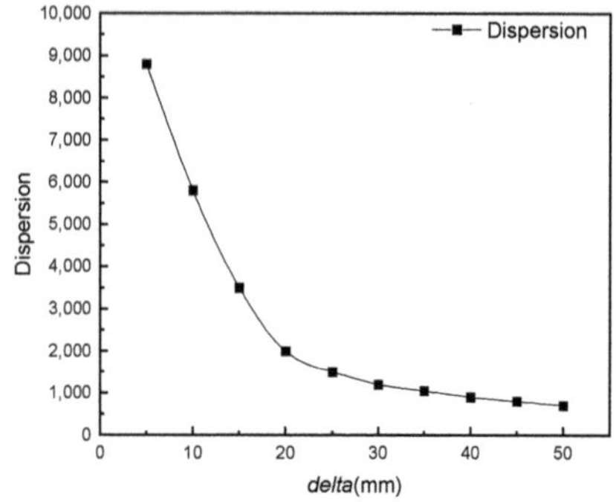

Figure 6. Relationship curve between dispersion and delta of the Pingnan Third Bridge.

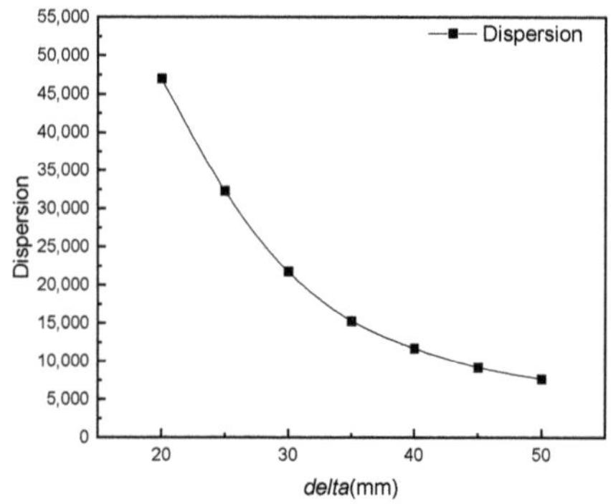

Figure 7. Relationship curve between dispersion and delta of the Tian'e Longtan Bridge.

4.3. Solution Results

The analysis results in Section 4.3 demonstrate that the displacement *delta* between the actual and target values of the cable closure for the Pingnan Third Bridge is 35 mm, while that of the Tian'e Longtan Bridge is 45 mm. The multi-objective arch rib alignment control technique was employed to evaluate the structural construction. The theoretical displacement analysis under the cable tension iteration's final value is reported in Tables 1 and 2.

Table 1. The theoretical displacement calculation results of Pingnan Third Bridge (mm).

Segment	The Target Displacement After Cable Closure	The Actual Displacement After Cable Closure	The Actual Displacement of the Current Tension Segment	The Actual Displacement of the Current Transverse Brace Segment	Relative Error of Actual Displacement to Target
1	−5	−14	0	−11	−9
2	−15	−38	8	−32	−23
3	−23	−49	9	−44	−26
4	−33	−58	106	−54	−25
5	−54	−71	−11	−80	−17
6	−77	−79	25	−87	−2
7	−100	−86	−97	−180	14
8	−119	−95	−51	−207	24
9	−137	−115	−191	−292	22
10	−152	−149	−84	−353	3
11	−162	−197	−322	−329	−35

Table 2. The theoretical displacement calculation results of Tian'e Longtan Bridge (mm).

Segment	The Target Displacement After Cable Closure	The Actual Displacement After Cable Closure	The Actual Displacement of the Current Tension Segment	The Actual Displacement of the Current Transverse Brace Segment	Percentage of Error Between Actual Displacement and Target Displacement
1	−6	−20	−5	−15	−14
2	−14	−41	−14	−35	−27
3	−28	−59	−31	−53	−31
4	−56	−85	−59	−69	−29
5	−98	−118	−105	−152	−20
6	−150	−156	−156	−180	−6
7	−208	−199	−216	−302	9
8	−270	−245	−277	−347	25
9	−333	−295	−339	−434	38
10	−384	−339	−388	−490	45
11	−415	−374	−417	−464	41
12	−423	−395	−420	−506	28

The segmental division of Pingnan Third Bridge is shown in Figure 8 below. The segmental division of Tian'e Longtan Bridge is shown in Figure 9 below.

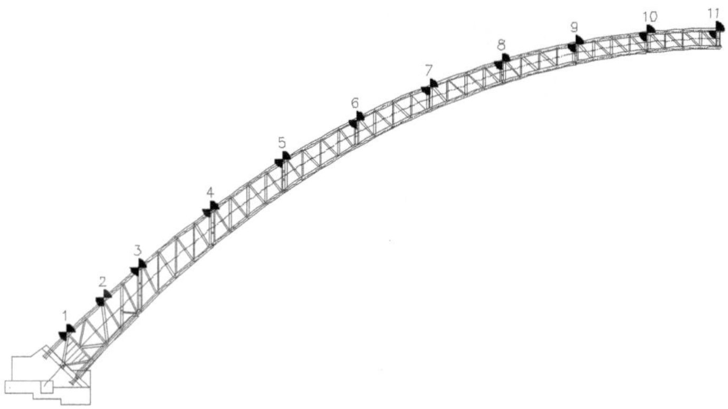

Figure 8. Arch rib segment division (Pingnan Third Bridge) (the number indicates the arch rib segment number).

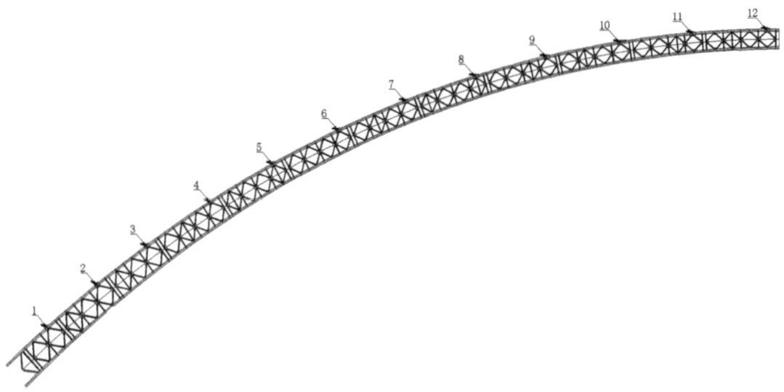

Figure 9. Arch rib segment division (Tian'e Longtan Bridge) (the number indicates the arch rib segment number).

The multi-objective arch rib alignment control method was analyzed and calculated to investigate the displacement change in different segments' cantilever end control points, as observed in Figures 10–13. The results revealed a more gradual displacement change in comparison to other methods. The control points of the two adjacent sections displayed no significant displacement change, indicating good continuity in construction alignment. After the cable closure of the Pingnan Third Bridge and the Tian'e Longtan Bridge, the error between the control point elevation and the target alignment was 10 mm and 30 mm, respectively. During the hoisting construction of the Tian'e Longtan Bridge, the error between the control point elevation and the target alignment was less than 45 mm. All of the errors observed in the study met the required specifications. These findings demonstrate the effectiveness of the multi-objective arch rib alignment control method in achieving precise alignment during construction.

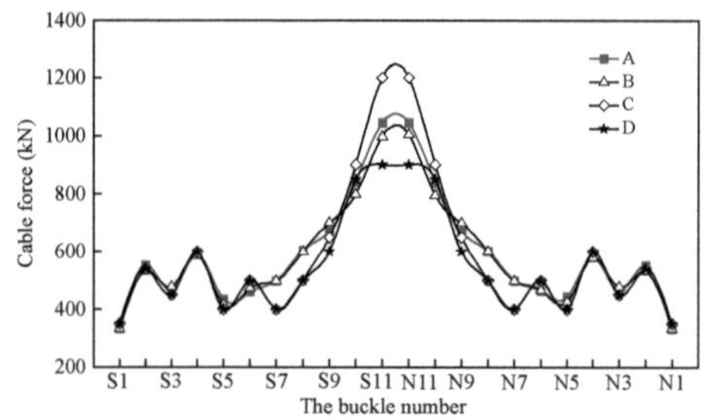

Figure 10. The comparison diagram of the initial iteration cable tension and optimized cable force of Pingnan Third Bridge (unit: kN). Note: S = South, N = North, A = Cable force after upstream optimization, B = Cable force after downstream optimization, C = Cable force before upstream optimization, D = Cable force before downstream optimization.

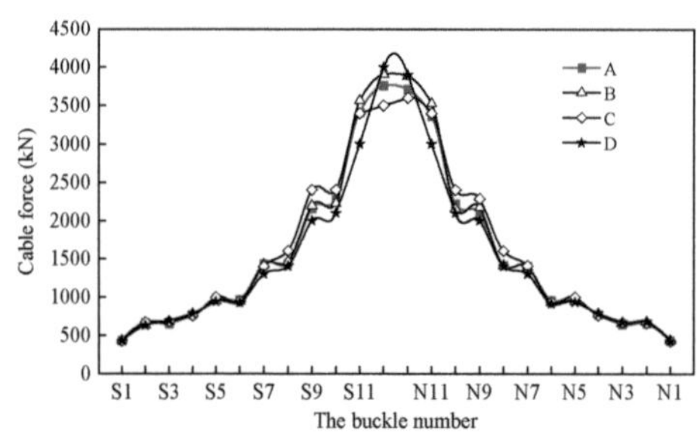

Figure 11. The comparison diagram of the initial iteration cable tension and optimized cable force of the Tian'e Longtan Bridge (unit: kN). Note: S = South, N = North, A = Cable force after upstream optimization, B = Cable force after downstream optimization, C = Cable force before upstream optimization, D = Cable force before downstream optimization.

4.4. *In Situ Measurement Comparison*

The comparison of in situ and theoretical data is presented using the Pingnan Third Bridge as an example, given that the Tian'e Longtan Bridge is currently under construction. During the in situ monitoring of the Pingnan Third Bridge, a one-time passive adjustment method for cable tension was utilized, and the optimization of cable tension was performed to achieve the convergence goal. Cable closure was carried out at a stable time of elevation and axis change, and the arch rib alignment was continuously monitored in situ. The displacement of the control point is presented in Figure 14 and analyzed using the multi-objective optimization method for arch rib alignment. Results show that the cable closure error of Pingnan Third Bridge is 3 mm. This precision adjustment method for arch rib alignment enabled control of each arch rib's alignment during the hoisting construction process, while ensuring uniform cable force that met safety requirements. This approach effectively resolved the issue of repeated cable adjustments and offered the advantages of high efficiency and convenience.

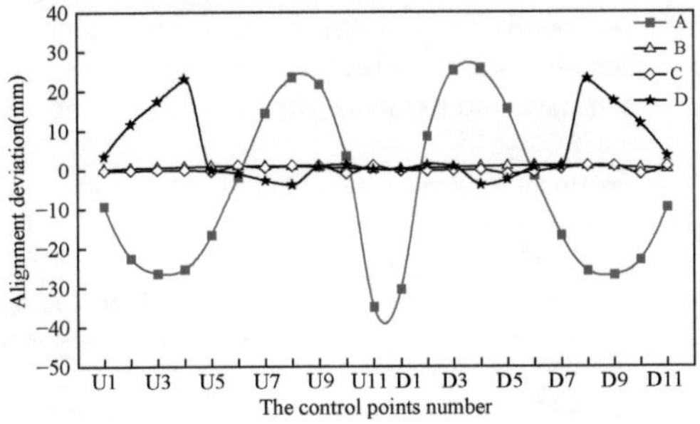

Figure 12. The displacement result diagram of the Pingnan Third Bridge. Note: U = Up, D = Down, A = The difference between the alignment of the after cable closure and the target displacement, B = The displacement difference between south and north after cable closure, C = The displacement difference between the upstream and downstream of the installed horizontal brace segment, D = The displacement difference between south and north in the current tension stage.

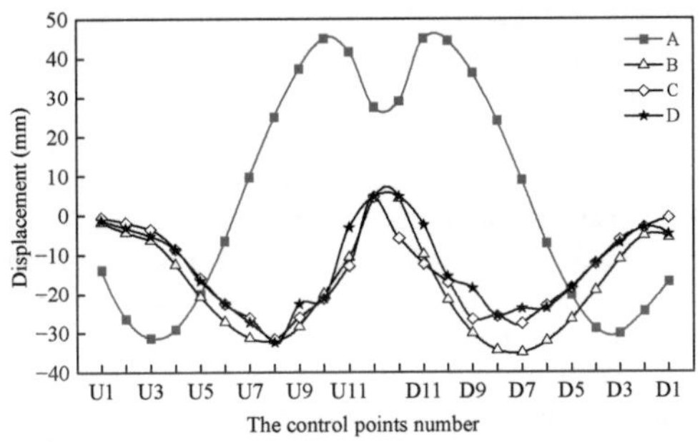

Figure 13. The displacement result diagram of the Tian'e Longtan Bridge. Note: U = Up, D = Down, A = The difference between the alignment of the after cable closure and the target displacement, B = The displacement difference between south and north after cable closure, C = The displacement difference between south and north in the current tension stage, D = The displacement difference between the upstream and downstream of the installed horizontal transverse brace segment.

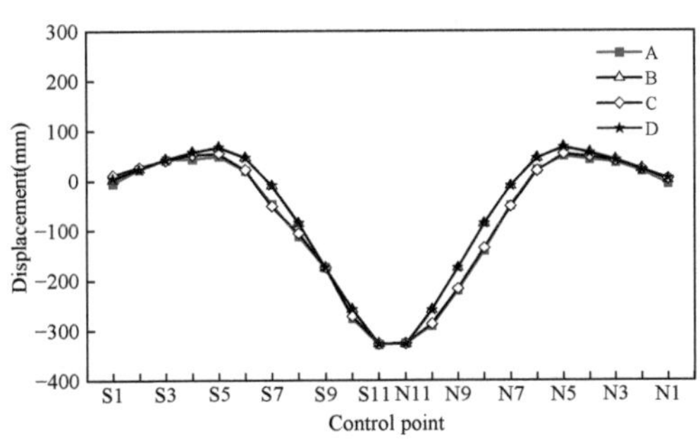

Figure 14. The displacement of in situ measured and theoretical of the Pingnan Third Bridge. Note: S = South, N = North, A = The upstream in situ measured displacement, B = The upstream theoretical displacement, C = The downstream in situ measured displacement, D = The downstream theoretical displacement.

5. Comparative Analysis of Each Optimization Scheme

According to the construction process of the Tian'e Longtan Bridge, the mid-span section of the exposed steel tube stress is the most concerning area. This observation indicates that the cable closure section of the CFST rigid skeleton arch bridge is the most critical and high-risk segment. Consequently, the division of construction stages commenced with the installation of the cable closure. To analyze the stress superposition of the exposed steel tube accurately, the construction stage and sequence were considered and are presented in Table 3.

Table 3. The construction stage division of the Tian'e Longtan Bridge.

Stage	Construction Stage Content
CS1	Install 12 segment rigid skeleton
CS2-CS6	Install cable closure segment and transverse braces
CS7	Dismantle cable
CS8	Dismantle interim transverse brace
CS9-CS24	Casting and activating concrete in the tube
CS25-CS26	Casting and activating permanent transverse braces
CS27-CS62	Bottom concrete encasement stage
CS63-CS68	Activate bottom stiffness
CS69-CS110	Web concrete encasement stage
CS111-CS115	Activate web stiffness
CS115-CS144	Top plate concrete encasement stage

In the process of constructing a rigid skeleton arch bridge, each stage of construction is characterized by creep, shrinkage, boundary constraints, and changes in structural stiffness due to the application of structural load step by step. The final structure's mechanical properties and stress state are closely related to the previous stages of construction. Specifically, in the concrete encasement process, the stress state of the exposed steel tube is affected by the previous stress superposition process. Therefore, optimizing the stress of the exposed steel tube requires considering the previous stress process. The construction process can be divided into two main stages: the installation of the rigid skeleton and the concrete encasement. The former is achieved through hoisting and cable-stayed buckle methods. At the same time, the latter is constructed using the multi-point balanced loading method, which ensures the stability of the structure by effectively balancing the stress of each section during construction. Therefore, optimization of the construction process focuses on the design of the rigid skeleton installation and concrete encasement casting schemes, as presented in Table 4.

Table 4. The optimization scheme of the Tian'e Longtan Bridge.

The Optimization Scheme	Optimization Stage	Optimization Content
A	Rigid skeleton installation stage	Optimize the cable closure scheme
B	Concrete encasement casting stage	Optimize the web casting sequence
C	Rigid skeleton installation stage Concrete encasement casting stage	Optimize the sequence of cable closure and web casting at the same time

During the installation phase of the rigid skeleton, the original plan for installing the upstream segments 11 and 12 involved first performing a unilateral cable closure followed by installing the downstream segments 11 and 12. However, an optimized scheme was developed to install the upstream segment 12 only after completing upstream and downstream segments 11 and 12 and to perform the downstream cable closure only after the upstream segment was installed. For the concrete encasement phase, the construction method adopted was the vertical ring and longitudinal section multi-surface casting, resulting in a uniform stress distribution and smooth deformation of the rigid skeleton. The exhaustive method was employed to adjust the sub-ring casting sequence, with the apex of the arch cast first to enhance the strength of the cross-section near the apex and reduce subsequent superimposed stress. Based on the optimized working surface and casting sequence, the stress and deformation process of the structure was calculated, as shown in Figure 15. The optimized cable closure and casting scheme significantly decreased the exposed steel tube's stress from the original 373 MPa to 316 MPa, meeting the specification requirements.

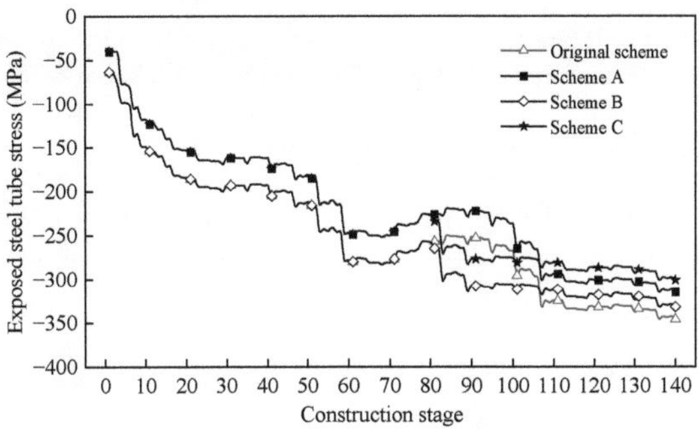

Figure 15. The comparison diagram of stress optimization schemes for the Tian'e Longtan Bridge.

6. Conclusions

The alignment control problem considered in this study can be summarized as cable force optimization, which is different from the existing cable force optimization, since the arch rib alignment, tower deviation, cable force uniformity, and safety during construction for long-span arch bridges' cable-stayed buckles are considered. This approach effectively addresses the defects of traditional monitoring and calculation methods. Additionally, this paper proposes a solution to address the impact of temperature by conducting daily measurements at low temperatures. From a construction perspective, it is essential to resolve the issue of arch rib positioning under varying temperature conditions. The following conclusions are drawn:

(1) The multi-objective cable force optimization method is an efficient and convenient approach for iterative calculation and engineering applications. It can analyze the relationship between delta and optimization dispersion to determine the difference between actual and target displacement after cable closure under the action of the bare arch.
(2) In constructing cable-stayed buckle long-span arch bridges, arch rib alignment, cable force uniformity, and safety are considered comprehensively. Optimizing the alignment of cable force during construction ensures that the cable force is uniform and meets specification requirements, effectively overcoming the traditional manual cable adjustment and cable force inhomogeneity issues.
(3) The alignment deviation in the construction process could affect the cable force. Still, the multi-objective cable force optimization method can ensure precision during arch segment installation, and the arch/bridge completion states meet specification requirements.
(4) The control cross-sectional stress of exposed steel tube in rigid skeleton arch bridge is related to the installation method and the concrete casting scheme. To achieve a good stress control effect, selecting the appropriate construction scheme to reduce construction risk is necessary.

Author Contributions: Conceptualization, X.Y. and M.Y.; methodology, L.W.; data curation, X.Y.; writing—original draft preparation, T.H. and N.D.; writing—review and editing, X.Y. All authors have read and agreed to the published version of the manuscript.

Funding: This study was partly sponsored by following funding programs: (1) National Natural Science Foundation of China (Code: 51738004); (2) National Natural Science Foundation of China (Code: 51868006); (3) The Scientific and Technological Project of the Science and Technology Department of Guangxi Province (grant number: 2021AC19125).

Data Availability Statement: The testing and analysis data used to support the findings of this study are included within the article.

Conflicts of Interest: Author Mengsheng Yu was employed by the company Guangxi Beitou Gulf Investment Group Co., Ltd. Author Mengsheng Yu, Xinyu Yao and Longlin Wang were employed by the company Guangxi Transportation Science and Technology Group Co., Ltd. Author Tianzhi Hao was employed by the company Guangxi Beitou Traffic Maintenance Technology Group Co., Ltd. The remaining authors declare that the research was conducted in the absence of any commercial or financial relationships that could be construed as a potential conflict of interest.

References

1. Zhao, R.; Zhang, Z. A Summary of Development of Concrete-Filled Steel Tube Framed Arch Bridges in China. *Bridge Constr.* **2016**, *46*, 45–50.
2. Fan, Y.; Xin, J.; Yang, L. Optimization method for the length of the outsourcing concrete working plane on the main arch rib of a rigid-frame arch bridge based on the NSGA-II algorithm. *Structures* **2024**, *59*, 105767. [CrossRef]
3. Tao, T.; Wang, H. Simulation of multivariate ergodic stochastic processes using adaptive spectral sampling and non-uniform fast Fourier transform. *Probabilistic Eng. Mech.* **2024**, *77*, 103669. [CrossRef]
4. Tao, T.; Wang, H. Efficient buffeting analysis of long-span bridges under non-stationary winds: A 2D interpolation enhanced approach. *J. Sound Vib.* **2023**, *559*, 117754. [CrossRef]
5. Yao, X.; Li, C.; Wang, L.; Yu, M.; Zhuo, X.; Hao, T.; Wang, X. A Practical Approach to Alignment and Error Feedback Control for Long-Span Arch Bridges. *Buildings* **2024**, *14*, 1995. [CrossRef]
6. Zhou, Q.; Feng, P.; Zhou, J.; Xin, J.; Wang, J. Analysis of concrete emptying in concrete-filled steel tube arch bridge under uneven temperature field. *J. Bridge Constr.* **2024**, *54*, 103–109. [CrossRef]
7. Mou, T.; Fan, B.; Zhao, Y.; Li, S. Application and Development of Steel Pipe Concrete Bridges in China. *Highway* **2017**, *62*, 161–165.
8. Zhou, Y.; Wang, Y.; Zhou, J. Arch calculation and control method of 500 m class steel pipe arch bridge. *China Highw.* **2022**, *35*, 60–72. [CrossRef]
9. Li, Y.; Li, Y.; Li, J. The calculation method of the whole process cable force optimization of the long-span concrete-filled steel tube arch bridge constructed by cable-stayed suspension method. *Prog. Build. Steel Struct.* **2019**, *21*, 33–39. [CrossRef]
10. Xie, K.; Wang, H.; Guo, X.; Zhou, J. Study on the safety of the concrete pouring process for the main truss arch structure in a long-span concrete-filled steel tube arch bridge. *Mech. Adv. Mater. Struct.* **2021**, *28*, 731–740. [CrossRef]
11. Xu, Y.; Shen, C.; Zhu, Y.; Wang, C. Improved Iteration Algorithm for Determination of Tension of Fastening Stays for Cantilever Construction of Arch Bridge. *Bridge Constr.* **2016**, *46*, 65–69.
12. Zhang, J.; Zheng, J.; Xiao, R. Optimization Calculation and Analysis of Concrete Filled Steel Tube Arch Bridge Hoisting Process. *China J. Highw. Transp.* **2005**, *2*, 40–44. [CrossRef]
13. Zhen, J.; Wang, J.; Feng, Z.; Han, Y.; Qin, D. Test on vacuum auxiliary filling technology of concrete-filled steel tube arch section. *China Highw. J.* **2014**, *27*, 44–50. [CrossRef]
14. Han, Y. Test and application of vacuum assisted filling of concrete in tubular arch bridge. *Bridge Constr.* **2015**, *45*, 19–25.
15. Zhou, D.; Deng, N.; Shi, T. Test and numerical simulation analysis of large scale hydration temperature field of concrete-filled steel tube arch bridge. *J. Guangxi Univ.* **2021**, *46*, 51–59. [CrossRef]
16. Sun, J.; Xie, J. Simulation analysis of the hydration heat of large diameter CFST arch and its effects on loading age. *Appl. Therm. Eng.* **2019**, *150*, 482–491. [CrossRef]
17. Zhou, Q.; Zhou, J.; Zhang, J.; Zhang, L. Self-regulating loading pouring method of long span CFST arch bridge. *J. Harbin Inst. Technol.* **2020**, *52*, 82–89. [CrossRef]
18. Zheng, J.; Wang, J. Chinese steel tube concrete arch bridge. *Engineering* **2018**, *4*, 306–331. [CrossRef]
19. Qin, D.; Zheng, J.; Du, H.; Han, Y.; Zheng, J.; Wei, L. Optimization Calculation Method for Stayed-Buckle Cable Force under One-Time Tension by Fastening Stay Method and Its Application. *China Railw. Sci.* **2020**, *41*, 52–60.
20. Han, Y.; Qin, D.; Zheng, J. Optimization calculation method for CFST arch bridge cable-stayed suspension construction. *Highway* **2018**, *63*, 100–104.
21. Zhou, J.; Liu, J.; Zhou, W.; Yan, R.; Yan, T. Analysis of the influence of temperature changes on the pre lifting value of cable-stayed buckle and the main arch ring shape of steel tube concrete arch bridges. *J. China Foreign Highw.* **2017**, *37*, 62–66. [CrossRef]
22. Chen, B.; Wei, J.; Zhou, J.; Liu, J.P. Application of concrete-filled steel tube arch bridges in China: Current status and prospects. *China Civ. Eng. J.* **2017**, *50*, 50.
23. Yu, M.; Deng, N.; Wang, L.; Hao, T.; Zhang, Z. Study on Sunshine Temperature Effect in Concrete-filled Steel Tubes Arch Rib of Extra-Large Arch Bridge. *Highw. Eng.* **2021**, *46*, 99–104. [CrossRef]

24. Liu, X.; Zhou, G.; Qiu, Y. Optimal Polynomial Time-varying Parameters Discrete Grey Model and Its Applications. *Stat. Decis.* **2022**, *38*, 31–36.
25. Li, X.; Wang, C. STI prediction based on metabolic GM (1, 1) Copula BP neural network. *Stat. Decis.* **2021**, *37*, 158–161. [CrossRef]
26. Vasques, J.F.; Gonalves, R.G.D.J.; Silva-Junior, A.J.D.; Martins, R.S.; Gubert, F.; Mendez-Otero, R. Gangliosides in nervous system development, regeneration, and pathologies. *Neural Regen. Res.* **2023**, *18*, 81–86. [CrossRef]
27. Li, C.; Huang, J.; Li, X. Consider the sunshine temperature field of concrete-filled steel tube arch. *Sino-Foreign Highw.* **2020**, *40*, 102–107.

buildings

MDPI

Article

A New Method for Correcting the Deviation of a Middle Pier Tower of a Long-Span Intermediate Arch Bridge

Xiaojin Dong and Dan Ye *

State Key Laboratory of Mountain Bridge and Tunnel Engineering, Chongqing Jiaotong University, Chongqing 400074, China; 18943995727@163.com
* Correspondence: yedan@cque.edu.cn

Abstract: To control the deviation of a long-span concrete-filled steel tube (CFST) arch bridge during construction monitoring, a practical method for controlling tower deviation is studied and established. The form of construction of this bridge is an intermediate double-arch bridge, which differs from conventional bridges, thus requiring the urgent resolution of the issue of unbalanced middle piers. Therefore, the mechanical characteristics and construction process of an intermediate long-span, dual-coupled steel pipe arch bridge are meticulously examined by using a 1:10 scale model, with particular focus on discussing the deflection of the buckle tower during the installation of the arch rib segments. Construction control is implemented using a novel tower deflection control method that addresses unilateral torsion problems and difficulties in controlling the deflection of the tower. The model results are compared with the finite element analysis output, demonstrating that this new approach can resolve unbalanced tower deviations by maintaining absolute values within 0.5 mm. After correcting these deviations, the measured results from the model bridge tower align with the calculated analytical results and even surpass the theoretical expectations for tower deviation. This remarkable new method accurately resolves real-world bridge tower deviations.

Keywords: bridge engineering; arch bridge; tower deviation control; control method

Citation: Dong, X.; Ye, D. A New Method for Correcting the Deviation of a Middle Pier Tower of a Long-Span Intermediate Arch Bridge. *Buildings* **2023**, *13*, 2498. https://doi.org/10.3390/buildings13102498

Academic Editor: Carmelo Gentile

Received: 10 August 2023
Revised: 20 September 2023
Accepted: 28 September 2023
Published: 30 September 2023

1. Introduction

With the wide application of steel pipe concrete arch bridges, its forms of construction are becoming more diversified [1]. The span of the bridge has developed from just over 100 m in the beginning to nearly 600 m at present [2,3]. The design of a concrete-filled steel tube (CFST) arch bridge is usually guided via finite element simulation, and the use of a scaled model test can not only verify the mechanical performance of the structure, but also preview and evaluate the actual control of the construction. The control of the accuracy of pylon deflection will affect the alignment of such long-span arch bridges [4], but there is little research available into pylon control (especially as related to the middle pier during cable-stayed suspension). At present, some Chinese researchers have studied the linear control of CFST arch bridges in different forms. Gu, Y [5] described arch rib construction technology to analyze arch rib installation control methods from the perspective of arch processing and the splicing of the steel tube based on the arch bridge characteristics and construction-control processes of CFST bridge types. Based on the grey prediction theory, Zhuo Xiaoli [6] established a grey model for line-shape prediction, considering the residual correction during the construction of main beams in the context of mid-span steel box arch bridges, and predicted the construction pre-elevation value of main beam segments through tower deviation. It can be seen that tower deflection and arch rib alignment in arch bridges interact [7–9]. The suspension assembly method is a common construction method for long-span arch bridges [10–12]. However, the studies on the suspension assembly method mainly focus on improving the accuracy of the cable forces [13], while the studies on tower deflection are few. At present, for a long-span steel truss arch bridge, suspension tower construction is the most difficult and risky part of the construction process. The steel truss arch erection side span of the Jianghan Seventh Bridge is installed by using the support method, and the middle span is installed via the cantilever method. The suspension tower plays an important role in the cantilever erection construction of the steel truss arch. The suspension tower is one of the key structures in the cantilever construction of the steel truss arch bridge, and it connects the cantilever end of the main arch with the balance end of the side beam using the suspension cable, so that the steel truss arch bridge can maintain overall stability during the construction of the main girder [14]. The application of monitoring technology to realize the construction-line monitoring of the suspension tower can not only meet the installation accuracy requirements of the suspension tower members, but also improve the safety and accuracy of subsequent main beam closure [15]. This issue has also been investigated by some foreign researchers in succession. In the large-span concrete arch bridge constructed via the cantilever pouring method [16], path control is key to ensuring construction quality, and the deflection of the buckle tower often affects the arch elevation of each segment. The influence of the deflection of the buckle tower on the elevation of each section during construction was analyzed via a geometric analysis method, and the finite element model was established in Midas software to verify the results. The results show that the deflection of the buckle tower has little effect on the elevation of the arch foot, but it has a greater effect on the elevation of the section near the arch top, and the section at the arch top is more sensitive to the change in the buckle-tower height. According to Jiang Wei [17], the structural state of the main cable of the hoisting system changes during the hoisting process. Based on the engineering background of a large bridge, this paper obtained a high-precision calculation method of the main cable sag and cable force during the hoisting process through theoretical deduction, and estimated the influence of vehicle traffic on the main cable force by numerical simulation. The results show that the proposed method for calculating the sag of the main cable has high accuracy and can be applied in practice. The main cable force increases gradually as the vehicle moves toward the mid-span and the main cable force at the pylon end reaches its maximum when the vehicle is within the span. Some scholars have also shown how to prevent the occurrence of tower deviations by monitoring via three-dimensional scanning, but they have not avoided the occurrence of tower deviations from the root.

In terms of the method of the tower deflection theory, Lin T.M. [18] used the suspension cable element method to derive a formula giving the deflection of the hinged tower of the cable hoisting system where the pressure tower is set, so that the calculation of the cable hoisting construction stage and the tower deflection calculation were unified. A feasibility control system of intelligent tower deviation was verified by theoretical analysis, but it has not been applied in engineering practice. Deng JiangMing [19] derived an analytical formula of the influence of the tower deflection with an integrated button-back on the linear arch rib by using a geometric analysis method. The feasibility of these proposed control systems as applied to intelligent tower deviation was validated by theoretical analysis, but the deformation of the tower is not considered.

In terms of calculation methods, Zhen-guo C et al. [20] considered the influence of the deflection of the buckle tower during the calculation of the cable force, and ascertained the that optimal buckle cable force matrix can be used to reduce the influence of buckle-tower deflection on the alignment of the arch rib. Ke-Jian Y [21] studied the sensitivity of each section of the buckle of the lower part of the buckle-tower to the vertical displacement of the closing hole, and found that too small a horizontal inclination would weaken the influence of the section buckle force on the vertical displacement of the closing hole, and they solved the problem of the low adjustment efficiency of the long buckle cable by calculating an optimal cable force via the formal installation iterative method. However, these two methods tend to be those viewed as lacking in innovation: if the number of iterations is excessive, the calculation will be slow. Xue-Tao D et al. [22] studied the influence of the main cable slip on tower deflection in the design of the cable hoisting system, and found that the main cable slip had a significant influence on the cable hoisting system, which could significantly change the deviation of the top of the tower and the stress thereon. Wei S et al. [23] found that the displacement of the top of the tower in the integrated construction of cables and buckle towers exerts a significant influence on the accuracy of the arch-rib splicing, and the key to construction control is to reduce the deflection of the cable and the buckle tower. Therefore, the initial tension of the wind rope of the top cable should be adjusted while observing the deviation of the tower during the construction of the arch rib assembly, thereby reducing the horizontal cable force difference and keeping the cable and buckle tower in a vertical state as far as possible. The method is novel and unique in its technical aspects, but this method lacks theoretical support.

The safety and construction accuracy of the tower has been studied widely; taking the Changshou Yangtze River Bridge as the background, Gao Z [24] introduced the construction technology of a 60 m high, single-story, suspension tower from the design, construction, and to demolition. Taking the Nanjing Dashingguan Bridge as the background, Sheng Zhiping [25] innovatively laid a suspension tower with three layers of cable stays on the outside of the two-span steel beams, with a height of 68.5 m, and adopted flat cable construction technology on the inside, which overcame the problems of the internal force and alignment adjustment difficulty during the erection of steel beams. Zhi-Hu Z et al. [26], taking the Jianghan Seventh Bridge as the background, conducted the stress analysis of various working conditions of the 89.6-m high sling tower during the assembly process. Renbo F et al. [27] studied the application of 39-m high suspension tower in flexible arch of a continuous steel girder in the Guanhe Super-Bridge on the Lianyan Railway. This long-span bridge has certain advantages and broad prospects for application. Due to its wide range of application, especially in long-span bridges, cable hoisting construction has become the main method of the erection of arch bridges [28–30]. When cables are used to lift arch ribs, the processes of each span affect the deformation of the tower, induce deviations, and have linking effects on the arch ribs suspended on the tower. When the linkage effect gradually accumulates, assembly accuracy and tower deviation will be affected if no adjustment is made [31,32]. In the present work, a set of new methods and theories are used to control tower deviation, limiting the occurrence of unbalanced tower deviation and the corresponding torsion at the root, which is different from the methods of other scholars. The problem of tower deviations is solved, and the accuracy of tower deviations is controlled to within the theoretical value, thereby achieving the ideal control effect.

2. Project Overview and Model Design

2.1. Project Overview

The Longtai Expressway is a section of China's national expressway network planning, which is one of the important links in the main skeleton expressway network planned across the Yangtze River, and is also the main channel from the central region to the south-east coast. The whole line adopts six-lane expressway standard construction. The Tianyun Bridge is one of the controlled, landscape-type, projects of the overall project. Its structure consists of two arch circles, three arch seats, middle piers and bridge panels. The main bridge adopts the main span of 2 × 260 m of a CFST to design the overall subgrade width of 16 m, the design load is Grade I (highway), and the structural safety grade is Grade I. The design speed is 80 km/h, and the bridge structure design life is 100 years. In the middle pier, there are buckle cables on both sides. Using the middle arch seat, sections are constructed on both sides at the same time, and the buckle cables are pulled up to complete its construction. The column on the arch adopts a row-type rectangular section structure, and the column is connected by a cross-beam. The span of the two-span continuous CFST intermediate truss arch is larger in the same type of bridge, the bridge continuously crosses the river, the bridge location on both sides of the mountains makes for difficult access to traffic, and its construction is difficult necessitating the present research into the key technology of design and construction. The general layout of the Tianyun Bridge is shown in (Figure 1).

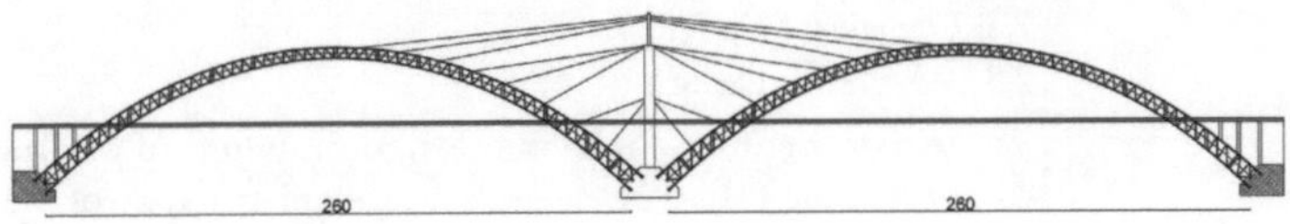

Figure 1. Layout of the Tianyun Bridge (unit: m).

2.2. Similarity Theory

For structural tests, it is most accurate to use the original actual structure for the simulation, so a certain proportion of model tests are often used, and the force state of the original structure is calculated through the use of the similarity ratio. To simulate the construction process of the Tianyuan Bridge, it is necessary to study the method used to control the tower deflection. To obtain more accurate test results and data, the geometric scaling ratio of the main arch ring, arch base and junction pier is proposed to be 1:10 on the basis of comprehensive consideration of the test content, model materials, production accuracy, and component volume, as well as considering time, economic cost, and data fidelity. The scale ratio of stress is 1:1, and the interface pier is designed in scale according to the ratio of pushing stiffness of 1:10.

From the scale ratio of geometry and force, the similarity coefficients are obtained according to the similarity theory as follows (Table 1):

Table 1. Model test similarity relationship.

Physical Quantity	Prototype	Model	Model Similarity Coefficient
Length	Lp	Lm = Lp (1/n)	1/10
Cross area	Ap	Am = Ap·(Ep/Em) (1/n2) (1/m)	1/100
Bending moment of inertia	Ip	Im = Ip·(Ep/Em) (1/n4) (1/m)	1/10,000
Torque moment of inertia	Jp	Jm = Jp·(Gp/Gm) (1/n4) (1/m)	1/10,000
Modulus of elasticity	Ep	Em	1
stress	σp	σm = σp (Ep/Em)	1
strain	εp	εm = εp	1
Linear displacement	δp	δm = δp·(1/n)	1/10
Angular displacement	αp	αm = αp	1
counterweight	Wp	Wm = Wp (1/n2) (1/m)	1/100
Concentrated load	Fp	Fm = Fp (1/n2) 1/m)	1/100
Shearing force	Qp	Qm = Qp (1/n2) (1/m)	1/100
reaction	Rp	Rm = Rp (1/n2)·(1/m)	1/100
Bending moment	Mbp	Mbm = Mbp (1/n3) (1/m)	1/1000
torque	MTp	MTm = MTp (1/n3)·(1/m)	1/1000

Note: In the table, n is the geometric scaling ratio (set to 10), and m denotes the scaling ratio of stress (set to 1).

2.3. Component Design

2.3.1. Arch Rib Design

According to the relevant design drawings of the Tianyun Bridge and the existing conditions, the values of the geometric similarity ratio of the main structural dimension are as per (Table 2).

Table 2. List of arch ring components (unit: mm) 30 September 2023.

Position	Original Bridge Dimension	Prototype Size	Model Size	Compressive Stiffness Ratio	Resistance to In-Plane Stiffness Ratio
Main chord 1	φ1600 × 35	φ80 × 1.7	φ36 × 4	1.02	1.04
Main chord 2	φ1600 × 32	φ80 × 1.6	φ35 × 4	1.2	1.11
Main chord 3	φ1200 × 26	φ70 × 1.3	φ35 × 3	1.1	1.12
Web rod 1	φ800 × 20	φ40 × 1	φ14 × 4	1.1	/
Web rod 2	φ600 × 16	φ30 × 0.8	φ15 × 3	1.2	/
Parallel connection 1	φ800 × 20	φ40 × 1	φ14 × 4	1.13	/
Parallel connection 2	φ800 × 22	φ40 × 1.1	φ12 × 3	1.15	/
Parallel connection 3	φ800 × 16	φ40 × 0.8	φ12 × 3	1.16	/
Cross-brace 1	φ800 × 22	φ40 × 1.1	φ14 × 4	1.13	/
Cross-brace 2	φ800 × 16	φ35 × 1.1	φ14 × 4	1.30	/
Cross-brace 3	φ700 × 22	φ40 × 0.8	φ12 × 3	1.09	/
Cross-brace 4	φ700 × 16	φ35 × 0.8	φ12 × 3	1.4	/
Fork stay	500 × 16	/	φ12 × 2.6	1.6	/

2.3.2. Arch Design

Due to this scaling model study, there are almost half spans on both sides. To make the scale model simulate the internal force and deformation of the real bridge, the arch design is strictly scaled down to ensure that the model undergoes no displacement in the construction and test process, which is aligned with reality. The arch seat diagram is shown in Figure 2.

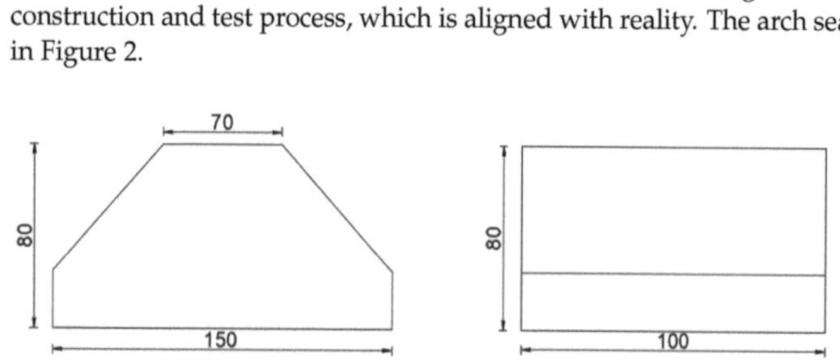

Figure 2. Arch abutment (unit: mm).

2.3.3. Pier Design

The height of the pier proposed in this design is 4.5 m + 1.5 m. The model test uses a steel tower pier (Figure 3).

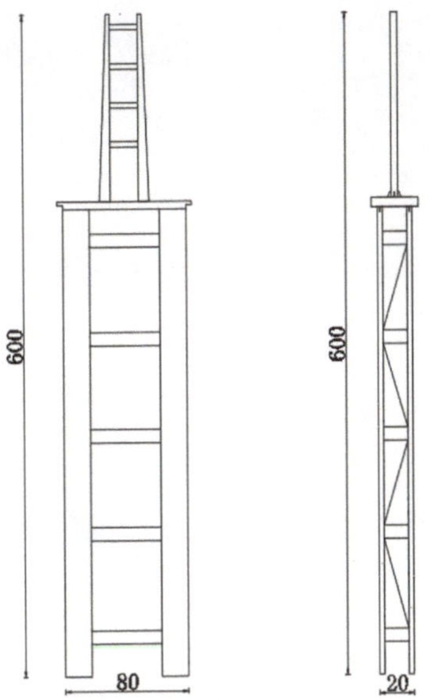

Figure 3. Intermediate pier (unit: mm).

3. Finite Element Analysis

3.1. Finite Element Model

The Midas finite element model is established based on the 1:10 scale model test. There are 2035 nodes and 3776 units in the whole bridge, among which the cable is a tension-only truss unit, and the main arch and tower are beam units. The simulation is conducted according to the hoisting sequence of the actual bridge during the construction stage.

For the convenience of description, the following is the partition diagram (Figure 4), for example: A2 is the second section of Side A.

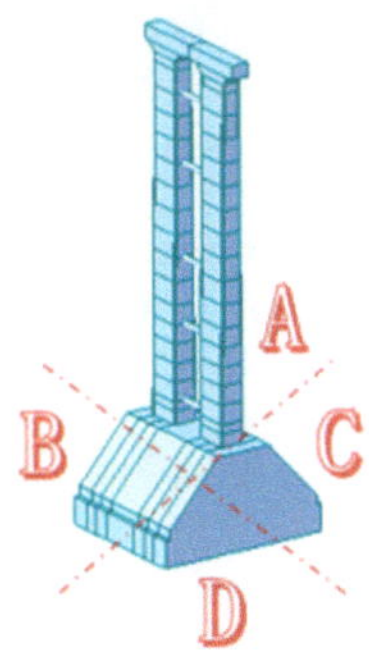

Figure 4. Azimuth zoning scheme.

3.2. Tower Deviation Calculation

The focus of the present research is the deviation of the middle pier tower. The deviation of the tower during normal construction is calculated: there are eight sections of the arch bridge, and some key sections are listed here, such as Section 2, Section 6, and Section 8. The outputs from the finite element analysis model are as follows:

Working condition 1: Hoisting arch rib A side of Section 2.

Working condition 2: Hoisting arch rib B side of Section 2.

In condition A2 (Figure 5), the unbalanced horizontal thrust is generated by unilateral lifting, with a 0.1 mm displacement at the bottom of the pier, a 0.88 mm displacement at the mid-point of the pier, and a 1.6 mm displacement at the top of the pier. In condition B2 (Figure 6), due to the balanced horizontal thrust generated by lifting on both sides, there is almost no displacement at the top of the pier. The maximum stress on the bottom of the tower is 52 MPa.

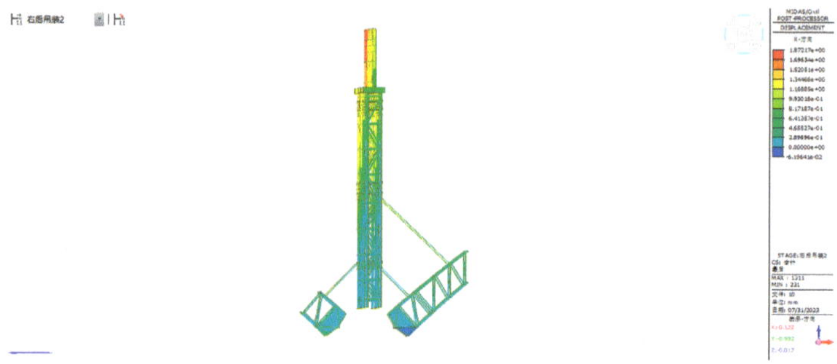

Figure 5. The second stage tower deviation (A2).

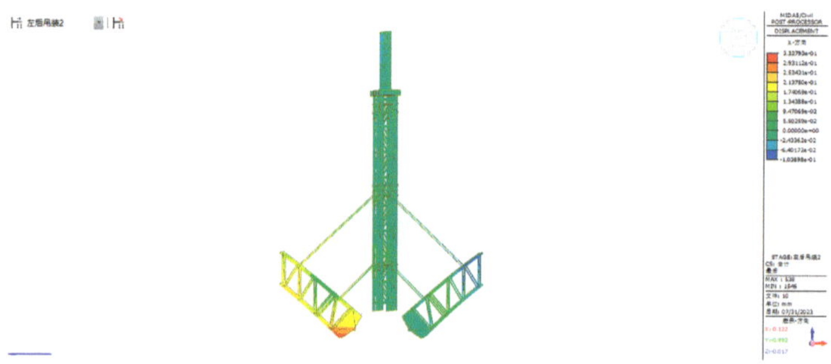

Figure 6. The second stage tower deviation (B2).

Working condition 3: Lifting Section 6 on side D of the arch rib.

Working condition 4: Lifting Section 6 on side C of the arch rib.

In condition D6 (Figure 7), the unbalanced horizontal thrust is generated by unilateral lifting. The displacement at the bottom of the pier is 0.1 mm, the displacement at the middle of the pier is about 0.9 mm, and the displacement at the top of the pier is 2.51 mm. In condition C6 (Figure 8), due to the balanced horizontal thrust generated by lifting on both sides, there is almost no displacement at the top of the pier. The maximum stress on the bottom of the tower is 82 MPa.

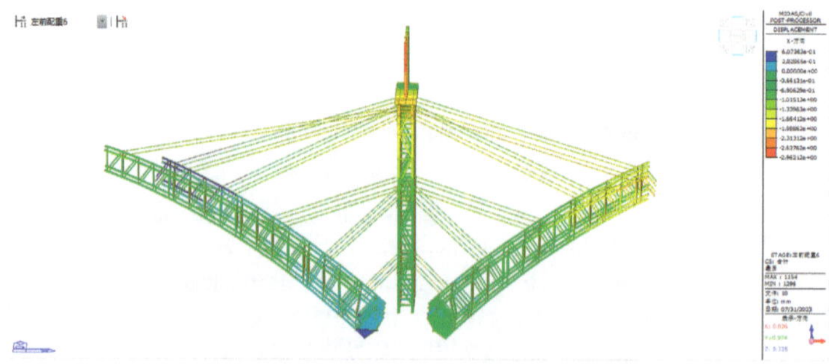

Figure 7. The sixth stage of tower deviation (D6).

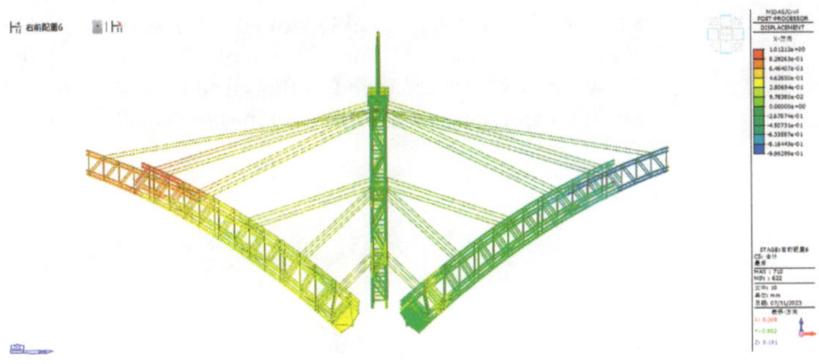

Figure 8. The sixth stage of tower deviation (C6).

Working condition 5: Lifting Section 8 on side A of the arch rib.

Working condition 6: Lifting Section 8 on side B of the arch rib.

In condition A8 (Figure 9), the unbalanced horizontal thrust is generated by unilateral lifting. The displacement at the bottom of the pier is 0.1 mm, the displacement at the middle of the pier is about 0.78 mm, and the displacement at the top of the pier is 2.85 mm. In condition B | 8 (Figure 10), due to the balanced horizontal thrust generated by lifting on both sides, there is almost no displacement at the top of the pier. The maximum stress on the bottom of the tower is 118 MPa.

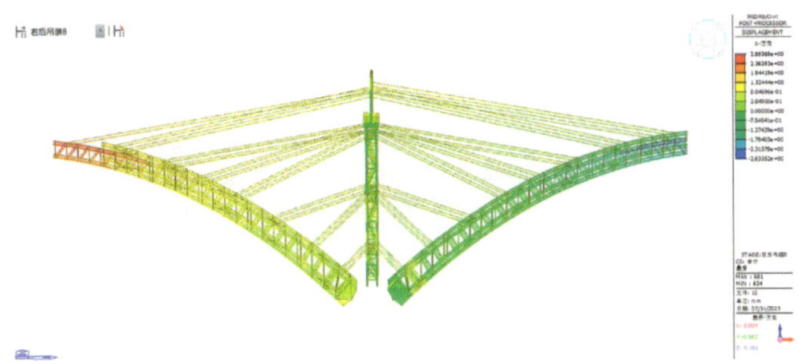

Figure 9. The eighth stage of tower deviation (A8).

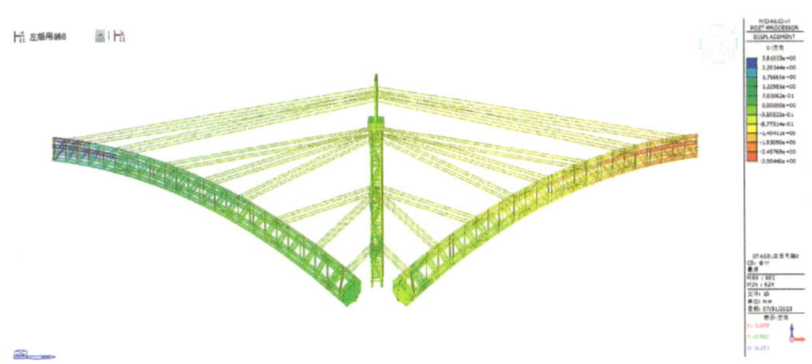

Figure 10. The eighth stage of tower deviation (B8).

The column deviation data under a normal construction sequence are listed below; these lay a foundation for the comparison of the data of new construction methods.

4. New Methods of Control Theory

4.1. Control Method

In this method, the single-side segment and the arch rib segment on the symmetrical side are hoisted first, the cable is not initially tensioned, the unbalanced bending moment is borne via the existing cable-stayed cable system, and the rear tensioned cable is used as a protective device. The aforementioned steps are repeated, and the two rear ends are symmetrically tensioned at the same time, which can ensure the reduction in tower deviation and the torsion caused by the single-side asymmetric hoisting problem (Figure 11)

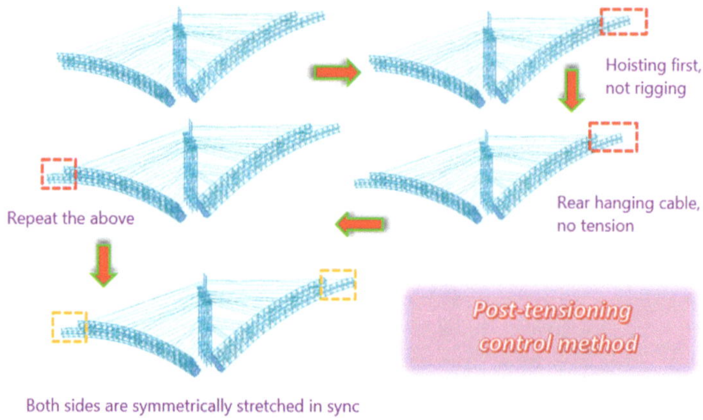

Figure 11. The method used to control tower deviation.

4.2. *Tower Deviation Analysis*

The following experimental results can be obtained: the first seven sections of the experiment are constructed in the form of the unilateral installation and tightening of segment buckle cables, and the other side is the same. Section 8 adopts the construction method of lifting on both sides first, not tightening the cable, and pulling the cable on both sides at the same time after the symmetrical segment is installed. The following table shows the deviation data for the traditional hoisting of segments 2 and 6, and the deviation data for segment 8 using the new method. T1 is the measurement point at the bottom of the tower, T2 is the measurement point in the middle of the tower, and T3–T6 are the four measurement points at the top of the tower (Figure 12).

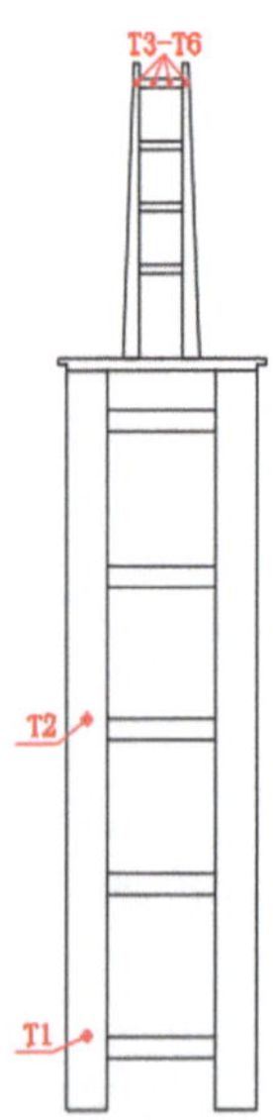

Figure 12. Measuring points T1 to T6.

In Tables 3 and 4, Δx is the longitudinal deviation of the bridge from the tower, Δy is the transverse deviation of the bridge from the tower, and Δz is the elevation change, indicating the change in the tower deviation on the hoisting side of the segment. $\Delta x'$ is the longitudinal deflection of the bridge towards the tower, $\Delta y'$ is the transverse deflection of the bridge towards the tower, and $\Delta z'$ is the elevation change, and the change in deflection of the mounting buckle tower on the other side of the symmetric segment. Measurements are made via total station. Tables 3 and 4 list the traditional tensioning column deviation data.

Table 3. Section 2 tower deviation (unit: mm).

	Δx	Δy	Δz	$\Delta x'$	$\Delta y'$	$\Delta z'$
T1	0.0006	0.0002	0.0004	0.0017	0.0003	−0.0002
T2	0.0018	0.0006	0.0014	0.0010	−0.0003	0.0002
T3	0.0048	0.0005	0.0021	0.0009	−0.0005	0.0008
T4	0.0031	0.0008	0.0005	0.0000	0.0002	−0.0002
T5	0.0021	0.0004	0.0005	−0.0010	−0.0001	−0.0011
T6	0.0009	0.0002	0.0001	−0.0026	−0.0003	−0.0004

Table 4. Section 6 tower deviation (unit: mm).

	Δx	Δy	Δz	Δx′	Δy′	Δz′
T1	−0.0001	0.0003	−0.0004	0.0003	0.0002	−0.0002
T2	−0.0005	−0.0003	−0.0003	0.0005	0.0006	0.0002
T3	−0.0020	−0.0005	0.0006	0.0009	0.0005	0.0008
T4	−0.0025	0.0002	−0.0001	0.0010	0.0008	−0.0002
T5	−0.0027	−0.0001	−0.0012	0.0014	0.0004	−0.0011
T6	−0.0047	−0.0003	−0.0007	0.0016	0.0002	−0.0004

Tables 5 and 6 list the column deviation data of the eighth stage after adopting the new tensioning method.

Table 5. Section 8 tower deviation (unit: mm).

	Δx	Δy	Δz	Δx′	Δy′	Δz′
T1	−0.0002	0.0003	−0.0004	0.0003	0.0002	0.0002
T2	−0.0010	−0.0009	−0.0003	0.0005	−0.0003	0.0004
T3	−0.0042	−0.0006	0.0006	0.0009	0.0008	0.0006
T4	−0.0037	−0.0005	−0.0001	0.0010	−0.0002	−0.0002
T5	−0.0030	0.0002	−0.0012	0.0010	0.0009	−0.0009
T6	−0.0025	−0.0008	−0.0007	0.0011	0.0008	−0.0004

Table 6. Section 8 tower deviation (unit: mm).

	Δx	Δy	Δz	Δx′	Δy′	Δz′
T1	0.0003	0.0002	0.0002	−0.0002	−0.0002	−0.0002
T2	0.0006	0.0005	0.0005	−0.0004	0.0003	0.0002
T3	0.0011	0.0004	0.0016	−0.0007	−0.0004	0.0008
T4	0.0009	0.0005	0.0006	−0.0006	−0.0002	−0.0002
T5	0.0007	0.0006	0.0001	−0.0007	0.0002	−0.0011
T6	0.0005	0.0003	0.0000	−0.0006	0.0005	−0.0004

5. New Methods of Optimization Data Analysis

Using data comparison, it can be found that the transverse deviation of the tower is insignificant, and most of the deviation is within 0.5 mm, so there is little scope for the description and calculation of the transverse deviation therein. The working conditions of

A2, B2, and Section 6 are given in Figure 13, and the data pertaining to both the simulation and experiment are compared. To improve practicability, the maximum stress of the arch foot of the two methods corresponding to the section is listed (Figure 13).

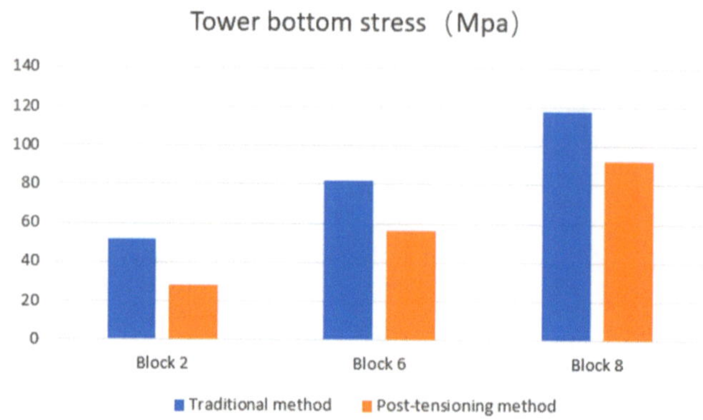

Figure 13. Comparison of arch-foot stress between the two methods.

It can be seen above that the method of the back cable can reduce the stress on some arch feet, and that on the eighth section, it can be reduced by about 25%. The following is a comparison of the actual and theoretical data of each section, and the last two charts allow the comparison of the data of the post-cable and traditional methods (Figures 14–19).

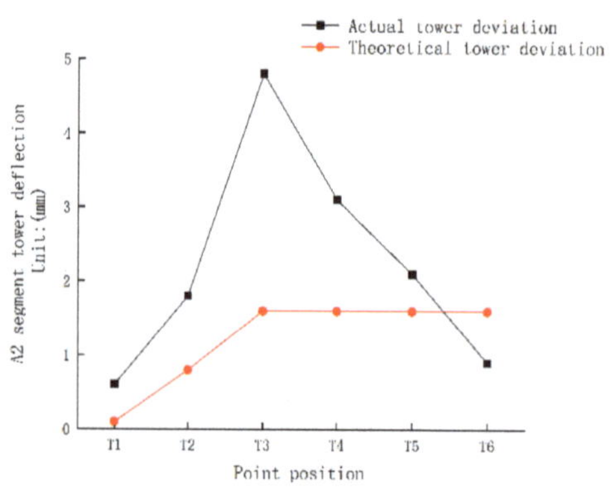

Figure 14. Section 2 tower bias comparison.

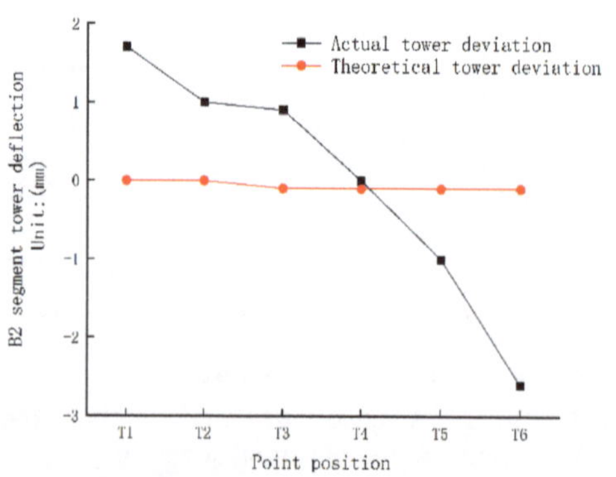

Figure 15. Section 2 tower bias comparison.

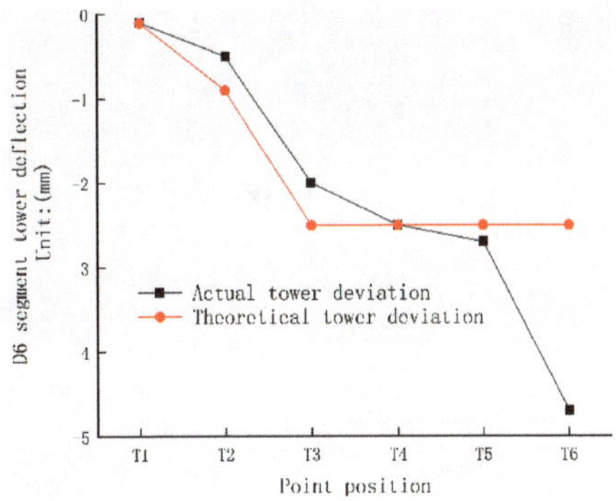

Figure 16. Section 6 tower bias comparison.

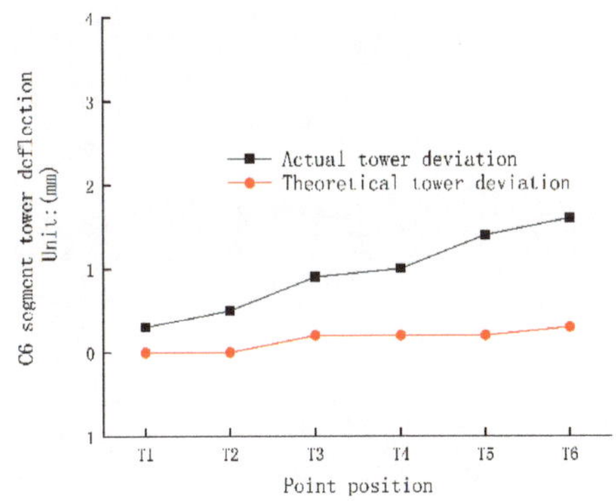

Figure 17. Section 6 tower bias comparison.

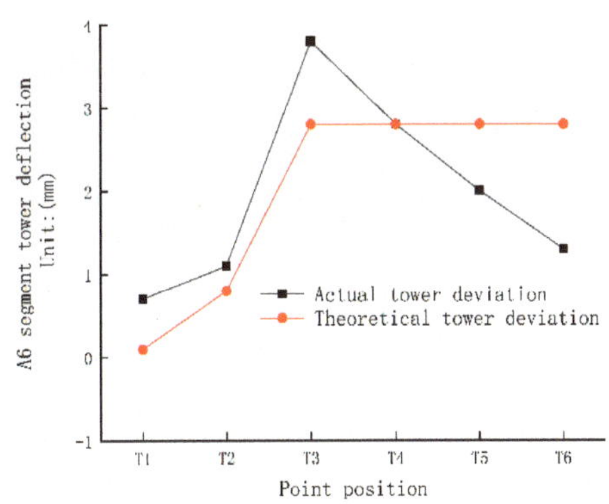

Figure 18. Section 6 tower bias comparison.

It can be seen that the traditional method induces asymmetric column deviation control and torsion problems. The actual data indicate that T3–T6 is the control point at the top of the buckle tower, and the actual data differ greatly, so the effect of lifting the unilateral rear cable is poor.

The following new method is applied to Section 8: after hoisting two symmetrical segments, both sides are tensioned at the same time, and the data are compared with that collected when hoisting one side segment at a time(Figures 20 and 21).

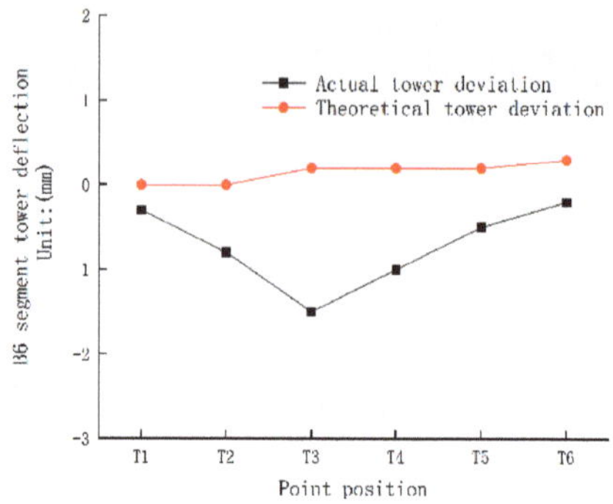

Figure 19. Section 6 tower bias comparison.

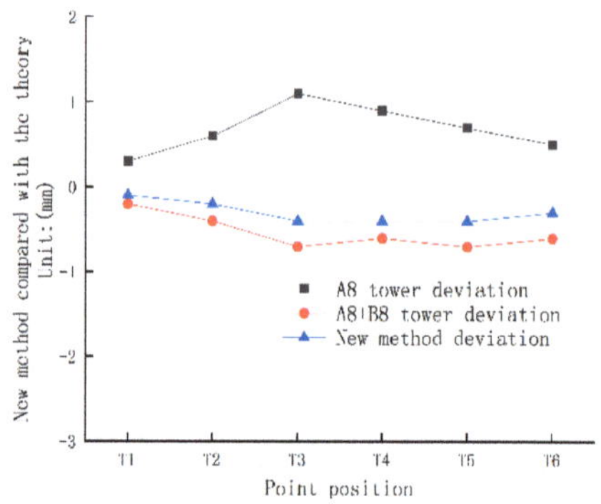

Figure 20. The new method is compared with the traditional method (A8 + B8 sides).

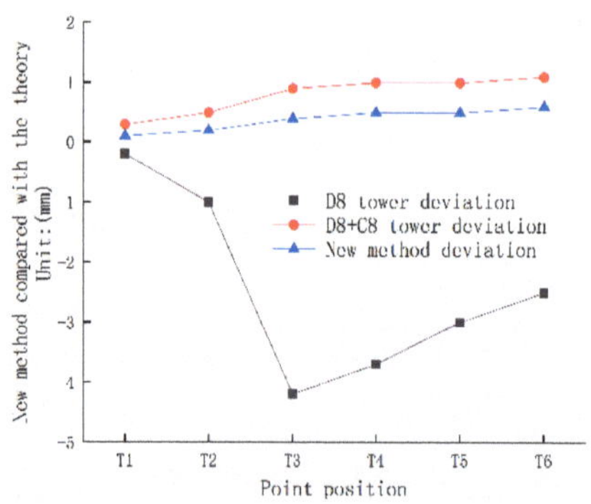

Figure 21. The new method is compared with the traditional method (C8 + D8 sides).

The comparison of the pictures shows that this method can actively and timeously correct the displacement deviation of the tower top and the torsion on the symmetric segment during the cable hoisting process, reducing the deviation. The maximum deviation of the tower pier can be controlled, in real time, to within the range of $\pm$ 0.5 mm.

6. Conclusions

To study the effect and operation of the new method of tower deflection control on the deflection of the buckle tower, and to apply it to the universality of similar bridge projects, the following conclusions can be drawn by establishing a 1:10 scale model:

(1) The deflection of the buckle tower increases with the increase in the number of arch-rib lifting segments and the increase in pier elevation, from the bottom to the top of the tower;
(2) The traditional way of controlling the deviation of the tower body is to pull a lot of steel wire and jack manually, which wastes a lot of materials and manpower. The new proposed method relies on the simultaneous tensioning technique, thereby greatly reducing the time cost, and obviating the need to adjust the offset;
(3) After adopting the new method, the influence of adjusting the tower deviation on the structure members is rendered insignificant. This means that the adjustment has a significant controlling effect on the longitudinal displacement of the tower, and can reduce the stress on the arch foot; each section analyzed benefitted from a stress reduction of about 30 MPa compared with the previous method;
(4) This method can actively and timeously correct the displacement deviation of the top of the tower and the torsion in the symmetric segment in the process of cable hoisting, reducing the deviation. The maximum deviation of the tower pier can be controlled, in real time, within the range of ±0.5 mm. Compared with the traditional methods and techniques, the results can provide a reference for the deviation control of the middle pier tower of similar types of bridge.

Author Contributions: X.D., methodology and finite element analysis, drawing and translation; D.Y., editing and modification. All authors have read and agreed to the published version of the manuscript as submitted.

Funding: This research received no external funding.

Data Availability Statement: The data are available on request only due to commercial restrictions.

Conflicts of Interest: The authors declare no conflict of interest.

References

1. Yazhou, X. Structural Study of Arch Bridge With a Span of 600 m Part I: Trial Design. *Appl. Mech. Mater.* **2011**, *138*, 300–304.
2. Zhang, F. Study on Filling Sequence of Long-Span Concrete-Filled Steel Tube Arch Ribs and Hydration Heat of Concrete in Pipes. Master's Thesis, Chongqing Jiaotong University, Chongqing, China, 2020.
3. Ma, Y. Application and development of concrete-filled steel tube arch bridge. *Eng. Res.* **2019**, *4*, 212–213.
4. Wang, F.; Zheng, W.; Liu, H. Determination and Control of Alignment in Construction of CFST Arch Bridge. *Technol. Highw. Transp.* **2021**, *37*, 98–104+110.
5. Gu, Y.; Li, Y.D.; Liu, S.Z. Research on Construction Control of Long-span CFST Arch Bridge. *Appl. Mech. Mater.* **2015**, *777*, 88–93. [CrossRef]
6. Zhuo, X. Research on Construction Alignment Control of Main Girder of Half-Through Steel Box Arch Bridge Based on Grey Theory. *J. Highw. Eng.* **2021**, *46*, 26–32+67.
7. Zheng, J.L.; Wang, J.J. Concrete-Filled Steel Tube Arch Bridges in China. *Engineering* **2017**, *4*, 143–155. [CrossRef]
8. Chen, B.C.; Wang, T.L. Overview of Concrete Filled Steel Tube Arch Bridges in China. *Pract. Period. Struct. Des. Constr.* **2009**, *14*, 70–80. [CrossRef]
9. Han, L.H.; Li, W.; Bjorhovde, R. Developments and advanced applications of concrete-filled steel tubular (CFST) structures: Members. *J. Constr. Steel Res.* **2014**, *100*, 211–228. [CrossRef]
10. Gao, X.; Liu, L.J.; Yao, S.K.; Yang, J.W.; Li, Y. Commonly Encountered Damages in Cable Members of CFST Arch Bridge and Detection Methods. *Key Eng. Mater.* **2012**, *619*, 71–80. [CrossRef]
11. Zhang, Z.-C.; Ye, G.-R.; Cheng, H.-Z. Calculation of SBC forces for concrete-filled steel tubular arch bridge with long span in process of lifting. *J. Zhejiang Univ. Eng. Sci.* **2004**, *38*. [CrossRef]
12. Zhou, Q.; Zhou, J.T.; Ma, H.; Li, X.-G.; Zhang, L. Improved algorithm of cable force for one time cable tensioning on steel tube arch ribs with segmental hoisting. *J. Traffic Transp. Eng.* **2020**, *20*, 92–101.
13. Zhou, Q.; Zhou, J.T.; Chen, J.W.; Feng, Y. Improved algorithm of cable force optimization of one—Time tensioning for arch bridge. *J. Fuzhou Univ. Nat. Sci. Ed.* **2019**, *47*, 412–416.
14. Liang, D.; Zhao, K. Virtual Steel Pylon Segments Assembly Method Based on 3D Laser Scanning. *Bridge Constr.* **2019**, *51*, 62–71.
15. Tian, W.; Liu, D. Closure Technique for Steel Truss Arch and Girder of the Second Hengqin Bridge in Zhuhai. *World Bridges* **2016**, *44*, 31–35.
16. Wu, P.; Pan, R.; Guo, J. Study on Influences of Phlon Displacement on Arch Rib Elevation of Concrete Arch Bridge Using Cantilever Casting Method. *J. Railw. Eng. Soc.* **2020**, *254*, 154–158.

17. Jiang, W.; Li, X. High Precision Calculation of Structural State of Main Cable during Cable Lifting Construction of Steel Tube Arch Bridge. *J. China Foreign Highw.* **2020**, *252*, 123–127.
18. Lin, T.M.; Zhong, Q.S.; Li, S.R. Segmental catenary method of calculating the cable curve of suspension bridge. *J. China Railw. Soc.* **2003**, *25*, 87–91.
19. Deng, J. Geometrical Analysis of Arch Rib Elevation Change Caused by Tower Deviation during Cable-Hoisting Construction. *J. Chongqing Jiaotong Univ. Nat. Sci.* **2009**, *3*, 505–508.
20. Cheng, Z.G.; Bai, Y.H.; Liu, C.T. The Control of Tower Deviation on Arch Installation in Xiaohe Grand Bridge. *Technol. Econ. Areas Commun.* **2010**, *3*, 87–89.
21. Yang, K.J.; Li, F.Q.; Zhang, Y.L. Research on the Design of Large Span Deck Steel Tube Concrete Arch Bridge. *J. Railw. Eng. Soc.* **2008**, *25*, 66–71.
22. Deng, X.T.; Zeng, D.R.; Zhou, Y.Q. Finite Element Analysis of Stability of Cable-Hoisting Erection System of Long-Span Arch Bridge. *J. Chongqing Jiaotong Univ. Nat. Sci.* **2007**, *26*, 1–5.
23. Shen, W.; Zhang, Y.Z.; Wen, L.U. Construction Control Method of Long-Span Steel Box Tied Arch Bridge. *West. China Commun. Sci. Technol.* **2009**, *12*, 85–89.
24. Gao, Z. Traveling wave effect of multi-span through concrete-filled steel tubular arch bridge. *J. Vibroengineering* **2022**, *24*, 19. [CrossRef]
25. Sheng, Z. Construction of Fastening Stay Tower at Pier No. 8 of Dashengguan Changjiang River Bridge in Nanjing. *Bridge Constr.* **2010**, *2*, 5–9.
26. Zhu, Z.H.; Lun, Y.I.; Gao, Z.Y. Analysis of Mechanical Behavior and Study of Construction Control Measures of Triple Main Truss Structure of Dashengguan Changjiang River Bridge in Nanjing. *Bridge Constr.* **2009**, *3*, 28–32.
27. Feng, R.; Zhao, H.; Tailei, C. Key Construction Techniques of Main Span Steel Truss Girder Flexible Arch of Guanhe Super-Long Bridge on Lianyungang-Yancheng Railway. *Railw. Eng.* **2017**, *8*, 47–49.
28. John, J.S. Construction of the Hoover Dam Bypass. *Concr. Int.* **2010**, *33*, 30–35.
29. Wei, J.; Chen, B. Application and research advancement of long span concrete arch bridges abroad. *World Bridge* **2009**, *2*, 4–8.
30. Shi, W.Z.; Ko, J.M.; Zhao, C.Y.; Chen, L. A bridge structural health monitoring information system based on GIS and DBMS. In Proceedings of the International Society for Optics and Photonics, San Diego, CA, USA, May 2005.
31. Yong, H.E.; Miao, Y.F.; Liu, H. Based on Beidou/GPS precise displacement monitoring technology in the application of the bridge monitoring. *J. Yunnan Univ. Nat. Sci. Ed.* **2016**, *38*, 35–39.
32. Wu, J.-C.; Wang, Z.-C.; Shi, X.-Y. Application of Beidou Satellite Communication and Positioning Technology to the Field Geological Survey in Northwestern China. *J. Geomech.* **2012**, *18*, 282–287.

sensors

MDPI

Article

Evolution Law of Concrete Interface Stress of Rigid-Frame Arch under Construction and Its Impact on Ultimate Load-Bearing Capacity

Yonghui Fan, Chao Luo *, Yin Zhou, Ligui Yang, Xinglin Li and Jinlong Liao

State Key Laboratory of Mountain Bridge and Tunnel Engineering, Chongqing Jiaotong University, Chongqing 400074, China; fyh1995@mails.cqjtu.edu.cn (Y.F.); yin_zhou935215@163.com (Y.Z.); yangligui@cqjtu.edu.cn (L.Y.); 622220970001@mails.cqjtu.edu.cn (X.L.); jinlong_liao@mails.cqjtu.edu.cn (J.L.)
* Correspondence: luochao@mails.cqjtu.edu.cn; Tel.: +86-186-8082-8616

Abstract: To study the evolution of stress on the ring and segment interfaces during the construction process of the concrete encapsulation of the main arch ring in a rigid-frame arch bridge, alongside its impact on the ultimate load-bearing capacity of the main arch ring, a 1:10 scale model experiment was conducted by taking the 600 m Tian'e Longtan Bridge as the prototype. The key cross-section concrete strain data were collected during the entire construction process of the main arch ring via fiber-optic strain sensors, which were used to investigate the stress evolution at ring and segment interfaces. ANSYS APDL was employed to simulate the ultimate bearing capacity under various loading conditions of two different finite element models, which were, respectively, formed segmentally and by single pouring. The results revealed that (1) after the closure of the concrete encapsulation of the main arch ring, the concrete stress in the cross-section exhibits significant stress disparities. At the same cross-section, the level of the web concrete stress can reach 76% of the floor concrete stress, while the roof concrete stress level is less than 20% of the floor concrete stress. (2) At the junction of two adjacent work planes, there are considerable differences in the stress levels of the concrete on both sides. After the closure of the main arch ring, the intersegment stress ratios of the floor, web, and roof concrete are 60~70%, 40~60%, and 0~5%, respectively. (3) Loading conditions remarkably affected the ultimate bearing capacity of the main arch ring. Under mid-span loading and 1/4 span symmetrical loading conditions, compared to single-pour concrete encapsulation, the ultimate bearing capacity of the main arch ring with concrete encapsulated by segmented and ring-divided pouring decreased by 19.16% and 5.23%, respectively, compared to single-pour concrete encapsulation. This suggests that the non-uniformity of stress distribution in the concrete sheath can lead to reductions in the ultimate bearing capacity of the arch ring.

Keywords: fiber-optic strain sensors; rigid-frame arch bridge; stress evolution; interface between rings and intersegment; ultimate load-bearing capacity

Citation: Fan, Y.; Luo, C.; Zhou, Y.; Yang, L.; Li, X.; Liao, J. Evolution Law of Concrete Interface Stress of Rigid-Frame Arch under Construction and Its Impact on Ultimate Load-Bearing Capacity. *Sensors* **2023**, *23*, 6868. https://doi.org/10.3390/s23156868

Academic Editor: Francesco Fabbrocino

Received: 30 June 2023
Revised: 23 July 2023
Accepted: 31 July 2023
Published: 2 August 2023

1. Introduction

The rigid-frame arch bridge is an advanced structural design featuring a strong steel truss as its backbone, whose concrete is poured into internal tubes and then covered with externally cast concrete in ring-segmented patterns [1]. Such a structure recently attracted increasing attention owing to its increased load-bearing capacity, attractive appearance, and lower long-term maintenance costs [2,3]. According to data collected, for rigid-frame arch bridges with a main span exceeding 300 m, casting the main arch ring's external concrete is typically performed in relation to segmented rings, also known as the multi-working planes balanced casting method [4]. This construction method results in significant temporal disparities in structural loading between the floor concrete, the web concrete, and the roof concrete of the main arch ring's external concrete, leading to substantial stress differences and compromising the load-bearing performance of the structure [5–7].

Comprehensive research on the stress evolution laws at the interfaces between the floor, web, and roof concrete of the main arch ring (hereafter referred to as inter-rings interface) and adjacent work plans interfaces (hereafter referred to as intersegments interface), as well as the influence of stress distribution on the ultimate load-bearing capacity, is scientifically and practically significant. It aids in clarifying the transmission mechanism of concrete at different ages and stress levels and optimizing the loading state of the main arch ring.

Currently, scholarly studies on the stress evolution during the construction process of external concrete encapsulation of the main arch ring in a rigid-frame arch bridge and its influence on the structural ultimate load-bearing capacity can be categorized into three types.

The first category focuses on the impact of the longitudinal segments' quantity and segmentation location of the external concrete encapsulation on the structural stress [8,9]. These studies mainly use numerical simulation methods to compare the structure's stress levels under different concrete encapsulation settings. For instance, Tong et al. [10] used the Nanpan River Bridge as an example and investigated the effect of the casting sequence of the longitudinal external concrete encapsulation on the arch ring structure's instantaneous and permanent stresses based on numerical analysis. Similarly, Yang et al. [11] employed the Beipanjiang Bridge as a representative instance to conduct a comparative analysis of diverse exogenous concrete construction methodologies, distinctly from the vantage points of horizontal ring segmentation and longitudinal partitioning, whose study focused on the force-bearing state of the arch ribs during the execution of the construction process and the corresponding controlling determinants. Lin et al. [12] studied the process of pouring exogenous concrete for the rigid frame of the main arch ring using the six working plane method and the combination method of inclined stay cable plus multiple working planes. They analyzed the patterns of instantaneous stress and permanent stress changes when the two methods were applied during construction. Such research primarily focuses on summarizing patterns through simulation results under different construction plans without proposing measures to reduce the stress levels of the structures.

The second research category optimized the concrete encapsulation construction scheme based on controlling the stress and the line shape of the main arch ring [13–15]. For example, Au et al. [16] used the influence matrix method to adjust the cable force, controlling the stress during the construction process of the concrete encapsulation of the main arch ring within a reasonable range. Lin et al. [17] proposed a casting method for the main arch ring concrete, which sets four work plans symmetrically along the entire arch. This method adjusts the casting length and sequence of the concrete by fitting a continuous function on the ascending and descending sections of the control stress process line, thereby reducing the instantaneous stress of the rigid frame. Tong et al. [18] focused their study on a rigid-frame arch bridge with a main span of 600 m by comparing the stress and deformation results of the main arch ring obtained from Midas Civil computations, and they optimized the segmented ring-by-ring concrete encapsulation scheme for the main arch ring of this bridge. However, these studies optimize the structural stress from the perspective of algorithms and simulations without delving into the impact of stress reduction on the force-bearing performance of the structure.

The third category of research investigates the influence of various factors on the stress-bearing performance of the main arch ring from the perspective of limit-bearing capacity [19–21]. For example, Liu et al. [22] conducted a 1:16 scale experiment based on the Washiwo Bridge with a span of 95 m, studying the failure modes of catenary arches under the loading conditions of the arch crown and 1/4 span. Wang et al. [23] proposed a double-layer corrugated steel–concrete composite arch structure and studied its bearing capacity through a single-point loading experiment at the mid-span. Yang et al. [24] studied the change in the structure's limit load-bearing capacity after the arch bridge's reinforcement with ultra-high-performance concrete (UHPC). Hu et al. [25] studied the impact of the initial defects, such as the main arch ring's fabrication and construction errors, on the structure's ultimate load-bearing capacity. This type of study was mainly concentrated on the influence of factors such as arch axis line shape, structural form, materials, and initial defects on the bearing performance of the structure, with less research focusing on the stress state after construction completion.

In summary, the majority of studies focusing on the stress evolution of the main arch ring tend to spotlight the summary of patterns. Although some scholars have proposed stress optimization methods, most of them rely heavily on numerical simulations and accomplish their research through trial-and-error selection, lacking experimental support from measured data. Furthermore, the research regarding the impact of the stress state of the main arch ring after construction completion on bearing performance remains a gap in the literature.

Focusing on the dual issues of the lack of experimental data support for the stress evolution rules during the construction process of the main arch ring and the absence of studies on the impact of structural stress states on the ultimate load-bearing capacity, a 1:10 scale model experiment of the main arch ring was conducted based on the world's largest rigid-frame arch bridge—the Tian'e Longtan Bridge—with a main span of 600 m. This 1:10 scale model has surpassed similar scale tests conducted on bridges worldwide previously [26,27]. ANSYS APDL was utilized to establish a finite element model of the bridge, and the model was validated by comparing the measured stress of the concrete at the key cross-sections and the simulation results. Ultimately, by comparing the ultimate load-bearing capacity using different construction methods as well as under different loading conditions, the influence of the structural stress state on the ultimate load-bearing capacity after the completion of the main arch ring construction is examined.

2. Engineering Example

The Tian'e Longtan Bridge, with a main span of 600 m and a height of 125 m (height-to-span ratio = 1/4.8), is located within Guangxi Province, China. The main arch is composed of three parts: a rigid steel skeleton, concrete-filled steel tube, and the external concrete forming a box-shaped cross-section; the size information of the external concrete at different sections is shown in Figure 1 [28]. The rigid steel skeleton adopts a Q420 material and is designed as a steel truss arch structure. During construction, it is installed using the cable-stayed pulling and hanging method. The concrete-filled steel tube is made of C80 concrete, and the construction process involves pumping through a lift-up pouring method. The external concrete uses C60 concrete and is constructed using the multi-working planes balanced casting method, as shown in Figure 2, the red part represents the floor and roof concrete, and the blue part represents the web concrete, the colours from deep to light and numbers both represent the sequence of concrete pouring. In the transverse direction, the external concrete is divided into three rings: the floor, web, and roof concrete. In the longitudinal direction, each concrete ring is further divided into eight working planes, and each working plane is poured multiple times to achieve a balanced pouring.

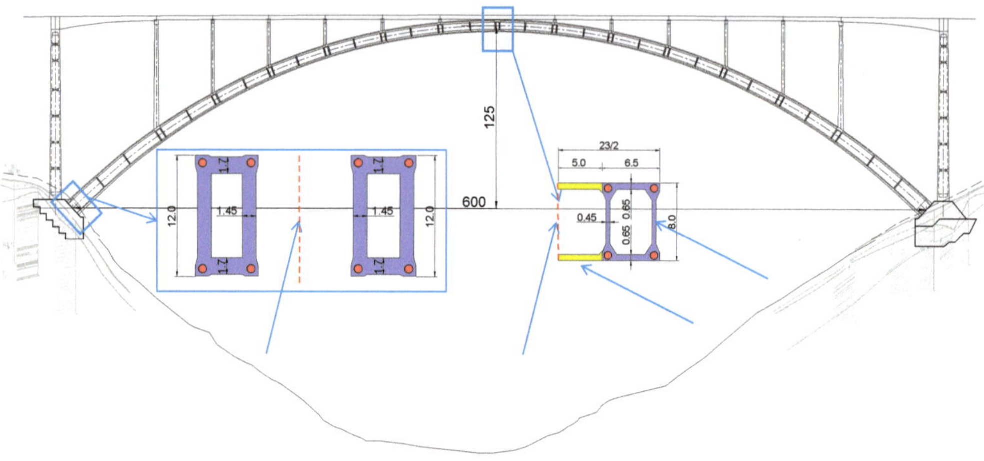

Figure 1. Elevation view of a concrete-filled steel tube stiff skeleton arch bridge (unit: m).

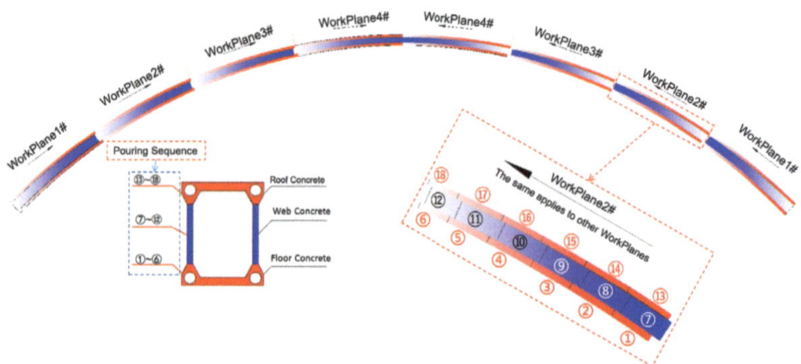

Figure 2. Schematic of segmental and ring-based concrete encapsulation on the primary arch ring.

3. Model Test Methodology

It should be pointed out that the load effect will vary with the change in scale, and a reasonable scale can improve the accuracy of inferring the original bridge state from the model test results. In addition, a smaller scale can lead to a smaller structure size, increasing the difficulty of welding the steel skeleton and the risk of pipe blockage during the pouring process of concrete inside the pipe. Additionally, thinner concrete will reduce the thickness of the protective layer of the steel bar. On the contrary, a larger scale will increase the demand for experimental space in the experiment. Considering multiple factors such as the load effect, availability of materials, the feasibility of steel structure fabrication, experimental space, and structural safety during construction, a scale of 1:10 was chosen for this experiment.

In accordance with the principle of stress equivalence [29,30], a main arch ring model experiment with a main span of 60 m was conducted. In order to minimize the influence of scale effects [31], strict scaling ratios were used in the geometric design, including overall dimensions such as span, the height of mid-span position, and linear forms. In terms of materials, a substantial number of concrete mix ratio experiments were carried out to ensure that the concrete of the model arch and the original bridge were essentially identical in parameters such as material strength and fluidity [32].

Notably, due to spatial constraints in the laboratory, the model experiment focused on a single arch ring. This focus made the issue of lateral stability of the main arch ring more prominent. To enhance the safety of the structure during construction, a limiting bracket along the longitudinal direction of the bridge every 10 m was arranged, as shown in Figure 3.

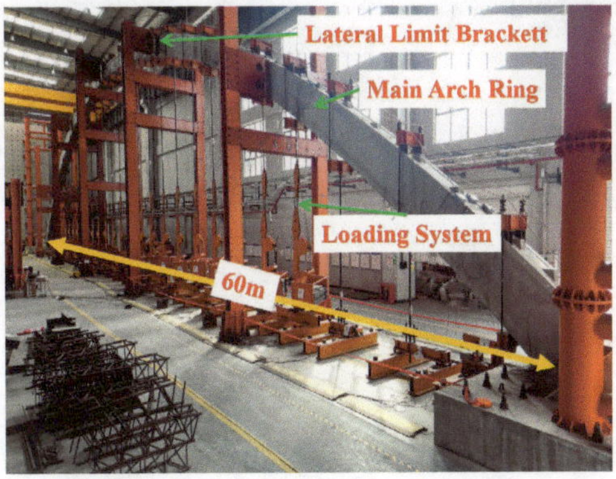

Figure 3. Picture of the model test.

To ensure stress equivalence between the model and the original bridge, the construction process required nine times the self-weight of the completed structure to be applied as a permanent load at the corresponding load points. The model arch is set with a loading point every 2 m along the longitudinal direction of the bridge, and a total of 29 loading points are set for the entire bridge. To enhance the local stability near the loading points, transverse partitions with a thickness of 50 mm are also arranged every 2 m along the length of the bridge [28]. During the casting process of the external concrete, the construction method involved using the permanent load balance weight before casting the concrete.

Considering the primary arch ring segmentation, the external concrete was cast in 18 stages. Considering the erection of the rigid steel skeleton, the pouring of inner tube concrete, and the application of the permanent load balance weight, the primary arch ring was divided into 40 construction phases, as illustrated in Table 1. Figure 4 shows pictures of some key construction nodes of the primary arch ring of the model bridge.

Table 1. Main arch ring construction stage division table.

Construction Stage	Construction Content
1# [1]	Closure of rigid steel frame.
2#	Pouring concrete into the pipe.
3#	Apply 9 times the self-weight of the rigid steel frame on the corresponding loading point.
4#	Apply 9 times the self-weight of the concrete inside the pipe on the corresponding loading point.
5#	Apply the first part of concrete self-weight counterweight on 8 working planes of the floor concrete on the corresponding loading point.
6#	Pour the first part of concrete on 8 working planes of the floor concrete.
7~14#	Apply the required load at each loading point (6 #, 8 #, 10 #, and 12 #) and pouring concrete (7 #, 9 #, 11 #, and 13 #) for the second to fifth part of the 8 working planes of the floor concrete.
15#	Apply the last part of concrete self-weight counterweight on 8 working planes of the floor concrete.
16#	Closure of the floor concrete.
17~26#	Apply the required load at each loading point (16#, 18#, 20#, 22#, and 24#) and pouring concrete (17#, 19#, 21#, 23#, and 25#) for the first to fifth part of the 8 working planes of the web concrete.
27#	Apply the last part of concrete self-weight counterweight on 8 working planes of the web concrete.
28#	Closure of the web concrete.
29~38#	Apply the required load at each loading point (28#, 30#, 32#, 34#, and 36#) and pouring concrete (29#, 31#, 33#, 35#, and 37#) for the first to fifth part of the 8 working planes of the roof concrete.
39#	Apply the last part of concrete self-weight counterweight on 8 working planes of the roof concrete.
40#	Closure of the roof concrete.

[1] The number + symbol "#" indicates the construction stage number to avoid confusion with other numbers in the text.

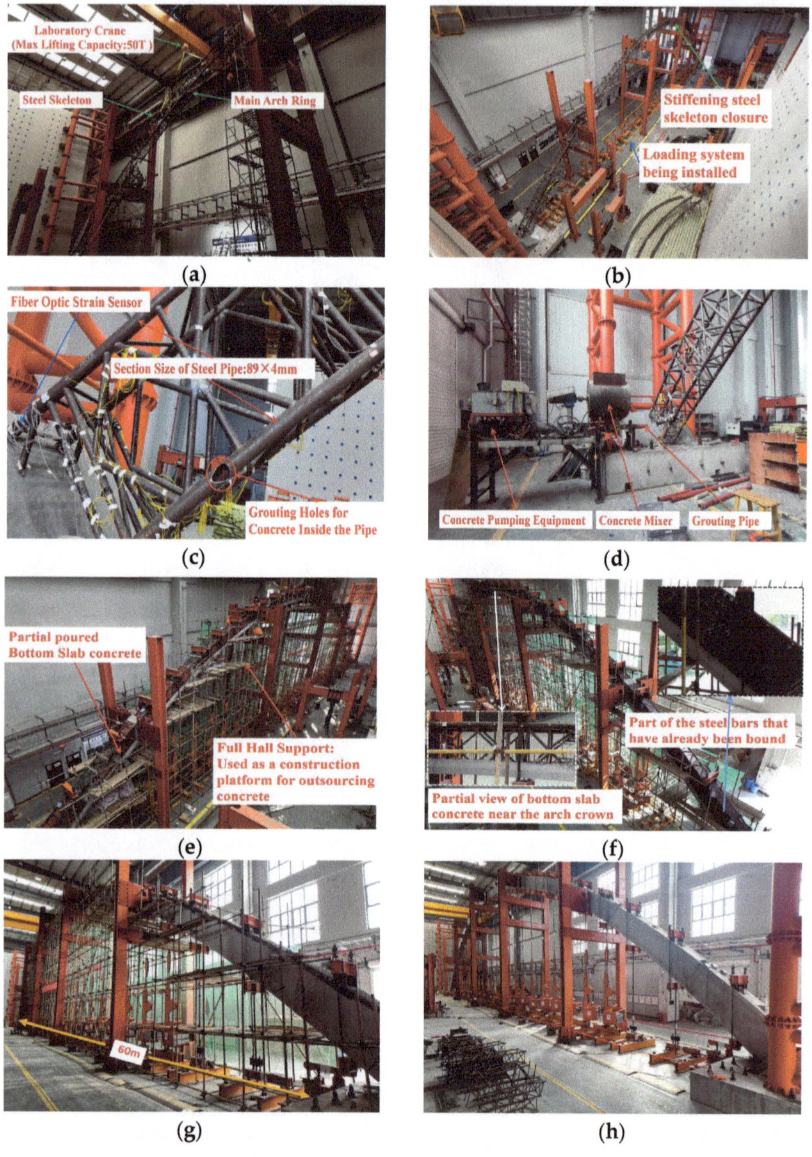

Figure 4. Display diagrams of the vital construction stages of the main arch ring of the model bridge. (**a**) Install the rigid steel skeleton using the large section lifting method; (**b**) closure of rigid steel skeleton (construction stage 1#); (**c**) concrete inlet in the pipe; (**d**) concrete grouting equipment inside the pipe (construction stage 2#); (**e**) layout of full framing; (**f**) closure of the floor concrete (construction stage 16#); (**g**) closure of the web concrete (construction stage 28#); (**h**) closure of the roof concrete (construction stage 40#).

4. Strain Testing Scheme for Externally Wrapped Concrete

The strain testing of externally wrapped concrete includes inter-ring strain tests at the 17 evenly distributed cross-sectional locations and strain tests near the section locations of the working plane of the concrete.

4.1. Concrete Inter-Rings Interface Strain Testing Scheme

The inter-rings interface refers primarily to the junction of the floor, web, and roof concrete, forming the three rings of the concrete structure, as depicted in Figure 5. Due to the varying sequence of pouring these three rings of concrete, their participation in the overall structural load-bearing varies, resulting in an uneven distribution of stress levels in these three rings of concrete after the primary arch ring is completed. This heterogeneity differs significantly from the stress distribution of arch rings formed through a single pouring operation and directly impacts the load transmission within the arch concrete, potentially reducing the limit load-bearing capacity of the arch rings below the original design value. Consequently, it is necessary to monitor the strain of the abovementioned panels throughout the construction process to capture the differences in interface stress [33,34]. Based on the model experiment, longitudinal strain sensors are evenly placed at 17 points along the span of the main arch ring specimen, as shown in Figure 6.

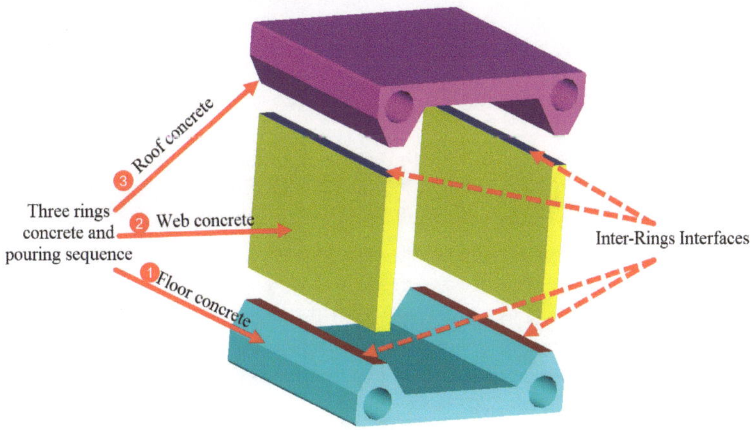

Figure 5. Schematic diagram of the force transfer surface and pouring sequence for the concrete rings wrapped around the main arch ring.

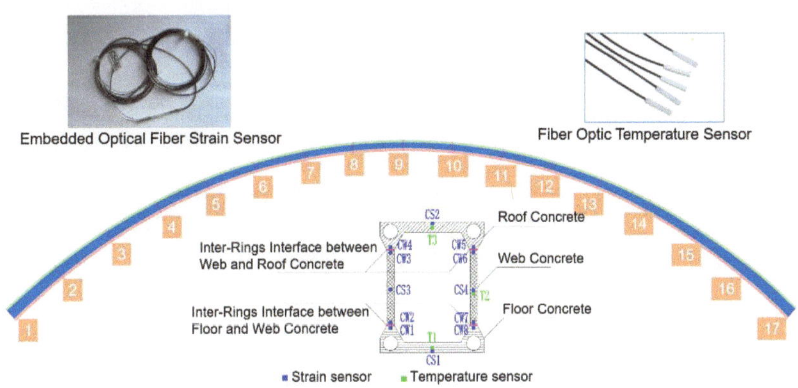

Figure 6. Diagram of the stress testing scheme for the externally wrapped concrete ring.

In Figure 6, 17 sections for strain testing were evenly distributed along the longitudinal direction of the main arch ring. Each section was equipped with 15 fiber-optic strain sensors [35] located at varying heights on the 17 sections; the relevant parameters of the sensor are shown in Table 2. Sensors CS1–CS4 are embedded and used to collect strain data of the floor, web, and roof concrete, while CW1–CW8 were used for testing strain data near the inter-rings interface between the floor and web concrete, and the inter-rings interface between the web and roof concrete.

Table 2. Sensor parameters table.

Sensor Type	Range	Accuracy	Wavelength Conversion Coefficient
Optical Fiber Strain Sensor	±2500 με	0.1% (2.5 με)	0.83
Optical Temperature Sensor	0~100 °C	0.1% (0.1 °C)	0.083

The structural strain can be calculated using Equation (1) and based on sensor wavelength data:

$$\varepsilon = (l_c - l_0)c_w \times 10^3 \tag{1}$$

In the equation, l_0 represents the initial wavelength at the time of sensor installation, l_c represents the current wavelength of the fiber-optic sensor, and c_w is the wavelength conversion coefficient, which is taken according to Table 2. ε represents the total strain that occurs at the current moment of the structure relative to the time when the sensor is installed.

To eliminate the influence of environmental temperature during the construction process on strain data, temperature sensors T1–T3 are embedded within the externally wrapped concrete, evenly distributed at three section heights, and the corresponding calculation method is shown in Equation (2):

$$\varepsilon_f = \varepsilon - (l_c - l_0)c_w\alpha_t \times 10^3 \tag{2}$$

In the equation, for this experiment, ε_f represents the strain of the structure at the measurement point position at the current moment relative to the sensor installation time under self-weight and counterweight load, and α_t represents the linear expansion coefficient of the structure.

4.2. Intersegments Interface Strain Testing Scheme

The intersegments interface primarily refers to the interface between two adjacent concrete sections within the same ring. When constructing externally wrapped concrete, a multi-plane balanced pouring method was used, and there is a particular age difference between the concrete on both sides of the intersegments interface. This age difference reaches its maximum when the interface is located between two adjacent working planes. To some extent, this difference reflects the timing of the concrete's participation in structural loading and the duration of stress accumulation; hence, the stress difference between two sections of concrete often correlates positively with the age difference. Based on these considerations, three intersegment interfaces were selected along the main arch ring in the longitudinal direction. Two test sections were set up on either side of each interface, as shown in Figure 7.

In Figure 7, the numbers encircled in brackets (①, ⑥, ⑦, ⑫, ⑬, and ⑱) represents the concrete pouring sequence, and the A*i*-*j*, B*i*-*j*, and C*i*-*j* (*i* =1,2,3,4; *j* =1,2, . . . ,6) respectively represent the concrete in the *j*th part of the *i*th workplane of the floor, web, and roof concrete. The three intersegments interfaces were located, respectively, at the junctions between work planes 1 and 2, 2 and 3, and 3 and 4. A test section, numbered from 1 to 6, was set up for each interface, with the sensor layouts for each section shown in Figure 8.

In Figure 8, each cross-section is equipped with 15 fiber-optic sensors. Among these, sensors CS1–CS2, CS3–CS4, and CS5–CS10 were strain gauges used to gather stress data from the concrete structure's roof, floor, and web. To eliminate the potential influence of environmental temperature and exothermic hydration reactions within the concrete on the strain data, temperature sensors T1–T3 were strategically positioned at three different heights along the intersegment interfaces, following the same arrangement as in the case of the annular space. The configuration allowed for accurate thermal correction and ensured the strain measurements' integrity.

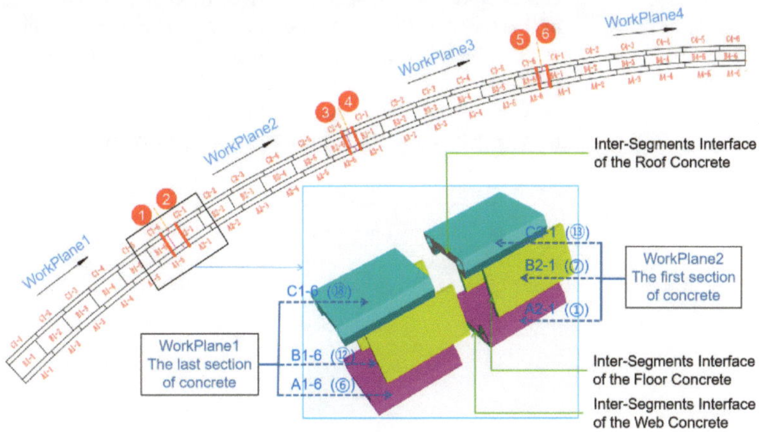

Figure 7. Distribution diagram of sectional interfaces in externally wrapped concrete.

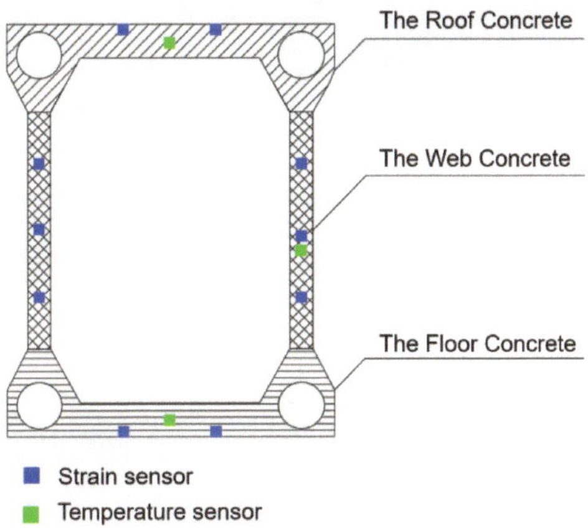

Figure 8. Schematic diagram of sensor layout at sectional interfaces.

5. Interfacial Stress Evolution in the Encapsulated Concrete Annular Space

5.1. Intersegments Interface Stress Evolution

To quantitatively analyze the evolution of interfacial stress in the main arch ring during the construction process of the encapsulated concrete, Hooke's Law was used to calculate the stress in the arch foot section, $\frac{1}{4}$ span section, and mid-span section. The stresses in the floor, web, and roof concrete at various construction stages were extracted from these sections. The derived stress profiles are illustrated in Figure 9.

Figure 9 shows that the stress level in the tri-ring concrete generally presented a pattern where the floor concrete bore the highest stress, followed by the web concrete, and the roof concrete had the least stress. After the encapsulated concrete was poured, the stress ratios of the tri-ring concrete in the arch foot section, 1/4-span section, and mid-span section were 1:0.67:0.20, 1:0.38:0.08, and 1:0.76:0.02, respectively. In Figure 9a,b, as the construction stage advances, the compression stress level of the tri-ring concrete exhibited an overall upward trend. Each ring's compression stress developed rapidly before the current ring closure. After the closure, it formed a unified structure with the rigid steel frame and participated in the overall structural load, which subsequently slowed the stress development. In Figure 9c, the compressive stress of the floor concrete at the mid-span position decreases from 8.10 MPa to 6.64 MPa (Area A in Figure 9c) during the web concrete closure stage and 9.51 MPa to 8.62 MPa (Area B) during the roof concrete closure stage, decreasing by 18.02% and 9.36%, respectively. This reduction was due to the increased negative moment at this location caused by the poured concrete in the web and roof concrete near the arch crown, thereby reducing the compressive stress in the floor concrete. Furthermore, as the moment of inertia at the arch crown section gradually increased with the closure of each ring of concrete, the negative moment's effect in reducing the compressive stress in the floor concrete was more pronounced in the early stages of construction.

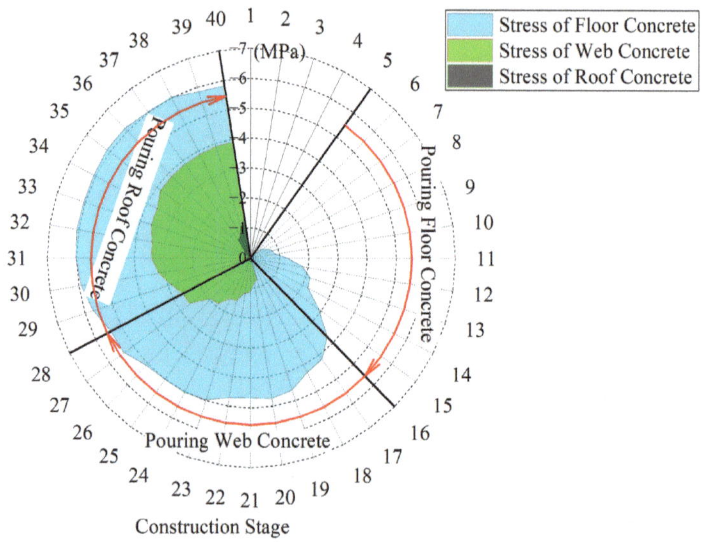

(a)

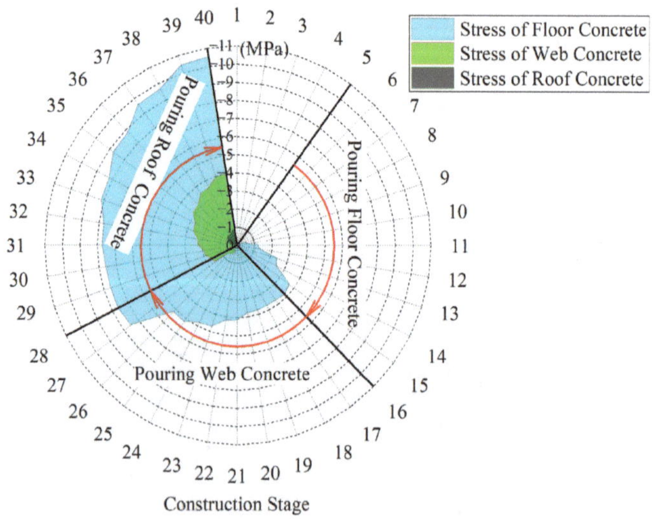

(b)

Figure 9. *Cont.*

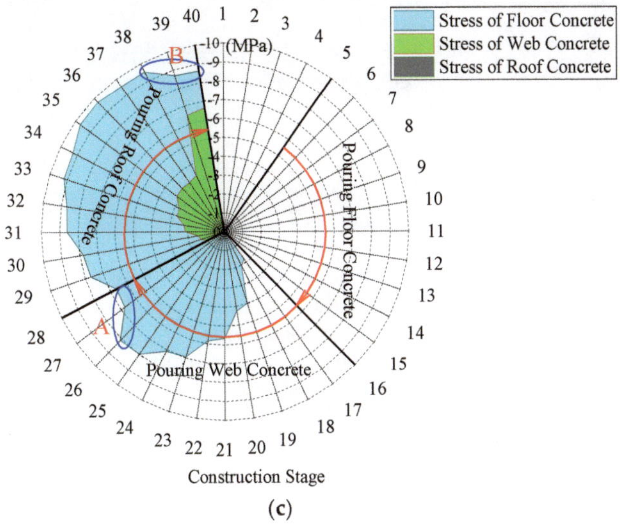

(c)

Figure 9. The stress curve of each main section in the construction stage of the main arch ring. (**a**) Section of arch foot; (**b**) 1/4 span section; (**c**) mid-span section.

To understand the changes in the relative stress state of the tri-ring concrete during the construction process of the main arch ring, we have calculated the stress ratios of the tri-ring concrete in the arch foot section, 1/4 span section, and arch crown section during the construction process, as shown in Figure 10.

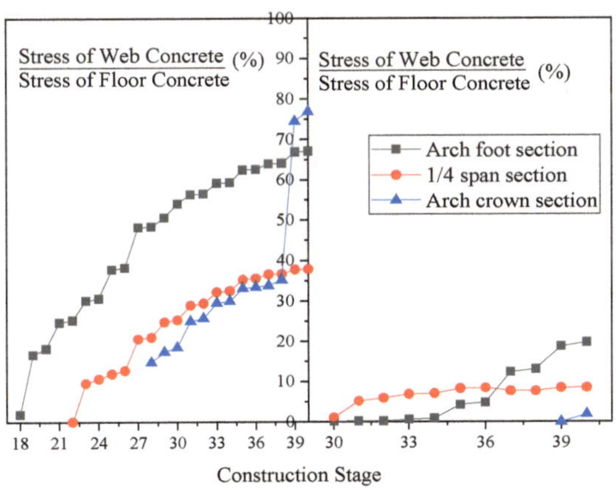

Figure 10. Three-ring stress ratio diagram of concrete encased in the main arch ring.

In Figure 10, all cross-sections present a characteristic where the stress level of the three rings' concrete gradually converged as the construction stage progressed. This was mainly manifested as an overall increasing trend in the ratio of the web concrete stress to the floor concrete stress and the roof concrete stress to the floor concrete stress at the same cross-section. After the closure of the main arch ring, the maximum web concrete stress at the same cross-section could reach 76.94% of the floor concrete, while the roof concrete accounted for only 19.73%. It should be pointed out that for the arch foot cross-section, the ratio of web concrete stress to floor concrete stress increased rapidly at the 19# construction stage. This was due to the large inclination angle of the arch axis and the horizontal plane at this location, resulting in a larger axial component of the self-weight load of the encapsulated concrete along the main arch ring.

However, when pouring the roof concrete, there was no similar phenomenon as in the web concrete. There were two main reasons for this. Firstly, the roof concrete had a smaller volume than the web concrete, which resulted in a smaller load component along the arch axis due to the self-weight of the former. Secondly, after the roof concrete of the arch foot cross-section hardened, it participated together with the main steel tube and the concrete within the tube in the local load-bearing of the structure. They jointly bore the axial component of the self-weight load from the newly poured roof concrete.

5.2. Analysis of Intersegments Interface Stress Evolution

During the pouring of the encapsulated concrete in the main arch ring, each ring was divided into multiple work surfaces in the longitudinal direction, and each work surface was completed in multiple stages of pouring. Figure 11 is a schematic diagram showing the location of the intersegment force transfer surfaces in the encapsulated concrete of the main arch ring. Before the closure of each ring of concrete, the axial forces between the work surfaces cannot be directly transmitted. Once a ring of concrete achieves closure and forms strength, the load-bearing capacity and stability of the main arch ring are enhanced. The axial forces generated in subsequent construction stages can be transmitted across work surfaces through the intersegment force transfer surfaces.

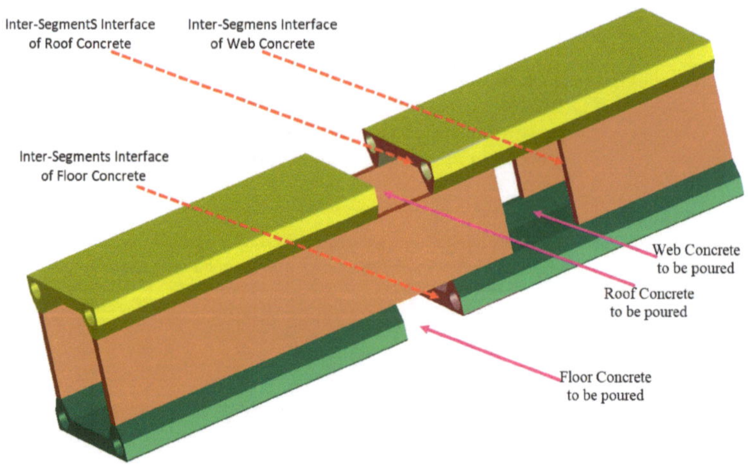

Figure 11. Schematic diagram of force transmission position between concrete sections wrapped around the main arch ring.

The construction sequence of the encapsulated concrete is shown in Figure 12, the dashed arrow in the figure indicates the direction of concrete pouring. The pouring sequence of each segment of concrete is denoted as $t_i(i = 1, 2, 3, \cdots, 18)$, and the interface between the Nth and the $(N + 1)$th work planes of each ring is represented as CP_N (N = 1, 2, 3). The concrete on either side of the interface is denoted as C_{Nf} (concrete of the earlier poured segment) and C_{Nl} (concrete of the later poured segment), according to the pouring sequence, and their respective stresses are denoted as σ_{Nf} and σ_{Nl}. P_{FN}, P_{WN}, and P_{RN} are used to represent the stress ratios of the concrete on either side of the intersegment interface in the floor, web, and roof concrete, respectively. Thus, we have:

$$P_{FN}, P_{WN}, P_{RN} = \frac{\sigma_{Nf}}{\sigma_{Nl}}, (N = 1, 2, 3) \tag{3}$$

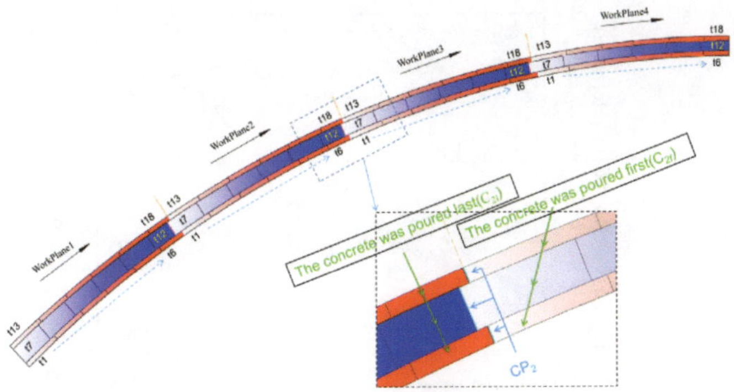

Figure 12. Construction sequence diagram of main arch ring wrapped with concrete.

The stress results for the intersegments interfaces during the construction process of the encapsulated concrete in the main arch ring have been collated, as illustrated in Figure 13.

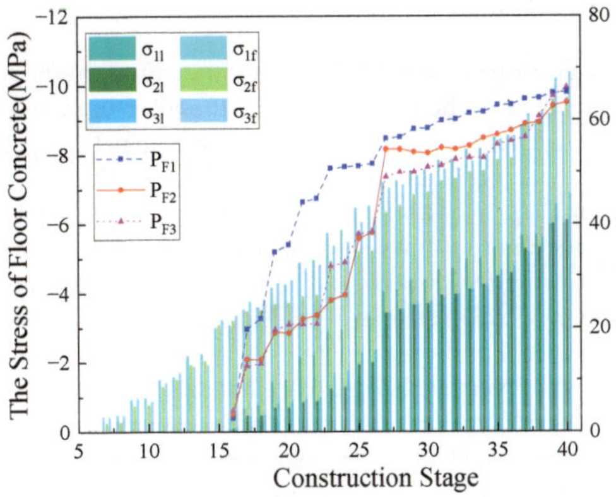

(a)

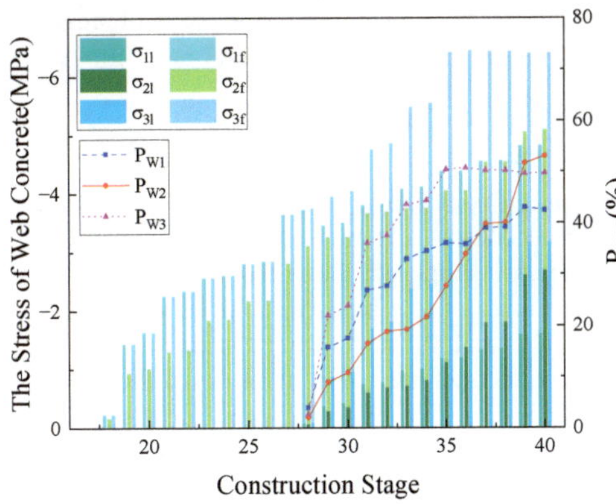

(b)

Figure 13. *Cont.*

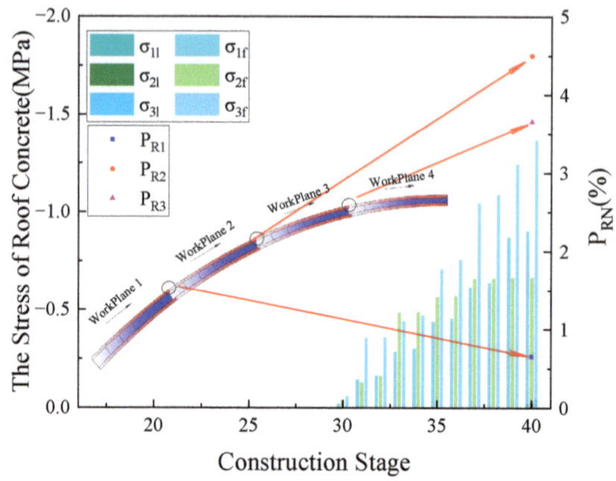

(c)

Figure 13. Concrete stress results of main arch ring wrapped with concrete in longitudinal section. (**a**) Stress results between concrete segments of floor concrete; (**b**) stress results between concrete segments of web concrete; (**c**) stress results between concrete segments of roof concrete.

As seen from Figure 13a,b, with the increase in the construction stage of the encapsulated concrete in the main arch ring, the intersegment stress ratio of the floor and web concrete gradually increases. After the construction of the encapsulated concrete was completed, the intersegment stress ratio P_{FN} of the floor concrete is within the range of 60~70%, and P_{F1}, P_{F2}, and P_{F3} are essentially equal; the intersegment stress ratio P_{WN} of the web concrete is within the scope of 40~60%. Figure 13c circles the longitudinal segment positions during the pouring of the roof concrete. The intersegment stress ratios P_{WN} for each segment are all less than 5%, which is significantly lower than the intersegment stress ratios of the floor and web concrete at this construction stage. This is because the pouring time of the roof concrete is later than that of the floor and web concrete, and the time it participates in structural load-bearing is also later. Compared with the floor and web concrete, the roof concrete lacked a process where the stress ratio increased with the construction stage. A large part of its stress was caused by its own gravitational load, and the mass of the roof concrete was smaller than that of the floor and web concrete, resulting in a smaller gravitational load. This eventually leads to the phenomenon where the intersegment stress ratio of the roof concrete is less than that of the first two rings of concrete.

6. Finite Element Model

6.1. Model Introduction

In the pursuit of a comprehensive study on the influence of uneven stress distribution in the external concrete ring and segmental joints, following the closure of the main arch of a rigid-frame arch bridge, a finite element model of the main arch ring has been constructed. This model utilizes the robust capabilities of the ANSYS APDL 18.0 [36], offering an extensive analysis of the structural behavior under different stress conditions. Figure 14 clearly represents the finite element model of the main arch ring.

Our approach, encompassing numerical simulation and advanced computational mechanics, aims to reveal potential vulnerabilities within the structural integrity of arch bridges. By focusing on the uneven distribution of stress after the arch ring closure, we can better comprehend its effects on the overall load-bearing performance of these structures. In turn, the acquired insights can significantly contribute to improving bridge design and safety strategies and provide a robust foundation for further academic studies in this field.

In Figure 14, the global coordinate system's X, Y, and Z directions correspond to the main arch ring's longitudinal, vertical, and lateral directions, respectively. The primary steel tubes, upper chord rods, lower chord rods, web rods of the rigid steel skeleton, and concrete-filled tubes are simulated using Beam188 elements. In contrast, the externally wrapped concrete is modeled using Shell181 elements. The complete bridge model comprises 6660 elements in total, including 3888 Beam188 elements and 2772 Shell181 elements.

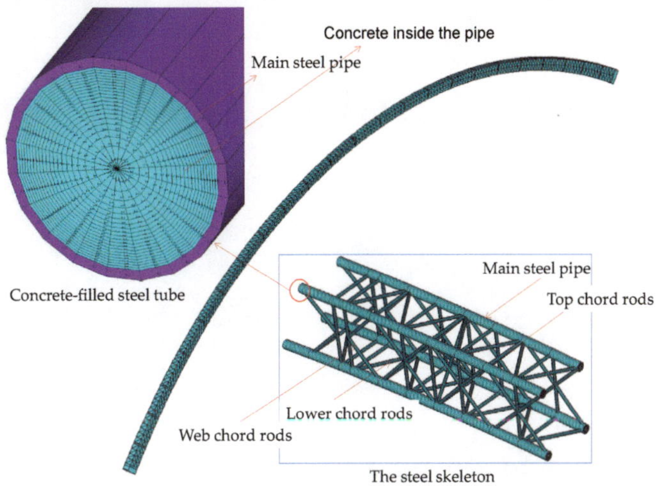

Figure 14. The finite model of main arch ring.

During calculations, displacement and rotation in all directions are restrained at every node on the surfaces where the arch feet are erected. Load conditions consist of the self-weight of the structure and the actual counterweights at the model bridge's loading points. The construction process of the main arch ring is simulated using birth-and-death elements.

6.2. Finite Element Model Validation

To validate the correctness of the finite element model, we have compared the stress results measured and simulated in the rigid steel skeleton, the concrete inside the tubes, and the externally wrapped concrete at some key cross-sections during various construction stages. These comparisons are graphically presented in Figure 15.

Our comparison across multiple stages ensures a rigorous validation of our finite element model, thus enhancing its credibility and potential application for future research. This method facilitates a better understanding of the stress distribution and performance of different bridge elements, making a significant contribution to the overall field of structural engineering.

As illustrated in Figure 15, each component's measured stress and simulation results during the construction process generally align well, exhibiting a consistent overall trend. Among them, the maximum relative error between the simulation results and experimental results for the main steel tube of the rigid steel skeleton throughout the construction process is 14.42 MPa, with the maximum relative error being 11.35%. These occur during the casting of the first segment of the web concrete stage, located at the top-chord main steel tube of the arch cross-section.

During the casting of externally wrapped concrete, the maximum relative error between the simulated and experimental results of the concrete inside the tubes is 4.09 MPa, which occurs during the casting of the third segment of the floor concrete stage. The maximum relative error between the simulated and experimental results of the floor concrete throughout the construction process is 1.44 MPa.

Throughout the construction process of the main arch ring, the stress results of the concrete inside the tubes and the externally wrapped concrete are significantly smaller compared to the main steel tube. Hence, the absolute error better represents the consistency between the ANSYS results and the experimental outcomes.

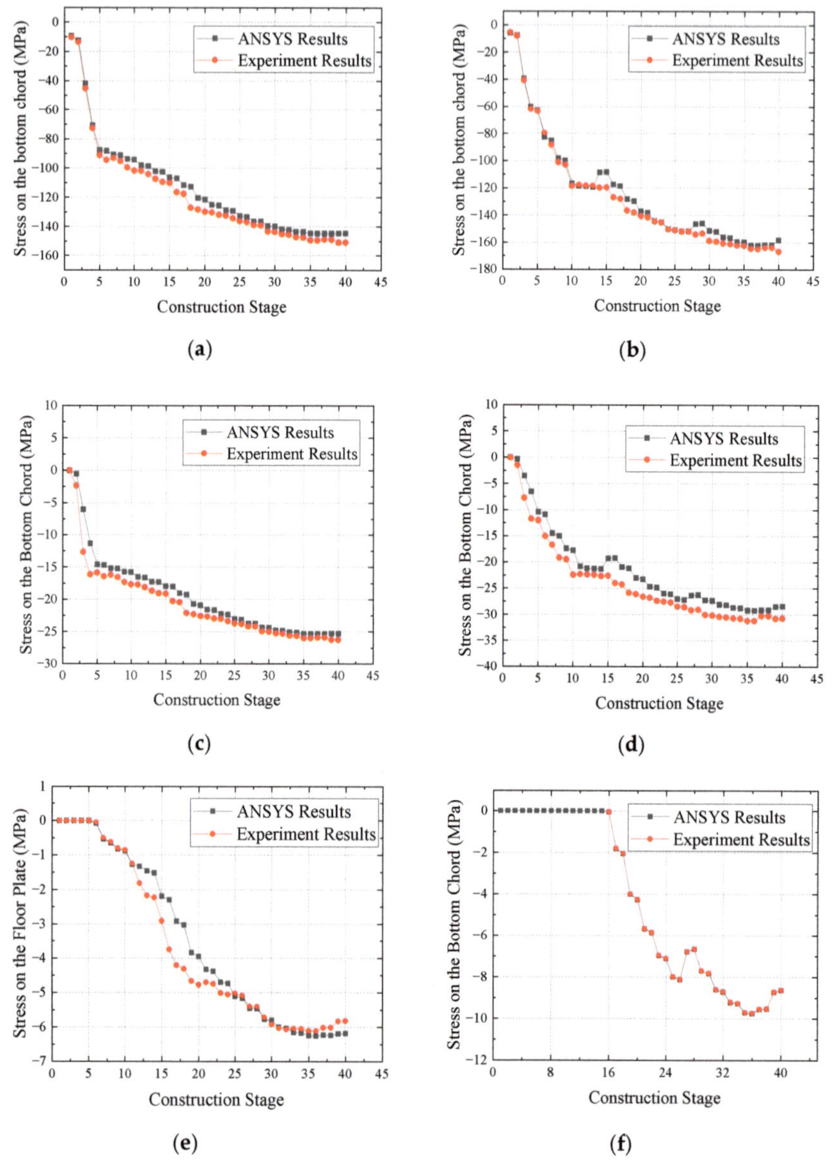

Figure 15. Comparison of simulated and measured results of each component throughout the construction process. (**a**) Stress results of the main steel pipe at the arch foot section; (**b**) stress results of the main steel pipe in the arch crown section; (**c**) stress results of concrete inside the arch foot section pipe; (**d**) stress results of concrete inside the arch crown section pipe; (**e**) stress results of floor concrete at arch foot section; (**f**) stress results of floor concrete at arch crown section.

Overall, the error between the simulated and experimental results falls within an acceptable range. These comparative results substantiate the finite element model's accuracy, thereby supporting the rationality of employing this finite element model for subsequent calculations.

7. Ultimate Load-Bearing Capacity Analysis

7.1. Introduction to the Computational Model

Experimental results demonstrate that when using a segmented ring-based construction method for externally concreted rigid-frame arch bridges, significant stress differentials exist between the tri-ring concrete segments and longitudinally adjacent concrete sections. In order to investigate the impact of this phenomenon on the structural performance of the main arch ring after its construction, we apply the 'birth-and-death' element method based on the validated finite element model in Section 6. This is achieved by adjusting the groups of activated elements under different construction stages, enabling a single pour for all the external concrete. This model, hereafter referred to as Model 2, demonstrates the construction sequence of the main arch ring, as shown in Table 3. The finite element model established following the conventional construction sequence will be designated as Model 1 for comparative purposes.

Table 3. Construction sequence table of the main arch ring in finite element model 2.

Construction Stage	Construction Content
1#	Closure of rigid steel frame.
2#	Pouring concrete into the pipe.
3#	Apply 9 times the self-weight of the rigid steel frame on the corresponding loading point.
4#	Apply 9 times the self-weight of the concrete inside the pipe on the corresponding loading point.
5#	Pour all external concrete at once.
6#	Apply the required counterweight at each loading point.

7.2. Ultimate Load-Bearing Capacity: Loading Scenarios and Results

In Model 2, the external concrete is activated simultaneously, meaning that all ringed and segmented concrete sections engage in structural stress simultaneously. After the completion of the sixth construction phase, the load-bearing distribution between and within the external concrete rings becomes more uniform, avoiding noticeable stress discontinuities. To quantitatively analyze the effect of uneven stress distribution between and within the rings of external concrete on the structure's load-bearing performance, we referenced other scholars' research methods in the arch bridge's ultimate load-bearing capacity [37–42]. We separately calculated the ultimate load-bearing capacity of Models 1 and 2 under four loading scenarios: mid-span loading, quarter-span single-point loading, symmetric loading at quarter-span, and full-span uniformly distributed load (as shown in Figure 16). The calculations consider the impact of geometric non-linearity and material non-linearity. Gradual loading during calculations was implemented through time step increments, with each adjacent time step having a load increment of 2000 N. The calculation results are presented in Table 4.

Table 4. Ultimate bearing capacity results of different models (unit: ton).

	Mid-Span Loading	1/4 Span Single-Point Loading	1/4 Span Symmetric Loading	Full-Span Uniformly Distributed Loading
Model 1	67.1	120.6	279.3	1098.6
Model 2	83.0	120.1	294.7	1307.6
Model 1/Model 2	80.84%	100%	94.77%	84.02%

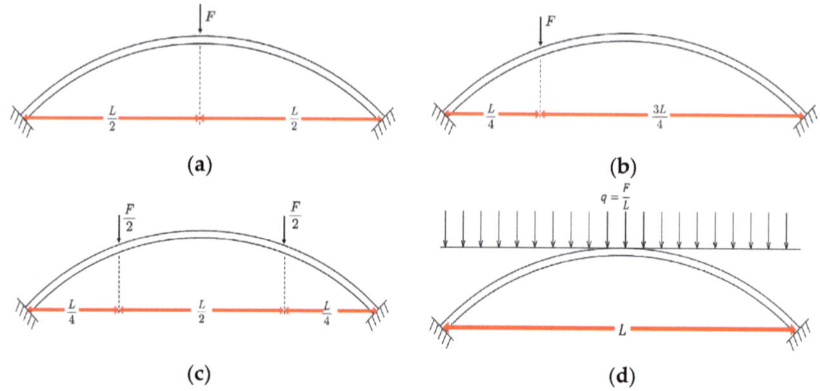

Figure 16. Schematic diagram of the loading conditions for the ultimate bearing capacity of the finite element model of the model bridge. (**a**) Single-point loading at the mid-span position; (**b**) single-point loading at 1/4 span position; (**c**) symmetrically loaded at 1/4 span position; (**d**) apply uniform load throughout the entire span.

The rigid steel frame of the main arch ring is made of Q420 steel. An ideal elastic–plastic model is used and simulated using the bilinear kinematic hardening model (BKIN) in the finite element model. As shown in Figure 17a, the relationship between δ and ε can be expressed using Equation (4):

$$\sigma = \begin{cases} E\varepsilon, \varepsilon \leq \sigma_y \\ \sigma_y, \sigma_y < \varepsilon \leq \varepsilon_u \end{cases} \tag{4}$$

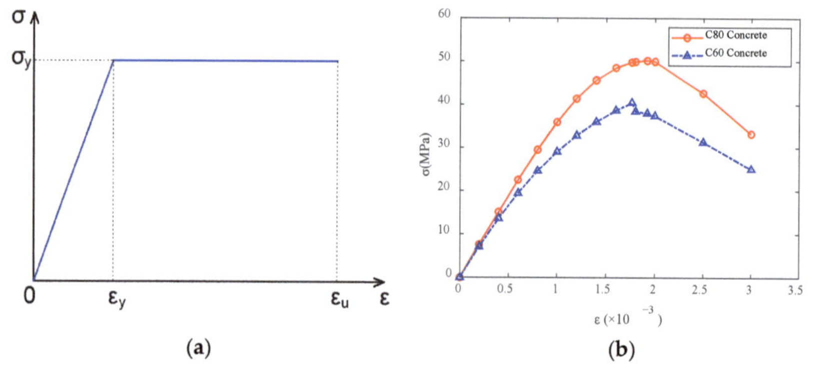

Figure 17. Stress–strain curve for main arch ring material. (**a**) Stress–strain curve for Q420 steel material; (**b**) stress–strain curve for concrete material.

In the equation, E represents the elastic modulus of Q420 steel, with $E = 2.06 \times 10^5$ MPa; σ_y represents the yield strength of the material, with $\sigma_y = 420$ MPa; ε_y represents the strain of the material when it reaches yield strength.

For the concrete material, the reference is "Code for Design of Concrete Structures" (GB50010-2010), [43]. It is assumed that both C80 and C60 concrete meet the following conditions:

$$\sigma = (1 - d_c)E_c\varepsilon \tag{5}$$

where d_c is the parameter for uniaxial compressive damage evolution of concrete, which can be determined using Equation (6):

$$d_c = \begin{cases} 1 - \frac{\rho_c n}{n-1+x^n}, x \leq 1 \\ 1 - \frac{\rho_c}{\alpha_c(x-1)^2+x}, x > 1 \end{cases} \tag{6}$$

In the equation, α_c is the parameter value of the descending segment of the uniaxial compressive stress–strain curve of concrete, while ρ_c, n, and x can be determined using Equations (7) to (9):

$$\rho_c = \frac{f_{c,r}}{E_c \varepsilon_{c,r}} \tag{7}$$

$$n = \frac{E_c \varepsilon_{c,r}}{E_c \varepsilon_{c,r} - f_{c,r}} \tag{8}$$

$$x = \frac{\varepsilon}{\varepsilon_{c,r}} \tag{9}$$

In these Equations, $f_{c,r}$ represents the representative value of the uniaxial compressive strength of the concrete; $\varepsilon_{c,r}$ indicates the peak compressive strain of the concrete corresponding to $f_{c,r}$. Ultimately, the stress–strain curves for the C80 concrete inside the main arch ring and the C60 concrete encasing are obtained, as shown in Figure 17b.

Table 4 shows a significant difference in the ultimate load-bearing capacity of the main arch ring under different loading conditions. The ultimate load-bearing capacities of both Model 1 and Model 2 do not exceed 100 tons when loaded at the mid-span. However, under uniform loading conditions, the ultimate load-bearing capacities of both models exceed 1000t, indicating a significant influence of loading conditions on the ultimate load-bearing capacity post-arch ring closure. In addition, with the exception of the 1/4 span single-point loading condition, the ultimate load-bearing capacity of Model 1 under all other loading conditions is less than Model 2. Compared to Model 2, the ultimate load-bearing capacities of Model 1 under mid-span and $\frac{1}{4}$ span symmetric loading conditions decreased by 19.16% and 5.23%, respectively. This suggests that an uneven stress distribution in the encased concrete can reduce the arch ring's ultimate load-bearing capacity after the completion of the main arch ring construction.

It is worth mentioning that for the 1/4 span single-point loading condition, the ultimate load-bearing capacities of Model 1 and Model 2 are nearly equal. This is because, under this loading condition, the ultimate load-bearing capacity of the main arch ring is controlled by the yield strength of the material near the loading point. During loading, vertical deflection occurs at the loading point. As shown in Figure 18, the concrete of the roof concrete at this location is compressed. Since the compressive stress of Models 1 and 2 are similar after the closure of the main arch ring, and the subsequent loading position and load step length are completely consistent, the roof concrete stress near the loading point of models 1 and 2 simultaneously reaches the material yield strength, causing structural instability.

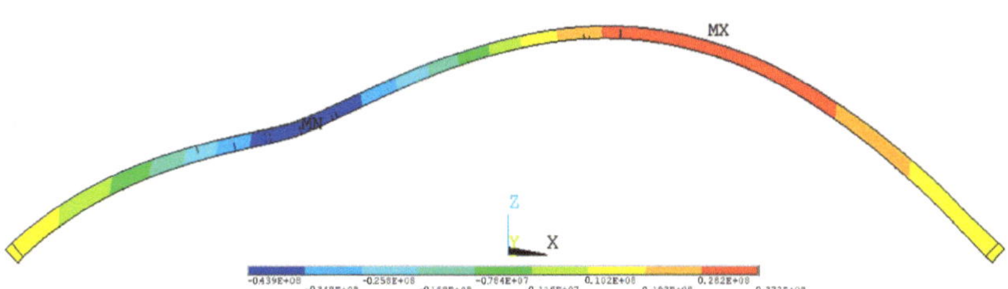

(a)

Figure 18. *Cont.*

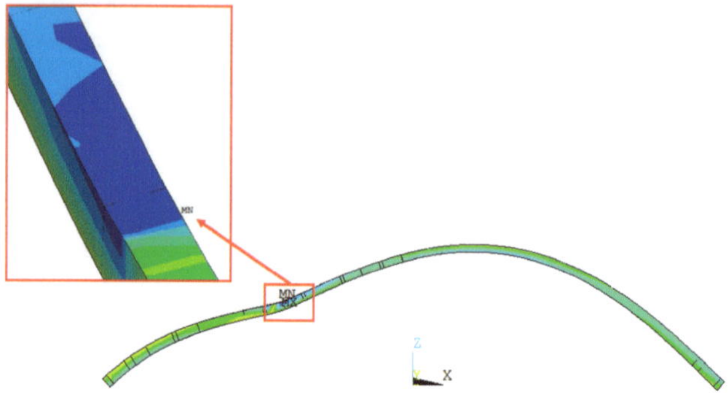

(b)

Figure 18. Instability state of the main arch ring under 1/4 span single-point loading condition. (a) Deformation diagram of the main arch ring at the point of instability (unit: Pa); (b) stress diagram of the main arch ring at the point of instability.

8. Conclusions

In order to investigate the stress evolution of inter-ring and intersegment interfaces during the separate ring and segment casting processes of concrete encasing the main arch ring of a rigid-frame arch bridge, and its impact on the ultimate load capacity, a 1:10 scaled model experiment was conducted on the main arch ring of the Tian'e Longtan Bridge with a main span of 600 m. Structural strain data at key cross-sections throughout the construction process were collected, and the stress evolution laws of inter-ring and intersegment interfaces were analyzed. The ANSYS simulation results and field data were compared, and the effect of uneven stress distribution of encased concrete on the ultimate load capacity under different loading conditions was calculated. The main conclusions are as follows:

(1) The construction order of the three rings of the main arch ring leads to differences in the timing of the floor, web, and roof concrete participating in the overall structural loading. After the closure of the main arch ring encased in concrete, the three rings of concrete exhibit noticeable stress differences. At the same cross-section, the stress level of the web concrete can reach 76% of the floor concrete stress, and the stress level of the roof concrete is less than 20% of the floor concrete stress.

(2) The three rings of concrete exhibit a characteristic of converging stress levels as the construction stage progresses, mainly manifested as an increase in the inter-ring stress ratio of the encased concrete at the same cross-section as the construction stage progresses.

(3) Due to the negative bending moment at the arch crown, casting concrete near the mid-span can significantly reduce the compressive stress level of the floor concrete. When casting the mid-span closure section web and roof concrete, the floor concrete compressive stress, respectively, decreases by 1.46 MPa and 0.89 MPa, with a decrease of 18.02% and 9.36%.

(4) For the same ring of concrete, due to the casting order, there is a difference in the timing of the concrete participating in local structural loading. At the junction of the work planes, there is a certain difference in the stress levels of the concrete on both sides of the interface. After the main arch ring is closed, the intersegment stress ratios of the floor, web, and roof concrete are 60~70%, 40~60%, and within 5%.

(5) ANSYS APDL was used to establish the finite element model of the main arch ring of the model bridge and the stress data of the top chord, inner concrete, and floor concrete on the rigid steel skeleton at the arch crown cross-section and the arch foot cross-section during the entire construction process were compared. The results show

that each component's measured stress and simulation results during the construction process generally match well, and the overall trend is basically consistent. Among them, the maximum relative errors of the simulation results and the measured results for the main steel tube, inner concrete, and floor concrete of the rigid steel skeleton throughout the construction process are 14.42 MPa, 4.09 MPa, and 1.44 MPa, respectively, verifying the correctness of the finite element model.

(6) To study the effect of the uneven stress distribution of encased concrete on the ultimate load capacity of the main arch ring, ANSYS APDL was used to calculate the ultimate load capacity of the two finite element models of the main arch ring encased in concrete under normal construction (Model 1) and encased in concrete in one pour (Model 2) under four loading conditions: mid-span loading, 1/4 span single-point loading, 1/4 span symmetrical loading, and full-span uniform loading. The results show that the loading conditions significantly affect the ultimate load capacity after the closure of the main arch ring. Under mid-span loading and 1/4 span symmetrical loading conditions, compared to Model 2, the ultimate load capacity of Model 1 decreased by 19.16% and 5.23%, respectively, indicating that an uneven stress distribution of encased concrete will reduce the ultimate load capacity after the closure of the main arch ring.

The next steps in the research will be to conduct related segment tests to clarify the stress evolution laws of the inter-ring and intersegment interfaces of ultra-large-span rigid-frame arch bridges under long-term service conditions, to define the load transfer paths between inter-ring and intersegment interfaces, and to conduct parametric analysis to study the impact of uneven stress on the ultimate bearing capacity of the main arch ring external concrete. Simultaneously, in conjunction with multi-objective optimization algorithms, we aim to minimize both the stress difference in the concrete of the floor, web, and roof concrete at the same section and the stress difference between adjacent segments of each ring of concrete. These are considered objective functions. With the maximum compressive stress of the concrete during the entire construction process as the constraint and the length of each working face of each ring of concrete as design variables, we plan to carry out research on the segmental optimization method for the external concrete of rigid-frame arch bridges. This is the task we are about to undertake.

Author Contributions: Conceptualization, Y.F. and Y.Z.; methodology, Y.F. and C.L.; software, Y.F.; validation, L.Y., X.L. and J.L.; formal analysis, L.Y.; investigation, Y.Z.; resources, C.L.; data curation, L.Y.; writing—original draft preparation, Y.F. and X.L.; writing—review and editing, J.L.; visualization, C.L.; supervision, Y.Z.; project administration, X.L.; funding acquisition, Y.F. All authors have read and agreed to the published version of the manuscript.

Funding: This work was supported by Guangxi Key Research and Development Plan Project (grant no. GuikeAB22036007); Special Key Project for Technological Innovation and Application Development in Chongqing (grant no. CSTB2022TIAD-KPX0205); Science and Technology Project of Chongqing Municipal Transportation Bureau (grant no. KJXM2021-0966); Independent Research and Development Project of the State Key Laboratory of Mountain Bridge and Tunnel Engineering (grant no. SKLBT-YF2103); Project of Science and Technology Program of Department of Transport, Hubei Province (grant no. 2020-186-1-6) and Chongqing Jiaotong University Postgraduate Scientific Research Innovation Project (grant no. CYB23247).

Institutional Review Board Statement: Not applicable.

Informed Consent Statement: Not applicable.

Data Availability Statement: The data presented in this study are available from the first and corresponding author upon request. The data is not publicly available due to the policy of the data provider.

Conflicts of Interest: The authors declare no conflict of interest. The sponsors had no role in the design, execution, interpretation, or writing of the study.

References

1. Zhao, R.; Zhang, Z. Overview of the development of concrete filled steel tubular rigid skeleton arch bridges in China. *Bridge Constr.* **2016**, *46*, 45–50.
2. Wang, Y.; Liao, P.; Jia, Y.; Zhao, R. Effects of Cyclic Temperature on Time-Dependent Deformation Behavior of Long-Span Concrete Arch Bridge. *Bridge Constr.* **2019**, *49*, 57–62.
3. Ma, K.; Xiang, T.; Zhao, R.; Xu, Y.; Xie, H. Stochastic analysis of long-term deformation of reinforced concrete arch bridges for high-speed railways. *China Civ. Eng. J.* **2012**, *45*, 141–146.
4. Lin, C.; Zheng, J. Pouring method of four working faces of main arch ring concrete of rigid skeleton arch bridge. *J. Transp. Eng.* **2020**, *20*, 82–89.
5. Zeng, Y.; He, H.; Qu, Y.; Sun, X.; Tan, H.; Zhou, J. Numerical Simulation of Fatigue Cracking of Diaphragm Notch in Orthotropic Steel Deck Model. *Materials* **2023**, *16*, 467. [CrossRef] [PubMed]
6. Chen, B.; Liu, J. Summary of world arch bridge construction and technology development. *J. Transp. Eng.* **2020**, *20*, 27–41.
7. Zeng, Y.; Qiu, Z.; Yang, C.; Su, H.; Xiang, Z.; Zhou, J. Fatigue experimental study on full-scale large sectional model of orthotropic steel deck of urban rail bridge. *Adv. Mech. Eng.* **2023**, *15*, 1–15. [CrossRef]
8. Zheng, J. New development trend of long-span concrete arch bridge in China. *J. Chongqing Jiaotong Univ.* **2016**, *35*, 8–11.
9. Hu, N.; Dai, G.; Yan, B.; Liu, K. Recent development of design and construction of medium and long span high-speed railway bridges in China. *Eng. Struct.* **2014**, *74*, 233–241. [CrossRef]
10. Tong, J.; Deng, N.; Lin, C.; Guo, X. Study on the pouring sequence of external concrete of long-span concrete-filled steel tubular rigid skeleton arch bridge. *Highway* **2020**, *65*, 92–97.
11. Yang, G.; Xu, Y.; Huang, Y. Optimization of concrete encasement scheme for long-span rigid skeleton arch bridge. *J. Railw. Eng.* **2017**, *34*, 50–54+82.
12. Lin, C.; Zheng, J. Construction analysis of arch ring concrete cable-stayed buckle method of Nanpanjiang super large bridge. *Bridge Constr.* **2016**, *46*, 116–121.
13. Kang, C.; Schneider, S.; Wenner, M.; Marx, S. Development of design and construction of high-speed railway bridges in Germany. *Eng. Struct.* **2018**, *163*, 184–196. [CrossRef]
14. Deng, N.; Li, C.; Guo, X.; Zhou, D.; Tong, J. Progress in construction method of main arch ring of concrete filled steel tubular rigid skeleton arch bridge. *Sci. Technol. Eng.* **2021**, *21*, 6132–6139.
15. Chen, B.; Wang, T. Overview of concrete filled steel tube arch bridges in China. *Pract. Period. Struct. Des. Constr.* **2009**, *14*, 70–80. [CrossRef]
16. Au, F.; Wang, J.; Liu, G. Construction Control of Reinforced Concrete Arch Bridges. *J. Bridge Eng.* **2003**, *8*, 39–45. [CrossRef]
17. Lin, C.; Zheng, J.; Li, X.; Lin, C.; Wang, C. Adjustment of structural stress of arch ring of Nanpanjiang bridge by cable-stayed buckle. *J. Guangxi Univ.* **2017**, *42*, 274–283.
18. Tong, J.; Deng, N.; Lin, C.; Li, H.; Li, C.; Zhou, D. Encased concrete pouring scheme of 600 m rigid skeleton arch bridge. *J. Guilin Univ. Technol.* **2021**, *41*, 791–796.
19. Liu, A.; Huang, Y.; Fu, J.; Yu, Q.; Rao, R. Experimental research on stable ultimate bearing capacity of leaning-type arch rib systems. *J. Constr. Steel Res.* **2015**, *114*, 281–292. [CrossRef]
20. Yang, L.; Xie, W.; Zhao, Y.; Zheng, J. Linear elastic iteration technique for ultimate bearing capacity of circular CFST arches. *J. Constr. Steel Res.* **2020**, *172*, 106135. [CrossRef]
21. Zhang, Y.; Liu, A.; Huang, Y.; Yang, J.; Fu, J.; Yu, Y.; Zeng, X. Experimental investigation of in-plane ultimate bearing capacity of parabolic high strength concrete-filled-steel-tubular arch. *Thin Walled Struct.* **2023**, *183*, 110348. [CrossRef]
22. Liu, Z.; Wang, S.; Xin, J.; Zhou, J.; Li, Y. Experimental study on failure of single-pipe arch of concrete-filled steel catenary pipe. *J. Railw. Sci. Eng.* **2022**, 1–11.
23. Wang, G.; Xia, Q.; Wang, Y.; Liu, C.; Sun, D. Research on static performance of double deck corrugated steel concrete composite arch. *J. Build. Struct.* **2021**, *42*, 358–364.
24. Yang, J.; Chen, R.; Zhang, Z.; Zou, Y.; Zhou, J.; Xia, J. Experimental study on the ultimate bearing capacity of damaged RC arches strengthened with ultra-high performance concrete. *Eng. Struct.* **2023**, *279*, 115611. [CrossRef]
25. Hu, J.; Yan, Q.; Yu, X.; Zheng, H. Ultimate bearing capacity analysis of concrete filled steel tubular arch considering the initial defects' influence. *Adv. Mater. Res.* **2012**, *446*, 1248–1251. [CrossRef]
26. Tang, X.; Xie, J.; Zhao, S.; Dong, W.; Pu, Q.; Yang, Y.; Weng, H. Experimental study on reinforced concrete arch model of Wanxian Yangtze River Bridge. *J. Southwest Jiaotong Univ.* **1994**, *4*, 362–367.
27. Zhang, S.; Zhao, R.; Jia, Y.; Wang, Y.; Xie, H. Experimental study on a large-span high-speed railway rigid skeleton concrete arch bridge model. *J. Southwest Jiaotong Univ.* **2017**, *52*, 1088–1096.
28. Fan, Y.; Zhou, J.; Luo, C.; Yang, J.; Xin, J.; Wang, S. Research on Loading Scheme for Large-Scale Model Tests of Super-Long-Span Arch Bridge. *Buildings* **2023**, *13*, 1639. [CrossRef]
29. Li, X.; Wang, M.; Xiao, J.; Zou, Q.; Liu, D. Experimental study on aerodynamic characteristics of high-speed train on a truss bridge: A moving model test. *J. Wind Eng. Ind. Aerod.* **2018**, *179*, 26–38. [CrossRef]
30. Zhang, C.; Jiang, G.; Su, L.; Lei, D.; Liu, W.; Wang, Z. Large-scale shaking table model test on seismic performance of bridge-pile-foundation slope with anti-sliding piles: A case study. *B Eng. Geol. Environ.* **2020**, *79*, 1429–1447. [CrossRef]

31. Gusella, F.; Orlando, M. Analysis of the dissipative behavior of steel beams for braces in three-point bending. *Eng. Struct.* **2021**, *244*, 112717. [CrossRef]
32. He, Y. *Preparation and Performance Study of C60 Fine-Stone Self-Compacting Concrete Encased in Steel Tube Arch*; Chongqing Jiaotong University: Chongqin, China, 2022.
33. Xin, J.; Zhou, C.; Jiang, Y.; Tang, Q.; Yang, X.; Zhou, J. A signal recovery method for bridge monitoring system using TVFEMD and encoder-decoder aided LSTM. *Measurement* **2023**, *214*, 112797. [CrossRef]
34. Zhang, H.; Li, H.; Zhou, J.; Tong, K.; Xia, R. A Multi-Dimensional Evaluation of Wire Breakage in Bridge Cable Based on Self-Magnetic Flux Leakage Signals. *J. Magn. Magn. Mater.* **2023**, *566*, 170321. [CrossRef]
35. Tan, X.; Abu-Obeidah, A.; Bao, Y.; Nassif, H.; Nasreddine, W. Measurement and Visualization of Strains and Cracks in CFRP Post-Tensioned Fiber Reinforced Concrete Beams Using Distributed Fiber Optic Sensors. *Autom. Constr.* **2021**, *124*, 103604. [CrossRef]
36. ANSYS. *Structural Finite Element Analysis (FEA) Software*; ANSYS Inc.: Canonsburg, PA, USA, 2023.
37. Du, C. *Study on Mechanical Properties of Concrete Filled Steel Tubular and Reinforced Concrete*; Tianjin University: Tianjin, China, 2011.
38. Wang, S.; Li, Y.; Liu, Z.; Cheng, T. Calculation model of concrete-filled steel tube arch bridges based on the arch effect. *Front. Mater.* **2023**, *9*, 1084999. [CrossRef]
39. Zheng, Z.; Chen, B. Nonlinear plane finite element analysis of reinforced concrete hingeless arch. *Appl. Mech.* **1988**, *4*, 72–77.
40. Tang, X.; Zhang, J.; Li, C.; Xu, F.; Pan, J. Damage analysis and numerical simulation for failure process of a reinforced concrete arch structure. *Comput. Struct.* **2005**, *83*, 2609–2631. [CrossRef]
41. Chen, B.; Qin, Z. Equivalent beam-column method for calculation of the ultimate load-carrying capacity of concrete filled steel tube (single) arch under in-plane loads. *J. China Railw. Soc.* **2006**, *28*, 99–104.
42. Wei, J.; Chen, B.; Wu, Q. Equivalent beam-column method to calculate nonlinear critical load for CFST arch under compression and bending. *Eng. Mech.* **2010**, *27*, 104–109.
43. *GB 50010-2010*; Ministry of Construction of the China, Code for Design of Concrete Structures. China Construction Industry Press: Beijing, China, 2015; pp. 210–211.

buildings

MDPI

Article

Effect of Shear Keys on the Quasi-Isolated Behavior of Small-to-Medium-Span Girder Bridges

Kefeng Yue [1], Lueqin Xu [1,2,*], Lei Fan [1], Jie Liu [1] and Hao Luo [3]

1 School of Civil Engineering, Chongqing Jiaotong University, Chongqing 400074, China; ykf204732@163.com (K.Y.); 2420705341@163.com (L.F.); 630782447@163.com (J.L.)
2 State Key Laboratory of Mountain Bridge and Tunnel Engineering, Chongqing Jiaotong University, Chongqing 400074, China
3 Greentown Architecture Design, Chongqing 401121, China; haoluo000@outlook.com
* Correspondence: xulueqin@163.com

Abstract: Small-to-medium-span girder bridges equipped with shear keys play a significant role in the Chinese highway bridge system. However, shear key failure was observed during the 2008 Wenchuan earthquake, which resulted in excessive superstructure displacements and even catastrophic span collapse. For this, six refined bridges were investigated for the quasi-isolated behaviors under different shear key strengths by using the Pushover and IDA methods. Results indicate that the bridges exhibit two distinct damage states upon the shear key strengths. The shear key failure and bearing sliding create a natural quasi-isolated mechanism, with the following damage sequence: shear key failure → bearing sliding → pier undamaged or slight damage. Quasi-isolated behavior leads to higher displacement demands for beams, especially when the peak ground acceleration (PGA) exceeds 0.45 g. By selecting suitable shear key strength, below 9% for 20 m piers and 30% for 10 m piers, quasi-isolated damage is expected to occur in bridges. The study offers a fresh perspective on the concept of seismic design for highway girder bridges in China.

Keywords: small-to-medium-span girder bridges; shear key; pushover analysis; incremental dynamic analysis (IDA); quasi-isolated behavior

Citation: Yue, K.; Xu, L.; Fan, L.; Liu, J.; Luo, H. Effect of Shear Keys on the Quasi-Isolated Behavior of Small-to-Medium-Span Girder Bridges. *Buildings* **2023**, *13*, 2246. https://doi.org/10.3390/buildings13092246

Academic Editors: Simon X. Yang, Jingzhou Xin, Yan Jiang and Hong Zhang

Received: 3 July 2023
Revised: 25 August 2023
Accepted: 30 August 2023
Published: 4 September 2023

1. Introduction

Small-to-medium-span girder bridges are the mainstay of China's highway bridges, many of which traverse high-intensity earthquake zones. The superstructures of these bridges are simply supported on piers or abutments by unanchored rubber bearings. To restrict the lateral displacement of the girder, reinforced concrete shear keys are installed on both sides of the cap beam and abutment. According to the Chinese seismic design code [1], girder bridges are designed to be ductile, utilizing predesignated plastic hinges in piers to withstand seismic forces. In the 2008 Wenchuan earthquake, a lateral seismic damage survey conducted on 1432 small-to-medium-span girder bridges revealed that, 16.6% of the total bearings suffered severe damage and 19.5% of the total bridge spans experienced lateral movement, including 10 spans that fell off from the supports. Interestingly, the damage proportion of piers was lower than anticipated [2,3]. Furthermore, damage to shear keys was also evidenced in the 1978 Tangshan earthquake [4], the 1994 Northridge earthquake [5], the 1999 Chi-Chi earthquake [6], and the 2010 Chile earthquake [7], revealing similar failure patterns of bridges compared with that in Wenchuan earthquake. The failure of shear keys resulted in the girder losing restraint, ultimately isolating the piers [3]. The isolated failure pattern contradicts the initial intent of the ductile design, highlighting the disparity between China's current seismic concept and engineering practices.

When it comes to the concrete shear keys, there has been a controversial issue associated with their role in the seismic behavior of bridges. While the role of the shear key in bridges has not been well-defined. In China, shear keys are classified as either a structural measure or secondary component, usually simplified and that disregard the constraining effect or assume the shear key to be rigid body [8,9]. Some codes, such as Caltrans [10], Goe [11] and Savelson [12], also tacitly accept these simplifications. After the Wenchuan earthquake, scholars have since realized that the shear key not only limits displacement but also plays a role in force transmission with digital models of bridges [13,14]. Various types shear keys have been proposed, including the double-layer seismic shear key [15], the sacrifice shear key [16,17], the resettable shear key [18], the wedge shear key [19], the shearing energy dissipation shear key [20], and the X-shaped steel plate shear key [21]. Bi and Hao [22] developed a detailed 3-dimensional finite element model to capture the shear key responses. The results indicated that neglecting the engagement of shear keys might lead to an inaccurate prediction of bridge seismic demands. Están et al. [23] evaluated the influence of external sacrificial shear keys on the seismic behavior of Chilean highway bridges, which concluded that the most vulnerable bridges were those without external shear keys, regardless of the seismic hazards and soil types. These studies concluded that the role of shear keys should not be overlooked. However, the strength of shear keys is frequently either excessive or insufficient due to lack of a design method for shear keys in the Chinese code, leading to significant uncertainties in the seismic response of bridges.

Additionally, the sliding of the superstructure on the unanchored bearing after the shear key failure has also attracted attention, since the sliding acts as isolation [24,25]. Tobias [26] proposed a so-called quasi-isolated bridge system to improve the Illinois Earthquake Resistant System (ERS). The AASHTO [27] stated that the sacrifice of connecting members could effectively weaken the superstructure–substructure connections, thereby limiting seismic forces transferred down to substructures and foundations. The IDOT [28] believed that the essence of the quasi-isolated mechanism lay in achieving a specific fusing mechanism at the connection members between the superstructure–substructure structures, which involved bearing sliding and shear key failure. Filipov [29,30] conducted experimental research on three types of US bearings, and found that bearing sliding results in higher residual displacement. Steelman [31] reached a similar conclusion by utilizing polytetrafluoroethylene bearings to achieve a more reliable quasi-isolated mechanism. Luo [32,33] discovered that achieving quasi-isolated behavior came at the expense of bearing sliding or even beam falling. Wang [34] and Li [35] noted for the study of girder bridges in China that the shear key should be designed as a sacrificial member to achieve the isolation. Nonetheless, there is still a lack of research on the coordination between bearing sliding and beam falling for typical small-to-medium-span girder bridges and are regarded as crucial research needs for further refining quasi-isolated bridge design methodology.

This paper investigated the role of shear keys on the quasi-isolated behavior in typical small-to-medium-span highway girder bridges, taking into account the effects of bearing sliding and shear key degradation. Six typical types of highway girder bridges were investigated for the influence of shear key strength on the transverse quasi-isolated behavior using the Pushover and IDA methods. The quasi-isolated behavior was researched as the failure of the shear key and the slippage of the bearing. The findings provide a basis for the establishment of quasi-isolated concepts.

2. Bridge Case Selection and Research Scope

This research primarily targets typical small-to-medium-span girder bridges in China. Skew bridges, curved bridges, and single-column pier bridges are not within the scope of this study. A typical single-span three-lane highway bridge was selected as the prototype basic bridge case. The bridge spans a U-shaped valley, and the superstructure is a 5 m × 25 m prestressed concrete simple T-beam, as shown in Figure 1. Each T-beam is equipped with a plate-type rubber bearing at both ends, with the bearing type being GJZ250 × 300 × 74 mm. The substructure adopts a round double-column pier, with each pier being 10 m and 1.5 m in diameter. The cap beam and the abutment are equipped with reinforced concrete shear keys. The position of the shear key on the gap beam is shown in Figure 1c, with a 20 mm gap between the shear key and T-beam. The T-beam is constructed using C40, while the pier, cap beam, and shear key are all made of C30. The bridge site features shallow bedrock and favorable geological sound with a rigid foundation.

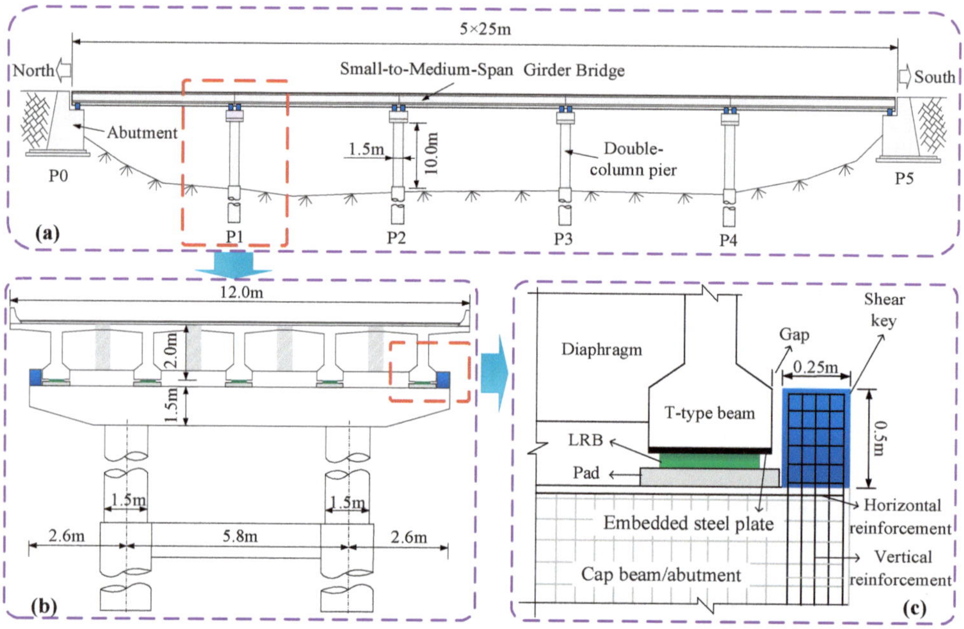

Figure 1. Bridge prototype: (**a**) typical girder bridge; (**b**) cross-sectional diagram; (**c**) shear key installation diagram.

Five other bridge cases are presented to make the research genera; they, are obtained from the above basic bridge case. Parameters include the pier height (10 m or 20 m), span diameter (25 m or 35 m) and foundation form (Rigid or Soft), as shown in Table 1. In addition, the shear key strengths of the different bridge cases were normalized by reference to [36] The shear key strength ratio is defined as the ratio of the shear key strength to the pier or abutment support reaction force, ranging from 0% to 100%. The M10S25Z bridge case introduces a soft soil foundation, which is modeled in the following analysis of pile–soil interaction. The M10S35 signifies that the pier height, pier form, and span of the bridge is 10 m, double-column pier (S), and 35 m, respectively. Table 1 also provides the basic transverse natural vibration period for the M10S35~M10S25Z.

Table 1. Basic information of bridge cases.

Bridge Cases	Pier Forms	Pier Heights	Spans	Foundations	Basic Periods	Illustration Diagrams
M10S25	1.5 m-Double column pier	10 m	5 × 25 m	Rigid	1.09 s	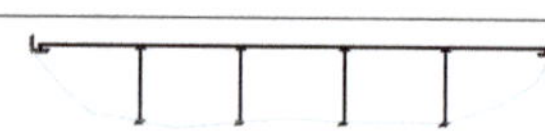
M10S35	1.5 m-Double column pier	10 m	5 × 35 m	Rigid	1.29 s	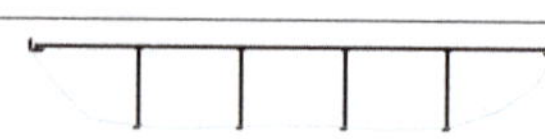
M20S25	1.5 m-Double column pier	20 m	5 × 25 m	Rigid	1.63 s	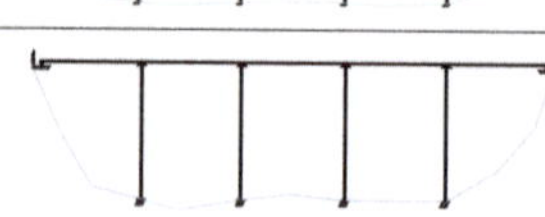

Table 1. *Cont.*

Bridge Cases	Pier Forms	Pier Heights	Spans	Foundations	Basic Periods	Illustration Diagrams
M1221S25	1.5 m-Double column pier	10-20-20-10 m	5 × 25 m	Rigid	1.38 s	
M2112S25	1.5 m-Double column pier	20-10-10-20 m	5 × 25 m	Rigid	1.47 s	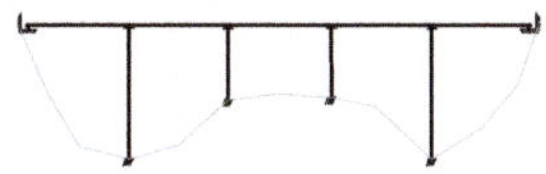
M10S25Z	1.5 m-Double column pier	10 m	5 × 25 m	Soft	2.14 s	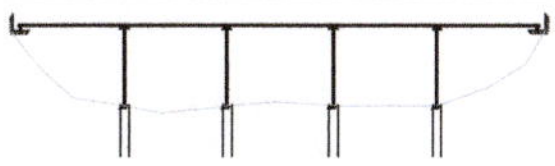

3. Model Construction and Verification

3.1. Bearing Sliding Analysis Model

This research primarily targets typical small-to-medium-span girder bridges in China. There is a large amount of literature available for reference on experimental research into the sliding mechanism of the bearing [37,38]. This paper proposes a simplified analysis model based on this, as shown in Figure 2. The bearing undergoes elastic shear deformation before sliding, with the shear stiffness being K_{bH}, and starts to slide when it reaches the critical friction force (F_{bH}). The sliding phenomenon is simulated by the smooth sliding unit in OpenSEES [39], and according to the Coulomb friction theory, the friction coefficient μ = 0.25 is assumed to remain constant during the sliding process. In the formula, G_b is the shear modulus of the rubber, taken as 1200 kN/m^2 according to the regulations [40]; A_b is the area of the bearing rubber plate; t is the total thickness of the rubber layer; and N is the axial force of the bearing. The sliding dynamic constitutive curve of the bearing in the horizontal direction can be calculated, as seen in Figure 2c.

$$K_{bH} = G_b A_b / \sum t \tag{1}$$

$$F_{bH} = \mu N \tag{2}$$

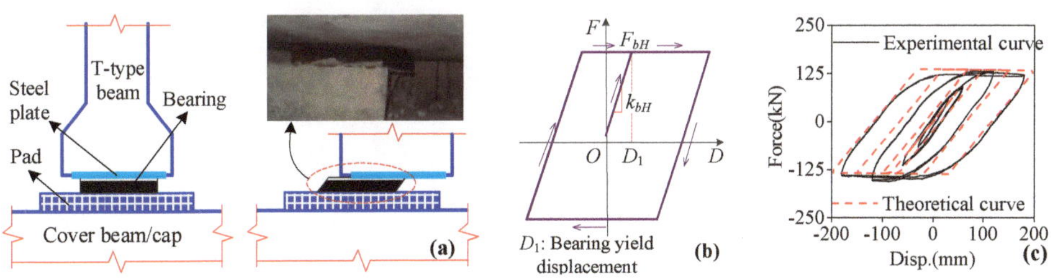

Figure 2. Bearing sliding analysis model: (**a**) simplified analysis of bearing sliding damage; (**b**) mechanical model; (**c**) verification analysis.

3.2. Shear Key Mechanical Analysis Model

Earthquake damage [2,3] and experimental research [16,17] show that reinforced concrete shear keys often suffer oblique section shear failure, as shown in Figure 3a,b. The force–deformation curve of the shear key under cyclic load can be decomposed using a two-term nonlinear spring, as shown in Figure 3c. The calculation method of the key node performance parameters corresponding to the two-term spring can be found in the literature [16]. The force–deformation relationship curve of the shear key calculated according to this model fits well with the measured curve, as shown in Figure 3b. There is generally an installation gap of about 20 mm between the shear key and the main beam in actual engineering, leading to a collision phenomenon between the main beam and the shear key. This paper uses a compression-only gap unit to simulate the shear key, as shown in Figure 3c. It is worth mentioning that the model is also widely adopted, such as Goel [11], Omrani [13], Wu Gang [24], etc.

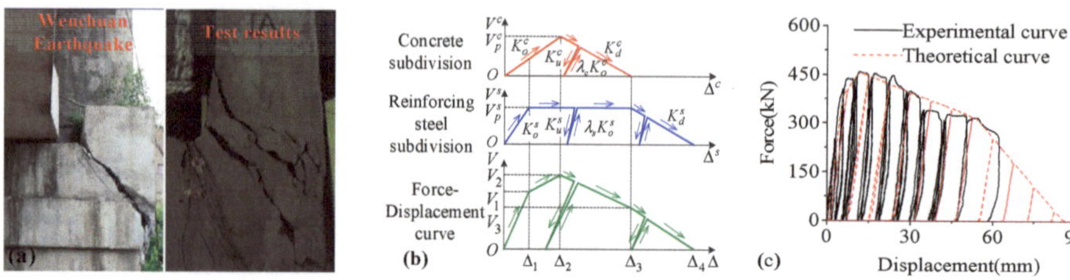

Figure 3. Shear key analysis model: (**a**) comparison of seismic damage and test phenomena; (**b**) simplified analysis; (**c**) verification analysis.

3.3. Pile-Soil Interaction Analysis Model

The pile–column foundation is a foundation form commonly used in the lower structure of girder bridges in China. This paper refers to the pile–soil–bridge pier nonlinear numerical analysis model [41] and establishes a pile–soil analysis model considering silt, clay, and sand soil, as is shown in Figure 4. The pile in the model is simulated by an elastoplastic fiber unit, with a pile length of 20 m, a pile diameter of 1.5 m, and a unit length of 1 m. The pile–soil lateral friction, vertical friction, and pile tip–soil interaction are simulated using the zero-length units of the P-y spring, T-z spring, and Q-z spring in the OpenSEES, respectively. The P-y curve suggested by Matlock is used for clay [42], and the P-y curve suggested by the American API specification is used for sandy soil [43]. The soil-related properties are shown in Figure 4, where φ represents the soil body friction angle; γ represents the soil body bulk density; c_u is the soil body undrained shear strength; ε_{50} represents the strain value when the soil body's 0.5 times the limit principal stress; and C_d represents the ratio of the pile side lateral resistance to the limit soil resistance.

Figure 4. Pile—soil interaction analysis model.

3.4. Whole Bridge Analysis Model

The whole bridge OpenSEES finite element analysis model is shown in Figure 5. The main beam and cap beam are simulated by elastic beam units. The pier column is simulated by elastoplastic fiber units, with concrete and steel bars using the built-in Kent–Scott–Park model and the Giuffré–Menegotto–Pinto model, respectively [39]. The boundary conditions at the abutment are modeled according to the Caltrans simplified method [4]. The prototype bridge case is located in a bedrock, where the nonlinearity of the soil body is not prominent. To avoid introducing too many complex factors in the analysis, the bridge cases of M10S25~M2112S25 all use rigid foundation modeling, with only the bridge case of M10S25Z considering the pile–soil interaction soft foundation modeling. The whole bridge model is detailed in Figure 5.

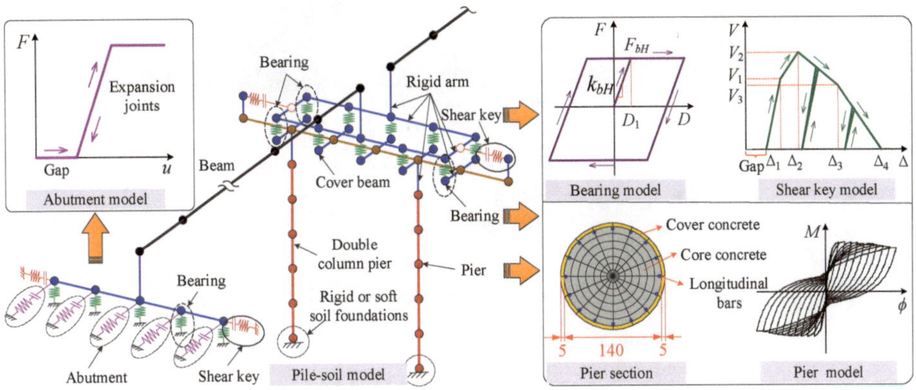

Figure 5. Analytical model of the whole bridge.

4. Analysis Method and Seismic Input

The static elastoplastic analysis method (Pushover method) is used to derive all bridge cases, establishing the damage state derivation curves. The study considers the first-order mode transversely to carry out the derivation analysis, ignoring the contribution of the high-order modes to the structural response [32], with the derivation analysis process shown in the literature [44]. Subsequently, incremental dynamic time history analysis (IDA method) is used for further analysis of the cases. Taking the design response spectrum corresponding to the type II site conditions in the regulations as a benchmark, based on the seismic magnitude $M \geq 6.5$, the epicentral distance $R \geq 20$ km, and the average shear wave velocity of the 30 m surface soil layer $250 \text{ m/s} < V_{se} \leq 500 \text{ m/s}$, 10 seismic waves were selected and amplified from 0.1 g to 1.0 g, as shown in Table 2. The horizontal acceleration response spectrum of the earthquake motion after the PGA = 0.4 g amplification is shown in Figure 6. Among them, T_t represents the first-order natural period of the basic case in the transverse direction. The earthquake motion is only input along the transverse direction, with the average results of the 10 seismic motions taken as its representative value.

Table 2. Ground motion recording parameters.

No.	Earthquake Records	Years	Magnitudes	Distance (km)	PGA (g)
1	Wenchuan	2008	8.0	54.5	0.840
2	Imperial Valley	1979	6.5	34.9	0.469
3	Loma Prieta	1989	6.9	76.8	0.103
4	Loma Prieta	1989	6.9	27.3	0.139
5	Landers	1992	7.3	69.2	0.154
6	Landers	1992	7.3	154.3	0.188
7	Northridge	1994	6.7	41.1	0.568
8	Northridge	1994	6.7	57.3	0.432
9	Chi-Chi	1999	7.6	24.9	0.468
10	Chi-Chi	1999	7.6	40.4	0.431

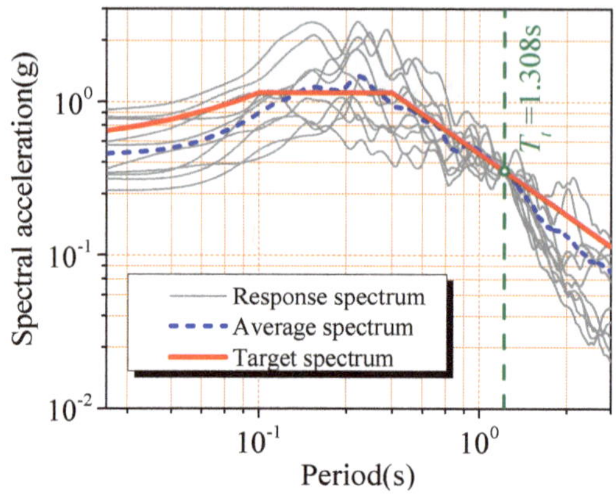

Figure 6. Scaling method of ground motions.

5. Definition of Critical States

5.1. Components Critical States

Components closely related to the quasi-isolated behavior include bearings, shear keys, and bridge piers or abutments. Given that the stiffness and strength of gravity abutments are generally large, the abutments are ignored. The pushover method is used to establish the force–displacement curve of the piers, including 10 m piers under rigid foundation($d_{10m,rigid}$), 20 m piers under rigid foundation($d_{20m,rigid}$) and 10 m piers under soft foundation($d_{10m,soft}$). According to the literature [45], the performance characteristic points corresponding to the first longitudinal reinforcement yielding, cover concrete crushing, core concrete crushing and equivalent yielding are determined. The equivalent bilinearization curve is carried out using the equal energy principle, obtaining the critical yielding displacement (d_y) and critical damage displacement (d_u), as shown in the outer and inner brackets of Figure 7.

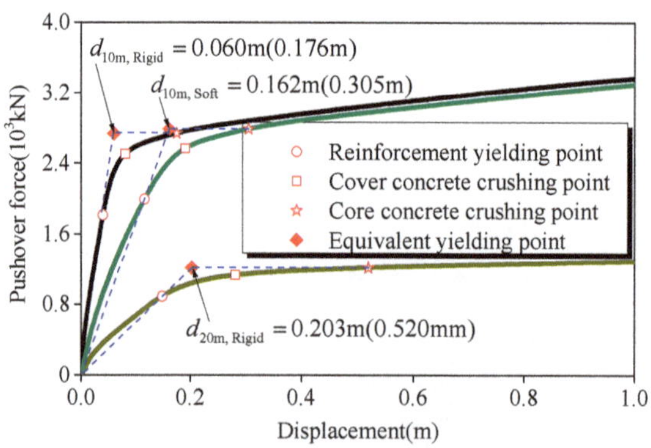

Figure 7. Definition of the pier critical states.

The critical states of the bearing are defined based on existing research results [21]. For the bearing, these are the critical sliding (D_d) and the critical unseating (D_u). The former is equal to the elastic shear deformation (D_1) in Figure 8, which can be calculated by the formulas (1) and (2). The latter is equal to the distance (d_0) plus elastic shear deformation (D_1), in which the distance (d_0) is the maximum distance the support can slip before unseating. The critical values are taken as shown in Table 3.

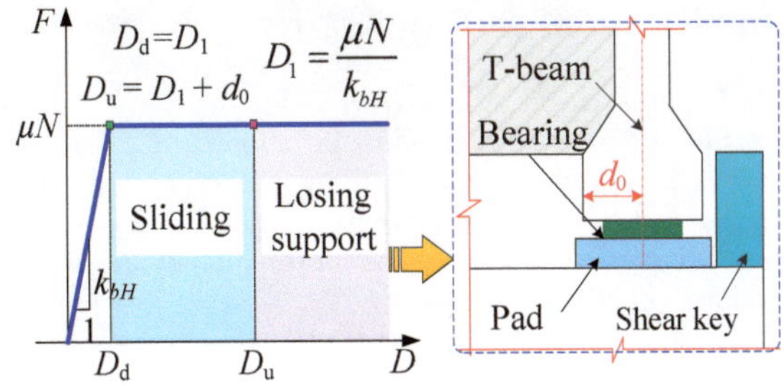

Figure 8. Definition of the bearing critical states.

Table 3. Definition of critical states.

Critical Points	Shear Key (m)	Bearing (m)	Pier (m)		
			$d_{10m,rigid}$	$d_{10m,soft}$	$d_{20m,rigid}$
A	Elasticity (RE)	0.02 (Gap)			
B	0.05 (Δ_d)	Elasticity (SE)			
C	Degradation (RD)	0.053 (D_d)	Elasticity (PE)	Elasticity (PE)	Elasticity (PE)
D	0.158 (Δ_u)	Sliding (SS)			
E		0.353 (D_u)			
F	Damage (RQ)	Losing support (SR)	0.060 (d_y)	0.162 (d_y)	0.203 (d_y)
			Yielding (PF)	Yielding (PF)	Yielding (PF)
G			0.176 (d_u)	0.305 (d_u)	0.520 (d_u)
			Damage (PG)	Damage (PG)	Damage (PG)

The damage critical states of shear keys have been categorized into intensive and degradation phases based on mechanical models [21] in Figure 3. The critical intensive degradation (Δ_d) and critical degradation (Δ_u) are defined in Figure 9. The intensive (Δ_d) and degradation phase (Δ_u) refer to the process of increasing and decreasing resistance provided by the shear key, respectively. Subsequently, the shear key is completely damaged, and fails to restrain the superstructures. The critical values are taken as shown in Table 3.

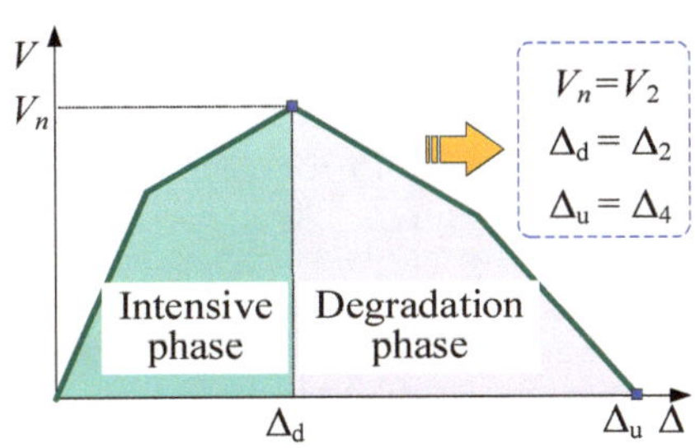

Figure 9. Definition of the shear key critical states.

5.2. System Critical States

Two typical structure systems are established under different shear key strengths by pushover analysis in Figure 10, including the quasi-isolated structural system (QS) and the ductile system (DS). The two systems are based on the relationship of pushover force (i.e., pier foundation force) and displacement. Combining the critical state indicators of each component given in the previous sections, the system critical state of the quasi-isolated structural system (QS) is divided into five critical points, namely: A~E. The damage state of the quasi-isolated system structure is divided into five states (RE, SE, RD, RQ/SS, SR). In addition, the ductile system (DS) that undergoes ductile failure is divided into two key critical points: F~G. The damaged state of the ductile system (DS) is divided into three states (PE, PF, PG). The definitions of A~G and RE~PG are shown in Figure 10 and Table 3. Table 3 lists the systems and components critical states, and provides judgment criteria and critical values. These critical states will be represented by the initial letter abbreviations shown in the following text.

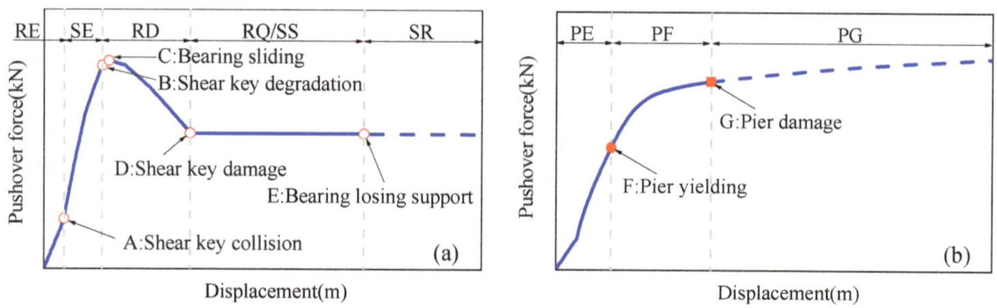

Figure 10. System critical states: (**a**) quasi-isolated system; (**b**) ductility system.

6. Pushover Results

6.1. *Quasi-Isolated Behavior of M10S25*

Six bridge cases were examined across a range of shear key strengths, varying from 0% to 100%. This ratio signifies the shear key's potency relative to the total supporting reaction force on one side of the pier or abutment. To elucidate, Figure 11 illustrates the system and component responses for M10S25 at shear key strengths of 20% and 60%. At 20%, the shear key strength equates to 0.2 times the support reaction force (2094 kN), approximately 418 kN, and at 60%, it stands at 1254 kN. A0# abutment, 1# pier, and 2# pier are selected for illustration, taking advantage of the symmetry. The pushover curves exhibit distinct trends due to the different shear key strengths. For shear key strengths below 20%, the pushover force (base resistance force) initially increases and subsequently decreases rapidly, reaching a relatively stable range. During this process, the shear keys, bearings, and bridge piers at the A0# abutment and 1# pier maintain an elastic state (B). At 2# pier, the superstructure collides with the shear key (A, B), resulting in bearing slippage (C) and rapid degradation of the shear key (D). After the failure of the shear key (D), bearing friction takes over as the only force transfer mechanism between the superstructure and piers, and the total pier resistance drastically drops. As the displacement keeps increasing, sliding of the bearings initiates (E) and the force transferred to the pier is capped by the kinematic friction. As the lateral force transfer path from the superstructure down to the pier is cut off by shear key failure, resulting in a rapid decrease in base resistance drops substantially. Shear key damage and bearing slipping effectively mitigate the damage to the structural system. This process exemplifies a typical quasi-isolated behavior. Conversely, with shear key strengths surpassing 60%, base resistance force rises proportionally with displacement. In this setup, the shear key exerts robust control over the superstructure, maintaining the bearings in an elastic state. This concurs with the ductile design of bearings as safeguarded components. At 1# pier, elasticity prevails, yielding a notably larger base force than in the quasi-isolated system; yielding (F) and damage (G) manifest at 2# pier. The bridge system then transgresses into a state of severe impairment (PF, PF). This sequence exemplifies a typical ductile behavior. Clearly, an excessive shear key strength renders bridges susceptible to untimely damage and even destruction.

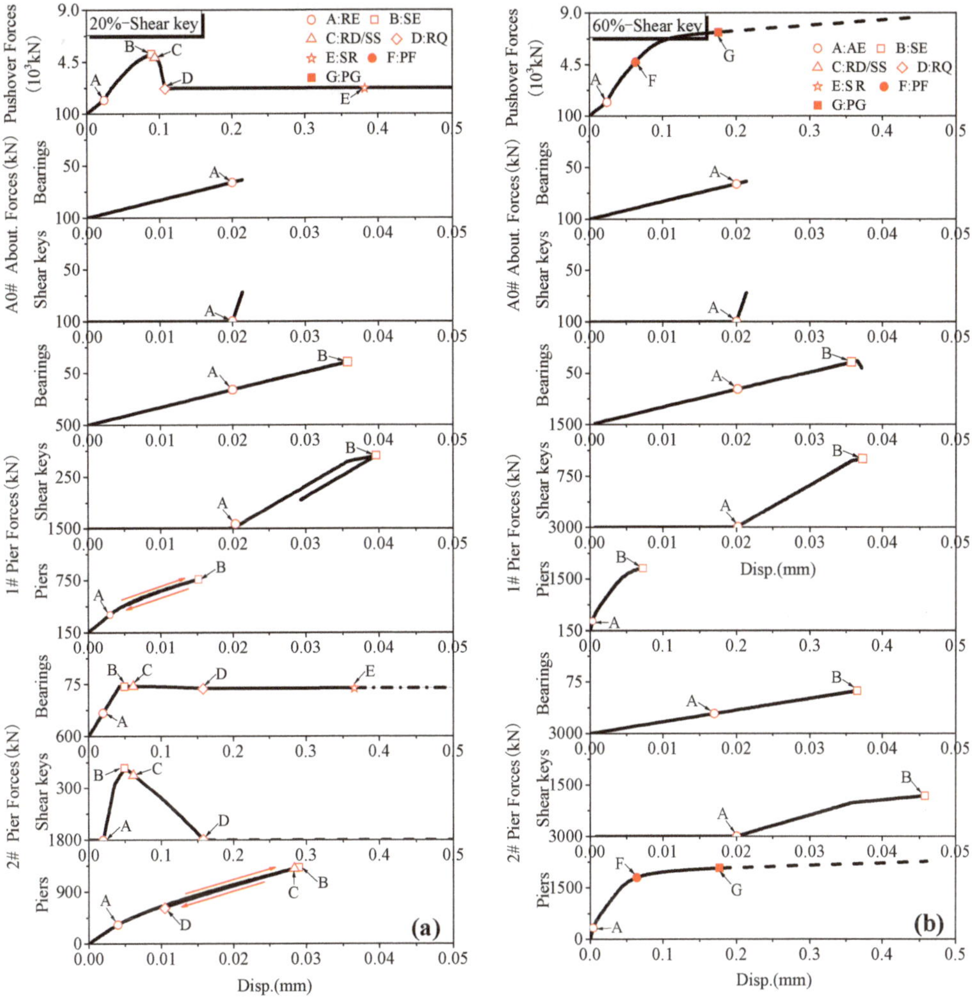

Figure 11. Quasi-isolated behavior of M10S25: (**a**) at 20%; (**b**) at 60%.

6.2. Quasi-Isolated Behavior of M1221S25

Figure 12 illustrates the quasi-isolated behavior curves of M1221S25 across different shear key strengths. Evidently, the pushover curves of M1221S25 exhibit a trend similar to M10S25, with alterations in both component and system behaviors as shear key strengths increase. At low shear key strengths (below 9%), a quasi-isolated behavior prevails. Here, the shear key fractures (b), the bearing slips (c), and the pier remains undamaged (d), leading to a sudden decline in the system's resistance (a). Conversely, when shear key strengths exceed 50%, the shear key and bearing primarily maintain an elastic state, while the pier experiences rapid damage (d), indicating ductile damage. Notably, in comparison to M10S25, M1221S25 demonstrates quasi-isolated behavior with a shear key strength ranging from 3% to 9%.

6.3. Quasi-Isolated Behavior of Other Cases

Figure 13 illustrates the system pushover curves of M10S35, M20S25, M20S25, and M10S25Z. The internal forces and the suitable range of shear key strengths for each bridge case are provided in Table 4. Notably, selecting the appropriate shear key strength can lead to a reduction of up to 70% in base shear forces. Quasi-seismic behavior can be achieved across all bridge cases, although the required shear key strength varies. For bridges with a 10 m pier, a shear key strength below 30% is effective, while those with a 20 m pier necessitate a relatively smaller shear key strength, specifically below 9%, to realize quasi-isolated behavior. In summary, the strategic selection of shear key strengths can facilitate a shift in bridge damage modes from ductile to quasi-isolated.

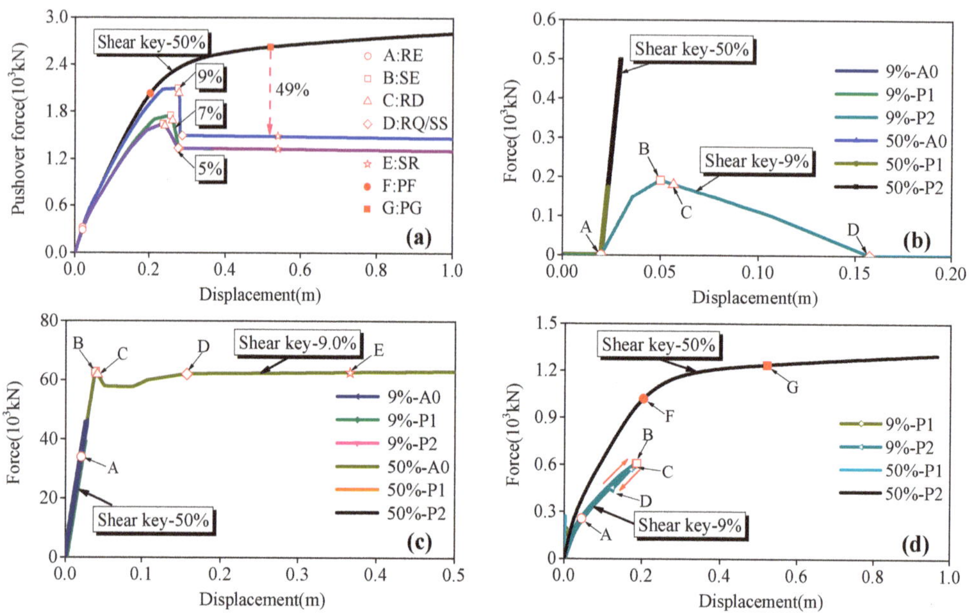

Figure 12. Quasi-isolated behavior of M1221S25: (**a**) system; (**b**) shear key; (**c**) bearing; (**d**) pier.

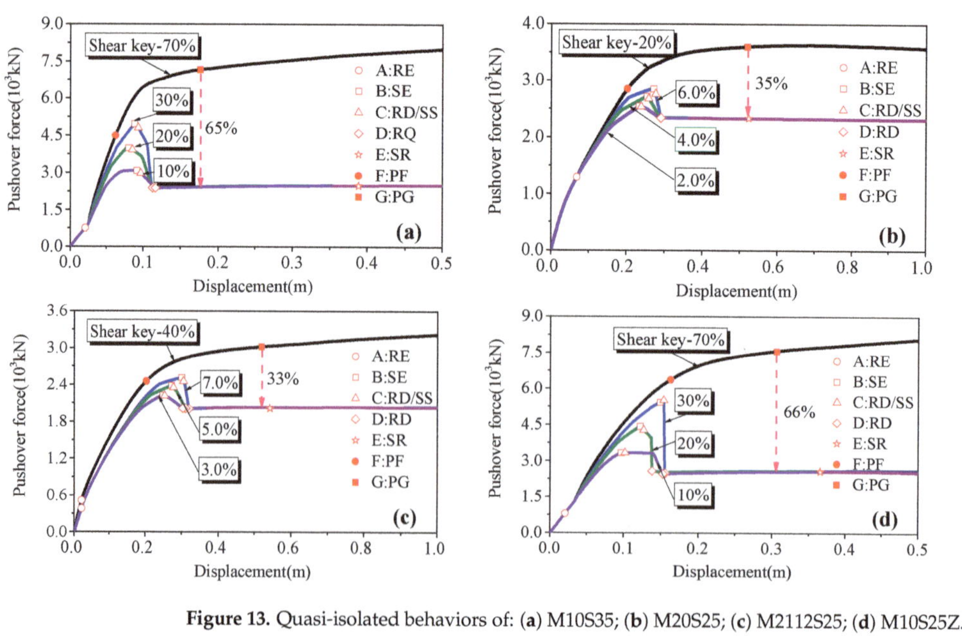

Figure 13. Quasi-isolated behaviors of: (**a**) M10S35; (**b**) M20S25; (**c**) M2112S25; (**d**) M10S25Z.

Table 4. Statistics under different shear key strengths of each bridge.

Bridge Cases	Shear Key Strength (%)		Base Shear Force (kN)			Pier Shear Force (kN)			Bearing Disp. (m)	
	DS	QS	DS	QS	Rates	DS	QS	Rates	DS	QS
M10S25	≥60%	10~30%	7320.6	2205.3	70%	2075.4	627.1	70%	0.0366	0.035
M10S35	≥70%	10~30%	7164.9	2482.0	65%	2075.5	800.1	61%	0.0366	0.049
M20S25	≥20%	2~6%	3601.3	2336.6	35%	1114.2	665.7	40%	0.0374	0.366
M2112S25	≥40%	3~7%	3021.6	2019.8	33%	1340.4	740.3	45%	0.0043	0.366
M1221S25	≥50%	5~9%	2630.3	1332.0	49%	1298.8	617.7	52%	0.0295	0.366
M10S25Z	≥70%	10~30%	7578.9	2580.9	66%	1833.3	802.3	56%	0.0256	0.055

7. IDA Results

7.1. Component Time–History Response

Figure 14 depicts the temporal evolution curves of bearing displacement, beam acceleration, and pier displacement for M10S25 under PGA = 0.1 g and PGA = 0.6 g. At lower PGA levels, the seismic responses show minimal discrepancies. In contrast, under higher PGA levels, the peak bearing displacement under 20% shear key strength reaches 0.479 m, which significantly surpasses the value under 60% strength. Concerning the beam acceleration response, a lower shear key strength can effectively limit the peak acceleration reaction of the beam, regardless of the PGA magnitude. Under increased PGA levels, the pier subjected to 20% shear key strength is approaching yielding, whereas the pier with 60% strength has already sustained substantial damage. A more robust shear key exerts a more adverse influence on the seismic behavior of the substructure, especially under high PGA conditions. Excessive shear key strength can lead to premature damage of the bridge.

7.2. Component Damage States

Figures 15–20 present the damage states of components for the 2# pier in each case, considering both lower and higher shear key strengths. The critical accelerations at which components transition to distinct states are indicated in parentheses. Overall, with an increase in PGA, components progressively enter distinct damage states, revealing diverse sequences of component damage. In the case of M10S25, under shear key strengths of 20% and 60%, the former remains undamaged, whereas the latter has already yielded at accelerations of 0.282 g and undergone significant damage at 0.65 g. Under 60% shear key strength, the pier experiences the initial component damage, while under 20% shear key strength, bearing slippage triggers an isolation phenomenon, deferring the damage onset and reducing the extent of damage. A similar quasi-isolated phenomenon can also be observed in other cases of M10S35~M10S25Z. The piers of these bridges exhibit yielding under two shear key strengths. In M10S35, the damage onset of the pier occurs earlier at 70% than at 20%, with accelerations of 0.228 g and 0.721 g, respectively. Moreover, the damage extent at 70% is also more severe than at 20%, resulting in significant damage to the piers. In summary, lower shear key strengths delay the damage and reduce the extent of damage, representing acceptable quasi-isolated behavior.

7.3. System Damage States

Figure 21 illustrates the damage state curves for each bridge system, established using the total pier base shear forces. The minimum acceleration at which all components transition to the critical state is indicated within parentheses. Furthermore, the figure depicts the rate of change of basal shear force at the point (G) of severe damage. The alteration in the system's base shear force is significant, with a maximum reduction of 37.3%. This alteration is observed in the figure to occur subsequent to the damage (D) of the shear key. In terms of the system's damage states, with higher shear key strength, predominant damage occurs on the piers (F, G), while the shear key and bearing remain intact. Conversely, when shear key strength is low, the shear key is more susceptible to damage, followed by the bearing and then the pier. However, the order in which damage occurs does not strictly adhere to a specific pattern. Notably, the piers of M10S25 consistently remained undamaged. In the cases of M20S25, M2112S25, and M10S25Z, the piers yielded earlier than the bearing, albeit generally with minor damage.

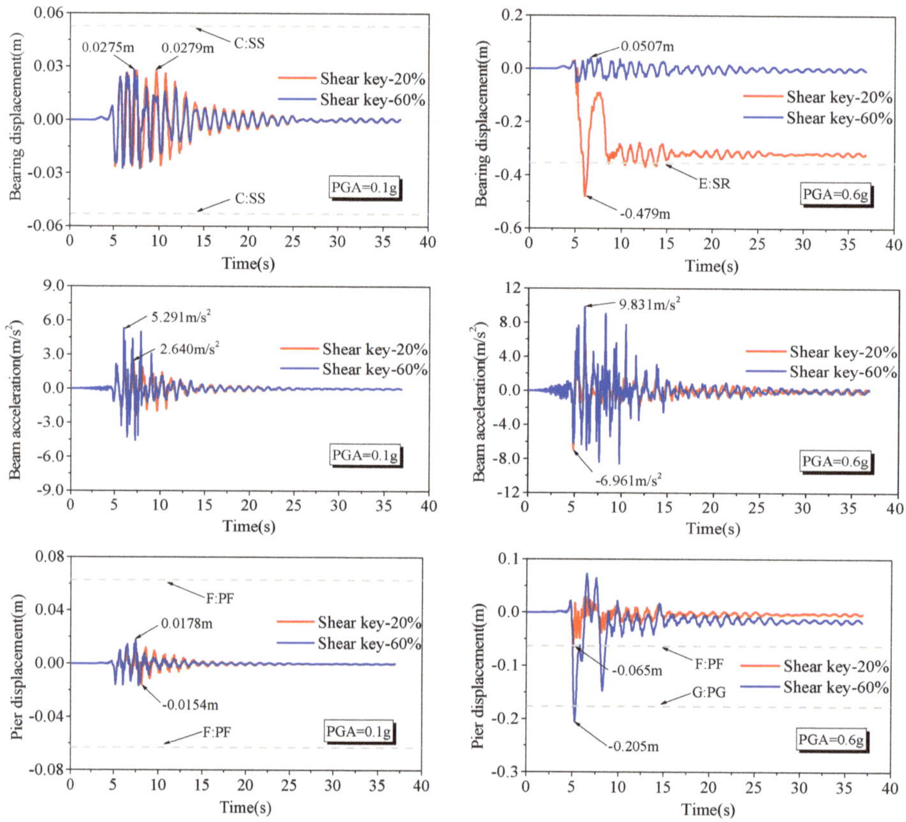

Figure 14. Time-history curves of M10S25 under PGA = 0.1 g and PGA = 0.6 g.

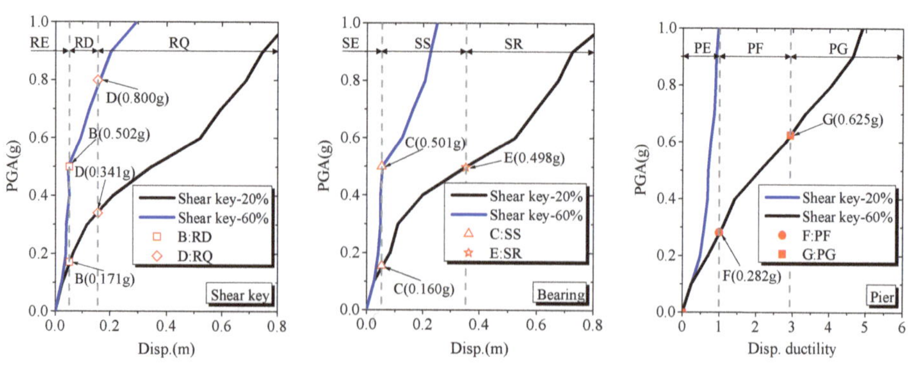

Figure 15. Components damage states of M10S25.

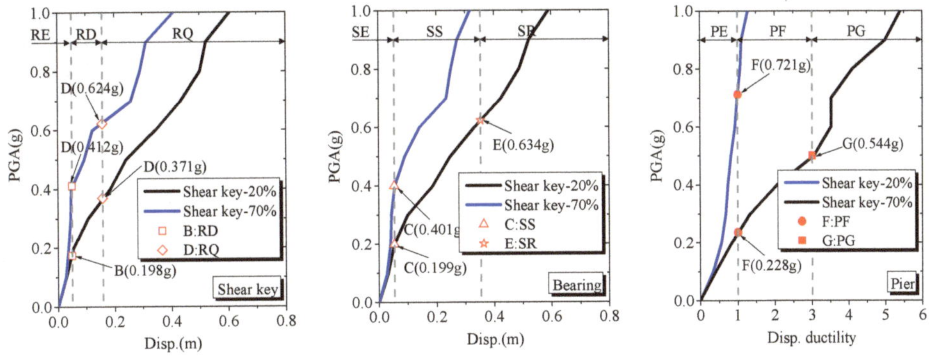

Figure 16. Components damage states of M10S35.

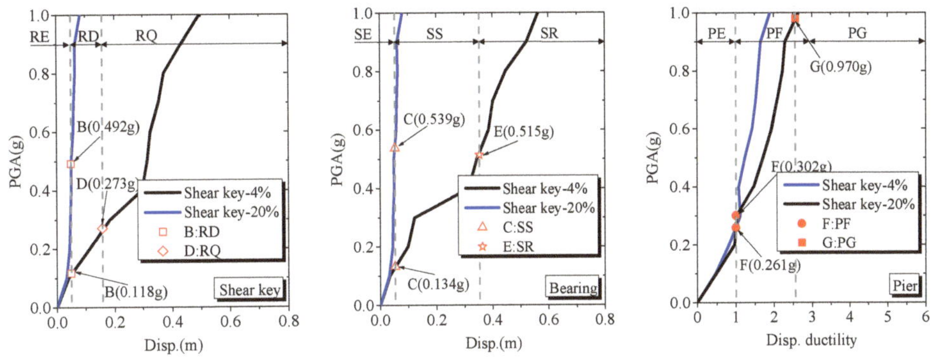

Figure 17. Components damage states of M20S25.

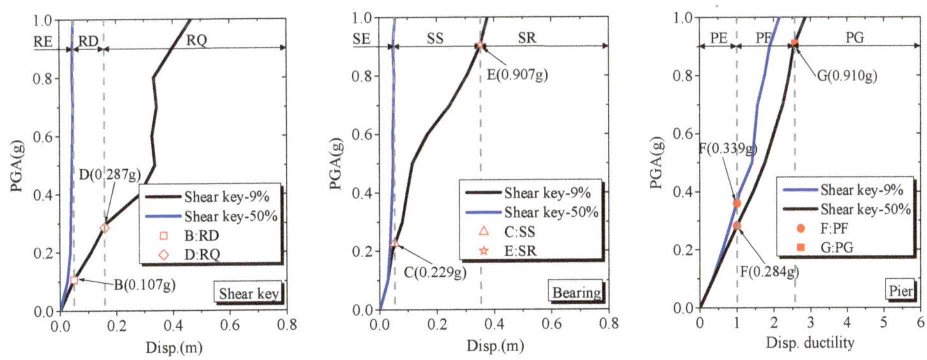

Figure 18. Components damage states of M1221S25.

7.4. Beam Displacement Demands

Figure 22 shows the displacement demand of the main beam, which is represented by the peak displacements of the bearings at different PGAs. Differentiating among colors, green, yellow, and orange correspond to the bearing states of elasticity, slip (0.053 m), and unseating (0.353 m) correspondingly. Darker shades indicate an escalated potential for bearing unseating. The demand for bearing displacement increases with reduced shear key strength, exceeding that at higher shear key strength, consequently resulting in inevitable bearing slippage. The bearing displacement under lower shear key strength (solid points) consistently surpasses that, signifying a marked residual displacement during quasi-isolated behavior. Notably, there is a risk that the main beam will lose support under a certain acceleration. The likelihood of bearing unseating is inconsequential when the ground motion probability is below 0.45 g. Nevertheless, within earthquake zones characterized by the PGA of ground motion exceeding 0.45 g, this risk cannot be ignored and a deliberation arises concerning the equilibrium between permitting bearing sliding and averting bearing unseating.

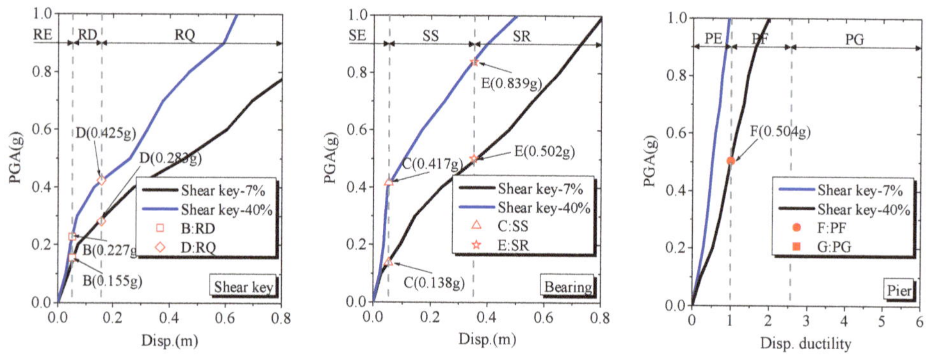

Figure 19. Components damage states of M2112S25.

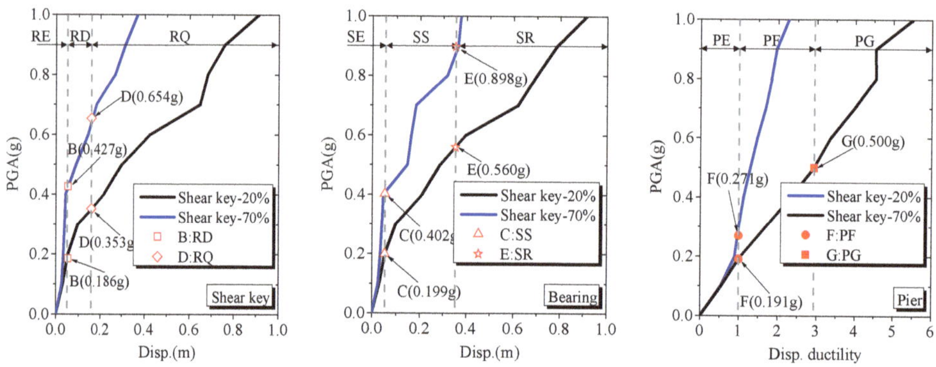

Figure 20. Components damage states of M10S25Z.

7.5. Damage Sequencs

Figure 23 presents a comparison of the damage sequence for each bridge under the two shear key strengths. The progression of each component through specific damage states is indicated. Let us take M10S25 as an example. At 60% shear key strength: 1PF@1#Pier → 2PF@2#Pier → 3SS@2#Pier → 4SS@A0#Abutment → 5SS@2#Pier → 6PG@2#Pier → 7SR@2#Pier → 8PG@1#Pier → 9SR@A0#Abutment. At 20% shear key strength: 1SS@2#Pier → 2SS@1#Pier → 3SS@A0#Abutment → 4SR@2#Pier → 5SR@1#Pier → 6SR@A0#Abutment. The observation reveals that, at 60% strength, the substructure (PF, PG) is primarily affected initially. In contrast, at 20% strength, the superstructures experience the initial impact, involving shear key failure (SR) and bearing slip (SS), with no significant effect on the piers. This similar damage sequence is observed in other bridge cases, which can be classified into two types: quasi-isolated damage and ductile damage. Quasi-isolated damage sequence: shear key damage → bearing slip → pier undamaged or slight damage. Ductile damage sequence: pier damage → bearing slip → pier serious damage. In summary, the most favorable damage sequence is the one aligned with quasi-isolated behavior, where the bridge undergoes minimal structural impairment.

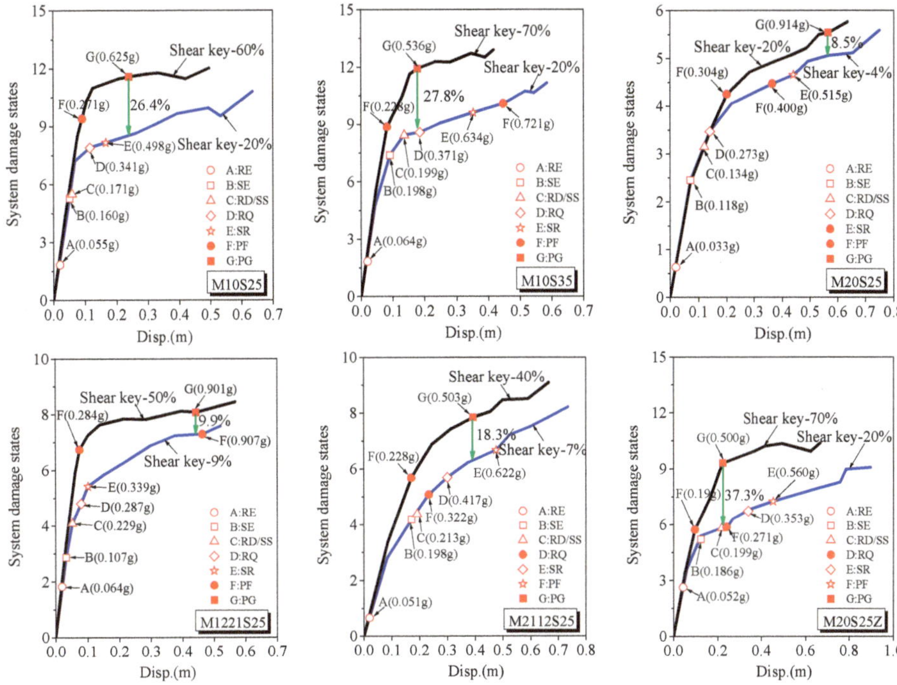

Figure 21. System damage states.

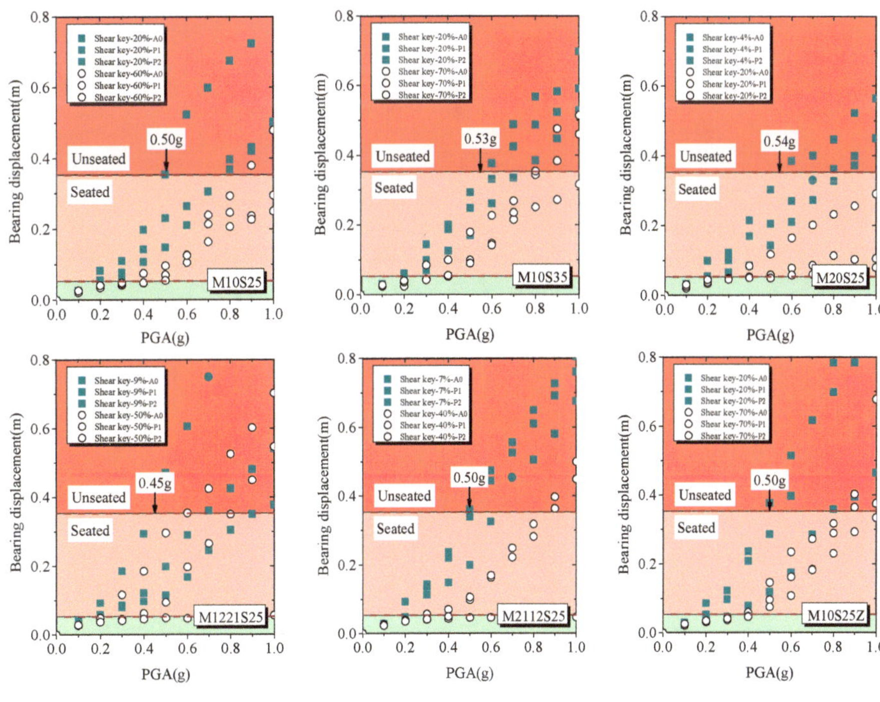

Figure 22. Beam displacement demands.

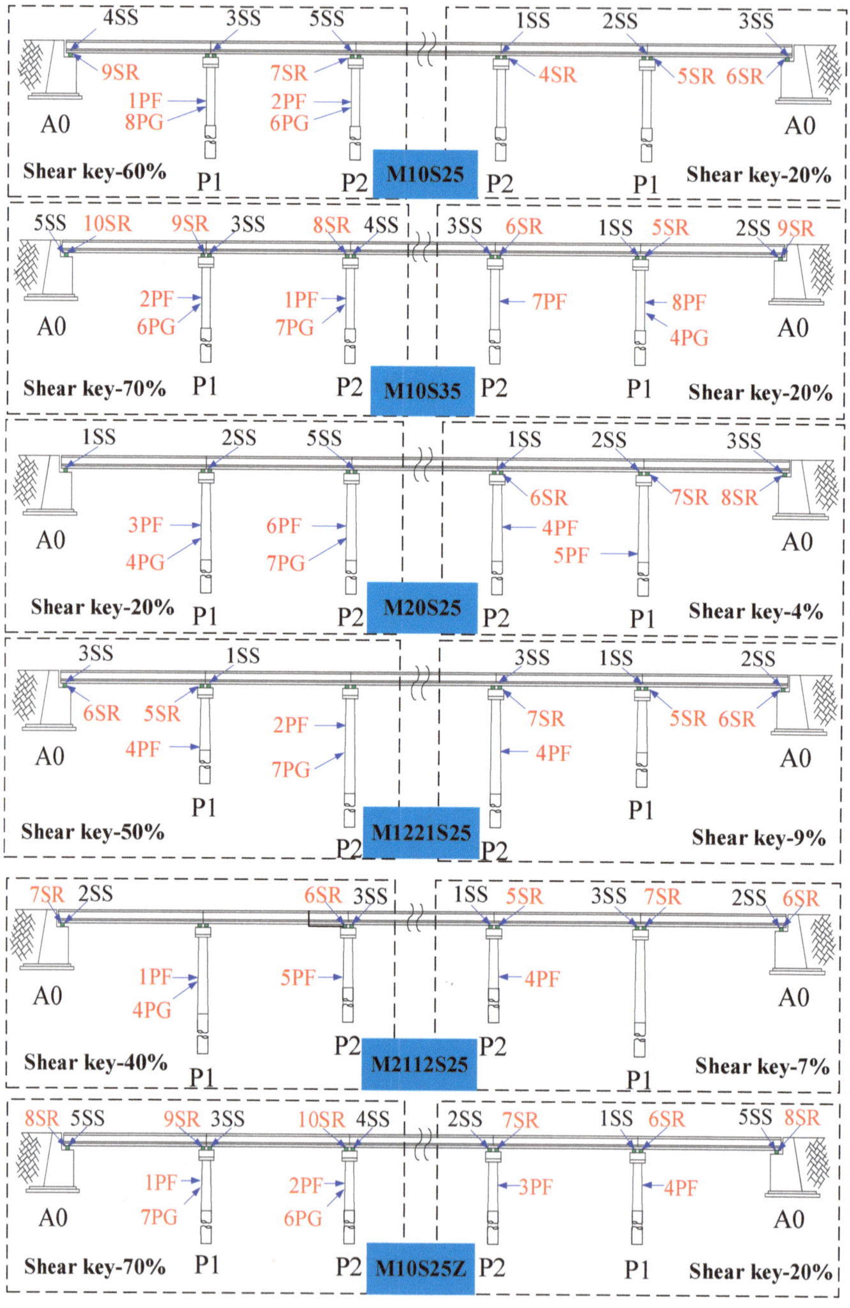

Figure 23. Bridge damage sequences.

7.6. Damage Patterns

Figure 24 provides the damage patterns of each bridge case. Notably, when subjected to a larger shear key strength (black line), bridges tend to enter the yellow and orange zones earlier, unlike those with a smaller shear key strength (red line). These observations hold despite the bridges having varying spans, pier heights, combination forms, and foundation types. Quasi-isolated behavior has been observed across these variations, even though they exhibit distinct damage sequences. The damage sequence of a bridge is not obligated to follow a specific pattern, aligning with the principles of the quasi-isolated strategy. An optimal quasi-isolated damage sequence involves: shear key damage → bearing slip → pier undamaged or slight damage. By opting for a smaller shear key strength, the damage patterns of bridges can transition from ductile failure to quasi-isolated failure, effectively mitigating premature and excessive structural damage.

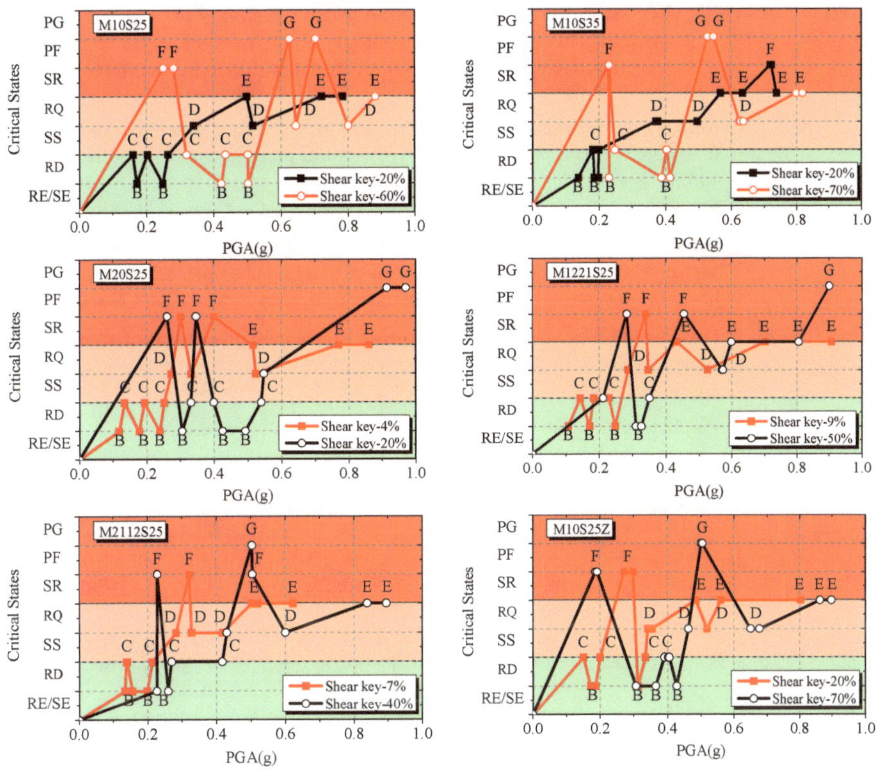

Figure 24. Bridge damage patterns.

8. Conclusions

This study employs the Pushover and the IDA methods to comparatively analyze the influence of shear keys on the quasi-isolated behavior of typical small-to-medium-span girder bridges. The key conclusions are summarized as follows:

(1) The bridge exhibits two distinct damage states upon the shear key strengths. At larger shear key strength, the bridge is damaged quickly. In cases where the shear key strength is lower, the base shear force declines, up to 70%. The bridges remain undamaged or slight damage due to the failure of shear key, which is a typical quasi-isolated behavior.

(2) In quasi-isolated bridges, the superstructure of bridges tends to generate larger displacement demands, which increases the risk of bridge unseating and, potentially, beams falling off. This concern becomes particularly pronounced in seismically active regions where the peak ground acceleration (PGA) exceeds 0.45 g.

(3) The damage sequence can be classified into two types: quasi-isolated damage and ductile damage. The quasi-isolated damage sequence is: shear key failure → bearing

slip → pier undamaged or slight damage. The damage sequence is not required to adhere to any specific pattern, which is permissible under the quasi-isolated strategy.

(4) By selecting suitable shear key, the response of the bridge during seismic events will shift from ductile damage to quasi-isolation damage. Optimal recommendations suggest maintaining the shear key strength ranging from 10% to 30% for bridges with 10 m piers, while for bridges with 20 m piers, a more suitable range lies between 2% and 9%.

(5) For typical highway girder bridges in China, selecting a lower and suitable shear key strength from the above range can effectively prevent premature damage to the bridges. This quasi-isolated strategy, which involves sacrificing auxiliary components to achieve isolated, offers a significant advantage in seismic fortification.

Author Contributions: Software, K.Y.; validation, K.Y., L.F. and J.L.; investigation, K.Y.; resources, L.X.; data curation, K.Y.; writing—original draft preparation, K.Y. and L.X.; writing—review and editing, K.Y. and L.X.; visualization, H.L.; supervision, L.X.; project administration, L.X.; funding acquisition, L.X. All authors have read and agreed to the published version of the manuscript.

Funding: This research was funded by the National Natural Science Foundation of China, grant number 51978113, Chongqing Talent Plan Project, grant number cstc2022ycjh-bgzxm0133, and Chongqing Jiaotong University Postgraduate Research Innovation Project (Grant No. CYB21213).

Data Availability Statement: Data available on request due to restrictions eg privacy or ethical. The data presented in this study are available on request from the corresponding author. The data are not publicly available due to [insert reason here].

Conflicts of Interest: The authors declare no conflict of interest.

References

1. *JTG/T 2231-01-2020*; Code for Seismic Design of Highway Bridges. China Communications Press: Beijing, China, 2020. (In Chinese)
2. Zhuang, W.L.; Liu, Z.Y.; Jiang, J.S. Analysis and countermeasures for the earthquake damage of highway bridges in Wenchuan, 5.12. *Southwest Highw.* **2009**. (In Chinese)
3. Du, X.L.; Han, Q.; Li, Z.X.; Li, L.Y.; Chen, S.F.; Zhao, J.F. Seismic damage of mountain highway bridges in the 5.12 Wenchuan earthquake and its inspiration. *J. Beijing Univ. Technol.* **2008**, *34*, 1270–1279.
4. Ding, J.; Jiang, S.; Bao, F. Review of bridge damage caused by Tangshan earthquake. *World Seismol. Eng.* **2006**, *22*, 68–71. (In Chinese)
5. Broderick, B.M.; Elnashai, A.S. Analysis of the failure of interstate 10 freeway ramp during the Northridge earthquake of 17 January 1994. *Earthq. Eng. Struct. Dyn.* **1995**, *24*, 189–208. [CrossRef]
6. Lu, C.H.; Liu, K.Y.; Chang, K.C. Seismic performance of bridges with rubber bearings: Lessons learnt from the 1999 Chi-Chi earthquake. *J. Chin. Inst. Eng.* **2011**, *34*, 889–904. [CrossRef]
7. Schanack, F.; Valdebenito, G.; Alvial, J. Seismic damage to bridges during the 27 February 2010 magnitude 8.8 Chile earthquake. *Earthq. Spectra* **2012**, *28*, 301–315. [CrossRef]
8. Nie, L.Y.; Li, J.Z.; Fan, L.C. Selection of pounding analysis parameters and its effects on structure under earthquake. *Eng. Mech.* **2005**, *22*, 142–146. (In Chinese) [CrossRef]
9. Liu, P.; Zheng, K.F.; Yang, L.; Wu, Z.W. Numerical method for bridge seismic response considering beam-shear key pounding effect. *Northwest. Seismol. J.* **2011**, *33*, 336–341. (In Chinese) [CrossRef]
10. SDC. *Caltrans Seismic Design Criteria Version 1.7*; California Department of Transportation: Sacramento, CA, USA, 2013.
11. Goel, R.K.; Chopra, A.K. Role of shear keys in seismic behavior of bridges crossing fault-rupture zones. *J. Bridge Eng.* **2008**, *13*, 398–408. [CrossRef]
12. Salveson, M.W.; Fell, B.V. Effect of abutment shear keys on the seismic response of bridges. In Proceedings of the Structures Congress 2011, Las Vegas, NV, USA, 14–16 April 2011. [CrossRef]
13. Omrani, R.; Mobasher, B.; Sheikhakbari, S. Variability in the predicted seismic performance of a typical seat-type California bridge due to epistemic uncertainties in its abutment backfill and shear-key models. *Eng. Struct.* **2017**, *148*, 718–738. [CrossRef]
14. Nicoletti, V.; Martini, R.; Carbonari, S.; Gara, F. Operational Modal Analysis as a Support for the Development of Digital Twin Models of Bridges. *Infrastructures* **2023**, *8*, 24. [CrossRef]
15. Wang, K.H.; Hui, Y.X.; Wu, G. Study on seismic performance of seismic double-layer stopper. *Earthq. Eng. Eng. Dyn.* **2014**, *34*, 505–510. (In Chinese)
16. Xu, L.Q.; Li, J.Z. Design and experimental investigation of a new type sliding shear key and its efficacy in seismic fortification. *Eng. Mech.* **2016**, *33*, 111–118. (In Chinese) [CrossRef]

17. Xu, L.Q.; Li, J.Z. A two-level design method for sacrificeable seismic blocks. *Chin. J. Highw.* **2015**, *28*, 59–66. (In Chinese)
18. Han, Q.; Zhou, Y.; Ou, Y.; Du, X. Seismic behavior of reinforced concrete sacrificial exterior shear keys of highway bridges. *Eng. Struct.* **2017**, *139*, 59–70. [CrossRef]
19. Zhao, J.; Xiang, C.S.; Zhou, Y.; Liu, G.S. Effect of wedge block on the transverse seismic performance of small-and-medium-span girder bridges. *China Earthq. Eng. J.* **2020**, *42*, 22–31. (In Chinese)
20. Deng, K.; Pan, P.; Su, Y.; Ran, T.; Xue, Y. Development of an energy dissipation restrainer for bridges using a steel shear panel. *J. Constr. Steel Res.* **2014**, *101*, 83–95. [CrossRef]
21. Xiang, N.L.; Alam, M.S.; Li, J. Yielding Steel Dampers as Restraining Devices to Control Seismic Sliding of Laminated Rubber Bearings for Highway Bridges: Analytical and Experimental Study. *J. Bridge Eng.* **2019**, *24*, 04019103.1–04019103.15. [CrossRef]
22. Bi, K.; Hao, H. Modelling of shear keys in bridge structures under seismic loads. *Soil Dyn. Earthq. Eng.* **2015**, *74*, 56–68. [CrossRef]
23. Están, J.J.W.; Santa María, H.; Riddell, R.; Arrate, C. Influence of the use of external shear keys on the seismic behavior of Chilean highway bridges. *Eng. Struct.* **2017**, *147*, 613–624. [CrossRef]
24. Wu, G.; Wang, Q.L.; Wang, K.H.; Zhang, P.P. Transverse seismic response analysis for bridges considering performance degradation of bearings and stoppers. *J. Vib. Shock* **2018**, *37*, 189–196. (In Chinese)
25. Li, Y.; Li, C.; Li, X. Effect of sliding of plate rubber bearings on seismic performance of small and medium span girder bridges under seismic action. *J. Civ. Eng.* **2014**, *47* (S1), 124–129. (In Chinese)
26. Tobias, D.H.; Anderson, R.E.; Hodel, C.E.; Kramer, W.M.; Wahab, R.M.; Chaput, R.J. Overview of earthquake resisting system design and retrofit strategy for bridges in Illinois. *Pract. Period. Struct. Des. Constr.* **2008**, *13*, 147–158. [CrossRef]
27. AASHTO. *AASHTO LRFD Bridge Design Specifications*; American Association of State Highway and Transportation Officials: Washington, DC, USA, 2012.
28. IDOT. *Illinois Department of Transportation. Standard Specification for Road and Bridge Construction*; IDOT: Springfield, IL, USA, 2007.
29. Filipov, E.T.; Fahnestock, L.A.; Steelman, J.S.; Hajjar, J.F.; LaFave, J.M.; Foutch, D.A. Evaluation of quasi-isolated seismic bridge behavior using nonlinear bearing models. *Eng. Struct.* **2013**, *49*, 168–181. [CrossRef]
30. Filipov, E.T.; Revell, J.R.; Hnestock, L.A.; Lafave, J.M.; Hajjar, J.F.; Foutch, D.A. Seismic performance of highway bridges with fusing bearing components for quasi-isolated. *Earthq. Eng. Struct. Dyn.* **2013**, *42*, 1375–1394. [CrossRef]
31. Steelman, J. Sacrificial Bearing Components for Quasi-Isolated Response of Bridges Subject to High-Magnitude, Low-Probability Seismic Hazard. Ph.D. Thesis, University of Illinois at Urbana-Champaign, Urbana, IL, USA, 2013. Available online: https://hdl.handle.net/2142/45455 (accessed on 22 August 2013).
32. Jie, L.; Fahnestock, L.A.; Lafave, J.M. Nonlinear Static Pushover and Eigenvalue Modal Analyses of Quasi-Isolated Highway Bridges with Seat-Type Abutments. *Structures* **2017**, *12*, 145–167. [CrossRef]
33. Luo, J.; Fahnestock, L.A.; Lafave, J.M. Seismic Performance Assessment of Quasi-Isolated Highway Bridges with Seat-Type Abutments. *J. Earthq. Eng.* **2019**, *25*, 2285–2324. [CrossRef]
34. Wang, K.H.; Wei, H.; Li, X.; Li, R. Seismic design concepts for small and medium span highway bridges. *J. Civ. Eng.* **2012**, *45*, 115–121. (In Chinese)
35. Li, J.Z.; Tang, H.; Guan, Z.G. A new isolation system for small and medium span bridges on laminated rubber bearings. *China J. Highw. Transp.* **2015**, *28*, 35–43. (In Chinese)
36. Xiang, N.; Li, J. Effect of exterior concrete shear keys on the seismic performance of laminated rubber bearing-supported highway bridges in China. *Soil Dyn. Earthq. Eng.* **2018**, *112*, 185–197. [CrossRef]
37. Steelman, J.S.; Fahnestock, L.A.; Filipov, E.T. Shear and friction response of nonseismic laminated elastomeric bridge bearings subject to seismic demands. *J. Bridge Eng.* **2013**, *18*, 612–623. [CrossRef]
38. Xiang, N.; Li, J. Experimental and numerical study on seismic sliding mechanism of laminated-rubber bearings. *Eng. Struct.* **2017**, *141*, 159–174. [CrossRef]
39. Mazzoni, S.; McKenna, F.; Scott, M.H. *Open System for Earthquake Engineering Simulation: OpenSees Command Language Manual*; Pacific Earthquake Engineering Center: Berkeley, CA, USA, 2007. [CrossRef]
40. *JT-T 4-2019*; Plate-Type Elastomeric Pad Bearings for Highway Bridges. China Communication Press: Beijing, China, 2019. (In Chinese)
41. Zhang, X.Y.; Tang, L.; Ling, X.C.; Su, L.; Liu, C.H.; Gao, X. Characterization of pile-soil dynamic interaction p-y curves for bridges at liquefaction sites. *J. Disaster Prev. Mitig. Eng.* **2014**, *5*, 619–625. (In Chinese)
42. API Recommended Practice 2A-WSD (RP 2A-WSD). *Recommended Practice for Planning, Designing, and Constructing Fixed Offshore Platforms-Working Stress Design*; American Petroleum Institute: Washington, DC, USA, 2002.
43. Lymon, C. *Single Pile and Pile Groups under Lateral Loading*; CRC Press: Boca Raton, FL, USA, 2010.
44. Manoukas, G.; Athanatopoulou, A.; Avramidis, I. Multimode pushover analysis for asymmetric buildings under biaxial seismic excitation based on a new concept of the equivalent single degree of freedom system. *Soil Dyn. Earthq. Eng.* **2012**, *38*, 88–96. [CrossRef]
45. Hwang, H.; Jernigan, J.B.; Lin, Y.W. Evaluation of Seismic Damage to Memphis Bridges and Highway Systems. *J. Bridge Eng.* **2000**, *5*, 322–330. [CrossRef]

buildings

MDPI

Article

Study on the Influence of Wind Fairing Parameters on the Aerodynamic Performance of Long-Span Double-Deck Steel Truss Suspension Bridge

Yang Yang [1], Long Li [1], Gang Yao [1,*], Bo Wu [2], Dawu Wang [3], Hui Yu [3] and Hao Qu [3]

1 School of Civil Engineering, Chongqing University, Chongqing 400045, China; 20121601009@cqu.edu.cn (Y.Y.); 202216021032@stu.cqu.edu.cn (L.L.)
2 School of Civil Engineering, Chongqing Jiaotong University, Chongqing 400074, China; bo.wu@cqjtu.edu.cn
3 Chongqing Railway Group Co., Ltd., Chongqing 400015, China; 15023482432@163.com (D.W.); 15310239541@163.com (H.Y.); 15859073223@163.com (H.Q.)
* Correspondence: yaogang@cqu.edu.cn; Tel.: +86-13608361176

Abstract: A long-span double-deck steel truss suspension bridge is easy to produce vortex-induced vibration (VIV) at low air velocity, which affects bridge service life. Additional aerodynamic measures play a role in suppressing VIV by changing the aerodynamic shape, which is a common control method. As the main aerodynamic measure to suppress the VIV response, wind fairing is widely used in engineering practice. In order to obtain the optimal additional position and shape parameters of the fairing, Huangjuetuo Yangtze River Bridge is the research target. Through the combination of a wind tunnel test and numerical simulation, the VIV response of the original and fairing section is studied. Based on data analysis, it is revealed that these additional fairings to the upper chord can significantly reduce the VIV response. When the shape parameters of the fairing are h/D = 1/4 and l/D = 1, the VIV inhibition efficiency is the highest, which can reach 65.51%. By analyzing the flow distribution, it can be seen that VIV is caused mainly by vortex separation in the upper bridge board area. Although this wind fairing does not change the original vortex shedding forms, it changes the first separation point and movement direction of the airflow, making the vortex scale generated by the airflow smaller and the vorticity lower, thus effectively suppressing VIV.

Keywords: double-deck steel truss girder; aerodynamic performance; wind fairing; wind tunnel test; numerical simulation

Citation: Yang, Y.; Li, L.; Yao, G.; Wu, B.; Wang, D.; Yu, H.; Qu, H. Study on the Influence of Wind Fairing Parameters on the Aerodynamic Performance of Long-Span Double-Deck Steel Truss Suspension Bridge. *Buildings* **2024**, *14*, 2255. https://doi.org/10.3390/buildings14072255

Academic Editor: Francesco Ricciardelli

Received: 13 June 2024
Revised: 12 July 2024
Accepted: 19 July 2024
Published: 22 July 2024

1. Introduction

Long-span bridges' modal frequency is always low, and vortex shedding is easy to occur when gas flows around the surface of the main girder, resulting in VIV. VIV has a limiting nature, although it cannot cause serious consequences such as bridge collapse caused by a divergent flutter of the bridge [1,2]. However, VIV has the characteristics of a low starting wind speed, large amplitude response and high frequency, which can easily cause the fatigue failure of a bridge's structure and affect structural safety. Also, a large VIV can easily cause social panic [3]. For example, China's Xihoumen Bridge [4] and Humen Bridge [5], Denmark's Great Belt Bridge [6], and Canada's Lions' Gate Bridge [7] all have VIV phenomena, so it is necessary to conduct special research on the VIV of long-span bridges.

The double-deck steel truss girder is a common girder form of a long-span suspension bridge, which can not only save line resources but also has excellent traffic capacity [8,9]. Studies have shown that when wind stream flows through a bridge's cross-section, it will generate obvious vortex generation, merging and shedding, resulting in periodic aerodynamic force [10–13]. Therefore, compared with the general single-deck main girder form, the aerodynamic shape of the more complex double-deck steel truss girder section has a greater hidden danger of VIV [14]. Right now, studies on the VIV of long-span double-deck steel truss suspension bridges have been comparatively scarce, but the number of long-span bridges with double-layer deck truss girders as the main girder is increasing, and the problem of VIV is also becoming increasingly prominent [15,16]. Therefore, explorations on double-deck steel truss suspension bridge VIV should be given more attention.

A wind fairing is an additional aerodynamic measure widely used in bridge engineering. This additional aerodynamic measure is a kind of bridge VIV control method. Its principle is to change the aerodynamic shape by adding an additional structure so as to improve aerodynamic performance. Bai et al. [17] studied the inhibition effect of the angle and eccentric position of a fairing on the VIV of a π-type composite girder. Regarding a small-size fairing, the results show that its VIV inhibition effectiveness is better than that of a large-size fairing. Li et al. [18] explored how impact affects the fairing angle on the VIV performance of a streamlined trapezoidal box girder by numerical simulation. According to the simulation's outcome, changing the fairing angle can significantly affect VIV performance, and the maximum amplitude of VIV decreases with a decrease in the fairing angle.

In the existing research, more attention has been paid to the impact of fairings on the aerodynamic characteristics of a box girder as well as a π-girder, and many valuable conclusions have been obtained. However, there are relatively few studies on the optimization of the fairing parameters of long-span double-deck steel truss suspension bridges, and most of them only consider the influence of $\pm 3^\circ$ and 0° conventional wind attack angles, and the case of a large $\pm 5^\circ$ angle is rarely considered. Due to the large differences in the structural system, section form and dynamic performance of an actual long-span double-deck truss bridge, it is necessary to carry out a wind tunnel test and numerical simulation research when there is a big difference between the bridge terrain and the bridge structure system. In order to effectively suppress VIV, Huangjuetuo Yangtze River Bridge is chosen as the research target, and we carry out a wind tunnel test and numerical simulation to test aerodynamic performance under different fairing parameters at 0°, $\pm 3^\circ$ and $\pm 5^\circ$ so as to obtain the optimal aerodynamic optimization measures and the vibration inhibition mechanism of the fairing. The research results can provide an effective reference for the design of the same type of bridge in the future.

2. Implementation of Vortex-Induced Vibration

2.1. Parameters of Wind Fairing

In order to find out the optimal aerodynamic shape of a fairing section and analyze how impact affects the aerodynamic performance of a double-deck steel truss bridge, a wind fairing is studied and analyzed according to different layout positions and shape parameters. Firstly, the fairing is set at different positions, and the optimal fairing arrangement position is obtained through an performance evaluation index. Then, based on this, different shapes of the fairing are designed, and the optimal fairing parameters are obtained by comparing them with each other. There are six working conditions of C1~C6 designed according to the position of the fairing, as shown in Figure 1, which are an upper chord wind fairing, web rod wind fairing, lower chord wind fairing, upper chord wind fairing + lower chord wind fairing, upper chord wind fairing + web rod wind fairing, and web rod wind fairing + lower chord wind fairing.

When setting the shape parameters of the fairing, the height, D, of the truss section is taken as the quantitative variable, and the horizontal size, l, of fairing and the relative height, h, of the fairing are taken as the variables. A total of nine working conditions of F1~F9 are set, in which the h/D value is 3/4, 1/2, and 1/4, and the l/D value is 1/2, 2/3, and 1. In Figures 2 and 3, the specific parameters of the fairing and its transverse section can be seen.

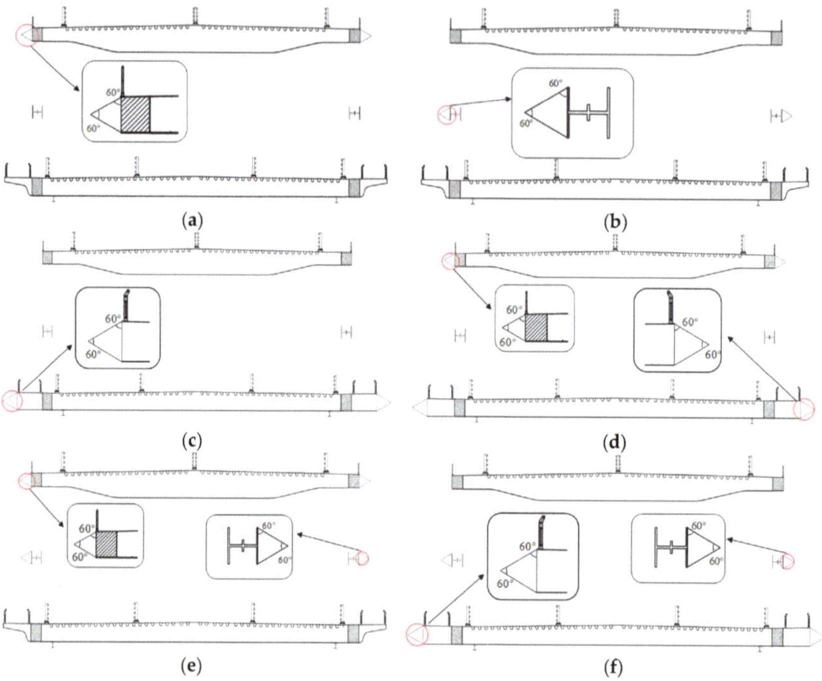

Figure 1. (**a**) Upper chord wind fairing; (**b**) web rod wind fairing; (**c**) lower chord wind fairing; (**d**) upper chord + lower chord wind fairing; (**e**) upper chord + web rod wind fairing; (**f**) web rod + lower chord wind fairing.

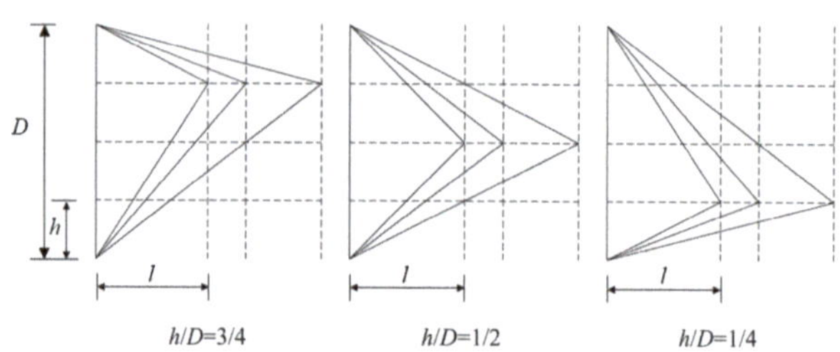

Figure 2. Nine types of wind fairing working conditions.

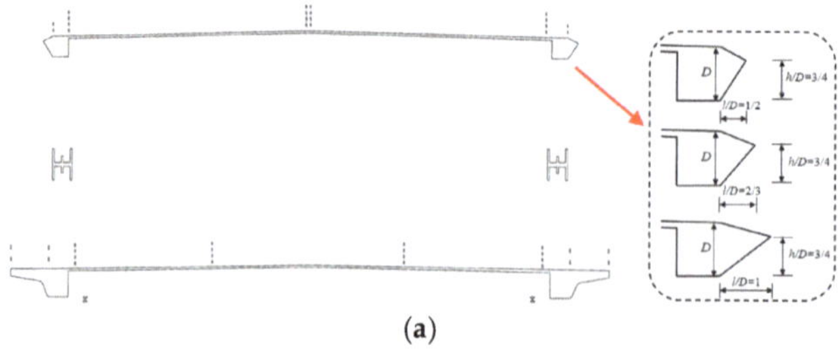

Figure 3. *Cont.*

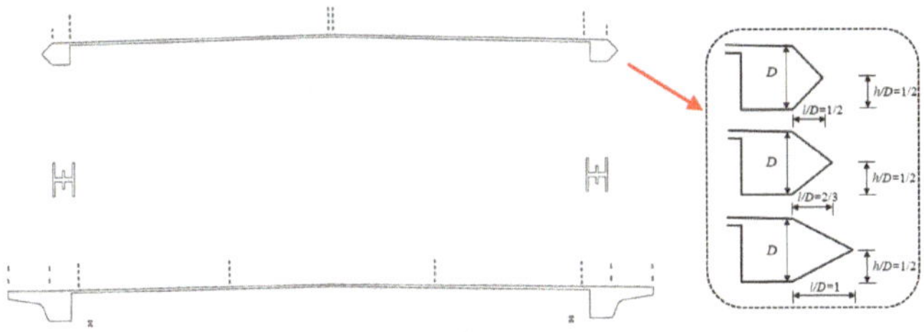

(**b**)

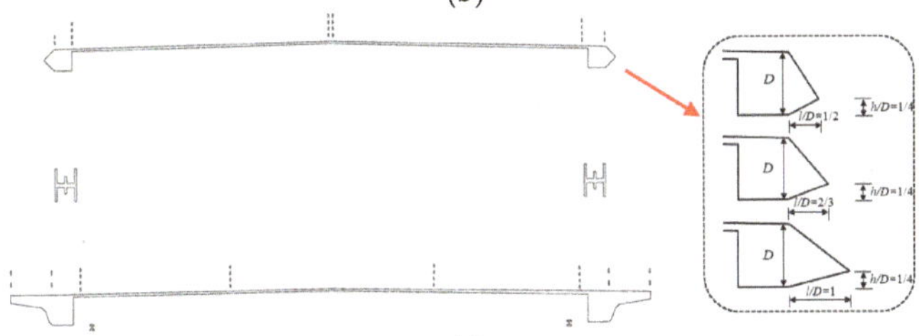

(**c**)

Figure 3. (**a**) *h/D* = 3/4; (**b**) *h/D* = 1/2; (**c**) *h/D* = 1/4.

2.2. VIV Evaluation Indicator

The setting of wind fairing measures changes the aerodynamic shape. Different cross-section aerodynamic shapes affect bridge VIV response, resulting in peak points of different heights in the VIV curve. To more intuitively evaluate the inhibition effect of different fairing parameters on VIV, we introduce parameter η to represent the inhibition effect of VIV. The expression of the VIV inhibition efficiency η is as follows:

$$\eta = \frac{y_0 - y_i}{y_0} \times 100\% \; or \; \eta = \frac{\theta_0 - \theta_i}{\theta_0} \times 100\% \quad (1)$$

In the formula, y_0 and θ_0 represent the maximum VIV amplitude of the original section, respectively, while y_i and θ_i represent the maximum VIV amplitude of the section with the set fairing, respectively. If η is positive, this indicates that the setting of the fairing has an inhibitory effect on VIV. If η is negative, this means that the setting of the fairing increases the VIV response. The larger the η, the better the fairing inhibition effect.

2.3. Aerodynamic Performance Optimization Process Based on Wind Fairing Parameters

By means of a wind tunnel test and numerical simulation, the optimization process of the aerodynamic shape of the main girder section with different fairing parameters as variables is shown in Figure 4. Firstly, a three-dimensional finite element model is established and its structural dynamic characteristics are calculated. The natural frequency, equivalent mass, vibration mode and other information are obtained. In accordance with this, the VIV test is designed, and the VIV response curves corresponding to different wind attack angles are obtained. In order to obtain the optimal position of the fairing, the vibration test is carried out under different working conditions with the position of the fairing as the variable, and the corresponding vibration inhibition efficiency is calculated. At the optimal fairing arrangement position, a variety of working conditions are further designed with different fairing size parameters for the numerical simulation. The vibration inhibition mechanism of the fairing is analyzed and the optimal fairing parameters are selected so as to obtain the optimal aerodynamic performance under fairing measures.

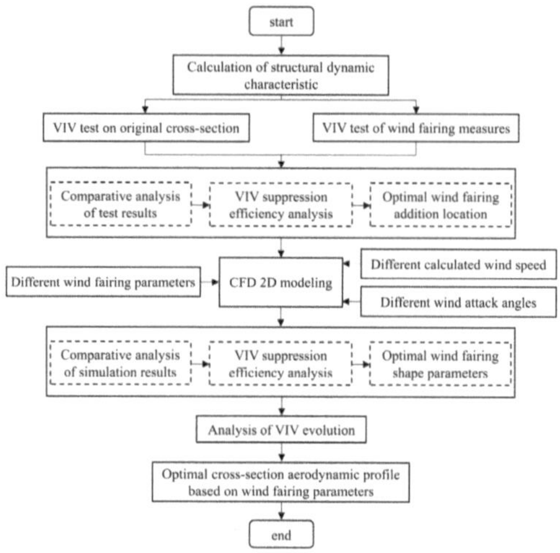

Figure 4. Aerodynamic shape optimization process for main girder section.

3. Calculation about Dynamic Characteristics of Bridge Structure

3.1. Simple Bridge Description

As the world's largest spanning highway–railway suspension bridge under construction [19], the Huangjuetuo Yangtze River Bridge is located in a subtropical humid monsoon climate zone and is greatly affected by the monsoon climate. The bridge is located in the Jiangbei District of Chongqing, China, and spans the Yangtze River. Due to the large span of the bridge and the harsh conditions of the wind on the river, the flow of the wind attack angle is easily affected. Therefore, it is particularly important to carry out wind resistance design. The bridge facade is shown in Figure 5. In addition to the web member, along the bridge axis, the transverse section of the bridge experiences almost no change. The bridge's typical transverse section can be seen in Figure 6.

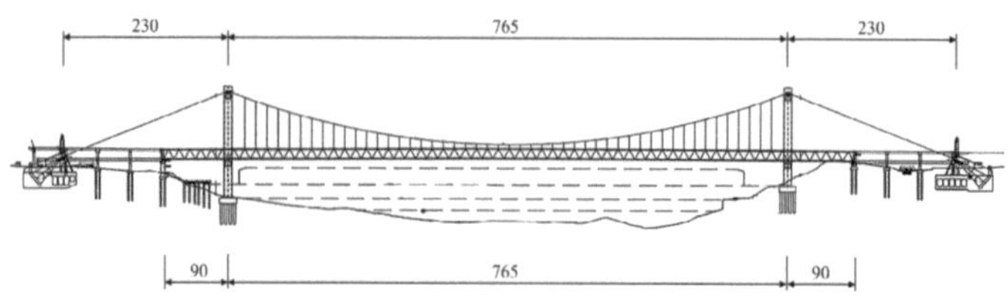

Figure 5. Bridge facade (unit: m).

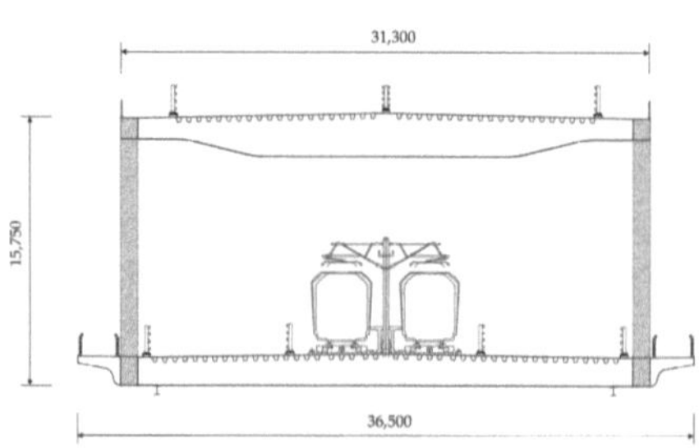

Figure 6. Bridge's typical transverse section (unit: mm).

3.2. Calculation of Structural Dynamic Characteristics

So as to ensure that the segment model can accurately simulate the vibration of the actual bridge under wind load, it is necessary to obtain accurate structural dynamic characteristics such as the natural frequency, equivalent mass and vibration mode of the bridge. Therefore, it is necessary to establish a three-dimensional finite element model that can fully reflect the dynamic effect of the structure. In Figure 7, the finite element model can be seen. For the benefit of clarifying the load and position of the loading point during the calculation, the stiffening girder is equivalent to 'single main girder' [20–23] by the principle of stiffness equivalence. The specific parameters of the main modes of the bridge are shown in Table 1, and the main modes are shown in Figure 8.

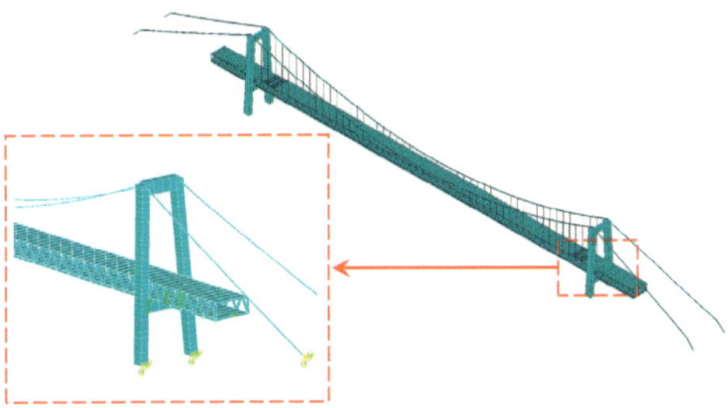

Figure 7. Finite element bridge model.

Table 1. Major mode parameters of bridge.

Order	Frequency	Mass	Moment of Inertia	Vibration Mode
3	0.2108 Hz	6.845×10^4 kg/m	/	first-order symmetric vertical bend
4	0.2368 Hz	7.008×10^4 kg/m	/	first-order objection vertical bend
11	0.4777 Hz	/	1.309×10^7 kg·m^2/m	first-order symmetric torsion
25	0.7055 Hz	/	1.228×10^7 kg·m^2/m	first-order objection torsion

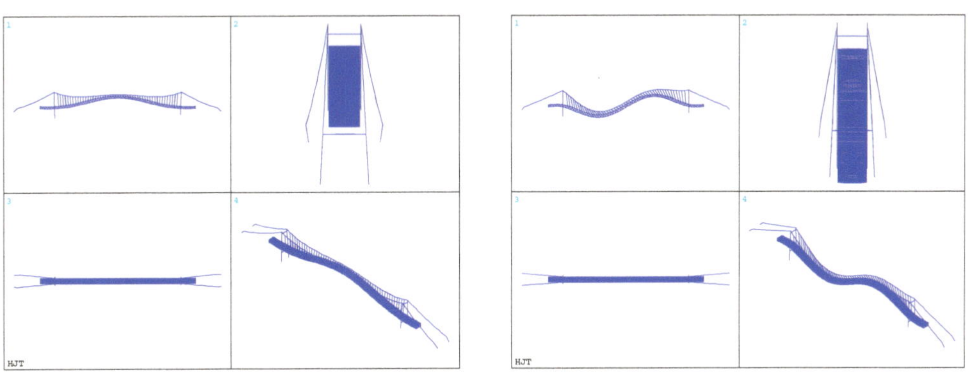

(a) (b)

Figure 8. *Cont.*

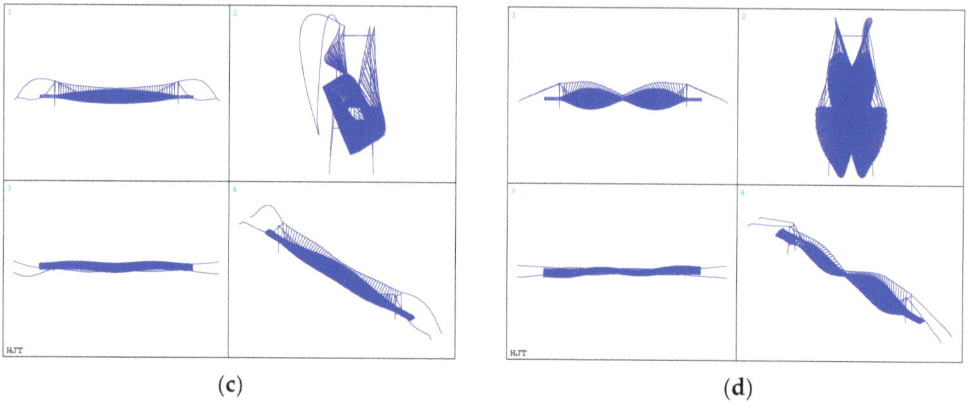

Figure 8. (**a**) First-order symmetric vertical bend; (**b**) first-order objection vertical bend; (**c**) first-order symmetric torsion; (**d**) first-order objection torsion.

4. Wind Tunnel Test of Main Girder Section with Wind Fairing Measures

After the preliminary preparation, the wind tunnel test is made in the DC wind tunnel laboratory of Chongqing University. This wind tunnel layout can be seen in Figure 9. It is 15 m long, 2.4 m wide, and 1.8 m high at the wind tunnel test section. In addition, the incoming flow in the wind tunnel laboratory is a uniform flow with low turbulence, and the wind speed adjustment range is 0.5 m/s~35 m/s, and turbulence, velocity field deviation, and other indicators meet the needs of this test.

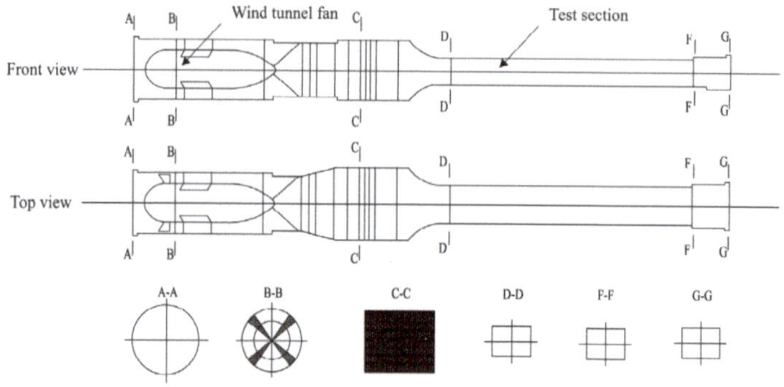

Figure 9. Wind tunnel layout.

4.1. *Vortex-Induced Vibration Test of Original Section*

Considering the laboratory conditions, segment model size and quality, the model scale ratio is selected as 1:55. In the VIV test, through the spring suspension system, the segment model can move vertically and rotate around the cross-section axis. A three-dimensional fluctuating wind speed measuring instrument is used to accurately measure the real-time wind speed of the flow field. A laser displacement sensor is used to measure and record the dynamic response of the bridge segment model. A slope measuring instrument is used to check the wind attack angle of the segment model. A high-frequency force balance is used to measure the three-component force of the segmental model. The specific test model is shown in Figure 10.

(**a**)

(**b**)

Figure 10. (**a**) Integral segment model; (**b**) local enlargement of the segment model.

The vibration test of the original cross-section model at five wind attack angles (α) of 0°, ±3°, and ±5° is carried out. The wind attack angle is the angle between the wind direction and the longitudinal axis of the bridge. The angle that increases lift is positive and the opposite is negative. The test wind speed is 0~10 m/s, and the corresponding real bridge wind speed is 0~44.8 m/s. The wind speed growth step is 0.08 m/s, and the corresponding wind speed of the real bridge is 0.36 m/s. When VIV may occur, the sampling frequency is encrypted, and the wind speed growth step is reduced to 0.04 m/s, corresponding to the actual bridge wind speed of 0.18 m/s. Each time, the displacement of model is recorded after the vibration is stable. The sampling time is 30 s, and the amplitude is the mean value of the vibration time history.

Through the vibration test, a variety of curves of VIV amplitude with the original section (C0) can be seen in Figure 11.

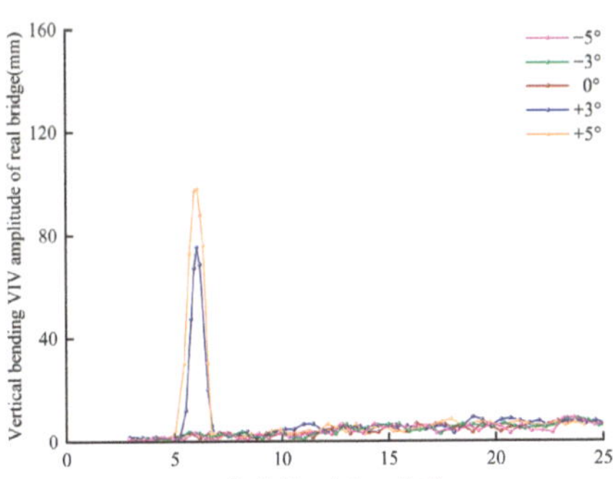

(a)

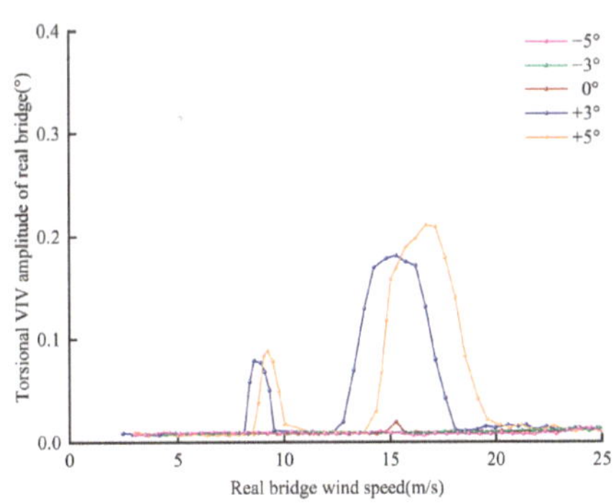

(b)

Figure 11. (**a**) Vertical bending VIV response; (**b**) torsional VIV response.

The specific test values are shown in Table 2. Compared with +3°, the maximum amplitudes of the vertical bending and torsional VIV increase significantly at +5°. The maximum amplitude of vertical bending VIV increased by 30.0%, while the maximum amplitude of torsional VIV at high air velocity and low air velocity increased by 16.6% and 12.7%, respectively. In addition, the onset air velocity of vertical bending VIV is slightly advanced, while its locking interval extent is evidently increased. The onset air velocity of torsional VIV is evidently delayed, while its locking interval extent is also slightly increased.

According to the test result, the wind attack angle plays a significant role in VIV response. The transverse section is more prone to VIV with a larger amplitude and longer locking range at a large attack angle.

Table 2. Major VIV modes of bridge.

α	VIV Type	Maximum Amplitude	Corresponding Real Speed	Frequency
+3°	Vertical bending	75.54 mm	6.05 m/s	2.227 Hz
	Torsional	0.079°, 0.181°	8.60 m/s, 15.28 m/s	4.289 Hz
+5°	Vertical bending	98.23 mm	6.06 m/s	2.227 Hz
	Torsional	0.089°, 0.211°	9.22 m/s, 16.67 m/s	4.289 Hz

4.2. Vortex-Induced Vibration Test of Cross-Section with Wind Fairing Measures

The 60° equilateral triangle fairing is selected as the research object, and the influence of the fairing's location on the VIV response is studied. According to the inhibition efficiency of VIV under various conditions, the optimal location of the fairing on the main girder section is obtained. The wind attack angle is set to +3° and +5°, and the test air velocity range is 0~10 m/s. According to the specific test conditions, the fairing test model is set as shown in Figure 12.

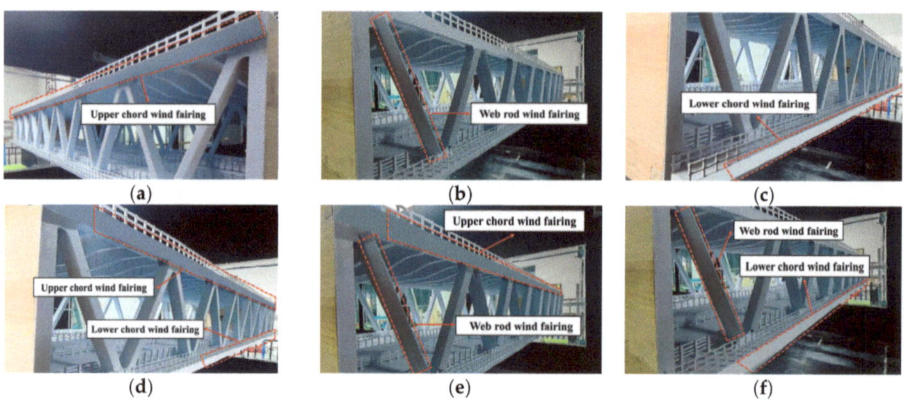

Figure 12. (**a**) C1; (**b**) C2; (**c**) C3; (**d**) C4; (**e**) C5; (**f**) C6.

Through the VIV inhibition test of the wind fairing, various VIV curves of six working conditions under two angles are obtained, as shown in Figures 13 and 14, separately (the wind speed and amplitude are transformed into real bridge values).

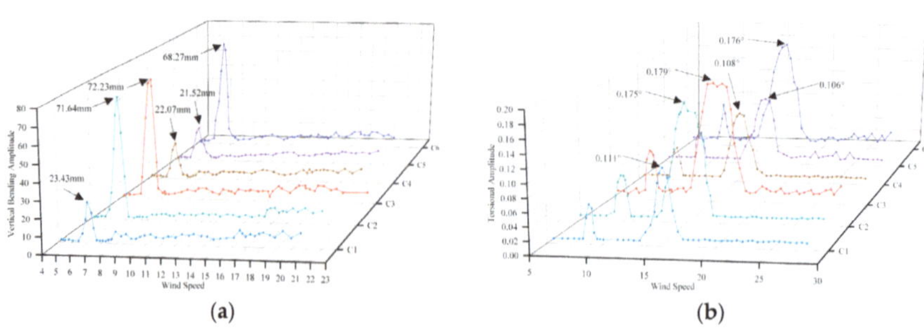

Figure 13. (**a**) Vertical bending VIV at +3°; (**b**) torsional VIV at +3°.

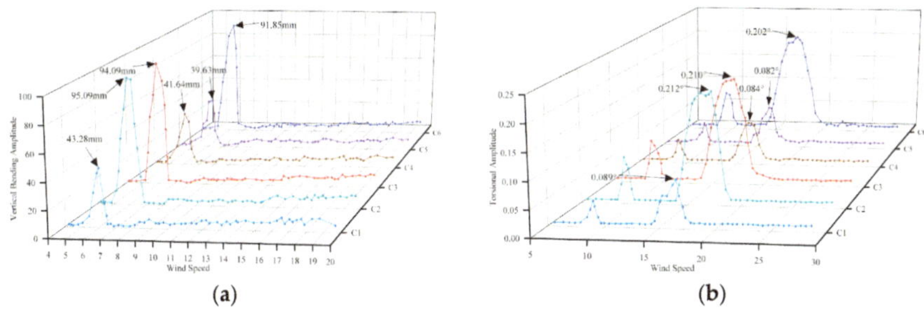

Figure 14. (**a**) Vertical bending VIV at +5°; (**b**) torsional VIV at +5°.

The VIV inhibition efficiency is shown in Figure 15 (the inhibition efficiency of torsional VIV is the average of the inhibition efficiency of the main and minor locking intervals).

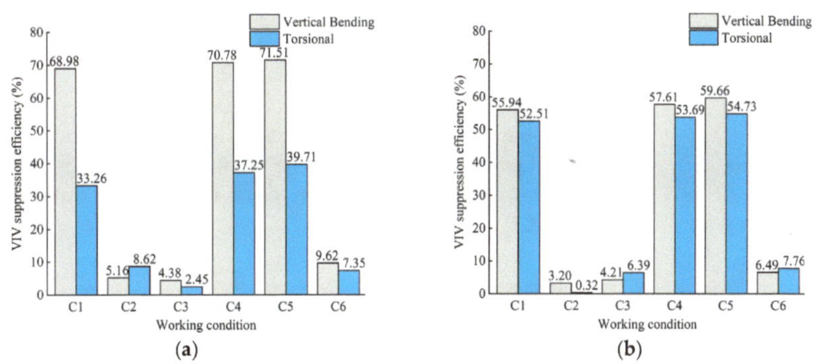

Figure 15. (**a**) VIV inhibition efficiency at +3°; (**b**) VIV inhibition efficiency at +5°.

Upon conducting a comprehensive comparison of the six fairing conditions, it can be seen that the influence of the fairing measures on the vibration inhibition of the vertical bending VIV is much greater than the influence on the torsional VIV. The highest inhibition efficiency of vertical bending VIV is 71.51%, the lowest is 3.20%, and the average is 34.80%. The highest inhibition efficiency of torsional VIV is 54.73%, the lowest is 0.32%, and the average is 25.34%. C1, C4, and C5 conditions are able to efficiently inhibit vertical bending and torsional VIV responses, but compared with the C1 condition, the VIV inhibition efficiency on the C4 and C5 conditions is not significantly improved. Therefore, the C1 working condition is the optimal working condition, that is, the upper chord is the optimal fairing addition position. The average inhibition efficiency of vertical bending VIV at two angles is 62.46%, and the average inhibition efficiency of torsional VIV is 42.89%. Regarding the vertical bending VIV, it is speculated that the vortex zone of the upper board dominates.

5. Influence of Wind Fairing Parameters on Vortex-Induced Vibration

For the benefit of finding out optimal wind fairing parameters, based on the optimal position of the fairing (upper chord), the vertical bending VIV response as well as the fairing influence mechanism on the main girder under different parameters are studied by means of numerical simulation.

5.1. Establishment of 2D CFD Model

The CFD geometric model uses the same scale ratio as the test model. The geometric model of main girder cross-section with wind fairing measures can be seen in Figure 16.

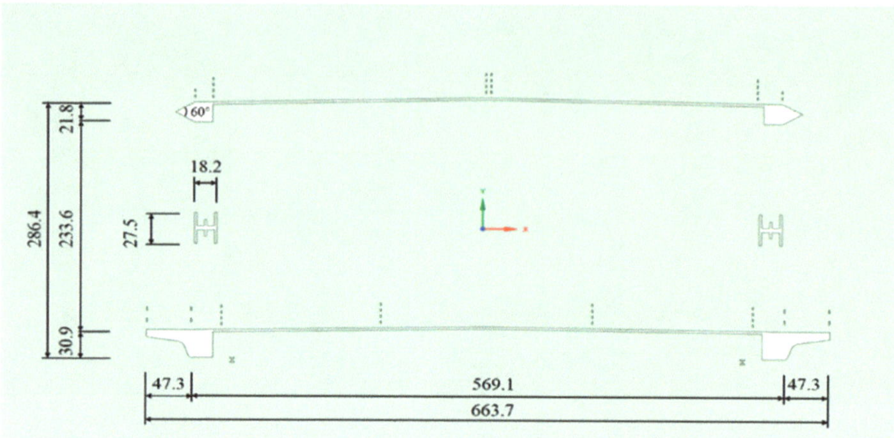

Figure 16. Geometric model of the main girder section setting fairing.

This numerical simulation domain is composed of a near-wall rigid motion region + a dynamic grid region + a background grid region. According to Figure 17, boundary conditions are strictly set. The turbulence model uses an SST *k*-*ω* model. The model blocking rate in the calculation region is 0.5%. It meets the permitted model requirement of 5% [24,25].

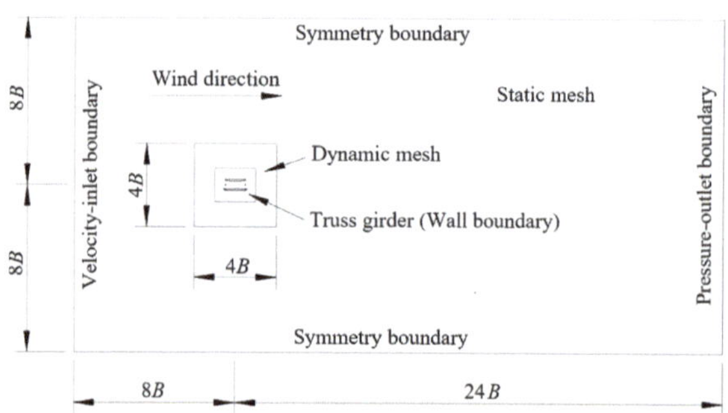

Figure 17. Calculation domain dimension and boundary condition setting.

Reasonable meshing is the key to the success of a numerical simulation [26]. The total number of grids of the two-dimensional section model generated in the simulation is about 78.5×10^4, and the specific grid division is shown in Figure 18.

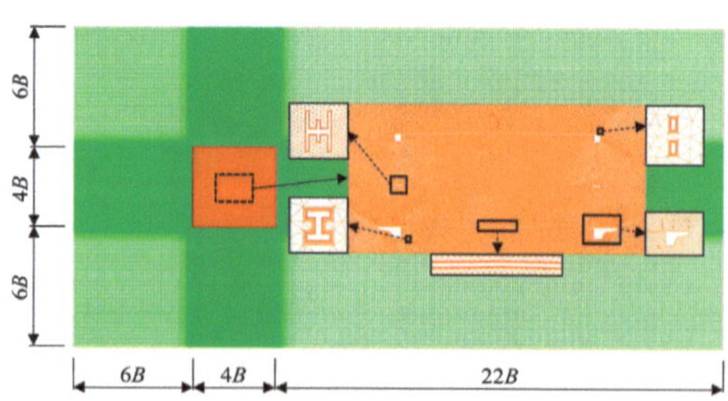

Figure 18. Mesh division of calculation domain.

5.2. Irrelevant Verification

For the purpose of ensuring the veracity of the CFD calculation outcome, the computational time step, the mesh quantity and the position independence of the web member are verified for the original two-dimensional model [27–32].

(1) Time step independence verification

Five time steps of 1×10^{-3} s, 5×10^{-4} s, 1×10^{-4} s, 5×10^{-5} s, and 1×10^{-5} s are used to calculate the drag, lift, and torque coefficients (C_D, C_L, and C_M) of the model at +3° and 10.0 m/s real bridge wind speed. The independence of the time step is verified by the average value ($\overline{C}$) and standard deviation (σ) of the three-force coefficient [33–37]. The specific results are shown in Table 3.

Table 3. Time step independence verification calculation results.

Computational Time Step (s)		0.001	0.0005	0.0001	0.00005	0.00001
mean value	$\overline{C_D}$	0.6741	0.6954	0.7221	0.7232	0.7250
	$\overline{C_L}$	0.8137	0.8254	0.8456	0.8451	0.8501
	$\overline{C_M}$	0.0181	0.0198	0.0285	0.0289	0.0290
root mean square error	σ_D	0.0345	0.0352	0.0205	0.0201	0.0174
	σ_L	0.0754	0.0711	0.0458	0.0421	0.0418
	σ_M	0.0222	0.0187	0.0119	0.0118	0.0125

The mean and standard deviation of 1×10^{-3} and 5×10^{-4} are significantly higher than those of the other three groups. On the basis of 1×10^{-4}, the time step is further reduced, and the calculation results have no significant change. Therefore, in the numerical simulation, the time step is 1×10^{-4} s, the calculation step is 1×10^5 steps, and the total calculation time is 10 s.

(2) Grid number independence verification

Four grid numbers of 39.3×10^4, 78.5×10^4, 98.0×10^4 and 117.6×10^4 are selected to calculate the three-force coefficients of the model under +3° and 10.0 m/s real bridge wind speed, and the independence of the grid number is verified. The specific results are shown in Table 4.

Table 4. Mesh number independence verification calculation results.

Mesh Number ($\times 10^4$)		39.3	78.5	98.0	117.6
y+		0.1215	0.0751	0.0749	0.0742
average value	$\overline{C_D}$	0.7091	0.7221	0.7243	0.7265
	$\overline{C_L}$	0.8217	0.8456	0.8502	0.8514
	$\overline{C_M}$	0.0219	0.0285	0.0288	0.0290
standard deviation	σ_D	0.0421	0.0205	0.0265	0.0224
	σ_L	0.0945	0.0458	0.0471	0.0431
	σ_M	0.0391	0.0119	0.0108	0.0134

On the basis of 78.5×10^4 grids, the number of grids is further increased, and the calculation results have no significant change. Increasing the number of grids improves the calculation accuracy but greatly increases the amount of calculation, so the number of grids for the numerical simulation is controlled at about 78.5×10^4.

(3) The position independence of the web member section verification

The three cross-sections of F1, F2, and F3 are intercepted, and the three-force coefficients of the model at +3° and 10.0 m/s real bridge wind speed are calculated, respectively, and the independence of the web member is verified by $\overline{C}$ and σ. The transverse sections of F1, F2, and F3 are shown in Figure 19, where d is the interval from the web member core to the upper chord core. These verification results of irrelevance are shown in Table 5.

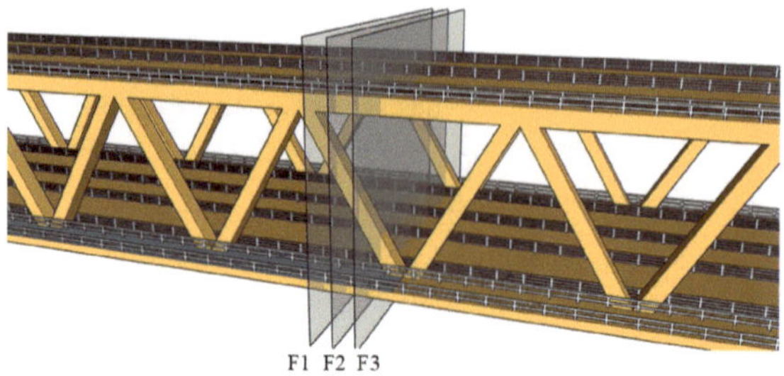

(a)

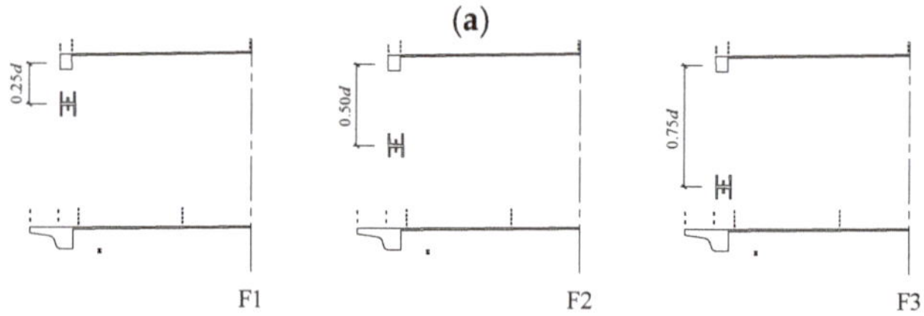

(b)

Figure 19. (**a**) Schematic representation of cross-section selection; (**b**) half-section of main girder.

Table 5. Web position independence verification calculation results.

Cross-Section Number		F1	F2	F3
average value	$\overline{C_D}$	0.7186	0.7221	0.7255
	$\overline{C_L}$	0.8245	0.8456	0.8512
	$\overline{C_M}$	0.0279	0.0285	0.0291
standard deviation	σ_D	0.0212	0.0205	0.0198
	σ_L	0.0525	0.0458	0.0478
	σ_M	0.0103	0.0119	0.0133

The difference in the static three-force coefficients under the three web sections obtained by numerical simulation is within 5%. Therefore, the F2 transverse section is selected in the numerical simulation.

5.3. Simulation of Vortex-Induced Vibration under Different Wind Fairing Parameters

In order to explore the optimal shape parameters of the fairing, 18 different wind speeds, two different wind attack angles, and nine different cross-section parameters of the fairing are numerically simulated. The wind speed range is 1.00~1.68 m/s, and the wind speed step is 0.04 m/s at +3° and +5°.

Through numerical simulation, various curves of different working conditions are obtained as shown in Figure 20. The corresponding vibration inhibition efficiency of wind fairing conditions is shown in Table 6.

From the overall working conditions, when the ratio of h to D is fixed, the greater the ratio of l to D, the higher the vertical bending VIV inhibition efficiency. On the contrary, when the ratio of l to D is fixed, the larger the ratio of h to D, the lower the vertical bending VIV inhibition efficiency is. The VIV inhibition efficiency at +3° is higher than that at +5°. The VIV inhibition efficiency of D9 is the highest, with an average of 63.01%. The VIV inhibition efficiency of D3 is the lowest, with an average of 34.16%.

To explore the mechanism of a wind fairing to inhibit VIV, these upper board stream area changes at the original section and the F3 cross-section at +5° are analyzed as seen in Figure 21.

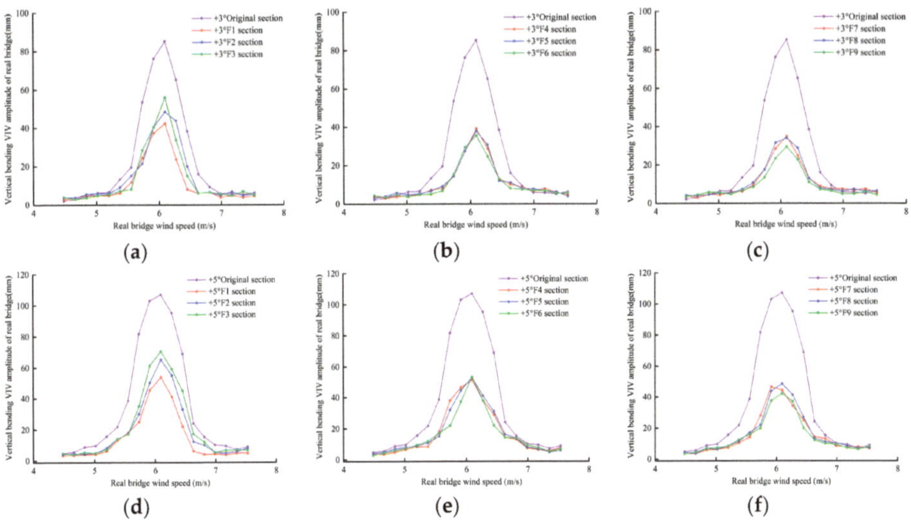

Figure 20. (**a**) $h/D = 3/4$, $\alpha = +3°$; (**b**) $h/D = 1/2$, $\alpha = +3°$; (**c**) $h/D = 1/4$, $\alpha = +3°$; (**d**) $h/D = 3/4$, $\alpha = +5°$; (**e**) $h/D = 1/2$, $\alpha = +5°$; (**f**) $h/D = 1/4$, $\alpha = +5°$.

Table 6. Wind fairing simulation results.

Section Number	Maximum Vertical Bending VIV Amplitude (+3°/+5°)	η
Original section	85.36 mm/107.14 mm	/
F1 section	42.38 mm/54.22 mm	50.35%/49.39%
F2 section	48.52 mm/65.37 mm	43.16%/38.99%
F3 section	56.13 mm/70.64 mm	34.24%/34.07%
F4 section	39.13 mm/51.62 mm	54.16%/51.92%
F5 section	37.93 mm/53.34 mm	55.56%/50.21%
F6 section	35.61 mm/53.71 mm	58.28%/49.87%
F7 section	34.83 mm/46.52 mm	59.20%/56.28%
F8 section	33.98 mm/48.53 mm	60.19%/57.70%
F9 section	29.44 mm/42.32 mm	65.51%/60.50%

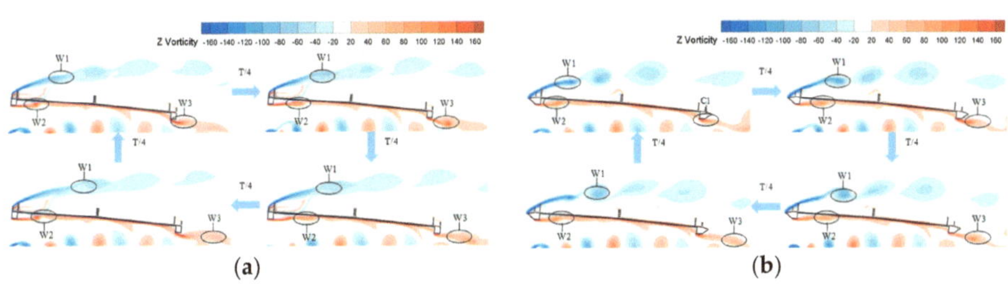

Figure 21. (**a**) Original section; (**b**) F3 section.

Compared with the original section, the first separation point of this air stream in the F3 section moves forward to the tip of the wind fairing. Subsequently, the airflow is rectified by the fairing during the flow to the upper bridge board, thereby significantly reducing the energy of the shear layer. The upper airflow is attached to the upper surface of the fairing, and a small part of the shunt is dispersed into a small-sized vortex after cross over the anti-collision railing slit, while most of the shunt occurs via the secondary separation on the upper side of the railing, resulting in a smaller W1 vortex than the original section scale. Meanwhile, this lower-side airflow produces vortex separation along the lower surface of the fairing. Not only the scale but also the vorticity of W2 vortex are significantly reduced comparable to the original cross-section, and airflow as well as the attachment point on the lower side of the upper bridge board are significantly moved backward compared to the original cross-section. Airflow moves slowly along the lower side of the upper bridge board to the leeward side. Accompanying this, it produces a W3 vortex which is basically as same as the original section scale but has a lower vorticity. Although this wind fairing does not change the vortex shedding forms of the original section, it changes the first separation point of the airflow, as well as movement direction, so that the vortex scale generated by the airflow is smaller and the vorticity is lower, thus effectively suppressing the VIV.

6. Conclusions

Making Huangjuetuo Yangtze River Bridge a research target, the VIV performance of the main girder section with wind fairing measures is tested by the section model wind tunnel test, and the optimal fairing arrangement position is found. Combined with CFD numerical simulation, the vibration inhibition efficiency and influence mechanism of different fairing parameters on VIV are analyzed, and the influence of fairing parameters on the aerodynamic performance of the long-span double-deck steel truss suspension bridge is realized. From the aspects of VIV response and flow field distribution, the following conclusions are drawn:

(1) Through the wind tunnel test, at +3° and +5° wind attack angles, with the addition of wind speed, at the original section of the main girder, what first occurs is the vertical bending VIV, then two torsional VIV with different maximum amplitudes occur, and the peak value of VIV at +5° is higher than that at +3°.

(2) By the VIV test, how impact affects the wind fairing position on VIV is studied. The upper chord is determined to be the optimal position of the fairing. Under two wind attack angle conditions, the average inhibition efficiency of vertical bending VIV is 62.46%, while the average inhibition efficiency of torsional VIV is 42.89%. It can be seen that the inhibition effect of the fairing on vertical bending VIV is greater than that of torsional VIV. Among them, the maximum inhibition efficiency of vertical bending VIV is 71.51%, the minimum is 3.20%, and the average is 34.80%. The highest inhibition efficiency of torsional VIV is 54.73%, the lowest is 0.32%, and the average is 25.34%.

(3) Based on the optimal position of the wind fairing, the height, D, of the truss section is taken as the quantitative value, and the horizontal size, l, of the fairing and the relative height, h, of the fairing are taken as the variables. A total of nine kinds of fairing conditions are set up for numerical simulation. When $h/D = 1/4$ and $l/D = 1$, the optimal fairing parameters are attained. From the airflow field distribution, these vortex zones around the upper board dominate vertical bending VIV. Therefore, the upper chord fairing weakens the energy of the shear layer by advancing the airflow separation point, breaking the vortex and reducing the vortex size to suppress VIV so as to achieve the purpose of optimizing the aerodynamic performance of the main girder.

Author Contributions: Conceptualization, G.Y. and Y.Y.; methodology, G.Y. and Y.Y.; software, L.L.; validation, Y.Y.; formal analysis, L.L.; writing—original draft preparation, L.L., B.W., D.W., H.Y. and H.Q.; writing—review and editing, G.Y. and Y.Y.; visualization, L.L.; supervision, G.Y.; funding acquisition, G.Y. and Y.Y. All authors have read and agreed to the published version of the manuscript.

Funding: This research was funded by National Natural Science Foundation of China (52108435), Central Universities' Basic Research Operating Expenses (2023CDJXY-031) and Chongqing Urban Rail Express Line Full Life Cycle CIM Technology Application Research and Demonstration Research Project (S20220413).

Data Availability Statement: The data presented in this study are available on request from the corresponding author.

Acknowledgments: The authors thanks the Innovation and Talent Recruitment Base of High-performance Wind Energy System and the Effective Operation of Wind Farms in Chongqing University and the Chongqing Key Laboratory of Wind Engineering and Wind Energy Utilization in Chongqing University.

Conflicts of Interest: Authors Dawu Wang, Hui Yu, and Hao Qu were employed by the company Chongqing Railway Group Co., Ltd. The remaining authors declare that the research was conducted in the absence of any commercial or financial relationships that could be construed as a potential conflict of interest.

References

1. Gang, Y.; Yang, Y.; Wu, B.; Liu, L.; Zhang, L.-L. Aerodynamic admittance influence on buffeting performance of suspension bridge with streamlined deck. *J. Vibroengineering* **2019**, *21*, 198–214. [CrossRef]
2. Wu, B.; Zhou, J.; Li, S.; Xin, J.; Zhang, H.; Yang, X. Combining active and passive wind tunnel tests to determine the aerodynamic admittances of a bridge girder. *J. Wind. Eng. Ind. Aerodyn.* **2022**, *231*, 105180. [CrossRef]
3. Zhao, L.; Cui, W.; Shen, X.; Xu, S.; Ding, Y.; Ge, Y. A fast on-site measure-analyze-suppress response to control vortex-induced-vibration of a long-span bridge. *Structures* **2022**, *35*, 192–201. [CrossRef]
4. Yang, Y.; Zhang, J.; Cao, F.; Ge, Y.; Zhao, L. Evaluation and improvement of wind environment and vehicle safety on long-span bridge deck under strong crosswind. *J. Wind Eng. Ind. Aerodyn.* **2022**, *228*, 105089. [CrossRef]
5. Ge, Y.; Zhao, L.; Cao, J. Case study of vortex-induced vibration and mitigation mechanism for a long-span suspension bridge. *J. Wind Eng. Ind. Aerodyn.* **2022**, *220*, 104866. [CrossRef]
6. Xiao, H.; Liu, Z.; Chen, Z.; Liu, Z.; Yan, A.; Yang, O. Experimental and numerical study on vibration response, VIV mechanism and aerostatic coefficients of triple-separated decks under different vertical distances. *J. Wind Eng. Ind. Aerodyn.* **2024**, *250*, 105782. [CrossRef]
7. He, M.; Liang, P.; Zhang, Y.; Wang, Y.; Wang, K. Identification, tracking and warning of vortex induced vibration using k-means clustering method. *Struct. Infrastruct. Eng.* **2022**, *20*, 380–393. [CrossRef]
8. Kim, S.-J.; Shim, J.-H.; Kim, H.-K. How wind affects vehicles crossing a double-deck suspension bridge. *J. Wind Eng. Ind. Aerodyn.* **2020**, *206*, 104329. [CrossRef]
9. He, X.-H.; Li, H. Review of aerodynamics of high-speed train-bridge system in crosswinds. *J. Cent. South Univ.* **2020**, *27*, 1054–1073. [CrossRef]
10. Xu, Z.; Wang, H.; Xing, C.; Tao, T.; Mao, J.; Liu, Y. Physics guided wavelet convolutional neural network for wind-induced vibration modeling with application to structural dynamic reliability analysis. *Eng. Struct.* **2023**, *297*, 117027. [CrossRef]
11. Irwin, P.A. Bluff body aerodynamics in wind engineering. *J. Wind Eng. Ind. Aerodyn.* **2008**, *96*, 701–712. [CrossRef]
12. Bruno, L.; Coste, N.; Fransos, D. Simulated flow around a rectangular 5:1 cylinder: Spanwise discretisation effects and emerging flow features. *J. Wind Eng. Ind. Aerodyn.* **2012**, *104–106*, 203–215. [CrossRef]
13. Bruno, L.; Salvetti, M.V.; Ricciardelli, F. Benchmark on the Aerodynamics of a Rectangular 5:1 Cylinder: An overview after the first four years of activity. *J. Wind Eng. Ind. Aerodyn.* **2014**, *126*, 87–106. [CrossRef]
14. Zhong, Y.; Li, M.; Ma, C. Aerodynamic admittance of truss girder in homogenous turbulence. *J. Wind Eng. Ind. Aerodyn.* **2018**, *172*, 152–163. [CrossRef]
15. Yao, G.; Wu, L.; Yang, Y.; Liu, M.; Chen, Y.; Du, H. Study on the Suppression Effect on Vortex-Induced Vibration of Double-Deck Truss Girder by the Spatial Position of the Deflector Plate. *Appl. Sci.* **2023**, *13*, 9764. [CrossRef]
16. Yao, G.; Chen, Y.; Yang, Y.; Zheng, Y.; Wu, L.; Du, H. Research on Mechanism of Vortex-Induced Vibration Railing Effect of Double-Deck Large-Span Suspension Bridge. *Appl. Sci.* **2023**, *13*, 9314. [CrossRef]
17. Bai, H.; Li, R.; Xu, G.; Kareem, A. Aerodynamic performance of Π-shaped composite deck cable-stayed bridges including VIV mitigation measures. *J. Wind Eng. Ind. Aerodyn.* **2021**, *208*, 104451. [CrossRef]
18. Li, Y.; Chen, X.; Yu, C.; Togbenou, K.; Wang, B.; Zhu, L. Effects of wind fairing angle on aerodynamic characteristics and dynamic responses of a streamlined trapezoidal box girder. *J. Wind Eng. Ind. Aerodyn.* **2018**. [CrossRef]
19. Yao, G.; Chen, Y.; Yang, Y.; Zheng, Y.; Du, H.; Wu, L. Investigation of Vortex-Induced Vibration of Double-Deck Truss Girder with Aerodynamic Mitigation Measures. *J. Mar. Sci. Eng.* **2023**, *11*, 1118. [CrossRef]
20. Li, P.; Wang, H. A novel strategy for the crossarm length optimization of PSSCs based on multi-dimensional global optimization algorithms. *Eng. Struct.* **2021**, *238*, 112238. [CrossRef]
21. Li, P.; Chen, G.; Qiu, J.; Qian, J.; Ding, D.; Jian, B.; Zhang, X.; Xiong, G. Experimental and numerical investigation into the load-carrying capacity of aluminium alloy H-sectional stocky columns under axial compression. *J. Build. Eng.* **2024**, *87*, 108777. [CrossRef]
22. Chen, X.; Zhu, C.; Ma, C.; Li, Q. Vortex-induced vibration of twin-box girder and its aerodynamic characteristics study based on high-order dynamic decomposition. *J. Wind Eng. Ind. Aerodyn.* **2024**, *249*, 105719. [CrossRef]
23. Zhao, C.-X.; Liu, J.-N.; Li, B.-Q.; Ren, D.; Chen, X.; Yu, J.; Zhang, Q. Multiscale Construction of Bifunctional Electrocatalysts for Long-Lifespan Rechargeable Zinc–Air Batteries. *Adv. Funct. Mater.* **2020**, *30*, 2003619. [CrossRef]

24. Khosravi, M.; Mosaddeghi, F.; Oveisi, M.; khodayari-b, A. Aerodynamic drag reduction of heavy vehicles using append devices by CFD analysis. *J. Cent. South Univ.* **2015**, *22*, 4645–4652. [CrossRef]
25. Wang, C.; Hua, X.; Huang, Z.; Tang, Y.; Chen, Z. Post-critical behavior of galloping for main cables of suspension bridges in construction phases. *J. Fluids Struct.* **2021**, *101*, 103205. [CrossRef]
26. Sarwar, M.W.; Ishihara, T. Numerical study on suppression of vortex-induced vibrations of box girder bridge section by aerodynamic countermeasures. *J. Wind Eng. Ind. Aerodyn.* **2010**, *98*, 701–711. [CrossRef]
27. Xiao, H.; Liu, Z.; He, F.; Chen, Z.; Liu, Z.; Yan, A. Aerodynamic Interference Effects on Vortex-Induced Vibration of Two Parallel Railway– and Highway–Cable-Stayed Bridges with Triple-Separated Decks. *J. Bridge Eng.* **2024**, *29*, 04023112. [CrossRef]
28. Haohong, L.; Liang, L.Z.; Bo, W. Effects of the angle of attack on the aerodynamic characteristics of a streamlined box girder. *Adv. Struct. Eng.* **2021**, *24*, 2090–2104. [CrossRef]
29. Lei, W.; Wang, Q.; Zhang, Y.; Li, Z. Study on VIV performance of streamlined steel box girder of a sea-crossing cable-stayed bridge. *Ocean Eng.* **2024**, *295*, 116897. [CrossRef]
30. Hu, C.; Zhao, L.; Wang, X.; Ge, Y. Spatio-temporal distribution of aerodynamic forces and their associated vortex drift patterns around a closed-box girder during torsional vortex-induced vibration. *Eng. Struct.* **2024**, *302*, 117459. [CrossRef]
31. Sun, H.; Zhu, L.-D.; Tan, Z.-X.; Zhu, Q.; Meng, X.-L. Spanwise layout optimization of aerodynamic countermeasures for multi-mode vortex-induced vibration control on long-span bridges. *J. Wind Eng. Ind. Aerodyn.* **2024**, *244*, 105616. [CrossRef]
32. Liu, Z.; He, F.; Yan, A.; Liu, Z.; Yin, T.; Chen, Z. Mechanism and Aerodynamic Countermeasures of Vortex-Induced Vibration of a Cable-Stayed Bridge with Narrow Π-Shaped Girder Sections. *J. Bridge Eng.* **2023**, *28*, 04023090. [CrossRef]
33. He, H.; Wang, J.; Wang, F. Investigation of the Effects of Vortex Isolation Plates with Different Opening Ratios and Sizes on Vortex-Induced Vibration. *Appl. Sci.* **2024**, *14*, 3863. [CrossRef]
34. Wang, C.; Huang, Z.; Hua, X.; He, X.; Zhou, S.; Chen, Z. Aerodynamic mechanism of triggering and suppression of vortex-induced vibrations for a triple-box girder. *J. Wind Eng. Ind. Aerodyn.* **2022**, *227*, 105051. [CrossRef]
35. Guo, S.; Jiang, Y.; Zhang, W.; Zeng, Y. Study on Dynamic Characteristics of Long-Span Highway-Rail Double-Tower Cable-Stayed Bridge. *Buildings* **2024**, *14*, 1733. [CrossRef]
36. Wang, Y.; Feng, H.; Xu, N.; Zhong, J.; Wang, Z.; Yao, W.; Jiang, Y.; Laima, S. A Data-Driven Model for Predictive Modeling of Vortex-Induced Vibrations of a Long-Span Bridge. *Appl. Sci.* **2024**, *14*, 2233. [CrossRef]
37. Zhang, Y.-M.; Wang, H.; Bai, Y.; Mao, J.-X.; Chang, X.-Y.; Wang, L.-B. Switching Bayesian dynamic linear model for condition assessment of bridge expansion joints using structural health monitoring data. *Mech. Syst. Signal Process.* **2021**, *160*, 107879. [CrossRef]

materials

MDPI

Article

Comparison of Dynamic Amplification Factor of Deflection and Bending Moment of Highway Continuous Box-Girder Bridges by Mode Superposition

Yelu Wang [1,2], Yongjun Zhou [1,*], Yang Zhao [3], Yu Zhao [1], Yuxin Xue [1] and Wei Feng [4]

[1] School of Highway, Chang'an University, Xi'an 710064, China; 2017021032@chd.edu.cn (Y.W.)
[2] Key Laboratory of Bridge Detection Reinforcement Technology Ministry of Communications, Xi'an 710064, China
[3] Shaanxi Provincial Transport Planning Design and Research Institute Co., Ltd., Xi'an 710065, China
[4] Xi'an Municipal Engineering Design and Research Institute Co., Ltd., Xi'an 710064, China
* Correspondence: zyj@chd.edu.cn; Tel.: +86-029-8233-4872

Abstract: There are differences between the dynamic deflection and bending moment (strain) in the same section of continuous girder bridges. However, the selection of the response for calculating dynamic amplification factors (DAFs), which are essential for bridge health monitoring and safety assessment, remains controversial. Modes may play a role in the relationship between the deflection DAF and the bending moment DAF in both numerical analysis and field tests. To investigate the distinctions between the DAFs of the deflection and bending moment in a continuous girder bridge, functional expressions of the DAFs were derived, taking into account multi-factor coupling under concentrated forces. The interaction effects of the mode and road surface condition (RSC), vehicle speed, bridge span length, and span number on the deflection DAF, the bending moment DAF, and the ratio of the deflection DAF to the bending moment DAF (RDM) of precast continuous box-girder bridges were analyzed using vehicle-bridge interaction. To ensure the accuracy of the DAF in numerical computations and experimental tests, two types of accuracy indexes and the corresponding cut-off modes were provided. Validation was conducted by performing dynamic load tests on two field bridges. The results indicate that different modes have a significant effect on the RDM of the mid-span section of a bridge. When considering multiple factors, the deflection DAF and bending moment DAF of the mid-span section increased rapidly with the considered modes and then stabilized. Statistically, the RDM of all nine bridges ranged from 1.00 to 1.12, indicating that the deflection DAF was greater than the bending moment DAF. The suggested cut-off modes can be utilized for efficient and accurate calculation of the DAF and response signal fidelity.

Keywords: bridge engineering; dynamic amplification factor; mode superposition; deflection; bending moment; cut-off mode

Citation: Wang, Y.; Zhou, Y.; Zhao, Y.; Zhao, Y.; Xue, Y.; Feng, W. Comparison of Dynamic Amplification Factor of Deflection and Bending Moment of Highway Continuous Box-Girder Bridges by Mode Superposition. *Materials* **2024**, *17*, 1041. https://doi.org/10.3390/ma17051041

Academic Editor: Yuri Ribakov

Received: 25 January 2024
Revised: 18 February 2024
Accepted: 20 February 2024
Published: 24 February 2024

1. Introduction

Small- and medium-span bridges are extensively used, with their live load effects accounting for a significant proportion of the bridge effects, making the vehicle-bridge interaction (VBI) effect significant [1–3]. The effect of the increased vertical dynamic load produced by vehicles on bridges is usually characterized by a dynamic amplification factor (DAF) [4]. Accurate evaluation of the DAF plays a critical role in the design and evaluation of bridges. A reasonable DAF leads to safer and more economical designs of new bridges and provides valuable information for the condition assessment and management of bridges in service [5,6].

Since the 1950s, significant efforts have been made to study the DAFs on bridges due to vehicle loading. The maximum bending moment and maximum deflection in a static state tend to occur together near the mid-span, so the corresponding DAFs have received much attention. Many researchers have found differences in DAFs obtained from different dynamic responses of a bridge through field tests or numerical simulations. Notably, there have been different views on the relationship between the deflection DAF and the bending moment (strain) DAF obtained at the same position. When the beam is kept in an elastic state, there is a correspondence between the strain and the bending moment. The AASHO road test [7,8] showed that the impact factor (IM, IM = DAF-1) obtained from measured deflection was always higher than that obtained from measured strains. Similar findings were also reported by other researchers. Mohseni [9] formulated the DAFs for bending moment and deflection based on an extensive numerical study involving 125 composite slab-on-girder prototype bridges with simple support. Szurgott [10] conducted experimental tests on a reinforced concrete simply supported girder bridge and found that the DAFs derived from deflection were significantly larger and less trustworthy than those recorded from strains. The deflection was clearly overestimated, including the actual deflection of the girders, the bearing displacements, and the contraction of the bridge piles. Li [11] examined a concrete type-III girder bridge and found that the IMs of deflection were greater than those of the bending moment at mid-span, especially for light trucks. The difference between the two decreased as the weight of the vehicle increased. Huang [12] utilized numerical methods to analyze seven three-span single-cell steel–concrete composite continuous girder bridges, concluding that the IMs of deflection usually exceed those of vertical bending. Majka [13] analyzed the dynamic characteristics of simply supported railway girder bridges through an efficient numerical model and stated that the DAFs of deflection might be conservative compared with those of strain in some cases. Yang [14] analyzed the DAFs of deflection and bending moments for both simply supported beams and continuous beams and showed that the IMs of deflection exceed those of bending moments at the mid-span.

Nonetheless, other researchers reported the opposite conclusion. Aluri [8] demonstrated, using test data from three FRP deck bridges, that the IMs obtained from deflection were lower than those acquired through strain measurements. Fafard [15] researched a cracked continuous concrete skewed girder bridge, and the numerical results showed that the DAF calculated with deflections was smaller than that estimated with strains and reactions. Senthilvasan [16] conducted an experimental study on a continuously curved concrete bridge, showing that strain has a greater dynamic amplification than deflection.

Through careful examination of the above studies, possible reasons for the deviations in the conclusions were identified: (1) whether the bridges are made of the same material or not: the relationship between strain and deflection may be different for steel, concrete, and FRP materials; (2) whether the structure is in an elastic state or not: cracks may change the structural response; (3) different methods of analyzing the structure: the accuracy of measurements made in the field test methods is affected by the performance of sensors and the environment; (4) different structural forms: straight bridges, curved bridges, and straight and skewed bridges may affect the results; (5) whether the measurements are taken at the same point or not: it was observed in some studies in the literature that the deflection and strain measurement points were not located at the edge of the section (upper or lower edge); (6) girder support conditions: girder deflections may include the compression of bearings or piers.

Since the available field test data are limited and tests are costly, numerous theoretical derivations and numerical simulations have been carried out to calculate the bridge responses to vehicle loads [17]. A plethora of scholars have extensively investigated the factors influencing DAFs, including the bridge characteristics, vehicle speed, bridge deck conditions, vehicle loads, etc. [12,17–20]. Existing studies have also reported the influence of modes on the DAFs owing to the prevalent use of the mode superposition technique [18,21,22]. Through theoretical extrapolation, Humar [23] deduced that the DAFs between the bending moment and deflection of simply supported bridges are equivalent when considering only the first-order modes. However, their precise relationship is contingent upon the contribution of higher-order modes, though they are closely approximated.

Li et al. [11] found that all the modes of the bridge are excited when a vehicle is traveling on one side of the bridge, and the first, third, and fifth modes contribute the most to the deflection response, which could be identified from the experiment data. Zhou et al. [24] performed a numerical analysis using a three-axis vehicle model. They found that for simply supported beams, the strain at the bottom of the beam at mid-span is significantly influenced by the higher-order modes, while the deflection is primarily influenced only by the first few modes. Zou et al. [25] showed that the lower-order modes dominate the deflection of the bridge, whereas bending and acceleration are more sensitive to the higher-order modes, necessitating the incorporation of higher-order bridge modes to ensure accurate results for bridge bending moments. Similar findings were also reported by other researchers [22,25–27]. In dynamic load tests, the acquired signals are usually filtered and de-noised, resulting in the removal of higher-order signals [16,28]. Currently, there is no consensus on the cut-off modes for calculating the DAFs of continuous girder bridges [29]. The above studies indicated that modes may potentially contribute to the relationship between the deflection DAF and bending moment DAF in numerical analyses and field tests.

It is, therefore, clear that more research is needed to understand the potential relationship between the deflection DAF, the bending moment DAF, and modes for continuous girder bridges. This will enable their more effective utilization in engineering practice. The primary scope of this study is (1) to clarify the magnitude relationship between the deflection and bending moment of straight precast concrete continuous girder bridges by theoretical and numerical methods; (2) to analyze the effect of vibration modes on the ratio of the deflection DAF to the bending moment DAF, including the coupled effects of some parameters (e.g., vehicle parameters, road surface condition, bridges); and (3) to propose different levels of cut-off modes to reduce the DAF errors caused by modes in numerical analysis or experimental tests. The outline of the paper is as follows: Section 2 presents the theory of the relationship between the DAFs of deflection and bending moments for two-span continuous beams. In Section 3, the development and verification process of the mode superposition method program for VBI analysis is presented. Section 4 analyzes the interactive effects of sensitive parameters and modes on the DAFs of the deflection and bending moment. Section 5 identifies the main vibration modes of the deflection and bending moment and provides the cut-off modes for different computational accuracies. Finally, Section 6 presents an experimental validation, and Section 7 concludes the study.

2. Theoretical Analysis

Take a two-span continuous beam as an example, as shown in Figure 1; assuming that the span is l and the velocity and displacement of the bridge are zero at the initial time, the vibration equation of the bridge subjected to a concentrated force F can be expressed as follows:

$$EI\frac{\partial^4 y_d(x,t)}{\partial x^4} + \overline{m}\frac{\partial^2 y_d(x,t)}{\partial t^2} + c\frac{\partial y_d(x,t)}{\partial t} = F(x,t) \tag{1}$$

$$F(x,t) = F\delta(x - vt) \tag{2}$$

where v denotes the load speed; t represents time; y_d is the vertical displacement of the beam; $\delta(x - vt)$ is the Dirac function; F is a moving load; EI is the vertical flexural stiffness of the beam.

According to the principle of mode superposition [30], the deflection and bending moment of the beam can be expressed as follows:

$$y_d(x,t) = \sum_{n=1}^{N} \varphi_n(x) q_n(t) \quad n = 1,\ 2,3.. \tag{3}$$

$$M_d(x,t) = -EI\frac{\partial^2 y(x,t)}{\partial x^2} = -EI\sum_{n=1}^{N} \frac{\partial^2 \varphi_n(x)}{\partial x^2} q_n(t) \tag{4}$$

where $\varphi_n(x)$ denotes the vibration mode function; $q_n(t)$ denotes the generalized modal coordinates; $M_d(x,t)$ is the bending moment of the beam at position x at time t.

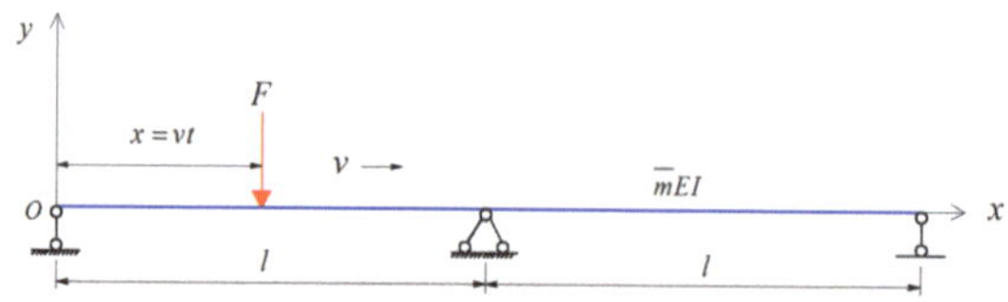

Figure 1. Schematic of a two-span continuous beam.

The vibration mode $\varphi_n(x)$ of the beam is divided into two cases.

When the mode order is odd, i.e., n = 1, 3, 5..., the mode function of the beam can be described as follows:

$$\varphi_n(x) = \sin\frac{(n+1)\pi x}{2l} \tag{5}$$

From Equations (1) to (3) and Equation (4), the following equations can be obtained:

$$\omega_n = \frac{(n+1)^2\pi^2}{4}\sqrt{\frac{EI}{\overline{m}l^4}} \tag{6}$$

$$q_n(t) = \frac{F}{\overline{m}l\omega_n^2}H_2(t) \tag{7}$$

$$\begin{aligned} H_2(t) = & \frac{1}{(1-\beta_n^2)^2+(2\zeta_n\beta_n)^2}\cdot[\mathrm{e}^{-\zeta_n\omega_n t}\frac{\zeta_n\omega_n(2\zeta_n\beta_n)-(1-\beta_n^2)\Omega_n}{\omega_D}\cdot\sin(\omega_\mathrm{D}t)+ \\ & (1-\beta_n^2)\sin(\Omega_n t)+\mathrm{e}^{-\zeta_n\omega_n t}(2\zeta_n\beta_n)\cos(\omega_\mathrm{D}t)] \end{aligned} \tag{8}$$

$$\Omega_n = \frac{(n+1)\pi v}{2l} \tag{9}$$

where $\overline{m}$ is the mass of the beam section; $\beta_n = \frac{\Omega_n}{\omega_n}$ is the frequency ratio, which is the ratio of the nth generalized excitation frequency Ω_n to the flexural frequency ω_n of the beam, $\omega_n = a_n^2\sqrt{\frac{EI}{\overline{m}l^4}}$; $\omega_D = \omega_n\sqrt{1-\xi_n^2}$; ξ_n is the damping ratio of the nth-order vibration mode.

When the mode order is even, i.e., n = 2, 4, 6..., the mode function of the beam must be given as a piecewise function.

When $0 \le x \le l$, the mode function is

$$\varphi_n(x) = \sin(\frac{\alpha_n x}{l}) - \frac{\sin(\alpha_n)}{\sinh(\alpha_n)}\cdot\sinh(\frac{\alpha_n x}{l}) \tag{10}$$

where $\alpha_n = 3.927 + (\frac{n}{2}-1)\cdot\pi$.

From the combination of Equations (1) to (3) and Equation (9), the following equation can be obtained:

$$q_n(t) = \frac{F}{M_n\omega_D}\cdot\int_0^t\left\{\left[\sin(\frac{\alpha_n vt}{l}) - \frac{\sin(\alpha_n)}{\sinh(\alpha_n)}\sinh(\frac{\alpha_n vt}{l})\right]\cdot\mathrm{e}^{-\zeta_n\omega_n(t-\tau)}\sin\omega_\mathrm{D}(t-\tau)\right\}\mathrm{d}\tau \tag{11}$$

where $M_n = \int_0^l \varphi_n(x)^2\overline{m}\,\mathrm{d}x$; τ is the integration factor related to time.

When $l \le x \le 2l$, the mode function is

$$\varphi_n(x) = \sin(\alpha_n\frac{2l-x}{l}) - \frac{\sin(\alpha_n)}{\sinh(\alpha_n)}\cdot\sinh(\alpha_n\frac{2l-x}{l}) \tag{12}$$

From the combination of Equations (1) to (3) and Equation (11), the following equation can be obtained:

$$q_n(t) = \frac{F}{M_n\omega_D} \cdot \int_{\frac{l}{v}}^{t} \left\{ \left[\int \sin\left(\alpha_n \frac{2l-x}{l}\right) - \frac{\sin(\alpha_n)}{\sinh(\alpha_n)} . \sinh\left(\alpha_n \frac{2l-x}{l}\right) \right] \cdot e^{-\xi_n\omega_n(t-\tau)} \sin\omega_D(t-\tau) \right\} d\tau \quad (13)$$

where $M_n = \int_0^l \varphi_n(x)^2 \overline{m}\, dx$; $\alpha_n = 3.927 + \left(\frac{n}{2} - 1\right) \cdot \pi$.

The static deflection and bending moment of the beam are

$$\begin{cases} y_s(x) = (\lambda_1(x)Fl^3)/EI \\ M_s(x) = \lambda_2(x)Fl \end{cases} \quad (14)$$

where $\lambda_1(x)$ and $\lambda_2(x)$ are the static load factors of the deflection and bending moment related to the location of the section, respectively.

When considering the first K order modes of the beam, the DAFs for the deflection and bending moment of the beam can be derived as follows:

$$DAF_u = \frac{y_d(x,t)}{y_s(x)} = g_y(\lambda_1, \frac{x}{l}, \frac{v}{l}, \xi_1 \ldots \xi_K, \beta_1 \ldots \beta_K, K) \quad (15)$$

$$DAF_M = \frac{M_d(x,t)}{M_s(x)} = f(\lambda_2, \frac{x}{l}, \frac{v}{l}, \xi_1 \ldots \xi_K, \beta_1 \ldots \beta_K, K) \quad (16)$$

where N is the total number of modes of the beam considered.

From Equations (15) and (16), the DAF of the two-span continuous beam is a function of the position coefficient x/l, frequency ratio β_n, static load factor λ_i, damping ratio ξ_i, and v/l. The damping ratio ξ_i of a beam ranges from 0.02 to 0.05 and therefore has a minor effect on the DAF. Once the section position is determined, λ_i and x/l are constant values. The number of modes considered could affect the accuracy of the DAFs. The more modes are chosen, the more accurate the DAF is.

For two equal-span continuous beams, the ratio of the deflection DAF to the bending moment DAF (RDM) at $l/2$ is

$$\mathrm{RDM}(\frac{l}{2}) = \frac{\mathrm{DAF_u}}{\mathrm{DAF}_M} = -\frac{312}{23l^2} \times \frac{\sum\limits_{n=1}^{N} [q_n(t)\varphi_n(x)]}{\sum\limits_{n=1}^{N} \left[q_n(t)\frac{\partial^2\varphi_n(x)}{\partial x^2}\right]} \quad (17)$$

The RDM at $l/2$ is obtained when only the first-order mode is considered as follows:

$$\mathrm{RDM}(\frac{l}{2}) = \frac{\mathrm{DAF_u}}{\mathrm{DAF}_M} = \frac{312}{23\pi^2} \approx 1.37 \quad (18)$$

From Equation (18), it can be seen that for two equal-span continuous beams, the deflection DAF is larger than that of the bending moment when only the first-order modes are considered, and the RDM is close to 1.37. When more modes are considered, the results of Equation (15) are dependent on factors such as the span and stiffness, necessitating the use of numerical solution methods. Nevertheless, it is clear that the RDM varies with the number of modes K.

3. Vehicle-Bridge Interaction Models

By the previous theoretical derivation, the functional relationship equation between the DAF of the bridge and the related factors subjected to concentrated force is obtained. The bridge modes have a significant effect on the RDM of continuous girder bridges. To further clarify the regularity pattern of DAFs with modes, the influence of real vehicles should be considered. At the same time, possible differences caused by vehicle parameters, the road surface condition (RSC), and bridge types need to be considered.

3.1. Vehicle and Bridge Vibration Equations

Vehicle vibration equations can be obtained by the direct balance method and Lagrange equation method [31]. The vehicle vibration equations are established based on Lagrange equations and reorganized in matrix form as follows:

$$[M_V]\{\ddot{Z}\} + [C_V]\{\dot{Z}\} + [K_V]\{Z\} = \{F_{Vb}\} \tag{19}$$

where $[M_v]$, $[C_v]$, and $[K_v]$ are the mass matrix, damping matrix, and stiffness matrix of the vehicle, respectively; $\{Z\}$, $\{\dot{Z}\}$, and $\{\ddot{Z}\}$ are the displacement, velocity, and acceleration vectors of the vehicle; $\{F_{Vb}\}$ is the force of the bridge on the vehicle.

A dynamic model of the bridge can be obtained through finite element modeling. The vibration equation of a bridge can be characterized as

$$[M_b]\{\ddot{Y}\} + [C_b]\{\dot{Y}\} + [K_b]\{Y\} = \{F_{bv}\} - \{F_g\} \tag{20}$$

where $[M_b]$, $[C_b]$, and $[K_b]$ are the mass matrix, damping matrix, and stiffness matrix of the bridge; $\{Y\}$, $\{\dot{Y}\}$, and $\{\ddot{Y}\}$ are the displacement, velocity, and acceleration vectors of the bridge structure; $\{F_{bv}\}$ is the interaction force vector of the vehicle on the bridge; $\{F_g\}$ is the load vector of the wheels due to self-weight.

3.2. Road Surface Condition

Existing studies suggest that the RSC is one of the main factors affecting bridge vibration and DAFs [32,33]. It can be assumed to be a smooth stochastic process with zero mean for each state history and can be expressed by a power spectral density (PSD) function. In the current study, the multi-path RSC was generated using the inverse Fourier transform, and the following power spectral density (PSD) function [6] was used:

$$G(n) = G(n_0)\left(\frac{n}{n_0}\right)^{-2} \tag{21}$$

where $G(n)$ denotes the roughness coefficient (m^3/cycle); n is the spatial frequency; $n_0 = 0.1\ m^{-1}$. Three types of RSCs of 'good', 'medium', and 'poor' represented by $G(n)$ as per ISO-8608 (1995) [34] were considered; the roughness coefficients were $8 \times 10^{-6}\ m^3$/cycle, $64 \times 10^{-6}\ m^3$/cycle, and $128 \times 10^{-6}\ m^3$/cycle for 'good', 'medium', and 'poor' RSCs, respectively.

RSCs can be generated by a summation of a series of harmonics as given below:

$$y(t) = \sum_{i}^{N} \sqrt{2G(n_i)\Delta n}\cos(2\pi n_i \beta t + \phi_i) \tag{22}$$

where ϕ_i is a random phase angle uniformly distributed from 0 to 2π; β represents a constant velocity; t represents a time corresponding to a particular location at which the roughness is to be found; N is the total number of terms used to build up the RSC sample in Equation (22).

The randomness of the RSC interferes with the results of VBI analysis [35], and in this paper, 10 sets of samples were selected to eliminate its effect. A sample of the RSC is shown in Figure 2.

3.3. Vehicle-Bridge Interaction and Numerical Solution

The vehicle's ith tire is subjected to a force transmitted by the bridge as shown in Figure 3, and the relationship between the two is as follows:

$$F_{vbi} = -k_{ti}\Delta_i - c_{ti}\dot{\Delta}_i \tag{23}$$

$$\Delta_i = Z_i(t) - Y_i(x,t) - r_i(x) \tag{24}$$

$$\dot{\Delta}_i = \dot{Z}_i(t) - \dot{Y}_i(x,t) - Y_i'(x,t)v - r_i(x)'v \tag{25}$$

where F_{vbi} is the force of the bridge on the ith wheel; k_{ti} is the stiffness of the ith wheel; c_{ti} is the damping of the ith wheel; Δ_i is the spring compression of the ith wheel; $Z_i(t)$ and $\dot{Z}_i(t)$ are the vertical displacement and velocity of the ith wheel, respectively; $Y_i(x,t)$ and $\dot{Y}_i(x,t)$ are the vertical displacement and velocity of the bridge at the ith wheel; $r_i(x)$ is the road surface condition at the ith wheel; $Y_i(x,t)'$ denotes the derivative of $Y_i(x,t)$ with respect to the longitudinal distance x; $r_i(x)'$ denotes the derivative of $r_i(x)$ with respect to the longitudinal distance x.

The mode superposition method can be used to solve the dynamic equations of the bridge subsystem. When the mode superposition method is used, Equation (23) should be transformed into the following form:

$$F_{\mathrm{vb}i} = -k_{ti}(Z_i - r_i - \langle N_{\mathrm{i}}\rangle \boldsymbol{\Phi}\{Q\}) - c_{ti}(\dot{Z}_i(t) - \langle N_{\mathrm{i}}\rangle' \boldsymbol{\Phi}\{Q\}v - \langle N_i\rangle \boldsymbol{\Phi}\{\dot{Q}\} - r_i'v) \tag{26}$$

where $\langle N_i\rangle$ is the shape function of the element at the contact point between the wheel and the bridge deck; $\boldsymbol{\Phi}$ denotes the vertical mode; $\{Q\}$ denotes the modal coordinate [30].

$$F_{vbi} = -F_{bvi} \tag{27}$$

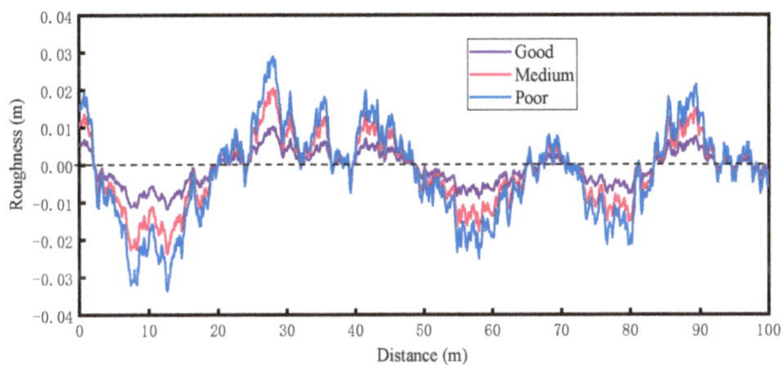

Figure 2. A sample of the road surface condition.

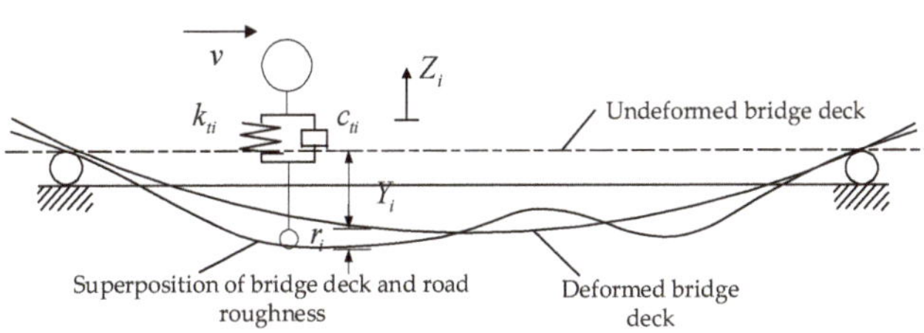

Figure 3. Displacement relationship between the wheel and bridge.

Assuming that the vehicle has n wheels, the force of the bridge on the vehicle can be expressed as follows:

$$\{F_{Vb}\} = \begin{Bmatrix} 0 \\ \vdots \\ F_{vb1} \\ \vdots \\ 0 \end{Bmatrix} + \begin{Bmatrix} 0 \\ \vdots \\ F_{vb2} \\ \vdots \end{Bmatrix} + \ldots + \begin{Bmatrix} 0 \\ \vdots \\ F_{vbn} \end{Bmatrix} = \begin{Bmatrix} 0 \\ \vdots \\ F_{vb1} \\ \vdots \\ F_{vbn} \end{Bmatrix} \tag{28}$$

After coordinate transformation, the final vibration equations of the vehicle-bridge coupled system are shown in the following:

$$\begin{bmatrix} M_b \\ M_v \end{bmatrix}\begin{Bmatrix} \ddot{Q} \\ \ddot{Z} \end{Bmatrix}+\begin{bmatrix} C_b+C_{b-b} & C_{b-v} \\ C_{v-b} & C_v \end{bmatrix}\begin{Bmatrix} \dot{Q} \\ \dot{Z} \end{Bmatrix}+\begin{bmatrix} K_b+K_{b-b} & K_{b-v} \\ K_{v-b} & K_v \end{bmatrix}\begin{Bmatrix} Q \\ Z \end{Bmatrix}=\begin{Bmatrix} F_b \\ F_v+F_G \end{Bmatrix} \quad (29)$$

where C_{b-b}, C_{b-v}, C_{v-b}, K_{b-b}, K_{b-v}, K_{v-b}, F_b, and F_v are the vehicle-bridge interaction coupling parameters.

The above vehicle-bridge coupled system may be solved using an uncoupled method [36,37]. Mode superposition makes it possible to separate the bridge modal analysis from the vehicle-bridge coupled model. In consequence, the number of calculations in Equation (29) and the complexity of the entire procedure are greatly reduced. The vibration equations of the vehicle-bridge coupled system are assembled automatically and solved in time history using the Newmark-β method. The program flow is shown in Figure 4. To verify the correctness of the vehicle-bridge coupling program, the vehicle parameters and bridge parameters from reference [38] were used; the calculation results of this paper's program were compared and are shown in Figure 5. In the figure, the trend of the time course curve and the values of the mid-span displacement response are highly consistent with the reference [38]. The maximum value of the displacement response of the self-programmed program differs from the reference by 0.05 mm, with an error of only 0.6%, which fully demonstrates the accuracy of the vehicle-bridge coupling analysis method and the numerical calculation results in this paper.

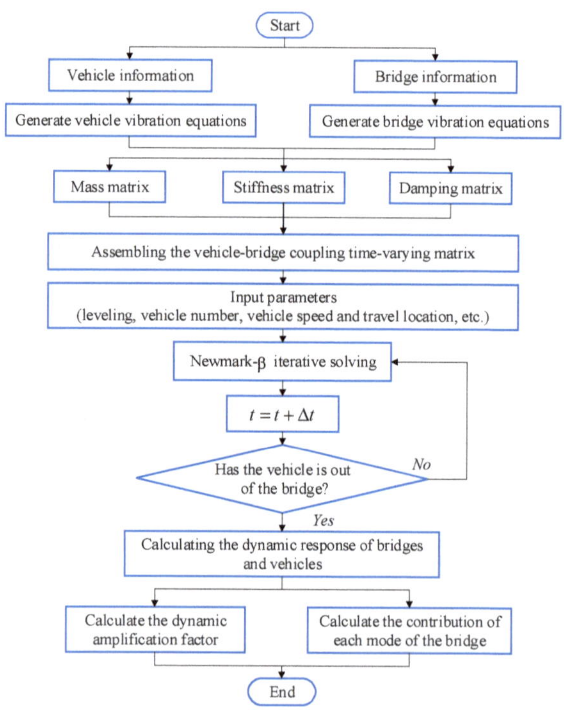

Figure 4. Flow of the vehicle-bridge coupling procedure.

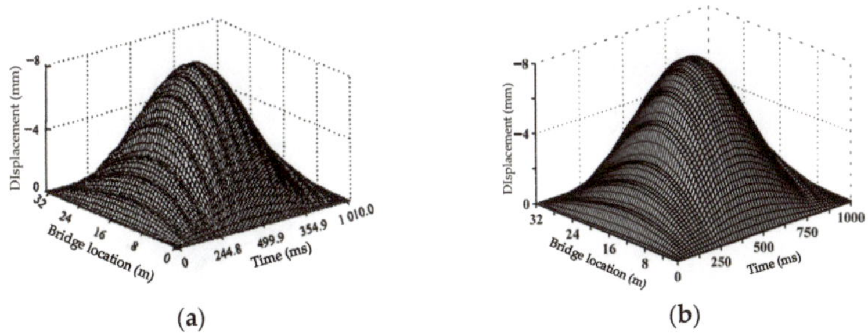

Figure 5. Comparison of the displacement responses at a speed of 40 km/h. (**a**) Reference [38]; (**b**) Self Programming.

4. Dynamic Analysis and Results

4.1. Prototype Bridges and Vehicle

A parametric investigation of precast continuous concrete box-girder bridges was conducted by considering the factors of span length, span number, road conditions, and vehicle speed. The levels of each influence factor analyzed by VBI are listed in Table 1. The corresponding beam heights for precast concrete continuous girder bridges with spans of 20 m, 30 m, and 40 m are 1.2 m, 1.8 m, and 2 m, respectively, and the other parameters are shown in Figure 6. The bridge was simplified to a grillage model [18,21,39] by ANSYS Software with cross-sectional properties assigned to each beam element. The roller support was modeled by releasing the horizontal movement. However, the hinged support was constrained from any horizontal movement. All supports were constrained in the vertical direction but allowed to rotate around the support line. Figure 7 shows a typical FEM applied for the dynamic analyses of the bridges. Three-node 3D beam elements, each with six degrees of freedom, were used to model all concrete beams. The modulus of elasticity E_c of the concrete was 3.45×10^4 MPa, the density ρ was 2500 kg/m^3, and the structural damping ratio was taken as 0.05. The 4 cm of concrete pavement and 18 cm of slab concrete formed a cross-section assigned to the girder elements. The rest of the 4 cm of concrete pavement and 10 cm of asphalt pavement were assigned as a uniform mass to the girder elements.

Table 1. Bridge and vehicle information.

Bridge Type	Span Length (m)	Span Number	Road Surface Condition	Vehicle Speed (km/h) (km/h)
Precast concrete	20	2	Smooth	40
continuous	30	3	Good	80
box-girder	40	4	Medium	120
bridge	—	—	Poor	—

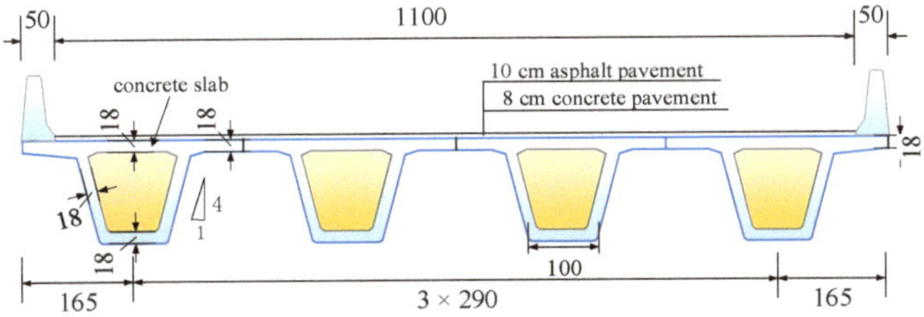

Figure 6. Typical cross-section of the bridge (unit: cm).

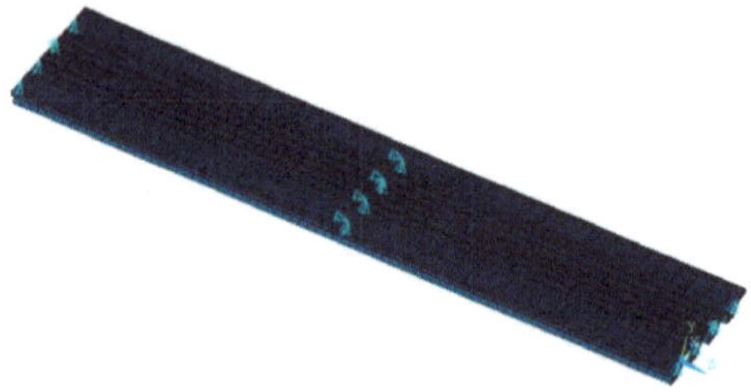

Figure 7. FEM of a 2 × 20 m continuous girder bridge.

The following assumptions were applied for the finite element modeling of the bridge: (1) the mass, stiffness, and damping characteristics of the bridge structure are uniformly distributed along the longitudinal direction; (2) the interaction between the bearing and the foundation is not taken into account; and (3) the torsional and distortional deformation of the beam are not taken into account.

The five-axle 3D vehicle with 15 degrees of freedom shown in Figure 8 was utilized for VBI analysis. Li et al. [40] stated the importance of 3D vehicle modeling in VBI simulations, which provides more realistic results compared to 2D models. The vehicle model consists of several rigid bodies connected by springs and dampers. More details of the vehicle can be found in [41].

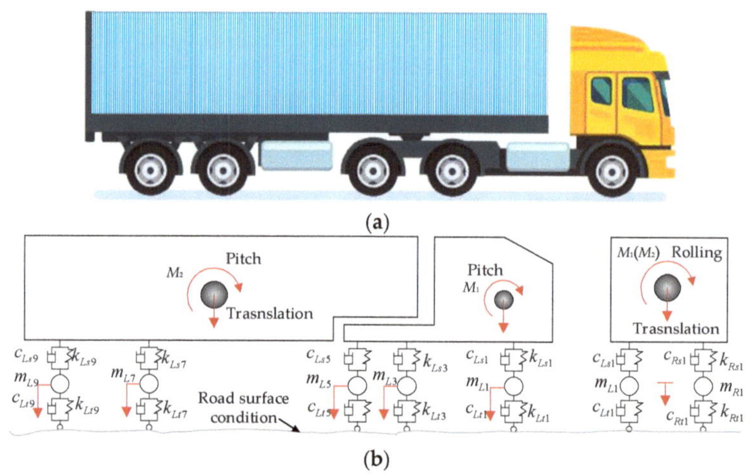

Figure 8. A 3D vehicle model with five axles: (**a**) prototype vehicle; (**b**) mechanical model.

As shown in Figure 8b, M is the mass of the vehicle body, K is the spring stiffness, C is the damping, m is the suspension mass, L represents the left-side wheel, R represents the right-side wheel, s represents the upper suspension, and t represents the lower tire.

The first fifty vibration orders of modes and corresponding frequencies of the bridge were obtained using the Lanczos method as shown in Figure 9. The bridge frequencies become denser as the span number and length increase; bridges with more spans or a longer span require a higher-order mode to achieve the same level of frequency. Then, the mode superposition method was used to obtain the total dynamic response and the dynamic components of the bridge in each mode. Figure 10 shows the dynamic deflection decomposition process of the first seven orders of modes of a 3 × 20 m continuous girder bridge. The percentage of the effect associated with each mode out of the total effect is defined as the modal contribution MPS_k. The calculation process is shown in the following equation [29]:

$$MPS_k = \frac{\sum\limits_{j=1}^{n_d} \sqrt{\sum\limits_{m=1}^{n_t} y_k^j(t)^2 / n_t}}{\sum\limits_{k=1}^{n_m} \sum\limits_{j=1}^{n_d} \sqrt{\sum\limits_{m=1}^{n_t} y_k^j(t)^2 / n_t}} \quad k = 1 \ \ to \ \ n_m, j = 1 \ \ to \ \ n_d \tag{30}$$

where y_k^j is the vertical modal displacement at structural node j for mode k; n_t is the total number of time steps; n_m is the number of selected modes; n_d is the number of structural nodes; k is the time step.

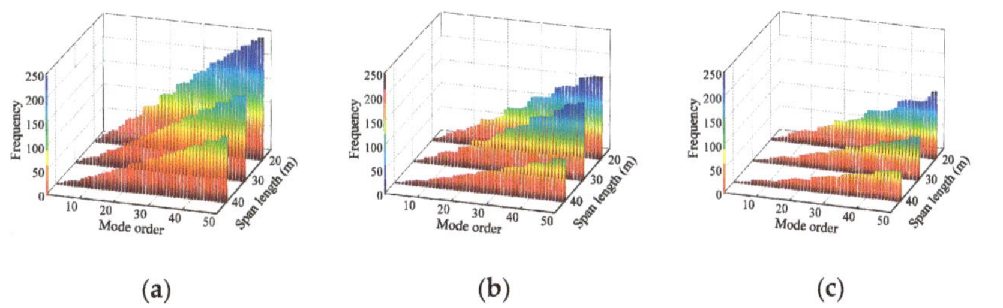

Figure 9. Bridge frequency: (**a**) two-span continuous girder bridges; (**b**) three-span continuous girder bridges; (**c**) four-span continuous girder bridges.

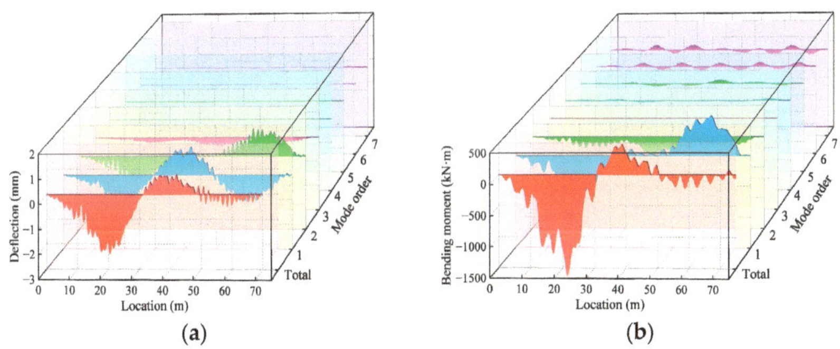

Figure 10. Decomposition of the total response: (**a**) deflection decomposition; (**b**) bending moment decomposition.

4.2. Comparison of Dynamic Amplification Factors

(1) *Road surface condition*

The RSC is an important factor affecting the VBI and has been studied intensively [30,42,43]. This section investigates the effect of the RSC and bridge mode coupling on the DAFs of the deflection and bending moment. Four RSCs of 'smooth', 'good', 'medium', and 'poor' were used in the VBI analysis. The vehicle passed over the three continuous box-girder bridges (2 × 20 m, 3 × 20 m, and 4 × 20 m) at a constant speed of 80 km/h, and the dynamic responses (deflection and bending moment) of the bridges were obtained by considering different numbers of modes. Finally, the deflection DAF, bending moment DAF, and RDM were determined by Equations (15) to (16).

Figures 11 and 12 present the variation in the DAFs of the deflection and bending moment with the modes and RSCs. The DAFs of the deflection and bending moment both increased with the RSC. For the three bridges, the maximum difference in the deflection DAF under the four RSCs is 0.32, while that in the bending moment DAF is 0.31. Therefore, periodic maintenance of the road surface can reduce the impact effect of vehicle loads on the bridge structure and improve vehicle occupants' traveling comfort and safety on the bridge structure. In Figure 13, the RDMs of the three bridges all first decrease and then stabilize with the modes under the influence of RSCs. The convergent RDMs of the three bridges under the influence of RSCs range from 1.02 to 1.06, 1.00 to 1.06, and 1.00 to 1.05, respectively, indicating that the deflection DAF is greater than the bending moment DAF at the same location.

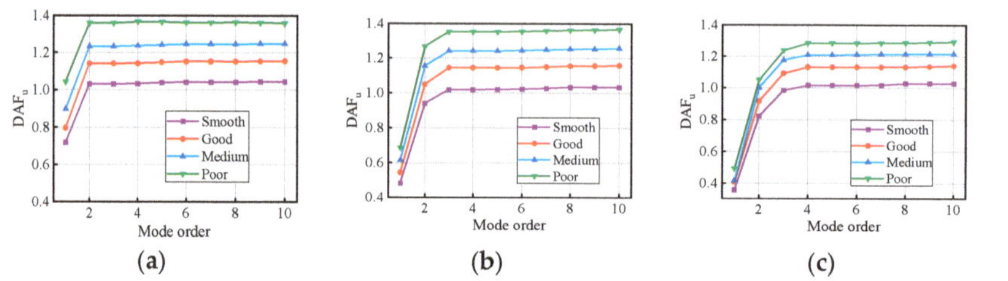

Figure 11. Deflection DAFs of the continuous girder bridges: (**a**) 2 × 20 m continuous girder bridge; (**b**) 3 × 20 m continuous girder bridge; (**c**) 4 × 20 m continuous girder bridge.

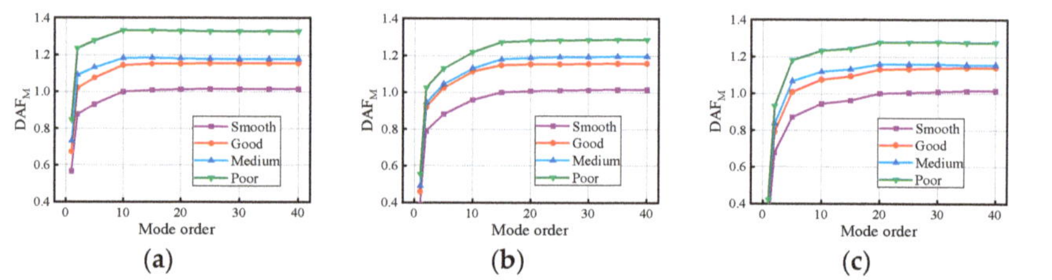

Figure 12. Bending moment DAFs of the continuous girder bridges: (**a**) 2 × 20 m continuous girder bridge; (**b**) 3 × 20 m continuous girder bridge; (**c**) 4 × 20 m continuous girder bridge.

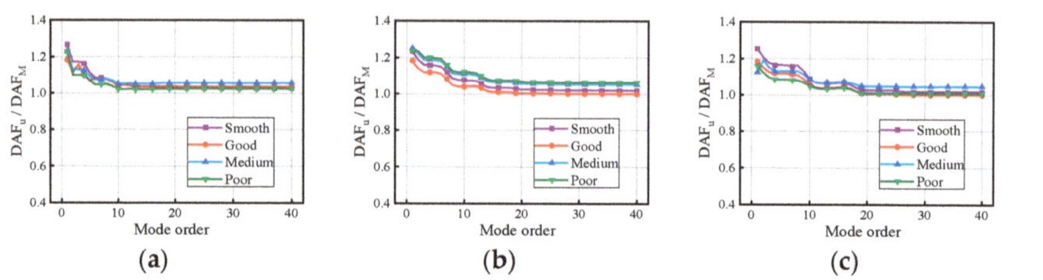

Figure 13. RDMs of the continuous girder bridges: (**a**) 2 × 20 m continuous girder bridge; (**b**) 3 × 20 m continuous girder bridge; (**c**) 4 × 20 m continuous girder bridge.

For the same RSC case, the deflection DAFs of the three bridges increase rapidly and then gradually stabilize with the vertical modes considered. The convergence mode (CM) is defined as the highest mode considered when the DAF first reaches stability. From Figures 11 and 12, it can be seen that the maximum CM for the deflection DAF is 10, while that for the bending moment DAF is 40 for all cases. The average contribution of the first two modes to the deflection DAF is more than 95%, while their average contribution to the bending moment DAF is only 90%. When the modes considered were increased from 10 to 20, the bending moment DAFs and deflection DAFs were increased by 4.1% to 6.7% and 0.4% to 0.6%, respectively. It can be seen that the higher-order modes have a greater influence on the dynamic moments compared to the dynamic deflections for all continuous girder bridges. A small bending moment DAF may be obtained when modes are poorly considered. The differences in mode contributions to the DAFs of the deflection and bending moment in the same bridge under 'smooth', 'good', 'medium', and 'poor' RSCs are all within 5%. The above results show that the RSC does not significantly change the contribution of each mode in the dynamic response, or the CM, but it significantly affects the dynamic responses and DAFs.

(2) *Vehicle Speed*

The effect of vehicle speed on DAFs has been demonstrated. Although many researchers have investigated the effect of vehicle speed on bridge response, this section focuses on the effect of vehicle speed on the CM and attempts to investigate the variation in the relationship between the deflection DAF and the bending moment DAF under the coupling of the vehicle speed and bridge modes. In the VBI analysis, the vehicle passed over the three bridges (2 × 20 m, 3 × 20 m, and 4 × 20 m continuous box-girder bridges) with a 'medium' RSC at a constant speed of 40 km/h, 80 km/h, and 120 km/h. The DAFs obtained from the dynamic deflection and dynamic bending moment are shown in Figures 14 and 15, respectively. Furthermore, the RDMs achieved by the combination of Figures 14 and 15 are presented in Figure 16.

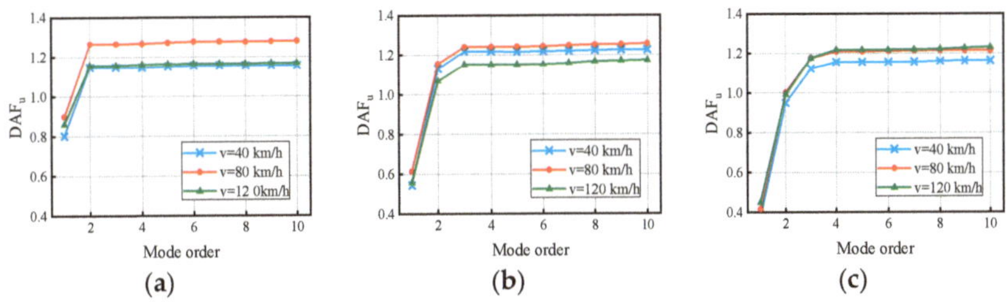

Figure 14. Deflection DAFs of the continuous girder bridges: (**a**) Results of the 2 × 20 m continuous girder bridge; (**b**) Results of the 3 × 20 m continuous girder bridge; (**c**) Results of the 4 × 20 m continuous girder bridge.

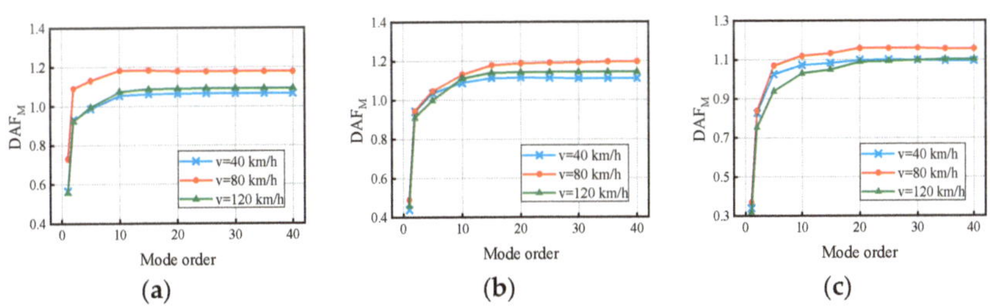

Figure 15. Bending moment DAFs of the continuous girder bridges: (**a**) Results of the 2 × 20 m continuous girder bridge; (**b**) Results of the 3 × 20 m continuous girder bridge; (**c**) Results of the 4 × 20 m continuous girder bridge.

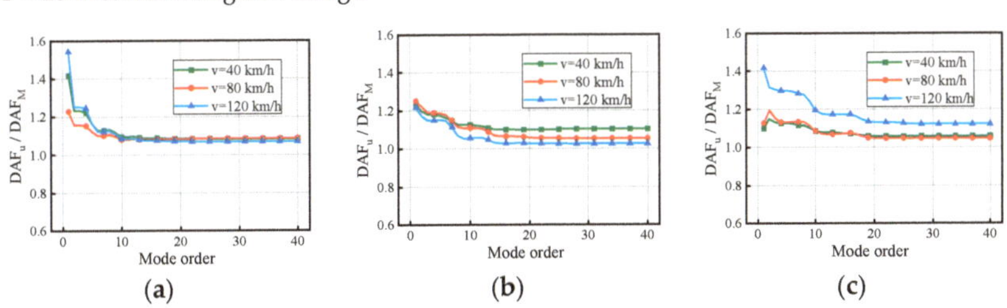

Figure 16. RDMs of the continuous girder bridges: (**a**) Results of the 2 × 20 m continuous girder bridge; (**b**) Results of the 3 × 20 m continuous girder bridge; (**c**) Results of the 4 × 20 m continuous girder bridge.

In Figures 15 and 16, the convergent deflection DAFs, bending moment DAFs, and RDMs of the bridges do not increase or decrease with vehicle speed. The differences in the deflection DAFs after considering the CM for the three bridges under the influence of vehicle speed are 0.18, 0.2, and 0.13, respectively, while those of the bending moment are 0.12, 0.08, and 0.07, respectively. The convergent RDMs of the three bridges under the influence of the RSCs range from 1.07 to 1.09, 1.03 to 1.10, and 1.05 to 1.12, respectively, indicating that the deflection DAF is greater than the bending moment DAF at the same location. The mechanism of influence between the vehicle speed and DAF has been proven to be complicated, and most scholars have attempted to interpret that in regard to the critical speed [19,30,44]. However, this is not the focus of discussion in this paper, so the mode-related results are discussed further.

For the same vehicle speed, the deflection DAFs and bending moment DAFs of the three bridges increase rapidly and then gradually stabilize with the vertical modes considered. Conversely, the RDMs of the three bridges first decrease and then stabilize with the modes under the effect of vehicle speed. The maximum CM of the deflection DAF influenced by vehicle speed for the three bridges is 10, while that of the bending moment DAF is 40. The average contribution of the first two modes to the deflection DAF is more than 95%, while the average contribution to the bending moment DAF is only 88% for all cases. When the modes considered were increased from 10 to 20, the bending moment DAFs and deflection DAFs were increased by 4.1% to 6.7% and 0.4% to 0.6%, respectively. For all continuous girder bridges, the higher-order modes have a greater influence on the dynamic moments compared to the dynamic deflections even when influenced by the vehicle speed. The maximum difference in the contribution of the first-order modes for the deflection DAF induced by various vehicle speeds is 6%, while that for the bending moment DAF is 11%. The above results indicate that the vehicle speed affects the contribution of each mode in the DAF to some extent, which, in turn, alters the DAF. Through Equations (15) and (16), it can be concluded that v/l and β_n are the major factors affecting the DAFs of the deflection and bending moment; vehicle speed exerts an impact on Ω_n, consequently influencing β_n, ultimately resulting in alterations to the contribution of each mode.

(3) *Span Length*

Many countries' design codes have defined the DAF as solely a function of the span length, such as the U.S. (1992), China (1989), New Zealand (2013), Europe (2003), and Japan (1996) [19]. The span is a significant structural factor affecting the DAFs. In the analysis, the vehicle passed over nine continuous box-girder bridges (2 × 20 m, 2 × 30 m, 2 × 40 m, 3 × 20 m, 3 × 30 m, 3 × 40 m, 4 × 20 m, 4 × 30 m, and 4 × 40 m) with a 'medium' RSC at a constant speed of 80 km/h. Bridge span changes can exacerbate the randomness of road surface conditions. For each bridge, 10 RSC samples were considered for analysis, and the final averaged DAFs and RDMs are plotted in Figure 17.

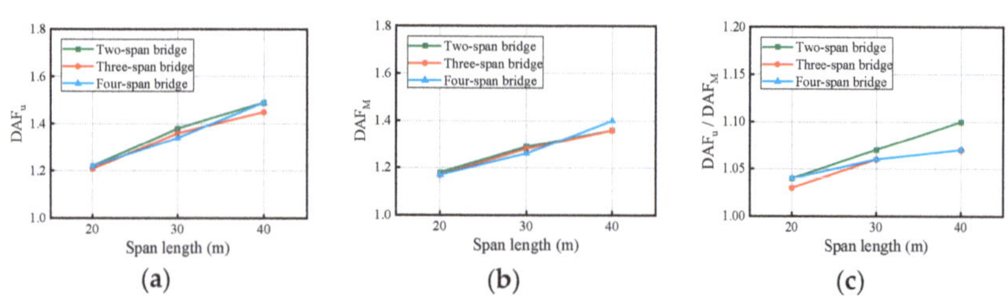

Figure 17. Bridge DAFs and RDMs influenced by span length: (**a**) deflection DAFs; (**b**) bending moment DAFs; (**c**) RDMs.

In Figure 17, the deflection DAF, bending moment DAF, and RDM all increase with the span length. The bridge RDMs for all cases range from 1.03 to 1.10, indicating that the deflection DAF is greater than the bending moment DAF.

Figures 18 and 19 show the variation in the deflection DAF and bending moment DAF with the mode for one of the samples, while Figure 20 shows the results for the corresponding RDMs. The deflection DAF and bending moment DAF of bridges with the same span numbers all converge gradually with the modes, while the bridge RDMs decrease with the mode. The effect of span length on the first-order modal contribution is the maximum at 13.9%, and its effect gradually decreases as the mode order increases. The maximum CM for the deflection DAF is 10, while that for the bending moment DAF is 40 for all cases; the average contribution of the first two modes to the deflection DAF is more than 80%, while the average contribution to the bending moment DAF is only 69.5%. When the modes considered were increased from 10 to 20, the bending moment DAFs and deflection DAFs were increased by 1.1% to 4.5% and 0.0% to 0.4%, respectively. This indicates that the higher-order modes have a greater influence on the bending moment compared to the dynamic deflection for all continuous girder bridges.

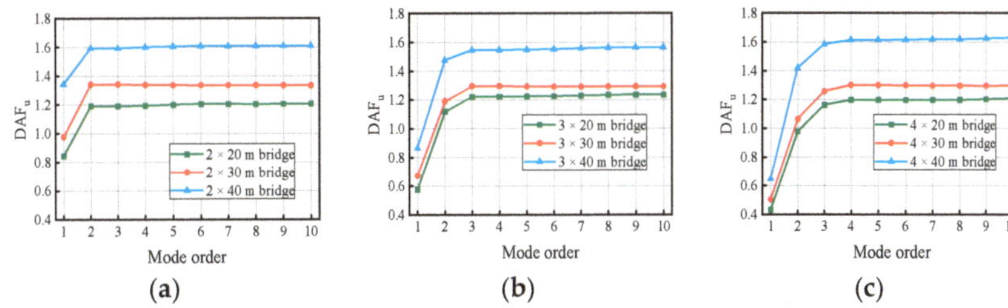

Figure 18. Deflection DAFs of the bridges: (**a**) DAFs for two-span continuous girder bridges; (**b**) DAFs for three-span continuous girder bridges; (**c**) DAFs for four-span continuous girder bridges.

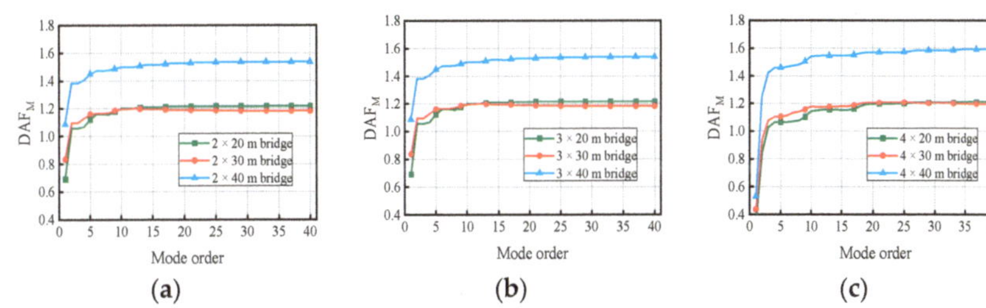

Figure 19. Bending moment DAFs of the bridges: (**a**) DAFs for two-span continuous girder bridges; (**b**) DAFs for three-span continuous girder bridges; (**c**) DAFs for four-span continuous girder bridges.

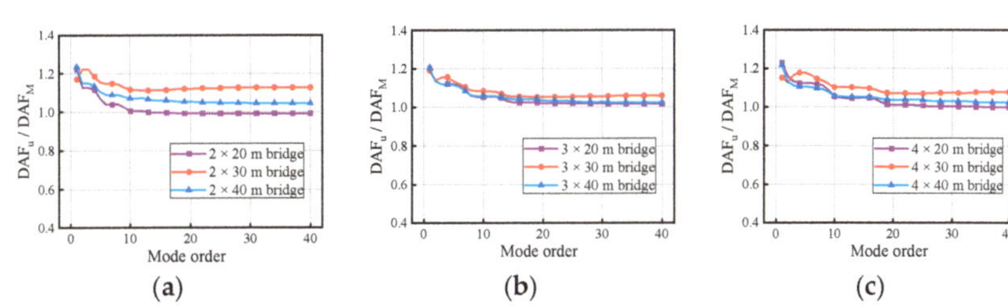

Figure 20. RDMs of the bridges: (**a**) RDMs for two-span continuous girder bridges; (**b**) RDMs for three-span continuous girder bridges; (**c**) RDMs for four-span continuous girder bridges.

(4) *Span Number*

Differences in the support conditions or span number could lead to differences in the response in the bridge even when subjected to the same loads. The data for bridges with the same spans in Figures 17–20 were further summarized to examine the interactive effects of span number and mode on bridge DAFs and the RDMs.

In Figure 21, the deflection DAF, bending moment DAF, and RDM show no clear relationship with the span number. The bridge deflection DAFs are all larger than the bending moment DAFs under the influence of the span number.

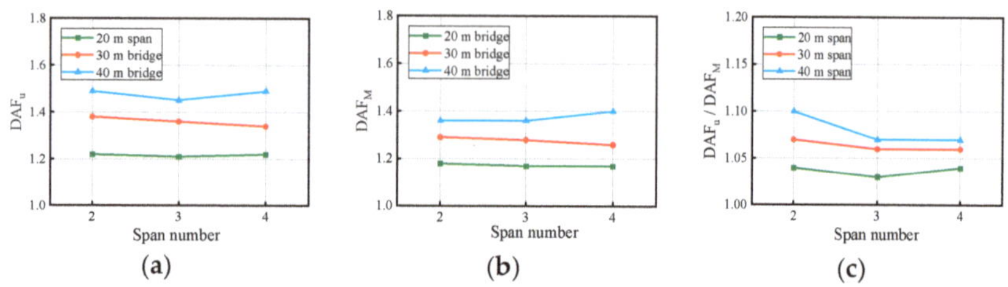

Figure 21. Bridge DAFs and RDMs influenced by span number: (**a**) deflection DAFs; (**b**) bending moment DAFs; (**c**) RDMs.

Figures 22 and 23 show the variation in the deflection DAF and bending moment DAF with the mode and span number for one of the samples. For the deflection DAF, two-span bridges, three-span bridges, and four-span bridges increase rapidly in the first two, first three, and first four orders of modes, respectively. In contrast, the bending moment DAFs of the bridges with two, three, and four spans increase rapidly in the first 10, 15, and 20 orders of mode, respectively, and then gradually stabilize. The CM of the deflection DAF gradually increases with the span number, e.g., the CMs of two-span, three-span, and four-span continuous girder bridges are of the 5th, 8th, and 10th orders, respectively. Meanwhile, for continuous girder bridges of four spans or less, the necessary CM of the bending moment DAF is 40.

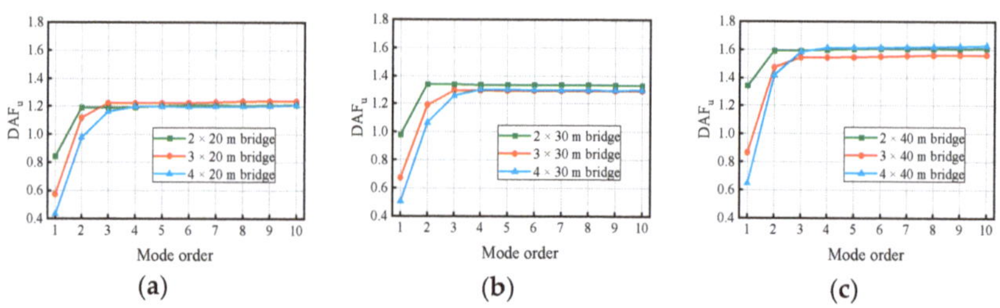

Figure 22. Deflection DAFs of the bridges: (**a**) DAFs for continuous girder bridges with 20 m spans; (**b**) DAFs for continuous girder bridges with 30 m spans; (**c**) DAFs for continuous girder bridges with 40 m spans.

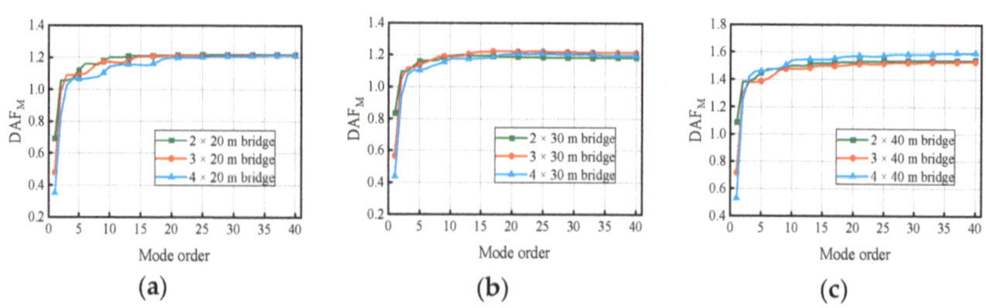

Figure 23. Bending moment DAFs of the bridges: (**a**) DAFs for continuous girder bridges with 20 m spans; (**b**) DAFs for continuous girder bridges with 30 m spans; (**c**) DAFs for continuous girder bridges with 40 m spans.

In addition, there is a 43.7% difference in the contribution of the first-order mode to the deflection DAF for the two-span bridge compared to the four-span bridge. The effect of span number on the other order modal contributions gradually decreases as the mode order increases. A phenomenon similar to the above was found for the bending moment DAF. The various spectral characteristics of the bridges are the major reason for the differences in the above conclusions. The contribution of the first three modes to the bending moment DAF is much smaller compared to that for the deflection DAF in continuous girder bridges, indicating that the bending moment DAF is influenced by higher-order modes. The RDM results in Figure 24, which are affected by the span number, have been discussed in terms of the span length factor.

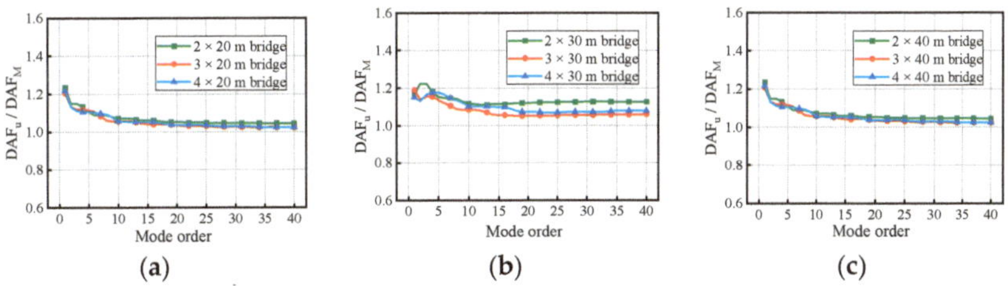

Figure 24. RDMs of the bridges: (**a**) RDMs for continuous girder bridges with 20 m spans; (**b**) RDMs for continuous girder bridges with 30 m spans; (**c**) RDMs for continuous girder bridges with 40 m spans.

5. Cut-Off Modes for the Deflection and Bending Moment

Four RSCs of 'smooth', 'good', 'medium', and 'poor', and three speeds of 40 km/h, 80 km/h, and 120 km/h were considered in the VBI analysis. Each bridge contains twelve cases. A total of 216 dynamic responses containing the deflections and bending moments of the nine precast continuous concrete box-girder bridges were obtained. After statistical analysis for these bridges, it was found that the RDMs were in the range of 1.00 to 1.12, and the deflection DAFs were always greater than the bending moment (strain) result. It is possible to obtain safer DAFs by adopting the deflection instead of the bending moment (strain) in the evaluation of a bridge's dynamic characteristics or the development of DAF criteria.

The contribution of each mode for each bridge was averaged over the twelve cases and visualized in Figure 25. It can be observed that the contribution of each mode to the deflection DAF or bending moment DAF varies depending on the type of bridge, and the span number also influences the dominant vibration modes. The cumulative contribution of modes to the deflection response for two-span, three-span, and four-span bridges can exceed 95% when considering the first two, first three, and first four orders of modes, respectively. By considering modes beyond the 10th order, all modal dynamic effects can be accounted for, enabling accurate results for the deflection DAF. It is worth noting that for an equivalent level of cumulative modal contribution, bending moments consistently require a greater number of modes.

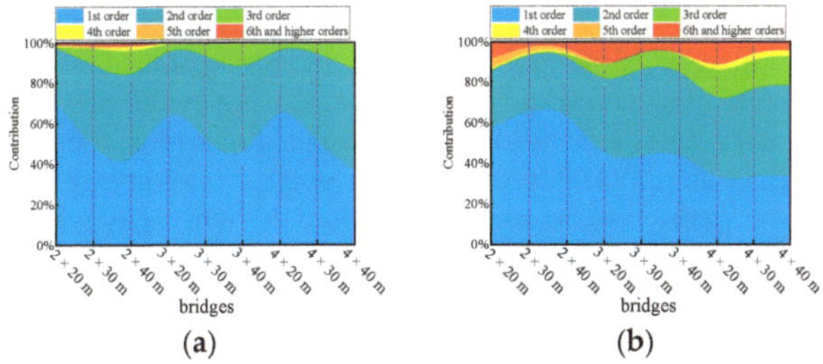

Figure 25. Mode participation for the continuous girder bridges: (**a**) mode participation for deflection; (**b**) mode participation for the bending moment.

To summarize, the dominant vibration modes of the deflection and bending moment for the mid-span section of a continuous girder bridge are of the first three orders. Currently, many national code provisions utilize the fundamental frequency to calculate DAFs [44–46]. However, when determining the deflection DAF or bending moment DAF for continuous girder bridges, it is crucial to consider the impact of modal contributions from the second and third orders, in addition to the fundamental frequency. In addition, two types of accuracy indexes are proposed in combination with practical engineering application requirements. The first type of accuracy index, P_1, is defined as a modal contribution of 95% and has a required minimum number of modes known as the cut-off mode K_1. The second type of accuracy index, P_2, is based on a modal contribution of 99% and has a cut-off mode of K_2. The first type of accuracy index is used for data filtering and noise elimination in field measurements without distortion. The second type of accuracy index is applied in VBI analysis to enhance solution efficiency.

After the statistical analysis, Table 2 provides the required cut-off modes for the accuracy indexes of the different bridge responses. The first type of accuracy index could be applied for data filtering and noise elimination in field measurements without distortion. It is worth noting that the cut-off mode can only eliminate high-frequency noise signals, while noise signals below the cut-off mode cannot be processed. The second type of accuracy index is utilized in VBI analysis to improve the efficiency of the solution. The mode superposition method, which does not require validation, is characterized by increased efficiency in VBI analysis when fewer modes are considered. Additional experimentation is required to validate the usefulness and effectiveness of the cut-off frequency.

Table 2. Cut-off mode orders for the DAFs of the deflection and bending moment.

Cut-Off Mode Order	Deflection			Bending Moment		
	Two-Span	Three-Span	Four-Span	Two-Span	Three-Span	Four-Span
K_1	3	3	4	7	8	9
K_2	6	7	8	15	20	25

6. Experimental Investigation

Two prestressed concrete continuous box-girder bridges, measuring 4 × 30 m and 4 × 20 m, were chosen for dynamic load tests on the Anlan Expressway in Shaanxi Province, China. The width of both bridges is 8.5 m. Details of the bridge sections can be found in the *Chinese General Bridge Atlas*. Strain measurements were performed using resistive, full-bridge circuit strain gauges with a double cantilever configuration. The measuring gauge length was 80 mm. These sensors have the advantages of a stable measurement value, high precision, high sensitivity, strong anti-interference ability, self-temperature compensation, wide adaptability, and easy installation and disassembly. Dial gauges with a resolution of 0.01 mm were used for deflection measurement. For the first bridge, strain gauges were applied to the bottom plate of the center girder at the mid-span of the side span, while the second bridge had strain gauges placed on the secondary side span. A visual inspection of the concrete in the bottom slab of the main girder was carried out before the measurement, and no apparent cracks were found. Suspension hammer systems including dial gauges were additionally installed near the strain gauges to measure the dynamic deflection. Furthermore, a piezoelectric accelerometer was attached to the bridge deck in the vicinity of the mid-span section; the sensitivity was 200 mV/g. A 36-ton truck with three axles passed over the bridge at a constant speed along the center of the bridge. Data were recorded using DASYLAB software. The sampling frequency was 200 Hz, and the measurement details are shown in Figure 26.

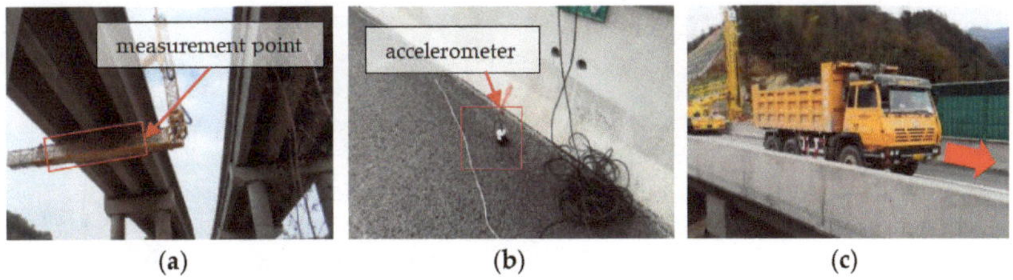

Figure 26. Field photographs: (**a**) deflection and strain sensor installation; (**b**) accelerometer arrangement; (**c**) dynamic test.

The static deflection and static bending moment were obtained by applying low-pass filtering [47,48]. In Figures 27 and 28, the calculated deflection DAFs for the two bridges were 1.13 and 1.11, while the strain (bending moment) DAFs were 1.09 and 1.06, respectively. The RDMs for the two bridges were 1.04 and 1.05, respectively, indicating that the deflection DAF at the mid-span location of the bridges was greater than the strain (bending moment) results. Only the first three orders of modes were obtained from the acceleration signals and were used to modify the finite element model. The higher modes can be extracted from the modified FE model, such as cut-off modes. To filter out high-frequency noise, a low-pass filter with a cut-off mode of K_1 was applied. For deflection, the K_1 order modal frequencies of the two bridges are 5.73 Hz and 9.8 Hz, while the frequencies for the bending moments are 16.5 Hz and 27.8 Hz, respectively. The deflection DAFs of the two bridges increased by 0.1% and 0.3% after noise elimination, while the strain DAF remained unchanged. Although the effect of noise elimination is not apparent due to the presence of a small noise component in the measured signal, it does highlight the effectiveness of the cut-off modal frequency in preserving signal fidelity.

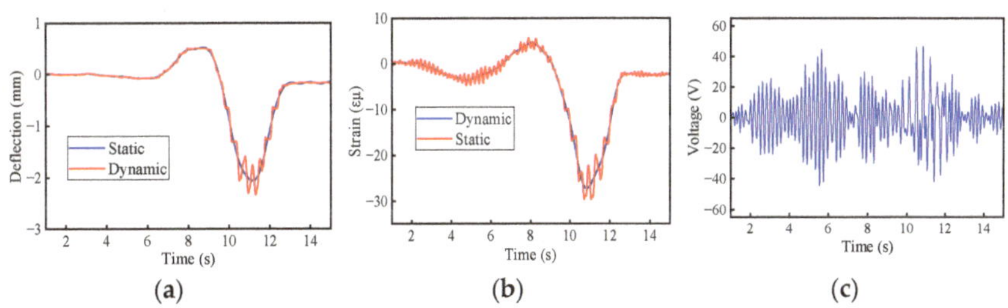

Figure 27. Measured dynamic responses for the 4 × 30 m bridge: (**a**) deflection; (**b**) bending moment; (**c**) acceleration.

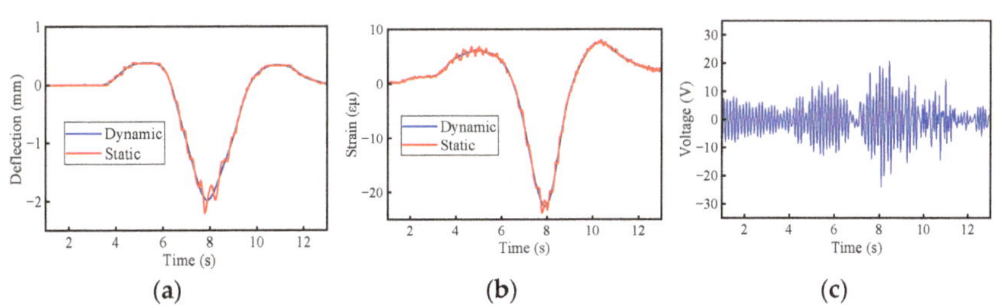

Figure 28. Measured dynamic responses for the 4 × 20 m bridges: (**a**) deflection; (**b**) bending moment; (**c**) acceleration.

7. Conclusions

In this paper, the association of the deflection DAF, bending moment DAF, and RDM with modes under multi-factor influence was investigated by numerical analyses. Two types of DAF accuracy indexes were proposed for VBI analysis and signal noise elimination. Finally, the relationship between the deflection DAF and bending moment DAF was verified by an experimental study, and the utility of cut-off modes in signal processing was demonstrated. The following conclusions were obtained: The deflection DAF and bending moment DAF of a continuous beam are related to the cross-sectional position and static load factor and are affected by the frequency ratio, damping ratio, vehicle speed, and span length. The number of modes also affects the DAF accuracy; the more modes are considered, the closer the result is to the analytical solution.

(1) Under varying conditions of travel speed, RSC, span length, and span number, the deflection DAF and positive bending moment DAF of the mid-span section of the continuous girder bridge increased rapidly with the modes and then stabilized. Under the influence of the above factors, the RDM decreased gradually with the number of modes but was greater than 1, indicating that the deflection DAF is always greater than the bending moment DAF.
(2) The span length, RSC, vehicle speed, and span number have strong, strong, complicated, and weak associations with DAF, respectively. The RSC did not significantly change the contribution of each mode to the DAF and CM, but it had a significant effect on the bridges' dynamic response and DAF. The vehicle speed altered the dynamic response, the DAF, and the contribution of each mode to the response. The span length had a greater effect on the modal contribution than did the span number.
(3) The dominant vibration modes of continuous girder bridges were of the first two to four orders, and it is unreasonable to calculate the bridge deflection DAF and bending moment DAF by considering only the fundamental frequency. For precast continuous concrete box-girder bridges with four spans or less, the deflection DAF and bending moment DAF could be obtained with 95% accuracy after considering the first four orders and the first nine orders of modes, respectively.
(4) In the field experiment, the RDMs of the mid-span section for the 4 × 30 m and 4 × 20 m continuous bridges were 1.04 and 1.05, respectively, demonstrating that the deflection DAF of the mid-span section of the bridges was greater than the strain (bending moment) results. After filtering at the cut-off mode frequency corresponding to the 95% accuracy index was considered, the maximum variation of the deflection DAF and bending moment DAF was 0.3%, proving the effectiveness of the cut-off mode in preserving signal fidelity.
(5) In this paper, box-girder bridges were investigated. For precast bridges with other cross-sections (T-beam or hollow slab), the above conclusions are also applicable when the modes exhibit similar characteristics. Bridges with various structural types (skewed or curved) need further research. A comparative analysis was conducted on various parameters under the dynamic load test conditions of a single vehicle on the bridge. The vehicle number and driving position may have a greater influence on the results of the cut-off frequency compared to the bridge deck width, girder section, and girder number and, therefore, also need to be further investigated.

Author Contributions: Conceptualization, Y.W.; Methodology, Y.W. and Y.Z. (Yang Zhao); Software, Y.W.; Validation, Y.X.; Formal analysis, Y.W. and Y.Z. (Yang Zhao); Investigation, Y.Z.(Yongjun Zhou) and Y.X.; Resources, Y.Z. (Yongjun Zhou) and W.F.; Data curation, Y.W., Y.Z. (Yang Zhao) and W.F.; Writing—original draft preparation, Y.W.; Writing—review and editing, Y.Z. (Yongjun Zhou), Y.Z. (Yu Zhao) and Y.X.; Visualization, Y.W. and W.F.; Supervision, Y.Z. (Yongjun Zhou) and Y.Z. (Yu Zhao); Project administration, Y.Z. (Yongjun Zhou) and Y.Z. (Yu Zhao); Funding acquisition, Y.Z. (Yongjun Zhou) and Y.W. All authors have read and agreed to the published version of the manuscript.

Funding: This research was funded by the National Natural Science Foundation of China, grant number [51978063, 52278138]. And The APC was funded by [51978063].

Institutional Review Board Statement: Not applicable.

Informed Consent Statement: Not applicable.

Data Availability Statement: Data are contained within the article.

Conflicts of Interest: Author Yang Zhao was employed by the company Shaanxi Provincial Transport Planning Design and Research Institute Co., Ltd. and Author Wei Feng was employed by the company Xi'an Municipal Engineering Design and Research Institute Co., Ltd. The remaining authors declare that the research was conducted in the absence of any commercial or financial relationships that could be construed as a potential conflict of interest.

References

1. Pieraccini, M.; Miccinesi, L.; Abdorazzagh Nejad, A.; Naderi Nejad Fard, A. Experimental dynamic impact factor assessment of railway bridges through a radar interferometer. *Remote Sens.* **2019**, *11*, 2207. [CrossRef]
2. Zhou, Y.J.; Ma, Z.J.; Zhao, Y.; Shi, X.W.; He, S.H. Improved definition of dynamic load allowance factor for highway bridges. *Struct. Eng. Mech.* **2015**, *54*, 561–577. [CrossRef]
3. Simoncelli, M.; Aloisio, A.; Zucca, M.; Venturi, G.; Alaggio, R. Intensity and location of corrosion on the reliability of a steel bridge. *J. Constr. Steel Res.* **2023**, *206*, 107937. [CrossRef]
4. González, A.; Cantero, D.; OBrien, E.J. Dynamic increment for shear force due to heavy vehicles crossing a highway bridge. *Comput. Struct.* **2011**, *89*, 2261–2272. [CrossRef]
5. Pourzeynali, S.; Zhu, X.; Ghari Zadeh, A.; Rashidi, M.; Samali, B. Comprehensive study of moving load identification on bridge structures using the explicit form of Newmark-β method: Numerical and experimental studies. *Remote Sens.* **2021**, *13*, 2291. [CrossRef]
6. Yu, Y.; Deng, L.; Wang, W.; Cai, C.S. Local impact analysis for deck slabs of prestressed concrete box-girder bridges subject to vehicle loading. *J. Vib. Control* **2017**, *23*, 31–45. [CrossRef]
7. O'Connor, C.; Pritchard, R.W. Impact studies on small composite girder bridge. *J. Struct. Eng.* **1985**, *111*, 641–653. [CrossRef]
8. Aluri, S.; Jinka, C.; Gangarao, H.V. Dynamic response of three fiber reinforced polymer composite bridges. *J. Bridge Eng.* **2005**, *10*, 722–730. [CrossRef]
9. Mohseni, I.; Ashin, A.; Choi, W.; Kang, J. Development of dynamic impact factor expressions for skewed composite concrete-steel slab-on-girder bridges. *Adv. Mater. Sci. Eng.* **2018**, *2018*, 4313671. [CrossRef]
10. Szurgott, P.; Wekezer, J.; Kwasniewski, L.; Siervogel, J.; Ansley, M. Experimental assessment of dynamic responses induced in concrete bridges by permit vehicles. *J. Bridge Eng.* **2011**, *16*, 108–116. [CrossRef]
11. Li, H.; Wekezer, J.; Kwasniewski, L. Dynamic response of a highway bridge subjected to moving vehicles. *J. Bridge Eng.* **2008**, *13*, 439–448. [CrossRef]
12. Huang, D.Z. Dynamic analysis of steel curved box girder bridges. *J. Bridge Eng.* **2001**, *6*, 506–513. [CrossRef]
13. Majka, M.; Hartnett, M. Effects of speed, load and damping on the dynamic response of railway bridges and vehicles. *Comput. Struct.* **2008**, *86*, 556–572. [CrossRef]
14. Yang, Y.B.; Liao, S.S.; Lin, B.H. Impact formulas for vehicles moving over simple and continuous beams. *J. Struct. Eng.* **1995**, *121*, 1644–1650. [CrossRef]
15. Fafard, M.; Laflamme, M.; Savard, M.; Bennur, M. Dynamic analysis of existing continuous bridge. *J. Bridge Eng.* **1998**, *3*, 28–37. [CrossRef]
16. Senthilvasan, J.; Thambiratnam, D.; Brameld, G. Dynamic response of a curved bridge under moving truck load. *Eng. Struct.* **2002**, *24*, 1283–1293. [CrossRef]
17. Huang, P.; Wang, J.; Han, W.S.; Yuan, Y.G. Study on impact factors of small-and medium-span bridges under the special-purpose vehicle load. *Structures* **2022**, *43*, 606–620. [CrossRef]
18. Zeng, H.; Bert, C.W. Dynamic amplification of bridge/vehicle interaction: A parametric study for a skewed bridge. *Int. J. Struct. Stab. Dyn.* **2003**, *3*, 71–90. [CrossRef]
19. Deng, L.; Yu, Y.; Zou, Q.L.; Cai, C.S. State-of-the-art review of dynamic impact factors of highway bridges. *J. Bridge Eng.* **2014**, *20*, 04014080. [CrossRef]
20. Kemal, A.; Talukdar, S. Evaluation of impact factor parameters for T-girder concrete bridge considering multi-lane vehicle movement. *Innov. Infrastruct. Solut.* **2022**, *7*, 220. [CrossRef]
21. Law, S.S.; Zhu, X.Q. Bridge dynamic responses due to road surface roughness and braking of vehicle. *J. Sound Vibr.* **2005**, *282*, 805–830. [CrossRef]
22. Yang, Y.B.; Wu, Y.S. A versatile element for analyzing vehicle-bridge interaction response. *Eng. Struct.* **2001**, *23*, 452–469. [CrossRef]
23. Humar, J.L.; Kashif, A.M. Dynamic response of bridges under travelling loads. *Can. J. Civ. Eng.* **1993**, *20*, 287–298. [CrossRef]
24. Zhou, Y.J.; Xue, Y.X.; Gao, X.J.; Li, R.R.; Wang, Y.L.; Zhao, Y. Research on dynamie amplification factor of highway simply supportedgirder bridge based on modal superposition method. *J. Traffic Transp. Eng.* **2023**, *23*, 146–155.
25. Zou, Q.; Deng, L.; Guo, T.; Yin, X. Comparative study of different numerical models for vehicle-bridge interaction analysis. *Int. J. Struct. Stab. Dyn.* **2016**, *16*, 1550057. [CrossRef]
26. Li, S.; Ren, J. Investigation on three-directional dynamic interaction between a heavy-duty vehicle and a curved bridge. *Adv. Struct. Eng.* **2018**, *21*, 721–738. [CrossRef]

27. Deng, L.; He, W.; Shao, Y. Dynamic impact factors for shear and bending moment of simply supported and continuous concrete girder bridges. *J. Bridge Eng.* **2015**, *20*, 04015005. [CrossRef]
28. Wang, Y.L.; Tian, J.; Zhou, Y.J.; Zhao, Y.; Feng, W.; Mao, K.Q. Assessing dynamic load allowance of the negative bending moment in continuous girder bridges by weighted average method. *Coatings* **2022**, *12*, 1233. [CrossRef]
29. Shao, Y.F.; Brownjohn, J.M.W.; Miao, C.Q.; Wang, M.M. A precise time-integration linear vehicle-bridge interaction method and dynamic sensitivity analysis. *Structures* **2021**, *33*, 4596–4603. [CrossRef]
30. Shi, X.; Cai, C.S.; Chen, S. Vehicle induced dynamic behavior of short-span slab bridges considering effect of approach slab condition. *J. Bridge Eng.* **2008**, *13*, 83–92. [CrossRef]
31. Huang, D.Z. Vehicle-induced vibration of steel deck arch bridges and analytical methodology. *J. Bridge Eng.* **2012**, *17*, 241–248. [CrossRef]
32. Park, Y.S.; Shin, D.K.; Chung, T.J. Influence of road surface roughness on dynamic impact factor of bridge by full-scale dynamic testing. *Can. J. Civ. Eng.* **2005**, *32*, 825–829. [CrossRef]
33. Broquet, C.; Bailey, S.F.; Fafard, M.; Brühwiler, E. Dynamic behavior of deck slabs of concrete road bridges. *J. Bridge Eng.* **2004**, *9*, 137–146. [CrossRef]
34. Parida, S.; Talukdar, S. An insight to the dynamic amplification factor for steel truss girder bridge. *Int. J. Steel Struct.* **2020**, *20*, 1341–1354. [CrossRef]
35. Aloisio, A.; Alaggio, R. Probabilistic comparative analysis of vehicle-bridge interaction models for predicting bridge response under moving vehicles. *J. Eng. Mech.* **2024**, *150*, 04023121. [CrossRef]
36. Ma, L.; Zhang, W.; Han, W.S.; Liu, J.X. Determining the dynamic amplification factor of multi-span continuous box girder bridges in highways using vehicle-bridge interaction analyses. *Eng. Struct.* **2019**, *181*, 47–59. [CrossRef]
37. Mo, S.; Chen, K.; Huang, Z.; Zhang, W. Study on the spatial system dynamics of long-span arch bridge interacting with vehicles subjected to multiple vehicle loads. *J. Vib. Eng. Technol.* **2023**, *6*, 1–15. [CrossRef]
38. Shen, H.M.; Xiao, X.B. Numerical method for vehicle-bridge coupled vibrations. *J. Southwest Jiao Tong Univ.* **2003**, *38*, 658–662.
39. Tan, G.H.; Brameld, G.H.; Thambiratnam, D.P. Development of an analytical model for treating bridge-vehicle interaction. *Eng. Struct.* **1998**, *20*, 54–61. [CrossRef]
40. Li, J.A.; Feng, D. A comparative study of vehicle-bridge interaction dynamics with 2D and 3D vehicle models. *Eng. Struct.* **2023**, *292*, 116493. [CrossRef]
41. Deng, L.; Duan, L.L.; He, W.; Ji, W. Study on vehicle model for vehicle-bridge coupling vibration of highway bridges in China. *China J. Highw. Transp.* **2018**, *31*, 92–100.
42. Gagarin, N.; Goulias, D.; Mekemson, J. Condition rating of bridge decks with fuzzy sets modeling for SF-GPR surveys. *Remote Sens.* **2023**, *15*, 3631. [CrossRef]
43. Wang, B.; Xia, C.; Min, Q.; Zhu, S.; Li, Y. Dynamic amplification factors of an arch bridge under random traffic flows. *Int. J. Struct. Stab. Dyn.* **2023**, *3*, 2450064. [CrossRef]
44. Moghimi, H.; Ronagh, H.R. Impact factors for a composite steel bridge using non-linear dynamic simulation. *Int. J. Impact Eng.* **2008**, *35*, 1228–1243. [CrossRef]
45. Yin, X.; Liu, Y.; Kong, B.; Chen, S. Experimental and numerical studies of a damaged bridge considering stochastic traffic flows and road roughness. *Int. J. Struct. Stab. Dyn.* **2017**, *17*, 1750089. [CrossRef]
46. Ling, T.; Deng, L.; He, W.; Cao, R.; Zhong, W. Determination of dynamic amplification factors for small-and medium-span highway bridges considering the effect of automated truck platooning loads. *Mech. Syst. Signal Proc.* **2023**, *204*, 110812. [CrossRef]
47. Ashebo, D.B.; Chan, T.H.; Yu, L. Evaluation of dynamic loads on a skew box girder continuous bridge Part II: Parametric study and dynamic load factor. *Eng. Struct.* **2007**, *29*, 1064–1073. [CrossRef]
48. Beben, D. Dynamic amplification factors of corrugated steel plate culverts. *Eng. Struct.* **2013**, *46*, 193–204. [CrossRef]

remote sensing

Article

High-Precision Monitoring Method for Bridge Deformation Measurement and Error Analysis Based on Terrestrial Laser Scanning

Yin Zhou [1], Jinyu Zhu [2], Lidu Zhao [2], Guotao Hu [2], Jingzhou Xin [2,*], Hong Zhang [1] and Jun Yang [1]

1 State Key Laboratory of Mountain Bridge and Tunnel Engineering, Chongqing Jiaotong University, Chongqing 400074, China; zhouyin@cqjtu.edu.cn (Y.Z.); hongzhang@cqjtu.edu.cn (H.Z.); yangjun@cqjtu.edu.cn (J.Y.)
2 School of Civil Engineering, Chongqing Jiaotong University, Chongqing 400074, China; jinyuzhu@mails.cqjtu.edu.cn (J.Z.); zhaolidu@cqjtu.edu.cn (L.Z.); guotaohu@mails.cqjtu.edu.cn (G.H.)
* Correspondence: xinjz@cqjtu.edu.cn

Abstract: In bridge structure monitoring and evaluation, deformation data serve as a crucial basis for assessing structural conditions. Different from discrete monitoring points, spatially continuous deformation modes provide a comprehensive understanding of deformation and potential information. Terrestrial laser scanning (TLS) is a three-dimensional deformation monitoring technique that has gained wide attention in recent years, demonstrating its potential in capturing structural deformation models. In this study, a TLS-based bridge deformation mode monitoring method is proposed, and a deformation mode calculation method combining sliding windows and surface fitting is developed, which is called the SWSF method for short. On the basis of the general characteristics of bridge structures, a deformation error model is established for the SWSF method, with a detailed quantitative analysis of each error component. The analysis results show that the deformation monitoring error of the SWSF method consists of four parts, which are related to the selection of the fitting function, the density of point clouds, the noise of point clouds, and the registration accuracy of point clouds. The error caused by point cloud noise is the main error component. Under the condition that the noise level of point clouds is determined, the calculation error of the SWSF method can be significantly reduced by increasing the number of points of point clouds in the sliding window. Then, deformation testing experiments were conducted under different measurement distances, proving that the proposed SWSF method can achieve a deformation monitoring accuracy of up to 0.1 mm. Finally, the proposed deformation mode monitoring method based on TLS and SWSF was tested on a railway bridge with a span of 65 m. The test results showed that in comparison with the commonly used total station method, the proposed method does not require any preset reflective markers, thereby improving the deformation monitoring accuracy from millimeter level to submillimeter level and transforming the discrete measurement point data form into spatially continuous deformation modes. Overall, this study introduces a new method for accurate deformation monitoring of bridges, demonstrating the significant potential for its application in health monitoring and damage diagnosis of bridge structures.

Keywords: deformation monitoring; laser scanning; point cloud; error modeling; deformation modes

Citation: Zhou, Y.; Zhu, J.; Zhao, L.; Hu, G.; Xin, J.; Zhang, H.; Yang, J. High-Precision Monitoring Method for Bridge Deformation Measurement and Error Analysis Based on Terrestrial Laser Scanning. *Remote Sens.* **2024**, *16*, 2263. https://doi.org/10.3390/rs16132263

Academic Editor: Susana Lagüela López

Received: 17 May 2024
Revised: 17 June 2024
Accepted: 18 June 2024
Published: 21 June 2024

1. Introduction

Bridge deformation is an important index to evaluate the safety of structures in bridge structural health monitoring, which includes the response to structural degradation and even damage. Extensive evidence supports that the stiffness decay degree of a structure can be evaluated, and the finite element model can be updated by periodically monitoring the deformation amplitude at key points [1,2]. Structural damage can be further located and quantified based on high-precision deformation modes [3,4]. Therefore, the efficient and accurate deformation mode identification of a structure is an important objective of bridge structural health monitoring, including identifying the deformation amplitude and deformation form at any position [3].

A linear variable differential transducer and clock gauge are commonly used sensors in bridge displacement monitoring, with a measurement accuracy of 0.001 mm. However, their use in long-span bridges is limited, owing to the requirement of stable support, such as temporary support [5]. The total station method is the most commonly used noncontact precision measurement method in engineering surveying. The measurement accuracy of the total station can reach the millimeter level, such as in the case of Leica TS30 with a measurement accuracy of 1 mm + 1 ppm. However, this requires the installation of reflective markers, such as prisms and reflective sheets, on the measuring points, thereby limiting the number of measuring points. Leveling is a method with higher accuracy than total station deformation measurement. The widely used static leveling-based connected pipe method has submillimeter accuracy for elevation deformation measurement, but the measuring points are fixed [6,7]. Moreover, the methods for fixed measuring points include the Global Navigation Satellite System (GNSS), acceleration data, and strain displacement mapping. GNSS provides reliable absolute displacement monitoring, with a sample frequency of up to 100 Hz, and is frequently used in long-span flexible bridges [8–10]. The dynamic accuracy of GNSS is at the centimeter level, whereas its static accuracy is at the millimeter level [11]. The acceleration integration method integrates the acceleration of bridge vibration into the displacement, which is an effective nonreference dynamic displacement monitoring method. Nevertheless, this method is not suitable for low-frequency and static deformation [12–14]. For bridges with simple cross-sections, a clear theoretical mapping relationship exists between strain and deflection [15,16]. However, the material and size of the actual bridges may be different from the design, and modifying the parameters in the initial monitoring period is necessary. Recently, a novel study used strain combined with acceleration to reconstruct the low- and high-frequency vibration displacements of bridges [17]. However, previous studies have used monitoring methods with fixed measuring points, and obtaining deformation modes composed of dense measuring points requires installing multiple sensors.

The vision-based method is a potential noncontact deformation monitoring technology, including the image-assisted total station method [18] and the photogrammetry method [19]. This method has the advantages of low cost and flexible work. The monitoring accuracy is mainly determined by the spatial dimensions corresponding to individual pixels [19]. The image coverage and monitoring accuracy in this method are contradictorily limited by camera hardware. Therefore, vision-based methods are mainly oriented toward the deformation of local areas or key points in a long-span bridge [20–24]. The structure should be divided into multiple monitoring windows for simultaneous monitoring to achieve geometric continuity through image stitching to achieve a high monitoring accuracy and improve the monitoring range [25]. The large size of the actual bridge structure results in a considerable disparity in the mapping relationship among pixels in different monitoring windows and physical objects [26]. Therefore, the image-based global deformation monitoring method requires establishing a mapping relationship in advance, which is very difficult in practical application. In addition, vision-based methods mainly monitor two-dimensional displacements in the plane because deformations in the depth direction are almost imperceptible in the image at long distances.

In recent years, three-dimensional (3D) laser scanning technology, which has emerged in the field of intelligent construction, has demonstrated the ability to capture 3D shapes [27]. Terrestrial laser scanning (TLS) is an important branch of 3D laser scanning, which is used to obtain high-precision 3D coordinates by placing the scanner on the ground. TLS is widely used in construction progress supervision, asset management, and reverse modeling owing to its flexible and fast data acquisition capabilities [28–31]. In terms of rapid and smart construction, TLS has been developed for checking the size of prefabricated components, spatial interference inspection, and virtual trial assembly [32–34]. Recently, TLS (Leica P50) has been used in the state assessment of long-span suspension bridges, showing great potential for engineering applications [35,36]. TLS is used to monitor the local deformation, tilt, and stability of historical buildings during deformation monitoring. This method can attain subcentimeter accuracy in local deformation monitoring [37,38]. Lohmus compared the test results for TLS (Leica C10) and the precision levels in the load tests of two small bridges, Loobu and Tartu. The error values of TLS in the two bridges were 3.4 and 0.8 mm, respectively [39]. Soni et al. used TLS (Focus 3D) to monitor the displacement of a stone arch bridge and compared it with close-range photogrammetry, with a deviation of 2.5 mm between the two different methods [40]. Truong-Hong et al. used TLS (ToF scanner) to monitor the dynamic load deflection of an underground passageway bridge, achieving a monitoring accuracy of 1 mm [41]. Park et al. developed a framework for TLS monitoring deformation to evaluate structural stress, and the deformation accuracy reached within 1 mm in an indoor experimental test [42]. The existing methods of using TLS to monitor bridge deformation with a small span can reach the millimeter level [43]. However, further research is needed to improve the accuracy of deformation monitoring using TLS to the submillimeter level in long-span bridges.

Structural deformation from point clouds can be calculated using three main data processing methods: (1) direct analysis based on point clouds, (2) feature extraction methods, and (3) model reconstruction methods. The basic principle of the direct analysis method is to calculate the spatial distance between neighborhoods from two epoch data. The advantage of this method is that feature extraction or model reconstruction are not necessary, making this method suitable for any scene. However, considering the presence of noise, the deformation obtained by this method also contains considerable noise [44]. The method based on feature extraction is widely applicable to structures with typical geometric features, such as frame buildings and bridges. Geometric corners, edges, and centerlines are the main extracted features [45]. Standard templates, such as right angles and circles, are usually used to detect and extract features [46]. Considering construction errors, differences may exist between the actual cross-section of the components and the standard cross-section, which can lead to errors in the extracted features. The model reconstruction method involves first reconstructing a 3D model of the structural surface based on point clouds and then calculating the distance between faces. The surface reconstruction methods can be further divided into the triangular mesh method [47] and the surface fitting method [48]. The triangular mesh method is simple and efficient, but it cannot filter out the noise of point clouds. Meanwhile, the surface fitting method uses highly adaptive surface functions, such as NURBS, to approximate the shape of the real surface [49]. This method may have the problem of random overfitting owing to the influence of noise, and the computational cost of data processing is very high. In the process of surface fitting, the elimination of noise and outliers is very important. M-estimation, represented by the least square method, is the most commonly used method, which is suitable for noise with symmetrical distribution [50]. Msplit estimation is an advanced variant of M-estimation, which can independently estimate multiple subsets of a point cloud [51,52]. Common bridge structures have simple geometric shapes, but they are large in scale and require extremely high accuracy in deformation. Therefore, further research is needed on high-precision deformation calculation methods applicable to bridges.

The main objectives of this research are to improve the monitoring accuracy of deformation to submillimeters and obtain spatially continuous deformation modes based on TLS point clouds. This study initially proposes a bridge deformation mode calculation method based on a combination of sliding windows and surface fitting, which effectively solves the influence of point cloud noise and overfitting problems in surface reconstruction. This part is displayed in Section 2. Then, a systematic analysis of the error composition of bridge deformation monitoring based on TLS is conducted, which provides a theoretical basis for improving the accuracy of deformation monitoring to the submillimeter level. This part is displayed in Section 3. In Section 4, microdeformation monitoring experiments at different measurement distances are conducted to verify the accuracy level of the proposed deformation monitoring method. In Section 5, real bridge testing experiments are conducted to test the effectiveness of the proposed deformation mode calculation method in real bridge environments.

2. Deformation Mode Monitoring Method

Typically, bridges have beams whose axial length significantly exceeds their cross-sectional dimensions. Consequently, the deformation modes of interest in bridges are usually spatial curves. The X-axis is defined along the spanwise direction of the bridge, the Y-axis along the horizontal direction perpendicular to the spanwise direction, and the Z-axis in the vertical upward direction described as the deformation direction. Deformation due to gravitational effects manifests as a two-dimensional curve in the XZ plane. When factors such as lateral foundation slip and uneven illumination of sunlight are considered, the deformation mode of the bridge becomes a 3D curve. However, this mode can still be decomposed into two-dimensional curves in the XY and XZ planes. Computing the deformation curves in the XY and XZ planes separately from the point cloud can sufficiently capture the spatial deformation mode of a bridge using TLS. The deformation curves in the XY plane are derived from the point clouds at the bottom or top surfaces of the beams, whereas the deformation curves in the XZ plane are computed from the point clouds on the sides of the beams. The general steps of the TLS-based spatial continuous deformation monitoring method for bridges proposed in this study are shown in Figure 1.

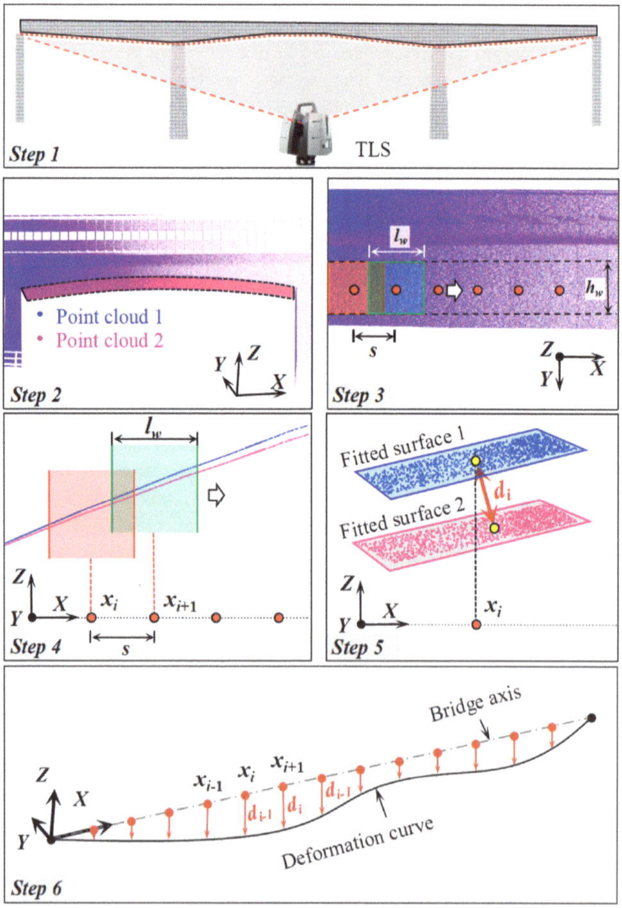

Figure 1. Key steps of bridge deformation mode monitoring method.

The proposed deformation mode calculation method integrates the sliding window and surface fitting methods, and is referred to as the SWSF method. The sliding window is used to separate point cloud blocks with the same boundary from two epoch point clouds. The sliding window method has previously been used for object detection and image filtering [53,54]. The surface function is used to fit the local surface of the bridge corresponding to the two epoch point cloud blocks separately. Due to the noise in the bridge point cloud following a Gaussian distribution, the M-estimation method based on least squares is used for surface fitting. The noise distribution in the bridge point cloud is described in detail in Section 3.2.3. The geometric deviation between the fitted surfaces is considered as the deformation value of the bridge within the sliding window area. By sliding the window, the deformation value at any position on the bridge axis can be obtained. The specific steps of the method are described as follows:

Step 1: TLS is used to scan the bridge in different periods, and the point clouds in different periods are registered in the same coordinate system.

Step 2: The point cloud for the region of interest is truncated from the two-phase point cloud that has been harmonized in the coordinate system. Figure 1 shows the bottom surface of the main girder. Given that the region of interest may be different in each bridge, this step needs to be performed manually in the preprocessing software of the point cloud.

Step 3: A window with a length of l_w and a width of h_w is constructed. This window slides along the axis of the region of interest. The step size s is the distance between the centers of adjacent windows, which determines the distance between measurement points on the deformation curve.

Step 4: The X-coordinate of the center of the sliding window i, which is defined as the X-coordinate x_i of deformation measurement point i, is calculated.

Step 5: A low-order polynomial surface function is used to fit the two point cloud blocks before and after deformation in sliding window i, resulting in two fitted surfaces: Surfaces 1 and 2. The coordinates (x_i, z_i) at x_i on Surface 1 are initially calculated, and the vertical distance d_i from point (x_i, z_i) to Surface 2 is calculated. (x_i, d_i) is a measurement point on the bridge deformation curve.

Step 6: The window is moved to the next position, such as $x_{i+1} = x_i+s$, and Steps 4 and 5 are repeated to obtain the deformation value d_{i+1}. (x_{i+1}, d_{i+1}) is another measurement point on the bridge deformation curve. By continuously sliding the window to obtain dense measurement points, a deformation curve is formed.

In *Step 5*, notably, the deformation calculated by the proposed method represents the normal distance between Surfaces 1 and 2, which is suitable for describing the deformation of beams and tower components with constant cross-sections. However, for components with variable cross-sections, an angle exists between the surface's normal direction and the deformation direction of the component, such as the bottom surface of the main beam shown in Figure 1. When the bridge's vertical deformation mode needs to be ascertained, the vertical deformation d_{zi} of measurement point i can be calculated using the following formula:

$$d_{zi} = d_i / \cos\theta, \quad (1)$$

where θ is the angle between the surface of the bridge at measurement point i and the deformation direction.

The proposed SWSF deformation calculation method belongs to the model reconstruction method. In terms of the precision of surface reconstruction, the SWSF method is between the triangular mesh and NURBS surface methods. The triangular mesh method directly establishes a triangular plane from three adjacent points, and the noise of the point cloud produces significant and random surface reconstruction errors. The NURBS surface method aims to reconstruct continuous and smooth surfaces for the entire point cloud. The reconstruction process is complex, and the surface has many control parameters. The proposed SWSF method uses a sliding window technique to achieve coverage at any position. Within the sliding window range, low-order surface fitting is implemented to reduce the influence of random noise in the point cloud while avoiding overfitting problems and making the algorithm simple.

The TLS-based bridge deformation measurement error accumulates gradually in the above steps, and the final deformation error is influenced by many complex factors, such as point cloud noise, point cloud registration error, and surface fitting error. In addition, the reasonable size of the sliding window in Step 3 and the choice of fitting surface function in Step 5 directly affect the surface fitting error. Therefore, the deformation measurement error based on TLS and SWSF needs to be systematically modeled and analyzed.

3. Error Analysis of Deformation Measurement Based on TLS

3.1. Error Model

In this study, the point clouds within a sliding window are analyzed, with the pre-deformation point cloud designated as "point cloud 1" and the post-deformation point cloud as "point cloud 2". Figure 2 illustrates the geometric relationships in the deformation process when utilizing point clouds from a sliding window. Bridge point clouds are typically 3D; however, for a straightforward illustration, Figure 2 restricts its depiction to the functional relationships within the xz plane. In Figure 2a, S_1 and S_2 represent the actual shape functions of the bridge surface before and after deformation, respectively. Similarly, F_1 and F_2 denote the fitted shape functions of the bridge surface based on point clouds 1 and 2, respectively. Figure 2b introduces d_S as the actual deformation function of the bridge and d_F as the calculated deformation function. The method for quantifying deformation is defined as follows:

$$d_S(x,y) = S_2(x,y) - S_1(x,y). \tag{2a}$$

$$d_F(x,y) = F_2(x,y) - F_1(x,y). \tag{2b}$$

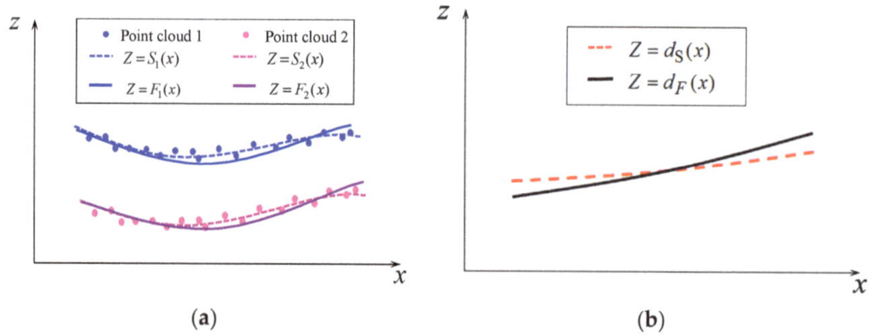

Figure 2. Geometric relationships in the deformation process. (**a**) The shape functions and (**b**) the deformation function.

The set of point coordinates in point cloud 1 can be defined as a set, which conforms to the following relationship:

$$\{x_{1i}, y_{1i}, z_{1i}\} = \{x_{1i}, y_{1i}, S_1(x_{1i}, y_{1i}) + \Delta z_{1i}\}, \tag{3}$$

where $S_1(x_{1i}, y_{1i})$ is the value of true shape function S_1 at (x_{1i}, y_{1i}). Δz_{1i} is the measurement noise of point cloud 1 in the z-direction, which may follow a particular statistical distribution law, such as a Gaussian distribution [55]. $\{x_{1i}, y_{1i}, z_{1i}\}$ can be considered a superposition of the point sets $\{x_{1i}, y_{1i}, S_1(x_{1i}, y_{1i})\}$ and $\{x_{1i}, y_{1i}, \Delta z_{1i}\}$ in the z-direction. $\{x_{1i}, y_{1i}, S_1(x_{1i}, y_{1i})\}$ strictly satisfies the actual surface function S_1, and $\{x_{1i}, y_{1i}, \Delta z_{1i}\}$ is the measurement noise of the point cloud for point cloud 1.

An approximate surface function is utilized to perform surface fitting for point cloud 1, and the resulting fitted surface function is defined as follows:

$$F_1(x,y) = f_{S_1}(x,y) + f_{E_1}(x,y), \tag{4}$$

where F_1 consists of two parts: $f_{S1}(x, y)$, which is the fitting function for the set of points $\{x_{1i}, y_{1i}, S_1(x_{1i}, y_{1i})\}$ following the true shape function S_1, and $f_{E1}(x, y)$, which is the fitting function for the measurement noise $\{x_{1i}, y_{1i}, \Delta z_{1i}\}$ of the point cloud.

Deformation of the bridge structure owing to damage or external loading occurs, and the true shape function after deformation is as follows:

$$S_2(x,y) = S_1(x,y) + d_S(x,y), \tag{5}$$

where $d_S(x, y)$ is the actual deformation function of the bridge.

The pseudo-deformation caused by the point cloud registration error is as follows:

$$Z = d_R(x,\ y). \tag{6}$$

The set of point coordinates of point cloud 2 can be considered as $\{x_{2j}, y_{2j}, z_{2j}\}$, which satisfies the following relation:

$$\{x_{2j},\ y_{2j}, z_{2j}\} = \{x_{2j},\ y_{2j},\ S_1(x_{2j},\ y_{2j}) + d_S(x_{2j},\ y_{2j}) + d_R(x_{2j},\ y_{2j}) + \Delta z_{2j}\}, \tag{7}$$

where z_{2j} is divided into four parts: $S_1(x_{2j}, y_{2j})$, $d_S(x_{2j}, y_{2j})$, $d_R(x_{2j}, y_{2j})$, and Δz_{2j}. $\{x_{2j}, y_{2j}, S_1(x_{2j}, y_{2j})\}$ satisfies the true shape function S_1 before deformation, $\{x_{2j}, y_{2j}, d_S(x_{2j}, y_{2j})\}$ satisfies the true deformation function d_S, $\{x_{2j}, y_{2j}, d_R(x_{2j}, y_{2j})\}$ satisfies the registration error function d_R, and $\{x_{2j}, y_{2j}, \Delta z_{2j}\}$ is the measurement noise of point cloud 2.

An approximate surface function is employed to conduct surface fitting to point cloud 2, with the corresponding fitted surface function defined as

$$F_2(x,y) = f^*_{S_1}(x,y) + f_{d_S}(x,y) + f_{d_R}(x,y) + f_{E_2}(x,y), \tag{8}$$

where F_2 consists of four components. $f_{S1}{}^*(x, y)$ is the fitting function of the component matrix $\{x_{2j}, y_{2j}, S_1(x_{2j}, y_{2j})\}$. Although $\{x_{1i}, y_{1i}, S_1(x_{1i}, y_{1i})\}$ and $\{x_{2j}, y_{2j}, S_1(x_{2j}, y_{2j})\}$ satisfy the shape function S_1, their coordinate values are not the same. Therefore, a difference exists in the fitting functions $f_{S1}(x, y)$ and $f_{S1}{}^*(x, y)$. $f_{dS}(x, y)$ is the fitting function of the component matrix $\{x_{2j}, y_{2j}, d_S(x_{2j}, y_{2j})\}$. $f_{dR}(x, y)$ is the fitting function of the component matrix $\{x_{2j}, y_{2j}, d_R(x_{2j}, y_{2j})\}$. $f_{E2}(x, y)$ is the fitting function of the component matrix $\{x_{2j}, y_{2j}, \Delta z_{2j})\}$.

Subtracting Equation (8) from Equation (4) yields the following computational deformation function of the structure:

$$\begin{aligned} d_F(x,y) &= F_2(x,y) - F_1(x,y) \\ &= f_{d_S}(x,y) + f_{d_R}(x,y) + [f^*_{S_1}(x,y) - f_{S_1}(x,y)] + [f_{E_2}(x,y) - f_{E_1}(x,y)] \end{aligned}. \tag{9}$$

By subtracting the actual deformation function from the calculated deformation function (Equation (9)), the error of the deformation measurement is obtained as follows:

$$\begin{aligned} E &= d_F(x,y) - d_S(x,y) \\ &= [f_{d_S}(x,y) - d_S(x,y)] + [f'_{S_1}(x,y) - f_{S_1}(x,y)] + [f_{E_2}(x,y) - f_{E_1}(x,y)] + f_{d_R}(x,y) \\ &= E_d + E_s + E_N + E_R \end{aligned}, \tag{10}$$

where the error of deformation measurement consists of four items: E_d, E_S, E_N, and E_R. E_d is the structural deformation fitting error, which is mainly generated by the inconsistency between the selected fitting function of the point cloud and the actual structural deformation function. The relative fitting error of the surface shape is E_S. When the selected fitting function is inconsistent with the actual shape function, the fitting results of the different point clouds (e.g., $\{x_{1i}, y_{1i}, S_1(x_{1i}, y_{1i})\}$ and $\{x_{2j}, y_{2j}, S_1(x_{2j}, y_{2j})\}$) with the same shape function are different; E_N is the relative fitting error of the measurement noise ($\{x_{1i}, y_{1i}, \Delta z_{1i})\}$ and $\{x_{2j}, y_{2j}, \Delta z_{2j})\}$), although $\{x_{1i}, y_{1i}, \Delta z_{1i})\}$ and $\{x_{2j}, y_{2j}, \Delta z_{2j})\}$

may follow the same noise distribution rule. Randomness is observed in the fitting results of the measurement noise components of the point clouds in different periods owing to the randomness of the samples. E_R is the fitted value of the registration bias component when the point clouds in two periods are registered to the same coordinate system.

3.2. *Error Analysis*

3.2.1. Error E_d

In common bridge structures, the cross-section of components is mainly polygonal and quasi-circular. Therefore, within the sliding window range, the local surface shape of the bridge is a simple plane or curved surface. The first- or second-order fitting function can accurately approximate the local surface shape. In addition, the overall deformation shape of the bridge may be a complex surface function owing to the small range of the sliding window. However, the deformation inside the sliding window can be regarded as a rigid translation of the local surface of the bridge. That is, the deformation function within the sliding window range usually satisfies a first-order plane function. The aforementioned inference is a fundamental characteristic of common bridges, which serves as the basis for analyzing the error E_d in the following text.

Taking bridges with circular cross-sections as the analysis object, we elaborate in detail on the deformation error component E_d. Given that the span of the bridge significantly exceeds its cross-sectional dimensions, its deformation can be conceptualized as an overall translation of the cross-section, as depicted in Figure 3a. A second-order surface function can be used to approximate the shape of the local surface area, as shown in Figure 3b. The deformation shape of the local surface can be accurately fitted with a first-order plane function, as demonstrated in Figure 3c. These fitting results also confirm the correctness of the aforementioned inference. Notably, the higher-order surface function encompasses elements of the first-order surface function. Therefore, when a higher-order surface is used to fit the point cloud of a bridge's local surface, the deformation function is automatically and precisely included in the surface fitting result. In conclusion, for bridge deformation analysis, the selection of any first-order or higher-order fitting function will not result in a deformation error E_d.

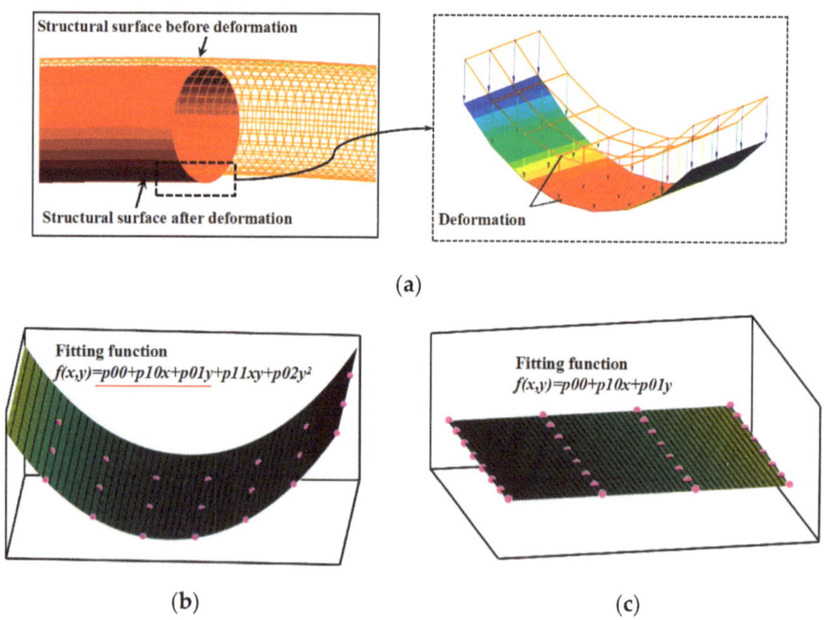

Figure 3. Relationship between local surface function and local deformation function. (**a**) Forms of deformation of bridges, (**b**) local surface fitting, and (**c**) local deformation fitting.

3.2.2. Error E_S

When the chosen fitting function accurately describes the local surface shape of the bridge, the fitting result is accurate, and the error E_S is equal to 0. The chosen function is only an approximate fit to the actual shape. Moreover, as long as the input point cloud coordinate matrix changes, the fitting result will also change. In real scanning scenarios, even in the absence of any measurement noise, the point cloud coordinate matrix is not the same for the same area scanned at different times. The reason is that keeping the coordinate system and the starting angle of the two scans exactly the same is difficult. Therefore, deviation exists in the results of the two fits because the positions of the point clouds from the two scans are not the same, and this deviation is the error E_S.

Further study showed that the E_S is mainly related to the point distance of the point cloud and the type of fitting function. Manually generating point clouds 1 and 2 is necessary to investigate the characteristics of E_S. First, two different point clouds were obtained by performing two random samples on the same surface function, which were free of any measurement noise. Cylinders were used as surface functions for this study. Then, two localized point cloud blocks were truncated from the two generated point clouds using the same sliding window, as shown in Figure 4a. Figures 4–6 demonstrate the fitting effects of choosing different fitting functions for different point cloud densities. Figure 4 shows an example of fitting a cylindrical surface point cloud using a second-order surface function in a dense point cloud, whereas Figure 5 shows an example of fitting a second-order surface function in a sparse point cloud scenario. Then, Figure 6 shows an example of fitting a cylindrical surface point cloud using a first-order plane function. Practically, the fitted surfaces in Figures 4–6 all deviate from the original point cloud owing to the inconsistency between the chosen fitting function and the actual shape function of the structure. However, in each example, the difference between the fitting results of point clouds 1 and 2, that is, E_S, is very small. Figure 7 demonstrates this pattern, showing E_S versus the point distance of the point clouds. In the figure, E_S is calculated as the maximum gap between the fitted surfaces of point clouds 1 and 2. Given that randomness exists in the location of the point cloud in each scan, the data in Figure 7 show the average value of the results of 1000 previous simulation experiments that were counted. As shown in Figure 7, E_S increases nonlinearly with increasing point distance. Hence, the following quantitative conclusions can be drawn. For the first-order planar function, E_S is less than 0.1 mm when the point spacing is less than 20 mm, and less than 1 mm when the point distance is less than 50 mm. For the second-order surface function, E_S is less than 0.1 mm when the point distance is less than 40 mm, and less than 1 mm when the point distance is less than 100 mm. Therefore, in a dense point cloud environment where the point spacing is less than 20 mm, E_S can be kept at 0.1 mm even if a planar function with a large difference from the actual shape is used as the fitting function. The point distance can be increased from 20 mm to 40 mm while keeping E_S at 0.1 mm by increasing the order of the fitting function from the first to the second order, which means that the effective scanning distance can be increased by a factor of 2. Taking the Leica P50 scanner as an example, its finest scanning mode has point spacing of 0.8 mm at a distance of 10 m [35]. The maximum allowable range when keeping E_S at 0.1 mm is up to 500 m. The smaller the size of the sliding window, the simpler the local surface shape it captures, and the more accurate the fitting results are. In the case shown in Figure 7, with unfavorable considerations, the center angle of the circle corresponding to the range extracted by the sliding window reaches 60°. In this extreme situation, E_S can still easily remain within 0.1 mm. Therefore, E_S is negligible in a general bridge scenario.

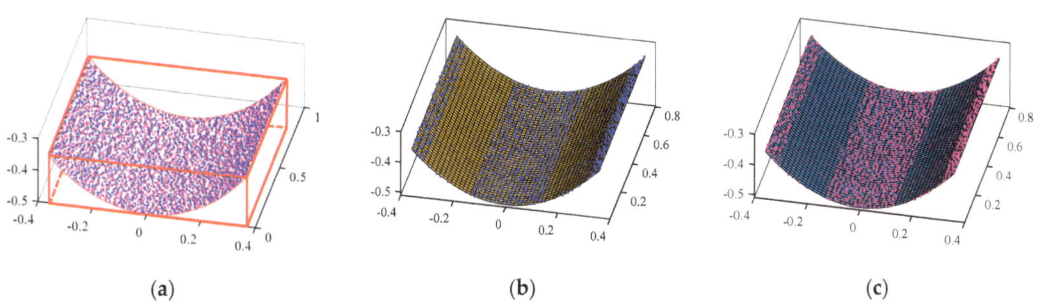

Figure 4. Dense point cloud high-order polynomial fitting surface. (**a**) Sliding window range, (**b**) point cloud 1, and (**c**) point cloud 2.

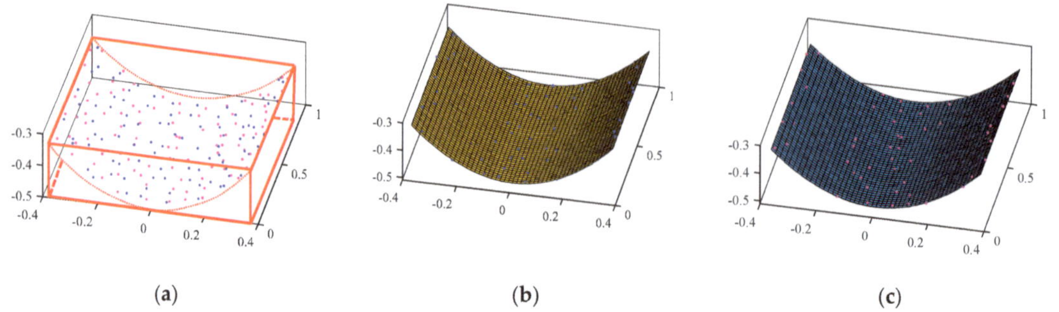

Figure 5. Sparse point cloud high-order polynomial fitting surface. (**a**) Sliding window range, (**b**) point cloud 1, and (**c**) point cloud 2.

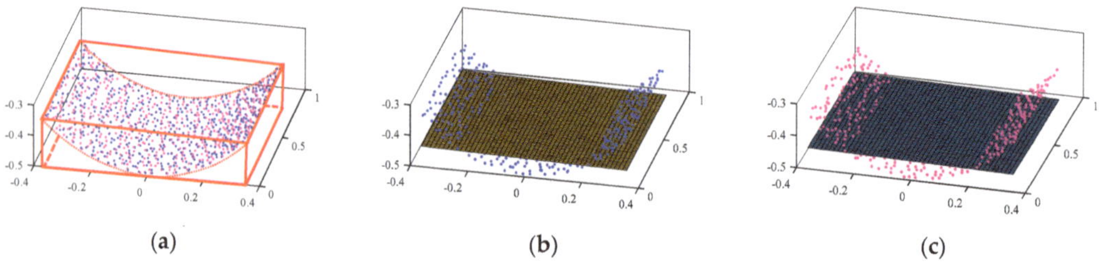

Figure 6. First-order polynomial fitting surface results. (**a**) Sliding window range, (**b**) point cloud 1, and (**c**) point cloud 2.

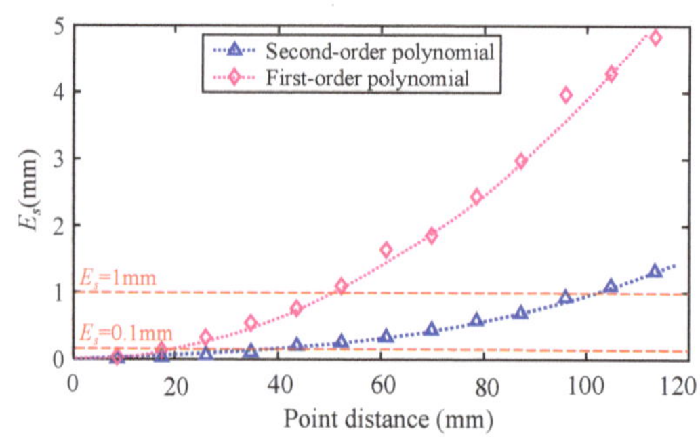

Figure 7. Error E_s.

3.2.3. Error E_N

The measurement noise of a scanner can be considered to satisfy a Gaussian distribution [55]. To further study the relationship between scanning point cloud noise and range, a point cloud noise survey experiment was conducted using a Leica P50 scanner. The experiment was conducted in an open square with set ranging distances of 30, 60, 100, 200, 300, and 400 m. The probability density of point cloud noise was statistically analyzed. The point cloud noise in all six tests met the Gaussian distribution $N(0, \sigma^2)$. Figure 8 shows the variation in the standard deviation σ of the measured point cloud noise with the range. Overall, the measured values of σ showed a linear relationship with the range. When the range was 100 m, σ was 1.2 mm; when the range was 400 m, σ was 3.0 mm. For every 100 m increase in range, the average increase in σ was 0.6 mm.

Considering that the scanning measurement noise satisfies the Gaussian distribution (N~$(0, \sigma^2)$), $\{x_{1i}, y_{1i}, \Delta z_{1i}\}$ and $\{x_{2j}, y_{2j}, \Delta z_{2j}\}$ are planar point clouds with Gaussian fluctuations near the plane of z = 0. According to Equation (9), E_N denotes the geometric difference between the fitted plane of the measurement noise $\{x_{1i}, y_{1i}, \Delta z_{1i}\}$ and that of the measurement noise $\{x_{2j}, y_{2j}, \Delta z_{2j}\}$. Therefore, E_N can be equated to the difference between the mean values of $\{\Delta z_{2j}\}$ and $\{\Delta z_{1i}\}$.

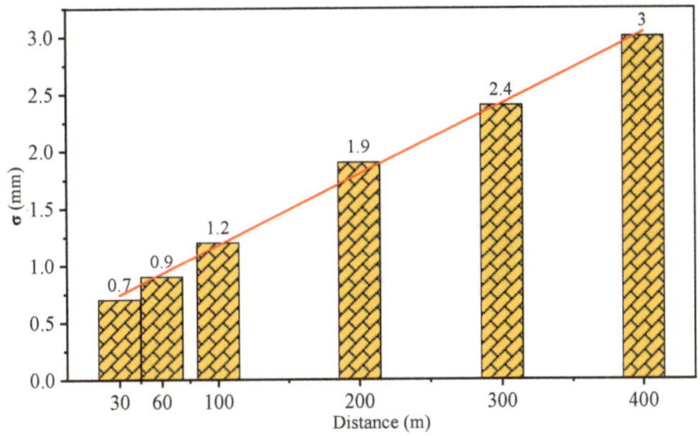

Figure 8. Relationship between the standard deviation of point cloud noise and range.

Clearly, E_N exhibits random variability and is significantly influenced by the number of points n involved in the computation and the noise standard deviation σ. This study conducted extensive numerical simulation experiments to explore the probabilistic statistical relationship between E_N, n, and σ. Figure 9 shows the results. Given that E_N is a random term, it cannot be directly quantified. The standard deviation of E_N obtained from multiple simulation experiments is defined as Sd(E_N). Sd(E_N) is used to describe E_N. In this study, each Sd(E_N) was statistically obtained from the results of 1000 random experiments. Figure 9a demonstrates the trend in Sd(E_N) as the number of samples n increases at σ = 1 mm. Similarly, Figure 9b shows the trend in Sd(E_N) with an increasing number of samples n at σ = 2 mm. The data indicate that Sd(E_N) decreases rapidly and then stabilizes as the number of points n involved in the calculation increases. Comparing Figure 9a and b, the variation pattern of Sd(E_N) with n is completely the same. Therefore, the variation in the standard deviation of point cloud noise will not change the variation pattern of Sd(E_N) and n. When the number of points used in the calculation reaches approximately 100, E_N reduces to one-tenth of the noise standard deviation level. Additionally, a point cloud with σ = 2 mm requires at least 100 points, whereas a point cloud with σ = 2 mm necessitates a minimum of 500 points to maintain Sd(E_N) within 0.1 mm.

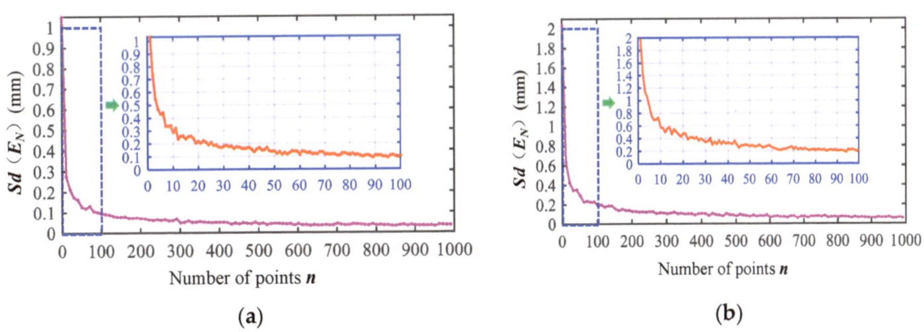

Figure 9. Relationship between Sd(E_N) and number of points n. (**a**) σ = 1 mm and (**b**) σ = 2 mm.

Figure 10a depicts the relationship between Sd(E_N) and the number of points n and the noise standard deviation σ within the range of 0.5–3 mm. Overall, when the number of points n is fixed, Sd(E_N) exhibits a linear relationship with σ. Twice the value of Sd(E_N) is adopted as the characterization value of E_N to further quantify the numerical magnitude of E_N. According to the basic characteristics of Gaussian distribution, this characterization value can achieve a reliability of 95%, which meets the general requirements of engineering measurement. When Sd(E_N) = 0.05 mm, the characterization error of E_N is 0.1 mm, and the relationship between the number of points n and noise σ is shown in Figure 10b. When the point cloud noise σ = 1, 2, and 3 mm, the sliding window point cloud needs at least 400, 1600, and 3500 points, respectively. When Sd(E_N) = 0.5 mm, the characterization error of E_N is 1 mm, and the relationship between the number of points n and noise σ is shown in Figure 10c. When the point cloud noise σ = 1, 2, and 3 mm, the sliding window point cloud needs at least 6, 17, and 38 points, respectively.

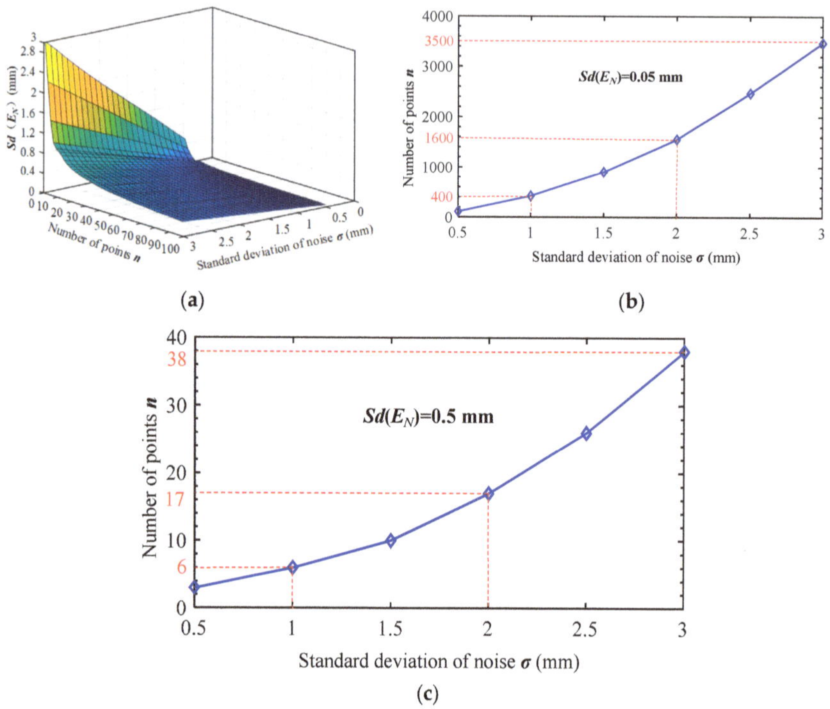

Figure 10. Regularity diagram of E_N. (**a**) E_N versus the number of points n and noise σ, (**b**) relationship between the number of points n and noise σ when Sd(E_N) = 0.05 mm, and (**c**) relationship between the number of points n and noise σ when Sd(E_N) = 0.5 mm.

In summary, the error E_N is a random term that is significantly related to the noise of the point cloud and the number of points in the sliding window point cloud. Overall, when the standard deviation of the point cloud noise is in the range of 0–3 mm, to make the error E_N less than 0.1 mm, the number of points n in the sliding window point cloud should be greater than 300. To make the error E_N less than 1 mm, the number of points n in the sliding window point cloud should be greater than 38. The number of points n in the sliding window point cloud is related to the size of the sliding window and the density of the point cloud. n can be calculated according to the following formula:

$$n = \frac{l_w h_w}{d^2}, \tag{11}$$

where d is the point spacing of the sliding window point cloud, which is related to the resolution setting mode of the scanner.

Therefore, in the SWSF method proposed in Section 2, the size of the sliding window mainly depends on the number of points in the sliding window. The width h_w of the sliding window is the width of the deformation-concerned region, and it is recommended that the maximum value is used as much as possible. In the example shown in Figure 1, the sliding window width h_w can be equal to the width of the bottom surface of the main beam. On the basis of the determined minimum number of points n in the sliding window, the point spacing d, and the sliding window width h_w, the length l_w of the sliding window can be calculated using Equation (10).

3.2.4. Error E_R

Deformation monitoring can be divided into two types: continuous and periodic measurements. In continuous measurement scenarios, which typically require a short interval between two deformation measurements, a scanning method with a fixed instrument position is employed, eliminating the registration error associated with two-phase point clouds. For periodic observations, however, the instrumentation often needs to be reset, necessitating multiple point cloud registrations into a consistent coordinate system. This case requires the use of precise registration methods to ensure minimal registration error. The iterative closest point (ICP) method and its variants are currently widely used, but they are all data-driven approaches, which are not suitable in the presence of deformations [56]. The black and white targets and spherical targets developed by the Leica High-definition Surveying (HDS) system are currently the most commonly used target-based registration methods. According to the user manual of the Leica scanning system, the registration error of HDS targets can reach the millimeter level. Recently, Zhou et al. introduced a registration method based on triangular pyramid targets [34]. The experimental results of this study indicate that the accuracy of triangular pyramid registration surpasses that of ICP and spherical markers, with registration errors ranging between 0.2 and 0.8 mm. Therefore, for continuous and periodic monitoring, the deformation monitoring error of the proposed method framework can reach the submillimeter level.

3.2.5. Summary of Measurement Errors

For the proposed method of bridge deformation monitoring, when deformation is calculated within a sliding window using fitting plane or surface subtraction methods, the components of the deformation measurement error include three parts: error E_S, error E_N, and error E_R. Error E_S arises from the discrepancy between the chosen fitting function and the actual surface function of the bridge. This error can be maintained within 0.1 mm in a suitable point cloud density environment (point spacing < 40 mm). Error E_N is primarily induced by scanning point cloud noise and varies with the point cloud noise and the number of points in the sliding window. Error E_R is largely dependent on the point cloud registration method and can currently reach submillimeter levels. In summary, the proposed deformation measurement method has been demonstrated to achieve submillimeter accuracy. In continuous monitoring scenarios without registration errors, the monitoring error of the bridge deformation mode based on TLS and SWSF mainly originates from E_N. Therefore, by reasonably setting the size of the sliding window, the number of points within the sliding window can meet the requirements of the corresponding monitoring accuracy level, and the deformation measurement accuracy can reach the level of 0.1 mm. Table 1 shows the summary of each error.

Table 1. Error summary table.

Error Term	Influence Factor	Key Parameter	Error Level
E_d	Type of fitting function	First- or second-order polynomial	0
E_S	Point distance and fitting function order	First-order function and point distance < 20 mm; Second-order function and point distance < 40 mm	<0.1 mm
E_N	Point cloud noise σ and number of points n	$\sigma \leq 1$ mm, $n > 400$; $\sigma \leq 2$ mm, $n > 1600$; $\sigma \leq 3$ mm, $n > 3500$	<0.1 mm
E_R	Registration method	Triangular pyramid targets [34]	>0.2 mm

4. Precision Verification Experiment

To verify the submillimeter accuracy level of deformation monitoring based on TLS and SWSF, a high-precision experimental device was designed, as shown in Figure 11. The experimental device is mainly composed of a fixed base, a sliding bracket, a concrete plate, and an adjustable bolt. The concrete plate is poured using steel formwork commonly used in practical bridges. Therefore, the concrete plate is similar to the local surface of the real bridge. The concrete plate is used to simulate the bridge surface within the sliding window region in the SWSF method. The size of the concrete plate is 0.6 m × 0.4 m. The fixed base is fixed on the ground and kept strictly motionless during the experiment. The sliding bracket is bound to the concrete plate and placed on the base surface. The movement of the sliding support is controlled by fine-tuning the bolts on the base plate to simulate the small deformation of the bridge. Multiple dial gauges are placed behind the concrete slab to measure accurately its true deformation value. The Leica P50 scanner is used in the experiment. Two scanners with different accuracy levels are used, Leica P50 and Leica RTC360.

The experiment involved conducting 10 scanning tests in 100 m and 200 m ranging conditions. Since the maximum range of Leica RTC360 is 130 m, Leica RTC360 was only used to conduct the tests with a 100 m range. By fine-tuning the bolts, the concrete plate was deformed by 0.1 mm relative to its initial position. During the experiment, the position of the scanner remained unchanged. Therefore, no registration error existed. The scanning points on the concrete plate in each test are not fewer than 3000 points. When a scan cannot obtain enough points, the scan must be repeated to increase the number of points on the concrete plate. By sampling the point cloud of the concrete slab to simulate the point clouds with different numbers of points in the sliding window, the calculation of deformation in the sliding window used a first-order function fitting. The details of the experimental scheme are shown in Table 2. The experimental results are shown in Figure 12. The deformation calculation results for different ranges and scanners are shown in Figure 12a–c. The mean square error of 10 deformation tests is shown in Figure 12d. Figure 12a–c indicate that the deformation calculation results tend to be stable with the increase in the number of points in the sliding window. The change in range and the difference in scanners did not affect the experimental results. When the number of points was 10, the deformation calculation results fluctuated significantly and could not effectively identify the 0.1 mm deformation. When the number of points was above 1000, the deformation calculation results stabilized around 0.1 mm and effectively identified the 0.1 mm deformation. Figure 12d indicates that the mean square error of the deformation test results decreases significantly with the increase in the number of points in the sliding window. This result was consistent for different ranges and different scanners. In addition, when the numbers of points in the sliding window were 10 and 100, the mean square error was affected by the noise standard deviation of the point cloud. The standard deviation of point cloud noise was 1.2 mm at 100 m range for the scanner Leica P50, 2.0 mm at 100 m range for the scanner Leica RTC360, and 1.9 mm at 200 m range for the scanner Leica P50. When the standard deviation of point cloud noise was larger, the mean square error of the deformation test results was relatively larger. However, when the number of points in the sliding window exceeded 1000, the mean square error of the deformation test results was less than 0.1 mm, and the point cloud noise was not sensitive to the mean square error of the deformation test results. The experimental results show that the measurement accuracy of deformation mode monitoring based on TLS and SWSF can be improved by increasing the number of points in the sliding window. This result is consistent with the simulation results shown in Figure 10. At the same time, under the condition of a range of 200 m, the deformation monitoring error based on TLS and SWSF can be measured to be less than 0.1 mm.

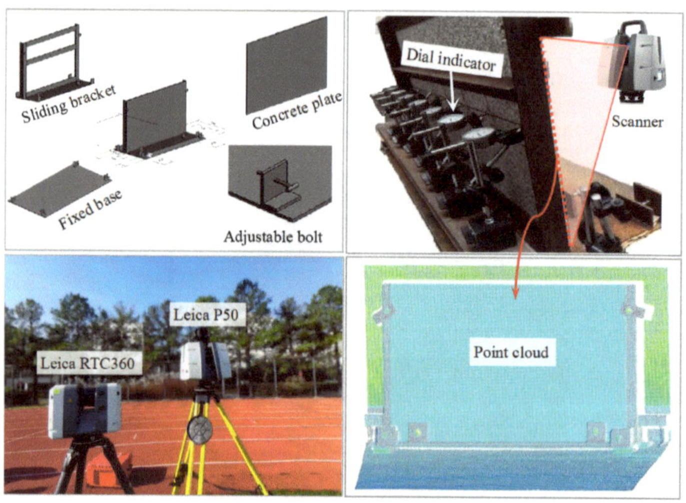

Figure 11. Deformation accuracy verification experiment.

Table 2. Overview of experimental scheme.

Range (m)	Scanner	Standard Deviation of Noise (mm)	Number of Points in Sliding Window	Number of Tests
100 m	Leica P50	1.2	10, 100, 1000, 3000	10
	Leica RTC360	2.0		
200 m	Leica P50	1.9		

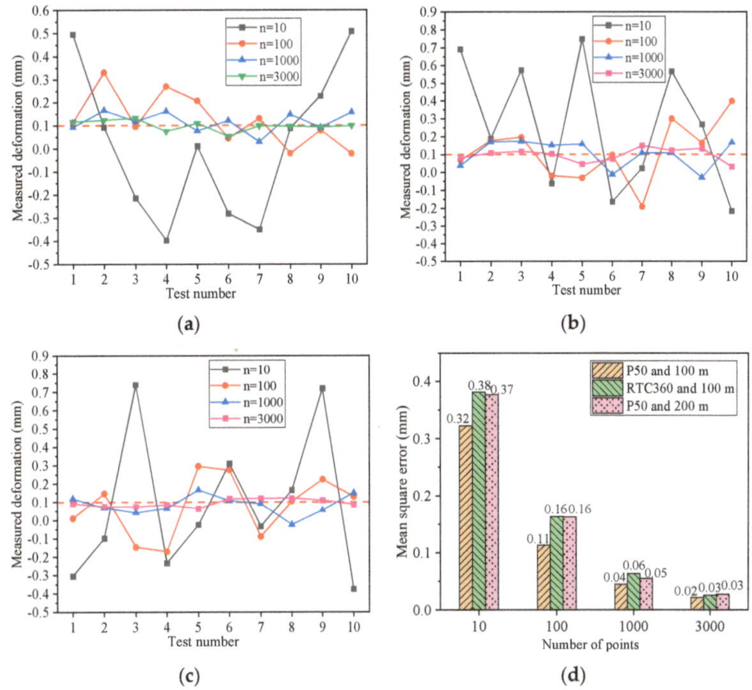

Figure 12. Deformation monitoring results under different conditions: (**a**) Leica P50 and range of 100 m, (**b**) Leica RTC360 and range of 100 m, (**c**) Leica P50 and range of 200 m, and (**d**) mean square error of deformation monitoring error.

5. Real Bridge Testing

5.1. Introduction to Bridge

Figure 13 shows a continuous rigid-frame bridge in Chongqing, China, spanning 40 m + 65 m + 40 m. This test bridge is nearing completion and will soon be open to traffic. Prior to its official operation, the bridge must undergo a load test to verify if its static and dynamic characteristics align with expectations. Deformation is a critical indicator of the static properties of a bridge. However, given that this bridge is a highly rigid rail bridge, the expected deformation under the test load is only a few millimeters, necessitating deformation monitoring with submillimeter accuracy. The bridge's main girder surface is enclosed by a sound barrier, making the use of a precision level impractical. Additionally, with a clearance of 50 m beneath the main girder, the use of dial indicators is not feasible owing to the lack of support. Therefore, a high-precision, noncontact method for deformation monitoring is necessary.

Figure 13. Site plan of a rail bridge.

5.2. Test Scheme

The proposed method was implemented in the deformation monitoring of the load tests of this bridge. At the same time, the traditional monitoring method of total station and prism combination, which is the most commonly used method in bridge monitoring, was adopted in parallel. Three prisms were installed on the side of the main girder using a lift truck positioned at L/4, L/2, and 3L/4 of the bridge's midspan. The scanner used was a Leica P50, set up beside the pier of the midspan of the test bridge, allowing complete observation of the left side- and center-span areas. A scanning resolution of 3.1 mm @10 m was used, and the single measurement time was approximately 3 min. The distance between the farthest point of the midspan and the scanner was 65 m, and the corresponding noise standard deviation of the point cloud was about 1 mm. The total station used was Leica TS30, which was set up next to the pier of the side span. The distance between the farthest prism and the total station was 85 m. The automatic prism identification accuracy of the Leica TS30 was 1 mm. In terms of single-point measurement accuracy, the selected Leica P50 and Leica TS30 had comparable accuracy. The layout plan of measuring instruments and loading trains is shown in Figure 14. The scanner and total station maintained their positions throughout the test, indicating that the error E_R described in Section 3.2 did not exist. The load was applied using a fully loaded train stationed at three preset loading positions. Given the varying positions of the train load, different deformations were induced on the test bridge.

(a)

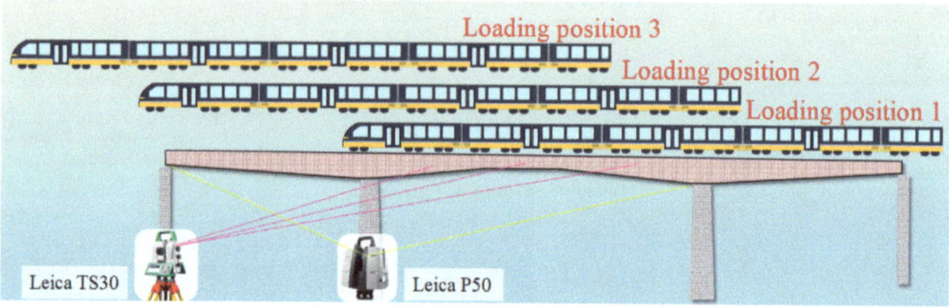

(b)

Figure 14. Test scheme. (**a**) Instrument and (**b**) loading scheme for trains.

5.3. Test Results

Figure 15 presents the point cloud model of the test bridge. The point clouds of the bottom surfaces of the left side span and the center span were completely captured. A total of five sets of data were collected throughout the testing process, including the initial state, three loading states, and the state after unloading. The point cloud of the bridge bottom surface was used to calculate the deformation. In the calculation process using the proposed method, the sliding window width was set to 3 m and the length to 1 m. The numbers of points in the sliding window on the left and farthest sides of the midspan were 120,000 and 2500, respectively. According to the numerical simulation results in Figure 10b and the experimental results in Figure 12d, under the conditions of a point cloud noise standard deviation of 1 mm and a sliding window point count of 2500, the deformation calculation accuracy of the sliding window could reach 0.1 mm. The average number of points in the point cloud within the sliding window was 8000. The step distance for sliding windows was 0.2 m. Figure 16 shows the results of the deformation test. The results include deformations after three loading states and after unloading. The test results demonstrate that the proposed SWSF method successfully captures the deformation modes of the side and center spans of the test bridge. The deformation curves were extremely smooth without fluctuating noise, indicating the strong robustness of the SWSF method in real bridge environments. This result indicates that the proposed method is robust indoors and outdoors. When comparing the test results of the proposed method with those of the prisms, the deformation results of the prisms fluctuate around the deformation curves obtained by the proposed method, with a maximum positive deviation of 1.1 mm and a maximum negative deviation of −1.0 mm. Given that the total station monitors bridge deformation by monitoring the changes in the prism, the automatic prism identification error is the main source of error in deformation monitoring. The automatic prism identification error of the Leica TS30 was 1 mm, which is consistent with the fluctuation in the measured data. Furthermore, the unloading deformation test results of the proposed method are analyzed. The unloading deformation is the difference between the state data when the train completely exits the bridge and the unloaded initial state data. Given that the test bridge is newly constructed, it should have no damage under normal conditions. Thus, the bridge is in the elastic phase with loading within the design's allowable range, and the theoretical deformation after unloading is 0 mm. The unloaded deformations in the side- and center-span regions obtained using the proposed method were close to 0 mm, with a maximum residual deformation of 0.2 mm. The residual deformation values of the three measuring points of the total station were −0.3, 0.9, and 0.6 mm. Clearly, the residual deformation indirectly proves that the proposed method accurately captures the deformation modes of the bridge. The monitoring results of the deformation mode also demonstrate the absence of initial structural damage in this test bridge.

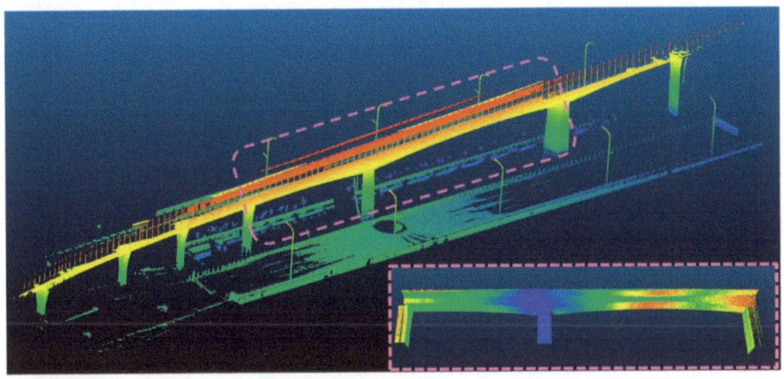

Figure 15. Point cloud model of the test bridge.

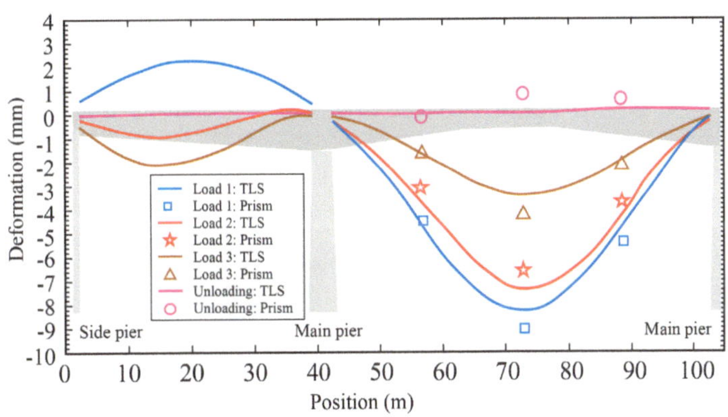

Figure 16. Results for the test bridge.

The practical application results show that the proposed method is a noncontact, high-precision monitoring technique that does not require the presetting of any reflective markers. The method can accomplish comprehensive coverage data collection within the visible range in just a few minutes. In comparison with the commonly used total station method, the proposed method improves the deformation monitoring accuracy from the millimeter level to the submillimeter level and transforms the discrete measurement point data format into a spatially continuous deformation mode.

6. Conclusions

In this study, a high-precision deformation monitoring method for bridges based on TLS is introduced. The contributions of this research are significant in two main aspects: (1) the establishment of a TLS-based method for calculating deformation modes of bridges, characterized by its substantial noise-resistant capability; and (2) the development of a scanning point cloud-based deformation measurement error model that is applicable for general bridges, accompanied by a quantitative analysis of each error component. The research and experimental verification presented in this study lead to the following conclusions:

1. The proposed bridge deformation monitoring method based on TLS and SWSF can obtain spatially continuous and smooth deformation curves of bridges. The fitting of point clouds within the sliding window only requires the use of first- or second-order polynomial functions.
2. The deformation measurement method, which is based on the differential of the fitted shape of sliding window point clouds, mainly comprises four components of error. These components are associated with the choice of fitting function, the density of the point cloud, the noise within the point cloud, and the accuracy of point cloud registration. Among these, the errors that cannot be neglected are those caused by the noise in the point cloud and the registration error.
3. The deformation measurement error caused by point cloud noise is related to the noise level and the number of points in the sliding window. At a certain level of point cloud noise, the deformation calculation error can be reduced to the submillimeter level, up to 0.1 mm, by simply increasing the number of points in the sliding windows. This conclusion was verified in experiments with ranges of 100 m and 200 m.
4. In comparison with the total station method, the proposed method does not require the preinstallation of reflective targets or sensors, and it has the advantages of high accuracy, no contact, high efficiency, and wide monitoring range. These advantages were verified in a real bridge test. The proposed method is also applicable to structures with simple surface geometries, such as buildings and dams.

In this study, registration error was eliminated by keeping the scanner's position constant, making the method ideal for scenarios where the instrument is permanently fixed or in short-term monitoring. In periodic monitoring scenarios where the scanner needs to be moved, reducing the registration error of two-phase point clouds remains a significant research focus. In addition, for flexible bridges in operation, structural vibrations owing to external excitation, which will appear as a significant point cloud noise, cannot be ignored. Developing strategies to eliminate these vibration noises is crucial for extending the applicability of the proposed method to flexible bridges.

Author Contributions: Methodology: Y.Z., J.Z. and J.X.; software: G.H. and L.Z.; validation: J.Y. and H.Z.; writing—original draft preparation: Y.Z., J.Z. and G.H.; writing—review and editing: Y.Z., J.X. and J.Y.; visualization: H.Z. and L.Z.; funding acquisition: J.X. All authors have read and agreed to the published version of the manuscript.

Funding: This research is funded by the ChongQing Postdoctoral Science Foundation (CSTB2022NSCQ-BHX0741), ChongQing Science Fund for Distinguished Young Scholars (CSTB2023NSCQ-JQX0029), Chongqing Natural Science Foundation of China (CSTB2022TIAD-KPX0205), and Science and Technology Project of Sichuan Provincial Transportation Department (2023-ZL-03).

Data Availability Statement: Data are contained within the article.

Conflicts of Interest: The authors declare no conflicts of interest.

References

1. Feng, D.; Feng, M.Q. Model Updating of Railway Bridge Using In Situ Dynamic Displacement Measurement under Trainloads. *J. Bridge Eng.* **2015**, *20*, 04015019. [CrossRef]
2. Moreu, F.; Jo, H.; Li, J.; Kim, R.E.; Cho, S.; Kimmle, A.; Scola, S.; Le, H.; Spencer, B.F.; LaFave, J.M. Dynamic Assessment of Timber Railroad Bridges Using Displacements. *J. Bridge Eng.* **2015**, *20*, 04014114. [CrossRef]
3. Feng, D.; Feng, M.Q. Output-only damage detection using vehicle-induced displacement response and mode shape curvature index. *Struct. Control Health Monit.* **2016**, *23*, 1088–1107. [CrossRef]
4. Tang, Q.; Xin, J.; Jiang, Y.; Zhang, H.; Zhou, J. Dynamic Response Recovery of Damaged Structures Using Residual Learning Enhanced Fully Convolutional Network. 2024. Available online: https://www.worldscientific.com/doi/10.1142/S0219455425500087 (accessed on 16 June 2024).
5. Sarthak, J.; Shrikant Madhav, H. Linear Variable Differential Transducer (LVDT) & Its Applications in Civil Engineering. *Int. J. Transp. Eng. Technol.* **2017**, *3*, 62–66. [CrossRef]
6. Liu, Y.; Deng, Y.; Cai, C.S. Deflection monitoring and assessment for a suspension bridge using a connected pipe system: A case study in China. *Struct. Control Health Monit.* **2015**, *22*, 1408–1425. [CrossRef]
7. Zhu, S.F.; Zhou, Z.X.; Wu, H.J. Application of semi-closed connected pipe differential pressure sensor in bridge deflection measurement. *Transducer Microsyst. Technol.* **2014**, *33*, 150–153.
8. Yu, J.; Meng, X.; Yan, B.; Xu, B.; Fan, Q.; Xie, Y. Global Navigation Satellite System-based positioning technology for structural health monitoring: A review. *Struct. Control Health Monit.* **2020**, *27*, e2467. [CrossRef]
9. Moschas, F.; Stiros, S. Measurement of the dynamic displacements and of the modal frequencies of a short-span pedestrian bridge using GPS and an accelerometer. *Eng. Struct.* **2011**, *33*, 10–17. [CrossRef]

10. Vazquez, B.G.E.; Gaxiola-Camacho, J.R.; Bennett, R.; Guzman-Acevedo, G.M.; Gaxiola-Camacho, I.E. Structural evaluation of dynamic and semi-static displacements of the Juarez Bridge using GPS technology. *Measurement* **2017**, *110*, 146–153. [CrossRef]
11. Zhao, L.; Yang, Y.; Xiang, Z.; Zhang, S.; Li, X.; Wang, X.; Ma, X.; Hu, C.; Pan, J.; Zhou, Y.; et al. A Novel Low-Cost GNSS Solution for the Real-Time Deformation Monitoring of Cable Saddle Pushing: A Case Study of Guojiatuo Suspension Bridge. *Remote Sens.* **2022**, *14*, 5174. [CrossRef]
12. Lee, H.S.; Hong, Y.H.; Park, H.W. Design of an FIR filter for the displacement reconstruction using measured acceleration in low-frequency dominant structures. *Int. J. Numer. Methods Eng.* **2010**, *82*, 403–434. [CrossRef]
13. Kim, K.; Choi, J.; Koo, G.; Sohn, H. Dynamic displacement estimation by fusing biased high-sampling rate acceleration and low-sampling rate displacement measurements using two-stage Kalman estimator. *Smart Struct. Syst.* **2016**, *17*, 647–667. [CrossRef]
14. Kim, J.; Kim, K.; Sohn, H. Autonomous dynamic displacement estimation from data fusion of acceleration and intermittent displacement measurements. *Mech. Syst. Signal Process.* **2014**, *42*, 194–205. [CrossRef]
15. Wang, Z.-C.; Geng, D.; Ren, W.-X.; Liu, H.-T. Strain modes based dynamic displacement estimation of beam structures with strain sensors. *Smart Mater. Struct.* **2014**, *23*, 125045. [CrossRef]
16. Shen, S.; Wu, Z.; Yang, C.; Wan, C.; Tang, Y.; Wu, G. An Improved Conjugated Beam Method for Deformation Monitoring with a Distributed Sensitive Fiber Optic Sensor. *Struct. Health Monit.* **2010**, *9*, 361–378. [CrossRef]
17. Ma, Z.; Chung, J.; Liu, P.; Sohn, H. Bridge displacement estimation by fusing accelerometer and strain gauge measurements. *Struct. Control Health Monit.* **2021**, *28*, e2733. [CrossRef]
18. Ehrhart, M.; Lienhart, W. Accurate measurements with image-assisted total stations and their prerequisites. *J. Surv. Eng.* **2017**, *143*, 04016024. [CrossRef]
19. Kohut, P.; Holak, K.; Uhl, T.; Ortyl, Ł.; Owerko, T.; Kuras, P.; Kocierz, R. Monitoring of a civil structure's state based on noncontact measurements. *Struct. Health Monit.* **2013**, *12*, 411–429. [CrossRef]
20. Xu, Y.; Brownjohn, J.; Kong, D. A non-contact vision-based system for multipoint displacement monitoring in a cable-stayed footbridge. *Struct. Control Health Monit.* **2018**, *25*, e2155. [CrossRef]
21. Yoon, H.; Shin, J.; Spencer, B.F., Jr. Structural Displacement Measurement Using an Unmanned Aerial System. *Comput.-Aided Civ. Infrastruct. Eng.* **2018**, *33*, 183–192. [CrossRef]
22. Mathew, M.; Ellenberg, A.; Esola, S.; McCarthy, M.; Bartoli, I.; Kontsos, A. Multiscale deformation measurements using multispectral optical metrology. *Struct. Control Health Monit.* **2018**, *25*, e2166. [CrossRef]
23. Wu, T.; Tang, L.; Zhang, X.; Liu, Y.; Li, X.; Zhou, Z. An Improved Structural Displacement Monitoring Approach by Acceleration-Aided Tilt Camera Measurement. *Struct. Control Health Monit.* **2023**, *2023*, 6247516. [CrossRef]
24. Jiang, S.; Zhang, J.; Gao, C. Bridge Deformation Measurement Using Unmanned Aerial Dual Camera and Learning-Based Tracking Method. *Struct. Control Health Monit.* **2023**, *2023*, 4752072. [CrossRef]
25. Shao, S.; Zhou, Z.; Deng, G.; Du, P.; Jian, C.; Yu, Z. Experiment of Structural Geometric Morphology Monitoring for Bridges Using Holographic Visual Sensor. *Sensors* **2020**, *20*, 1187. [CrossRef] [PubMed]
26. Deng, G.; Zhou, Z.; Shao, S.; Chu, X.; Jian, C. A Novel Dense Full-Field Displacement Monitoring Method Based on Image Sequences and Optical Flow Algorithm. *Appl. Sci.* **2020**, *10*, 2118. [CrossRef]
27. Aryan, A.; Bosché, F.; Tang, P. Planning for terrestrial laser scanning in construction: A review. *Autom. Constr.* **2021**, *125*, 103551. [CrossRef]
28. El-Omari, S.; Moselhi, O. Integrating 3D laser scanning and photogrammetry for progress measurement of construction work. *Autom. Constr.* **2008**, *18*, 1–9. [CrossRef]
29. Tang, P.; Huber, D.; Akinci, B.; Lipman, R.; Lytle, A. Automatic reconstruction of as-built building information models from laser-scanned point clouds: A review of related techniques. *Autom. Constr.* **2010**, *19*, 829–843. [CrossRef]
30. Qin, G.; Zhou, Y.; Hu, K.; Han, D.; Ying, C. Automated Reconstruction of Parametric BIM for Bridge Based on Terrestrial Laser Scanning Data. *Adv. Civ. Eng.* **2021**, *2021*, 8899323. [CrossRef]
31. Yang, L.; Cheng, J.C.P.; Wang, Q. Semi-automated generation of parametric BIM for steel structures based on terrestrial laser scanning data. *Autom. Constr.* **2020**, *112*, 103037. [CrossRef]
32. Kim, M.-K.; Wang, Q.; Park, J.-W.; Cheng, J.C.P.; Sohn, H.; Chang, C.-C. Automated dimensional quality assurance of full-scale precast concrete elements using laser scanning and BIM. *Autom. Constr.* **2016**, *72*, 102–114. [CrossRef]
33. Kim, D.; Kwak, Y.; Sohn, H. Accelerated cable-stayed bridge construction using terrestrial laser scanning. *Autom. Constr.* **2020**, *117*, 103269. [CrossRef]
34. Zhou, Y.; Han, D.; Hu, K.; Qin, G.; Xiang, Z.; Ying, C.; Lidu, Z.; Hu, X. Accurate Virtual Trial Assembly Method of Prefabricated Steel Components Using Terrestrial Laser Scanning. *Adv. Civ. Eng.* **2021**, *2021*, 9916859. [CrossRef]
35. Park, H.S.; Lee, H.M.; Adeli, H.; Lee, I. A New Approach for Health Monitoring of Structures: Terrestrial Laser Scanning. *Comput.-Aided Civ. Infrastruct. Eng.* **2007**, *22*, 19–30. [CrossRef]
36. Mosalam, K.M.; Takhirov, S.M.; Park, S. Applications of laser scanning to structures in laboratory tests and field surveys. *Struct. Control Health Monit.* **2014**, *21*, 115–134. [CrossRef]
37. Sánchez-Rodríguez, A.; Riveiro, B.; Conde, B.; Soilán, M. Detection of structural faults in piers of masonry arch bridges through automated processing of laser scanning data. *Struct. Control Health Monit.* **2018**, *25*, e2126. [CrossRef]

38. Zhou, Y.; Xiang, Z.; Zhang, X.; Wang, Y.; Han, D.; Ying, C. Mechanical state inversion method for structural performance evaluation of existing suspension bridges using 3D laser scanning. *Comput.-Aided Civ. Infrastruct. Eng.* **2022**, *37*, 650–665. [CrossRef]
39. Liang, D.; Zhang, Z.-S.; Sui, C.-L.; Pu, J.; Su, L.-C.; Zhao, K. Performance assessment of self-anchored suspension footbridge using 3D laser scanning. *Struct. Control Health Monit.* **2022**, *29*, e2978. [CrossRef]
40. Lõhmus, H.; Ellmann, A.; Märdla, S.; Idnurm, S. Terrestrial laser scanning for the monitoring of bridge load tests—Two case studies. *Surv. Rev.* **2017**, *50*, 270–284. [CrossRef]
41. Soni, A.; Robson, S.; Gleeson, B. Structural monitoring for the rail industry using conventional survey, laser scanning and photogrammetry. *Appl. Geomat.* **2015**, *7*, 123–138. [CrossRef]
42. Truong-Hong, L.; Laefer, D.F. Using Terrestrial Laser Scanning for Dynamic Bridge Deflection Measurement. 2014. Available online: http://hdl.handle.net/10197/7495 (accessed on 13 August 2014).
43. Shen, N.; Wang, B.; Ma, H.; Zhao, X.; Zhou, Y.; Zhang, Z.; Xu, J. A review of terrestrial laser scanning (TLS)-based technologies for deformation monitoring in engineering. *Measurement* **2023**, *223*, 113684. [CrossRef]
44. Girardeau-Montaut, D.; Roux, M.; Marc, R.; Thibault, G. Change detection on point cloud data acquired with a ground laser scanner. *Int. Arch. Photogramm. Remote Sens. Spat. Inf. Sci.* **2005**, *36*, W19.
45. Yu, Y.; Guan, H.; Li, D.; Jin, S.; Chen, T.; Wang, C.; Li, J. 3-D Feature Matching for Point Cloud Object Extraction. *IEEE Geosci. Remote Sens. Lett.* **2020**, *17*, 322–326. [CrossRef]
46. Si, H.; Wei, X. Feature extraction and representation learning of 3D point cloud data. *Image Vis. Comput.* **2024**, *142*, 104890. [CrossRef]
47. Meneses Diego, M.; Zheng, L.; Guo, Q. Identification and Quantification of Surface Depressions on Grassy Land Surfaces of Different Topographic Attributes Using High-Resolution Terrestrial Laser Scanning Point Cloud and Triangulated Irregular Network. *J. Hydrol. Eng.* **2023**, *28*, 04023004. [CrossRef]
48. Khameneifar, F.; Feng, H.-Y. Establishing a balanced neighborhood of discrete points for local quadric surface fitting. *Comput.-Aided Des.* **2017**, *84*, 25–38. [CrossRef]
49. Dimitrov, A.; Gu, R.; Golparvar-Fard, M. Non-Uniform B-Spline Surface Fitting from Unordered 3D Point Clouds for As-Built Modeling. *Comput.-Aided Civ. Infrastruct. Eng.* **2016**, *31*, 483–498. [CrossRef]
50. Dąbrowski, P.; Zienkiewicz, M. Impact of cross-section centers estimation on the accuracy of the point cloud spatial expansion using robust M-estimation and Monte Carlo simulation. *Measurement* **2022**, *189*, 110436. [CrossRef]
51. Janicka, J.; Rapiński, J.; Błaszczak-Bąk, W.; Suchocki, C. Application of the Msplit estimation method in the detection and dimensioning of the displacement of adjacent planes. *Remote Sens.* **2020**, *12*, 3203. [CrossRef]
52. Duchnowski, R.; Wyszkowska, P. Msplit Estimation Approach to Modeling Vertical Terrain Displacement from TLS Data Disturbed by Outliers. *Remote Sens.* **2022**, *14*, 5620. [CrossRef]
53. Dong, Q.; Wang, S.; Chen, X.; Jiang, W.; Li, R.; Gu, X. Pavement crack detection based on point cloud data and data fusion. *Philos. Trans. R. Soc. A* **2023**, *381*, 20220165. [CrossRef]
54. Gao, R.; Park, J.; Hu, X.; Yang, S.; Cho, K. Reflective Noise Filtering of Large-Scale Point Cloud Using Multi-Position LiDAR Sensing Data. *Remote Sens.* **2021**, *13*, 3058. [CrossRef]
55. Zahs, V.; Winiwarter, L.; Anders, K.; Williams, J.G.; Rutzinger, M.; Höfle, B. Correspondence-driven plane-based M3C2 for lower uncertainty in 3D topographic change quantification. *ISPRS J. Photogramm. Remote Sens.* **2022**, *183*, 541–559. [CrossRef]
56. Zhao, X.; Kargoll, B.; Omidalizarandi, M.; Xu, X.; Alkhatib, H. Model Selection for Parametric Surfaces Approximating 3D Point Clouds for Deformation Analysis. *Remote Sens.* **2018**, *10*, 634. [CrossRef]

buildings

MDPI

Article

Bridge Condition Deterioration Prediction Using the Whale Optimization Algorithm and Extreme Learning Machine

Liming Jiang [1], Qizhi Tang [1,*], Yan Jiang [1,2,*], Huaisong Cao [3] and Zhe Xu [4]

1 State Key Laboratory of Mountain Bridge and Tunnel Engineering, Chongqing Jiaotong University, Chongqing 400074, China; jlm1065155897@163.com
2 College of Engineering and Technology, Southwest University, Chongqing 400715, China
3 Chongqing Yuhe Expressway Co., Ltd., Chongqing 400799, China; 19912422134@163.com
4 Guangxi Nanbai Expressway Co., Ltd., Nanning 530029, China; xz18844090147@163.com
* Correspondence: tangqizhi1015@163.com (Q.T.); xnjtjiangyan@163.com (Y.J.)

Abstract: To address the problem in model computations and the limited accuracy of current bridge deterioration prediction methods, this paper proposes a novel bridge deterioration prediction method using the whale optimization algorithm and extreme learning machine (WOA-ELM). First, we collected a dataset consisting of 539 sets of bridge inspection data and determined the necessary influencing factors through correlation analysis. Subsequently, the WOA-ELM algorithm was applied to establish a nonlinear mapping relationship between each influencing factor and the bridge condition indicators. Furthermore, the extreme learning machine (ELM), back-propagation neural network (BPNN), decision trees (DT), and support vector machine (SVM) were employed for comparison to validate the superiority of the proposed method. In addition, this paper provides further substantiation of the model's exceptional predictive capabilities across diverse bridge components. The results demonstrate the accurate predictive capability of the proposed method for bridge conditions. Compared with ELM, BPNN, DT, and SVM, the proposed method exhibits significant improvements in predictive accuracy, i.e., the correlation coefficient is increased by 4.1%, 11.4%, 24.5%, and 33.6%, and the root mean square error is reduced by 7.3%, 18.0%, 14.8%, and 18.1%, respectively. Moreover, the proposed method presents considerably enhanced generalization capabilities, resulting in the reduction in mean relative error by 11.6%, 15.3%, 6%, and 16.2%. The proposed method presents a robust framework for proactive bridge maintenance.

Keywords: bridge engineering; inspection data; deterioration condition prediction; whale optimization algorithm; extreme learning

Citation: Jiang, L.; Tang, Q.; Jiang, Y.; Cao, H.; Xu, Z. Bridge Condition Deterioration Prediction Using the Whale Optimization Algorithm and Extreme Learning Machine. *Buildings* **2023**, *13*, 2730. https://doi.org/10.3390/buildings13112730

Academic Editor: Andrea Benedetti

Received: 2 September 2023
Revised: 21 October 2023
Accepted: 26 October 2023
Published: 29 October 2023

1. Introduction

The number of highway bridges in China exceeded 960,000 in 2023. With the passage of time, these bridge structures inevitably experience performance degradation, caused by the coupled effects of external service environment factors and internal material deterioration [1–3]. Accurately predicting bridge performance evolution holds great theoretical and practical importance for scientific maintenance and extending structural service life [4–6].

Extensive research has been conducted by a considerable number of scholars on the establishment of bridge condition prediction models, which are broadly categorized into deterministic and probabilistic models [7]. The former category assumes a fixed and deterministic degradation trend for bridge condition, and utilizes historical periodic inspection data to perform regression fitting of predefined deterioration decay equations. Through this process, the degradation rate of bridge performance under different conditions is estimated. For instance, Yang et al. [8] employed detection data from 398 reinforced concrete bridges to fit a bridge condition degradation model that reveals the changing characteristics of bridge performance during different operational phases. Similarly, Sahar et al. [9] established a degradation regression model for the superstructure of bridges based on extensive bridge inspection data. Subsequently, the model considered eight influencing factors, including service time, span, and traffic volume, and the researchers also conducted sensitivity analysis on these factors. While deterministic models are relatively straightforward to construct, they can be adjusted and updated for different bridges. However, they still struggle to account for the stochastic nature of bridge degradation and demand high-quality historical inspection data. In contrast, the second type of model considers the degradation rate as a random variable and utilizes the theory of stochastic processes to simulate the deteriorating trends of bridge structural conditions. Moreover, most probability models have implemented Markov processes [7]. For example, Zhang et al. [10] used five years of continuous bridge inspection data from 445 hollow slab bridges and developed a bridge condition degradation prediction model based on a multi-stage Markov chain. Furthermore, their findings revealed that the state transition probability matrix in the model closely matched actual conditions. Additionally, Wellalage et al. [11] proposed a Metropolis–Hastings optimized Markov chain Monte Carlo method to calculate the state transition matrix of the typical components of railway bridges. Moreover, they compared it with regression and Bayesian models to demonstrate its superiority. In addition, Thanh et al. [12] addressed data insufficiency issues in Markov processes by incorporating a physical-empirical model. The results indicated that the model could derive the state transition matrix using the least squares method. Nevertheless, probabilistic models remain dependent on subjective engineering judgments and necessitate ongoing updates, constraining their potential for further optimization to ensure predictive efficacy.

In recent years, owing to the rapid development and extensive application of machine learning, various scholars have started adopting the extreme learning machine (ELM) model to enhance the predictive performance of models. For instance, Jiang et al. [13] utilized the ELM model to indirectly predict the remaining lifespan of lithium batteries, achieving an error control within 5%. Furthermore, He et al. [14] demonstrated that the ELM model could achieve a 94.44% accuracy in circuit fault prediction in just one millisecond, highlighting the significant advantages of ELM in predictive performance. Subsequently, researchers discovered that optimizing the initial weights and thresholds could further improve ELM's predictive capabilities.

The whale optimization algorithm (WOA) is a novel swarm intelligence and bio-inspired optimization algorithm proposed by Mirjalili et al. [15] in 2016. It draws inspiration from the hunting behavior of humpback whales and simulates their unique spiral bubble-net hunting strategy, aiming to achieve optimization for complex problems. Additionally, WOA incorporates three independent population update mechanisms: search for prey, encircling prey, and spiral update. Effectively, WOA eliminates the need for manually setting various control parameter values. Consequently, this approach significantly enhances algorithm efficiency and reduces application complexity. In this context, Lu et al. [16] established a microgrid fault analysis model using the whale optimization algorithm-enhanced extreme learning machine (WOA-ELM). When compared to the backpropagation neural network (BPNN), radial basis function neural network, and conventional ELM, WOA-ELM demonstrated faster learning speed, stronger generalization capability, and higher recognition accuracy. Similarly, Li et al. [17] conducted experiments comparing various optimization algorithms for ELM prediction models, and their findings revealed that WOA-ELM outperformed ELM, genetic algorithm-optimized ELM, cuckoo search-optimized ELM, and dandelion algorithm-optimized ELM in terms of predictive performance. Certainly, there are many other effective metaheuristic algorithms. For instance, Nadimi-Shahraki et al. [18] analyzed the performance of the MTDE algorithm based on the multi-trial vector-based differential evolution method for problems such as the pressure vessel, welded beam, tension/compression spring, and three-bar truss. Their analysis demonstrated that the MTDE algorithm exhibits improved performance and high precision in finding the optimal solution. Liu et al. [19] conducted research on the agricultural drone route-planning problem using the grey wolf optimization algorithm. They discovered that this algorithm can effectively generate drone trajectories that meet agricultural operational requirements, and exhibits improved performance and high precision in finding optimal solutions.

Considering the high data quality requirements, limited applicability, and the subjective influence on model updates in existing methods, this paper proposes a bridge deterioration prediction model based on WOA-ELM. To enhance the model's applicability beyond a single bridge, the study utilizes a dataset comprising 539 sets of diverse bridge inspection data. It establishes nonlinear mapping relationships between 11 influencing factors, including time, bridge type, span, and others, and the indicators representing the bridge's condition. By comparing the results of this paper's approach with various machine learning prediction models, its superiority is confirmed.

2. Extreme Learning Machine

The ELM [20–22] is a machine learning algorithm based on the construction of a feedforward neural network. Its fundamental principle involves randomly generating connection weights between the input layer and the hidden layer, as well as thresholds for the hidden layer nodes, and then obtaining the optimal solution through straightforward matrix computations. Compared to traditional feedforward neural network algorithms, ELM exhibits strong learning capabilities, superior generalization performance, and simplicity in parameter configuration. The network structure of ELM is illustrated in Figure 1.

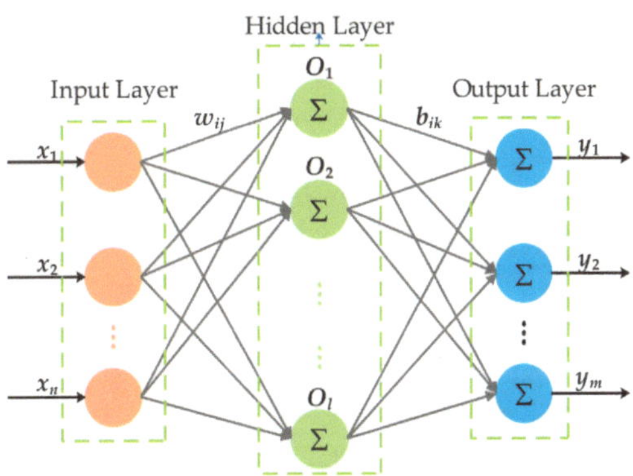

Figure 1. Extreme learning machine network structure.

In consideration of the single-hidden-layer ELM network architecture illustrated in Figure 1, we assume the existence of n arbitrary samples (X_i, Y_i). In this context, $X_i = [x_{i1}, x_{i2}, \ldots, x_{in}]^T \in R^n$ represents the input matrix, and $Y_i = [y_{i1}, y_{i2}, \ldots, y_{im}]^T \in R^m$ signifies the output matrix. The input layer of the ELM comprises n nodes, while the hidden layer consists of l nodes, and the output layer encompasses m nodes. The mathematical representation of the ELM can be succinctly articulated as follows:

$$\sum_{i=1}^{l} \beta_i g(\omega_i x_j + b_j) = y_j, j = 1, 2, \ldots, n \tag{1}$$

In this equation, the symbol $\omega_i = [\omega_{i1}, \omega_{i2}, \ldots, \omega_{in}]^T$ signifies the input weight matrix, while $\beta_i = [\beta_{i1}, \beta_{i2}, \ldots, \beta_{im}]^T$ represents the output weight matrix. Additionally, the notations $g(x)$ and b_i are employed to denote the activation function and bias of the hidden layer neurons correspondingly.

The primary objective of learning in a single-hidden-layer neural network is to minimize the output error. This objective necessitates identifying distinctive values for β_i, ω_i, and b_i that satisfy the requirements in Equation (2). Consequently, these values can be succinctly represented using matrices as follows:

$$H\beta = Y \tag{2}$$

$$H(\omega_1, \ldots, \omega_l, x_1, \ldots, x_l, b_1, \ldots, b_l) = \begin{bmatrix} g(\omega_1 x_1 + b_1) & \ldots & g(\omega_l x_1 + b_l) \\ \vdots & \ldots & \vdots \\ g(\omega_1 x_N + b_1) & \ldots & g(\omega_l x_N + b_l) \end{bmatrix}_{N \times l} \tag{3}$$

In this expression, $\beta_{l \times m} = [\beta_1, \beta_2, \ldots, \beta_l]^T$ represents the output weight matrix, and $T_{N \times m} = [T_1, T_2, \ldots, T_N]^T$ corresponds to the desired output matrix.

The learning process of ELM can be approximated as solving a nonlinear optimization problem. When the activation function is infinitely differentiable, the input weights and biases of ELM are stochastically determined. Simultaneously, the output matrix of the hidden layer becomes uniquely determined throughout the training process. Consequently, the ELM learning process is analogous to finding the least squares solution, which can be mathematically expressed as follows:

$$\hat{\beta} = H^{+} Y \tag{4}$$

In this equation, H^{+} stands for the Moore–Penrose pseudoinverse of the hidden layer output matrix.

3. Optimization of the WOA Algorithm

Given the unknown nature of the optimal initial weights and thresholds for ELM, the present study employs WOA for optimization. By considering the optimal position of an individual within the initial whale population as the best candidate set for the target position, two essential steps are achieved. Firstly, the optimal individual's position serves as a reference for identifying the best candidate set. Secondly, the remaining individuals progressively converge towards this identified best candidate set, while iterative position updates are conducted to iteratively approach the optimal solution [23]. The mathematical expression for this stage is presented below:

$$\begin{aligned} X(t+1) = X^{*}(t) - A \cdot D,\ A = 2a \cdot r - a \\ D = |C \cdot X^{*}(t) - X(t)|,\ C = 2r,\ a = 2(1 - t/T) \end{aligned} \tag{5}$$

In this equation, $X(t+1)$ represents the position vector of the individual whale after iterative updates at the current iteration. $X(t)$ denotes the position vector of the individual whale at the current iteration. $X^{*}(t)$ signifies the position vector of the optimal individual within the current whale population. D stands for the random distance vector between the whale and the target. A and C are coefficient vectors. a represents the linearly decreasing attenuation coefficient of $2 \rightarrow 0$. r is *a* random value between 0 and 1. t represents the current iteration number, and T denotes the maximum number of iterations.

During each iteration update, a random individual whale is selected, and the distance between this whale and the target is calculated. Subsequently, a spiral equation is established between the individual whale and the target based on this distance calculation. The mathematical expression for this spiral equation is as follows:

$$X(t+1) = X^{*}(t) + D' \cdot e^{bl} \cdot \cos(2\pi l);\ D' = |X^{*}(t) - X(t)| \tag{6}$$

In the provided equation, D' represents the distance vector between the current individual whale and the target. The constant b defines the logarithmic spiral curve, and l is a random variable between 0 and 1.

In this study, a probability threshold p is set to simultaneously achieve both of the aforementioned approaches. This ensures that the sperm whale randomly selects either of the two models with equal probabilities. The final mathematical expression for this stage is presented below:

$$X(t+1) = \begin{cases} X^*(t) - A \cdot D & p < 0.5 \\ X^*(t) + D' \cdot e^{bl} \cdot \cos(2\pi l) & p \geq 0.5 \end{cases} \tag{7}$$

In this equation, p is a random variable between −1 and 1.

To enhance the search capability further, this study calculates the magnitude of A in real time during each iteration. When $|A|$ is below the threshold 1, the selection of the optimal individual from the whale population as the target for position updates, in accordance with Equation (7), exemplifies the algorithm's local search capability. Additionally, when the magnitude of $|A|$ surpasses a specific threshold of 1, the algorithm enhances its global search capability. This enhancement is achieved by adopting a strategy that involves randomly selecting a whale individual as the target position and updating the positions of other individuals accordingly. The mathematical expression for this process is provided below:

$$X(t+1) = X_{rand}(t) - A \cdot D; D = |C \cdot X_{rand}(t) - X(t)| \tag{8}$$

In this formula, the symbol $X_{rand}(t)$ represents the position vector of a randomly selected whale individual from the current whale population.

4. Bridge Condition Deterioration Prediction Model Based on WOA-ELM

The objective of this study is to establish a data-centric bridge health prediction model. In this context, a substantial amount of bridge inspection data are utilized within this study. However, the ELM model's initial weights and thresholds exhibit randomness and lack of consistency, thereby constraining the further enhancement of model accuracy. To address this challenge, WOA is introduced as a remedy. The initial phase entails data preprocessing. During this phase, 539 sets of bridge inspection data are organized into a dataset to facilitate training and testing with WOA-ELM. The subsequent phase involves leveraging the training set to calculate optimal hyperparameters for the ELM predictive model and establish the network topology. Following that, WOA is applied, utilizing the training set and network topology, to compute the most suitable model weights for this dataset. In the ultimate phase, optimized initial weights and thresholds are incorporated into the ELM model for training, culminating in the development of a bridge deterioration prediction model. Figure 2 illustrates the workflow of the regional bridge deterioration prediction model based on WOA-ELM.

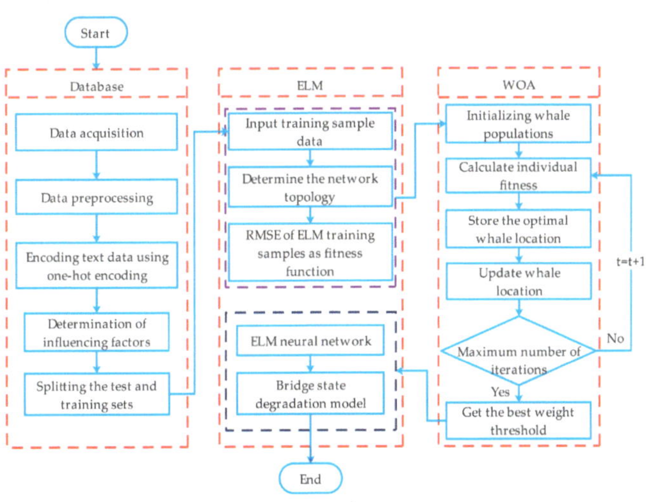

Figure 2. Construction process of bridge state deterioration prediction model based on WOA-ELM.

5. Performance Metrics for Evaluation

To assess the performance of the model, this study employs the mean absolute error (MAE), mean relative error (MRE), root mean square error (RMSE), and correlation coefficient (R) as evaluation metrics for the predictive model [24,25]. The computations for each performance evaluation metric are as follows:

$$MAE = \frac{1}{N}\sum_{i=1}^{N}|y_i - \hat{y}_i| \tag{9}$$

$$MRE = \frac{1}{N}\sum_{i=1}^{N}\left|\frac{y_i - \hat{y}_i}{y_i}\right| \tag{10}$$

$$RMSE = \sqrt{\frac{1}{N}\sum_{i=1}^{N}(y_i - \hat{y}_i)^2} \tag{11}$$

$$R = \frac{N\sum_{i=1}^{N} y_i\hat{y}_i - \sum_{i=1}^{N} y_i \sum_{i=1}^{N} \hat{y}_i}{\sqrt{\left[N\sum_{i=1}^{N} y_i^2 - \left(\sum_{i=1}^{N} y_i\right)^2\right]\left[N\sum_{i=1}^{N} \hat{y}_i^2 - \left(\sum_{i=1}^{N} \hat{y}_i\right)^2\right]}} \tag{12}$$

where *MAE* measures the absolute deviation between the predicted values and the expected values. It effectively addresses the issue of error cancellation and serves as an indicator of the model's generalization capability. *MRE* measures the relative deviation between the predicted and expected values, providing valuable insights into the model's generalization ability. *RMSE* assesses the fluctuations in deviation between the predicted and expected values, thus serving as a reflection of the model's accuracy. *R* signifies the goodness of fit and predictive accuracy of the model. Here, y_i denotes the actual value of the *i*th sample, and $\hat{y}_i$ represents the model's predicted value for the same sample. The variable N corresponds to the total number of samples, and $|R| < 0.4$ denotes a low-degree linear correlation. A correlation coefficient value within the range $0.4 < |R| < 0.7$ indicates a moderate correlation, whereas a value within the range $0.7 < |R| < 1$ signifies a high-degree linear correlation.

6. Method Validation

6.1. Dataset

Establishing a comprehensive bridge condition database serves as an essential foundation for investigating the evolutionary patterns of bridge deterioration. This article gathers documentary materials, including sets of periodic inspection reports, construction design drawings, repair drawings, and maintenance records, for 539 urban bridges situated from 2011 to 2019.

6.1.1. Data Preprocessing

When dealing with multi-source data, a key-value pair collection approach is employed to represent and store each individual bridge entity. The former specifies the specific features, while the latter assigns corresponding data to these features [26]. Bridge entity attributes should encompass information from three aspects: the route level, bridge level, and component level. Different attributes have varying data formats, such as text-based or numerical data. Text-based data typically represent the name or label of the subject under study, while numerical data indicate quantitative relationships between data points.

However, the integrated raw database cannot be directly used for subsequent analyses. This is primarily due to disparities in the quality of historical raw data, such as limited preservation time for paper-based records and instances of missing data. These issues significantly affect the continuity and traceability of data representation. Additionally, the storage of electronic data is constrained by technical standards and management practices, resulting in inconsistent storage formats, information gaps, and inconsistencies.

Therefore, it is necessary to preprocess the initial raw data to maximize the elimination of data noise's impact on subsequent evaluations and ensure the reliability and applicability of the database. The construction process of the regional bridge condition database proposed in this paper is illustrated in Figure 3. Common data cleaning methods include deletion, analogy filling, and mean filling.

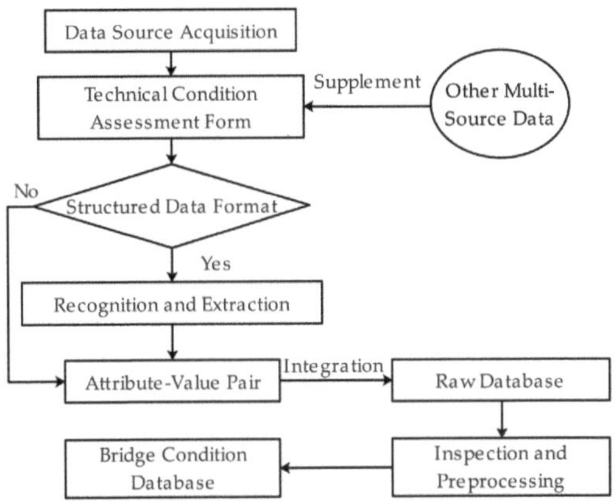

Figure 3. Bridge condition database construction process.

6.1.2. Mathematical Representation of Bridge Condition Data

The bridge condition database encompasses diverse data types. The input variables adopted for modeling encompass numerical variables, such as bridge age and length. Moreover, textual variables are included in region and bridge type. Additionally, Boolean variables are employed to signify maintenance conditions. To ensure numerical stability and avert gradient explosions, the normalization technique will be applied to standardize the numerical variables, and eliminate their influence on the model. Concerning the textual variables, there is no explicit correlation or ordinal relationship among distinct values. Therefore, one-hot encoding will be utilized to transform them into $1 \times N$ binary vectors, where N denotes the number of categories for each variable.

For instance, in the context of bridge type, predefined categories will be used for encoding. These categories include hollow slab beam, T-beam, and box girder, which will be encoded as (1, 0, 0), (0, 1, 0), and (0, 0, 1), respectively, thereby serving as input neurons. Similarly, the area variable will be divided into three major regions within the primary urban area: Region 1, Region 2, and Region 3. These regions possess diverse climatic, spatial, geographical, and environmental characteristics. They will be encoded using the same methodology as previously mentioned. Furthermore, for the Boolean variable maintenance condition, a value of 1 will signify a year with maintenance performed, and a value of 0 will represent a year without maintenance.

6.1.3. Establishment of Bridge Condition Database

Secondary encoding, such as minimum-maximum normalization and one-hot encoding, will be performed on the data to prepare the input variables suitable for network training. The dataset will be randomly divided into a training set comprising 80% of the samples (430 cases) and a test set with 20% of the samples (109 cases). The ranges of values for each variable are presented in Table 1. The complete dataset is shown in Appendices A and B.

Table 1. Prediction model variable representation and value ranges.

Variable	Value Ranges
Bridge Condition Index: y	$[0, 100]$
The service life of the bridge t/years	$[0, 20]$ rounding
The regions {Region 1, Region 2, Region 3} T_a	$\{1,0,0\}, \{0,1,0\}, \{0,0,1\}$
The bridge types {Slab beam, T-beam, Box girder} T_b	$\{1,0,0\}, \{0,1,0\}, \{0,0,1\}$
The structural forms {Simply supported beam, Continuous beam} T_c	$\{0, 1\}$
Crossing l/m	$[6, 54]$
Bridge Length L/m	$[8, 1650]$
Bridge Width B/m	$[7, 44]$
Number of Lanes N	$[1, 8]$ rounding
Maintenance Condition {No maintenance, Maintenance performed} M	$\{0, 1\}$

6.2. Data Normalization

For numerical variables, there are significant differences in value ranges among different features [27]. Consequently, unequal weights are assigned to the variables, which leads to the phenomenon of feature dilution in variables with smaller value ranges. To address this, the minimum-maximum normalization method is employed to map the feature values of variables into the [0, 1] interval, as shown in Equation (13):

$$x_i^{j\prime} = \frac{x_i^j - x_{i\min}}{x_{i\max} - x_{i\min}} \quad i = 1,2,\ldots,m \ \ j = 1,2,\ldots,n \tag{13}$$

where m denotes the number of variables; n represents the number of samples for each variable; x_i^j is the original value of the jth sample for the ith variable; $x_i^{j\prime}$ is the normalized value corresponding to the variable after processing; and $x_{i\max}$ and $x_{i\min}$ are the lower and upper bounds, respectively, for the variable values.

6.3. Determination of Influencing Factors

Selecting the appropriate input variables is pivotal for ensuring accurate predictions of BCI. Inadequate input variables cannot adequately capture the core aspects of bridge degradation issues, while an excess of variables can lead to challenges like overfitting and model incongruity, simultaneously elevating the model's computational complexity [28]. Consequently, this study adheres to the methodology described in references [29–31], utilizing statistical analysis and mutual information correlation coefficients for the meticulous selection of input variables customized for diverse bridge components. Because whether the bridge structure has undergone maintenance is a crucial influencing factor in this study, we will focus on selecting variables from the remaining factors in the variable selection process.

The scatterplot distribution of sample feature variables and BCI established in this study is depicted in Figure 4.

Subsequently, an exploration of the relationships between various variables and the output was conducted. Appropriate correlation coefficients were chosen for statistical correlation analysis. We employed coefficients to determine the correlation between two categorical variables, used the Pearson correlation coefficient for computing the correlation between two interval variables, and typically applied eta-squared coefficients for the correlation between mixed variables involving both categorical and interval data. These findings are presented in Figure 5.

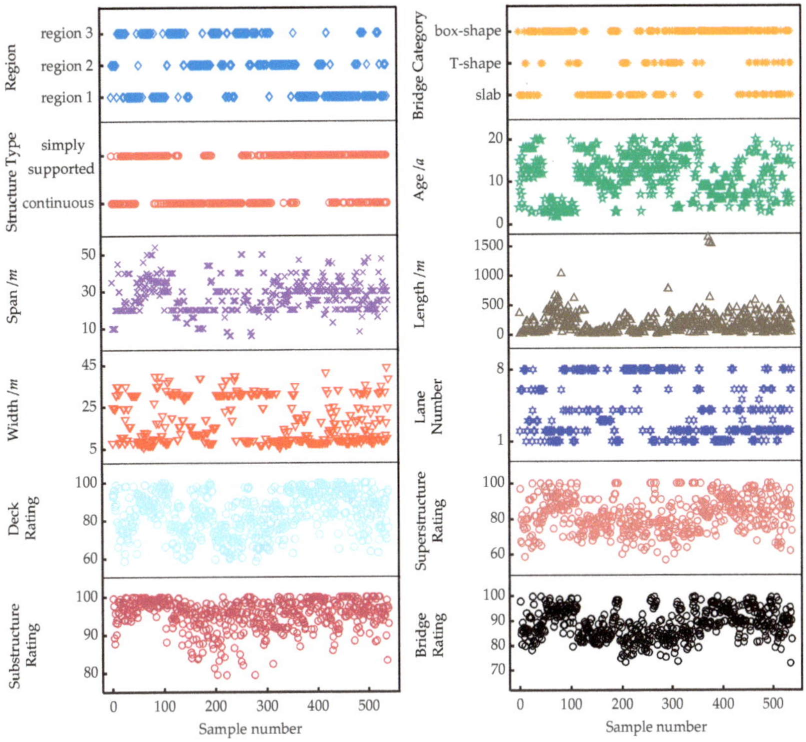

Figure 4. Scatter plot of dataset samples.

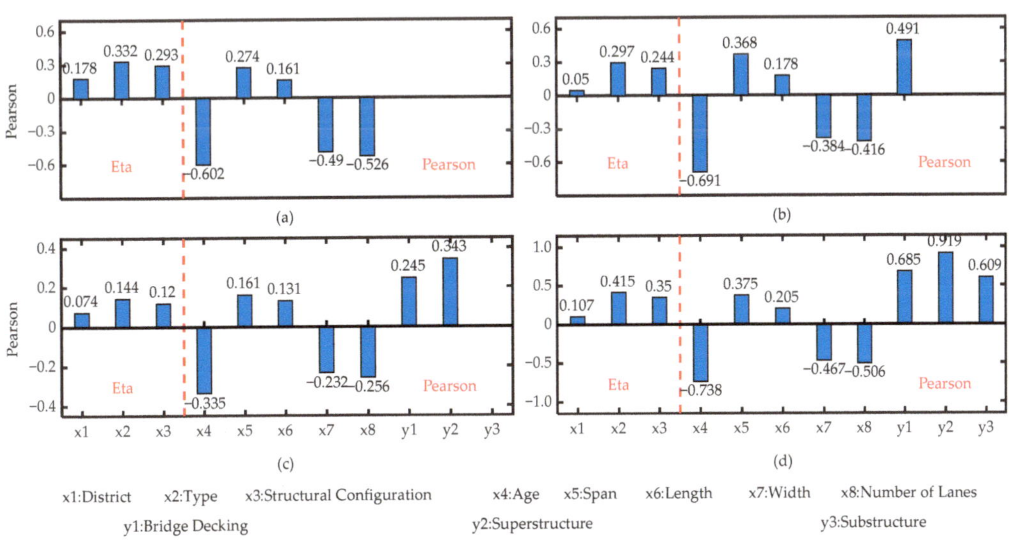

Figure 5. Bridge rating and variable correlation degree based on statistical correlation coefficients. (**a**) Bridge decking rating and correlation with variables; (**b**) superstructure rating and correlation with variables; (**c**) substructure rating and correlation with variables; (**d**) overall structural rating and correlation with variables.

Figure 5 depicts the correlation between bridge ratings, determined using statistical correlation coefficients, and various variables. Notably, a strong correlation is observed between the bridge deck, superstructure, and substructure variables and the overall structural rating. This correlation aligns with the bridge assessment methodology that takes into account the component weights. When examining the correlations between bridge decking, superstructure, and substructure, the highest correlations are observed in the following order: [bridge decking, superstructure], [superstructure, substructure], and [bridge decking, substructure]. This pattern emerges because, during the operational phase, the bridge decking and superstructure act as a unified entity, sharing the load, which results in the highest correlation among all combinations. Conversely, the superstructure and substructure are typically connected through supports, characterized by load transmission rather than simultaneous loading, resulting in slightly lower correlation coefficients. Additionally, the bridge decking and substructure lack a direct mutual interaction relationship, which accounts for the lower observed correlation. These findings reflect the inherent structural characteristics and common interaction patterns in bridges. Furthermore, it is worth noting that the overall structural rating is significantly affected by its service life. This is evident from correlation coefficients exceeding 0.6 for all variables, except for the substructure. Bridge age emerges as the primary factor influencing structural performance degradation. Additionally, the bridge type and structural form contribute to the rating to varying degrees. Effective combinations of hollow box girders, T-beams, box girders, simple supports, and continuous beams can enhance the bridge's service life throughout its full lifespan and provide sufficient performance reserves. Notably, the overall structural rating decreases with increasing bridge width and lane count, indirectly reflecting the unique contribution of traffic volume to the bridge deterioration process.

The aforementioned correlation analysis reveals the interdependencies between bridge age, region, bridge type, structural form, span, bridge length, bridge width, lane count, and the ratings for the bridge decking, superstructure, substructure, and overall structural performance. It highlights the unique contributions of each variable to the deterioration of bridge performance, forming the basis for extrapolation and prediction in long-term structural performance assessment. However, it is worth noting that some of the correlation coefficients are relatively low, indicating either weak or very weak correlations. This can be attributed, in part, to variations in rating data stemming from differences in inspection personnel's habits and cognitive levels in describing bridge defects. This leads to data inaccuracies and instability, resulting in pronounced data variability. Furthermore, Pearson correlation coefficients are inherently sensitive to linear data and may struggle to quantify and identify potential nonlinear relationships between variables. This may account for the lower correlation coefficients observed in certain cases.

The discrete distribution characteristics of the bridge inspection sample data are illustrated in Figure 6, which serves as an example of visualizing inherent relationships among variables. It is evident that a pronounced linear relationship exists between bridge age, span, superstructure, and the BCI of overall bridge structure, as indicated by robust Pearson correlation coefficients. Conversely, for bridge type and bridge length, the data exhibit a scattered and disorganized distribution, reflecting some nonlinear relationships. These nonlinear associations can be effectively identified and quantified using maximum mutual information coefficients, resulting in relatively high correlation coefficients.

Furthermore, taking into consideration the diverse correlation patterns existing among the data, we delve deeper into the intrinsic characteristics and associations among bridge data. Based on joint statistical analysis and mutual information correlation coefficients, we identify the degree of correlation between variables. Employing a correlation threshold of 0.2, variables with correlation coefficients lower than 0.2 are excluded from the subsequent modeling of bridge degradation states. This screening process results in a candidate feature set that exhibits higher correlation with bridge ratings. This serves to reduce the dimensionality and computational complexity in subsequent model training and prediction, thereby minimizing unnecessary computations. The comprehensive results of the correlation analysis, considering various correlation patterns, are presented in Figure 7. The variable candidate sets, ranked by correlation coefficient magnitude, are as follows: [bridge decking: bridge age, lane count, bridge type, structural form, bridge width, span, bridge length, region], [Superstructure: bridge age, lane count, bridge type, structural form, bridge width, span, bridge length], [substructure: bridge age, bridge type, structural form, lane count, bridge width, region], and [overall structure: bridge age, lane count, bridge width, bridge type, structural form, span, bridge length, region].

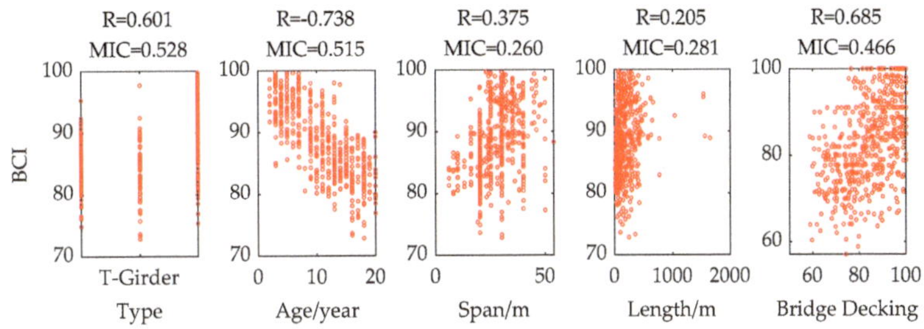

Figure 6. Discrete distribution characteristics of correlation analysis variables.

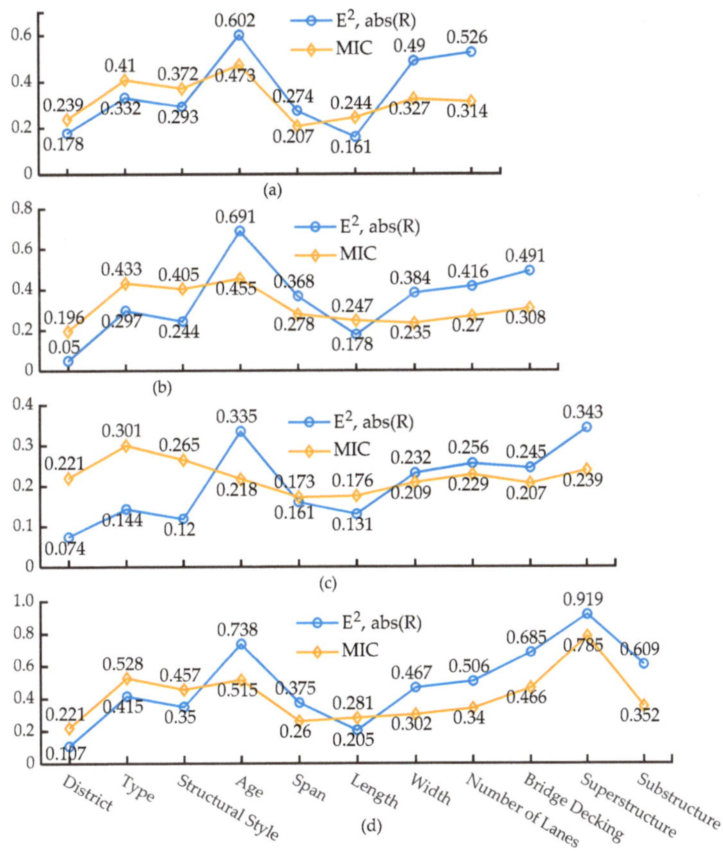

Figure 7. Discrete distribution characteristics of correlation analysis variables. (**a**) Bridge decking rating and correlation with variables; (**b**) superstructure rating and correlation with variables; (**c**) substructure rating and correlation with variables; (**d**) overall structural rating and correlation with variables.

To further substantiate the robustness of the aforementioned analysis, we have employed the quantile–quantile plot to visualize the predictive outcomes, both before and after the selective removal of specific input variables in Figure 8.

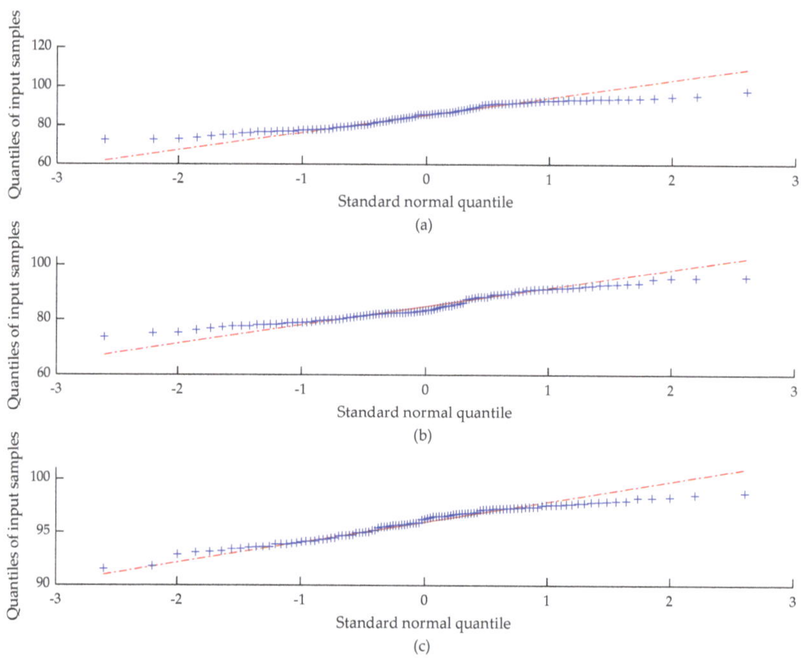

Figure 8. Quantile–quantile plot of BCI values. (**a**) Quantile–quantile plot of BCI values for bridge deck; (**b**) quantile–quantile plot of BCI values for superstructure; (**c**) quantile–quantile plot of BCI values for substructure.

From our examination of the plots, we can draw the following conclusions: for the predictions of BCI values pertaining to the bridge deck, superstructure, and substructure, data points exhibit a predominantly linear distribution along the diagonal line. This observation signifies that the predictive outcomes from the model remain largely consistent in terms of their distribution characteristics, irrespective of the inclusion or exclusion of specific input variables. More precisely, in the context of bridge deck BCI predictions, the mean difference between predictions with and without the exclusion of specific input variables is 2.1%, with a corresponding standard deviation variance of 3.7%. Similarly, for superstructure BCI value predictions, the mean difference stands at 2.7%, with a standard deviation variance of 2.0% after selective input exclusion. In the case of substructure BCI value predictions, the mean difference is 2.1%, accompanied by a standard deviation variance of 4.0% following the removal of specific inputs. These findings serve to further underline the rationality behind the removal of these specific input variables.

6.4. Establishment of the WOA-ELM Model

6.4.1. Selection of Activation Functions

Activation functions play a pivotal role in neural networks. Here are some of the most prevalent activation functions in neural networks, including the sigmoid function, the sine function, and the hardlim function [32,33], with specific details provided below:

$$Sigmoid(x) = \frac{1}{1+e^{-x}} \tag{14}$$

$$Sine(x) = Sin(x) \tag{15}$$

$$Hardlim(x) = \begin{cases} 0 & x < 0 \\ 1 & x \geq 0 \end{cases} \tag{16}$$

In light of the findings in Figure 9, utilizing the sigmoid function as the activation function in the extreme learning machine model demonstrates improved learning efficiency and enhanced fitting precision. Following that, the performance of the sine function is observed, while among these three activation functions, the hardlim function displays inferior performance. Therefore, we have opted for the sigmoid function as the activation function for subsequent modeling.

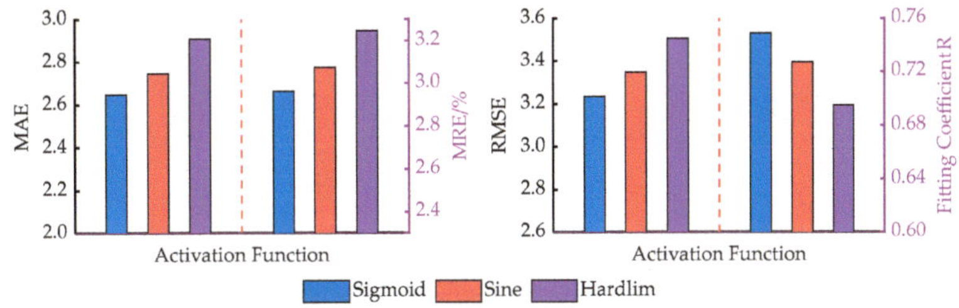

Figure 9. Different Activation Functions' ELM Prediction Accuracy.

6.4.2. Determining the Number of Neurons in the Neural Network

In this study, we evaluated the predictive accuracy of the ELM model using the sigmoid activation function. The assessment was conducted for various numbers of neurons in the hidden layer, namely, 5, 10, 15, 20, 25, and 30. The obtained accuracy results exhibited a parabolic distribution, and it was observed that the highest accuracy was achieved when the number of neurons was set to 20 in the hidden layer. Consequently, we determined the number of neurons in the hidden layer as 20 for our model. The results are illustrated in Figure 10.

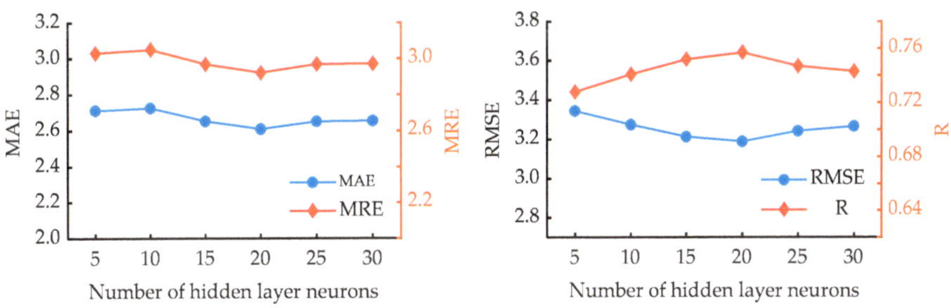

Figure 10. ELM prediction accuracy of different numbers of hidden layer neurons.

6.5. Forecasting Results and Comparisons

6.5.1. Prediction of BCI of Overall Bridge Structure

Before each training and testing process, a set of 430 data points was randomly chosen as training samples. Additionally, 109 data points were reserved as testing samples. Based on the previous research, the activation function of the ELM algorithm is set to the sigmoid function, with 20 hidden-layer neurons; the WOA population size is set to 20 individuals, and the maximum number of iterations is 50 times.

To mitigate the influence of random sample allocation on predictive results, the best outcome from 10 runs was adopted as the final result. Furthermore, a comparative analysis was conducted to validate the superiority of the proposed method. The analysis involved

WOA-ELM, ELM, BPNN, decision trees (DT) and support vector machine (SVM) as the subjects of comparison.

The predictive model's performance was evaluated using several metrics, including MAE, MRE, RMSE, and R. The comparison between the predicted and expected BCI is illustrated in Figure 11. The input layer of the model comprised 11 neurons, while the output layer consisted of 1 neuron.

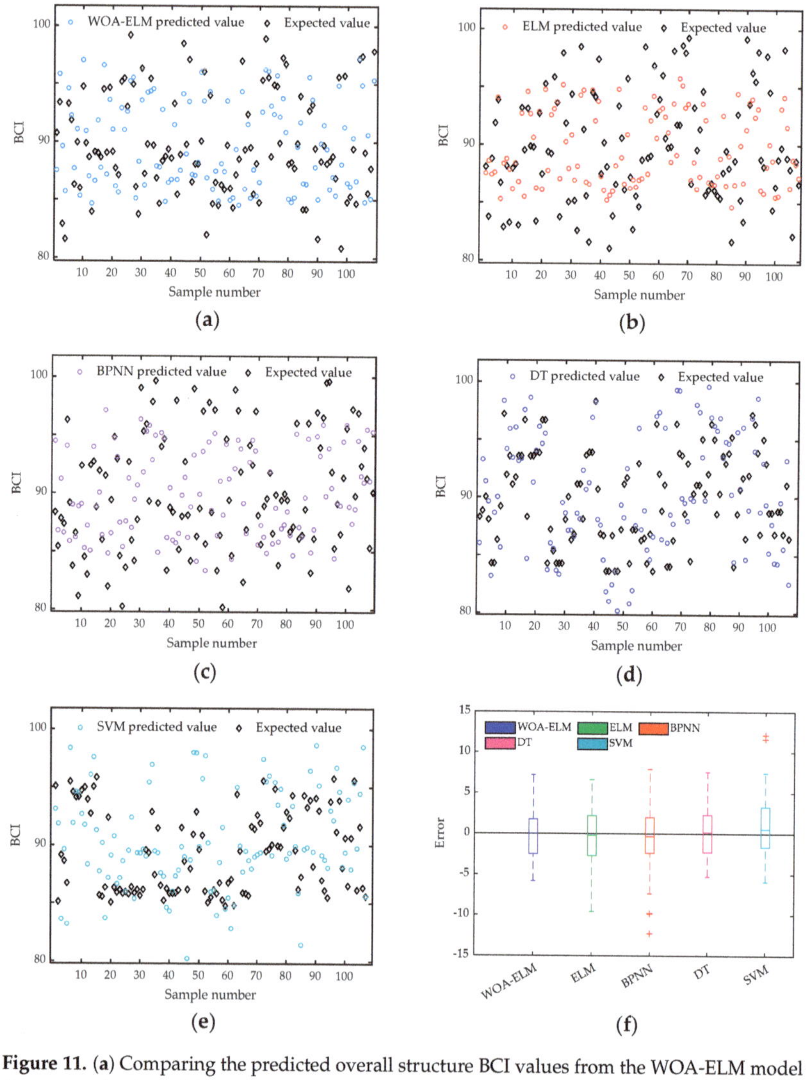

Figure 11. (**a**) Comparing the predicted overall structure BCI values from the WOA-ELM model with the expected values; (**b**) comparing the predicted overall structure BCI values from the ELM model with the expected values; (**c**) comparing the predicted overall structure BCI values from the BPNN model with the expected values; (**d**) comparing the predicted overall structure BCI values from the DT model with the expected values; (**e**) comparing the predicted overall structure BCI values from the SVM model with the expected values; (**f**) box plot of model error.

Figure 11 reveals differences between the predictions of BCI values for the technical condition of bridges by five different algorithms and their expected values, resulting in a substantial overlap among certain data points. In practice, the predictive models based on the WOA-ELM algorithm and the ELM algorithm demonstrate remarkable proficiency in capturing sample characteristics, and exhibit especially favorable proficiency for forecasting trends in the technical state of bridges. Accurately predicting changes in bridge conditions will aid managers in taking preemptive measures to maintain bridge performance and enhance safety. Additionally, upon closer examination of the prediction error distribution for BCI values among the five algorithms, the WOA-ELM algorithm manages the magnitude of prediction errors significantly, with overall error fluctuations primarily concentrated around values near zero. Following this, the ELM algorithm exhibits relatively lower error levels. However, the SVM and BP neural networks, among the five algorithms, exhibit the most significant overall deviations in prediction errors and possess a larger number of outliers.

Figure 12 portrays scatter plots that exhibit the forecasted and anticipated BCI values, employing five algorithms: WOA-ELM, ELM, BPNN, DT, and SVM. The data points from all five models extend outward around the diagonal line. Additionally, five is a dense distribution near the diagonal, indicating a high degree of agreement between the predicted and expected values. Furthermore, the probability density distribution curves of the BCI prediction errors for the five algorithms are compared. The WOA-ELM algorithm exhibits a smaller extension range and a higher concentration of data points around the peak of the density curve. These observations indicate its superior ability to track the expected values. Moreover, the ELM and DT algorithms show a slightly lower capturing effect on the expected values, and the BPNN and SVM algorithms display a larger dispersion in the distribution of prediction data, indicating poorer control over error fluctuations.

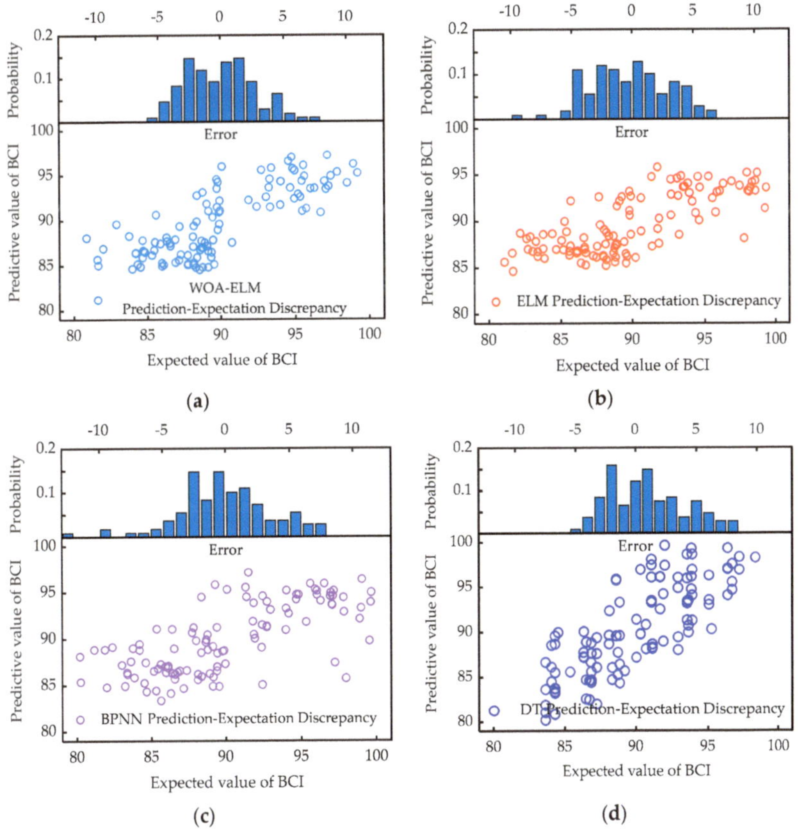

Figure 12. *Cont.*

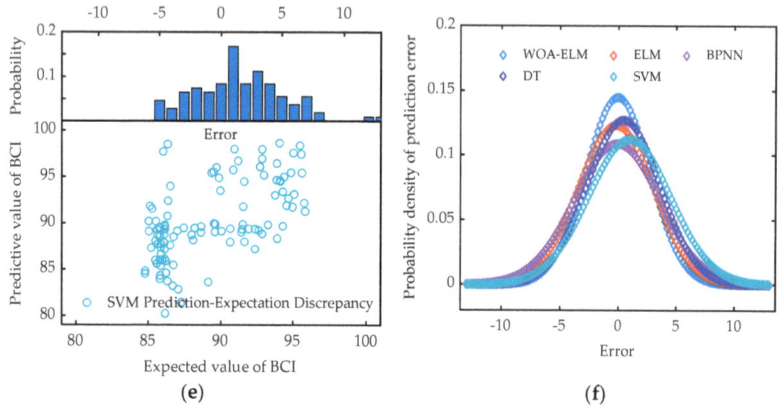

Figure 12. (**a**) Comparative analysis of deviation between predicted and expected BCI using the WOA-ELM model; (**b**) comparative analysis of deviation between predicted and expected BCI using the ELM model; (**c**) comparative analysis of deviation between predicted and expected BCI using the BPNN model; (**d**) comparative analysis of deviation between predicted and expected BCI using the DT model; (**e**) comparative analysis of deviation between predicted and expected BCI using the SVM model; (**f**) probability density analysis of prediction errors among the WOA-ELM, ELM, BPNN, DT, and SVM models.

The results depicted in Table 2 reveal significant advancements attained by the WOA-ELM model when compared to the ELM, BPNN, DT, and SVM models. In particular, the WOA-ELM model exhibits significant improvements in R, with enhancements of 4.1%, 11.4%, 24.5%, and 33.6%. Additionally, the model demonstrates reductions in MAE by 9.9%, 13.6%, 5.4%, and 15.7%, and MRE by 11.6%, 15.3%, 6%, and16.2%, respectively. Furthermore, the RMSE is lower by 7.3%, 18.0%, 14.8%, and 18.1% for the corresponding evaluations. The obtained results establish a remarkable level of consistency between the proposed WOA-ELM model and the other four models concerning the predictions of BCI values. When compared to other models, the proposed model exhibits superior performance in various aspects. It shows superiority in terms of both absolute and relative deviations in predictions. Additionally, the proposed model demonstrates better performance in deviation fluctuation and goodness of fit measures. The proposed model exhibits the lowest level of deviation between predicted and expected values. This signifies its high prediction accuracy and remarkable generalization capability. As a result, the model enables the effective forecasting of bridge states under various time points and feature variables. These achievements can be primarily attributed to the optimization of the initial weights and thresholds of the extreme learning machine through the whale optimization algorithm. Following this, the network undergoes training and testing on the dataset with the optimized initial weights and thresholds. As a consequence, commendable performance is achieved in both accuracy and generalization capability.

Table 2. Summary of predictive performance indicators of different models.

Category	Prediction Model	MAE	MRE (%)	RMSE	R
Overall structure	WOA-ELM	2.412	2.632	3.003	0.768
	ELM	2.676	2.978	3.241	0.738
	BPNN	2.791	3.106	3.661	0.689
	DT	2.549	2.800	3.151	0.616
	SVM	2.862	3.14	3.667	0.575

6.5.2. Prediction of BCI of Bridge Components

Similarly, employing the WOA-ELM model with consistent parameters, predictions were made for the degradation scores of the bridge deck, superstructure, and substructure. A comparative analysis between the predicted BCI and the expected values is presented in Figures 12–14. Specifically, the bridge deck model incorporates an input layer consisting of nine neurons, the superstructure model features an input layer comprising eight neurons, and the substructure model is equipped with an input layer comprising seven neurons. Each of these models has an output layer consisting of one neuron.

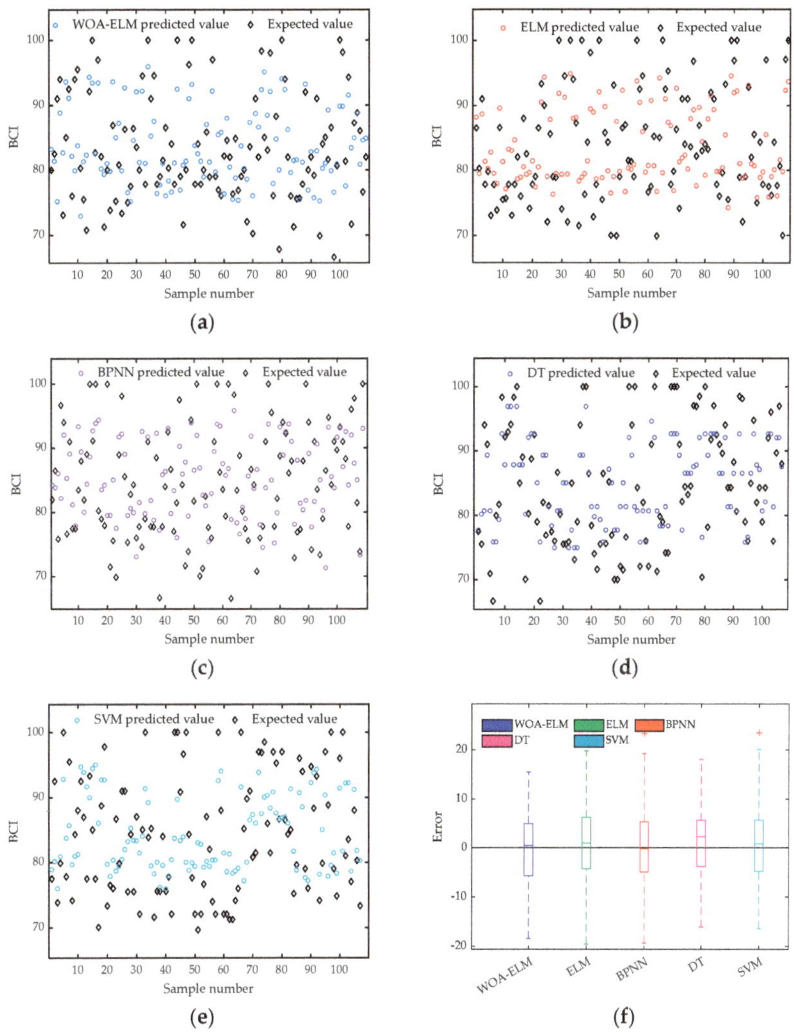

Figure 13. (**a**) Comparing the predicted bridge deck BCI values from the WOA-ELM model with the expected values; (**b**) comparing the predicted bridge deck BCI values from the ELM model with the expected values; (**c**) comparing the predicted bridge deck BCI values from the BPNN model with the expected values; (**d**) comparing the predicted bridge deck BCI values from the DT model with the expected values; (**e**) comparing the predicted bridge deck BCI values from the SVM model with the expected values; (**f**) box plot of model error in the bridge deck BCI ratings.

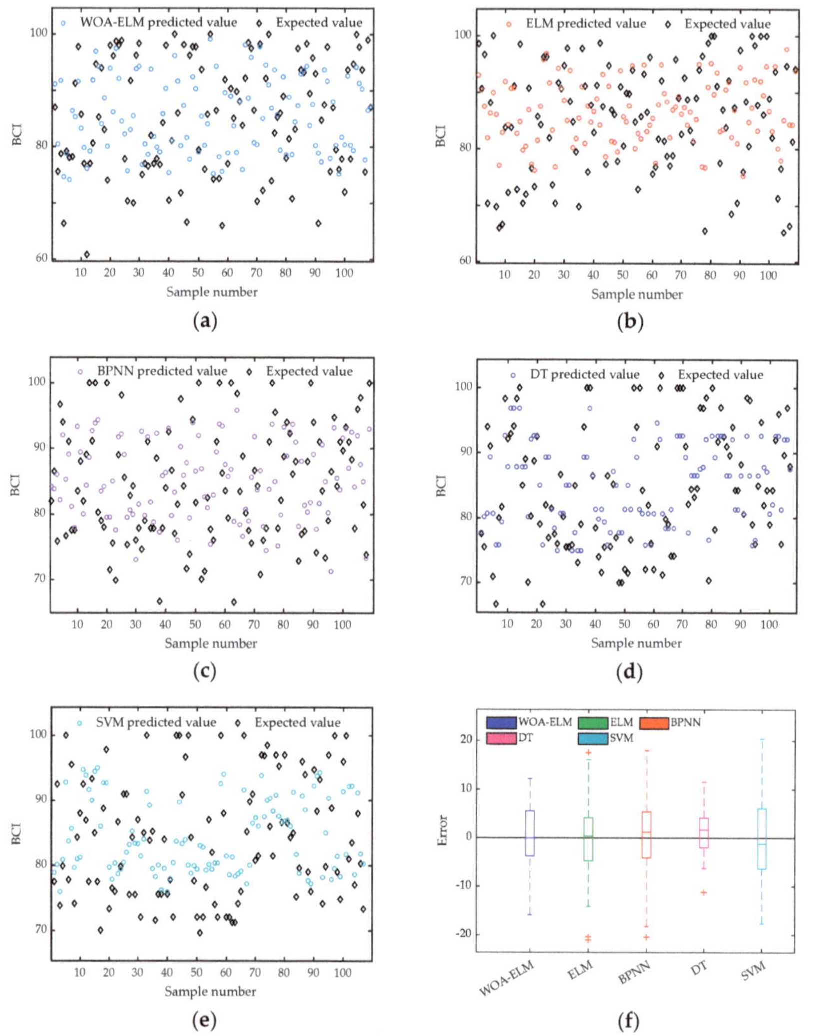

Figure 14. (**a**) Comparing the predicted superstructure BCI values from the WOA-ELM model with the expected values; (**b**) comparing the predicted superstructure BCI values from the ELM model with the expected values; (**c**) comparing the predicted superstructure BCI values from the BPNN model with the expected values; (**d**) comparing the predicted superstructure BCI values from the DT model with the expected values; (**e**) comparing the predicted superstructure BCI values from the SVM model with the expected values; (**f**) box plot of model error in the superstructure BCI ratings.

According to Figures 13–15, the WOA-ELM algorithm demonstrates excellent generalization performance and fitting accuracy in the evaluation of various components and overall ratings of bridges. It also exhibits a minimal presence of outliers and high applicability across diverse data features. While the DT model shows low error variance and dispersion in predicting BCI values for upper structural elements, there is room for improvement in other performance aspects. It is worth noting that the model exhibits relatively lower fitting accuracy in the evaluation of lower structural components. This is primarily due to the fact that, within the entire regional road network, most bridge substructures remain in good condition, with no localized minor damage or deterioration. As a result, BCI values tend to be uniformly high, and the variability is low, making the algorithm susceptible to noise interference and, thus, challenging in accurately capturing the relationship between structural states and feature variables. Especially taking bridge decks as an example, the WOA-ELM model demonstrates a significant advantage in terms of R-squared, with improvements of 9.5%, 12.7%, 16.7%, and 19.7%, respectively. Furthermore, the model exhibits reductions of 4.3%, 6.5%, 1.7%, and 12.5% in terms of MAE, and reductions of 6.4%, 6.7%, 2.8%, and 13.3% in terms of MRE. Additionally, the corresponding RMSE evaluations show reductions of 5.4%, 9.0%, 4.3%, and 12.7%, respectively.

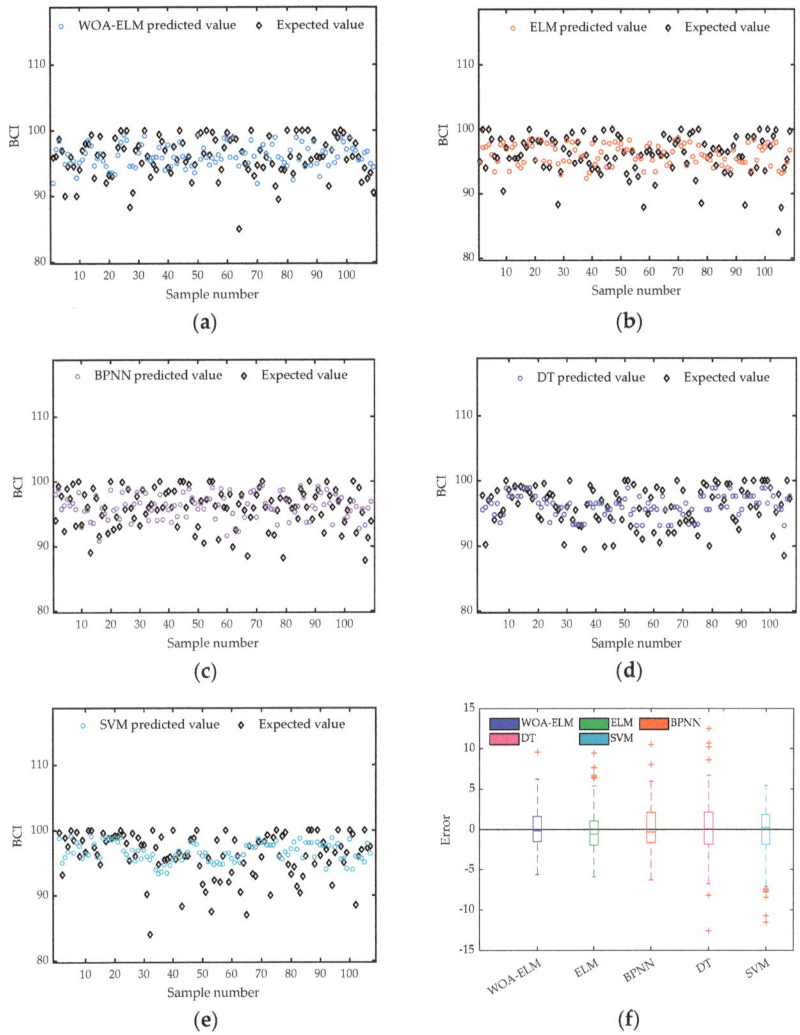

Figure 15. (**a**) Comparing the predicted substructure BCI values from the WOA-ELM model with the expected values; (**b**) comparing the predicted substructure BCI values from the ELM model with the expected values; (**c**) comparing the predicted substructure BCI values from the BPNN model with the expected values; (**d**) comparing the predicted substructure BCI values from the DT model with the expected values; (**e**) comparing the predicted substructure BCI values from the SVM model with the expected values; (**f**) box plot of model error in the substructure BCI ratings.

Given that the randomness of the data can affect the model to some extent, we conducted ten random splits of the dataset and generated a box plot of the errors between predicted values and expected values, as illustrated below.

According to Figure 16, after ten runs, the WOA-ELM algorithm demonstrated that the median errors in both the overall bridge structure and component ratings were primarily within proximity to zero, reaffirming the model's outstanding predictive performance. Additionally, WOA-ELM exhibited the lowest number of outliers among the five models, indicating strong generalization capabilities. While the BPNN and SVM models had relatively good error distributions, the presence of too many outliers was deemed unacceptable. Furthermore, despite the ELM and DT models showcasing solid generalization and accuracy under ideal conditions, an overall examination of ten runs revealed more scattered error distributions and a higher occurrence of outliers, indicative of the lower stability of these two models.

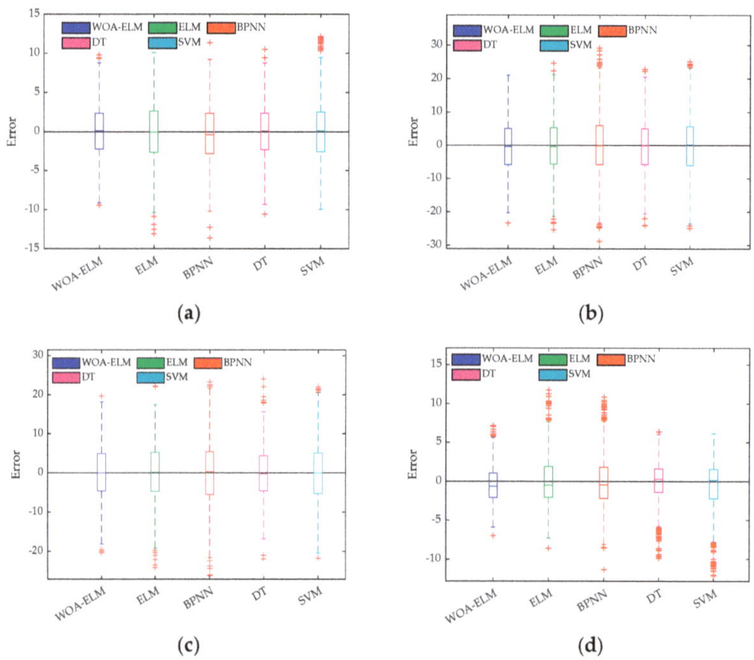

Figure 16. (**a**) Box plot of model error in the overall structure BCI ratings; (**b**) box plot of model error in the bridge deck BCI ratings; (**c**) box plot of model error in the superstructure BCI ratings; (**d**) box plot of model error in the substructure BCI ratings.

7. Conclusions

In this study, a comprehensive dataset of bridge condition deterioration was meticulously collected by the aggregation of extensive bridge inspection data. The dataset encompasses 11 significant input features, including bridge service time, region, and bridge type. Leveraging this comprehensive dataset, a data-driven WOA-ELM bridge condition deterioration prediction model was successfully constructed. To ensure the reliability and robustness of the proposed method, it was subjected to rigorous validation using a substantial volume of inspection data. The extensive validation process has enabled us to draw the following concluding insights:

(1) The foundation of this study rests upon a comprehensive dataset of bridge inspection data, which facilitates the establishment of complex nonlinear connections between essential features and bridge conditions. This research effectively harnesses and explores the inherent data patterns within the long-term bridge inspection data. Setting

itself apart from other prediction models that focus solely on individual bridges, the proposed model demonstrates exceptional proficiency in precisely forecasting the states of diverse bridges within the region.

(2) In this study, we conducted correlation analysis, taking into consideration several key factors, including the bridge's age, lane count, bridge width, bridge type, structural form, span, bridge length, geographic location, and maintenance status. The aim was to enhance the accuracy of BCI prediction. Furthermore, the WOA-ELM prediction model proposed in this paper outperforms the ELM, BPNN, DT, and SVM models in terms of sample fitting capability and accuracy. Specifically, the model presented in this paper exhibited improvements in R-values by 4.1%, 11.4%, 24.5%, and 33.6%, reductions in RMSE by 7.3%, 18.0%, 14.8%, and 18.1%, decreases in MAE by 9.9%, 13.6%, 5.4%, and 15.7%, and reductions in MRE by 11.6%, 15.3%, 6%, and 16.2%. These results clearly demonstrate a significant enhancement in the model's performance for bridge condition prediction.

(3) In the context of predicting the BCI for bridge components, we utilized Pearson correlation analysis and mutual information theory to identify the critical influencing factors that need to be taken into account for each specific component. For instance, within the realm of the bridge superstructure, it was imperative to consider variables such as bridge age, lane count, bridge width, bridge type, structural form, span, bridge length, geographic location, and maintenance status. When undertaking the prediction BCI for various bridge components, our proposed method consistently demonstrated remarkable advantages, surpassing the performance of the ELM, BPNN, DT, and SVM models in terms of predictive accuracy. Specifically, with regard to bridge decking components, our method resulted in significant improvements in R-values, with increases of 9.5%, 12.7%, 16.7%, and 19.7%. Additionally, there were reductions in RMSE by 5.4%, 9.0%, 4.3%, and 12.7%, decreases in MAE by 4.3%, 6.5%, 1.7%, and 12.5%, and reductions in MRE by 6.4%, 6.7%, 2.8%, and 13.3%. These outcomes prominently underscore the exceptional predictive prowess of our methodology across diverse bridge component conditions. The significance of this research lies in the provision of a more dependable technical assessment tool for bridge management and maintenance, poised to assume a pivotal role in practical applications.

Although our proposed method exhibits significant advantages over other models, there is still room for further improvement in the predictive performance of our model (e.g., enhancing the optimization precision and learning efficiency of meta-learning). Therefore, we will focus on exploring meta-learning algorithms to further enhance the performance of our model in the field of BCI prediction in future research.

Author Contributions: L.J. contributed to methodology, conceptualization, data curation, formal analysis, investigation, and writing. Q.T. was involved in data curation, methodology, validation and the preparation of the original draft. Y.J. contributed to conceptualization, investigation and supervision. H.C. was involved in conceptualization, the preparation of the original draft, and data collection. Z.X. participated in validation, reviewing and data collection. All authors have read and agreed to the published version of the manuscript.

Funding: This work was supported by the National Natural Science Foundation of China (Grant No. 52278292), the Chongqing Outstanding Youth Science Foundation (Grant No. CSTB2023NSCQ-JQX0029), the Chongqing Science and Technology Project (CSTB2022TIAD-KPX0205), the Chongqing Transportation Science and Technology Project (Grant No. 2022-01), the Science and Technology Project of Guizhou Department of Transportation (Grant No. 2023-122-001), the China Postdoctoral Science Foundation (Grant No. 2023M730431), the Special Funding of Chongqing Postdoctoral Research Project (Grant No. 2022CQBSHTB2053), and the Research and Innovation Program for Graduate Students in Chongqing (Grant No. CYS23477).

Data Availability Statement: The data presented in this study are available on request from the corresponding author.

Conflicts of Interest: The authors declare no conflict of interest.

Appendix A

Table A1. Dataset.

NO.	A	B	C	D	E	F	G	H	I	J	K	NO.	A	B	C	D	E	F	G	H	I	J	K
1	2	1	4	3	35	365	8	2	0	0	0	271	1	2	13	1	8	20	31	8	0	0	0
2	1	2	15	1	10	20	25	6	0	1	0	272	1	2	19	1	6	8.3	31	8	1	0	0
3	1	2	10	1	10	20	25	6	1	0	1	273	1	2	13	1	6	8.3	31	8	0	0	0
4	1	2	8	1	10	20	25	6	0	0	0	274	2	2	19	3	21	59	9	1	1	0	0
5	1	2	12	1	10	21	24.5	6	1	1	0	275	2	2	15	3	21	59	9	1	0	0	0
6	1	2	7	1	10	21	24.5	6	0	0	0	276	1	3	16	1	20	39.5	33	8	0	1	1
7	1	2	5	1	10	21	24.5	6	0	0	0	277	1	3	13	1	20	39.5	33	8	1	0	0
8	1	2	16	1	20	85	31	8	0	1	0	278	1	3	10	1	20	39.5	33	8	0	0	0
9	1	2	14	1	20	85	31	8	0	0	0	279	1	3	16	1	20	33	31.5	8	1	0	0
10	1	2	18	1	20	52	35	8	0	1	0	280	1	3	13	1	20	33	31.5	8	0	1	0
11	1	2	13	1	20	52	35	8	0	0	0	281	1	3	10	1	20	33	31.5	8	0	0	0
12	1	2	11	1	20	52	35	8	0	0	0	282	1	3	16	2	20	41	31.5	8	1	0	0
13	2	1	4	3	45	45	8	1	0	0	0	283	1	3	13	2	20	41	31.5	8	1	0	0
14	1	3	18	2	40	180	31.5	8	1	1	0	284	1	3	16	1	20	35	31.5	8	0	1	0
15	1	3	18	2	30	230	8.5	2	1	0	0	285	1	3	13	1	20	35	31.5	8	1	0	0
16	1	3	13	2	30	230	8.5	2	0	0	0	286	1	3	10	1	20	35	31.5	8	0	0	0
17	1	3	18	1	20	66	11	3	0	0	0	287	2	2	18	3	19	93	7	1	1	1	0
18	1	3	13	1	20	66	11	3	0	0	0	288	2	2	14	3	19	93	7	1	0	0	0
19	1	3	11	1	20	66	11	3	0	0	0	289	1	2	18	3	24	29	7	1	1	0	0
20	2	3	18	3	20	65	11	3	0	0	0	290	1	2	14	3	24	29	7	1	0	0	0
21	2	3	13	3	20	65	11	3	0	0	0	291	2	2	18	3	24	176	7	1	1	0	0
22	2	3	11	3	20	65	11	3	0	0	0	292	2	2	14	3	24	176	7	1	0	0	0
23	1	3	18	1	20	57	24.5	6	1	1	0	293	1	2	13	2	50	768	24.5	6	1	0	0
24	1	3	13	1	20	57	24.5	6	1	0	0	294	1	2	10	2	50	768	24.5	6	0	0	0
25	2	1	3	3	23	73	8	1	0	0	0	295	1	3	12	2	40	379	32.6	8	1	0	1
26	1	3	18	1	25	84	19.3	4	0	1	0	296	1	3	9	2	40	379	32.6	8	0	0	0
27	2	1	3	3	40	289	35	8	0	0	0	297	1	3	7	2	40	379	32.6	8	0	0	0
28	2	3	18	3	24	236	9.5	2	1	1	0	298	2	2	11	3	30	350	10	2	0	0	1
29	2	3	13	3	24	236	9.5	2	0	0	0	299	2	2	8	3	30	350	10	2	0	0	0
30	2	3	11	3	24	236	9.5	2	0	0	0	300	1	2	19	2	30	234	31	8	0	0	0
31	1	3	18	1	20	29	24.5	6	1	0	0	301	1	2	14	2	30	234	31	8	0	0	0
32	1	3	13	1	20	29	24.5	6	1	0	0	302	1	3	16	2	20	59	31.5	8	1	1	0
33	2	1	20	3	40	100	16	3	0	0	0	303	1	3	16	2	20	125	31.5	8	1	1	0
34	2	1	15	3	40	100	16	3	1	0	0	304	2	2	20	3	20	111	8	2	0	1	0
35	2	1	14	3	40	100	16	3	1	0	0	305	1	2	20	1	16	192	7	1	0	0	0
36	2	1	12	3	40	100	16	3	0	0	0	306	1	3	16	2	20	95	31.5	8	1	1	
37	1	1	20	1	20	35	31	6	1	0	0	307	1	3	13	2	20	95	31.5	8	0	0	0
38	1	1	15	1	20	35	31	6	0	0	0	308	1	3	11	2	20	95	31.5	8	0	0	0
39	1	1	14	1	20	35	31	6	0	0	0	309	2	1	4	3	25	194	9	2	1	0	0
40	1	1	12	1	20	35	31	6	0	0	0	310	2	1	4	3	24	57	9	2	0	0	0
41	2	1	3	3	23	73	8	1	0	0	0	311	1	3	16	2	20	94	31.5	8	0	1	1
42	1	1	20	3	20	54	31	6	0	1	0	312	1	3	13	2	20	94	31.5	8	1	0	0
43	1	1	15	3	20	54	31	6	1	0	0	313	2	2	4	3	28	62	9	2	0	0	0
44	1	1	12	3	20	54	31	6	0	0	0	314	2	2	4	3	34	263	9	2	1	0	0
45	1	1	20	2	25	38	31	6	0	0	0	315	2	2	4	3	29	236	9	2	0	0	0
46	1	1	15	2	25	38	31	6	1	0	0	316	2	2	4	3	35	122	9	2	0	0	0
47	1	1	14	2	25	38	31	6	1	1	0	317	2	2	4	3	36	52	9	2	0	0	0
48	1	1	12	2	25	38	31	6	0	0	0	318	2	2	4	3	35	265	9	2	0	0	0
49	2	1	4	3	35	275	8	1	0	0	0	319	2	2	18	2	20	94	31.5	8	0	0	0
50	2	1	3	3	30	42	9	2	0	0	0	320	2	2	15	2	20	94	31.5	8	1	0	0
51	2	1	3	3	35	111	9	2	0	0	0	321	2	2	18	3	28	294	7	1	0	0	0
52	2	3	6	3	27	207	8	2	0	0	0	322	2	2	15	3	28	294	7	1	1	0	0
53	2	1	3	3	33	72	9	2	0	0	0	323	2	2	18	3	38	288	9	2	0	0	0
54	2	1	3	3	24	420	9	2	0	0	0	324	2	2	15	3	38	288	9	2	1	0	0
55	2	2	3	3	20	90	9	2	0	0	0	325	2	2	14	3	35	133	7	1	1	0	0
56	2	1	3	3	37.5	373	11.5	2	0	0	0	326	2	2	14	3	33	153	5.5	1	0	0	0
57	2	1	3	3	24	68	7	1	0	0	0	327	2	2	18	3	24	176	7	1	0	1	0
58	2	1	3	3	27	407	9	2	0	0	0	328	2	2	15	3	24	176	7	1	1	0	0
59	2	1	4	3	35	465	9	2	0	0	0	329	2	2	18	3	33	262	9	2	0	1	0
60	2	3	6	3	37	150	8	2	0	0	0	330	2	2	15	3	33	262	9	2	0	0	0

Table A1. *Cont.*

NO.	A	B	C	D	E	F	G	H	I	J	K	NO.	A	B	C	D	E	F	G	H	I	J	K
61	2	3	4	3	37	491	6	1	0	0	0	331	2	2	18	3	23	204	9.5	2	1	0	0
62	2	3	6	3	46	171	11.7	2	0	0	0	332	2	2	15	3	23	204	9.5	2	0	0	0
63	2	1	5	3	32	443	13	3	0	0	0	333	2	2	14	3	23	204	9.5	2	1	0	0
64	2	3	6	3	30	142	8	2	0	0	0	334	2	2	18	3	23	140	14.5	3	0	0	0
65	2	3	2	3	33	391	7	1	0	0	0	335	2	2	18	3	28	294	9.5	2	0	0	0
66	2	3	6	3	27	118	9	1	0	0	0	336	1	2	18	3	24	29	13	2	0	0	0
67	2	3	6	3	29	289	9	2	0	0	0	337	2	2	18	3	38	288	9	2	0	1	0
68	2	3	6	3	38	238	12.8	2	0	0	0	338	2	2	18	3	28	294	7	1	1	0	0
69	2	3	4	3	40	526	7	1	0	0	0	339	2	2	18	3	24	176	7	1	1	0	0
70	2	3	4	3	50	660	7	1	0	0	0	340	2	2	5	3	25	157	8	1	0	0	0
71	2	3	2	3	39	228	7	1	0	0	0	341	2	2	5	3	26	270	8	1	0	0	0
72	2	3	4	3	39	400	7	1	0	0	0	342	2	2	5	3	28	171	8	1	0	0	0
73	2	3	2	3	36	530	7	1	0	0	0	343	2	2	5	3	25	25	13.5	3	0	0	0
74	2	3	4	3	39	618	7	1	0	0	0	344	2	2	5	3	30	243	8	1	0	0	0
75	2	3	2	3	32	370	7	1	0	0	0	345	2	2	5	3	31	301	8	1	0	0	0
76	2	3	2	3	38	645	9	2	0	0	0	346	2	2	5	3	28	233	8	1	0	0	0
77	2	3	2	3	50	580	9	2	0	0	0	347	2	2	11	3	20	292	9.5	2	0	0	0
78	2	3	2	3	30	238	7	1	0	0	0	348	2	2	12	3	20	109	9	2	0	0	0
79	2	3	4	3	36	147	7	1	0	0	0	349	1	2	11	2	40	430	24.5	6	0	0	0
80	2	3	4	3	50	465	9.5	2	0	0	0	350	1	2	7	2	40	430	24.5	6	0	0	0
81	2	1	4	3	36	350	10	2	0	0	0	351	1	1	20	1	20	58	31	6	1	0	0
82	2	1	3	3	32	452	27	6	0	0	0	352	1	1	17	1	20	58	31	6	0	0	0
83	2	1	5	3	32	1038	9	2	0	0	0	353	1	2	19	1	20	31	31	8	0	0	0
84	1	1	3	3	30	46	40	8	0	0	0	354	1	2	16	1	20	31	31	8	1	0	0
85	2	3	6	3	54	285	9.5	2	0	0	0	355	1	2	14	1	20	31	31	8	0	1	0
86	2	1	4	3	35	115	17.5	4	0	0	0	356	2	3	4	3	30	71	35	6	0	0	0
87	1	1	4	3	36	151	35	8	0	0	0	357	1	2	17	1	20	70	16	4	0	0	0
88	2	1	4	3	40	168	35	8	0	0	0	358	2	2	17	3	30	150	9.5	2	0	0	0
89	2	1	4	3	40	168	35	8	0	0	0	359	2	2	9	3	32	220	19.5	4	0	0	0
90	2	1	4	3	35	325	35	8	0	0	0	360	2	2	9	3	40	311	19.5	4	0	0	0
91	2	1	4	3	35	180	17.5	4	0	0	0	361	1	2	9	2	35	437	20	4	0	0	0
92	1	1	4	3	36	396	40	8	0	0	0	362	2	2	9	3	30	120	9.5	2	0	0	0
93	2	1	3	3	35	259	37	8	0	0	0	363	2	2	9	3	33	258	19.5	4	0	0	0
94	2	1	3	3	28	340	10	2	0	0	0	364	2	1	9	3	22	27	9	2	0	0	0
95	2	1	3	3	35	145	11.5	2	0	0	0	365	2	1	9	3	47	148	9	2	0	0	0
96	2	1	3	3	31	475	11.5	2	0	0	0	366	2	1	9	3	32	126	9	2	0	0	0
97	2	1	3	3	42	130	11.5	2	0	0	0	367	2	1	9	3	30	400	9	2	0	0	0
98	1	1	3	2	40	296	21	4	0	0	0	368	2	1	9	3	33	182	9	2	0	0	0
99	2	1	3	3	30	367	9	2	0	0	0	369	2	1	9	3	35	345	9	2	1	1	0
100	2	1	3	3	31	72	9	2	0	0	0	370	2	1	9	3	35	170	9	2	0	0	0
101	2	1	3	3	35	321	9	2	0	0	0	371	2	1	9	3	30	106	9	2	0	0	0
102	2	1	3	3	35	328	9	2	0	0	0	372	2	1	9	3	30	1650	18.5	4	0	0	0
103	1	3	20	2	20	162	15.8	4	1	1	0	373	2	1	9	3	30	353	18.5	4	1	0	0
104	2	1	5	3	30	124	8	1	0	0	0	374	2	1	9	3	30	1560	18.5	4	0	0	0
105	2	1	5	3	35	115	33	8	0	0	0	375	2	1	9	3	33	626	18.5	4	1	0	0
106	2	1	5	3	32	128	8	1	0	0	0	376	2	1	7	3	38	113	11.5	2	1	0	0
107	2	1	4	3	30	120	8	1	0	0	0	377	2	1	4	3	38	113	11.5	2	0	0	0
108	1	1	4	3	36	617	37	8	0	0	0	378	2	1	7	3	40	1537	21.5	4	0	0	0
109	2	1	4	3	35	285	8	1	0	0	0	379	2	1	4	3	40	1537	21.5	4	0	0	0
110	2	2	3	3	30	197	9	2	0	0	0	380	2	1	7	3	40	245	11.5	2	1	0	0
111	2	2	3	3	30	373	9	2	0	0	0	381	2	1	4	3	40	245	11.5	2	0	0	0
112	2	2	3	3	31	70	38	8	0	0	0	382	2	1	7	3	30	190	8	1	1	0	0
113	1	3	18	2	40	260	31.5	8	0	1	0	383	2	1	13	3	30	343	10	1	1	0	0
114	1	3	13	2	40	260	31.5	8	1	0	0	384	2	1	7	3	29	143	8	1	1	0	0
115	1	3	11	2	40	260	31.5	8	0	0	0	385	2	1	13	3	23	155	8	1	1	0	0
116	1	3	14	1	20	28	20	4	1	0	0	386	2	1	13	3	22	276	8.5	1	1	0	0
117	1	3	12	1	20	28	20	4	0	0	0	387	2	1	13	3	30	310	8	1	1	0	0
118	1	3	18	2	40	400	31.5	8	0	0	0	388	2	1	9	3	23	61	8	1	1	0	0
119	1	3	13	2	40	400	31.5	8	0	1	0	389	2	1	6	3	23	61	8	1	0	0	0
120	1	3	11	2	40	400	31.5	8	0	0	0	390	2	1	9	3	25	180	8	1	1	0	0
121	1	3	18	1	20	20	31	8	0	0	0	391	2	1	6	3	25	180	8	1	0	0	0
122	1	3	13	1	20	20	31	8	1	1	0	392	2	1	9	3	33	171	8	1	1	0	0
123	1	3	11	1	20	20	31	8	0	0	0	393	2	1	6	3	33	171	8	1	0	0	0
124	1	3	18	1	16	27	32.5	8	1	0	0	394	2	1	9	3	25	74	8	1	0	0	0
125	1	3	13	1	16	27	32.5	8	1	0	0	395	2	1	6	3	25	74	8	1	0	0	0
126	1	3	11	1	16	27	32.5	8	0	0	0	396	2	1	9	3	25	92	8	1	1	0	0
127	2	3	18	3	24	403	9.5	2	0	0	0	397	2	1	6	3	25	92	8	1	0	0	0

Table A1. *Cont.*

NO.	A	B	C	D	E	F	G	H	I	J	K	NO.	A	B	C	D	E	F	G	H	I	J	K
128	2	3	18	3	24	419	9.5	2	0	0	0	398	2	1	9	3	25	191	8	1	1	0	0
129	1	3	18	1	16	73	31.5	8	0	0	0	399	2	1	6	3	25	191	8	1	0	0	0
130	2	3	18	3	24	138	15.8	4	1	0	0	400	2	1	9	3	30	278	8	1	1	0	1
131	2	3	13	3	24	138	15.8	4	0	0	0	401	2	1	6	3	30	278	8	1	0	0	0
132	2	3	11	3	24	138	15.8	4	0	0	0	402	2	1	9	3	25	100	8	1	0	0	0
133	2	1	10	3	25	100	12.5	2	0	0	0	403	2	1	6	3	25	100	8	1	0	0	0
134	1	3	17	1	20	37	31	8	1	0	1	404	2	2	11	3	33	90	7.8	1	0	0	0
135	1	3	12	1	20	37	31	8	0	0	0	405	2	2	11	3	30	90	7.8	1	0	0	0
136	1	3	10	1	20	37	31	8	0	0	0	406	2	2	11	3	30	30	7.8	1	0	0	0
137	1	3	17	1	30	74	19	4	1	0	0	407	2	2	11	3	30	40	17.8	4	0	0	0
138	1	3	12	1	30	74	19	4	0	0	0	408	2	2	11	3	30	40	17.8	4	0	0	0
139	1	3	10	1	30	74	19	4	0	0	0	409	2	2	11	3	35	110	7.8	1	0	0	0
140	1	3	18	1	25	30	29	8	1	0	0	410	2	2	11	3	33	350	7.8	1	0	0	0
141	1	3	13	1	25	30	29	8	0	1	0	411	2	2	11	3	30	30	7.8	1	0	0	0
142	1	3	11	1	25	30	29	8	0	0	0	412	2	2	7	3	40	374	9.5	2	0	0	0
143	1	2	18	1	20	20	30	8	1	0	0	413	2	2	7	3	38	440	9.5	2	1	0	0
144	1	2	13	1	20	20	30	8	0	1	0	414	2	2	4	3	38	440	9.5	2	0	0	0
145	1	2	11	1	20	20	30	8	0	0	0	415	2	2	7	3	18	70	8	1	0	0	0
146	1	3	18	1	13	13	31	8	1	0	1	416	2	2	7	3	17	38	8	1	0	0	0
147	1	3	13	1	13	13	31	8	0	0	0	417	2	1	11	3	35	80	41.5	8	0	0	0
148	1	3	11	1	13	13	31	8	0	0	0	418	2	1	9	3	30	105	33	8	0	0	0
149	1	1	16	1	20	50	31	8	1	1	0	419	2	1	6	3	20	40	8	2	0	0	0
150	1	1	14	1	20	50	31	8	0	0	0	420	2	1	9	3	30	252	8	2	0	0	0
151	1	1	19	1	20	50	16	4	1	0	0	421	2	3	9	3	25	100	31.5	8	1	0	0
152	1	1	14	1	20	50	16	4	0	1	1	422	2	3	9	3	31	98	12.5	2	1	1	0
153	1	1	11	1	20	50	16	4	0	0	0	423	2	1	6	3	45	305	24.5	6	0	0	0
154	1	1	19	1	20	50	16	4	0	0	1	424	2	1	6	3	45	90	14	2	0	1	0
155	1	1	14	1	20	50	16	4	1	0	0	425	2	1	6	3	46	97	14	2	0	0	0
156	1	2	15	1	16	60	12.3	3	0	0	1	426	1	1	7	3	25	125	9	2	1	0	0
157	1	2	10	1	16	60	12.3	3	0	1	0	427	2	1	13	3	30	90	22	4	0	0	0
158	1	2	9	1	16	60	12.3	3	0	0	0	428	2	1	12	3	33	450	13	2	0	0	0
159	1	2	15	1	16	24	12.3	3	0	0	0	429	2	2	14	3	30	78	13	2	1	0	0
160	1	2	10	1	16	24	12.3	3	1	0	0	430	2	2	14	3	30	210	14	2	1	0	0
161	1	2	9	1	16	24	12.3	3	0	0	0	431	2	1	11	3	45	255	20	4	0	0	0
162	1	2	12	1	20	80	12.3	3	0	0	1	432	1	1	13	1	13	40	25	6	1	0	0
163	1	2	7	1	20	80	12.3	3	0	0	0	433	1	1	15	3	27	60	17	4	0	1	0
164	1	2	6	1	20	80	12.3	3	0	0	0	434	1	1	15	3	27	60	17	4	0	0	0
165	1	2	15	1	20	109	12.3	3	0	0	1	435	2	1	11	3	33	298	17	4	0	0	0
166	1	2	10	1	20	109	12.3	3	1	1	0	436	1	1	11	3	30	170	17	4	0	0	0
167	1	2	9	1	20	109	12.3	3	0	0	0	437	1	1	11	1	20	20	10	2	0	0	0
168	1	2	15	1	10	20	12.3	3	1	0	1	438	1	1	4	3	26	286	23	5	0	0	0
169	1	2	10	1	10	20	12.3	3	0	0	0	439	1	1	4	3	25	25	14	3	0	0	0
170	1	2	9	1	10	20	12.3	3	1	0	0	440	2	1	4	3	30	186	9	2	0	0	0
171	1	2	15	1	10	20	39	8	1	1	0	441	1	1	15	1	30	111	33	6	0	0	0
172	1	2	10	1	10	20	39	8	1	1	0	442	1	1	15	1	20	40	33	6	1	0	0
173	1	2	9	1	10	20	39	8	0	0	0	443	2	1	7	3	38	417	9	2	1	0	0
174	1	2	14	1	20	120	12.3	3	1	0	1	444	2	1	5	3	38	417	9	2	0	0	0
175	1	2	9	1	20	120	12.3	3	0	1	0	445	2	1	7	3	35	370	9	2	0	0	0
176	1	2	8	1	20	120	12.3	3	0	0	0	446	2	1	7	3	35	259	9	2	0	0	0
177	1	2	9	1	20	32	12.3	3	0	1	0	447	2	1	5	3	35	259	9	2	0	0	0
178	1	2	15	1	10	20	12.3	3	1	1	0	448	2	1	7	3	35	182	9	2	0	0	0
179	1	2	10	1	10	20	12.3	3	0	0	0	449	2	1	5	3	35	182	9	2	0	0	0
180	1	3	13	3	20	40	8.5	1	0	0	0	450	2	1	7	3	38	340	9	2	0	0	0
181	1	3	9	3	20	40	8.5	1	0	0	1	451	2	1	7	3	30	246	9	2	0	0	0
182	2	2	13	3	20	50	19	4	0	0	0	452	2	1	7	3	25	167	9	2	0	0	0
183	2	2	9	3	20	50	19	4	0	0	0	453	1	1	17	1	20	152	11	2	1	0	0
184	2	2	13	3	20	40	13	2	0	0	1	454	1	1	17	2	30	150	11	2	1	0	0
185	2	2	9	3	20	40	13	2	0	0	0	455	1	1	17	2	30	485	17	4	0	0	0
186	2	2	8	3	20	40	13	2	0	0	0	456	1	1	17	1	20	248	17	4	0	0	0
187	2	2	7	3	44	141	8	1	0	0	0	457	2	1	7	3	30	90	7.8	1	0	0	0
188	2	2	3	3	44	141	8	1	0	0	0	458	2	1	7	3	32	172	7.8	1	0	0	0
189	1	2	7	3	25	25	8	1	0	0	0	459	2	1	7	3	30	161	7.8	1	0	0	0
190	1	2	3	3	25	25	8	1	0	0	0	460	2	1	9	3	30	120	15	4	0	0	0
191	2	2	7	3	44	141	8	1	0	0	0	461	2	1	9	3	27	75	7.8	1	0	0	0
192	2	2	3	3	44	141	8	1	0	0	0	462	2	1	9	3	30	318	7.8	1	0	0	0
193	1	2	13	3	35	45	8	1	0	0	1	463	2	1	9	3	30	110	7.8	1	0	0	0
194	1	2	9	3	35	45	8	1	0	0	0	464	2	1	9	3	27	100	10.5	1	0	0	0
195	2	2	13	3	36	260	8	1	0	0	0	465	1	1	17	1	20	573	21	4	0	0	0

Table A1. *Cont.*

NO.	A	B	C	D	E	F	G	H	I	J	K	NO.	A	B	C	D	E	F	G	H	I	J	K
196	2	2	9	3	36	260	8	1	0	0	0	466	1	1	18	1	20	393	21	4	0	0	0
197	1	3	18	1	20	132	15	4	0	1	1	467	1	1	20	2	20	420	11	2	1	1	0
198	1	3	14	1	20	132	15	4	0	0	0	468	1	1	7	3	20	25	8.8	2	0	0	0
199	1	3	12	1	20	132	15	4	0	0	0	469	2	1	17	1	20	172	9	2	0	0	0
200	1	3	17	1	20	146	15.5	4	1	1	0	470	2	1	17	1	20	172	9	2	0	0	0
201	1	3	13	1	20	146	15.5	4	0	0	0	471	2	1	9	3	30	180	9	2	0	0	0
202	1	3	11	1	20	146	15.5	4	0	0	0	472	1	1	19	2	20	91	32	5	1	0	0
203	1	3	16	2	30	180	31	8	0	0	1	473	2	1	5	3	32	125	8	2	0	0	0
204	1	3	12	2	30	180	31	8	0	1	0	474	2	1	4	3	25	300	11	2	0	0	0
205	1	3	10	2	30	180	31	8	0	0	0	475	2	1	7	3	38	225	17.8	4	0	0	0
206	1	3	16	2	30	347	31	8	0	1	1	476	2	1	4	3	30	100	9	1	0	0	0
207	1	3	12	2	30	347	31	8	0	0	0	477	2	2	3	3	35	233	8	1	0	0	0
208	1	3	10	2	30	347	31	8	0	0	0	478	2	1	11	3	30	190	9	2	0	1	0
209	1	3	16	2	30	190	31	8	0	1	0	479	2	1	7	3	40	442	18.5	4	0	0	0
210	1	3	12	2	30	190	31	8	0	0	0	480	1	2	10	1	20	32	24.5	6	0	1	0
211	1	3	10	2	30	190	31	8	0	0	0	481	1	2	7	2	30	254	24.5	6	0	0	0
212	1	3	16	2	40	288	31	8	0	0	0	482	1	1	12	1	20	51	9	2	0	0	0
213	1	3	12	2	40	288	31	8	0	0	0	483	1	1	13	1	20	50	31.5	6	0	0	0
214	1	3	10	2	40	288	31	8	0	0	0	484	1	2	16	1	8	20	31	8	0	0	0
215	1	2	16	1	25	50	35	8	1	0	0	485	2	1	12	3	23	53	10	2	0	0	0
216	1	2	12	1	25	50	35	8	0	0	0	486	2	1	7	3	34	196	8	1	0	0	0
217	1	2	17	1	30	240	31	8	1	0	1	487	1	3	16	2	30	70	19	4	0	0	0
218	1	2	13	1	30	240	31	8	0	0	0	488	2	3	16	1	25	84	19	4	0	1	0
219	1	2	11	1	30	240	31	8	0	0	0	489	1	3	13	1	16	60	31.5	8	0	0	0
220	1	2	19	1	16	45	23	4	0	0	0	490	2	3	13	3	24	403	9.5	2	0	0	0
221	1	2	17	1	20	74	38	8	0	1	1	491	1	3	13	2	40	180	31.5	8	1	0	0
222	1	2	13	1	20	74	38	8	1	1	0	492	2	3	13	3	24	419	9.5	2	0	0	0
223	1	2	11	1	20	74	38	8	1	0	0	493	2	1	3	3	46	97	14	2	0	0	0
224	1	1	20	1	8	19	31	8	0	1	0	494	2	1	3	3	46	305	24	4	0	0	0
225	1	1	16	1	8	19	31	8	1	0	0	495	2	3	6	3	25	100	35.5	6	0	0	0
226	1	1	14	1	8	19	31	8	0	0	0	496	2	1	3	3	45	90	14	2	0	0	0
227	1	3	20	1	30	60	31	8	1	0	0	497	2	1	5	3	28	100	9.5	2	0	0	0
228	1	3	16	1	30	60	31	8	1	0	0	498	1	1	13	1	20	50	31	6	0	0	0
229	1	3	14	1	30	60	31	8	0	1	0	499	1	1	12	1	27	91	32.5	6	0	0	0
230	1	1	20	1	6	9.3	24	6	0	1	0	500	1	1	13	1	20	40	16	3	0	0	0
231	1	1	16	1	6	9.3	24	6	1	0	0	501	2	2	8	3	20	50	9.75	2	0	0	0
232	1	1	14	1	6	9.3	24	6	0	1	0	502	1	1	10	3	48	190	9	2	0	0	0
233	1	2	18	3	20	27	18.5	4	0	0	0	503	2	1	11	3	23	53	10	2	0	0	0
234	1	2	16	3	20	27	18.5	4	0	0	0	504	2	1	19	2	20	280	13	2	1	1	0
235	1	2	18	1	20	65	31	8	1	0	0	505	2	1	9	3	30	90	22	4	0	1	0
236	1	2	14	1	20	65	31	8	0	0	1	506	1	1	5	3	20	25	8.8	2	0	0	0
237	1	2	12	1	20	65	31	8	0	1	0	507	2	1	5	3	32	172	7.8	1	0	0	0
238	1	1	16	1	25	43	38	8	1	1	0	508	2	1	5	3	30	90	7.8	1	0	0	0
239	1	1	7	3	25	125	40	8	0	0	0	509	2	1	15	1	23	300	9	2	0	1	0
240	1	1	3	3	25	125	40	8	0	0	0	510	2	1	15	1	22	172	9	2	0	1	0
241	1	1	16	3	25	150	9	2	0	1	0	511	2	1	5	3	30	161	7.8	1	0	0	0
242	1	1	12	3	25	100	9	2	0	0	0	512	2	1	15	1	23	152	11	2	0	1	0
243	1	1	17	1	30	111	33	8	1	0	0	513	1	1	15	2	30	150	11	2	0	0	0
244	1	3	20	1	20	28	32	8	1	1	0	514	2	1	15	1	19	193	17	4	1	1	0
245	1	3	16	1	20	28	32	8	0	0	0	515	2	1	15	1	20	248	17	4	0	0	0
246	1	3	14	1	20	28	32	8	0	0	0	516	1	1	18	1	20	420	11	2	0	1	0
247	1	3	17	2	40	140	31	8	0	0	1	517	1	1	15	1	20	573	24	4	0	0	0
248	1	3	13	2	40	140	31	8	1	1	0	518	1	1	16	1	20	393	24	4	0	1	0
249	1	3	11	2	40	140	31	8	1	0	0	519	2	3	11	3	24	419	9.5	2	0	1	0
250	1	3	17	2	40	150	31	8	0	0	1	520	1	3	11	1	16	60	31.5	8	0	0	0
251	1	3	13	2	40	150	31	8	1	0	0	521	2	3	11	1	24	403	9.5	2	0	1	0
252	1	3	11	2	40	150	31	8	1	0	0	522	1	3	11	2	40	180	31.5	8	0	1	0
253	1	3	17	2	50	280	31.5	8	1	1	1	523	1	3	15	2	25	84	19	4	0	0	0
254	1	3	13	2	50	280	31.5	8	0	0	0	524	1	3	15	2	30	71	19	4	0	0	0
255	1	3	11	2	50	280	31.5	8	0	0	0	525	1	2	15	1	30	40	32	8	0	0	0
256	2	3	13	3	24	116	9.25	2	0	0	0	526	1	2	15	1	25	70	35	8	0	0	0
257	2	3	9	3	24	116	9.25	2	0	0	0	527	2	1	7	3	48	158	9	2	0	0	0
258	2	3	7	3	24	116	9.25	2	0	0	0	528	2	1	5	3	32	182	9	2	0	0	0
259	2	3	16	3	28	110	8.5	1	1	1	0	529	2	1	5	3	35	345	9	2	0	0	0
260	2	3	12	3	28	110	8.5	1	0	0	0	530	2	1	5	3	35	170	9	2	0	0	0
261	2	3	10	3	28	110	8.5	1	0	0	0	531	2	1	5	3	30	106	9	2	0	0	0
262	2	3	13	3	25	278	8.5	1	1	0	0	532	2	1	7	3	30	400	9	2	0	0	0
263	2	3	9	3	25	278	8.5	1	0	0	0	533	1	1	15	3	30	30	32	8	0	0	0

Table A1. *Cont.*

NO.	A	B	C	D	E	F	G	H	I	J	K	NO.	A	B	C	D	E	F	G	H	I	J	K
264	2	3	7	3	25	278	8.5	1	0	0	0	534	2	2	15	1	20	95	11.5	2	0	1	1
265	2	3	13	3	25	159	8.5	1	0	0	0	535	2	1	7	3	32	298	17.5	4	0	0	0
266	2	3	9	3	25	159	8.5	1	0	0	0	536	1	2	18	2	30	240	44	6	0	1	1
267	2	3	16	2	30	347	8.5	1	0	0	0	537	2	2	6	3	30	109	10	2	0	0	0
268	2	3	14	2	30	347	8.5	1	0	0	0	538	1	2	5	2	30	254	24.5	6	0	0	0
269	1	3	16	1	20	179	31.5	8	0	0	0	539	1	1	14	1	25	42	38.5	8	0	0	0
270	1	2	19	1	8	20	31	8	1	1	0	-	-	-	-	-	-	-	-	-	-	-	-

In this Appendix, A represents Structural Style, B represents District, C represents Age of Bridge, D represents Type of Bridge, E represents Span, F represents Length of Bridge, J represents Width of Bridge, H represents Number of Lanes, I represents Maintenance of Bridge Deck, J represents Maintenance of Superstructure, K represents Maintenance of Substructure. In column A, 1 represents Simply Supported Beam, 2 represents Continuous Beam. In column B, 1, 2, and 3 represent three different regions, respectively. In column D, 1 represents Plate Girder, 2 represents T-Girder, 3 represents Box Girder. In columns I, J, and K, 0 represents no maintenance, and 1 represents maintenance conducted during that year.

Appendix B

Table A2. BCI Values Set.

NO.	BCI	NO.	BCI	NO.	BCI	NO.	BCI	NO.	BCI	NO.	BCI	NO.	BCI	NO.	BCI
1	97.82	69	92.68	137	87.93	205	82.14	273	83.84	341	96.06	409	96.72	477	92.19
2	88.92	70	94.68	138	90.75	206	86.07	274	87.26	342	97.41	410	94.88	478	96.71
3	89.16	71	95.94	139	90.78	207	82.68	275	89.88	343	93.59	411	93.06	479	89.76
4	80.57	72	91.35	140	86.72	208	83.22	276	88.89	344	99.66	412	97.85	480	88.36
5	94.78	73	94.52	141	87.24	209	82.96	277	87.65	345	99.29	413	94.14	481	87.74
6	86.05	74	95.39	142	81.61	210	80.2	278	90.22	346	95.01	414	93.46	482	87.59
7	85.64	75	94.74	143	87.18	211	81.17	279	85.5	347	92.69	415	89.71	483	85.46
8	93.29	76	91.76	144	86.65	212	89.02	280	85.99	348	89.96	416	89.78	484	83.66
9	83.57	77	98.57	145	80.87	213	89.13	281	83.57	349	87.96	417	91.77	485	87.62
10	85.95	78	97.18	146	87.9	214	89.37	282	84.29	350	89.47	418	88.87	486	94.65
11	85.16	79	92.71	147	81.76	215	83.09	283	86.13	351	81.45	419	97.08	487	88.98
12	86.88	80	91.71	148	85.31	216	87.55	284	86.57	352	84.63	420	89.37	488	88.18
13	93.12	81	95.6	149	89.13	217	85.17	285	84.2	353	85.94	421	93.95	489	95.3
14	94.29	82	92.42	150	83.24	218	84.82	286	80.21	354	86.74	422	93.25	490	89.42
15	88.88	83	92.51	151	88.1	219	86.56	287	83.61	355	87.74	423	94.73	491	93.98
16	85.42	84	85.07	152	89.4	220	83.18	288	82.97	356	98.39	424	93.41	492	87.98
17	81.73	85	88.36	153	88.64	221	89.29	289	85.65	357	89.51	425	90.3	493	98.13
18	91.38	86	93.45	154	87.18	222	83.78	290	87.15	358	88.72	426	89.32	494	97.38
19	91.86	87	88.31	155	86.75	223	84.32	291	82.23	359	89.56	427	89.43	495	97.08
20	83.63	88	91.49	156	88.64	224	86	292	84.26	360	88	428	89.95	496	99.33
21	91.38	89	96.31	157	88.15	225	88.18	293	84.47	361	89.96	429	82.38	497	95.43
22	91.22	90	94.82	158	86.43	226	86.63	294	88.8	362	89.92	430	84.49	498	86.59
23	86.54	91	93.97	159	85.97	227	88.28	295	87.04	363	89.03	431	95.73	499	88.88
24	81.59	92	96.07	160	91.47	228	88.15	296	82.51	364	89.34	432	85.89	500	85.51
25	99.75	93	93.18	161	88.78	229	92.32	297	84.39	365	92.39	433	90.61	501	90.6
26	88.13	94	94.81	162	89.55	230	87.36	298	91.87	366	93.97	434	90.8	502	94.36
27	96.63	95	93.38	163	88.34	231	84.77	299	91.85	367	93.29	435	89.72	503	85.17
28	89.64	96	97.58	164	92.4	232	87.88	300	84.66	368	87.63	436	90.75	504	87.46
29	85.59	97	89.28	165	89.46	233	87.57	301	86.71	369	89.74	437	88.65	505	89.24
30	86.64	98	84.71	166	85.38	234	87.5	302	90.17	370	89.23	438	96.12	506	92.94
31	83.22	99	94.94	167	82.9	235	85.43	303	89.41	371	88.7	439	90.18	507	99.2
32	88.66	100	99.05	168	87.23	236	86.72	304	82.9	372	88.89	440	89.56	508	95.61
33	89.29	101	98.64	169	83.93	237	84.66	305	83.96	373	89.4	441	81.44	509	85.25
34	89.03	102	91.3	170	83.37	238	86.3	306	87.14	374	89.23	442	84.7	510	86.89
35	87.14	103	87.94	171	89.34	239	89.88	307	83.01	375	90.17	443	88.95	511	97.84

Table A2. *Cont.*

NO.	BCI	NO.	BCI	NO.	BCI	NO.	BCI	NO.	BCI	NO.	BCI	NO.	BCI	NO.	BCI
36	92.97	104	89.29	172	88.91	240	92.52	308	86.64	376	98.07	444	89.69	512	87.72
37	85.63	105	93.42	173	82.67	241	89.02	309	97.08	377	97.44	445	89.84	513	87.73
38	84.71	106	94.07	174	89.31	242	88.74	310	97.41	378	95.52	446	96.92	514	89.79
39	86.52	107	95.25	175	87.4	243	88.11	311	86.41	379	96.01	447	98.5	515	84.39
40	88.59	108	95.31	176	84.21	244	86.73	312	85.58	380	98.51	448	96.44	516	84.4
41	99.75	109	96.14	177	86.32	245	80.25	313	90.9	381	97.12	449	96.85	517	88.92
42	88.69	110	94.67	178	89.65	246	80.85	314	95.05	382	93.54	450	95.05	518	89.19
43	88	111	95.55	179	86.5	247	85.55	315	94.65	383	89.1	451	98.9	519	94.49
44	89.43	112	98.92	180	89.07	248	83.02	316	96.06	384	93.62	452	94.08	520	89.86
45	90.04	113	88.79	181	92.35	249	82	317	90.08	385	89.15	453	89.66	521	95.46
46	91.9	114	86.72	182	91.26	250	88.75	318	94.66	386	86.92	454	88.09	522	94.79
47	91.84	115	87.76	183	93.5	251	87.6	319	84.19	387	89.78	455	89.48	523	84.33
48	85.68	116	84.19	184	90.7	252	86.69	320	86.39	388	97.26	456	89.5	524	89.6
49	98.36	117	83.75	185	91.7	253	92.28	321	86.12	389	99.63	457	97.16	525	85.47
50	92.04	118	81.47	186	97.25	254	82.04	322	88.54	390	89.73	458	98.16	526	89.6
51	96.55	119	86.07	187	91.28	255	84.05	323	80.66	391	96.98	459	98.64	527	94.24
52	97.05	120	84.67	188	98.24	256	97.99	324	85.04	392	88.76	460	96.9	528	91.83
53	94.85	121	81.64	189	95.43	257	97.96	325	87.06	393	96.27	461	94.92	529	90.54
54	95.24	122	90.21	190	97.98	258	98.99	326	87.34	394	93.21	462	93.2	530	88.42
55	91.71	123	85.58	191	96.94	259	89.89	327	91.56	395	95.9	463	96.22	531	89.8
56	92.96	124	83.71	192	98.28	260	86.71	328	93.25	396	89.88	464	95.18	532	90.8
57	97.92	125	83.38	193	89.5	261	87.63	329	84.52	397	94	465	89.17	533	88.32
58	98.98	126	89.56	194	88.12	262	96.34	330	85.46	398	95.84	466	88.79	534	98.5
59	91.91	127	88.68	195	82.04	263	96.03	331	81.66	399	95.46	467	89.58	535	95.33
60	94.18	128	88.9	196	89.35	264	99.55	332	82.84	400	96.93	468	92.71	536	85.6
61	94.21	129	88.12	197	87.63	265	95.74	333	84.67	401	93.97	469	87.51	537	93.27
62	96.3	130	88.67	198	82.16	266	96.59	334	85.72	402	89.76	470	88.11	538	90.12
63	93.84	131	87.17	199	87.33	267	85.36	335	84.3	403	96.85	471	90.73	539	82.58
64	94.6	132	87.82	200	84.61	268	97.71	336	88.37	404	89.69	472	83.75	-	-
65	93.38	133	87.2	201	81.88	269	86.9	337	86.56	405	94.26	473	95.71	-	-
66	87.73	134	89.54	202	81.05	270	88.54	338	84.82	406	93.51	474	92.71	-	-
67	94.52	135	83.36	203	82.49	271	83.86	339	87.75	407	96.06	475	93.02	-	-
68	89.9	136	89.13	204	83.63	272	86.41	340	99.33	408	94.6	476	91.49	-	-

References

1. Xin, J.; Zhou, C.; Jiang, Y.; Tang, Q.; Yang, X.; Zhou, J. A signal recovery method for bridge monitoring system using TVFEMD and encoder-decoder aided LSTM. *Measurement* **2023**, *214*, 112797.
2. Tang, Q.; Jiang, Y.; Xin, J.; Liao, G.; Liu, S.; Zhou, J.; Yang, X. A novel method for the recovery of continuous missing data using multivariate variational mode decomposition and fully convolutional networks. *Measurement* **2023**, *220*, 113366. [CrossRef]
3. Wang, D.; Yang, S.X. Intelligent feature extraction, data fusion and detection of concrete bridge cracks: Current development and challenges. *Intell. Robot.* **2022**, *2*, 391–406. [CrossRef]
4. Tang, Q.; Xin, J.; Jiang, Y.; Zhou, J.; Li, S.; Fu, L. Fast identification of random loads using the transmissibility of power spectral density and improved adaptive multiplicative regularization. *J. Sound Vib.* **2022**, *534*, 117033. [CrossRef]
5. Xin, J.; Zhou, J.; Zhou, F.; Yang, S.X.; Zhou, Y. Bearing capacity model of corroded RC eccentric compression columns based on hermite interpolation and fourier fitting. *Appl. Sci.* **2019**, *9*, 24. [CrossRef]
6. Tao, T.; He, J.; Wang, H.; Zhao, K. Efficient simulation of non-stationary nonhomogeneous wind field: Fusion of multi-dimensional interpolation and NUFFT. *J. Wind Eng. Ind. Aerod.* **2023**, *236*, 105394. [CrossRef]
7. Martinez, P.; Mohamed, E.; Mohsen, O. Comparative study of data mining models for prediction of bridge future conditions. *J. Perform. Constr. Fac.* **2020**, *34*, 04019108. [CrossRef]
8. Yang, L.; Sun, L. Analysis of performance decay characteristics of reinforced concrete bridges. *Highway* **2015**, *60*, 132–135.
9. Hasan, S.; Elwakil, E. Stochastic regression deterioration models for superstructure of prestressed concrete bridges in California. *J. Struct. Integr. Maint.* **2019**, *4*, 97–108. [CrossRef]
10. Zhang, Y.; Huang, Y.; Ren, C. Multi-stage degradation model of bridge technical condition based on inspection and evaluation big data. *Highway* **2018**, *63*, 87–91.
11. Wellalage, N.K.; Zhang, T.; Dwight, R. Calibrating Markov chain-based deterioration models for predicting future conditions of railway bridge elements. *J. Bridge Eng.* **2015**, *20*, 04014060. [CrossRef]

12. Thanh, N.L.; Hackl, J.; Adey, B. Determination of Markov transition probabilities to be used in bridge management from mechanistic-empirical models. *J. Bridge Eng.* **2017**, *22*, 01017063.
13. Jiang, Y.; Liu, Z.; Luo, H. ELM indirect prediction method for the remaining life of lithium-ion battery. *J. Electron. Meas. Instrum.* **2016**, *30*, 179–185.
14. He, X.; Wang, H.; Lu, J. Analog circuit fault diagnosis method based on preferred wavelet packet and ELM. *Chin. J. Sci. Instrum.* **2013**, *34*, 2614–2619.
15. Mirjalili, S.; Lewis, A. The whale optimization algorithm. *Adv. Eng. Softw.* **2016**, *95*, 51–67. [CrossRef]
16. Lu, X.; Li, C.; Wu, Z. Microgrid fault diagnosis based on extreme learning machine optimized by whale algorithm. *Smart Power* **2022**, *50*, 15–67.
17. Li, L.; Sun, J.; Tseng, M. Extreme learning machine optimized by whale optimization algorithm using insulated gate bipolar transistor module aging degree evaluation. *J. Expert Syst. Appl.* **2019**, *127*, 58–67. [CrossRef]
18. Nadimi-Shahraki, M.H.; Taghian, S.; Mirjalili, S.; Faris, H. MTDE: An effective multi-trial vector-based differential evolution algorithm and its applications for engineering design problems. *J. Appl. Soft. Comput.* **2020**, *97*, 106761. [CrossRef]
19. Liu, X.; Li, G.; Yang, H.; Zhang, N.; Wang, L.; Shao, P. Agricultural UAV trajectory planning by incorporating multi-mechanism improved grey wolf optimization algorithm. *J. Expert Syst. Appl.* **2023**, *233*, 120946. [CrossRef]
20. Huang, G.; Zhu, Q.; Siew, C.K. Extreme learning machine: A new learning scheme of feedforward neural networks. In Proceedings of the IEEE International Joint Conference on Neural Networks, Budapest, Hungary, 25–29 July 2004; Volume 2, pp. 985–990.
21. Cao, J.; Zeebaree, D.Q.; Chen, Q. Breast cancer diagnosis using hybrid AlexNet-ELM and chimp optimization algo-rithm evolved by Nelder-mead simplex approach. *Biomed. Signal Process. Control.* **2023**, *85*, 105053.
22. Wang, Y.; He, Q.; Zhang, D. Improving Li-ion battery health: Predicting remaining useful life using IWBOA-ELM algorithm. *J. Energy Storage* **2023**, *72*, 108547. [CrossRef]
23. Wu, L.; Chen, E.; Guo, Q. Smooth exploration system: A novel ease-of-use and specialized module for improving exploration of whale optimization algorithm. *Knowl.-Based Syst.* **2023**, *272*, 110580. [CrossRef]
24. Li, S.; Xin, J.; Jiang, Y. Temperature-induced deflection separation based on bridge deflection data using the TVFEMD-PE-KLD method. *J. Civ. Struct. Health Monit.* **2023**, *13*, 781–797. [CrossRef]
25. Tong, K.; Zhang, H.; Zhao, R. Investigation of SMFL monitoring technique for evaluating the load-bearing capacity of RC bridges. *Eng. Struct.* **2023**, *293*, 116667. [CrossRef]
26. Liu, S.; Jiang, Y.; Qiao, K.; Peng, L.; Liu, D. Record-based simulation of three-component long-period ground motions: Hybrid of surface wave separation and multivariate empirical mode decomposition. *Soil. Dyn. Earthq. Eng.* **2023**, *172*, 108037. [CrossRef]
27. Liu, M.; Peng, L.; Huang, G.; Yang, Q.; Jiang, Y. Simulation of stationary non-Gaussian multivariate wind pressures using moment-based piecewise Hermite polynomial model. *J. Wind Eng. Ind. Aerod.* **2020**, *196*, 104041. [CrossRef]
28. Jiang, Y.; Zhao, N.; Peng, L.; Zhao, L.; Liu, M. Simulation of stationary wind field based on adaptive interpolation-enhanced scheme. *J. Wind Eng. Ind. Aerod.* **2019**, *195*, 104001. [CrossRef]
29. Liang, J.; Feng, C.; Song, P. A Survey on Correlation Analysis of Big Data. *Chin. J. Comput.* **2016**, *39*, 18.
30. Sun, L.; Shang, Z.; Xia, Y. Development and Prospect of Bridge Structural Health Monitoring in the Context of Big Data. *China J. Highw. Transp.* **2019**, *32*, 1–20.
31. Reshef, D.N.; Reshef, Y.A.; Finucane, H.K. Detecting novel associations in large data sets. *Science* **2011**, *334*, 1518–1524. [CrossRef]
32. Zhou, K.; Wang, X.; Wang, K. Research on Motor Short—Circuit Fault Prediction Based on Grey Limit Learning Machine. *Comput. Simul.* **2021**, *38*, 488–492.
33. Zhang, N.; Li, Q.; Li, C. Rock and soil catastrophe early warning research based on extreme learning machine and entropy method. *Sci. Technol. Eng.* **2019**, *19*, 251–258.

sensors

MDPI

Article

Bridge Model Updating Based on Wavelet Neural Network and Wind-Driven Optimization

Haifang He [1], Baojun Zeng [2], Yulong Zhou [1], Yuanyuan Song [3], Tianneng Zhang [3], Han Su [3,*] and Jian Wang [4,*]

[1] National Engineering Laboratory of Bridge Safety and Technology (Beijing), Research Institute of Highway Ministry of Transport, Beijing 100088, China
[2] Anhui Provincial Highway Management Service Center, Hefei 230022, China; lhy02027082@163.com
[3] School of Civil Engineering, Beijing Jiaotong University, Beijing 100044, China; 22125918@bjtu.edu.cn (Y.S.)
[4] School of Civil Engineering, Sun Yat-sen University, Guangzhou 510275, China
* Correspondence: hansu@bjtu.edu.cn (H.S.); wangj833@mail.sysu.edu.cn (J.W.)

Abstract: Aging, corrosive environments, and inadequate maintenance may result in performance deterioration of civil infrastructures, and finite element model updating is a commonly employed structural health monitoring procedure in civil engineering to reflect the current situation and to ensure the safety and serviceability of structures. Using the finite element model updating process to obtain the relationship between the structural responses and updating parameters, this paper proposes a method of using the wavelet neural network (WNN) as the surrogate model combined with the wind-driven optimization (WDO) algorithm to update the structural finite element model. The method was applied to finite element model updating of a continuous beam structure of three equal spans to verify its feasibility, the results show that the WNN can reflect the nonlinear relationship between structural responses and the parameters and has an outstanding simulation performance; the WDO has an excellent ability for optimization and can effectively improve the efficiency of model updating. Finally, the method was applied to update a real bridge model, and the results show that the finite element model update based on WDO and WNN is applicable to the updating of a multi-parameter bridge model, which has practical significance in engineering and high efficiency in finite element model updating. The differences between the updated values and measured values are all within the range of 5%, while the maximum difference was reduced from −10.9% to −3.6%. The proposed finite element model updating method is applicable and practical for multi-parameter bridge model updating and has the advantages of high updating efficiency, reliability, and practical significance.

Keywords: finite element model updating; wavelet neural network; wind-driven optimization; surrogate model; bridge

Citation: He, H.; Zeng, B.; Zhou, Y.; Song, Y.; Zhang, T.; Su, H.; Wang, J. Bridge Model Updating Based on Wavelet Neural Network and Wind-Driven Optimization. *Sensors* **2023**, *23*, 9185. https://doi.org/10.3390/s23229185

Academic Editor: Filippo Ubertini

Received: 20 October 2023
Revised: 6 November 2023
Accepted: 11 November 2023
Published: 14 November 2023

1. Introduction

Aging, corrosive environments, cracks caused by seismic activities, and inadequate maintenance may result in performance deterioration of civil infrastructures [1–4]; thus, employment of a structural health monitoring system to reflect the current situation is necessary to ensure both the safety and serviceability of structures. Finite element model updating is among the most common procedures in the structural health monitoring of civil engineering [5–7]. The traditional finite element model updating method utilizes the finite element model for calculation in each iteration when iteratively optimizing the updating parameters. The process is time-consuming, and the amount of calculation is huge; thus, it does not apply to engineering practice. Additionally, finite element models of complex structures require many parameters to be updated, the computational workload is large, and the traditional method is difficult to implement [8]. The method [9,10] based on surrogate models replaces the structural finite element model with a surrogate model with strong fitting ability, simple structure, and broad applicability and is combined with an optimization algorithm with strong global search ability and high efficiency to iteratively optimize the updating parameters. Thus, this method simplifies the model updating process and improves the efficiency of the updating work, and it is the most widely used method in the field of model updating [11–14].

The surrogate models commonly used in structural finite element model updating mainly include the polynomial response surface model [15–17], radial basis function (RBF) model [18,19], Kriging model [20], and neural network model [21–25]. Xu [26] proposed a bridge structure finite element model updating method based on the sparrow search algorithm (SSA) and the polynomial response surface method, and by updating the model of the high-dimensional locally damaged cantilever beam, the feasibility of the proposed method was verified. To more reasonably determine the number of node points of the Kriging surrogate model and to improve the prediction accuracy of the Kriging model for the extreme value region of the objective function, Qin [27] proposed an adaptive Kriging model and then applied it to the finite element model updating of bridge structures; the first-order frequency response surfaces of standard Kriging and adaptive Kriging fitting were established, and the finite element model updating of the two models was compared: The results show that the adaptive Kriging model can obtain better updating results. Qin [28] also combined the Kriging surrogate model with the improved gravity search algorithm (GSA), updated the initial finite element model using the load test data of the Nanzhonghuan Bridge, and obtained a reliable structural benchmark finite element model. Yang [29] employed the weighted response surface method to modify the single-girder finite element model of a bridge and adopted the central composite design experiment and variance analysis to screen out the parameters with high significance, thereby establishing a weighted response surface surrogate model; the objective function was constructed from the test load test data, and the structural parameters were identified by combining the chaotic particle swarm optimization algorithm; the obtained results are promising.

Among the models mentioned above, the neural network model has outstanding learning ability and widespread applicability [30], and it can directly update the structural parameters without solving the complex sensitivity matrix; in addition, it has a powerful nonlinear mapping function and strong robustness, which can fit the implicit function relationship between structural parameters and structural responses and better deal with data noise and ensure that the fitting results are not distorted [31,32]. Wavelet neural network (WNN) [33] is a product of the perfect combination of wavelet analysis theory and neural network theory. Based on the theory of neural networks, the wavelet function is used as the excitation function of the WNN. It inherits the superiority of wavelet analysis and artificial neural network, which has excellent approximation and fault tolerance, and is widely used in many engineering fields. However, the WNN lacks application in the finite element model updating of structures, and the related literature is scarce, but its excellent function approximation ability enables the network to be used as a surrogate model. Therefore, using the WNN as the surrogate model to update the finite element model has both research and application significance. This paper focuses on model updating and proposes a WNN combined with the WDO algorithm as the surrogate model of the finite element model.

In this paper, firstly, the theories of WNN and WDO and their realization are introduced; then, a structural finite element model updating method using a WNN as the surrogate model and WDO algorithm is proposed, and the method is implemented on the finite element model updating of Ningbo Bund Bridge; the updating parameters are studied and selected, and then, the updating procedure is carried out. It is demonstrated that the model updating method is applicable to the updating of the multi-parameter bridge model, which has practical significance in engineering and high efficiency in finite element model updating.

2. Wavelet Neural Network

In this section, the inception and concepts of wavelet neural network (WNN) are introduced, the structure of WNN is discussed, the training process of WNN is demonstrated conceptually, and finally, the shortcomings of WNN and measures to improve WNN are discussed.

Backpropagation (BP) neural network is the training process of feeding error rates back through a neural network to make it more accurate. The excitation function selected by the BP neural network is generally the Sigmoid function, which makes it difficult to guarantee the uniqueness of the solution, and the Sigmoid functions do not interact with each other; therefore, the learning convergence is slow. The RBF neural network needs to determine the center and width of the radial basis, and then, the linear optimization method is employed to determine the parameters, which are difficult to solve, and the radial basis function is non-orthogonal, which cannot guarantee the uniqueness of the approximation function. Wavelet neural network (WNN) was first proposed in 1992 [34] and was gradually developed into a mathematical modeling analysis method. WNN originated from the BP neural network [35]; however, it replaces the original excitation function Sigmoid function of the BP neural network with the wavelet function and obtains a new neural network similar to RBF neural network [36]. The excitation function of WNN is orthogonal or approximately orthogonal, which guarantees fast convergence and high fitting accuracy [37].

In addition, WNNs have the following characteristics [38]:

(1) The WNN can select different excitation functions for different fitting functions to improve the fitting;

(2) WNN can describe the function mutation stepwise, resulting in the function's better fitting performance;

(3) Determining wavelet function and the whole network structure has a reliable theoretical foundation, which avoids blindness in network design.

2.1. WNN Structure

The WNN is consistent with the BP neural network structure, which consists of three layers: input layer, hidden layer, and output layer [39], as shown in Figure 1. The function of the input layer is to input sample data, the hidden layer uses the wavelet function $\Psi(t)$ as the excitation function, and the output layer can use the Sigmoid function and the linear Purline function as the excitation function.

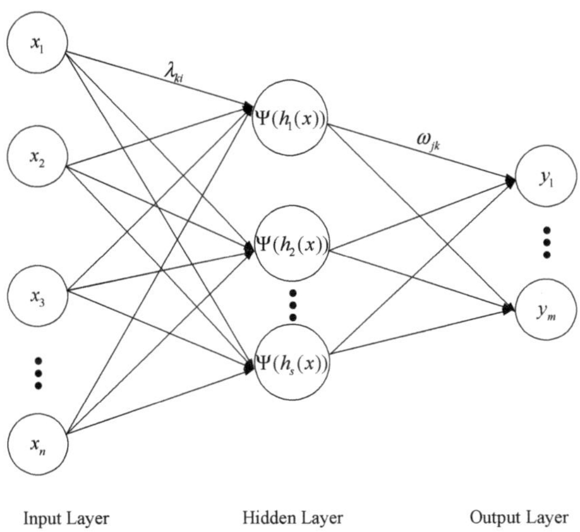

Figure 1. Network structure of WNN.

Let $\{x_1, x_2, \ldots, x_i, \ldots, x_n\}$ be the network input, $\{y_1, y_2, \ldots, y_i, \ldots, y_n\}$ the network output, n the number of samples, and m the number of outputs; thus, the hidden layer can be defined as in Equations (1)–(4):

$$\begin{aligned}\Psi(h_k(x)) &= \Psi\left(\frac{\sum_{i=1}^{n} \lambda_{ki}x_i - b_k}{a_k}\right) \\ &= \Psi\left(\sum_{i=1}^{n} g_{ki}x_i + r_k\right)\end{aligned} \tag{1}$$

$$\begin{aligned}h_k(x) &= \frac{\sum_{i=1}^{n} \lambda_{ki}x_i - b_k}{a_k} \\ &= \sum_{i=1}^{n} g_{ki}x_i + r_k\end{aligned} \tag{2}$$

$$g_{ki} = \frac{\lambda_{ki}}{a_k} \tag{3}$$

$$r_k = -\frac{b_k}{a_k} \tag{4}$$

where $k = 1, 2, \ldots, s$, s is the number of hidden layer nodes, λ_{ki} is the connection coefficient between the input layer and hidden layer, a_k is the kth scaling factor, and b_k is the kth translation factor.

The output layer can be defined as Equation (5):

$$\begin{aligned}y_j &= f\left(\sum_{k=1}^{S} \omega_{jk}\Psi(h_k(x)) + \delta_j\right) \\ &= f\left(\sum_{k=1}^{S} \omega_{jk}\Psi\left(\sum_{i=1}^{n} g_{ki}x_i + r_k\right) + \delta_j\right)\end{aligned} \tag{5}$$

where $f(x)$ is the output layer excitation function, ω_{jk} is the connection coefficient between the hidden layer and the output layer, and δ_j is the neural network threshold. Similarly, for multiple samples, the hidden layer and output layer of the WNN can be defined as Equations (6)–(8):

$$\Psi(h_{kt}(x)) = \Psi\left(\sum_{i=1}^{n} g_{ki}x_{it} + r_k\right) \tag{6}$$

$$h_{kt}(x) = \sum_{i=1}^{n} g_{ki}x_{it} + r_k \tag{7}$$

$$y_{jt} = f\left(\sum_{k=1}^{S} \omega_{jk}\Psi\left(\sum_{i=1}^{n} g_{ki}x_{it} + r_k\right) + \delta_j\right) \tag{8}$$

where $t = 1, 2, \ldots, T$, T is the number of samples, x_{it} is the ith element of the tth input samples, and y_{jt} is the jth neural network output, which corresponds to the tth input samples.

The matrix method is used to simplify the expression of hidden layer and output layer:

$$\mathbf{X}(n) = \begin{bmatrix} x_{11} & \cdots & x_{1T} \\ \vdots & \ddots & \vdots \\ x_{n1} & \cdots & x_{nT} \end{bmatrix} \tag{9}$$

$$\mathbf{I}(s) = \begin{bmatrix} g_{11} & \cdots & g_{1n} & r_1 \\ \vdots & \ddots & \vdots & \vdots \\ g_{s1} & \cdots & g_{sn} & r_s \end{bmatrix} \tag{10}$$

$$V(m) = \begin{bmatrix} \omega_{11} & \cdots & \omega_{1s} \\ \vdots & \ddots & \vdots \\ \omega_{m1} & \cdots & \omega_{ms} \end{bmatrix} \tag{11}$$

$$W(m) = \begin{bmatrix} \omega_{11} & \cdots & \omega_{1s} & \delta_1 \\ \vdots & \ddots & \vdots & \vdots \\ \omega_{m1} & \cdots & \omega_{ms} & \delta_m \end{bmatrix} \tag{12}$$

where $\boldsymbol{X}(n)$ is the input sample matrix; $\boldsymbol{I}(s)$ is the coefficient matrix consisting of g_{ki} and r_k; $\boldsymbol{W}(m)$ is the coefficient matrix consisting of ω_{jk} and δ_j; $\boldsymbol{V}(m)$ is the 1~n columns of $\boldsymbol{W}(m)$, which is the coefficient matrix consisting of ω_{jk}. Thus, we obtain the following:

$$\boldsymbol{H}(s) = \begin{bmatrix} h_{11}(x) & \cdots & h_{1T}(x) \\ \vdots & \ddots & \vdots \\ h_{s1}(x) & \cdots & h_{sT}(x) \end{bmatrix} = \boldsymbol{I}(s) \begin{bmatrix} \boldsymbol{X}(n) \\ \boldsymbol{E}_{1\times T} \end{bmatrix} \tag{13}$$

$$\boldsymbol{R}(s) = \Psi(\boldsymbol{H}(s)) \tag{14}$$

$$\boldsymbol{Y}(m) = \begin{bmatrix} y_{11} & \cdots & y_{1T} \\ \vdots & \ddots & \vdots \\ y_{m1} & \cdots & y_{mT} \end{bmatrix} = f\left(\boldsymbol{W}(m) \begin{bmatrix} \boldsymbol{R}(s) \\ \boldsymbol{E}_{1\times T} \end{bmatrix} \right) \tag{15}$$

where $\boldsymbol{H}(s)$ is composed of $h_{kt}(x)$, $E_{1\times T}$ is the identity matrix of 1 row and T columns, $\boldsymbol{R}(s)$ is the output matrix of hidden layer, and $\boldsymbol{Y}(m)$ is the output matrix of neural network output layer.

2.2. WNN Training

As a mature learning algorithm, gradient descent is often used to train neural networks. According to the idea of gradient descent and matrixing and introducing a momentum factor to accelerate the learning process, the iterative equations of coefficient matrices $\boldsymbol{W}(m)$ and $\boldsymbol{I}(s)$ are given:

$$\boldsymbol{W}(m)_{p+1} = \boldsymbol{W}(m)_p - \eta \frac{\partial U_p}{\partial \boldsymbol{W}(m)_p} + a\left(\boldsymbol{W}(m)_p - \boldsymbol{W}(m)_{p-1}\right) \tag{16}$$

$$\boldsymbol{I}(s)_{p+1} = \boldsymbol{I}(s)_p - \eta \frac{\partial U_p}{\partial \boldsymbol{I}(s)_p} + a\left(\boldsymbol{I}(s)_p - \boldsymbol{I}(s)_{p-1}\right) \tag{17}$$

where p is the number of iterations of the gradient descent method; $\boldsymbol{W}(m)_p$ and $\boldsymbol{I}(s)_p$ are the matrices obtained by the p-th iteration of $\boldsymbol{W}(m)$ and $\boldsymbol{I}(s)$; U_p is the matrix of total error function of the neural network after the p-th iteration, that is, the performance function of the neural network; η is the learning rate; a is the momentum factor. To speed up the network training process and avoid falling into local optimum, the learning rate η needs to be adjusted, that is, the adaptive learning rate method. Using the adaptive learning rate to adjust the network can not only effectively improve the convergence rate but also improve the stability of the network learning process. The adjustment mechanism is as in Equation (18):

$$\eta = \begin{cases} 1.05\eta & U_{p+1} < U_p \\ 0.95\eta & U_{p+1} > 1.04\, U_p \\ \eta & otherwise \end{cases} \tag{18}$$

The learning rate is adjusted by comparing the network's total error function size before and after the update. When the error function increases, it indicates that the updating direction is wrong, or the updating amount is too large, and the learning rate needs to be reduced; when the error function decreases, it indicates that the direction is correct, and the learning rate should be increased.

Let $\boldsymbol{D}(m) = \begin{bmatrix} d_{11} & \cdots & d_{1T} \\ \vdots & \ddots & \vdots \\ d_{m1} & \cdots & d_{mT} \end{bmatrix}$ be the actual output sample; the mean squared error (MSE) function can be used as the network performance function:

$$U_p = \text{MSE} = \frac{1}{T}\left\| \boldsymbol{D}(m) - \boldsymbol{Y}(m)_p \right\|_2^2 = \frac{1}{T}\sum_{t=1}^{T}\sum_{j=1}^{m}\left(\text{d}_{jt} - \text{y}_{jt,p}\right)^2 \tag{19}$$

Thus, we obtain the following:

$$\frac{\partial U_p}{\partial \boldsymbol{W}(m)_p} = \left(\boldsymbol{D}(m) - \boldsymbol{Y}(m)_p\right) \odot f'\left(\boldsymbol{W}(m)_p \begin{bmatrix} \boldsymbol{R}(s)_p \\ \boldsymbol{E}_{1\times T} \end{bmatrix}\right) \tag{20}$$

$$\frac{\partial U_p}{\partial \boldsymbol{I}(s)_p} = \Psi'\left(\boldsymbol{H}(s)_p\right) \odot \left[\boldsymbol{V}(m)_p{}^{\text{T}}\left(\left(\boldsymbol{Y}(m)_p - \boldsymbol{D}(m)\right) \odot f'\left(\boldsymbol{W}(m)_{p+1}\begin{bmatrix} \boldsymbol{R}(s)_p \\ \boldsymbol{E}_{1\times T} \end{bmatrix}\right)\right)\begin{bmatrix} \boldsymbol{X}(n) \\ \boldsymbol{E}_{1\times T} \end{bmatrix}\right] \tag{21}$$

where $y_{jt,p}$ is the element of $\boldsymbol{Y}(m)$ in the matrix $\boldsymbol{Y}(m)_p$ after the p-th iteration; $\boldsymbol{V}(m)_p{}^{\text{T}}$ is the inversion of $\boldsymbol{V}(m)$ after the p-th iteration; $\boldsymbol{R}(s)_p$ is the matrix obtained after the p-th iteration of $\boldsymbol{R}(s)$.

2.3. *Improved WNN*

The WNN still has shortcomings. For example, with the increase of the dimension of the problem, the number of training samples required by the network also increases sharply, which makes the network structure huge, and the learning and training convergence speed decreases; the number of hidden layer nodes of the network is difficult to determine, lacking scientific theoretical guidance; however, whether the selection of the number of hidden layer nodes is appropriate is directly related to the fitting performance and convergence accuracy of the network. The selection of the initial coefficient matrices $\boldsymbol{W}(m)$ and $\boldsymbol{I}(s)$ has a significant impact on the establishment and performance of the network, as an inappropriate initial coefficient matrix may lead to slow convergence of network errors, low convergence accuracy, or even non-convergence, resulting in an ill-conditioned network.

As for the problem of the dramatic increase of required training samples, it can be dealt with by sampling. The uniform design method (UDM) is employed to ensure the quality of the sample [40], and the sample experimental design approach is implemented at the expense of the minimal sample size to reduce the required number of samples.

As for the problem that the number of hidden layer nodes is difficult to determine since there is no conventional selection method for the number of hidden layer nodes, the empirical formula with mature application and universal applicability is employed to determine the number of hidden layer nodes, as in Equation (22) [41]:

$$\text{s} = \sqrt{m+n} + \gamma \tag{22}$$

where γ is a constant in the range of 1 to 10.

To solve the problem of the selection of initial coefficient matrices $\boldsymbol{W}(m)$ and $\boldsymbol{I}(s)$, the wind-driven optimization (WDO) algorithm with strong robustness and high efficiency is used to select the initial coefficient matrix, thus improving the performance of the network. The combined use of the wind-driven algorithm and WNN is shown in Figure 2.

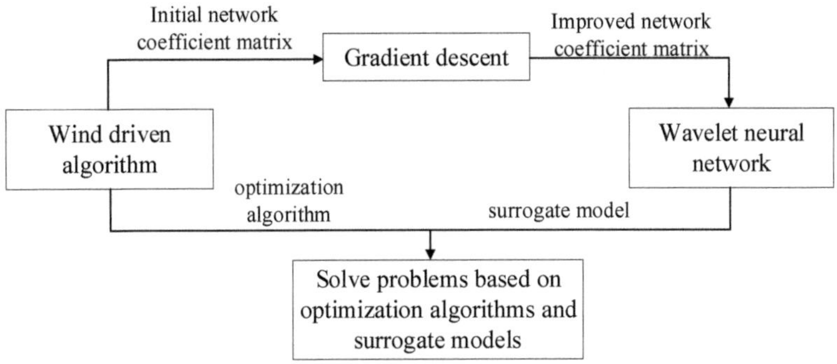

Figure 2. Combination of WDO and WNN.

3. Wind-Driven Optimization Algorithm

In this section, the fundamentals of the algorithm of wind-driven optimization (WDO) are introduced, the update equations are derived, and finally, the realization of WDO is demonstrated.

WDO [42–44] is a novel-group iterative heuristic global optimization algorithm. The algorithm has the advantages of strong robustness and high optimization efficiency and is suitable for dealing with nonlinear, multi-extremal, and non-differentiable complex problems. It was first proposed in 2010 and has been rapidly developed and applied recently.

3.1. Fundamentals of the Algorithm

The motion of the atmosphere inspires WDO; that is, the atmosphere is considered to be composed of many air particles with a certain mass and volume, and the air particles satisfy the ideal gas law, so the motion state of each air particle can be described by Newton's second law [45]. The fundamentals of the WDO algorithm are derivations of velocity and position update equations during the movement of air particles. Newton's second law and ideal gas law are as Equations (23) and (24):

$$\rho \boldsymbol{a} = \sum_i \boldsymbol{F}_i \tag{23}$$

$$\boldsymbol{Q} = \rho RT \tag{24}$$

where $\boldsymbol{a}$ is the acceleration of air particles; ρ is the density of air particles in the atmosphere; $\boldsymbol{F}_i$ is the different forces acting on the air particle; Q is the atmospheric pressure; R is the ideal gas constant; T is the atmospheric temperature.

The atmosphere moves on the earth's surface and has a certain speed relative to the earth, usually called wind speed. When air particles move in a straight line with the wind speed vector in a rotating system, their inertia will produce a linear motion offset relative to the rotating system. Since Newton's law is established in the inertial coordinate system, when dealing with the problem of the non-inertial rotating coordinate system, an imaginary force, the Coriolis force, needs to be introduced to describe the linear offset phenomenon in the rotating system. Combined with the physics of atmospheric motion, it is assumed that air particles are in a state of hydrostatic equilibrium, and the atmospheric motion is simulated by the four main forces acting on air particles, which are as follows:

(1) The pressure gradient force F_p that makes the gas move from the high-pressure region to the low-pressure region is calculated by Equation (25):

$$F_p = -\Delta \boldsymbol{Q} \times \delta V = -\Delta \rho RT \times \delta V; \tag{25}$$

(2) The friction force F_u that prevents the gas from flowing due to the pressure gradient force is calculated by Equation (26):

$$F_u = -\rho\gamma \boldsymbol{u}; \tag{26}$$

(3) The gravity force F_g which is perpendicular to the center of the earth is calculated by Equation (27):

$$F_g = \rho \times \boldsymbol{g} \times \delta V; \tag{27}$$

(4) The Coriolis force F_c describing the linear offset of air particles is calculated by Equation (28):

$$F_c = -2\boldsymbol{\Omega} \times \boldsymbol{u} \tag{28}$$

where $-\Delta \boldsymbol{Q}$ is the pressure gradient, and the negative sign indicates the direction of descent along the gradient; δV is the finite volume of air particles; γ is the friction coefficient; $\boldsymbol{u}$ is the wind velocity vector; $\boldsymbol{g}$ is the gravitational acceleration vector; $\boldsymbol{\Omega}$ is the earth's rotation angular velocity vector. The acceleration of air particles can be expressed as $\boldsymbol{a} = \Delta \boldsymbol{u}/\Delta t$, which can be obtained by substituting the four main forces into Newton's second law as Equation (29):

$$\begin{aligned} \rho \tfrac{\Delta \boldsymbol{u}}{\Delta t} &= F_p + F_u + F_g + F_c \\ &= (-\Delta \boldsymbol{Q} \times \delta V) + (-\rho\gamma \boldsymbol{u}) + (\rho \times \boldsymbol{g} \times \delta V) + (-2\boldsymbol{\Omega} \times \boldsymbol{u}) \end{aligned} \tag{29}$$

3.2. Derivation of Update Equations

To derive the velocity update equation, Equation (29) is simplified; let the air particle volume be $\delta V = 1$, $\Delta t = 1$. Both sides of the equation are divided by ρ, and then it is converted to $\rho = \boldsymbol{Q}/RT$ according to the ideal gas law, which leads to the following:

$$\Delta \boldsymbol{u} = \frac{-\Delta \boldsymbol{Q} \times RT}{\boldsymbol{Q}} - \gamma \boldsymbol{u} + \boldsymbol{g} - \frac{2\boldsymbol{\Omega} \times \boldsymbol{u} \times RT}{\boldsymbol{Q}} \tag{30}$$

In Equation (30), the gravity vector $\boldsymbol{g}$ in the three-dimensional coordinate system centered at the earth will change with the distance between the air particle and the center of the earth, and the WDO algorithm replaces the earth-centered three-dimensional coordinate system with a $[-1, 1]$-type one-dimensional coordinate system with 0 as the origin and assumes that the relationship between the vector $\boldsymbol{g}$ and the current position $\boldsymbol{x}_{cur}$ of the air particle is $\boldsymbol{g} = |\boldsymbol{g}| \cdot (0 - \boldsymbol{x}_{cur})$. Therefore, the position of the air particle is also limited to the range of $[-1, 1]$ so that the optimization range of the algorithm is $[-1, 1]$.

Meanwhile, in the one-dimensional coordinate system, the direction of the pressure gradient force F_p is from the high-pressure region to the low-pressure region, and the pressure value at the current position $\boldsymbol{x}_{cur}$ of the air particle is defined as $\boldsymbol{Q}_{cur}$, and the current minimum pressure value in the air particle swarm is $\boldsymbol{Q}_{opt}$, which is the optimal pressure value, corresponding to the current optimal position $\boldsymbol{x}_{opt}$ in the air particle swarm. Then, it can be assumed that the relationship between the pressure gradient $-\Delta \boldsymbol{Q}$ and the position of air particle is $-\Delta \boldsymbol{Q} = \left|\boldsymbol{Q}_{opt} - \boldsymbol{Q}_{cur}\right| \cdot (\boldsymbol{x}_{opt} - \boldsymbol{x}_{cur})$. The gravity force F_g and the pressure gradient force F_p are shown in Figure 3.

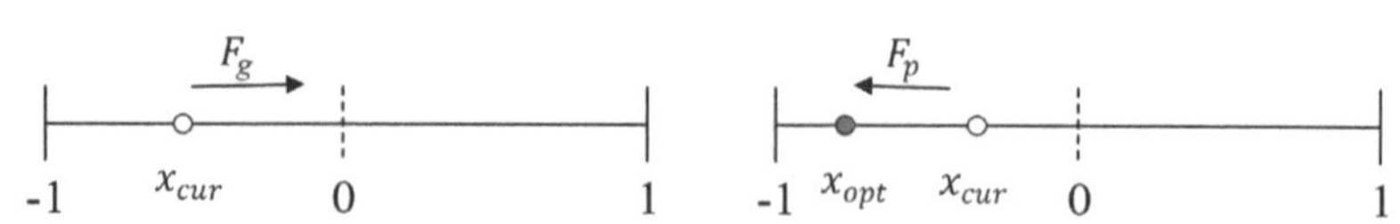

Figure 3. The friction force (F_g) and pressure gradient force (F_p) in the one-dimensional coordinate system.

In Equation (30), the velocity variable $\Delta \boldsymbol{u}$ can be expressed as $\Delta \boldsymbol{u} = \boldsymbol{u}_{new} - \boldsymbol{u}_{cur}$, where $\boldsymbol{u}_{cur}$ is the current velocity of air particle, and $\boldsymbol{u}_{new}$ is the updated velocity of air particle. The friction force F_u is affected by the current velocity; therefore, $\boldsymbol{u} = \boldsymbol{u}_{cur}$ is the friction force term, but the Coriolis force F_c describes the influence of inertia on current velocity, and thus, the velocity $\boldsymbol{u}_{cur}^{other\ dim}$ is used to describe the velocity of the current dimension that is affected by other dimensions; that is, $\boldsymbol{u} = \boldsymbol{u}_{cur}^{other\ dim}$ is the Coriolis force term, and this velocity ensures the balance between the algorithm's exploration ability and development ability. Thereafter, Equation (30) can be further derived:

$$\begin{aligned} \boldsymbol{u}_{new} = (1-\gamma)\boldsymbol{u}_{cur} - RT(\boldsymbol{x}_{opt} - \boldsymbol{x}_{cur})\frac{|Q_{opt}-Q_{cur}|}{Q_{cur}} \\ +g\boldsymbol{x}_{cur} - \frac{2\Omega RT\boldsymbol{u}_{cur}^{other\ dim}}{Q_{cur}} \end{aligned} \tag{31}$$

To further improve Equation (31), let $= -2|\boldsymbol{\Omega}|RT$. In addition, to make the WDO algorithm operational, the algorithm uses the ratio of air particle sorting numbers to replace the actual air pressure ratio; that is, $\frac{Q_{opt}}{Q_{cur}} = \frac{1}{k}$, where k is the sorting number obtained by sorting the air particles in ascending order according to the distance between the current position $\boldsymbol{x}_{cur}$ and the current optimal position $\boldsymbol{x}_{opt}$, and the sorting number for $\boldsymbol{x}_{opt}$ is 1. Hence, the velocity update equation can be obtained as Equation (32):

$$u_{\text{new}} = (1-\gamma)u_{\text{cur}} - g\ x_{\text{cur}} + RT\left|\frac{1}{k} - 1\right|(x_{\text{opt}} - x_{\text{cur}}) + \frac{cu_{\text{cur}}^{\text{other dim}}}{k} \tag{32}$$

where the recommended value ranges of correlation coefficients γ, g, c, and RT are shown in Table 1 [45].

Table 1. Range of parameters [45].

Correlation Coefficient	γ	g	c	RT
Range	[0.8, 0.9]	[0.6, 0.7]	[0.05, 3.6]	[1.0, 2.0]

Then, the position update equation of air particles is derived, and the relationship between position and velocity is as Equation (33):

$$x_{new} = x_{cur} + u_{new} \times t \tag{33}$$

where $\boldsymbol{x}_{new}$ is the update speed of air particles. Let t = 1; the equation can be simplified as in Equation (34):

$$\boldsymbol{x}_{new} = \boldsymbol{x}_{cur} + \boldsymbol{u}_{new} \tag{34}$$

Each updated air particle position is limited to the range of $[-1, 1]$ due to the theoretical limit of the WDO algorithm. Meanwhile, its velocity must be limited to prevent the air particle position from moving too much. The limited update speed should be within the range of $[-u_{max}, u_{max}]$, where $-u_{max}$ and u_{max} are upper and lower bounds of air particle velocity.

The WDO algorithm takes the state of air particles as the research object and is an optimization algorithm established in a one-dimensional coordinate system, but the algorithm can map one-dimensional problems into N-dimensions to solve multi-dimensional optimization problems.

3.3. *Realisation of WDO*

A relational expression between the two needs to be established to relate the updating parameters of the structure to the position of the air particle. Let the updating range of the i-th updating parameter p_{i} be $\lambda_{1i}\% \sim \lambda_{2i}\%$; then, the range of p_{i} is $[p_{0i}(1+\lambda_{1i}\%), p_{0i}(1+\lambda_{1i}\%)]$, where $p_{0\mathrm{i}}$ is the initial value of p_{i}. Suppose the position of an air particle (in N dimension) is $(x_1, x_2, \ldots, x_N)$, and the updating parameter of the structure is p_i:

$$\lambda_{1i}\% \leq \left(\frac{\lambda_{2i}+\lambda_{1i}}{2} + \frac{\lambda_{2i}-\lambda_{1i}}{2} \times x_i\right)\% \leq \lambda_{2i}\% \tag{35}$$

$$p_i = p_{0i} \times \left(1 + \left(\frac{\lambda_{2i}+\lambda_{1i}}{2} + \frac{\lambda_{2i}-\lambda_{1i}}{2} \times x_i\right)\%\right) \tag{36}$$

The WDO algorithm calculation flow is shown in Figure 4.

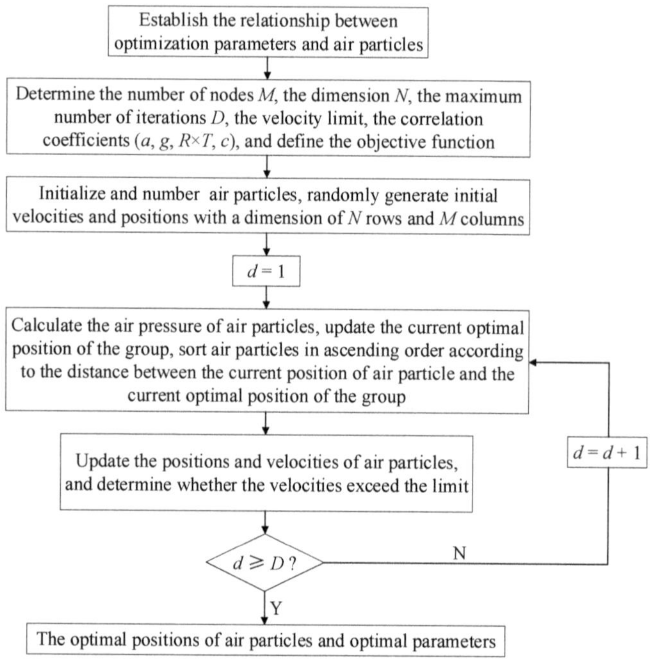

Figure 4. Calculation flow of WDO.

4. Verification of the Proposed Method

In this section, firstly, an initial finite element model that needs to be modified is established, and the model is regarded as the initial structure; secondly, the physical parameters in the model are adjusted appropriately, and the modified parameters are used as the physical parameters of the structure to establish a new finite element model, and the model is regarded as the actual structure; then, based on the difference in static and dynamic responses of the two models, the model is modified based on the WDO and CNN; finally, the updating results are compared and analyzed: If the difference between updated and actual values is slight, the model updating is completed.

4.1. Finite Element Model

The numerical model is a continuous beam structure of three equal spans, and the length of each span is 4 m. Since the elastic modulus and density of materials could change along with time, and their changes could influence structural behavior significantly, these data were therefore selected for verification and demonstration purposes. The elastic modulus of the material of the beam is $E = 3.25 \times 10^4$ MPa, and the density is $\rho = 25\ \text{kN/m}^3$. The beam's cross-section is a rectangular section of 0.2 m × 0.15 m. The finite element model was established by finite element analysis software Midas Civil 2020, and the model was divided into 12 beam elements. The elastic moduli of each span of the three-span continuous beam are denoted as E_1, E_2, and E_3 and the midpoint of each span as C_1, C_2, and C_3. The finite element model is shown in Figure 5.

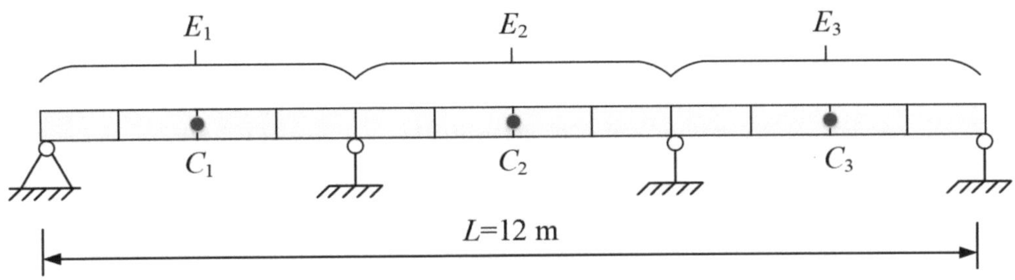

Figure 5. The structural representation of the finite element model.

In practice, the material density of the structure may deviate, and the stiffness may change during service time. Therefore, for illustrative purposes, the material density ρ is assumed to be increased by 10%, and the elastic moduli E_1, E_2, and E_3 are assumed to be decreased by 10%, 15%, and 20%, respectively, as the parameters of the actual structure. Values of these deviated parameters are called the actual values of the structure for simplicity, and the static and dynamic responses calculated by these deviated values are called the actual values of responses. The initial and actual values of parameters are shown in Table 2, where the difference is: (initial value − actual value)/actual value.

Table 2. Initial value and actual value of physical parameters.

Parameter	E_1 ($\times 10^4$ MPa)	E_2 ($\times 10^4$ MPa)	E_3 ($\times 10^4$ MPa)	ρ ($\times 10^1$ kg/m^3)
Initial value	3.250	3.250	3.250	2.500
Actual value	2.925	2.762	2.600	2.750
Difference/%	11.11	17.67	25.00	−9.09

The finite element model updating is carried out using the static and dynamic response of the structure, in which the dynamic response adopts the first six natural frequencies of the structure, and the static response is obtained by applying a vertical downward force of 30 kN at the two points C1 and C3 to obtain displacements at C1, C2, and C3. The initially calculated values obtained from the responses of the initial finite element model and the actual values of the finite element model with updated parameters are shown in Table 3. The difference in the table is: (initial value − actual value)/ actual value.

Table 3. The initial value and actual value of the static and dynamic responses of the structure. (The positive sign indicates downwards displacement, and the negative sign indicates upwards displacement.)

Responses		Initial Value	Actual Value	Difference (%)
Natural frequency/Hz	1st order	15.151	13.295	13.96
	2nd order	19.383	17.040	13.75
	3rd order	28.203	24.788	13.78
	4th order	59.869	52.429	14.19
	5th order	67.678	59.568	13.61
	6th order	81.533	71.722	13.68
Displacement/mm	C1	−17.033	−19.243	−11.48
	C2	9.843	11.589	−15.07
	C3	−17.033	−20.926	−18.60

It can be seen from Table 3 that the initial values of the initial finite element model show considerable differences from actual values, and the differences in natural frequencies and displacements are both larger than 11%; therefore, the initial finite element model needs to be updated. According to the basic process of model updating, the four parameters of the initial model, E_1, E_2, E_3, and ρ were updated, and for illustrative purposes, the objective function was constructed without conducting sensitivity analysis to screen the updating parameters.

Combined with static and dynamic responses of the finite element model, the objective function was constructed. The objective function takes the form of the residual sum of squares and is defined as Equation (37):

$$U = \sum_{i=1}^{3}\left(\frac{c_i^t - c_i^a}{c_i^t}\right)^2 + \sum_{j=1}^{6}\left(\frac{f_j^t - f_j^a}{f_j^t}\right)^2 \quad (37)$$

where c_i^t is the actual displacement at the midpoint C_i of the i-th span, $i = 1, 2, 3$; c_i^a is the initial displacement value; f_j^t is the actual value of the jth order natural frequency, $j = 1, 2, \ldots, 6$; f_j^a is the initial value of the natural frequency.

The four parameters to be updated were divided into 21 levels in the range of -10% to 10% with 1% as the step size, and the uniform design table [46] $U_{21}(21^4)$ was used for experimental design. The parameter values of the 21 levels were sequentially brought into the initial finite element model, and the corresponding model responses were obtained. Taking 21 sets of parameter test values as input samples and taking the corresponding model responses as output samples, the input and output samples were normalized and used for WNN training.

4.2. Establishment of WNN

WNN parameters were selected tentatively by experience for verification purposes. The number of nodes in the network's input layer is four, and the number of nodes in the output layer is nine. According to the empirical formula, the number of nodes in the network's hidden layer was selected as 11. Meanwhile, the learning rate η of the network was selected as 0.01, and the momentum factor a is 0.5. In addition, the excitation function of the hidden layer adopts the Morlet wavelet function, and the excitation function of the output layer adopts the tansig function.

When establishing the network, the WDO algorithm and the normalized test sample data are used to optimize the connection coefficient matrix of the WNN initially. Secondly, the WNN learning and training are carried out, and the MSE function is used as the network performance function. The learning and training process of the WNN is shown in Figure 6.

It can be seen from the figure that the WNN obtains a small network error in the process of 100 times of learning and training, and the MSE value is only 4.7×10^{-6}. The small network error indicates that the WNN as a surrogate model of the finite element model has good fitting accuracy.

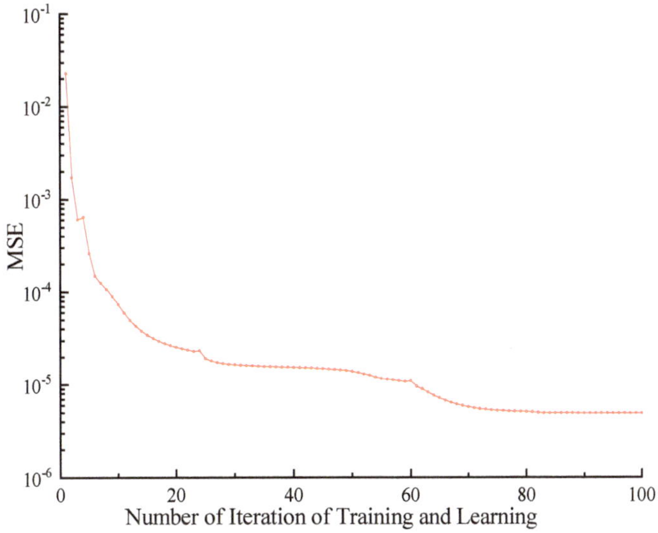

Figure 6. The training process of WNN.

4.3. Results of Algorithm Optimization

Combined with the WNN, the WDO algorithm optimizes parameters E_1, E_2, E_3, and ρ. The four updating parameters correspond to the four dimensions of air particles, and 100 air particles are used for optimization calculation. The velocity limit is $[-1, 1]$, the maximum number of iterations is 100, and the coefficients a, g, RT, and c in the velocity update equation are 0.85, 0.65, 0.5, and 2, respectively. The iterative curve of the objective function is shown in Figure 7.

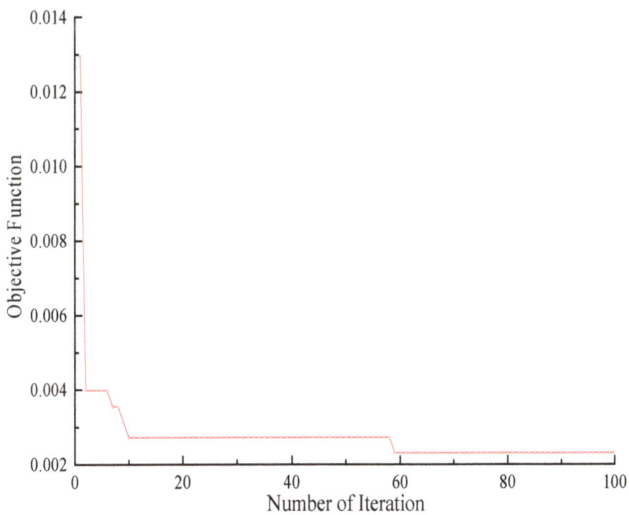

Figure 7. The iterative curve of WDO.

It can be seen from the figure that the objective function converges at 58 iterations, and the objective function value is as small as 2.3×10^{-3}, which shows that the optimization effect of the algorithm is obvious, and it is suitable for multi-parameter optimization. The corrected values of physical parameters are shown in Table 4.

Table 4. The optimization results of WDO.

Parameter	Optimal Position of Air Particles	Corrected Value
$E_1/\times10^4$ MPa	−0.444	2.889
$E_2/\times10^4$ MPa	−0.576	2.782
$E_3/\times10^4$ MPa	−0.674	2.702
$\rho/\times10^3$ kg/m^3	0.470	2.849

The differences in updated and actual values are shown in Table 5. Based on the parameters, the WNN and the updated finite element model were used to calculate the structural response, respectively, to obtain the network value and the updated value of the structural response and then compared with actual responses. The comparison is shown in Table 6. Among them, the difference I is: (network value − actual value)/actual value; difference II is: (updated value − actual value)/actual value.

Table 5. Differences in physical parameters.

Parameter	E_1 ($\times10^4$ MPa)	E_2 ($\times10^4$ MPa)	E_3 ($\times10^4$ MPa)	ρ ($\times10^3$ kg/m^3)
Updated value	2.890	2.782	2.702	2.849
Actual value	2.925	2.762	2.600	2.805
Difference/%	−1.22	0.74	3.92	1.57

Table 6. Comparison of structural responses.

Responses		Network Value	Updated Value	Actual Value	Difference I (%)	Difference II (%)
Natural frequency/Hz	1st order	13.415	13.274	13.295	0.90	−0.16
	2nd order	17.174	16.996	17.040	0.78	−0.26
	3rd order	24.973	24.714	24.788	0.75	−0.30
	4th order	52.996	52.416	52.429	1.08	−0.02
	5th order	59.955	59.377	59.568	0.65	−0.32
	6th order	72.214	71.453	71.722	0.69	−0.37
Displacement /mm	C1	−19.263	−19.361	−19.243	0.10	0.61
	C2	11.367	11.468	11.589	−1.91	−1.04
	C3	−20.082	−20.305	−20.926	−4.03	−2.97

It can be seen from Table 5 that after updating parameters E_1, E_2, E_3, and ρ, the differences become smaller, and the maximum difference is reduced from 25.00% to 3.92%. It can be seen from Table 6 that the values of difference I and difference II are both small, and the network value, updated value, and actual value of the static and dynamic responses are relatively close; thus, the model updating is proven to be effective. Therefore, the WNN can approximately reflect the nonlinear relationship between structural response and parameters and has good simulation performance; using the WNN as a surrogate model can effectively update the finite element model and improve work efficiency.

5. Finite Element Model Updating of Ningbo Bund Bridge

In this section, the proposed is implemented on the model updating of Ningbo Bund Bridge, updating parameters are selected based on sensitivity analysis, WNN is established, and model updating is then carried out based on static and dynamic responses of the structure.

5.1. Introduction of Ningbo Bund Bridge

The Ningbo Bund Bridge is a cable-stayed bridge made of steel with a single tower and four cable planes, as shown in Figure 8.

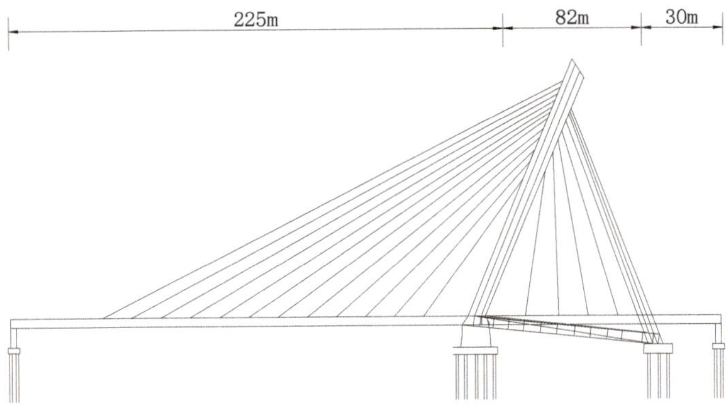

Figure 8. The elevation of Ningbo Bund Bridge.

The bridge's total length is 337 m, and the span arrangement is (225 + 82 + 30) m from west to east. The main girder of the bridge adopts a separate steel box girder, which is connected by the cross-beam. The main girder adopts a varying cross-section along the longitudinal direction. The maximum beam height is 2.4 m, designed as a two-way six-lane bridge, and the standard width of the one-way is 21.4 m.

There are 64 stay cables in the bridge, including 46 main-span stay cables and 18 side-span stay cables. The stay cables are anchored on the head part of the front tower column and the inner and outer webs on both sides of the girders. The stay cables are anchored to the tower and inner and outer webs of the steel girder. The bridge tower adopts a triangular inclined tower structure composed of front tower columns, rear-inclined rods, horizontal rods, and other parts. The elevation of the bridge is shown in Figure 8.

5.2. Finite Element Model of Ningbo Bund Bridge

According to design drawings and engineering construction practice, the finite element model of the bridge was established by the finite element analysis software Midas Civil 2020. The main girder and triangular inclined tower structure were simulated by the spatial beam element, the stay cable was simulated by the truss element, and the cable force was input by the measured cable force. The corresponding element mass adopts the equivalent density method, which is equivalent to the mass density of the element according to the mass of the segment, so that the structural mass of the finite element model is close to the actual mass of the structure. The tower and the bottom of the pier were consolidated to constrain all the degrees of freedom; the beam end restraint was simulated with spring elements to restrain the vertical and longitudinal displacements of the bridge. The established finite element model has a total of 309 nodes, 64 truss elements, and 193 beam elements, as shown in Figure 9. In the model, the triangular inclined tower structure is shown in Figure 10. The design load of the bridge is vehicle load highway-level I and city-level A, and pedestrian load is 2.4 kN/m^2.

Figure 9. Finite element model of Ningbo Bund Bridge.

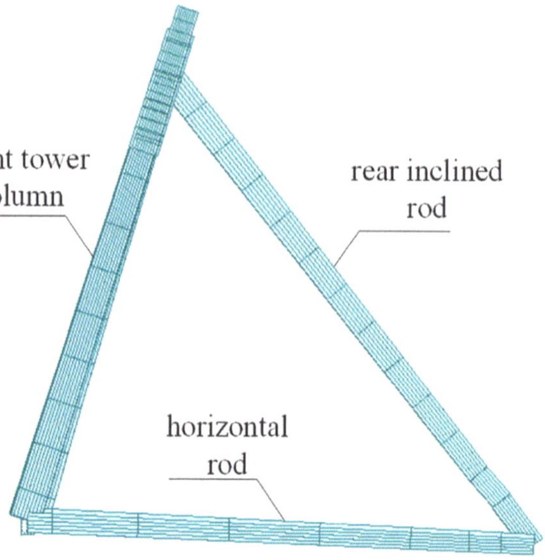

Figure 10. The schematic diagram of the triangular inclined tower structure.

An ambient vibration test was carried out on the Ningbo Bund Bridge to obtain the structure's measured frequencies and mode shapes. The ultra-low frequency (minimum sampling frequency up to 0.05 Hz) 991-type accelerometer was used in the test, the operation details in the experimental modal analysis are detailed in the literature [47], and a low-speed gear was selected for signal acquisition for low-frequency analysis. The measuring points were arranged in the main beam's vertical, horizontal, and longitudinal direction, and the INV306D data signal acquisition and analysis system was used for data acquisition and modal analysis.

Meanwhile, the bridge's initial finite element analysis was conducted to obtain the initial calculation results of the frequency and mode shape. The first three-order mode shapes are shown in Figure 11, and the first eight-order natural frequencies are shown in Table 7.

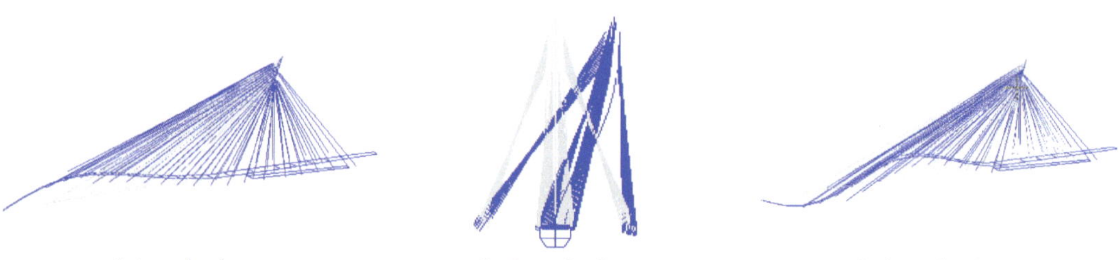

1st mode shape 2nd mode shape 3rd mode shape

Figure 11. First three-order mode shapes of the bridge.

Table 7. Measured and calculated values of natural frequencies.

Order of Natural Frequency	Measured Value (Hz)	Calculated Value (Hz)	Difference (%)	Description of Mode Shape
1	0.55	0.49	−10.9%	225 m span main girder vertical bending vibration
2	0.73	0.78	6.8%	Lateral vibration of the bridge tower
3	0.86	0.88	2.3%	225 m span main girder vertical bending vibration
4	1.26	1.21	−3.9%	Lateral bending vibration of the main beam
5	1.34	1.41	5.2%	Torsional vibration of the main beam
6	1.40	1.46	4.3%	Vertical bending vibration of the main beam
7	1.64	1.76	7.3%	Lateral vibration of the bridge tower
8	1.98	2.09	5.6%	Torsional vibration of the main beam

It can be seen from Table 6 that differences exist between the initially calculated value of the structural natural frequency and the measured value: The difference in first-order natural frequency is more than 10%, and the differences in the second-, fifth-, seventh-, and eighth-order natural frequencies are all above 5%; thus, the initial finite element model needs to be updated.

5.3. Selection of Updating Parameters

The selection of correction parameters is mainly based on engineering and modeling experience and sensitivity analysis methods. The Ningbo Bund Bridge has a novel structure and complex form and has many structural design parameters. When establishing the initial finite element model, the physical parameters of the structure, such as the elastic modulus of the material, mass density, geometric parameters, boundary conditions, and others, were mainly determined based on the construction design drawings, and these physical parameters are often different from the actual situation.

The Ningbo Bund Bridge is a steel bridge, and most components were prefabricated and assembled. Assembly may adversely affect the structural rigidity, and some detailed components may contribute to the structural rigidity to a certain extent. The influence of assembly and the contribution of detailed components to structural rigidity should be considered when establishing the model. Therefore, based on engineering experience, the elastic modulus of the main beam E_1, the elastic modulus of the front tower column E_2, the elastic modulus of the rear-inclined rod E_3, the elastic modulus of the horizontal rod E_4, the elastic modulus of the stay cable E_5, and the elastic modulus of the cross-beam E_6 were selected as updating parameters. Meanwhile, the equivalent density method was used in the modeling to equalize the mass of the detailed components, bridge deck pavement, and bridge deck auxiliary structures to the main structure, and there may be deviations in the mass conversion process. Therefore, the girder density ρ_1, the front tower column density ρ_2, the rear-inclined rod density ρ_3, the horizontal rod density ρ_4, the stay cable density ρ_5, and the cross-beam density ρ_6 were also selected as updating parameters. Additionally, the cable force of the bridge was adjusted according to the construction acceptance standard of the cable-stayed bridge, and a small cable force error was obtained; the error between the cable force of the completed bridge and the designed cable force is within 5% [48,49]. The whole bridge has a total of 64 cables, and to simplify the model updating process, as the cable force error is small, the actual measured cable force of the bridge was used as the input rather than selected as updating parameters. Furthermore, the error between the actual size and the design size of the prefabricated components of the bridge is very small, and the element section of the finite element model was established according to design drawings; thus, the geometric parameters of the sections were not considered. The boundary constraints of the bridge are few and all controlled in an ideal state, and the partially consolidated piers have little effect on the natural frequencies of the structure; hence, the boundary conditions were also not considered.

When conducting parameter sensitivity analysis [50], each of the 12 updating parameters was individually increased by 1% and brought into the finite element model to approximately obtain the sensitivity values of each parameter corresponding to each order of natural frequency, as shown in Figure 12.

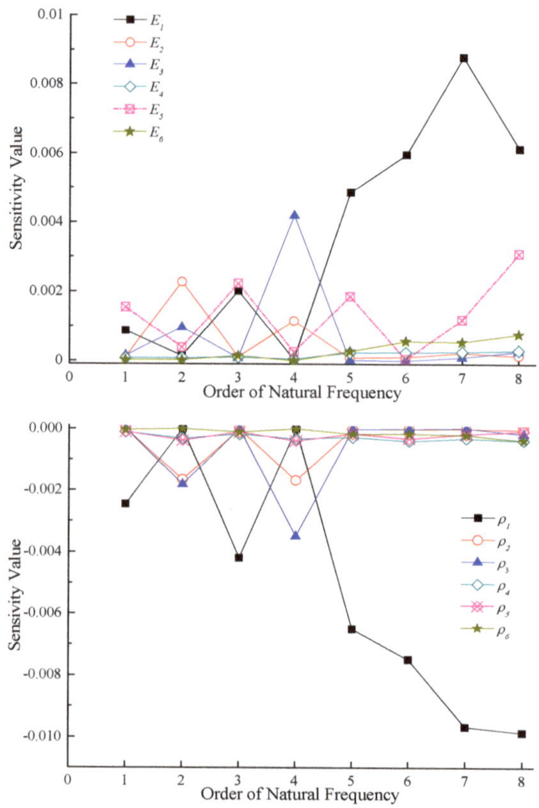

Figure 12. Sensitivity analysis of updating parameters.

Through sensitivity analysis of 12 updating parameters, 7 parameters with large sensitivity values were finally selected as updating parameters, which include the elastic modulus of the main beam E_1, the elastic modulus of the front tower column E_2, the elastic modulus of the rear-inclined rod E_3, the elastic modulus of the stay cable E_5, the density of the main beam ρ_1, the density of the front tower column ρ_2, and the density of the rear-inclined rod ρ_3. The initial values of the final updating parameters in the finite element model are shown in Table 8.

Table 8. Initial values of updating parameters.

E_1 ($\times 10^4$ MPa)	E_2 ($\times 10^4$ MPa)	E_3 ($\times 10^4$ MPa)	E_5 ($\times 10^4$ MPa)	ρ_1 ($\times 10^3$ kg/m^3)	ρ_2 ($\times 10^3$ kg/m^3)	ρ_3 ($\times 10^3$ kg/m^3)
2.06	2.06	2.06	2.06	12.59	8.05	8.05

In the table, the elastic modulus of the main beam E_1, the elastic modulus of the front tower column E_2, the elastic modulus of the rear-inclined rod E_3, and the elastic modulus of the stay cable E_5 are all the elastic modulus values of steel: 2.06×10^4 MPa; the density of the main beam ρ_1, the density of the front tower column ρ_2, and the density of the rear-inclined rod ρ_3 are the values after mass conversion by the equivalent density method; hence, their densities are all greater than the density of steel, which is 7.85×10^3 kg/m^3.

5.4. Establishment of WNN

Seven updating parameters were divided into 31 levels within the range of −15% to 15% with a step size of 1%, and the uniform design table U31(31^7) was employed for experimental design to obtain input samples. The parameters were brought into the initial finite element model of the bridge in turn, and the corresponding response output samples were obtained. Then, the input and output samples were, respectively, normalized and used as training samples of the neural network.

The number of nodes in the input layer of the network was set to seven and that of nodes in the output layer to eight; according to Equation (22), the number of nodes in the hidden layer of the network was selected as 15; the learning rate η of the network was selected as 0.01, and the momentum factor a was chosen as 0.5. In addition, the excitation function of the hidden layer adopts the Morlet wavelet function, and the excitation function of the output layer adopts the tansig function.

When the network is established, the WDO algorithm and the normalized training samples are used to optimize the connection coefficient matrix of the WNN preliminarily. Secondly, the MSE function is adopted as the network performance function to train the WNN further. The learning and training process of the WNN is shown in Figure 13.

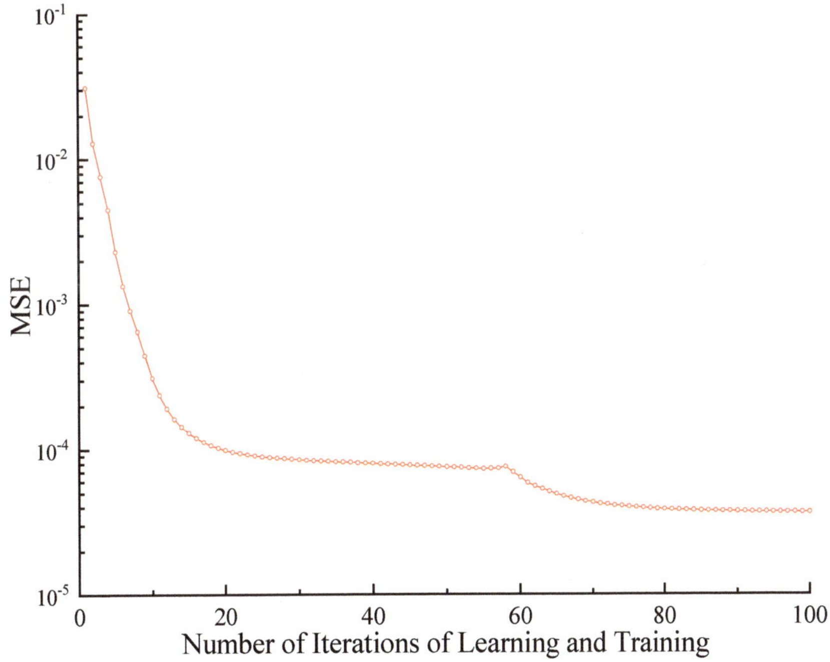

Figure 13. The training process of WNN.

It can be seen from the figure that the WNN obtained a small network error after 100 times of training and learning, and the MSE value is only 3.7×10^{-5}. The network training is efficient, and the network error is small; thus, it can be used as the surrogate model for the initial finite element of the bridge.

5.5. Results and Discussion

Combined with the well-trained WNN and WDO algorithm, seven parameters were updated. The seven updating parameters correspond to the seven dimensions of air particles. One hundred air particles were used for optimization calculation, and the velocity limit is $[-1, 1]$, the maximum number of iterations is 100, and the coefficients a, g, RT, and c are, respectively, 0.85, 0.65, 0.1, and 2. The optimization process of the WDO algorithm is shown in Figure 14.

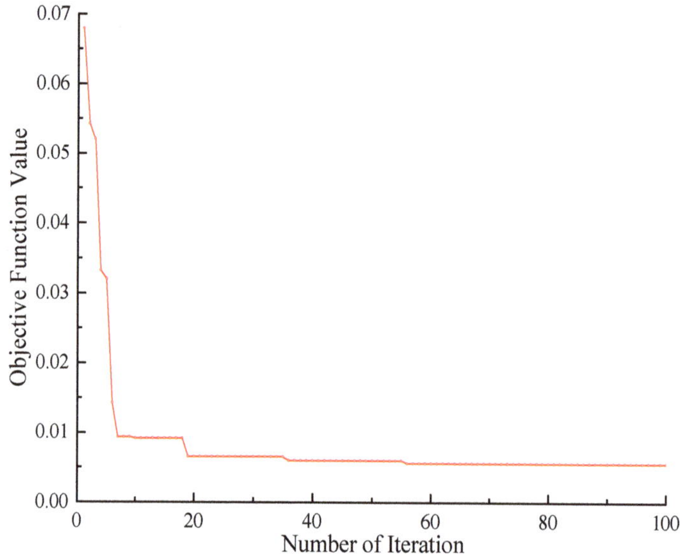

Figure 14. The iterative curve of WDO.

The objective function value reaches convergence in 57 iterations, and the convergence error is 5.6×10^{-3}; thus, the optimization effect is obvious. The comparison between updated values and initial values is shown in Table 9. In the table, the difference is: (updated value − initial value)/initial value.

Table 9. Comparison between updated values and initial values.

Parameter	Updated Value	Initial Value	Difference (%)
$E_1/\times10^4$ MPa	1.95	2.06	−5.3%
$E_2/\times10^4$ MPa	1.92	2.06	−6.8%
$E_3/\times10^4$ MPa	2.09	2.06	1.5%
$E_5/\times10^4$ MPa	2.15	2.06	4.4%
$\rho_1/\times10^3$ kg/m^3	12.68	12.59	0.7%
$\rho_2/\times10^3$ kg/m^3	7.88	8.05	−1.9%
$\rho_3/\times10^3$ kg/m^3	8.43	8.05	4.7%

It can be seen from the table that the updating ranges of seven parameters are small, which are all within the range of −10~10%. Thus, it can be inferred that the decrease in values of E_1 and E_2 may be caused by the adverse effects of the structural assembly; the increase of E_3 and E_4 might be caused by the favorable effects of auxiliary structures and detailed components; the increase in the value of ρ_1 is likely to be caused by the addition of bridge deck pavement load and auxiliary structures or the deviation caused by the mass conversion method during modeling; the decrease in the value of ρ_2 may be caused by considering the extra mass of the detailed components in the initial finite element model or the deviation of the mass conversion method during modeling; the increase in the value of ρ_3 may be caused by ignoring part of the mass of detailed components or the deviation of the quality conversion method during modeling. The selected seven updating parameters have reasonable changes in value, and the updated values of ρ_1, ρ_2, and ρ_3 are all greater than 7.85×10^3 kg/m^3, which still lie in reasonable ranges.

The seven updated parameters were brought into the initial finite element model to obtain the updated value of the dynamic response of the structure. The comparison results with initial values are shown in Table 10. In the table, the difference is: (updated value − measured value)/measured value.

Table 10. Comparison of structural responses.

Structural Response		Updated Value	Measured Value	Difference (%)
Natural frequency/Hz	1st order	0.53	0.55	−3.6%
	2nd order	0.75	0.73	2.7%
	3rd order	0.87	0.86	1.1%
	4th order	1.22	1.26	3.2%
	5th order	1.35	1.34	0.7%
	6th order	1.44	1.40	2.9%
	7th order	1.69	1.64	3.0%
	8th order	2.01	1.98	1.5%

It can be seen from the table that after updating the selected seven parameters, the differences between the updated values and measured values all fell within the range of 5%, and the maximum difference was reduced from −10.9% to −3.6%; thus, the updating is proven to be effective. The finite element model updating method based on the WDO algorithm and WNN has proven to be applicable and practical for multi-parameter bridge model updating.

6. Conclusions

Finite element model updating makes a significant contribution to the healthy operation and damage detection of structures. This paper mainly focuses on model updating and proposes a WNN combined with the WDO algorithm as the surrogate model of the finite element model. The following conclusions are drawn:

(1) A structural finite element model updating method using a WNN as the surrogate model and WDO algorithm is proposed. The finite element model of the structure was updated based on static and dynamic responses of structures. The updated results show that the WNN can reflect the nonlinear relationship between the structural response and the parameters and has good simulation performance; the wind-driven optimization algorithm has the advantages of high optimization efficiency and fast convergence speed, and the method can effectively update the finite element model;

(2) The selection of updating parameters is important for improving the quality and efficiency of model updating. To avoid selecting unnecessary parameters, in this paper, the elastic modulus and mass density of the main beam and the elastic modulus and mass density of the triangular tower were preliminarily selected as updating parameters according to the engineering and modeling experience; then, sensitivity analysis was conducted, and the seven updating parameters were finally determined;

(3) The finite element model updating of the Ningbo Bund Bridge was studied. Combined with measured dynamic responses of the structure, the initial model was updated based on the WDO algorithm and WNN. The results show that the differences between measured dynamic responses and dynamic responses calculated by the updated finite element model were greatly reduced: The maximum difference was reduced from −10.9% to −3.6%. It can be concluded that the finite element model updating method based on the WDO algorithm and WNN is applicable and practical for multi-parameter bridge model updating and has the advantages of high updating efficiency, reliability, and practical significance.

For future research, as evidenced in Section 2 of the paper, since the fitting accuracy and stability of WNNs are affected by the initial value of coefficient matrix, the number of hidden layer nodes, the selected excitation function, the network learning method, and the number of hidden layers, currently, the information and scale of the network are often determined based on experience and trial calculation. How to correctly select the wavelet function, the number of hidden layers, and the number of hidden layer nodes still needs research.

Author Contributions: Methodology, H.H.; software, Y.Z.; validation, B.Z.; investigation, T.Z.; data curation, Y.S.; funding acquisition, H.S. and J.W. All authors have read and agreed to the published version of the manuscript.

Funding: This work was financially sponsored by the National Key Research and Development Program of China (Grant No. 2022YFC3005305), and the Open Project of National Engineering Laboratory of Bridge Safety and Technology (Grant No. 003-04150201122, 76140-42990014).

Institutional Review Board Statement: Not applicable.

Informed Consent Statement: Not applicable.

Data Availability Statement: Data is available upon request.

Conflicts of Interest: The authors declare no conflict of interest.

References

1. Chrysanidis, T. Experimental and numerical research on cracking characteristics of medium-reinforced prisms under variable uniaxial degrees of elongation. *Eng. Fail. Anal.* **2023**, *145*, 107014. [CrossRef]
2. Chrysanidis, T.A.; Panoskaltsis, V.P. Experimental investigation on cracking behavior of reinforced concrete tension ties. *Case Stud. Constr. Mater.* **2022**, *16*, e00810. [CrossRef]
3. Chrysanidis, T.; Mousama, D.; Tzatzo, E.; Alamanis, N.; Zachos, D. Study of the Effect of a Seismic Zone to the Construction Cost of a Five-Story Reinforced Concrete Building. *Sustainability* **2022**, *14*, 10076. [CrossRef]
4. Gong, C.; Wang, Y.; Peng, Y.; Ding, W.; Lei, M.; Da, Z.; Shi, C. Three-dimensional coupled hydromechanical analysis of localized joint leakage in segmental tunnel linings. *Tunn. Undergr. Space Technol.* **2022**, *130*, 104726. [CrossRef]
5. Razavi, M.; Hadidi, A. Structural damage identification through sensitivity-based finite element model updating and wavelet packet transform component energy. *Structures* **2021**, *33*, 4857–4870. [CrossRef]
6. Aloisio, A.; Pasca, D.P.; Battista, L.D.; Rosso, M.M.; Cucuzza, R.; Marano, G.C.; Alaggio, R. Indirect assessment of concrete resistance from FE model updating and Young's modulus estimation of a multi-span PSC viaduct: Experimental tests and validation. *Structures* **2022**, *37*, 686–697. [CrossRef]
7. Liu, C.; Huang, X.; Zhao, Y.; Miao, J. Vibration analysis of concrete T-beam at elevated temperatures based on modified finite element model. *J. Build. Eng.* **2022**, *52*, 104381. [CrossRef]
8. Cheng, X.X.; Song, Z.Y. Modal experiment and model updating for Yingzhou Bridge. *Structures* **2021**, *32*, 746–759. [CrossRef]
9. Li, J. Study of Surrogate Model Approximation and Surrogate-Enhanced Structural Reliability Analysis. Ph.D. Thesis, Shanghai Jiao Tong University, Shanghai, China, 2013. (In Chinese)
10. Jin, R.; Chen, W.; Simpson, T.W. Comparative studies of metamodelling techniques under multiple modelling criteria. *Struct. Multidiscip. Optim.* **2001**, *23*, 1–13. [CrossRef]
11. Zhou, Q.R.; Ou, J.P. Finite element model updating of long-span cable-stayed bridge based on the response surface method of radial basis function. *China Railw. Sci.* **2012**, *33*, 8–15. (In Chinese)
12. Wan, H.P.; Ren, W.X.; Wei, J.H. Structural finite element model updating based on Gaussian process response surface methodology. *J. Vib. Shock* **2013**, *31*, 82–87. (In Chinese)
13. Yang, C.; Xie, Y.M.; Long, Q. Sheet metal forming springback prediction based on agent model of radius basis function. *Mach. Tool Hydraul.* **2013**, *41*, 8–11. (In Chinese)
14. Peng, L.; Liu, L.; Long, T. Satellite bus truss structure optimization using radial basis function. *J. Nanjing Univ. Aeronaut. Astronaut.* **2014**, *46*, 475–480. (In Chinese)
15. Deng, L.; Cai, C.S. Bridge model updating using response surface method and genetic algorithm. *J. Bridge Eng.* **2010**, *15*, 553–564. [CrossRef]
16. Zhong, J.; Gou, H.; Zhao, H.; Zhao, T.; Wang, X. Comparison of several model updating methods based on full-scale model test of track beam. *Structures* **2022**, *35*, 46–54. [CrossRef]
17. Zhao, Y.; Zhang, J.; Li, D.; Zhou, D.; Xin, D. Finite Element Model Updating of Bridge Structures Based on Improved Response Surface Methods. *Struct. Control Health Monit.* **2023**, *2023*, 2488951. [CrossRef]
18. Yang, H.; Yao, W.X. 2D surrogate model of wing lift distribution based on radial basis function. *Chin. J. Comput. Mech.* **2008**, *25*, 797–802. (In Chinese)

19. Chen, S.Z.; Zhong, Q.M.; Hou, S.T.; Wu, G. Two-stage stochastic model updating method for highway bridges based on long-gauge strain sensing. *Structures* **2022**, *37*, 1165–1182. [CrossRef]
20. Krige, D.G. A statistical approach to some basic mine valuation problems on the Witwatersrand. *J. South. Afr. Inst. Min. Metall.* **1951**, *52*, 119–139.
21. Levin, R.I.; Lieven, N.A.J. Dynamic finite element model updating using neural networks. *J. Sound Vib.* **1998**, *210*, 593–607. [CrossRef]
22. Park, Y.S.; Kim, S.; Kim, N.; Lee, J.J. Finite element model updating considering boundary conditions using neural networks. *Eng. Struct.* **2017**, *150*, 511–519. [CrossRef]
23. Nejad, R.M.; Sina, N.; Moghadam, D.G.; Branco, R.; Macek, W.; Berto, F. Artificial neural network based fatigue life assessment of friction stir welding AA2024-T351 aluminum alloy and multi-objective optimization of welding parameters. *Int. J. Fatigue* **2022**, *160*, 106840. [CrossRef]
24. Masoudi, N.R.; Sina, N.; Ma, W.; Liu, Z.; Berto, F.; Gholami, A. Optimization of fatigue life of pearlitic Grade 900A steel based on the combination of genetic algorithm and artificial neural network. *Int. J. Fatigue* **2022**, *162*, 106975. [CrossRef]
25. Han, F.; Yang, J. A New Multi-Resolution Wavelet Neural Network for Bridge Health Monitoring. *Intell. Autom. Soft Comput.* **2010**, *16*, 771–776. [CrossRef]
26. Xu, Z.; Xin, J.Z.; Tang, Q.Z.; Xiao, W.N.; Li, S.J. Finite element model updating based on response surface method and sparrow search algorithm. *Sci. Technol. Eng.* **2021**, *21*, 9094–9101. (In Chinese)
27. Qin, S.Q.; Liao, S.P.; Huang, C.L.; Tang, J. Adaptive Kriging model based finite element model updating of a cable-stayed pedestrian bridge. *Acta Sci. Nat. Univ. Sunyatseni* **2021**, *60*, 43–53. (In Chinese)
28. Qin, S.Q.; Gan, Y.W.; Kang, J.T. Model updating of Nanzhonghuan Bridge based on improved gravitational search algorithm. *J. Vib. Shock* **2021**, *40*, 116–124+136. (In Chinese)
29. Yang, Y.X.; Yang, F.L.; Yu, H.B.; Chai, W.H. Single beam finite element model modification based on a weighted response surface method and chaos particle swarm optimization algorithm. *J. Lanzhou Univ.* **2021**, *57*, 823–829. (In Chinese)
30. Zhou, Z.; Zhang, J.; Gong, C. Hybrid semantic segmentation for tunnel lining cracks based on Swin Transformer and convolutional neural network. *Comput.-Aided Civ. Infrastruct. Eng.* **2023**, *38*, 2491–2510. [CrossRef]
31. Gou, H.; Zhao, T.; Qin, S.; Zheng, X.; Pipinato, A.; Bao, Y. In-situ testing and model updating of a long-span cable-stayed railway bridge with hybrid girders subjected to a running train. *Eng. Struct.* **2022**, *253*, 113823. [CrossRef]
32. Ponsi, F.; Bassoli, E.; Vincenzi, L. A multi-objective optimization approach for FE model updating based on a selection criterion of the preferred Pareto-optimal solution. *Structures* **2021**, *33*, 916–934. [CrossRef]
33. Xiao, S.Z. *Wavelet Neural Network Theory and Application*; Northeastern University Press: Shenyang, China, 2006. (In Chinese)
34. Zhang, Q.H.; Benveniste, A. Wavelet networks. *IEEE Trans. Neural Netw.* **1992**, *3*, 889–898. [CrossRef] [PubMed]
35. Durbin, R.; Rumelhart, D.E. Product units: A computationally powerful and biologically plausible extension to backpropagation networks. *Neural Comput.* **1989**, *1*, 133–142. [CrossRef]
36. Billings, S.A.; Zheng, G.L. Radial basis function network configuration using genetic algorithms. *Neural Netw.* **1995**, *8*, 877–890. [CrossRef]
37. Xiao, S.M.; Yan, Y.J.; Jiang, B.L. Damage identification for bridge structures based on the wavelet neural network method. *Appl. Math. Mech.* **2016**, *37*, 149–159. (In Chinese)
38. Ulagammai, M.; Venkatesh, P.; Kannan, P.S.; Padhy, N.P. Application of bacterial foraging technique trained artificial and wavelet neural networks in load forecasting. *Neurocomputing* **2007**, *70*, 2659–2667. [CrossRef]
39. Alexandridis, A.K.; Zapranis, A.D. Wavelet neural networks: A practical guide. *Neural Netw.* **2013**, *42*, 1–27. [CrossRef]
40. Zhao, Z.; Zhao, Y.G.; Li, P.P. Efficient approach for dynamic reliability analysis based on uniform design method and Box-Cox transformation. *Mech. Syst. Signal Process.* **2022**, *172*, 108967. [CrossRef]
41. Tong, F. A Maritime Accident Forecast Based on BP Neural Network and Carried out by MATLAB. Master's Thesis, Wuhan University of Technology, Wuhan, China, 2005. (In Chinese)
42. Bayraktar, Z.; Komurcu, M.; Werner, D.H. Wind Driven Optimization (WDO): A novel nature-inspired optimization algorithm and its application to electromagnetics. In Proceedings of the 2010 IEEE Antennas and Propagation Society International Symposium, Toronto, ON, Canada, 11–17 July 2010; pp. 1–4.
43. Ibrahim, I.A.; Hossain, M.J.; Duck, B.C.; Nadarajah, M. An improved wind driven optimization algorithm for parameters identification of a triple-diode photovoltaic cell model. *Energy Convers. Manag.* **2020**, *213*, 112872. [CrossRef]
44. Zhong, L.; Zhou, Y.; Luo, Q.; Zhong, K. Wind driven dragonfly algorithm for global optimization. *Concurr. Comput. Pract. Exp.* **2021**, *33*, e6054. [CrossRef]
45. Ren, Z.L.; Zhang, R.J.; Tian, Y.B. Wind driven optimization algorithm. *J. Jiangsu Univ. Sci. Technol. (Nat. Sci. Ed.)* **2015**, *29*, 153–158. (In Chinese)
46. Fang, K.T. *Uniform Design and Uniform Design Table*; Science Press: Beijing, China, 1994. (In Chinese)
47. Komarizadehasl, S.; Huguenet, P.; Lozano, F.; Lozano-Galant, J.A.; Turmo, J. Operational and analytical modal analysis of a bridge using low-cost wireless Arduino-based accelerometers. *Sensors* **2022**, *22*, 9808. [CrossRef]
48. Hu, Z.R.; Wang, Y.W. Analysis on construction control technology of single-tower and four-cable plane space special-shaped cable-stayed bridge. *J. China Foreign Highw.* **2015**, *35*, 155–160. (In Chinese)

49. Zapico, J.L.; Gonzalez, M.P.; Friswell, M.I.; Taylor, C.; Crewe, A. Finite element model updating of a small scale bridge. *J. Sound Vib.* **2003**, *268*, 993–1012. [CrossRef]
50. Yuan, A.M. Studies on Several Key Problems of Finite Element Model Updating Based on the Sensitivity Analysis. Ph.D. Thesis, Southeastern University, Nanjing, China, 2006. (In Chinese)

materials

Article

Lumped Plasticity Model and Hysteretic Performance of Ultra-High-Performance Concrete Rocking Pier

Haifang He, Yulong Zhou *, Shoushan Cheng and Hongyi Liu

National Engineering Laboratory of Bridge Safety and Technology (Beijing), Research Institute of Highway Ministry of Transport, Beijing 100088, China; hf.he@rioh.cn (H.H.); ss.cheng@rioh.cn (S.C.); hongyi.liu@ctvic.cn (H.L.)
* Correspondence: zhouyulong4554@163.com

Abstract: Rocking piers using ultra-high-performance concrete (UHPC) have high damage-control capacity and self-centering characteristics that can limit the post-earthquake recovery time of bridges. To study the hysteretic behavior of UHPC rocking piers, a lumped plasticity model is proposed that comprises two parallel rotational springs and which can accurately calculate their force-displacement hysteretic behavior. Three states of the rocking piers, decompression, yield, and large deformation, are considered in this study. The model is verified based on existing experimental results, and the hysteretic characteristics of the UHPC rocking piers, such as strength, stiffness, and energy dissipation, are further analyzed. The research results show that the lumped plasticity analysis model proposed in this study can predict the force-displacement hysteretic behavior of the rocking piers accurately. Moreover, the hysteretic performance of the UHPC rocking piers is better than that of rocking piers using normal-strength concrete. An increase in the energy dissipation reinforcement ratio, pre-stressed tendon ratio, and initial pre-stress improves the lateral stiffness and strength of the UHPC rocking piers. However, the increase in the pre-stressed tendon ratio and initial pre-stress reduces their energy-dissipation capacity.

Keywords: rocking pier; ultra-high-performance concrete; lumped plasticity model; hysteretic performance; parametric analysis

Citation: He, H.; Zhou, Y.; Cheng, S.; Liu, H. Lumped Plasticity Model and Hysteretic Performance of Ultra-High-Performance Concrete Rocking Pier. *Materials* **2023**, *16*, 6515. https://doi.org/10.3390/ma16196515

Academic Editor: Sérgio Manuel Rodrigues Lopes

Received: 22 August 2023
Revised: 17 September 2023
Accepted: 26 September 2023
Published: 30 September 2023

1. Introduction

Numerous post-earthquake investigations have shown that the traffic interruption caused by earthquake damage to bridges significantly affects the post-disaster rescue and reconstruction operations in the stricken areas. This results in indirect losses that are difficult to estimate and far exceed social expectations [1,2]. Therefore, an important concept in the future seismic design of bridge structures is to effectively control earthquake damage and shorten the recovery time after an earthquake. Currently, the ductile seismic design method is primarily used in the seismic design of bridges in China; it experiences serious plastic damage and large residual displacements during earthquakes, which is highly unfavorable to the recovery of bridge capacity after earthquakes. Therefore, in recent times, a rocking structure with strong earthquake recovery ability has been widely focused on as a new seismic system for bridge structures.

Recently, the seismic concept of rocking bridges has been applied to bridge engineering [3–7]. In 1963, Housner [8] proposed for the first time the seismic response of a structure that was weakened by the rocking of available structures during earthquakes and presented a model to analyze the rocking behavior of rigid bodies. Subsequently, the Housner rigid-body rocking model was improved, and its applicability became possible [9–13]. Mander et al. [14] conducted a quasi-static experimental study on a rocking pier with only unbonded pre-stressed steel bars and proposed a simplified calculation formula for its force–displacement relationship. Pampanin et al. [15] and Palermo et al. [16] proposed the combined application of built-in energy-dissipating reinforcements and unbonded pre-stressed technology to improve the energy-dissipation capacity of a rocking pier. Guo et al. [17,18] adopted fiber-reinforced materials and modified alloys to improve the durability of pre-stressed tendons and energy-dissipating devices in corrosive environments. At present, seismic studies on rocking piers are still in the stage of structural research and basic theoretical development. Although the advantages of the seismic toughness of rocking piers have been confirmed via experiments and numerical methods, the hysteretic and post-seismic resilience of rocking piers using new materials, such as ultra-high-performance concrete (UHPC), remain to be further studied. UHPC is a reasonable combination of an ultra-fine admixture, ultra-high-strength steel fibers, an efficient water-reducing agent, and small particle aggregates [19,20]. Compared with ordinary concrete, UHPC has the characteristics of high strength, ductility, and durability. The seismic damage of a rocking pier is mainly the compressive damage of the column toe (concrete at the edge of the rocking interface). The application of UHPC to the plastic hinge area of a rocking pier (as shown in Figure 1) can significantly improve the compressive bearing capacity of its column toe, which allows for the limitation of seismic damage and failure of the bridge pier to a greater degree. At present, the multi-axial spring model is widely used to simulate the hysteretic behavior of rocking piers, but it is complex to determine the stiffness of the multi-axial springs. Therefore, a simple and clear analysis model of a UHPC rocking pier needs to be developed.

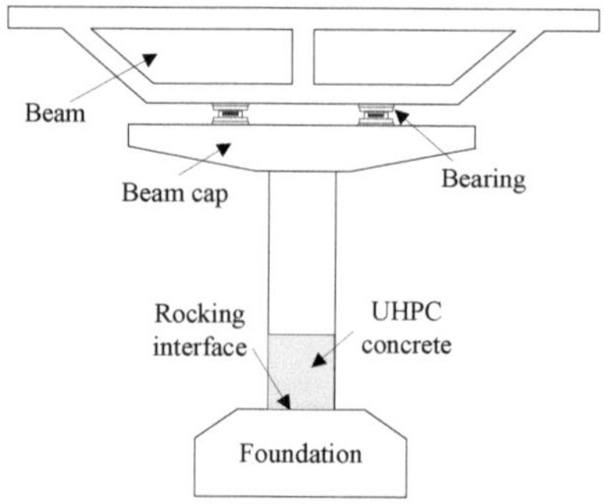

Figure 1. Schematic diagram of UHPC rocking pier.

In this study, a UHPC rocking pier is taken as the research object, and the relationship between the height of the compression zone on the rocking interface and the stress states of the pre-stressed tendon and energy-dissipating steel bar are carefully considered. The section analysis of the rocking pier is divided into the analyses of self-centering and energy-dissipating components, and then a double-plastic hinge analysis model of the rocking pier is established. The hysteretic properties of the UHPC rocking bridge pier are analyzed. The influence of the energy dissipation reinforcement ratio, pre-stressed reinforcement ratio, and initial pre-stress ratio on the strength, stiffness, and energy-dissipation characteristics of UHPC rocking bridge piers is investigated. The obtained results are compared with those of a normal rocking bridge analysis. Based on the above, the research findings can provide reference for the seismic design of bridge structures and foundations.

2. Double-Plastic Hinge Analysis Model

2.1. Mechanical Mechanism

A rocking bridge pier is placed directly on the top of a bearing platform to form a rocking interface, and the bridge pier and the bearing platform are connected as a whole by an unbonded pre-stressed tendon and an energy-dissipating device. As shown in Figure 2, under the action of an earthquake, the rocking pier, as the main compression member, is lifted and closed on the cap. Therefore, the plastic deformation of the pier is concentrated on the rocking interface, avoiding damage to the pier body. The rocking of the pier causes the unbonded pre-stressed tendon and the energy-dissipating device to provide the functions of self-centering and dissipation of seismic energy, respectively. The bending moments applied by the energy-dissipating and self-centering components and the height of the compression zone on the rocking interface change with the change in the rotation angle of the rocking interface during the deformation of the rocking pier. The bending moment–rotation and height of the compression zone in a typical deformation process are shown in Figure 2.

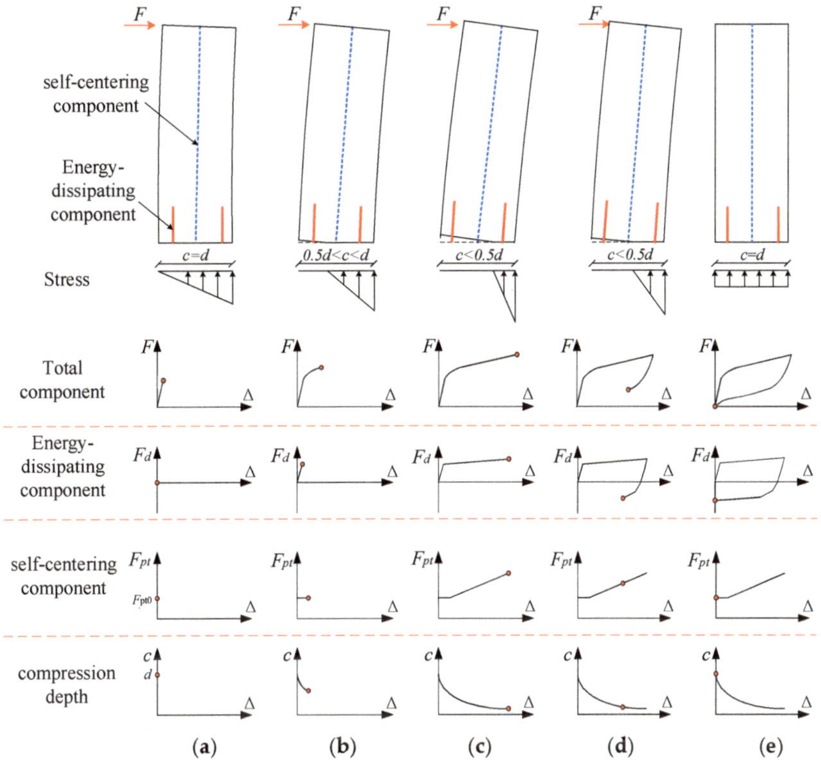

Figure 2. Deformation mode of rocking pier (c is the compression depth, d is the width of the column, θ is the rotation of the rocking interface. F and Δ are the lateral force and displacement of the rocking pier, respectively. F_d and F_{pt} are the lateral forces of the energy-dissipating component and self-centering component, respectively. F_{pt0} is the initial pre-stress force.). (**a**) Decompression stage, (**b**) Yield stage, (**c**) Large deformation stage, (**d**,**e**) Re-centering stage.

The stress stage of the rocking pier can be divided into four parts. In the first stage, named the decompression stage, as shown in Figure 2a, the stress of the concrete at the edge of the rocking pier is zero. In the second stage, named the yield stage, as shown in Figure 2b, openings appear at the rocking interface, and the assumption of a flat section is not satisfied. Therefore, introducing a displacement coordination condition to conduct a moment-angle analysis of the rocking interface is necessary [21]. In the third stage, named the large deformation stage, as shown in Figure 2c, the displacement coordination condition of the rocking pier is similar to that of the equivalent concept of the yield state. In the fourth stage, named the reset stage, as shown in Figure 2d, the lateral displacement of the rocking pier and the angle of the rocking interface decrease until the elastic displacement of the rocking pier and the angle of the rocking interface are zero (as shown in Figure 2e).

2.2. Analytical Model

Based on the force mechanism described in the above section, the moment-rotation relationship of a rocking pier is decomposed into a self-centering component composed of superstructure gravity and a pre-stress tendon and an energy-dissipating component composed of an energy-dissipating reinforcement. Considering the pressure release, yield, and large deformation of a rocking pier, the compressive strains at the outermost edge of the rocking interface in the three stress states are calculated.

$$\varepsilon_c = \begin{cases} \frac{2(N+F_{pt,0})}{E_c BD} & \theta = 0 \\ \left(\frac{3\theta}{l_{cant}} + \varphi_{dec}\right)\cdot c & 0 < \theta \le \frac{1}{3}(\varphi_y - \varphi_{dec})l_{cant} \\ \left[\frac{\frac{3\theta}{l_{cant}} - (\varphi_y - \varphi_{dec})}{\frac{3l_p}{l_{cant}}\left(1 - \frac{l_p}{2l_{cant}}\right)} + \varphi_y\right]\cdot c & \theta > \frac{1}{3}(\varphi_y - \varphi_{dec})l_{cant} \end{cases}, \tag{1}$$

where N is the superstructure gravity, $F_{pt,0}$ is the initial tension of the pre-stressed tendon, E_c is the elastic modulus of concrete, B is the width of the rocking interface, D is the height of the rocking interface, θ is the rotation angle of the rocking interface, l_{cant} is the calculated height of the rocking pier, φ_{dec} is the pressure relief limit state curvature of the rocking pier, C is the compression depth of the rocking pier, Y is the yield state curvature of the rocking pier, and L_p is the equivalent plastic hinge length of the rocking pier.

The strain of the ith layer pre-stressed tendon ($\varepsilon_{pt,i}$) can be calculated using the following formula:

$$\varepsilon_{pt,i} = \varepsilon_{pt,0} + \frac{\theta(d_{pt,i} - c)}{l_{pt}} = \frac{F_{pt,0}}{E_{pt}\cdot A_{pt,i}} + \frac{\theta(d_{pt,i} - c)}{l_{pt}} \tag{2}$$

where $\varepsilon_{pt,0}$ is the initial tensile strain of the pre-stressed tendon, $d_{pt,i}$ is the distance between the ith layer pre-stressed tendon and the edge of the compression zone on the rocking interface, l_{pt} is the unbonded length of the pre-stressed tendon, E_{pt} is the elastic modulus of the pre-stressed tendon, and $A_{pt,i}$ is the section area of the ith layer pre-stressed tendon.

If the energy-dissipating reinforcement strain of the rocking pier is lower than the yield strain, ε_y, the energy-dissipating reinforcement strain of the ith layer ($\varepsilon_{ms,i}$) can be calculated using the following formula:

$$\varepsilon_{ms,i} = \frac{\theta_{imp}(d_{ms,i} - c)}{l_{ms} + \frac{4}{3}l_{sp}}, \tag{3}$$

where $d_{ms,i}$ is the distance between the energy-dissipating device of the ith root and the edge of the compression zone of the rocking interface, l_{sp} is the strain distribution length, and l_{ms} is the unbonded length of the energy-dissipating reinforcement.

If the strain of the energy-dissipating steel bar exceeds the yield strain (ε_y), its strain in the ith layer ($\varepsilon_{ms,i}$) is calculated using the following formula:

$$\varepsilon_{ms,i} = \frac{\theta_{imp}(d_{ms,i} - c)}{l_{ms} + 2l_{sp}}, \tag{4}$$

When the strain is defined on the entire interface, the stresses of the self-centering and energy-dissipating components can be calculated to satisfy the force equilibrium condition. The bending moment bearing capacity of the rocking interface is typically calculated according to the centroid of the concrete compression zone. In this process, the bending moments generated by the self-centering and energy-dissipating components of the rocking pier can be calculated.

A lumped plasticity model as shown in Figure 3 is built using the OpenSees finite element analysis platform (Version 2.2.2). An elastic beam-column element is applied to simulate the lateral elastic stiffness of the pier. Two zero-length elements are used to simulate the moment-rotation hysteretic behaviors of the self-centering component and the energy-dissipating component, respectively. Then, the two zero-length elements are connected in parallel to the elastic beam-column element. The moment-rotation hysteretic behavior of the self-centering component is simulated using the ElasticMultiLinear material model. The moment-rotation hysteretic behavior of the energy-dissipating component is simulated using the RambergOsgoodSteel material model. Based on the mechanical mechanism, the moment-rotation hysteretic behaviors of the self-centering component and energy-dissipating component can be obtained using the deformation mode. In addition, the compression depth of the rocking pier and the rotation can be calculated using an iterative method proposed in the literature [21].

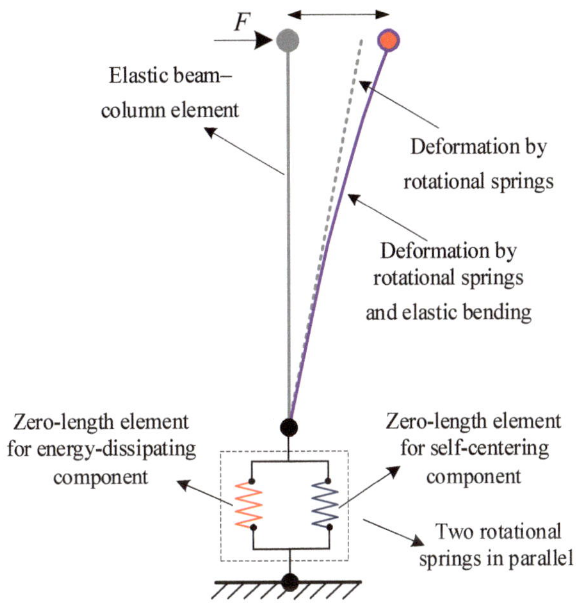

Figure 3. Lumped plasticity analysis model.

2.3. Model Verification

Based on the studies by Marriot et al. [21–23], 1:3-scaled quasi-static test results of a rocking pier are used to verify the validity of the double-plastic hinge analysis model. The calculated height of the test pier is 1.6 m, and the width of the rectangular section is 0.35 m. The section layouts of the selected rocking pier specimens—HBD1, HBD2, HBD4, and HBD5—are shown in Figure 4, in which the energy-consuming steel bars of HBD1 and HBD2 are internal, whereas those of HBD4 and HBD5 are external. The quasi-static loading regime is consistent with that used in the literature [21]. Also, the parameters used for the modeling of the rocking pier specimens HBD1, HBD2, HBD4, and HBD5 can be seen in the literature [21]. The force-displacement hysteretic curves of the four specimens are calculated using the double-plastic hinge analysis model proposed in the above section and compared with the test results. Based on the comparison shown in Figure 4, the double-plastic hinge analysis model can better simulate the hysteretic behavior of the rocking pier. Also, the equivalent viscous damping ratio is closer between the experimental and simulated results, as shown in Figure 4e. However, the analysis model overestimated the initial stiffness, due to the lack of protection of the pier rocking toe. Therefore, this analysis model is more suitable for a rocking pier with enough protection of the rocking toe. Also, the soil-structure interaction effects are not considered in this study; this is a worthy topic for future research [24].

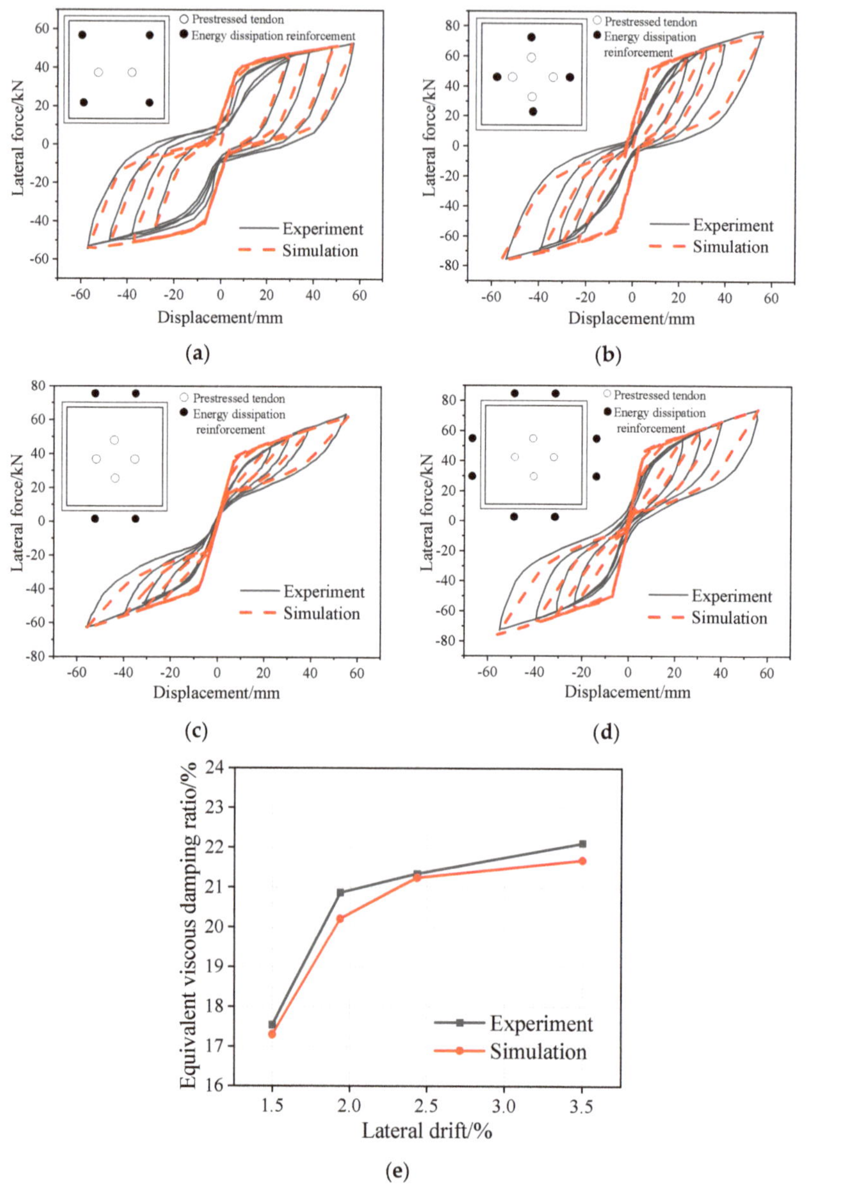

Figure 4. Comparisons between the experimental and simulated results of the hysteretic curves. (**a**) Hysteretic curves of HBD1, (**b**) Hysteretic curves of HBD2, (**c**) Hysteretic curves of HBD4, (**d**) Hysteretic curves of HBD5, (**e**) Equivalent viscous damping ratio of HBD1.

3. Hysteretic Performance and Parameter Analysis of UHPC Rocking Bridge Pier

3.1. Hysteretic Properties of UHPC Rocking Bridge Pier

In this section, the hysteretic properties of a UHPC rocking bridge pier are compared with those of the pier specimen HBD2. The stress-strain model of UHPC adopts that of concrete proposed by Scott et al. [25]. Wang et al. [26] confirmed the accuracy of the model proposed by Scott et al. [25] applied to unconstrained and constrained UHPC. For constrained UHPC, the peak stress ($\sigma_{con,pk}$), peak strain ($\varepsilon_{con,0}$), limit stress ($\sigma_{con,20}$), and limit strain ($\varepsilon_{con,20}$) can be calculated as follows:

$$\sigma_{con,pk} = \sigma_{un,pk}\left(1 + 1.8\rho_w f_{yv}/\sigma_{un,pk}\right), \quad (5)$$

$$\varepsilon_{con,0} = \varepsilon_{un,0}\left(1 + 5\rho_w f_{yv}/\sigma_{un,pk}\right), \quad (6)$$

$$\sigma_{con,20} = \lambda_{con,rs}\sigma_{con,pk}, \quad (7)$$

$$\varepsilon_{con,20} = 0.06(1 - \lambda_{con,rs}) + \lambda_{con,rs}\varepsilon_{con,0}, \quad (8)$$

where $\sigma_{un,pk}$ is the unconstrained peak stress of UHPC, $\varepsilon_{un,0}$ is the peak strain of unconstrained UHPC, ρ_w is the volume reinforcement ratio of the stirrup, f_{yv} is the yield stress of the stirrup, and $\lambda_{con,rs}$ is the strain coefficient of constrained UHPC, which is generally set at 0.2.

In this study, the concrete is replaced with UHPC based on the specimen HBD2, and then a UHPC rocking bridge pier is established. Wang et al. [26] showed that the compressive strength and elastic modulus of UHPC are 124.3 MPa and 48,000 MPa, respectively. A double-plastic hinge model of this UHPC rocking pier was established, the lateral cyclic load of the HBD2 rocking pier was applied to the double-plastic hinge model, and the corresponding hysteretic curve was obtained.

As shown in Figure 5a, the UHPC rocking pier has larger initial stiffness and lateral bearing capacity than HBD2. As shown in Figure 5b, with an increasing drift rate, the secant stiffness change trends of the specimens HBD2 and UHPC are the same; however, the secant stiffness of specimen UHPC is higher than that of specimen HBD2. For example, under a drift rate of 1.5%, the secant stiffness of UHPC is 7.7% higher than that of HBD2 (2.6 kN/mm). At a drift rate of 3.5%, the secant stiffness of UHPC is 9.6% higher than that of HBD2.

As shown in Figure 5c, with an increase in the drift rate, the variation trends of the energy dissipation moments of HBD2 and UHPC are the same; however, the energy dissipation moment of UHPC is larger than that of HBD2. For example, at a drift rate of 1.5%, the energy dissipation torques of HBD2 and UHPC are 1016.6 kN·mm and 1594.8 kN·mm, respectively. Thus, the energy dissipation torque of UHPC is 56.9% larger than that of HBD2. At a drift rate of 2.4%, the energy dissipation moment of UHPC is 23.3% larger than that of HBD2.

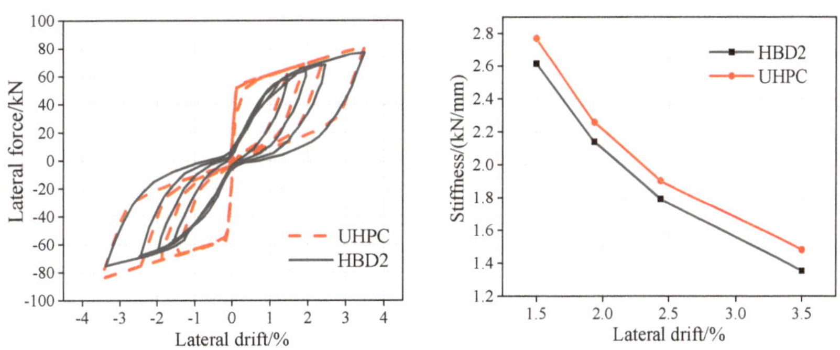

(a) (b)

Figure 5. *Cont.*

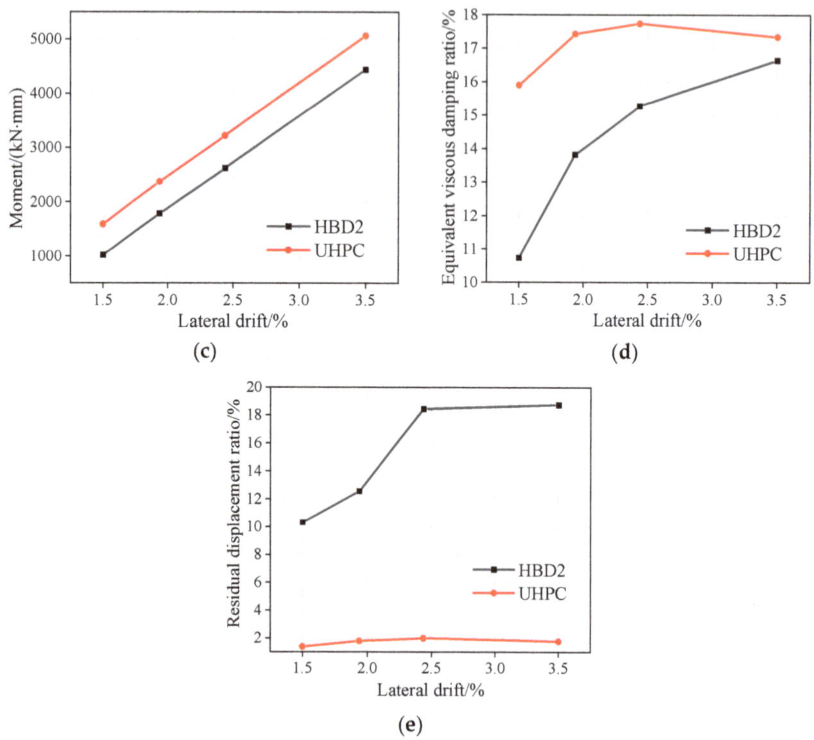

Figure 5. Hysteretic performance comparison of UHPC and HBD2 rocking piers. (**a**) Hysteretic curves, (**b**) Stiffness degradation curves, (**c**) Moment of energy-dissipating components, (**d**) Equivalent viscous damping ratio, (**e**) Residual displacement ratio.

As shown in Figure 5d, when the drift rate is lower than 3.5%, the initial stage of the equivalent viscous damping ratio of specimen UHPC is greater than that of specimen HBD2. Specifically, under a drift rate of 1.5%, the equivalent viscous damping ratios of specimens UHPC and HBD2 are 15.9% and 10.7%, respectively. At a drift rate of 3.5%, the corresponding values are 17.4% and 16.6%. The equivalent viscous damping, an index of energy-dissipation capacity, can be calculated using the following formula:

$$\xi_{eq} = \Delta W_i / (2\pi k u_{\max}^2) \tag{9}$$

where ΔW_i is the dissipated energy in one hysteretic loop, k is the secant stiffness, and $u_{\max}$ is the maximum displacement in a cycle.

In this study, the residual displacement ratio is used to quantitatively examine the self-centering ability of the specimens. A large residual displacement ratio implies a poor self-centering ability. As shown in Figure 5e, under a drift rate of 2.4%, the residual displacement ratios of HBD2 and UHPC are 2.0% and 18.4%, respectively. Therefore, UHPC has a small residual displacement ratio and good self-centering ability.

3.2. Parameter Analysis of UHPC Rocking Pier

The strength, stiffness, and energy-dissipation characteristics of the UHPC rocking pier were studied using the developed double-plastic hinge analysis model. The parameters mainly included the reinforcement ratio of the energy-dissipating steel bar, initial pre-stress force, and reinforcement ratio of the pre-stress tendon.

3.2.1. Reinforcement Ratio of the Energy-Dissipating Steel Bar

The reinforcement diameters of the rocking bridge pier specimens UHPC-E1, UHPC-E2, UHPC-E3, and UHPC-E4 are 12.5 mm, 16 mm, 18 mm, and 20 mm, respectively. The corresponding steel reinforcement ratios are 0.40%, 0.66%, 0.83%, and 1.03%. As shown in Figure 6a, with an increase in the energy consumption steel reinforcement ratio, the UHPC rocking lateral bearing capacity of the bridge pier increases. For example, under a drift rate of 5.44%, as the energy consumption steel reinforcement ratio increases from 0.40% to 1.03%, the lateral force increases by 43.43%. Before yielding, the pier is in the elastic stage, the slope of the curve increases gradually, and the initial stiffness increases gradually. After yielding, the curve slope is basically the same; therefore, the stiffness of the UHPC rocking pier is basically the same. As shown in Figure 6b, with an increase in the steel reinforcement ratio of the energy-dissipating steel bar, under the same drift, the energy consumption and equivalent viscous damping ratio of the bridge pier are remarkably improved. For example, at a drift rate of 4.31%, the equivalent viscous damping ratio increases by 5.22% when the reinforcement ratio of the energy-dissipating bar increases from 0.40% to 0.66%.

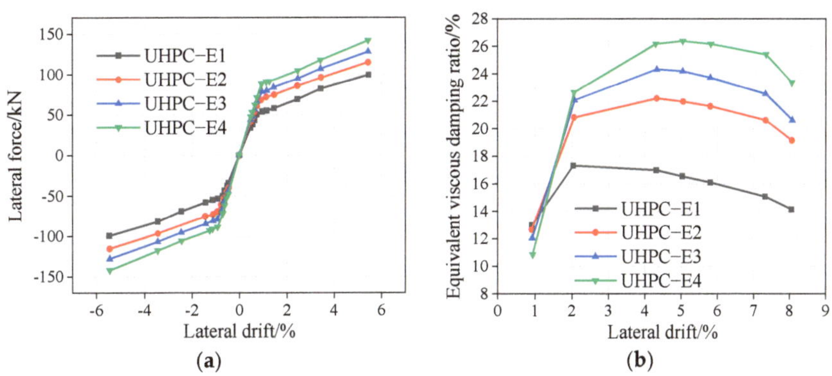

Figure 6. Influence of energy dissipation reinforcement ratio on UHPC rocking piers. (**a**) Envelop curves, (**b**) Equivalent viscous damping ratio.

3.2.2. Initial Pre-Stress Force

The initial pre-stress force of the tendons of the rocking pier specimens UHPC-T1, UHPC-T2, UHPC-T3, and UHPC-T4 are 75 kN, 112.5 kN, 150 kN, and 187.5 kN, respectively. As shown in Figure 7a, with an increase in the initial pre-stress force, the rocking UHPC lateral bearing capacity of the bridge pier increases. For example, under a drift rate of 5.44%, when the initial pre-stress force increases from 75 kN to 150 kN, the lateral force increases by 27.27%. Before yielding, the pier is in the elastic stage, and the slope of the curve increases gradually, indicating that the initial stiffness of the rocking pier can be improved by increasing the initial pre-stress. After yielding, the curve slope is basically the same; therefore, the stiffness of the UHPC rocking pier remains basically the same. As shown in Figure 7b, with an increase in the initial pre-stress force, under the same drift, the equivalent viscous damping ratio is significantly reduced due to the decrease of openings on the rocking interface. For example, at a drift rate of 4.31%, the equivalent viscous damping ratio decreases from 10.73% to 16.94% as the initial pre-stress force increases from 75 kN to 187.5 kN.

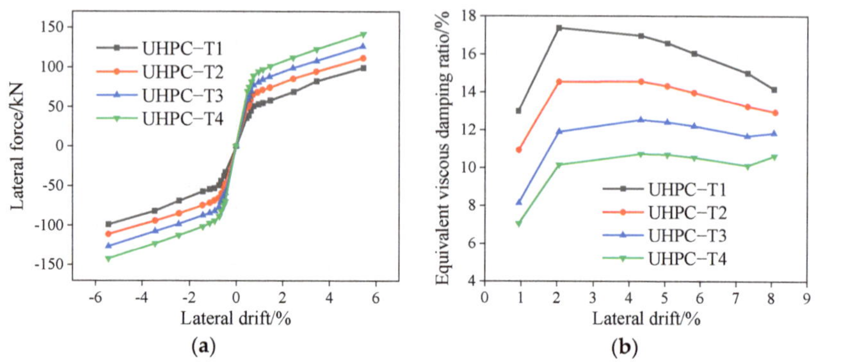

Figure 7. Influence of initial pre-stress on UHPC rocking piers. (**a**) Envelop curves, (**b**) Equivalent viscous damping ratio.

3.2.3. Reinforcement Ratio of Pre-Stressed Tendon

The pre-stressed tendon areas of the rocking pier specimens UHPC-P1, UHPC-P2, UHPC-P3, and UHPC-P4 are 99 mm^2, 198 mm^2, 297 mm^2, and 396 mm^2, respectively. The corresponding ratios of the pre-stressed tendons are 0.32%, 0.65%, 0.97%, and 1.29%. As shown in Figure 8a, with an increase in the pre-stress reinforcement ratio, the lateral bearing capacity of the UHPC rocking bridge pier increases. For example, under a drift rate of 5.44%, when the pre-stressed tendon ratio increases from 0.32% to 1.29%, the lateral force increases by 83.84%. Before yielding, the bridge pier is in the elastic stage, and the curve slope and the initial stiffness are basically unchanged. The slope of the curve increases gradually after yielding; therefore, the stiffness of the UHPC rocking pier increases gradually after yielding. As shown in Figure 8b, with an increase in the pre-stress reinforcement ratio, under the same drift ratio, the equivalent viscous damping ratios of the UHPC rocking bridge piers are significantly reduced. For example, at a drift rate of 5.81%, the equivalent viscous damping ratio decreases from 16.02% to 8.46% when the reinforcement ratio increases from 0.32% to 1.29%.

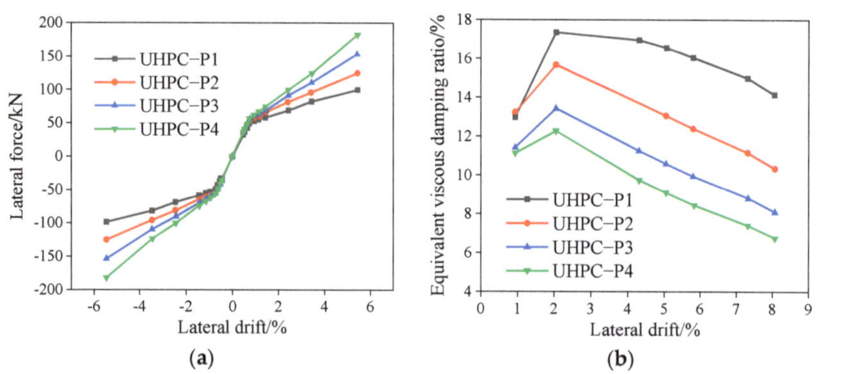

Figure 8. Influence of pre-stressed tendon ratio on UHPC rocking piers. (**a**) Envelop curves, (**b**) Equivalent viscous damping ratio.

4. Conclusions

In this study, a double-plastic hinge analysis model of a rocking pier is established based on its three stress states: compressive, yield, and large deformation. The hysteretic performance of a UHPC rocking pier is compared with that of an ordinary concrete rocking pier. The main conclusions of the parameter analyses of the UHPC rocking pier are as follows:

(1) The hysteretic behavior of the UHPC rocking bridge pier is studied based on the rocking bridge pier-loading mechanism. The dissipation yield pressures and large deformations of the three types of stress states of the rocking bridge pier are fully considered. Moreover, the bending moment–corner relations of the rocking bridge pier are divided into those of self-centering and energy-dissipating components. Accordingly, this paper proposes a double-plastic hinge analysis model that can finely calculate the force–displacement hysteretic behavior of the rocking bridge pier.
(2) The effectiveness of the model is verified using existing experimental results, and the hysteretic curves from the model simulation and experiments fit well. Moreover, the hysteretic properties, such as strength, stiffness, and energy dissipation of the UHPC rocking pier, are further analyzed.
(3) This study adopts the double-plastic hinge analysis model, which can accurately simulate the force–displacement hysteretic behavior of the rocking bridge pier. The rocking hysteretic performance of the UHPC rocking bridge piers is verified to be better than that of ordinary concrete rocking bridge piers. Moreover, compared to an ordinary concrete rocking bridge pier, the UHPC rocking bridge pier has higher lateral stiffness and strength and a higher energy-dissipation capacity and reset ability.
(4) The reinforcement ratio of the energy-dissipating reinforcement and an increase in the initial pre-stress can improve the lateral stiffness and strength of the UHPC rocking pier. For example, the lateral bearing capacity increases by approximately 30% when the initial pre-stress doubles. However, the equivalent viscous damping ratio decreases and the energy-dissipation capacity of the pier decreases with an increase in the reinforcement ratio of the pre-stress tendon and the initial pre-stress.

Author Contributions: Methodology, S.C.; Software, H.L.; Investigation, Y.Z.; Writing—original draft, H.H. All authors have read and agreed to the published version of the manuscript.

Funding: This research is funded by the Open Project of National Engineering Laboratory of Bridge Safety and Technology (Beijing) (Grant No. 2020-GJKFKT-8).

Data Availability Statement: Data sharing is not applicable to this article.

Acknowledgments: This research is funded by the Open Project of National Engineering Laboratory of Bridge Safety and Technology (Beijing) (Grant No. 2020-GJKFKT-8). This support is gratefully acknowledged. The results and conclusions presented in this paper are those of the authors and do not necessarily reflect the views of the sponsors.

Conflicts of Interest: The authors declare no conflict of interest.

References

1. Hsu, Y.T.; Fu, C.C. Study of damaged Wushi Bridge in Taiwan earthquake. *Pract. Period. Struct. Des. Constr.* **2000**, *5*, 166–171. [CrossRef]
2. Kawashima, K.; Takahashi, Y.; Ge, H.; Wu, Z.; Zhang, J. Reconnaissance report on damage of bridges in 2008 Wenchuan, China, earthquake. *J. Earthq. Eng.* **2009**, *13*, 965–996. [CrossRef]
3. Priestley, M.J.N.; Seible, F.; Calvi, G.M. *Seismic Design and Retrofit of Bridges*; John Wiley and Sons: New York, NY, USA, 1996.
4. Astaneh-Asl, A.; Shen, J.H. Rocking behavior and retrofit of tall bridge piers. In *Structural Engineering in Natural Hazards Mitigation*; American Society of Civil Engineers: Tustin, CA, USA, 1993; pp. 121–126.
5. Dowdell, D.J.; Hamersley, B.A. Lions' Gate Bridge North Approach: Seismic retrofit. In *Behaviour of Steel Structures in Seismic Areas*; Balkema: Rotterdam, The Netherlands, 2000; pp. 319–326.
6. Jones, M.H.; Holloway, L.J.; Toan, V.; Hinman, J. Seismic retrofit of the 1927 Carquinez Bridge by a displacement capacity approach. In *Bridges and Highways Progress in Research and Practice*; Federal Highway Administration: Washington, DC, USA, 1997; pp. 445–456.
7. Ingham, T.J.; Rodriguez, S.; Nader, M.; Taucer, F.; Seim, C. Seismic retrofit of the golden gate bridge. In *Progress in Research and Practice*; Federal Highway Administration: Washington, DC, USA, 1995; pp. 2021–2030.
8. Housner, G.W. The behavior of inverted pendulum structures during earthquakes. *Bull. Seismol. Soc. Am.* **1963**, *53*, 403–417. [CrossRef]
9. Tso, W.K.; Wong, C.M. Steady state rocking response of rigid blocks part 1: Analysis. *Earthq. Eng. Struct. Dyn.* **1989**, *18*, 89–106. [CrossRef]

10. Aslam, M.; Scalise, D.T.; Godden, W.G. Earthquake rocking response of rigid bodies. *J. Struct. Div.* **1980**, *106*, 377–392. [CrossRef]
11. Makris, N.; Konstantinidis, D. *The Rocking Spectrum and the Shortcomings of Design Guidelines*; Pacific Earthquake Engineering Research Center: Berkeley, CA, USA, 2001.
12. Makris, N.; Konstantinidis, D. The rocking spectrum and the limitations of practical design methodologies. *Earthq. Eng. Struct. Dyn.* **2003**, *32*, 265–289. [CrossRef]
13. Makris, N.; Zhang, J. Rocking response of anchored blocks under pulse-type motions. *J. Eng. Mech.* **2001**, *127*, 484–493. [CrossRef]
14. Mander, J.B.; Cheng, C.T. *Seismic Resistance of Bridge Piers Based on Damage Avoidance Design*; US National Center for Earthquake Engineering Research: New York, NY, USA, 1997.
15. Pampanin, S.; Priestley MJ, N.; Sritharan, S. Analytical modelling of the seismic behavior of precast concrete frames designed with ductile connections. *J. Earthq. Eng.* **2001**, *5*, 329–367. [CrossRef]
16. Palermo, A.; Pampanin, S.; Marriott, D. Design, modeling, and experimental response of seismic resistant bridge piers with posttensioned dissipating connections. *J. Struct. Eng.* **2007**, *133*, 1648–1661. [CrossRef]
17. Guo, T.; Cao, Z.; Xu, Z.; Lu, S. Cyclic load tests on self-centering concrete pier with external dissipators and enhanced durability. *J. Struct. Eng.* **2016**, *142*, 04015088. [CrossRef]
18. Antonellis, G.; Panagiotou, M. Seismic response of bridges with rocking foundations compared to fixed-base bridges at a near-fault site. *J. Bridge Eng.* **2014**, *19*, 04014007-1–04014007-14. [CrossRef]
19. Zamani, A.A.; Ahmadi, M.; Dalvand, A.; Aslani, F. Effect of single and hybrid fibers on mechanical properties of high-strength self-compacting concrete incorporating 100% waste aggregate. *J. Mater. Civ. Eng.* **2023**, *35*, 04022365. [CrossRef]
20. Sattarifard, A.R.; Ahmadi, M.; Dalvand, A.; Sattarifard, A.R. Fresh and hardened-state properties of hybrid fiber–reinforced high-strength self-compacting cementitious composites. *Constr. Build. Mater.* **2022**, *318*, 125874. [CrossRef]
21. Marriott, D.J. The Development of High-Performance Post-Tensioned Rocking Systems for the Seismic Design of Structures. Ph.D. Thesis, University of Canterbury, Christchurch, New Zealand, 2009.
22. Marriott, D.; Pampanin, S.; Palermo, A. Quasi-static and pseudo-dynamic testing of unbonded post-tensioned rocking bridge piers with external replaceable dissipaters. *Earthq. Eng. Struct. Dyn.* **2009**, *38*, 331–354. [CrossRef]
23. Marriott, D.; Pampanin, S.; Palermo, A. Biaxial testing of unbonded post-tensioned rocking bridge piers with external replacable dissipaters. *Earthq. Eng. Struct. Dyn.* **2011**, *40*, 1723–1741. [CrossRef]
24. Bulajić, B.D.; Todorovska, M.I.; Manić, M.I.; Trifunac, M.D. Structural health monitoring study of the ZOIL building using earthquake records. *Soil Dyn. Earthq. Eng.* **2020**, *133*, 106105. [CrossRef]
25. Scortt, B.D.; Park, R.; Priestly, M.J.N. Stress-Strain Behavior of Concrete Confined by Overlapping Hoops at Low and High Strain Rates. *ACI J.* **1982**, *79*, 13–27.
26. Wang, Z.; Wang, J.; Zhu, J.; Zhao, G.; Zhang, J. Energy dissipation and self-centering capacities of post-tensioning precast segmental UHPC bridge columns. *Struct. Concr.* **2020**, *21*, 517–532. [CrossRef]

buildings

MDPI

Article

A Stress Measurement Method for Steel Strands Based on Spatially Self-Magnetic Flux Leakage Field

Shangkai Liu [1], Cheng Cheng [2], Ruiqiang Zhao [3,*], Jianting Zhou [1] and Kai Tong [1]

1 State Key Laboratory of Mountain Bridge and Tunnel Engineering, Chongqing Jiaotong University, Chongqing 400074, China; skliu@mails.cqjtu.edu.cn (S.L.); jtzhou@cqjtu.edu.cn (J.Z.); kaitong@mails.cqjtu.edu.cn (K.T.)

2 Chongqing Wukang Technology Co., Ltd., Chongqing 404000, China; chengcheng@cmhk.com

3 School of Materials Science and Engineering, Chongqing Jiaotong University, Chongqing 400074, China

* Correspondence: rqzhao@cqjtu.edu.cn; Tel.: +86-153-2036-4035

Abstract: Metal Magnetic Memory (MMM) exhibits the advantage of not requiring embedded sensors or external excitation, making it suitable for inspecting ferromagnetic components in engineering structures. This study introduced MMM into stress detection of steel strands. Graded tensile tests were conducted on the steel strands to investigate the correlation between Self-Magnetic Flux Leakage (SMFL) signals and stress levels. Different spatial detection positions with varying Lift-Off Values (LOV) and Rotation Angle Values (RAV) were set to examine the distribution of spatial SMFL field under load. Furthermore, a magnetic characteristic parameter A_N was proposed to assess the stress level of the steel strands. The results indicate that the rate of change in the middle region of the SMFL curve was lower than that at the beginning and the end. Additionally, with increased applied load, the SMFL curve exhibited systematic variations, and the dispersion of the normal component curve gradually decreased. By utilizing the magnetic characteristic parameter A_N, the stress in the steel strands can be calculated, with the parameters determined based on LOV and RAV. This achievement expanded the nondestructive testing methods for steel strands and holds significant research value.

Keywords: steel strand; stress detection; metal magnetic memory; spatially SMFL field; detection position

Citation: Liu, S.; Cheng, C.; Zhao, R.; Zhou, J.; Tong, K. A Stress Measurement Method for Steel Strands Based on Spatially Self-Magnetic Flux Leakage Field. *Buildings* **2023**, *13*, 2312. https://doi.org/10.3390/buildings13092312

Academic Editor: Huiyong Ban

Received: 2 August 2023
Revised: 11 September 2023
Accepted: 11 September 2023
Published: 12 September 2023

1. Introduction

Steel strands are commonly used in critical load-bearing components of bridges, such as inclined cables, suspender rods, and main cables. Accurately understanding the stress state of steel strands can effectively ensure the safe operation of bridges [1,2].

Nondestructive testing is the basis of structural health monitoring and can improve the efficiency of structural reinforcement [3,4]. Common nondestructive testing methods for tension include: (i) the fiber Bragg grating-based method [5], which measures the strain of steel strands using the principle of fiber Bragg gratings, offering advantages of high-speed detection, high precision, and strong anti-interference capability, but with the drawback of high cost and complex installation and usage; (ii) the vibration frequency method [6], which detects tension based on the relationship between the natural frequency of steel strands and their stress state, providing advantages of rapid and efficient testing with good repeatability, but susceptible to environmental influences, limited applicability, and requiring specialized equipment; (iii) the magnetic flux method [7], which indirectly infers the stress state of steel strands by measuring the changes in magnetic flux, offering the advantages of high sensitivity and quantitative detection, but also affected by environmental factors and requiring complex equipment and specialized operation.

The common issue with the above methods is the need for embedded sensors or external excitation, marking the installation and application of the testing equipment relatively complex [8]. Metal Magnetic Memory (MMM) detection is an environmentally friendly, non-exciting, nondestructive testing method. It features lightweight detection instruments and does not require embedded sensors, making it of significant engineering significance when applied to the stress detection of steel strands [9,10].

The foundation of MMM detection lies in the Self-Magnetic Flux Leakage (SMFL) effect [11], which refers to local magnetic field changes inside magnetic materials without an external magnetic field due to defects or stress concentrations [12–14]. MMM detection determines defects in structures by measuring the SMFL field changes of ferromagnetic materials, and it is suitable for detecting microscopic defects [15,16], making it widely applied in fields such as pipeline transportation, power facilities, and industrial manufacturing [17–19].

Meanwhile, scholars have conducted extensive research on this technology, proposing magneto-mechanical models [20], energy conservation laws [21], magnetic dipole models [22], and other theories. They have also conducted experiments on various components, including reinforcing bars [23], steel structures [24,25], and welds [26]. Numerous studies have been conducted in terms of evaluation indicators for MMM detection. Roskosz measured the Residual Magnetic Field (RMF) signal on the surface of specimens under load and found that the tangential component parallel to the direction of the load is most correlated with the stress level [27], proposing a residual stress evaluation method based on RMF gradient [28]. Zhou measured the Self-Magnetic Flux Leakage (SMFL) signal of corroded reinforced concrete and found that the normal component $H_S(Z)$ of the SMFL signal can qualitatively determine the location and degree of reinforcement corrosion. The $H_S(Z)$ gradient is approximately linearly correlated with the flexural strength loss rate of the beam [29]. Huang proposed using the maximum gradient K_{max} of the normal component $H_p(y)$ of the magnetic model to represent the degree of stress concentration [30]. Yao proposed using the peak-to-peak amplitude of the normal and tangential gradient curves of local magnetic signals, along with horizontal distances $W(X)_{p\text{-}p}$, $W(Y)_{p\text{-}p}$, $S(X)_{p\text{-}p}$, and $S(Y)_{p\text{-}p}$, to characterize the degree and scope of plastic deformation [31].

To sum up, research on metal magnetic memory has mainly focused on the microscopic damage, stress state, and stress concentration detection in ferromagnetic materials [12,32]. Stress detection was primarily conducted on standard specimens and reinforcing bars with distinct elastic and plastic stages [33–36]. However, in the case of steel strands, this technology was mainly utilized for detecting corrosion damage, and there is limited research on stress detection in steel strands and the impact of spatial detection positions on the results [37,38].

Therefore, this experiment subjected steel strands to graded loading to investigate the SMFL field distribution under different load conditions. A stress measurement method for steel strands based on the spontaneous SMFL field is proposed. The influence of LOV (the vertical distance between the sensor surface and the tested specimen) and RAV (the clockwise rotation angle of the sensor with the initial position as the origin and the tested specimen as the center) on the detection results was discussed.

This study presents a method to measure the working stress of steel strands using the magnetic characteristic parameters A_N (normalized mean absolute deviation). Section 1 introduces the basic principles and research status of Metal Magnetic Memory (MMM) detection and its application to stress detection in steel strands. Section 2 describes the theoretical derivation of the relationship between SMFL signals and stress using the Jiles–Atherton model. Section 3 presented the experimental design for collecting SMFL signals. Section 4 analyzes the distribution of the original magnetic field and SMFL field on the surface of the specimens under graded loading. Section 5 discusses the relationship between the normal component of the SMFL field and applied loads for different sampling points or spatial detection positions. The study proposes a calculation method for steel strand stress based on the magnetic characteristic parameter A_N and further investigates the effects of LOV and RAV on the detection results.

2. Theoretical Background

2.1. Effect of Applied Stress on Magnetic Properties of Materials

In MMM detection, components exhibit the phenomenon of maximum tangential component and zero-crossing point of the normal component of the SMFL field in the damaged or stress-concentrated areas. However, research on the magnetic domain is required for components with magnetic field redistribution caused by axial tension, where the stress-concentrated area is not clearly defined.

A magnetic domain refers to regions in ferromagnetic materials with the same magnetization direction at the microscopic scale. Without an external magnetic field, magnetic domains in ferromagnetic materials are randomly distributed. The applied stress is equivalent to applying an external magnetic field to the magnetic domains, resulting in a directional distribution of the magnetic domains and altering the material's magnetism, ultimately manifesting as an additional magnetic field on the material's surface, as shown in Figure 1.

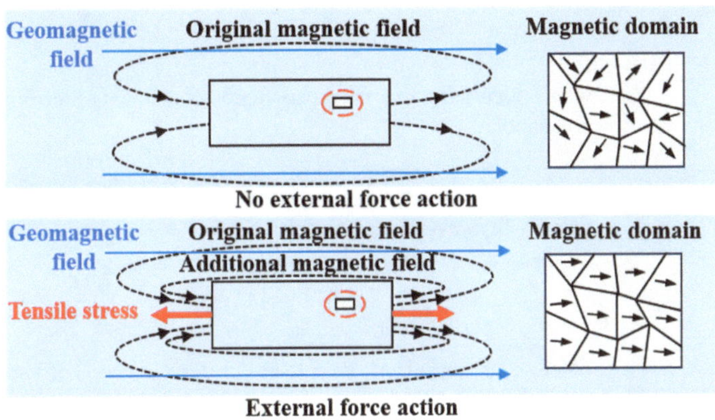

Figure 1. Magnetic field distribution and internal magnetic domain changes of components under external stress.

2.2. Jiles–Atherton Model

To further analyze the influence of stress on the distribution of the leakage magnetic field and explore the magneto-mechanical coupling mechanism, it is necessary to apply the Jiles–Atherton (J-A) model [39,40]. The J-A model was derived from Jiles and Atherton's theory of ferromagnetization. These scholars pointed out that uniaxial stress can change the internal effective field through magnetostriction. Under an external magnetic field, the magnetization of a ferromagnetic material was composed of reversible magnetization caused by domain wall bending and reversible magnetization caused by domain wall movement.

The impact of applied stress on magnetization is equivalent to adding a magnetic field H_σ to the magnetic domain structure. This additional field is described by the energy A along the reversible non-hysteresis magnetization curve [41,42].

$$A = \mu_0 HM + \frac{H_0}{2}\alpha M^2 + \frac{3}{2}\alpha\lambda + TS \tag{1}$$

where T is temperature, S is entropy, H is the external magnetic field, M is the magnetization intensity, α is the domain coupling coefficient, σ is the applied stress, and λ is the magnetostriction coefficient.

The effective field H_e can be expressed as the derivative of energy A concerning magnetization intensity M.

$$H_e = \frac{1}{M_0}\frac{dA}{dM} = H + \alpha M + \frac{3}{2}\frac{\sigma}{\mu_0}\frac{d\lambda}{dM} \tag{2}$$

when the applied stress σ_0 is not in the same direction as λ and M, the stress used in Equation (2) represents the component of stress applied along that direction. For isotropic materials:

$$\sigma = \sigma_0\left(\cos^2\theta - v\sin^2\theta\right) \tag{3}$$

where θ is the angle between the applied stress σ_0 and the direction of the magnetic field strength H, and V is the Poisson's ratio. The effective field component H_σ caused by the applied stress is calculated as follows:

$$H_\sigma = \frac{3}{2}\frac{\sigma}{\mu_0}\left(\frac{d\lambda}{dM}\right)_\sigma = \frac{3}{2}\frac{\sigma_0}{\mu_0}\left(\frac{d\lambda}{dM}\right)_\sigma\left(\cos^2\theta - v\sin^2\theta\right) \tag{4}$$

Before calculating H_σ, it is necessary to determine the magnetostriction coefficient λ of the material, which is related to the magnetization intensity M [39]:

$$\lambda = \sum_{i=0}^{\infty}\gamma_i M^{2i} \tag{5}$$

when calculating the magnetostriction coefficient of ferromagnetic materials, $i = 2$ is more reasonable:

$$\lambda = \gamma_1 M^2 + \gamma_2 M^4 \tag{6}$$

The relationship between the magnetostriction coefficient and stress is represented using a Taylor series expansion:

$$\gamma_i(\sigma) = \gamma_i(0) + \sum_{n=1}^{\infty}\frac{\sigma^n}{n!}\gamma_i^n(0) \tag{7}$$

where $\gamma_i(\sigma)$ represents the nth-order derivative γ_i corresponding to $\sigma = 0$. For ferromagnetic materials, $n = 1$, and $\gamma_i(\sigma)$ is substituted into Equations (2) and (4):

$$H_\sigma = \frac{3\sigma}{2H_0}\left[\left(2\gamma_1(0) + 2\gamma'_1(0)\sigma\right)M + \left(4\gamma_2(0) + 4\gamma'_2(0)\sigma\right)\right]M^3 \tag{8}$$

$$H_e = H + \alpha M + \frac{3\sigma}{2H_0}\left[\left(2\gamma_1(0) + 2\gamma'_1(0)\sigma\right)M + \left(4\gamma_2(0) + 4\gamma'_2(0)\sigma\right)\right]M^3 \tag{9}$$

From this, it can be observed that if $H_\sigma > 0$, the application of stress strengthens the existing magnetic field, while if $H_\sigma < 0$, the effect of stress weakens the existing magnetic field. The application of stress significantly influences the distribution of the existing magnetic field, and this effect shows non-linear changes with increasing stress.

3. Experimental Material and Methods

To explore the stress detection method based on the SMFL effect, this study conducted experiments on 1860-grade steel strands using a universal testing machine to clarify the surface leakage magnetic field distribution of steel strands under different stress states. The steel strands were subjected to uniform tension at a constant speed and maintained at the design tension value. The sensor was scanned along the axial direction of the specimen at the designated positions.

3.1. Experimental Specimen and Platform

The steel strand used in this experiment was of type Φs15.24(1×7), with a nominal diameter of 15.2 mm, composed of 7 steel wires, each having a diameter of 5 mm. The material parameters are shown in Table 1.

Table 1. Chemical composition and mechanical properties of steel strand test piece.

Chemical Composition/%					Mechanical Properties			
C	Si	Mn	S	P	Tensile Strength/MPa	Modulus of Elasticity/GPa	Hardness	Coefficient of Friction
0.75~0.85	0.12~0.32	0.60~0.90	<0.025	<0.025	1860	193.9	HRC 40~45	0.15~0.25

The steel strands were transformed into test specimens of 80 cm in length. Before testing, the PE protective sheath and anticorrosive grease were removed from the surface. The specimens were clamped and stretched using the universal testing machine, with a 15 cm clamping length on both sides. Aluminum Anti-skid Spacers were inserted in the clamping area to prevent slippage between the specimen and the fixture.

In practical engineering scenarios, pre-magnetizing steel specimens is often not feasible. Therefore, a single steel wire was cut into multiple segments for testing in this experiment, and no pre-magnetization was applied to the specimens. Additionally, no external magnetic field source was introduced. Consequently, the initial magnetization strength of the specimens falls between demagnetized and saturated magnetization. This approach closely simulates the real-world usage conditions of steel strands, laying a solid foundation for the practical application of this technology.

The relationship between initial magnetization strength and initial magnetic field strength is as follows:

$$M_i = \mu_0 H_i(\mu_r - 1) \tag{10}$$

where μ_r represents the relative magnetic permeability, μ_0 represents the vacuum magnetic permeability, taken as $4\pi \times 10^{-7}$ N/A^2, M_i represents the initial magnetization strength, and H_i represents the initial magnetic field strength.

The HMR2300 magnetic probe with an accuracy of 70 microgauss was utilized to collect the SMFL signals. The data were transmitted via a serial port server and stored on a computer, as depicted in Figure 2. A coordinate system was established with the magnetic probe as the origin. The relative position between the probe and the specimen determined that: (i) the *y*-axis was parallel to the specimen axis, (ii) the *z*-axis was perpendicular to the specimen axis, and (iii) the *x*-axis was perpendicular to the *y*–*z* plane.

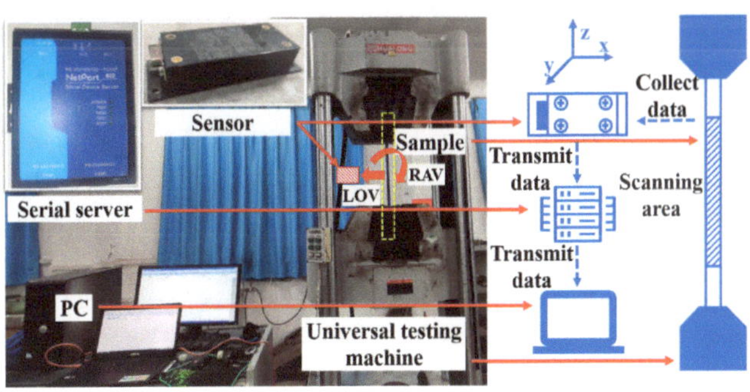

Figure 2. Schematic diagram of the experimental platform.

The position of the sensor is determined by the scanning area, lift-off value (LOV), and Rotation Angle Values (RAV). The scanning area is parallel to the test piece, indicating the moving range of the sensor during one scan. LOV is the vertical distance between the sensor surface and the tested specimen, and RAV is the clockwise rotation angle of the sensor with the initial position as the origin and the tested specimen as the center.

3.2. *Experimental Procedures and Data Acquisition Method*

In this study, the experimental phenomena were described by the applied tension, which could be converted into stresses in the relevant results. The test specimens were pulled up at a constant speed of 2 kN/s using the universal testing machine, with the loading range varying from 0 to 160 kN. At every 10 kN increase in tension, the machine paused to maintain the current load. This process was repeated for a total of 17 loading levels. The SMFL signals were collected by the sensor at the middle section of the specimen, specifically from 25 cm to 55 cm (y = 0~30 cm). Data were collected at every 1 cm interval, resulting in 31 sampling points. After completing the scan under the current load, the machine was loaded to the next level, and the scanning process continued.

To investigate the spatial SMFL field distribution of steel strands under different tension levels, a graded loading test was designed using three Lift-Off Values (LOV) (H_1 = 0.5 cm, H_2 = 1.5 cm, H_3 = 3 cm) and three Rotation Angle Values (RAV) ($\theta_1 = 0°$, $\theta_2 = 120°$, $\theta_3 = 240°$). Sensors were used to scan the SMFL field on the surface of the specimen at a total of 9 detection positions within the range of y = 0~30 cm. Refer to Table 2 and Figure 3 for specific details.

Table 2. Arrangement of detection positions.

	$\theta_1 = 0°$	$\theta_2 = 120°$	$\theta_3 = 240°$
H_1 = 0.5 cm	P_1	P_2	P_3
H_2 = 1.5 cm	P_4	P_5	P_6
H_3 = 3 cm	P_7	P_8	P_9

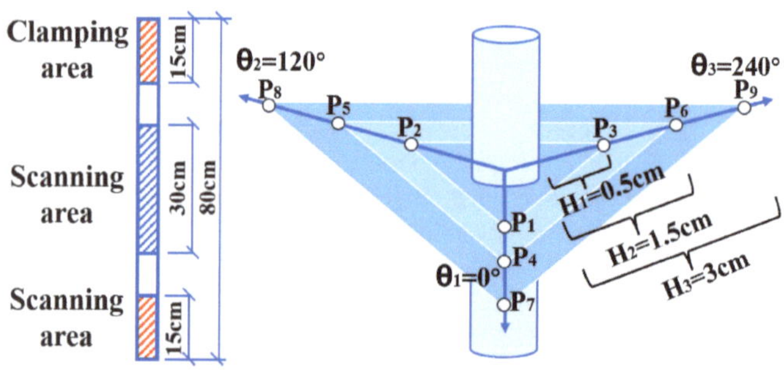

Figure 3. Schematic diagram of the detection area and detection positions.

4. Experimental Results

The SMFL field is a weak magnetic field, resulting in lower flux density in the measured signals. At the same detection location, we measured the leakage magnetic field in the presence and absence of the specimen, referred to as the scanning and external fields, respectively. As discussed in Section 2, ferromagnetic materials generate an additional magnetic field on their surface when subjected to applied stress. This phenomenon means that the scanning magnetic field consists of the external and SMFL fields under tension. The SMFL field discussed in this manuscript is obtained by subtracting the external field from the scanning field. This process is considered as shielding against the Earth's magnetic field and other steel objects' proximity, effectively eliminating a significant amount of noise.

4.1. Magnetic Scanning Curve Distribution

The distribution of the scanning magnetic field at detection position P_1 (H_1 = 0.5 cm, $\theta_1 = 0°$) is illustrated in Figure 4, where By represents the tangential component of the magnetic field, Bz represents the normal component of the magnetic field, and Bx represents the magnetic field component perpendicular to the *By*–*Bz* plane. The distribution patterns of the scanning magnetic field are as follows:

(1) The Bx curve gradually increased along the scanning path, with a significantly lower rate of change in the middle region (y = 10 cm~20 cm) compared to the two ends. As the tension in the specimen changed, the Bx curve shifted as a whole, with similar

trends and a relatively high degree of coincidence. In the y = 0~16 cm region, the Bx curves corresponding to F = 80 kN and F = 60 kN were located above and below the overall curve, respectively;

(2) The By curve showed a quadratic-like shape, decreasing first and then increasing along the scanning path, reaching the minimum value at y = 14cm. Similar to the Bx curve, the rate of change of the By curve in the middle region was lower than at the two ends. The shape of the By curve remained similar with varying applied stress, and the magnetic field curves intersected at y = 3 cm and y = 28 cm for different tension levels;

(3) The Bz curve followed a cubic-like shape, with significant differences in the magnetic field curves under different tension levels. When F = 0 kN, the Bz curve first increased, then decreased, and finally increased again along the scanning path. There were two distinct peaks at y = 8 cm and y = 25 cm, and as the tension in the steel strand increased, these peaks gradually flattened. Similar to the other two components, the Bz curve also showed a lower rate of change in the middle region compared to the two ends.

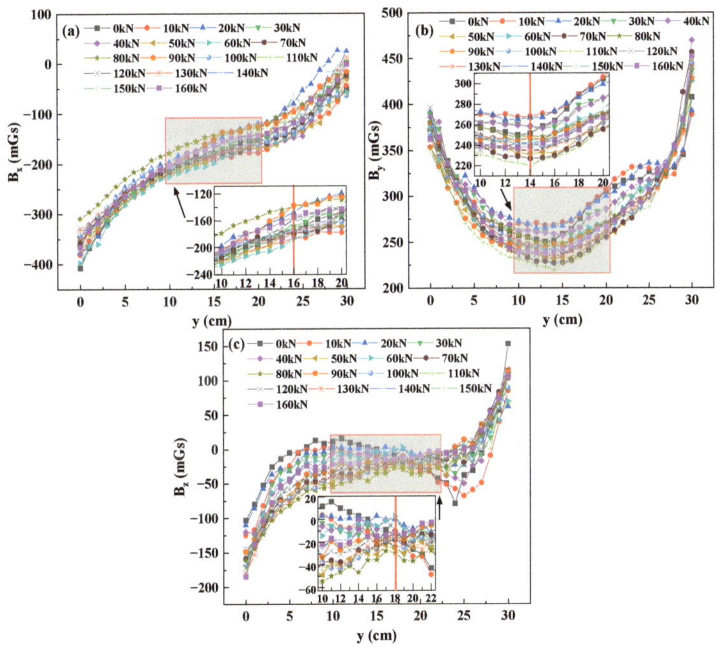

Figure 4. Scanning magnetic field distribution at P_1. (**a**) Bx component of Scanning magnetic field; (**b**) By component of Scanning magnetic field; (**c**) Bz component of Scanning magnetic field.

4.2. SMFL Field Curve Distribution

The distribution of the external magnetic field at detection position P_1 is depicted in Figure 5a. By subtracting the external magnetic field from the scanning magnetic field discussed in Section 4.1, the SMFL field of the specimen was obtained, as shown in Figure 5b–d. The distribution patterns of the SMFL field were as follows:

(1) The Bx curve decreased along the scanning path. When y < 15 cm, the curves for different loads were approximately parallel and linearly declined. The curve bulged upward between y = 15 cm and 26 cm, and some curves intersected. When y > 26 cm, the curves exhibit severe fluctuations, and at y = 30 cm, the curve showed a significant drop, reaching the minimum value;

(2) The By curve exhibited drastic changes at both ends of the scanning region and remained relatively flat in the middle area (y = 10 cm~20 cm). When y < 11 cm, the curve showed a significant upward trend along the scanning path, and the curves for different loads were approximately parallel. Between y = 11 cm to 27 cm, the curve's variation was relatively small, with the most significant change of 39.1 mGs for the 0 kN curve. When y > 27 cm, the curves converged and increased dramatically, reaching the maximum value at y = 30 cm;

(3) The Bz curve was similar to the By curve, with more significant changes at the ends than in the middle region. When y < 6 cm, the curves rapidly decreased and were relatively close to each other. Between y = 6 cm to 20 cm, the curves entered the "gentle region" with more minor variations than the previous region. The magnitude of the curve changes decreased further with increasing load, causing the distance between different curves to first increase, then fall, and finally intersect at y = 20 cm. When y > 20 cm, the distance between the curves increased again, and the curve's trend changed around y = 25 cm.

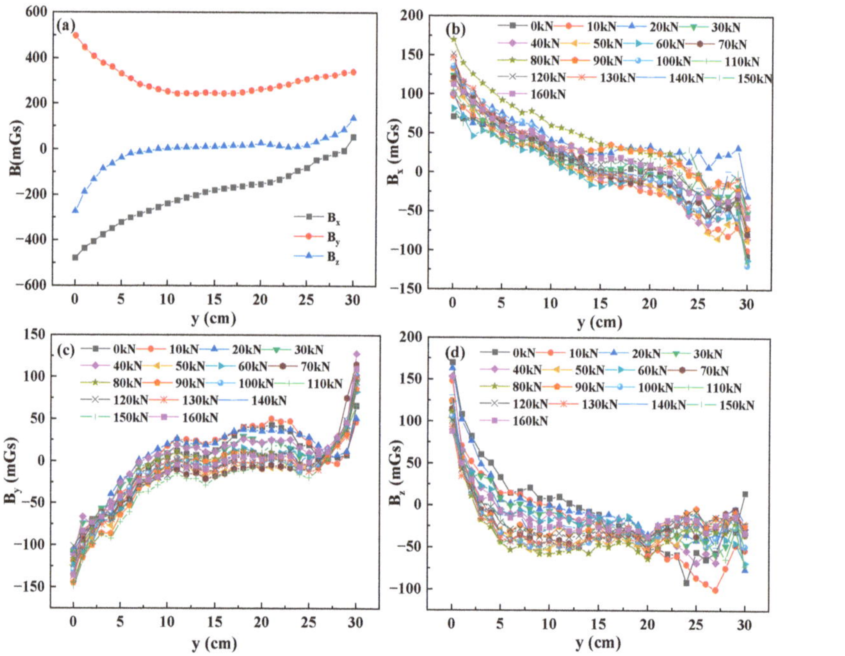

Figure 5. Distribution curves of external magnetic field and SMFL field at P_1. (**a**) External magnetic field; (**b**) Bx component of SMFL field; (**c**) By component of SMFL field; (**d**) Bz component of SMFL field.

Upon analyzing the scanning magnetic field and SMFL field curves, it was evident that the three components of the two magnetic fields differ significantly in magnitude and shape. However, they shared a common characteristic of exhibiting a lower rate of change in the middle region compared to the two ends. This phenomenon could be attributed to the influence of the fixture's magnetic field in the areas near the ends. The curves for different loads tended to converge and become nearly parallel at the ends, while they dispersed in the middle region. However, the curves of the other components, besides the tangential component Bx, exhibited clear intersection points. This phenomenon could be attributed to transforming the internal magnetic domains from a random distribution to an oriented distribution under increasing load, resulting in regular changes on the specimen's surface.

Compared to the other two components, the normal component Bz of the scanning magnetic field and leakage magnetic field had two distinct intersection points at y = 18 cm and 20 cm, respectively. These curves showed different features and exhibited regular changes under stress. Thus, further analysis of the normal component Bz of the SMFL field is necessary in the next section. The effect of load on the SMFL field intensity at various sampling points on the Bz curve was discussed, and the distribution of the spatial SMFL field under load was analyzed. A magnetic feature parameter was proposed to characterize the stress state of the steel strand, resulting in a stress measurement method based on the spatial SMFL distribution.

5. Discussion

5.1. Influence of Load on SMFL Intensity at Each Sampling Point

To investigate the stress characterization method for steel strands based on magnetic feature parameters, it was necessary to explore the influence of loads on the SMFL field at various sampling points. The relationship between the leakage magnetic field and the applied loads is shown in Figure 6. The 31 curves in the graph represent the 31 sampling points along the scanning path, and the horizontal axis represents the loads applied on the specimen (0~160 kN).

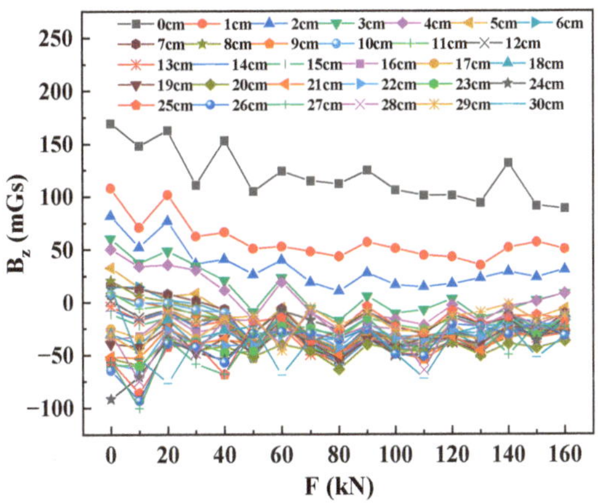

Figure 6. Magnetic force curves of each sampling point on the scanning path.

During the loading process, there were significant variations in the magnetic field intensity at different positions. The magnetic force curves for the sampling points near the ends (such as points 1–5 and 26–30) exhibited more significant fluctuations and were located on the outer side of all curves, far from other curves. This observation was consistent with the inference mentioned in Section 4.1 that the fixture influenced magnetic fields more at the ends.

When F was 0~50 kN, the Bz curve changed rapidly and showed an overall downward trend. When F exceeded 50 kN, the Bx values fluctuated within a small range. The range from F = 0 to 50 kN was referred to as Region 1; the remaining range was Region 2. There were distinct differences in the curve distributions between these two regions. The range of data in Region 1 (R_1) was generally more extensive than in Region 2 (R_2), and their difference was calculated as $\Delta = R_1 - R_2$. The results are shown in Figure 7.

It can be observed that, except at points y = 16 cm and 29 cm, the range of data in Region 1 was generally greater than that in Region 2 ($\Delta > 0$) within the scope of y = 14~20 cm. Moreover, the range values within this range were relatively small, indicating that the values vary within a narrow range.

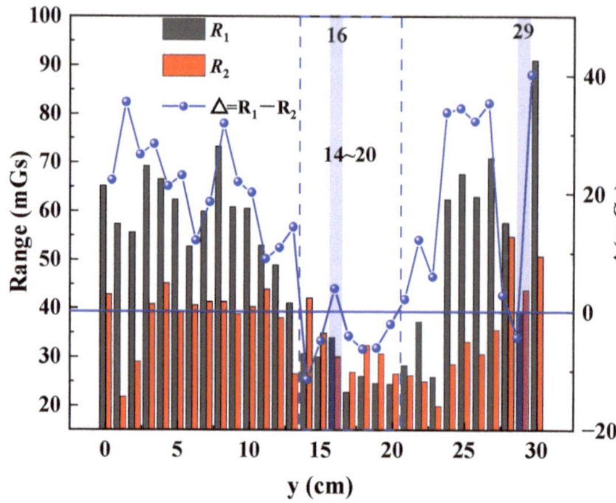

Figure 7. The range of data in Region 1 and Region 2.

5.2. *Influence of Load on the Spatial SMFL Distribution*

Based on the analysis conducted in Sections 3.1 and 5.1, it was evident that under axial loading, the magnetic force curves at different sampling points on the steel strand exhibited significant variations in numerical values and diverse patterns of change. Such a phenomenon made identifying stress concentration points difficult. Evaluating the stress on the specimen directly based on the SMFL intensity or extreme values and differences in the magnetic force curves from a single sampling point was also challenging. Therefore, it was necessary to extract SMFL curve characteristics to characterize the stress state of the specimen effectively.

Regarding the normal component of the leakage magnetic field (Bz), when the load was small (e.g., F = 0 kN), its corresponding curve rapidly decreased along the scanning path and started to rise after y > 25 cm. As the load increased, a "gentle zone" appeared in the SMFL curve, where the change was significantly smaller than in the first and last areas. When the load was large (e.g., F = 160 kN), its corresponding curve only dropped rapidly at y < 6 cm and the subsequent curve exhibited slight fluctuations within a specific range.

Based on the above analysis, it was observed that as the applied load on the specimen increased, the variation in the normal component of the SMFL field (Bz) became gentler, with reduced slope, range, and fluctuation. It was necessary to identify a parameter that could evaluate the scope and fluctuation of the data. AVEDEV is a function commonly used to assess data dispersion and is denoted here as *A*. The formula for calculating AVEDEV for the Bz curve of the leakage magnetic field component is given by Equation (11):

$$Avedev = \frac{1}{n-1}\sum_{i=1}^{n-1}\left|B_{z_{i+1}} - \overline{B_z}\right| \tag{11}$$

In this experiment, the parameter *A* for each load level curve was calculated using Equation (11), resulting in the A–F curves for each of the nine detection positions. Each curve contained 17 data points, representing a 17-level load, as shown in Figure 8. This figure provided preliminary insight into the correspondence between the SMFL signals and the applied load on the specimen.

From Figure 8, it is evident that *A* decreased monotonically with the increase in load F, indicating that the dispersion of the Bz curve reduced as the tensile force on the specimen increased. However, there were significant differences in the distribution of A–F curves for different detection positions. For example, the curves for P_3 and P_6 had larger ranges and more extreme values than others. In contrast, the curves for P_1, P_5, and P_8 exhibited a more pronounced change in range in Region 1 compared to Region 2, and the other curves showed relatively uniform changes, presenting an excellent linear relationship.

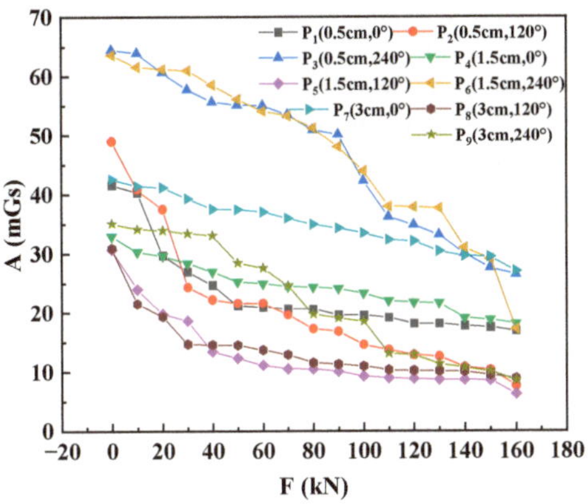

Figure 8. A–F curve of leakage magnetic field direction component Bz.

To achieve stress measurement based on the spatial SMFL field distribution in the steel strand, it was necessary to understand how the spatial detection positions influence the distributions of A–F curves. These positions were determined by two factors: the LOV and the RAV. The curves corresponding to the same factors were plotted together in Figures 9 and 10.

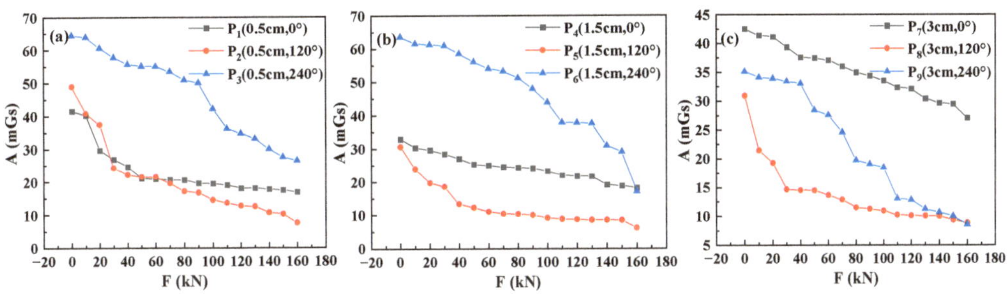

Figure 9. A–F curves of the same LOV: (**a**) $H_1 = 0.5$ cm; (**b**) $H_2 = 1.5$ cm; (**c**) $H_3 = 3$ cm.

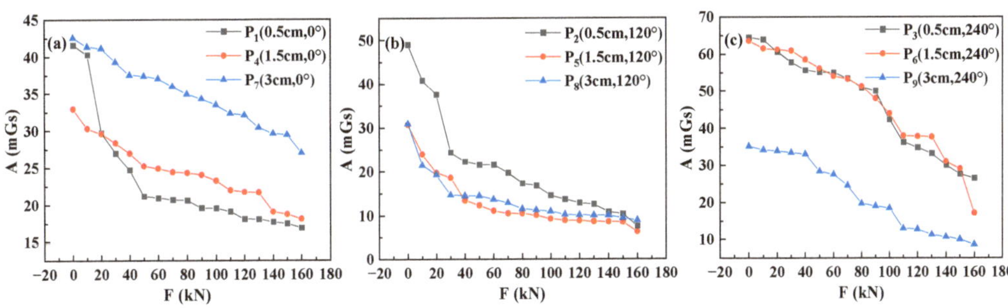

Figure 10. A–F curves with the same RAV: (**a**) $\theta_1 = 0°$; (**b**) $\theta_2 = 120°$; (**c**) $\theta_3 = 240°$.

From Figures 9 and 10, it can be observed that the RAV affected both the range and rate of change of the A–F curves, while the LOV mainly influenced the numerical values of the curves, with minimal impact on their shape. Except when $H_1 = 0.5$ cm, the range of the curves for $\theta_3 = 240°$ was smaller than $\theta_2 = 120°$, the range of the curves decreased in the order of θ_3, θ_2, and θ_1, and any parameter *A* for θ_2 was smaller than those for θ_3 and θ_1. Moreover, with an increase in load value, the rate of change for the A–F curves corresponding to $\theta_1 = 0°$ and $\theta_3 = 240°$ did not exhibit significant differences. In contrast, the rate of change for the curve corresponding to $\theta_2 = 120°$ gradually decreased.

Based on the above analysis, it is evident that as the tensile force on the specimen increases, the dispersion of the SMFL curves decreases. The RAV affected the range and rate of change of the dispersion, while the LOV influenced the numerical values. To achieve stress measurement based on the spatial SMFL fields in the steel strand, it was necessary to establish a relationship between parameter A and the applied external load. However, due to significant differences in the range of A–F curves for different detection positions, the further processing of parameter A was needed.

5.3. Stress Characterization Method of Steel Strand Based on Magnetic Characteristic Parameter

To better quantify the dispersion of the Bz curve, the parameter A was normalized to obtain A_N through Equation (12).

$$A_N = \frac{A_i - A_{min}}{A_{max} - A_{min}} \quad (12)$$

Figure 11 shows that the A_N values decreased with the increase in F after normalization, and an excellent linear relationship between A_N and F was observed. Some A_N–F curves exhibit a concave trend, indicating that the dispersion changes more rapidly in the early loading stage compared to the later stage. This phenomenon was attributed to the transition of scattered magnetic domains to a more regular distribution during the initial loading, significantly affecting the magnetic field distribution. As the loading progresses, the magnetic domains tend to have a more regular distribution, leading to a gradual reduction in their influence on the variation of the SMFL field.

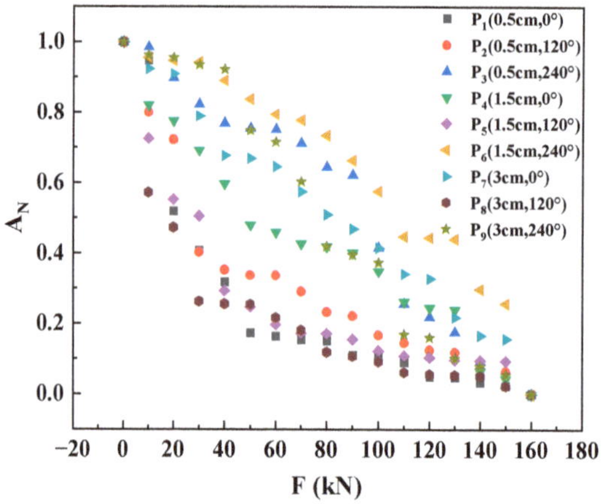

Figure 11. Normalized result of parameter A.

A linear fitting was performed on the A_N–F data to establish the correspondence between A_N and F, as shown in Figure 9. The fitting results are summarized in Figure 12 and Table 3.

Table 3. Linear fitting results of A_N–F curve of each detection position.

		$\theta_1 = 0°$		$\theta_2 = 120°$		$\theta_3 = 240°$
$H_1 = 0.5$ cm	P_1	$y = -0.00506x + 0.6576$ $R^2 = 0.7016$	P_2	$y = -0.00499x + 0.7168$ $R^2 = 0.8179$	P_3	$y = -0.00667x + 1.0721$ $R^2 = 0.9594$
$H_2 = 1.5$ cm	P_4	$y = -0.00546x + 0.8652$ $R^2 = 0.9574$	P_5	$y = -0.00454x + 0.6369$ $R^2 = 0.7304$	P_6	$y = -0.00559x + 1.095$ $R^2 = 0.9364$
$H_3 = 3$ cm	P_7	$y = -0.00589x + 0.9809$ $R^2 = 0.9887$	P_8	$y = -0.00419x + 0.5579$ $R^2 = 0.6923$	P_9	$y = -0.00722x + 1.0834$ $R^2 = 0.9668$

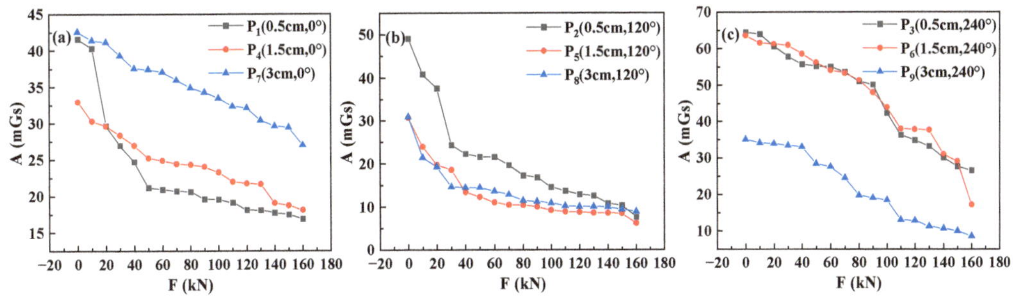

Figure 12. Linear fitting result of A_N–F curve with the same LOV: (**a**) $H_1 = 0.5$ cm; (**b**) $H_2 = 1.5$ cm; (**c**) $H_3 = 3$ cm.

The linear fitting curve formula is:

$$A_N = K \cdot F + B \tag{13}$$

The formula for calculating the stress of the specimen can be obtained from this:

$$\sigma = \frac{F}{S} = \frac{A_N - B}{K \cdot S} \tag{14}$$

where *S* is the cross-sectional area of the specimen, and the values of *K* and *B* were as follows:

From Table 4, it can be observed that the values for *K* and *B* changed depending on the detection position. At $\theta_1 = 0°$, the absolute values of *K* and B increased with LOV (*H*). At $\theta_2 = 120°$, *K* and *B*'s absolute values decreased as H increases. At $\theta_3 = 240°$, the relationship between the absolute values of *K*, *B*, and *H* was more complicated. That is why choosing the appropriate parameters based on the detection position was essential when calculating the specimen's stress using Equation (14).

Table 4. The values of the parameters *K* and *B* at each detection position.

		$\theta_1 = 0°$			$\theta_2 = 120°$			$\theta_3 = 240°$	
		K	*B*		*K*	*B*		*K*	*B*
$H_1 = 0.5$ cm	P_1	−0.00506	0.6576	P_2	−0.00499	0.7168	P_3	−0.00667	1.0721
$H_2 = 1.5$ cm	P_4	−0.00546	0.8652	P_5	−0.00454	0.6369	P_6	−0.00559	1.095
$H_2 = 1.5$ cm	P_7	−0.00589	0.9809	P_8	−0.00419	0.5579	P_9	−0.00722	1.0834

Figures 12 and 13 and Table 4 show that the linear fitting results of A_N–F curves for different detection positions exhibited significant differences, and both the LOV and RAV can influence the fitting results. At $\theta_1 = 0°$, the fitting function's linear correlation coefficient (R^2) increased with the increasing H, and all LOVs except $H_1 = 0.5$ had an R^2 greater than 0.95, which means good linearity. At $\theta_2 = 120°$, the R^2 decreased with increased H and was less than 0.82, indicating poor linearity. At $\theta_3 = 240°$, the R^2 only slightly changed with H and was always greater than 0.93, indicating good linearity.

Based on the above analysis, it can be concluded that the dispersion of the normal component of the SMFL field in the steel strand decreased as the stress on the specimen increased. The degree of dispersion *A* was normalized for each detection position as a feature of value $A_{N,}$ and linear regression was performed to establish a relationship between the distribution of the SMFL field and the stress on the specimen. The results indicate that the LOV and RAV could influence the final fitting curves. Among the three rotation angle values, $\theta_3 = 240°$ exhibited the best linearity and stability in the fitting results, while $\theta_2 = 120°$ showed the worst linearity, with $\theta_1 = 0°$ lying in between them.

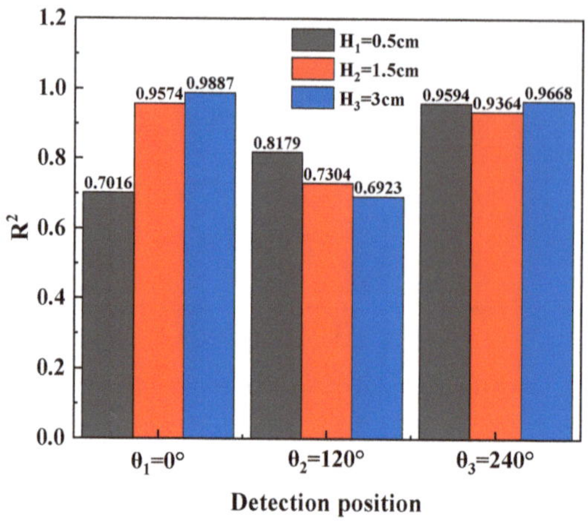

Figure 13. Linear fitting correlation coefficient of each detection position.

6. Conclusions

This study aimed to develop a stress measurement method for steel strands based on the spatial distribution of the SMFL field. A graded loading test for steel strands was designed to achieve this, and nine detection positions with different LOV and RAV were set to detect the spatial SMFL field. The original magnetic field and SMFL field distribution on the specimen surface under graded loading, the numerical values, the range of variation, and the curve shape of the normal component of the leakage magnetic field (Bz) were discussed. The study proposed a stress calculation method for steel strands based on the magnetic characteristic parameter A_N and investigated the influence of LOV and RAV on the detection results. The following conclusions were drawn:

(1) External magnetic fields and fixture-induced magnetic fields significantly impact the distribution of the SMFL field. Regardless of whether the influence of the external magnetic field was excluded, the curves of all components of the SMFL field exhibited a "three-segment" pattern, with drastic changes at the beginning (Y = 0~5 cm) and end (Y = 25~30 cm) of the curves and relatively smooth changes in the middle;

(2) When the external load was in the range of F = 0~50 kN, the variation range of the SMFL field intensity at the sampling points was greater than in the subsequent loading stages. There were noticeable differences in the SMFL field intensity values at different sampling points along the same scanning path during loading. However, there was no clear correspondence between the SMFL field intensity and the stress level of the specimen. Consequently, evaluating the stress level of the specimen solely based on the SMFL field intensity from a single detection point proved challenging;

(3) As the external load increased, the dispersion of the normal component of the SMFL field curve (Bz) on the surface of the steel strand decreased gradually. The magnetic characteristic parameter A_N could be used to evaluate the dispersion of the Bz curve, and there was a good linear relationship between A_N and F. By performing linear fitting, a formula for calculating the stress of the steel strand based on the magnetic characteristic parameter A_N was obtained;

(4) The LOV and RAV significantly affected the results of the linear fitting, and the specific detection positions determined the parameters of the stress calculation formula. Therefore, it is essential to conduct experiments at multiple spatial locations to optimize the formula parameters in the future.

Author Contributions: Conceptualization, S.L. and C.C.; methodology, S.L. and R.Z.; software, S.L.; validation, R.Z., J.Z. and K.T.; formal analysis, S.L.; investigation, J.Z.; resources, R.Z.; data curation, C.C.; writing—original draft preparation, S.L.; writing—review and editing, S.L.; visualization, S.L.;

supervision, K.T.; project administration, J.Z.; funding acquisition, C.C. All authors have read and agreed to the published version of the manuscript.

Funding: This work was supported by the National Natural Science Foundation of China (U20A20314), the Chongqing Natural Science Foundation of China (CSTB2022NSCQ-LZX0006, cstc2022ycjh-bgzxm0086), and Major Scientific Research Projects of China Railway Group Co., Ltd. (2022-ZD-11).

Data Availability Statement: The data presented in this study are available from the first and corresponding author upon request. The data are not publicly available due to the policy of the data provider.

Conflicts of Interest: The authors declare no conflict of interest.

References

1. Minaei, A.; Daneshjoo, F.; Goicolea, J.M. Experimental and numerical study on cable breakage equivalent force in cable-stayed structures consisting of low-relaxation seven-wire steel strands. *Structures* **2020**, *27*, 595–606. [CrossRef]
2. Zhang, Y.; Fang, Z.; Jiang, R.; Xiang, Y.; Long, H.; Lu, J. Static Performance of a Long-Span Concrete Cable-Stayed Bridge Subjected to Multiple-Cable Loss during Construction. *J. Bridge Eng.* **2020**, *25*, 04020002.1–04020002.16. [CrossRef]
3. Xin, J.; Zhou, C.; Jiang, Y.; Tang, Q.; Yang, X.; Zhou, J. A signal recovery method for bridge monitoring system using TVFEMD and encoder-decoder aided LSTM. *Measurement* **2023**, *214*, 112797. [CrossRef]
4. Yang, J.; Chen, R.; Zhang, Z.; Zou, Y.; Zhou, J.; Xia, J. Experimental study on the ultimate bearing capacity of damaged RC arches strengthened with ultra-high performance concrete. *Eng. Struct.* **2023**, *279*, 115611. [CrossRef]
5. Yao, Y.; Yan, M.; Bao, Y. Measurement of cable forces for automated monitoring of engineering structures using fiber optic sensors: A review. *Autom. Constr.* **2021**, *126*, 103687. [CrossRef]
6. Fu, Z.; Ji, B.; Wang, Q.; Wang, Y. Cable force calculation using vibration frequency methods based on cable geometric parameters. *J. Perform. Constr. Facil.* **2017**, *31*, 04017021. [CrossRef]
7. Ni, Y.; Zhang, Q.; Xin, R. Magnetic flux detection and identification of bridge cable metal area loss damage. *Measurement* **2021**, *167*, 108443. [CrossRef]
8. Zhang, L.; Qiu, G.; Chen, Z. Structural health monitoring methods of cables in cable-stayed bridge: A review. *Measurement* **2021**, *168*, 108343. [CrossRef]
9. Wang, Z.D.; Yao, K.; Deng, B.; Ding, K.Q. Quantitative study of metal magnetic memory signal versus local stress concentration. *NDT E Int.* **2010**, *43*, 513–518. [CrossRef]
10. Dubov, A.A. A study of metal properties using the method of magnetic memory. *Met. Sci. Heat Treat.* **1997**, *39*, 401–405. [CrossRef]
11. Wang, Z.D.; Yao, K.; Deng, B.; Ding, K.Q. Theoretical studies of metal magnetic memory technique on magnetic flux leakage signals. *NDT E Int.* **2010**, *43*, 354–359. [CrossRef]
12. Shi, P.; Su, S.; Chen, Z. Overview of researches on the nondestructive testing method of metal magnetic memory: Status and challenges. *J. Nondestruct. Eval.* **2020**, *39*, 43. [CrossRef]
13. Liu, B.; Fu, Y.; Xu, B. Study on metal magnetic memory testing mechanism. *Res. Nondestruct. Eval.* **2015**, *26*, 1–12. [CrossRef]
14. Wang, Z.D.; Gu, Y.; Wang, Y.S. A review of three magnetic NDT technologies. *J. Magn. Magn. Mater.* **2012**, *324*, 382–388. [CrossRef]
15. Bao, S.; Jin, P.; Zhao, Z.; Fu, M. A review of the metal magnetic memory method. *J. Nondestruct. Eval.* **2020**, *39*, 11. [CrossRef]
16. Pospisil, K.; Manychova, M.; Stryk, J.; Korenska, M.; Matula, R.; Svoboda, V. Diagnostics of reinforcement conditions in concrete structures by GPR, impact-echo method and metal magnetic memory method. *Remote Sens.* **2021**, *13*, 952. [CrossRef]
17. Dubov, A.; Dubov, A.; Kolokolnikov, S. Application of the metal magnetic memory method for detection of defects at the initial stage of their development for prevention of failures of power engineering welded steel structures and steam turbine parts. *Weld. World* **2014**, *58*, 225–236. [CrossRef]
18. Shi, M.; Liang, Y.; Zhang, M.; Huang, Z.; Feng, L.; Zhou, Z. Pipeline damage detection based on metal magnetic memory. *IEEE Trans. Magn.* **2021**, *57*, 1–15. [CrossRef]
19. Zhao, B.; Yao, K.; Wu, L.; Li, X.; Wang, Y.S. Application of metal magnetic memory testing technology to the detection of stress corrosion defect. *Appl. Sci.* **2020**, *10*, 7083. [CrossRef]
20. Shi, P. Magneto-mechanical model of ferromagnetic material under a constant weak magnetic field via analytical anhysteresis solution. *J. Appl. Phys.* **2020**, *128*, 115102. [CrossRef]
21. Yang, X.; He, C.; Pu, H.; Chen, L. An extended magnetic-stress coupling model of ferromagnetic materials based on energy conservation law and its application in metal magnetic memory technique. *J. Magn. Magn. Mater.* **2022**, *544*, 168653. [CrossRef]
22. Xia, R.; Zhou, J.; Zhang, H.; Zhou, D.; Zhang, Z. Experimental study on corrosion of unstressed steel strand based on metal magnetic memory. *KSCE J. Civ. Eng.* **2019**, *23*, 1320–1329. [CrossRef]
23. Pang, C.; Zhou, J.; Zhao, R.; Ma, H.; Zhou, Y. Research on internal force detection method of steel bar in elastic and yielding stage based on metal magnetic memory. *Materials* **2019**, *12*, 1167. [CrossRef] [PubMed]
24. Zhao, X.; Su, S.; Wang, W.; Zhang, X. Metal magnetic memory inspection of Q345B steel beam in four point bending fatigue test. *J. Magn. Magn. Mater.* **2020**, *514*, 167155. [CrossRef]

25. Liu, B.; Zeng, Z.; Wang, H. Study on the early fatigue damage evaluation of high strength steel by using three components of metal magnetic memory signal. *NDT E Int.* **2021**, *117*, 102380. [CrossRef]
26. Roskosz, M. Metal magnetic memory testing of welded joints of ferritic and austenitic steels. *NDT E Int.* **2011**, *44*, 305–310. [CrossRef]
27. Roskosz, M.; Gawrilenko, P. Analysis of changes in residual magnetic field in loaded notched samples. *NDT E Int.* **2008**, *41*, 570–576. [CrossRef]
28. Roskosz, M.; Bieniek, M. Analysis of the universality of the residual stress evaluation method based on residual magnetic field measurements. *NDT E Int.* **2013**, *54*, 63–68. [CrossRef]
29. Zhou, J.; Qiu, J.; Zhou, Y.; Zhou, Y.; Xia, R. Experimental study on residual bending strength of corroded reinforced concrete beam based on micromagnetic sensor. *Sensors* **2018**, *18*, 2635. [CrossRef]
30. Huang, H.; Jiang, S.; Yang, C.; Liu, Z. Stress concentration impact on the magnetic memory signal of ferromagnetic structural steel. *Nondestruct. Test. Eval.* **2014**, *29*, 377–390. [CrossRef]
31. Yao, K.; Deng, B.; Wang, Z.D. Numerical studies to signal characteristics with the metal magnetic memory-effect in plastically deformed samples. *NDT E Int.* **2012**, *47*, 7–17. [CrossRef]
32. Zhang, H.; Li, H.; Zhou, J.; Tong, K.; Xia, R. A multi-dimensional evaluation of wire breakage in bridge cable based on self-magnetic flux leakage signals. *J. Magn. Magn. Mater.* **2023**, *566*, 170321. [CrossRef]
33. Wang, H.; Dong, L.; Wang, H.; Ma, G.; Xu, B.; Zhao, Y. Effect of tensile stress on metal magnetic memory signals during on-line measurement in ferromagnetic steel. *NDT E Int.* **2021**, *117*, 102378. [CrossRef]
34. Leng, J.; Liu, Y.; Zhou, G.; Gao, Y. Metal magnetic memory signal response to plastic deformation of low carbon steel. *NDT E Int.* **2013**, *55*, 42–46. [CrossRef]
35. Ren, S.; Ren, X. Studies on laws of stress-magnetization based on magnetic memory testing technique. *J. Magn. Magn. Mater.* **2018**, *449*, 165–171. [CrossRef]
36. Tong, K.; Zhang, H.; Zhao, R.; Zhou, J.; Ying, H. Investigation of SMFL monitoring technique for evaluating the load-bearing capacity of RC bridges. *Eng. Struct.* **2023**, *293*, 116667. [CrossRef]
37. Qu, Y.; Zhang, H.; Zhao, R.; Liao, L.; Zhou, Y. Research on the method of predicting corrosion width of cables based on the spontaneous magnetic flux leakage. *Materials* **2019**, *12*, 2154. [CrossRef]
38. Liu, S.; Sun, Y.; Jiang, X.; Kang, Y. Comparison and analysis of multiple signal processing methods in steel wire rope defect detection by hall sensor. *Measurement* **2021**, *171*, 108768. [CrossRef]
39. Jiles, D.C. Theory of the magnetomechanical effect. *J. Phys. D Appl. Phys.* **1995**, *28*, 1537. [CrossRef]
40. Jiles, D.C.; Atherton, D.L. Theory of ferromagnetic hysteresis. *J. Magn. Magn. Mater.* **1986**, *61*, 48–60. [CrossRef]
41. Li, J.; Xu, M. Modified Jiles-Atherton-Sablik model for asymmetry in magnetomechanical effect under tensile and compressive stress. *J. Appl. Phys.* **2011**, *110*, 8. [CrossRef]
42. Li, J.; Xu, M.; Leng, J.; Xu, M. Modeling plastic deformation effect on magnetization in ferromagnetic materials. *J. Appl. Phys.* **2012**, *111*, 1. [CrossRef]

buildings

MDPI

Article

Research on an Intelligent Identification Method for Cable-Stayed Force with a Damper Based on Microwave Radar Measurements

Xiaogang Li [1,2,*], Xiangsheng Huang [1], Peng Ding [3], Qiansong Wang [1] and Qin Wang [1]

1 School of Civil Engineering and Architecture, Chongqing University of Science & Technology, Chongqing 400074, China; 2022206095@cqust.edu.cn (X.H.); 13618333005@163.com (Q.W.); 2023206039@cqust.edu.cn (Q.W.)
2 Chongqing Key Laboratory of Energy Engineering Mechanics & Disaster Prevention and Mitigation, Chongqing University of Science & Technology, Chongqing 400074, China
3 T. Y. Lin International Engineering Consulting (China) Co., Ltd., Chongqing 401121, China; dingpeng@tylin.com.cn
* Correspondence: 2021058@cqust.edu.cn; Tel.: +86-135-9409-4526

Abstract: Aiming at the shortcomings of traditional contact cable force monitoring technology in accuracy, efficiency, and applicability, an assessment method based on microwave radar measurements is proposed to measure a sloping cable with a damper for lengths greater than 200 m in this study. A formula for calculating the cable-stayed force with a damper is derived, and an intelligent cable force monitoring platform is developed based on cloud technology. Based on the Chongqing Nanjimen Railway Bridge, a real bridge test was carried out. It was indicated that the microwave radar method could be used to freely adjust the measurement angle and possessed high applicability and penetration. It significantly improved the measurement accuracy and efficiency of cables with a damper for lengths greater than 200 m. It has good application value for the solution of the problems of complicated operation and high costs in the monitoring of cables with a damper. The formula for calculating the cable force with a damper was proven to be reliable and accurate when compared to the results of direct calculation and the equivalent cable length method. It was able to significantly reduce the calculation error of the cable force caused by the influence of the damper. Additionally, the intelligent cable force monitoring platform was utilized to enhance the level of digitization, providing technical support for the scientific management and maintenance of bridges.

Keywords: microwave radar; damper; cable-stayed bridge; cable force monitoring; smartization

Citation: Li, X.; Huang, X.; Ding, P.; Wang, Q.; Wang, Q. Research on an Intelligent Identification Method for Cable-Stayed Force with a Damper Based on Microwave Radar Measurements. *Buildings* **2024**, *14*, 568. https://doi.org/10.3390/buildings14030568

Academic Editor: George Morcous

Received: 17 January 2024
Revised: 14 February 2024
Accepted: 17 February 2024
Published: 20 February 2024

1. Introduction

The cable is a vital component of a cable-stayed bridge, as it directly impacts the bridge's load-bearing capacity, shape, and structural safety [1–3]. Accurate measurement of the cable force and its variations is crucial for various purposes, including construction control, health monitoring, safety assessment, and damage diagnosis of cable-stayed bridges [4].

Currently, there are four common methods for cable force monitoring: the tension jack method, pressure sensor method, vibration frequency method, and magnetic flux method [5–7]. However, each method has its limitations and advantages. The tension jack method [8,9] is primarily used during the tensioning process, but it cannot provide continuous monitoring of the cable force once a bridge is constructed. The pressure sensor method [10] offers high measurement accuracy and simplicity of operation. However, a pressure sensor needs to be embedded during the cable's construction stage, and if the sensor is damaged, it cannot be easily replaced. The magnetic flux method [11] has limitations in terms of measurement efficiency and applicability, and it is still in a critical stage of research and development. When using the vibration frequency method [12,13] to measure a cable-stayed cable with a damper, the vibration sensor is more susceptible to the influence of the damper. Thus, a specific operating platform or measurement technique is often required for the accurate identification of the natural vibration frequency of the cable-stayed cable.

Traditional cable force measurement methods have some disadvantages, such as their limited range, low efficiency, insufficient accuracy, and applicability under certain conditions. Therefore, the development of a convenient and efficient cable force measurement method is of great practical significance for the development of long-span bridges. In view of the methods of cable force measurement, scholars have conducted much related research. Jo, Hyeon Cheol et al. [14] proposed a tension and tension range evaluation method based on multiple digital images that was able to obtain the cable force more economically and efficiently than traditional methods could. Xiaofeng Liu et al. [15] proposed a vibration-based elastic–magnetic (VBEM) method that had a good ability to identify the natural frequencies of each order of steel strand. Hongbo Liu [16] conducted dynamic response tests on closed cables and high vanadium cables under prestress and proposed a deep learning model that could intelligently identify the cable force for construction based on test data. Wen-Hwa Wu et al. [17] considered the specific arrangement of the boundary constraints of a suspender of an arch bridge and were able to simply determine the tension of the suspender by concentrating multiple measurements at one end. Marco Bonopera et al. [18] studied the free transverse vibration of a post-reinforced thin-walled steel box girder and found that the dynamic force of the beam largely depended on the contact between the cable and the surrounding section.

There are two common methods for calculating the cable-stayed force with a damper (length $\geq$ 200 m): the equivalent cable length method and the energy method [19–21]. Xiongjun He et al. [19] used the finite-element equivalent cable length instead of the calculated cable length in the traditional vibration frequency method to measure the cable force, which effectively improved the accuracy of the cable force test. Shenghua Tang et al. [20] constructed a formula for calculating the equivalent cable length based on the differential equation of cable vibration and the energy principle. Based on the derivation of cable force formulas under different constraint conditions with the energy method, Xianyu Wang et al. [21] conducted a systematic analysis from three perspectives: the optimization of the cable force at the boom design stage, the measured cable force at the bridge completion stage, and the influence of the cable force on the bending stiffness at the operation stage. Gong Junh [22] took a concrete cable-stayed bridge with a main span containing a 260 m high-speed railway as an example, and they designed the bridge structure according to the functional needs and site conditions; then, they established a three-dimensional finite-element model and adopted the minimum bending energy method and the influence matrix method to determine the reasonable construction state of the cable-stayed bridge. However, the equivalent cable length method only considers the damper as the anchoring section and reduces the length of the cable. It fails to effectively consider the influence of the damper. The energy method only considers the damping of the damper when studying the shape function of the cable, ignoring its stiffness. However, the damper has an equivalent stiffness K that affects the shape function of the cable. This results in an insufficient calculation of the cable force [23,24]. Thus, it is essential to adjust the shape function to align with reality and enhance the accuracy of calculations.

In summary, the field of cable force monitoring with a damper (length $\geq$ 200 m) presents a significant contradiction. Traditional cable force measurement methods are easily affected by the damper, resulting in low measurement accuracy, efficiency, and applicability. This limitation makes it difficult to meet actual engineering needs. Furthermore, the influence of the damper is not fully considered in the calculation of cable force, which requires improved accuracy. The emergence of big data cloud technology has opened up new opportunities for intelligent monitoring. An intelligent cable force monitoring platform that integrates monitoring, evaluation, and early warning [24] can reduce costs, increase efficiency, and provide technical support for the scientific management and maintenance of bridges.

This study utilizes microwave radar as a tool for detecting the vibration frequency of a cable. The energy method is modified to incorporate the presence of a damper, allowing for the description of the cable's shape function. Then, a new formula for calculating the cable force with a damper is derived by combining the equivalent stiffness formula of the damper with the modified energy method. The data collected by the microwave radar are then uploaded to an intelligent cable force monitoring platform. The platform automatically processes, analyses, and stores the data, providing cable force monitoring values through the derived calculation formula. An evaluation and early warning system is implemented to assess the cable's operational status, facilitating intelligent monitoring, evaluation, and early warning of the cable force. This approach provides a scientific basis for ensuring the safe operation of cables.

2. Working Principle

2.1. Microwave Radar Differential Interferometry Technique

Microwave radar differential interferometry [25–28] is a technique that captures the subtle vibrations of an object at different moments using microwave radar. This technique relies on the characteristics of microwaves, which can be absorbed, reflected, and projected. The phase sum of two observations of an inclined cable is obtained using the differential interferometry technique to determine its deformation over time. A working diagram of microwave radar (Figure 1) and the deformation calculation formula are presented below.

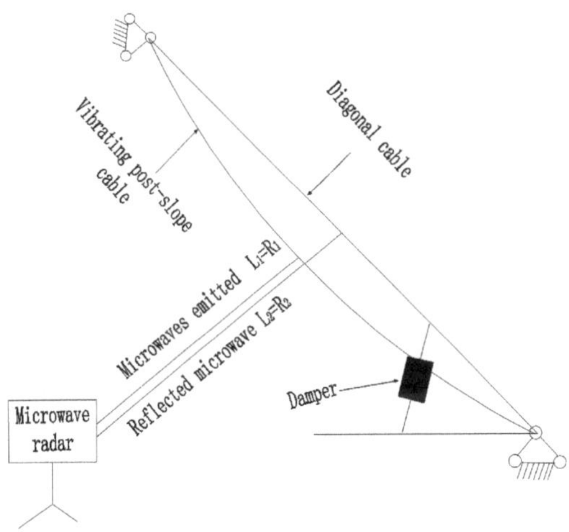

Figure 1. Schematic diagram of the operation of microwave radar.

The sum of the two observations is used to obtain the phase:

$$\Delta\varphi = \varphi_2 - \varphi_1 = \frac{4\pi}{\lambda} \cdot (R_2 - R_1) \tag{1}$$

The form variables can be expressed as follows:

$$d = R_2 - R_1 = \frac{\lambda}{4\pi} \cdot \Delta\varphi \tag{2}$$

2.2. Microwave Radar Frequency Conversion

After the microwave radar collects the time-domain signals of the two vibrations before and after the diagonal cable, the frequency-domain signals are obtained through a Fourier transformation: $d(t_s)$ [29]. $d(t_s)$ can be calculated with Equation (3):

$$X(f_d) = \int_0^{T_x} d(t_s)\exp(-j\pi f_d t_s)dt_s \tag{3}$$

where

T_x—the deformation observation time;

f_d—the frequency resolution, which is inversely proportional.

3. Calculation Formula for a Diagonal Cable with a Damper

3.1. Principle

Firstly, the mode function of the cable is determined, followed by the integration of its kinetic and potential energy over the entire length at a specific time. Then, based on the law of conservation of energy, the relationship between the cable force S and the angular frequency ω is established. Finally, the equation for the relationship between the cable force and frequency is derived by solving and simplifying the equations [30–32]. The equivalent stiffness K in the calculation formula can be eliminated by calculating different order frequencies. However, this approach increases the workload of monitoring and calculation. The damper's information can typically be obtained through design data, field information, or the nameplate. By calculating the equivalent stiffness K based on the known information of the damper, the workload can be significantly reduced, the monitoring efficiency can be improved, and future tests can be facilitated. An equivalent stiffness calculation method was used to modify the existing energy calculation formula given that the damper information was available. It resulted in a more convenient, concise, and efficient formula for calculating cable force with dampers.

3.2. Boundary Conditions

Under the action of dampers, the cable boundary conditions become complicated. According to [33–35], the dynamic model was reasonably simplified in this research, and the damper was simplified as the equivalent stiffness K. The cable was connected to the bridge tower and the main beam through an anchor, and its boundary condition was between articulation and consolidation. However, when the cable reached a certain length, the boundary condition of articulation or solid connection had little influence on the calculation of the cable force. Therefore, this study assumes that the cable's boundary conditions were hinged, so it was assumed that the boundary condition of the cable was articulation.

3.3. Calculation Process

According to [36], the sagging of a cable has little influence on high-order frequencies (4th and above). Therefore, this study took the fourth-order mode as an example for the calculation, without considering the influence of the cable's dead weight, sagging, and out-of-plane vibration.

When a cable vibrates, the existence of a stagnation point significantly influences the vibration mode of the span with the damper, but it has little influence on the other spans. Consequently, it is assumed that the shape function of the other spans is still a sine function with only one constant difference, and the span with the damper is superimposed with an anti-symmetric higher-order term. Therefore, this study only corrected the calculation of this cross-shape function with the damper, as shown in Figure 2 below.

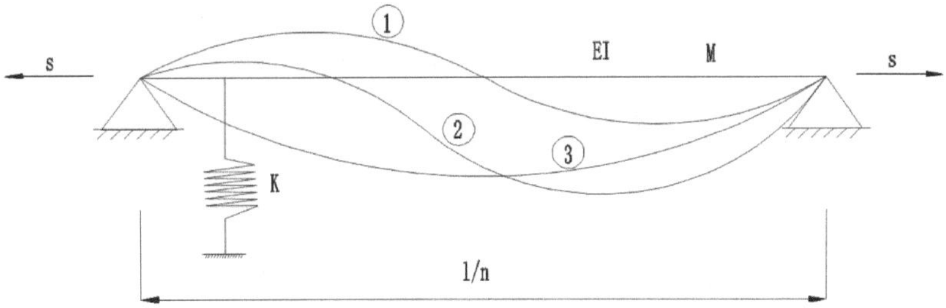

Figure 2. Shape diagram of the diagonal cable (i = 4).

Here,

$$\text{①}-v(x,t)=\frac{\sin(2n\pi x/l)}{\sin(n\gamma\pi)}z(t),$$
$$\text{②}-v(x,t)=\frac{\sin(n\pi x/l)}{\sin(n\gamma\pi)}z(t),$$
$$\text{③}-v(x,t)=\psi_1(x)z(t)=\frac{A\sin(n\pi x/l)+B\sin(2n\pi x/l)}{A\sin(n\gamma\pi)}z(t).$$

The shape function that is superimposed can be expressed as

$$\psi_1(x)=\frac{A\sin\left(\frac{\pi xn}{l}\right)+B\sin\left(\frac{2\pi xi}{l}\right)}{A\sin(n\gamma\pi)} \tag{4}$$

where

A, B—constants to be determined;

γ—distance from the damper to the end.

The displacement of points on the diagonal cable at any instant t is described as follows:

$$v(x,t)=\psi_1(x)z(t)=\psi_1\sin(\omega t+\alpha) \tag{5}$$

The kinetic energy (T) and potential energy (V) of the lasso at any given moment (t) are determined as follows:

$$T=m\omega^2L(A^2+B^2)\cos^2(\omega t+a)/(4A^2n\sin^2\gamma\pi) \tag{6}$$

$$\begin{aligned}V=&[0.5EI\tfrac{l}{n}(0.5A^2\tfrac{\pi^4n^4}{l^4}+8B^2\tfrac{\pi^4n^4}{l^4})+0.5K(A\sin n\pi\gamma+B\sin 2n\pi\gamma)^2]\\&+[0.5S\tfrac{l}{n}(0.5A^2\tfrac{\pi^2n^2}{l^2}+2B^2\tfrac{\pi^2n^2}{l^2})]\tfrac{\sin^2(\omega t+\alpha)}{A^2\sin^2 n\gamma\pi}\end{aligned} \tag{7}$$

According to the law of conservation of energy, the maximum kinetic energy is equal to the maximum potential energy:

$$T_{\max}=V_{\max} \tag{8}$$

Finally, the force of the rope is obtained:

$$\begin{aligned}S=&\tfrac{l^2}{8n^2\pi^2}\{-\left\{\tfrac{2Kn}{l(\sin^2(n\pi\gamma)+4\sin^2(2n\pi\gamma))}+20\tfrac{n^4\pi^4}{l^4}EI-5m\omega^2\right\}+[(2Kn/l(\sin^2(n\pi\gamma)-\\&4\sin^2(2n\pi\gamma))+12\tfrac{n^4\pi^4}{l^4}EI+3m\omega^2)^2+64K^2\sin^2(n\pi\gamma)\sin^2(2n\pi\gamma)n^2/l^2]^{1/2}\}\end{aligned} \tag{9}$$

The term $64K^2\sin^2(n\pi\gamma)\sin^2(2n\pi\gamma)n^2/l^2]^{1/2}$ in Equation (9) is small and negligible compared to the other terms, so it can be simplified to

$$S=\frac{4mf_n^2l^2}{n^2}-\frac{n^2\pi^2}{l^2}EI-\frac{2Kl}{n\pi^2}\sin^2(n\pi\gamma) \tag{10}$$

Compared with the direct calculation method, the modified energy method includes an additional term ($\frac{2Kl}{n\pi^2}\sin^2(n\pi\gamma)$) in the calculation formula for the cable force with dampers, which increases the effect of the dampers.

According to the "Technical Regulations for Energy Dissipation and Shock Absorption in Buildings (JGJ297-2013)" [37], the equivalent stiffness K of the damper is calculated based on the parameter information.

$$\mu = \frac{x}{x_1} \tag{11}$$

$$K = \frac{(1-\alpha)k_1}{\mu} + \alpha k_1 = \frac{F}{x} \tag{12}$$

where

K—the damper's equivalent stiffness;
F—the damper's maximum damping force;
x—the damper's maximum stroke;
x_1—the yield displacement;
α—the post-yield stiffness ratio;
k_1—the initial elastic stiffness of the damper.

By substituting Equations (11) and (12) into Equation (10), we obtain

$$S = \frac{4mf_i^2 l^2}{i^2} - \frac{\pi^2 i^2}{l^2}EI - \frac{2Fl}{n\pi^2 x}\sin^2(n\pi\gamma) \tag{13}$$

When the flexural stiffness is disregarded, Equation (14) takes the following form:

$$S = \frac{4mf_i^2 l^2}{i^2} - \frac{2Fl}{n\pi^2 x}\sin^2(n\pi\gamma) \tag{14}$$

3.4. Consideration of the Correction of Bending Stiffness

There is a correlation between the cable force S calculated with Equation (14) and the actual cable force S_0, which can be corrected with the correlation coefficient ζ: $S_0 = \zeta S$. For different cable lengths l, the relationship between the measured cable force and the actual cable force is nonlinear. Therefore, through the nonlinear fitting of the correlation coefficient, the functional relationship between the correlation coefficient ζ and the cable length l can be obtained [5].

$$\zeta = 1.00038 - \left\{ 0.53036 \times \left[1 + \left(\frac{l}{4.29473} \right)^{1.82264} \right]^{-1} \right\} \tag{15}$$

4. Evaluation and Early Warning through Cable Monitoring Based on Cloud Technology

An intelligent cable force monitoring platform that integrated monitoring, evaluation, and early warning was established using big data cloud technology [38–44]. The platform used microwave radar as a sensor to collect cable vibration data, which were then transmitted to the data platform through an interface protocol. After processing and analyzing the data, the cable's status was evaluated using an objective evaluation and early warning system, and corresponding management measures were provided. The platform comprised the following components and the structure of the intelligent cable force monitoring platform is shown in Figure 3:

1. Sensor and data acquisition and transmission system: First, microwave radar was used to collect the cable vibration data in real time, and then the data were transmitted to the processing, analysis, and storage system through a data acquisition module and data transmission module.

2. Data processing, analysis, and storage system: After the data were transmitted to the platform, the calculation formula for the cable force with a damper was used to obtain the monitoring values for professional analysis.
3. Evaluation and early warning system: After the monitoring values were obtained, the following evaluation mechanism was established according to the "Technical specification for operation monitoring of urban rail transit facilities Part 2: Bridges (GB/T 39559.2-2020) [S]" [45]:

$$\Delta = |S_\sigma| - \beta|M_\sigma| \leq 0, \text{ normal} \tag{16}$$

$$\Delta = |S_\sigma| - \beta|M_\sigma| > 0, \text{ early warning} \tag{17}$$

where

$|S_\sigma|$—the monitoring value of the cable force;

$|M_\sigma|$—the corresponding theoretical calculated value of the cable force;

β—the discount factor of the theoretical calculated value of the cable force, $\beta \in [0.95, 1.00]$.

After the evaluation system, the following determinations were made: (1) The status of the cable was normal according to the routine management; (2) there was a cable status warning, which should be analyzed through research and judgment to put forward solutions, including the need for special inspections, train speed limits, and other recommendations.

4. User front-end: Instructions were issued and data management functions were performed.

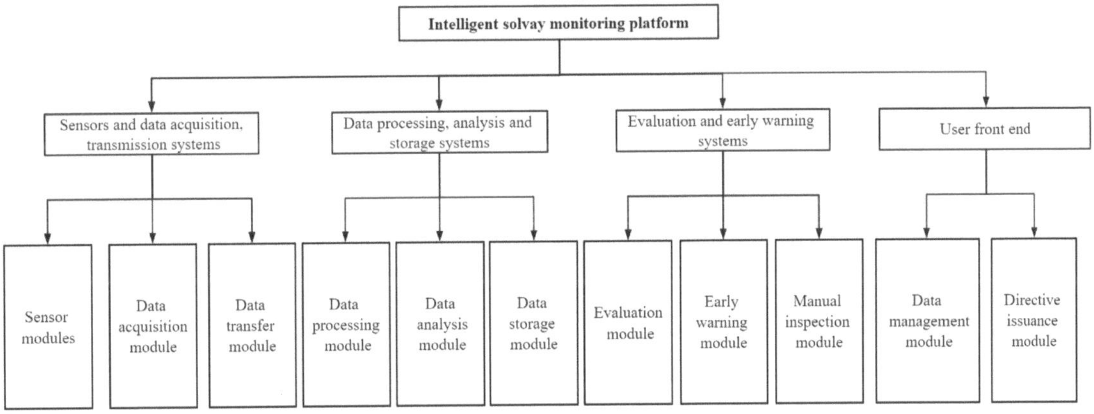

Figure 3. Schematic structure of the intelligent cable force monitoring platform.

5. Engineering Applications

5.1. Project Overview

The Chongqing Nanjimen Railway Bridge is currently the world's largest dedicated cable-stayed bridge for railway tracks. Its structural form is that of a five-span high- and low-tower double-cable-plane semi-floating system, and it has a span arrangement of 34.5 m + 180.5 m + 480 m + 215.5 m + 94.5 m. The main beams of the main bridge are all in the form of steel box composite beams, and the bridge tower is in the form of a portal bridge tower. The low bridge tower has a total height of 158 m, while the high bridge tower has a total height of 227 m. The diagonal cable used is a Φ 7.0 mm galvanized high-strength low-relaxation parallel-steel-wire HDPE-sheathed and finished cable, with a tensile strength of not less than 1770 MPa and a tensile elastic modulus of not less than 195,000 MPa. The cable-stayed steel wire is coated with zinc −5% aluminum/mixed rare earth alloy; the steel wire is arranged in a compact concentric twist, and the outermost steel wire is twisted at an angle of 3.5° with respect to the ground with a left turn. The outer layer of the cable is a high-density polyethylene sheath, and a double-helix coil pipe jacket is used to resist wind and rain vibrations. The high tower's side has 27 pairs of permanent cables—NMC1–NMC27(north mid-span cable) and NSC1–NSC27 (north side-span cable)—while the low tower's side has 16 pairs of diagonal cables: SSC1–SSC16 (south side-span cable) and SMC1–SMC16 (south mid-span cable). A linear viscous damper with a length greater than 200 m is installed at both ends of the diagonal cables and connected to the main beam. The bridge's layout is shown in Figures 4–9.

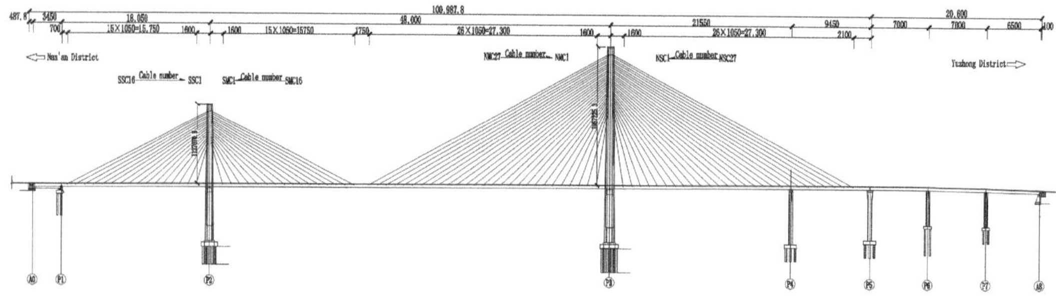

Figure 4. Layout of the whole bridge (unit: cm).

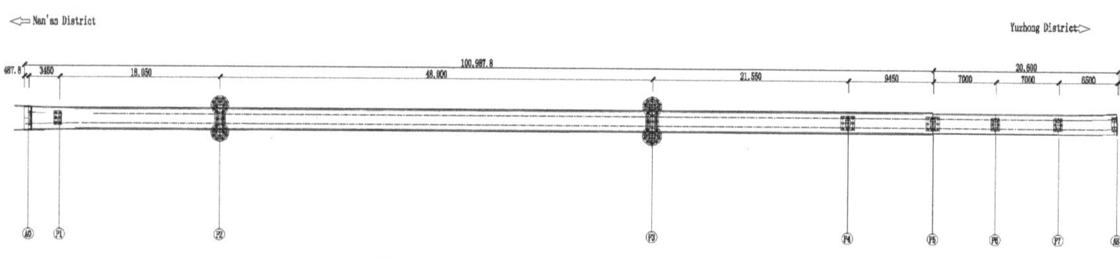

Figure 5. General layout of the bridge (unit: cm).

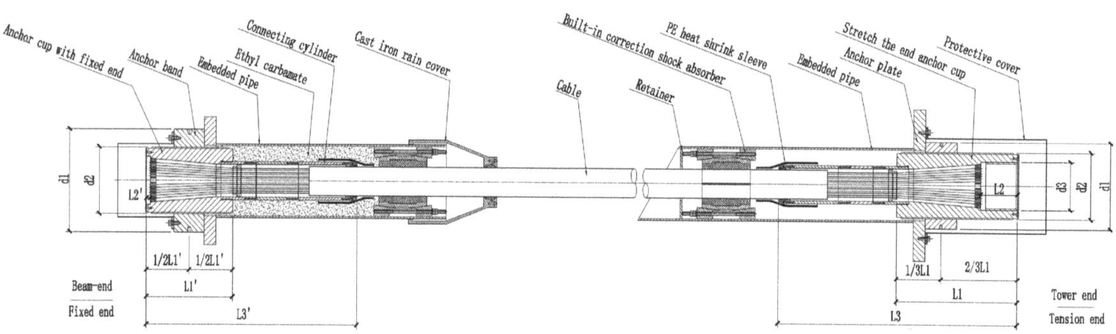

Figure 6. Diagram of the diagonal cables' construction.

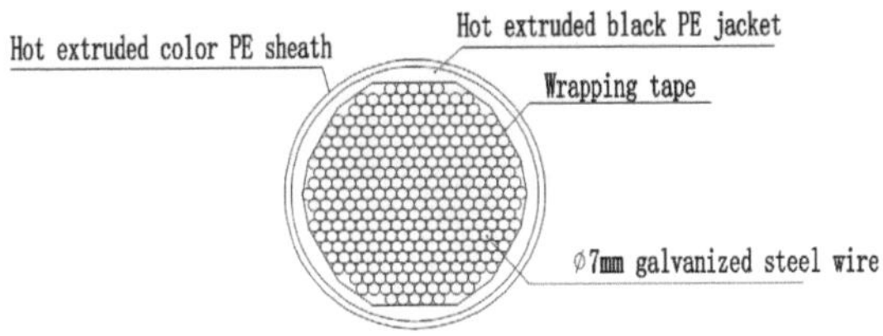

Figure 7. Cross-sectional drawing of cables.

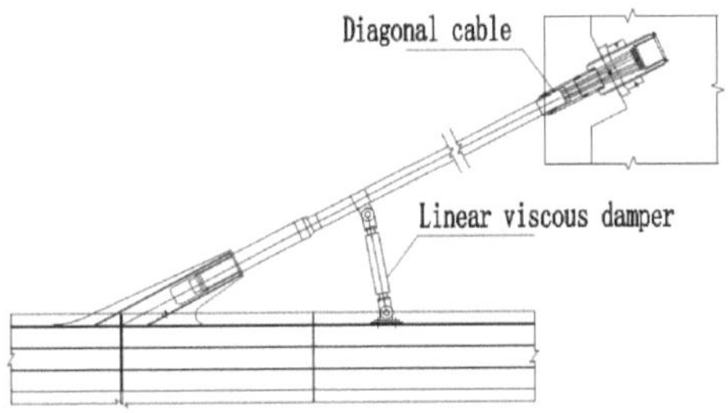

Figure 8. Schematic diagram of the damper connection.

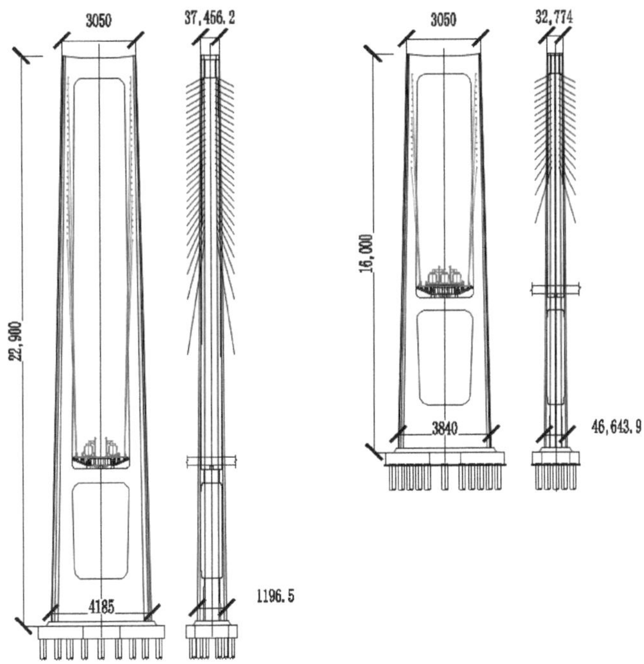

Figure 9. Cross-sectional view of the main tower (unit: cm).

5.2. Health Monitoring System

5.2.1. System Overview

The long-term health monitoring of the Chongqing Nanjimen Railway Bridge includes load and environmental monitoring, overall static and dynamic response monitoring, and local response monitoring. The overall layout of the health monitoring points of the whole bridge and the sensor information are shown in Figure 10 and Table 1: ①: displacement of the expansion joint; ②: angle of the beam end; ③: video surveillance; ④: vertical deformation of the main beam; ⑤: main beam stress; ⑥: main beam temperature; ⑦: steel structure fatigue; ⑧: horizontal displacement of the pier top; ⑨: main beam vibration (transverse); ⑩: main beam vibration (longitudinal); ⑪: GNSS deformation system; ⑫: ship collision (lateral); ⑬: ship collision (vertical); ⑭: ship collision (longitudinal); ⑮: wind speed and direction; ⑯: temperature and humidity; ⑰: water level; ⑱: cable force monitoring; ⑲: rainfall; ⑳: main tower vibration (transverse); ㉑: main tower vibration (longitudinal); ㉒: main tower stress; ㉓: main tower temperature.

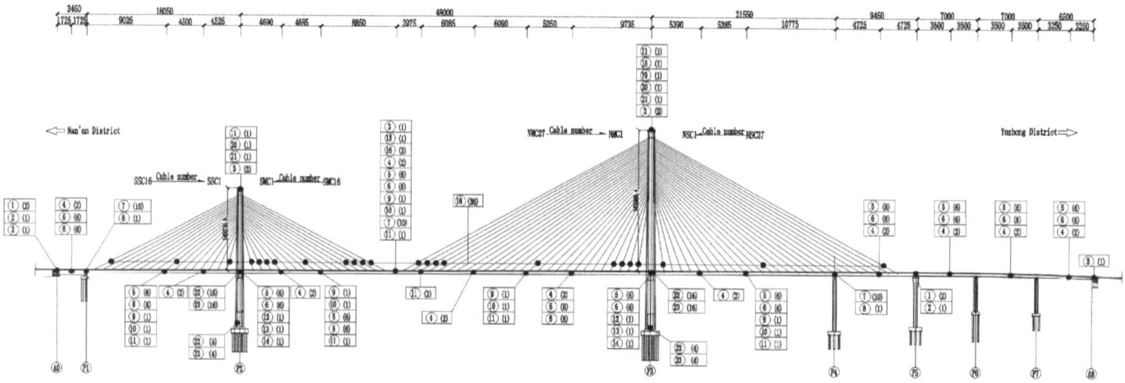

Figure 10. General layout of the health monitoring points (unit: cm).

Table 1. Health monitoring equipment information.

Monitoring Item	Equipment Category	Monitoring Frequency
①	Fiber grating displacement meter	1 Hz
②, ⑧	Inclination sensor	1 Hz
④	Differential pressure—static level	1 Hz
⑤, ⑦, ㉒	Fiber grating strain gauge	1 Hz
⑥, ⑯, ㉓	Hygrograph	1/600 Hz
⑨, ⑩, ⑳, ㉑	Acceleration sensor	1/120 Hz
⑪	GNSS	1 Hz
⑫, ⑬, ⑭	Acceleration sensor	1 Hz
⑮	Mechanical anemometer	1 Hz
⑰	Radar water level gauge	1/600 Hz
⑱	Cable force acceleration sensor	1/120 Hz
	Transcore pressure sensor	1/60 Hz
⑲	Rain gauge	1/60 Hz

5.2.2. Microwave Radar Application

The microwave radar method, vibration frequency method, and pressure sensor method were used to measure SMC1, SMC16, NMC1, and NMC27 of the Chongqing Nanjimen Railway Bridge. The schematic diagrams of the field measurements are shown in Figure 11, and the test data are shown in Figure 12 and Tables 2 and 3. The deviation rate of the cable force measured with the three test methods—namely, the vibration frequency method, pressure sensor method, and microwave radar method—was equivalent to the deviation ratio of the theoretical cable force: $\delta = \frac{S_2 - S_1}{S_1} \times 100\%$; here, δ is the deviation rate of the cable force, S_1 is the theoretical cable force, and S_2 is the measured cable force.

(a)

(b)

Figure 11. *Cont.*

(c)

Figure 11. Schematic diagram of the field measurements. (**a**) Microwave radar method. (**b**) Vibration frequency method. (**c**) Pressure transducer method.

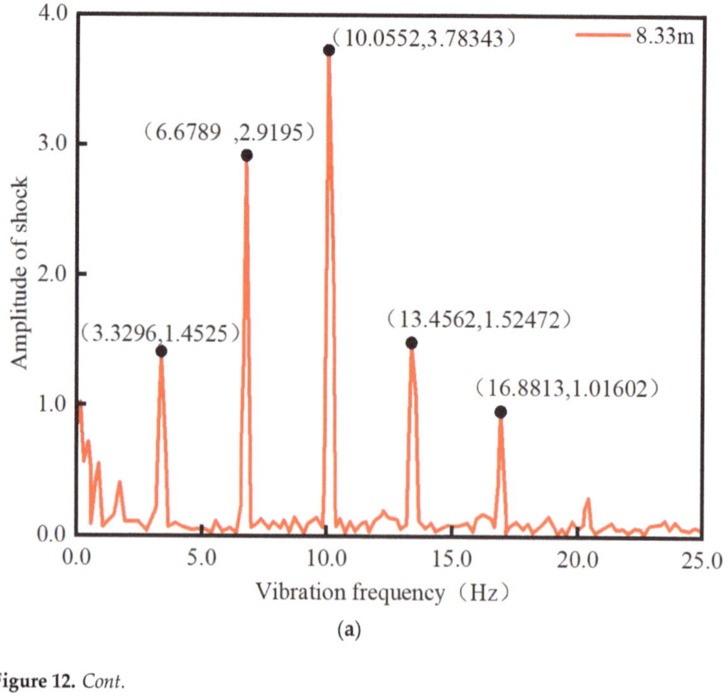

(a)

Figure 12. *Cont.*

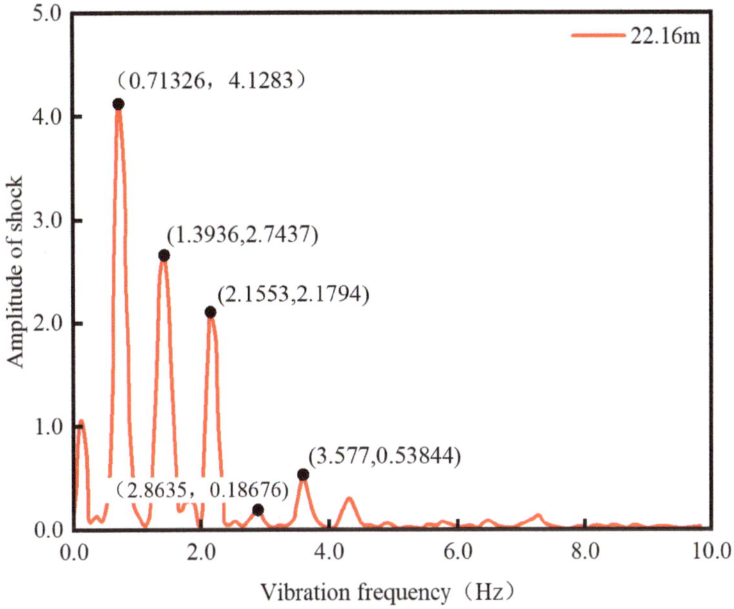

(**b**)

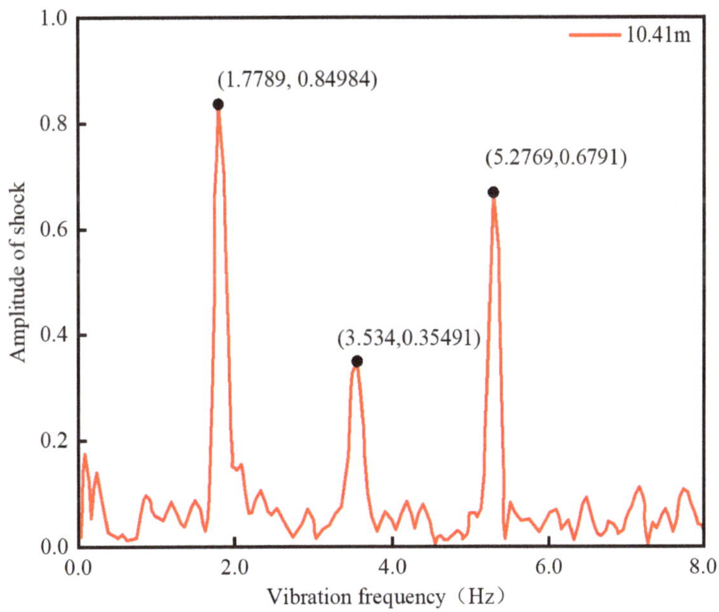

(**c**)

Figure 12. *Cont.*

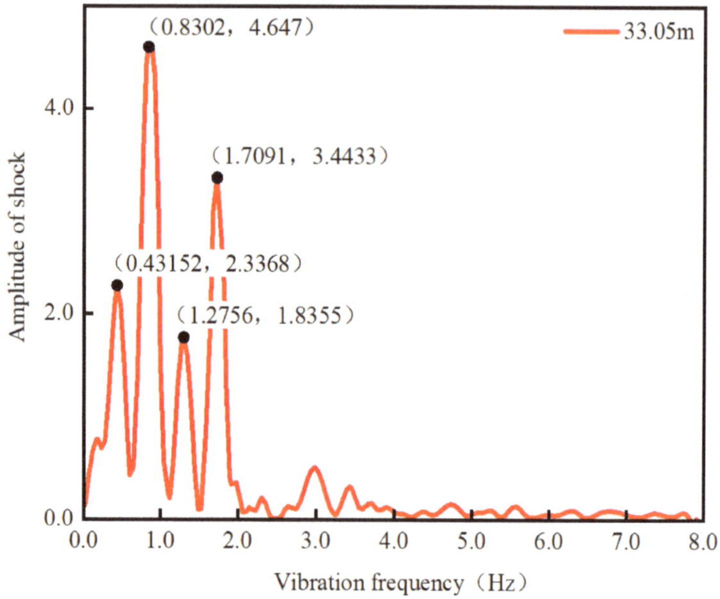

(d)

Figure 12. Spectral diagram of the SMC1, NMC16, NMC1, and NMC27 cables. (**a**) Base frequency diagram for SMC1. (**b**) Base frequency diagram for SMC16. (**c**) Base frequency diagram for NMC1. (**d**) Base frequency diagram for NMC27.

Table 2. Measurement results of the vibration frequency method, pressure transducer method, and microwave radar method for an inclined cable with a damper.

Number	Fundamental Frequency	Theoretical Cable Force S_1 (kN)	Measured Cable Force S_2 (kN)			Deviation Rate of Cable Forceδ		
			Vibration Frequency Method	Pressure Sensor Method	Microwave Radar Method	Vibration Frequency Method	Pressure Sensor Method	Microwave Radar Method
NMC27	0.43	6902	6550	6262	6860	−5.1%	−9.3%	−0.6%

Table 3. Measurement results of the vibration frequency method, pressure sensor method, and microwave radar method for an inclined cable without a damper.

Number	Fundamental Frequency	Theoretical Cable Force S_1 (kN)	Measured Cable Force S_2 (kN)			Deviation Rate of Cable Forceδ		
			Vibration Frequency Method	Pressure Sensor Method	Microwave Radar Method	Vibration Frequency Method	Pressure Sensor Method	Microwave Radar Method
SMC1	3.33	4139	4010	3901	4060	−3.1%	−5.8%	−1.9%
SMC16	0.71	6853	6403	6302	6638	−6.6%	−8.0%	−3.1%
NMC1	1.78	3695	3903	3609	3804	5.6%	−2.3%	2.9%

Table 2 demonstrates that for the NMC27 cable with a damper longer than 200 m, the deviation rate of the force measured with microwave radar was only 0.6%. This rate was significantly lower than those of the vibration frequency and pressure sensor methods, indicating that the microwave radar method was highly accurate and reliable. Table 3 shows that the cables without dampers, namely, SMC1, SMC16, and NMC1, which had a length of less than 200 m, had a maximum deviation rate for the cable force of 3.1% when measured using the microwave radar method. This was lower than the deviation rates obtained using the vibration frequency method (6.6%) and the pressure sensor method (8.0%). The results confirmed the excellent applicability of the microwave radar method in meeting the required specifications.

In this study, three methods were applied to measure the cable force of a cable-stayed cable of the Chongqing Nanjimen Railway Bridge. The following were the results: Microwave radar can be used to directly set up freely adjustable equipment, and the cable force measurement can be completed by only two people. The vibration frequency is influenced by the damper, and a ladder needs to be set up. The pressure sensor method is more complex and requires embedding in advance for measurement. These two methods need 3–4 people to complete the work. The vibration frequency method and pressure sensor method do not support multi-target measurement, as only one cable can be measured at a time. In contrast, the microwave radar method can be used to measure the cable force of more than 10 cables simultaneously. This method is based on the principle of non-contact measurement, and it provides better stability and efficiency than contact measurement methods do. Additionally, it can penetrate the HDPE sheath, resulting in more accurate measurement results. Compared to traditional contact measurement methods, the microwave radar method offers significant advantages in terms of measurement accuracy and efficiency for stayed cables with dampers. Table 4 demonstrates that the microwave radar method has significant advantages over traditional contact measurement methods in terms of measurement accuracy and efficiency for an inclined cable with a damper.

Table 4. Comparison of the advantages and disadvantages of the vibration frequency method, pressure sensor method, and microwave radar method.

Project	Vibration Frequency Method	Pressure Sensor Method	Microwave Radar Method
Labor cost	3–4 people	3–4 people	2 people
Multi-object measurement	Nonsupport	Nonsupport	Up to 10 cables can be measured simultaneously
Precision	The deviation of cable force is large	The cable force deviation is large	Cable force deviation is low
Efficiency	One at a time	One at a time	More than 10 at a time
Measurement parameter	Frequency	Stress	Displacement
Measurement mode	Contact measurement	Contact measurement	Non-contact measurement
How tedious the preparation is	A ladder or an elevator is necessary	The sensor must be embedded in advance	Can be directly set up and freely adjusted
Penetration characteristic	No	No	Penetrable HDPE/PE sheath

5.2.3. GNSS Application

The Chongqing Nanjimen Railway Bridge relies on a GNSS deformation system to monitor the displacement of the main beam and bridge tower. The monitoring items include the transverse, longitudinal, and vertical spatial deformation data of the main beam, as well as the transverse, longitudinal, and vertical spatial deformation data of the top bridge tower. This system provides an important basis for the assessment of the state of the bridge structure. The on-site installation diagram is shown in Figure 13.

Figure 13. Photo of the special GNSS antenna used in the field.

Based on the application of the GNSS deformation monitoring system, the maximum vertical deformation, transverse deformation, and longitudinal deformation of the main beam of the Chongqing Nanjimen Railway Bridge were monitored from 1 January 2023 to 31 December 2023. It was found that the deformation of the main beam at the measuring point was not abnormal within the theoretically calculated deformation envelope. The interval of the change in the vertical GNSS deformation of the upper 10 upstream sides of the main beam monitoring section was [−411.4, 88.6], and the amount of change was 500.0 mm.

5.2.4. Application of an Environmental Monitoring System

A temperature and humidity sensor was used to measure the changes in the ambient temperature of the bridge, and the changes in the wind speed and direction at the bridge site were measured with a wind speed and direction instrument, which provided a basis for the analysis of the working environment of the bridge and an evaluation of the driving safety. The ambient temperature measurement points were arranged at the closing section of the main span of the main bridge, and the wind speed and direction measurement points were arranged at the top of tower P3 and the closing section of the main span of the main bridge. The on-site installation diagram is shown in Figures 14 and 15.

Figure 14. Photo of the hygrograph site.

Figure 15. Field photo of the anemometer.

Based on the ambient temperature monitoring system and the wind speed and direction monitoring system, the temperature and wind speed and direction of Chongqing Nanjimen Railway Bridge were monitored from 1 January 2023 to 31 December 2023. It was found that the maximum wind speed at the measurement point of the middle span of the main bridge was about 17.3 m/s, the daily maximum temperature was 39.3 °C, and the daily minimum temperature was 4 °C.

5.3. Application of the Derived Equation for the Calculation of the Cable Force with Dampers

5.3.1. Error Analysis of the Formula for Calculating the Cable Force of a Cable-Stayed Cable with a Damper

The direct calculation method, the equivalent cable length method, and the derived formula for the calculation of the force of a cable-stayed cable with a damper were used to calculate the force of a cable-stayed cable (NMC16-NM27) of the Chongqing Nanjimen Railway Bridge with a length of over 200 m, as shown in Figure 16; the detailed parameters of the damper are shown in Table 5; the results were analyzed and verified, and they are shown in Table 6.

Equation (14) of this paper was used to calculate the cable force of the cable. As shown in Table 6 and Figure 16, for the cable with a damper, the maximum error of the cable force calculated with the direct calculation method with respect to the theoretical cable force was 5.5%, and the total cable force was too large. The maximum error of the equivalent cable length method was −7.2%, and the total calculated cable force value was small, which proved that using the equivalent cable length method to calculate the cable force could not effectively solve the influence of the damper on the cable force. The maximum error of the cable force obtained by using the derived formula of the cable with a damper was 3.9%, which was lower than that of the equivalent cable length method and the direct calculation method. It was proved that the cable calculation formula was effective, and the calculation accuracy was higher than that of the direct calculation method and the equivalent cable length method.

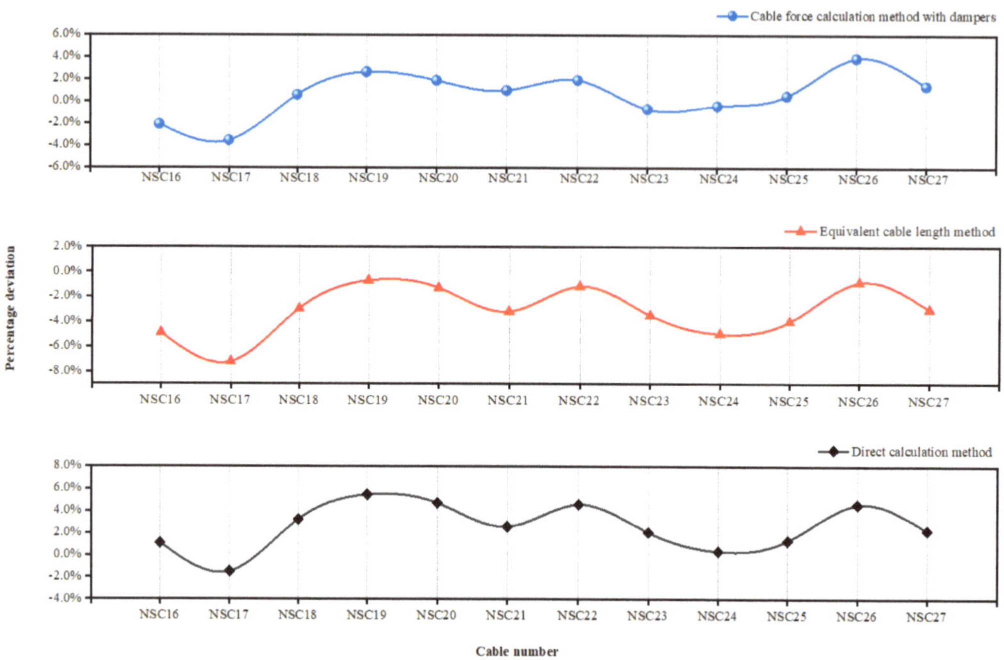

Figure 16. Comparison of the deviation values of the three calculation methods for the cable force.

Table 5. Information on the damper parameters.

Types	Maximum Damping Force (kN)	Damping Index α	Damping Factor C $KN/(m/s)^{0.4}$	Maximum Stroke x (mm)
Linear viscous dampers	2700	0.3	3000	±650

Table 6. Comparison of the calculation results of the three methods.

Number	Designed Cable Force (kN)	Actual Calculated Cable Force (kN)			Error Relative to the Theoretical Force		
		Direct Calculation Method	Equivalent Cable Length Method	Cable Force Calculation Method with Dampers	Direct Calculation Method	Equivalent Cable Length Method	Cable Force Calculation Method with Dampers
NSC16	3106	3281	3086	3178	5.6%	−0.6%	2.3%
NSC17	5075	5217	4915	5109	2.8%	−3.2%	0.7%
NSC18	5052	5366	5046	5230	6.2%	−0.1%	3.5%
NSC19	5006	5360	5047	5218	7.1%	0.8%	4.2%
NSC20	4370	4592	4330	4469	5.1%	−0.9%	2.3%
NSC21	5022	5136	4848	5008	2.3%	−3.5%	−0.3%
NSC22	5103	5285	4995	5152	3.6%	−2.1%	1.0%
NSC23	5124	5175	4897	5037	1.0%	−4.4%	−1.7%
NSC24	5141	5115	4845	5077	−0.5%	−5.8%	−1.3%
NSC25	5019	5058	4795	5018	0.8%	−4.5%	0.0%
NSC26	6406	6677	6333	6636	4.2%	−1.1%	3.6%
NSC27	5336	5454	5177	5411	2.2%	−3.0%	1.4%

5.3.2. Analysis of the Influence of Bending Stiffness on the Calculation

In this study, Equation (14) was used to calculate the cable force of a cable-stayed cable, and the influence of the bending stiffness was not considered in the calculations. In order to reduce the deviation between the calculated cable force and the measured cable force, the bending stiffness of the 12 cables mentioned above was corrected according to Equation (15) in Section 3.4, and the calculations are shown in Table 7.

Table 7. Influence of the flexural stiffness of the cable.

Cable Number	Original Cable Length (m)	S (EI = 0 (kN))	S_0 (EI $\neq$ 0 (kN))	ζ
NSC16	208	3178	3178	1.00
NSC17	218	5109	5109	1.00
NSC18	228	5230	5230	1.00
NSC19	238	5218	5218	1.00
NSC20	248	4469	4469	1.00
NSC21	259	5008	5008	1.00
NSC22	269	5152	5152	1.00
NSC23	279	5037	5037	1.00
NSC24	289	5077	5077	1.00
NSC25	300	5018	5018	1.00
NSC26	310	6636	6636	1.00
NSC27	321	5411	5411	1.00

After the coefficient correction, it was found that when the cable was long, the influence of the bending stiffness on the accuracy of the measured cable force was almost zero and could be ignored. Therefore, when calculating the cable force of a long and thin cable, the influence of flexural stiffness should not be considered.

5.4. Intelligent Cloud Platform Application

To improve the scientific management of bridges and monitor the operational status of the cable-stayed cables in real time, this study established an intelligent cable force monitoring platform using the Chongqing Nanjimen Railway Bridge as a case study. An evaluation and early warning mechanism was established based on Equations (16) and (17), as shown in Figure 17.

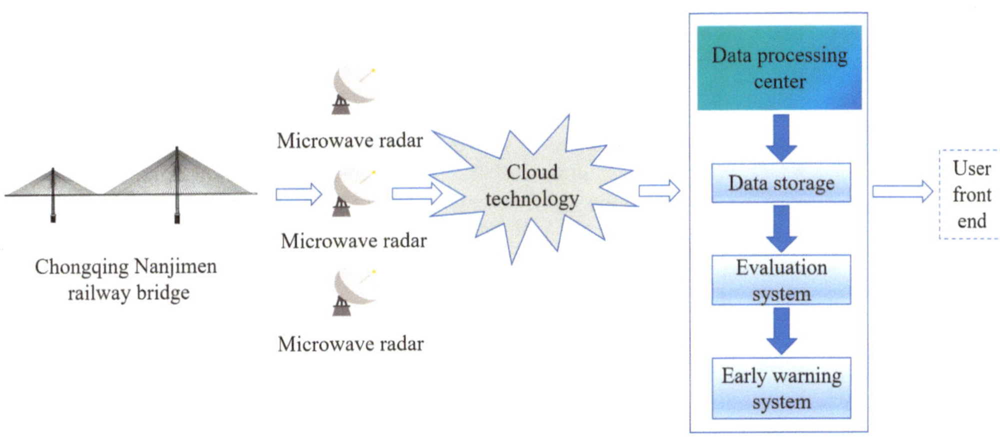

Figure 17. Schematic diagram of the operation of the intelligent cable monitoring platform.

Figure 17 shows that the use of sensors and data acquisition and transmission systems to collect real-time data in the field can greatly improve the collection efficiency and accuracy and can save labor costs. Then, the data were uploaded to the data processing, analysis, and storage system through cloud technology, analysis, and storage, the complex data were refined, and the cable force monitoring value was obtained after calculation and analysis. Finally, the cable force monitoring value was entered into the evaluation and early warning system. After systemic research and judgment, the real-time operation status of the cable could be presented to the user's front end simply and clearly. The intelligent cable force monitoring platform had the functions of fine analysis and efficient data processing, which could effectively improve the level of digitization of bridge monitoring and provide technical support for the scientific management of bridges.

6. Conclusions and Prospects

6.1. Conclusions

Relying on the microwave radar method, a derived formula for calculating the cable force with a damper (length $\geq$ 200 m), and an intelligent cable force monitoring platform for the Chongqing Nanjimen Railway Bridge after a real bridge test, this study obtained the following conclusions:

1. Measuring the force of a cable with a damper based on microwave radar is an efficient and reliable method. The actual engineering experiments showed that the deviation rate of the cable force measurements was only 0.6% for a cable with a damper over 200 m in length. At the same time, the measurement angle could be freely adjusted, and the measurement penetration was stronger than that of the traditional method. It was proved that the measurement accuracy, efficiency, and applicability of microwave radar for cable force measurement were superior to those of traditional monitoring methods. This has wide application prospects in the field of cable force monitoring for slender cables with dampers.
2. Considering the problem of calculating the cable force of cable-stayed cables with dampers, the presented formula is feasible and can effectively improve the accuracy of the calculation of the cable force. Three calculation methods were used to calculate the force of 12 cable-stayed cables with dampers in the mid-span of the north side of the Chongqing Nanjimen Railway Bridge. The maximum cable force error of the calculation formula for the force of a cable-stayed cable with dampers was only 3.9%, which was lower than 7.1% for the direct calculation method and 5.8% for the equivalent cable length method, which proved that the derived calculation formula was reliable, effective, and accurate. It has reference value in the field of cable force calculation for cable-stayed cables with dampers.
3. The intelligent cable monitoring platform integrating monitoring, evaluation, and early warning has the functions of real-time monitoring, real-time analysis, and real-time early warning, and it can provide technical support for the scientific management and maintenance of bridges.
4. However, microwave radar is susceptible to electromagnetic interference in the environment, resulting in measurement errors. In the presence of occlusions or space constraints, the measurement accuracy will be reduced. Therefore, it should not be used in an environment with strong electromagnetic interference, occlusions, or space restrictions. The measurement distance and frequency should be paid attention to in the measurement process to ensure the penetration of the signal and the accuracy of the measurement results.

6.2. Prospects

In this study, an assessment method for cable force measurement based on microwave radar and a calculation method for the cable force with a damper were studied to some extent, but there are still the following deficiencies:

(1) The research results in this study showed the achievement of efficient and high-precision measurement of long cables with dampers, but the influence of the

cable diameter was not taken into account, and the measurement of short cables with dampers was not studied; further research will be carried out in follow-up work.

(2) The calculation formula for a stayed cable with a damper derived in this study fully considers the influence of the damper, and the calculation results were in line with the expectations, but the influence of the damper was ignored when solving the model. The influence of the damper will be further studied in subsequent work.

Author Contributions: Conceptualization, X.L. and Q.W. (Qin Wang); methodology, P.D.; software, P.D.; validation, X.L., Q.W. (Qin Wang) and P.D.; formal analysis, Q.W. (Qin Wang); investigation, Q.W. (Qin Wang) and X.H.; resources, P.D. and Q.W. (Qin Wang); data curation, P.D. and Q.W. (Qin Wang); writing—original draft preparation, X.H.; writing—review and editing, X.L. and Q.W. (Qin Wang); visualization, X.H.; project administration, X.L.; formal analysis, Q.W. (Qiansong Wang). All authors have read and agreed to the published version of the manuscript.

Funding: This work was supported by the Open Fund of Chongqing Key Laboratory of Energy Engineering Mechanics and Disaster Prevention and Mitigation (EEMDPM2021201) and the Chongqing Institute of Science and Technology Research Grant Program (CKRC2021074 and YKJCX2220644).

Data Availability Statement: All of the data and models that were generated or used during the study are available from the corresponding author by request.

Conflicts of Interest: Author Peng Ding was employed by the company T. Y. Lin International Engineering Consulting (China) Co., Ltd. The remaining authors declare that the research was conducted in the absence of any commercial or financial relationships that could be construed as a potential conflict of interest.

References

1. Fang, R.; Wu, Y.; Wei, W.; Na, L.; Biao, Q.; Jiang, P.; Yang, Q. An Improved Static Residual Force Algorithm and Its Application in Cable Damage Identification for Cable-Stayed Bridges. *Appl. Sci.* **2022**, *12*, 2945. [CrossRef]
2. Dai, J.; Qin, F.; Geng, J.; Chen, Y. Review on Cable Force Optimization Method for Cable-staved Bridge in Completed Bridge State. *China J. Highw. Transp.* **2019**, *32*, 17–37.
3. Liu, X.; Yi, X.; Wang, B.; Liu, Y. Condition Assessment of Grouped Cable Forces of Cable-Stayed Bridge Using Deflection Data. *Buildings* **2023**, *13*, 472. [CrossRef]
4. Zhou, Y.; Huang, Z.-H.; Wang, Y.; Xiang, Z.-F.; Zhou, J.-T.; Zhang, X.-S. Measurement Method of Stay Cable Force Based on Measured Cable Shape Using Laser Scanning. *China J. Highw. Transp.* **2021**, *34*, 91–103.
5. Wang, S.-R.; Wang, B.; Cheng, C.-S.; Zhou, J.-T.; Qu, J.-Q.; Tan, H. Research on Influencing Factors of Cable Force Measurement Methods Based on Alignment Recognition. *China J. Highw. Transp.* **2023**, *36*, 154–165.
6. Zhang, L.; Qiu, G.; Chen, Z. Structural health monitoring methods of cables in cable-stayed bridge: A review. *Measurement* **2021**, *168*, 108343. [CrossRef]
7. Zhang, Z.-J.; Wang, R.-H.; Zhen, X.-X.; Liu, X.-H. Evaluation of Measurement Methods for Tension of Parallel Steel Strand Stay Cables. *Bridge Constr.* **2016**, *46*, 42–47.
8. Zhang, R.; Yang, Z.; Zhu, X.; Liang, Q.; Xu, R. Practical Formulas to Calculate Suspender Tension Based on Frequency. *J. Southwest Jiaotong Univ.* **2015**, *50*, 823–829.
9. Li, G.-p.; Li, J.-x. Tension Strategy Derived by Non-stress Length of Parallel Steel Strand Stay Cables. *China J. Highw. Transp.* **2017**, *30*, 48–56.
10. Shen, Q.; Yuan, D.-R.; Chen, C.-X. A New Monitoring Device for the Force States of the Pressure Dispersed Anchor Cable. In Proceedings of the 2015 International Conference on Advanced Engineering Materials and Technology, Guangzhou, China, 22–23 August 2015; pp. 836–841.
11. Liu, X.; Wu, D.; He, C.; Feng, H.; Wu, B. Comparison of AC and pulsed magnetization-based elasto-magnetic methods for tensile force measurement in steel strand. *Measurement* **2018**, *117*, 410–418. [CrossRef]
12. Ji, B.; Cheng, M.; Fu, Z.; Chen, X.; Sun, Y. Influential factors in cable force measurement of cable-stayed bridges based on vibration frequency method. *J. Cent. South Univ. (Sci. Technol.)* **2015**, *46*, 2620–2625.
13. Fang, Z.; Wang, J.-q. Practical Formula for Cable Tension Estimation by Vibration Method. *J. Bridge Eng.* **2012**, *17*, 161–164. [CrossRef]
14. Jo, H.C.; Kim, S.H.; Lee, J.; Sohn, H.-G.; Lim, Y.M. Sag-based cable tension force evaluation of cable-stayed bridges using multiple digital images. *Measurement* **2021**, *186*, 110053. [CrossRef]
15. Liu, X.; Chen, Y.; Hu, H.; Feng, S.; Feng, Z. Measurement method of natural frequencies and tension forces for cables based on elasto-magnetic sensors calibrated by frequencies. *AIP Adv.* **2022**, *12*, 015301. [CrossRef]

16. Liu, H.; Wang, L.; Guo, L.; Zhang, F.; Chen, Z. Intelligent identification of cable force for short and thick building cables based on deep learning. *J. Build. Struct.* **2023**, *44*, 204–213.
17. Wu, W.-H.; Chen, C.-C.; Chen, Y.-C.; Lai, G.; Huang, C.-M. Tension determination for suspenders of arch bridge based on multiple vibration measurements concentrated at one end. *Measurement* **2018**, *123*, 254–269. [CrossRef]
18. Bonopera, M.; Chang, K.-C.; Tullini, N. Vibration of prestressed beams: Experimental and finite-element analysis of post–tensioned thin-walled box-girders. *J. Constr. Steel Res.* **2023**, *205*, 107854. [CrossRef]
19. He, X.-J.; Yang, Y.-C.; Xiao, X.; Wang, J.-J. A vibration frequency method based on finite element equivalent cable length. *Bridge Constr.* **2016**, *46*, 40–44.
20. Tang, S.-H.; Fang, Z. Equivalent cable length calculation method considering influence of intermediate supports. *J. Vib. Shock* **2013**, *32*, 82–87.
21. Wang, X.; Wang, X.; Liang, X.; Zhou, M.; Du, Y. Analysis of suspension cable force of large-span steel tube concrete tied arch bridge. *J. Vib. Shock* **2023**, *42*, 306–313.
22. Gong, J. Design and research of long-span concrete cable-stayed bridge for high-speed railway. *J. Railw. Sci. Eng.* **2020**, *17*, 1611–1619.
23. Sohaib, M. Study on the Influence of Dampers on Cable-Stayed Bridge Cable Force Estimation. Master's Thesis, Harbin Institute of Technology, Harbin, China, 2019.
24. Yang, X. Research on Calculation Method and Influencing Factors of Cable Force of Asymmetric Cable-Stayed Bridge. Master's Thesis, Lanzhou Jiaotong University, Lanzhou, China, 2020.
25. Liang, X.; Hui, D.; Zhou, J.; Zhong, J.; Yang, J.; Xiao, Y. Construction and application of health monitoring and evaluation system for long-span cable-stayed Bridges. *J. China Foreign Highw.* **2016**, *36*, 92–97.
26. *T/CCTAS 68-2023*; Technical Specification for the Detection of Bridge Structure Deformation by millimeter Wave Radar (T/CCTAS 68-2023) Group Standard. China Transportation Association: Beijing, China, 2023.
27. Wang, J.; Wang, X.; Zhou, Z. Nano millimeter wave radar for bridge cable tension measurement. *J. Natl. Univ. Def. Technol.* **2022**, *44*, 118–122.
28. Wang, T.; Zhang, J.; Jiang, S.; Liu, W.; Chen, Y. Time-varying cable force identification method based on radar vibration measurement technique. *J. Vib. Shock* **2023**, *42*, 205–212.
29. Liu, Y.; Xie, J.-Z.; Tafsirojjaman, T.; Yue, Q.-R.; Tan, C.; Che, G.-J. CFRP lamella stay-cable and its force measurement based on microwave radar. *Case Stud. Constr. Mater.* **2022**, *16*, e00824. [CrossRef]
30. Xu, W.; Du, D.; Kasai, K.; Wang, S. Optimization layout of viscous dampers and its application in retrofitting of a high-rise steel structure. *J. Build. Struct.* **2022**, *43*, 32–44.
31. Duan, Y.-F.; Huang, J.-S.; Deng, N.; Wang, S.-M.; Ying, Z.-G.; He, W. Vector form intrinsic finite element based simulation on parametric vibration of cables. *J. Vib. Eng.* **2023**, *36*, 188–195.
32. Main, J.A.; Jones, N.P. Vibration of tensioned beams with intermediate damper. II: Damper near a support. *J. Eng. Mech.* **2007**, *133*, 379–388. [CrossRef]
33. Jiang, L.L.; Chen, Q.Z. Research on calculation formula of frequency faso force test. *Highway* **2018**, *63*, 164–166.
34. Dan, D.; Zhao, Y.; Yang, T. Vibration measurement based cable trnsion identifcation method for cable-damper system. *J. Vib. Shock* **2013**, *32*, 123–127+204.
35. Huang, W.; Wu, Q.; Chen, B. Force measurement of stay cable considering damper influence. *J. Guangxi Univ. (Nat. Sci. Ed.)* **2018**, *43*, 416–424.
36. Fang, Z.; Zhang, Z.Y. Test of Cable Tension in Cable-stayed Bridges. *China J. Highw. Transp.* **1997**, *1*, 51–58.
37. *JGJ297-2013*; Technical Specification for Seismic Energy Dissipation of Buildings. Ministry of Housing and Urban-Rural Development of the People's Republic of China: Beijing, China, 2013.
38. *JTG/T3650-01-2022*; Technical Regulations for Highway Bridge Construction Monitoring. Ministry of Transport of the People's Republic of China: Beijing, China, 2022.
39. Teng, G.; Cao, X.; Gai, W.; Jiang, R. An Experimental Study on Intelligent Bearing System for Bridge Health Monitoring. In Proceedings of the International Conference on Innovative Solutions in Hydropower Engineering and Civil Engineering, Berlin, Germany, 25–27 December 2022; pp. 319–324.
40. He, Z.; Li, W.; Salehi, H.; Zhang, H.; Zhou, H.; Jiao, P. Integrated structural health monitoring in bridge engineering. *Autom. Constr.* **2022**, *136*, 104168. [CrossRef]
41. Li, X.; Luo, H.; Ding, P.; Chen, X.; Tan, S. Prediction Study on the Alignment of a Steel-Concrete Composite Beam Track Cable-Stayed Bridge. *Buildings* **2023**, *13*, 882. [CrossRef]
42. Sun, L.-M.; Shang, Z.-Q.; Xia, Y. Development and Prospect of Bridge Structural Health Monitoring in the Context of Big Data. *China J. Highw. Transp.* **2019**, *32*, 1–20.
43. Zhou, Y.; Sun, L. Insights into temperature effects on structural deformation of a cable-stayed bridge based on structural health monitoring. *Struct. Health Monit.* **2018**, *18*, 778–791. [CrossRef]

44. Tong, K.; Zhang, H.; Zhao, R.; Zhou, J.; Ying, H. Investigation of SMFL monitoring technique for evaluating the load-bearing capacity of RC bridges. *Eng. Struct.* **2023**, *293*, 116667. [CrossRef]
45. *GB/T 39559.2-2020*; Specifications for Operational Monitoring of Urban Rail Transit Facilities—Part 2: Bridge. Ministry of Transport of the People's Republic of China: Beijing, China, 2021.

buildings

MDPI

Article

Research on the Method of Absolute Stress Measurement for Steel Structures via Laser Ultrasonic

Hongsong Tian [1], Yujiang Kong [1], Bin Liu [1], Bin Ouyang [1], Zhenfeng He [2,3,*] and Leng Liao [2,3]

1 Guizhou Bridge Construction Group Co., Ltd., Guiyang 550001, China; thongsong96@163.com (H.T.); kyj2969@163.com (Y.K.); liubin-xinguang@126.com (B.L.); gqzgb@gzql.cn (B.O.)
2 State Key Laboratory of Mountain Bridges and Tunnel Engineering, Chongqing Jiaotong University, Chongqing 400074, China
3 School of Civil Engineering, Chongqing Jiaotong University, Chongqing 400074, China
* Correspondence: 631501010724@mails.cqjtu.edu.cn

Abstract: Accurate measurement of the stress in steel structures is crucial for structural health monitoring. To achieve this goal, a novel technique, the laser ultrasonic technique, was used in absolute stress measurement in this study. The feasibility of this technique has been verified through theoretical analysis and finite element (FE) analysis. A stress measurement experiment in steel specimens was conducted and the relationship between ultrasonic relative wave velocity and stress was explored. The results revealed that there is a similar linear correlation between the ultrasonic relative wave velocity and absolute stress. The stress can be obtained based on ultrasonic relative wave velocity. According to the stress measurement results, it was found that the absolute error between the measured stress and theoretical stress was largest when the stress level was low, and that the measured error of stress gradually decreased with increases in the applied stress. The relative error between the measured stress and the theoretical stress was within 10% when the stress was higher than 100 MPa. This further verifies the reliability of the laser ultrasonic technique under high-stress conditions. Additionally, the impact of temperature and surface roughness on stress measurement was analyzed. The stress error in stress measurement increased similarly linearly with the increase in temperature and increased non-linearly with the increase in roughness. The corresponding compensation methods were proposed to effectively improve the accuracy of measurement.

Keywords: absolute stress measurement; steel structure; laser ultrasonic; influence factors

Citation: Tian, H.; Kong, Y.; Liu, B.; Ouyang, B.; He, Z.; Liao, L. Research on the Method of Absolute Stress Measurement for Steel Structures via Laser Ultrasonic. *Buildings* **2024**, *14*, 602. https://doi.org/10.3390/buildings14030602

Academic Editor: Harry Far

Received: 29 January 2024
Revised: 21 February 2024
Accepted: 21 February 2024
Published: 23 February 2024

1. Introduction

Steel structures are renowned for their robustness and superior seismic capabilities, making them a popular choice in constructing large-span bridges; however, they are vulnerable to significant damage from loads such as earthquakes, leading to potential safety risks. To maintain their operational integrity and prevent accidents, it is crucial to regularly assess the stress state of these bridges, as this directly reflects their structural health [1–4]. Traditional stress measurement techniques fall into two categories: destructive and non-destructive testing. Destructive testing, which can harm the integrity of the bridge, is not ideal for regular checks [5,6]. Consequently, the non-destructive testing/dynamic monitoring approaches of stress in steel structures, including ultrasonic and magnetic flux leakage testing, along with the corresponding artificial intelligence algorithms, have become the primary focus of research [7–10]. Ultrasonic testing, in particular, has widespread application, but its effectiveness is hampered by the need for direct contact and couplants, which can affect measurement sensitivity and accuracy [11,12]. In addition, the size of the transducers further limits the precision and resolution of this method. An innovative solution to these challenges is the laser ultrasonic technique. This technique employs lasers to generate ultrasonic waves without physical contact, bypassing the limitations associated with couplants and transducers. Suitable for use in high-temperature or corrosive settings, the laser ultrasonic technique offers enhanced precision and spatial resolution, positioning it as a superior alternative in assessing the health of steel bridges.

White [13] et al. demonstrated the feasibility of exciting ultrasonic waves in solids using pulsed lasers, by constructing a one-dimensional theoretical model of the response of a material to laser excitation. Subsequent studies on laser ultrasonic technology used in stress measurement have achieved a series of developments. Zhan et al. [14–16] determined the acoustoelastic coefficients for TC4 titanium alloy and 7075 aluminum alloy specimens using a prestress loading method, and calculated the transverse and longitudinal residual stresses of the test pieces using the acoustoelastic equation, discussing the impact of welding parameters on residual stresses. Xue et al. [17] conducted experiments on stress detection in aluminum alloy under different tensile loads using laser ultrasonics. The experimental results were similar to those obtained using the X-ray method, demonstrating the high precision and accuracy of the laser ultrasonic detection method in measuring stress. Ye et al. [18] proposed a method for measuring residual stress on the surface/subsurface of welded components. Pan et al. [19] explored the significant quantitative relationship between laser ultrasonic velocity and residual stress in high-temperature alloys and verified the method using X-ray detection results, realizing the quantitative detection of alloy stress. Ji et al. [20,21] found a positive correlation between the relative velocity of laser ultrasonic guided waves and tension, and a negative correlation with group velocity, through experiments. Qian et al. [22] conducted a detection on the variation of residual stress on the shot-peened surface of the K417G turbine disk, using laser ultrasonic technology and X-ray methods. The results showed that the changes detected by both methods were consistent, verifying that laser ultrasonic technology can effectively identify the trend of residual stress changes. The research of He et al. [23] indicated that ultrasonic surface waves propagating in the direction of the principal stress are most reasonable for assessing the residual principal stress in 45# steel, and that there is a significant negative correlation between the peak amplitude of the ultrasonic wave and the stress. Santhakumar et al. [24] proposed a theoretical calculation model for the distribution of residual stress based on the laser ultrasonic detection method, achieving accurate detection of the residual stress distribution in titanium alloys. These studies have shown the great potential of laser ultrasonic detection technology in stress measurement. In summary, laser ultrasonic detection technology is primarily used to characterize residual stress in metals, but there are few studies on the detection of absolute stress in steel structures. Therefore, in this work, a method for detecting absolute stress in steel structures based on laser ultrasonics is proposed, drawing on the application of the method in the detection of metal residual stress. The laser ultrasonic wave speed variation characteristics of loading steel structures were explored based on theory, simulation, and experiments. On this basis, the effective detection of absolute stress in steel structures based on laser ultrasonics was achieved. The workflow is shown in Figure 1.

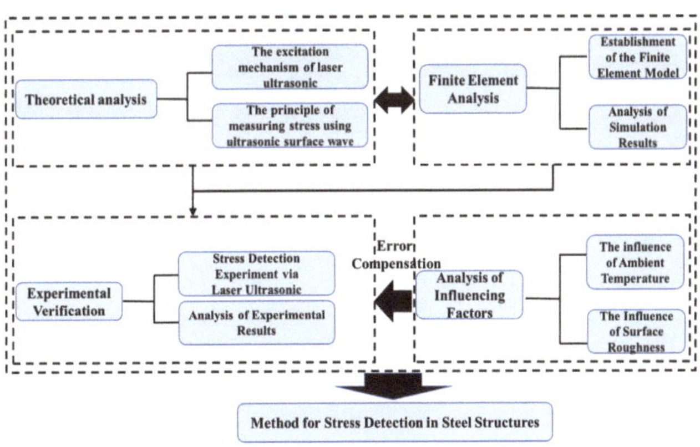

Figure 1. Schematic diagram of the workflow.

2. Theoretical Background

2.1. The Excitation Mechanism of Laser Ultrasonics

When the surface of a solid material is irradiated by a pulsed laser, the absorbed laser energy is converted into thermal energy. A portion of the energy is absorbed by the material and propagates from the surface into the interior, converting into thermal energy. This process rapidly increases the surface temperature of the material through heat conduction, forming a transient, uneven temperature field. Thus, the area of the material exposed to the laser radiation undergoes thermal expansion, which in turn generates thermal stress. The thermal stress propagates to the surrounding medium in the form of a transient pulse so that the ultrasonic waves are excited on the surface of the solid material. The principle of the excitation with laser ultrasonics is shown in Figure 2.

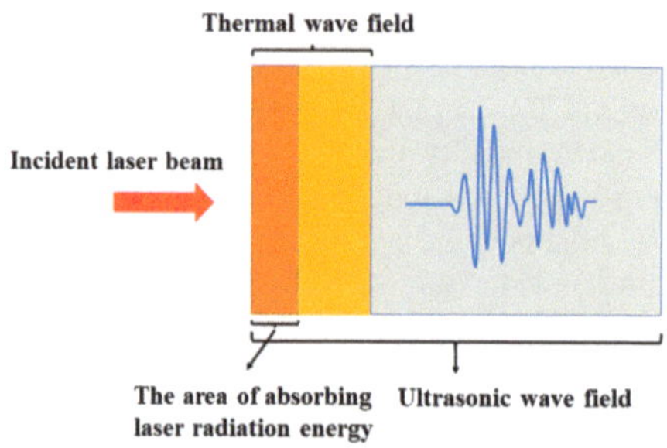

Figure 2. The principle of the excitation with laser ultrasonics.

The theory of laser ultrasonics is based on the thermoelastic mechanism. If the incident laser energy increases the temperature of the material without causing a phase change, this results in thermal expansion and, consequently, the generation of thermoelastic stress. This stress excites ultrasonic waves at the surface or within the material. The thermoelastic stress and strain induced by temperature changes can be solved using the principles of thermoelasticity. There are three fundamental equations in thermoelasticity, which are as follows [25]:

Equation of motion:

$$\sigma_{ij,j} + f_i = \rho \frac{\partial u_i^2}{\partial t} \tag{1}$$

Strain displacement equation:

$$\varepsilon_{ij} = \frac{1}{2}\left(u_{i,j} + u_{j,i}\right) \tag{2}$$

Constitutive equation for isotropic materials:

$$\sigma_{ij} = \lambda\delta_{ij}\varepsilon_{kk} + 2\mu\varepsilon_{ij} - \alpha(3\lambda + 2\mu)\delta_{ij}T \tag{3}$$

where $i,j = 1,2,3$, u_i represents the displacement component, ε_{ij} represents the component of the strain tensor, σ_{ij} represents the component of the stress tensor, ρ is the density of materials, f_i is the external force applied, T represents the increase in temperature, λ and μ are the Lamé constants, δ_{ij} is the unit tensor, α is the coefficient of thermal expansion.

In an isotropic solid, by combining the motion equation shown in Equation (1), the strain-displacement equation shown in Equation (2), and the constitutive equation shown in Equation (3), the Navier–Stokes equation can be derived as follows:

$$\mu U_{i,kk} + (\lambda + \mu)U_{k,ki} - \alpha(3\lambda + 2\mu)T_i + f_i = \rho\ddot{U}_i \tag{4}$$

For isotropic materials, the thermoelastic coupling equations can be simplified when the external forces and internal heat sources in the material are neglected:

$$\mu\nabla^2 U + (\lambda+\mu)\nabla\nabla\cdot U - \alpha(3\lambda+2\mu)\nabla T = \rho\ddot{U} \tag{5}$$

where U represents the displacement vector and T denotes the increase in temperature. Considering the scenario where a pulsed laser with a Gaussian distribution irradiates the surface of an isotropic solid, the boundary conditions on the upper and lower surfaces of the material are free in the normal direction, as shown in Equation (6):

$$n\cdot[\sigma - \alpha(3\lambda+2\mu)T(r,z,t)I] = 0 \tag{6}$$

where I represents the identity tensor, σ denotes the stress tensor, and n is the unit vector perpendicular to the surface of the material. Additionally, the following initial conditions should be satisfied:

$$U(r,z,t)|_{t=0} = \left.\frac{\partial U(r,z,t)}{\partial t}\right|_{t=0} = 0 \tag{7}$$

Finally, the ultrasonic wave displacement field at each point within the material can be obtained by simultaneously solving Equation (5) through Equation (7).

2.2. The Principle of Measuring Stress Using Ultrasonic Surface Wave

Due to the nature of the ultrasonic surface waves propagating solely along the surface of the material, their amplitude is minimally affected by the propagation distance, resulting in a high signal-to-noise ratio. The ultrasonic surface waves are highly effective in measuring absolute stress by using this characteristic. For isotropic solids, Hughes and Kelly measured the third-order elastic constants of an object based on the nonlinear elasticity theory proposed by Murnaghan [26], and the relationship between ultrasonic wave velocity and stress in isotropic materials under stress was formulated.

For an isotropic solid under stress, the principal stress directions can be represented using a coordinate system, as shown in Figure 3. The relationship between the velocity of different modes of ultrasonic waves and stress is given by Equation (8) [27].

$$\begin{cases} \rho_0 V_{11}^2 = \lambda+2\mu+\frac{\sigma_{11}}{3\lambda+2\mu}\left[\frac{\lambda+\mu}{\mu}(4m+4\lambda+10\mu)+\lambda+2l\right] \\ \rho_0 V_{12}^2 = \rho_0 V_{13}^2 = \mu+\frac{\sigma_{11}}{3\lambda+2\mu}\left[m+\frac{\lambda n}{4\mu}+4\lambda+4\mu\right] \\ \rho_0 V_{22}^2 = \rho_0 V_{33}^2 = \lambda+2\mu+\frac{\sigma_{11}}{3\lambda+2\mu}\left[2l-\frac{2\lambda}{\mu}(m+\lambda+2\mu)\right] \\ \rho_0 V_{21}^2 = \rho_0 V_{31}^2 = \mu+\frac{\sigma_{11}}{3\lambda+2\mu}\left[m+\frac{\lambda n}{4\mu}+\lambda+2\mu\right] \\ \rho_0 V_{23}^2 = \rho_0 V_{32}^2 = \mu+\frac{\sigma_{11}}{3\lambda+2\mu}\left[m-\frac{\lambda n}{2\mu}-\frac{n}{2}-2\lambda\right] \end{cases} \tag{8}$$

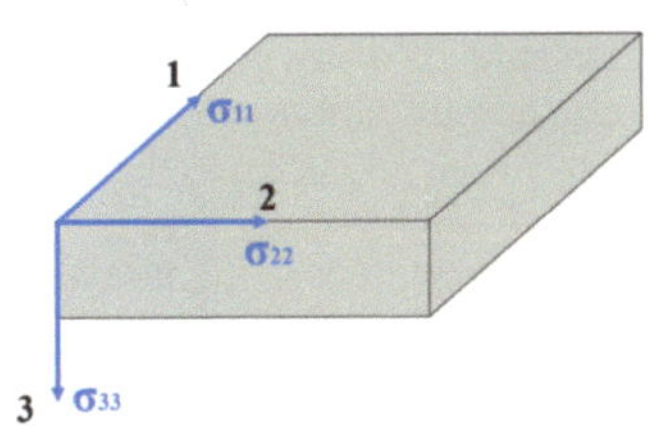

Figure 3. The direction of principal stress.

In this equation, ρ_0 represents the initial density of the material in the stress-free state and V_{ij} denotes the velocity of the ultrasonic wave (where i,j = 1,2,3; the direction of wave propagation is indicated by the first subscript i, and the polarization direction of the wave is indicated by the second subscript j. When $i = j$, V_{ij} is the longitudinal wave velocity V_L. When $i \neq j$, V_{ij} is the shear wave velocity V_S). λ and μ are the Lamé constants. l, m, and n

are the Murnaghan constants. σ_{11} is the normal stress in direction 1. Similarly, σ_{22} and σ_{33} are the stresses in directions 2 and 3, respectively.

When the material is in a plane stress state, it is assumed that σ_{33} = 0, and σ_{11} and σ_{22} are not zero. This condition reflects a scenario where stress is applied in two dimensions (directions 1 and 2), with no stress in the third dimension. For shear waves, the change in wave velocity caused by stress is relatively small. Therefore, it can be assumed that [28]:

$$V_{12} \cong V_{13} \cong V_{S0} \text{ but } V_{12} - V_{13} \neq 0 \tag{9}$$

where V_{S0} is the initial wave velocity of transverse waves without initial stress. According to Equation (9), Equation (8) can be simplified into two equations that are used to represent the relationship between transverse and longitudinal waves and stress.

$$\frac{(V_{12} - V_{13})}{V_{S0}} = \frac{n + 4\mu}{4\mu^2}(\sigma_{22} - \sigma_{11}) = K_S(\sigma_{22} - \sigma_{11}) \tag{10}$$

$$\frac{V_{11} - V_{L0}}{V_{L0}} = \frac{\mu l - \lambda(m + \lambda + 2\mu)}{\mu(3\lambda + 2\mu)(\lambda + 2\mu)}(\sigma_{22} + \sigma_{11}) = K_L(\sigma_{22} + \sigma_{11}) \tag{11}$$

where K_S is the acoustic elastic coefficient of the transverse wave, and K_L is the acoustic elastic coefficient of the longitudinal wave.

The propagation of ultrasonic surface waves can be considered a composite of longitudinal and transverse motions. The vibration displacement of the particle in these waves aligns with the propagation characteristics of the transverse waves. Therefore, the acoustoelastic coefficients of ultrasonic surface waves can be equated to those of the transverse waves. Under uniaxial stress, the velocity of the surface waves parallel to the stress direction is more sensitive to stress changes. The stress in the perpendicular direction can be considered negligible. The values of σ_{22} = σ and σ_{11} = 0 can be set, and the relationship between the Rayleigh wave velocity and stress is then expressed as shown in Equation (12) [29].

$$\frac{V_R - V_{R0}}{V_{R0}} = K_R\sigma = K_S\sigma = \frac{n + 4\mu}{8\mu^2}\sigma \tag{12}$$

where V_{R0} represents the velocity of the ultrasonic surface waves under the stress-free state, V_R is the velocity of ultrasonic surface waves when stress is present, and K_R is the acoustoelastic coefficient for the ultrasonic surface waves. The absolute stress in the steel structure under the current conditions can be determined using Equation (12) according to the acoustoelastic coefficient and the velocities of the ultrasonic surface waves in both the stress-free and stressed states, which are measured through experimentation.

3. Finite Element Modeling

The FE analysis of laser ultrasonics is a complex problem involving thermo-mechanical coupling. Based on the fundamental theory of laser ultrasonics, the laser energy absorbed by the material creates a transient, non-uniform temperature field through thermal conduction effects around the area of laser interaction. Local thermal expansion under the action of a non-uniform temperature field will produce thermal stress which, in turn, excites the ultrasonic waves. The heat conduction is treated as the essential mechanism for laser-induced ultrasonic excitation in the FE model. Based on the principles of heat conduction and thermoelastic coupling, the process of stress measurement through laser-induced ultrasonic can be simulated using the software, COMSOL Multiphysics version 6.1.

3.1. Setting of Model Parameters

In the FE model, the material was modeled as isotropic Q235 steel, characterized by dimensions of 30 mm in length and 10 mm in width. The thermal and mechanical properties pertinent to this steel type are systematically outlined in Table 1. To streamline the computational process, the process of laser excitation of ultrasound on the surface of a solid was equivalent to a two-dimensional plane strain model. This approximation facilitated a more manageable analysis while capturing the essential aspects of the laser-material interaction. In the model, the laser pulse was directed to impinge upon the central area of the surface of the material. The configuration and details of this FE model are depicted in Figure 4.

Table 1. Physical parameters of Q235.

Density/(kg/m^3)	Elastic Modulus/Pa	Poisson's Ratio	Thermal Expansion Coefficient/(1/K)	Thermal Conductivity/(W/(m·K))	Heat Capacity at Constant Pressure/(J/(kg·K))
7850	2.06 × 1011	0.25	0.000012	52.34	502

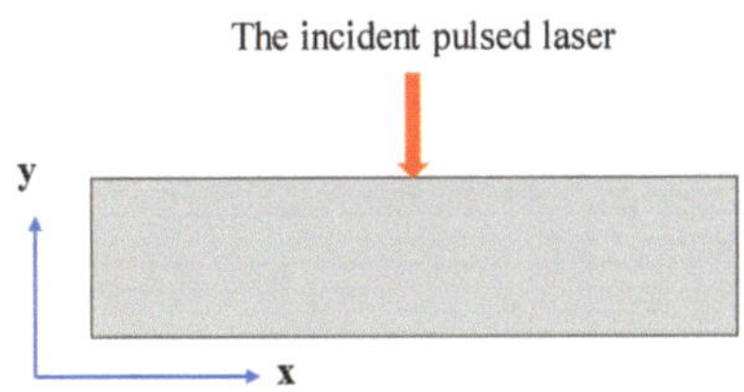

Figure 4. FE model.

The incident laser was executed as a boundary load in the FE model. This load adheres to a Gaussian distribution, integral to the formulation of the laser ultrasonic thermo-mechanical coupling model. The thermal source distribution function of the laser with Gaussian distribution in time and space is as follows [30]:

$$Q(I_0, r, z, t) = I_0 A(T) f(r) g(t) \tag{13}$$

In the equation, I_0 represents the energy density of the incident pulsed laser, measured in watts per square meter (W/m^2), and $A(T)$ denotes the absorptivity of the surface of the material to the laser. For simplification in calculations, it was assumed that the pulsed laser energy was entirely absorbed by the material, hence $A(T) = 1$. $f(x)$ is the Gaussian function describing the spatial distribution of the laser, while $g(t)$ is the Gaussian function that characterizes the temporal distribution of the laser.

In the FE model depicted in Figure 5, the pulsed laser was effectively represented by a Gaussian thermal source. The upper boundary of the model incorporated a heat flux equation that corresponded to this defined thermal source. For the analysis of the temperature field, the baseline temperature of the model was established at ambient room temperature. The boundaries of the metal material that were not subjected to laser irradiation, specifically the lower and lateral surfaces, were characterized as adiabatic, preventing heat transfer through these surfaces.

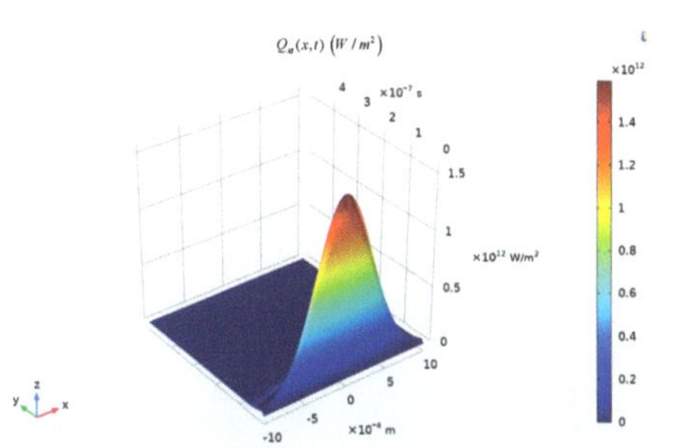

Figure 5. The Gaussian heat source.

The initial displacement and initial stress of the model were both set to zero. The upper surface of the model, directly exposed to the laser, was designated with a free boundary condition. To mitigate the influence of the reflected waves, the model incorporated low-reflection boundary conditions on the three surfaces not directly exposed to the laser irradiation. External stress or pre-stress and strain were applied to the linear elastic material within the model, serving as the external load of the model. Finally, the multi physics coupling of solid heat transfer and solid mechanics was carried out to obtain the acoustic parameters of the laser-induced ultrasound under different stresses.

Transient research was used in this study. For the solution of the transient models, their accuracy was limited by the resolution of the grid in space for waves and the resolution of the time step for temporal changes [31]. Depending on the actual needs of the model, there were different standards for dividing the grid size. In this study, the grid size L was controlled to be less than one-quarter of the ultrasonic wavelength λ [32].

Based on the relationship between ultrasonic wave speed, wavelength, and frequency, the wavelength can be determined through the ultrasonic speed and frequency. After expressing the pulsed laser using a Gaussian function, the maximum frequency of the ultrasonic wave excited by the laser was represented by a specific formula: Equation (14) [33]. By substituting $C = \lambda f$ into Equation (14), the minimum wavelength of the surface acoustic wave excited by the laser was obtained.

$$f_{\max} = \frac{\sqrt{2} C_R}{\pi r_0} \tag{14}$$

$$\lambda_{\min} = \frac{\pi r_0}{\sqrt{2}} \tag{15}$$

In the equation, r_0 represents the radius of the laser spot and C_R is the wave speed of the surface acoustic wave, which can be calculated using the material parameters according to Equation (16) [34]:

$$C_R = \frac{0.87 + 1.12\nu}{1 + \nu} \sqrt{\frac{\hat{E}}{2\rho(1 + \nu)}} \tag{16}$$

where $\hat{E}$ represents the elastic modulus of the material, ρ is the density of the material, and ν is the Poisson's ratio of the material. In the FE model, the radius of the laser spot, r_0 was set to 0.5 mm. It was found that the minimum wavelength of the surface wave excited by the laser on the material surface was approximately 1.11 mm, using the calculation. To ensure the accuracy and reliability of the FE calculation, the grid width needed to be less than 0.28 mm. The mesh types and quantities are shown in Table 2.

Table 2. The sizes and quantity of mesh.

Finite Element Type	Quantity in Model	Mesh Size(mm)
Quad4	18,750	0.25
Tri3	9979	0.25

In the FE analysis of the laser-generated ultrasound, the time resolution needed to be sufficiently fine to ensure the reliability of the FE calculation results. Therefore, the time step, Δt should be minimized [35].

$$\Delta t = \frac{1}{180 f_{\max}} \tag{17}$$

Based on Equations (14), (16) and (17), the time step Δt can be calculated as approximately 2.67 ns. Considering the time resolution and the complexity of the computation, the time step Δt was set to 1 ns in the FE analysis and the total time duration was set to 10,000 ns.

3.2. Analysis of Finite Element Results

Upon the incidence of the pulsed laser on the material surface, a transient temperature field was generated, subsequently exciting the production of the ultrasonic waves. First, the temperature field of the model in a stress-free state was analyzed. The temperature distribution across the entire field is shown in Figure 6, which represents the temperature field when the temperature at the laser irradiation center reached its highest, at $t = 70$ ns. In the figure, it can be observed that the pulse laser irradiation to the material surface caused a temperature rise only in a very small area around the laser irradiation center; the temperature of the rest of the material remained unchanged. In Figure 6, it is shown that the transient temperature field produced after laser irradiation was limited in scope, with instantaneous temperature rises occurring only within a very small area near the irradiation center; this area was only in the micron (μm) scale. Beyond this range, the change in the material surface temperature was minimal or almost non-existent. It was precisely the temperature difference between the transient temperature field and the steady temperature field that caused thermal expansion on the material surface, thereby forming the ultrasonic waves. In addition, it should be noted that the characteristics of materials under high stress change as the stress level increases, which has a certain impact on the generation of ultrasonic waves. Figure 7 shows the distribution of the full-field waveform at a particular moment. From the figure, it is evident that the incidence of the pulsed laser on the material surface excites longitudinal, transverse, and surface waves, each propagating at their respective ultrasonic wave velocities. In this study, the surface waves for stress detection were employed and the acoustic characteristics of the surface waves were analyzed.

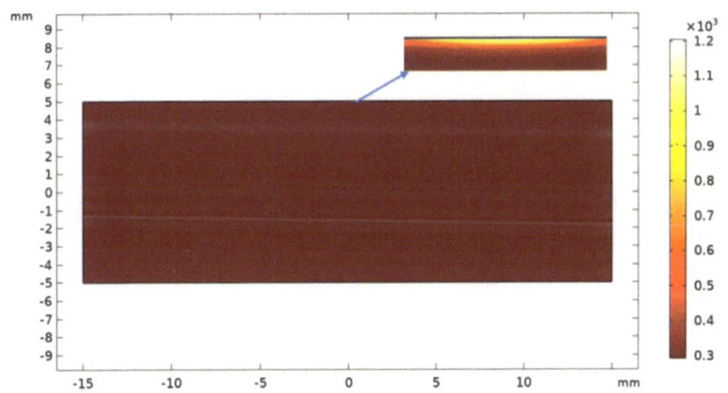

Figure 6. The temperature distribution of the whole field when t is 70 ns.

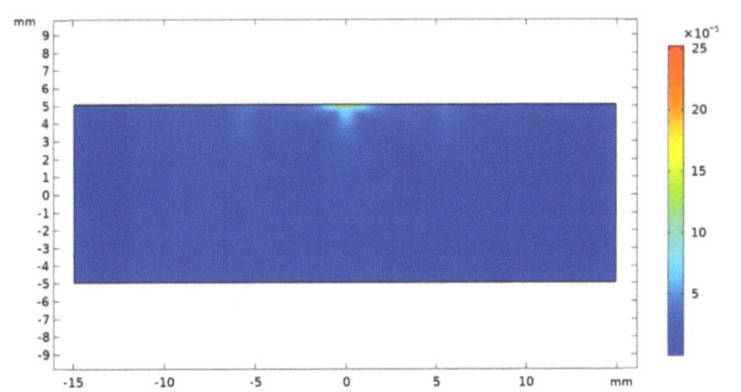

Figure 7. The waveform distribution of the entire field.

In the FE model, the signal reception points were designated at locations 2 mm, 3 mm, 4 mm, and 5 mm to the right of the laser irradiation center on the upper surface. The time-domain waveforms of the ultrasonic waves received at these points are shown in Figure 8. As observed in Figure 8, the pulsed laser incident on the material surface generates ultrasonic waves, which propagate outward in all directions at the time $t = 0$. There were similar variation characteristics for waveforms of signals at the different reception points. As the propagation distance of the ultrasonic waves increased, the amplitude of the time-domain waveforms gradually attenuated, and the wave packet widened. There were delays for the time-domain signals of the ultrasonic surface waves. Additionally, it was indicated that the flight time of the ultrasonic waves also increased progressively, consistent with the propagation characteristics of the surface waves. When the propagation distances of the ultrasonic waves were 2 mm, 3 mm, 4 mm, and 5 mm, the corresponding flight times of the ultrasonic waves were 0.7065 μs, 1.0300 μs, 1.4295 μs, and 1.8295 μs, respectively. The velocity of the ultrasonic waves was determined by linearly fitting the different propagation distances and corresponding flight times at the four signal reception points. The calculated velocity of the ultrasonic waves was 2647.06 m/s, which is close to the general propagation speed of ultrasonic waves in metals. Through the analysis of the propagation characteristics and velocity of the ultrasonic waveforms, this study identified the ultrasonic waves extracted from the upper surface of the model, at a distance to the right of the laser irradiation center, as ultrasonic surface waves. These were used in the subsequent analysis of the correlation between wave speed and stress.

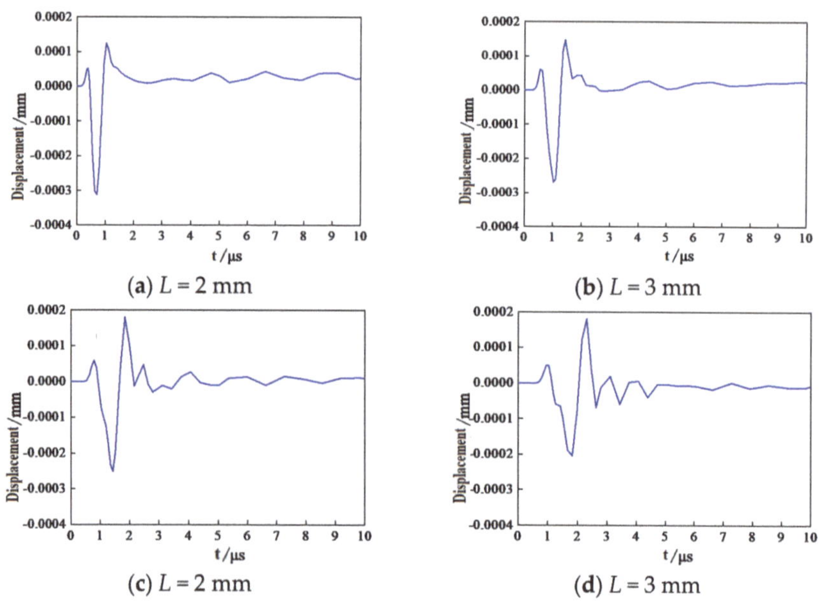

Figure 8. Ultrasonic time-domain signal at different locations.

The model was subjected to stress levels ranging from 20 MPa to 200 MPa, and the time-domain signals of the ultrasonic waves on the surface of the material under different stress levels were measured. To enhance the time resolution, the interpolation method from the multi-rate sampling technique was employed. Based on the distance between each reception point and the laser excitation point, as well as the flight time of the ultrasonic waves, the average ultrasonic wave velocity under various stress levels was determined using the method of differences. The relative wave velocity was defined as $(V - V_0)/V_0$, which can be represented as $\Delta V/V_0$. The relationship between the relative velocity of ultrasonic waves and stress is shown in Figure 9. The observation of the variation trend of relative wave velocity with stress in Figure 9 revealed that the relative wave velocity gradually decreased with the increase of stress. This trend was consistent with the theory mentioned in the literature [36], which states that stress causes a reduction in wave velocity. Furthermore, there exists an approximately linear relationship between relative wave velocity and stress, aligning with the predictions of acoustoelastic theory. Therefore, the ultrasonic surface waves generated by pulsed laser excitation can be effectively used to measure the absolute stress on metal surfaces.

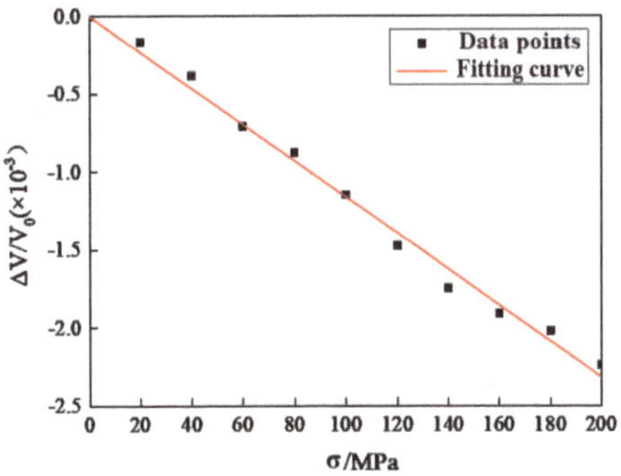

Figure 9. Relationship between the relative wave velocity of ultrasound and stress.

4. Experimental Details

4.1. Specimens Preparation

To align with the practical engineering scenarios, Q235 steel, which is a material commonly used in engineering projects, was selected as the test material in this experiment. The dimensions of the test specimens were 60 mm × 60 mm × 60 mm, and there were a total of nine specimens, as shown in Figure 10. Since the experiment involved the uniaxial compression of cubic Q235 steel specimens, and the laser excitation point and ultrasonic signal reception point were located on the same side of the specimen, it was necessary to ascertain whether the stress at the surface measurement points of the specimen was equivalent to the axially applied stress. A three-dimensional cubic model, identical in size to the test specimens, was established using the software. This model simulated the uniaxial compression process based on the loading method of the pressure testing machine. Constraints were added to the upper surface of the specimen. The stress variation on one side of the specimen is depicted in Figure 11, when a stress of 100 MPa was applied to the bottom surface. As observed in Figure 11, the stress increased near the constraints when the top surface of the specimen was constrained and the bottom surface was subjected to stress loading. However, the rest of the side surface of the specimen still experienced the magnitude of the uniaxial stress. Therefore, this experiment ensured consistency between the applied stress and the actual measured stress.

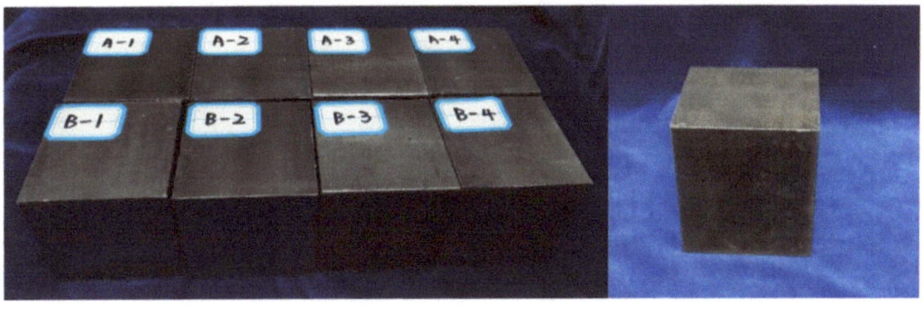

Figure 10. Q235 steel specimens.

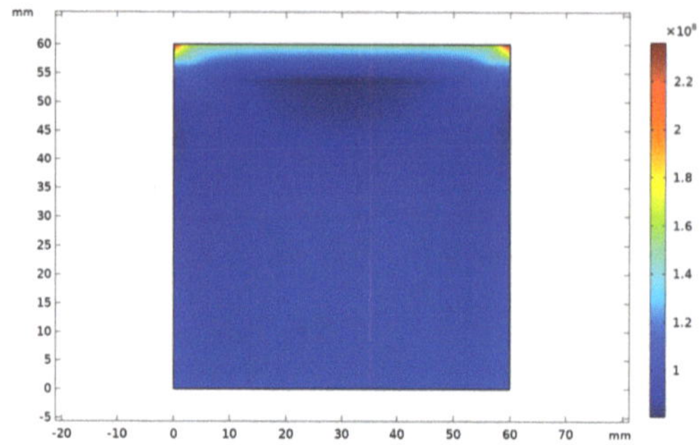

Figure 11. The distribution of stress on the side surface under uniaxial stress.

4.2. Experiment Device and Detection Scheme

To facilitate the laser ultrasonic detection of stress in steel structures, an experimental laser ultrasonic system was established, as shown in Figure 12. The experimental system comprised both laser excitation and reception devices, as well as a loading apparatus. The system included a 3D scanning platform with a displacement accuracy of 1 mm, an Nd:YAG pulsed laser with a wavelength of 1064 nm and a repetition rate of 10 Hz, a total reflection mirror, a lens, an interferometric pickup (vibrometer) with a frequency bandwidth of DC~24 MHz and a displacement resolution of 0.1 pm, an oscilloscope, a PC terminal that was used to control the sampling frequency and sampling direction parameters, and a 300-ton fully automatic pressure testing machine. Since the stress on the side surfaces of the specimen was the same, one side was chosen to perform the stress detection. Both the laser excitation device and the ultrasonic wave receiver were located on the same side of the specimen.

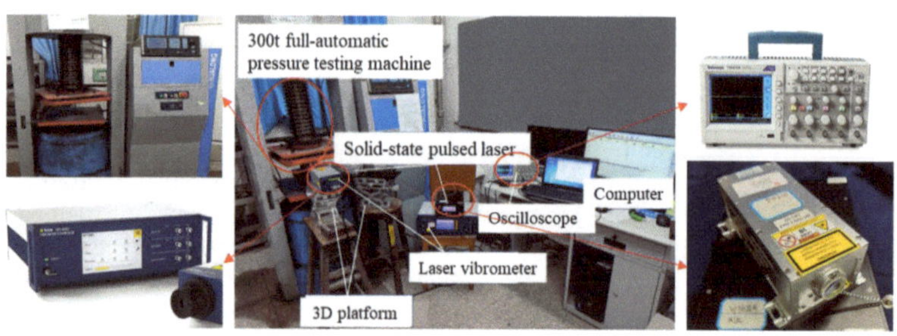

Figure 12. The laser ultrasonic testing system.

The nine Q235 steel specimens were divided into three groups for measurement, as shown in Table 2. To enhance the signal-to-noise ratio of the received ultrasonic signals, the surface of the material was polished, and aluminum foil with a thickness of about 1 mm was adhered at the signal reception points to amplify the signal strength. All tests were conducted at room temperature, which is approximately 20 °C. Group 1 was used to measure the velocity of the ultrasonic surface waves under stress-free conditions. The signal collection points were located 10 mm from the left edge of the specimen; the upper and lower edges were 30 mm each. The laser excitation points were located at different positions to the right of the collection points, with a 5 mm interval between each point. Scanning from left to right according to the path shown in Figure 13 allowed for the collection of time-domain signals of the ultrasonic waves at different positions. Groups 2 and 3 were used for the loading tests. Taking specimen A1 as an example, it was placed between the upper and lower platens of the pressure testing machine. The collection point was fixed at the center of the side surface of the specimen, and the laser excitation point was located below the collection point. The distance between the laser excitation point and the signal collection point was distributed as shown in Table 3. The ultrasonic wave propagation direction was aligned with the principal stress direction, as illustrated in Figure 14.

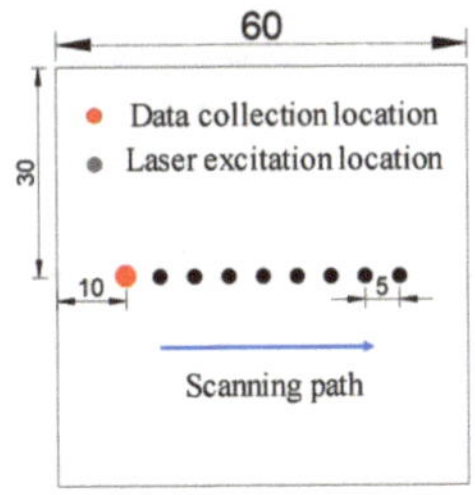

Figure 13. The scanning path of the specimen under stress-free conditions/mm.

Table 3. The situation for sample grouping.

Group	The Number of Specimens	Distance between Incentive Point and Acquisition Point *L*/mm	The Purpose
1	SD-1	5—40	Analyzing the variation of Rayleigh wave velocity under stress-free conditions
2	A1 A2 A3 A4	15 15 10 20	Detecting the variation of wave velocity with stress at the same/different positions of the test block
3	B1 B2 B3 B4	15 15 10 20	Repeating group for Group 2

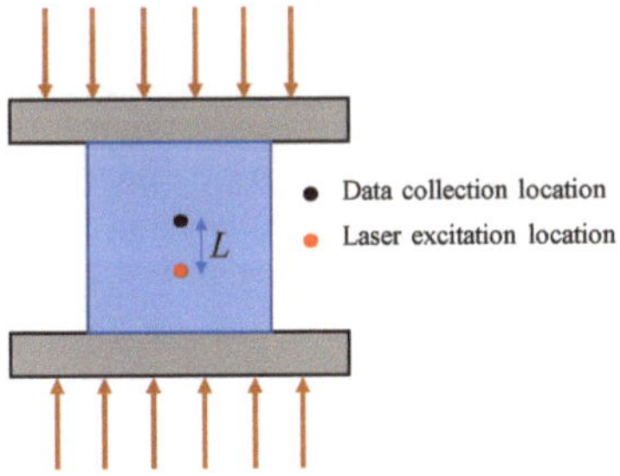

Figure 14. The scanning path of the specimen under stress conditions/mm.

After completing the ultrasonic wave measurements under stress-free conditions, the test machine was controlled to apply the load to the specimens according to a preset program, with a loading increment of 20 MPa, and continuing up to a maximum of 200 MPa. At each stress level during loading, the testing machine was stabilized in the load-holding state and the ultrasonic wave signals received by the oscilloscope were saved. Due to the vibrations of the pressure testing machine, the specimens placed on it would also vibrate. This caused random changes in the relative position between the laser vibrometer and the specimen, and led to a deviation of the signal collection point from its original position. Therefore, the signals at the measurement points were collected multiple times and averaged to reduce the errors caused by the operational characteristics of the testing machine.

4.3. Analysis and Discussion

During the experimental process, although the oscilloscope sampling filtered out some noise, further filtering of the specific frequency bands from the signal was still required to ensure accurate analysis and processing of the ultrasonic wave signals. For the ultrasonic signals, there is often more information for low-frequency components, whereas high-frequency components are typically associated with noise. In this study, the Sym8 wavelet was used to decompose the experimentally obtained ultrasonic wave signals. The decomposition was performed over five levels, and a soft thresholding function [37] was chosen to achieve this purpose. A comparison of the ultrasonic surface wave signals before and after wavelet denoising is illustrated in Figure 15.

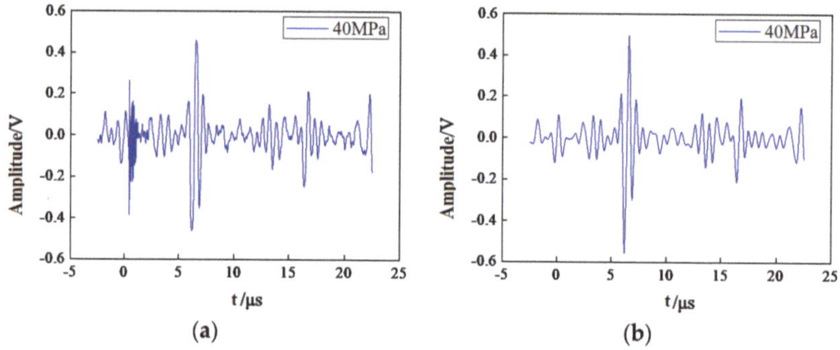

Figure 15. Ultrasonic time-domain signal for B1. (**a**) Ultrasonic time—domain signal before wavelet denoising; (**b**) Ultrasonic time-domain signal after wavelet denoising.

Upon comparing the time-domain signals before and after denoising, it was observed that the noise components within the signal were effectively removed, leading to an enhanced signal-to-noise ratio and a smoother time-domain curve. Additionally, the denoised ultrasonic signals were further processed using a cubic spline interpolation method to improve the resolution of the time difference measurements. A comparison of the ultrasonic surface wave signals before and after interpolation is depicted in Figure 16. While the overall trend of the time-domain signal curve remained unchanged, the resolution was enhanced and the precision in extracting time differences was improved post interpolation. The waveforms of the ultrasonic signals received at various stress levels were similar, with a consistent frequency and other characteristics, and differed only in the arrival time of the ultrasonic waves. The velocity of the ultrasonic surface waves was determined by reading the time delay in the ultrasonic wave time-domain waveforms and combining it with the distance over which the ultrasonic wave was propagated.

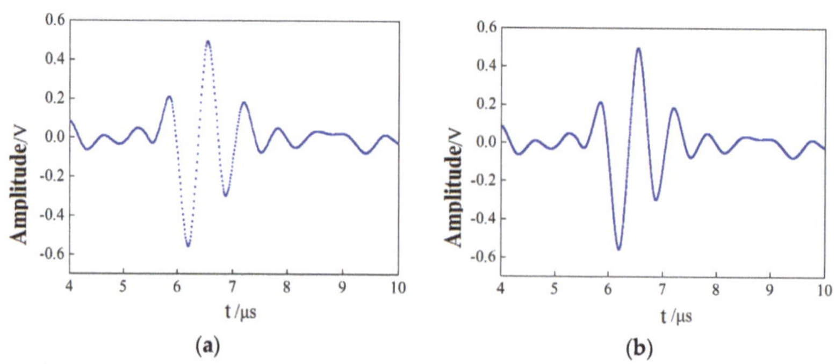

Figure 16. Ultrasonic time-domain signal for B1. (**a**) Ultrasonic time-domain signal before interpolation (**b**) Ultrasonic time-domain signal after interpolation.

In the context of the Q235 steel structural specimens subjected to uniaxial loading, a subtle expansion occurs on the side surfaces of the specimens due to compression. This phenomenon leads to an extended propagation path for the surface waves, as depicted in Figure 17. While the magnitude of the surface expansion during the elastic phase of the specimen was relatively minor, the influence of the distance alterations on the variations in wave velocity was notably significant. Consequently, it became imperative to account for the effects of surface expansion of the specimen when conducting laser ultrasonic testing. This consideration was crucial for ensuring the accuracy and reliability of the ultrasonic measurements, particularly in stress analysis and material characterization applications.

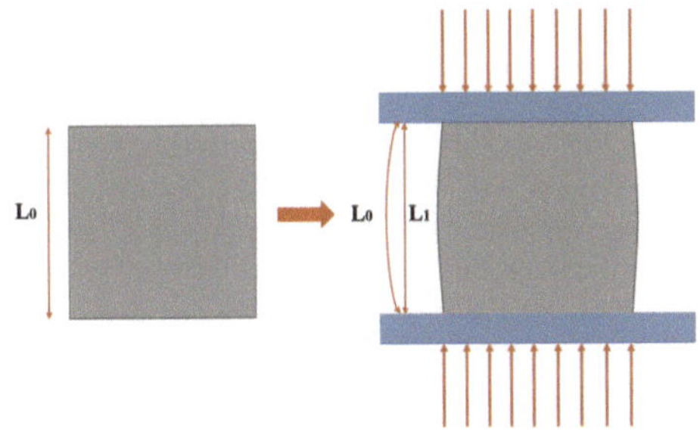

Figure 17. The surface changes of the specimen during the loading process.

For specimens subjected to uniaxial compression, the acoustoelastic coefficient is a key factor in calculating stress. The acoustoelastic coefficient can be expressed as follows:

$$C_R = \frac{v_R - v_{R0}}{v_{R0}\sigma} \tag{18}$$

The calculated velocities of the ultrasonic surface waves under both stress-free and stress-loaded conditions can be represented as follows:

$$v_{R0} = \frac{L}{t_0}, v_R = \frac{L_\Delta}{t_\sigma} \tag{19}$$

In the equation, L denotes the theoretical distance over which the ultrasonic wave propagates. $L_\triangle$ denotes the actual distance over which the ultrasonic wave propagates. The variables t_0 and t_σ represent the travel times of the ultrasonic wave for this theoretical propagation distance under two different conditions: t_0 is the travel time when there is no applied stress, and t_σ is the travel time when the specimen is subjected to a stress of σ. Taking into account the surface expansion of the specimen leads to an increase in the propagation distance (L), as used in the computation of the Rayleigh wave velocity (V_R) in Equation (19). Under identical stress and wave travel time conditions, this adjustment resulted in a higher calculated wave velocity. Subsequently, as per the formulation in Equation (18), there is a consequential increase in the acoustoelastic coefficient so that there are some errors in the calculation of the stress.

To eliminate the impact of the lateral surface expansion of the specimen, the wave velocity was calculated using the actual propagation distance $L_\triangle$ of the ultrasonic wave after determining the time delay, as illustrated in Figure 18. During the uniaxial compression process, the expansion curve of the specimen was approximately parabolic. A Cartesian coordinate system was established with the midpoint of the parabola as the origin. The arc length direction was set as the *x*-axis and the direction perpendicular to the arc length was set as the *y*-axis. The origin of the coordinate system was spatially parallel to the signal reception point and can be considered as the signal reception point itself.

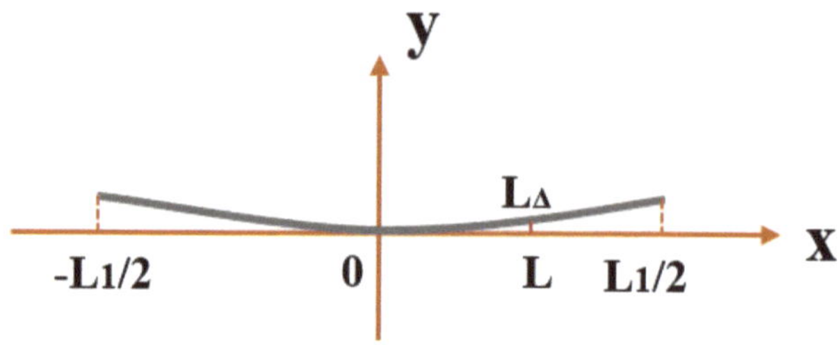

Figure 18. Changes of specimen during loading process.

The original side length of the specimen was L_0. When compressed and expanded, the length of the side itself remained unchanged, and it can be assumed that the arc length of the expanded parabola was L_0. At this time, the distance between the upper and lower platens of the testing machine was L_1. The stress applied to the specimen by the testing machine was achieved by moving the lower platform upwards, so that L_1 is equal to $L_0 - \Delta x$. Δx is the displacement of platform of the testing machine. It can be obtained from the control interface of the testing machine. As shown in Figure 18, the horizontal coordinates of the two ends of the parabola were $\pm L_1/2$. L is the theoretical propagation distance of the ultrasonic wave, and L_Δ is the actual propagation distance of the ultrasonic wave. It is necessary to calculate the arc length of the parabola from the horizontal coordinate zero to L. The solution expression is as follows:

$$L_\Delta = \int_0^L \sqrt{1 + \frac{1}{p^2} x^2 dx} \tag{20}$$

where p can be obtained through the parabolic equation. The calculation results for the actual propagation distance of the surface waves on the specimen are shown in Tables 4 and 5. From these tables, it is evident that the farther the distance between the laser excitation point and the signal reception point, the greater the change in the actual propagation distance of the surface waves during compression expansion. For a theoretical propagation distance of 20 mm, a stress of 200 MPa results in a change of approximately 0.1 to 0.2 mm in the propagation distance, leading to a change of approximately 1% in the ultrasonic wave velocity. Since the acoustoelastic effect is a weak effect, it is essential to consider the changes in wave velocity caused by variations in the propagation distance. Subsequently, calculations of ultrasonic wave velocity were performed using the expanded propagation distance. The relationship curves between the relative wave velocity and stress for specimens A1 and B1, from the stress-free state to bearing 200 MPa, are shown in Figure 19.

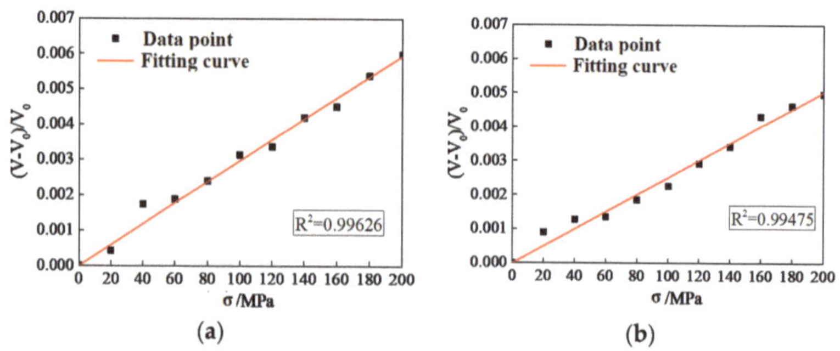

Figure 19. The relationship between the relative wave velocity and stress. (**a**) The relationship for A1 (**b**) The relationship for B1.

Table 4. The actual propagation distance of the surface waves for Group 2.

A1		A2		A3		A4	
σ/MPa	**Actual Propagation Distance/mm**	**σ/MPa**	**Actual Propagation Distance/mm**	**σ/MPa**	**Actual Propagation Distance/mm**	**σ/MPa**	**Actual Propagation Distance/mm**
0	15	0	15	0	10	0	20
20	15.0166	20	15.0096	20	10.0047	20	20.0136
40	15.0248	40	15.0182	40	10.007	40	20.0285
60	15.0322	60	15.0251	60	10.0089	60	20.0428
80	15.039	80	15.0317	80	10.0108	80	20.0566
100	15.0453	100	15.0378	100	10.0126	100	20.0691
120	15.0514	120	15.0437	120	10.0143	120	20.0809
140	15.0577	140	15.0496	140	10.016	140	20.0935
160	15.0642	160	15.0555	160	10.0177	160	20.1061
180	15.0706	180	15.0612	180	10.0194	180	20.1187
200	15.077	200	15.0669	200	10.021	200	20.1312

Table 5. The actual propagation distance of the surface waves for Group 3.

B1		B2		B3		B4	
σ/MPa	**Actual Propagation Distance/mm**	**σ/MPa**	**Actual Propagation Distance/mm**	**σ/MPa**	**Actual Propagation Distance/mm**	**σ/MPa**	**Actual Propagation Distance/mm**
0	15	0	15	0	10	0	20
20	15.0095	20	15.0156	20	10.0033	20	20.0469
40	15.0205	40	15.0331	40	10.0072	40	20.0857
60	15.028	60	15.0478	60	10.0098	60	20.1176
80	15.0342	80	15.0588	80	10.012	80	20.1445
100	15.0398	100	15.0676	100	10.0141	100	20.1655
120	15.045	120	15.0751	120	10.016	120	20.1835
140	15.0499	140	15.086	140	10.018	140	20.2004
160	15.0548	160	15.0886	160	10.0198	160	20.2165
180	15.0595	180	15.0952	180	10.0216	180	20.2319
200	15.0641	200	15.1015	200	10.0359	200	20.2475

Figure 19 presents the relationship between the relative wave speed and stress. From Figure 19, it is indicated that there is an approximate linear relationship between relative wave speed and stress, consistent with the simulation analysis. To clearly characterize the linear relationship between the relative wave speed and stress, the curve has been linearly fitted. The fitting goodness R^2 was above 0.99, indicating a strong linear relationship between the relative wave velocity and stress, which is in line with the requirements of the acoustoelastic equation. In the process of stress solving, the acoustoelastic coefficient C_R is a key parameter for the solution. Based on Equation (18), the acoustoelastic coefficient can be obtained through the linear relationship between the ultrasonic relative wave speed and stress. The slope of the fitted curve is the acoustoelastic coefficient. The obtained acoustoelastic coefficients for the various specimens fluctuated between 2.51×10^{-5} and 3.21×10^{-5}, which was minimal variance among different specimens. Therefore, the acoustoelastic coefficient of specimen A1, 2.97×10^{-5}, was chosen as the optimal coefficient for calculating the absolute stress of the specimens.

Based on the acoustoelastic coefficient obtained, the absolute stress for specimens was calculated. The results of the stress measurement and their theoretical values are shown in Figures 20–22. Due to the surface roughness of the steel specimens used in the experiment and the fact that the material uniformity cannot be guaranteed to be completely consistent, the acoustoelastic coefficients of the different specimens are different. In addition, the measurement results are also influenced not only by material characteristics but also by the stress level and the distance of ultrasonic wave propagation. Therefore, there will be errors for the calculated absolute stress. Figures 20–22 show the results of stress detection at the different ultrasonic propagation distances. From the figures, it can be observed that the absolute error between the measured stress and the theoretical stress of the specimens was the largest when the stress level was low. When the ultrasonic propagation distance was 15 mm, the maximum absolute errors between the measured stress and the theoretical stress of specimens A1, B1, A2, and B2 were 18.48 MPa, 16.07 MPa, 21.92 MPa, and 15.35 MPa, respectively, with average errors of 5.18 MPa, 7.46 MPa, 11.45 MPa, and 7.89 MPa. When the ultrasonic propagation distance was 10 mm, the maximum absolute errors between the measured stress and the theoretical stress of specimens A3 and B3 were 21.92 MPa and 33.52 MPa, respectively, with average errors of 11.45 MPa and 20.66 MPa. When the ultrasonic propagation distance was 20 mm, the maximum absolute errors between the measured stress and the theoretical stress of specimens A4 and B4 were 18.46 MPa and 15.38 MPa, respectively, with average errors of 7.03 MPa and 8.12 MPa. Overall, compared with the other specimens, the stress measurement error was the largest when the ultrasonic propagation distance was 10 mm. The main reason is that the distance between the laser excitation point, and the signal collection point was too close, causing the laser spot energy to interfere with the beam of the laser vibrometer, leading to greater data volatility. Therefore, it is necessary to ensure that the excitation point and the measurement point are sufficiently far apart when measuring the stress.

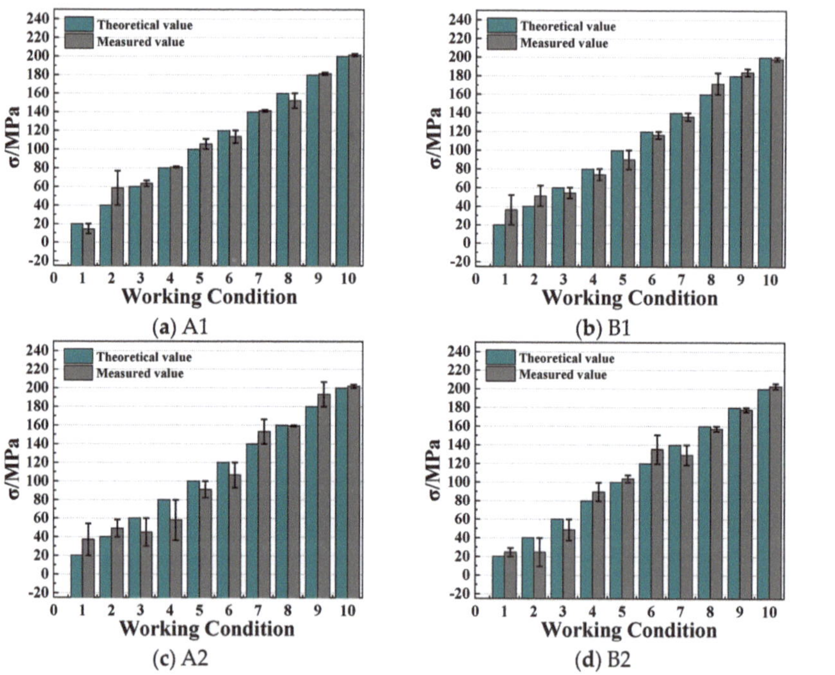

Figure 20. Stress calculation results for theoretical propagation distance at 15 mm.

Figure 23 presents the relative errors in the experimental results at different stress levels. As the stress level continued to increase, the relative error between the calculated stress and the theoretical stress of the specimens gradually decreased. When the stress of the steel component was 20 MPa, the relative error of the measurement results was the largest, with a maximum relative error of 85.3%. When the stress of the steel component was 200 Mpa, the relative error of the measurement results was the smallest, with a minimum relative error of 0.4%. In addition, it was found that the relative errors were generally within 10% when the stress was 100 Mpa and above. Overall, the laser ultrasonic method demonstrates good reliability and accuracy in measuring the absolute stress of steel structures, especially at higher stress levels. This makes this method a viable option for engineering measurements. In practical engineering applications, if there is a need to measure the absolute stress of a steel structure, it is only necessary to calibrate the test specimens of the steel structure to obtain the acoustoelastic coefficient. The absolute stress of the steel components can be determined through the acoustoelastic formula, using the laser ultrasonic method to detect the relative velocity of the ultrasonic surface waves. This approach offers a practical and efficient solution for assessing the stress state in steel structures, which is crucial for ensuring their structural integrity and performance.

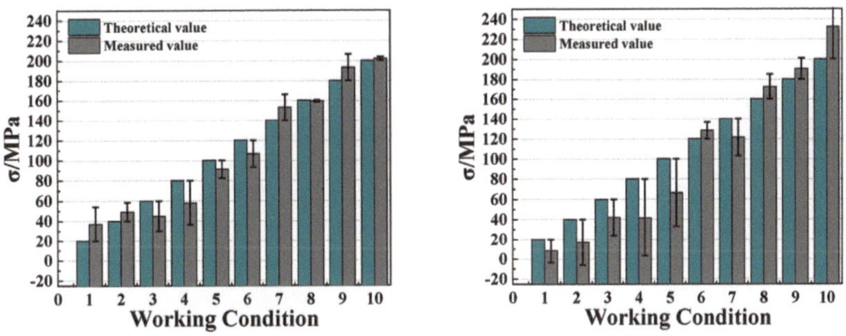

(a) A3 (b) B3

Figure 21. Stress calculation results for theoretical propagation distance at 10 mm.

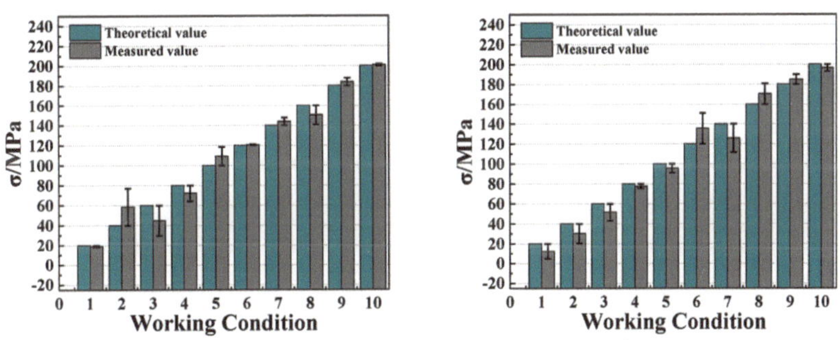

(a) A4 (b) B4

Figure 22. Stress calculation results for theoretical propagation distance at 20 mm.

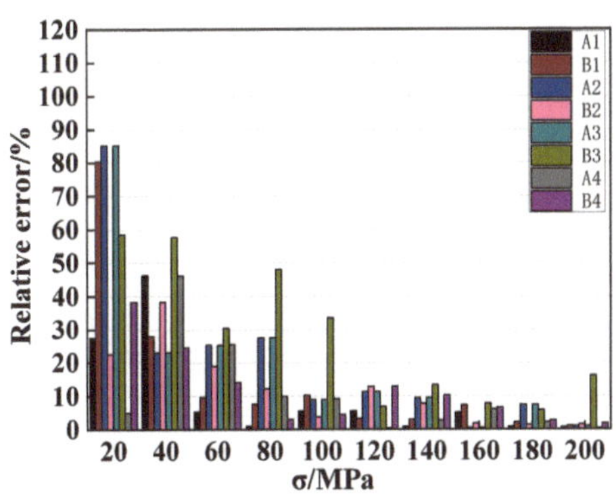

Figure 23. Relative error of stress calculation results.

5. Analysis of the Influencing Factors for Laser Ultrasonic Technology

5.1. Sensitivity of Stress Detection to Environmental Temperature

When using the laser ultrasonic method to detect the absolute stress in steel structures, changes in environmental temperature will affect the propagation of the ultrasonic waves in the following ways:

(1) Change in the propagation speed of the ultrasonic surface waves

According to the literature [34], it is known that the velocity of the surface waves in an isotropic solid under no stress is primarily related to the elastic modulus of the material, the density of the material, and Poisson's ratio, and is directly proportional to the elastic modulus. Changes in temperature generally cause changes in the elastic modulus and Poisson's ratio of steel structural materials.

According to the literature [38], the variation of the elastic modulus of several commonly used steels in engineering with temperature is usually described in detail, as shown in Figure 24. Since the change in Poisson's ratio within 100 °C is negligible, temperature mainly causes changes in the elastic modulus of the material. The increase in temperature leads to a decrease in the elastic modulus of metal materials, as shown in the Equation (21) [34].

$$E_T = E_0(1 - \eta \Delta T) \tag{21}$$

where E_T is the elastic modulus of the material when the temperature reaches, E_0 is the initial elastic modulus of the material, and η is the temperature variation coefficient of the elastic modulus of the metal. This coefficient is difficult to determine, but it is related to the metal material's linear expansion coefficient α. According to the results calculated using the experimental values in reference [39], η is approximately 25 times α. Therefore, the surface wave velocity, which is directly proportional to the elastic modulus, also decreases with the increase in temperature. The change in the velocity of ultrasonic surface waves under the influence of temperature can be expressed using the following equation:

$$V_{RT1} = V_{R0}(1 - \beta \Delta T) \tag{22}$$

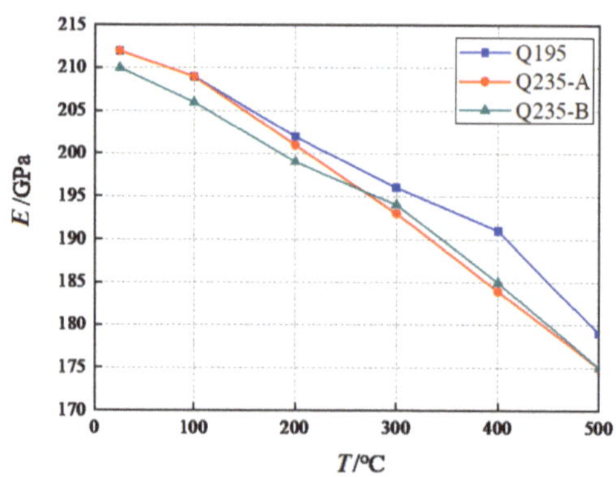

Figure 24. The elastic modulus at different temperatures.

In this equation, V_{RT1} is the wave velocity of the ultrasonic surface wave corresponding to the change in the elastic modulus when the temperature reaches T; V_{R0} is the wave velocity of the ultrasonic surface wave at the initial temperature, with no stress in the steel structure; β is the temperature coefficient of the change in the surface wave velocity with temperature, and ΔT is the change in temperature.

(2) Change in the propagation distance of the ultrasonic surface waves

Due to the thermal expansion and contraction of the materials with temperature changes, the propagation distance of ultrasonic waves is also affected. As the temperature rises, the propagation distance of the ultrasonic waves increases.

$$L_T = L_0(1 + \alpha \Delta T) \tag{23}$$

where L_0 is the initial ultrasonic wave propagation distance, L_T is the propagation distance of the ultrasonic waves at temperature T, α is the thermal expansion coefficient of the metal materials, and ΔT is the temperature change amount. The velocity of the ultrasonic surface wave after the temperature increase can be represented using the following equation:

$$V_{RT2} = \frac{L_0(1 + \alpha \Delta T)}{t_T} \tag{24}$$

In this equation, V_{RT2} is the wave velocity of the ultrasonic surface wave corresponding to the change in the elastic modulus when the temperature reaches T; t_T is the ultrasonic wave flight time when the temperature reaches T.

For metal materials, the increase in temperature leads to a decrease in the ultrasonic wave velocity, which is caused by both a decrease in the elastic modulus and an increase in the ultrasonic wave propagation distance. Therefore, the ultrasonic wave velocity when the temperature increases to T is as follows:

$$V_{RT} = V_{RT1} - (V_{R0} - V_{RT2}) \tag{25}$$

According to the literature [40], the change in the linear expansion coefficient α is relatively small compared with the change in the temperature coefficient β and can be ignored. Combining Equations (22) and (24), the change in the relative wave velocity with temperature can be expressed as follows:

$$\frac{V_{RT} - V_{R0}}{V_{R0}} = -\beta \Delta T + \frac{t_0}{t_T} - 1 \tag{26}$$

where t_0 is the flight time of the ultrasonic surface wave measured at the initial temperature. When the temperature increases, t_0 is less than t_T. From Equation (26), it can be established that the relative wave speed decreased approximately linearly with the increase in temperature.

Based on the above analysis, the effect of temperature on ultrasonic wave propagation characteristics is simulated by changing the elastic modulus. Taking 25 °C as the initial temperature, the results obtained from the model are shown in Figures 25 and 26.

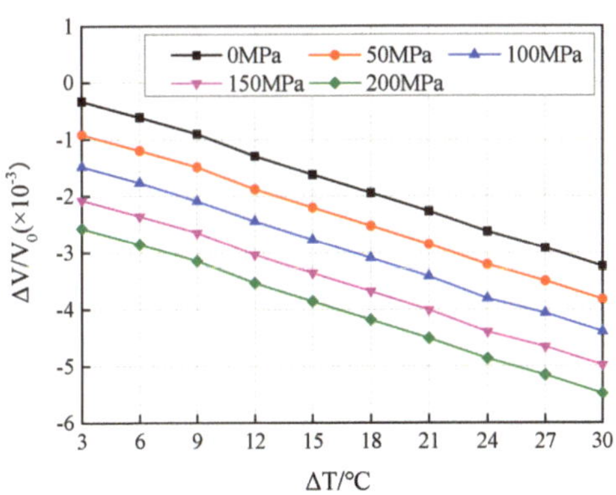

Figure 25. The variation in the relative wave velocity with temperature under different stress levels.

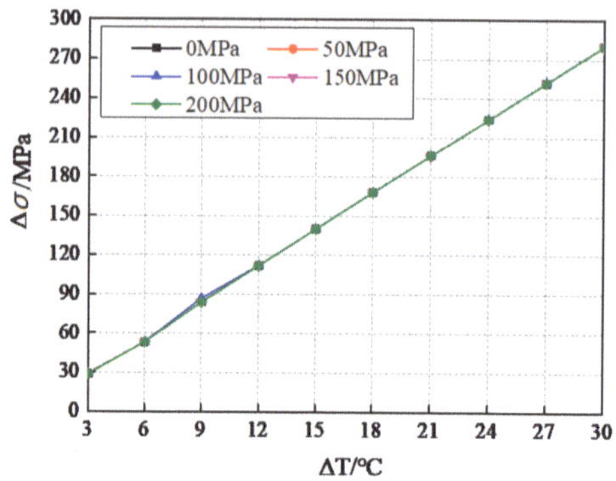

Figure 26. The variation in the stress differential with temperature under different stress levels.

In Figure 25, the relative wave speed of the ultrasonic wave decreased approximately linearly at the same stress level as the temperature increased. The change in stress affected the change in relative wave speed but did not alter the trend of change with temperature. Under the same temperature change, the higher the stress level, the lower the relative wave speed of the ultrasound. To more clearly understand the impact of ambient temperature on the stress calculation results, the deviation of calculated stress at different stress levels with temperature change is shown in Figure 26. The results indicate that an increase in temperature caused a significant error in the stress calculation. Under the same temperature change, the stress error values at various stress levels were approximately consistent. The stress deviation exceeded 270 MPa when the environmental temperature changed by 30 °C. At this time, the error far exceeded the measured value, making the stress measurement results no longer of reference value. To reduce the impact of temperature changes on stress detection, it is necessary to conduct the detection in an environment where the temperature is relatively stable.

When temperature changes are inevitable, it is necessary to propose compensation methods for the temperature variations. According to Equation (26), the temperature coefficient β is fitted based on the relationship between relative wave speed and the change in temperature. Temperature errors are compensated based on the temperature coefficient. The temperature compensation equation obtained using Figure 25 is as follows:

$$\frac{V_{RT} - V_{R0}}{V_{R0}} = -1.0816 \times 10^{-4}\Delta T + 8.4286 \times 10^{-5} \tag{27}$$

According to Equation (27), the stress calculation results after compensation are shown in Figure 27. It can be seen from Figure 27 that the stress after temperature compensation is close to those at the initial temperature, with a maximum error of 3.68 MPa. This indicates the effectiveness of the temperature compensation equation.

5.2. *Sensitivity of Stress Detection to Material Surface Roughness*

The surfaces of metal materials, such as steel structures, are not absolutely smooth. Scratches and other uneven undulations may exist on the surfaces of materials, which are defined as the surface roughness of the material. The inevitable random roughness of the material surface will undoubtedly affect the propagation of the surface waves when using the laser ultrasonic method to detect the absolute stress in steel structures; therefore, the measurement accuracy of stress will be affected.

To study the impact of the material surface roughness on the propagation characteristics of ultrasonic waves, it is necessary to establish a relevant model for targeted analysis. Typically, the creation of a random rough surface is performed using the Monte Carlo method. A random rough surface is composed of a superposition of harmonics with different frequencies. The height of each point on the rough surface represents the amplitude of these superimposed harmonics. After randomly generating a two-dimensional rough surface using the Monte Carlo method, the rough surface curve was discretized into sample point data. The contour arithmetic mean deviation, *Ra*, was used to determine the degree of surface roughness and ten sets of rough surfaces with *Ra* values ranging from 0.3 μm to 3 μm were selected. The FE model is shown in Figure 28.

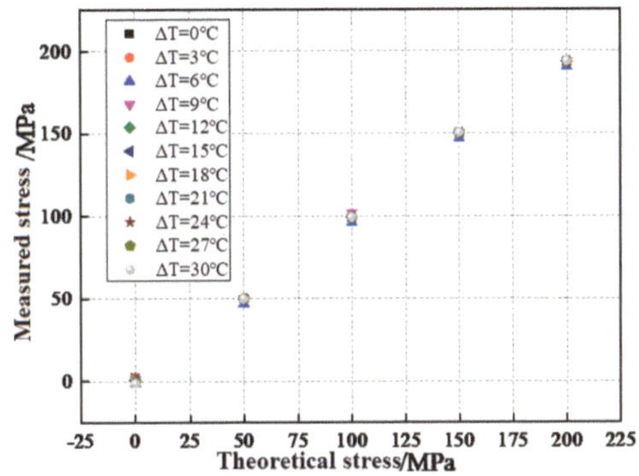

Figure 27. Measured results after temperature compensations.

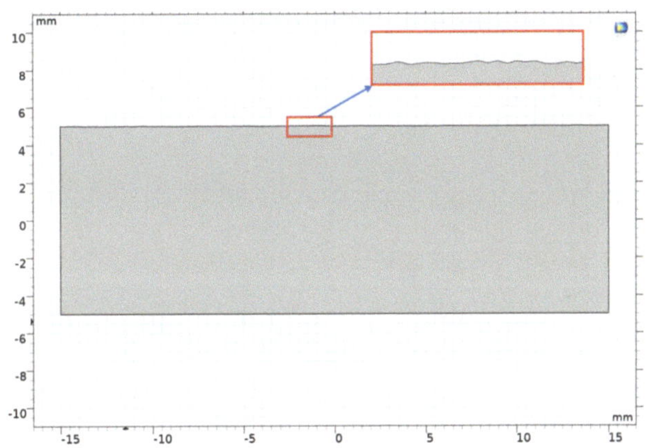

Figure 28. The FE model with a rough surface.

The curve showing the change in relative wave speed of the ultrasonic surface waves with different roughness at different stress levels is illustrated in Figure 29. From Figure 29, it can be observed that the relative wave speed decreased nonlinearly, as the roughness of the material surface increased at the same stress level. At different stress levels, the non-linear relationship between relative wave speed and material surface roughness was similar. At the same time, the deviation of calculated stress at different stress levels with changes in material roughness is represented in Figure 30. From Figure 30, it can be seen that the stress error showed a non-linear growth trend as the increase in surface roughness for the steel specimen was at the same stress level. The variation trend of stress error with changes in roughness remains essentially consistent at different stress levels. When the roughness is approximately 3 μm, the difference in stress exceeds 50 MPa. At this time, the measurement results are no longer of reference value. In summary, it is necessary to polish the surface of the test piece with fine sandpaper or a grinder before the detection, due to the impact of material surface roughness on stress detection. This will ensure the accuracy of the detection results.

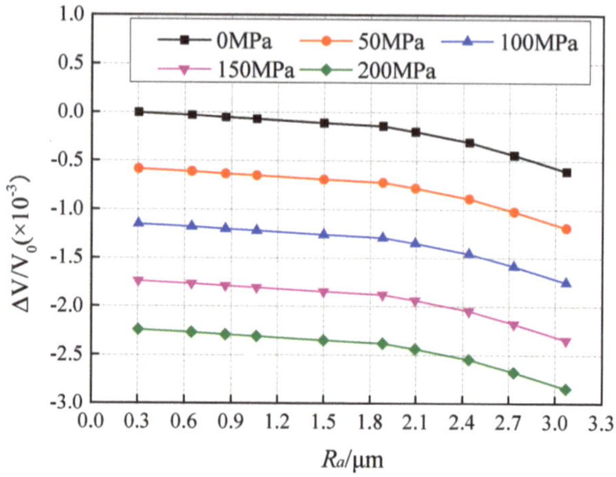

Figure 29. The variation in the relative wave velocity with *Ra* under different stress levels.

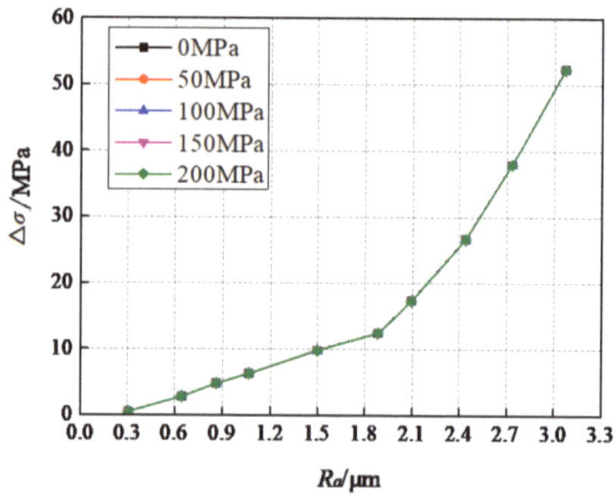

Figure 30. The variation in the stress differential with *Ra* under different stress levels.

6. Conclusions

In this work, laser ultrasonic technology was used for the measurement of absolute stress in steel structures. The velocity variation of laser-induced ultrasonic waves caused by stress variations in steel structures was analyzed based on the experiment and FE simulation. The experimental and FE results are consistent and indicate that the relative wave velocity of the ultrasonic wave is approximately linearly related to the applied stress, and that the stress can be obtained based on the linear relationship. The results confirm the feasibility of using laser ultrasonics to measure the absolute stress of steel structures.

The experimental results show that the absolute error between the measured stress and the theoretical stress of the specimens is largest when the stress level is low, and that the relative error of measured stress gradually decreased as the stress increased. When the stress experienced by the specimen was above 100 MPa, the relative error was generally within 10%. It has been proven that this technology for stress measurement has good reliability and accuracy at a high stress level.

Considering the temperature and material surface roughness as the two main influencing factors of stress detection, the sensitivity for stress detection to these factors was analyzed based on theoretical analysis and FE. The analysis results indicate that the error in stress measurement increased similarly linearly with the increase in temperature and increased non-linearly with the increase in roughness. Consequently, error compensation methods tailored for different influencing factors were proposed, effectively enhancing the accuracy of absolute stress detection methods for steel structures.

The research on laser ultrasonic stress measurement in this paper focuses on the surface stress of specimens under uniaxial stress loading. However, steel structures in actual service are subject to complex stresses; therefore, the study of laser ultrasonic detection methods in actual steel structures is necessary. Additionally, the nonlinear geometrical effects along the members will be experimentally considered in the future.

Author Contributions: Conceptualization, Z.H.; methodology, L.L., Z.H. and H.T.; Software, Y.K.; Validation, B.L.; investigation, B.O. and B.L.; resources, Z.H.; data curation, H.T. and Y.K.; writing—original draft, H.T. and Z.H.; writing—review and editing, L.L.; supervision, L.L.; funding acquisition, L.L. All authors have read and agreed to the published version of the manuscript.

Funding: This study was supported by the National Natural Science Foundation of China (U20A20314, 52278146, 52278291), the Natural Science Fund for Distinguished Young Scholars of Chongqing (cstc2020jcyj-jqX0006), the Chongqing Natural Science Foundation of China (cstc2019jcyj-cxttX0004, cstc2020yszx-jscxX0003), and the Science and Technology Project of Guizhou Provincial Transportation Department (2020-123-017).

Data Availability Statement: The data presented in this study are available from the first and corresponding author upon request. The data are not publicly available due to the policy of the data provider.

Conflicts of Interest: Authors Hongsong Tian, Yujiang Kong, Bin Liu and Bin Ouyang were employed by the company Guizhou Bridge Construction Group Co., Ltd. The remaining authors declare that the research was conducted in the absence of any commercial or financial relationships that could be construed as a potential conflict of interest.

Abbreviations

u	the displacement component;
ε_{ij}	the component of the strain tensor;
ρ	the density of materials;
f_i	the external force applied;
T	the increase in temperature;
λ, μ	Lamé constant;
δ_{ij}	the unit tensor;
α	the coefficient of thermal expansion;
U	the displacement vector;
I	the identity tensor;
σ	the stress tensor;
n	the unit vector perpendicular to the material's surface;
V_L	the longitudinal wave velocity;
V_S	transverse wave velocity;
l, m, n	the Murnaghan constants;
K_S	the acoustic elastic coefficient of the transverse wave;
K_L	acoustic elastic coefficient of the longitudinal wave;
V_{S0}	the initial wave velocity of transverse waves without initial stress;
V_{R0}	the velocity of ultrasonic surface waves under the stress-free state;
K_R	the acoustoelastic coefficient for ultrasonic surface waves;
λ	the ultrasonic wavelength;
C_R	the wave speed of the surface acoustic wave;
$\hat{E}$	the material's elastic modulus;
ν	the material's Poisson's ratio;
r_0	the radius of the laser spot;
E_T	the elastic modulus of the material when the temperature reaches T;
E_0	the initial elastic modulus of the material;

η	the temperature variation coefficient of the metal's elastic modulus;
α	the metal material's linear expansion coefficient;
β	the temperature coefficient of the change in surface wave velocity with temperature;
V_{RT}	the wave velocity of the ultrasonic surface wave corresponding to the change in elastic modulus;
L_T	the propagation distance of ultrasonic waves at temperature T;

References

1. Zhang, Z.; Xie, K.; Xie, W. Analysis of Existing Steel Structure Bridge Diseases and Research on Protection Technology. *North. Transp.* **2021**, *9*, 14–17.
2. Yue, K.; Xu, L.; Liu, J. Energy dissipation evaluation of butterfly-shaped steel plate shear keys under different loading protocols. *Structures* **2024**, *60*, 105893. [CrossRef]
3. Yue, K.; Xu, L.; Liu, J. Seismic performance of an energy dissipating shear key for highway bridges using butterfly-shaped steel plates. *Eng. Struct.* **2023**, *295*, 116885. [CrossRef]
4. Tong, K.; Zhang, H.; Zhao, R.; Zhou, J.; Ying, H. Investigation of SMFL monitoring technique for evaluating the load-bearingcapacity of RC bridges. *Eng. Struct.* **2023**, *293*, 116667. [CrossRef]
5. Hagara, M.; Trebuňa, F.; Pástor, M. Analysis of the Aspects of Residual Stresses Quantification Performed by 3D DIC Combined with Standardized Hole-drilling Method. *Measurement* **2019**, *137*, 238–256. [CrossRef]
6. Li, R.; Xu, L.; Zhang, S. Application of Residual Stress High-Speed Drilling Method in Thick-Walled Pipeline Steel Pipes. *Phys. Test.* **2019**, *37*, 1–5.
7. Zhang, H.; Ma, X.; Jiang, H. Grading Evaluation of Overall Corrosion Degree of Corroded RC Beams via SMFL Technique. *Struct. Control Health Monit.* **2023**, *2023*, e6672823. [CrossRef]
8. Zhang, H.; Qi, J.; Zheng, Y. Characterization and grading assessment of rebar corrosion in loaded RC beams via SMFL technology. *Constr. Build. Mater.* **2024**, *411*, 134484. [CrossRef]
9. Rainieri, C.; Fabbrocino, G. Development and validation of an automated operational modal analysis algorithm for vibration-based monitoring and tensile load estimation. *Mech. Syst. Signal Process.* **2015**, *60*, 512–534. [CrossRef]
10. Bonopera, M. Stress Evaluation in Axially Loaded Members of Masonry Buildings and Space Structures: From Traditional Methods to Combinations with Artificial Intelligence Approaches. *Buildings* **2023**, *13*, 2097. [CrossRef]
11. Wang, G.; Liao, L.; Zhang, S. Laser Ultrasonic Detection of Concrete Stress under the Influence of Stress History. *Sci. Technol. Eng.* **2021**, *21*, 14734–14741.
12. Liu, H.; Liu, T.; Li, Y. Ultrasonic Measurement Method of Residual Stress on Metal Material Surfaces. *J. Mech. Eng.* **2021**, *57*, 118–125.
13. White, R.M. Generation of Elastic Waves by Transient Surface Heating. *J. Appl. Phys.* **1963**, *34*, 3559–3567. [CrossRef]
14. Zhan, Y.; Zhang, E.; Ge, Y. Residual Stress in Laser Welding of TC4 Titanium Alloy Based on Ultrasonic laser Technology. *Appl. Sci.* **2018**, *8*, 1997. [CrossRef]
15. Zhan, Y.; Liu, C.; Zhang, J. Measurement of residual stress in laser additive manufacturing TC4 titanium alloy with the laser ultrasonic technique. *Mater. Sci. Eng.* **2019**, *762*, 138093. [CrossRef]
16. Zhan, Y.; Li, Y.; Zhang, E. Laser ultrasonic technology for residual stress measurement of 7075 aluminum alloy friction stir welding. *Appl. Acoust.* **2019**, *145*, 52–59. [CrossRef]
17. Xue, M. Research on Non-Destructive Detection of Residual Stress in Metal Surfaces Based on Laser Ultrasonics. Master's Thesis, Zhejiang University, Hangzhou, China, 2018.
18. Ye, C.; Ume, I.C.; Zhou, Y. Inspection of the residual stress on welds using laser ultrasonic supported with finite element analysis. *Manuf. Rev.* **2019**, *6*, 3. [CrossRef]
19. Pan, Z.; Song, J.; Gao, Y. Residual Stress Measurement of GH4169 Superalloy Based on Laser Ultrasonic Method. *Chin. Q. Mech.* **2021**, *42*, 98–107.
20. Ji, B.; Cao, J.; Huang, G. Research on Laser Ultrasonic Guided Wave Detection Method for Internal Stress in Steel Plates. *Chin. J. Lasers* **2022**, *49*, 7–15.
21. Ji, B. Research on Laser Ultrasonic Detection Method for the Composite Quality of Metal Laminated Plates. Ph.D. Thesis, University of Science and Technology Beijing, Beijing, China, 2022.
22. Qian, Z.; Han, X.; Wang, G. Laser Ultrasonic Detection of Residual Stress on the Surface of K417G Turbine Disk after Shot Peening. *Appl. Laser* **2022**, *42*, 91–95.
23. He, X. Research on Nondestructive Testing of Residual Stress in Metal Components Based on Laser Ultrasonics. Master's Thesis, North University of China, Taiyuan, China, 2024.
24. Sampath, S.; Zhang, Z.; Tham, Z.; Chen, Y.; Seng, D.; Gantala, T.; Zhang, L. Non-contact measurements of residual stress distribution and grain size in titanium alloys with laser ultrasonic system. *Int. J. Mech. Sci.* **2024**, *264*, 108809. [CrossRef]
25. Wang, Y. Numerical Study on Laser Ultrasonic Detection of Surface Cracks on Shafts. Master's Thesis, Dalian University of Technology, Dalian, China, 2019.
26. Truesdell, C. Book Review: Finite deformation of an elastic solid. *Bull. Am. Math. Soc.* **1952**, *58*, 577–580. [CrossRef]

27. Bompan, K.F.; Haach, V.G. Ultrasonic tests in the evaluation of the stress level in concrete prisms based on the acoustoelasticity. *Constr. Build. Mater.* **2018**, *162*, 740–750. [CrossRef]
28. Xu, S. Stress Identification of Steel Components Based on Ultrasonics and Deep Learning. Master's Thesis, Hunan University, Changsha, China, 2022.
29. Wang, G.; Liao, L.; Zhang, Z. Laser Wakefield Electron Acceleration and Novel Radiation Sources. *Sci. Technol. Eng.* **2021**, *21*, 14734–14741.
30. Yu, C.; Qin, Z.; Liu, C. Laser Wakefield Electron Acceleration and Novel Radiation Sources. *Chin. J. Lasers* **2024**, *51*, 0101002.
31. Adin, M.Ş.; Adin, H.; Ergün, R.K. Finite Element Analysis of Safety Pin in Snowplow Equipment. *Eur. J. Tech. (EJT)* **2022**, *12*, 89–92. [CrossRef]
32. Dong, L.; Feng, S.; Zhu, L.; Jiang, H. Research on the Localization and Quantification of Internal Crack Defects in Aluminum Plates Based on Laser Ultrasonics. *Appl. Laser* **2022**, *42*, 102–111.
33. Bai, X. Calculation of Surface Stress Field of Objects Based on Laser Grating Method. Master's Thesis, Beijing University of Chemical Technology, Beijing, China, 2023.
34. Tian, J. Research on Full Focal Imaging Detection of Laser Cladding Layer Defects Using Ultrasonic Phased Array Surface Waves. Master's Thesis, Changsha University of Science & Technology, Changsha, China, 2022.
35. Liang, C. Research on the Interaction between Laser Ultrasound and Metal Defects Based on COMSOL. Master's Thesis, Beijing Jiaotong University, Beijing, China, 2020.
36. Shi, Y.; Shen, Z.; Ni, X. Laser-Excited Rayleigh Wave Measurement of Residual Stress in Aluminum Alloy Welds. *Chin. J. Lasers* **2008**, *10*, 1627–1631.
37. Li, L. Research on the Application of Ultrasonic Nondestructive Testing in Bolt Flaw Detection. Master's Thesis, North China Electric Power University, Beijing, China, 2022.
38. Zeng, Z. *Practical Handbook of Steel Materials*; Mechanical Industry Press: Beijing, China, 2015.
39. Xu, Z.; Fan, Z. Phenomenological Interpretation of the Variation Law of Elastic Modulus of Metal Materials with Temperature. *J. Southwest Jiaotong Univ.* **1993**, 2, 92–97.
40. Luo, L. Testing of Internal Plane Stress in Metal Materials by Acoustoelastic Effect. Master's Thesis, Jiangsu University of Science and Technology, Zhenjiang, China, 2010.

materials

MDPI

Article

Active-Passive Joint Acoustic Emission Monitoring Test Considering the Heterogeneity of Concrete

Zhehan Chen [1], Tianjiao Miao [1], Tao Liu [1], Xuandong Chen [1,2] and Aiping Yu [1,2,*]

[1] College of Civil and Architectural Engineering, Guilin University of Technology, Guilin 541004, China; 2120210725@glut.edu.cn (Z.C.); 1020230752@glut.edu.cn (T.M.); 2120230888@glut.edu.cn (T.L.); 6616051@glut.edu.cn (X.C.)
[2] Guangxi Key Laboratory of Geomechanics and Geotechnical Engineering, Guilin 541004, China
* Correspondence: apyu@glut.edu.cn; Tel.: +86-0773-5896112

Abstract: The heterogeneity of concrete is a major challenge for acoustic emission monitoring. A method of active-passive joint acoustic emission monitoring considering the heterogeneity of concrete is presented herein, and the time-frequency-space multi-parameter response characteristics of active and passive acoustic emission signals were studied in relation to the damage evolution of concrete. This method provides an idea of evaluating the damage state of concrete more actively and quantitatively than traditional methods. The results show that the microscopic damage model of concrete based on the acoustic emission penetrating wave velocity and amplitude is in agreement with the damage process of concrete. The standard deviation of the wave velocity up to 1000 m/s and the change rate of the amplitude up to −0.66 can be adopted as two signs that the load of concrete reached 70% of the ultimate load. The time-of-arrival localization based on variable velocity was used to correct the acoustic emission localization results, and the localization accuracy was increased by 44.74%. The damage process of concrete undergoes diverse changes; that is, the distribution of damage changes from heterogeneous to homogeneous and then back to heterogeneous. Hence, it is necessary for researchers to consider the heterogeneity of concrete when using acoustic emission monitoring. The active-passive joint acoustic emission monitoring is an effective method.

Keywords: concrete; active-passive joint acoustic emission monitoring; wave velocity; amplitude; time-of-arrival localization

Citation: Chen, Z.; Miao, T.; Liu, T.; Chen, X.; Yu, A. Active-Passive Joint Acoustic Emission Monitoring Test Considering the Heterogeneity of Concrete. *Materials* **2023**, *16*, 7694. https://doi.org/10.3390/ma16247694

Academic Editors: Sergei Tarasov and DIMOS A. Triantis

Received: 4 November 2023
Revised: 29 November 2023
Accepted: 13 December 2023
Published: 18 December 2023

1. Introduction

The damage evolution of concrete has always been of great concern. Traditional strategies rely on the continuity of materials and often treat the concrete material as homogeneous. Although this method can produce the desired results under certain conditions, the findings are often inconsistent, and some meaningful information about the heterogeneity of concrete is easily ignored [1,2]. Fang et al. [3] established a heterogeneous concrete model and found that the fracture evolution process of the heterogeneous model was more consistent with the results of a CT test than that of the homogeneous model. Zima et al. [4] found that Lamb waves propagate with different speeds in homogeneous and heterogeneous concrete models.

Acoustic emission (AE) technology is a passive monitoring technology, which can collect the instantaneous elastic waves released during concrete fracture in real time [5–7]. The fracture information carried by elastic waves is used to provide valuable clues for the dynamic damage of concrete [8–10]. A variety of AE characteristic parameters (energy, ringing count, hit count, etc.) related to concrete microcracks are often directly used to describe the accumulation, development, and acceleration of microcracks [11,12]. AE energy is the area of the AE waveform envelope, AE ringing count is the number of AE wave peaks, and AE hit is the number of waveforms that are collected. Chen et al. [13] found that the time-varying curves of cumulative AE hits and ringing counts have two obvious inflection points, which are related to the peak load and the boundary effect of concrete. In addition, cumulative AE energy can represent the changing trend of concrete fracture energy under different loading rates. Song et al. [14] thought that P-wave speed can be used as the precursor of concrete failure under cyclic tensile load. Some methods in signal processing (B-value, RA-AF, frequency-domain analysis, etc.) have been proposed and used to explain the fracture mechanism of concrete and identify the types of cracks [15–20].

The above research proved the feasibility of AE in identifying and diagnosing the damage of concrete, but the following two main problems still exist:

(1) The heterogeneity of concrete was not fully considered. AE waves are attenuated, dispersed, and diffracted in concrete, resulting in their distortion [21–25]. The complex propagation behavior of AE waves makes it difficult to consider the heterogeneity of concrete. In the current literature, numerical methods have been used to investigate the effect of heterogeneity on the propagation characteristics of AE waves. It was usually related to the size of specimen [26–28]. The effect of heterogeneity on the propagation characteristics of AE waves in small samples is generally considered to be negligible [29]. However, it is difficult to explain some experimental phenomena because the heterogeneity of concrete usually leads to uneven damage distribution, which further affects the propagation behavior of AE waves, resulting in greater errors. It is not conducive to the formation of a unified AE damage evaluation index. Mpalaskas et al. [30] and Tayfur et al. [31] emphasized that the heterogeneity of concrete can lead to incorrect AE results. For the same specimen, changing the position of AE sensors may result in different results. Li et al. [32] found that free water in a concrete medium influences its AE characteristics and ultrasonic propagation behavior. Obviously, the development of an active AE method considering the heterogeneity of concrete is of great significance for AE monitoring.

(2) The existing AE monitoring method is too dependent on the presence of loads. It is difficult for existing AE methods to actively detect the damage state of concrete under no load and constant load, and it is also difficult to quantitatively characterize the damage state of concrete. On the other hand, most of the AE microscopic damage models developed in the laboratory are dependent on the cumulative AE parameters [33–35]. Yu et al. [2] established a damage evaluation model of concrete using cumulative AE ringing counts and AE energy uniaxial compression. Gu et al. [36] established a damage constitutive model by AE energy, considering critical damage based on Weibull-distributed. To obtain the cumulative AE parameters, it is necessary to monitor the entire loading history of concrete. This makes AE monitoring too passive and time-consuming. Unexpected interruptions in testing are very detrimental to AE monitoring. This limits the application of AE. It is very important to find a method to evaluate the damage state of concrete actively, quantitatively, and timely.

This paper aims to propose an active-passive joint AE monitoring method considering the heterogeneity of concrete. Existing passive AE methods are the usual means of qualitatively evaluating the real-time developmental state of the damage of concrete. In contrast, active AE methods can quantitatively evaluate the damage state of concrete at a certain moment or stress level. Therefore, the active-passive joint AE monitoring method provides a more comprehensive understanding of the damage evolution mechanism of concrete. While exploring the active AE method, the influence of the heterogeneity of concrete in small scale samples on AE testing is discussed. The fundamental basis for the active AE detection method is that pores and cracks in concrete affect the propagation behavior of AE waves, and thus affect the AE information [37–39]. The artificially excited AE waves propagate through the sample, and the signal characteristics change, thus conveying information about the heterogeneity and damage state of concrete. Moreover, concrete samples with different water–cement ratios (w/c ratio) were tested by step-loading, and the active-passive joint AE tests were carried out under different loading levels in this paper. The time-varying and frequency-varying characteristics of active signals and passive signals are analyzed. The damage process of concrete is quantitatively explained by AE wave velocity and amplitude. The results indicate that the effect of concrete heterogeneity on AE characteristics should be considered even in samples of small size, and the wave velocity of active signals is used to correct the AE localization results, which turns out to be an effective approach.

2. Materials and Methods

2.1. Experiment Design

In the experiment, three groups of samples, A, B, and C, were prepared, and the w/c ratios were 0.58, 0.48, and 0.38, respectively. There were 6 samples in each group, numbered from 1 to 6. The size of the specimen was 150 mm $\times$150 mm $\times$ 150 mm. The mix ratios of the samples are shown in Table 1.

Table 1. Proportions of concrete mix.

Group Code	W/C Ratio	Cement (kg/m^3)	Water (kg/m^3)	Sand (kg/m^3)	Macadam (kg/m^3)
A	0.58	458.72	266.06	534.05	991.80
B	0.48	500.30	240.10	545.60	1013.20
C	0.38	550.10	209.00	559.40	1038.80

The cement was ordinary Portland cement (OPC). Fine aggregate is medium sand, and coarse aggregate is gravel with a maximum particle size of 20 mm. Cement, sand, and gravel were all produced in Guilin Xiangjiu Sand Field. All samples were made at the same time. After demolding, they were maintained under standard curing conditions for 28 days.

2.2. Testing System

The test system is composed of mechanical, AE, and AWG systems, as shown in Figure 1. The mechanical system is SUNS universal testing machine. The AE system is Sensor Highway 3 AE device and is used to collect AE signals. AE sensor is PK15I narrow-band resonance sensor, and the operating frequency range is 100–450 kHz. The lower limit of frequency filtering is 20 kHz, and the upper limit is 1 MHz. The sampling frequency is 1 MHz. The AWG system is an arbitrary waveform generator (AWG), and the model is ARB-1410. The software WaveGen 13.0 can control the AWG to transmit AE signals of different frequencies. After initial testing, AWG signals of 120 kHz were relatively stable. Therefore, the signal of this frequency was selected for active AE test. Three AE and one AWG measuring points were set up on the left side of the sample, and four AE measuring points were set up on the right side of the sample. The field ambient noise and mechanical noise were less than 35 dB, so the AE threshold value was set to 40 dB.

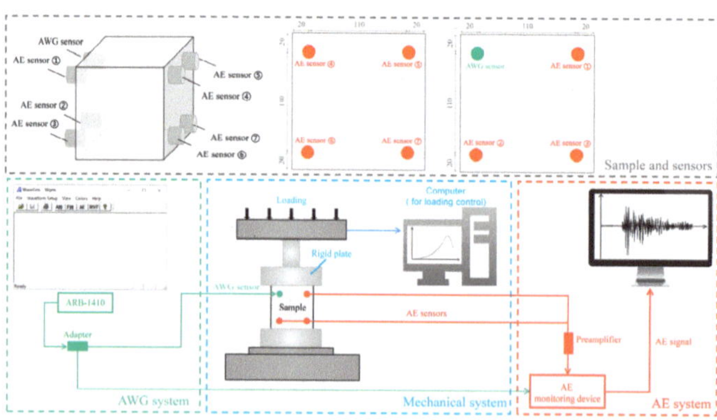

Figure 1. Test system diagram.

2.3. Test Program

As shown in Figure 2, the test adopted the step-loading method. The loading process included Rise phase (*RP*) and Invariant phase (*IP*). The loading rate of *RP* was 0.1 MPa/s, and the *RP* lasted for 30 s. AE damage signals were collected throughout the *RP*, which is called passive monitoring. During the *IP*, which lasted for 400 s, the load remained unchanged. The samples were loaded to failure according to the above loading program. To eliminate the mutual interference between active and passive tests, the active test was conducted during the last 300 s of *IP*. The AWG was used to excite active signals, and the signals were received by 4#~7# AE sensors, which is called active detection.

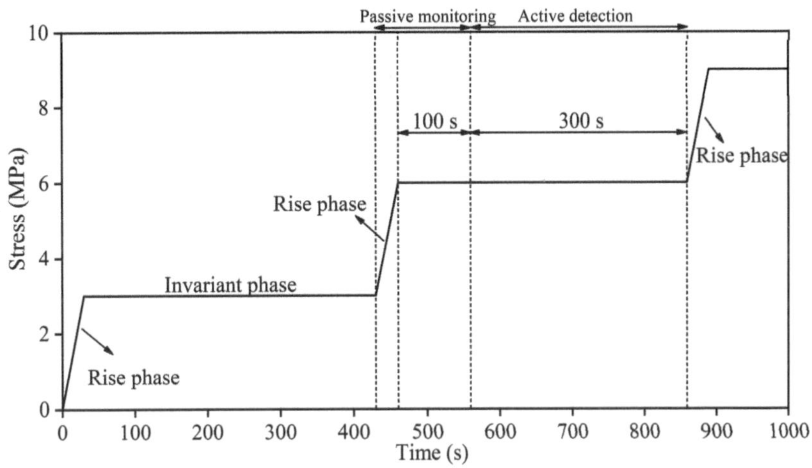

Figure 2. Test program diagram.

2.4. Test Procedure

With an increase in load, microcracks in concrete will gradually develop. However, AE waves cannot propagate directly through cracks. When a crack is encountered, the propagation path of AE waves will change. The AE waves produced by AWG are also affected by cracks, and their characteristic parameters and frequency-domain characteristics will change. If this change can be captured, it can provide a basis for AE active detection. It is worth noting that the propagation mechanism of AE waves on the surface and inside the concrete is different. The microcracks are generated and developed inside the concrete. Microcracks have a more pronounced effect on AE waves propagating through the interior of the concrete (defined as penetrating waves). Therefore, it is more reliable to study the effect of microcracks on AE waves by using penetrating waves.

To further clarify these descriptions, AWG signals are called active signals, and the damage signals of samples are called passive signals.

2.5. Test Theory

2.5.1. Microscopic Damage Model of Concrete

AE waves are generally assumed to propagate uniformly in a straight line through the material. Based on this assumption, the wave velocity of AE waves can be obtained using the following formula:

$$v = \frac{l}{t_1 - t_0}, \quad (1)$$

where v is the average wave velocity of AE wave; l is the propagation distance of the wave; and t_0 and t_1 represent the arrival times of the transmitted and received waves, respectively.

Liu et al. [40] defined the damage variable of concrete as:

$$D = 1 - \frac{\widetilde{E}}{E_0},\tag{2}$$

where D is the damage variable, E_0 is the elastic modulus of concrete in the initial state, and $\widetilde{E}$ is the elastic modulus of concrete after damage.

The wave velocity of AE wave in medium is:

$$v_p = \sqrt{\frac{E(1-\mu)}{\rho(1+\mu)(1-2\mu)}},\tag{3}$$

where v_p is the AE wave velocity, E is the elastic modulus of the medium, μ is the Poisson ratio of the medium, and ρ is the density of the medium.

Ignoring the changes of μ and ρ during the damage process of concrete, the microscopic damage model of concrete based on AE wave velocity can be obtained:

$$D_v = 1 - \left(\frac{\widetilde{v}_p}{v_{p0}}\right)^2,\tag{4}$$

where v_{p0} and $\widetilde{v}_p$ are the initial AE wave velocity and damaged AE wave velocity, respectively.

In fact, even for the same sample, the spatial evolution law of cracks observed at different measuring points is different. This makes the variation law of AE wave velocity also reflect a certain discreteness. This divergence can be demonstrated by using standard deviation. The standard deviation of the velocity of AE penetrating waves is obtained using Formula (5):

$$S = \sqrt{\frac{\sum_{i=1}^{N}(v_i - \overline{v})^2}{N}},\tag{5}$$

where S is the standard deviation of wave velocity, v_i is the AE penetrating wave velocity, $\overline{v}$ is the average wave velocity, and N is the number of samples of penetrating wave velocity (N = 24).

Further, the microscopic damage model of concrete based on amplitude is obtained by using the method similar to Formula (6):

$$D_A = 1 - \left(\frac{\widetilde{A}_p}{A_{p0}}\right)^2,\tag{6}$$

where A_{p0} and $\widetilde{A}_p$ are initial AE amplitude and damaged AE amplitude, respectively.

The change rate of amplitude can be used to describe the trend of amplitude change. The change rate of amplitude is defined as:

$$P = \frac{A_i - A_j}{\sigma_i - \sigma_j},\tag{7}$$

where A_i and A_j are the amplitudes obtained by the i and j measurements, respectively; σ_i and σ_j are the load on the sample at the i and j measurements, respectively.

2.5.2. Time-of-Arrival Localization Based on Variable Velocity

The basic idea of traditional time-of-arrival localization method is to estimate the location of AE source by the difference of arrival time of AE signal from different sensors and AE wave velocity. As shown in Figure 3, when an AE signal is received by multiple

AE sensors (R_1, R_2, R_3), AE wave velocity can be calculated using Formula (1). Then the source location is calculated using Formula (8):

$$\|X_S - X_{R,i}\| - \|X_S - X_{R,j}\| = v \cdot (t_i - t_j), \forall i, j \in [1, 2, \ldots, n], \tag{8}$$

where X_S, $X_{R,i}$, and $X_{R,j}$ represent the AE source location, the location of sensor i, and the location of sensor j, respectively.

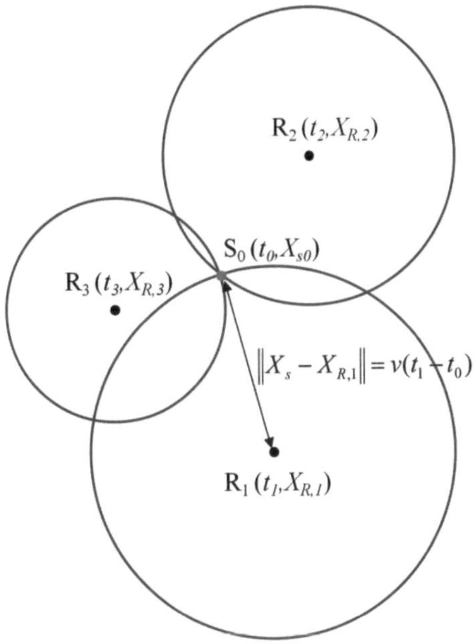

Figure 3. Diagram of time-of-arrival localization.

In the traditional method, the wave velocity calculated using Formula (1) is considered to be constant. The results of this paper show that the development of cracks has a great influence on the AE penetrating wave velocity. Traditional methods do not consider this phenomenon, so it is easy to produce large errors. This paper proposes to correct the error of wave velocity related to the time-varying characteristics of AE wave velocity. By substituting the variable velocity into Formula (8), the localization results can be corrected. The above method is called the time-of-arrival localization based on variable velocity.

3. Results

3.1. Time-Varying Characteristics of AE Parameters

3.1.1. Wave Velocity

As mentioned above, the initiation and development of microcracks will change the propagation path of AE waves. Therefore, the wave velocity calculated using Formula (1) will change. It should be noted that the wave velocity is usually constant in the same medium. The wave velocity described in this paper is the relative wave velocity calculated to reflect the effect of cracks on the AE wave propagation path (hereinafter referred to as wave velocity). The wave velocity is the average of the test results of 4#~7# sensors. Figure 4 shows the time-domain joint response relationship among AE wave velocity, passive signals, and stress in the damage process of the concrete.

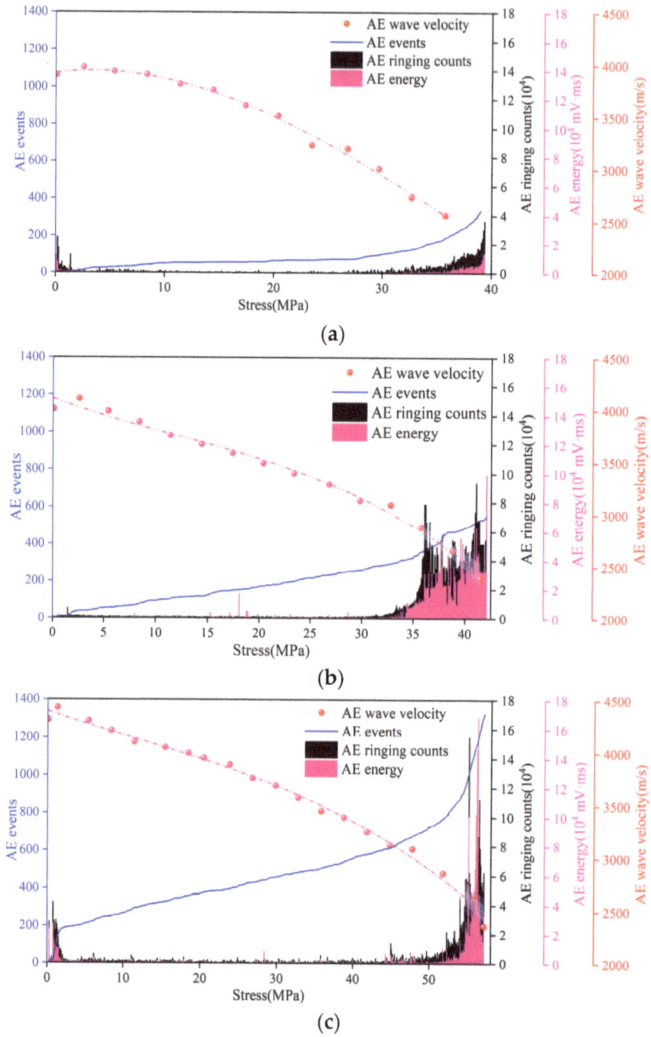

Figure 4. Wave velocity fitted curve and the relationship between stress and AE parameters for (**a**) Group A, (**b**) Group B, and (**c**) Group C.

It can be seen in Figure 3 that the concrete damage evolution process can be divided into three stages according to the time-varying characteristics of AE parameters:

- Compaction stage (stress level 0–10%): The initial microcracks within the concrete were closed under compression, resulting in AE events with certain ringing counts and energy. In this stage, the penetrating wave velocity increased slightly because the propagation distance of the penetrating wave was shortened to some extent by the closure of the original microcrack. The initial penetrating wave velocities of Groups A, B, and C were approximately 3899 m/s, 4044 m/s, and 4311 m/s, respectively. The peak penetrating wave velocities of Groups A, B, and C were approximately 3975 m/s, 4103 m/s, and 4425 m/s, respectively. The penetrating wave velocities of Groups A, B, and C were increased by 3.01%, 1.46% and 2.64%, respectively.
- Stable stage (stress level 10–70%): The closed initial microcracks opened under pressure and began to develop. New cracks started in the matrix and developed continuously. In this stage, the number of AE events continued to accumulate, and the penetrating

wave velocity gradually decreased, indicating that the cracks were developing continuously. The AE ringing count and energy were low, indicating that the development of cracks was not drastic. Therefore, this stage was the AE quiet period.

- Unstable stage (stress level 70–100%): There was a qualitative change in the development of microcracks. The speed of crack development was accelerating. Many cracks crossed and aggregated and finally formed macroscopic cracks. The AE events increased sharply and carried high ringing counts and energies, indicating that the crack development was very intense. The penetrating velocity dropped sharply. The final penetrating wave velocity measurements of Groups A, B, and C were 2561 m/s, 2386 m/s, and 2359 m/s, respectively. Compared with the peak velocity, it was reduced by 35.57% (for Group A), 41.86% (for Group B), and 46.69% (for Group C). The smaller the w/c ratio, the larger the wave velocity loss ratio.

By fitting the wave velocity in Figure 4, the relationship between stress σ and wave velocity v can be obtained:

Group A:

$$v = 3948.72801 + 3.02704\sigma - 1.18115\sigma^2, R^2 = 0.99215, \tag{9}$$

Group B:

$$v = 4055.42762 - 13.45897\sigma - 0.57768\sigma^2, R^2 = 0.98677, \tag{10}$$

Group C:

$$v = 4354.65118 - 11.21413\sigma - 0.36486\sigma^2, R^2 = 0.98549 \tag{11}$$

The correlation coefficients R^2 of Formulas (9)–(11) are all greater than 0.98, and the accuracy of the results is high. The fitting curve is shown by the red dashed line in Figure 4.

3.1.2. Microscopic Damage Model of Concrete Based on Wave Velocity

As mentioned in Section 3.1, the AE wave velocity first increased and then decreased during the damage process of the concrete. When the AE wave velocity increased, D_v was negative, which was unreasonable. To avoid this contradiction, we took the maximum AE wave velocity during the damage process of the concrete as v_{p0}. The AE wave velocity used for analysis here was the average of the data obtained from each set of six samples and four measuring points.

As shown in Figure 5, S and D_v in each group showed similar changes. D_v represents the initial damage of the concrete before loading. During stage I (stress level 0–10%), the degree of damage of the concrete was reduced. D_v decreased to 0, and there was little change in S. The maximum S values of groups A, B, and C were 123 m/s, 98 m/s, and 104 m/s, respectively. During stage II (stress level 10–40%), microcracks began to develop, and D_v and S began to increase. D_v and S grew slowly during this stage because the crack development was still very slow at this time. The maximum S values of groups A, B, and C were 235 m/s, 233 m/s and 222 m/s, respectively. During stage III (stress level 40–70%), the crack development accelerated. D_v and S also accelerated, and their curves became steeper. The maximum S values of Groups A, B, and C were 1019 m/s, 1301 m/s, and 1532 m/s, respectively. At the end of this phase, the S reached its peak value. During stage IV (stress level 70 to 100%), crack development was very active, and D_v increased at a faster rate until the sample failed. However, S decreased significantly during this stage. In the last test, S in each group decreased by 62.01% (for group A), 39.43% (for group B), and 26.79% (for group C) from the peak values. In summary, S reaching 100 m/s can be used as a harbinger of the beginning of microfracture development. When S reaches 1000 m/s or drops significantly, the concrete is about to be damaged.

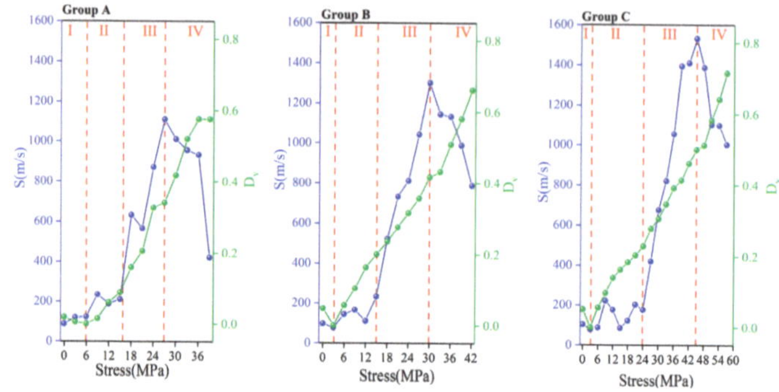

Figure 5. Concrete damage model D_v and standard deviation of wave velocity.

3.1.3. Amplitude

The amplitude can reflect the strength of the signal. The frequent occurrence of high-amplitude signals may be related to the macroscopic cracking of concrete. To facilitate observation, the amplitude of all passive signals was normalized.

Figure 6 shows the amplitude density distribution of passive signals and 50 points with the lowest density. Signals in the low amplitude range of 0.4–0.6 occupied the largest proportion. The density of low-amplitude signals increased with the increase in stress and reached a peak density near peak stress. The high-amplitude signals with amplitudes above 0.8 were mainly distributed in the later part of the loading process. At this point, it entered the unstable phase. The AE response associated with macroscopic cracking of the sample was very dramatic. This mirrors the AE ringing and energy surge described in Section 3.1.1 because, in general, the ringing count and energy of the high-amplitude signals was generally higher. The low-density distribution of high-amplitude signals was due to the small number of such signals. Compared with the low-amplitude signals throughout the whole damage process of the concrete, the number of high-amplitude signals was not dominant.

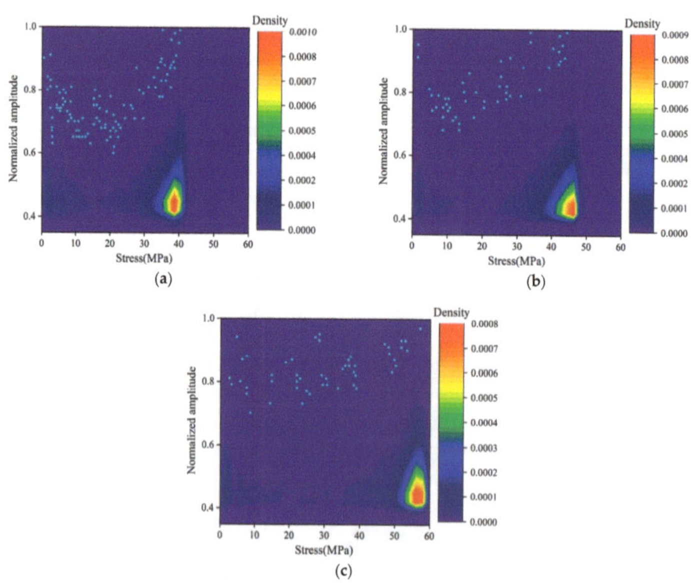

Figure 6. Amplitude density distribution of the passive signals for (**a**) Group A, (**b**) Group B, and (**c**) Group C.

Figure 7 shows the evolution of the amplitude of active signals. Similar to wave velocity, the amplitude showed a trend of first increasing and then decreasing. Before loading, the amplitude of the samples with a low w/c ratio was higher because the samples with a low w/c ratio had smaller porosity. The propagation distance of the penetrating wave was relatively short, and the attenuation degree was small. During the compaction stage, the amplitudes of groups A, B, and C increased by 4.88%, 2.68% and 1.44%, respectively. Unlike the wave velocity, the amplitude dropped more slowly during the late loading period than during the middle loading period. According to the quality factor Q theory, there is a negative exponential relationship between the amplitude and the propagation distance [41]. Therefore, the amplitude attenuation in this test resulted in a trend from fast to slow.

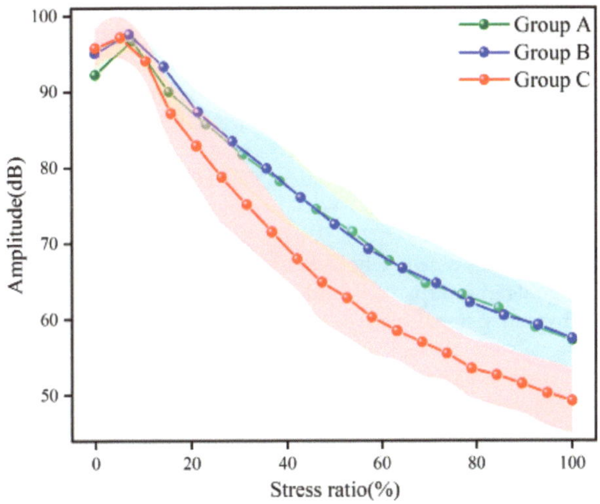

Figure 7. The amplitude of active signals. (The background color is error band).

3.1.4. Microscopic Damage Model of Concrete Based on Amplitude

Similarly, to avoid the unreasonable situation where D_V is negative, the maximum AE amplitude during the loading process was taken as A_{p0}. As mentioned in Section 3.1.3, the attenuation of the amplitude showed a tendency from fast to slow.

As shown in Figure 8, D_A underwent a change of first decreasing and then increasing, which is similar to the change in D_v in Section 3.1.2. The difference is that the ascending section of D_A showed a trend from steep to gentle. This corresponded with a change in P. When P was greater than 0, the amplitude increased. At this time, the sample was in the compaction stage, the damage degree was slightly reduced, and the D_A was reduced to 0. Subsequently, the crack began to develop, and P plummeted below 0, indicating that the amplitude began to decrease. The D_A began to rise, indicating that the damage was deepening. With the increase in load, P gradually increased, which means that the attenuation of the amplitude gradually slowed down. When the P reached −1.00000 (for group A), −0.66917 (for group B), and −0.65667 (for group C), the samples entered the unstable stage (Figure 8). This indicated that the sample was about to be destroyed. In summary, the changes in D_A and P were consistent with the damage process of the concrete.

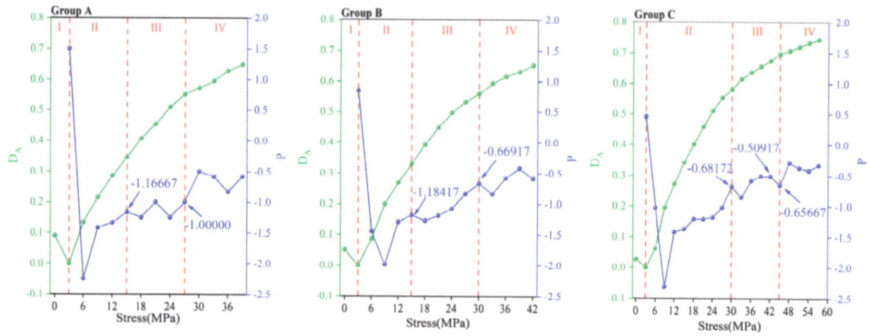

Figure 8. Microscopic damage model of concrete D_A and change rate of amplitude P.

3.2. Frequency-Varying Characteristics of AE Signals

Fast Fourier transform (*FFT*) was performed on the active and passive signals, and it was found that the spectrum characteristics of the two signals were obviously different.

Figure 9 shows the main frequency for all passive signals. The passive signals had a wide frequency range within 0–500 kHz. Overall, the passive signals can be divided into low-frequency signals (0–200 kHz), medium-frequency signals (200–300 kHz), and high-frequency signals (300–500 kHz). The low-frequency and medium-frequency signals were almost all over the whole damage process of concrete. The proportions of low-frequency signals in groups A, B, and C were 90.62%, 96.18%, and 95.54%, respectively. The low-frequency signals occupied the dominant position regarding quantity. The high-frequency signals mainly existed during the late period of the loading process because the cracks during this period were very active, and it was easy to produce high-frequency signals. This corresponded to a surge of passive AE characteristic parameters in the later stage of loading process, as shown in Figure 3. It can be inferred that the high-frequency signals were related to the macroscopic cracking of the concrete. This feature can be used as a harbinger of concrete failure.

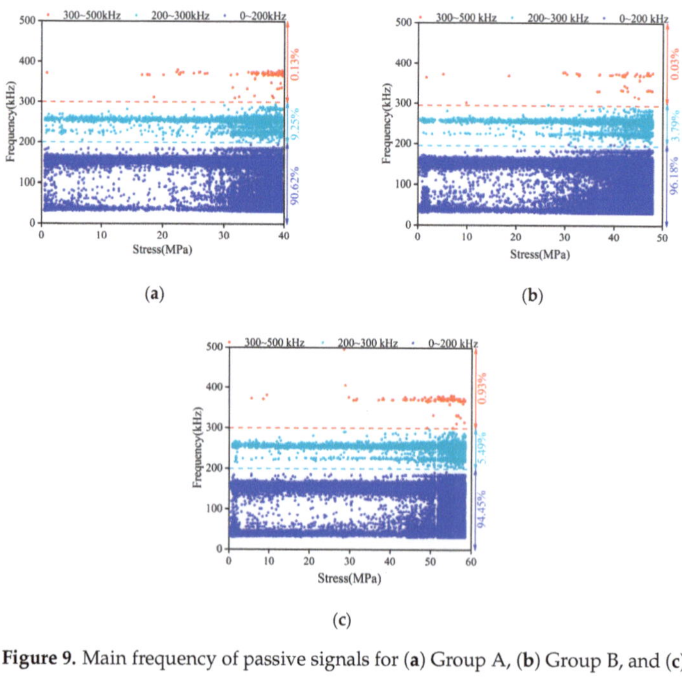

Figure 9. Main frequency of passive signals for (**a**) Group A, (**b**) Group B, and (**c**) Group C.

Further, taking group *A* as an example, the frequency spectrum signature of the signals in different damage stages were analyzed. Overall, the amplitude of the active signal was high, and the energy was concentrated near the main frequency. There was basically no energy distributed at the high-frequency bands. The main frequency and secondary frequency of the passive signals were messier, and there were even multiple secondary frequencies. The amplitude of the passive signal was lower, but the energy distribution range was significantly wider.

During stage I, the active signal consisted of multiple peaks, and the microcracks were closed under pressure (Figure 10). With the increase in load, the initial damage of the sample decreased, and the load was not enough to produce nascent cracks. Therefore, the main frequency and the amplitude of each frequency band of the passive signal increased. At this stage, the attenuation of the AE signals decreased. The spectrum energy of the active signals shifted from 138.7 kHz to 149.2 kHz.

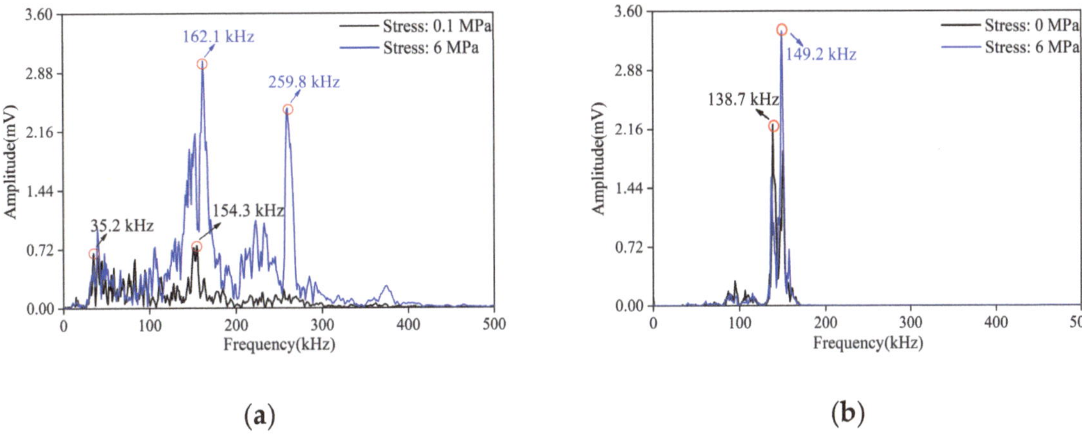

Figure 10. Spectrum diagram of active signals (**a**) and passive signals (**b**) during stage I.

As shown in Figure 11, when the stress reached 12 MPa (stage II), initial cracks developed, and new cracks started. At this time, the fracture scale was small, and the development was slow. The spectrum energy of the active signal was still concentrated around 150.4 kHz, but the amplitude was reduced. When the stress reached 24 MPa (stage III), the fracture development accelerated. The main frequency of the passive signal moved from low frequency to medium frequency and the spectrum energy of the passive signal was concentrated in the middle-frequency band. The spectrum energy of the active signal shifted to 137.7 kHz. When the stress reached 39 MPa (stage IV), the sample was close to failure. The spectrum energy distribution of the passive signal was more complex, and the number of secondary frequencies increased. Due to the existence of many cracks, the attenuation of the AE signals increased and the amplitude of the active signal decreased.

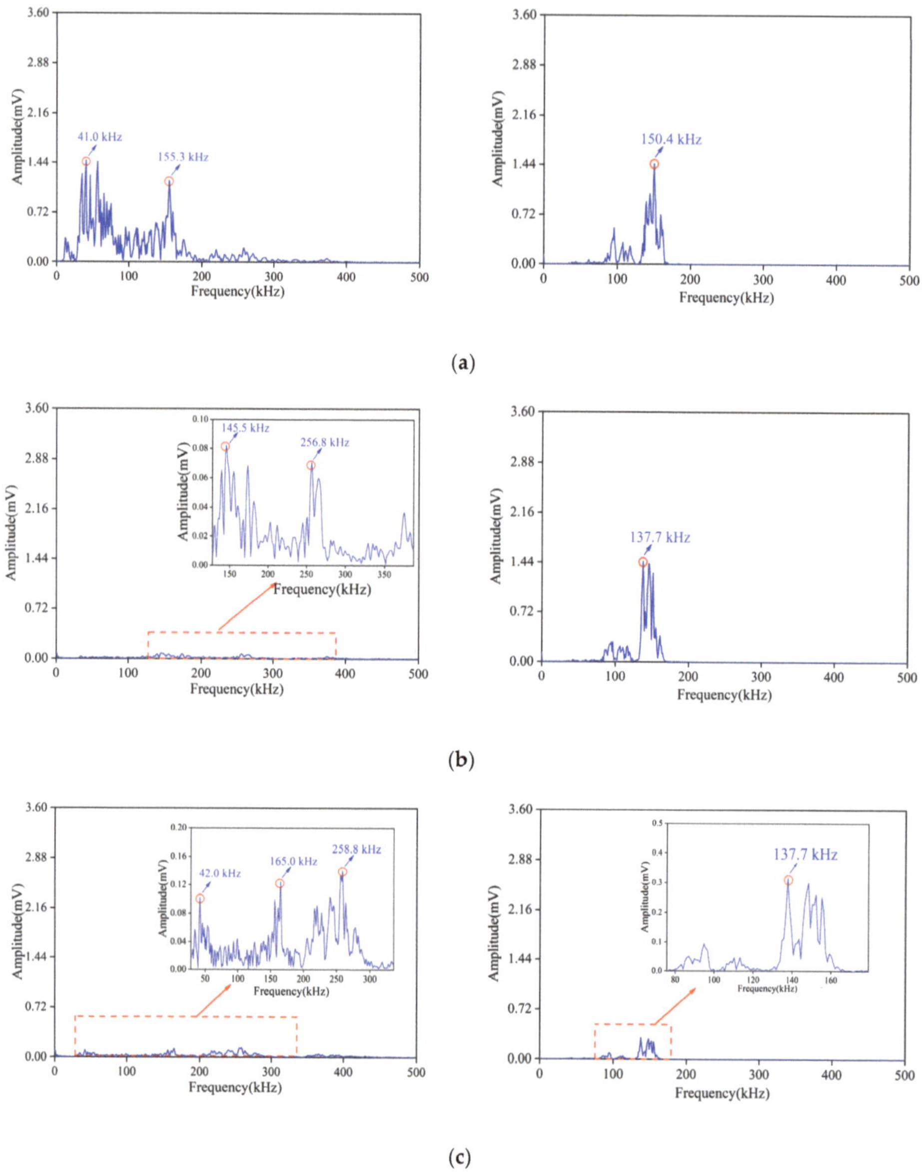

Figure 11. Spectrum diagram of active signals (**left**) and passive signals (**right**) during (**a**) stage II, (**b**) stage III, and (**c**) stage IV.

3.3. Spatial Joint Response of Active and Passive Signals

In this paper, a two-dimensional localization method was adopted. Sample A_1 was taken as an example to illustrate the correction process of the localization results. Figure 12a shows the crack morphology of the front façade when the sample is damaged. Taking the lower left corner of the facade of the sample as the origin, a plane coordinate system was established, and a crack morphology diagram was drawn, as shown in Figure 12b. The coordinate planes were divided into 10 longitudinal bands numbered 1#–10#. The bands where the cracks were located were defined as the crack bands. The AE localization points within crack bands were considered accurate.

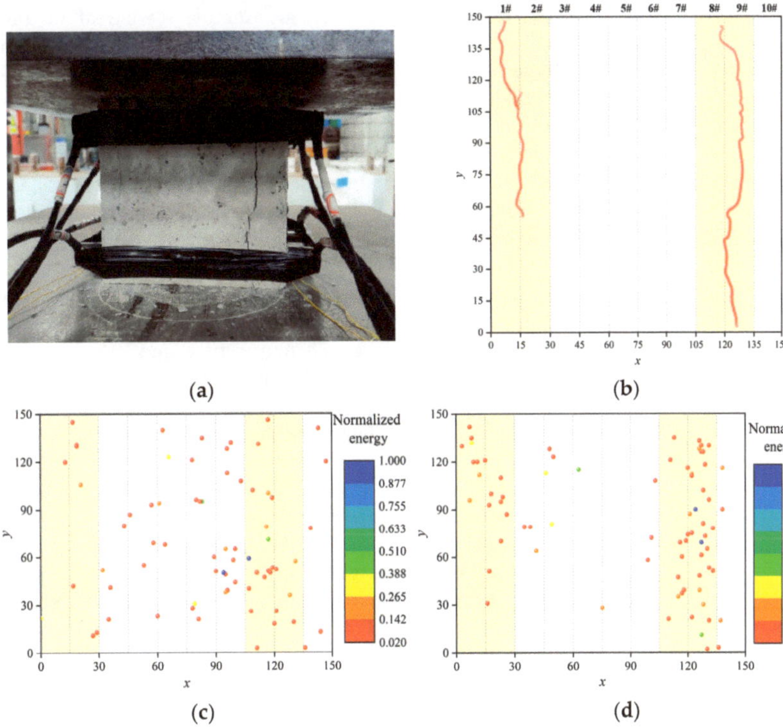

Figure 12. (**a**) Cracks in front facade of sample A_1. (**b**) Crack bands. (**c**) Localization results before correction. (**d**) Corrected localization results.

The 467 AE events generated during the loading process of sample A_1 were screened. The AE energy was used as a screening index because it was very sensitive to the macroscopic cracking of concrete. The AE energy of the 467 AE events was normalized, and the 76 AE localization points with an energy greater than 0.020 were analyzed. The localization results based on the traditional method are shown in Figure 12c. The AE localization points were dispersed in the plane. There were 30 AE events that were located in the crack bands with an accuracy of 39.47%. The AE localization results calculated using the time-of-arrival localization method based on the variable velocity proposed in this paper are shown in Figure 12d. The distribution of AE localization points in the plane was significantly changed. Most of the points converged towards the crack bands. There were 64 AE events located in the crack bands with an accuracy of 84.21%. The accuracy of the localization results increased by 44.74%. From the localization results, the new method proposed in this paper was closer to the actual location of the cracks. This shows that it is feasible to improve the accuracy of an AE source location by using the time-of-arrival localization method based on variable velocity.

4. Discussion

Generally, microcracks start and develop in heterogeneous layers. With the increase in microcrack density, microcracks interconnect with each other and form macroscopic fractures. Linear elastic fracture mechanics cannot fully explain this complex process [42,43]. In this study, the time-varying and frequency-varying characteristics of AE signals were used to visualize the damage evolution process of concrete. These characteristics provide a reference for AE to detect concrete actively. In addition, it was found that the damage evolution process of concrete is a changing process from heterogeneous to homogeneous and back to heterogeneous.

Take the wave velocity for example. If we treat the concrete sample as a homogeneous material, then the wave velocity captured by the 4#~7# sensors should be the same when there is no load. However, the 4#~7# sensors were different. As the load increased, cracks developed in the homogeneous layer. This development was random and had different effects on the sensors in different locations. Therefore, with the increase in load, the wave velocity captured by the different sensors will have a significant dispersion. This dispersion will increase with an increase in load. In short, if the concrete is regarded as a homogeneous material, the wave velocity will not be discrete in the initial state. As the load increases, the dispersion of the wave speed should increase.

However, in our experiments, we observed a different pattern. We used the standard deviation S of the wave velocity to measure the dispersion of the wave velocity and found that the dispersion of the wave velocity exists all the time. In the initial state, the S of groups A, B, and C were 88 m/s, 98 m/s, and 104 m/s, respectively. This showed that even in the initial state, the initial damage of concrete samples at different locations was different. This reflected the non-uniformity of the concrete sample. Even for the same sample, the wave velocity at different measuring points also showed discreteness. This is a satisfying detail because AE equipment usually calculates the AE source location based on wave velocity and then defines AE characteristics. The dispersion of the wave velocity showed the non-negligible effect of the heterogeneity of concrete on AE monitoring.

During the compaction stage, the wave velocity of the AE signal from different sensors increased by different degrees, whereas S decreased by different degrees. This indicated that the internal damage of the concrete sample was reduced during the compaction stage. The sample changed from heterogeneous to homogeneous. Similarly, with the increase in stress, the wave velocity decreased and S increased significantly. This indicated that the sample changed from homogeneous to heterogeneous.

To sum up, it is one-sided to regard small concrete samples as homogeneous materials, which was the case in previous tests. The heterogeneity of concrete should be considered in the exploration of concrete failure precursors and damage assessment methods and evaluated using AE technology. This also reflects the necessity of active-passive joint AE monitoring.

5. Conclusions and Prospects

The main conclusions can be summarized as follows:

1. During stage I, D_A and D_V decreased. Wave velocity and amplitude of the active signals increased. The main frequency and amplitude of the passive signals increased. During stage II, D_A and D_V began to rise slowly. The wave velocity and amplitude of the active signals decreased. The energy of the passive signals decreased. During stage III, D_A and D_V increased rapidly. The wave velocity of the active signals decreased significantly. The passive signals shifted to an intermediate frequency. During stage IV, the spectrogram of the active signals showed multimodal characteristics. The number of passive signals increased sharply, and the number of spectral peaks increased.
2. The microscopic damage model of concrete established in this paper based on the varying characteristics of the wave velocity and amplitude are in agreement with the damage evolution process of concrete. The standard deviation of the wave velocity reached 1000 m/s, and the change rate of amplitude reached −0.66; these changes can be used to signal that the load on the concrete has reached 70% of the ultimate load. Based on the relationship between the penetrating wave velocity and stress, the localization accuracy of the concrete was improved by 44.74%.
3. The damage evolution process of concrete is a changing process from heterogeneous to homogeneous and then back to heterogeneous. The dispersion of the penetrating wave velocity decreases first and then increases, and it is necessary to consider the heterogeneity of concrete even when monitoring samples of small size.

Future research could investigate the following:

1. Other factors that contribute to the heterogeneity of concrete, such as age, moisture content, aggregate size, and sand rate should also be considered.
2. The AE response of concrete under different work conditions should be considered. For example, it is necessary to analyze the damage sequence of concrete using fatigue mode, which is conducive to further improving the feasibility of proposed active-passive joint AE monitoring methods. A discussion of the Kaiser effect is indispensable during this step. We will complete this task in the next study.
3. This paper is a preliminary exploration of active-passive joint AE monitoring. In the future, this method will be used for the assessment of greater structures to enhance its applicability. Moreover, the application conditions of the active-passive joint AE monitoring method in RC structures should be further considered.
4. In addition, some non-contact detection methods, such as X-ray computed tomography and digital image correlation, can be combined with AE monitoring. Establishing the relationship between crack morphology, surface displacement, and the characteristics of active AE signals can further reveal the damage mechanism of concrete and improve the accuracy of AE results.

Author Contributions: Conceptualization, A.Y. and Z.C.; methodology, Z.C.; investigation, T.M. and T.L.; data curation, A.Y. and Z.C.; writing—original draft preparation, Z.C.; writing—review and editing, A.Y. and X.C.; supervision, A.Y.; project administration, A.Y. and Z.C. All authors have read and agreed to the published version of the manuscript.

Funding: This research was funded by the National Natural Science Foundation of China, grant number 51968014, the Innovation Project of Guangxi Graduate Education, grant number YCSW2023367, and the Guangxi Natural Science Foundation, grant number 2022GXNSFAA035553.

Institutional Review Board Statement: Not applicable.

Informed Consent Statement: Not applicable.

Data Availability Statement: The data presented in this study are available on request from the corresponding author.

Acknowledgments: The first author would like to acknowledge all those who helped him during the writing of this dissertation.

Conflicts of Interest: The authors declare no conflict of interest.

References

1. Friedrich, L.F.; Cezar, É.S.; Colpo, A.B.; Tanzi, B.N.R.; Sobczyk, M.; Lacidogna, G.; Niccolini, G.; Kosteski, L.E.; Iturrioz, I. Long-Range Correlations and Natural Time Series Analyses from Acoustic Emission Signals. *Appl. Sci.* **2022**, *12*, 1980. [CrossRef]
2. Yu, A.; Chen, Z.; Zhang, L.; Li, X.; Shi, J.; Feng, F. Study on AE characteristics of concrete with different w/c ratio under uniaxial compression. *Structures* **2023**, *58*, 105443. [CrossRef]
3. Fang, J.; Yuan, Z.; Liang, J.; Li, S.; Qin, Y. Research on compressive damage mechanism of concrete based on material heterogeneity. *J. Build. Eng.* **2023**, *79*, 107740. [CrossRef]
4. Zima, B.; Kędra, R. Numerical Study of Concrete Mesostructure Effect on Lamb Wave Propagation. *Materials* **2020**, *13*, 2570. [CrossRef] [PubMed]
5. Zhang, H.; Pan, L.Y.; Pan, Q.; Jin, C.J.; Jin, S.S.; Liu, X.Y. Experimental Study on Splitting Tensile Damage Characteristics and Discrete Element Simulation of Concrete Based on Acoustic Emission Technology. *J. Mater. Civ. Eng.* **2023**, *35*, 04022385. [CrossRef]
6. Oneschkow, N.; Timmermann, T. Influence of the composition of high-strength concrete and mortar on the compressive fatigue behaviour. *Mater. Struct.* **2022**, *55*, 83. [CrossRef]
7. Wang, Y.; Zhang, T.; Yan, C.; Wang, N.; Yao, F.; Chen, L.; Gu, J. Effect of different loading systems on acousto-ultrasonic characteristics of concrete under axial compression. *Mater. Test.* **2019**, *61*, 591–599. [CrossRef]
8. Burud, N.; Kishen, J. Investigation of long memory in concrete fracture through acoustic emission time series analysis under monotonic and fatigue loading. *Eng. Fract. Mech.* **2023**, *277*, 108975. [CrossRef]
9. Yang, K.; Li, D.; He, Z.; Zhou, H.; Li, J. Study on Acoustic Emission Characteristics of Low-Temperature Asphalt Concrete Cracking Damage. *Materials* **2021**, *14*, 881. [CrossRef]
10. Goyal, P.; Sharma, S.; Kwatra, N. Acoustic emission monitoring of steel fiber reinforced beams under simultaneous corrosion and sustained loading. *Eur. J. Environ. Civ. Eng.* **2023**, *27*, 1535–1560. [CrossRef]

11. Singh, P.; Yogesh, R.; Bhowmik, S.; Kishen, J.M.C. Insights into the fracturing process of plain concrete under crack opening. *Int. J. Fract.* **2023**, *241*, 153–170. [CrossRef]
12. Yang, L.; Xie, H.; Fang, S.; Huang, C.; Yang, A.; Chao, Y.J. Experimental study on mechanical properties and damage mechanism of basalt fiber reinforced concrete under uniaxial compression. *Structures* **2021**, *31*, 330–340. [CrossRef]
13. Chen, C.; Fan, X.; Chen, X. Experimental investigation of concrete fracture behavior with different loading rates based on acoustic emission. *Constr. Build. Mater.* **2020**, *237*, 117472. [CrossRef]
14. Song, Z.; Frühwirt, T.; Konietzky, H. Fatigue characteristics of concrete subjected to indirect cyclic tensile loading: Insights from deformation behavior, acoustic emissions and ultrasonic wave propagation. *Constr. Build. Mater.* **2021**, *302*, 124386. [CrossRef]
15. Tra, V.; Kim, J.Y.; Jeong, I.; Kim, J.M. An Acoustic Emission Technique for Crack Modes Classification in Concrete Structures. *Sustainability* **2020**, *12*, 6724. [CrossRef]
16. Ju, S.; Li, D.; Jia, J. Machine-learning-based methods for crack classification using acoustic emission technique. *Mech. Syst. Signal Process.* **2022**, *178*, 109253. [CrossRef]
17. Bai, J.; Su, H.; Yin, B.; Cai, Y. Mechanical properties and damage mechanisms of concrete under four temperature gradients combined with acoustic emission method. *J. Build. Eng.* **2022**, *57*, 104906. [CrossRef]
18. Lv, Z.; Jiang, A.; Jin, J. Influence of ultrafine diatomite on cracking behavior of concrete: An acoustic emission analysis. *Constr. Build. Mater.* **2021**, *308*, 124993. [CrossRef]
19. Li, X.; Chen, X.; Jivkov, A.; Hu, J. Assessment of damage in hydraulic concrete by gray wolf optimization-support vector machine model and hierarchical clustering analysis of acoustic emission. *Struct. Control Health Monitoting* **2022**, *29*, e2909. [CrossRef]
20. Chen, C.; Chen, X.; Xu, W.; Cheng, X. Fracture behavior and crack mode of steel slag pervious concrete using acoustic emission technique. *Struct. Control Health Monitoting* **2021**, *28*, e2796. [CrossRef]
21. Aggelis, D.G.; Shiotani, T.; Papacharalampopoulos, A.; Polyzos, D. The influence of propagation path on elastic waves as measured by acoustic emission parameters. *Struct. Health Monitoting-Int. J.* **2012**, *11*, 359–366. [CrossRef]
22. Ospitia, N.; Hardy, A.; Larbi, A.S.; Aggelis, D.; Tsangouri, E. Size Effect on the Acoustic Emission Behavior of Textile-Reinforced Cement Composites. *Appl. Sci.* **2021**, *11*, 5425. [CrossRef]
23. Tayfur, S.; Alver, N. A 3D parameter correction technique for damage assessment of structural reinforced concrete beams by acoustic emission. *Constr. Build. Mater.* **2019**, *215*, 148–161. [CrossRef]
24. Verbis, J.T.; Kattis, S.E.; Tsinopoulos, S.V.; Polyzos, D. Wave dispersion and attenuation in fiber composites. *Conmputational Mech.* **2001**, *27*, 244–252. [CrossRef]
25. Aggelis, D.; Tsinopoulos, S.; Polyzos, D. An iterative effective medium approximation (IEMA) for wave dispersion and attenuation predictions in particulate composites, suspensions and emulsions. *J. Acoust. Soc. Am.* **2004**, *116*, 3443–3452. [CrossRef]
26. Schubert, F.; Koehler, B. Three-dimensional time domain modeling of ultrasonic wave propagation in concrete in explicit consideration of aggregates and porosity. *J. Comput. Acoust.* **2001**, *9*, 1534–1560. [CrossRef]
27. Bai, Y.; Liu, Y.; Gao, G.; Su, S. Acoustic Emission Monitoring of Concrete Plate-like Structures Using Deep Transfer Learning and Beamforming. *Instrum. Exp. Tech.* **2023**, *66*, 156–167. [CrossRef]
28. Gollob, S.; Kocur, G. Analysis of the wave propagation paths in numerical reinforced concrete models. *J. Sound Vib.* **2021**, *494*, 115861. [CrossRef]
29. Zhang, F.; Pahlavan, L.; Yang, Y. Evaluation of acoustic emission source localization accuracy in concrete structures. *Struct. Health Monitoting-Int. J.* **2020**, *19*, 2063–2074. [CrossRef]
30. Mpalaskas, A.C.; Thanasia, O.V.; Matikas, T.E.; Aggelis, D.G. Mechanical and fracture behavior of cement-based materials characterized by combined elastive wave approaches. *Constr. Build. Mater.* **2014**, *50*, 649–658. [CrossRef]
31. Tayfur, S.; Zhang, T.; Mahdi, M.; Issa, M.; Ozevin, D. Cluster-based sensor selection framework for acoustic emission source localization in concrete. *Measurement* **2023**, *219*, 113293. [CrossRef]
32. Li, H.; Meng, S.; Shi, D.; Wei, Q.; Xu, Z.; Zhao, W. Influence of moisture on ultrasonic propagation, acoustic emission activity, and failure mechanism in concrete media. *Constr. Build. Mater.* **2023**, *386*, 131499. [CrossRef]
33. Suzuki, T.; Nishimura, S.; Shimamoto, Y.; Shiotani, T.; Ohtsu, M. Damage estimation of concrete canal due to freeze and thawed effects by acoustic emission and X-ray CT methods. *Constr. Build. Mater.* **2020**, *245*, 118343. [CrossRef]
34. Geng, J.; Sun, Q.; Zhang, Y.; Cao, L.; Zhang, W. Studying the dynamic damage failure of concrete based on acoustic emission. *Constr. Build. Mater.* **2017**, *149*, 9–16. [CrossRef]
35. Men, J.; Wang, J.; Guo, L.; Wang, K. Acoustic emission behavior and damage evaluation of recycled aggregate concrete under compression. *Struct. Health Monitoting-Int. J.* **2020**, *27*, e2612. [CrossRef]
36. Gu, Q.; Ma, Q.; Tan, Y.; Jia, Z.; Zhao, Z.; Huang, D. Acoustic emission characteristics and damage model of cement mortar under uniaxial compression. *Constr. Build. Mater.* **2019**, *213*, 377–385. [CrossRef]
37. Li, D.; Yang, K.; He, Z.; Zhou, H.; Li, J. Acoustic Emission Wave Velocity Attenuation of Concrete and Its Application in Crack Localization. *Sustainability* **2020**, *12*, 7405. [CrossRef]
38. Wu, X.; Yan, Q.; Hedayat, A.; Wang, X. The influence law of concrete aggregate particle size on acoustic emission wave attenuation. *Sci. Rep.* **2021**, *11*, 22685. [CrossRef]
39. Tayfur, S.; Alver, N. Attenuation and Frequency Characteristics of Acoustic Waves in Steel and Synthetic Fiber-Reinforced Concrete: 3D-PCT and Unsupervised Pattern Recognition. *Appl. Sci.* **2022**, *12*, 12976. [CrossRef]

40. Liu, X.; Wang, X.; Wang, E.; Liu, Z.; Xu, X. Study on Ultrasonic Response to Mechanical Structure of Coal under Loading and Unloading Condition. *Shock Vib.* **2017**, *2017*, 7643451. [CrossRef]
41. Toksoz, M.; Johnston, D.; Timur, A. Attenuation of seismic waves in dry and saturated rocks: I. laboratory measurements. *Geophysics* **1979**, *4*, 681–690. [CrossRef]
42. Nanjo, K.Z. A fiber-bundle model for the continuum deformation of brittle material. *Int. J. Fract.* **2017**, *204*, 225–237. [CrossRef]
43. Wang, X.; Wang, E.; Liu, X. Damage characterization of concrete under multi-step loading by integrated ultrasonic and acoustic emission techniques. *Constr. Build. Mater.* **2019**, *221*, 678–690. [CrossRef]

sensors

MDPI

Article

Experimental Analysis of the Magnetic Leakage Detection of a Corroded Steel Strand Due to Vibration

Hong Zhang [1,2,*], Yaxi Ding [1,2], Ye Yuan [1,2], Runchuan Xia [1,2] and Jianting Zhou [1,2]

1 State Key Laboratory of Mountain Bridge and Tunnel Engineering, Chongqing Jiaotong University, Chongqing 400074, China; 622210086048@mails.cqjtu.edu.cn (Y.D.); yuanye202307@163.com (Y.Y.); rcxia@mails.cqjtu.edu.cn (R.X.); jtzhou@cqjtu.edu.cn (J.Z.)
2 School of Civil Engineering, Chongqing Jiaotong University, Chongqing 400074, China
* Correspondence: hongzhang@cqjtu.edu.cn

Abstract: The self-magnetic flux leakage (SMFL) detection technique has great potential in the corrosion detection of bridge stay cables due to its advantages of small testing equipment, high accuracy, and fast testing rate. However, the vibration effect in the cable's SMFL detection is unclear. To address this, the influence of vibration on the magnetic field distribution of cable structure is analyzed theoretically. According to the theoretical model, the effect of vibration on SMFL detection primarily manifests as displacement changes (displacement-added magnetic field) and defect shape changes (deformation-added magnetic field). SMFL detection experiments are conducted on steel strands. The results demonstrate that the displacement-added magnetic field exhibits statistical characteristics in the form of a normal distribution, fluctuating around the zero value. The impact of the deformation-added magnetic field on SMFL is linearly correlated with the corrosion ratio c. Moreover, a corrosion characterization index A was proposed and has an excellent linear fit with the corrosion ratio c. The index A effectively improves the accuracy of corrosion detection and provides early warning for the maintenance of cable structures.

Keywords: stranded steel; cable structure; vibration; corrosion; self-magnetic flux leakage

Citation: Zhang, H.; Ding, Y.; Yuan, Y.; Xia, R.; Zhou, J. Experimental Analysis of the Magnetic Leakage Detection of a Corroded Steel Strand Due to Vibration. *Sensors* **2023**, *23*, 7130. https://doi.org/10.3390/s23167130

Academic Editor: Yuan Yao

Received: 13 July 2023
Revised: 5 August 2023
Accepted: 9 August 2023
Published: 11 August 2023

1. Introduction

1.1. Motivation

In recent years, cable-stayed bridges were rapidly promoted and applied worldwide due to their beautiful appearance, high-crossing capacity, and relatively low construction cost [1]. As the primary force-transmitting component of cable-stayed bridges, the performance of the cable directly affects the reliability of the bridge's structural system [2–5]. Due to the prolonged exposure of cables to external conditions, they are susceptible to various environmental factors, such as ultraviolet radiation, high temperature and humidity, acid rain, as well as the combined effect of cyclic loads. As a result, different degrees of damage, such as protective coating erosion, water penetration, and corrosion or breakage of internal steel wires, can occur in the cable system [6–9]. Existing research and engineering practices showed that the performance degradation of a cable often begins with sheath breakage and corrosion of high-strength steel wires [10,11]. Therefore, the research on diagnostic techniques for assessing the service condition of cables was a focal point of attention in both the engineering and academic communities [12].

Non-destructive testing methods became the fastest-growing inspection method due to their ability to assess structural conditions without causing direct damage [13]. The main non-destructive testing methods for detecting defects in cable structures include ultrasonic-guided wave testing [14], acoustic emission testing [15], magnetostrictive-guided wave testing [16], and magnetic leakage testing [17]. In addition, the cables' internal steel wires/strands are ferromagnetic materials, and the magnetic leakage phenomenon occurs in the damaged areas. The magnetic leakage signal in these areas is not hindered by the PE sheath of the cable and exhibits a strong correlation with cable corrosion and wire breakage.

Therefore, among the various non-destructive testing methods for cable structures, the self-magnetic flux leakage (SMFL) testing technique is increasingly recognized and valued by experts and scholars due to its advantages of small testing equipment, high accuracy, and fast testing rate [18,19]. However, this method is prone to interference from spatial magnetic fields, the PE sheath, and external environmental factors [20]. Therefore, it is of great scientific significance and potential engineering application prospects to research the spontaneous magnetic leakage testing method for detecting wire corrosion in cable structures under the coupling effect of multiple factors in natural bridge conditions [21–23].

1.2. Literature Review

In the past few decades, SMFL testing for wire corrosion in cables received significant attention from experts and became a research hotspot. The research focus was primarily on the characterization mechanism of magnetic leakage, diagnostic methods for wire corrosion, magnetic signal acquisition, analysis methods, as well as studies on SMFL imaging and inversion methods [24]. For example, Shi [25] reviewed the mechanisms of the magnetic memory effect, influencing factors of magnetic signals, and evaluation methods for magnetic memory damage over the past decade. Qiu [26] derived the theoretical framework for SMFL testing of parallel steel wires/strands, based on the magnetic dipole model. Xia [27] simplified the defective region as a sector shape for the case of continuous corrosion of multiple wires in the outer layer of parallel wire ropes. They derived a three-dimensional spatial model for spontaneous magnetic leakage by combining magnetic charge density theory and the magnetic dipole model [28].

Experts and scholars have conducted numerous basic experiments while conducting spontaneous magnetic leakage theory characterization research. For example, Fernandes found that the corrosion region evaluated with the SMFL typically matched the actual corrosion observed by visual inspection. They used SMFL to detect the remaining cross-sectional area of a corroded prestressed steel strand [29]. Qu [30] proposed a method of quantitatively evaluating the corrosion width of a steel bar using the intersection point of SMFL curves. They offered a novel way of predicting corrosion width with better accuracy. Singh [31] developed a flexible giant magnetoresistance (GMR) sensor array for the magnetic leakage detection of steel wire ropes, and conducted preliminary binary image imaging exploration of wire break defects. Ren [32] studied the effect of initial magnetization on the magnetic memory signal for the quantitative testing of SMFL detection. Xia [33–35] developed a logistic model to analyze the eigenvalue of the SMFL signal and thus quantitatively assessed the corrosion of steel strands. Shi [36] determined the effects of the load, defect size and lifting height on the magnetic signals. A theoretical model was applied to solve stress concentration problems in demonstrating the feasibility of the model in early diagnosis. Le [37] proposed a method of simulating the magnetic field distribution in magnetic flux leakage detection using a deep neural network. In terms of quantitative characterization of corrosion morphology, Xu [38] developed an indirect method for describing the corrosion morphology by injecting 60 mL of red ink into the cracked sheath of the old cable removed from a bridge and observing the distribution of the red ink within the cable body after two days. Additionally, Li [39] utilized a high-resolution three-dimensional contour measurement device to obtain geometric point cloud images of the steel wire surface, and through digital image processing techniques they derived relevant statistical features of the corroded pits on the steel wire surface. This method provides a feasible and reliable approach for accurately quantifying the corrosion morphology.

Inspired by the literature review, the diagnosis of corrosion and broken wire status in tensioned suspension cables based on the spontaneous leakage magnetic detection technique is feasible, and significant progress is made in existing research. While some challenges and issues still need to be addressed, the above research on detecting the corrosion of cable structures using SMFL technology mainly focused on the testing technique.

However, the effects of the environment on natural structures were ignored. The high accuracy of SMFL detection technology, for instance, is easily affected by the external environment [40,41]. The cable undergoes continuous vibrations due to alternating loads, including the vibrations caused by bridge structures, vehicle loads, wind-induced vibrations [42,43], and rain-wind-induced vibrations [44]. The deformation of the cable structure differs from its static state, which inevitably influences corrosion measurement [45], as well as the stability and accuracy of the data. Precisely measuring the corrosion of bridge cables has become an urgent task for making maintenance decisions. Analyzing the impact of vibration on the measurement is also of critical scientific research significance. Therefore, the paper aims to study the influence of the vibration on the SMFL detection of cable corrosion, analyze its influencing mechanism, and find effective methods to reduce the impact of the vibration.

1.3. Contribution

In this study, the vibration characteristics of the corrosion distribution of cable structures using SMFL technology were theoretically interpreted. Both vibration simulations and SMFL tests were performed on corroded steel strands through experimental analysis. The characteristics of the magnetic field distribution under vibration conditions and the effect of vibration on SMFL detection were analyzed. Finally, a characterization index for cable structure corrosion was proposed based on the theoretical model and experimental results. The main contributions of this paper concerning the published literatures are summarized as follows:

(1) In terms of theory, the influence of vibration was analyzed based on the existing two-dimensional magnetic dipole model for rectangular corroded defects. The two-dimensional magnetic dipole model under vibration conditions was derived, and a theoretical model for the additional magnetic field under vibration was established. The original theoretical model was optimized.
(2) In terms of application, this study goes beyond the previous ideal experimental conditions by considering the cable's vibration effects generated by the coupling action between the environmental and alternating loads. Through empirical analysis, the impact of this vibration effect on the detection of leakage magnetic flux was assessed. This assessment lays the foundation for the accurate diagnosis of cable corrosion and wire breakage under varying environmental and load conditions.
(3) In terms of diagnostic methods, an evaluation index A for cable corrosion under vibration conditions was proposed. This index effectively reduces the influence of vibration and improves the accuracy of cable corrosion diagnosis.

1.4. Organization of the Paper

The rest of this paper is organized as follows. In Section 2, the principle of spontaneous leakage magnetic flux detection is introduced, and optimizations are made based on the original two-dimensional magnetic dipole model. In Section 3, experimental tests on the spontaneous leakage magnetic flux detection of corroded steel strands under vibration conditions are conducted. In Section 4, the impact of vibration on the corrosion-induced leakage magnetic flux signals is analyzed from multiple perspectives. The evaluation index suitable for assessing the corrosion of steel strands under vibration conditions are proposed. The last section outlines some main conclusions.

2. Theoretical Background

According to the principle of SMFL detection, a galvanized steel strand, serving as a ferromagnetic material, produces a magnetic leakage field on its corroded surface. This field can be understood as the vector sum of several types of magnetic fields [46], as illustrated in Figure 1. When the cable is undamaged and stationary, the magnetic field measured by the magnetic detection instrument is only the vector sum of the environmental magnetic field $\boldsymbol{B}_{\mathrm{E}}$ and the self-induced magnetic field $\boldsymbol{B}_{\mathrm{I}}$. Hence, the measured magnetic induction intensity $\boldsymbol{B}_{\mathrm{S}}$ is expressed as Equation (1). However, when the cable experiences radial vibrations, the displacement variation leads to an increase in the magnetic field of the measured signal, Consequently, the magnetic induction intensity measured during vibration can be expressed as $\boldsymbol{B}_{\mathrm{V}}$, as shown in Equation (2). Once a corrosion defect occurs within the cable structure, a corrosion SMFL emerges near the defect and continuously changes under the influence of vibrations. The magnetic induction strength at rest, $\boldsymbol{B}_{\mathrm{S\text{-}C}}$, can be defined as presented in Equation (1), whereas the magnetic induction strength during vibration, $\boldsymbol{B}_{\mathrm{V\text{-}C}}$, is expressed in Equation (2).

$$\boldsymbol{B}_{\mathrm{S}} = \boldsymbol{B}_{\mathrm{E}} + \boldsymbol{B}_{\mathrm{I}} \rightarrow \boldsymbol{B}_{\mathrm{S-C}} = \boldsymbol{B}_{\mathrm{S}} + \boldsymbol{B}_{\mathrm{C}}(c) \quad (1)$$

$$\boldsymbol{B}_{\mathrm{V}} = \boldsymbol{B}_{\mathrm{S}} + \Delta\boldsymbol{B}_{\mathrm{V}}(0) \rightarrow \boldsymbol{B}_{\mathrm{V-C}} = \boldsymbol{B}_{\mathrm{S}} + \Delta\boldsymbol{B}_{\mathrm{V}}(c) + \boldsymbol{B}_{\mathrm{C}}(c) \quad (2)$$

where $\boldsymbol{B}_{\mathrm{E}}$ is the environmental magnetic field. $\boldsymbol{B}_{\mathrm{I}}$ is the structural self-induced magnetic field. $\boldsymbol{B}_{\mathrm{C}}(c)$ is the corrosion self-magnetic flux leakage. $\boldsymbol{B}_{\mathrm{S}}$, $\boldsymbol{B}_{\mathrm{S\text{-}C}}$ are the magnetic induction at rest. $\boldsymbol{B}_{\mathrm{V}}$, $\boldsymbol{B}_{\mathrm{V\text{-}C}}$ are the magnetic induction during vibration, corresponding to the intact structure and the corroded structure. $\Delta\boldsymbol{B}_{\mathrm{V}}(0)$, $\Delta\boldsymbol{B}_{\mathrm{V}}(c)$ are the magnetic field increments affected by the vibration effect (c is the corrosion ratio).

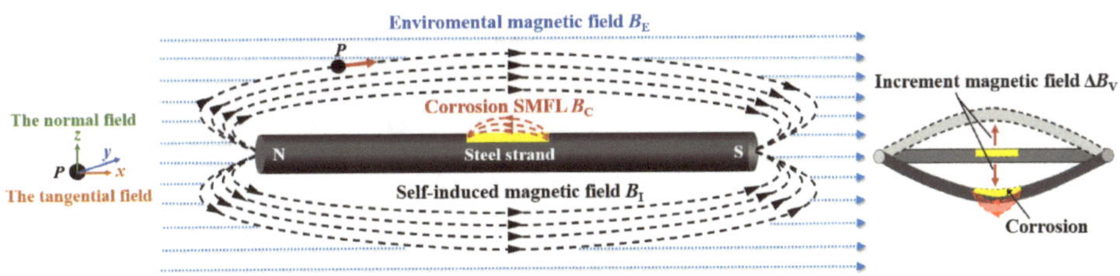

Figure 1. Magnetic field distribution of cable structure.

Figure 1 shows the different degrees of displacement and defect shape changes during the vibration of the steel strand, which will cause the change in magnetic induction intensity. $\Delta\boldsymbol{B}_{\mathrm{V}}(0)$ and $\Delta\boldsymbol{B}_{\mathrm{V}}(c)$ are the increments of the magnetic field under the influence of vibration. To analyze the effect of vibration on the magnetic field and its varying behavior, two types of magnetic field increments are defined: One type is the magnetic field increment caused by displacement change, denoted as $\Delta\boldsymbol{B}^{\mathrm{dis}}$ and described by Equation (3). This increment is termed the displacement-added magnetic field. The other type is the magnetic field increment caused by defect deformation, represented as $\Delta\boldsymbol{B}^{\mathrm{def}}$ and outlined in Equation (4). This increment is referred to as the deformation-added magnetic field.

$$\Delta\boldsymbol{B}^{\mathrm{dis}} = \boldsymbol{B}_{\mathrm{V}} - \boldsymbol{B}_{\mathrm{S}} = \Delta\boldsymbol{B}_{\mathrm{V}}(0) \quad (3)$$

$$\Delta\boldsymbol{B}^{\mathrm{def}} = \boldsymbol{B}_{\mathrm{V-C}} - \boldsymbol{B}_{\mathrm{S-C}} - \Delta\boldsymbol{B}^{\mathrm{dis}} = \Delta\boldsymbol{B}_{\mathrm{V}}(c) - \Delta\boldsymbol{B}^{\mathrm{dis}} \quad (4)$$

2.1. Displacement-Added Magnetic Field

The steel strand vibrating under external load excitation can be regarded as an Euler beam, which only considers the bending deformation and ignores the effects of shear deformation and moment of inertia. Assuming the displacement function $u(x,t)$ of the strand vibration response is known, Point P along the axial position x_1 experiences an external magnetic induction $\boldsymbol{B}_{\mathrm{p}}$. A two-dimensional overall coordinate system, xoz, is established in this plane. Additionally, a local coordinate system, x_1oz_1, is established, where the x_1 axial aligns with the axial direction of the strand, and the z_1 axis is perpendicular to it, as shown in Figure 2.

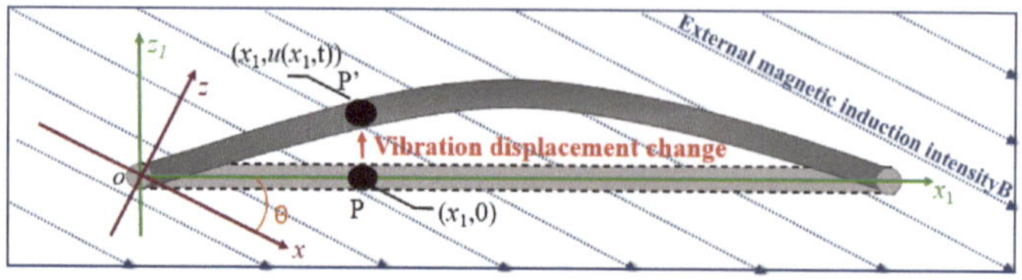

Figure 2. Schematic diagram of displacement-added magnetic field.

The angle between the local coordinate system and the overall coordinate system is denoted as θ. Consequently, $\boldsymbol{B}_{\mathrm{p}}$ can be expressed as the vector sum of the two magnetic fields in the x-axis and z-axis directions. Assuming that the magnetic field component $\boldsymbol{B}_{\mathrm{p}x}$ in the x-direction follows a specific functional relationship $\boldsymbol{B}_{\mathrm{p}x} = f_{\mathrm{m}}(x)$, and the magnetic field component $\boldsymbol{B}_{\mathrm{pz}}$ in the z-direction satisfies another specific functional relationship $\boldsymbol{B}_{\mathrm{pz}} = f_{\mathrm{n}}(y)$, the external magnetic induction $\boldsymbol{B}_{\mathrm{P}}$ at the point P(x_1,0) in the local coordinate system x_1oz_1 is described by Equation (5). During the vibration of the steel strand, point P moves to P′ at time t, where the location of P′ becomes (x_1, u (x_1, t)). The magnetic induction strength at P′, denoted as $\boldsymbol{B}_{\mathrm{P}'}$, can be determined as Equation (6).

$$\boldsymbol{B}_{\mathrm{p}} = \begin{bmatrix} f_{\mathrm{m}}(x_1 \times \cos\theta) \\ f_{\mathrm{n}}(x_1 \times \sin\theta) \end{bmatrix} \tag{5}$$

$$\boldsymbol{B}_{\mathrm{p}'} = \begin{bmatrix} f_{\mathrm{m}}(x_1 \cdot \cos\theta - u(x_1,t) \cdot \sin\theta) \\ f_{\mathrm{n}}(x_1 \cdot \sin\theta + u(x_1,t) \cdot \cos\theta) \end{bmatrix} \tag{6}$$

$$\Delta\boldsymbol{B}^{\mathrm{dis}} = \boldsymbol{B}_{\mathrm{p}'} - \boldsymbol{B}_{\mathrm{p}} = \begin{bmatrix} \Delta\boldsymbol{B}_x^{\mathrm{dis}} \\ \Delta\boldsymbol{B}_z^{\mathrm{dis}} \end{bmatrix} = \begin{bmatrix} f_{\mathrm{m}}(x_1 \cdot \cos\theta - u(x_1,t) \cdot \sin\theta) \\ f_{\mathrm{n}}(x_1 \cdot \sin\theta + u(x_1,t) \cdot \cos\theta) \end{bmatrix} - \begin{bmatrix} f_{\mathrm{m}}(x_1 \cdot \cos\theta) \\ f_{\mathrm{n}}(x_1 \cdot \sin\theta) \end{bmatrix} \tag{7}$$

The displacement-added magnetic field $\Delta\boldsymbol{B}^{\mathrm{dis}}$ can be approximated by Equation (7). It can be seen that the increment of the magnetic induction intensity $\Delta\boldsymbol{B}^{\mathrm{dis}}$ caused by the change of the vibration displacement of the steel strand is mainly related to the magnetic field strength around the steel strand and the vibration response displacement function. Combined with the knowledge of structural dynamics, it is known that the vibration response displacement function u (x, t) of the steel strand is a periodic function. Therefore, in a relatively stable geomagnetic field, the displacement-added magnetic field will fluctuate due to the periodicity of the displacement function.

2.2. Deformation-Added Magnetic Field

$\Delta\boldsymbol{B}^{\mathrm{def}}$ reveals the defective deformation effect of steel strands during vibration, and the characteristic law of $\Delta\boldsymbol{B}^{\mathrm{def}}$ under the vibration state is analyzed based on the magnetic dipole model. Suppose there is a rectangular defect with a width of $2a$ and a depth of h_0 at position x_0 of the strand at rest, as shown in Figure 3, then a right-angle coordinate system, xoz, is established at the center of the defect surface. On both sides of the defect, positive and negative magnetic charges are uniformly distributed. At a particular point Q in space, the magnetic field magnitudes $\boldsymbol{H}_{\mathrm{A}}$ and $\boldsymbol{H}_{\mathrm{B}}$ generated by all the magnetic charge microelements on surfaces A and B can be expressed as Equations (8) and (9). According to the magnetic field superposition principle, the magnetic induction intensity at point Q can be expressed as Equation (10).

$$\boldsymbol{H}_{\mathrm{A}} = \begin{pmatrix} \boldsymbol{H}_{\mathrm{A}x} \\ \boldsymbol{H}_{\mathrm{Az}} \end{pmatrix} = \frac{\mathrm{M}(\sigma, H_1^0)}{2\pi} \begin{pmatrix} \arctan\frac{-h}{x+a} + \arctan\frac{h+h_0}{x+a} \\ \frac{1}{2}\ln\frac{(x+a)^2+(h+h_0)^2}{(x+a)^2+h^2} \end{pmatrix} \tag{8}$$

$$H_B = \begin{pmatrix} H_{Bx} \\ H_{Bz} \end{pmatrix} = \frac{M(\sigma, H_1^0)}{2\pi} \begin{pmatrix} \arctan\frac{-h}{-x+a} + \arctan\frac{h+h_0}{-x+a} \\ \frac{1}{2}\ln\frac{(x-a)^2+h^2}{(x-a)^2+(h+h_0)^2} \end{pmatrix} \tag{9}$$

$$B_Q = \mu_0\mu_r H_Q = \mu_0\mu_r \begin{pmatrix} H_{Qx} \\ H_{Qz} \end{pmatrix} = \mu_0\mu_r \begin{pmatrix} H_{Ax} + H_{Bx} \\ H_{Az} + H_{Bz} \end{pmatrix} \tag{10}$$

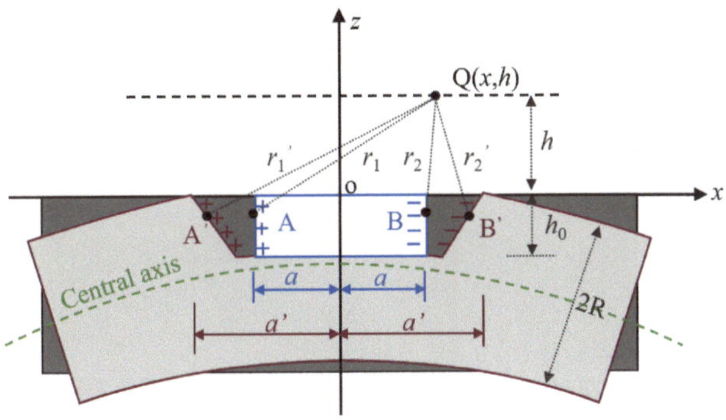

Figure 3. Schematic diagram of the shape change of the defect.

When the strand vibrates at a certain moment t, the depth of the rectangular defect remains constant, while the corrosion width expands to $2a'$, as depicted in Figure 3. The defect cross-section still remains perpendicular to the central axis and satisfies the assumption of flat cross-section. The coordinate transformation of the magnetic charge unit of A′ and B′ during the vibration process is obtained by Equation (11). Their respective magnetic field strengths $H_{A'}$ and $H_{B'}$ generated at point Q can be calculated by Equation (12). The total magnetic induction strength $B_{Q'}$ at point Q is calculated by integrating Equation (13).

R is the radius of the strand. M_0 is the bending moment at the damage. EI is the bending stiffness of the section at the damage. k is the curvature of the central axis. μ_0 and μ_r are the vacuum permeability and relative permeability of the material, respectively.

$$\begin{aligned} A'(x_1, z_1) &\Rightarrow A'\left(-a - \frac{aM_0(R+z_1)}{EI}, z_1\right) \Rightarrow A'\left(a(R+z_1)\frac{\partial^2 u(x_0,t)}{\partial x^2} - a, z_1\right) \\ B'(x_2, z_2) &\Rightarrow B'\left(a + \frac{aM_0(R+z_2)}{EI}, z_2\right) \Rightarrow B'\left(a - a(R+z_2)\frac{\partial^2 u(x_0,t)}{\partial x^2}, z_2\right) \end{aligned} \tag{11}$$

$$\begin{aligned} H_{A'} &= \begin{pmatrix} H_{A'x} \\ H_{A'z} \end{pmatrix} = \int \frac{Mds}{2\pi (r_1')^2}\vec{r_1'} = \frac{M}{2\pi}\int_{-h_0}^{0} \frac{\sqrt{1+\left(\frac{1}{ak}\right)^2}dy \begin{pmatrix} x+a-ak(R+z_1) \\ h-z_1 \end{pmatrix}}{[x+a-ak(R+z_1)]^2+(h-z_1)^2} \\ H_{B'} &= \begin{pmatrix} H_{B'x} \\ H_{B'z} \end{pmatrix} = \int \frac{Mds}{2\pi (r_2')^2}\vec{r_2'} = \frac{M}{2\pi}\int_{-h_0}^{0} \frac{\sqrt{1+\left(\frac{1}{ak}\right)^2}dy \begin{pmatrix} x-a+ak(R+z_2) \\ h-z_2 \end{pmatrix}}{[x-a+ak(R+z_2)]^2+(h-z_2)^2} \end{aligned} \tag{12}$$

$$B_{Q'} = \mu_0\mu_r H_{Q'} = \mu_0\mu_r \begin{pmatrix} H_{Q'x} \\ H_{Q'z} \end{pmatrix} = \mu_0\mu_r \begin{pmatrix} H_{A'x} + H_{B'x} \\ H_{A'z} + H_{B'z} \end{pmatrix} \tag{13}$$

$$\Delta B^{def} = B_{Q'} - B_Q = \mu_0\mu_r \begin{pmatrix} H_{Q'x} - H_{Qx} \\ H_{Q'z} - H_{Qz} \end{pmatrix} = \begin{pmatrix} \Delta B_x^{def} \\ \Delta B_z^{def} \end{pmatrix} \tag{14}$$

The difference between the measured magnetic induction intensity of the Q-point in the vibrating and stationary states is defined as the deformation-induced additional magnetic field ΔB^{def}, see Equation (14). Equations (8)–(14) demonstrate that the self-leakage magnetic field generated by corroded defects is mainly related to the measurement distance h, defect depth h_0, and defect width a. To further analyze the influence of h, h_0, and a on the leakage magnetic field of defects, MATLAB 2021a software was used for numerical simulation of the defect leakage magnetic field $\boldsymbol{B}_{Q'}$ in the vibrating state. The results of these simulations are presented in Figure 4.

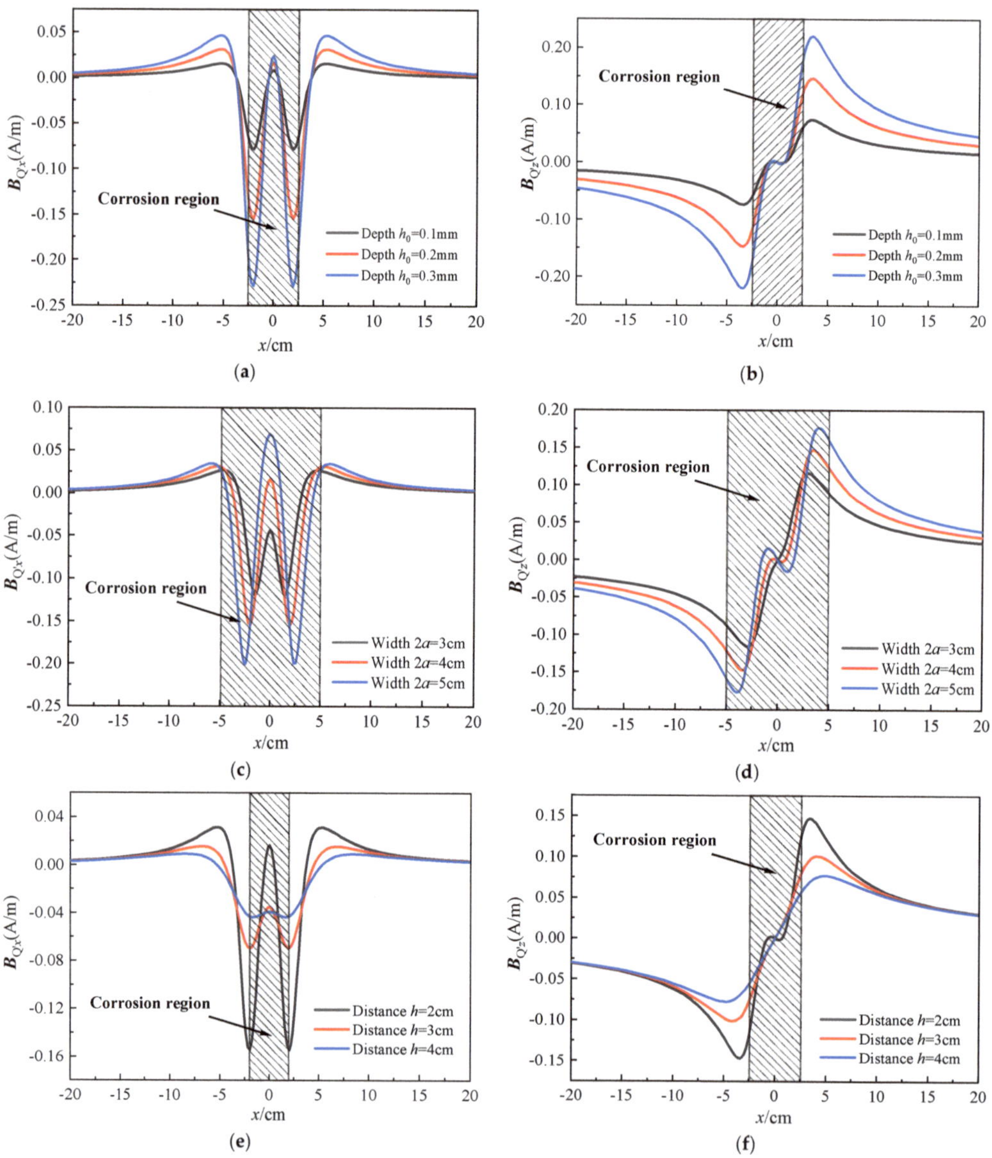

Figure 4. Theoretical distribution law of $\boldsymbol{B}_{Q'}$: (**a**,**b**) are the x-axial component and z-axial component under various depths; (**c**,**d**) are the x-axial component and z-axial component under various widths; (**e**,**f**) are the x-axial component and z-axial component under various distances.

Figure 4 demonstrates that both components of $\boldsymbol{B}_{Q'x}$ and $\boldsymbol{B}_{Q'z}$ possess symmetry, and the extremum and zero-crossing phenomena in the curve conform to the general distribution law of the SMFL method. Figure 4a–d indicate that the magnetic field intensity is positively correlated with defect depth h_0 and defect width a. Figure 4e,f indicate that the magnetic field intensity is negatively correlated with the measurement distance h. h, h_0, and a do not alter the distribution pattern of the magnetic field, but only affect the extremum. In fact, different degrees of h_0 and a represent varying section loss rates of the steel strand, indicating different levels of corrosion. Therefore, in the subsequent experimental study, the degree of corrosion in the steel strand (denoted as c) is used to represent the differences in defect width a and depth h_0. This theoretical model also serves as a guiding principle for the subsequent experimental research.

In summary, the alterations in magnetic field distribution within the strand during vibration can be categorized into two main aspects. First, the displacement of the strand leads to incremental magnetic induction intensity ($\Delta \boldsymbol{B}^{dis}$) at the measurement point, which is primarily influenced by the stability of the surrounding magnetic field. Under relatively stable geomagnetic conditions, $\Delta \boldsymbol{B}^{dis}$ will exhibit fluctuating changes. Second, the defect size undergoes varying degrees of change due to the vibration, resulting in an increase in corrosion self-magnetic flux leakage. This increase, denoted as $\Delta \boldsymbol{B}^{def}$, is primarily determined by the defect's size. The above simplified theoretical model effectively explains the impact of vibration on SMFL and provides the theoretical foundation for the following tests. Moreover, the model highlights the significance of both the stability of the surrounding magnetic field and the size of the defect in defining the incremental changes observed in the magnetic field distribution during vibration.

3. Methodology and Experiment

The experimental objective is to investigate the characteristics of SMFL in corroded steel strands under vibration and examine the effects of vibration on SMFL detection. As shown in Figure 5, experiments were conducted for tension, corrosion, vibration, and signal acquisition. The signals were collected subsequent to the corrosion process.

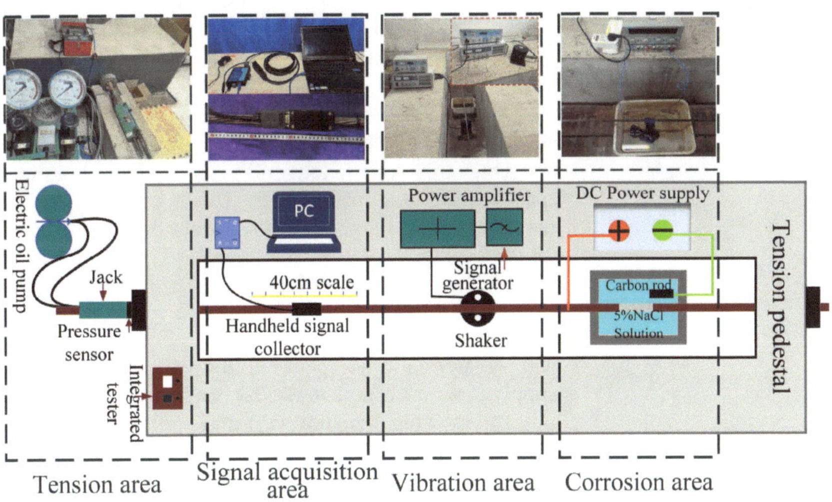

Figure 5. Schematic of the test work platform.

3.1. Preparation and Tension

The experiment material was a smooth steel strand (1 × 7Φ5). The strand had a nominal diameter of 15.20 mm and a single length of 7 m. Table 1 provides the chemical components and material properties of steel strand specimens. The relative permeability of steel strand μ = 2000, and the saturation magnetization intensity $M_s = 1 \times 10^6$ A/m.

Table 1. Chemical components and material properties of steel strand specimens.

Material Proportion	Tensile Strength	Nominal Diameter	Elasticity Modulus	Weight	Elongation	Yield Load Fp0.2
Values	1860 MPa	15.20 mm	195 GPa	1.101 kg/m	≥3.5%	≥229 kN
Chemical compositions	C	Si	Mn	P	S	Cu
Proportion	0.8–0.85%	0.12–0.32%	0.6–0.9%	<0.025%	<0.025%	<0.2%

Four groups of experiments were performed in Table 2; 1# specimen was conducted to explore the effect of measuring distance on SMFL; 2# and 3# specimens were conducted to investigate the effect of corrosion degree on SMFL under different excitation conditions; and 4 # specimen was not corroded, and the purpose is to explore the effect of the change in displacement of the strand on the magnetic measurement in the vibration state.

Table 2. Specimen grouping.

Number	Corrosion Loss Ratio *c*	Frequency *f*/Hz	Current *I*/A	Measuring Distance *d*/cm
1#	4%, 8%	5	10	1, 2, 3
2#	2%, 4%, 6%, 8%	10	2, 3, 4, 5	1
3#	3%, 6%, 9%, 12%, 15%	5, 8, 10	5	1
4#	0%	5, 8, 10	2, 3, 4, 5	1

To simulate real-life bridge conditions, the strand was intentionally corroded after being subjected to a tensioning force. Given that the length of the tensioning platform is 6 m, each strand was set to a length of 7 m to ensure an ample tensioning area. The team's tensioning platform, demonstrated in Figure 5 as the "Tension area", was primarily responsible for the tensioning work. The tensioning force is set to 70 kN.

3.2. Specimen Corrosion

Due to the long length and prior tensioning of the strand specimen, weighing the specimen itself was inconvenient. Therefore, in this experiment, the degree of corrosion of the steel strands was controlled using electrochemical corrosion and the empirical formula for the theoretical corrosion rate, denoted as c. Presented as the "Corrosion area" in Figure 5, the specimen to be corroded was placed in 5% NaCl electrolyte solution. A water-absorbent cloth was placed at the position to be corroded and immersed in the electrolyte solution. Moreover, the positive pole of an external power supply was connected at the specimen end. A carbon rod was placed in the solution to connect the negative pole of the power supply to form a complete electrochemical corrosion closed circuit.

By applying Equation (15) and setting the constant DC power supply's current $I = 0.4$ A, the corrosion degree c of the specimen could be determined. The actual corrosion effect of the steel strand specimen is shown in Figure 6. It is found that with the increase of corrosion time, the area of the corrosion section of the steel strand was significantly reduced.

$$c = \frac{\Delta m_{Fe}}{m} = \frac{M_{Fe} It}{nF\lambda l} = 2.63 \times 10^{-7} \frac{It}{l} \tag{15}$$

where I is the current size, t is the corrosion time, l is the strand corrosion width. M_{Fe} is the atomic molar mass of Fe, $M_{Fe} = 55.85 \times 10^{-3}$ kg/mol. n is the transfer of electrons in the anodic reaction, $n = 2$. F is the Faraday constant, F = 96,485 C/mol. λ is the theoretical weight of the strand per meter, $\lambda = 1.101$ kg/m.

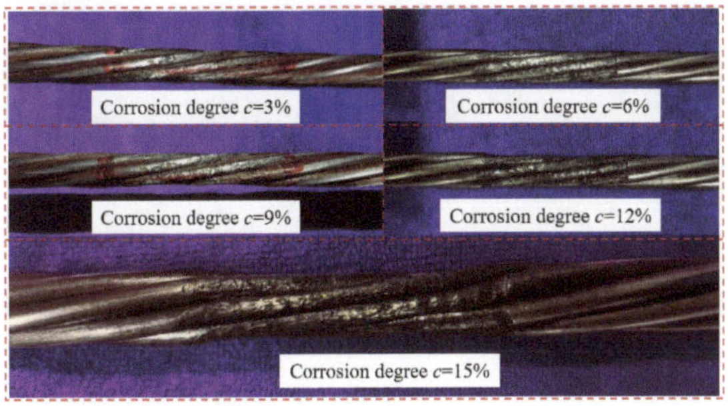

Figure 6. The actual corrosion effect of the specimen.

3.3. Specimen Vibration and Magnetic Signal Acquisition

A modal exciter (Chongqing, China, Sa-jz005) was placed at the middle of the specimen to excite the steel strand to simulate the cable vibration generated by external loading during service. A low-frequency signal generator was used to simulate the sinusoidal signal of the cable vibration. A power amplifier drove the exciter. The modal exciter was a permanent magnet electric modal one. When the power amplifier supplied variable-frequency current to the moving coil, the exciting force is shown in Equation (16).

$$F = BIL \tag{16}$$

where B is the average magnetic induction strength, I is the instantaneous value of the current supplied by the power amplifier, L is the effective length of the coil wire. Therefore, the output frequency of the signal generator controlled the vibration speed, whereas the output current of the power amplifier controlled the exciting force.

In order to distinguish the magnetic leakage signal before and after the vibration of the steel strand, three types of magnetic signal were recorded in the experiments: initial magnetic signal $\boldsymbol{B}_{\mathrm{S}}$ (without corrosion), stationary magnetic signal $\boldsymbol{B}_{\mathrm{S\text{-}C}}$ (The corrosion rate is c), and vibrating magnetic signal $\boldsymbol{B}_{\mathrm{V\text{-}C}}$ (The corrosion rate is c). When the corrosion degree $c = 0$, the vibrating magnetic signals $\boldsymbol{B}_{\mathrm{V\text{-}C}}(0) = \boldsymbol{B}_{\mathrm{V}}$.

The magnetoresistive sensor (HMR 2300) was used to obtain the magnetic field distribution of the specimens. The resolution of the sensor was 67μGs with 10–157 sampling points per second, and the output ranged from −2Gs–2Gs. In the magnetic field measurement system, the magnetic sensor was carried by the positioning device to move freely on the structure's surface. The measurement area has a length of 40 cm, divided into 41 measurement points, each point is 1 cm away from the other, as shown in Figure 7. The magnetic sensor has three directional components (abbreviated as $\boldsymbol{B}_x$, $\boldsymbol{B}_y$, and $\boldsymbol{B}_z$) and only the tangential ($\boldsymbol{B}_x$) and normal ($\boldsymbol{B}_z$) components are needed for the measurement. The lift-off height d is the distance between the magnetic sensor and the surface of the specimen and the measurement heights of this experiment are 1 cm, 2 cm and 3 cm. During the vibration, the magnetic field data were collected three times at each data point. The Scanning scheme of magnetic field signals is shown in Figure 7.

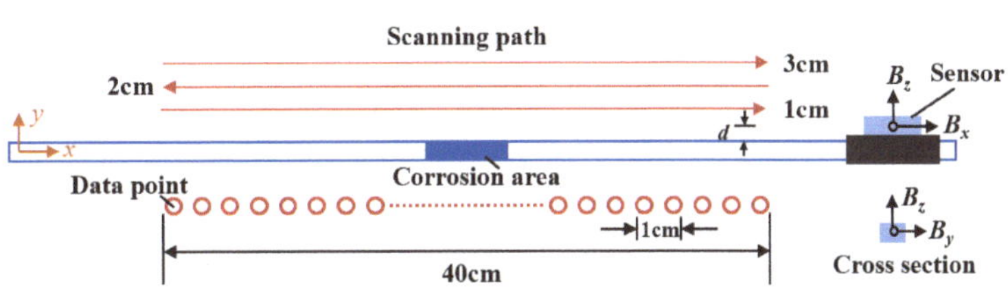

Figure 7. Scanning scheme of magnetic field signals.

4. Results and Discussion

The experiment results are analyzed to investigate the difference in the SMFL of the corroded steel strand between vibration and static states, while the interference of the external magnetic and geomagnetic fields needed to be eliminated. To achieve this, Equation (17) is applied to eliminate the effects caused by the environmental magnetic field and the structural self-induced magnetic field. Subsequently, 1# specimen (corrosion ratio c = 8%) is taken as an example and its distribution laws of the corrosion SMFL are shown in Figure 8.

$$\boldsymbol{B}_{\mathrm{C}} = \boldsymbol{B}_{\mathrm{S-C}} - \boldsymbol{B}_{\mathrm{S}} \tag{17}$$

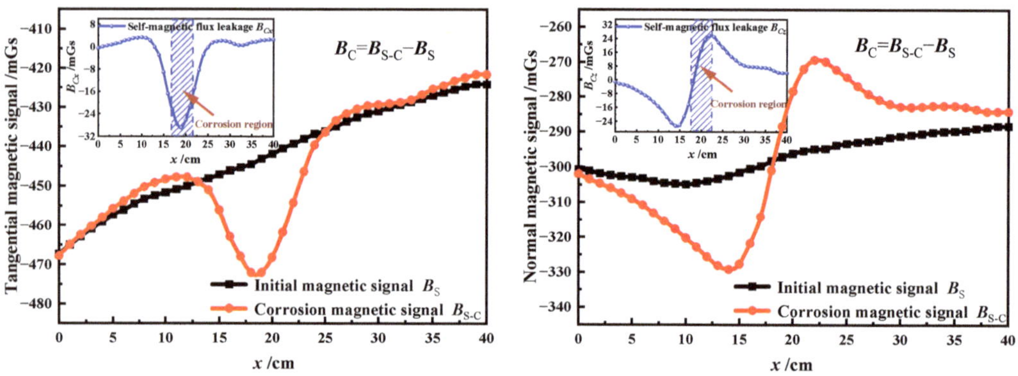

Figure 8. Calculation of the SMFL of corrosion damage: (**a**) tangential magnetic signal; (**b**) normal magnetic signal.

It can be found from Figure 8 that the tangential component curve has a minimal value in the center of the corrosion region. On the other hand, the normal component curve crosses the zero point and has two extreme values in the corrosion region. There is a notable correlation between the extremum points and the corrosion area, which aligns with the outcomes obtained from the magnetic dipole model [26].

4.1. SMFL in Vibration and Stationary States

In order to investigate the distribution characteristics and evolution law of cable structures' self-magnetic flux leakage (simplified as "corrosion SMFL") under vibration and stationary states, the effects caused by the environmental magnetic field and the structural self-induced magnetic field are eliminated by Equation (17). The following paper analyzes the characteristics of the corroded SMFL under vibration and stationary states from two aspects: different lift-off heights and corrosion degrees. Where, $\boldsymbol{B}_{\mathrm{C}}$ and $\boldsymbol{B}_{\mathrm{C}}^{\mathrm{V}}$ corresponded to the stationary state and vibration state of cable structure. $\boldsymbol{B}_{\mathrm{C}x}$ and $\boldsymbol{B}_{\mathrm{C}z}$ corresponded to the tangential and normal components of the corrosion SMFL curve in stationary state. $\boldsymbol{B}_{\mathrm{C}x}^{\mathrm{V}}$ and $\boldsymbol{B}_{\mathrm{C}z}^{\mathrm{V}}$ corresponded to the tangential and normal components of the corrosion SMFL curve in vibration state.

4.1.1. Different Lift-Off Heights

Considering the spatial nature of the self-magnetic flux leakage (SMFL) field generated by the defect, it is essential to note that the signal strength of this field is affected by the sensing locations. The theoretical analysis indicates that the SMFL of corrosion damage obtained by different lift-off heights of the testing equipment is different. Therefore, the 1# specimen is taken as an illustration. The distribution curves of SMFL field under the same corrosion condition but in different measuring lift-off heights are analyzed and shown in Figure 9.

As shown in Figure 9, on the one hand, the corrosion SMFL in the vibration state has similar characteristics to the stationary state. The SMFL curves for various lift-off heights all demonstrate signal extremes near the corrosion area (highlighted by the gray line box). In the B_{Cx} and B_{Cx}^{V} curves, each signal curve under different lift-off heights displays consistent behavior, where the signal strength of the magnetic field changed reciprocally to the distance. Specifically, The higher the lift-off height, the lower the signal strength. In the B_{Cz} and B_{Cz}^{V} curves, the magnetic signals are also inversely proportional to the lift-off heights. Two extreme points emerge on each side of the corrosion area, while one intersection point is identified at the center of the corrosion area. The experimental results are consistent with the numerical simulation results presented in the previous theoretical section.

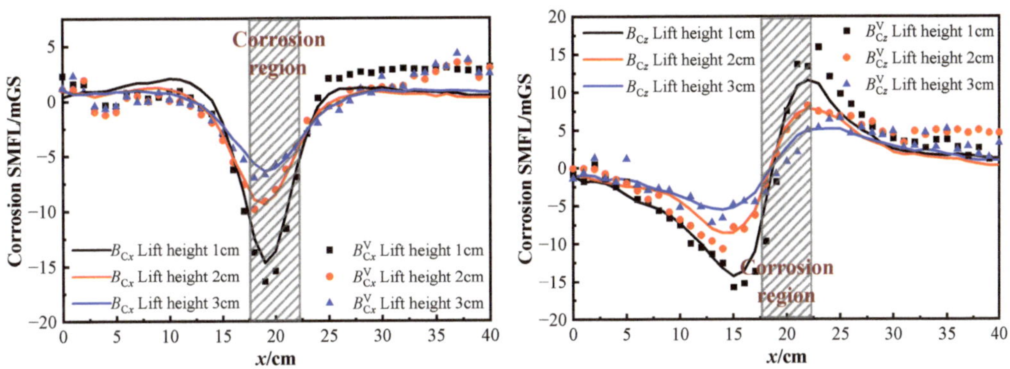

Figure 9. Corrosion SMFL with different lift-off heights.

On the other hand, although the corrosion SMFL in the vibration state has similar characteristics to the stationary state, there are still varying degrees of fluctuations which are particularly significant at the curve's inflection point. With the increase of the measurement lift-off heights, the B_C and B_C^V curves tend to be flatter, resulting in a smaller extreme value near the corrosion area. This feature indicates that the influence of the measurement lift-off heights on the SMFL curve is similar for both vibration and stationary states. To mitigate the impact of the measurement lift-off heights on the SMFL detection, the sensor should be as close as possible to the measured part during the detection and choose the appropriate measurement distance.

4.1.2. Different Corrosion Degrees

To explore the characteristics of the SMFL in different corrosion degrees of the strand under vibration and stationary states, the 2# specimen is taken as an example and the results of the analysis are shown in Figure 10. B_{Cx} and B_{Cz} corresponded to the tangential and normal components of the corrosion SMFL curve.

From Figure 10, it is evident that the corrosion SMFL signals in the vibration state fluctuate around the stationary SMFL curve and follow a similar trend with it under different corrosion degrees. The changes in the SMFL curves under different corrosion degree are obvious. As the corrosion degree increases, the peak of the tangential and normal components of the corrosion SMFL curve increases significantly, indicating that the extreme value of the curve is positively correlated with the corrosion degree. The normal component B_{Cz} and B_{Cz}^{V} curves display double polar points and cross the zero threshold, while the B_{Cz}^{V} signals have an overall upward shift based on the B_{Cz} curve. In the B_{Cx} and B_{Cx}^{V} curves, the SMFL signals with different corrosion degrees have obvious intersection points near the corrosion area.

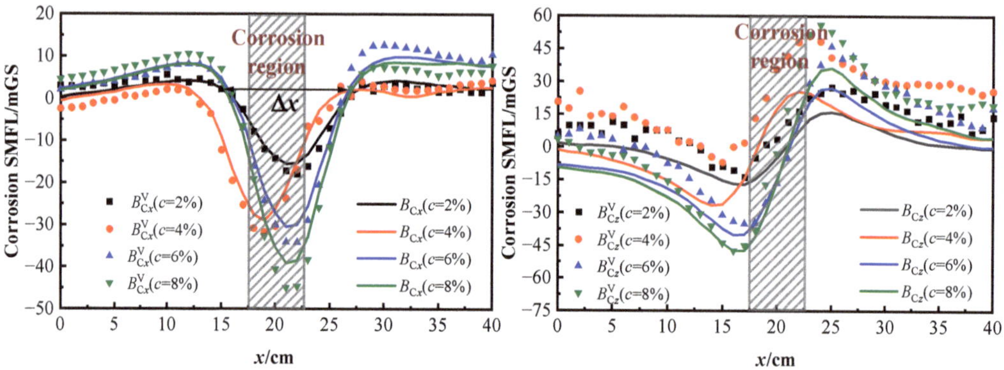

Figure 10. Corrosion SMFL with different corrosion degrees.

The width Δx of the two intersection points is greater than the width of the corrosion area, which is about twice the corrosion width [33]. Through the location and width of Δx, it can be qualitatively determined regarding the location and width of the corrosion area. The extreme value point is positively correlated with the degree of corrosion, which can be used to characterize the degree of corrosion. The relationship between the degree of corrosion of steel wire ropes and their self-induced magnetic leakage signals also agrees with the results of numerical simulation.

In summary, the extreme point of the SMFL curve is negatively correlated with the measured lift-off height and positively correlated with the corrosion degree. The corrosion SMFL signals in the vibration state fluctuate around the stationary SMFL curve and have the same trend and characteristics. In order to analyze the fluctuation characteristics of the SMFL signals in the vibration state and to explore the impact of vibration on the quantification and localization of the corrosion area, the two aspects will be discussed later: vibration displacement effect and vibration deformation effect.

4.2. Analysis of the Vibration Effects on SMFL

4.2.1. Displacement Effects ($\Delta \boldsymbol{B}^{dis}$)

Theoretical analysis demonstrates that the uncorroded strand is only affected by displacement change during vibration. Therefore, the 4# specimen is taken as an example and its displacement-added magnetic field $\Delta \boldsymbol{B}^{dis}$ can be derived from Equation (3). Figure 11 presents the statistical characteristics of normal ($\Delta \boldsymbol{B}_z^{dis}$) and tangential ($\Delta \boldsymbol{B}_x^{dis}$) components of the $\Delta \boldsymbol{B}^{dis}$ signals under different excitation currents, represented as a violin diagram. In this diagram, the white vertical line represents the magnetic signal range from 1/4 to 3/4. The yellow vertical line represents the average value and the area of the violin represents the distribution interval and size of this set of data. It can be seen from Figure 11 that the curves of the $\Delta \boldsymbol{B}^{dis}$ signals fluctuate up and down with no apparent change pattern and 50% of the magnetic signals are concentrated within −1–1 mGs. The overall signals exhibit a "high in the middle and low on both sides" pattern, remaining within the ±5 mGs range. This feature indicates that the fluctuation of the $\Delta \boldsymbol{B}^{dis}$ signals is slight because the external magnetic induction intensity around the steel strand in the experiment chamber is relatively stable. Therefore, the variation in the $\Delta \boldsymbol{B}^{dis}$ signals is mainly affected by the vibration amplitude. The displacement-added magnetic field only fluctuates up and down in a small range when the vibration amplitude of the steel strand is not significant.

Figure 11 reflects that the form of $\Delta \boldsymbol{B}_x^{dis}$ x is "narrow and high" and the shape of $\Delta \boldsymbol{B}_z^{dis}$ is "wide and flat", indicating that the normal component of the ΔB^{dis} signals is more discrete than the tangential component. To assess the fluctuation of the data set, the variance s^2 of 4# specimen is calculated by Equation (18). The results are depicted in Figure 12. Figure 12a–c represent the variance results of the excitation frequency of 10 Hz, 8 Hz, 5 Hz. It can be shown from Figure 12 that the variances of $\Delta \boldsymbol{B}_z^{dis}$ are approximately

0.2 mGs^2 greater than those of ΔB_x^{dis}, indicating that the degree of fluctuation of the normal component is greater than that of the tangential component. This phenomenon can be attributed to the experimental setup, which applies radial vibration to the strand in the same direction as the normal component of the magnetic signals. Consequently, the normal displacement of the strand is larger than the tangential displacement. The results show that the strand vibration has a more pronounced impact on the normal component of the magnetic signal.

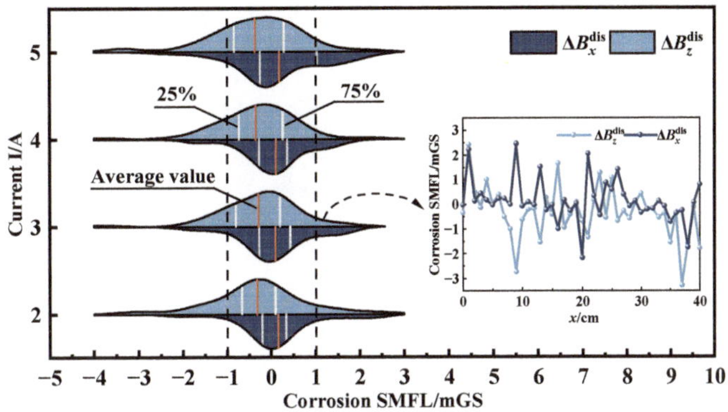

Figure 11. Displacement-added magnetic field of 4# specimen: f = 8 Hz.

$$s^2 = \frac{\sum_{i=1}^{n} (x_i - \overline{x})^2}{n} \quad (18)$$

where x_i is the value of tangential and normal components of ΔB^{dis}, $\overline{x}$ is the mean of the set of data, n is the quantity of data in the group.

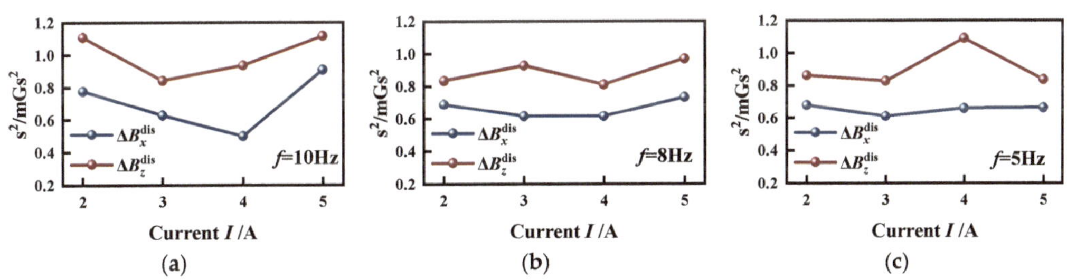

Figure 12. Variances of ΔB_x^{dis} and ΔB_z^{dis} of 4# specimen under different excitation conditions: (**a**) f = 10 Hz; (**b**) f = 8 Hz; (**c**) f = 5 Hz.

Considering the fluctuation of ΔB^{dis}, Figure 13 analyzes the statistical distribution law of ΔB^{dis} under other excitation conditions in the form of a histogram. Where β denotes the range of ΔB^{dis}, each data group is divided into intervals of 0.1 mGs and F is the frequency of each interval. Similar to Figure 11, the overall magnetic signals are "high in the middle and low on both sides", mainly distributed around the zero value. The statistical distribution of the data at different excitation frequencies has no obvious variation pattern, but all show characteristics of normal distribution. The maximum frequency corresponds to the data distribution interval β_{max} around "0", which indicates that the data's fluctuation range is small.

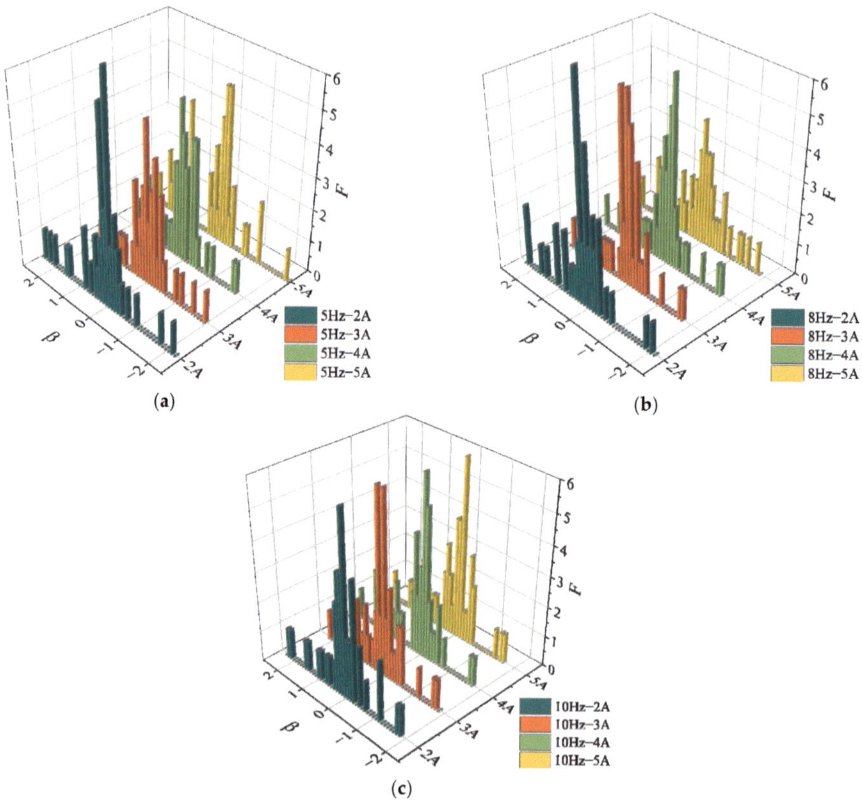

Figure 13. Statistical histogram of ΔB^{dis} of 4# specimen under different excitation conditions: (**a**) f = 10 Hz I = 2A–5A; (**b**) f = 8 Hz I = 2A–5A; (**c**) f = 5 Hz I = 2A–5A.

When the external magnetic induction intensity around the strand is stable, the magnitude of the displacement-added magnetic field is mainly influenced by the vibration amplitude. The ΔB^{dis} signals fluctuate up and down around the zero value with the statistical characteristics of a normal distribution. The effects of displacement on SMFL are small, but the normal component fluctuated more. Therefore, the subsequent analysis would be carried out only for the tangential component.

4.2.2. Corrosion Defect Deformation Effects (ΔB^{def})

Figure 1 shows that the corrosion SMFL is generated when the strand is corroded. It is shown from Figure 3 that the defect width changes when the strand vibrates, which affects the magnitude of the SMFL. To investigate the influence law of defect deformation on the SMFL during vibration, the 2# and 3# specimens are taken as examples. The analysis is carried out only for the tangential component of SMFL because the stability of the tangential component is better. Figure 14a,c show the SMFL at different corrosion degrees, where $B_{\text{C}x}^{V}$ and $B_{\text{C}x}$ corresponded to the vibration and stationary states of the strand. Figure 14b,d are the results of the vibrating magnetic signals minding the static magnetic signals, which included the ΔB_x^{dis}, as shown in Equation (19). The absolute value of $\Delta B_{\text{V}x}$ poles is shown in Equation (20).

$$\Delta B_{\text{V}x} = B_{\text{C}x}^{\text{V}} - B_{\text{C}x} = \Delta B_x^{\text{dis}} + \Delta B_x^{\text{def}} \tag{19}$$

$$\eta = |\Delta B_{\text{V}x}(\min)| \tag{20}$$

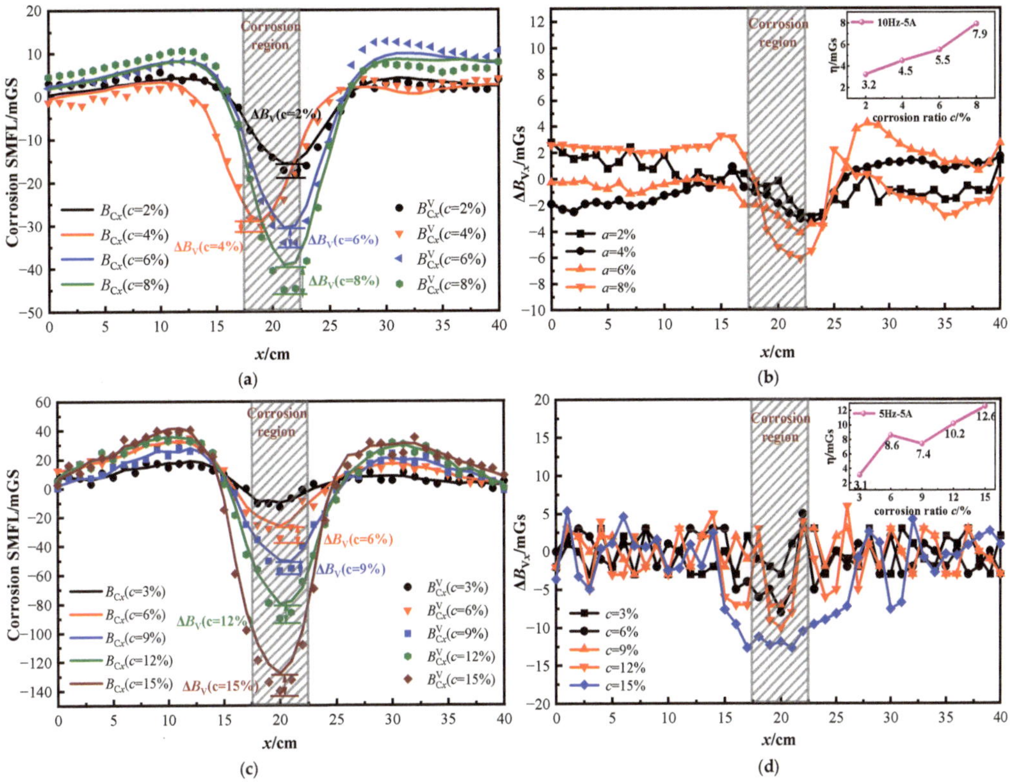

Figure 14. Corrosion SMFL and ΔB_{Vx} of 2# and 3# specimens under different corrosion degrees: (**a**) and (**b**) are 2# specimens; (**c**,**d**) are 3# specimens.

From Figure 14a,c, it is evident that the SMFL curves of the strand exhibit a strong correlation with the corrosion degree. The vibration magnetic signals are consistent with the variation pattern of the stationary magnetic signal, fluctuating around the stationary SMFL curve. Figure 13b,d show that the ΔB_{Vx} signals oscillate near the zero value in the uncorroded region spanning from 0–15 cm and from 25–40 cm. This indicates that the displacement changed mainly occur in the uncorroded region, consistent with the characteristics of the ΔB^{dis}. In the vicinity of the corroded region, ΔB_{Vx} displays a more prominent variation with obvious extreme values, indicating an increase in defect width during vibration. From the relationship graph of η-*c*, it is seen that the extreme value of ΔB_{Vx} is larger as the corrosion degree increases and has a linear correlation. It means that the greater the degree of corrosion, the greater the effect of vibration on SMFL.

On the other hand, Figure 13b,d also reveal that when the degree of corrosion of the steel strand rope ranges between 2% and 10%, the peak value of ΔB_{Vx} does not exceed 10 mGs, and in the non-corroded area, ΔB_{Vx} fluctuates within the range of −5 to 5 mGs. ΔB_{Vx} represents the difference in magnetic signal between vibration and static station, indicating the influence of vibration. These experimental results demonstrate that the impact of vibration on magnetic flux leakage detection is correlated with the degree of corrosion. The greater the degree of corrosion, the greater the influence of vibration, primarily reflected in the fact that the deformation-added magnetic field is greater than the displacement-added magnetic field.

The relationship between η and *c* for the remaining excitation conditions of specimens 2# and 3# is illustrated in Figure 15a–e. There is an excellent linear relationship between η and c for different excitation conditions, demonstrating that η increased with the increase of corrosion degree *c*. Due to some specific errors in the experiments, certain working conditions do not confirm this linear relationship. The results of all conditions are fitted to Figure 15f. From the fitting result, the linear relationship between η and c is well fitted with a fitted regression rate $R^2 = 0.87$. The corresponding linear equation is shown in Figure 15f. Considering that η represents the difference between vibration and rest at the extremes of the tangential component of the SMFL curve, it follows that this difference increases with the corrosion ratio. The extreme value of the SMFL curve is often used to characterize the corrosion degree of the strand, and the greater the corrosion degree, the greater the error in describing the corrosion in the vibration state.

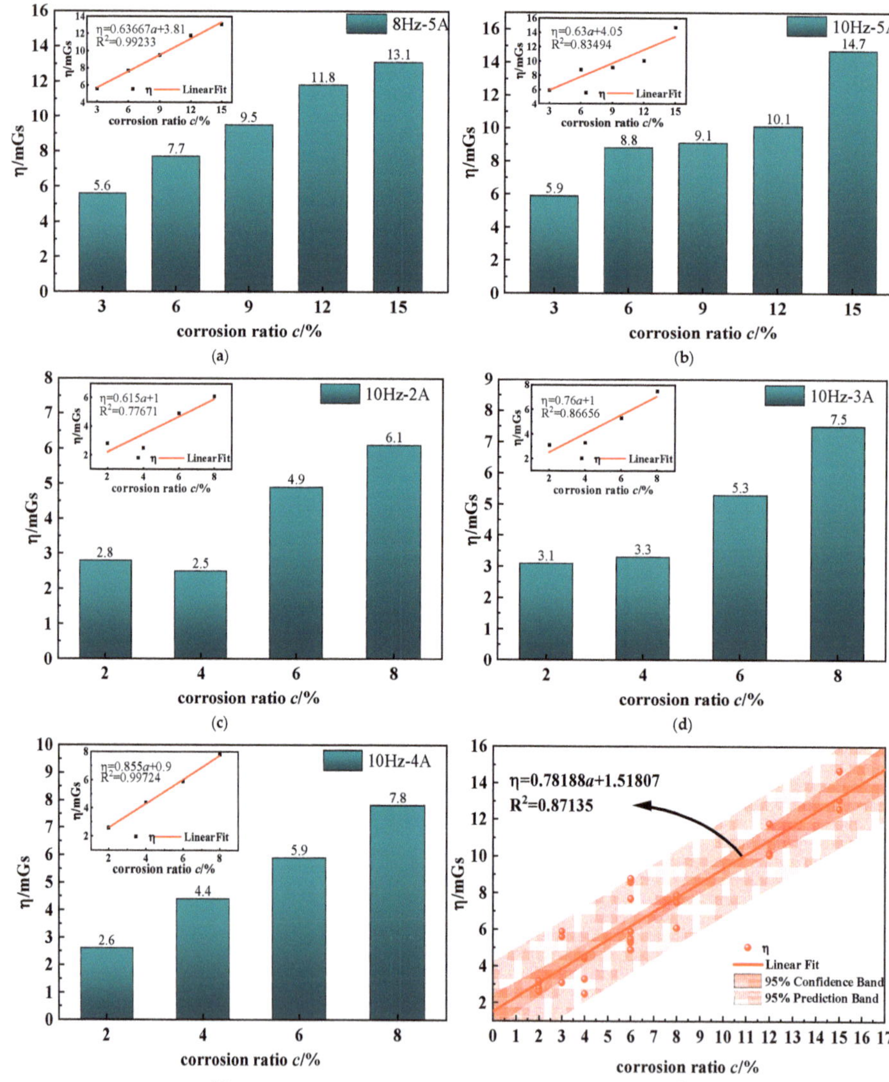

Figure 15. Relationship of η and c for the remaining excitation conditions of specimens 2# and 3#: (**a**,**b**) are 2# specimens; (**c**–**e**) are 3# specimens; (**f**) is the fitting result of η and c.

In summary, the effect of vibration deformation on SMFL is more significant than that of vibration displacement and increases with the increase of corrosion degree. If the extreme point of the SMFL curve is used as the characteristic value to characterize the corrosion degree, there will be a large error. Therefore, it is necessary to reduce the influence of vibration on corrosion characterization as much as possible and to seek the characteristic value suitable for corrosion characterization under the vibration state.

4.3. Characterization Index of SMFL under Vibration

According to the analysis of the vibration effects on SMFL in Section 4.2, it is evident that the tangential component of SMFL exhibits higher stability. Consequently, the characterization index for corrosion is selected from the tangential component curve. As shown in Figure 16, M, N and P are the extreme value points on the SMFL curve. The experiment results show that points M, N are located on both sides of the corrosion area, with the width of the two intersection points (Δx) being approximately twice the width of the corrosion. Furthermore, the extreme value point P is positioned at the center of the corrosion area and exhibits a linear correlation with the degree of corrosion. Considering the vibration effects on the minima P of the SMFL curve, the three characteristic points (M, N, P) are used as cut-off points to analyze the characterization value of the corrosion degree.

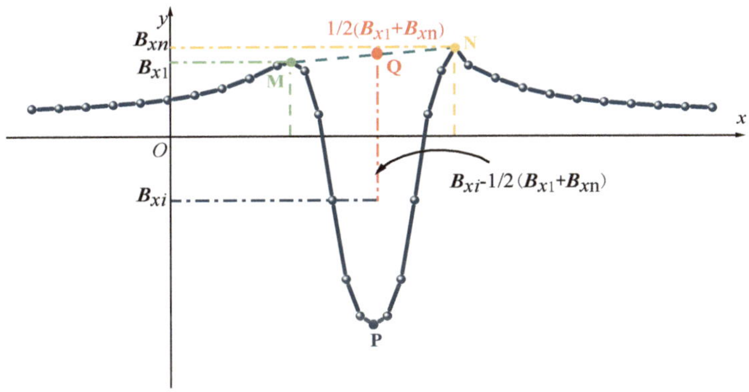

Figure 16. Schematic diagram of magnetic characteristics parameter.

In previous studies on the characterization model for corrosion in cables, only the magnetic signal values at points M, N, and P were used as reference points to construct indicators [20]. However, the theoretical and experimental research results mentioned earlier found that the magnetic signals between points M and N fluctuate to varying degrees due to vibration. Furthermore, as the degree of corrosion increases, the fluctuation at the extremum point P also becomes larger. Therefore, considering only the magnetic signals at these three reference points can lead to significant errors in determining the degree of corrosion. Therefore, this study incorporates the magnetic signals between points M and N into the construction of the characterization indicator. The average value is taken to reduce the influence of signal fluctuations, resulting in the construction of the characterization indicator A as expressed in Equation (21).

In Figure 16, $\boldsymbol{B}_{\mathrm{M}} = \boldsymbol{B}_{x1}$, $\boldsymbol{B}_{\mathrm{N}} = \boldsymbol{B}_{x\mathrm{n}}$, Q is the midpoint of M and N,$\boldsymbol{B}_{\mathrm{Q}} = 1/2(\boldsymbol{B}_{x1}+ \boldsymbol{B}_{x\mathrm{n}})$. The magnitude of the magnetic signal at any point between M and N is $\boldsymbol{B}_{x\mathrm{i}}$. The index A is constructed and its expression is presented in Equation (21).

$$A = \frac{\sum\limits_{\mathrm{i}=1}^{n} \boldsymbol{B}_{xi} - \frac{1}{2}(\boldsymbol{B}_{x1} + \boldsymbol{B}_{xn})}{n} \tag{21}$$

To verify the correlation between index *A* and corrosion degree *c*, the 2# and 3# specimens are taken as examples and the index *A* under different corrosion degrees is calculated by Equation (21). At the same time, the indicators mentioned in document [20] are also calculated using the vibration state magnetic signals of specimens 2# and 3# in this study. These calculated indicators are denoted as A′. The purpose is to compare whether the indicators A proposed in this paper are more accurate than those in previous research. The results are shown in Figure 17. Figure 17a,b are the calculation results of the characteristic index of specimen 3#. Figure 17c,d are the calculation results of the characteristic index of specimen 2#.

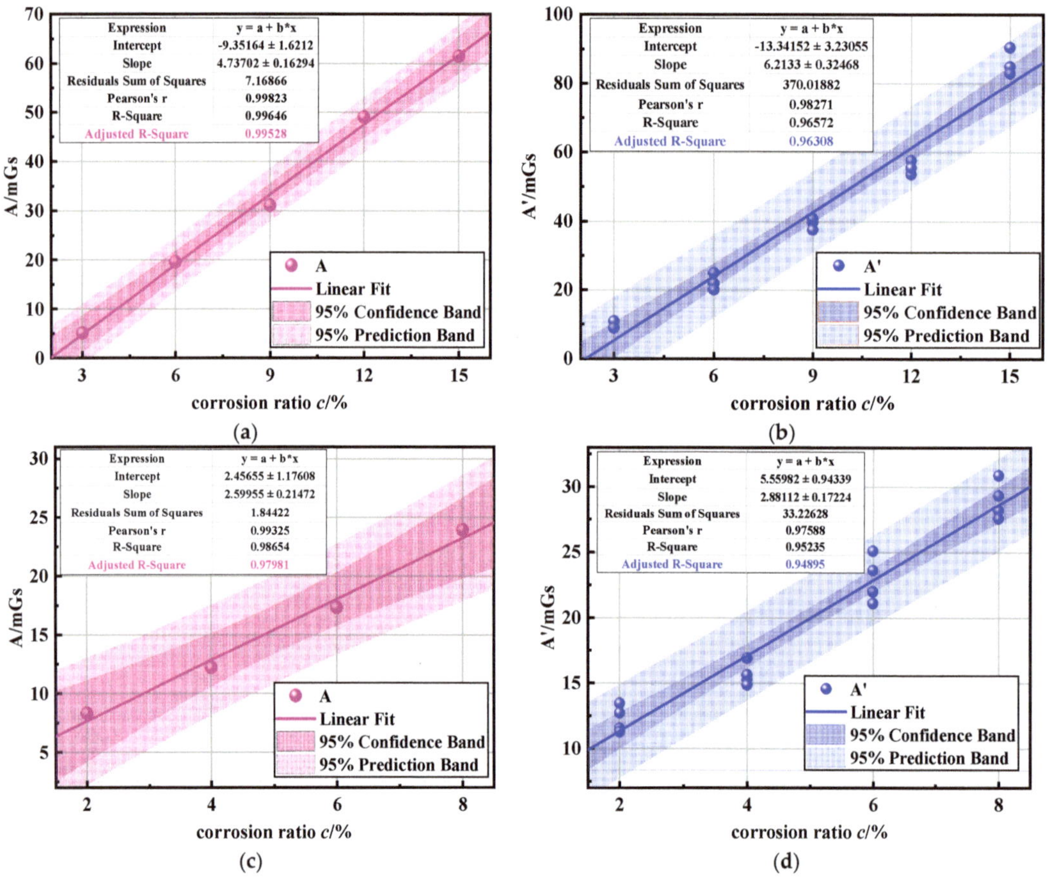

Figure 17. Correlation between parameters *A* and corrosion ratio *c*: (**a**,**b**) were 3# specimens; (**c**,**d**) were 2# specimens.

It is shown from Figure 17 that the relationship between the index *A* and the corrosion degree *c* can be approximately regarded as a linear relationship, satisfying the equation $A = a + b * c$, where a is the slope of the fitted straight line and b is the intercept of the fitted straight line. To assess the potential of index A for corrosion detection in cable structures under vibration, the fitting effect of the equation is analyzed. This analysis includes both the significance test and the fitting test. The significance test of the regression equation is usually the *F* test. The principle is to decompose the total sum of squares of deviations into regression sums of squares and residual squares and see how many regression sums of squares and residual sums of squares are included in the total sum of squares of deviations. Its specific expression was determined by Equation (22).

$$F = \sum_{i=1}^{n} (\hat{y}_i - \overline{y})^2 \times \left(\frac{\sum_{i=1}^{n} (y_i - \hat{y}_i)^2}{n-2}\right)^{-1} \quad (22)$$

where $\sum_{i=1}^{n} (\hat{y}_i - \overline{y})^2$ is the regression sum of squares, $\sum_{i=1}^{n} (y_i - \hat{y}_i)^2$ is the residual sum of squares.

$F_\alpha(1, n-2)$ is the critical value, when $F > F_\alpha$, the regression equation is significant. To illustrate the fitting effect of the regression line, the goodness of fit R^2 of the regression equation is further calculated. The results are shown in Table 3. Figure 17 and Table 3 show that the four groups of data all satisfied $P < 0.05$, so the regression equation is significant. In fact, for the univariate linear regression, the significance of the regression equation is equivalent to the test of the regression coefficient. R^2 values of all samples are more significant than 0.95, indicating that the data in the working stage with cracks had a higher linear relationship.

Table 3. Linear fitting parameters.

Number	F	P	R^2
3#-A	854.155	0	0.99
3#-A′	366.208	0	0.96
2#-A	146.569	0.0067	0.98
2#-A′	664.068	0	0.95

Furthermore, the calculation results of R^2 in Table 3 indicate that the calculated results of indicator A in specimens 2# and 3# exhibits a superior performance compared to A′. This indicates that indicator A possesses a stronger correlation with corrosion level. Indicator A can reduce the influence of vibrations to some extent. Therefore, using the characteristic parameter A for corrosion characterization is feasible and it can provide more accurate determination results.

5. Conclusions

Taking the corrosion of steel strand under vibration as the focus of this study, the change of magnetic leakage signal of corroded steel strand under vibration was analyzed theoretically and the concept of additional magnetic field was put forward. The existence of additional magnetic field was verified by experiments and its variation characteristics were further analyzed, which provided a new idea for cable corrosion detection in practical engineering. Based on the test results, the correlation between magnetic characteristic index and corrosion degree was analyzed, and the main conclusions were as follows:

(1) The corrosion SMFL signal in the vibration state fluctuated around the stationary SMFL curve, exhibiting similar trends and characteristics. The extreme point of the SMFL curve was negatively correlated with the measured lift-off height and positively correlated with the corrosion degree.

(2) The influence of vibration on magnetic field distribution of cable structure was defined as the displacement-added magnetic field ($\Delta \boldsymbol{B}^{\text{dis}}$) and the deformation-added magnetic field ($\Delta \boldsymbol{B}^{\text{def}}$). The tangential components of both have good data stability. $\Delta \boldsymbol{B}^{\text{dis}}$ followed a statistical normal distribution, fluctuating around zero value, and had no obvious relationship with excitation current and frequency. The effect of $\Delta \boldsymbol{B}^{\text{def}}$ was mainly related to the degree of corrosion. The larger the degree of corrosion, the greater the impact of $\Delta \boldsymbol{B}^{\text{def}}$. When the external magnetic induction intensity was stable, the increment of magnetic field intensity change was mainly determined by $\Delta \boldsymbol{B}^{\text{def}}$.

(3) Combined with the characteristics of SMFL curve under vibration and the influence of $\Delta \boldsymbol{B}^{dis}$ and $\Delta \boldsymbol{B}^{def}$ on the SMFL curve, a corrosion index A was proposed. Moreover, through testing and analyzing, the index A exhibited a strong linear fit with the degree of corrosion (with R^2 values greater than 0.97 in all cases) and can reduce the influence of vibrations to some extent. Using index A for corrosion characterization was feasible and can provide more accurate determination results.

In summary, considering the influence of vibration effect on magnetic leakage detection, this study proposed a suitable characterization index for cable corrosion detection under vibration. This approach effectively improved the accuracy of cable corrosion detection. However, this method is also affected by environmental magnetic field, magnetization difference of specimens and other factors and the influence of temperature on magnetic leakage detection was not taken into account. These questions would be the focus of future research.

Author Contributions: Conceptualization, Y.D.; methodology, H.Z. and Y.D. and R.X.; Software, H.Z.; Validation, Y.Y.; investigation, H.Z. and R.X.; resources, Y.Y.; data curation, Y.D. and Y.Y.; Writing—original draft, H.Z. and Y.D.; writing—review and editing, H.Z.; supervision, R.X.; funding acquisition, J.Z. All authors have read and agreed to the published version of the manuscript.

Funding: This research was funded by the National Natural Science Foundation of China (52278291, U20A20314, 52208301), the Postgraduate Research Innovation Project of Chongqing Jiaotong University (2022S0003); and the Chongqing Natural Science Foundation of China (cstc2022ycjh-bgzxm008, CSTB2022NSCQ-LZX0006, CSTB2022NSCQ-BHX0036).

Institutional Review Board Statement: Not applicable.

Informed Consent Statement: Not applicable.

Data Availability Statement: The data presented in this study are available from the first and corresponding author upon request. The data is not publicly available due to the policy of the data provider.

Conflicts of Interest: The authors declare no conflict of interest. The sponsors had no role in the design, execution, interpretation, or writing of the study.

References

1. Huang, W.; Pei, M.; Liu, X.; Wei, Y. Design and Construction of Super-Long Span Bridges in China: Review and Future Perspectives. *Front. Struct. Civ. Eng.* **2020**, *14*, 803–838. [CrossRef]
2. Guzmán, S.; Gálvez, J.C.; Sancho, J.M. Cover Cracking of Reinforced Concrete Due to Rebar Corrosion Induced by Chloride Penetration. *Cem. Concr. Res.* **2011**, *41*, 893–902. [CrossRef]
3. Qiu, J.; Zhang, H.; Zhou, J.; Ma, H.; Liao, L. Experimental Analysis of the Correlation between Bending Strength and SMFL of Corroded RC Beams. *Constr. Build. Mater.* **2019**, *214*, 594–605. [CrossRef]
4. Fernandes, B.; Nims, D.; Devabhaktuni, V. Computer Aided Modeling of Magnetic Behavior of Embedded Prestressing Strand for Corrosion Estimation. *J. Nondestruct. Eval.* **2013**, *32*, 124–133. [CrossRef]
5. Chernin, L.; Val, D.V. Prediction of Corrosion-Induced Cover Cracking in Reinforced Concrete Structures. *Constr. Build. Mater.* **2011**, *25*, 1854–1869. [CrossRef]
6. Yuan, Y.; Liu, X.; Pu, G.; Wang, T.; Zheng, D. Temporal and Spatial Variability of Corrosion of High-Strength Steel Wires within a Bridge Stay Cable. *Constr. Build. Mater.* **2021**, *308*, 125108. [CrossRef]
7. Mangual, J.; Elbatanouny, M.; Ziehl, P.; Matta, F. Corrosion Damage Quantification of Prestressing Strands Using Acoustic Emission. *J. Mater. Civ. Eng.* **2013**, *25*, 1326–1334. [CrossRef]
8. Du, P.; Xu, D.; Huang, S.; Cheng, X. Assessment of Corrosion of Reinforcing Steel Bars in Concrete Using Embedded Piezoelectric Transducers Based on Ultrasonic Wave. *Constr. Build. Mater.* **2017**, *151*, 925–930. [CrossRef]
9. Ben Seghier, M.E.A.; Corriea, J.A.F.O.; Jafari-Asl, J.; Malekjafarian, A.; Plevris, V.; Trung, N.-T. On the Modeling of the Annual Corrosion Rate in Main Cables of Suspension Bridges Using Combined Soft Computing Model and a Novel Nature-Inspired Algorithm. *Neural Comput. Applic.* **2021**, *33*, 15969–15985. [CrossRef]
10. Qiu, G.; Cai, Y.; Li, Z. Multiscale Investigation of Magnetic Field Distortion on Surface of Ferromagnetic Materials Caused by Stress Concentration for Metal Magnetic Memory Method. *Comput. Mater. Sci.* **2022**, *209*, 111353. [CrossRef]
11. Ismail, M.; Muhammad, B.; Ismail, E.G. Compressive Strength Loss and Reinforcement Degradations of Reinforced Concrete Structure Due to Long-Term Exposure. *Constr. Build. Mater.* **2010**, *24*, 898–902. [CrossRef]

12. Ma, Y.; Wang, G.; Guo, Z.; Wang, L.; Jiang, T.; Zhang, J. Critical Region Method-Based Fatigue Life Prediction of Notched Steel Wires of Long-Span Bridges. *Constr. Build. Mater.* **2019**, *225*, 601–610. [CrossRef]
13. Li, S.; Xin, J.; Jiang, Y.; Wang, C.; Zhou, J.; Yang, X. Temperature-Induced Deflection Separation Based on Bridge Deflection Data Using the TVFEMD-PE-KLD Method. *J Civ. Struct. Health Monit.* **2023**, *13*, 781–797. [CrossRef]
14. Dong, B.; Fang, G.; Liu, Y.; Dong, P.; Zhang, J.; Xing, F.; Hong, S. Monitoring Reinforcement Corrosion and Corrosion-Induced Cracking by X-Ray Microcomputed Tomography Method. *Cem. Concr. Res.* **2017**, *100*, 311–321. [CrossRef]
15. Liu, S.; Ma, H. Estimation of the Stress Level on a Cross Section of a Reinforced Concrete Beam via Acoustic Emission Intensity Distribution (AID) Analysis. *Constr. Build. Mater.* **2018**, *164*, 463–476. [CrossRef]
16. Xu, J.; Wu, X.J. Nondestructive Testing of Bridge Cables Using Magnetostrictive Guided Wave Technique. *AMM* **2011**, *130–134*, 2015–2018. [CrossRef]
17. Ni, Y.; Zhang, Q.; Xin, R. Magnetic Flux Detection and Identification of Bridge Cable Metal Area Loss Damage. *Measurement* **2021**, *167*, 108443. [CrossRef]
18. Zhang, H.; Liao, L.; Zhao, R.; Zhou, J.; Yang, M.; Xia, R. The Non-Destructive Test of Steel Corrosion in Reinforced Concrete Bridges Using a Micro-Magnetic Sensor. *Sensors* **2016**, *16*, 1439. [CrossRef]
19. Zhou, J.; Qiu, J.; Zhou, Y.; Zhou, Y.; Xia, R. Experimental Study on Residual Bending Strength of Corroded Reinforced Concrete Beam Based on Micromagnetic Sensor. *Sensors* **2018**, *18*, 2635. [CrossRef]
20. Xia, R.; Zhang, H.; Zhou, J.; Liao, L.; Zhang, Z.; Yang, F. Probability Evaluation Method of Cable Corrosion Degree Based on Self-Magnetic Flux Leakage. *J. Magn. Magn. Mater.* **2021**, *522*, 167544. [CrossRef]
21. Yang, J.; Chen, R.; Zhang, Z.; Zou, Y.; Zhou, J.; Xia, J. Experimental Study on the Ultimate Bearing Capacity of Damaged RC Arches Strengthened with Ultra-High Performance Concrete. *Eng. Struct.* **2023**, *279*, 115611. [CrossRef]
22. Xin, J.; Zhou, C.; Jiang, Y.; Tang, Q.; Yang, X.; Zhou, J. A Signal Recovery Method for Bridge Monitoring System Using TVFEMD and Encoder-Decoder Aided LSTM. *Measurement* **2023**, *214*, 112797. [CrossRef]
23. Liu, S.; Jiang, Y.; Qiao, K.; Peng, L.; Liu, D. Record-Based Simulation of Three-Component Long-Period Ground Motions: Hybrid of Surface Wave Separation and Multivariate Empirical Mode Decomposition. *Soil Dyn. Earthq. Eng.* **2023**, *172*, 108037. [CrossRef]
24. Wu, L.; Yao, K.; Shi, P.; Zhao, B.; Wang, Y.-S. Influence of Inhomogeneous Stress on Biaxial 3D Magnetic Flux Leakage Signals. *NDT E Int.* **2020**, *109*, 102178. [CrossRef]
25. Shi, P.; Su, S.; Chen, Z. Overview of Researches on the Nondestructive Testing Method of Metal Magnetic Memory: Status and Challenges. *J. Nondestruct. Eval.* **2020**, *39*, 43. [CrossRef]
26. Qiu, J.; Zhang, H.; Zhou, J.; Zhang, W. An SMFL-Based Non-Destructive Quantification Method for the Localized Corrosion Cross-Sectional Area of Rebar. *Corros. Sci.* **2021**, *192*, 109793. [CrossRef]
27. Xia, R.; Zhang, H.; Zhou, J.; Liao, L.; Di, H.; Zhang, Z. Studies on the Spatial Location Method of Corrosion Defect in Bridge Cable Considering Self-Magnetic Flux Leakage Effect. *J. Test. Eval.* **2021**, *49*, 4526–4539. [CrossRef]
28. Han, W.; Xu, J.; Wang, P.; Tian, G. Defect Profile Estimation from Magnetic Flux Leakage Signal via Efficient Managing Particle Swarm Optimization. *Sensors* **2014**, *14*, 10361–10380. [CrossRef]
29. Fernandes, B.; Titus, M.; Nims, D.K.; Ghorbanpoor, A.; Devabhaktuni, V.K. Practical Assessment of Magnetic Methods for Corrosion Detection in an Adjacent Precast, Prestressed Concrete Box-Beam Bridge. *Nondestruct. Test. Eval.* **2013**, *28*, 99–118. [CrossRef]
30. Qu, Y.; Zhang, H.; Zhao, R.; Liao, L.; Zhou, Y. Research on the Method of Predicting Corrosion Width of Cables Based on the Spontaneous Magnetic Flux Leakage. *Materials* **2019**, *12*, 2154. [CrossRef]
31. Singh, W.S.; Rao, B.P.C.; Thirunavukkarasu, S.; Jayakumar, T. Flexible GMR Sensor Array for Magnetic Flux Leakage Testing of Steel Track Ropes. *J. Sens.* **2012**, *2012*, 129074. [CrossRef]
32. Ren, S.; Ren, X.; Duan, Z.; Fu, Y. Studies on Influences of Initial Magnetization State on Metal Magnetic Memory Signal. *NDT E Int.* **2019**, *103*, 77–83. [CrossRef]
33. Xia, R.; Zhang, H.; Zhou, J.; Liao, L.; Yang, W.; Li, Y. Corrosion Non-Destructive Testing of Loaded Steel Strand Based on Self-Magnetic Flux Leakage Effect. *Nondestruct. Test. Eval.* **2022**, *37*, 56–70. [CrossRef]
34. Xia, R.; Zhou, J.; Zhang, H.; Zhou, D.; Zhang, Z. Experimental Study on Corrosion of Unstressed Steel Strand Based on Metal Magnetic Memory. *KSCE J. Civ. Eng.* **2019**, *23*, 1320–1329. [CrossRef]
35. Xia, R.; Zhou, J.; Zhang, H.; Liao, L.; Zhao, R.; Zhang, Z. Quantitative Study on Corrosion of Steel Strands Based on Self-Magnetic Flux Leakage. *Sensors* **2018**, *18*, 1396. [CrossRef]
36. Shi, P.; Jin, K.; Zheng, X. A Magnetomechanical Model for the Magnetic Memory Method. *Int. J. Mech. Sci.* **2017**, *124–125*, 229–241. [CrossRef]
37. Le, M.; Pham, C.-T.; Lee, J. Deep Neural Network for Simulation of Magnetic Flux Leakage Testing. *Measurement* **2021**, *170*, 108726. [CrossRef]
38. Xu, J.; Chen, W. Behavior of Wires in Parallel Wire Stayed Cable under General Corrosion Effects. *J. Constr. Steel. Res.* **2013**, *85*, 40–47. [CrossRef]
39. Xu, Y.; Li, H.; Li, S.; Guan, X.; Lan, C. 3-D Modelling and Statistical Properties of Surface Pits of Corroded Wire Based on Image Processing Technique. *Corros. Sci.* **2016**, *111*, 275–287. [CrossRef]
40. Machine Learning-Assisted Probabilistic Fatigue Evaluation of Rib-to-Deck Joints in Orthotropic Steel Decks. *Eng. Struct.* **2022**, *265*, 114496. [CrossRef]

41. GPR-Assisted Evaluation of Probabilistic Fatigue Crack Growth in Rib-to-Deck Joints in Orthotropic Steel Decks Considering Mixed Failure Models. *Eng. Struct.* **2022**, *252*, 113688. [CrossRef]
42. Cheng, S.; Fan, X.; Tao, H. Wind-Induced Vibration Analysis of a Cable-Stayed Bridge during Erection by a Modified Time-Domain Method. *J. Sound Vib.* **2007**, *303*, 330–342. [CrossRef]
43. Matsumoto, M.; Shirato, H.; Yagi, T.; Goto, M.; Sakai, S.; Ohya, J. Field Observation of the Full-Scale Wind-Induced Cable Vibration. *J. Wind Eng. Ind. Aerodyn.* **2003**, *91*, 13–26. [CrossRef]
44. Ni, Y.Q.; Wang, X.Y.; Chen, Z.Q.; Ko, J.M. Field Observations of Rain-Wind-Induced Cable Vibration in Cable-Stayed Dongting Lake Bridge. *J. Wind Eng. Ind. Aerodyn.* **2007**, *95*, 303–328. [CrossRef]
45. Ahmad, J.; Cheng, S. Effect of Cross-Link Stiffness on the in-Plane Free Vibration Behaviour of a Two-Cable Network. *Eng. Struct.* **2013**, *52*, 570–580. [CrossRef]
46. Zhang, H.; Li, H.; Zhou, J.; Tong, K.; Xia, R. A Multi-Dimensional Evaluation of Wire Breakage in Bridge Cable Based on Self-Magnetic Flux Leakage Signals. *J. Magn. Magn. Mater.* **2023**, *566*, 170321. [CrossRef]

buildings

MDPI

Article

Propagation Effect Analysis of Existing Cracks in Box Girder Bridges Based on the Criterion of Compound Crack Propagation

Yong-Jian Chen [1], Xian-Cheng Huang [1], Shi-Zheng Chen [1], Li-Yuan Wang [1,*], Ji-Wei Zhong [2] and Hui Cheng [2]

[1] College of Civil Engineering, Fuzhou University, Fuzhou 350108, China; chenyongjian@fzu.edu.cn (Y.-J.C.); hxc935857710@gmail.com (X.-C.H.); a946822490@gmail.com (S.-Z.C.)

[2] Railway Bridge Science Research Institute, Ltd., Wuhan 430034, China; 19835521657@163.com (J.-W.Z.); 18359391658@163.com (H.C.)

* Correspondence: eyuan369@163.com

Abstract: Cracking in concrete box girder bridges will have a significant impact on the safety and durability of the structure, and many box girder bridges which are in service have undergone varying degrees of cracking. Currently, the safety design of actual bridge projects place an emphasis on the stress or the load value of a cross section at the limit value specified in the code for safety control. This design method assumes that the member itself is of uniform and continuous material and is internally undamaged. However, the bridge structure is more or less cracked to varying degrees during the period from casting to construction to operation of the concrete members. In this paper, a finite element computational model of a three-span prestressed concrete box girder bridge with existing cracks is established based on the fracture mechanics theory, and the critical parameters of crack extension are introduced to evaluate the extension state of cracks. At the same time, the extended stability of the existing cracks of the box girder bridge is analyzed by considering the temperature effect, vehicle loading, and prestressing loss, and the sensitivity of crack extension under each working condition is investigated. The results show that, with the increase in crack length and depth, the crack expansion is promoted, but the effect is relatively small, and the maximum stress intensity factor is only 6.48 MPa $mm^{1/2}$. Under the multi-factor coupling effect, the cracks show a composite crack expansion dominated by type I cracks, the longitudinal cracks of the existing base plate are in a stable state, the maximum value of the crack expansion critical parameter of the vertical cracks of the webs reaches 1.087, and there is a tendency to expand locally. The maximum value of the critical parameter for crack extension of the vertical crack in the web plate reaches 1.087, and there is a tendency towards local expansion. The crack extension evaluation criteria proposed in this paper have a certain reference value for crack extension research on similar concrete box girder bridges and provide a scientific basis for the optimized design of similar bridges.

Keywords: prestressed concrete box girder; finite element analysis; fracture mechanics; fracture propagation

Citation: Chen, Y.-J.; Huang, X.-C.; Chen, S.-Z.; Wang, L.-Y.; Zhong, J.-W.; Cheng, H. Propagation Effect Analysis of Existing Cracks in Box Girder Bridges Based on the Criterion of Compound Crack Propagation. *Buildings* **2024**, *14*, 2958. https://doi.org/10.3390/buildings14092958

Academic Editors: Simon X. Yang, Fabrizio Gara, Jingzhou Xin, Yan Jiang and Hong Zhang

Received: 19 August 2024
Revised: 9 September 2024
Accepted: 16 September 2024
Published: 19 September 2024

1. Introduction

Prestressed concrete box girder bridges are subjected to a variety of operating conditions in the long-term operation process; there are generally different degrees of concrete cracking problems, and the existence of cracks reduces the effective cross section of the concrete structure, weakening the stiffness of the concrete box girder and making the mid-span deflection increase, thus reducing the safety and durability of the bridge [1]. In the past, the bridge collapsed before the bridge reached its standard strength and was damaged, often due to the existence of cracks in the fatigue damage caused by the expansion of instability. Therefore, it is particularly important to investigate whether the existing cracks in concrete box girder bridges further expand during operation [2].

The problem of cracking in concrete arises from the initial defects inherent in its being a non-homogeneous material, and these small cracks will gradually expand to form macroscopic cracks when subjected to different loads. The traditional strength theory assumes the concrete material to be continuous and homogeneous, which has limitations [3]. However, the application of fracture mechanics theory presupposes the existence of macroscopic cracks in the structure, which can make up for the shortcomings of the traditional strength theory. Fracture mechanics theory can more accurately describe the crack behavior and expansion law of concrete when subjected to external forces and provide a theoretical basis for the design, assessment, and maintenance of existing cracked concrete structures.

Scholars at home and abroad have carried out relevant studies on concrete cracking based on the theory of fracture mechanics. Walsh [4] investigated precast notched concrete beams using LEFM, focusing on the effect of concrete beam dimensions on the fracture parameters, and pointed out that the size of the member has a significant effect on the extrapolation of the actual structural fracture parameters, i.e., there is a size effect. Li et al. [5] used resistance strain gauges to measure the initiation loads of three-point bending beams with different heights and deduced the initiation fracture toughness and the destabilization fracture toughness under the double K model based on the crack opening displacement curves. Xu Shilong collected experimental data on the fracture toughness of concrete three-point bending specimens and conducted a probabilistic modeling study [6]. The results showed that the fracture toughness of concrete conformed to the Weibull distribution, revealing the phenomenon that the fracture toughness of concrete is affected by size effect. Combined with the reliability theory, an expression for the statistical distribution of the fracture toughness of concrete specimens was proposed and the relative failure probability of a structural fracture was calculated. Feng Bolin studied the fracture mechanics analysis method of arch dam cracks under temperature loading. He concluded that it is reasonable to apply the theory and method of fracture mechanics to the analysis and calculation of crack extension in dams as well as to practical engineering [7]. Yanhua Zhao proposed the double G criterion for concrete crack extension from the perspective of energy, combining the linear elastic fracture mechanics and the viscous cohesion distribution in the virtual crack zone with the energy release rate as a parameter [8]. Wu Zhimin proposed the corresponding crack extension criterion to calculate the whole process of I–II composite crack extension based on the experimentally measured critical K_I–K_{II} curves of I–II composite cracks in concrete materials [9]. Fan Guogang used the interaction integral method to study the fracture parameters of different crack patterns in concrete dams and used the cohesive crack model to analyze the effect of crack patterns on the bearing capacity of concrete structures [10]. Chen Xingda used a variable stiffness crack model to investigate the dynamic characteristics of an existing cracked T-beam under vehicular loading and evaluated the crack extension state using a three-dimensional dynamic stress intensity factor [11]. Mi Zhengxiang proposed the concept of fracture strength by combining the traditional strength theory with fracture mechanics theory for the problem of concrete cracking in dams, which brought about new ideas for the design of concrete structures [12]. Li-Qiong Ren used the cohesive crack model to study the subcritical crack expansion stage of concrete before complete failure and used fracture mechanics-related parameters to assess the mechanism of subcritical crack expansion -n concrete [13].

Based on fracture mechanics, most of the cracking problems of concrete by the above scholars focus on the test members, and there is a lack of research on the crack propagation of practical projects. In this paper, the finite element modeling of a prestressed concrete box beam bridge with existing cracks is carried out and the composite crack expansion evaluation criteria based on fracture mechanics is introduced. Under the comprehensive consideration of temperature effect, vehicle load, prestress loss, and other factors, the expansion stability of existing cracks is analyzed, which provides references for the design of and research on similar bridge structures.

2. Basic Theory of Fracture Mechanics and Extended Evaluation Criteria

In fracture mechanics, according to the stress type and displacement characteristics of cracks, the fracture morphology is divided into three types: open (type I) cracks, in-plane shear (type II) cracks, and anti-plane shear (type III) cracks, as shown in Figure 1 [14].

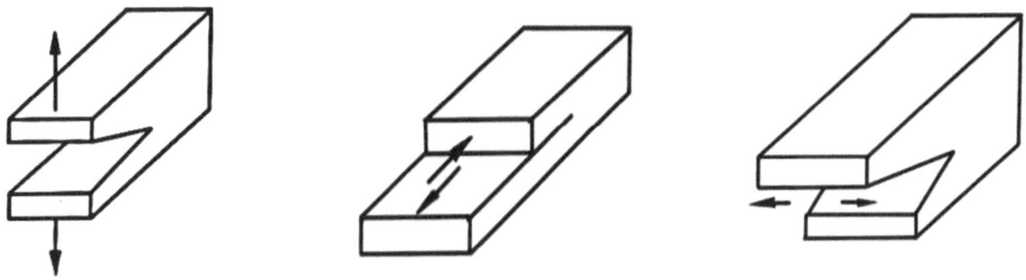

(**a**) Open crack (**b**) In-plane shear crack (**c**) anti-plane shear cracks

Figure 1. Basic morphology of cracks.

In practical engineering, because of the action form of loads, the different characteristics of materials and the development direction of cracks, cracks are often presented in a composite form, and people are concerned about the expansion conditions and directions of cracks, that is, the critical state and angle of composite crack expansion [15]. Therefore, it is necessary to seek a reasonable and reliable crack propagation criterion for judgment. Currently, the common composite crack propagation criteria include the maximum circumferential tensile stress intensity factor criterion, the maximum energy release rate criterion, and the K judgment criterion based on composite critical curves. The first two types of criteria are mostly used for metal materials. Due to the particularity of concrete materials, it has certain limitations, so this paper adopts the K judgment criterion of composite critical curve, which is quoted in the data of fracture test conducted on concrete I–II composite cracks in Yu Xiaozhong [16] and gives the curve expression of the empirical fracture criteria of I–II composite cracks as follows:

$$\left(\frac{K_I}{K_{IC}}\right)^2 + 4.2\left(\frac{K_{II}}{K_{IC}}\right)^2 = 1 \tag{1}$$

According to the above expression, the critical parameter of crack propagation φ is introduced in this paper to describe whether existing cracks enter the expanding state under the multi-factor coupling action of concrete box beams, as shown in Equation (2):

$$\varphi = \frac{K_I^2 + 4.2K_{II}^2}{K_{IC}^2} \tag{2}$$

According to Equations (1) and (2), when the critical parameter of crack propagation is $\varphi = 1$, the crack is in a critical state of propagation. $\varphi > 1$, the cracks were in a spreading state; $\varphi < 1$, the cracks were stable. However, it should be emphasized that, although $\varphi < 1$ may indicate that the fracture is stable at the macro level, in some cases the fracture propagation trend is still affected by the stress intensity factor at the crack tip. Therefore, in addition to the evaluation φ, it is necessary to comprehensively consider the stress field of the crack tip under the action of different factors and the influence on the stress intensity factor to judge the potential expansion behavior of the existing stable crack.

In the determination of the fracture toughness of concrete structures K_{Ic}, it has been observed that the values measured by small size specimens are affected by the size effect and are not applicable to large-size concrete structures. Through the fracture testing of large-size concrete specimens, Wu Zhimin [17] found that, when the height of the specimen was above 800 mm, the measured value K_{Ic} had nothing to do with the size of the specimen However, for concrete box girder bridges, it is very difficult to carry out large-scale fracture tests of the same size or scale as the structure. In view of this, based on the method proposed by Mai Jiaxuan [18] and Xu Shi-lang [6], this paper deduces the fracture toughness suitable for large-size specimens from the measured values K_{Ic} of small-size specimens. The corresponding size effect equation is as follows:

$$K_{IC} = K_{ICs}\sqrt{\frac{h}{h_s}}\left(\frac{V_s}{V}\right)^{\frac{1}{\alpha}} \tag{3}$$

where, K_{ICs}, h_s, V_s are corresponding to the fracture toughness, height, and volume of small-size specimens; K_{IC}, h, V are derived the fracture toughness, height, and volume of large-size specimens, wherein $h \le 800$ mm; and α is the Weibull modulus, which is related to the coefficient of variation in the experimental value of fracture toughness K_{Ic} measured, with the specific value being referred to in the literature.

The above equation is used to calculate the fracture toughness of specimens of different sizes in the literature; the relative error between the results and the test values is less than 10% and the agreement is high. Therefore, this paper refers to the wedge splitting test conducted by Gao Hongbo [19]; the specimen size is 300 mm × 300 mm × 200 mm and the concrete strength grade is C50. The test results show that the cubic compressive strength, tensile strength, and elastic modulus of the concrete specimen are 50.2 MPa, 3.95 Mpa, and 34.8 Gpa, and the average fracture toughness of the concrete specimen is 1.147 Mpa$\sqrt{\text{m}}$. Based on the small size specimen, the fracture toughness of the prestressed concrete continuous box girder bridge is 0.8781 Mpa$\sqrt{\text{m}}$ by Equation (3). This result is consistent with the interval range of size-independent fracture toughness of C50 concrete measured by Xu Shi-lang [20].

3. Finite Element Model

This paper takes a prestressed concrete continuous box girder bridge as the engineering background. The total length of the box girder bridge is 869 m with nine connections. Among them, the crack phenomenon of the third joint is the most prominent, so this paper takes the third joint box girder bridge as the research object. The span arrangement of the joint is 35 m + 35 m + 35 m. The prestressed concrete box girder structure with a single box and four chambers is adopted, and the beam height is 1.8 m. The thickness of the top plate and bottom plate of the box girder is 0.25 m, while the top plate thickness of the cantilever part is 0.2 m and the length of the cantilever is 2.5 m. The horizontal width of the inclined web is 0.433 m. The width of the middle beam is 2 m and the width of the end beam is 1.5 m. In addition, the width of the box girder top plate is 20 m, and the width of the bottom plate is 15 m. The general layout of the bridge and the section layout of the box girder are shown in Figures 2 and 3, respectively.

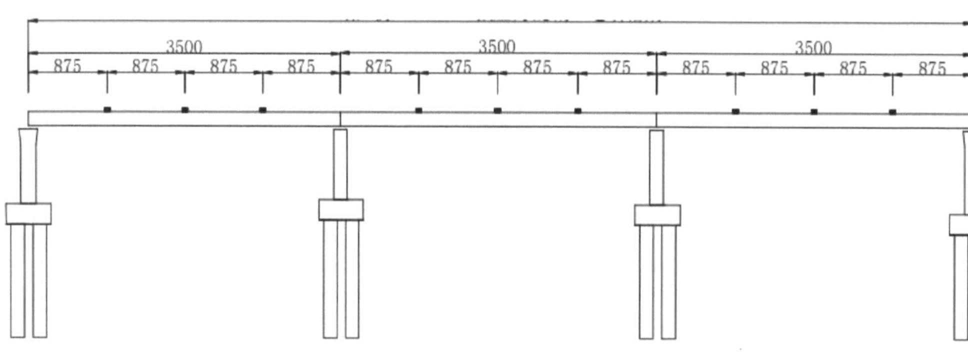

Figure 2. General layout of bridge (unit: cm).

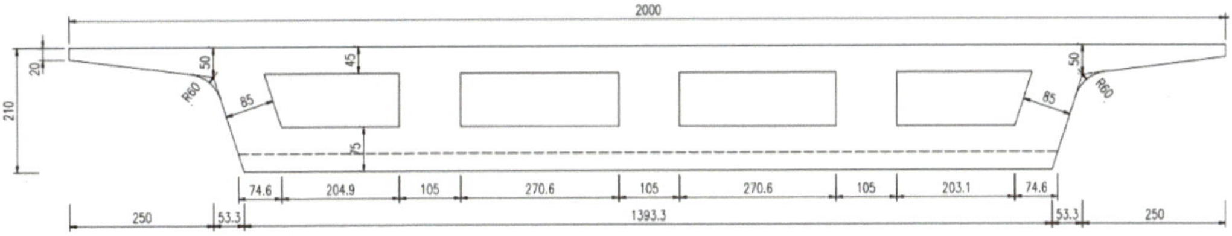

Figure 3. Box girder section diagram (unit: cm).

In this paper, the three-dimensional finite element model is established by ANSYS Workbench 16.0 and the concrete and rebar units are simulated by SOLID186 and BEAM188 units, respectively. The hexahedral grid division method is used to grid the concrete solid model, and the rebar is meshed separately. The initial stress is applied to the prestressed steel bar by the cooling method, several nodes of the concrete unit are selected by CEINTF command, and the constraint equation is established with the nodes of the steel bar in the tolerance range to realize the coupling between the prestressed steel bar and the concrete node. According to the actual situation of the bridge, the boundary conditions are set and the external load is simulated. The finite element model of prestressed concrete continuous box girder bridge has a total of 520,460 nodes and 939,448 units. The overall wire frame and grid division are shown in Figures 4 and 5.

Figure 4. Three-dimensional wireframing of box girder.

Figure 5. Box girder grid division diagram.

The main objective of the bridge static load test is to assess whether the actual working condition of the bridge structure is consistent with the design expectation by measuring the deformation, stress, and strain responses of the bridge structure under static loading. The test not only directly and effectively verifies the mechanical properties of the bridge structure but also provides key data to verify that the finite element model accurately simulates the initial stress state of the bridge.

(1) Test control section

The static loading test takes into account the distribution characteristics of bridge cracks, structural characteristics, and the actual situation on site, and it selects the maximum positive bending moment cross section of the third span and the maximum positive bending moment cross section of the second span of the link as the test control cross sections.

(2) Loading method

According to the current situation of the bridge structure, this static load test adopts the automobile-loading method.

(3) Layout of test points

The selected test control sections include the A-A section of the side span, the B-B section subjected to negative moments, and the C-C section of the center span. The layout of the test sections is shown in Figure 6, and the arrangement of the measurement points for each section is shown in Figure 7.

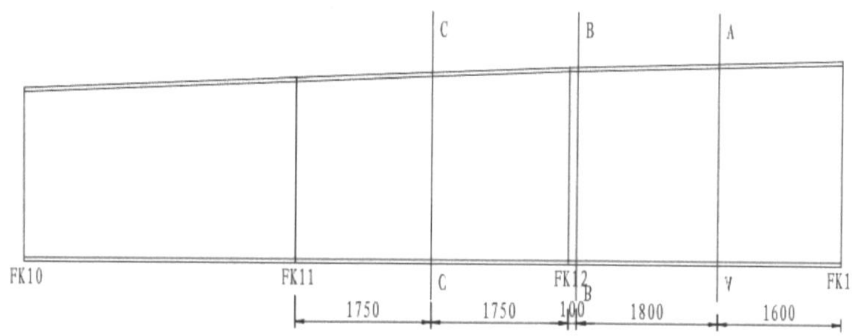

Figure 6. Layout of test control sections (unit: cm).

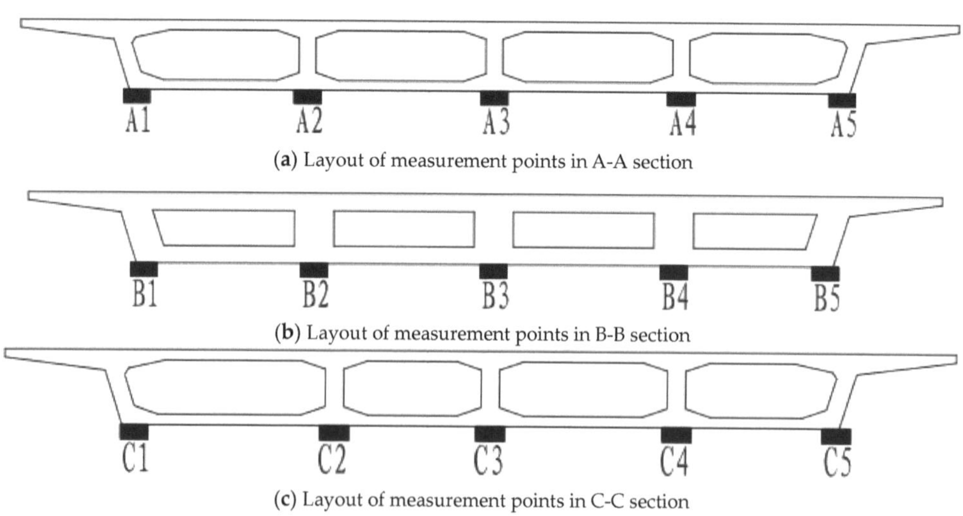

(**a**) Layout of measurement points in A-A section

(**b**) Layout of measurement points in B-B section

(**c**) Layout of measurement points in C-C section

Figure 7. Arrangement of measurement points in each section.

In order to enable the solid finite element model to accurately reflect the stress state of the prestressed concrete continuous box girder bridge under the action of multiple factors, the static load test data of the bridge with the maximum positive bending moment control section at the mid-span and side spans were compared and analyzed. The related measured and calculated values of deflection and strain are detailed in Tables 1 and 2, respectively.

Table 1. Side span maximum positive moment control section measured point values.

Measurement Point	Deflection/mm		Relative Error	Strain/με		Relative Error
	Measured Value	Calculated Value		Measured Value	Calculated Value	
A1	1.75	2.74	56.57%	11	17	54.55%
A2	2.45	3.37	37.56%	9	28	211.11%
A3	3.92	4.70	19.90%	30	41	36.67%
A4	5.93	6.97	17.54%	47	58	23.40%
A5	7.30	8.86	21.37%	22	83	277.27%

Table 2. Mid-span maximum positive moment control section measured point values.

Measurement Point	Deflection/mm		Relative Error	Strain/με		Relative Error
	Measured Value	Calculated Value		Measured Value	Calculated Value	
C1	1.65	2.11	27.88%	9	14	55.56%
C2	2.38	2.65	11.34%	12	26	116.67%
C3	3.46	3.86	11.56%	25	39	56%
C4	5.04	5.85	16.07%	86	61	29.07%
C5	5.85	7.09	21.19%	55	79	43.64%

By comparing the finite element simulation results in the above table with the static load test data of the bridge, the deflection and strain values of the corresponding measurement points are more consistent with the measured data, and the rule of change is consistent. However, some measurement points still show large errors [21,22]. These errors are mainly attributed to the existence of new cracks in the actual structure, resulting in the distortion of the strain values at some measurement points. In addition, the discrepancy between the inhomogeneity of the actual concrete material and the ideal material properties assumed in the model, coupled with the deviation of the eccentric or locally concentrated force of the actual bridge load test from the model setting and the actual complexity of the bearing and boundary conditions not accurately reflected in the model, may lead to the discrepancy between the actual stress state and the simulation results [23]. Although the model shows some errors due to the above factors, the overall error is still within the acceptable range, indicating that the model can effectively simulate the initial static load-bearing behavior of the bridge under the basic working conditions and can be used for the subsequent analysis of the expansion behavior of existing cracks in box girders [24].

4. Expansion Effect Analysis of Box Girder Bridges with Existing Cracks

According to the spatial stress distribution of the box girder and the location where the actual cracks are concentrated, the cracks are selected based on the longitudinal cracks in the bottom plate of the middle span and the vertical cracks in the web plate of the side span. According to the actual bridge test, the most unfavorable loading arrangement is selected for the vehicle load and the most unfavorable loading is arranged for the middle and side spans; the specific crack parameters and arrangement are shown in Table 3 and Figure 8.

Table 3. Set basic crack parameters.

Fracture Morphology	Length (mm)	Depth (mm)
Longitudinal cracks in mid-span floor	800∼1600 (Δ200)	10∼50 (Δ10)
Vertical crack of side span web	300∼500 (Δ50)	

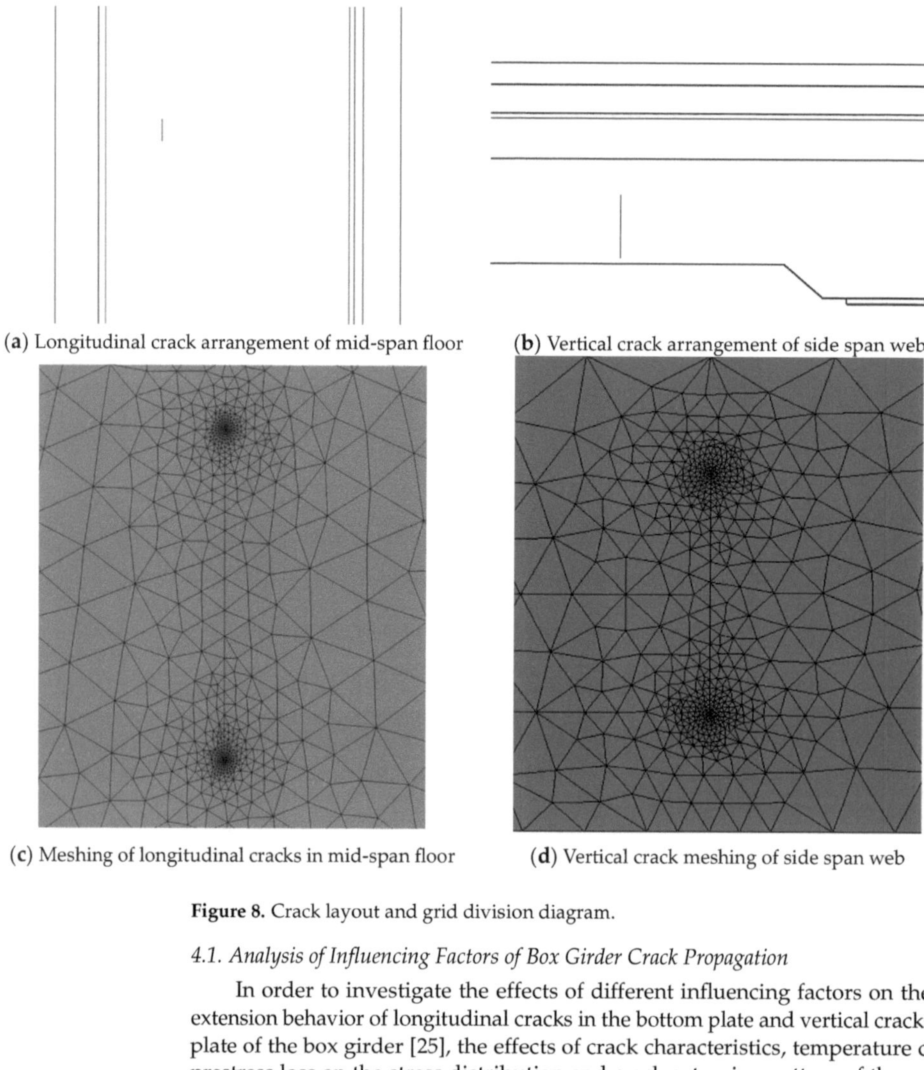

Figure 8. Crack layout and grid division diagram.

4.1. Analysis of Influencing Factors of Box Girder Crack Propagation

In order to investigate the effects of different influencing factors on the stress and extension behavior of longitudinal cracks in the bottom plate and vertical cracks in the web plate of the box girder [25], the effects of crack characteristics, temperature change, and prestress loss on the stress distribution and crack extension pattern of the crack tips are investigated under the combined action of the box girder's self-weight, the second-phase constant load, and the prestressing stress as the basic working condition [26–28].

The calculation working condition is set up in combination with the field bridge static load test in order to more accurately respond to the stress state of the prestressed concrete continuous box girder bridge; the finite element model of the load is applied in accordance with the relevant provisions of the JTG D60-2015 [29] general specification for the design of highway bridges to load, and the analysis of the overall spatial cracking effect of the box girder is mainly considered for the following influencing factors:

(1) Phase I dead load: Box girder bridge dead weight.
(2) Phase II dead load: Bridge deck pavement layer, anti-collision guardrail and other facilities, a total of 129.5 kN/m.
(3) Vehicle load: The bridge road grade is a double-width eight-lane first-level highway. According to the regulations, the lateral reduction coefficient should be considered when the multi-lane loading is carried out in the overall analysis and calculation of the bridge. The model in this paper is a single four-lane road with a reduction coefficient of 0.67. qk = 10.5 kN/m, Pk = 360 kN. The loading schematic for single-lane loading is shown in Figure 9.
(4) Temperature gradient: According to the provisions of the general specification for highway bridge design, the vertical temperature gradient is selected for loading, as shown in Figure 10. The specifications specify T1 and T2 of different paving types. The paving layer thickness of the box girder bridge in this paper is 10 cm asphalt concrete, so the vertical positive temperature gradient T1 = 14 °C and T2 = 5.5 °C. The vertical negative temperature gradient of the concrete superstructure is a positive temperature gradient multiplied by −0.5, i.e., negative temperature gradient T1 = −7 °C, T2 = −2.75 °C.

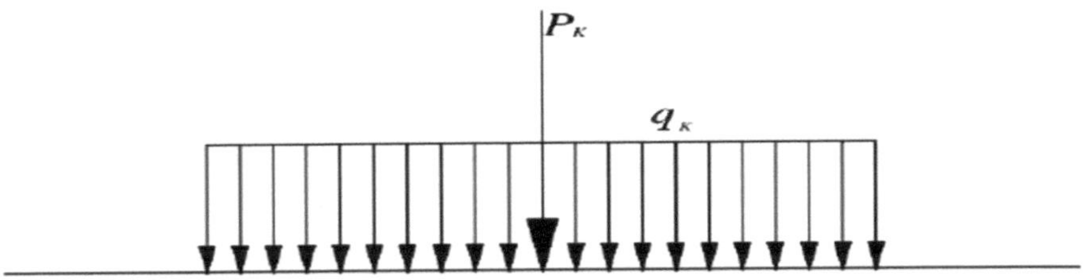

Figure 9. Loading diagram of lane load.

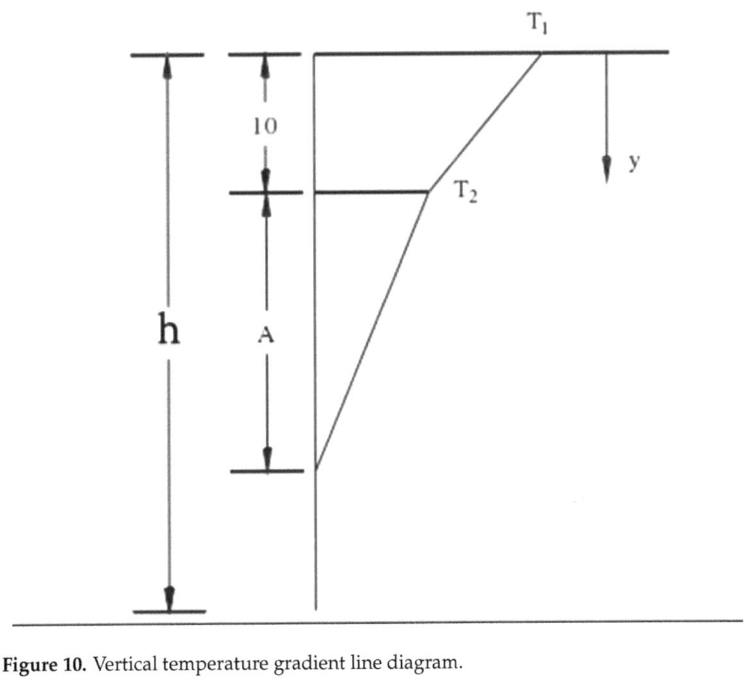

Figure 10. Vertical temperature gradient line diagram.

4.1.1. Effect of Fracture Length on Fracture Propagation

(1) Longitudinal crack of floor

Taking the mid-span longitudinal crack of the bottom plate with a crack depth of 30 mm as the reference, five working conditions with crack lengths of 800 mm, 1000 mm, 1200 mm, 1400 mm, and 1600 mm were considered, respectively, to extract the maximum stress intensity factor of the crack tip and draw the integrated path value into a graph, as shown in Table 4 and Figure 11.

Table 4. Maximum value of stress intensity factor in longitudinal crack length of different floor.

Longitudinal Crack Length (mm)	800	1000	1200	1400	1600
K_I (MPa·mm$^{1/2}$)	−0.919	−0.235	−0.889	0.209	1.227
K_{II} (MPa·mm$^{1/2}$)	0.508	3.01	2.21	3.22	6.48
K_{III} (MPa·mm$^{1/2}$)	0.377	1.03	−1.38	−1.73	−3.52

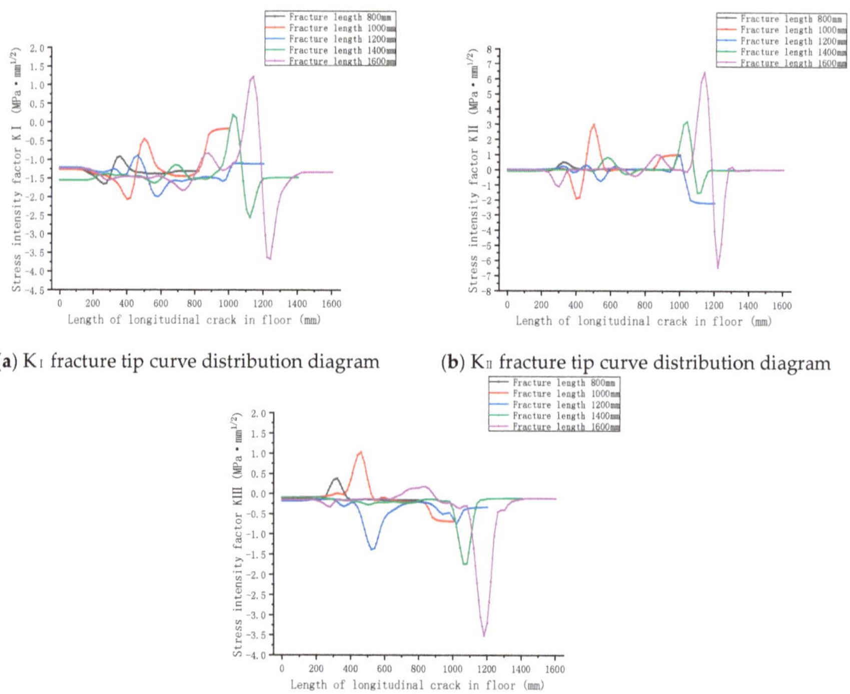

(**a**) K_I fracture tip curve distribution diagram

(**b**) K_{II} fracture tip curve distribution diagram

(**c**) K_{III} fracture tip curve distribution diagram

Figure 11. Subdistribution of stress intensity at longitudinal crack tip of floor.

It can be seen from Table 4 and Figure 11 that, under the basic working conditions, the stress intensity factor changes in the longitudinal crack tip of the bottom plate in a similar way, increasing first and then decreasing from the two ends of the crack to the middle. This trend becomes more obvious with the increase in fracture length, and the peak value tends to appear in the middle of the fracture, while the value at both ends of the fracture is relatively stable and does not change significantly with the increase in fracture length. The maximum value of the stress intensity factor increases with the increase in crack length, and the maximum value is 6.48 MPa·mm$^{1/2}$. K_{II} is larger than K_I and K_{III}, and, with the increase in fracture length, the mixture of the types of fractures, mainly type II, is the main factor affecting the longitudinal fracture propagation of the floor.

(2) Vertical cracks in web

Taking the vertical crack of the side span web with a crack depth of 30 mm as the basis, five working conditions of crack lengths of 300 mm, 350 mm, 400 mm, 450 mm, and 500 mm were considered, respectively, the extreme value of the stress intensity factor of the crack tip under each working condition was extracted, and the integral path value was plotted as a graph, as shown in Table 5 and Figure 12.

Table 5. The maximum value of stress intensity factor for different vertical crack lengths of web.

Vertical Crack Length of Web (mm)	300	350	400	450	500
K_I (MPa·mm$^{1/2}$)	−79.157	−80.282	−81.152	−81.832	−82.659
K_{II} (MPa·mm$^{1/2}$)	−0.532	0.513	0.566	0.613	0.727
K_{III} (MPa·mm$^{1/2}$)	−1.67	−1.63	−1.654	−1.824	−1.803

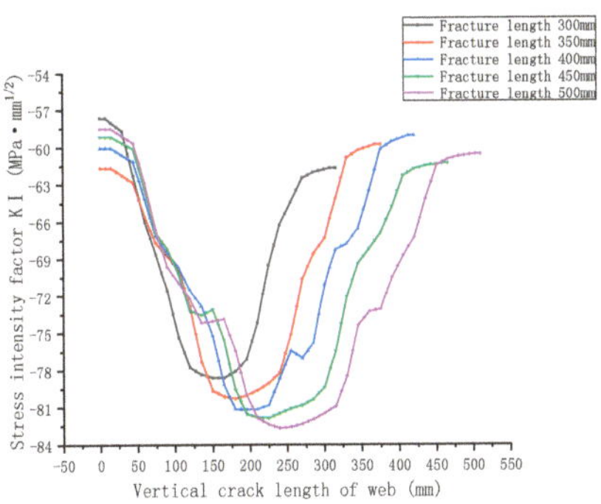

(**a**) K_I fracture tip curve distribution diagram

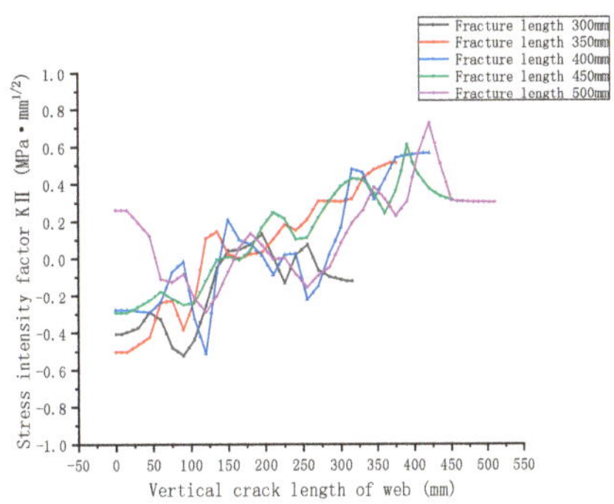

(**b**) K_{II} fracture tip curve distribution diagram

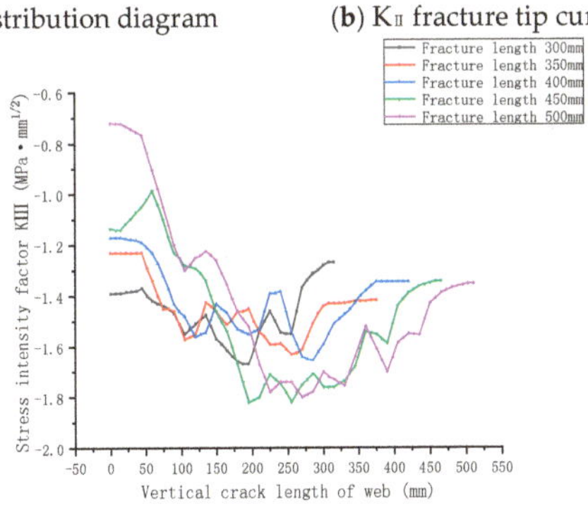

(**c**) K_{III} fracture tip curve distribution diagram

Figure 12. Subdistribution of stress intensity at vertical crack tip of web.

As can be seen from Table 5 and Figure 12, with the increase in the vertical crack length of the web, the K_I value increases somewhat at both ends of the crack, but the increase is small; meanwhile, the extreme value in the middle of the crack decreases continuously, showing a large negative value as a whole, with the maximum negative value being −82.659 MPa·mm$^{1/2}$. It is shown that the presence of large prestress in the longitudinal direction of the web has a significant inhibitory effect on the expansion of type I cracks. For K_{II} and K_{III}, the maximum value increases with the increase in crack length, and the fracture end away from the floor side shows a higher value. The results show that the propagation of type I cracks is inhibited under the basic working conditions, and type II and type III cracks are the main factors affecting the vertical crack propagation of the web.

4.1.2. Effect of Fracture Depth on Fracture Propagation

(1) Longitudinal crack of floor

Taking the mid-span longitudinal crack of the bottom plate with the crack length of 1200 mm as the basis, five working conditions with crack depths of 10 mm, 20 mm, 30 mm, 40 mm, and 50 mm were considered, respectively. The maximum stress intensity factor of the crack tip under each working condition was extracted, and the integrated path value was plotted as a graph, as shown in Table 6 and Figure 13.

Table 6. The maximum value of stress intensity factor at different longitudinal crack depth of floor.

Longitudinal Crack Depth (mm)	10	20	30	40	50
K_I (MPa·mm$^{1/2}$)	−0.984	−1.744	−1.995	−2.438	−2.612
K_{II} (MPa·mm$^{1/2}$)	−0.676	−0.704	−2.217	−1.392	−1.436
K_{III} (MPa·mm$^{1/2}$)	−0.607	−0.993	−1.381	−1.573	−1.992

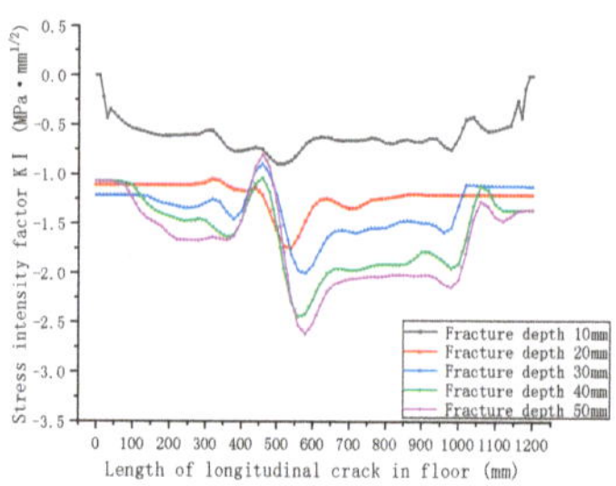

(**a**) K_I fracture tip curve distribution diagram

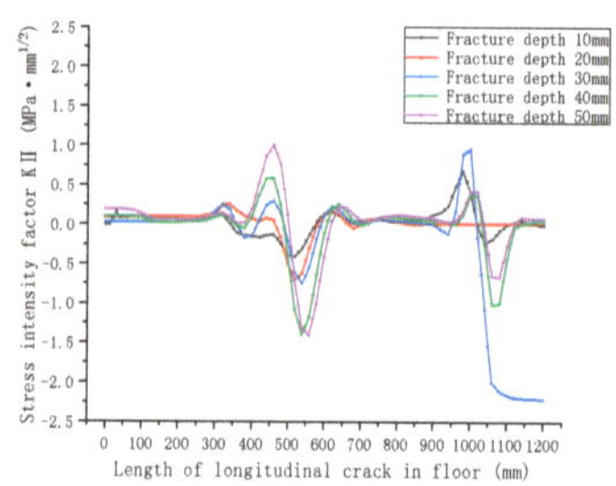

(**b**) K_{II} fracture tip curve distribution diagram

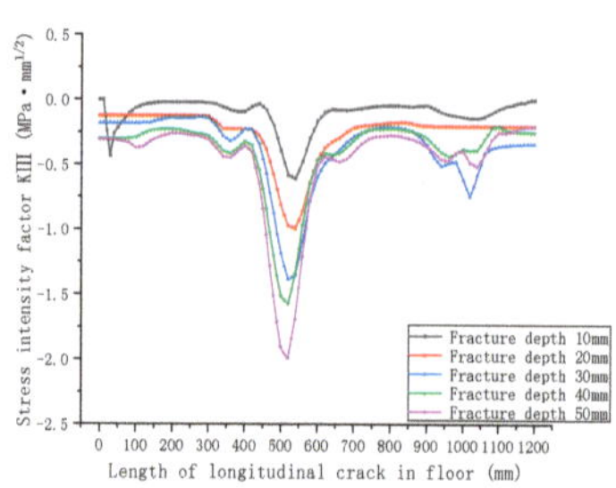

(**c**) K_{III} fracture tip curve distribution diagram

Figure 13. Subdistribution of stress intensity at longitudinal crack tip of floor.

As can be seen from Table 6 and Figure 13, with the increase in crack depth, the change in stress intensity factor presents a trend in first increasing, then decreasing, and then increasing. A peak value of −2.217 MPa·mm$^{1/2}$ occurs at the crack tip when the depth is 30 mm and stabilizes with the increasing crack depth. The distribution of K_I and K_{III} at the fracture tip is stable, and their maximum values appear in the middle of the fracture and increase with the increase in fracture depth. All K_I values are negative, which indicates that the influence of typeIcrack propagation decreases with the increase in crack depth, and type II and III cracks are the main factors affecting the longitudinal crack propagation of the floor.

(2) Vertical cracks in web

Taking the vertical crack of the side span web with a crack length of 400 mm as the basis, five working conditions with crack depths of 10 mm, 20 mm, 30 mm, 40 mm, and 50 mm were considered, respectively, and the maximum stress intensity factor of the crack tip under each working condition was extracted, while the integral path value was plotted as a graph, as shown in Table 7 and Figure 14.

Table 7. The maximum value of stress intensity factor at different vertical crack depths of web.

Vertical Crack Depth of Web (mm)	10	20	30	40	50
K_I (MPa·mm$^{1/2}$)	−21.556	−65.692	−81.152	−91.443	−98.956
K_{II} (MPa·mm$^{1/2}$)	−0.816	−0.388	0.541	0.503	1.187
K_{III} (MPa·mm$^{1/2}$)	−0.324	−1.552	−1.568	−1.891	−3.169

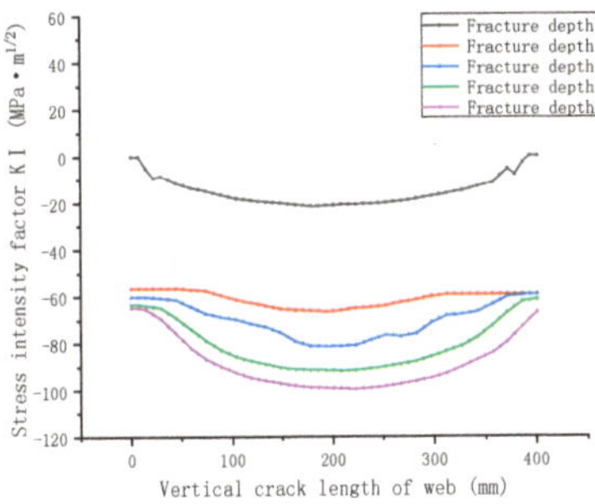

(**a**) K_I fracture tip curve distribution diagram

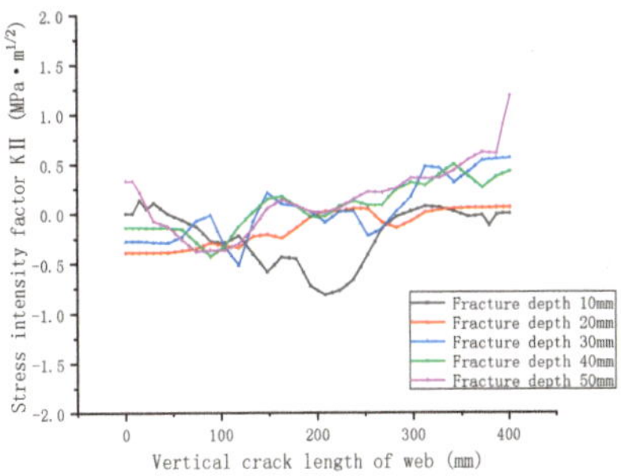

(**b**) K_{II} fracture tip curve distribution diagram

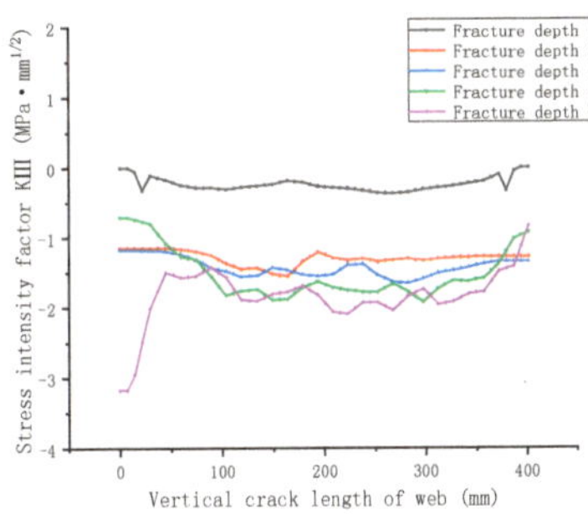

(**c**) K_{III} fracture tip curve distribution diagram

Figure 14. Subdistribution of stress intensity at vertical crack tip of web.

As can be seen from Table 7 and Figure 14, with the increase in vertical crack depth, the maximum value of K_I continues to increase with a large range of changes, all of which are negative. When the depth reaches 30 mm, the change tends to be stable. K_{II} showed a trend in first decreasing and then increasing, and its overall distribution had no obvious rule, while the maximum value mainly appeared at the crack tip far away from the bottom plate. When the crack depth is 50 mm, the stress intensity factor can reach 1.187 MPa·mm$^{1/2}$. K_{III} increases with depth, but the overall change is relatively gentle. The value of K_I is much higher than that of K_{II} and K_{III} and is sensitive to the change in fracture depth. The results show that, with the increase in fracture depth, the propagation of type I fractures is inhibited to a large extent, while the effect on the propagation of type II and type III fractures is small.

4.2. *Effect of Prestress Loss on Crack Propagation*

The loss of prestress is unavoidable during bridge operation, and, with the loss of prestress, the crack resistance of the bridge decreases. In order to investigate the effect of prestress loss on the crack extension performance of box girders, the effects of longitudinal cracks in the bottom plate and vertical crack extension in the web plate were analyzed under five working conditions, with no prestress loss, 10%, 20%, 30%, and 40% prestress loss being considered, respectively.

(1) Longitudinal crack of floor

Taking the longitudinal crack length of 1200 mm and depth of 30 mm in the base plate as the reference, the maximum value of the stress intensity factor at the crack tip under different degrees of prestress loss is extracted and the integral path value is plotted as a graph; the results are shown in Table 8 and Figure 15.

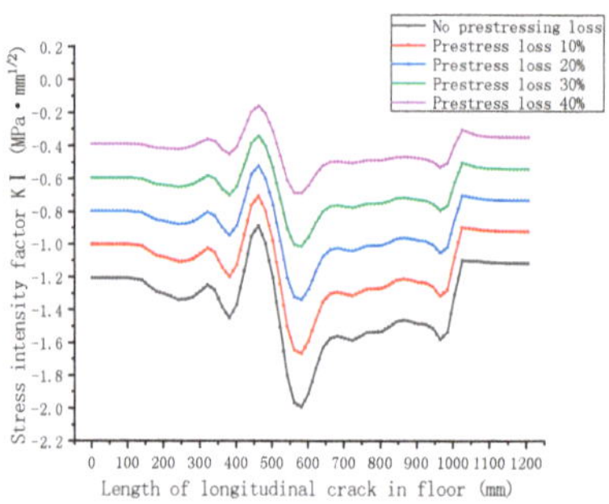

(**a**) K_I fracture tip curve distribution diagram

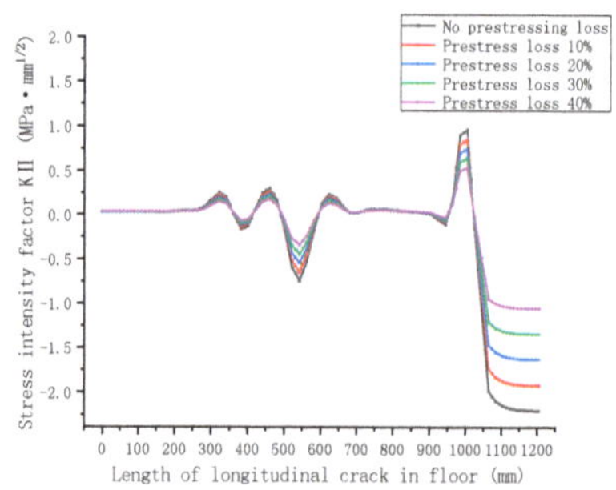

(**b**) K_{II} fracture tip curve distribution diagram

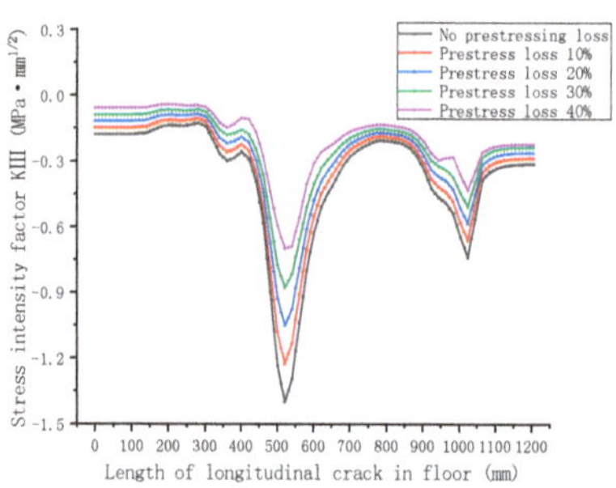

(**c**) K_{III} fracture tip curve distribution diagram

Figure 15. Subdistribution of stress intensity at longitudinal crack tip of bottom plate.

Table 8. Maximum value of stress intensity factor of longitudinal crack in bottom plate under different prestress losses.

Loss of Prestress	0	10%	20%	30%	40%
K_I (MPa·mm$^{1/2}$)	−0.889	−0.703	−0.520	−0.338	−0.156
K_{II} (MPa·mm$^{1/2}$)	−2.205	−1.917	−1.629	−1.340	−1.052
K_{III} (MPa·mm$^{1/2}$)	−1.401	−1.225	−1.052	−0.879	−0.699

From Table 8 and Figure 15, it can be seen that the maximum values of the stress intensity factor K_I for longitudinal cracks in the base plate all tend to increase with the increase in prestress loss, while the maximum values of K_{II} and K_{III} show a decreasing trend. When the prestress loss reaches 40%, the increase in K_I maxima is as high as 82.45%.

The K_{II} maxima appear at the crack ends and the K_{III} maxima appear in the middle of the cracks, which show overall negative values. In the case of larger prestress loss, it promotes the expansion of type I cracks and inhibits the expansion of type II and III cracks to some extent.

(2) Vertical cracks in web

Taking a web vertical crack length of 400 mm and a depth of 30 mm as the reference, the maximum value of the stress intensity factor at the crack tip under different degrees of prestress loss is extracted and the integral path value is plotted as a graph; the results are shown in Table 9 and Figure 16.

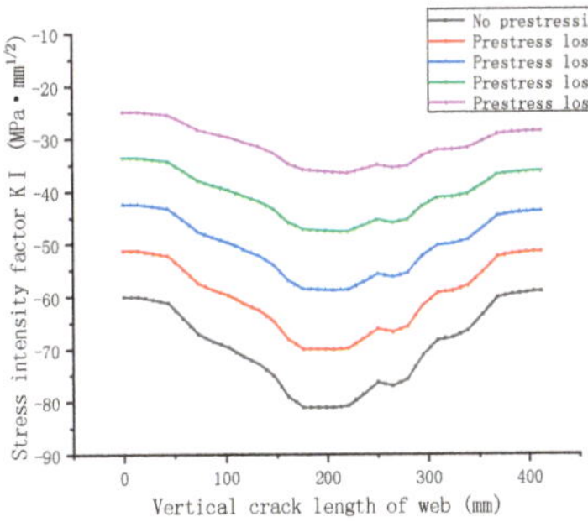

(a) K_I fracture tip curve distribution diagram

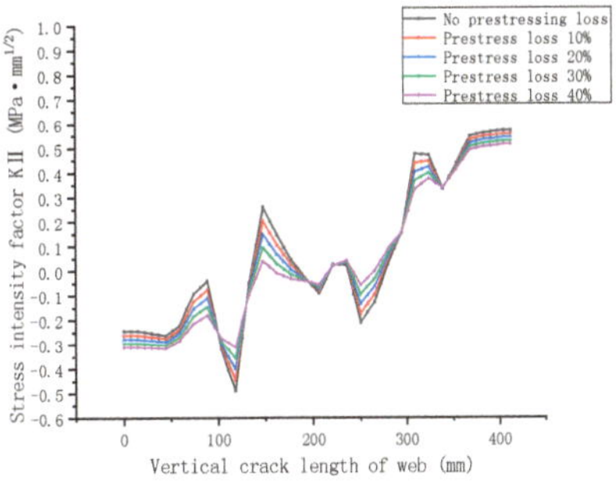

(b) K_{II} fracture tip curve distribution diagram

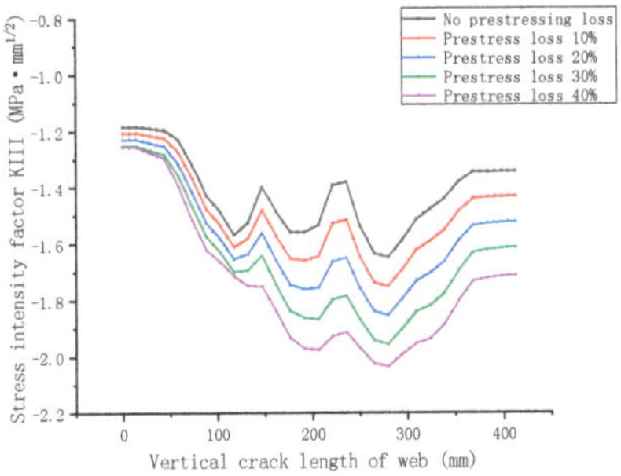

(c) K_{III} fracture tip curve distribution diagram

Figure 16. Subdistribution of stress intensity at vertical crack tip of web.

Table 9. Maximum stress intensity factors of vertical cracks of web under different prestress losses.

Loss of Prestress	0	10%	20%	30%	40%
$K_I(MPa \cdot mm^{1/2})$	−60.049	−51.246	−42.443	−33.641	−24.838
$K_{II}(MPa \cdot mm^{1/2})$	0.574	0.559	0.544	0.529	0.515
$K_{III}(MPa \cdot mm^{1/2})$	−1.649	−1.751	−1.853	−1.956	−2.035

As can be seen from Table 9 and Figure 16, the expansion of type I cracks was suppressed due to the presence of prestressing, which resulted in a large negative value of K_I as a whole. However, with the increase in prestress loss, the maximum value of the stress intensity factor K_I of the vertical cracks in the web increases more significantly, and it increases by 35.21 when the prestress loss reaches 40%. In contrast, the overall values of K_{II} and K_{III} do not change much, but the trend in change is the opposite. Under the influence of prestress loss, it mainly promotes the expansion of type I and III cracks and has less influence on the expansion of type II cracks.

4.3. *Stability Analysis of Longitudinal Cracks in Existing Floor*

In order to investigate the influence of multi-factors on the expansion performance and force state of existing longitudinal cracks in the base plate under the operation stage, combined with the results of the analysis of the factors affecting the expansion of the cracks in the box girder, the longitudinal cracks in the base plate with a length of 1200 mm and a depth of 30 mm are selected as the cracking conditions, the most unfavorable bias loading arrangement of the spanning span in the automobile load is taken into account, and the combined effect of the factors is considered comprehensively, while the specific conditions are set up as follows (Table 10):

Table 10. Working condition setting.

	Combination of Operating Conditions
A	Dead weight of box girder + second stage dead load + no prestressed loss + automobile load + temperature gradient
B	Dead weight of box girder + second phase dead load + no prestress loss + 1.5 times automobile load + temperature gradient
C	Dead weight of box girder + second stage dead load + prestress loss 20% + car load + temperature gradient
D	Dead weight of box girder + second stage dead load + prestress loss 20% + 1.5 times automobile load + temperature gradient

In order to study the stability of longitudinal cracks in the floor during operation, the combined loads that may produce the most adverse effects are first analyzed to determine the distribution trend and maximum value of the stress intensity factors so as to judge the type of crack expansion. At the same time, based on the empirical fracture criterion of Equation (2), whether the existing fracture is in the expanding state is judged.

The maximum value of the stress intensity factor of the crack tip under various working conditions and the integral path value are drawn into a curve, as shown in Figure 17.

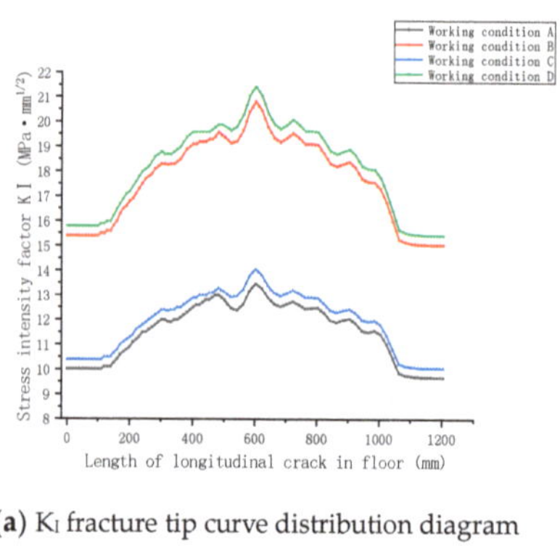

(**a**) K_I fracture tip curve distribution diagram

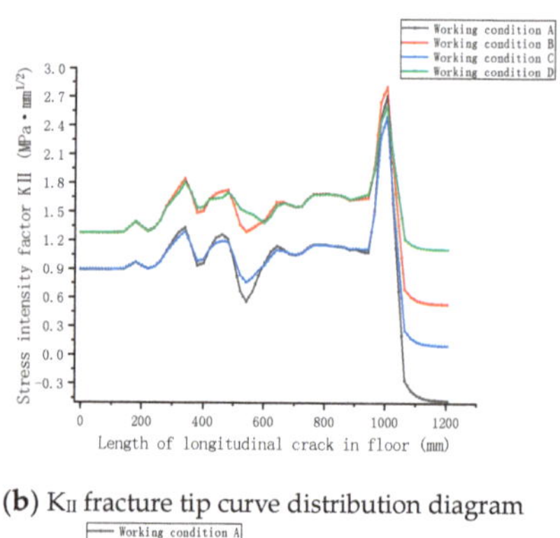

(**b**) K_{II} fracture tip curve distribution diagram

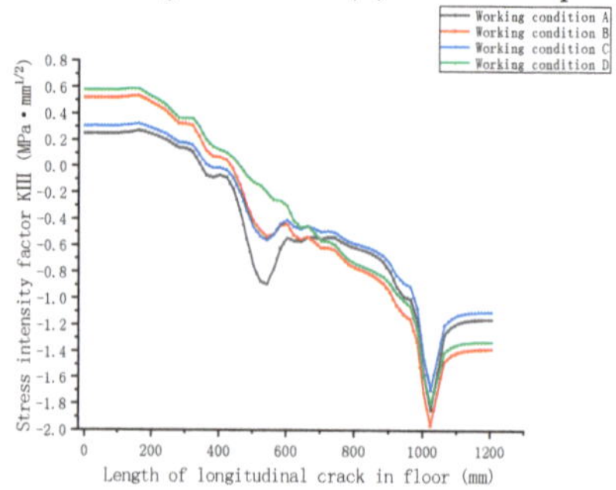

(**c**) K_{III} fracture tip curve distribution diagram

Figure 17. Subdistribution of stress intensity at longitudinal crack tip of floor.

As can be seen from Figure 13, under the action of automobile load, the maximum value of the stress intensity factor at the crack tip is positive and K_I is much larger than K_{II} and K_{III}. According to the distribution of the curve diagram, the distribution trend in K values for different types of cracks is basically the same, with the larger value of K_I mainly appearing in the middle of the crack, and the value of the two ends of the crack is relatively small. Comparing the working conditions, it can be seen that the effect of automobile load is obviously greater than the loss of prestressing force, and the influence on K_I is the largest, with an overall increase of more than 50% in the case of overloading, while the change in value is more significant. In contrast, although the overall value of K_{II} shows an upward trend, the increase is only about 30% and the overall value of K_{III} does not change much. The above analysis shows that the automobile load is the main factor affecting the stability of longitudinal crack expansion in the base plate during the operation stage, which mainly shows the expansion of mixed type I–II cracks, with type I cracks as the main type, and it tends to expand to the middle part of the cracks instead of both ends of the cracks.

In order to further explore the stability of longitudinal crack growth in the floor, the maximum critical parameter of crack growth under various working conditions is extracted and the curve of critical parameter of crack growth is drawn according to the empirical fracture criteria of composite cracks, as shown in Table 11 and Figure 18:

Table 11. The maximum value of critical parameter φ of crack propagation under different working conditions.

Working Condition	A	B	C	D
φ	0.240	0.572	0.469	0.607

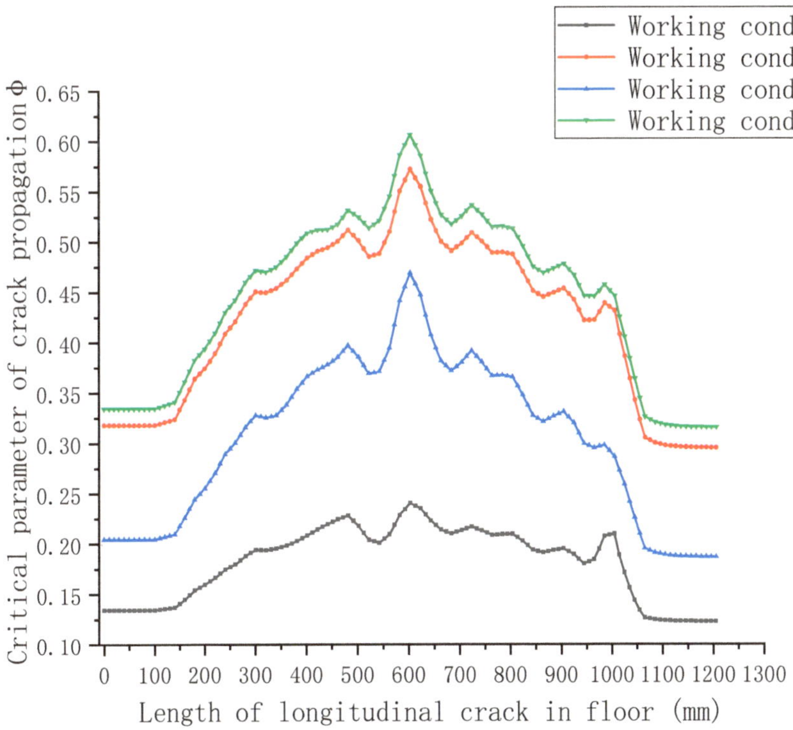

Figure 18. Distribution of critical parameters of longitudinal crack propagation in floor under different working conditions.

It can be seen from Table 11 and Figure 18 that prestress loss and overload are the main factors affecting the increase in critical parameter φ for crack expansion. When prestress loss reaches 20%, the maximum value of the corresponding parameter increases nearly twice, while, under overload, the maximum value increases more than twice. Under different working conditions, the maximum value of critical parameter φ of crack propagation is less than 1, indicating that the longitudinal crack of the floor is in a stable state and has not entered the expansion state.

4.4. Stability Analysis of Vertical Cracks of Existing Web

As shown in Section 4.3, the maximum stress intensity factor of the crack tip under working conditions is plotted as a curve, as shown in Figure 19.

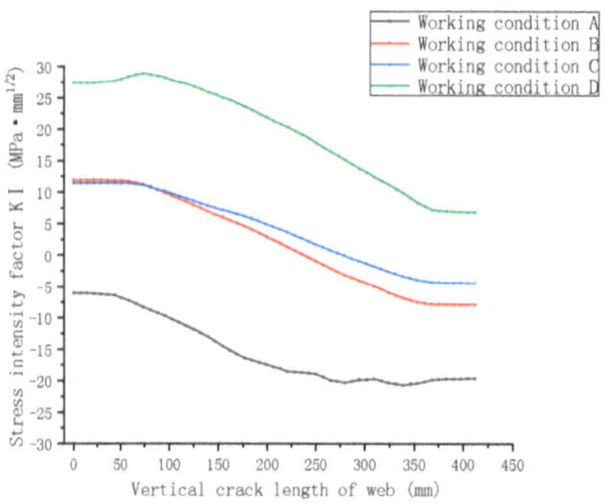

(**a**) K_I fracture tip curve distribution diagram

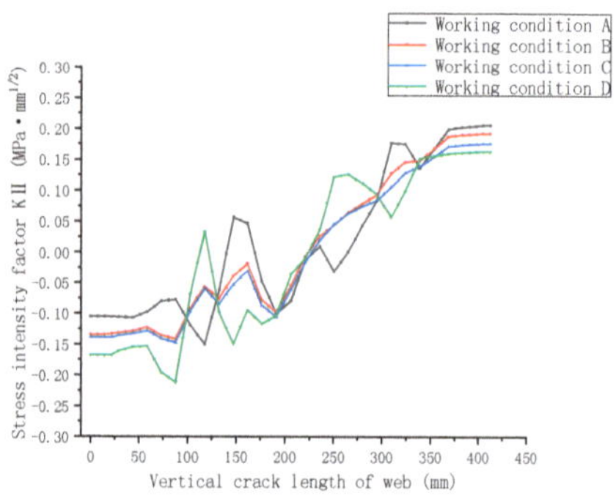

(**b**) K_{II} fracture tip curve distribution diagram

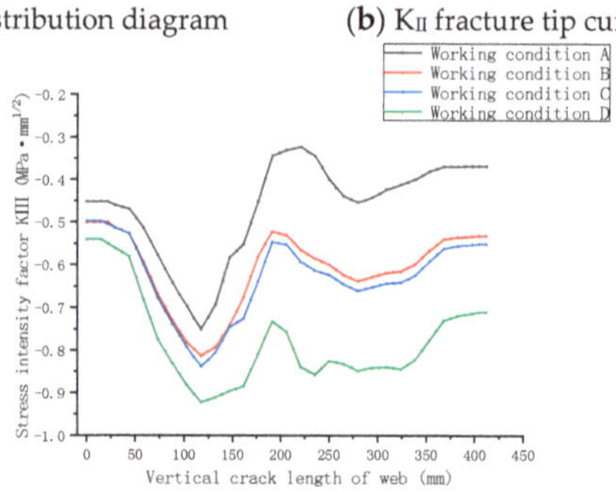

(**c**) K_{III} fracture tip curve distribution diagram

Figure 19. Subdistribution of stress intensity at vertical crack tip of web.

It can be seen from Figure 19 that the maximum value of K_I changes most significantly in various working conditions, especially when prestressing loss and overload are considered. When only automobile load is considered, the value of K_I is negative, which indicates that the tensile stress at the crack tip is small and the crack propagation is inhibited. However, under the influence of prestress loss and overload, the increase of K_I is large and positive, and large values appear at both ends of the crack. The maximum values of K_{II} and K_{III} do not change much overall, and K_I is much larger than K_{II} and K_{III}. The above analysis shows that prestress loss and overload have a great influence on the expansion of type I cracks in the operation stage, the influence of the two is not a simple linear superposition, and their combined effect will be far more than the sum of the single influencing factors. During operation, the vertical crack propagation of the web may show a trend in regard to compound cracks, mainly type I crack propagation along both ends of the crack.

In order to further explore the stability of the vertical crack propagation of the web, the maximum critical parameter of crack propagation under various working conditions is extracted and the curve φ of critical parameter of crack propagation is drawn, as shown in Section 4.3. In this case, the stress intensity factor at the crack tip is basically negative under working condition A, the crack propagation is inhibited, and the φ value is 0, as shown in Table 12 and Figure 20.

Table 12. The maximum value of critical parameters of crack propagation under different working conditions.

Working Condition	A	B	C	D
φ	0	0.186	0.171	1.087

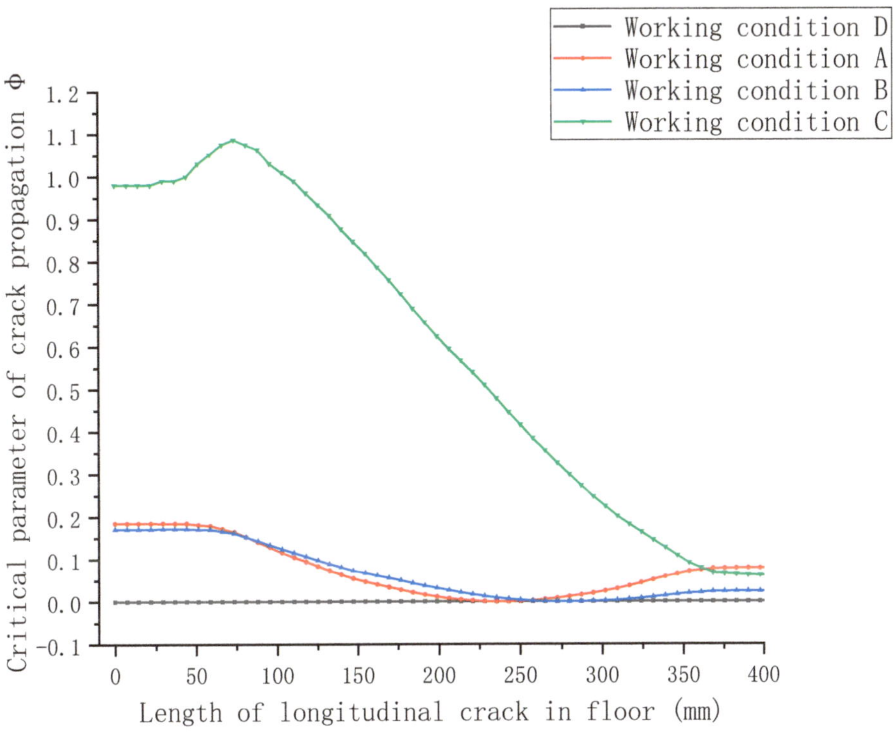

Figure 20. Distribution of critical parameters of vertical crack propagation of web under different working conditions.

It can be seen from Table 12 and Figure 20 that the critical parameter φ of crack propagation is less than 1 only when the vehicle load and prestress loss act separately, which indicates that the vertical cracks of the web are in a stable state and have not yet entered the expanding state. However, under the combined influence of prestress loss and overload, the critical parameter φ of crack propagation appears to be greater than 1 on the side of the crack tip near the box girder bottom plate, which means that the vertical cracks of the web plate may be in an unstable state and have a tendency to expand towards the box girder bottom plate.

5. Conclusions

In this paper, based on the crack distribution characteristics of actual prestressed concrete continuous box girder bridges, the longitudinal cracks in the bottom slab of the mid-span and the vertical cracks in the webs of the side spans are selected as the research objects and the effects of different crack lengths, depths, temperature gradients, prestress losses, and vehicle loads on the crack extension of box girders are investigated. The crack extension critical parameters are introduced to assess the crack extension stability, the crack extension performance is quantitatively analyzed, and the following conclusions are obtained:

1. Under the action of basic working conditions, the results of the analysis of the stress intensity factor at the tips of cracks of different lengths and depths show that the longitudinal cracks of the bottom plate mainly show the expansion of composite cracks dominated by type II cracks as the lengths of the cracks increase. Meanwhile, for the increase in the vertical crack length of the web plate, it mainly shows a type II and III crack extension dominated by the type II and III cracks. When the crack depth increases, the common characteristic of both is to promote the expansion of II and III type cracks while restricting the development of I type cracks.
2. Under the effect of prestress, the expansion of longitudinal cracks in the bottom plate and vertical cracks in the web plate showed obvious differences. The reduction in prestress promoted the expansion of longitudinal cracks of type I in the bottom plate and inhibited the expansion of type II and III cracks to some extent. In contrast, the expansion of vertical cracks in the web is more sensitive to the prestress reduction, which promotes the expansion of type I, II, and III cracks but mainly affects the expansion of type I cracks.
3. The analysis results of the extension stability of the existing longitudinal cracks on the base plate show that the automobile load is the main factor affecting the extension stability of the longitudinal cracks on the base plate, which mainly shows the composite crack extension dominated by the type I cracks.
4. The results of the extended stability analysis of the existing longitudinal cracks in the web plate show that the loss of prestress and overloading significantly affect the expansion of the type I cracks. In the case of no overloading, the vertical cracks of the web plate will not be extended under the action of prestressing, while, under the joint influence of the loss of prestressing and overloading, the critical parameter φ of the crack extension locally appear to be greater than 1, which indicates that the vertical cracks of the web plate are in an unstable state and that the possibility of destabilization and expansion is present.

The causes of cracks in continuous prestressed concrete box girder bridges, as well as the crack patterns themselves, are complex and diverse, and there are still many deficiencies in this paper that deserve further study.

1. The analysis of existing crack extensions in box girders in this paper only considers the longitudinal cracks of the bottom plate and the vertical cracks of the web plate, and it is a single crack, while the actual box girder needs to be further analyzed for the combination of different crack patterns and multiple cracks.
2. This paper is limited to only analysing crack extensions under static loads [30,31], and stability analyses of existing crack extensions under dynamic loads should be considered in later studies.

Author Contributions: Writing—original draft, Y.-J.C. and X.-C.H.; Writing—review & editing, Y.-J.C., S.-Z.C., L.-Y.W., J.-W.Z. and H.C. All authors have read and agreed to the published version of the manuscript.

Funding: Analysis of causes and treatment measures for cracks in prestressed concrete continuous box-girder bridges, Grant Number: 2022032903; research on real-time monitoring system of important bridges, Grant Number: 0050229916.

Data Availability Statement: The datasets presented in this article are not readily available due to technical and time limitations. Requests to access the datasets should be directed to the corresponding author.

Conflicts of Interest: Authors Jiwei Zhong and Hui Cheng were employed by the company Railway Bridge Science Research Institute, Ltd. The remaining authors declare that the research was conducted in the absence of any commercial or financial relationships that could be construed as a potential conflict of interest.

References

1. Chung, W.S. *A Cracked Concrete Material Model for the Nonlinear Finite Element Analysis of Slab-on-Girder Bridges*; Purdue University: West Lafayette, IN, USA, 2003.
2. Wang, G.; Xie, J.; Fu, Y. Investigation on Crack of Long-span Prestressed Concrete Box Girder Bridges in Service. *J. Highw. Transp. Res. Dev.* **2009**, *4*, 71–76. [CrossRef]
3. Täljsten, B. CFRP Strengthening and Monitoring of a Box Girder Bridge. In *Sensing Issues in Civil Structural Health Monitoring*; Springer: Dordrecht, The Netherlands, 2005; pp. 373–382.
4. Walsh, P.F. Crack initiation in plain concrete. *Mag. Concr. Res.* **1976**, *28*, 37–41. [CrossRef]
5. Li, X.D.; Dong, W.; Wu, Z.M.; Chang, J.R. Experimental study on double-K fracture parameters of small size concrete specimens. *Eng. Mech.* **2010**, *2*.
6. Shi-lang, X.; Guo-Fan, Z. Study on probability model of fracture toughness of concrete. *J. Civ. Eng.* **1988**, *4*, 9–23.
7. Feng, B.L.; Xu, D.Y.; Zhou, X.G.; Liu, Y.P. Fracture mechanics analysis of arch dam cracks under temperature load. *J. Hefei Univ. Technol.* **2000**, *6*, 1003–1008.
8. Zhao, Y.H.; Xu, S.H.; Wu, Z.M. Double G criterion for crack propagation of concrete structures. *J. Civ. Eng.* **2004**, *10*.
9. Wu, Z.M.; Dong, W.; Xu, Q. Criteria for the growth of I–II composite cracks in concrete and numerical simulation of the whole process. *J. Hydraul. Eng.* **2009**, *2*, 180–187.
10. Fan, G.G. Research on Numerical Calculation Method of Failure Process of Concrete Dam Based on Fracture Mechanics. Master's Thesis, Dalian University of Technology, Dalian, China, 2013.
11. Xing-Da, C. Research on Dynamic Characteristics of Cracked Reinforced Concrete Beams under Vehicle Load. Master's Thesis, Tianjin University, Tianjin, China, 2017.
12. Mi, Z.X.; Hu, Y.; Li, Q.B. Fracture strength of dam concrete and its determination method. *J. Hydraul. Power Gener.* **2019**, *6*, 19–28.
13. Liqiong, R. Application of J-Integral to Simulation of Concrete Subcritical Crack growth. Master's Thesis, Lanzhou University, Lanzhou, China, 2023.
14. Xu, S.; Zhao, Y.H. Fracture process criteria and analysis of concrete crack propagation. *Eng. Mech.* **2008**, *S2*, 20–33.
15. Kaplan, M.F. Crack propagation and the fracture of concrete. *J. Proc.* **1961**, *58*, 591–610.
16. Yu, X.Z. Lecture on Concrete fracture Mechanics (3)—Concrete fracture toughness K_(IC) and fracture criterion of composite cracks. *Hydroelectricity* **1988**, *3*, 49–53.
17. Zhi-Min, W.; Guo-fan, Z. Fracture toughness of large-size concrete specimens. *J. Hydraul. Eng.* **1997**, *6*, 67–70.
18. Jiaxuan, M.; Hui-Juan, L.; Wen-Lin, P. Research on temperature cracks of concrete surface using fracture mechanics method. *J. Hydraul. Power Gener.* **2002**, *2*, 31–36.
19. Gao, H.B. Research on the Determination of Type I and Type II Fracture Parameters of Concrete. Ph.D. Thesis, Dalian University of Technology, Dalian, China, 2009.
20. Xu, S.-L. Calculation theory and standardized test method of double K fracture parameters of concrete. *J. China Three Gorges Univ.* **2002**, *1*, 1–8.
21. Liu, C.; Xu, D. Influence of cracking on deflections of concrete box girder bridges. *Balt. J. Road Bridge Eng.* **2012**, *7*, 104–111. [CrossRef]
22. Peiheng, L. Numerical Analysis Method for Cracking of prestressed concrete box girder Bridges. Ph.D. Thesis, Tongji University, Shanghai, China, 2005.
23. Jinsong, Z.; Yifeng, Z.; Xing-Da, C. Crack propagation of beam body under moving vehicle load. *J. Southeast Univ.* **2018**, *4*, 678–686.
24. Wells, G.N.; Sluys, L.J. A new method for modelling cohesive cracks using finite elements. *Int. J. Numer. Methods Eng.* **2001**, *50*, 2667–2682. [CrossRef]
25. Dwairi, H.M.; Wagner, M.C.; Kowalsky, M.J.; Zia, P. Behavior of instrumented prestressed high performance concrete bridge girders. *Constr. Build. Mater.* **2010**, *24*, 2294–2311. [CrossRef]
26. Chen, Q.; Andrawes, B. Plasticity modeling of concrete confined with NiTiNb shape memory alloy spirals. *Structures* **2017**, *11*, 1–10. [CrossRef]
27. Azadeh, P.; Janet, R. Internal and external reinforcement of concrete members by use of shape memory alloy and fiber reinforced polymers under cyclic loading—A Review. *Polymers* **2018**, *10*, 376. [CrossRef]
28. Kinam, H.; Sugyu, L.; Yeongmo, Y.; Jung, K. Flexural response of reinforced concrete beams Strengthened with Near-Surface-Mounted Fe-Based Shape-Memory Alloy Strips. *Int. J. Concr. Struct. Mater.* **2018**, *12*, 45.
29. *JTG D60-2015*; General Specification for Design of Highway Bridges and Culverts. China Communication Press: Beijing, China, 2015.

30. Aberson, J.A.; Anderson, J.M.; King, W.W. Mechanics of Fracture 4. In *Elastodynamic Crack Problems*; Sih, G.C., Ed.; Springer: Berlin/Heidelberg, Germany, 1977.
31. Barre, F.; Bisch, P.; Danièle, C.; Cortade, J.; Coste, J.F.; Dubois, J.P.; Erlicher, S.; Gallitre, E.; Labbé, P.; Mazars, J.; et al. *Control of Cracking in Reinforced Concrete Structures (Research Project CEOS.fr)*; Cracking in Walls; John Wiley & Sons: Hoboken, NJ, USA, 2016; pp. 75–104.

sensors

MDPI

Article

A Lightweight and Efficient Method of Structural Damage Detection Using Stochastic Configuration Network

Yuanming Lu [1], Di Wang [1,*], Die Liu [2] and Xianyi Yang [2]

[1] School of Information Science and Engineering, Chongqing Jiaotong University, Chongqing 400074, China; diaodiaolu@gmail.com

[2] School of Civil Engineering, Chongqing Jiaotong University, Chongqing 400074, China; liudie111@163.com (D.L.); xianyi@yahoo.com (X.Y.)

* Correspondence: diwang@cqjtu.edu.cn

Abstract: With the advancement of neural networks, more and more neural networks are being applied to structural health monitoring systems (SHMSs). When an SHMS requires the integration of numerous neural networks, high-performance and low-latency networks are favored. This paper focuses on damage detection based on vibration signals. In contrast to traditional neural network approaches, this study utilizes a stochastic configuration network (SCN). An SCN is an incrementally learning network that randomly configures appropriate neurons based on data and errors. It is an emerging neural network that does not require predefined network structures and is not based on gradient descent. While SCNs dynamically define the network structure, they essentially function as fully connected neural networks that fail to capture the temporal properties of monitoring data effectively. Moreover, they suffer from inference time and computational cost issues. To enable faster and more accurate operation within the monitoring system, this paper introduces a stochastic convolutional feature extraction approach that does not rely on backpropagation. Additionally, a random node deletion algorithm is proposed to automatically prune redundant neurons in SCNs, addressing the issue of network node redundancy. Experimental results demonstrate that the feature extraction method improves accuracy by 30% compared to the original SCN, and the random node deletion algorithm removes approximately 10% of neurons.

Keywords: sensors; multi-sensors; stochastic configuration networks; structural health monitoring; convolutional neural network; neural network pruning

Citation: Lu, Y.; Wang, D.; Liu, D.; Yang, X. A Lightweight and Efficient Method of Structural Damage Detection Using Stochastic Configuration Network. *Sensors* **2023**, *23*, 9146. https://doi.org/10.3390/s23229146

Academic Editor: Luca De Marchi

Received: 11 September 2023
Revised: 9 November 2023
Accepted: 10 November 2023
Published: 13 November 2023

1. Introduction

In engineering structures, structural damage represents an intrinsic impairment, shaped by both environmental and mechanical factors, with the potential to spread internally. This type of damage accelerates the aging process and diminishes the intended design lifespan of the structure, highlighting the pivotal importance of diligent monitoring. In the field of structural health monitoring (SHM), monitoring techniques have evolved from visual inspection to more advanced methods for structural health monitoring and damage detection. Various technologies have been developed for the detection, localization, and quantification of damage in structures [1–3]. Among them, vibration-based damage detection techniques have been extensively researched and successfully applied. These techniques assess damage in structures by recording and analyzing their vibration responses, enabling informed decisions regarding structural health [4]. Over the course of several decades, significant advancements have been achieved in this field [5–7]. The principle of vibration-based damage detection is based on observing changes in the vibration response characteristics of structures when subjected to external excitation. By employing suitable sensors to record the structural vibration data and applying signal processing and pattern recognition techniques, potential damage within the structure can be detected.

Damage refers to changes in the geometric shape or material properties of a structure, which have adverse effects on its performance, safety, reliability, and service life [8]. Unlike complete failure, damage refers to the deterioration of structural functionality, resulting in a decrease in performance [9,10]. If left untreated, damage accumulates gradually and eventually leads to structural failure. The type of damage determines the mode of failure, some of which develop gradually, while others occur suddenly [11]. For instance, damage caused by corrosion or fatigue typically develops slowly, whereas destruction resulting from earthquakes or fires may lead to rapid failure [12]. This difference is attributed to variations in the mode of action and the extent of impact of different types of damage.

In order to monitor the structural health condition, it is necessary to deploy a monitoring system on the monitored object. SHM systems have been applied in various engineering fields, such as mechanical and civil engineering [13–15]. The monitoring system conducts long-term measurements on the monitored object itself and its surrounding environment [16] and analyzes the data collected by sensors to locate, identify, and quantify damage [17]. A typical damage identification system is illustrated in Figure 1. A number of sensors are installed on the monitored object, with different sensors serving different purposes. For example, some are used for monitoring the vehicle load [18], while others are used for anomaly detection [19]. The monitored object can be a bridge, tunnel, building, mechanical equipment, etc. The sensors include vibration sensors, acceleration sensors, pressure sensors, temperature sensors, and so on. These sensors collect relevant data from the object under investigation, and, after undergoing operations such as data storage, transmission, and management by the data system, the data are sent to the control center. The control center performs relevant analysis using algorithms on the processed structured data [20].

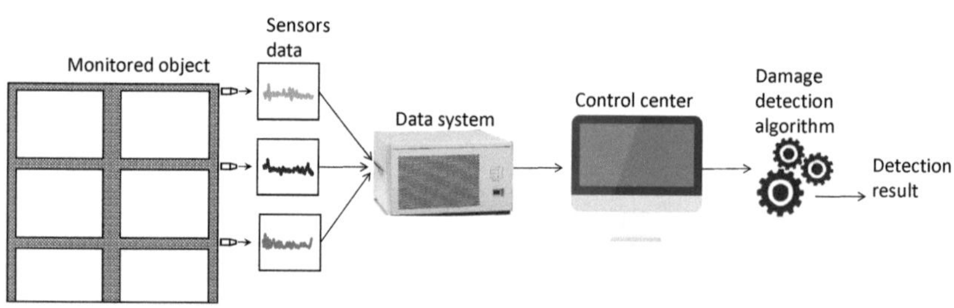

Figure 1. Structure diagram of damage detection systems.

With the advancement of sensor technology and machine learning techniques, data-driven monitoring methods have gained considerable attention. The development of sensors has enabled the collection of large amounts of real-time monitoring data, while machine learning techniques have provided various neural network algorithms. Leveraging machine learning, these monitoring methods can extract features from extensive real-time monitoring data and accomplish the immediate assessment and monitoring of engineering structure conditions. These methods not only automate the monitoring process, improving its accuracy and reliability, but also provide timely alerts and warnings, assisting engineers in taking appropriate maintenance and repair measures to ensure structural safety and reliability. A large number of traditional machine learning methods have been applied to structural health monitoring [21]. Traditional machine-learning-based identification methods mainly included support vector machines [22], neural networks [23,24], clustering [25,26], and principal component analysis [27]. However, these methods still rely on manual feature extraction, although they alleviate the issue of low monitoring efficiency.

In recent years, deep learning has rapidly developed, and various deep learning methods have been introduced into structural damage identification, alleviating the inefficiency of traditional machine learning methods in feature extraction [28]. Among these methods, convolutional neural networks (CNNs) and recurrent neural networks (RNNs) have been widely used [29–31]. Some researchers have combined CNNs and RNNs to capture the spatio-temporal features of monitoring data [32,33]. Furthermore, to enhance the attention of the model to the relationships among sensors, attention mechanisms have been introduced by some researchers [34].

Although deep-learning-based methods have improved the accuracy of damage detection, these methods overlook the operational efficiency of the entire monitoring system. In a monitoring system, numerous algorithms are integrated, and some algorithms are deployed on embedded devices with limited computational capabilities, posing significant challenges to algorithm efficiency. In addition to neglecting monitoring efficiency, the definition of neural network structures is subjective, experience-dependent, and static. Typically, a fixed network structure is defined in advance and trained on fixed data. This approach presents some issues. If the network is defined too complexly, there is a risk of overfitting and wastage of computational resources. Conversely, if the network is defined too simply, there is a risk of underfitting. For different monitoring tasks, traditional neural networks require different structure definitions based on past experience. Once a network is defined, its structure is generally not subject to alteration, and a static network configuration may not be suitable for monitoring dynamic changes in data.

Data randomization learning is a fast modeling method. Pao et al. proposed a random vector functional link network (RVFL) [35], but the RVFL does not have universal approximation capabilities in certain cases [36,37]. To address this limitation, Wang et al proposed a stochastic configuration network (SCN) [38], which is an emerging incremental learning method that dynamically increases network nodes based on previous errors and data. With the introduction of the SCN, several outstanding studies related to SCNs have gradually emerged. The strategy of "block learning" was introduced from the perspective of parallel learning by Dai et al. [39], which accelerated the training process of the SCN. Zhao et al. [40] introduced the chaotic sparrow algorithm to search for suitable hyperparameters, making the training process of SCNs more stable. Liu et al. [41] utilized statistical characteristics of vibration signals as inputs for the SCN, enabling the diagnosis of mechanical equipment faults. Li et al. [42] combined convolution with an SCN to automatically grade vegetables and fruits.

In terms of the dynamic nature of the structure, a randomly configured network can alleviate the shortcomings of previous methods in dynamic environments. However, due to its inherent nature as a fully connected neural network and the insufficient updating of network nodes compared to gradient descent, it still has its drawbacks. Firstly, in relation to the monitoring task, its input format is not user-friendly as it requires data inputs to be one-dimensional vectors. In practical scenarios, multiple sensors [18,32,43] are often deployed on the monitored object, resulting in the majority of monitoring data being two-dimensional matrices. If these matrices are directly converted into one-dimensional vectors and fed into the SCN, it can lead to loss of internal information within the data, resulting in a degraded performance of the SCN. Secondly, there are redundant nodes within the network. These extra nodes increase the number of network parameters and slow down the inference speed, which is unfavorable for monitoring systems.

To address the first issue, inspired by extreme learning machines (ELMs) [44] and random convolutional kernel transform (Rocket) [45], a feature extraction method based on random mapping is proposed. Specifically, a large number of random convolutional kernels are employed to extract features, which are independent of the backpropagation algorithm, and the parameters of the convolutional kernels are not updated. The extracted features serve as mappings in a high-dimensional space and are then fed into the SCN for recognition. To tackle the second issue, this paper presents a random node deletion algorithm that can prune redundant neurons in the SCN. Through simulation analysis, the effectiveness of the proposed algorithm is verified. Specifically, the feature extraction method proposed in this paper is capable of achieving a 30% performance enhancement for the SCN on two monitoring datasets, thereby addressing the inherent deficiency of the SCN in handling temporal data. The node pruning algorithm introduced in this paper can remove 10% of neurons within an acceptable margin of error.

2. Methods

In the context of structural health monitoring systems, a multitude of monitoring models are deployed, necessitating consideration not only of the accuracy of these models in monitoring tasks but also of their inferential speed. Some neural-network-based models, such as recurrent neural networks [46], while capable of accurately predicting time series monitoring data, exhibit a large number of parameters and require the manual predefinition of network architecture based on prior experience. Defining a conventional neural network that is too large results in a wasteful allocation of computational resources, whereas defining one that is too small fails to meet the performance requirements. In contrast, SCNs represent an incremental neural network approach wherein the network structure does not require manual definition but is generated on-the-fly during the training process.

2.1. Review of SCN

The SCN is a neural network that has emerged in recent years that differs from traditional gradient-based neural networks. It does not rely on gradient descent algorithms and instead incrementally constructs the network structure by randomly configuring suitable network nodes. The specific structure is depicted in the accompanying Figure 2. The nodes in the intermediate hidden layer are incrementally constructed, starting with the generation of the first node, followed by the second node, and so on until the N-th node is generated. The solid lines represent already configured neurons, while the dashed lines represent yet-to-be-configured and generated neurons. In the following review, the main aspects of SCNs will be revisited and summarized. First, the general approximation properties of an SCN will be introduced. Given a target function $f : \mathbb{R}^d \rightarrow \mathbb{R}^m$, suppose that a network has been configured with $L-1$ hidden nodes, denoted as $f_{L-1} = \sum_{l=1}^{L-1} \beta_l \Phi_l(w_l^T x + b_l)(L = 1, 2, \ldots; f_0 = 0)$, where $\beta_l = [\beta_{l,1}, \beta_{l,2}, \ldots, \beta_{l,m}]^T$ and $\Phi_l(w_l^T x + b_l)$ is the activation function of the l-th node. The residual error is defined as $e_{L-1}^* = f - f_{L-1} = [e_{L-1,1}^*, \ldots, e_{L-1,m}^*]$.

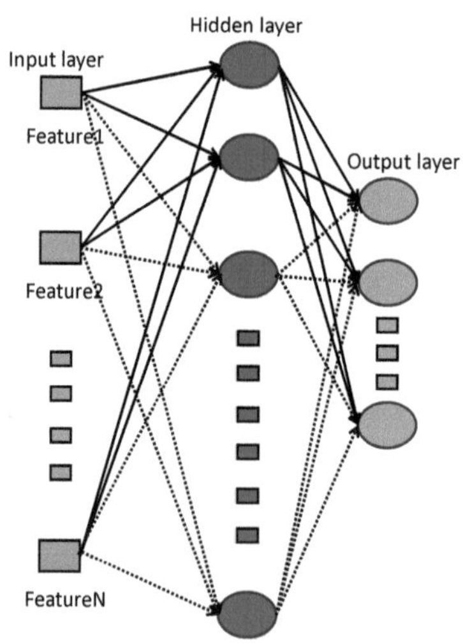

Figure 2. Structure diagram of stochastic configuration network.

Let $\Gamma = \{\Phi_1, \Phi_2, \Phi_3, \ldots\}$ be the set of real-valued functions and span(Γ) be the space spanned by Γ. $L_2(D)$ represents all Lebesgue measurable functions $f = [f_1, f_2, \ldots, f_m] : \mathbb{R}^d \rightarrow \mathbb{R}^m$ defined on $D \subset \mathbb{R}^d$. The L_2 norm is defined as

$$\|f\| = (\sum_{q=1}^{m} \int_D |f_q(x)|^2 dx)^2. \quad (1)$$

The inner product of f and $\theta = [\theta_1, \theta_2, \ldots, \theta_m]$ is defined as

$$\langle f, \theta \rangle = \sum_{q=1}^{m} \langle f_q, \theta_q \rangle = \sum_{q=1}^{m} \int_D f_q(x)\theta_q(x)dx. \tag{2}$$

Suppose that span(Γ) is dense in the L_2 space and holds for $\Phi \in \Gamma, 0 < \|\Phi\| < b_\Phi$. Given the inequality $0 < r < 1$, as well as a non-negative sequence $\{\mu_L\}$ with limit $lim_{L \to \mu_L} = 0$ and $\mu_L \leq (1-r)$, for $L = 1,2,3,\ldots$, define $\delta_L = \sum_{q=1}^{m} \delta_{L,q}, \delta_{L,q} = (1 - r - \mu_L)\|e_{L-1,q}\|$. If the random basis functions satisfy the inequality

$$\langle e_{L-1,q}, \Phi_L \rangle^2 \geq b_\Phi^2 \delta_{L,q}, q = 1,2,\ldots,m, \tag{3}$$

and the output weights of the hidden layer are

$$\beta = [\beta_1, \beta_2, \ldots, \beta_L] = \text{argmin}_\beta \|f - \sum_{j=1}^{L} \beta_j \Phi_j\|, \tag{4}$$

then $\lim_{L\to\infty}\|f - f_L\| = 0$. After the introduction of the universal approximation property, the subsequent section presents the generation process of an SCN. Consider a training set consisting of N sample pairs $\{(x_n, y_n), n = 1,2,\ldots,N\}$, where $x_n \in \mathbb{R}^d, y_n \in \mathbb{R}^m$. Let $X \in \mathbb{R}^{N\times d}$ and $Y \in \mathbb{R}^{N\times m}$ represent the inputs and outputs, respectively, corresponding to these N samples in the network. Furthermore, let $e_{L-1}(X) \in \mathbb{R}^{N\times M}$ denote the residual error matrix, where each column $e_{L-1,q}(X) = [e_{L-1,q}(x_1), \ldots, e_{L-1,q}(x_N)]^T \in \mathbb{R}^N$, $q = 1,2,\ldots,m$. The output vector of X at the L-th node, denoted on Φ_L, is defined as

$$h_L(X) = [\Phi(w_L^T x_1 + b_L), \ldots, \Phi(w_L^T x_N + b_L)] \tag{5}$$

Then, the output matrix of the hidden layer in f_L can be represented as $H_L = [h_1, h_2, \ldots, h_L]$. Let

$$\xi_{L,q} = \frac{(e_{L-1,q}^T(X) \cdot h_L(X))^2}{h_L^T(X) \cdot h_L(X)} - (1 - r - \mu_L)e_{L-1,q}^T(X)e_{L-1,q}(X), q = 1,2,\ldots,m \tag{6}$$

be defined as an intermediate variable. Perform T_{max} configurations on node Φ_L, selecting the candidate node parameters that maximize $\xi_L = \sum_{q=1}^{m} \xi_{L,q} \geq 0$ as the parameters for the L-th node. Substitute the label matrix Y for the objective function in Equation (4), and then calculate Equation (4) using the least squares method.

$$\beta = \text{argmin}_\beta \|f - \sum_{j=1}^{L} \beta_j \Phi_j\| = H^+ Y. \tag{7}$$

Here, H^+ represents the generalized inverse matrix of H. If the predicted results do not meet the tolerance error or the number of nodes does not reach the predetermined maximum number of nodes, the above operations are continued to generate new nodes.

As neural network node deletion is involved in this paper, the following formula is provided for calculating the number of parameters in a single-hidden-layer fully connected neural network. Assuming that the input layer has a dimension of I, the output layer has a dimension of O, and the number of neurons in the intermediate hidden layer is L, with each neuron having a bias term, the total number of parameters is denoted as P. The calculation of the parameter of network count is given by

$$P = (I + O + 1)L. \tag{8}$$

In order to investigate the computational cost of the SCN, floating-point operations per second (FLOPs) are introduced for a quantitative assessment of the SCN's computational performance. An SCN is fundamentally a single-hidden-layer fully connected neural network with an activation function. The activation function employed in this study is the Sigmoid activation function, expressed as $S(z) = \frac{1}{1+exp(-z)}$. Through straightforward derivation, the SCN's FLOPs can be represented as

$$FLOPs = (2L * (I + O) + 3L) * bs, \tag{9}$$

where bs represents the batch size, denoting the number of input samples.

2.2. Feature Extraction Method Based on Randomly Parameterized Rectangular Convolution

The monitoring data for structural health monitoring can be defined using multivariate time series from multiple sensors. In structural health monitoring, the focus is typically on the state and performance of structures such as bridges and buildings. To obtain information about the state of the structure, multiple sensors are commonly used to measure different physical quantities, forming a monitoring dataset. For example, N acceleration sensors may be used to collect signals at a frequency f for a duration of T, and these collected data are combined to form a dataset based on acceleration signals. The input of the SCN is a one-dimensional vector, while monitoring data are typically two-dimensional multivariate time series data. If the monitoring data are directly flattened into a one-dimensional format (as shown in Figure 3), the performance of the model trained on such data will be poor. This is because flattening the data disrupts or even loses the information in the temporal and channel domains. Therefore, a way needs to be found to ensure that the features of the monitoring data can be effectively extracted.

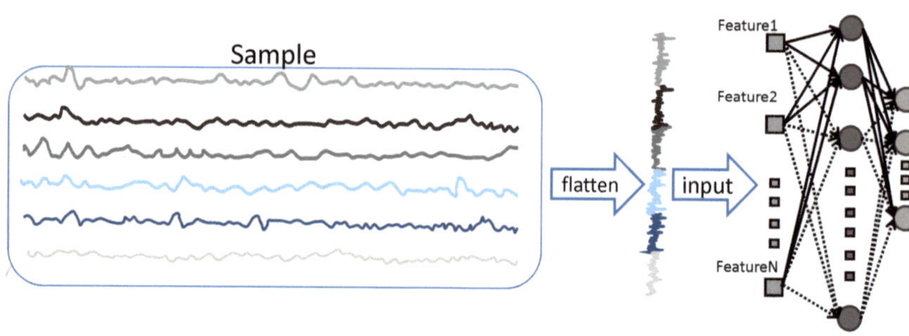

Figure 3. Flattening of multivariate time series: dimension mixing and information loss.

An ELM [44] is a simple machine learning approach. Initially, the ELM randomly initializes the input weights of the hidden layer. Then, it linearly maps the data using these weights to a high-dimensional space. Finally, it updates the weights between the hidden layer and the output layer using the least squares method. Rocket [45] is a simple yet effective method for single-variable time series classification. Its basic idea is to use a large number of randomly generated convolutional kernels to extract features from the time series, which are then used for subsequent modeling tasks. Rocket argues that time series data are not as complex as images or natural language, and therefore do not require complex neural networks or tedious training processes. Rocket has been validated on single-variable time series data. However, the original Rocket method uses one-dimensional convolutional kernels, which may not capture the channel-domain information effectively for multivariate time series monitoring data. Inspired by ELMs and Rocket, this study introduces a feature extraction approach based on random two-dimensional convolutional kernels. Experimental results demonstrate that extending the one-dimensional kernels in Rocket to two-dimensional kernels leads to performance improvements. Traditional two-dimensional convolutions use square kernels with odd side lengths, which may be more suitable for images than monitoring data. Therefore, it is worth emphasizing that this study selects random two-dimensional convolutional kernels that are more suitable for monitoring data, with the width and height chosen from a given range of positive integers. The experimental results also confirm the effectiveness of using random rectangular kernels. The feature generation methods based on Rocket, square convolutional kernels, and rectangular convolutional kernels are illustrated in Figure 4. In all three methods, the convolutional kernels move from left to right and from top to bottom, and the features are selected by directly choosing the maximum value in the feature maps. The forward figure of the entire framework is shown in Figure 5. Given a sample, features are constructed using the approach depicted in Figure 4. N convolutional kernels are randomly generated, with each kernel generating a feature. Thus, a sample corresponds to N features. These features, generated through random convolutions, are used as inputs for the SCN. For the hyperparameters of the convolutional kernel (kernel size, input–output channel numbers, stride, padding, dilation factor, number of convolution groups, bias), a range is given. In the experiments, the convolutional kernels are randomly generated based on these specified ranges. The weight parameters of the convolutional kernels are initialized randomly using the *Kaiming uniform* [47] method. The specific configurations of the random convolutional kernels used in the experiments are provided below.

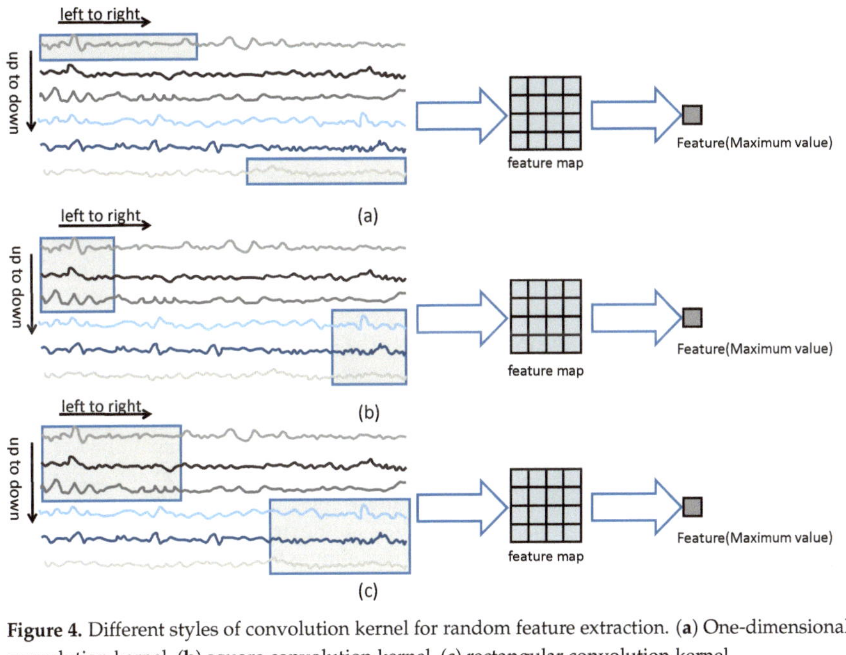

Figure 4. Different styles of convolution kernel for random feature extraction. (**a**) One-dimensional convolution kernel, (**b**) square convolution kernel, (**c**) rectangular convolution kernel.

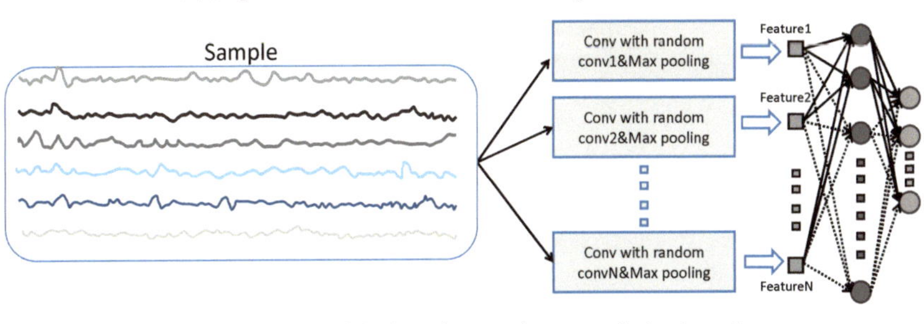

Figure 5. A framework for training SCN based on random convolution kernel.

Specifically, the side length of the three convolution kernels is randomly determined by selecting integers from 1 to 10, with each integer having an equal probability of selection. The number of channels in the output feature map is set to 1. The stride, zero-padding, and dilation factors are equally and randomly determined from integers within the range of 1 to 3. The bias of the convolution kernels is randomly drawn from a uniform distribution between $(-1, 1)$. The weights are initialized using PyTorch, a deep learning framework's built-in weight initialization method. The input features of the SCN are represented by the maximum values of the feature maps. The specific setting methods for parameters of the convolution kernel are shown in Table 1.

Table 1. The range of values for convolutional kernel parameters.

Parameters	$1 \times K$	$K \times K$	$K_1 \times K_2$
Kernel size	randint [1, 10]	randint [1, 10]	randint [1, 10]
I/O channels	1	1	1
Stride	randint [1, 3]	randint [1, 3]	randint [1, 3]
Padding	randint [1, 3]	randint [1, 3]	randint [1, 3]
Dilation	randint [1, 3]	randint [1, 3]	randint [1, 3]
Groups	1	1	1
Bias	$u(-1, 1)$	$u(-1, 1)$	$u(-1, 1)$
Weight	$Kaiming\ uniform$	$Kaiming\ uniform$	$Kaiming\ uniform$

2.3. Random Node Deletion Algorithm

Given a pre-trained SCN with L neurons, the node deletion problem can be considered as an approximate dynamic programming problem. The original problem is defined as follows: the accuracy of the network is maximized by deleting a certain number of neurons. The state of this problem, referred to as $states[n]$, represents the maximum accuracy achieved by deleting n nodes from the total of L nodes. In other words, it is necessary to compute the combination of nodes, calculate the accuracy for each combination, and select n nodes from the L nodes for deletion, resulting in C_L^n possible combinations. After the nodes are deleted, the accuracy of the network is calculated and the accuracies are stored in a $List$. The state transition equation can be defined as follows:

$$states[n] = \max(states[n-1], \max(List)) \tag{10}$$

The current state n represents the maximum accuracy achieved among all models with n nodes removed, compared to the maximum accuracy achieved in the previous state. However, due to the involvement of combinatorial calculations, the time complexity is considerably high. To mitigate this, a randomized node deletion approach is employed as an approximation, aiming to reduce the time cost. It is worth emphasizing that the n removed nodes are deleted randomly, and the probability of removal is equal for each node to be deleted.

An acceptable accuracy threshold, denoted as ϵ, is set as the pruning criterion. In this study, B is defined as $\epsilon = \alpha - 0.01$, where α represents the accuracy of the original network. This indicates that the accuracy of all pruned networks generated during the pruning process is expected to be no more than 1% lower than the accuracy of the original network.

The following pseudo-code (Algorithm 1) presents the algorithm for node deletion:

Algorithm 1. Node Random Pruning Algorithm.

Given the parameters to be deleted, $\boldsymbol{W} = [w_1, \ldots, w_L]$, $b = [b_1, \ldots, b_L]$, $\boldsymbol{\beta} = [\beta_1, \ldots, \beta_L]$, where L is the number of node; the initial accuracy of the network, α; test data X, Y; setting an acceptable accuracy $\epsilon = \alpha - 0.01$; and the maximum number of nodes that can be deleted, $N_{max} = \lfloor 0.2 * L \rfloor$.

1. Initialize $states \leftarrow zeros(N_{max} + 1)$, $states[0] \leftarrow \alpha$, $diclist \leftarrow [\ ]$
2. **For** $n = 1$ **To** $N_{max} + 1$ **Do**
3. $num = C_L^n$, $accuracy\_and\_indices \leftarrow \{\ \}$, $accuracy_{l}ist \leftarrow [\ ]$
4. **If** $num > 10000$
5. $num = 10000$
6. **For** $k = 1$ **To** num **Do**
7. Randomly choice *indices* of n nodes from original nodes.
8. Remove the corresponding nodes from the original list based on the *indices* of the nodes to be deleted.
9. $\boldsymbol{W}^* =$delete$(\boldsymbol{W}, indices)$
10. $b^* =$delete$(b, indices)$
11. $\boldsymbol{\beta}^* =$delete$(\boldsymbol{\beta}, indices)$
12. Calculate the accuracy of the model after removing the nodes.
13. $accuracy =$accuracy_score$(\boldsymbol{W}^*, b^*, \boldsymbol{\beta}^*, X, Y)$
14. **If** $accuracy \geq \epsilon$
15. $accuracy_and_indices \leftarrow$append$\{accuracy : indices\}$
16. $accuracy_list \leftarrow$append$[accuracy]$
17. **End For** (corresponds to **Step 4**)
18. $diclist \leftarrow$append$[accuracy_and_indices]$
19. $states[n] =$max$(state[n-1]$,max$(accuracy_list))$
20. **End For** (corresponds to **Step 2**)
21. **Return** $states, diclist$.

In the pseudo-code, the parameter "*num*" represents a hyperparameter that can be appropriately adjusted as the problem scale increases. In addition to returning the approximate state matrix, the pseudo-code also returns a list that stores the model accuracy and node indices. This list can be utilized to select an appropriate model based on the stored data. For instance, in a monitoring environment with a high tolerance for errors, models with a minimal number of parameters can be preferred. Conversely, in a monitoring environment with a low tolerance for errors, models with relatively higher accuracy are prioritized.

3. Explanation of Experimental Dataset

To validate the effectiveness of the proposed method in this paper, experiments will be conducted using a scaled-down mode of the Heichonggou Extra Large Bridge located in Yunnan, China (hereinafter referred to as the scaled-down mode) [32], as well as benchmark public data [43]. The following section provides an overview of the basic information and loading conditions of the data.

The scaled-down mode is divided into two halves, with the scaled-down mode (left half) serving as the prototype for constructing a scaled-down mode. Figure 6a shows the original image of the bridge, Figure 6b depicts a scaled image at a ratio of 20:1, and Figure 6c represents the scaled-down mode built in the laboratory. The basic structural parameters of the bridge are as follows: the superstructure of the main bridge consists of three-span prestressed concrete continuous rigid frame bridges with lengths of (98 + 180 + 98 m), and the total length of the bridge is 397 m. The box girder adopts a single-box, single-cell structure with vertical web plates. The top width of the box girder is 12.5 m, and the width of the box is 6.5 m. To simulate the structural damage and deterioration of the prototype of the Heichonggou Grand Bridge, excitation was applied at the mid-span position of the scaled-down mode based on the stress characteristics of the bridge structure. During three different static loadings applied to the mid-span of the scaled-down mode, varying degrees of cracks were observed in the structure. In the first excitation, a single transverse crack with a width of 0.06 mm and a length of 14.3 cm appeared in the mid-span bottom plate of the scaled-down mode after applying an external force of 2.889 KG. In the second excitation, two transverse cracks formed in the mid-span bottom plate of the scaled-down mode after applying an external force of 6.097 KG. The crack widths were measured to be in the range of (0.11–0.13) mm and (0.02–0.04) mm, while the crack length remained the same. In the third excitation, two transverse cracks in the mid-span bottom plate of the scaled-down mode further developed. The crack widths increased to 0.12 mm and (0.06–0.08) mm, respectively, while the crack length remained unchanged. The specific conditions for each working condition are shown in Table 2.

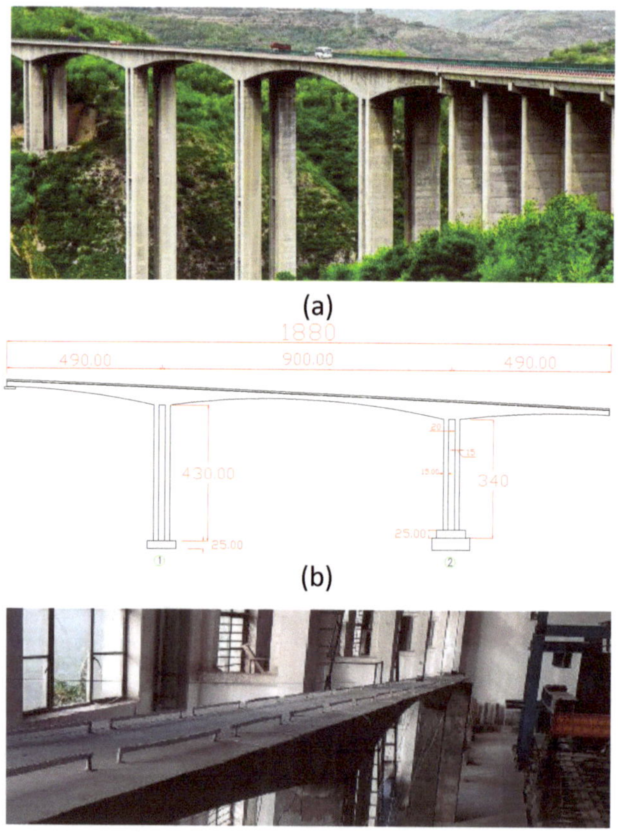

Figure 6. Schematic diagram of the scaled-down model. (**a**) Real scene image of the bridge; (**b**) Actual dimensions of the bridge; (**c**) Concrete model of the bridge scaled down to 1:20 ratio.

Table 2. Description of scaled-down bridge dataset damage states.

Case	Descriptions
D1	No damage in the bridge structure.
D2	One crack with width of 0.06 mm.
D3	Two cracks in the mid-span of scale model with widths of (0.11–0.13) mm and (0.02–0.04) mm, respectively.
D4	Two cracks in the mid-span of scale model with widths of 0.12 mm and (0.06–0.08) mm, respectively.

The second dataset used in this study is the IASC-ASCE benchmark finite element model dataset. The structural damage scenarios of the model are denoted as D1, D2, D3, D4, D5, D6, and D7, representing structural damage scenarios one to seven, respectively. The corresponding details of each structural damage scenario are presented in the Table 3 below.

Table 3. Description of IASC-ASCE benchmark finite element model dataset damage states.

Case	Descriptions
D1	No damage.
D2	Remove all diagonal supports on the first floor.
D3	Remove all diagonal supports on the first and second floors.
D4	Remove one diagonal support on the first floor.
D5	Remove a diagonal support on the first and third floors.
D6	D4 + Weaken the left side of the 18 element (first layer beam element).
D7	Reserve 2/3 of an oblique support area on the first floor.

The IASC-ASCE benchmark finite element model consists of six different structural damage scenarios and one initial structural condition. Each scenario is associated with 16 collected accelerometer data, with 8 sensors along the x-axis and 8 sensors along the y-axis, distributed across a four-story frame structure. Each accelerometer is used to capture the vibration signals of the model. Different structural damage states are simulated by removing supporting units or loosening bolt connections within the structure.

As shown in Figure 7, the damage conditions for scenarios D2 to D7 are depicted, with each damage scenario corresponding to the conditions listed in the table. The damaged regions are highlighted with red circles and red dashed lines. Due to the simultaneous sampling of vibration signals by the 16 sensors for each scenario, the acceleration vibration data for each scenario form a multivariate time series with 16 dimensions.

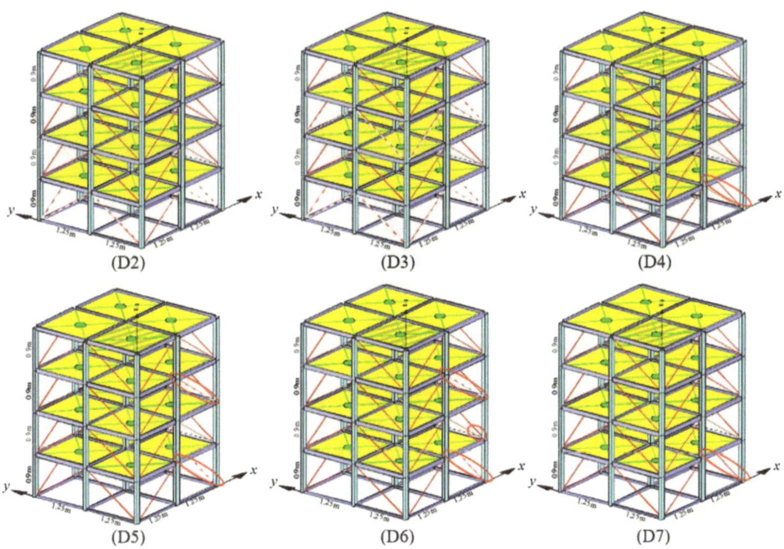

Figure 7. The benchmark framework.

In general, there are multiple sensors on the monitored object, which collect monitoring information of different dimensions; that is to say, the monitoring data usually consist of multiple dimensions, and it is necessary to analyze the correlation between different dimensions of data. From the perspective of strong correlation between sensors, a one-dimensional convolutional kernel cannot simultaneously scan the data of multiple sensors, which may result in the extracted features ignoring the similarity between them. From the perspective of weak correlation between sensors, this may also lead to the convolutional kernel neglecting the differential information between different sensors. The correlation matrices of sensors for the scaling model and the benchmark model are provided below. From the perspective of the quantitative analysis of correlation coefficients, it is necessary to choose a two-dimensional convolutional kernel.

The correlation matrix of sensors for the scaled-down mode is shown in Figure 8. According to the rule of correlation strength assessment, if the correlation coefficient falls between 0.3 and 0.5, it indicates a moderate level of correlation. It can be observed that there are moderate correlations between sensors 3 and 4, 1 and 6, 6 and 7, and so on. If the correlation coefficient ranges from 0.5 to 1, it signifies a strong correlation between variables. Strong correlations can be observed between sensors 1 and 3, 7 and 9, 9 and 11, and so forth.

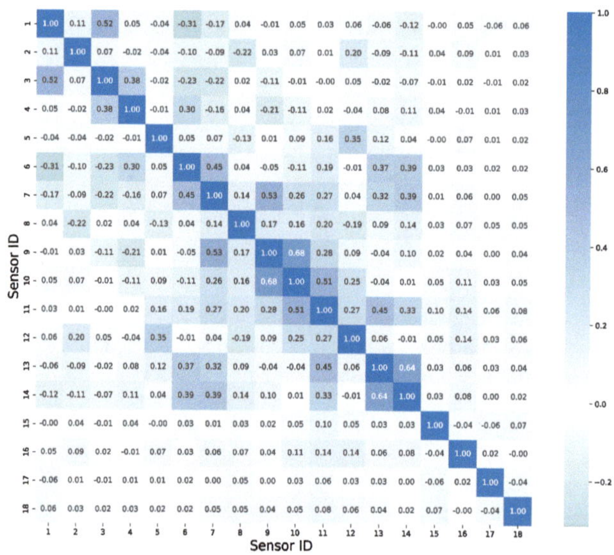

Figure 8. Spearman correlation matrix of sensors in scaled-down mode.

The correlation matrix of sensors for the benchmark model is shown in Figure 9, providing quantitative measures of correlation. Due to the smaller size of the benchmark model compared to the scaling model, the sensors are more densely distributed, and they exhibit a higher degree of correlation. Whether it is a monitoring object with densely distributed sensors or loosely distributed sensors, there are always sensors within these objects that exhibit strong correlation. Therefore, when extracting features, it is important to consider extracting features from multiple sensors simultaneously. Although the correlation matrix graph provides specific values of correlation coefficients, it may not be user-friendly for engineers who are interested in identifying sensors with strong correlations. To address this, a novel network graph approach for determining correlations is presented below. Figure 10 illustrates a network graph depicting the correlations of benchmark among sensors of benchmark model. The black nodes with assigned numbers represent the corresponding sensor IDs. Researchers can rapidly and qualitatively assess the correlation between sensors based on the thickness and length of the connecting edges. The shorter and thicker the lines in the figure, the stronger the correlation among the sensors indicated. Referring to Figure 9, taking the strongest correlation between sensor 8 and sensor 6 as an example, the two most strongly associated points can be quickly identified in the graph. Without the assistance of Figure 10, this identification process would be considerably slower. Therefore, the introduction of such a network graph is deemed necessary.

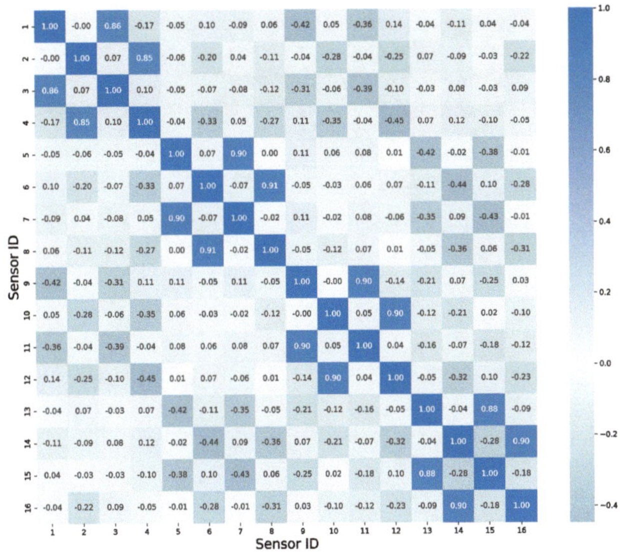

Figure 9. Spearman correlation matrix of sensors in benchmark model.

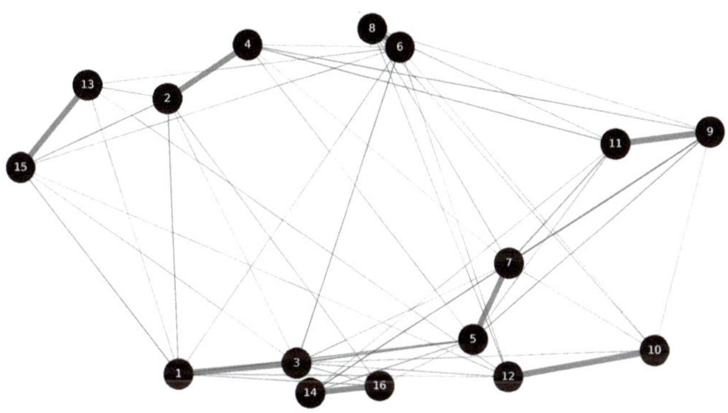

Figure 10. Network graph of node correlations.

4. Experiments

In the benchmark, an arrangement of 16 accelerometer sensors was set up within the structural framework. Thus, the dimensionality of the monitoring data collected by these 16 sensors is 16. A random selection of data, using a window of length 60, was performed for each working condition, collecting 500 samples for each. In total, 3500 samples were collected. Experimental validation indicated that both longer sample lengths and a greater number of samples (beyond 500) had negligible effects on the model's performance, as excessively long samples could slow down the training of the model. The 3500 collected samples were divided into training and testing sets in a 7.5:2.5 ratio. For the scaled-down model with 18 sensors, samples of length 60 were also taken, but there were 1000 samples for each category, resulting in a total of 4000 samples. These were similarly divided into training and testing sets at a 7.5:2.5 ratio.

4.1. Experiments on the Random Feature Extraction Algorithm

To examine the impact of the number of network nodes on model performance, four scenarios were selected with node numbers of 150, 200, 250, and 300, while controlling the number of features, NK, to be 150, 200, 250, and 300, respectively. In other words, in this section of the experiment, there are three variables: maximum node number, number of convolutional kernels, and types of convolutional kernels. By manipulating these variables, the accuracy of the model was observed (based on the average of 20 repetitions). The experiments were conducted on two datasets, and the results are shown in Figures 11 and 12. The subplots in the figures represent the accuracy of different models trained with varying numbers of convolutional kernels.

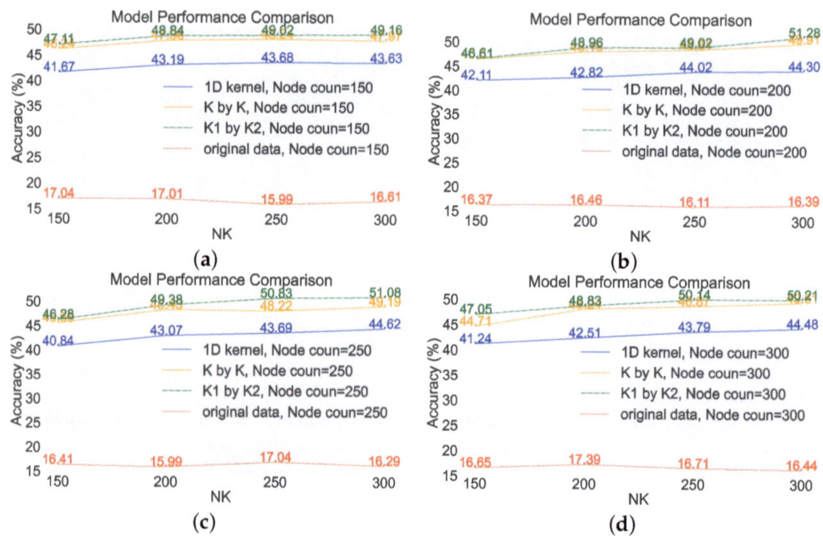

Figure 11. Analyzing the impact of the count of nodes on the performance of the model on the benchmark dataset. (**a**) Node count: 150; (**b**) node count: 200; (**c**) node count: 250; (**d**) node count: 300.

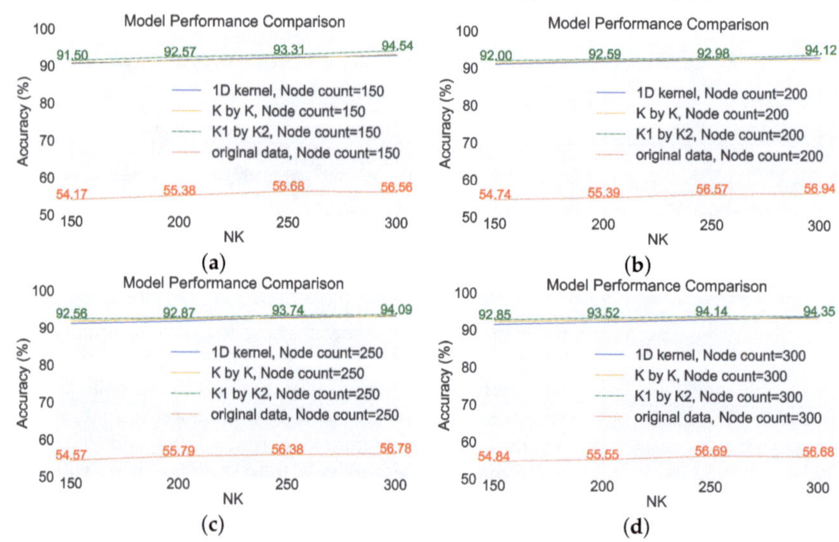

Figure 12. Analyzing the impact of the count of nodes on the performance of the model on the scaled-down model dataset. (**a**) Node count: 150; (**b**) node count: 200; (**c**) node count: 250; (**d**) node count: 300.

Taking subplot (a) as an example, the number of convolutional kernels, NK, is set to 150, which means that the input dimension of the network is 150. The red line represents the data without any transformations directly fed into the model, while the lines of other colors represent the data after feature extraction before being fed into the model. It can be observed that directly concatenating the raw sensor data and feeding them into the network for training yields poor results, with an accuracy of only about ten percent. However, after feature extraction using random convolutional kernels, the overall model performance improved by approximately 30%. In general, as long as random convolutional kernels are used, the model performance will be improved compared to the original SCN.

Specifically, among the three feature extraction methods, the model using rectangular convolutional kernels exhibited the best performance, followed by the model using square convolutional kernels with a slight decrease in performance compared to the rectangular kernel model. The model using one-dimensional convolutional kernels showed a larger decrease in performance compared to the rectangular kernel model.

Examining the red lines in the four subplots, these lines fluctuate around 16% without any clear trend. For a seven-classification model, achieving a performance of only 16% after training indicates that the model performs poorly in the prediction task. The main reasons for the low model performance are insufficient model complexity, poor data quality, inadequate feature extraction, and overfitting. The number of nodes is varied, i.e., the complexity of the network changes, but the overall model performance does not show a changing trend, which suggests that the low predictive performance is not due to insufficient model complexity.

Based on the performance of the model on the training set during training, overfitting is not evident, as the accuracy of the model on the training set remains consistently low. The reliability of the benchmark task data can be ensured, but directly inputting the raw data into the SCN leads to a deterioration in data quality. From the lines of different colors, it can be seen that after introducing feature extraction methods, the performance of the SCN improves significantly, and the accuracy of the model can increase with increasing complexity. From this perspective, it can also be concluded that the main reason for the poor performance of the SCN on monitoring data is its inability to effectively extract features from the monitoring data.

Therefore, for the original SCN, changing the number of nodes has little impact on performance. This is because the operation of directly concatenating the raw data together as input leads to the loss of information between different sensors. Once the distribution of data is disrupted, different classes of samples will lack separability, and the compactness among samples of the same class will decrease. However, for models that incorporate feature extraction methods, the performance shows an improvement of around 30% in terms of accuracy, and, in terms of trend, the performance of model can increase with an increase in the number of nodes.

In order to investigate the influence of the hyperparameter K on the model, four scenarios were selected with the number of features, K, set to 150, 200, 250, and 300. Similar to previous experiments, four sets of experiments were conducted with the number of nodes set to 150, 200, 250, and 300, respectively. The validation was performed on both the benchmark dataset and the scaled-down model, as shown in Figures 13 and 14. From the two plots, it can be observed that, under the same value of K, the performance of the proposed random rectangular convolutional kernel is the best. On the test data of the scaled-down model, the detection accuracy increases as K increases.

However, the feature-based SCN is more sensitive to the hyperparameter K. Although the accuracy does not increase with the increase in K, it has been proven to be a hyperparameter that can alter the performance of the model.

Furthermore, on the scaled-down dataset, the performance of the original SCN exhibits a clear trend with the variation in K. Overall, after introducing the feature extraction method, the accuracy of the SCN is no longer in a nearly constant state regardless of how the hyperparameters are changed. As K increases, the performance of the model also improves.

The accuracy obtained from 20 training runs was utilized to construct a box plot for analysis, as shown in Figure 15, depicting the results on the benchmark dataset. This further confirms the influence of introducing feature extraction methods on the accuracy of the model. It can be observed that the box representing the accuracy of models trained on raw data is relatively low, indicating a lower level of fluctuation in accuracy compared to models trained on feature-extracted data. Regardless of the model type, the achieved accuracy is not considered high, suggesting that the benchmark task is relatively challenging. In terms of performance improvement, this approach offers a solution to enhancing the low accuracy of SCNs in complex monitoring tasks.

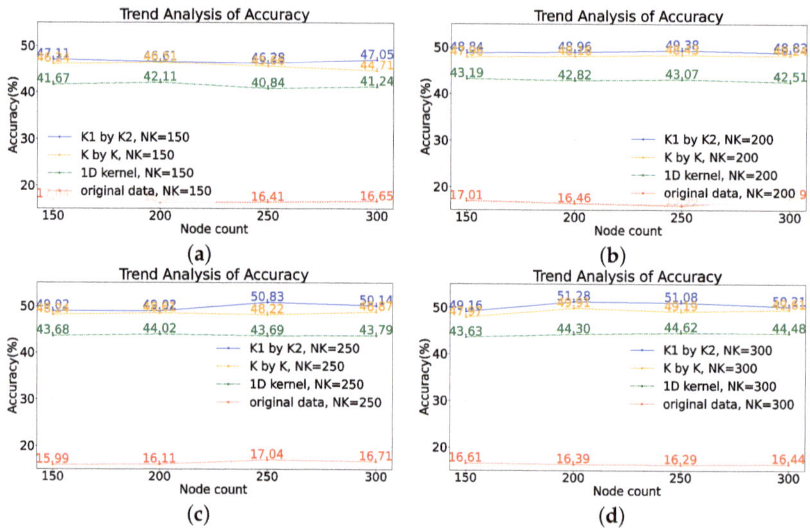

Figure 13. Influence of K number (NK) on model performance in benchmark dataset. (**a**) NK: 150; (**b**) NK: 200; (**c**) NK: 250; (**d**) NK: 300.

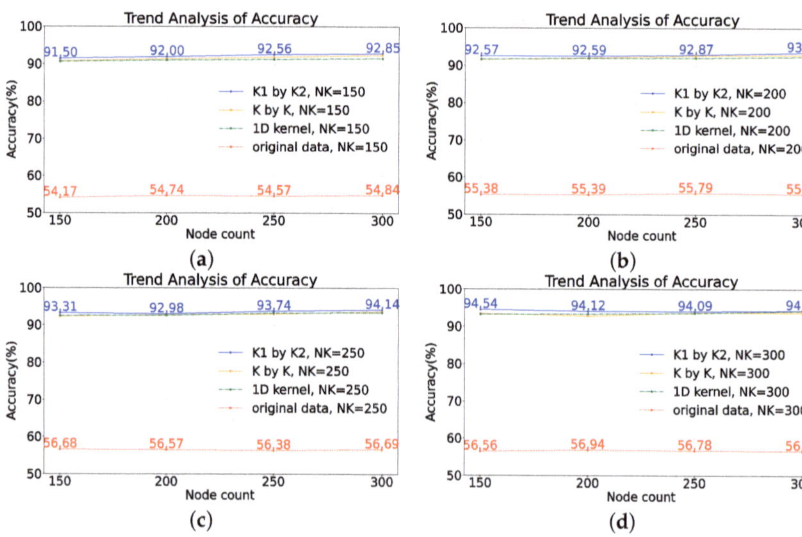

Figure 14. Influence of K number (NK) on model performance in scaled-down bridge dataset. (**a**) Node count: 150; (**b**) node count: 200; (**c**) node count: 250; (**d**) node count: 300.

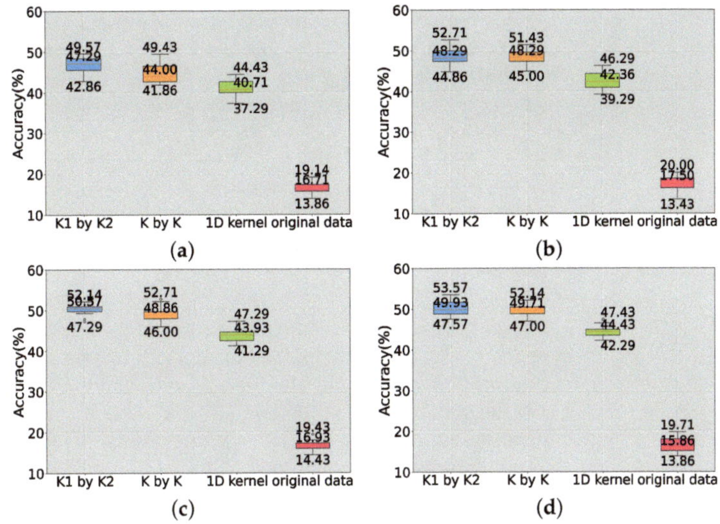

Figure 15. Box plot of benchmark accuracy. (**a**) The number of K is 150; (**b**) the number of K is 200; (**c**) the number of K is 250; (**d**) the number of K is 300.

The box plot on the scaled-down model is shown in Figure 16. It can be observed that the introduction of feature extraction methods on the scaled-down model data leads to a more stable training process and a smaller fluctuation range in model performance compared to the original SCN. In comparison to the benchmark task, the proposed method demonstrates stability on the monitoring data of the scaled-down model. It is evident that the accuracy of the model exceeds 90% when using the proposed method. In situations where the performance of the model is high, training stability is crucial. A large fluctuation in the model can easily lead to local optima. The defect of the tendency to converge to local optima has led to an increase in the training cost required for the model to achieve optimal performance.

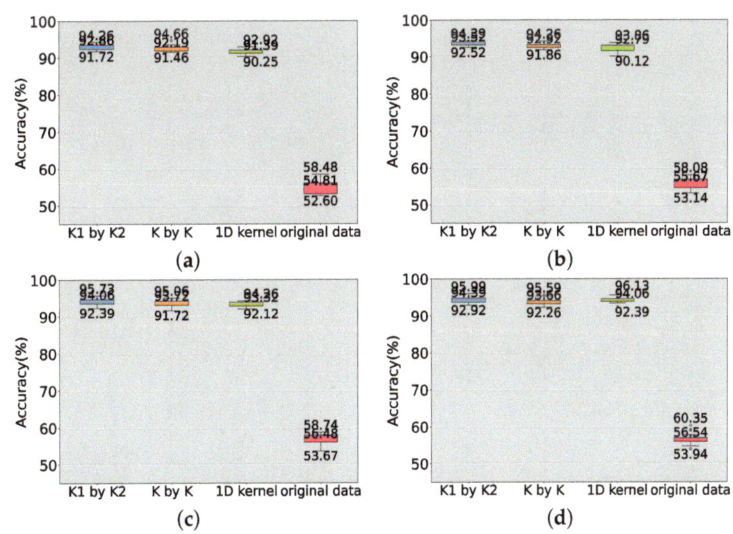

Figure 16. Box plot of scaled-down model accuracy. (**a**) The number of K is 150; (**b**) the number of K is 200; (**c**) the number of K is 250; (**d**) the number of K is 300.

In summary, this section validates the effectiveness of the proposed method on different types of datasets. The proposed method maximizes model performance in more challenging monitoring tasks. Additionally, in easier monitoring tasks, the proposed method not only improves performance but also enhances training stability.

4.2. Experiments on the Random Node Deletion Algorithm

In the experiments of this section, the convolutional kernel pattern was set as rectangular kernels, with 200 features and 250 neurons. Based on the aforementioned settings, experiments were conducted on two datasets.

4.2.1. Experiments on the Random Node Deletion Algorithm in the Benchmark Dataset

Figure 17 presents a statistical analysis of the frequency of deleted nodes in the model when applying the random deletion algorithm. The x-axis represents the node numbers in the hidden layer, while the y-axis represents the corresponding frequency of nodes being deleted. It can be observed that some nodes have a high frequency of being deleted, indicating the presence of relatively direct redundant neurons in the network. Further experiments revealed that removing these highly deleted neurons individually does not significantly affect the performance of the network and may even lead to improvements. Figure 18 presents the statistics of the threshold-satisfied models in the random node deletion algorithm. The x-axis represents the accuracy of the model, while the y-axis represents the number of neurons deleted relative to the original model. The title indicates that the accuracy of the original model before node deletion is 48.29%. It can be observed that during the execution of the entire random node deletion algorithm, the highest accuracy reaches 51.71%, with a deletion of 10 neurons relative to the original network. The accuracy improvement of 3.42% relative to the original accuracy is achieved. Considering the input dimension of 200, neuron count of 250, and output dimension of 7, the reduction in network parameters is calculated as 2080 based on Equation (8).

In the algorithm, the upper limit for neuron deletion is set to 20% of the original network, allowing for a maximum deletion of 50 neurons. Given an acceptable accuracy, among the models that meet the ϵ, as Figure 18 shows, over half of them have deleted at least 30 neurons, and over one-third have deleted at least 40 neurons. When using an SCN for damage identification on this challenging benchmark dataset, a considerable number of redundant neurons are present. This method effectively eliminates these redundant nodes. Many models have deleted more than 45 neurons, indicating that the limit of algorithm is not restricted to 20%. Hence, when performing node deletion on an SCN exhibiting high redundancy, it is possible to set a larger value for the upper limit.

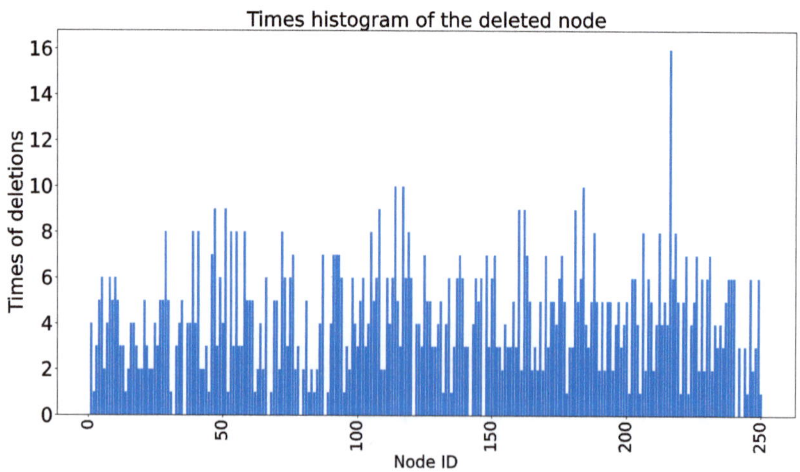

Figure 17. The frequency of deleted nodes (Experiments based on the Benchmark dataset).

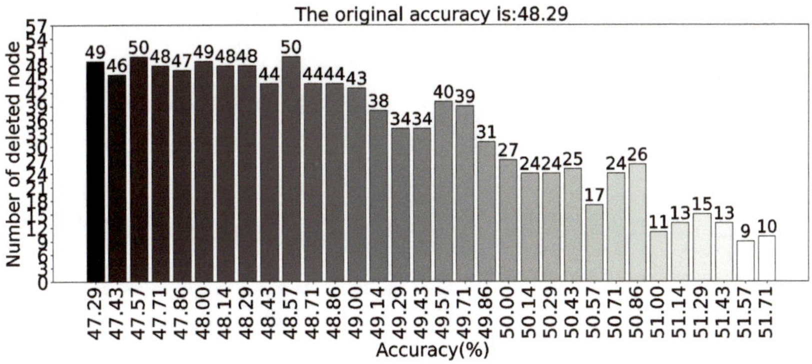

Figure 18. The relationship between the accuracy and the number of deleted nodes after the completion of the random node deletion algorithm (Experiments based on the Benchmark dataset).

4.2.2. Experiments on the Scaled-Down Model

The frequency of deleted nodes in the model during the execution of the random deletion algorithm is presented in Figure 19. The horizontal axis represents the node index in the hidden layer, and the vertical axis represents the number of times the corresponding node was deleted. Compared to Figure 17, Figure 19 appears to be more sparse. In other words, in the experiments conducted on the scaled-down model, the degree of node redundancy is not higher than that of the models trained on the benchmark dataset. This is because the data in the scaled-down model are relatively easier to learn compared to the benchmark dataset, resulting in a higher effectiveness of nodes generated by the SCN and lower node redundancy. Despite the lower node redundancy, the random node deletion algorithm still removes some neurons while ensuring the accuracy of the model. Figure 20 shown displays the accuracy of the selected models and the corresponding number of deleted nodes after executing the random node deletion algorithm. Without any node being deleted, the original accuracy of the model is 93.06%. In the case of maximum accuracy optimization, five neurons are deleted, resulting in a reduction of 1040 parameters. With a 0.13% improvement in accuracy, 17 nodes are deleted, and the parameter count is reduced by 3536. On average, the SCN in the experiments on the scaled-down model can delete 10 neurons.

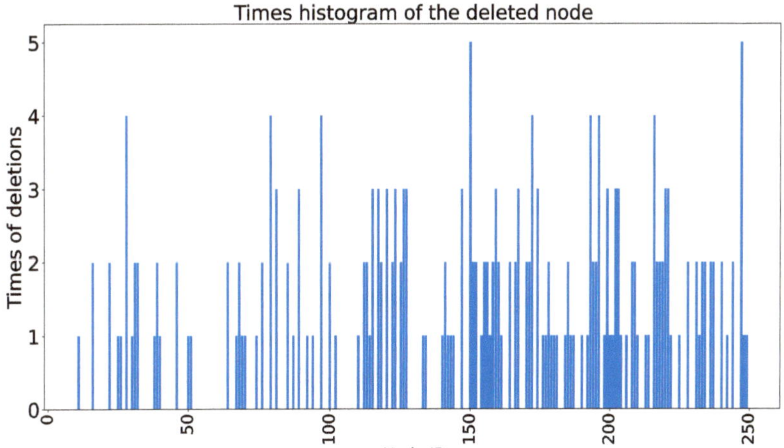

Figure 19. The frequency of deleted nodes (Experiments based on the scaled-down model dataset).

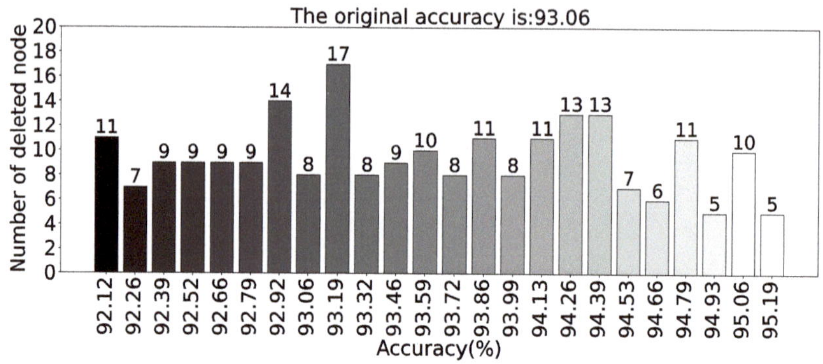

Figure 20. The relationship between the accuracy and the number of deleted nodes after the completion of the random node deletion algorithm (Experiments based on the scaled-down model dataset).

4.3. FLOPs Analysis of SCN

In accordance with the FLOPs computation Formula (9) for the SCN, the influence of node removal on FLOPs is visualized. In general, the dimensions of the input layer and output layer of the SCN are kept constant, i.e., the values of I and O remain unchanged, while the number of neurons in the hidden layer, denoted as L, is varied. Given that this study encompasses two datasets, each dataset identifies a different number of operating conditions. First, an analysis of FLOPs is conducted based on the benchmark dataset. Specifically, the input layer length of the SCN is set to 200, the number of neurons in the hidden layer is set to 250, and the output layer length is set to 7, with a batch size of 100. To observe the changes in FLOPs before and after the removal of 10% of neurons, a reduction of 10% in neurons is applied to the initial 250 neurons. As depicted in Figure 21, after removing 10% of the neurons, FLOPs decrease by approximately 10%. More precisely, after removal, the floating-point computations of the SCN are reduced by around one million. Subsequently, an analysis of FLOPs is carried out based on the downsampled model dataset, with the only alteration being the output layer length set to 4. All other settings remain consistent with the experiments based on the benchmark dataset. Analyzing Figure 22 leads to the same conclusion, where the removal of 10% of neurons results in a reduction of approximately one million floating-point computations.

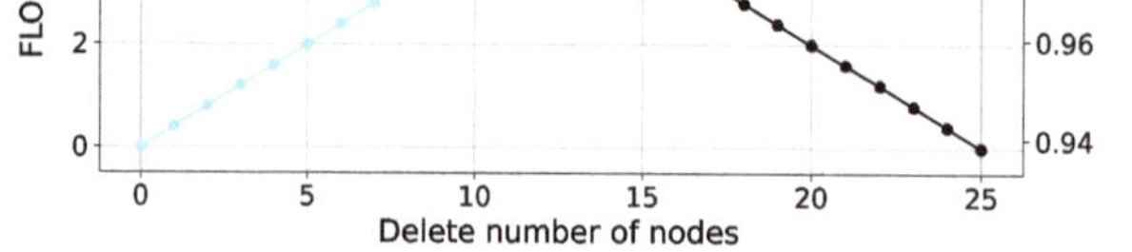

Figure 21. The relationship between the number of deleted nodes and SCN FLOPs (based on the benchmark dataset).

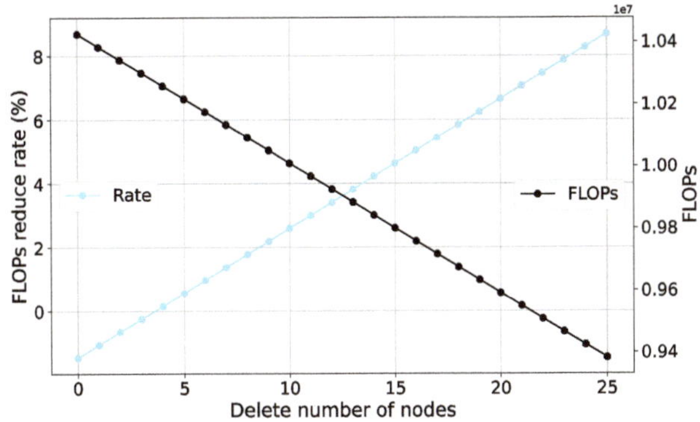

Figure 22. The relationship between the number of deleted nodes and SCN FLOPs (based on the scaled-down model dataset).

5. Conclusions

1. Key Findings: The study's primary findings can be summarized in three key points:
   a. Using raw multisensor data directly as input for the self-constructing network (SCN) is unsuitable.
   b. Employing randomly generated rectangular convolutional kernels for feature extraction is effective.
   c. SCNs contain redundant nodes, and the random node deletion algorithm efficiently eliminates them.
2. Feature Extraction Enhancement: In traditional deep learning, a fully connected network handles classification by extracting features. SCNs operate similarly but lack robust feature extraction capabilities for complex monitoring data. To address this, we introduce a convolution-based feature extraction approach inspired by deep learning, utilizing randomly generated rectangular convolutional kernels. This method is validated through experimental results.
3. Improved Model Performance: Concatenating raw data as input leads to information loss between sensors and low accuracy. The introduction of feature extraction using random convolutional kernels significantly improves model performance, with accuracy increasing by approximately 30%. Notably, rectangular convolutional kernels outperform one-dimensional kernels similar to Rocket.
4. Random Node Deletion: The experiments with the random node deletion algorithm demonstrate its potential for enhancing model performance, particularly when used on benchmark datasets. This approach achieves parameter reduction and improved performance.
5. Overall Implications: The experimental results emphasize the impact of employing random feature extraction, adjusting convolutional kernels, and node deletion on model performance. These findings are valuable for optimizing model design and parameter selection in monitoring data processing.
6. SCN for Multisensor Data: Conventional SCNs are not suitable for multisensor monitoring data. The study introduces random 2D convolutional kernels for feature extraction in this context. Future research should explore improved feature extraction methods and enhanced random mapping approaches.
7. Future Applications: While this paper primarily focuses on loss recognition, the framework has the potential to extend to other monitoring tasks, such as anomaly detection, remaining useful life prediction, and vehicle load modeling.
8. The method proposed in this paper is better suited for data with small mean fluctuations (such as acceleration data oscillating around zero), and is not applicable to data where mean values suddenly increase or decrease, such as deflection and strain [48]. When sampling from non-Gaussian or mixed probability distributions, it is relatively

easier to collect data with skewed characteristics, heavy tails, and the presence of outliers as compared to the Gaussian distribution. This poses a significant challenge for time series models. Subsequent research will consider more universally applicable models for data of different distributions.

Author Contributions: Y.L., methodology and writing—original draft preparation; D.W. methodology and reviewed the manuscript; D.L. and X.Y., data processing and figure plotting. All authors have read and agreed to the published version of the manuscript.

Funding: This research was funded by the National Natural Science Foundation of China (Grant No. 62103068, and 51978111), the Science and Technology Research Program of Chongqing Municipal Education Commission (Grant No. KJQN202100745, KJQN202301803).

Conflicts of Interest: The authors declare no conflict of interest.

References

1. Chesné, S.; Deraemaeker, A. Damage localization using transmissibility functions: A critical review. *Mech. Syst. Signal Process.* **2013**, *38*, 569–584. [CrossRef]
2. Amezquita-Sanchez, J.P.; Adeli, H. Signal processing techniques for vibration-based health monitoring of smart structures. *Arch. Comput. Methods Eng.* **2016**, *23*, 1–15. [CrossRef]
3. Meruane, V.; Heylen, W. An hybrid real genetic algorithm to detect structural damage using modal properties. *Mech. Syst. Signal Process.* **2011**, *25*, 1559–1573. [CrossRef]
4. Wu, R.T.; Jahanshahi, M.R. Data fusion approaches for structural health monitoring and system identification: Past, present, and future. *Struct. Health Monit.* **2020**, *19*, 552–586. [CrossRef]
5. Yang, J.; Xiang, F.; Li, R.; Zhang, L.; Yang, X.; Jiang, S.; Zhang, H.; Wang, D.; Liu, X. Intelligent bridge management via big data knowledge engineering. *Autom. Constr.* **2022**, *135*, 104118. [CrossRef]
6. Li, R.; Mo, T.; Yang, J.; Li, D.; Jiang, S.; Wang, D. Bridge inspection named entity recognition via BERT and lexicon augmented machine reading comprehension neural model. *Adv. Eng. Inform.* **2021**, *50*, 101416. [CrossRef]
7. Yang, J.; Huang, L.; Tong, K.; Tang, Q.; Li, H.; Cai, H.; Xin, J. A Review on Damage Monitoring and Identification Methods for Arch Bridges. *Buildings* **2023**, *13*, 1975. [CrossRef]
8. Cosenza, E.; Manfredi, G. Damage indices and damage measures. *Prog. Struct. Eng. Mater.* **2000**, *2*, 50–59. [CrossRef]
9. Kaouk, M.; Zimmerman, D.C. Structural damage assessment using a generalized minimum rank perturbation theory. *AIAA J.* **1994**, *32*, 836–842. [CrossRef]
10. Zimmerman, D.C.; Kaouk, M. Structural damage detection using a minimum rank update theory. *J. Vib. Acoust. Trans. ASME* **1994**. [CrossRef]
11. Farrar, C.R.; Worden, K. An introduction to structural health monitoring. *Philos. Trans. R. Soc. A Math. Phys. Eng. Sci.* **2007**, *365*, 303–315. [CrossRef] [PubMed]
12. Farrar, C.R.; Worden, K. *Structural Health Monitoring: A Machine Learning Perspective*; John Wiley & Sons: Hoboken, NJ, USA, 2012. [CrossRef]
13. Avci, O.; Abdeljaber, O.; Kiranyaz, S.; Hussein, M.; Inman, D.J. Wireless and real-time structural damage detection: A novel decentralized method for wireless sensor networks. *J. Sound Vib.* **2018**, *424*, 158–172. [CrossRef]
14. Chaabane, M.; Hamida, A.B.; Mansouri, M.; Nounou, H.N.; Avci, O. Damage detection using enhanced multivariate statistical process control technique. In Proceedings of the 2016 17th International Conference on Sciences and Techniques of Automatic Control and Computer Engineering (STA), Sousse, Tunisia, 19–21 December 2016; pp. 234–238. [CrossRef]
15. Catbas, F.N.; Celik, O.; Avci, O.; Abdeljaber, O.; Gul, M.; Do, N.T. Sensing and monitoring for stadium structures: A review of recent advances and a forward look. *Front. Built Environ.* **2017**, *3*, 38. [CrossRef]
16. Park, S.; Yun, C.B.; Roh, Y.; Lee, J.J. PZT-based active damage detection techniques for steel bridge components. *Smart Mater. Struct.* **2006**, *15*, 957. [CrossRef]
17. Farrar, C.R.; Doebling, S.W.; Nix, D.A. Vibration–based structural damage identification. *Philos. Trans. R. Soc. Lond. Ser. A Math. Phys. Eng. Sci.* **2001**, *359*, 131–149. [CrossRef]
18. Xia, Y.; Jian, X.; Yan, B.; Su, D. Infrastructure safety oriented traffic load monitoring using multi-sensor and single camera for short and medium span bridges. *Remote Sens.* **2019**, *11*, 2651. [CrossRef]
19. Hu, L.; Bao, Y.; Sun, Z.; Meng, X.; Tang, C.; Zhang, D. Outlier Detection Based on Nelder-Mead Simplex Robust Kalman Filtering for Trustworthy Bridge Structural Health Monitoring. *Remote Sens.* **2023**, *15*, 2385. [CrossRef]
20. Adewuyi, A.; Wu, Z. Vibration-Based Structural Health Monitoring Technique Using Statistical Features from Strain Measurements. *J. Eng. Appl. Sci.* **2006**, *4*, 3. Available online: https://www.researchgate.net/publication/237143477_Vibration-based_structural_health_monitoring_technique_using_statistical_features_from_strain_measurements (accessed on 11 September 2023).
21. Noori Hoshyar, A.; Rashidi, M.; Yu, Y.; Samali, B. Proposed Machine Learning Techniques for Bridge Structural Health Monitoring: A Laboratory Study. *Remote Sens.* **2023**, *15*, 1984. [CrossRef]

22. Li, J.; Yang, C.; Chen, J. Sound Damage Detection of Bridge Expansion Joints Using a Support Vector Data Description. *Sensors* **2023**, *23*, 3564. [CrossRef]
23. Osornio-Rios, R.A.; Amezquita-Sanchez, J.P.; Romero-Troncoso, R.J.; Garcia-Perez, A. MUSIC-ANN analysis for locating structural damages in a truss-type structure by means of vibrations. *Comput.-Aided Civ. Infrastruct. Eng.* **2012**, *27*, 687–698. [CrossRef]
24. Jayaswal, P.; Verma, S.N.; Wadhwani, A.K. Application of ANN, fuzzy logic and wavelet transform in machine fault diagnosis using vibration signal analysis. *J. Qual. Maint. Eng.* **2010**, *16*, 190–213. [CrossRef]
25. Diez, A.; Khoa, N.L.D.; Makki Alamdari, M.; Wang, Y.; Chen, F.; Runcie, P. A clustering approach for structural health monitoring on bridges. *J. Civ. Struct. Health Monit.* **2016**, *6*, 429–445. [CrossRef]
26. Alamdari, M.M.; Rakotoarivelo, T.; Khoa, N.L.D. A spectral-based clustering for structural health monitoring of the Sydney Harbour Bridge. *Mech. Syst. Signal Process.* **2017**, *87*, 384–400. [CrossRef]
27. Malhi, A.; Gao, R.X. PCA-based feature selection scheme for machine defect classification. *IEEE Trans. Instrum. Meas.* **2004**, *53*, 1517–1525. [CrossRef]
28. Wang Di, Y.S.X. Intelligent Feature Extraction, Data Fusion and Detection of Concrete Bridge Cracks: Current Development and Challenges. *Intell. Robot.* **2022**, *2*, 391–406. [CrossRef]
29. Tang, Q.; Zhou, J.; Xin, J.; Zhao, S.; Zhou, Y. Autoregressive model-based structural damage identification and localization using convolutional neural networks. *KSCE J. Civ. Eng.* **2020**, *24*, 2173–2185. [CrossRef]
30. Xu, Y.; Wei, S.; Bao, Y.; Li, H. Automatic seismic damage identification of reinforced concrete columns from images by a region-based deep convolutional neural network. *Struct. Control Health Monit.* **2019**, *26*, e2313. [CrossRef]
31. Abdeljaber, O.; Avci, O.; Kiranyaz, M.S.; Boashash, B.; Sodano, H.; Inman, D.J. 1-D CNNs for structural damage detection: Verification on a structural health monitoring benchmark data. *Neurocomputing* **2018**, *275*, 1308–1317. [CrossRef]
32. Yang, J.; Zhang, L.; Chen, C.; Li, Y.; Li, R.; Wang, G.; Jiang, S.; Zeng, Z. A hierarchical deep convolutional neural network and gated recurrent unit framework for structural damage detection. *Inf. Sci.* **2020**, *540*, 117–130. [CrossRef]
33. Yang, J.; Yang, F.; Zhou, Y.; Wang, D.; Li, R.; Wang, G.; Chen, W. A data-driven structural damage detection framework based on parallel convolutional neural network and bidirectional gated recurrent unit. *Inf. Sci.* **2021**, *566*, 103–117. [CrossRef]
34. Liao, S.; Liu, H.; Yang, J.; Ge, Y. A channel-spatial-temporal attention-based network for vibration-based damage detection. *Inf. Sci.* **2022**, *606*, 213–229. [CrossRef]
35. Pao, Y.H.; Takefuji, Y. Functional-link net computing: Theory, system architecture, and functionalities. *Computer* **1992**, *25*, 76–79. [CrossRef]
36. Gorban, A.N.; Tyukin, I.Y.; Prokhorov, D.V.; Sofeikov, K.I. Approximation with random bases: Pro et contra. *Inf. Sci.* **2016**, *364*, 129–145. [CrossRef]
37. Li, M.; Wang, D. Insights into randomized algorithms for neural networks: Practical issues and common pitfalls. *Inf. Sci.* **2017**, *382*, 170–178. [CrossRef]
38. Wang, D.; Li, M. Stochastic configuration networks: Fundamentals and algorithms. *IEEE Trans. Cybern.* **2017**, *47*, 3466–3479. [CrossRef]
39. Dai, W.; Li, D.; Zhou, P.; Chai, T. Stochastic configuration networks with block increments for data modeling in process industries. *Inform. Sci.* **2019**, *484*, 367–386. [CrossRef]
40. Zhang, C.; Ding, S. A stochastic configuration network based on chaotic sparrow search algorithm. *Knowl.-Based Syst.* **2021**, *220*, 106924. [CrossRef]
41. Liu, J.; Hao, R.; Zhang, T.; Wang, X. Vibration fault diagnosis based on stochastic configuration neural networks. *Neurocomputing* **2021**, *434*, 98–125. [CrossRef]
42. Li, W.; Tao, H.; Li, H.; Chen, K.; Wang, J. Greengage grading using stochastic configuration networks and a semi-supervised feedback mechanism. *Inf. Sci.* **2019**, *488*, 1–12. [CrossRef]
43. Johnson, E.A.; Lam, H.F.; Katafygiotis, L.S.; Beck, J.L. Phase I IASC-ASCE structural health monitoring benchmark problem using simulated data. *J. Eng. Mech.* **2004**, *130*, 3–15. [CrossRef]
44. Huang, G.B.; Zhu, Q.Y.; Siew, C.K. Extreme learning machine: Theory and applications. *Neurocomputing* **2006**, *70*, 489–501. [CrossRef]
45. Dempster, A.; Petitjean, F.; Webb, G.I. ROCKET: Exceptionally fast and accurate time series classification using random convolutional kernels. *Data Min. Knowl. Discov.* **2020**, *34*, 1454–1495. [CrossRef]
46. Zhao, H.W.; Ding, Y.L.; Li, A.Q.; Chen, B.; Wang, K.P. Digital modeling approach of distributional mapping from structural temperature field to temperature-induced strain field for bridges. *J. Civ. Struct. Health Monit.* **2023**, *13*, 251–267. [CrossRef]
47. He, K.; Zhang, X.; Ren, S.; Sun, J. Delving deep into rectifiers: Surpassing human-level performance on imagenet classification. In Proceedings of the IEEE International Conference on Computer Vision, Santiago, Chile, 7–13 December 2015; pp. 1026–1034. [CrossRef]
48. Zhao, H.; Ding, Y.; Meng, L.; Qin, Z.; Yang, F.; Li, A. Bayesian Multiple Linear Regression and New Modeling Paradigm for Structural Deflection Robust to Data Time Lag and Abnormal Signal. *IEEE Sens. J.* **2023**, *23*, 19635–19647. [CrossRef]

buildings

MDPI

Review

A Review on Damage Monitoring and Identification Methods for Arch Bridges

Jiafeng Yang [1,2], Lei Huang [3], Kai Tong [3], Qizhi Tang [3], Houxuan Li [3], Haonan Cai [3] and Jingzhou Xin [4,*]

1 School of Civil Engineering, Central South University, Changsha 410075, China; ctjtyjf@126.com
2 Chongqing Railway Investment Group Co., Ltd., Chongqing 400023, China
3 School of Civil Engineering, Chongqing Jiaotong University, Chongqing 400074, China; huanglei@mails.cqjtu.edu.cn (L.H.); kaitong@mails.cqjtu.edu.cn (K.T.); tangqizhi1015@163.com (Q.T.); hxli@mails.cqjtu.edu.cn (H.L.); 611210080005@mails.cqjtu.edu.cn (H.C.)
4 State Key Laboratory of Mountain Bridges and Tunnel Engineering, Chongqing Jiaotong University, Chongqing 400074, China
* Correspondence: xinjz@cqjtu.edu.cn

Abstract: The damage monitoring and identification of arch bridges provide an important means to ensure the safe operation of arch bridges. At present, many methods have been developed, and the applicability and effectiveness of these methods depend on the damage type, structural configuration and available data. To guide the practical application of these methods, a systematic review is implemented in this paper. Specifically, the damage monitoring and identification methods of arch bridges are divided into the damage monitoring of local diseases and damage identification of overall performance. Firstly, the research on the damage monitoring of the local diseases of arch bridges is reviewed. According to the disease type, it is divided into four categories, including suspender inspection, void monitoring, stress detection and corrosion detection. For each disease, this paper analyzes the principles, advantages and shortcomings of various methods. Then, the damage identification methods of the overall performance of arch bridges are reviewed, including masonry arch bridges, steel arch bridges, reinforced concrete arch bridges and concrete-filled steel tubular arch bridges. And the commonly used damage indexes of damage identification methods are summarized. This review aims to help researchers and practitioners in implementing existing damage detection methods effectively and developing more reliable and practical methods for arch bridges in the future.

Keywords: arch bridge; damage monitoring; damage identification

Citation: Yang, J.; Huang, L.; Tong, K.; Tang, Q.; Li, H.; Cai, H.; Xin, J. A Review on Damage Monitoring and Identification Methods for Arch Bridges. *Buildings* **2023**, *13*, 1975. https://doi.org/10.3390/buildings13081975

Academic Editor: Weixin Ren

Received: 29 June 2023
Revised: 25 July 2023
Accepted: 28 July 2023
Published: 2 August 2023

1. Introduction

Arch bridges have been widely used in transportation infrastructure construction because of their large structural stiffness, affordability and strong adaptability in mountainous areas. However, under the coupling effect of an external complex environment and internal material degradation, arch bridges inevitably face the risk of performance degradation [1–4], resulting in different degrees of damage to the structure. These damages will reduce the bearing capacity and service life of the arch bridge and even endanger the safety of the bridge. An unpredicted structural failure can be catastrophic not only in terms of life and economic losses but also in terms of the subsequent societal impacts [5]. Therefore, monitoring and identifying the damage state of the arch bridge in a timely manner can improve the structural safety and reduce the later maintenance cost, which is of great significance for ensuring the normal operation of the arch bridge and prolonging its service life.

Structural damage identification methods can be divided into two categories: local damage identification and overall damage identification [5]. The local damage monitoring method uses advanced sensors, non-destructive testing and other means to directly diagnose the local damage status of the structure. By contrast, the overall damage identification method uses data mining of structural response information to indirectly extract the characterization index of the structural damage status [6–12]. As a large and complex civil structure, arch bridges have many key components, and the damage monitoring and identification methods of arch bridges are complicated. In order to make research on the damage identification of arch bridges more systematic, this paper makes a clear classification of the existing methods, which is more practical and comprehensive. Considering the specific structural characteristics of a bridge, the existing methods are classified and then reviewed. Specifically, the damage monitoring and identification methods of arch bridges are divided into two categories: damage monitoring methods for local diseases of arch bridges and damage identification methods for the overall performance of arch bridges. The former mainly focuses on the detailed detection of a certain part or a component of the arch bridge structure, such as suspenders, arch rings, joints, etc., to find out whether there are local diseases such as corrosion, cracks and fractures; the latter mainly analyzes the overall response of the arch bridge structure, such as frequency, mode, stiffness, etc., to determine whether there is a global damage such as overall performance degradation or structural deformation.

Section 2 summarizes the damage monitoring for local diseases of arch bridges, including suspender inspection, void monitoring, stress detection and corrosion detection. In Section 3, the damage identification of the overall performance of arch bridges is described in detail. According to the development history of an arch bridge and the different construction materials used, it is divided into the following four types: masonry arch bridge, steel arch bridge, reinforced concrete (RC) arch bridge and concrete-filled steel tube (CFST) arch bridge.

2. Damage Monitoring for Local Diseases of Arch Bridge

The damage monitoring of local diseases for arch bridges is mainly realized by physical means, such as the ultrasonic method, magnetic memory method, thermal imaging method, etc. Considering the key bearing member, common disease problems and important safety performance indicators of various types of arch bridges, this paper divides the damage monitoring of local diseases into the following four categories: suspender inspection, void monitoring, stress detection and corrosion detection. Different structural objects and different diseases have their own detection methods. The following summarizes these methods, clarifies their detection principles, compares their advantages and shortcomings, and concludes the development status of each method.

2.1. Suspender Inspection

Among many types of arch bridges, a tied arch bridge is a special bridge type that combines the advantages of arch and beam. It supports the bridge deck load by the cables suspended at the top of the main arch, thereby reducing the span and height of the main arch and making the bridge light, economical and aesthetic [13]. The cable is the key bearing member of this kind of bridge to transmit the dead load and live load. It is composed of internal stress steel wires, an anchorage system and an external protection system (high-density polyethylene (HDPE) material sheath) (as shown in Figure 1). However, subjected to the bridge dead load, vehicles, wind and temperature, the protection system of the cable is prone to damage, which makes the internal steel wire of the cable rust or even break, seriously affecting the safety of the whole bridge [14]. In addition, because the cable is high above the ground, and the outer surface of the cable is wrapped by the HDPE protective sleeve, the damage of the cable is concealed and difficult to detect and prevent. The detection of the cable has attracted extensive attention from researchers [15–17]. With the development of non-destructive techniques, a variety of innovative detection methods based on acoustics, magnetism, electricity and machine vision have emerged in the detection of arch bridge cables.

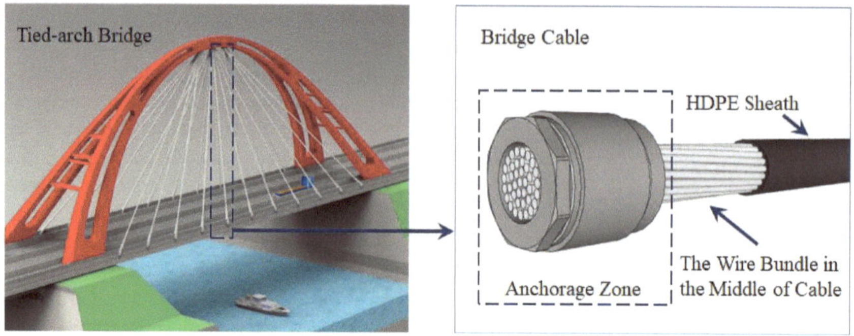

Figure 1. Cable structure diagram of tied arch bridge.

Acoustic emission testing (AET), as a long-developed non-destructive technique, has been widely used in the detection of suspenders. Its principle is to judge whether there are cracks or fractures inside the suspenders by analyzing the characteristics of acoustic signals. To reduce the influence of the environment on the detection effect, Zhu et al. [18] clarified the influence of the stress state of the suspender, the position of the measuring point, and the HDPE sheath on the AET effect and used AET to accurately determine the location of the suspender fracture. On this basis, Djeddi et al. [19] used AET to monitor the cracking process of the cable under the stress–corrosion coupling state, which filled the gap of acoustic emission in the suspender corrosion detection field. Some researchers also used AET for damage detection inside the suspender anchor head [20]. Sun et al. [21] determined the location and generation time of fatigue cracks in the steel wire inside the anchor head by analyzing the variation law of fatigue crack acoustic signals. The detection principle can be found in Figure 2. In addition, Kharrat and Gaillet [22] innovatively combined the acoustic emission principle with ultrasonic technology to realize the detection and classification of different damage degrees of anchor head.

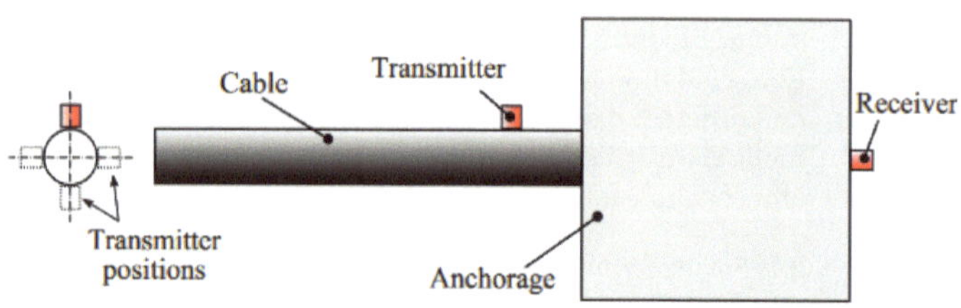

Figure 2. The principle of acoustic emission and the principle of ultrasonic technology to detect the damage of anchorage zone [21].

Magnetostrictive guided wave technology is a high sensitivity method for detecting broken wires of suspenders. The principle is that when the magnetic field induction coil generates an alternating magnetic field, a specific reflection signal will be generated at the damage of the suspender based on the magnetostrictive effect, which is converted into the information of the internal defects or damage of the suspender by signal processing technology (as shown in Figure 3). In the case of a broken wire in the suspender, Xu et al. [23] provided a method combining low-frequency wave detection and high-frequency wave detection, which can simultaneously determine the broken wire defects at different locations. In order to quantify the length of the broken wire, Chen et al. [24] studied the attenuation characteristics of the low-frequency longitudinal guided wave generated by the magnetostrictive sensor in the cable theoretically and experimentally and proposed an energy-based model, which can predict the energy attenuation well. Zhang et al. [25] designed a ring permanent-magnet-biased magnetostrictive longitudinal guided wave transducer, which improved the energy conversion efficiency of large-diameter bridge cable and promoted the practical engineering application of magnetostrictive guided wave technology.

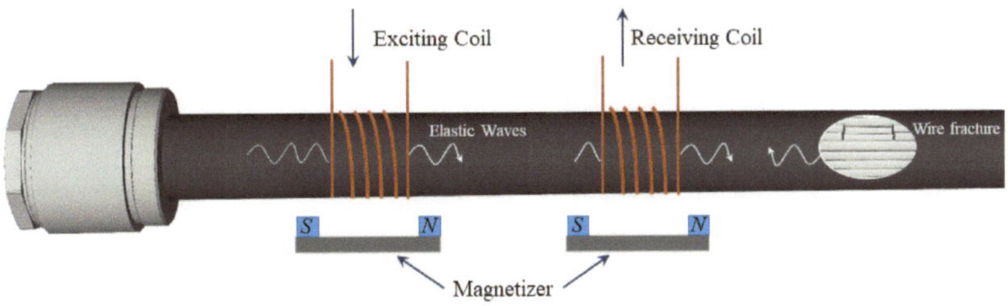

Figure 3. Magnetostrictive guided wave detection of steel wire fracture principle diagram.

Because the arch bridge suspender uses steel, which is a magnetic material, as the base material, magnetic flux leakage detection technology has also been widely used in suspender damage detection. When there are corrosion, cracks and broken wires in the suspender, the magnetic field probe (usually the Hall element) will detect the magnetic flux leakage signal on the surface of the suspender. By analyzing these magnetic flux leakage signals, the damage location and size inside the suspender can be determined (as shown in Figure 4).

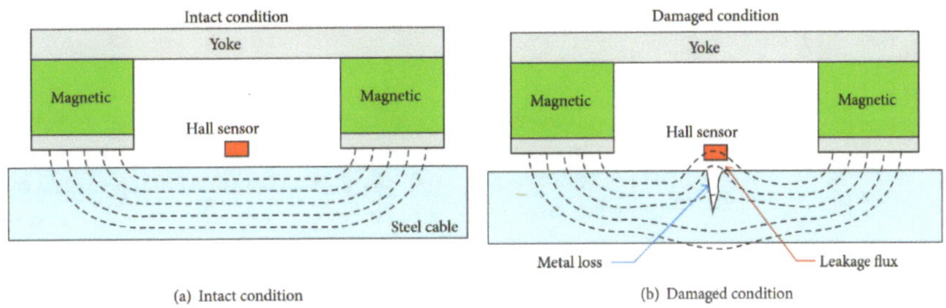

Figure 4. Principle of magnetic flux leakage detection technology [26].

At present, scholars have determined the fracture location of the steel wire inside the cable using magnetic flux leakage detection technology. However, quantifying the damage length has always been a research focus, along with the difficulty of magnetic flux leakage detection. Zhang and Xin [27] proposed an analysis model to quantify the influence of defect length, studied the influence of defect length for a magnetic flux leakage signal and verified the correctness of the model during experiments. Ni et al. [28] explored the influence of the suspender cross-section loss rate on the magnetic signal and the proposed defect length coefficient to correct the magnetic flux signal and improved the quantitative identification accuracy of the corrosion damage. To promote the practical engineering application of magnetic flux leakage detection technology, Xu et al. [29] developed the suspender climbing detection robot. The magnetic flux leakage detection system was used to detect the state of the suspender, and the filtering algorithm was used to reduce the interference signal of the environment.

The above method needs to magnetize the suspender via a strong magnetic field. The magnetization process requires heavy magnetization equipment (as shown in Figure 5), and the detection process is relatively complex [30]. Based on the principle of magnetic flux leakage detection technology, some scholars combine metal magnetic memory technology instead of artificial magnetic field excitation and analyze the magnetic flux leakage signal of suspender damage under the action of a geomagnetic field. This method has the advantages of simple operation, low cost and high efficiency [31]. In terms of suspender fracture damage detection, Qu et al. [32] proposed a cable circumferential broken wire damage detection method based on spontaneous magnetic flux leakage, which made up for the deficiency of the circumferential detection corrosion method along the suspender. In terms of suspender corrosion detection, Xia et al. [33] proposed a probabilistic evaluation method of the suspender corrosion degree based on a Bayesian model and proposed the evaluation standard of the suspender corrosion degree based on metal magnetic memory technology for the first time. In order to explore the influence of suspender stress on this method, Xia et al. [34] carried out corrosion detection on steel strand samples under different stress states and proposed a dimensionless index, X, for the quantitative detection of corrosion.

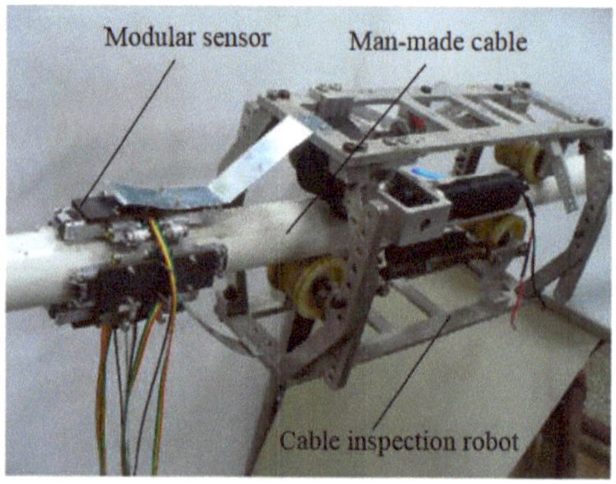

Figure 5. Suspender inspection robot based on magnetic flux leakage detection technology [30].

In recent years, piezoelectric sensors have received extensive attention from researchers due to their high sensitivity and long service life. Their principle is that the piezoelectric effect of crystals (such as quartz, strontium titanate, etc.) can be used to measure the strain of the suspender, and the anchoring efficiency of the suspender can be inferred via the relationship between stress and strain. As shown in Figure 6, Nguyen et al. [35] used the impedance-based damage estimation model to quantify the damage of the anchorage zone. The feasibility of the method was verified by the cable model experiment in the laboratory. Zhang et al. [36] monitored the loosening state of the wedge anchorage system based on the time reversal data of the sensor and proposed a method for detecting the anchoring efficiency of the suspender by using the patch piezoelectric sensor. This detection technique is the only method retrieved in this paper that can monitor the anchorage efficiency of the anchorage system.

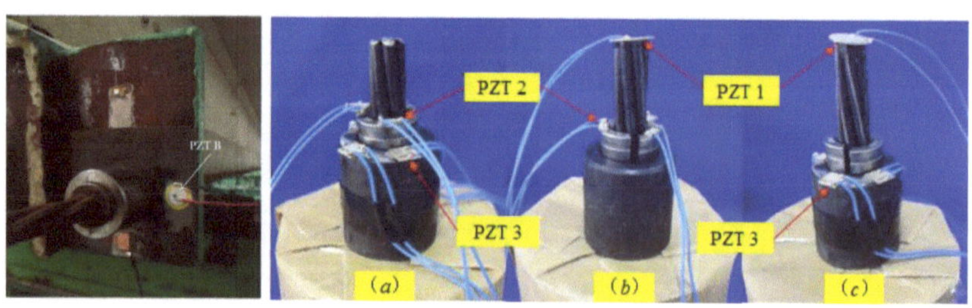

Figure 6. Piezoelectric sensor arrangement method for detecting the anchoring efficiency of suspenders: (a) contact surface between the wedge and barrel; (b) contact surface between the steel strand and wedge; (c) contact surface between the steel strand and barrel [35,36].

With the continuous development of machine vision technology, scholars have begun to use machine vision technology to detect the damage of an HDPE sheath on the surface of the suspender. This method first uses the camera or other image acquisition equipment to obtain the image or video of the suspender and then preprocesses the collected image, including image denoising, image enhancement, edge detection and other operations, to improve the quality and clarity of the image. Finally, the machine learning algorithm is used to classify and identify the damage features, thereby realizing the automatic damage detection of the HDPE sheath. Under the influence of wind or rain flow, the suspension cables of the arch bridge may suffer from vibrations. To reduce these vibrations, the surface of the suspension cables is equipped with an HDPE protective sheath designed with indentations or protruding axials. However, this design interferes with the effectiveness of machine vision inspection. To realize image acquisition, Ho et al. [37] developed a cable-climbing detection robot. The image enhancement technology was combined with the principal component analysis (PCA) algorithm to improve the overall quality of the image collected by the robot. The length of the damage crack and the aging damage was determined by the Mahalanobis square distance (as shown in Figure 7). Hou et al. [38] proposed a transfer learning method based on the cascaded mask region-based convolutional neural network (RCNN) model for the automatic detection of surface defects of suspenders and compared it with other algorithms via field tests. It is found that this method improves the detection accuracy. Due to the different view of each camera, a single image may only collect a part of the entire defect. In order to solve this problem, Li et al. [39] proposed an efficient scale invariant feature transform (SIFT) algorithm, which realized the multi-image stitching of partially overlapping regions in different defect images and provided ideas for the all-round detection of suspenders.

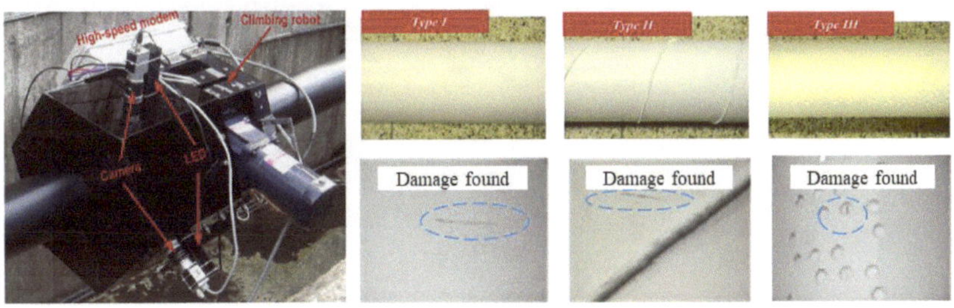

Figure 7. Climbing robot and the damage detection under three suspender surface types [37].

For the damage of an arch bridge, such as the cracking or aging of a suspender HDPE sheath, steel wire corrosion, steel wire crack or fracture, and anchorage efficiency decline, this paper summarizes the advantages, shortcomings and application scope of various detection methods, as shown in Table 1.

Table 1. Advantages and shortcomings of various detection methods for arch bridge suspenders.

Damaged Parts of Suspenders	Damage Type	Method	Advantages	Shortcomings	References
HDPE Sheath	Cracks or aging	Machine vision	The crack or aging damage of the HDPE sheath can be accurately located, and the length, width and area of the damage can be quantitatively measured.	In the early stage, a great amount of time is needed to train and debug the model. The detection effect is limited by the quality and quantity of the data set. It is impossible to judge whether the crack has penetrated into the internal steel wire.	[37–39]

Table 1. *Cont.*

Damaged Parts of Suspenders	Damage Type	Method	Advantages	Shortcomings	References
Internal steel wire of cable	Corrosion	Magnetic flux leakage testing	The quantitative measurement of the corrosion state can be realized.	The detection sensitivity of initial corrosion is poor.	[33,34]
	Fracture or crack	Acoustic emission technique	It has high sensitivity and high precision for the detection of steel wire fracture or crack position inside the suspender.	Due to the influence of the HDPE sheath, suspender length and the external HDPE sheath, the quantitative measurement of fracture length and fracture number cannot be realized, and the large area corrosion damage cannot be detected.	[18,19]
		Magnetostrictive guided wave detection technology	The testing equipment is portable and can quantitatively detect the fracture length of the internal steel wire.	The test results are affected by the length of the suspender.	[23–25]
		Magnetic flux leakage testing	The accurate positioning of steel wire fracture position and the quantitative measurement of fracture length and fracture number can be realized.	Affected by the environmental magnetic field, it is not sensitive to closed cracks.	[27–29,32]
Anchorage system	Anchoring efficiency decreased	Piezoelectric transducer	The detection cost is low. This is the only method that can monitor the anchorage efficiency of the anchorage system found in this paper.	Some sensors need to be moisture-proof, the output signal is weak, and sensors need to be deployed during the bridge construction period. There is no practical application yet.	[35,36]

2.2. Void Monitoring

The cooperative work of a steel tube and concrete is the key to ensure the bearing capacity of a CFST arch bridge [40,41]. Due to the load, temperature stress, creep shrinkage and construction stage bubble aggregation, among other reasons, the CFST arch bridge often suffers void disease. At present, many nondestructive means have been used in the void detection of a CFST, such as the ultrasonic method [42], impact–echo method [43], optical fiber sensor method [44,45] and infrared thermal imaging method [46].

Ultrasonic detection detects interface changes by detecting echo signals to achieve void detection. Researchers use the velocity of ultrasonic waves in different media to evaluate the defects of a CFST [47,48]. For example, Dong et al. [49] studied the motion path of ultrasonic waves in a CFST with void disease. Callejas et al. [50] showed that broadband ultrasound attenuation (BUA) has a linear relationship with the percentage of air in a CFST via experimental research. The void in a CFST can be detected effectively via BUA. Liu et al. [51] proposed an ultrasonic detection method for voids in CFSTs by combining travel time tomography (TTT) and inversion time migration (RTM) technology. This method can image the size and shape of voids in CFSTs with high resolution. Liu et al. [52] proposed an ultrasonic array imaging method based on full-matrix capture (FMC), which can semi-quantitatively characterize the voids of CFSTs. Liu et al. [52] used this method to detect the void in a CFST and verified the test results by core drilling method. The results show that the method has good detection accuracy and can adequately locate the void in a CFST. Although the ultrasonic method has been widely used in engineering to detect the void of CFST, the method still has the shortcomings of limited penetration and the influence of concrete aggregate on the detection results [53]. In addition, ultrasonic testing requires scanning the entire structure, which takes a long time, and, especially for large CFST arch bridge structures, the detection time will be longer.

The impact–echo method was first proposed by Sansalone and Carino of Cornell University in the United States [54]. This method uses the propagation and reflection of high-frequency sound waves generated by tapping the surface of the measured object in the concrete structure to detect defects. The waveform and spectrum of the impact elastic wave are analyzed to determine the location, size and boundary of the defect [55,56]. In recent years, due to the rapid development of machine learning, many scholars have used machine learning to optimize the impact–echo method [57,58]. Chen et al. [59] carried out the void detection test on a CFST based on the impact–echo method. The power spectral density (PSD) of the void area and the non-void area was extracted, and its features were used as the database of the support vector machine (SVM) model. The detection accuracy of the void detection model obtained by machine learning can reach 94.17%. The impact-echo method can well characterize the depth and shape of the void. However, the impact-echo detection is mainly affected by the impact error of the impact hammer and the installation error of the sensor. In actual use, it is difficult for the operator to accurately control the angle and amplitude of the impact. It is usually accompanied by a double hit in the impact process, resulting in a decrease in detection accuracy [60]. At the same time, the detection efficiency of the impact–echo method is very low, and it is required to detect the area to be measured point by point. For the CFST arch bridge, it is difficult to detect the whole arch rib.

The distributed optical fiber sensor has been widely used in civil engineering in recent years [61]. It uses the optical fiber itself as a sensor to continuously monitor the optical fiber along the length direction. The principle is to receive and analyze the scattered light by the optical time-domain reflectometer (OTDR) to obtain its attenuation waveform to determine the range, location and value of the void and crack [44,45]. Ding [44] obtained the semi-empirical formula of loss by a regression analysis of experimental data. At the same time, based on the force-light constitutive model, Ding [44] proposed a quantitative algorithm for the void–crack of a CFST and applied it in the Wuxia Yangtze River Bridge. Gong et al. [62] proposed a distributed optical fiber detection method with active heating function based on Brillouin reflection and clarified the relationship between the void depth and the equivalent heat absorption coefficient (EHAC). This method quantitatively detects the void by monitoring the surface temperature of a CFST. The optical fiber sensor has good accuracy and can accurately monitor the amount of void. However, the optical fiber material itself is low in strength and expensive. In practical engineering, it is necessary to embed the optical fiber sensor in the concrete structure, which greatly improves the requirements for construction technology and reduces the service life of the sensor.

The application of infrared thermal imaging technology in civil engineering has been developed for a long time. Because of its lack of contact and high efficiency, it has been widely studied in structural non-destructive testing. When there is a void in the CFST, the thermal conductivity of the air in the void area is quite different from that in the non-void area, so there is a certain temperature difference between the two areas. Because the steel tube has good thermal conductivity, the void can be detected by the temperature difference of the outer surface of the CFST. Wang et al. [63] used the characteristics of the high thermal efficiency of a rubber heating belt to wrap and heat CFST specimens. The results showed that the heating method could effectively detect small defects with a thickness of 5 mm. Guo and Yan [64] used eddy current thermal excitation method to explore the influencing factors of the internal hole detection of a CFST and established the multiple regression relationship between defect size, thickness of steel tube, heating time, heating power and heating rate. Jiang and Chen [65] used pulse heat source to heat the specimen, and the temperature difference between the surface of the specimen above the defect and other areas appeared via heat conduction. Due to the large pulse energy, the signal-to-noise ratio of the obtained heat map was improved. Because the detection results of the infrared thermal imaging method are finally presented as infrared thermal imaging images, many scholars have used image processing algorithms to improve the detection accuracy in recent years. Zhang et al. [66] used contrast limited adaptive histogram equalization (CLAHE) algorithm to solve the problem of gray loss and image patch in traditional algorithm, effectively reducing the excessive distortion of local area contrast and regional noise interference and making the image of the defect area and the non-defect area clearer and more complete after processing. Zheng and Tan [67] combined the time domain fitting with the CLAHE algorithm to enhance the image and processed the edge image of the internal defects of the concrete in detail so that the recognition effect of the defects was improved. The infrared thermal imaging method has the advantages of non-contact detection, large-area detection and rapid detection. However, the infrared thermal imaging method also has certain shortcomings. The existing numerical simulation [65] has demonstrated that the void monitoring of CFSTs using the thermal imaging method is influenced by various factors, including the geometric dimensions of defects and the thickness of the steel tubes. Thicker steel tubes lead to a more uniform surface temperature distribution, making void monitoring challenging in large-scale CFSTs. Furthermore, the surrounding environment also influences the surface temperature field of CFSTs, which in turn affects the monitoring process. Moreover, this testing approach requires specialized heat equipment and involves a complex testing system [68].

Different detection methods have their own advantages and shortcomings in practical engineering applications. Therefore, the detection principles, advantages and shortcomings of the above methods are sorted out in Table 2.

Table 2. Advantages and shortcomings of various methods.

Methods	Detection Principles	Advantages	Shortcomings
Ultrasonic method	The difference in propagation velocity of ultrasonic wave in different media is used for detection.	The equipment is simple, and the detection cost is low.	Contact point-by-point measurement has slow detection speed. The penetration of ultrasonic wave is limited and easily affected by factors such as aggregate size.
Impact–echo method	The spectrum analysis of the P wave generated by hammering is performed to determine whether there is a void in the detection area.	The detection accuracy is high, and the depth and shape of the void can be well characterized.	Using contact point-by-point measurement, the detection speed is slow. It is difficult for operators to accurately control the angle and amplitude of the impact, resulting in errors.
Optical fiber sensor method	The scattered light is received and analyzed to obtain the attenuation waveform to determine the value, location and range of the void and crack.	The fiber optic sensor has good accuracy and can accurately monitor the degree of voids	The optical fiber sensor is expensive and needs to be embedded in concrete. The sensor is fragile, and the durability is poor.
Infrared thermography	The temperature field of the void area is different from the temperature field of the non-void area to distinguish the void.	Using non-contact remote detection, the detection range is large, and the detection speed is fast.	Only the surface void can be detected, and the detection accuracy is not high due to the limitation of weather and excitation mode.

2.3. Stress Detection

The bearing capacity of an arch bridge structure has always been a hot issue in engineering and academia. Both reinforced concrete structures and steel structures are important bearing components, and their stress state is an important index to measure the safety performance of an arch bridge [69]. At present, the detection methods of steel structure stress are mainly divided into the destructive detection method and the non-destructive detection method [70].

Destructive detection mainly refers to the stress release method. The principle is to partially destroy the area to be measured to release the stress in the area [71]. The strain of the region before and after the stress release is different due to the elastic deformation, so the stress magnitude and distribution of the region can be obtained by measuring the strain.

The hole-drilling method is a relatively mature stress detection method, which is widely used in engineering [72]. When using the drilling method to measure stress, it is necessary to paste the strain flowers at the position to be measured first. When the stress has been released, the strain gauge can measure the strain generated at the position, and the stress in the area to be measured is obtained by using the correlation formula. Ju et al. [73] measured the residual stress distribution on the surface of 304 stainless steel via the drilling method, and the measurement results were consistent with the finite element simulation results. The stress was the smallest in the middle and gradually increased to both sides in both vertical and horizontal directions. Bobzin et al. [74] measured the residual stress distribution along the depth direction of the aluminum-based coating of the steel structure by the drilling method and found that the heat treatment had a relaxation effect on the residual stress level of the coating. Mohamed et al. [75] measured the residual stress of stainless steel coatings with different thicknesses by the drilling method. It was found that the residual stress of the coating at all thicknesses was tensile stress, and the bonding layer did not seem to affect the stress field. The advantage of drilling method is that the technology is mature, and the principle and operation are simple, but its measurement accuracy is easily affected by external test conditions, such as strain gauge sticking manners, drilling depth and so on, which will have a great impact on the accuracy of the measurement results [76].

The basic principle of the ring core method is to fully release the residual stress by machining an annular groove on the surface of the rotor. By pasting a strain gauge on the surface of the rotor, the electrical signal is converted into a strain value by the instrument, and finally, the stress size and direction are calculated [77,78]. In the experimental test research, Li and Ren [79] used the ring core method to detect the micro-indentation residual stress on the surface of steel structure specimens and combined it with ISSR technology to realize the distribution identification of residual stress. Hu et al. [80] used the ring core method combined with the three-dimensional digital image method to detect the residual stress on the surface of the steel plate, and the results were in good agreement with the theoretical model. Dang et al. [81] used the ring core method to evaluate the residual stress state of the double-layer film and obtained a feasible semi-destructive method.

The grooving method is an improvement based on the ring core method. The purpose is to facilitate cutting, and the release of stress is generated by cutting a linear or square groove on the material. The grooving method is easy to operate in practical engineering, causes less damage to the steel structure and is suitable for low-level stress measurement [82]. In the experimental test, Zhu et al. [83] studied the high temperature release law of residual stress on the surface of alloy materials by using the grooving method combined with an FIB-EB double-beam system. Shomali et al. [84] used the grooving method to test the shear performance of reinforced RC beams and carried out numerical analysis to verify the accuracy of the method. Yan et al. [85] carried out the grooving method for reinforced concrete specimens and analyzed the effects of different initial stresses, grooving methods and grooving depths on stress release efficiency. The detection principles, advantages and shortcomings of the above methods are sorted out in Table 3.

Table 3. Comparison of destructive detection methods.

Detect Methods	Advantages	Shortcomings	Detection Principles
Hole-drilling method	The technology is mature, and the principle and operation are simple.	The measurement accuracy is easily affected by many external factors.	Correlation between stress and strain
Ring core method	The accuracy and sensitivity of measurement are higher than those of the drilling method.	The operation is complex, and the processing process easily affects the material and stress.	
Grooving method	It is easy to operate and causes less damage to the steel structure.	The vibration of the cutting machine will disturb the strain test.	

Nondestructive testing has the advantages of not destroying structural integrity and being able to measure the same position multiple times to reduce measurement errors [86]. At present, the commonly used non-destructive testing methods for the stress state of key structures of arch bridges mainly include the resistance strain gauge method, X-ray diffraction method, magnetic measurement method and ultrasonic method.

The resistance strain gauge method is to paste the resistance strain gauge on the surface of the measured component. The sensitive gate of the strain gauge will change the resistance of the strain gauge due to the small deformation of the surface after the component is stressed [87,88]. According to the resistance strain effect, the strain at the strain gauge can be obtained, and then according to the elastic relationship between stress and strain, the purpose of obtaining the stress value of the component can be achieved. This method is common in test and measurement work. Kovačič et al. [89] verified the accurate characteristics of this method for bridge load test and measurement via signal processing examples measured by strain gauges, induction sensors and total stations. Turan et al. [90] used strain gauge to detect the residual stress at the groove on the surface of steel and effectively analyzed different types of stress. Ma et al. [91] proposed a displacement estimation technique based on a finite impulse response filter by using a resistance strain gauge and realized the accurate monitoring for the structural state of long-span bridges by integrating strain and acceleration measurements. The advantage of the resistance strain gauge method is that the measurement accuracy and sensitivity are high, and the stability in the measurement process is good. However, only the strain of a point on the surface of the component in a single direction can be measured, and the stress distribution cannot be reflected.

The basic principle of the X-ray diffraction method is that the lattice strain of the material caused by the stress is consistent with the macroscopic strain obtained by the elastic theory [92]. X-rays are incident on the metal surface under stress, and the lattice strain is calculated according to the Bragg diffraction equation, and the stress can be further calculated [93]. Monin et al. [94] measured the residual stress distribution of the welded joint by X-ray diffraction method and compared it with the residual stress state obtained by finite element analysis. Epp et al. [95] measured the radial residual stress intensity and distribution inside the steel hole extrusion strengthening parts using the X-ray method and analyzed the phase change characteristics of the crystal structure. Based on the inherent mean value of stress provided by X-ray diffraction, Morin et al. [96] inverted the linear system constructed by the average data set, developed a spatial deconvolution technique to reconstruct the local stress field and obtained the spatial distribution of stress. The X-ray diffraction method has high measurement accuracy and accurate measurement results, but it has high requirements for the surface finish of the component, and the detection equipment is expensive and complicated. In addition, the penetration depth of X-rays can only reach tens of microns, and the stress inside the structure cannot be measured. Crucially, the X-ray diffraction method may be harmful to human health, and specialized protective equipment and inspection techniques are required.

The basic principle of the weak magnetic detection method is to determine the magnitude of the stress by the coupling relationship between the stress of the magnetic material and the change in its internal magnetic properties [97]. According to the magnetic properties of ferromagnetic materials in arch bridge structures, the internal stress state is speculated by collecting and analyzing the variation characteristics of the surface magnetic field of ferromagnetic materials, only relying on the excitation of the geomagnetic field [98]. This method was first proposed by Russian scholar Dubov et al. [99]. In recent years, the research on the detection of steel bar stress state in bridge engineering has gradually increased. Huang et al. [100] and Chen et al. [101] carried out the axial tensile test of Q235 steel and explored the relationship between the magnetic signal and the steel fatigue load and the feasibility of magnetic flux leakage detection to predict structural damage. Tong et al. [102,103] measured the magnetic field data of HRB400 steel bar at different stress stages and obtained the coupling effect of elastic load and plastic deformation on magnetic memory signal. Zhang et al. [104] carried out the tensile–tensile fatigue test on a defective steel plate. It was found that the weak magnetic field strength can accurately detect the fatigue area and component life, and the results showed high detection sensitivity. At the same time, the results showed that there are many factors affecting the detection of weak magnetic field signals, such as the initial residual state of the material, the external magnetic field and the ambient temperature [105–108]. Therefore, the quantitative evaluation of the self-stress state of ferromagnetic components needs to be further studied.

The detection principle of the ultrasonic method is based on the acoustic elasticity theory; i.e., the change in stress state in elastic components will cause the change in ultrasonic propagation velocity [109]. In the stress detection of a steel structure of an arch bridge, the application of the ultrasonic method is very extensive. The commonly used ultrasonic types include transverse waves, longitudinal waves, LCR waves, surface waves, etc. According to the different detection objects, the appropriate ultrasonic type and method can be selected for detection. Pilarski and Rose [110] effectively detected the surface defects of steel joints according to the characteristics of ultrasonic transverse waves. Chaki et al. [111] used the interaction of longitudinal waves and transverse waves to effectively detect the tightening degree of bolts under the internal stress of a steel structure, and the reliability reached 95%. Liu et al. [112] optimized the existing LCR wave detection method and developed a new surface coupling agent, which improved the accuracy of LCR waves for steel stress measurement. Eremeyev et al. [113] used surface wave detection technology to detect and analyze the flatness and stress state of a steel tube cylindrical surface. The detection principles, advantages and shortcomings of the above methods are sorted out in Table 4.

Table 4. Comparison of nondestructive detection methods.

Nondestructive Means	Advantages	Shortcomings	Detection Principles
Resistance strain gauge method	The measurement accuracy and sensitivity are high, and the stability of the measurement process is good.	The strain gauge must be closely attached to the surface of the component, which cannot reflect the stress distribution.	Resistance strain effects
X-ray diffraction method	The measurement precision is high, and the result is accurate.	The surface finish of the component is required to be high, and the internal stress cannot be measured.	Bragg diffraction equation
Weak magnetic detection method	Non-contact measurement can be realized, and the detection speed is fast.	It is susceptible to external influences, and the measurement accuracy is poor.	Magneto-mechanical coupled effect
Ultrasonic method	This method is fast, convenient and suitable for on-line detection.	The transducer and coupling agent are needed, and the detection accuracy and spatial resolution are limited by the size of the transducer.	Acoustoelastic effect

2.4. Corrosion Detection

The durability of reinforced concrete is a difficult problem faced by the whole world. Steel corrosion, chemical erosion, concrete carbonization, freeze–thaw damage, alkali aggregate reaction and so on will cause insufficient structural durability, but the main reason for accelerating concrete damage is steel corrosion [114,115]. The corrosion of steel bars reduces the cross-sectional area and the strength of steel bars and leads to cracks on the surface of concrete [116], which leads to the deterioration of the structural performance of arch bridges and reduces the structural bearing capacity of arch bridges. At present, there are three main non-destructive testing methods for steel corrosion, which are the analytical method, electrochemical method and physical method [117–119].

The analytical method is to detect the degree of concrete cracks, the thickness of the protective layer, the type of steel bar, the initial chloride ion content and the carbonization degree of the concrete structure [120]; collect a large amount of information; and analyze the information. Finally, the mathematical model of steel bar corrosion is established to achieve the purpose of predicting the degree of steel bar corrosion. The crack backstepping method after concrete cracking is the main method of analysis. Since the corrosion rate of steel bars in reinforced concrete exceeds a critical corrosion rate, it will lead to cracking on the surface of concrete. Therefore, the degree of corrosion of internal steel bars is inversely deduced by the information of crack width after concrete corrosion expansion and cracking. Alonso et al. [121] carried out the corrosion test of reinforced concrete and analyzed the cracking law of the concrete protective layer under various water–cement ratios and thickness–diameter ratios. Liang and Wang [122] analyzed the thick-walled cylinder corrosion model and verified the model's prediction of the cracking time of the concrete cover using the measured data.

The process of concrete cover cracking caused by steel bar corrosion is quite complicated. The research on the process of concrete corrosion damage at home and abroad is mainly focused on the pre-cracking. Based on experimental research and theoretical analysis, the calculation formula of the corrosion rate of a steel bar section during cracking is given directly or indirectly. The development process of rust expansion and its influencing factors are simulated and analyzed by the nonlinear finite element method, but the relationship between the width of a rust expansion crack and the corrosion rate of a steel bar is rarely given. Using the rapid corrosion test, the calculation formula of the corrosion rate of the steel bar section based on the width of the corrosion expansion crack is given directly or indirectly. Throughout some of the existing research results, the Table 5 lists several theoretical calculation models for the loss rate of steel corrosion W loss and crack width w after concrete surface cracking.

Table 5. Theoretical calculation model of steel corrosion loss rate and crack width after concrete surface cracking.

Theoretical Model	Computational Expressions	Main Technical Indexes
Di model [123]	$W_{loss\ rate} = \begin{cases} 507 \cdot e^{0.007c} \cdot f_{cu}{}^{-0.09} \cdot d^{-1.76} \\ (w \leq 0.2 \text{ mm}) \\ 332 \cdot e^{0.008c} \cdot f_{cu}{}^{-0.567} \cdot d^{1.108} \\ (0.2 \leq w \leq 0.4 \text{ mm}) \end{cases}$	c: concrete cover depth f_{cu}: concrete strength d: bar diameter w: crack width
Hui model [124]	Class I round rebar at the corner: $W_{loss\ rate} = (32.43 + 0.303 f_{cu} + 0.65c + 27.45w)/d$ Threaded Class II rebar at the corner: $W_{loss\ rate} = (34.486w + 0.789 f_{cu} - 1.763)/d$ Grade I round steel bars located at the stirrup position: $W_{loss\ rate} = (276w + 1.07 f_{cu} + 59.45)/d$	c: concrete cover depth f_{cu}: concrete strength d: bar diameter w: crack width
Niu model [125]	$W_{loss} = 1.173 \cdot P_{RH} \cdot D_0 \cdot (t - t_{cr})$	P_{RH}: correction coefficient of environmental humidity D_0: oxygen diffusion coefficient
Wang model [126]	$W_{loss\ rate} = 4[(0.2345w + 0.0175)/d - (0.2345w + 0.0175)^2/d^2]$	d: bar diameter w: crack width
"Durability evaluation standard of concrete structure" GB/T51355-2019 [127]	Corner round steel bar: $W_{loss\ rate} = \begin{cases} 0.07w + 0.012c/d + 0.00084 f_{cu} + 0.08 \\ (w \geq 0.33 \text{ mm}) \\ 0.35w + 0.012c/d + 0.00084 f_{cu} - 0.013 \\ (w < 0.33 \text{ mm}) \end{cases}$ Angular deformed steel bar: $W_{loss\ rate} = \begin{cases} 0.086w + 0.008c/d + 0.00055 f_{cu} + 0.015 \\ (w \geq 0.10 \text{ mm}) \\ 0.35w + 0.008c/d + 0.00055 f_{cu} - 0.013 \\ (w < 0.10 \text{ mm}) \end{cases}$	c: concrete cover depth f_{cu}: concrete strength d: bar diameter w: crack width
Mao model [128]	$W_{loss\ rate} = 2.9598w + 1.5647$	w: crack width
Andrade model [129]	$W_{loss\ rate} = 4[(0.222w - 0.011)/d - (0.222w + 0.011)^2/d^2]$	d: bar diameter w: crack width

The corrosion of steel bars in concrete is an electrochemical process, so the electrochemical detection method can be used as a powerful means to reflect the degree of corrosion of steel bars [130]. The half-cell potential (HCP) method is the most widely used detection method in the electrochemical method. When the steel bar is corroded in the concrete, the primary battery is formed; that is, the anode area and the cathode area are formed on the surface of the steel bar. Between these regions, a micro-current will be generated inside the concrete, and the electrical activity of the steel bar and the concrete can be considered as a half-weak battery pack, where the steel bar acts as an electrode, and the concrete acts as an electrolyte [131]. The measurement method is to connect the steel bar to one end of the potentiometer, while the other end is connected with a reference electrode. The reference electrode moves on the surface of the concrete, and the relative potential difference between the steel bar electrode and the reference electrode in different regions is measured. As shown in Figure 8, the equipotential line on the surface of the concrete is drawn, and the equipotential zone is determined. The most negative part is the corrosion zone. The evaluation criteria of steel corrosion are as follows: when the potential is larger than −200 mV, the probability of no corrosion is greater than 90%; when the potential is −200 mV~−350 mV, the degree of corrosion is uncertain; when the potential is less than −350 mV, the probability of corrosion is larger than 90%. In previous research, Amiri et al. [132] used the HCP method to monitor the bridge deck corrosion of reinforced concrete bridges, and the method showed good accuracy and effectiveness. Adriman et al. [133] diagnosed the area and size of steel corrosion in reinforced concrete by combining the field data and simulation calculation of semi-unit potential measurement, and the accuracy rate reached 95.65%. Almashakbeh et al. [134] used the HCP method to investigate the effect of corrosion grade on the flexural capacity of reinforced concrete members and found a positive correlation between the measured value of semi-element potential and the flexural capacity of RC beams. And the flow chart of steel corrosion detection by half-cell potential method is shown in Figure 9.

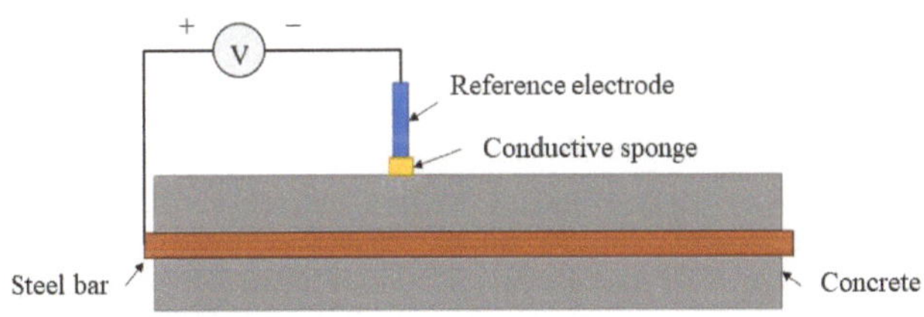

Figure 8. Half-cell potential method to detect steel corrosion schematic diagram.

Physical method tests are related to physical characterizations, including changes in electromagnetic, thermal, acoustic, and electrical resistance. And the changes are caused by steel corrosion [135]. The magnetic memory nondestructive testing method belongs to the magnetic field measurement method in the physical method, which is the research hotspot at the present stage. Under the combined action of a working load and the Earth's magnetic field, the ferromagnetic components of an arch bridge structure will undergo directional and irreversible reorientation of the magnetic domain structure with magnetostrictive properties in the corrosion concentration area and stress concentration area, resulting in the formation of a leakage magnetic field on the surface of the material [97]. By detecting the distribution of the leakage magnetic field on the surface of the ferromagnetic component, it can not only accurately judge the macroscopic defects that have been formed in the component but also detect the microscopic defects by other non-destructive testing methods such as ultrasound and eddy current. It can indicate the stress concentration position of the hidden defects and has the ability of early rapid warning [136]. The corrosion of steel bars in reinforced concrete is a prominent problem in the disease of arch bridges. Scholars have conducted a great amount of research on solving the problem of determining the corrosion area and corrosion degree of steel bars. Zhou et al. [137,138] first carried out the magnetic memory signal detection test of steel corrosion in the reinforced concrete of a bridge structure. It was found that the tangential magnetic induction intensity in the steel corrosion area had a large magnetic anomaly, and the magnetic flux leakage intensity curves of different lift-off heights had intersection points in the corrosion area, which provided a strong magnetic memory criterion for the detection of steel corrosion width. Xia et al. [139,140] optimized the magnetic dipole model, explored the change characteristics of the magnetic memory signal under the conditions of corrosion and wire breaking in an arch bridge suspender structure, and established the relationship between corrosion degree and magnetic memory signal. Su et al. [141,142] explored the mechanical properties of steel bars and steel plates under different degrees of corrosion and used magnetic memory testing to effectively analyze the influence of the corrosion of steel specimens on service life. The magnetic field distribution in the corrosion area of steel plate can be found in Figure 10.

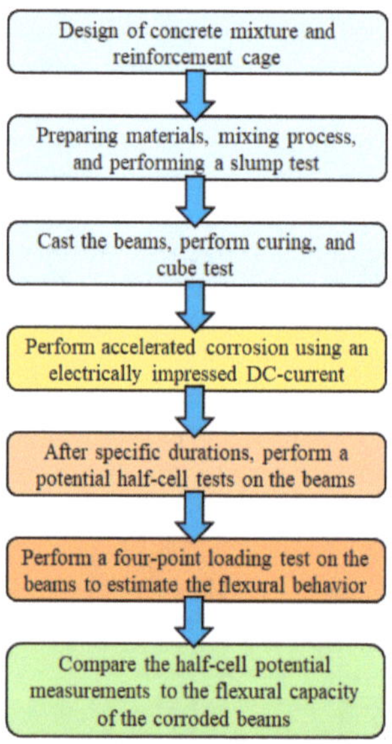

Figure 9. Flow chart of steel corrosion detection by half-cell potential method [134].

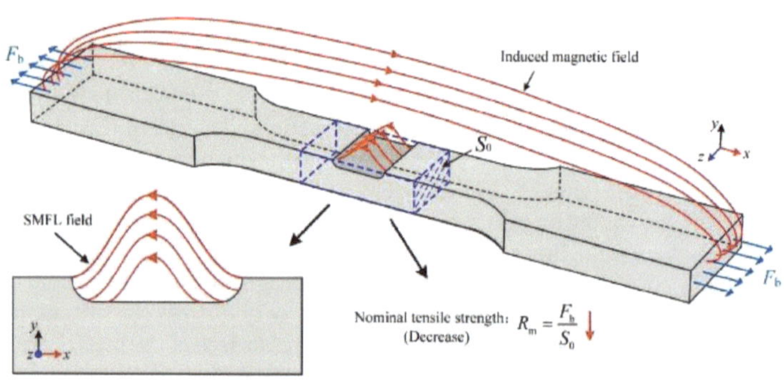

Figure 10. Magnetic field distribution in the corrosion area of steel plate [142].

3. Damage Identification for the Overall Performance of Arch Bridge

In addition to physical means (such as the ultrasonic method, magnetic memory method, etc.), the health monitoring system has also been widely used in the damage identification of arch bridges. Differently from the damage diagnosis method of arch bridges based on local detection data (such as acoustic wave, magnetic signal, etc.), the overall damage identification method is mainly based on the structural vibration response obtained from the monitoring system [143,144]. The characteristic indexes of structural damage are defined by means of modal analysis or mathematical statistics. The damage occurrence, damage location or quantitative damage degree of the structure can be identified via the change in damage index. According to different development histories and construction materials, arch bridges can be divided into the following four types: masonry arch bridge, steel arch bridge, RC arch bridge and CFST arch bridge. In order to show the differences between damage identification methods caused by different structural characteristics comprehensively, the damage identification methods of a masonry arch bridge, a steel arch bridge and an RC arch bridge have been reviewed in the existing literature [145]. On this basis, this paper analyzes the relevant literature on the damage identification of arch bridges in the past five years, and the introduction of the damage identification method of a CFST arch bridge is added. The characteristics of damage identification methods for different types of arch bridges are summarized to show the latest progress on arch bridge damage identification methods more comprehensively and systematically.

3.1. Masonry Arch Bridge Damage Identification Method

The masonry arch bridge has a long history because the materials used are mostly stone. The early structural forms are mainly circular arc and filled spandrel arch bridges, and the span is short. Stone arch bridges, including Zhaozhou Bridge, have become precious cultural heritage sites. However, due to the long-term erosion of the natural environment, a large number of structural diseases have occurred in masonry arch bridges, and the structural performance degradation is serious. In order to better maintain and retain such cultural heritage, scholars have tried various methods, and they carry out research on the damage identification of masonry arch bridges. Some scholars use the structural response data obtained by traditional sensing methods to realize damage identification by means of modal analysis or finite element model correction [146]. For example, Serra et al. [147] simulated the foundation displacement of a 1: 2 scale masonry arch bridge model in the laboratory and obtained the acceleration along the structure generated by hitting the bridge wall with a hammer. The coda wave interferometry was used to detect the change in the elastic properties of the medium. Furthermore, the evolution of the structural damage mechanism was analyzed. Emre [148] focused on the scale model of a single-span historical masonry arch bridge in the laboratory and used the enhanced frequency domain decomposition (EFDD) technology to identify the operating mode of the bridge. The ANSYS finite element of the structure was modified via response surface methodology to realize the approximate identification of the damage location. Civera et al. [149] proposed a fast relaxation vector fitting (FRVF) method for the fast, effective and reliable identification of structural systems based on vibration. The results show that this method can match other mature SI technologies well, such as the Eigensystem Realization Algorithm (ERA), and can be used for damage assessment. On this basis, they combined machine learning and operational modal analysis to propose a novel automatic OMA multi-stage clustering algorithm and applied it to the damage detection and damage degree assessment of masonry arch bridges [150]. Conde et al. [151] obtained the most probable distribution of the finite element model parameters by using a Bayesian method and the Markov chain Monte Carlo (MCMC) method based on the comparison between the finite element calculation results and the actual measured geometric data so as to reproduce the existing damage mode with the highest accuracy using the numerical model. The results showed that the proposed method could reasonably infer the original geometric structure of the bridge and the possible damage loading scene so as to obtain the crack mode, which was almost the same as the current damage state. Borlenghi et al. [152] analyzed the damage and operation mode of a long-standing masonry arch bridge, Olla bridge, via visual inspection and a field test, and modified the finite element model of the structure based on the measured data. Alexakis et al. [153] used acoustic emission sensors to study the deterioration of masonry structures under gradually increasing cyclic loading. The results showed that the acoustic emission results had a significant correlation trend with different damage stages. Based on this, a real-time warning system for early structural damage can be developed.

Other scholars use advanced measurement methods to obtain the key geometric parameters of masonry arch bridges in service and realize structure damage identification by modifying the finite element model. For example, Pepi et al. [154] combined advanced measurement procedures, such as close-range photogrammetry based on drones and environmental vibration testing, to propose a finite element model correction method for severely damaged historical masonry structures and to guide the structure damage assessment. Batar et al. [155] used terrestrial laser scanning (TLS) technology to determine the geometric shape of the bridge with high precision. Based on the obtained point cloud data, a three-dimensional CAD-based structural solid model was established. The CDP material model was used to describe the inelastic behavior of the homogeneous structure, and then the structural behavior was analyzed with high precision and reliability. Deng et al. [156] proposed an automatic identification method for the arch axis of a stone arch bridge based on 3D laser scanning and applied it to Lugou Bridge with a history of 800 years. The results show that the relative deformation of the upstream and downstream arch rings of the bridge and the relative deviation of the arch rise and span of the belly arches at symmetrical positions are significant.

In addition, relevant scholars have paid attention to the identification and evaluation of seismic damage of arch bridges. For example, Jara et al. [157] established a numerical model based on the disease data of bridges. By revealing the failure mechanism of different types of structural members, the effects of epicentral distance and seismic fault on the seismic vulnerability of bridges were evaluated. By performing a series of environmental vibration tests, the dynamic characteristics of five existing historical bridges were determined, which could be used to construct the vulnerability curve and determine the numerical model of bridge seismic vulnerability. In order to investigate the seismic performance of damaged masonry arch bridges, Kaya et al. [158] simulated different damage states using a shaking table test. Then, based on the environmental vibration test, the structural modal parameters under undamaged and damaged conditions were obtained. The experimental results were in good agreement with the numerical simulation results, which provided a reference for the damage assessment of masonry arch bridges after earthquakes. Considering the uncertainty of material properties and seismic input, Gönen et al. [159,160] used a 3D finite element model for probabilistic nonlinear analysis, combined with a Monte Carlo simulation and first-order reliability method (FORM), to propose a probabilistic evaluation method for the seismic performance of masonry arch bridges.

3.2. *Damage Identification Method of Steel Arch Bridge*

The steel arch bridge is sensitive to the stiffness change caused by damage, which will lead to changes in structural modal characteristics. Therefore, most of the damage identification methods of steel arch bridges are based on structural modal characteristics. Due to the certain impact of environmental factors on the modal characteristics of structures, the existing literature has studied variations in the modal characteristics of steel arch bridges under environmental influences, aiming to further improve the accuracy of damage identification by eliminating the impact of environmental factors. For example, Mu et al. [161] used Bayesian network to model the relationship between the long-term monitoring data (temperature, humidity, wind speed and traffic volume) and the modal frequency of Xinguang Bridge (a 782 m steel arch bridge). The results showed that the selected network structure adequately captured the relationship between modal frequency and multiple environmental factors. Anastasopoulos et al. [162,163] used fiber Bragg grating strain sensors to continuously monitor a steel arch bridge for two years, automatically identifying the fundamental frequency and vibration modes of the structure from the measured strain time history. The results showed that the temperature had a great influence on the fundamental frequency but only had a slight effect on some high-order modes. In addition, the finite element simulation confirmed that the local damage had a significant effect on the modal characteristics of the structure. Considering the bearing characteristics and spatial temperature distribution, Zhu et al. [164] discussed the relationship between the temperature change and temperature response of a long-span steel truss arch bridge with spherical bearing by using elastic beam theory and verified it using field monitoring data. The results showed that this relationship was determined by the structural parameters, especially the section characteristics and bearing characteristics of the bridge. Based on this, the criterion of bridge state identification could be established. Combined with correlation analysis, numerical simulation and neural network technology, Teng et al. [165] studied the mechanism of the influence of temperature on the structural frequency. The results showed that the modal frequency of the arch was mainly affected by the change in elastic modulus. Sunca et al. [166] used EFDD and the stochastic subspace identification method to identify the modal parameters of an Eynel steel arch bridge in the frequency domain and time domain, respectively. The results showed that the change in natural frequency was very limited, and there was no change between the vibration modes, which was in line with the actual situation of the bridge.

Some scholars have identified the occurrence of structural damage based on changes in the natural frequency, vibration modes or modal damping of steel arch bridges. For example, Chiaia et al. [167] used modal analysis technology and a data-driven stochastic subspace identification algorithm to extract the dynamic characteristics of the structure, including natural frequency, damping ratio and vibration modes, and then identify the damage of the structure. Dammika et al. [168] believed that structural damage would lead to changes in structural stiffness and damping characteristics, thereby changing the natural frequency, vibration mode and modal damping ratio of the structure. The energy-based damping model was introduced to estimate the damping parameters such as the equivalent loss factor of the structural members of the steel arch bridge, and the modal damping ratio of the bridge was analyzed by using the damping parameters. The results show that the method could identify the damping source of the steel bridge and its contribution to the modal damping ratio. Tran and Ozer [169] identified modal parameters based on the information matrix system implementation method. Then, the damage of the structure was identified from the stable modal poles collected from the stability diagram via the anomaly detection algorithm based on Gaussian distribution. Whelan et al. [170] monitored the environmental vibration of a tied arch bridge using a wireless sensor network and then estimated its modal parameters. In the parallel computing cluster, the integer-constrained genetic algorithm was used to optimize the objective function globally to update the cross-scale finite element model so as to evaluate the damage state of the structure.

Furthermore, some scholars define the damage sensitivity index to characterize the damage location or degree of the structure according to the structural modal parameters so as to realize the damage location and damage degree quantification of the steel arch bridge. For example, Jayasundara et al. [171] used the dual-criteria method based on modal flexibility and modal flexibility change to identify and locate the damage of steel arch bridges. The results showed that under certain damage conditions, the damage of arch ribs and longitudinal columns can be detected and located accurately via only a small number of early vibration modes. On this basis, Jayasundara et al. [172,173] proposed a vibration-based damage location and quantification method for arch bridge ribs. Based on the change of damage mode of frequency response functions (FRFs), the location and severity of structural damage are predicted by artificial neural networks (ANNs). Based on the nonlinear and non-stationary dynamic response of bridges, Delgadillo et al. [174] used Hilbert–Huang transform and instantaneous phase difference to construct damage sensitive features. Furthermore, the damage under traffic transient vibration was detected, located and quantified. The artificial damage of a real steel arch bridge was identified. The results showed that the proposed method had good robustness and sensitivity. Based on the seismic acceleration response signal, Kordestani et al. [175] calculated the energy level according to the Arias strength and normalized it to locate the damaged suspender.

Using the three-dimensional numerical simulation of the tied arch bridge under 16 different ground motions, the effectiveness of the method was verified.

3.3. Damage Identification Method of RC Arch Bridge

Similarly to the damage identification method of a steel arch bridge, the damage identification method of an RC arch bridge is mostly based on the modal parameters of the structure. For example, Magalhães and Cunha [176] proposed three effective output-only modal identification algorithms. By processing the test data of a concrete arch bridge with a span of 280 m, the steps of these identification procedures were clarified. At the same time, the operational modal analysis under the background of structural health monitoring was introduced. Furthermore, based on the environmental excitation response of an RC arch bridge, Magalhães et al. [177] used the automatic operation modal analysis algorithm to identify the modal parameters of the bridge online. By reducing the influence of environmental and operational factors on the natural frequency of the bridge. The evolution law of the modal characteristics of the bridge over time in the past two years is obtained, and the damage identification is carried out by using the control chart. The results showed that the proposed method could detect the real damage scene related to the frequency shift of about 0.2%. Ren et al. [178,179] proposed a new damage feature of a vibration system by using the system matrix based on covariance-driven stochastic subspace identification. In order to reduce the influence of modeling error, noise and environmental changes on the response of the measurement structure. The statistical pattern recognition paradigm was added to this method. By defining the damage index based on statistics, the Mahalanobis and Euclidean distance judgment functions of the damage feature vector were used. And the method was verified by the reinforced concrete beam tested in the laboratory and the full-scale arch bridge tested in the field. Based on multivariate statistical techniques, Comanducci et al. [180] processed the monitoring data of arch bridges and defined damage-sensitive features under changing environments and operating conditions. It was applied to the time history of the modal frequency of the bridge, and the control chart was used to evaluate its ability to reveal structural damage of varying severity. Ferrari et al. [181] identified the structural modal characteristics based on the collected experimental vibration data. Combined with Latin hypercube sampling, a self-fulfilling global optimization method for model updating was proposed. It had been successfully applied to a century-old reinforced concrete parabolic arch bridge with historical significance. Based on the Fourier amplitude spectrum (FAS) of acceleration response, Duan et al. [182] used a CNN to identify the damage of suspenders of tied arch bridges. The results showed that the CNN had good robustness under various observation noise levels and wind speeds.

In addition, scholars have also explored some novel damage identification methods for RC arch bridges, including data-driven methods and methods based on radar interferometry. For example, Zhou et al. [183] proposed a damage identification method for RC arch bridges based on monitoring data. The acceleration data were preprocessed by a Kalman filter and an autoregressive moving average (ARMA) model. The damage index was constructed, and the identification accuracy was further improved by the generalized autoregressive conditional heteroscedasticity (GARCH) model. Qin et al. [184] used a modern differential interferometric synthetic aperture radar (DInSAR) to improve the reliability of PT identification, thermal expansion separation and the structural risk assessment of arch bridges. Taking Rainbow Bridge and Lupu Bridge as examples, the effectiveness of the method was verified. And the experimental results were verified by horizontal benchmark verification, cross-sensor comparison and structural reliability evaluation.

3.4. Damage Identification Method of CFST Arch Bridge

The damage identification method of a CFST arch is different from that of other types of arch bridges. Due to the particularity of CFST arch bridge structure and material, its damage identification is mainly based on static means. For example, Wang et al. [185]

established the relationship between the static deflection change of the suspender and the cable force. Based on the static deflection difference, the damage identification index of the suspender was proposed. The numerical simulation and experimental results show that the method can accurately locate the damage location and give the relative damage degree. Li et al. [186] combined monitoring data and detection data to modify the finite element model from the component and global, respectively. Thus, several common defects of a CFST arch bridge were investigated and updated. Fu and Jiang [187] proposed a new intelligent damage detection data fusion system by combining a probabilistic neural network (PNN), data fusion technology and correlation fractal dimension (CFD). Single and multiple damage modes were identified by the numerical simulation of a two-span CFST arch bridge. The results showed that the hybrid data fusion system had good damage identification performance, anti-noise ability and robustness. Gui et al. [188] analyzed the ultimate bearing capacity modeling, limit threshold setting and limit state prediction of concrete-filled steel tubular members. A system maintenance decision model was proposed, which combined the relative monitoring state, the relative final state of numerical analysis and the relative residual life of degraded components. The results showed that these technologies had great potential for the life cycle performance evaluation of the existing concrete bridge structure health monitoring system.

In addition, scholars have also made some explorations on the damage identification methods of CFST arch bridges using modal characteristics. For example, Yan and Ren [189] took the idea that the power spectral density transmittance (PSDT) was independent of the external excitation and the transmitted output of the system poles. It was proposed to extract the working modal parameters of the structure based on PSDT, which was verified and analyzed by a half-arch bridge of CFST. The recognition results were compared with those of the peak picking method, random subspace identification and finite element analysis. The results showed that the working modal parameters identified by this method were in good agreement with other methods and had the potential to be applied to structural damage identification. Xiong et al. [190] used the global navigation satellite real-time kinematics system to monitor the vibration response of long-span CFST arch bridges under environmental excitation. The structural modal parameters were calculated by fast Fourier transform and a random decrement technique.

Based on the above study, the damage indexes of damage identification for the overall performance are summarized in Table 6.

Table 6. Damage indexes of damage identification for the overall performance.

Damage Index	Computational Expressions	Main Parameter
Modal flexibility damage index [171]	$MFDI_{ij} = \frac{\left[\sum_{i=1}^{m} \frac{1}{\omega_i^2}\phi_i\phi_i^T\right]_D - \left[\sum_{i=1}^{m} \frac{1}{\omega_i^2}\phi_i\phi_i^T\right]_H}{\left[\sum_{i=1}^{m} \frac{1}{\omega_i^2}\phi_i\phi_i^T\right]_H}$	m: the mode number ω_i: the natural frequency of the structure at mode i ϕ_i: ith mode shape vector D: the damage conditions H: the healthy conditions
Modal strain energy damage index [172]	$\beta_{ij} = \frac{k_j}{k_j^*} = \frac{\left[(\phi''^{*}_{i})^2 + \Sigma(\phi''^{*}_{i})^2\right]\left[\Sigma(\phi''_{i})^2\right]}{\left[(\phi''_{i})^2 + \Sigma(\phi''_{i})^2\right]\left[\Sigma(\phi''^{*}_{i})^2\right]}$	β_{ij}: Modal strain energy damage index for jth member at ith mode k_j: the bending stiffness of the beam $\phi_i(x)$: the mode shape of ith modal vector
Normalized arias intensity [175]	$I_A = \frac{\pi}{2g}\int_0^T a^2(t)\mathrm{d}t$ $\beta_i = \frac{I_{A-IN_i}}{\overline{I_{A-IN}}}$ $NI_{A_i} = \frac{I_{A-DM_i}}{\overline{I_{A-DM}}} \times \frac{100}{\beta_i} = \frac{I_{A-DM_i}}{\overline{I_{A-DM}}} \times \frac{\overline{I_{A-IN}}}{I_{A-IN_i}} \times 100$	I_A: arias intensity T: duration of acceleration a: acceleration data β: normalizing factor i: the node counter NI_A: normalized arias intensity I_{A-IN}: arias intensity for intact condition I_{A-DM}: arias intensity for damaged condition

Table 6. *Cont.*

Damage Index	Computational Expressions	Main Parameter
Statistics-based damage index [178]	$NI_{i,\gamma}^{E} = \frac{\overline{D}_{i,T}^{E}}{\overline{D}_{i,J}^{E}}$ or $NI_{i,\gamma}^{M} = \frac{\overline{D}_{i,T}^{M}}{\overline{D}_{i,J}^{M}}$	$\overline{D}_{i,T}^{E}$: The mean value of Euclidean of the state to be evaluated $\overline{D}_{i,J}^{E}$: The mean value of Euclidean of baseline state
Deflection-based identification index [185]	$DI = \Delta w(i+2) - 2\Delta w(i+1) + 2\Delta w(i-1) - \Delta w(i-2)$	$\Delta w(i+2)$, $\Delta w(i+1)$, $\Delta w(i-1)$ and $\Delta w(i-2)$ are the static deflection changes at nodes i + 2, i + 1, i − 1 and i − 2 before and after hanger damage, respectively
Transmissibility-based damage index [191]	$DI(T_{ud}, T_d) = \frac{\lvert\lvert\lvert\{T_{ud}\}\rvert - \lvert\{T_d\}\rvert\rvert\rvert}{\lvert\lvert\lvert\{T_{ud}\}\rvert + \lvert\{T_d\}\rvert\rvert\rvert}$	T_{ud}: transmissibility matrices from the slab response to the girder response in undamaged states T_d: transmissibility matrices from the slab response to the girder response in damaged states

4. Conclusions

This paper reviews the research progress of damage monitoring and identification methods for arch bridges, including damage monitoring for local diseases of arch bridges and damage identification for the overall performance of arch bridges. The main conclusions are as follows:

(1) In the monitoring of local disease damage, the principles of various methods and their advantages and shortcomings are summarized. According to the different damage parts, the different damage types (cracks or aging, corrosion, fractures or cracks, and anchorage efficiency decrease) and the commonly used methods (the machine vision method, magnetic flux leakage detection method, acoustic emission method, magneto strictive guided wave detection method, and piezoelectric sensor method) are summarized from three aspects of an HDPE sheath, the internal steel wire of the cable body and the anchorage system. At present, there are four commonly used methods for void monitoring: the ultrasonic method, impact–echo method, optical fiber sensor method and infrared thermal imaging method. The stress detection of the main bearing components of the arch bridge is divided into two categories: destructive detection and non-destructive testing. The destructive detection mainly includes the drilling method, ring core method and segmentation grooving method. The non-destructive testing mainly includes the resistance strain gauge method, X-ray diffraction method, weak magnetic detection method and ultrasonic method. There are three main non-destructive testing methods for steel corrosion detection, i.e., the analysis method, electrochemical method and physical method (electromagnetic method, heat conduction method, acoustic emission method and resistance method). In the Introduction and summary, it is found that non-destructive testing is the research focus of the local disease damage monitoring of arch bridges. Among them, the physical method can include most of the detection methods, and the other methods can be divided into machine vision methods and sensor methods.

(2) In the damage identification of the overall performance, the modal-based damage identification is still the main research direction. In addition to the static damage identification of the CFST arch bridge, the other three bridge types are mainly based on modal-based damage identification. In the masonry arch bridge, the traditional sensing method and the advanced measurement method are taken into account. In addition to the modal analysis, the finite element model correction method is also introduced to identify the damage. At the same time, the identification and evaluation of seismic damage have also attracted the attention of scholars. Most of the research on steel arch bridges is based on modal characteristics. In order to achieve high-accuracy identification, the variation law of the modal characteristics of steel arch bridges under environmental influence is studied to eliminate the influence of environmental factors. The damage sensitivity index is also proposed to realize the damage location and damage degree quantification of steel arch bridges. In addition to the modal research of RC arch bridges, scholars have also explored some novel methods, including data-driven methods and methods based on radar

interferometry. Although CFST arch bridges are mainly based on static means, scholars have also made some explorations on the modal-based damage identification method.

(3) It can be found that the applicability and effectiveness of different available techniques depend on the damage type, structural configuration and available data. However, a universal method that can identify all damage types of different structures remains to be developed. In addition, although some achievements have been made in damage monitoring and identification, it is still necessary to develop more accurate and reliable damage identification methods to realize the effective combination of local disease damage monitoring and overall performance damage identification. In particular, the research on the damage location method of the overall performance is imminent. Once the two-step strategy of "overall structural damage location" and "local disease damage monitoring" is realized, the efficiency of damage identification will be greatly improved, and the structure can also be accurately and timely repaired and maintained, thereby prolonging the service life of the structure.

Author Contributions: Conceptualization, J.Y. and J.X.; methodology, J.Y. and L.H.; investigation, K.T., Q.T. and H.L.; writing—original draft preparation, J.Y. and L.H.; writing—review and editing, J.Y. and J.X.; visualization, L.H., K.T. and H.C.; supervision, J.X.; funding acquisition, J.X. All authors have read and agreed to the published version of the manuscript.

Funding: The supports by National Natural Science Foundation of China (Grant No. 52278292), Chongqing Outstanding Youth Science Foundation (Grant No. CSTB2023NSCQ-JQX0029), Chongqing Science and Technology Project (CSTB2022TIAD-KPX0205), Chongqing Transportation Science and Technology Project (Grant No. 2022-01), Science and Technology Project of Guizhou Department of Transportation (Grant No. 2023-122-001), China Postdoctoral Science Foundation (Grant No. 2023M730431), and Special Funding of Chongqing Postdoctoral Research Project (Grant No. 2022CQBSHTB2053) are greatly acknowledged.

Data Availability Statement: Some or all data, models or code that support the findings of this study are available from the corresponding author upon reasonable request.

Conflicts of Interest: The authors declare no conflict of interest.

References

1. Zhang, H.; Li, H.; Zhou, J.; Tong, K.; Xia, R. A multi-dimensional evaluation of wire breakage in bridge cable based on self-magnetic flux leakage signals. *J. Magn. Magn. Mater.* **2023**, *566*, 170321. [CrossRef]
2. Xin, J.; Zhou, J.; Zhou, F.; Yang, S.; Zhou, Y. Bearing Capacity Model of Corroded RC Eccentric Compression Columns Based on Hermite Interpolation and Fourier Fitting. *Appl. Sci.* **2018**, *9*, 24. [CrossRef]
3. Chen, Z.; Xu, Y.; Hua, J.; Wang, X.; Huang, L.; Zhou, X. Mechanical Properties and Shrinkage Behavior of Concrete-Containing Graphene-Oxide Nanosheets. *Materials* **2020**, *13*, 590. [CrossRef] [PubMed]
4. Yang, J.; Chen, R.; Zhang, Z.; Zou, Y.; Zhou, J.; Xia, J. Experimental study on the ultimate bearing capacity of damaged rc arches strengthened with ultra-high performance concrete. *Eng. Struct.* **2023**, *279*, 115611. [CrossRef]
5. Hou, R.; Xia, Y. Review on the new development of vibration-based damage identification for civil engineering structures: 2010–2019. *J. Sound Vib.* **2021**, *491*, 115741. [CrossRef]
6. Wang, C.; Ansari, F.; Wu, B.; Li, S.; Morgese, M.; Zhou, J. LSTM approach for condition assessment of suspension bridges based on time-series deflection and temperature data. *Adv. Struct. Eng.* **2022**, *25*, 3450–3463. [CrossRef]
7. Tang, Q.; Xin, J.; Jiang, Y.; Zhou, J.; Li, S.; Fu, L. Fast identification of random loads using the transmissibility of power spectral density and improved adaptive multiplicative regularization. *J. Sound Vib.* **2022**, *534*, 117033. [CrossRef]
8. Zeng, Y.; He, H.; Qu, Y.; Sun, X.; Tan, H.; Zhou, J. Numerical Simulation of Fatigue Cracking of Diaphragm Notch in Orthotropic Steel Deck Model. *Materials* **2023**, *16*, 467. [CrossRef]
9. Xin, J.; Zhou, C.; Jiang, Y.; Tang, Q.; Yang, X.; Zhou, J. A Signal Recovery Method for Bridge Monitoring System Using TVFEMD and Encoder-Decoder Aided LSTM. *Measurement* **2023**, *214*, 112797. [CrossRef]
10. Xin, J.; Jiang, Y.; Zhou, J.; Peng, L.; Liu, S.; Tang, Q. Bridge Deformation prediction based on SHM data using improved VMD and conditional KDE. *Eng. Struct.* **2022**, *261*, 114285. [CrossRef]
11. Zeng, Y.; Qiu, Z.; Yang, C.; Haozheng, S.; Xiang, Z.; Zhou, J. Fatigue Experimental Study on Full-Scale Large Sectional Model of Orthotropic Steel Deck of Urban Rail Bridge. *Adv. Mech. Eng.* **2023**, *15*, 168781322311552. [CrossRef]
12. Li, S.; Xin, J.; Jiang, Y.; Wang, C.; Zhou, J.; Yang, X. Temperature-induced deflection separation based on bridge deflection data using the TVFEMD-PE-KLD method. *J. Civ. Struct. Health* **2023**, *13*, 781–797. [CrossRef]
13. Zheng, J.; Wang, J. Concrete-filled steel tube arch bridges in china. *Engineering* **2018**, *4*, 143–155. [CrossRef]

14. Zhang, L.; Qiu, G.; Chen, Z. Structural health monitoring methods of cables in cable-stayed bridge: A review. *Measurement* **2021**, *168*, 108343. [CrossRef]
15. Jafari, M.; Hou, F.; Abdelkefi, A. Wind-induced vibration of structural cables. *Nonlinear Dyn.* **2020**, *100*, 351–421. [CrossRef]
16. Abdullah, A.B.M.; Rice, J.A.; Hamilton, H.R. Wire breakage detection using relative strain variation in unbonded posttensioning anchors. *J. Bridge Eng.* **2015**, *20*, 04014056. [CrossRef]
17. Rizzo, P.; Enshaeian, A. Challenges in bridge health monitoring: A review. *Sensors* **2021**, *21*, 4336. [CrossRef]
18. Zhu, Z.; Wang, D.; Liu, C.; Wang, B. Research on wire-broken monitoring of bridge cable based on acoustic emission technique. In Proceedings of the 4th International Conference on Civil Engineering and Materials Science (ICCEMS 2019), Bangkok, Thailand, 17–19 May 2019; IOP Publishing Ltd.: Bristol, UK, 2019; Volume 652, p. 012065.
19. Djeddi, L.; Khelif, R.; Benmedakhene, S.; Favergeon, J. Reliability of acoustic emission as a technique to detect corrosion and stress corrosion cracking on prestressing steel strands. *Int. J. Electrochem. Sci.* **2013**, *8*, 8356–8370. [CrossRef]
20. Nair, A.; Cai, C.S. Acoustic emission monitoring of bridges: Review and case studies. *Eng. Struct.* **2010**, *32*, 1704–1714. [CrossRef]
21. Guohao, S.; Zhengxing, W.; Delian, K.; Mingqing, B.; Yang, Z.; Fengjing, X.; Jiazhen, P. Analysis on acoustic signal characteristics of the fatigue cracks in anchor structure of cable-stayed bridge tower. *Russ. J. Nondestr. Test* **2012**, *48*, 718–730. [CrossRef]
22. Kharrat, M.; Gaillet, L. Non-destructive evaluation of anchorage zones by ultrasonics techniques. *Ultrasonics* **2015**, *61*, 52–61. [CrossRef] [PubMed]
23. Xu, J.; Wu, X.; Sun, P. Detecting broken-wire flaws at multiple locations in the same wire of prestressing strands using guided waves. *Ultrasonics* **2013**, *53*, 150–156. [CrossRef]
24. Chen, X.; Zhu, J.; Lin, Y. Attenuation characteristics of low frequency longitudinal guided waves generated by magnetostrictive transducers in bridge cables. *Mech. Syst. Signal Proc.* **2022**, *164*, 108296. [CrossRef]
25. Zhang, Z.; Xu, J.; Guo, Z. A longitudinal mode guided wave transducer with the ring permanent biased magnet based on magnetostrictive effect for large diameter bridge cables. *IEEE Sens. J.* **2021**, *21*, 18544–18553. [CrossRef]
26. Park, S.; Kim, J.W.; Lee, C.; Lee, J.J. Magnetic flux leakage sensing-based steel cable NDE technique. *Shock Vib.* **2014**, *2014*, 929341. [CrossRef]
27. Zhang, Q.; Xin, R. The Defect-length effect in corrosion detection with magnetic method for bridge cables. *Front. Struct. Civ. Eng.* **2018**, *12*, 662–671. [CrossRef]
28. Ni, Y.; Zhang, Q.; Xin, R. Magnetic flux detection and identification of bridge cable metal area loss damage. *Measurement* **2021**, *167*, 108443. [CrossRef]
29. Xu, F.; Wang, X.; Wu, H. Inspection method of cable-stayed bridge using magnetic flux leakage detection: Principle, sensor design, and signal processing. *J. Mech. Sci. Technol.* **2012**, *26*, 661–669. [CrossRef]
30. Zhang, D.; Zhang, E.; Pan, S. A new signal processing method for the nondestructive testing of a steel wire rope using a small device. *NDT E Int.* **2020**, *114*, 102299. [CrossRef]
31. Xu, K.; Qiu, X.; Tian, X. Theoretical Investigation of Metal Magnetic Memory Testing Technique for Detection of Magnetic Flux Leakage Signals from Buried Defect. *Nondestruct. Test. Eval.* **2018**, *33*, 45–55. [CrossRef]
32. Qu, Y.; Zhou, J.; Liu, R.; Liao, L.; Zhao, Q. Research on the Detection of the Broken Wire Damage of a Cable in the Circumferential Directions Based on Self-Magnetic Flux Leakage. *KSCE J. Civ. Eng.* **2021**, *25*, 879–890. [CrossRef]
33. Xia, R.; Zhang, H.; Zhou, J.; Liao, L.; Zhang, Z.; Yang, F. Probability Evaluation Method of Cable Corrosion Degree Based on Self-Magnetic Flux Leakage. *J. Magn. Magn. Mater.* **2021**, *522*, 167544. [CrossRef]
34. Xia, R.; Zhang, H.; Zhou, J.; Liao, L.; Yang, W.; Li, Y. Corrosion Non-Destructive Testing of Loaded Steel Strand Based on Self-Magnetic Flux Leakage Effect. *Nondestruct. Test. Eval.* **2022**, *37*, 56–70. [CrossRef]
35. Nguyen, K.D.; Kim, J.T.; Park, Y.H. Multiscale Structural Health Monitoring of Cable-Anchorage System Using Piezoelectric PZT Sensors. *Int. J. Distrib. Sens. Netw.* **2013**, *9*, 254785. [CrossRef]
36. Zhang, X.; Zhang, L.; Liu, L.; Huo, L. Prestress Monitoring of a Steel Strand in an Anchorage Connection Using Piezoceramic Transducers and Time Reversal Method. *Sensors* **2018**, *18*, 4018. [CrossRef]
37. Ho, H.N.; Kim, K.D.; Park, Y.S.; Lee, J.J. An Efficient Image-Based Damage Detection for Cable Surface in Cable-Stayed Bridges. *NDT E Int.* **2013**, *58*, 18–23. [CrossRef]
38. Hou, S.; Dong, B.; Wang, H.; Wu, G. Inspection of Surface Defects on Stay Cables Using a Robot and Transfer Learning. *Autom. Constr.* **2020**, *119*, 103382. [CrossRef]
39. Li, X.; Gao, C.; Guo, Y.; He, F.; Shao, Y. Cable Surface Damage Detection in Cable-Stayed Bridges Using Optical Techniques and Image Mosaicking. *Opt. Laser Technol.* **2019**, *110*, 36–43. [CrossRef]
40. Han, L.; Li, W.; Bjorhovde, R. Developments and advanced applications of concrete-filled steel tubular (CFST) structures: Members. *J. Constr. Steel Res.* **2014**, *100*, 211–228. [CrossRef]
41. Chen, B.; Wang, T. Overview of concrete filled steel tube arch bridges in china. *Pract. Period. Struct. Des. Constr.* **2009**, *14*, 70–80. [CrossRef]
42. Iyer, S.; Sinha, S.K.; Tittmann, B.R.; Pedrick, M.K. Ultrasonic signal processing methods for detection of defects in concrete pipes. *Autom. Constr.* **2012**, *22*, 135–148. [CrossRef]
43. Hola, J.; Sadowski, L.; Schabowicz, K. Nondestructive identification of delaminations in concrete floor toppings with acoustic methods. *Autom. Constr.* **2011**, *20*, 799–807. [CrossRef]

44. Ding, R. Research on fiber sensing of health monitoring for steel tube-confined concrete arch bridge. *Chin. Civil. Eng. J.* **2005**, *38*, 69–147. (In Chinese)
45. Yan, Q.; Cheng, X.; Ding, R. Application of distributed optical technique in sensing interface disengaging and cracks of steel tube-confined concrete. In Proceedings of the 2011 International Conference on Electric Technology and Civil Engineering (ICETCE), Lushan, China, 22–24 April 2011; pp. 122–126.
46. Clark, M.R.; McCann, D.M.; Forde, M.C. Application of infrared thermography to the non-destructive testing of concrete and masonry bridges. *NDT E. Int.* **2003**, *36*, 265–275. [CrossRef]
47. Zhang, H.; Yu, Q.; Lu, Y. Application study of quality testing of arch-rib concrete of concrete filled steel tube arch bridge by ultrasonic transmission method. *Chin. Civil Eng. J.* **2004**, *37*, 50–53. (In Chinese)
48. Liu, Y.; Zhang, Y.; Zou, Z. Study of ultrasonic detection and evaluation for the deficiency of concrete-filled steel tube arch bridge. *J. Beijing Jiaotong Univ.* **2004**, *28*, 54–57. (In Chinese)
49. Dong, W.; Wu, Z.; Zhou, X.; Tan, Y. Experimental studies on void detection in concrete-filled steel tubes using ultrasound. *Constr. Build. Mater.* **2016**, *128*, 154–162. [CrossRef]
50. Callejas, A.; Palma, R.; Hernández-Figueirido, D.; Rus, G. Damage detection using ultrasonic techniques in concrete-filled steel tubes (CFSTs) columns. *Sensors* **2022**, *22*, 4400. [CrossRef] [PubMed]
51. Liu, H.; Xia, H.; Zhuang, M.; Long, Z.; Liu, C.; Cui, J.; Xu, B.; Hu, Q.; Liu, Q.H. Reverse Time Migration of Acoustic Waves for Imaging Based Defects Detection for Concrete and CFST Structures. *Mech. Syst. Signal Process.* **2019**, *117*, 210–220. [CrossRef]
52. Liu, H.; Chen, Z.; Liu, Y.; Chen, Y.; Du, Y.; Zhou, F. Interfacial debonding detection for cfst structures using an ultrasonic phased array: Application to the Shenzhen SEG building. *Mech. Syst. Signal Process.* **2023**, *192*, 110214. [CrossRef]
53. Büyüköztürk, O. Imaging of concrete structures. *NDT E Int.* **1998**, *31*, 233–243. [CrossRef]
54. Sansalone, M.; Carino, N.J.; Hsu, N.N. Flaws detection in concrete by frequency spectrum analysis of impact-echo waveforms. *J. Res. Natl. Bur. Stand.* **1987**, *7*, 279–290. [CrossRef]
55. Zhang, D.; Wang, Y. Study of impact-echo method detection in arch bridge of concrete-filled steel tube. *Chin. J. Eng. Geophys.* **2009**, *6*, 364–367. (In Chinese)
56. Liu, Y.; Liu, Z.; Lai, S.; Luo, L.; Dai, J. Debonding detection in the grouted joints of precast concrete shear walls using impact-echo method. *J. Nondestruct. Eval.* **2021**, *40*, 50. [CrossRef]
57. Yang, C.; Yu, Q. Placement and Size-Oriented Heat Dissipation Optimization for Antenna Module in Space Solar Power Satellite Based on Interval Dimension-Wise Method. *Aerosp. Sci. Technol.* **2023**, *134*, 108155. [CrossRef]
58. Yang, C.; Xia, Y. Interval Uncertainty-Oriented Optimal Control Method for Spacecraft Attitude Control. *IEEE Trans. Aerosp. Electron. Syst.* **2023**, 1–13. [CrossRef]
59. Chen, D.; Montano, V.; Huo, L.; Fan, S.; Song, G. Detection of subsurface voids in concrete-filled steel tubular (CFST) structure using percussion approach. *Constr. Build. Mater.* **2020**, *262*, 119761. [CrossRef]
60. Chen, H.; Nie, X.; Gan, S.; Zhao, Y.; Qiu, H. Interfacial imperfection detection for steel-concrete composite structures using ndt techniques: A state-of-the-art review. *Eng. Struct.* **2021**, *245*, 112778. [CrossRef]
61. Ansari, F. Practical implementation of optical fiber sensors in civil structural health monitoring. *J. Intell. Mater. Syst. Struct.* **2007**, *18*, 879–889. [CrossRef]
62. Gong, S.; Feng, X.; Zhang, G. A thermal-driven method based on brillouin fiber-optic sensors for the quantitative identification of subsurface cavities in concrete-filled steel tube structures. *J. Civ. Struct. Health Monit.* **2021**, *11*, 521–536. [CrossRef]
63. Wang, W.; Wang, G.; Wei, D.; Yang, H. Experimental study on detection of compactness of steel pipe concrete by rubber heating belt. *China Concr. Cem. Prod.* **2018**, 74–77. (In Chinese) [CrossRef]
64. Guo, X.; Yan, H. Analysis of influence factors for CFST arch bridge void based on eddy current thermal imaging. *Int. J. Robot. Autom.* **2021**, 36. [CrossRef]
65. Jiang, H.; Chen, L. Research status and development trend of pulsed infrared nondestructive testing technology. *Infrared Technol.* **2018**, *40*, 946–951+965. (In Chinese)
66. Zhang, S.; Tan, S.; Xu, L.; Chen, L. Thermoimage enhancement method for hollowing detection of steel tube concrete based on CLAHE. *J. Xihua Univ. Nat. Sci. Ed.* **2019**, *38*, 107–112. (In Chinese)
67. Zheng, D.; Tan, S. Research on internal defect detection of concrete based on infrared image enhancement algorithm. *J. Chongqing Jiaotong Univ. Nat. Sci. Ed.* **2021**, *40*, 114–117+127. (In Chinese)
68. Xu, R.; Jiang, D.; Lu, X. Research on the application of the infrared thermal image method in detection of concrete density of concrete-filled steel tube. *Appl. Mech. Mater.* **2012**, *166*, 998–1001. [CrossRef]
69. Sarhosis, V.; Santis, S.D.; de Felice, G. A Review of Experimental Investigations and Assessment Methods for Masonry Arch Bridges. *Struct. Infrastruct. E* **2016**, *12*, 1439–1464.
70. Gholizadeh, S. A Review of Non-Destructive Testing Methods of Composite Materials. *Procedia Struct. Integr.* **2016**, *1*, 50–57. [CrossRef]
71. Shang, R.; Zeng, B.; Rong, H.; Xu, Q.; Xu, M.; Xu, X. Research on Standardized Method of Stress Release Method for Measuring the Stress of Concrete. *Ind. Constr.* **2022**, *52*, 151–156. (In Chinese)
72. Pan, L. Example for calculating actual internal forces of concrete members by Hole Method of Reinforcement. *Build. Struct.* **2019**, *49*, 57–60. (In Chinese)

73. Ju, H.; Lin, C.; Zhang, J.; Liu, Z. Residual Stress Simulation and Measurement of Fe-Mn-Si Shape Memory Alloy Coating. *Infrared Laser Eng.* **2017**, *46*, 1017009. (In Chinese)
74. Bobzin, K.; Wietheger, W.; Knoch, M.A.; Schacht, A.; Reisgen, U.; Sharma, R.; Oster, L. Comparison of Residual Stress Measurements Conducted by X-Ray Stress Analysis and Incremental Hole Drilling Method. *J. Therm. Spray Tech.* **2020**, *29*, 1218–1228. [CrossRef]
75. Mohamed Karim, H.; Lamine, H.; Nadir, M. Assessment of Non-Uniform Residual Stress Field of the Thermal Sprayed Stainless Steel Coatings on Aluminium Substrates by the Integral Hole Drilling Method. *Proc. Inst. Mech. Eng. Part C J. Mech. Eng. Sci.* **2022**, *236*, 10701–10712. [CrossRef]
76. Razumovskii, I.A.; Usov, S.M. Development of the Hole-Drilling Method as Applied to the Study of Inhomogeneous Residual Stress Fields. *J. Mach. Manuf. Reliab.* **2021**, *50*, 727–734. [CrossRef]
77. Keil, S. Experimental determination of residual stresses with the ring-core method and an on-line measuring system. *Exp. Tech.* **2008**, *16*, 17–24. [CrossRef]
78. Chen, Z.; Sun, G.; Zhao, Y. Extension of Applicable Range of Measuring Residual Stress by Ring-Core Method. *Mater. Mech. Eng.* **2015**, *39*, 47–50+54. (In Chinese)
79. Li, K.; Ren, W. Application of Minature Ring-Core and Interferometric Strain/Slope Rosette to Determine Residual Stress Distribution With Depth—Part I: Theories. *J. Appl. Mech.* **2007**, *74*, 298–306. [CrossRef]
80. Hu, Z.; Xie, H.; Lu, J.; Zhu, J.; Wang, H. Residual Stresses Measurement by Using Ring-Core Method and 3D Digital Image Correlation Technique. *Meas. Sci. Technol.* **2013**, *24*, 085604. [CrossRef]
81. Dang, N.; Ku, W.; Wang, Z.; Lin, C.; Chen, T.; Lin, M. Incremental FIB-DIC Ring-Core Methods for the Residual Stress Measurement of Bilayer Thin Films. *Exp. Mech.* **2022**, *62*, 1489–1499. [CrossRef]
82. Sanginabadi, K.; Yazdani, A.; Mostofinejad, D.; Czaderski, C. RC Members Externally Strengthened with FRP Composites by Grooving Methods Including EBROG and EBRIG: A State-of-the-Art Review. *Constr. Build. Mater.* **2022**, *324*, 126662. [CrossRef]
83. Zhu, R.; Yin, Y.; Xie, H.; Liu, Y. On the Micro-Residlual Stress Measurement of Shot PeenedNickel-Based Alloy Material by using Slot-Milling Method. *J. Exp. Mech.* **2017**, *32*, 145–151. (In Chinese)
84. Shomali, A.; Mostofinejad, D.; Esfahani, M.R. Experimental and Numerical Investigation of Shear Performance of RC Beams Strengthened with FRP Using Grooving Method. *J. Build. Eng.* **2020**, *31*, 101409. [CrossRef]
85. Yan, Y.; Sun, Y.; He, L.; Zheng, L. Concrete Beam Stress Test Based on Grooving Method. *J. Chongqing Jiaotong Univ. (Nat. Sci.)* **2022**, *41*, 71–75+94. (In Chinese)
86. Dwivedi, S.K.; Vishwakarma, M.; Soni, A. Advances and Researches on Non Destructive Testing: A Review. *Mater. Today Proc.* **2018**, *5*, 3690–3698. [CrossRef]
87. Huang, Y.H.; Liu, L.; Sham, F.C.; Chan, Y.S.; Ng, S.P. Optical Strain Gauge vs. Traditional Strain Gauges for Concrete Elasticity Modulus Determination. *Optik* **2010**, *121*, 1635–1641. [CrossRef]
88. Zhao, Y.; Liu, Y.; Li, Y.; Hao, Q. Development and Application of Resistance Strain Force Sensors. *Sensors* **2020**, *20*, 5826. [CrossRef]
89. Kovačič, B.; Kamnik, R.; Štrukelj, A.; Vatin, N. Processing of Signals Produced by Strain Gauges in Testing Measurements of the Bridges. *Procedia Eng.* **2015**, *117*, 795–801. [CrossRef]
90. Turan, M.E.; Aydin, F.; Sun, Y.; Cetin, M. Residual Stress Measurement by Strain Gauge and X-Ray Diffraction Method in Different Shaped Rails. *Eng. Fail. Anal.* **2019**, *96*, 525–529. [CrossRef]
91. Ma, Z.; Chung, J.; Liu, P.; Sohn, H. Bridge Displacement Estimation by Fusing Accelerometer and Strain Gauge Measurements. *Struct. Control Health Monit.* **2021**, *28*, e2733. [CrossRef]
92. Takebayashi, S.; Kunieda, T.; Yoshinaga, N.; Ushioda, K.; Ogata, S. Comparison of the Dislocation Density in Martensitic Steels Evaluated by Some X-Ray Diffraction Methods. *ISIJ Int.* **2010**, *50*, 875–882. [CrossRef]
93. Macherauch, E. X-Ray Stress Analysis: Paper Covers Some Rather Important Developments during the Last Decade, Including New Experimental Results and Theoretical Aspects of X-Ray Stress Analysis. *Exp. Mech.* **1966**, *6*, 140–153. [CrossRef]
94. Monin, V.I.; Gurova, T.; Castello, X.; Estefen, S.F. Analysis of residual stress state in welded steel plates by x-ray diffraction method. *Rev. Adv. Mater. Sci.* **2009**, *20*, 172–175.
95. Epp, J. X-Ray Diffraction (XRD) Techniques for Materials Characterization. In *Materials Characterization Using Nondestructive Evaluation (NDE) Methods*; Elsevier: Amsterdam, The Netherlands, 2016; pp. 81–124.
96. Morin, L.; Braham, C.; Tajdary, P.; Seddik, R.; Gonzalez, G. Reconstruction of Heterogeneous Surface Residual-Stresses in Metallic Materials from X-Ray Diffraction Measurements. *Mech. Mater.* **2021**, *158*, 103882. [CrossRef]
97. Shi, P.; Su, S.; Chen, Z. Overview of Researches on the Nondestructive Testing Method of Metal Magnetic Memory: Status and Challenges. *J. Nondestruct. Eval.* **2020**, *39*, 43. [CrossRef]
98. Bao, S.; Jin, P.; Zhao, Z.; Fu, M. A Review of the Metal Magnetic Memory Method. *J. Nondestruct. Eval.* **2020**, *39*, 11. [CrossRef]
99. Dubov, A.; Dubov, A.; Kolokolnikov, S. Application of the Metal Magnetic Memory Method for Detection of Defects at the Initial Stage of Their Development for Prevention of Failures of Power Engineering Welded Steel Structures and Steam Turbine Parts. *Weld. World* **2014**, *58*, 225–236. [CrossRef]
100. Huang, H.; Jiang, S.; Wang, Y.; Zhang, L.; Liu, Z. Characterization of Spontaneous Magnetic Signals Induced by Cyclic Tensile Stress in Crack Propagation Stage. *J. Magn. Magn. Mater.* **2014**, *365*, 70–75. [CrossRef]
101. Chen, C.; Zeng, S.; Su, L. Strain Energy Based Method for Metal Magnetic Memory Effect of Tensile Tested Structures. *J. Nondestruct. Eval.* **2019**, *38*, 36. [CrossRef]

102. Tong, K.; Zhou, J.; Zhao, R.; Hu, W.; Qu, Y.; Cheng, C. Experimental Study on Rebar Stress Measurement Based on Force-Magnetic Coupling under Excited Magnetic Field. *Measurement* **2022**, *189*, 110620. [CrossRef]
103. Tong, K.; Zhou, J.; Zhao, R.; Ying, H.; Zhang, S. Quantitative Measurement of Stress in Steel Bars under Repetitive Tensile Load Based on Force-Magnetic Coupling Effect. *Measurement* **2022**, *202*, 111820. [CrossRef]
104. Zhang, W.; Qiu, Z.; Yu, X.; Sun, H. Feasibility Analyses of Improving Magnetic Memory Testing Sensibility by Strengthening Magnetic Excitation Field. *China Mech. Eng.* **2015**, *26*, 3375–3378.
105. Leng, J.; Xu, M.; Zhou, G.; Wu, Z. Effect of Initial Remanent States on the Variation of Magnetic Memory Signals. *NDT E Int.* **2012**, *52*, 23–27. [CrossRef]
106. Tong, K.; Zhou, J.; Ma, X.; Ying, H.; Zhao, R. Investigation of the Effect of Initial Magnetization State on the Force-Magnetic Coupling Effect of Rebars. *J. Magn. Magn. Mater.* **2023**, *569*, 170382. [CrossRef]
107. Hu, B.; Li, L.; Chen, X.; Zhong, L. Study on the Influencing Factors of Magnetic Memory Method. *Int. J. Appl. Electrom.* **2010**, *33*, 1351–1357. [CrossRef]
108. Liu, L.; Chen, W.; Zhang, P.; Tan, A. Effects of external magnetic field interference on tensile stress measurement with magnetic memory technology. *Chin. J. Sci. Instrum.* **2012**, *33*, 2194.
109. Acevedo, R.; Sedlak, P.; Kolman, R.; Fredel, M. Residual Stress Analysis of Additive Manufacturing of Metallic Parts Using Ultrasonic Waves: State of the Art Review. *J. Mater. Res. Technol.* **2020**, *9*, 9457–9477. [CrossRef]
110. Pilarski, A.; Rose, J.L. A Transverse-wave Ultrasonic Oblique-incidence Technique for Interfacial Weakness Detection in Adhesive Bonds. *J. Appl. Phys.* **1988**, *63*, 300–307. [CrossRef]
111. Chaki, S.; Corneloup, G.; Lillamand, I.; Walaszek, H. Combination of Longitudinal and Transverse Ultrasonic Waves for In Situ Control of the Tightening of Bolts. *J. Press. Vessel Technol.* **2007**, *129*, 383–390. [CrossRef]
112. Liu, H.; Li, Y.; Li, T.; Zhang, X.; Liu, Y.; Liu, K.; Wang, Y. Influence Factors Analysis and Accuracy Improvement for Stress Measurement Using Ultrasonic Longitudinal Critically Refracted (LCR) Wave. *Appl. Acoust.* **2018**, *141*, 178–187. [CrossRef]
113. Eremeyev, V.A.; Rosi, G.; Naili, S. Transverse Surface Waves on a Cylindrical Surface with Coating. *Int. J. Eng. Sci.* **2020**, *147*, 103188. [CrossRef]
114. Balonis, M.; Sant, G.; Burkan Isgor, O. Mitigating Steel Corrosion in Reinforced Concrete Using Functional Coatings, Corrosion Inhibitors, and Atomistic Simulations. *Cement Concrete Comp.* **2019**, *101*, 15–23. [CrossRef]
115. Goyal, A.; Ganjian, E.; Pouya, H.S.; Tyrer, M. Inhibitor Efficiency of Migratory Corrosion Inhibitors to Reduce Corrosion in Reinforced Concrete Exposed to High Chloride Environment. *Constr. Build. Mater.* **2021**, *303*, 124461. [CrossRef]
116. Kim, B.; Cho, S. Image-based Concrete Crack Assessment Using Mask and Region-based Convolutional Neural Network. *Struct. Control Health Monit.* **2019**, *26*, e2381. [CrossRef]
117. James, A.; Bazarchi, E.; Chiniforush, A.A.; Panjebashi Aghdam, P.; Hosseini, M.R.; Akbarnezhad, A.; Martek, I.; Ghodoosi, F. Rebar Corrosion Detection, Protection, and Rehabilitation of Reinforced Concrete Structures in Coastal Environments: A Review. *Constr. Build. Mater.* **2019**, *224*, 1026–1039. [CrossRef]
118. Rodrigues, R.; Gaboreau, S.; Gance, J.; Ignatiadis, I.; Betelu, S. Reinforced Concrete Structures: A Review of Corrosion Mechanisms and Advances in Electrical Methods for Corrosion Monitoring. *Constr. Build. Mater.* **2021**, *269*, 121240. [CrossRef]
119. Tešić, K.; Baričević, A.; Serdar, M. Non-Destructive Corrosion Inspection of Reinforced Concrete Using Ground-Penetrating Radar: A Review. *Materials* **2021**, *14*, 975. [CrossRef]
120. Amalia, Z.; Qiao, D.; Nakamura, H.; Miura, T.; Yamamoto, Y. Development of Simulation Method of Concrete Cracking Behavior and Corrosion Products Movement Due to Rebar Corrosion. *Constr. Build. Mater.* **2018**, *190*, 560–572. [CrossRef]
121. Alonso, C.; Andrade, C.; Rodriguez, J.; Diez, J.M. Factors Controlling Cracking of Concrete Affected by Reinforcement Corrosion. *Mater. Struct.* **1998**, *31*, 435–441. [CrossRef]
122. Liang, Y.; Wang, L. Prediction of Corrosion-Induced Cracking of Concrete Cover: A Critical Review for Thick-Walled Cylinder Models. *Ocean Eng.* **2020**, *213*, 107688. [CrossRef]
123. Di, X.; Zhou, Y. Durability design methods for concrete structures. *Archit. Sci.* **1997**, *1*, 16–20. (In Chinese)
124. Hui, Y.; Li, R.; Lin, Z.; Quan, M. Experimental study on the performance of steel bars in basic concrete components before and after corrosion. *Ind. Constr.* **1997**, *6*, 15–19+58. (In Chinese)
125. Niu, D.; Zhai, B.; Wang, K.; Di, X.; Wang, Q. Analysis of bearing capacity of corroded reinforced concrete beams. *Build. Struct.* **1999**, *29*, 23–25. (In Chinese) [CrossRef]
126. Wang, S.; Zhang, W.; Zhang, Y. Research on Predicting Steel Bar Section Loss from Longitudinal Crack Width of Concret. In Proceedings of the Engineering Safety and Durability—Proceedings of the 9th Annual Conference of the Chinese Society of Civil Engineering, Shanghai, China, 30 June–5 July 2019; pp. 361–364. (In Chinese).
127. Wang, Q. *Durability Evaluation Standard for Concrete Structures Shaanxi Province*; Xi'an University of Architecture and Technology: Xi'an, China, 2005. (In Chinese)
128. Zhang, J.; Mao, Y.; Zhang, J. Comparative analysis of steel bar corrosion prediction model experiments. *Highw. Transp. Technol. (Appl. Technol. Ed.)* **2016**, *12*, 223–225. (In Chinese)
129. Molina, F.J.; Alonso, C.; Andrade, C. Cover Cracking as a Function of Rebar Corrosion: Part 2—Numerical Model. *Mater. Struct.* **1993**, *26*, 532–548. [CrossRef]
130. Hu, J.; Zhang, S.; Chen, E.; Li, W. A Review on Corrosion Detection and Protection of Existing Reinforced Concrete (RC) Structures. *Constr. Build. Mater.* **2022**, *325*, 126718. [CrossRef]

131. Mei, K.; He, Z.; Yi, B.; Lin, X.; Wang, J.; Wang, H.; Liu, J. Study on Electrochemical Characteristics of Reinforced Concrete Corrosion under the Action of Carbonation and Chloride. *Case. Stud. Constr. Mat.* **2022**, *17*, e01351. [CrossRef]
132. Amiri, A.S.; Erdogmus, E.; Richter-Egger, D. A Comparison between Ultrasonic Guided Wave Leakage and Half-Cell Potential Methods in Detection of Corrosion in Reinforced Concrete Decks. *Signals* **2021**, *2*, 413–433. [CrossRef]
133. Adriman, R.; Bin, M.; Ibrahim, I.; Huzni, S.; Fonna, S.; Ariffin, A.K. Improving Half-Cell Potential Survey through Computational Inverse Analysis for Quantitative Corrosion Profiling. *Case. Stud. Constr. Mat.* **2022**, *16*, e00854. [CrossRef]
134. Almashakbeh, Y.; Saleh, E.; Al-Akhras, N.M. Evaluation of Half-Cell Potential Measurements for Reinforced Concrete Corrosion. *Coatings* **2022**, *12*, 975. [CrossRef]
135. Fan, L.; Bao, Y. Review of Fiber Optic Sensors for Corrosion Monitoring in Reinforced Concrete. *Cement Concrete Comp.* **2021**, *120*, 104029. [CrossRef]
136. Zhao, B.; Yao, K.; Wu, L.; Li, X.; Wang, Y. Application of Metal Magnetic Memory Testing Technology to the Detection of Stress Corrosion Defect. *Appl. Sci.* **2020**, *10*, 7083. [CrossRef]
137. Zhou, J.; Qiu, J.; Zhou, Y.; Zhou, Y.; Xia, R. Experimental Study on Residual Bending Strength of Corroded Reinforced Concrete Beam Based on Micromagnetic Sensor. *Sensors* **2018**, *18*, 2635. [CrossRef]
138. Qiu, J.; Zhang, H.; Zhou, J.; Ma, H.; Liao, L. Experimental Analysis of the Correlation between Bending Strength and SMFL of Corroded RC Beams. *Constr. Build. Mater.* **2019**, *214*, 594–605. [CrossRef]
139. Xia, R.; Zhou, J.; Zhang, H.; Liao, L.; Zhao, R.; Zhang, Z. Quantitative Study on Corrosion of Steel Strands Based on Self-Magnetic Flux Leakage. *Sensors* **2018**, *18*, 1396. [CrossRef]
140. Xia, R.; Zhou, J.; Zhang, H.; Zhou, D.; Zhang, Z. Experimental Study on Corrosion of Unstressed Steel Strand Based on Metal Magnetic Memory. *KSCE J. Civ. Eng.* **2019**, *23*, 1320–1329. [CrossRef]
141. Su, S.; Ma, X.; Wang, W.; Yang, Y.; Hu, J. Quantitative Evaluation of Cumulative Plastic Damage for Ferromagnetic Steel under Low Cycle Fatigue Based on Magnetic Memory Method. *Strain* **2021**, *57*, e12379. [CrossRef]
142. Yang, Y.; Ma, X.; Su, S.; Wang, W. Study on Corrosion Damage Characterization and Tensile Strength Evaluation for Locally Corroded Bridge Steel via Metal Magnetic Memory Method. *Measurement* **2023**, *207*, 112406. [CrossRef]
143. Yang, C.; Liang, K.; Zhang, X.; Geng, X. Sensor Placement Algorithm for Structural Health Monitoring with Redundancy Elimination Model Based on Sub-Clustering Strategy. *Mech. Syst. Signal Process.* **2019**, *124*, 369–387. [CrossRef]
144. Yang, C.; Xia, Y. A Novel Two-Step Strategy of Non-Probabilistic Multi-Objective Optimization for Load-Dependent Sensor Placement with Interval Uncertainties. *Mech. Syst. Signal Process.* **2022**, *176*, 109173. [CrossRef]
145. An, Y.; Chatzi, E.; Sim, S.; Laflamme, S.; Blachowski, B.; Ou, J. Recent Progress and Future Trends on Damage Identification Methods for Bridge Structures. *Struct. Control Health Monit.* **2019**, *26*, e2416. [CrossRef]
146. Yang, C.; Ouyang, H. A Novel Load-Dependent Sensor Placement Method for Model Updating Based on Time-Dependent Reliability Optimization Considering Multi-Source Uncertainties. *Mech. Syst. Signal Process.* **2022**, *165*, 108386. [CrossRef]
147. Serra, M.; Festa, G.; Vassallo, M.; Zollo, A.; Quattrone, A.; Ceravolo, R. Damage Detection in Elastic Properties of Masonry Bridges Using Coda Wave Interferometry. Struct. *Control Health Monit.* **2017**, *24*, e1976. [CrossRef]
148. Emre Alpaslan, Z.K. Response Surface-Based Model Updating to Detect Damage on Reduced-Scale Masonry Arch Bridge. *Struct. Eng. Mech.* **2021**, *79*, 9–22.
149. Civera, M.; Calamai, G.; Zanotti Fragonara, L. System Identification via Fast Relaxed Vector Fitting for the Structural Health Monitoring of Masonry Bridges. *Structures* **2021**, *30*, 277–293. [CrossRef]
150. Civera, M.; Mugnaini, V.; Zanotti Fragonara, L. Machine Learning-based Automatic Operational Modal Analysis: A Structural Health Monitoring Application to Masonry Arch Bridges. *Struct. Control Health Monit.* **2022**, *29*, e3028. [CrossRef]
151. Conde, B.; Eguía, P.; Stavroulakis, G.E.; Granada, E. Parameter Identification for Damaged Condition Investigation on Masonry Arch Bridges Using a Bayesian Approach. *Eng. Struct.* **2018**, *172*, 275–284. [CrossRef]
152. Borlenghi, P.; Saisi, A.; Gentile, C. ND Testing and Establishing Models of a Multi-Span Masonry Arch Bridge. *J. Civ. Struct. Health Monit.* **2023**. [CrossRef]
153. Alexakis, H.; Liu, H.; DeJong, M.J. Damage Identification of Brick Masonry under Cyclic Loading Based on Acoustic Emissions. *Eng. Struct.* **2020**, *221*, 110945. [CrossRef]
154. Pepi, C.; Cavalagli, N.; Gusella, V.; Gioffrè, M. An Integrated Approach for the Numerical Modeling of Severely Damaged Historic Structures: Application to a Masonry Bridge. *Adv. Eng. Softw.* **2021**, *151*, 102935. [CrossRef]
155. Batar, O.S.; Tercan, E.; Emsen, E. Ayvalıkemer (Sillyon) Historical Masonry Arch Bridge: A Multidisciplinary Approach for Structural Assessment Using Point Cloud Data Obtained by Terrestrial Laser Scanning (TLS). *J. Civ. Struct. Health Monit.* **2021**, *11*, 1239–1252. [CrossRef]
156. Deng, Y.; Jia, Y.; Li, Y.; Li, A. Structural Parameter Identification of Ancient Stone Arch Bridge via Three-Dimensional Laser Ranger Scanning. *J. Perform. Constr. Facil.* **2022**, *36*, 04022041. [CrossRef]
157. Jara, J.M.; López, J.I.; Olmos, B.A.; Martínez, G. Effect of Seismic Source Type on the Expected Behavior of Historic Arch Bridges. *Int. J. Archit. Herit.* **2022**, *16*, 789–815. [CrossRef]
158. Kaya, A.; Adanur, S.; Bello, R.A.; Genç, A.F.; Okur, F.Y.; Sunca, F.; Günaydin, M.; Altunişik, A.C.; Sevim, B. Post-Earthquake Damage Assessments of Unreinforced Masonry (URM) Buildings by Shake Table Test and Numerical Visualization. *Eng. Fail. Anal.* **2023**, *143*, 106858. [CrossRef]

159. Gönen, S.; Soyöz, S. Reliability-Based Seismic Performance of Masonry Arch Bridges. *Struct. Infrastruct. Eng.* **2022**, *18*, 1658–1673. [CrossRef]
160. Aytulun, E.; Soyoz, S.; Karcioglu, E. System Identification and Seismic Performance Assessment of a Stone Arch Bridge. *J. Earthq. Eng.* **2022**, *26*, 723–743. [CrossRef]
161. Mu, H.; Zheng, Z.; Wu, X.; Su, C. Bayesian Network-Based Modal Frequency–Multiple Environmental Factors Pattern Recognition for the Xinguang Bridge Using Long-Term Monitoring Data. *J. Low Freq. Noise Vib. Act. Control* **2020**, *39*, 545–559. [CrossRef]
162. Anastasopoulos, D.; Maes, K.; De Roeck, G.; Lombaert, G.; Reynders, E.P.B. Influence of Frost and Local Stiffness Variations on the Strain Mode Shapes of a Steel Arch Bridge. *Eng. Struct.* **2022**, *273*, 115097. [CrossRef]
163. Anastasopoulos, D.; De Roeck, G.; Reynders, E.P.B. One-Year Operational Modal Analysis of a Steel Bridge from High-Resolution Macrostrain Monitoring: Influence of Temperature vs. Retrofitting. *Mech. Syst. Signal Process.* **2021**, *161*, 107951. [CrossRef]
164. Zhu, Q.; Wang, H.; Spencer, B.F. Investigation on the Mapping for Temperature-Induced Responses of a Long-Span Steel Truss Arch Bridge. *Struct. Infrastruct. Eng.* **2022**, 1–18. [CrossRef]
165. Teng, J.; Tang, D.; Hu, W.; Lu, W.; Feng, Z.; Ao, C.; Liao, M. Mechanism of the Effect of Temperature on Frequency Based on Long-Term Monitoring of an Arch Bridge. *Struct. Health Monit.* **2021**, *20*, 1716–1737. [CrossRef]
166. Sunca, F.; Ergün, M.; Altunişik, A.C.; Günaydin, M.; Okur, F.Y. Modal Identification and Fatigue Behavior of Eynel Steel Arch Highway Bridge with Calibrated Models. *J. Civ. Struct. Health Monit.* **2021**, *11*, 1337–1354. [CrossRef]
167. Chiaia, B.; Marasco, G.; Ventura, G.; Zannini Quirini, C. Customised Active Monitoring System for Structural Control and Maintenance Optimisation. *J. Civ. Struct. Health Monit.* **2020**, *10*, 267–282. [CrossRef]
168. Dammika, A.J.; Kawarai, K.; Yamaguchi, H.; Matsumoto, Y.; Yoshioka, T. Analytical Damping Evaluation Complementary to Experimental Structural Health Monitoring of Bridges. *J. Bridge Eng.* **2015**, *20*, 04014095. [CrossRef]
169. Tran, T.T.X.; Ozer, E. Automated and Model-Free Bridge Damage Indicators with Simultaneous Multiparameter Modal Anomaly Detection. *Sensors* **2020**, *20*, 4752. [CrossRef]
170. Whelan, M.; Salas Zamudio, N.; Kernicky, T. Structural Identification of a Tied Arch Bridge Using Parallel Genetic Algorithms and Ambient Vibration Monitoring with a Wireless Sensor Network. *J. Civ. Struct. Health Monit.* **2018**, *8*, 315–330. [CrossRef]
171. Jayasundara, N.; Thambiratnam, D.; Chan, T.; Nguyen, A. Vibration-Based Dual-Criteria Approach for Damage Detection in Arch Bridges. *Struct. Health Monit.* **2019**, *18*, 2004–2019. [CrossRef]
172. Jayasundara, N.; Thambiratnam, D.P.; Chan, T.H.T.; Nguyen, A. Damage Detection and Quantification in Deck Type Arch Bridges Using Vibration Based Methods and Artificial Neural Networks. *Eng. Fail. Anal.* **2020**, *109*, 104265. [CrossRef]
173. Jayasundara, N.; Thambiratnam, D.P.; Chan, T.H.T.; Nguyen, A. Locating and Quantifying Damage in Deck Type Arch Bridges Using Frequency Response Functions and Artificial Neural Networks. *Int. J. Struct. Stab. Dyn.* **2020**, *20*, 2042010. [CrossRef]
174. Delgadillo, R.M.; Tenelema, F.J.; Casas, J.R. Marginal Hilbert Spectrum and Instantaneous Phase Difference as Total Damage Indicators in Bridges under Operational Traffic Loads. *Struct. Infrastruct. Eng.* **2023**, *19*, 824–844. [CrossRef]
175. Kordestani, H.; Xiang, Y.; Ye, X.; Yun, C.; Shadabfar, M. Localization of Damaged Cable in a Tied-arch Bridge Using Arias Intensity of Seismic Acceleration Response. *Struct. Control Health Monit.* **2020**, *27*, e2491. [CrossRef]
176. Magalhães, F.; Cunha, Á. Explaining Operational Modal Analysis with Data from an Arch Bridge. *Mech. Syst. Signal Process.* **2011**, *25*, 1431–1450. [CrossRef]
177. Magalhães, F.; Cunha, A.; Caetano, E. Vibration Based Structural Health Monitoring of an Arch Bridge: From Automated OMA to Damage Detection. *Mech. Syst. Signal Process.* **2012**, *28*, 212–228. [CrossRef]
178. Ren, W.; Lin, Y.; Fang, S. Structural Damage Detection Based on Stochastic Subspace Identification and Statistical Pattern Recognition: I. Theory. *Smart Mater. Struct.* **2011**, *20*, 115009. [CrossRef]
179. Lin, Y.; Ren, W.; Fang, S. Structural Damage Detection Based on Stochastic Subspace Identification and Statistical Pattern Recognition: II. Experimental Validation under Varying Temperature. *Smart Mater. Struct.* **2011**, *20*, 115010. [CrossRef]
180. Comanducci, G.; Magalhães, F.; Ubertini, F.; Cunha, Á. On Vibration-Based Damage Detection by Multivariate Statistical Techniques: Application to a Long-Span Arch Bridge. *Struct. Health Monit.* **2016**, *15*, 505–524. [CrossRef]
181. Ferrari, R.; Froio, D.; Rizzi, E.; Gentile, C.; Chatzi, E.N. Model Updating of a Historic Concrete Bridge by Sensitivity- and Global Optimization-Based Latin Hypercube Sampling. *Eng. Struct.* **2019**, *179*, 139–160. [CrossRef]
182. Duan, Y.; Chen, Q.; Zhang, H.; Yun, C.; Wu, S.; Zhu, Q. CNN-Based Damage Identification Method of Tied-Arch Bridge Using Spatial-Spectral Information. *Smart Struct. Syst.* **2019**, *23*, 507–520.
183. Zhou, S.; Jiang, Y.; Li, X.; Wu, Q. Vibration-based damage identification of reinforced concrete arch bridges using KALMAN–ARMA–GARCH model. *Int. J. Robot. Autom.* **2021**, *36*. [CrossRef]
184. Qin, X.; Zhang, L.; Yang, M.; Luo, H.; Liao, M.; Ding, X. Mapping Surface Deformation and Thermal Dilation of Arch Bridges by Structure-Driven Multi-Temporal DInSAR Analysis. *Remote Sens. Environ.* **2018**, *216*, 71–90. [CrossRef]
185. Wang, W.; Su, M.; Wang, C. Static Deflection Difference-Based Damage Identification of Hanger in Arch Bridges. *KSCE J. Civ. Eng.* **2022**, *26*, 5096–5106. [CrossRef]
186. Li, S.; Ou, J.; Wang, J.; Gao, X.; Yang, C. Level 2 Safety Evaluation of Concrete-filled Steel Tubular Arch Bridges Incorporating Structural Health Monitoring and Inspection Information Based on China Bridge Standards. *Struct. Control Health Monit.* **2018**, *26*, e2303. [CrossRef]
187. Fu, C.; Jiang, S. A Hybrid Data-Fusion System by Integrating CFD and PNN for Structural Damage Identification. *Appl. Sci.* **2021**, *11*, 8272. [CrossRef]

188. Gui, C.; Lin, W.; Huang, Z.; Xin, G.; Xiao, J.; Yang, L. A Decision-Making Algorithm for Concrete-Filled Steel Tubular Arch Bridge Maintenance Based on Structural Health Monitoring. *Materials* **2022**, *15*, 6920. [CrossRef]
189. Yan, W.; Ren, W. Operational Modal Parameter Identification from Power Spectrum Density Transmissibility. *Comput. Aided Civ. Infrastruct. Eng.* **2012**, *27*, 202–217. [CrossRef]
190. Xiong, C.; Yu, L.; Niu, Y. Dynamic Parameter Identification of a Long-Span Arch Bridge Based on GNSS-RTK Combined with CEEMDAN-WP Analysis. *Appl. Sci.* **2019**, *9*, 1301. [CrossRef]
191. Li, J.; Hao, H.; Xia, Y.; Zhu, H.-P. Damage detection of shear connectors in bridge structures with transmissibility in frequency domain. *Int. J. Struct. Stab. Dyn.* **2014**, *14*, 1350061. [CrossRef]

buildings

MDPI

Article

A Study on the Ultimate Span of a Concrete-Filled Steel Tube Arch Bridge

Yuexing Wu [1,2], Xiangchuan Wang [3], Yonghui Fan [2,*], Jun Shi [2], Chao Luo [2] and Xinzhong Wang [1]

1 School of Civil Engineering, Hunan City University, Yiyang 413000, China; wuyuexing@hncu.edu.cn (Y.W.); 622210970020@mails.cqjtu.edu.cn (X.W.)
2 State Key Laboratory of Mountain Bridge and Tunnel Engineering, Chongqing Jiaotong University, Chongqing 400074, China; 611200080015@mails.cqjtu.edu.cn (J.S.); luochao@mails.cqjtu.edu.cn (C.L.)
3 Huasheng Testing Technology Co., Ltd., Chongqing 400039, China; 622220970001@mails.cqjtu.edu.cn
* Correspondence: fyh1995@mails.cqjtu.edu.cn; Tel.: +86-17132308545

Abstract: In order to study the ultimate span of a concrete-filled steel tube (CFST) arch bridge, taking the structural strength, stiffness, and stability as the limiting conditions, the finite element analysis method is adopted to carry out research on the influence law of a single parameter of the pipe diameter, wall thickness, and cross-section height on the ultimate span of the arch axial shape. The result is used as a sample point to determine the ultimate span of the CFST arch bridge under multifactor coupling based on the response surface method. The finite element method is used to check the strength, stiffness, stability, number of segments and maximum lifting weight, steel content rate, and steel pipe concrete constraint effect coefficient of the CFST arch bridge under the ultimate span diameter. The results show that, when analyzed using a single parameter, the ultimate span diameter of the CFST arch bridge increases with the increase in the steel pipe diameter and the cross-section height, and then decreases. Moreover, it increases with the increase in the wall thickness of the steel pipe, and the CFST arch bridge reaches the ultimate span with the increase in the steel pipe wall thickness. When the pipe diameter is 1.38 m, the CFST arch bridge reaches the ultimate span; according to a multi-parameter coupling analysis, when the pipe diameter is 1.49 m, wall thickness is 37 mm, and cross-section height is 17 m, the CFST arch bridge reaches the ultimate span of 821 m, which meets all of the limiting conditions, and, at this point, the arch axial coefficient is 1.2. The results of the finite element calculation show that the structural strength, prior to the stiffness, stability, and other limitations, just reaches the critical value of the limiting conditions.

Keywords: concrete-filled steel tube arch bridge; ultimate span; strain energy minimization method; finite element analysis; response surface method

Citation: Wu, Y.; Wang, X.; Fan, Y.; Shi, J.; Luo, C.; Wang, X. A Study on the Ultimate Span of a Concrete-Filled Steel Tube Arch Bridge. *Buildings* **2024**, *14*, 896. https://doi.org/10.3390/buildings14040896

Academic Editor: Fabrizio Gara

Received: 21 February 2024
Revised: 19 March 2024
Accepted: 23 March 2024
Published: 26 March 2024

1. Introduction

With the increasing improvement of arch bridge design theory, the application of high-strength materials, and the advancement of cable-stayed buckling construction technology, large-span CFST arch bridges are becoming increasingly more prevalent [1]. At present, the largest CFST arch bridge in the world is the Pingnan Three Bridge in Guangxi, with a main span of 575 m. Zheng [2] pointed out that China has the ability to design and build 700 m class CFST arch bridges. However, no matter how developed the theory and construction technology of this type of bridge is, it is difficult to increase the span diameter endlessly. The limit to which the span diameter of CFST arch bridges can reach is no longer included in the program comparison, which is a matter of concern for bridge designers [3]. The study of the ultimate span diameter of CFST5 G arch bridges is of great scientific significance and engineering application value for the deep understanding of the mechanical behavior of the steel pipe and the concrete in the pipe, when they work in coordination, and for the improvement of the design methods and specifications of this type of bridge.

The current domestic and international research on the ultimate span diameter of arch bridges can be mainly divided into three categories.

The first category focuses on improving the span diameter of arch bridges by optimizing the structural form [4–6]. For example, Xie et al. [7] proposed a new bridge structural system known as the medium-bearing cable arch bridge; here, the main span and side spans of suspension bridges are set up in the arch rib and the main cables are anchored to the arch foot of the side arches, and the structural mechanical performance is improved by adjusting the yaw-to-span ratio. A study was carried out with a 700 m span as an example, and the results showed that this structural form improved the strength-bearing capacity and stability-bearing capacity by about 25% and 70%, respectively, compared with a continuous arch bridge under the same conditions, which lays a solid foundation for a breakthrough in the improvement of the arch bridge span. To address the problem that the large self-weight of arch bridges restricts the development of the spanning diameter, Huang et al. [8] put forward a new type of arch structure consisting of top concrete, bottom concrete, and steel web rods (Steel-Reinforced Concrete, SRC). The calculation results showed that the self-weight of the new type of arch rib was reduced by about 31% compared with the original bridge, and nearly one-third of the construction time of the arch rib pouring concrete could be saved, which has certain feasibility in design and construction technology. This kind of research improves the span diameter by optimizing the structural form but pays less attention to the theoretical span limit of this structural form of arch bridge.

The second type of research focuses on improving the span diameter of arch bridges by using high-strength materials [9,10]. Zhang et al. [11] used theoretical derivation to analytically determine the ultimate span diameter of arch bridges and studied the ultimate span diameter of the bridges when R200 activated fly ash concrete and Q690 high-strength steel were used. The researchers found that the theoretical ultimate span diameter of concrete arch bridges using R200 activated powder concrete can reach up to 2000 m, while the main span of steel arch bridges using Q690 high-strength steel can be greater than 2500 m. Shao et al. [12] carried out the conceptual design of a 1000 m class arch bridge based on steel–ultra-high-performance fiber-reinforced concrete (UHPFRC). The results of a finite element analysis showed that the design solution meets the structural stability, strength, and stiffness requirements of the code. This type of research increases the span diameter of the arch bridge by improving the strength of the material but neglects the sharp increase in the construction difficulty after this increase in span diameter [13].

The third category of research primarily investigates the mechanical performance of concrete-filled steel tube (CFST) composite members from the component level. For instance, Wei et al. [14] conducted dynamic performance tests on CFST composite columns, indicating that a higher axial compression ratio leads to earlier yield initiation and more severe final failure of the specimens. Additionally, Wang et al. [15] performed axial compression bearing capacity tests on 18 short steel tube concrete columns, analyzing the impact of basalt fiber length on their mechanical properties. Compared to conventional short steel tube concrete columns, the addition of basalt fibers significantly enhanced the axial compression-bearing capacity and ductility coefficient of the columns, with average increases of approximately 8.1% and 31.6%, respectively. Zhang et al. [16] conducted dynamic performance tests and numerical simulation analyses on eight Recycled Aggregate Concrete-Filled Circular Steel Tube (RACFCST) specimens, noting that the load-bearing capacity of these specimens decreased with an increase in the axial compression ratio and with a reduction in the beam–column linear stiffness ratio and yield bending moment ratio. Additionally, Zhang et al. [17] explored the dynamic performance of S-RACFST specimens and proposed a displacement-based seismic design method for S-RACFST frames, verifying that this method meets the requirements for "temporary use", "normal use", and "collapse prevention". Such studies mainly focus on the performance optimization and pattern analysis of specimens, failing to deeply explore the potential and efficacy of CFST composite members within the overall structure, particularly regarding their role and contribution to the design of large-span arch bridges. Although CFST composite members exhibit excellent performance in enhancing local mechanical properties, research on their application in overall structural design, construction technology, and long-term performance remains relatively scarce.

In summary, current research on the ultimate span of arch bridges primarily exhibits three issues: a focus on structural optimization while neglecting theoretical limits, an emphasis on the application of high-strength materials without considering construction difficulties, and a concentration on the mechanical performance optimization of specimens without a systematic study of structural aspects from design and construction to long-term performance. In view of these shortcomings, this study selects several key factors affecting the span diameter of large-span steel-tube concrete arch bridges, taking into account the feasibility of the structural design, construction programs, and other constraints to examine the influence of these factors on the ultimate span of the bridges. This approach is combined with the response surface method to comprehensively consider the coupling effect between a number of factors on the acquisition of the ultimate span of steel-tube concrete arch bridges, and to provide a reference for the promotion of the arch bridges for the development of a greater span diameter.

2. Study of Reasonable Parameters for the Limit Span

2.1. Parameter Selection

In conducting research on the ultimate span of CFST arch bridges, the rationality of parameter selection is crucial to the reliability of the research results. Considering several key factors, such as structural geometry, material mechanical properties, section-bearing capacity, and structural stability, four types of parameters, namely, arch axial shape, material strength, section height, and height-to-span ratio, are selected for this study [18–20].

2.1.1. Rationalization of the Arch Axis

The selection of the arch axis is the basis of arch bridge design. Under the action of external load, the main arch rib only produces axial pressure. Without bending moment and shear force, the arch axis and the pressure line completely coincide; such an arch axis is known as a reasonable arch axis [21], and its determination must follow the principle of the line shape being reasonable, aesthetically appealing, and convenient to construct.

According to statistics, large-span steel pipe concrete is mostly used in hollow-belly arch bridges [22]. The constant load set of this type of structure is distributed discontinuously along the arch ribs, and the suspension chain line is a curve that matches its pressure line. The authors of [23] investigated the proportion of suspension chain lines, parabolic lines, and spline curves as arch axes in CFST arch bridges in China, and pointed out that 66.67% of CFST arch bridges used suspension chain lines as arch axes. Currently, the Poseidon Yangtze River Bridge with a main span of 530 m, the Hejiang Yangtze River Highway Bridge with a main span of 507 m, and the Pingnan Third Bridge with a main span of 560 m [24] all use suspended chain lines as arch axes.

In summary, the suspension chain line is formulated as the arch axis line shape for the limit span study in this paper. It can be seen through the equation of the hanging chain line that the selection of the arch axis coefficient m of the hanging chain line is closely related to the load set on the arch, and the load sets of arch bridges under different spans are different, so the research process of the ultimate span should be carried out in accordance with the existing optimization method of the arch axis coefficient under the optimal arch axis coefficient.

2.1.2. Material Parameters

The steel pipe concrete structure is a combination of structures consisting of a steel pipe filled with concrete, and this creates a concrete hoop effect so that the steel pipe is in a three-direction compression state. This combined structure has the advantages of high strength, light weight, good ductility, fatigue resistance, impact resistance, and other good mechanical properties.

According to statistics [25], with the increase in the arch bridge span and the need for structural strength, the proportion of Q345 steel used in CFST arch bridges increases year by year, and C60 or even C70 concrete is also gradually being applied in large-span CFST arch bridges. In the 507 m Hejiang Yangtze River Bridge, and in the world's largest-spanning steel arch bridge, the Pingnan Bridge, the upper and lower chords are made of Q345 steel pipe and C70 concrete. Some builders of large-span arch bridges have begun to use Q420 steel [26]; the concrete in the tube is self-compacting compensatory concrete, and the strength grade is mostly C60~C80 [27]. Combined with the development trend of the material and the current application prospects, this paper adopts the combination of Q420 steel and C80 concrete for the subsequent ultimate span study.

2.1.3. Cross-Sectional Construction Parameters

The arch rib section is one of the most important aspects of arch bridge design. According to statistics relating to the arch rib section forms of 327 CFST arch bridges [28], the four-limbed truss section is used most often in large-span CFST arch bridges, so this research adopts the four-limbed truss section for the study of ultimate span.

For the truss section, the steel pipe concrete diameter, wall thickness, cross-section truss height, and variable cross-section mode are the key parameters of design, and the reasonableness of their values directly affects the structural stress performance. The values of some CFST arch bridge section parameters in China are summarized in Table 1.

Table 1. Values of cross-section parameters of some steel pipe arches in China.

Serial Number	Bridges	Span (m)	Pipe Diameter (m)	Wall Thickness (mm)	Section Height Ratio Foot/Top of Arch
1	Guangxi Pingnan Third Bridge	560	1.4	34	2.0 (17 m/8.5 m)
2	Sichuan Hejiang Bridge	530	1.3	30	2.0 (16 m/8.0 m)
3	Sichuan Hejiang Yangtze River Highway Bridge	507	1.32	32	2.0 (14 m/7.0 m)
4	Wushan Yangtze River Bridge	492	1.22	25	2.0 (14 m/7.0 m)
5	Guizhou Dazaijing Bridge	450	1.36	35	1.8 (14 m/8 m)
6	Yunnan Mengxin Highway Liangshuigou Bridge	430	1.2	35	2.0 (13 m/6.5 m)
7	Hunan Xiangtan Liancheng Bridge	388	0.85	24	1.8 (9 m/5 m)
8	Hunan Yiyang Maocaojie Bridge	368	1	28	2.0 (8 m/4 m)
9	Guizhou Zongxi River Bridge	360	1.2	35	1.8 (11 m/6 m)

From referencing the aforementioned information and taking into account the manufacturing capabilities of steel structure factories, the range for the pipe diameter in the limit span analysis process of this study is 1260 mm~1500 mm, the wall-thickness range is 32 mm~48 mm, the truss-height range is 13 m~24 m, and the ratio of the cross-section height of the foot of the arch to the top of the arch takes a value of 2.

2.1.4. Height-to-Span Ratio

The height-to-span ratio refers to the ratio of the span to the height of the main arch rib, the value of which affects the structural stress performance and the degree of coordination between the structure and the external environment.

For CFST arch bridges, the arch rib height-to-span ratio should be suitable in the range of 1/6~1/4, considering appearance, landscape, construction difficulty, etc. [29]. According to the statistical analysis of the height-to-span ratio of the constructed truss CFST arch bridges in China [30], most of the height-to-span ratios are concentrated at 1/5 or 1/4, accounting for as much as 66%, and taking a larger height-to-span ratio value within a reasonable range is conducive to the overall performance of the arch ribs; thus, a height-to-span ratio of 1/4 is proposed in this study to carry out the subsequent ultimate spanning diameter analysis.

2.2. Limit Indicators for Limit Span Analysis

In order to overcome the problem that the existing research cannot take into account the inadequacy of structural design, material strength, construction methods, and other factors, and the fact that the selected material strength, structural stiffness, stability, number of main arch segments, maximum lifting weight of the segments, steel content of the steel pipe concrete, steel pipe concrete constraint effect coefficients, and steel pipe diameter and thickness ratios are research process limitations in combination with the current specification requirements, the limitations of the indicators take the following values:

In terms of material strength, Q420 steel and C80 concrete are selected as the materials for the upper and lower chords of the steel pipe concrete, and the design values of the axial compressive strength and axial tensile strength of the C80 concrete are taken as 34.6 MPa and 2.14 MPa, respectively. The design values of the strength of the Q420 steel are taken as 320 MPa (wall thickness of 16–40 mm) or 305 MPa (wall thickness of 40–63 mm).

In terms of structural rigidity, the maximum vertical deflection of the main arch under lane loading should not be greater than L/1000 (L is the span diameter).

In terms of stability, the elastic overall stability coefficient of the main arch should not be less than 4.0 in the construction and operation use stage, and the stability coefficient of the local components should not be less than the elastic overall stability coefficient of the main arch.

Regarding the number of main arch sections and the maximum lifting weight of the sections, with reference to the equipment transfer capacity of the existing bridge construction process, the proposed number of single-rib section divisions shall not exceed 32 sections, and the maximum lifting weight of the sections shall not exceed 200 tons.

The steel content of the steel pipe concrete (the ratio of the steel pipe area to the concrete area in the cross-section) ranges from 0.04 to 0.20, the steel pipe concrete constraint effect coefficient (the ratio of the product of the area of the steel pipe and the design value of the steel strength to the product of the area of the concrete and the design value of the concrete strength in the cross-section) should not be less than 0.6, and the ratio of the steel pipe diameter to thickness should range from 40 to 90.

2.3. Parametric Coupling Analysis Methods

Finite element software is used to determine the influence of multiple parameters on the ultimate span of CFST arch bridges, which has a huge computational workload and can only obtain the analytical value rather than the functional relationship equation. Meanwhile, in the process of studying the ultimate span diameter, parameters such as material, construction parameters, arch axis coefficient, and height-to-span ratio are in a mutually coupled relationship, for which the response surface method can be used to analyze and solve. The response surface method [31] is a statistical method of experimental design and optimization used to study the effects of multiple factors on one or more response variables and to determine the optimal conditions, which has the advantages of being able to analyze multiple variables, reduce the number of tests, and be visualized [32]. In this paper, the response surface model is used to replace the complex finite element calculation model, and the combination of the design of experiments and the response surface method establishes a mathematical model of the limit span diameter, which can considerably improve the efficiency of the ultimate span determination process. The specific steps are as follows: First, an analysis of the impact of various parameters on the ultimate span of CFST (concrete-filled steel tube) arch bridges is conducted based on the finite element model. Subsequently, using the analysis results as sample data, a response surface function for the ultimate span of CFST arch bridges is obtained by fitting the sample data. This function is then subjected to an F-test and R^2 test. Following this, a mathematical optimization model for the ultimate span of CFST arch bridges is established, yielding the ultimate span under various constraints. Finally, the optimization results are verified through the finite element model.

3. Reasonable Arch Axis Analysis

3.1. Arch Axis Line Analysis

The arch axis coefficient m, as the linear characteristic value of the suspension chain line [33], depends on the arch axis coefficient to a large extent. In order to ensure that the deviation between the suspension chain line and the load pressure line is small, it is necessary to adopt a reasonable arch axis coefficient optimization method to solve this coefficient, and, with the change in the cross-section parameters, spanning diameter, and other parameters, the value of the reasonable arch axis coefficient is not certain; thus, the study of the ultimate spanning diameter needs to be based on the optimal arch axis coefficient.

3.2. Optimization Method for Reasonable Arch Axis Coefficients

The arch rib bending strain energy minimization method [34] is a commonly used arch axis coefficient optimization method for large-span arch bridges which controls the value of bending moments in each section of the arch rib by controlling the deviation of the arch axis from the load pressure line. The principle is as follows:

In the elastic range of the structure, the elastic deformation energy U of an elastomer due to deformation under external load is numerically equal to the work carried out W by the external load on the corresponding displacement:

$$U = W \tag{1}$$

where the elastic strain energy U can be expressed as

$$U = \int_l \frac{M(x)^2}{2EI} dx + \int_l \frac{Q(x)^2}{2GA} dx + \int_l \frac{N(x)^2}{2EA} dx \tag{2}$$

where $\int_l \frac{M(x)^2}{2EI} dx$ denotes the bending energy caused by the bending moment M, $\int_l \frac{Q(x)^2}{2GA} dx$ denotes the shear energy caused by the shear force Q, and $\int_l \frac{N(x)^2}{2EA} dx$ denotes the axial deformation energy caused by the axial force N; E and G denote the modulus of elasticity and shear modulus of the material, respectively; and I and A denote the moment of inertia and the area of the cross-section, respectively.

The magnitude of the strain energy can reflect the stress state of the structure. For compression structures such as arch bridges, the ideal main arch rib force form should only bear the axial force and not the bending moment and shear force. Therefore, for the design of the arch bridge structure, the bending deformation energy and shear deformation energy of the main arch rib need to be as small as possible so that the structure remains mostly unaffected by the bending moment and shear force. The shear deformation energy is generally only about 5% of the bending deformation energy [35], which may not be considered. Therefore, in the selection of arch-axis alignment for arch bridges, the bending strain energy minimum method can be considered for the optimization of arch-axis alignment:

$$U = \int_l \frac{M(x)^2}{2EI} dx \tag{3}$$

For discrete rod system units, Equation (3) can be modified to

$$U = \sum_{i=1}^{m} \frac{l_i}{4E_i I_i} \left(M_{Li}^2 + M_{Ri}^2 \right) \tag{4}$$

where m is the total number of structural units; L_i, E_i, and I_i are the element length, modulus of elasticity, and section bending moment of inertia of element i; and M_{L_i} and M_{R_i} are the left- and right-end bending moments of element i.

4. Analysis of the Effect of Different Parameters on the Ultimate Spanning Diameter

4.1. Methods for Analyzing the Effect of Different Parameters on the Ultimate Span Diameter

In this paper, the ultimate span study of the CFST arch bridge is based on the suspension chain line. As the arch axis line shape and the suspension chain line shape are controlled by the arch axis coefficient m as a single variable, the value of which is closely related to the set of loads on the arch, the ultimate span study process needs to determine the optimal arch axis coefficient under each parameter, such as the diameter of the steel pipe and the wall thickness.

According to the parameter value range formulated in Section 2.1, a steel pipe concrete arch bridge is taken as the engineering basis, and finite element analysis and calculations are carried out according to the established step length. In the process of calculation, firstly, according to the control variables to be studied, a set of initial state parameters for the structure is selected, the corresponding finite element model is established, and the optimal arch axis coefficient corresponding to the set of parameters is searched by minimizing the eccentricity distance of the cross-section; then, strength, stiffness, and stability analyses are carried out. If all of these factors are in the proposed limiting conditions, then the span diameter is increased and analyzed again according to the above steps until any one of the strength, stiffness, and stability analysis results exceeds the limiting value. In this case, the calculation is stopped, and the current span is recognized as the limit span under the group of parameters. Subsequently, according to the above steps, the values of the ultimate span of the sample points increase according to the proposed step length until each group of variables is obtained, and the trend of the changes in the ultimate span under different variables is also obtained. A flowchart of the study on the ultimate span of the CFST arch bridge is shown in Figure 1.

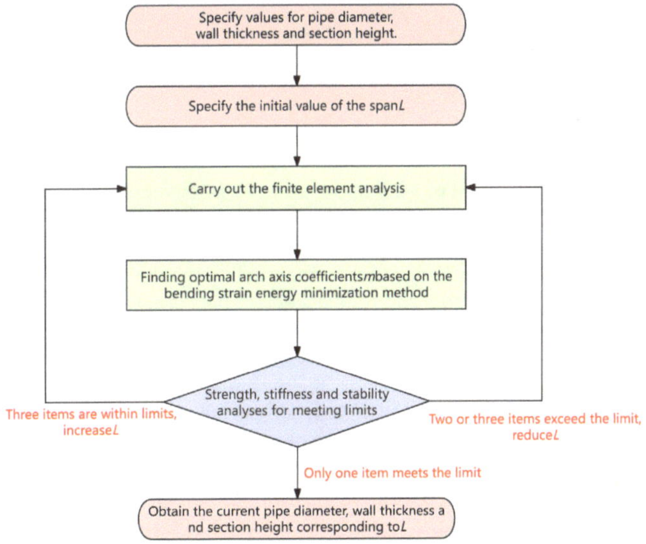

Figure 1. Flowchart for the study of the trend of the influence of single-parameter variation on the ultimate span diameter.

4.2. Finite Element Calculation Models

4.2.1. Project Overview

The Hejiang Yangtze River Highway Bridge is located in Hejiang, Luzhou City, Sichuan Province. The main bridge structure is the swallow-type steel pipe concrete tie arch bridge, and the main arch ribs use steel pipe concrete truss sections. The span arrangement is (80.5 + 507 + 80.5) m with a total length of 668 m, and the bridge is the world's largest-spanning arch bridge. The net height-to-span ratio of the main span is 1/4, the arch axis coefficient is 1.5, and the spacing between the arch ribs on both sides is 25.3 m. The radial height of the arch top cross-section is 7.0 m, and that of the arch foot cross-section is 14 m. The layout of the bridge is shown in Figure 2.

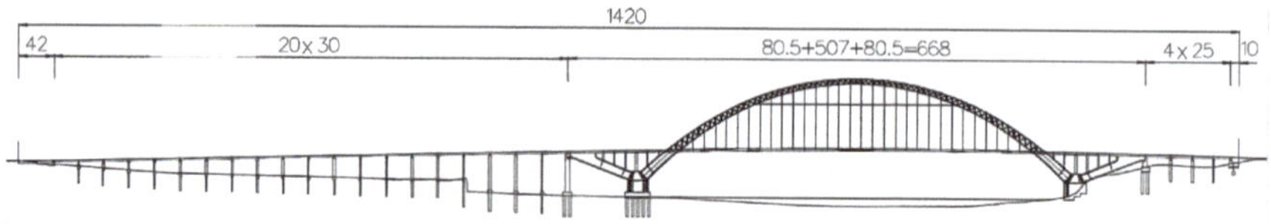

Figure 2. Bridge layout (unit: meter).

4.2.2. Finite Element Model

Combined with the structural characteristics of the Hejiang Yangtze River Highway Bridge, this study adopts Midas Civil V2.1 finite element analysis software to establish a three-dimensional rod system calculation model. The finite element model of the Hejiang Yangtze River Highway Bridge is shown in Figure 3.

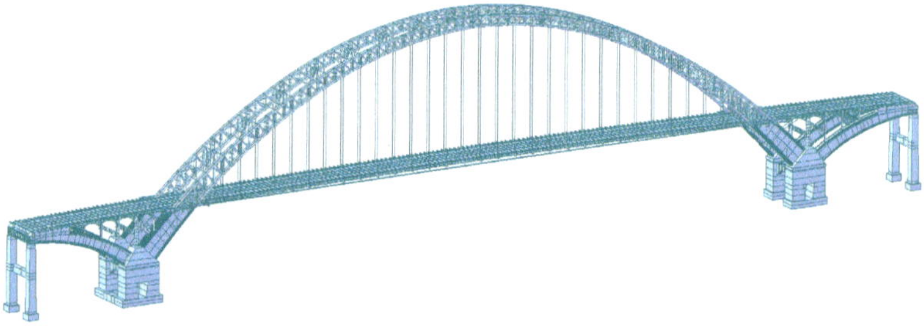

Figure 3. Finite element model of Hejiang Yangtze River Highway Bridge.

A total of 9607 nodes and 23,014 elements are established in the finite element simulation of the Hejiang Yangtze River Highway Bridge, and different element types are adopted according to different structural stress states and considerations. Among them, the suspension rods are simulated by the tensile truss elements, the stiffened plate at the foot of the arch and the steel bridge panels are simulated by the shell elements, and the rest are simulated by the beam elements. The steel pipe is made of Q345 steel: the modulus of elasticity is taken as 2.06×10^5 MPa and the Poisson's ratio is taken as 0.3. The concrete in the pipe is C70 concrete: the modulus of elasticity is taken as 3.7×10^4 MPa and the Poisson's ratio is taken as 0.2. When analyzing the structure, the displacements and rotations at the base of the arch seat and both abutments are constrained. Elastic connections are used between the main arch rib and the columns on the arch, and between the columns and the main beams, with geometric nonlinearities taken into account in the calculations, which only consider the structural deadweight loads.

4.2.3. Validation of Finite Element Models

In order to verify the correctness of the finite element model, the theoretical elevation and the measured elevation of the arch rib were compared after the main arch ring was merged and loosened, and the results are shown in Figure 4.

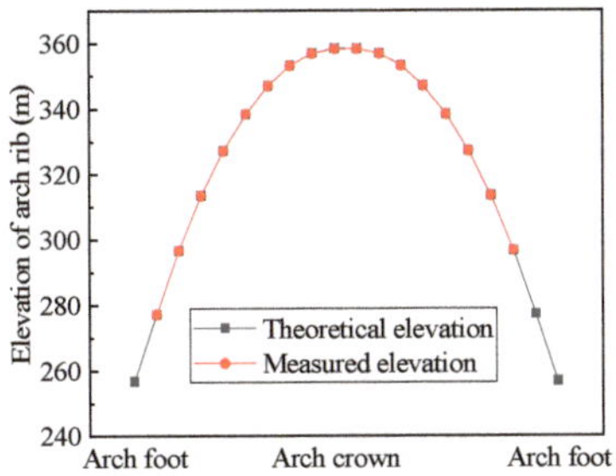

Figure 4. Comparison between theoretical and measured elevation of arch rib.

In Figure 4, the measured and theoretical elevation curves of the arch rib basically coincide with each other, and the maximum relative error is 34 mm, which occurs at the top of the arch, with a maximum relative error of 0.03%, verifying the correctness of the finite element model.

4.3. Analysis of the Effect of Pipe Diameter on the Ultimate Span

In order to study the change trend of the pipe diameter and ultimate span of the CFST arch bridge, the wall thickness of the steel pipe, height-to-span ratio, cross-section parameter, and material type were formulated as constants, and the change trend of the ultimate span of the CFST arch bridge under the corresponding parameter was studied by changing the pipe diameter. The optimal arch axis coefficients of the suspension chain line under the seven groups of different pipe diameter values for the main arch static analysis calculations, in order to meet the strength, stiffness, and stability of the corresponding parameter under the ultimate span value, are shown in Table 2, and an illustration of the pipe diameter and the trend of the ultimate span diameter is shown in Figure 5.

Table 2. Ultimate spans of steel pipe concrete arch bridges corresponding to different pipe diameters.

Pipe Diameter (m)	Wall Thickness (mm)	Height of Arch Foot Section (m)	Height of Vault Section (m)	Height-to-Span Ratio	Optimum Arch Axis Coefficient	Ultimate Span (m)
1.26					1.35	737
1.30					1.35	768
1.34					1.30	776
1.38	32	14.24	7.12	1/4	1.30	784
1.42					1.30	780
1.46					1.35	758
1.50					1.35	751

According to Figure 5, it can be concluded that, with the increase in the cross-section pipe diameter, the limit span value under the corresponding parameter shows a trend of increasing and then decreasing. At the peak apex of the curve, i.e., when the diameter of the steel–concrete pipe is taken as 1.38 m, the stress of the steel pipe of the lower chord under the optimal arch axial shape of the hanging chain line reaches the limit value, and the span cannot be further increased. On the left side of the peak point of the curve, the compressive and flexural stiffness of the concrete section of the steel pipe increases with the increase in the diameter of the steel pipe, which leads to the growth of the ultimate span; however, on the right side of the peak point of the curve, in the case of the constant wall thickness of the steel pipe, the proportion of the steel pipe that bears the constant load stress increases with the increase in the diameter of the steel pipe, and the proportion of the rate of increase is greater than that of the compressive stiffness of the concrete section of the steel pipe. Therefore, the stress of the steel pipe increases gradually and exceeds the allowable strength of the material, which results in a gradual decrease in the ultimate span under the corresponding parameter.

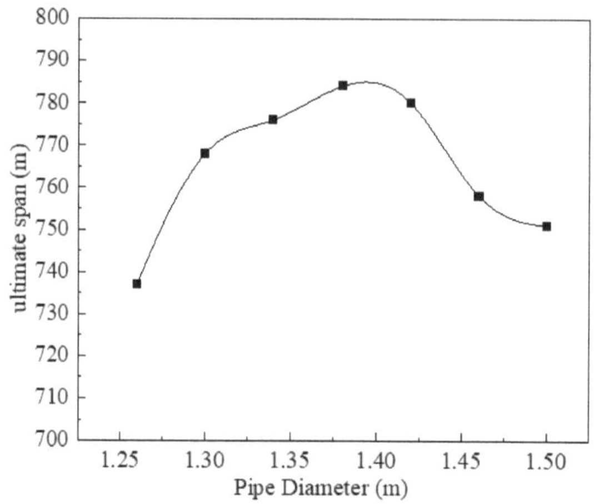

Figure 5. Effect on ultimate span under pipe diameter change.

4.4. Analysis of the Effect of Pipe Wall Thickness on Ultimate Span

In order to study the change trend of the wall thickness of the steel pipe in the main arch section and the ultimate span of the CFST arch bridge, the diameter of the steel pipe, concrete pipe, truss height, height-to-span ratio, cross-section parameters, and material type were formulated as constants, and the change in the ultimate span of the CFST arch bridge was studied by changing the wall thickness of the steel pipe concrete under the corresponding parameters. Based on the optimal arch axis coefficient of the hanging chain line, the main arch static analysis was calculated for five groups of different wall thickness values, and the ultimate span values were obtained under the corresponding parameters to satisfy the strength, stiffness, and stability. The values are shown in Table 3, and the trend of the wall thickness and ultimate span is shown in Figure 6.

Table 3. Ultimate span corresponding to different tube wall thicknesses.

Pipe Diameter (m)	Wall Thickness (mm)	Height of Arch Foot Section (m)	Height of Vault Section (m)	Height-to-Span Ratio	Optimum Arch Axis Coefficient	Ultimate Span (m)
	0.032				1.40	710
	0.036				1.30	780
1.5	0.040	20	10	1/4	1.25	850
	0.044				1.20	900
	0.048				1.20	960

As can be seen from Table 3 and Figure 6, the ultimate span of the CFST arch bridge increases approximately linearly with the increase in the wall thickness of the steel pipe, which is due to the fact that, with the increase in the steel content in the cross-section, the stresses in the steel pipe and in the concrete are gradually reduced, which allows the value of the ultimate span of the structure to be further increased. The steel content is 9.11% at a wall thickness of 32 mm, and increases to 10.33%, 11.58%, 12.85%, and 14.14% according to the proposed increase in step length, corresponding to an increase in the ultimate span of 8.94%, 16.03%, 21.06%, and 25.62%, which shows that an increase in the wall thickness increases the value of the ultimate span of the corresponding parameter.

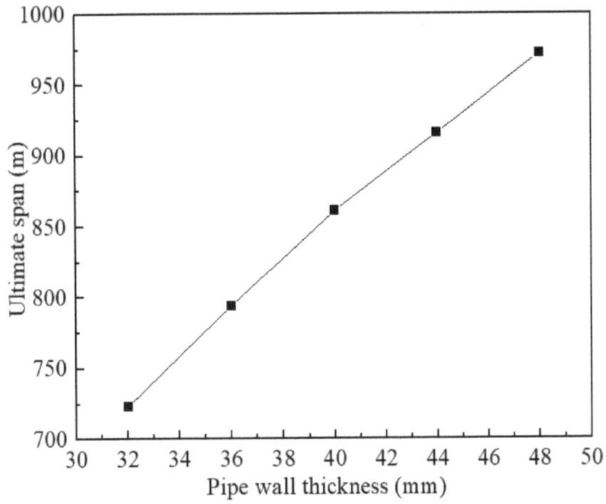

Figure 6. Ultimate spans of CFST arch bridges with different steel tube wall thicknesses.

4.5. Analysis of the Effect of Section Height on Ultimate Span Diameter

In order to study the trends for the main arch cross-section truss height and the ultimate span of the CFST arch bridge, the steel pipe concrete diameter, wall thickness, height-to-span ratio, cross-section parameter, and material type were set as constants, and the trend of the ultimate span of the CFST arch bridge under the corresponding parameter was investigated by changing the cross-section height. Based on the optimal arch axis coefficient of the hanging chain line, 11 groups of different truss heights were calculated for the static analysis of the main arch rib. The values of the ultimate span under the corresponding parameters of strength, stiffness, and stability are shown in Table 4, and the trend of the truss height and the ultimate span is plotted in Figure 7.

Table 4. Table of ultimate span parameters for main arch with different section heights.

Pipe Diameter (m)	Wall Thickness (mm)	Height of Arch Foot Section (m)	Height of Vault Section (m)	Height-to-Span Ratio	Optimum Arch Axis Coefficient	Ultimate Span (m)
1.5	40	14	7.0	1/4	1.30	800
		15	7.5		1.25	850
		16	8.0		1.20	885
		17	8.5		1.25	875
		18	9.0		1.25	865
		19	9.5		1.25	858
		20	10.0		1.25	850
		21	10.5		1.30	836
		22	11.0		1.30	820
		23	11.5		1.30	805
		24	12.0		1.30	810

According to Table 4 and Figure 7, with the increase in the cross-section truss height, the limit span value under the corresponding parameter shows a trend of increasing and then decreasing. At the peak of the curve, i.e., when the height of the foot of the arch truss is taken as 16 m, the stress of the steel pipe of the lower chord under the optimal arch axis shape of the suspension chain line reaches the limit value, and the span cannot be further increased. The bending stiffness of the section of the arch foot increases with the increase in the section height, but the stability coefficients of each order of the main arch ring first decrease slowly and then increase rapidly with the increase in the section height. This is due to the fact that the stability of the main arch ring is affected by both the section stiffness and the span diameter at the same time, and the favorable influence of the former is greater than the unfavorable influence of the latter when the section height is in the range of 14–16 m. The opposite is true when the section height is more than 16 m, which ultimately results in the increase and then the decrease in the ultimate span diameter of the CFST arch bridge with the increase in the section height.

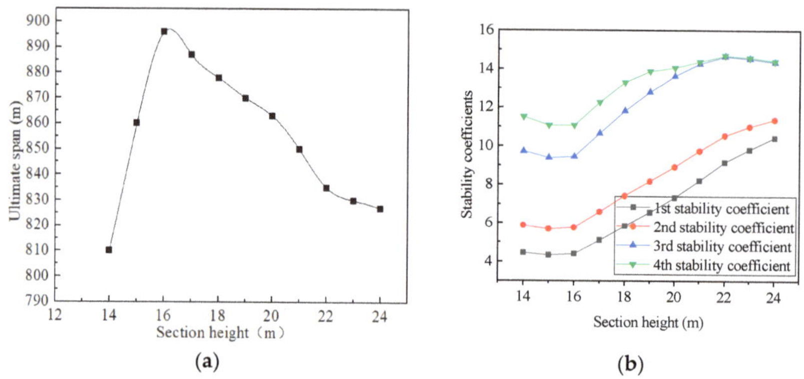

Figure 7. Ultimate span and stability coefficients of main arch ribs at different section heights. (**a**) Ultimate span at different section heights of the main arch rib; (**b**) stability coefficients of the first four orders of the main arch rib under different cross-section heights.

5. Ultimate Spanning Analysis Based on Response Surface Methodology

In the previous section, the effect of single-parameter changes, such as pipe diameter, wall thickness, and section truss height, on the ultimate span of a CFST arch bridge was investigated, and the analysis process only considered the three constraints of structural strength, stiffness, and stability. In this section, the number of section divisions and the maximum lifting weight of the sections, steel content rate, constraint effect coefficient, and diameter/thickness ratio are increased in order to study the ultimate span of a CFST arch bridge with four additional constraints.

However, through the change of a single variable and according to the proposed step size exhaustive calculation, the workload is huge, the iterative process is complicated, and only the numerical solutions can be obtained rather than the functional equation. This section describes the use of the response surface method for the ultimate spanning diameter; the solution process uses the results from Section 4 as the sample point data, uses the response surface model instead of the complex finite element computation model, and establishes the ultimate spanning diameter mathematical model, which can significantly improve the efficiency of the ultimate spanning diameter determination process.

5.1. Multiple Nonlinear Regression Analysis Based on Response Surface Methodology

The response surface method (RSM) analyzes the response surface and contour lines to seek the optimal parameters. This approach uses multiple quadratic regression equations to fit the response value and the factors of the functional relationship between the optimization of statistical methods. The basic idea is to test the system of the target response value as a function of a number of experimental factors, and such a functional relationship occurs through the derivation of the means of the optimization of the experimental design of the optimal parameters, i.e., the deterministic sample points are tested in order to achieve the goal of fitting a response surface to approximate the distribution of response values.

5.2. Mathematical Modeling of Response Surfaces for Extreme Spans

5.2.1. Identification of Sample Points and Sample Data

The study variables of steel pipe concrete diameter, wall thickness, and cross-section joist height are denoted as x_1, x_2, and x_3, with units of m, mm, and m, respectively, and the study range of the variables is formulated as $x_1 \in (1.26, 150), x_2 \in (32, 48)$, and $x_3 \in (14, 24)$. The study range of the variables is defined as follows.

The finite element response value y is the limiting span under the corresponding parameter, i.e., the sample point data, which are obtained using finite element analysis software calculations. Considering the interval distribution of the sample points and the fact that the appropriate increase in the number of sample points can improve the fitting accuracy, and based on the center combination test design method, a total of 41 groups of sample point data were extracted. The sample points and finite element response values are shown in Table 5.

Table 5. Sample points and corresponding ultimate spans of steel tube concrete arch bridges.

Finite Element Response Value y	Random Variable			Finite Element Response Value y	Random Variable		
Ultimate Span (m)	x_1	x_2	x_3	Ultimate Span (m)	x_1	x_2	x_3
737	1.26	32	16.76	760	1.26	32	16.00
768	1.30	32	16.84	735	1.26	32	19.00
776	1.34	32	16.92	700	1.26	32	24.00
784	1.38	32	17.00	790	1.34	32	16.00
780	1.42	32	17.08	755	1.34	32	19.00
758	1.46	32	17.16	730	1.34	32	24.00
751	1.50	32	17.24	790	1.42	32	16.00
800	1.50	40	14.00	760	1.42	32	19.00
850	1.50	40	15.00	735	1.42	32	24.00
885	1.50	40	16.00	795	1.26	36	16.00
875	1.50	40	17.00	830	1.26	36	19.00
865	1.50	40	18.00	812	1.26	36	24.00
858	1.50	40	19.00	825	1.34	36	16.00
850	1.50	40	20.00	833	1.34	36	19.00
836	1.50	40	21.00	810	1.34	36	24.00
820	1.50	40	22.00	835	1.42	36	16.00
805	1.50	40	23.00	850	1.42	36	19.00
810	1.50	40	24.00	825	1.42	36	24.00
710	1.50	32	20.00	850	1.50	40	20.00
780	1.50	36	20.00				

5.2.2. Limit Span Response Surface Fitting and Accuracy Tests

In this paper, the ultimate span study considers the parameter correlation and chooses the second-order polynomials containing cross terms as the basic equations of the response surface model:

$$y = a + \sum_{i=1}^{n} b_i x_i + \sum_{i=1}^{n} \sum_{j=1}^{n} c_{ij} x_i x_j \tag{5}$$

where a is a constant term, x_i, $x_i x_j$ is the basis function of the response surface model, n is the number of random variables, and a, b_i, c_{ij} are the coefficients of the basis function.

According to the 41 groups of sample points and sample data formulated in Section 5.2.1, the model of the response surface containing only significant terms is fitted based on stepwise regression to obtain the response surface function of the ultimate span of the steel pipe concrete arch bridge. The coefficients of the response surface function and the results of the accuracy test are shown in Tables 6 and 7 below.

Table 6. Response surface modeling function coefficients.

Basis Functions	Constant	x_1^2	x_1	x_2	x_3
Extreme Span Response Surface Modeling	−3182.68	−1887.04	4973.20	−0.43	41.10
Basis Function	x_2^2	x_3^2	x_1x_2	x_1x_3	x_2x_3
Limit Span Response Surface Model	−0.23	−0.89	15.59	−17.68	0.41

Table 7. Sample points and corresponding ultimate spans of steel tube concrete arch bridges.

Response Surface Modeling	F-Test	$R_{adj}{}^2/\%$
Extreme span response surface	124.42	95.40

The F-test mentioned in Table 7 is a statistical method used to determine that at least one of the coefficients in the model is significantly non-zero. In response to the surface methodology, it is primarily used to test the significance of the entire regression model. If the p-value of the F-test is less than a certain threshold, the entire model is considered significant, i.e., at least one of the predictor variables in the model has a significant effect on the response variable. The coefficient of determination R^2 is a statistic that measures the goodness of fit of a model, with a value between 0 and 1. R^2 describes the degree to which the independent variable in the model explains the variation in the response variable. An R^2 value close to 1 means that the model explains most of the variation in the response variable, and in response surface methodology, higher R^2 values usually indicate a better model fit to the data.

In practice, the F-test and R^2 are often used in conjunction to assess the statistical significance and goodness-of-fit of a model. The F-test tells us whether the model is significant overall, while the R^2 provides a quantitative measure of the model's ability to account for variation. Considering both statistics together can be used to assess the validity and fitness of a model.

The maximum value of the critical F value is 3.41 at a significance level of $\alpha = 0.05$. From Table 7, it can be seen that the F value of the extreme spanning response surface model is 124.42, which is much larger than the critical F value, indicating that the response surface model is significant. From Table 7, it can be seen that the $R_{adj}{}^2$ value is 95.4%, which indicates that 95.4% of the total deviation of the extreme spanning response surface model under the corresponding parameter is caused by the change in the random variable, i.e., the total response is not explained by the resulting response surface model, and, at most, 4.6% of the deviation cannot be explained by the resulting response surface model. The fitting accuracy of the response surface model is high, and the obtained response surface model is reliable and can be used for the subsequent solving of the limiting spanning values that satisfy each constraint.

5.3. 1:10 Optimization of the Response Surface Model for Extreme Spanning Solutions

5.3.1. Mathematical Model for Limit Span Optimization

The steel pipe concrete diameter, wall thickness, and section truss height are selected as design variables, which are labeled as x_1, x_2, and x_3, and the ultimate span diameter is selected as the objective function. Under the restriction indexes formulated in Section 2.2, the maximum value of the ultimate span of the steel pipe concrete arch bridge can be achieved, and then the ultimate span optimization problem can be expressed as follows:

$$\begin{cases} \max F \\ 1.26 \le x_1 \le 1.5 \\ 32 \le x_2 \le 48 \\ 14 \le x_3 \le 24 \\ 0.04 \le \frac{A_s}{A_c} = \frac{x_1^2}{(x_1-2x_2)^2} - 1 \le 0.2 \\ \xi = \frac{420}{50.2} \times \left(\frac{x_1^2}{(x_1-2x_2)^2} \right) \ge 0.6 \\ x_1/x_2 \ge 0.04 \end{cases} \tag{6}$$

where A_s/A_c denotes the cross-section steel content of the steel pipe concrete member; A_s and A_c denote the total area of steel reinforcement and the total area of concrete in the cross-section, respectively; and ξ denotes the coefficient of the confinement effect of the steel pipe concrete.

In order to more intuitively analyze the relationship between the study variables and the objective function as the ultimate spanning diameter, the response surface model for the ultimate spanning diameter was analyzed graphically and intuitively using the computational analysis software MATLAB R2018a.

The trend of the ultimate spanning response surface with respect to the study variables x_2 and x_3 when the study variable $x_1 = 1.4$ is shown in Figure 8, the trend of the ultimate spanning response surface with respect to the study variables x_1 and x_3 when the study variable $x_2 = 36$ is shown in Figure 9, and the trend of the ultimate spanning response surface with respect to the study variables x_1 and x_2 when the study variable $x_3 = 20$ is shown in Figure 10.

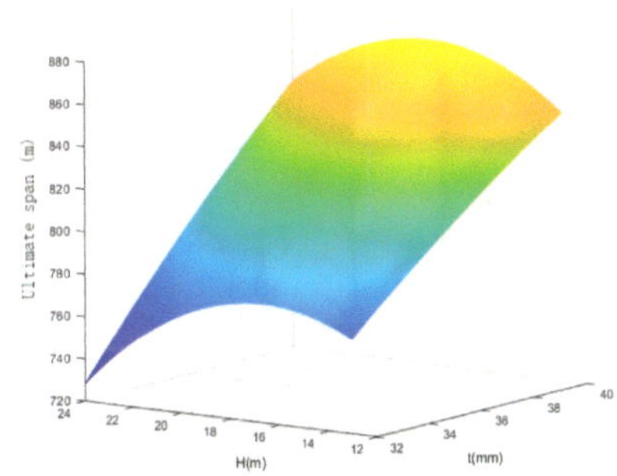

Figure 8. Limit span response surface $x_1 = 1.40$.

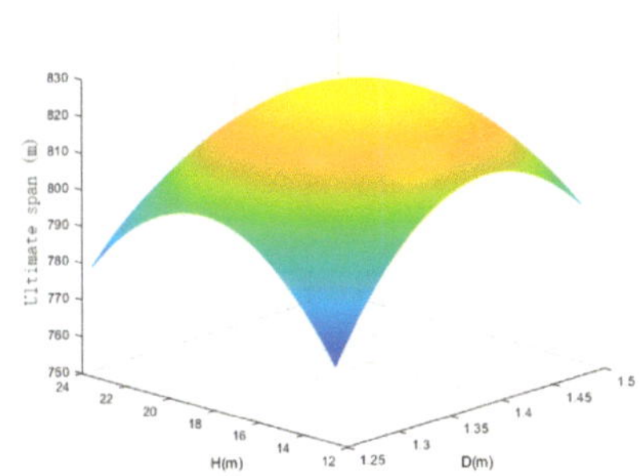

Figure 9. Limit span response surface $x_2 = 36$.

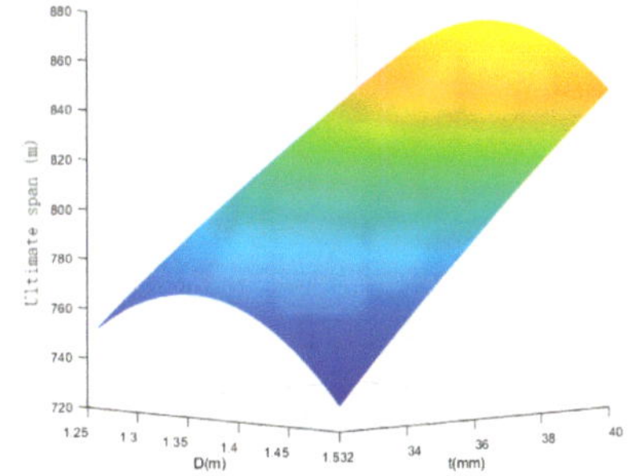

Figure 10. Limit span response surface $x_3 = 20$.

The above Figures 8–10 showing the variation surface of the limit span with the study variables clearly reflect the response values and the direction of finding the maximum limit span at different values of the variables. The change in color from blue to yellow in Figures 8–10 represents the increase in ultimate span from small to large.

According to the ultimate span response surface model and the optimization mathematical model, the optimization toolbox of MATLAB and the multivariate nonlinear optimization function were used to optimize and solve the problem and ultimately reached the optimum at $x_1 = 1.49, x_2 = 37$, and $x_3 = 17$, corresponding to the value of an ultimate span of 821 m and an optimal arch axis coefficient of 1.2.

This ultimate span exceeds the span of the current largest CFST arch bridge in the world, the Guangxi Pingnan Third Bridge, which has a main span of 575 m; this represents a span increase of 246 m (42.78%), indicating that CFST arch bridges still have significant potential for development in large-span bridges.

It should be additionally noted that, when an i5-11300H four-core CPU with 16G RAM is used to carry out the optimization calculations, the optimization time is no more than 1 min, which is much less than the time required using finite element analysis, verifying the high level of efficiency of the response surface method for determining the ultimate span of CFST arch bridges.

5.3.2. Finite Element Verification of Optimization Results

According to the above findings, in order to verify the correctness of the optimization results based on the response surface model, the maximum ultimate span and the corresponding structural parameters of the optimization solution were calculated using finite element analysis. The maximum ultimate span finite element model is shown in Figure 11.

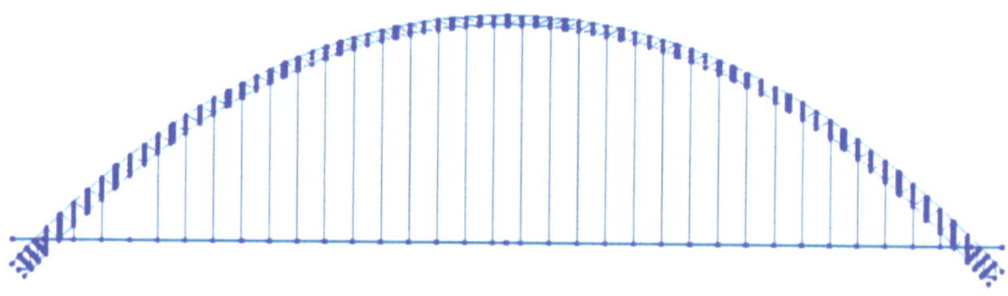

Figure 11. Finite element modeling of maximum ultimate span of CFST arch bridge.

The results of the strength calculations of the steel pipe and the concrete in the pipe under the combination of the basic effects of the ultimate load-carrying capacity state are shown in Figure 12.

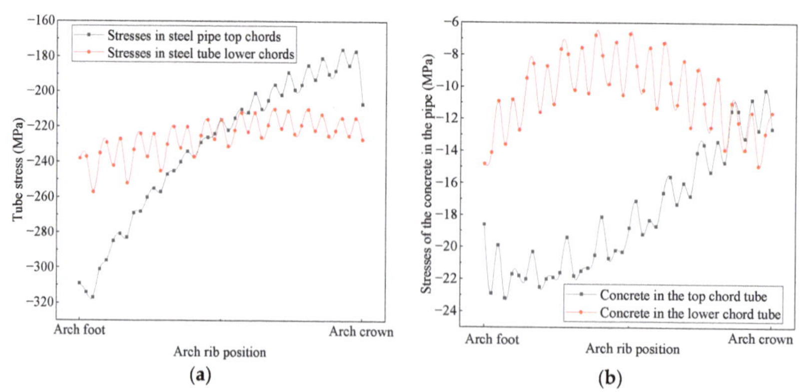

Figure 12. Main arch rib stress simulation results. (**a**) Stress distribution along the span of steel pipe stringers; (**b**) stress distribution along the span of the concrete in the pipe.

As can be seen from Figure 12, the maximum stresses of the steel pipe and concrete in the pipe appear at the foot of the arch, and the stresses of the steel pipe and concrete in the pipe in the upper chord are more uniformly distributed along the span, with a large increase in the middle of the span. The stresses of the steel pipe and concrete in the pipe in the lower chord gradually decrease along the distribution of the span, and the steel pipe and the concrete in the pipe are in a state of compression. The maximum stress values under load are as follows:

$$\begin{cases} (\sigma_s)_{\max} = |-317.81|\text{MPa} \le 320\ \text{MPa} \\ (\sigma_c)_{\max} = |-23.80|\text{MPa} \le 34.6\ \text{MPa} \end{cases} \tag{7}$$

According to the maximum stress value, it can be seen that the concrete stress reserve in the pipe is large, and the maximum stress value of the steel pipe is approaching the strength limit value. At this time, the limit span continues to increase with the strength of the steel pipe.

The maximum vertical deflection of the main arch of the CFST arch bridge under lane loads (excluding impact forces) is shown in Figure 13.

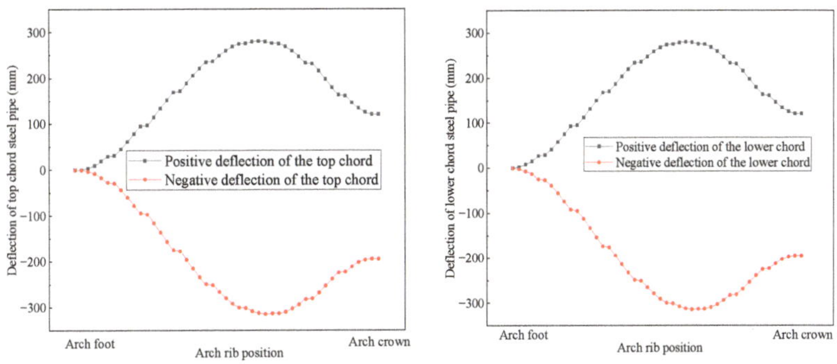

(a) (b)

Figure 13. Distribution of steel pipe deflection along the span diameter. (**a**) Distribution of deflection of top chord steel pipe along span diameter; (**b**) distribution of deflection of lower chord steel pipe along span diameter.

From Figure 13, it can be seen that the deflection extremes of the upper and lower chord bars under lane loading (not counting the impact force) both appeared at 1/4L, and the maximum vertical deflection value was as follows:

$$W_{\max} = 593.21\ \text{mm} \le \frac{\text{L}}{1000} = 829.00\ \text{mm} \tag{8}$$

An elastic buckling stability analysis was carried out for the overall and local stability of the steel pipe concrete main arch only. The buckling analysis was carried out using the Midas civil space finite element model with the moving load equivalent concentrated force as a variable load and other loads as a constant load. The results of the first four orders of instability modal calculations are shown in Figure 14.

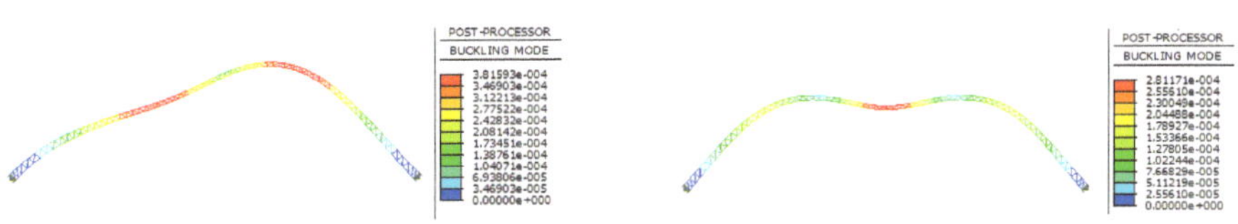

(a) (b)

Figure 14. *Cont.*

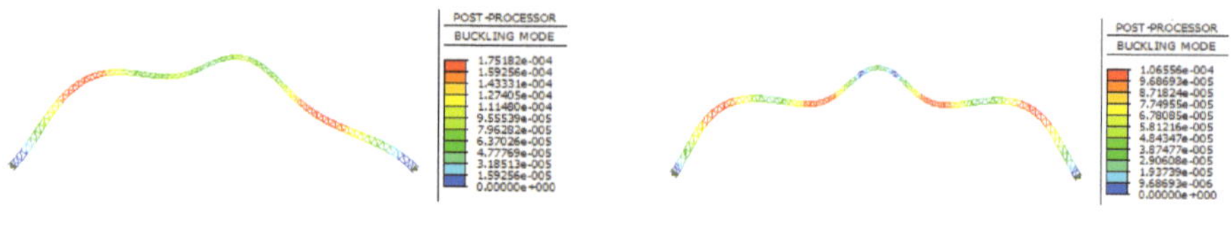

(c) (d)

Figure 14. Modal diagram of the first four orders of instability of the main arch rib. (**a**) The 1st instability modal diagram of the main arch rib; (**b**) 2nd instability modal diagram of the main arch rib; (**c**) 3rd instability modal diagram of the main arch rib; (**d**) 4th instability modal diagram of the main arch rib.

The first four orders of instability forms and elastic buckling stability coefficients during the operational phase of the completed bridge are shown in Table 8.

Table 8. Sample points and corresponding ultimate spans of CFST arch bridges.

Order	Destabilized Mode	Elastic Flexural Stability Coefficient
1st destabilization	In-plane antisymmetric instability	6.33
2nd destabilization	In-plane symmetric instability	9.07
3rd destabilization	In-plane antisymmetric instability	14.49
4th destabilization	In-plane symmetric instability	17.30

From the instability modal diagrams presented in Figure 14 and the elastic buckling stability coefficients shown in Table 8, it can be seen that the elastic buckling stability coefficients of all orders are greater than the allowable values.

The CFST arch bridge steel structure arch rib (excluding the transverse brace) weighs a total of 12,512 t. Single rib lifting can be divided into 32 segments, with the maximum lifting weight segment control of 200 t, to meet the proposed restrictions.

Steel pipe concrete contains steel at a rate α_s of:

$$0.04 \leq \alpha_s = \frac{A_s}{A_c} = \frac{x_1^2}{(x_1 - 2x_2)^2} - 1 = 0.108 \leq 0.2 \quad (9)$$

The steel content meets the limits drawn up in accordance with the specifications.

The steel pipe concrete constraint effect coefficient ξ is:

$$\xi = \frac{420}{50.2} \times \left(\frac{x_1^2}{(x_1 - 2x_2)^2} \right) = 0.89 \geq 0.6 \quad (10)$$

The constraint effect coefficients satisfy the constraints drawn up in accordance with the specification.

6. Conclusions

In order to study the ultimate span of a CFST arch bridge, taking the structural strength, stiffness, and stability as the limiting conditions, the finite element analysis method was adopted to examine the trend of the influence of the change in a single parameter of the pipe diameter, wall thickness, and truss height on the ultimate span on the basis of the optimal analysis of the arch axial shape. The response surface method was introduced into the optimization of the ultimate span to solve the problem, and then the value of the ultimate span of the CFST arch bridge was obtained. The main conclusions are as follows:

(1) The diameter of the steel pipe concrete and the corresponding parameters of the ultimate span diameter showed a trend of increasing and then decreasing. When the

diameter of the steel pipe concrete was taken as 1.38 m, the stress of the lower chord steel pipe reached the limiting value under the optimal arch axial shape of the suspension chain line; the wall thickness of the steel pipe concrete and the corresponding parameters of the ultimate span diameter showed a trend of continuous increase; and the truss height of the main arch cross-section and the corresponding parameters of the ultimate span diameter showed a trend of increasing and then decreasing.

(2) An optimization method based on the response surface method was proposed to solve the ultimate spanning diameter, and the response surface model of the ultimate spanning diameter was established by taking the pipe diameter, wall thickness, and truss height as the research variables, and the response surface function of the ultimate spanning diameter with high reliability was obtained. By establishing the ultimate spanning diameter optimization mathematical model and adopting the multivariate nonlinear optimization function for the optimization and solution, a maximum value of the ultimate spanning diameter of 821 m was reached when the pipe diameter was 1.49 m, the wall thickness t was 37 mm, and the truss height was 17 m, satisfying all the constraints. The optimal arch axis coefficient was 1.2.

(3) The verification analysis was carried out using finite element software, and the results showed that the structural strength just reached the limit value, and the remaining variables, such as the stiffness, stability, and the maximum lifting weight of the section, were all within the limit values.

Environmental factors, as important factors affecting the construction quality of large-span CFST arch bridges, restrict the span development of these bridges to a certain extent. Mountain canyon winds, ambient temperatures, construction schemes, etc., tend to directly affect the structural stability, deformation, and stress during the construction of arch bridges, and carrying out a study on the ultimate spans of CFST arch bridges that considers environmental factors and the actual construction scheme used will be the next step in our research.

Author Contributions: Conceptualization, Y.W. and X.W. (Xiangchuan Wang); methodology, Y.W. and Y.F.; software, Y.W.; validation, C.L., J.S. and X.W. (Xiangchuan Wang); formal analysis, Y.W.; investigation, X.W. (Xinzhong Wang); resources, Y.F.; data curation, C.L.; writing—original draft preparation, J.S.; writing—review and editing, X.W. (Xinzhong Wang); visualization, Y.W.; supervision, J.S.; project administration, C.L.; funding acquisition, X.W. (Xiangchuan Wang). All authors have read and agreed to the published version of the manuscript.

Funding: This work was supported by the National Natural Science Foundation of China (Grant No. 51908094) and the Project of the Education Bureau of Hunan Province of China (Grant No. 23B0732, No. 23A0559, and No. 22A0561).

Data Availability Statement: The data presented in this study are available from the first and corresponding author upon request. The data are not publicly available due to the policy of the data provider.

Conflicts of Interest: Author Xiangchuan Wang was employed by the company Huasheng Testing Technology Co., Ltd. The remaining authors declare that the research was conducted in the absence of any commercial or financial relationships that could be construed as a potential conflict of interest.

References

1. Liu, J.; Li, X.; Liu, H.; Chen, B. Recent Application and Development of Concrete-Filled Steel Tube Arch Bridges in China. In *Advances in Civil Engineering Materials, Proceedings of the 6th International Conference on Architecture and Civil Engineering (ICACE 2022), Kuala Lumpur, Malaysia, 18 August 2022*; Nia, E.M., Ling, L., Awang, M., Emamian, S.S., Eds.; Springer: Singapore, 2023; pp. 263–272.
2. Zheng, J.; Wang, J.; Mou, T.; Feng, Z.; Han, Y.; Qin, D. Feasibility Study on Design and Construction of Concrete Filled Steel Tubular Arch Bridge with a Span of 700 m. *Chin. Acad. Eng.* **2014**, *16*, 33–37.
3. Tong, K.; Zhang, H.; Zhao, R.; Zhou, J.; Ying, H. Investigation of SMFL Monitoring Technique for Evaluating the Load-Bearing Capacity of RC Bridges. *Eng. Struct.* **2023**, *293*, 116667. [CrossRef]
4. Yu, M.; Deng, N.; Chen, Q.; Hao, T. Refined Finite Element Analysis of Crack Causes in SRC Arch Rib Bridges Considering Multiple Factors. *Adv. Civ. Eng.* **2018**, *2018*, 2690951. [CrossRef]

5. Yang, Y.; Luo, X.; Tang, Y.; Wwei, J.; Chen, B. Research on the Ultimate Bearing Capacity and Simplified Design Formulae of Steel Truss Web-Steel Reinforced Concrete (SRC) Composite Arches. *Eng. Mech.* **2017**, *34*, 200–206.
6. Pouraminian, M.; Ghaemian, M. Shape Optimisation of Concrete Open Spandrel Arch Bridges. *Gradjevinar* **2015**, *67*, 1177–1185.
7. Xie, X.; Fu, Y.; Deng, N. Design of Half-Through Cable-Arch Bridge with 700 m Main Span. *J. Southwest Jiaotong Univ.* **2019**, *54*, 1162–1176. [CrossRef]
8. Huang, Q.; Wei, J.; Chen, B.; Wu, M. Trial-Design Research on 420m SRC-Steel Web Composite Box Arch Bridge. *J. Fuzhou Univ. Nat. Sci. Ed.* **2011**, *39*, 936–940.
9. Huang, Q.; Fu, Y.; Xu, C.; Chen, B. Trial Design Research on Super Long Span Reactive Powder Concrete Arch Bridge. *J. Nanchang Univ. Eng. Technol.* **2015**, *37*, 252–256+266. [CrossRef]
10. Shao, X.; He, G. Conceptual Design and Feasibility Study of an 800 m Scale Steel-UHPC Composite Truss Arch Bridge. *China J. Highw. Transp.* **2020**, *33*, 73.
11. Zhang, X.; Deng, Z.; Fang, G.; Ge, Y. Theoretical Analysis of Ultimate Main Span Length for Arch Bridge. *Sustainability* **2022**, *14*, 17043. [CrossRef]
12. Shao, X.; He, G.; Shen, X.; Zhu, P.; Chen, Y. Conceptual Design of 1000 m Scale Steel-UHPFRC Composite Truss Arch Bridge. *Eng. Struct.* **2021**, *226*, 111430. [CrossRef]
13. Zheng, J.; Du, H.; Mu, T.; Liu, J.; Qin, D.; Mei, G.; Tu, B. Innovations in Design, Construction, and Management of Pingnan Third Bridge—The Largest-Span Arch Bridge in the World. *Struct. Eng. Int.* **2022**, *32*, 134–141. [CrossRef]
14. Wei, J.; Ying, H.; Yang, Y.; Zhang, W.; Yuan, H.; Zhou, J. Seismic Performance of Concrete-Filled Steel Tubular Composite Columns with Ultra High Performance Concrete Plates. *Eng. Struct.* **2023**, *278*, 115500. [CrossRef]
15. Wang, X.; Li, L.; Xiang, Y.; Wu, Y.; Wei, M. The Influence of Basalt Fiber on the Mechanical Performance of Concrete-Filled Steel Tube Short Columns under Axial Compression. *Front. Mater.* **2024**, *10*, 1332269. [CrossRef]
16. Zhang, X.; Zhou, G.; Liu, X.; Fan, Y.; Meng, E.; Yang, J.; Huang, Y. Experimental and Numerical Analysis of Seismic Behaviour for Recycled Aggregate Concrete Filled Circular Steel Tube Frames. *Comput. Concr.* **2023**, *31*, 537.
17. Zhang, X.; Liu, X.; Zhang, S.; Wang, J.; Fu, L.; Yang, J.; Huang, Y. Analysis on Displacement-Based Seismic Design Method of Recycled Aggregate Concrete-Filled Square Steel Tube Frame Structures. *Struct. Concr.* **2023**, *24*, 3461–3475. [CrossRef]
18. Huang, W.; Lai, Z.; Chen, B.; Xie, Z.; Varma, A.H. Concrete-Filled Steel Tube (CFT) Truss Girders: Experimental Tests, Analysis, and Design. *Eng. Struct.* **2018**, *156*, 118–129. [CrossRef]
19. Lai, X.; Chen, B.; Zheng, J.; Huang, L. Experimental Study on Stability Capacity of CFT Columns Considering the Influence of Long-Term Load. *Build. Sci.* **2023**, *39*, 65–73.
20. Xu, L.; Lu, Q.; Chi, Y.; Yang, Y.; Yu, M.; Yan, Y. Axial Compressive Performance of UHPC Filled Steel Tube Stub Columns Containing Steel-Polypropylene Hybrid Fiber. *Constr. Build. Mater.* **2019**, *204*, 754–767. [CrossRef]
21. Shi, Z.; Hu, H.; Li, J. Axis Optimisation of Arch-Shaped Pylons for High-Speed Railway Cable-Stayed Bridges. *Eng. Struct.* **2021**, *227*, 111424. [CrossRef]
22. Zheng, J.; Wang, J. Concrete-Filled Steel Tube Arch Bridges in China. *Engineering* **2018**, *4*, 143–155. [CrossRef]
23. Lan, H. Investigation and Analysis of Development Situation of CFST Truss Arch Bridge in China. *West. China Commun. Technol.* **2018**, *7*, 134–138. (In Chinese) [CrossRef]
24. Chen, B.-W.; Han, L.-H.; Qin, D.-Y.; Li, W. Life-Cycle Based Structural Performance of Long-Span CFST Hybrid Arch Bridge: A Study on Arch of Pingnan Third Bridge. *J. Constr. Steel Res.* **2023**, *207*, 107939. [CrossRef]
25. Chen, B.; Liu, J.; Habib, T. Recent Research and Application of Arch Bridges in China. In *Proceedings of ARCH 2019, Proceedings of the 9th International Conference on Arch Bridges, Porto, Portugal, 2–4 October 2019*; Arêde, A., Costa, C., Eds.; Springer International Publishing: Cham, Switzerland, 2020; pp. 536–544.
26. Fan, Y.; Xin, J.; Yang, L.; Zhou, J.; Luo, C.; Zhou, Y.; Zhang, H. Optimization Method for the Length of the Outsourcing Concrete Working Plane on the Main Arch Rib of a Rigid-Frame Arch Bridge Based on the NSGA-II Algorithm. *Structures* **2024**, *59*, 105767. [CrossRef]
27. Fan, Y.; Zhou, J.; Luo, C.; Yang, J.; Xin, J.; Wang, S. Research on Loading Scheme for Large-Scale Model Tests of Super-Long-Span Arch Bridge. *Buildings* **2023**, *13*, 1639. [CrossRef]
28. Chen, B.; Liu, F.; Wei, J. Statistical Analysis of 327 Steel-Tube Concrete Arch Bridges. *J. China Foreign Highw.* **2011**, *31*, 96–103. (In Chinese) [CrossRef]
29. Ye, W.; Wang, L.; Sun, X.; Xu, D.; Wang, T. Key Parameter Influence Analysis of Half-through Concrete-Filled Steel Tubular Arch Bridge. *Bridge Constr.* **2023**, *53*, 79–86. (In Chinese) [CrossRef]
30. Chen, B.; Liu, J.; Wei, J. *Concrete-Filled Steel Tubular Arch Bridges*; Springer: Singapore, 2023.
31. Khuri, A.I.; Mukhopadhyay, S. Response Surface Methodology. *Wiley Interdiscip. Rev. Comput. Stat.* **2010**, *2*, 128–149. [CrossRef]
32. Ye, D.; Xu, Z.; Liu, Y. Solution to the Problem of Bridge Structure Damage Identification by a Response Surface Method and an Imperialist Competitive Algorithm. *Sci. Rep.* **2022**, *12*, 16495. [CrossRef] [PubMed]
33. Tang, Q.; Xin, J.; Jiang, Y.; Zhang, H.; Zhou, J. Dynamic Response Recovery of Damaged Structures Using Residual Learning Enhanced Fully Convolutional Network. *Int. J. Struct. Stab. Dyn.* **2024**. [CrossRef]

34. Temür, R. Analyses of Plane Stress and Plane Strain through Energy Minimization. *Structures* **2021**, *33*, 728–736. [CrossRef]
35. Juvinall, R.C.; Marshek, K.M. *Fundamentals of Machine Component Design*; John Wiley & Sons: Hoboken, NJ, USA, 2020.

sensors

MDPI

Article

An Evaluation Framework for Construction Quality of Bridge Monitoring System Using the DHGF Method

Jingzhou Xin [1], Chen Wang [1], Qizhi Tang [1,*], Renli Zhang [2] and Tao Yang [2]

[1] School of Civil Engineering, Chongqing Jiaotong University, Chongqing 400074, China; xinjz@cqjtu.edu.cn (J.X.); 18738259901@163.com (C.W.)
[2] Northeast Operation Branch of Chongqing Expressway Group Co., Ltd., Chongqing 401120, China; zhangrenli@cegc.com.cn (R.Z.); yangtao@cegc.com.cn (T.Y.)
* Correspondence: tangqizhi1015@163.com

Abstract: Aiming at comprehensively evaluating the status of a bridge monitoring system, an evaluation framework based on the improved Delphi, analytic Hierarchy process, Grey relations analysis and Fuzzy integrated evaluation (DHGF) is selected. Firstly, the evaluation indexes for the bridge monitoring system are determined by an anonymous group discussion and expert questionnaire using the improved Delphi method. Secondly, a comparison matrix of the evaluation indexes is constructed to determine the comprehensive weight via the analytic hierarchy process. Then, based on the gray relations analysis, the albino weight function is constructed, the evaluation gray class is determined, and the single-factor fuzzy evaluation matrix is obtained. Finally, the final evaluation result was obtained by the fuzzy comprehensive evaluation. The evaluation results of a real bridge monitoring system show that the evaluation level of the monitoring system was level II, and the proposed framework could better reflect the construction and operation status of the monitoring system.

Keywords: bridge health monitoring; monitoring system evaluation; DHGF; comprehensive evaluation

Citation: Xin, J.; Wang, C.; Tang, Q.; Zhang, R.; Yang, T. An Evaluation Framework for Construction Quality of Bridge Monitoring System Using the DHGF Method. *Sensors* **2023**, *23*, 7139. https://doi.org/10.3390/s23167139

Academic Editor: Ching-Tai Ng

Received: 28 June 2023
Revised: 26 July 2023
Accepted: 9 August 2023
Published: 12 August 2023

1. Introduction

With the long-term effect of natural factors and increasing traffic, the safety of bridge structures will inevitably deteriorate, and catastrophic accidents will occur in extreme cases. To ensure the safe operation of bridges, bridge monitoring systems have been widely established. At present, the goals of real-time monitoring, synchronous analysis, and data network sharing of monitoring systems have been gradually achieved [1–5]; however, the accurate diagnosis of bridge status, as the main functional goal of health-monitoring systems [6], is still difficult due to the complexity of the bridges, the randomness of external loads, and the uncertainty of the service environment. Therefore, most of the existing studies focus on the accumulation of historical data [7–13], the identification of external load and internal damage [14–16], and the assessment of overall safety [17]. For example, Jian et al. [18] automatically detected and classified a large number of fault monitoring data points by using the histogram of the relative frequency distribution of the data and the one-dimensional convolutional neural network and then comprehensively verified the selected acceleration data of the two long-span bridges. Hajializadeh et al. [19] propose a bridge damage recognition method based on deep learning, which can accurately and automatically identify the damage at different speeds and track unevenness and ambient noise. Zhang et al. [20] proposed a robustness index evaluation method based on bridge health monitoring data, which can effectively evaluate the structural state of bridges.

After the long period of operation of some monitoring facilities, there are phenomena such as a lack of basic data, damage to instruments and equipment, unstable system operation, and chaotic monitoring data. The monitoring system belongs to complex multidisciplinary cross-system engineering, and many factors, such as the coverage degree of sensor layout, the installation of monitoring equipment, and the quality of monitoring data, may cause large deviations in structural evaluation results. In this context, the evaluation of bridge monitoring systems (BHM) has become a new topic for scholars and engineers in the field. In terms of the operation state analysis of BHM systems, Li et al. [21] proposed a method for sensor fault detection using a generalized likelihood ratio and correlation coefficient and carried out manual fault sensor detection and classification verification tests on the Yangtze River Bridge, which can realize the online diagnosis of multiple sensor faults. Li et al. [22] proposed an abnormal signal detection method for the BHM system based on a two-segment deep convolutional neural network, which improved the recognition accuracy of abnormal signal patterns. Yu et al. [23] proposed a sensor state assessment and fault diagnosis method based on multi-dimensional information fusion, which can accurately obtain information from multiple sensors through the simultaneous interpretation of different sensor data. The above research focuses on the diagnosis of the monitoring status of a single sensor, and the evaluation model for the overall quality of the bridge monitoring system is rarely reported.

In view of the complexity of the operation status evaluation of BHM systems and the shortcomings of existing research, this paper proposes a DHGF-based operation status evaluation method for bridge monitoring systems and verifies the effectiveness of this method using a real-world system.

2. DHGF Algorithm

The DHGF comprehensive evaluation method integrates the successes of the improved Delphi method, analytic hierarchy process, gray relations analysis, and fuzzy comprehensive evaluation method [24]. The theoretical basis of the DHGF comprehensive evaluation method is the comprehensive, integrated method proposed by Qian [25] and the physics–affairs–human science (WSR) analysis method proposed by Gu, which combines disciplinary theory and human experience and knowledge and makes full use of the advantages of these two methods. To this end, the DHGF algorithm is introduced in the evaluation of bridge health monitoring systems to accurately grasp the operation status of bridge monitoring systems.

The basic idea of the DHGF comprehensive evaluation method is to use the improved Delphi method for anonymous discussion and statistics, optimize the evaluation indicators of the system to be evaluated, formulate the comprehensive evaluation index system, and establish a hierarchy of evaluation indicators. The analytic hierarchy process is used to calculate the combined weights of the underlying elements, and the evaluation value matrix of the evaluation index is given. The gray system theory is used to determine the evaluation gray class, and the gray evaluation coefficient is calculated to obtain the gray evaluation weight vector and weight matrix. According to the fuzzy integrated evaluation, the evaluation matrix is formed, and fuzzy calculation is carried out to obtain comprehensive evaluation results. The application of different theories and methods in different steps of the systematic evaluation can take advantage of their strengths and avoid their weaknesses, and the process of the DHGF algorithm for system state evaluation is shown in Figure 1.

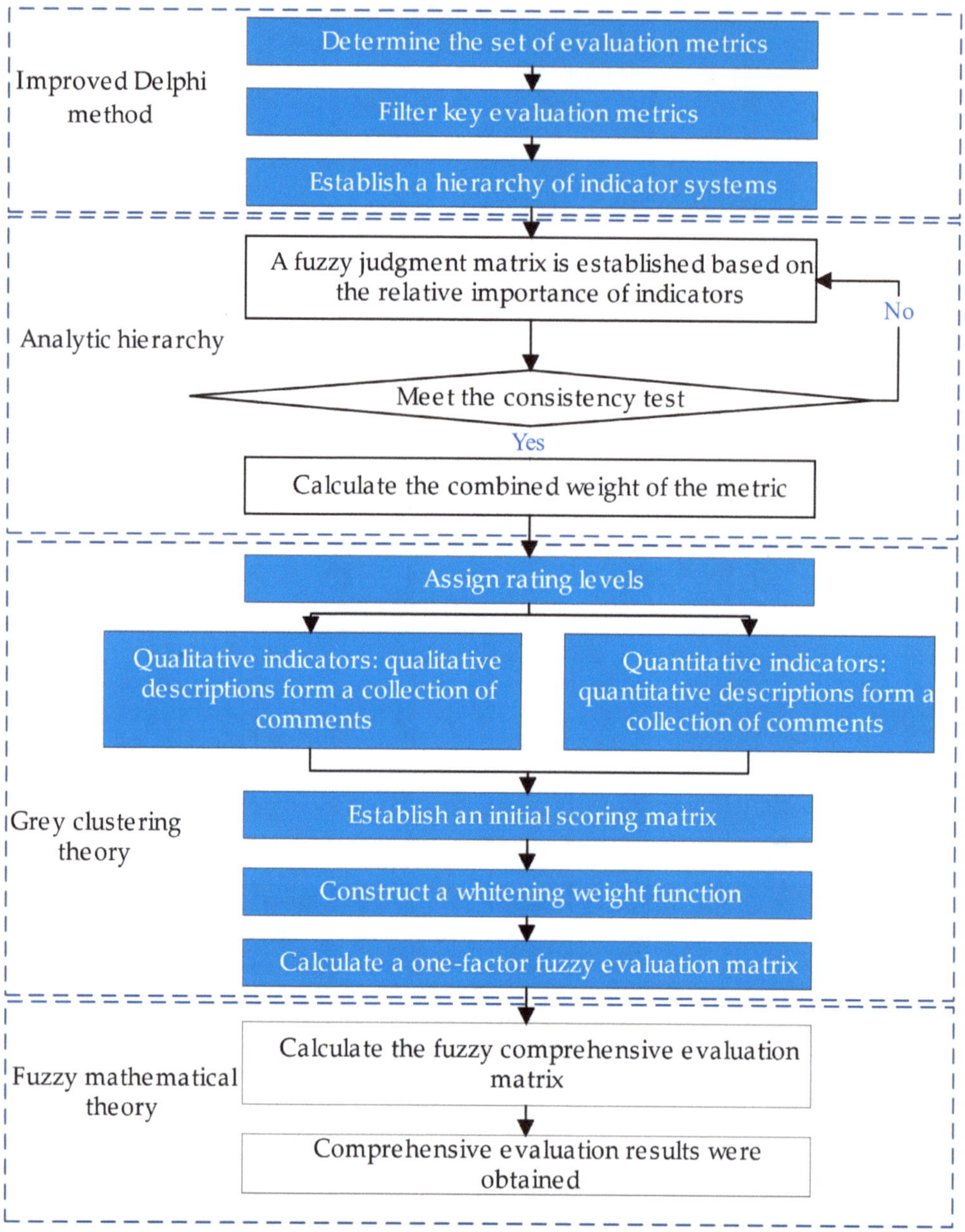

Figure 1. The holistic structure of the proposed method.

3. Construction of Bridge Monitoring System Evaluation Model

3.1. Selection of Evaluation Indicators

For the selection of evaluation indicators, the indicators are determined by means of expert group discussion and the current standard "Technical Code for Monitoring Highway Bridge Structures" (JT T 1037-2022). In order to ensure the independence of the indicators, principles based on scientificity, systematicness, accuracy, and operability are adopted. In addition, the completeness of the design function of the monitoring system, the quality control of the construction stage, the stability of the system operation status, and the timeliness of manual maintenance management are taken into account. On this basis, a hierarchical index system for the evaluation of the bridge monitoring system is constructed, and the established state evaluation index system is a three-level hierarchy, including criterion layer indicators ($B_1 - B_4$) and several index layer indicators ($B_{11} - B_{43}$), as shown in Figure 2. The evaluation of the BHM system is mainly divided into four stages (i.e., design, construction, operation, and maintenance) to evaluate the whole lifecycle of the BHM system.

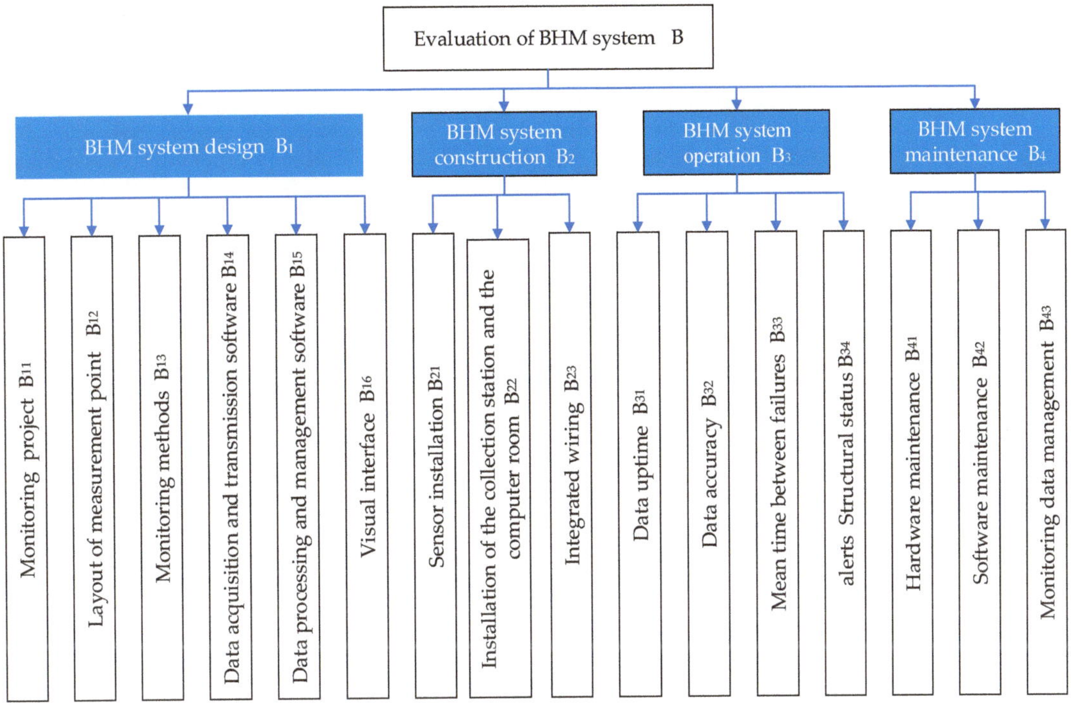

Figure 2. Evaluation index hierarchy of bridge monitoring system.

Assuming that r experts (E) participate in the evaluation and there are n evaluation indicators, d_{hi} represents the score of the h expert on the index i. Then, the evaluation data of r experts on n indicators constitute an evaluation sample size matrix:

$$D = (d_{hi})_{r\times n} = \begin{bmatrix} d_{11} & d_{12} & \dots & d_{1n} \\ d_{21} & d_{22} & \dots & d_{2n} \\ \dots & \dots & \dots & \dots \\ d_{r1} & d_{r2} & \dots & d_{rn} \end{bmatrix}. \quad (1)$$

3.2. The Evaluation Index Weights

The importance of each indicator in reflecting the operation status of the bridge monitoring system varies. Thus, it is necessary to empower each evaluation indicator before conducting a comprehensive evaluation. In this paper, the analytic hierarchy process [24] is used to construct two pairs of judgment matrices to determine the combined weights of each index, and the determination of the judgment matrix scale is determined in Table 1. The reciprocal of the scale (1/3, 1/5, 1/7, 1/9, etc.) indicates that the importance of the two indicators is opposite to the description in the table.

Table 1. Determining the scale and meaning of the matrix.

Scale	Meaning
1	Indicator *a* is just as important as indicator *b*
3	Indicator *a* is slightly more important than indicator *b*
5	Indicator *a* is more important than indicator *b*
7	Indicator *a* is important compared to indicator *b*
9	Indicator *a* is particularly important compared to indicator *b*
The median value	The median of the above scales (2, 4, 6, 8)

A consistency test is required for the obtained judgment matrix. The results of the consistency test can exhibit the logical rationality of the judgment matrix. By this test, an awkward situation can be avoided, i.e., the indicator a is more important than the indicator b, and the indicator b is more important than the indicator c, but the indicator c is more important than the indicator a. The consistency metric is defined below:

$$CI = \frac{\lambda_{\max} - m}{m - 1}, \tag{2}$$

where CR is the consistency ratio; CI is an average random consistency metric; RI is determined by the order of the matrix (see Table 2); $\lambda_{\max}$ is the maximum eigenvalue; and m represents the order of the judgment matrix.

Table 2. Reference table of average consistency indicators.

Matrix Rank	1	2	3	4	5	6	7	8	9	10
RI	0	0	0.58	0.90	1.12	1.24	1.32	1.41	1.45	1.49

In the process of consistency test, m is the only non-zero eigenvalues of the m-rank consensus matrix, and the maximum eigenvalues satisfy $\lambda_{\max} \geq m$; if and only if max = m, the matrix has complete consistency. Therefore, the greater the $\lambda_{\max}$ is than m, the more serious the inconsistency of the matrix. As a result, the value of $\lambda_{\max} - m$ can be used to measure the degree of inconsistency of the matrix. As for CI, the smaller the CI, the greater the consistency. It has complete consistency when $CI = 0$, while there is good consistency when CI is close to 0.

To measure the value of CI, the random consistency index RI is introduced, which is determined by the rank of the matrix. In general, the larger the rank of the matrix, the greater the possibility of random deviation of consistency. Table 2 presents the relationship between RI and matrix rank.

Considering that the deviation of consistency may be caused by randomness, it is also necessary to compare CI and the immediate agreement index to obtain the test coefficient CR when testing whether the matrix has satisfactory consistency. The formula is as follows:

$$CR = \frac{CI}{RI}. \tag{3}$$

When $CR < 0.1$, it is considered that the judgment matrix has passed the consistency test, otherwise it does not have satisfactory consistency, and the judgment matrix needs to be readjusted. By calculating each judgment matrix and finding the feature vector corresponding to the largest feature value as the weight set of each indicator, the weight set can be expressed by

$$W = \{W_1, W_2, \ldots, W_n\}, \tag{4}$$

where W_i is the weight corresponding to indicator i. The weight vector satisfies the normalization condition, and the normalization condition is as follows:

$$\sum_{i=1}^{n} W_i = 1\&0 < W_i < 1. \tag{5}$$

3.3. Determine the Evaluation Level of the BHM System

In this paper, the evaluation level of the operation status of the BHM system is set as I, II, III, and IV, i.e., $m = 4$, $V = (90, 70, 50, 20)^T$. Each rating corresponds to a state of the BHM system (Table 3):

- Level I indicates that the system is in an "excellent" state, where the BHM system fully meets the needs of the monitoring work and has performance beyond the requirements of "Technical Code for Monitoring Highway Bridge Structures" (JT T 1037-2022).
- Level II indicates that the system is in a "good" state, where the BHM system meets the needs of monitoring work, and the system meets the specifications of (JT T 1037-2022).
- Level III indicates that the system is in a "medium" state, where the BHM system has a few defects but basically meets the needs of monitoring work.
- Level IV means that the system is in a "poor" state, where the monitoring system defects are more obvious and cannot meet the needs of monitoring work.

Table 3. Evaluation grades and score divisions of BHM systems.

Evaluation Level	I	II	III	IV
Hierarchical	[90, 100]	[70, 90)	[50, 70)	[20, 50)

According to the "Technical Code for Monitoring Highway Bridge Structures" (JT T 1037-2022), the evaluation standards for each indicator are formulated, as shown in Table 4. Among the 16 secondary evaluation indicators, two quantitative evaluation indicators are included, and the calculation of the quantitative evaluation indicators is given below.

Table 4. Indicator evaluation sets.

Index	I [90–100]	II [70–90)	III [50–70)	IV [20–50)
Monitoring project	The monitoring project exceeds the specifications	Monitor whether the project meets the specifications	Most of the monitoring projects meet the specifications	The monitoring project basically does not meet the requirements of the specification
Layout of measurement point	The layout of the measurement point accurately meets the needs of monitoring work, and there are redundant measurement points in key parts	The layout of the measurement point meets the needs of monitoring work, and there are redundant measurement points in key parts	The layout of the measurement points basically meets the needs of monitoring work	The layout of the measurement points basically does not meet the needs of the monitoring work
Monitoring methods	All sensor parameters meet the requirements, the monitoring method is suitable, and the sampling frequency meets the requirements	More than 90% of the sensor parameters meet the requirements, the monitoring method is appropriate, and the sampling frequency meets the requirements	More than 80% of the sensor parameters meet the requirements, the monitoring method is basically suitable, and the sampling frequency basically meets the requirements	The sensor parameters meet the requirements, the monitoring methods are mostly inappropriate, and the sampling frequency mostly does not meet the requirements
Acquisition and transmission software	The indicators are fully functional and exceed the needs of monitoring work	The indicators are fully functional to meet the needs of monitoring	The indicators have basic functions and basically meet the needs of monitoring work	The indicators are not fully functional and do not meet the needs of monitoring

Table 4. *Cont.*

Index	I [90–100]	II [70–90)	III [50–70)	IV [20–50)
Data processing and management software	The indicators are fully functional and exceed the needs of monitoring work	The indicators are fully functional to meet the needs of monitoring	The indicators have basic functions and basically meet the needs of monitoring work	The indicators are not fully functional and do not meet the needs of monitoring
Visual interface	The layout of the visual interface is very clear and reasonable, which can intuitively reflect data changes, and the response time of the operation is timely, and mobile software is available	The layout of the visual interface is clear and reasonable, which can intuitively reflect data changes, and the response time of the operation is timely, and mobile software is available	The layout of the visual interface is reasonable, which can reflect data changes, and the response time of the operation is acceptable, and mobile software is not available	The layout of the visual interface is chaotic, which does not clearly reflect data changes, operation response times are too long, and mobile software is not available
Sensor installation	Each measurement point is stable and has excellent working environment and very good protection measures	Each measurement point is stable, has a good working environment, and good protection measures	Each measurement point is stable and has an acceptable working environment and general protection measures	Each measurement point is not in a stable position and has a poor working environment and no protective measures
Installation of the collection station and the computer room	The indicator has a reasonable position and very stable installation, which is very neat and beautiful, meets the process requirements, and the terminal contact is good	The indicator has a reasonable position and stable installation, which is neat and beautiful, in line with process requirements, and the terminal contact is good	The indicator has a reasonable position, and the installation is relatively stable, which is relatively neat and basically meets the process requirements, and the terminal contact is good	The indicator does not have a reasonable position, the installation is not stable, which does not meet the process requirements, and the terminal contact is poor
Integrated wiring	The wiring is specified and is clearly marked, beyond the requirements of specifications	The wiring is relatively specified, has a relatively clear mark, and satisfies the requirements of specifications	The wiring meets the requirements of use, which has marks but falls below the specifications	The wiring is cluttered and has no clear marks
Data uptime	$S \geq 95\%$	$90\% \leq S \leq 95\%$	$85\% \leq S \leq 90\%$	$S \leq 85\%$
Data accuracy	The BHM system data are in good agreement with the manual observation data	The BHM system data are consistent with the manual observation data	The BHM system data are basically consistent with the manual observation data	The BHM system data do not match the manual observation data
Mean time between failures	$MTBF \geq 99\%$	$99\% \leq MTBF \leq 95\%$	$90\% \leq MTBF \leq 95\%$	$MTBF \leq 90\%$
Structural status alerts	The upload, analysis, and submission of the collected data are very timely; structural safety alarms are very accurate, and almost no false positives	The upload, analysis, and submission of the collected data are timely, accurate alarms for structural safety issues, and the probability of false positives is within an acceptable range	The upload, analysis, and submission of the collected data are timely, the structural safety alarm is more accurate, and there are false positives, which do not affect the use of the system	The upload, analysis, and submission of the collected data are not timely, the alarm of structural safety is inaccurate, or the system's alarm does not work if there is a major security problem in the structure, and the probability of false positives is high

Table 4. *Cont.*

Index	I [90–100]	II [70–90)	III [50–70)	IV [20–50)
Hardware maintenance	The maintenance is very well and on time	The maintenance is well and on time	The maintenance is relatively on time	The maintenance is not on time
Software maintenance	The maintenance is very well and on time	The maintenance is well and on time	The maintenance is relatively on time	The maintenance is not on time
Monitoring data management	The monitoring data are regularly reported in detail, exceeding the requirements of the current norms	The monitoring data are regularly reported in detail, meeting the requirements of current norms	There is a slight delay in the formation of the monitoring report, and the content basically meets the requirements of the current norms	Reports are not produced regularly

The indicator of measurement point data integrity rate can be expressed as

$$S = \left(1 - \frac{\sum_{i=1}^{p} t_i}{P \times T}\right) \times 100\%, \quad (6)$$

where S is the integrity rate of the measurement point data; p is the number of fault measurement points; t_i is the failure time of the i-th fault measurement point (take days (d) as a unit); P is the total number of measurement points; and T is the check cycle time in days (d).

The indicator $MTBF$ is calculated by

$$MTBF = \frac{\sum_{i=1}^{n} T_i}{\sum_{i=1}^{n} r_i}, \quad (7)$$

where T_i is the normal working time of the unit i, r_i is the abnormal working time of unit i, and n is the total number of units in the data acquisition. Taking the integrity rate (S) of measurement point data as an example, the calculation method of the specific score of quantitative evaluation indicators is shown. From Table 4, if S is located in the interval $[p, q)$, the corresponding score interval is $[a, b)$, and the score (O) of the indicator is calculated by

$$O = a + (b - a) \times \frac{S - p}{q - p}. \quad (8)$$

3.4. Determine the Gray Class Assessment

To comprehensively calculate the overall evaluation level of the BHM system by combining the evaluation level of each indicator, it is necessary to construct a whitening weight function to quantitatively describe the importance of the evaluation indicator. The albino weight function corresponds to an evaluation gray class for each evaluation level. Using the lowest score of the evaluation level as the threshold, the gray category is set as I, II, III, and IV according to the defined four evaluation levels; the minimum threshold of each level is 20, 50, 70, and 90, which are the gray numbers of each evaluation gray class. The corresponding whitening weight function of each level can be obtained:

For the first gray class "I" (i.e., $j = 1$ and gray number $\otimes \in [0, 90, \infty)$), the albino weight function is written as

$$f_1(d_{hi}) = \begin{cases} \frac{d_{hi}}{90} & d_{hi} \in [0, 90) \\ 1 & d_{hi} \in [90, \infty) \\ 0 & d_{hi} \in (-\infty, 0) \end{cases}. \quad (9)$$

For the second gray class "II" (i.e., $j = 2$, and gray number $\otimes \in [0, 70, 140]$), the albino weight function is written as

$$f_2(d_{hi}) = \begin{cases} \frac{d_{hi}}{70} & d_{hi} \in [0, 70) \\ 2 - \frac{d_{hi}}{70} & d_{hi} \in [70, 140). \\ 0 & d_{hi} \in [140, \infty) \end{cases} \quad (10)$$

For the third gray class "III" (i.e., $j = 3$, and gray number $\otimes \in [0, 50, 100]$), the albino weight function is written as

$$f_3(d_{hi}) = \begin{cases} \frac{d_{hi}}{50} & d_{hi} \in [0, 50) \\ 2 - \frac{d_{hi}}{50} & d_{hi} \in [50, 100). \\ 0 & d_{hi} \in [100, \infty) \end{cases} \quad (11)$$

For the fourth gray class "IV" (i.e., $j = 4$, and gray number $\otimes \in [0, 20, 40]$), the albino weight function is written as

$$f_4(d_{hi}) = \begin{cases} 1 & d_{hi} \in [0, 20) \\ \frac{40-d_{hi}}{40-20} & d_{hi} \in [20, 40). \\ 0 & d_{hi} \in [40, \infty) \end{cases} \quad (12)$$

3.5. Gray Statistical Calculation

Based on the gray theory, the whitening function can find the weight $f_j(d_{hi})$ of each expert's score d_{hi} belonging to the j category evaluation gray class, which is denoted as n_{ij}. Therefore, the gray statistics for each indicator belonging to the evaluation gray category of the category j is calculated by

$$n_{ij} = \sum_{h=1}^{r} f_j(d_{hi}). \quad (13)$$

The total gray statistics for each evaluation indicator are calculated by

$$n_i = \sum_{j=1}^{k} n_{ij}. \quad (14)$$

The gray weights of the i-th evaluation factor and j gray category by the comprehensive r experts are calculated by

$$r_{ij} = \frac{n_{ij}}{n_i}. \quad (15)$$

The univariate gray clustering coefficient matrix composed of r_{ij} can be constructed by

$$R = \begin{bmatrix} r_{11} & r_{12} & \cdots & r_{1m} \\ r_{21} & r_{22} & \cdots & r_{2m} \\ \cdots & \cdots & \cdots & \cdots \\ r_{n1} & r_{n2} & \cdots & r_{nm} \end{bmatrix}. \quad (16)$$

3.6. Fuzzy All-Round Assessment Matrix

The multiplication of the univariate weighted matrix and the gray clustering coefficient matrix is carried out to obtain the fuzzy comprehensive evaluation matrix can be constructed below [25,26]:

$$B = [b_1, b_2, \ldots, b_m] = WR = [w_1, w_2, \ldots, w_n] \begin{bmatrix} r_{11} & r_{12} & \cdots & r_{1m} \\ r_{21} & r_{22} & \cdots & r_{2m} \\ \cdots & \cdots & \cdots & \cdots \\ r_{n1} & r_{n2} & \cdots & r_{nm} \end{bmatrix}, \quad (17)$$

where $b_j = \sum_{i=1}^{n} w_i r_{ij}$, $\sum_{j=1}^{m} b_j = 1$.

3.7. Calculation of Evaluation Results

The final evaluation result Z is calculated from the fuzzy all-round evaluation matrix and the evaluation grade matrix [27]:

$$Z = BV^T = (WR)V^T. \tag{18}$$

The final BHM system evaluation level can be obtained via Table 3.

4. Analysis of Example

4.1. Overview of the BHM System

A cable-stayed bridge with a total length of 1001 m is employed for analysis. The BHM system is mainly composed of an automatic perception subsystem, a data acquisition and transmission subsystem, a data processing and analysis subsystem, a structural early warning and evaluation subsystem, and a visual integration subsystem, which can realize early warning in abnormal conditions. There are a total of 288 measurement points. The layout of the measurement point is shown in Figure 3. The labels in Figure 3 correspond to the type of measurement point, as shown in Table 5, and the number in parentheses is the number of measurement points.

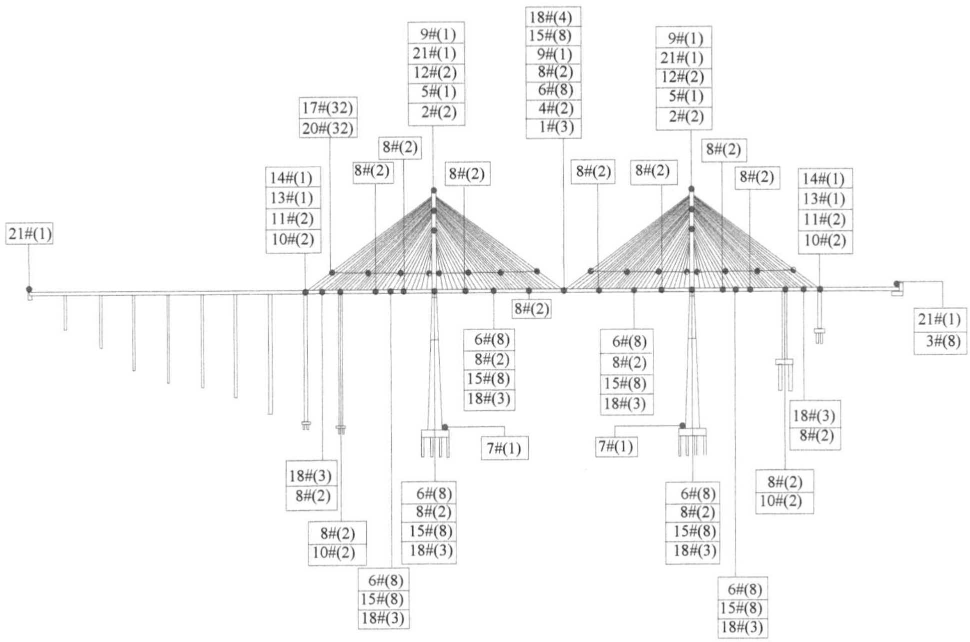

Figure 3. The layout of the bridge measurement point.

Table 5. Labels correspond to the types of measurement points.

Label of Measurement Point	Type of Measurement Point	Number of Measurement Points
1#	Ambient temperature and humidity measurement point	3
2#	Temperature and humidity measurement point in the Sota anchoring area	4
3#	Vehicle load measurement point	8
4#	Bridge deck wind speed and wind direction measurement point	2
5#	Wind speed and direction measurement point at the top of the tower	2

Table 5. *Cont.*

Label of Measurement Point	Type of Measurement Point	Number of Measurement Points
6#	Structural temperature measurement point	56
7#	Ground motion measurement point	2
8#	Main beam deflection measurement point	34
9#	Lateral displacement measurement point of main beam	1
10#	Seat shift measurement point	8
11#	Beam end displacement measurement point	4
12#	Offset measurement point at the top of the tower	4
13#	Horizontal angle measurement point at beam end	2
14#	Vertical angle measurement point of beam end	2
15#	Main beam strain measurement point	40
16#	Tower column strain measurement point	16
17#	Cable-stayed cable force measurement point	32
18#	Main beam vibration measurement point	28
19#	Vibration measurement point at the top of the tower	4
20#	Cable vibration measurement point	32
21#	Video surveillance measurement points	4

4.2. Comprehensive Evaluation

The operation quality of the above BHM system was evaluated using the evaluation model established in Section 2.

4.2.1. Building the Sample Matrix of Evaluation Quantities

Based on the evaluation, ten experts were invited to objectively evaluate the 16 indicators of the BHM system. The scoring range is 1–100 grades, and a sample matrix D with the size of 10×16 was obtained. The initial scores of the experts for each indicator are shown in Table 6 and Figure 4.

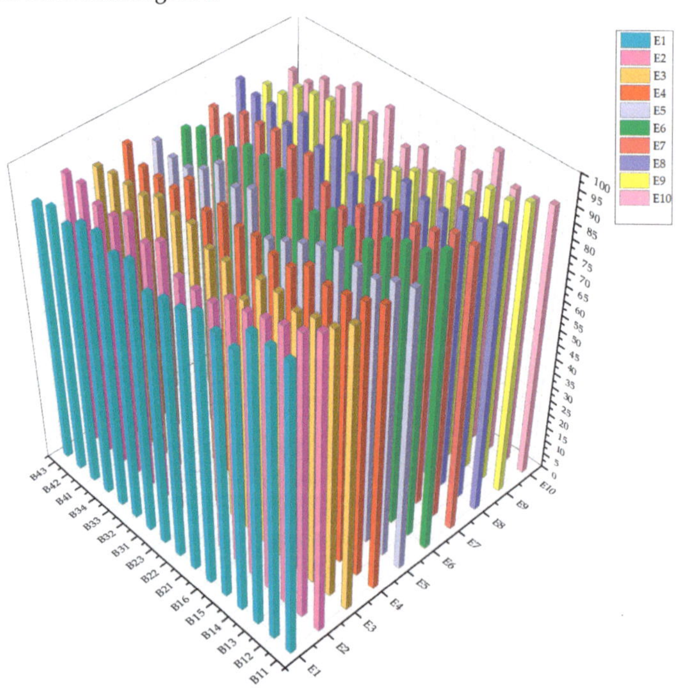

Figure 4. Three-dimensional view of the evaluation sample matrix.

Table 6. Initial expert scoring matrix.

d_{li}	B_{11}	B_{12}	B_{13}	B_{14}	B_{15}	B_{16}	B_{21}	B_{22}	B_{23}	B_{31}	B_{32}	B_{33}	B_{34}	B_{41}	B_{42}	B_{43}
E_1	98	98	98	89	90	92	89	88	86	92	90	93	92	88	90	88
E_2	100	96	94	92	90	90	85	85	85	92	88	93	89	89	92	92
E_3	96	91	90	88	90	90	80	88	88	92	91	93	90	90	90	89
E_4	96	93	91	90	92	88	88	90	89	92	87	93	87	87	87	91
E_5	95	93	90	90	91	89	86	83	80	92	89	93	88	85	85	87
E_6	100	96	95	92	88	88	90	86	86	92	93	93	89	89	89	86
E_7	96	96	93	91	91	90	86	82	86	92	91	93	92	92	88	88
E_8	96	94	94	90	90	92	83	86	84	92	86	93	87	90	90	92
E_9	98	95	95	90	90	90	87	84	83	92	89	93	92	91	86	86
E_{10}	92	90	90	98	88	92	81	86	83	92	87	93	89	89	85	86

4.2.2. The Weights of Evaluation Indicators

There are 16 indicators reflecting the operation status of the BHM system. To fully consider the relationship between all evaluation indicators, the weight of the evaluation indicators is calculated in hierarchical blocks [24], and the weight of each level of indicators is analyzed according to Figure 1.

The four indicators of the criterion layer are taken as examples to present the process of weights allocation: First, in comparing design (B_1) and construction (B_2), B_1 is more important than B_2, and the degree of importance is slightly more important, which is recorded as scale 2; in comparing B_1 and running (B_3), B_1 is more important than B_3 and the degree of importance is slightly more important, which is denoted as scale 2; in comparing B_1 and B_4, B_1 is more important than maintenance (B_4), and the degree of importance is clearly important, which is recorded as scale 4. Then, in comparing B_2 and B_3, they are of similar importance and are recorded as scale 1; in comparing with B_2 and B_4, B_2 is more important than B_4, and the degree of importance is slightly more important, which is recorded as scale 2. Finally, in comparing B_3 and B_4, B_3 is more important, and the degree of importance is slightly more important, which is denoted as scale 3. The judgment matrix between the indicators of the criterion layer is shown in Table 7. Based on Equations (2) and (3), the consistency ratio is calculated. According to Table 2, due to $CI = (4.02 - 4)/(4 - 1) = 0.0067$, $CR = 0.0067/0.89 = 0.0075 < 0.1$, it can be seen that there is no logical error in the judgment matrix, and the weights of the B_1, B_2, B_3, and B_4 layers are shown in the eigenvalue corresponding to the indicators in Table 7. Similarly, the final combination weights of the index layer are displayed in Table 8.

Then, the consistency for weights of the criterion layer was checked:

$$\sum_{i=1}^{4} W_i = 0.4515 + 0.2257 + 0.2507 + 0.1007 = 1.0,$$

for the weights of the index layer:

$$\sum_{i=1}^{16} W_i = 0.0939 + 0.0939 + 0.0325 + \ldots + 0.0258 + 0.0517 = 1.0.$$

Table 7. Criterion layer judgment matrix of BHM system.

Criterion Layer	BHM System Design B_1	BHM System Construction B_2	BHM System Operation B_3	BHM System Maintenance B_4	Eigenvalue	λ_{max}
BHM system design B_1	1	2	2	4	0.4515	4.02
BHM system construction B_2	1/2	1	1	2	0.2257	
BHM system operation B_3	1/2	1	1	3	0.2507	
BHM system maintenance B_4	1/4	1/2	1/3	1	0.1033	

Table 8. Evaluation index system and weight of BHM system.

Target Layer	Criterion Layer	Weights of Criterion Layer	Index Layer	Weights of Index Layer	Index Layer Combination Weights
BHM system evaluation	BHM system design B_1	0.4515	Monitoring project B_{11}	0.2080	0.0939
			Layout of measurement point B_{12}	0.2080	0.0939
			Monitoring methods B_{13}	0.0719	0.0325
			Data acquisition and transmission software B_{14}	0.1699	0.0767
			Data processing and management software B_{15}	0.1699	0.0767
			Visual interface B_{16}	0.1848	0.0834
	BHM system construction B_2	0.2257	Sensor installation B_{21}	0.5559	0.1255
			Installation of the collection station and the computer room B_{22}	0.3537	0.0798
			Integrated wiring B_{23}	0.0904	0.0204
	BHM system operation B_3	0.2507	Data uptime B_{31}	0.1033	0.0259
			Data accuracy B_{32}	0.2179	0.0887
			Mean time between failures B_{33}	0.2179	0.0546
			Structural status alerts B_{34}	0.4609	0.1155
	BHM system maintenance B_4	0.1007	Hardware maintenance B_{41}	0.2500	0.0258
			Software maintenance B_{42}	0.2500	0.0258
			Monitoring data management B_{43}	0.5000	0.0517

Obviously, the combined weights of the above standard layer and index layer meet the normalization test.

The weights of the index layer are shown in Figure 5. It can be seen that the five most important indicators accounting for the heaviest proportion are B_{21}, B_{34}, B_{11}, B_{12}, B_{32}, and B_{16}.

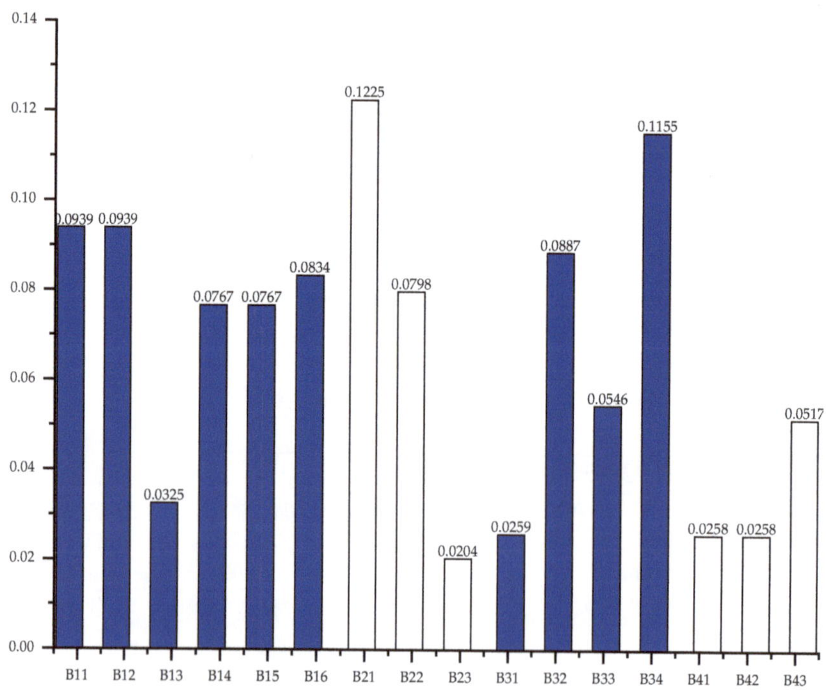

Figure 5. Index layer weight diagram.

4.2.3. Gray Statistical Calculation

Taking the weight of the BHM evaluation indicator B_{11} for the first gray group as an example, the gray group statistics are calculated from Equation (13):

$$n_{11,1} = f_1(98) + f_1(100) + f_1(96) + \ldots + f_1(98) + f_1(92) = 10.0000.$$

Similarly, the evaluation indicator B_{11} of gray statistics for gray groups 2, 3, and 4 is calculated $n_{11,2} = 6.1857$, $n_{11,3} = 0.6600$, and $n_{11,4} = 0$.

The gray statistics matrix calculated for the 16 secondary evaluation indicators belonging to each gray category is as follows:

$$n_{ij} = \begin{bmatrix} 10.0000 & 6.1857 & 0.6600 & 0.0000 \\ 10.0000 & 6.5429 & 1.1600 & 0.0000 \\ 10.0000 & 6.7143 & 1.4000 & 0.0000 \\ 9.9667 & 7.0000 & 1.8000 & 0.0000 \\ 9.9556 & 7.1429 & 2.0000 & 0.0000 \\ 9.9445 & 7.1286 & 1.9800 & 0.0000 \\ 9.5001 & 7.7857 & 2.9000 & 0.0000 \\ 9.5334 & 7.7429 & 2.8400 & 0.0000 \\ 9.4445 & 7.8571 & 3.0000 & 0.0000 \\ 10.0000 & 6.8571 & 1.6000 & 0.0000 \\ 9.8446 & 7.2714 & 2.1800 & 0.0000 \\ 10.0000 & 6.7143 & 1.4000 & 0.0000 \\ 9.8779 & 7.2143 & 2.1000 & 0.0000 \\ 9.8556 & 7.2857 & 2.2000 & 0.0000 \\ 9.7778 & 7.4000 & 2.3600 & 0.0000 \\ 9.7780 & 7.3517 & 2.3000 & 0.0000 \end{bmatrix}$$

The gray statistics of the indicator B_{11} is calculated by Equation (14):

$$n_{11} = n_{11,1} + n_{11,2} + n_{11,3} + n_{11,4} = 16.8457.$$

The evaluation indicator B_{11} for each gray class weight can be obtained:

$$r_{11,1} = \frac{n_{11,1}}{n_{11}} = 0.5936,\ r_{11,2} = \frac{n_{11,2}}{n_{11}} = 0.3672,$$
$$r_{11,3} = \frac{n_{11,3}}{n_{11}} = 0.0392,\ \mathrm{r}_{11,4} = 0$$

By analogy, the univariate gray clustering matrix R for all indicators of the index layer is as follows:

$$R = \begin{bmatrix} 0.5936 & 0.3672 & 0.0392 & 0 \\ 0.5649 & 0.3696 & 0.0655 & 0 \\ 0.5521 & 0.3707 & 0.0773 & 0 \\ 0.5311 & 0.3730 & 0.0959 & 0 \\ 0.5213 & 0.3740 & 0.1047 & 0 \\ 0.5219 & 0.3741 & 0.1039 & 0 \\ 0.4709 & 0.3857 & 0.1437 & 0 \\ 0.4739 & 0.3849 & 0.1412 & 0 \\ 0.4652 & 0.3870 & 0.1478 & 0 \\ 0.5418 & 0.3715 & 0.0867 & 0 \\ 0.5102 & 0.3768 & 0.1130 & 0 \\ 0.5521 & 0.3707 & 0.0773 & 0 \\ 0.5147 & 0.3759 & 0.1094 & 0 \\ 0.5096 & 0.3767 & 0.1137 & 0 \\ 0.5005 & 0.3788 & 0.1208 & 0 \\ 0.5031 & 0.3785 & 0.1183 & 0 \end{bmatrix}$$

4.2.4. Fuzzy Comprehensive Assessment Matrix

The fuzzy comprehensive evaluation matrix of the BHM system is calculated by combining the weight matrix and the univariate fuzzy evaluation matrix using Equation (17):

$$B = W \times R = (0.5161,\ 0.3719,\ 0.1013,\ 0)$$

The comprehensive clustering coefficients of the four gray groups of the BHM system are 0.5161, 0.3719, 0.1013, and 0.

4.2.5. Calculate the Results of the Overall Evaluation

The comprehensive evaluation results are calculated by the fuzzy all-around evaluation matrix and the evaluation grade matrix by Equation (18), and the final comprehensive evaluation score is Z.

The comprehensive evaluation results of the BHM system are

$$Z = BV^T = 83.9217 \in (70, 90)$$

From Table 3, the BHM system belongs to evaluation level "II".

5. Conclusions

To accurately grasp the operation status of the BHM system, this paper establishes an evaluation framework based on the DHGF comprehensive evaluation method. The feasibility of the proposed framework is verified by a real-world BHM system, and some main conclusions are summarized below:

The DHGF comprehensive evaluation method integrates the advantages of the improved Delphi method, the analytic hierarchy process, the gray theory, and the fuzzy comprehensive evaluation method, which can effectively meet the needs of qualitative and quantitative evaluation. Specifically, the improved Delphi method can comprehensively consider the opinions of multiple experts and avoid one-sidedness caused by personal subjective factors. Meanwhile, the analytic hierarchy process can comprehensively consider the logical relationship between the importance of multiple evaluation indicators when determining the weight of evaluation indicators. In addition, the gray theory can effectively divide the evaluation levels of each evaluation indicator according to the principle of the maximum gray clustering coefficient and effectively display the influence of each indicator on different evaluation levels; and the fuzzy comprehensive evaluation method can transform qualitative and quantitative analysis into the final quantitative evaluation results;

The evaluation of the operation status of the BHM system is a comprehensive evaluation problem involving multiple indicators. The improved Delphi method is employed to determine the evaluation indicators of the BHM system, and the evaluation indicator system is established, which can reflect the interrelationship between the indicators well. On this basis, the 16 proposed evaluation indicators can depict the operation status of the BHM system well, which is scientific, comprehensive, and feasible;

The case study shows that the proposed indicators are detailed and reasonable, and the comprehensive evaluation framework is effective and feasible. By this framework, the complex indicators of the BHM system can be integrated, and the qualitative and quantitative evaluation can be realized. Therefore, a comprehensive evaluation of the operation status of the BHM system is achieved, and the evaluation results accurately reflect the operation status of the BHM system.

This paper provides a framework for the evaluation of the BHM system and presents the evaluation process through a new BHM system. In this framework, the evaluation indicators (e.g., data quality) can be flexibly adjusted to evaluate the in-service BHM system to achieve more accurate evaluation results.

Author Contributions: J.X. conceived the framework of this paper and wrote this paper. C.W. and Q.T. advised the data analysis and paper organization. R.Z. and T.Y. were involved in paper writing. All authors have read and agreed to the published version of the manuscript.

Funding: This work was supported by National Natural Science Foundation of China (Grant No. 52278292), Chongqing Outstanding Youth Science Foundation (Grant No. CSTB2023NSCQ-JQX0029); Chongqing Science and Technology Project (CSTB2022TIAD-KPX0205); Chongqing Transportation Science and Technology Project (Grant No. 2022-01); Science and Technology Project of Guizhou Department of Transportation (Grant No. 2023-122-001); China Postdoctoral Science Foundation (Grant No. 2023M730431); Special Funding of Chongqing Postdoctoral Research Project (Grant No. 2022CQBSHTB2053).

Institutional Review Board Statement: Not applicable.

Informed Consent Statement: Not applicable.

Data Availability Statement: The data presented in this study are available on request from the corresponding author.

Conflicts of Interest: The authors declare no conflict of interest.

References

1. Xin, J.; Tang, H.; Wang, J.; Tang, Q.; Xin, D.; Qu, J. Automatic separation of temperature effects from bridge cable force data based on the VMD-PE-KLD algorithm. *Int. J. Robot. Autom.* **2023**, *38*, 247–258. [CrossRef]
2. Li, S.; Xin, J.; Jiang, Y.; Wang, C.; Zhou, J.; Yang, X. Temperature-induced deflection separation based on bridge deflection data using the TVFEMD-PE-KLD method. *J. Civ. Struct. Health Monit.* **2023**, *13*, 781–797. [CrossRef]
3. Xin, J.; Jiang, Y.; Zhou, J.; Peng, L.; Liu, S.; Tang, Q. Bridge deformation prediction based on SHM data using improved VMD and conditional KDE. *Eng. Struct.* **2022**, *261*, 114285. [CrossRef]
4. Jiang, Y.; Hui, Y.; Wang, Y.; Peng, L.; Huang, G.; Liu, S. A novel eigenvalue-based iterative simulation method for multi-dimensional homogeneous non-Gaussian stochastic vector fields. *Struct. Saf.* **2023**, *100*, 102290. [CrossRef]
5. Chen, Z.; Zhang, C.; Wang, X.; Ma, C. Wind Tunnel Measurements for Flutter of a Long-Afterbody Bridge Deck. *Sensors* **2017**, *17*, 335. [CrossRef] [PubMed]
6. Tang, Q.; Zhou, J.; Xin, J.; Zhao, S.; Zhou, Y. Autoregressive model-based structural damage identification and localization using convolutional neural networks. *KSCE J. Civ. Eng.* **2020**, *24*, 2173–2185. [CrossRef]
7. Tao, T.; Wang, H. Efficient Buffeting Analysis of Long-Span Bridges under Non-Stationary Winds: A 2D Interpolation Enhanced Approach. *J. Sound Vib.* **2023**, *559*, 117754. [CrossRef]
8. Liu, S.; Jiang, Y.; Qiao, K.; Peng, L.; Liu, D. Record-based simulation of three-component long-period ground motions: Hybrid of surface wave separation and multivariate empirical mode decomposition. *Soil Dyn. Earthq. Eng.* **2023**, *172*, 108037. [CrossRef]
9. Jiang, Y.; Zhao, N.; Peng, L.; Zhao, L.; Liu, M. Simulation of stationary wind field based on adaptive interpolation-enhanced scheme. *J. Wind Eng. Ind. Aerod.* **2019**, *195*, 104001. [CrossRef]
10. Zhao, N.; Jiang, Y.; Peng, L.; Chen, X. Fast simulation of nonstationary wind velocity fields by proper orthogonal decomposition interpolation. *J. Wind Eng. Ind. Aerod.* **2021**, *219*, 104798. [CrossRef]
11. Yang, J.; Chen, R.; Zhang, Z.; Zou, Y.; Zhou, J.; Xia, J. Experimental study on the ultimate bearing capacity of damaged RC arches strengthened with ultra-high performance concrete. *Eng. Struct.* **2023**, *279*, 115611. [CrossRef]
12. Jiang, Y.; Liu, S.; Peng, L.; Zhao, N. A novel wind speed prediction method based on robust local mean decomposition, group method of data handling and conditional kernel density estimation. *Energ. Convers. Manag.* **2019**, *200*, 112099. [CrossRef]
13. Xin, J.; Zhou, C.; Jiang, Y.; Tang, Q.; Yang, X.; Zhou, J. A signal recovery method for bridge monitoring system using TVFEMD and encoder-decoder aided LSTM. *Measurement* **2023**, *214*, 112797. [CrossRef]
14. Tang, Q.; Xin, J.; Jiang, Y.; Zhou, J.; Li, S.; Chen, Z. Novel identification technique of moving loads using the random response power spectral density and deep transfer learning. *Measurement* **2022**, *195*, 111120. [CrossRef]
15. Tang, Q.; Xin, J.; Jiang, Y.; Zhou, J.; Li, S.; Fu, L. Fast identification of random loads using the transmissibility of power spectral density and improved adaptive multiplicative regularization. *J. Sound Vib.* **2022**, *534*, 117033. [CrossRef]
16. Jiang, Y.; Zhao, N.; Peng, L.; Liu, S. A new hybrid framework for probabilistic wind speed prediction using deep feature selection and multi-error modification. *Energ. Convers. Manag.* **2019**, *199*, 111981. [CrossRef]
17. Wang, C.; Ansari, F.; Wu, B.; Li, S.; Morgese, M.; Zhou, J. LSTM approach for condition assessment of suspension bridges based on time-series deflection and temperature data. *Adv. Struct. Eng.* **2022**, *25*, 3450–3463. [CrossRef]
18. Jian, X.D.; Zhong, H.Q.; Xia, Y.; Sun, L.M. Faulty data detection and classification for bridge structural health monitoring via statistical and deep-learning approach. *Struct. Control Health Monit.* **2021**, *28*, e2824. [CrossRef]
19. Hajializadeh, D. Deep learning-based indirect bridge damage identification system. *Struct. Health Monit.-Int. J.* **2023**, *22*, 897–912. [CrossRef]
20. Zhang, L.; He, Y.; Cheng, S.; Wang, G.; Wang, B.; Chen, K.; Dong, L. Bayes method for bridge robustness assessment based on monitoring data. In Proceedings of the 10th International Conference on Bridge Maintenance, Safety and Management (IABMAS), Electr Network, Sapporo, Japan, 11–18 April 2021; pp. 2126–2230.
21. Li, L.; Liu, G.; Zhang, L.; Li, Q. Sensor fault detection with generalized likelihood ratio and correlation coefficient for bridge SHM. *J. Sound Vib.* **2019**, *442*, 445–458. [CrossRef]

22. Li, S.; Jin, L.; Qiu, Y.; Zhang, M.M.; Wang, J. Signal Anomaly Detection of Bridge SHM System Based on Two-Stage Deep Convolutional Neural Networks. *Struct. Eng. Int.* **2023**, *33*, 74–83. [CrossRef]
23. Yu, X.B. Research on Sensor State Evaluation and Fault Diagnosis based on Multi-dimensional Information Fusion. In Proceedings of the International Conference on Optoelectronic Materials and Devices (ICOMD), Guangzhou, China, 18–20 December 2020.
24. Qin, C.X.; Li, B.D.; Shi, B.; Qin, T.; Xiao, J.T.; Xin, Y.L. Location of substation in similar candidates using comprehensive evaluation method base on DHGF. *Measurement* **2019**, *146*, 152–158. [CrossRef]
25. Jin, D.Z.; Zhou, Y.T.; Liu, J.Y.; Han, M.G. Culture innovation of building enterprises based on WSR. *Tech. Technol. Educ. Manag.-Ttem* **2010**, *5*, 785–791.
26. Xu, X.Y.; Yu, F.S.; Pedrycz, W.; Du, X.B. Multi-source fuzzy comprehensive evaluation. *Appl. Soft Comput.* **2023**, *135*, 110042. [CrossRef]
27. Meng, M.; Zhao, P.; Bai, M.; Wang, J. Fuzzy comprehensive evaluation of performance on urban water supply system. In Proceedings of the 40th Chinese Control Conference, Shanghai, China, 26–28 July 2021; pp. 6679–6684.

remote sensing

MDPI

Article

An Integrated Multisource and Multiscale Monitoring Technique for Assessing the Health Status of High-Speed Railway Subgrade

Yuanxingzi He [1,2], Yongwei Li [1,2,*] and Linrong Xu [1,2]

1 School of Civil Engineering, Central South University, 68 Shaoshan Road, Changsha 410075, China; 224812359@csu.edu.cn (Y.H.); xulin_rong@163.com (L.X.)
2 National Engineering Research Center of High-Speed Railway Construction Technology, Central South University, Changsha 410075, China
* Correspondence: yongweili@csu.edu.cn; Tel.: +86-153-8750-6044

Abstract: The precise identification of railway subgrade defects remains a significant challenge for the railway industry globally. Due to the limitations of individual monitoring techniques, comprehensive information on subgrade defects cannot be obtained. In fact, the presence of subgrade defects can significantly increase the risk of traffic accidents during high-speed train operations, which may affect the safety of train operations and economic development. The monitoring of subgrade health status is used as a pre-disaster planning method that is urgently required to avoid accidents and guide the maintenance strategy. Therefore, a novel "integrated" holistic monitoring approach for subgrade structures is presented based on satellite remote sensing, a comprehensive inspection vehicle, and a ground-based testing technique. Additionally, the monitoring content is more clearly defined during the service life of the subgrade. The method is used to investigate the location, development trend, and the cause of subgrade defects on the Shanghai–Nanjing high-speed railway. Some new viewpoints are put forward: First, the monitoring content for assessing the health status of the subgrade should encompass the foundation settlement, the track geometry status, and the monitoring of deformation and defects within the subgrade. Second, the mileage points K235 and K299 of the subgrade, as well as K236 and K237 of the bridge–subgrade transition sections, are estimated to be locations with potential defects based on the differential InSAR and track quality index. Third, the result of settlement monitoring and ground-penetrating radar analysis illustrates that sections K235 +540 to +680 and K299 +680 to +750 are diagnosed as defect positions triggered by the rapid drop of water level and engineering activity, respectively. Fourth, the "integrated" holistic monitoring technique for subgrade service status might be expected to be a promising method that can be useful in developing maintenance plans and implementing fault recovery for railway infrastructure.

Keywords: health status of high-speed subgrade; monitoring content; "integrated" holistic monitoring approach; InSAR; comprehensive inspection vehicle; ground-based testing technique

Citation: He, Y.; Li, Y.; Xu, L. An Integrated Multisource and Multiscale Monitoring Technique for Assessing the Health Status of High-Speed Railway Subgrade. *Remote Sens.* **2024**, *16*, 1972. https://doi.org/10.3390/rs16111972

Academic Editor: Michele Saroli

Received: 26 March 2024
Revised: 10 May 2024
Accepted: 21 May 2024
Published: 30 May 2024

1. Introduction

By the end of 2022, the length of the operating railways in China had increased to a total of 155,000 km (including over 42,000 km of high-speed railways), and China had become the country with the largest operating mileage in the world [1]. The over-increased embankment and railway-cutting slope have led to the destruction of the original geological balance; thus, the railway is susceptible to disasters during operation in extreme climates. For example, in 2020, a train derailed at position K302 +500 of the Litang–Zhanjiang railway, caused by a subgrade collapse, which resulted in railway disruption of up to 41 h and a cost of CNY 14.46 million [2]. In 2020, the railway derailment accident of the Beijing–Guangzhou railway was also triggered by a subgrade collapse, resulting in 122 people injured and a loss of CNY 22.35 million [2]. As a result, the frequent accidents caused by subgrade defects are major impediments to the development of transportation and the economic efficiency of railways.

The Shanghai–Nanjing High-Speed Railway (SNHR) is mainly located in a typical coastal soft soil area that is often affected by typhoons and rainstorms after the "El Nino" event. In addition, the subgrade stability is also affected by soil foundation deformation, construction activities, and train load. Therefore, the number of subgrade defects is continuously increasing due to several factors, making the investigation of subgrade health status much more difficult. Detection or monitoring of subgrade service status is always regarded as a pre-disaster planning measure. Early warning and prevention activities are conducted to minimize the damage caused by subgrade defects. Therefore, accurate recognition of defects and railway health monitoring is crucial for the safe operation and disaster prevention of railways. Recently, the method of detection or monitoring of infrastructure status on railways has mainly contained four aspects according to technical type.

First, the evaluation of track irregularity mainly relies on the power spectral density based on amplitude and wavelength [3,4] and the evaluation method of track quality. In addition, the evaluation method of track quality is always represented by the standard deviation or maximum value of irregularity amplitude with specific track parameters, which is often used to guide the temporary repair and the management of speed limitations. The track quality index is typically determined by calculating the standard deviation of track parameters within a specific section. The Q-index takes into account the ratio of average standard deviations of level and alignment corresponding to threshold limits in the Swedish National Railway [5]. The W_5 track defectiveness index was proposed by the Australian Rail Track Corporation and is based on five parameters of track defectiveness (profile, alignment, gauge, cant, and twist) [6]. The TGI index of the Indian Railway focused on the weighted average of surface, twist, gauge, and alignment [7]. The J index in Poland was a weighted average of surface, alignment, warp, and gauge [8]. The P index adopted by the Japanese Railway was the percentage of mileage with track deviation exceeding 3 mm to the 100 m or 500 m lengths of track segments [9]. Every country chooses an appropriate track condition index to reflect the real state of the track based on its own experience of track maintenance. However, the Chinese Track Quality Index (TQI) was proposed by accumulating the standard deviation of seven geometric deviations in a 200 m railway unit, and its management standards of track irregularity for 250–350 km/h high-speed railways are used to maintain the railway structure, which is more stringent than the European norm [10]. The comprehensive inspection train of each country can obtain the track quality index at the same speed as a high-speed train and can consider the wheel–rail interaction effect in the real state. But, subgrade defects are only one of the factors that cause track geometric irregularities [11]. Thus, the track deviation cannot directly reflect the deterioration of the railway subgrade [5] and is also subjected to the impact of certain environmental factors [12].

Second, there is a ground-based monitoring technique. Precision leveling and total stations are often adopted to collect the elevations of the track vertical control network (CPIII), which are widely used in railway management departments in various countries [13]. The subgrade settlement with millimeter magnitudes is obtained at different periods by these methods. In addition, the sedimentation plate method, single-point settlement, and the magnetic ring-layered sedimentation method are also used to gain high-accuracy deformation information for subgrades. However, manual measuring techniques require a lot of time and labor costs. Fiber Bragg grating and GPS (Global Positioning System) have been utilized to automatically monitor the collapse and long-term deformation of the subgrade in recent years [14]. But, real-time monitoring techniques require many devices/sensors, which is impractical for covering the whole subgrade.

Third, InSAR (interferometric synthetic aperture radar) interferometry technology has the advantages of wide coverage, non-intrusive surveying, and remote monitoring [15], which is often used to monitor surface subsidence that is caused by external environmental factors and human engineering activities during the long-term service of the railway.

Various studies have highlighted that the deformation of railway infrastructure can be indirectly reflected by analyzing the influence of ground settlement around the railway on the subgrade, e.g., differential subsidence caused by the overexploitation of groundwater [16], karst collapse [17], mining [18], and engineering construction activities [19]. Even though displacements measured by differential interferometric InSAR are evaluated with millimeter-level precision, the positioning precision of scatterers is relatively low [20]. Moreover, the interferometric coherence of InSAR will be affected in areas with vegetation or complicated terrain topography [21,22].

Fourth, traditional manual patrols are still widely conducted in routine detection. However, if subgrade defects like cracking, large settlement, and mud pumping can be detected by visual inspections, then the degree of subgrade deterioration is very serious. In other words, the subgrade undergoing creep or internal failures is undetectable by surface visual inspection alone. But, the result of visual inspection is mainly dependent on the specialized knowledge of technicians. Non-destructive testing methods such as ground-penetrating radar (GPR) can detect information on the loose area, holes, water accumulation, and local distortion inside the subgrade [23]. Additionally, the development trend of subgrade deterioration can also be carried out by qualitative predictions. However, the methods of GPR image interpretation are considered to be time-consuming and require significant experience [24].

Great progress has been made in current studies, and the monitoring methods are constantly innovating and have multiplied with the development of science and technology. However, there are still three limitations that need to be further studied: First, the monitoring subjects are not comprehensive enough, and the technology system is not yet perfected in current practical applications of subgrade state monitoring. If infrastructure degradation or a worsening environment cannot be detected in a timely manner, it may threaten the safe operation of trains and cause traffic accidents. Second, the service status of the subgrade is almost imperceptible due to it being covered by an enclosed waterproof sealing structure. The standalone use of comprehensive inspection vehicles, leveling, and InSAR cannot be considered to be an impeccable solution for detecting subgrade defects on the whole railway line. Third, it is difficult to estimate subgrade defects accurately during railway operations due to the lack of comprehensive analysis ability for multisource data and the capabilities to retrieve key information from massive data.

The integrated monitoring approach has been implemented in many research fields like volcano [25,26], forest fire [27,28], and hazard monitoring [29,30]. However, the implementation of various monitoring technologies for the detection of subgrade defects is still at the exploratory stage. Therefore, this work aims to integrate multisource and multiscale monitoring technologies, leveraging the unique advantages of InSAR, comprehensive vehicle inspection, leveling, and GPR to monitor the health status of subgrades and investigate defects. In this work, the external hazards threatening the stability of the subgrade and monitoring the content of the subgrade structure are more clearly defined during the subgrade service process. An integrated technique of multimeasures is proposed to identify the potential defect locations, monitor the critical deformation areas, and investigate the cause of subgrade defects in detail. The method is applied in SNHR. The detailed procedure and the shortcomings of this integrated method are also described and discussed, respectively. It should be noted that this "integrated" holistic monitoring technique for subgrade service status may be expected to be a promising method that can provide valuable insights for the safe operation and disaster prevention of railways.

2. Study Area

The SNHR is in Shanghai and Jiangsu Province, with a total mileage of 300 km and a design speed of 350 km/h. About 33.6% of the railway structure consists of subgrade throughout the entire mileage. SNHR is one of the railways with the highest proportion of subgrade in China. The K228~K299 section of SNHR is selected as the study region in Nanjing City, Jiangsu Province (Figure 1). This study area has a humid subtropical climate with 1090 mm average annual rainfall [31] and has often suffered from typhoons and rainstorms. The study area focuses on the Yangtze River coast, which has a well-developed water network that has led to the widely distributed alluvial deposits of soft soil. Soft soil is characterized by high water content, low strength, and high compressibility [32]. As shown in Figure 1c, the region is extensively covered by Quaternary strata, which are primarily composed of sedimentary soils, including soft soil, silt, and clay formed by fluvial and alluvial deposition in river floodplains [33]. The pile–raft, and pile–net composite foundations were mainly used in the foundation treatment of SNHR. In addition, intensive human engineering activity has become a critical factor that causes ground subsidence during the period of urbanization in this study area, such as groundwater extraction, deep foundation pit engineering, and metro construction [34].

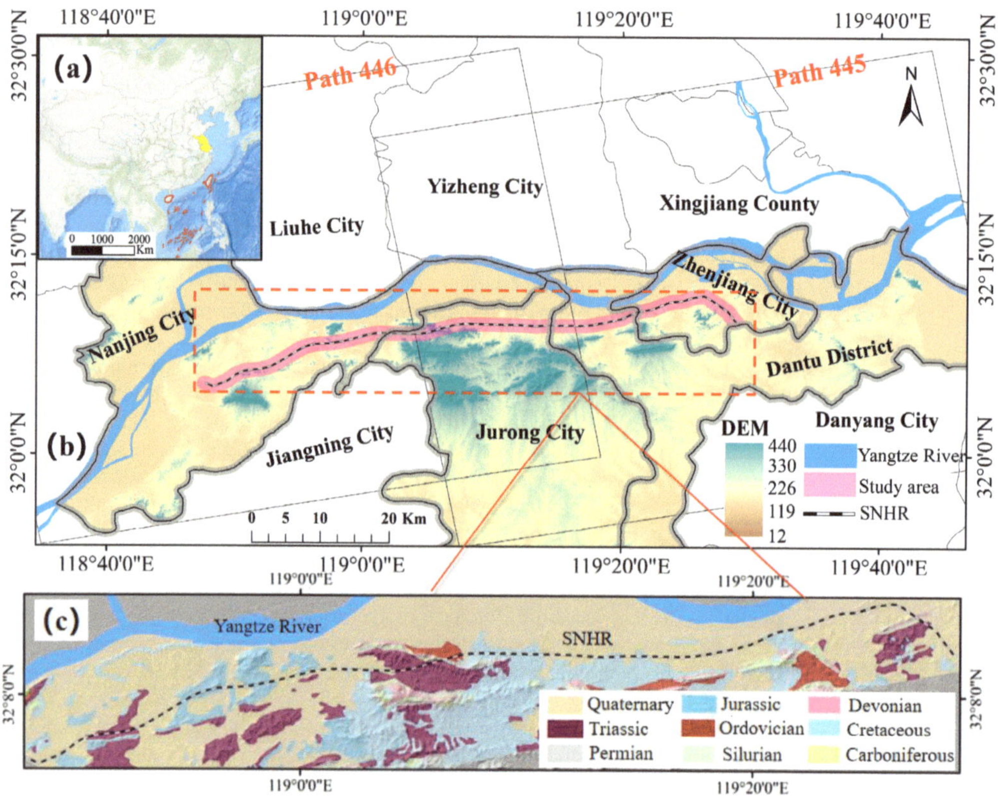

Figure 1. Location map of Shanghai–Nanjing High-Speed Railway (SNHR) in Jiangsu Province. (**a**) Map of China's administrative boundaries, and the yellow area is the location of Jiangsu province. (**b**) Map of digital elevation model (DEM) in this study area. (**c**) Lithology map in red dashed line.

3. Data

This study utilized five types of multisource and multiscale-monitoring data techniques. (1) ALOS PALSA operated in a radar carrier frequency of 1270 MHz (L band) with an incidence angle range of 8~46 degrees in dual polarization mode, which can penetrate the vegetation and reach the ground. Two ALOS PALSA data sets were prepared to obtain the deformation (Figure 1 and Table 1). The level 1.5 georeferenced amplitude image with 15 m spatial resolution was Single Look Complex (SLC) data in vertical transmit and vertical receive (VV) mode. The data were derived from the ASF (Alaska Satellite Facility) website. (2) Digital elevation model (DEM) (12.5 m $\times$ 12.5 m) data generated on 3 May 2008 was also obtained from the sensor of ALOS PALSAR. (3) Track Quality Index (TQI) indicated the track irregularity which was recorded by the comprehensive track inspection train of Shanghai Railway Bureau. It realistically represents the quality state within the 200 m track section. (4) A precise leveling method was used to measure the vertical settlement of the CPIII control network of subgrades in key railway sections. (5) Ground-penetrating radar was used to investigate subgrade internal structure information, detect structural cavities and hydraulic channels, and collect other subgrade defect information.

Table 1. Parameters of ALOS PALSA SAR data over the study area.

Data Sets	Date Acquired	Absolute Orbit	Time—Baseline	Spatial—Baseline	Path	Frame
first	24 September 2010 9 November 2010	24,441 25,112	46 days	421.84 m	446	630
second	26 August 2010 11 October 2010	24,864 25,535	46 days	−683.66 m	445	630

4. Methods

4.1. Differential Radar Interferometry (D–InSAR)

D–InSAR technique can detect minor surface deformation by differential interference of two SAR images obtained from two different positions and visit times. Two SAR images are coregistered by offset estimation of image matching. The initial interference phase can be acquired by pixels of the master image multiplied by the complex conjugate of the corresponding pixel of the slave image. The interferometric phase can be expressed as follows:

$$\varphi_{int} = \varphi_{flat} + \varphi_{topo} + \varphi_{atm} + \varphi_{defo} + \varphi_{noi} \tag{1}$$

where φ_{int} indicates the initial interference phase. φ_{flat} expresses the flat earth phase. φ_{topo} is the topographic phase. φ_{atm} donates the atmospheric delay phase. φ_{defo} is the deformation phase. φ_{noi} is the residual noise phase. The SAR images are processed with the assumption of a flat earth. Hence, the phase difference between the flat and actual earth should be considered, which can be expressed as follows:

$$\varphi_{flat} = \frac{4\pi}{\lambda \cdot B_{||}} \tag{2}$$

where the $B_{||}$ is the parallel baseline for each pixel. The topographic phase is the phase component of the terrain caused by the terrain fluctuation, which is proportional to the vertical baseline. As shown in Equation (3):

$$\varphi_{topo} = \frac{4\pi}{\lambda} \cdot \frac{B_{\perp}}{R \sin\theta \cdot \Delta z} \tag{3}$$

where the $B_{\perp}$ indicates the parallel baseline for each pixel. R is the distance between the target and the satellite. θ is the antenna look angle, and Δz is the topography above the reference ellipsoid (Figure 2). The topographic phase can be removed using the external DEM data. φ_{atm} is the atmospheric delay phase, which is estimated by external measurements or atmospheric models [35]. φ_{noi} indicates the interferometric phase noise, it is removed by filtering via the SAR imaging software of ENVI 5.3. The deformation interferometric phase is caused by the deformation of the target object as shown as follows:

$$\varphi_{defo} = \frac{4\pi}{\lambda \cdot \Delta R_{defo}} \tag{4}$$

where λ is the radar wavelength, ΔR_{defo} is the deformation value.

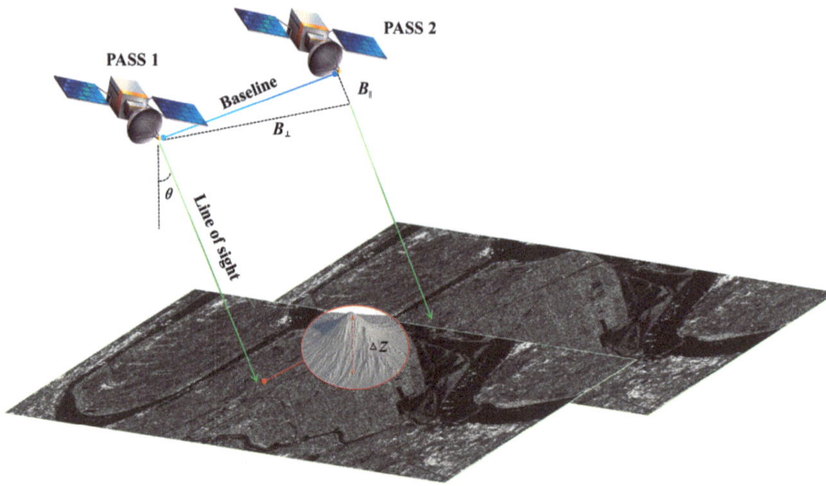

Figure 2. Repeat pass interferometry and geometric relationship of LOS (Line of sight) and radar slant plane.

To obtain the surface deformation, the interferogram results after filtering should be unwrapped first because it is wrapped around modulus 2π with amplitude images as background. The coherence value of each pixel is calculated at the same time. In this work, the minimum cost flow (MCF) is used to unwrap the phase by network planning, and the pixels whose coherence is lower than predefined are masked [36]. In addition, orbit refinement and reflattening are needed to correct satellite orbit and phase offset before generating the deformation map, which makes its spatial coordinate information more accurate. Finally, the phase is converted into the deformation value. However, the map is in the SAR image coordinate system, which should be converted to the geographic coordinate system using the geocoding method.

4.2. Settlement Monitoring and Track Quality Index (TQI)

Subgrade settlement is measured by precision leveling based on the difference of elevation control points of CPIII at different times. The round-trip mapping method is adopted to measure the subgrade settlement of SNHR. The measurement error of precision level is limited to ± 0.3 mm. The monitoring frequency of key railway sections depends on the degree of deterioration of the railway structure.

TQI is a management method of average value, which reflects the fluctuation of track geometric irregularity in a 200 m railway unit. It is also an accumulative value of the standard deviation of seven geometric deviations (longitudinal level of left and right, left and right alignment, cross-level, gauge, and track twist). The comprehensive inspection train was used to obtain the TQI data of SNHR every 30–60 days. TQI management threshold is 5 mm for high-speed railways with a maximum allowable speed of 250–350 km/h, and the TQI is given as follows:

$$TQI = \sum_{i=1}^{7} \delta_i \tag{5}$$

where δ_i is the deviation of a single index, including the left and right longitudinal level, left and right alignment, cross-level, gauge, and track twist.

$$\delta_i = \sqrt{\frac{1}{N}\sum_{j=1}^{N} \left(X_{ij} - \overline{x}_i\right)^2} \tag{6}$$

where N indicates the sampling number in the 200 m railway unit, $\overline{x}_i$ is the average value of the peak of single sampling X_{ij} (j = 1,2,3, ..., N) in each geometric parameter (i = 1, 2, ...,7).

4.3. Exploration Using Ground-Penetrating Radar (GPR)

GPR can survey the locations and sizes of different electrical layers and objects within materials. Various soil layers have different electrical properties, and the incident wave generates different reflected signals when it pulses into unique media, while the rest of the wave will continue to propagate. The antenna of the device receives reflected signals, and the reflected energy is displayed as a waveform, which shows amplitudes and time elapses between wave transmission and reflection. Hence, defect signal features inside the subgrade can be detected after data preprocessing like enhanced signal-to-noise ratio. The characteristics of railway subgrade defects in GPR images can be identified by visual interpretation. For example, the image of water accumulation typically exhibits large amplitude, reverse phase, and multiple reflections at the interface. Because the amplitude reflection of the water accumulation layer is strong, it filters the high-frequency electromagnetic waves and leaves low-frequency waves. The image characteristic of subsidence shows that the reflected signals of the interface between the surface and the bottom layer of the subgrade exhibit significant bending, discontinuity, and offset of the depth downwards [37]. Reflected signals of different intensities are represented by hexadecimal colors based on GPR analysis software. As shown in Figure 3, the waveforms of electromagnetic signals are transformed into reflection signals. The reflection intensities of the khaki color in the middle are relatively weak.

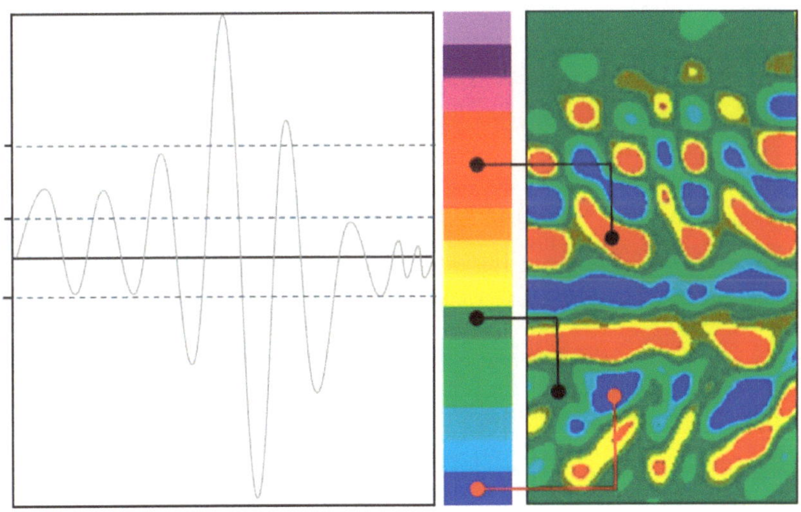

Figure 3. Waveform of electromagnetic signals and reflection signals.

5. Results

5.1. Land Subsidence along the Shanghai–Nanjing Railway

Due to the impact of overexploitation of groundwater, subway construction, and heavy loading of high-rise buildings, the problem of land subsidence occurs frequently in soft soil areas [38]. The land subsidence will harm the railway subgrade and affect the function of urban infrastructure and railway tracks. Therefore, the D-InSAR technique was adopted to obtain the land subsidence in the range of 1 km covering the region along SNHR after the railway operation (1 July 2010). Subsequently, the geological environment of the railway was monitored, and the areas of land settlement that may affect the smoothness of the subgrade were delineated. The value of ground displacement in the radar line-of-sight

(LOS) direction was measured (Figure 4). Positive LOS deformation (blue colors) indicates movement towards the satellite. Conversely, the red color area (negative LOS deformation) represents the movement away from the satellite (like land subsidence).

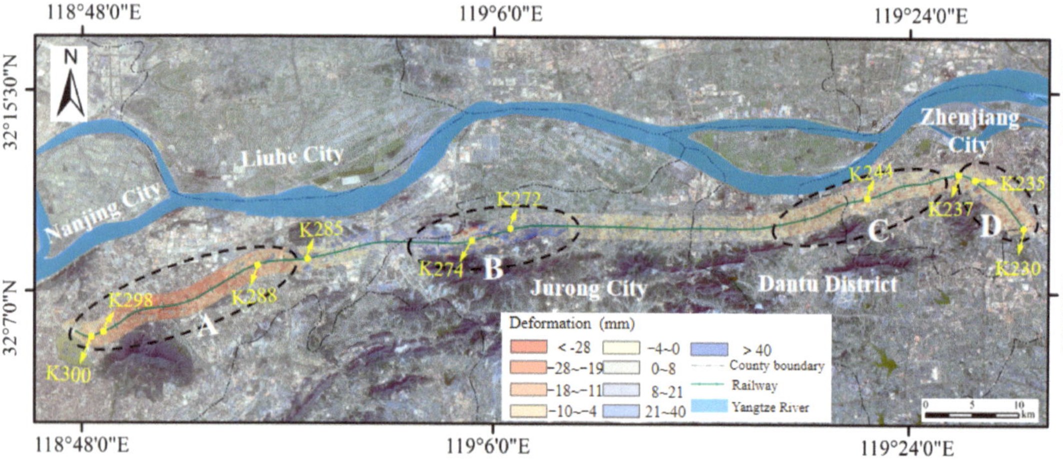

Figure 4. The deformation map along the K230 to K300 of the high-speed railway from radar line-of-sight (LOS) direction.

The differential interference result shows that the significant subsidence is mainly located in area A (black oval dotted line A in Figure 4) in the K288~K300 railway section. The maximum value of subsidence is over −28 mm near the railway in Nanjing City. The deformation in subgrade section K298~K300 is about −10~−15 mm. Moreover, the deformation in the range of −28~−10 mm is mostly located in section K288~K298 (10 km length), and the railway line in this section was mainly composed of bridges with 24~32 m prestressed concrete box beams. Although the surface settlement is relatively large, it has little impact on the bridge foundation structure. Meanwhile, the uplifted region is mainly distributed in the B section, located at the border area of Jurong City, where the landform is mainly characterized by hilly and mountainous terrain with an altitude of 300 m. There is a railway tunnel of about 2000 m, and the land uplift value is about 40~60 mm. The subsidence is comparatively low in the plain area of section C (Figure 4). The surface subsidence of the K237~K244 section is within ±4 mm in most areas, and the structure of the railway is constructed as a low embankment in this region. Conversely, another region of section C is characterized by depression with relatively low terrain. Therefore, bridges and high embankments are built to maintain the slope of the railway and meet the requirements of smoothness in this area. Furthermore, the surface subsidence is about −8~−6 mm at the K235 mileage point. The groundwater level of the subgrade may be vulnerable to the fluctuation of the river, as this area is close to the river (less than 100 m). This area should be monitored by the railway bureau in the dry season.

Above all, the significantly large land subsidence is evident at mileage point K235, and the K298~K300 section (2 km in total) is very likely to damage the subgrade sections. The deformation of the foundation triggered by regional subsidence will directly damage the structure of the subgrade, resulting in subsidence and other defects. However, the subgrade defects are exacerbated under the cyclic load of trains, which will affect the strength and smoothness of the railway subgrade, potentially leading to disruptions in rail traffic and operational accidents. Hence, the question of whether the subgrade has deteriorated due to environmental factors should be considered by combining the TQI data for further study. If irregularities arise in these areas of the railway line, indicating the deteriorating state or defect in these positions.

5.2. Track Irregularity Detection by Track Quality Index (TQI)

The critical factors affecting track smoothness are the abrasion of track components and the loosening of rail fastenings. Furthermore, the corrosion of rail and fasteners and cracks in sleepers will cause changes in track geometry, which are often aggravated by the dynamic load of the train. But, the irregular state of the railway will be improved by railway conservation using vibration machines, and there are not only monthly minor repairs, but there are also major repairs once every 6~12 months. Another reason for geometric irregularity is the uneven settlement of the subgrade. Hence, the TQI consistently exceeded the threshold management threshold (TQI $\geq$ 5) after tamping by the device, which may be caused by subgrade deterioration.

Five TQI data points are recorded for each kilometer mileage, with each data point representing 200 m of the unit segment. For example, the TQI data of the K235 mileage point contains units of K235 +200, +400, +600, +800 and K236 +000. Hence, the data from a total of 20 periods were used to analyze the geometric smoothness state along the K230 to K300 sections of the high-speed railway from 16 August 2010 to 20 February 2014 in this work. It is demonstrated that the TQI value of sections was relatively high from 2010 to 2014 in sections such as K235, K236, K237, K239, K299, and so on (Figure 5a,b). The highest value of TQI was 13.3 mm on 15 February 2011. The TQI value is about 5.16~7.7 mm from section K235 +200 to K236 +800 and about 5.08~10.3 mm in section K298 +800~K300. Notably, the TQI value of all these sections has exceeded the management threshold (TQI $\geq$ 5 mm). The railway mileage, number of TQI threshold overruns, maximum TQI value, and track formation type were collected in Table 2. In this research, the notation "Kxyz" is used to represent the railway mileage along the SNHR, where "K" represents "kilometer" and "xyz" indicates the specific mileage point. It was indicated that exceeding threshold frequency was 58, 46, and 31 times in mileage points K235, K299, and K237, respectively. The structural type of K235 and K299 is subgrade. Furthermore, the bridge–subgrade transition in track geometric irregularity is also prominent in mileage points K236, K239, and K241 due to different settlements caused by differences in the rigidity of the soil subgrade and concrete bridge.

In some mileage points where the threshold is only exceeded a few times, it may be caused by an error in the measurement of deviation or other defects of rail. Nevertheless, it will be improved after track maintenance. For instance, track irregularity occurred three times in the subgrade mileage point K246 (appearing at K246 +400) on 15 February 2011 and 22 May 2013. However, no irregularities were observed in other inspections. The same phenomenon occurred in the bridge–subgrade transition of mileage point K232 (specifically in K232 +800 to K233 +000). The smooth state of the track was improved after repair. Consequently, mileage points K235, K236, K237, and K299 should be considered to be potential defective areas because the track irregularity was still detected in these subgrade mileage points even after being repaired by the tamping machine and restored by the workers.

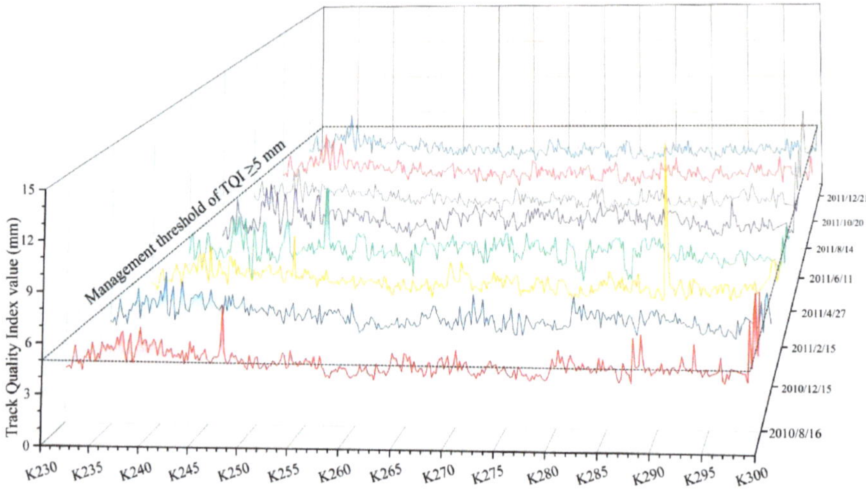

(**a**)

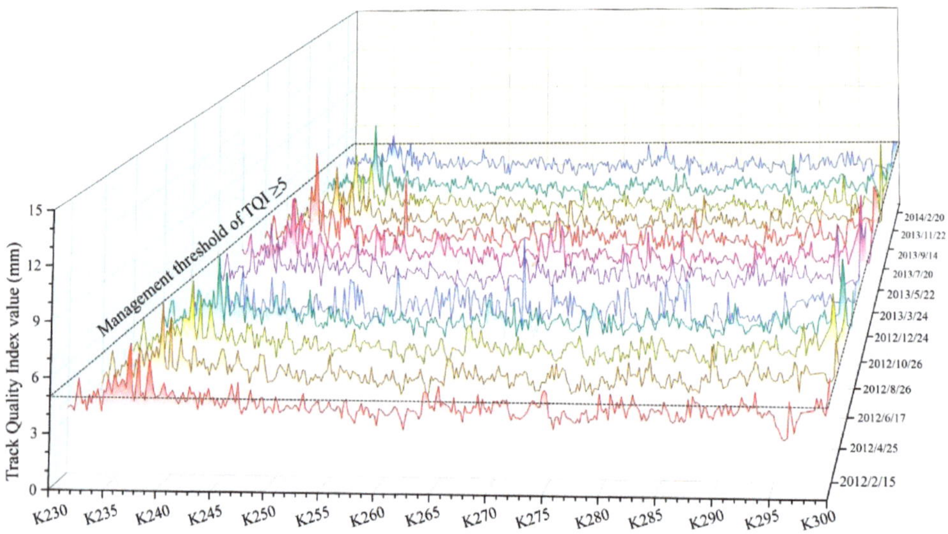

(**b**)

Figure 5. Track quality index (TQI) reflecting geometric irregularities acquired from comprehensive inspection trains along the K230 to K300 sections of the high-speed railway. (**a**) TQI data from 2010 to 2011, (**b**) TQI data from 2012 to 2014.

Table 2. Frequency of TQI above management threshold in various railway mileage points.

Ranking	Railway Mileage	Exceeded Times	Maximum Value	Structural Type
1	K235	58	9.2	Subgrade
2	K299	46	10.3	Subgrade
3	K237	31	6.28	Bridge–subgrade transition
4	K236	28	6.94	Bridge–subgrade transition
5	K233	14	5.64	Subgrade
6	K239	14	6.21	Bridge–subgrade transition
7	K298	13	7.63	Subgrade
8	K241	12	6.06	Bridge–subgrade transition
9	K234	11	5.44	Subgrade
10	K238	7	5.93	Subgrade
11	K230	6	5.32	Subgrade
12	K243	5	5.18	Subgrade
13	K263	4	5.18	Bridge–subgrade transition
14	K287	3	13.3	Subgrade
15	K279	3	5.38	Bridge–subgrade transition
16	K245	3	5.74	Subgrade
17	K246	3	7.42	Subgrade
18	K232	3	5.81	Bridge–subgrade transition

5.3. Settlement Monitoring and Ground-Penetrating Radar Survey

Using InSAR, an area of land subsidence was observed at mileage points K235, K298, and K299, which may affect the subgrade. Additionally, the geometric smoothness of the track, as indicated by TQI (Track Quality Index) data, can assist in locating potential defects. Therefore, the four mileage points (K235, K236, K237, and K299) with the highest frequency of TQI exceeded the threshold and were selected as the key observation areas, while the K298 mileage point was excluded. The tracks on these four mileage points were continuously found to be uneven. Among them, two mileage points, i.e., K235 and K299, were severely affected by ground settlement. These four key locations, spanning a 70 km stretch of the railway (from K230 to K300) across four cities, significantly reduce the workload for a detailed field survey. Therefore, the settlement monitoring data are obtained by measuring the CPIII control network for the ballastless track of the high-speed railway using a precise leveling technique. It is crucial to emphasize that the post-construction settlement of the subgrade is usually strictly controlled, ensuring that the deformation does not exceed 15 mm.

The cumulative vertical settlement of the railway cross-section in sections K235 +390 to K236 +335 and K279 +368 to K279 +992 was displayed by field investigation, as shown in Figure 6. It is observed that the settlement at positions K235 +587, K236 +050, and K236 +095 exceed the standard value of deformation (15 mm). The position K235 +587 was the most severely deformed location (up to −46 mm), and the displacement at sections K236 +050 and K236 +095 was −17.7 and −16.8 mm, respectively. Figure 7 shows the relationship between cumulative settlement and time variation at the specific position, and the settlement was found to increase with time from 2012 to 2014. It could be suggested that the settlement of the subgrade is progressively deteriorating, and this status cannot be improved by routine maintenance. The excessive settlement observed at these three positions has persisted for over a year. Therefore, the subgrades K235 +587, K236 +050, and K236 +095 were identified as the defect positions. TQI value exceeded the management threshold (5 mm) over 31 times at the K237 mileage point, but the settlement never exceeded the settlement control standards (15 mm). Thus, this site was not observed as the defect location. The differential settlement was observed between the bridge and subgrade due to the differences in the rigidity of the soil subgrade and concrete bridge. Subsequently, the track irregularity state is detected in the subgrade side based on TQI when the train passes through here. However, this irregular state can be repaired by tamping devices or by enhancing the stiffness of the subgrade or slab using grouting technology. Figure 8 illustrates the subgrade deformation at K299 +235 to K299 +800 in 2014. The observed deformations were minor and did not exceed 5 mm. It is observed that the subgrade is still in good working condition in terms of settlement.

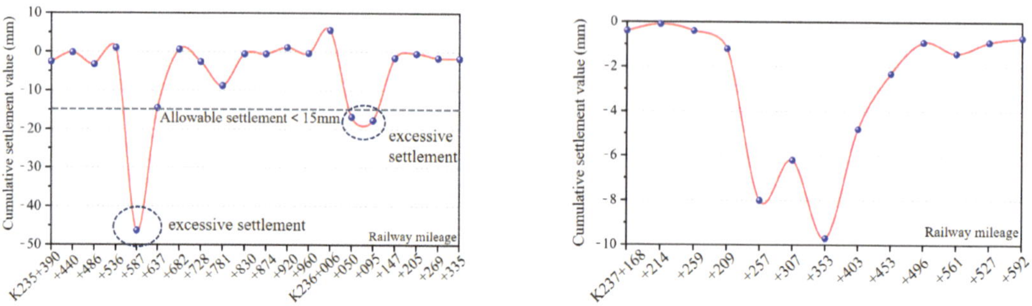

Figure 6. Monitoring data of accumulated settlement at different locations.

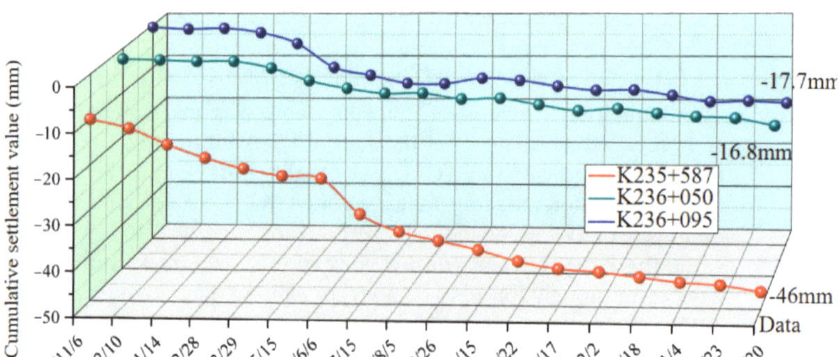

Figure 7. Time-series monitoring data of K235 +587, K236 +050, and K236 +095 position.

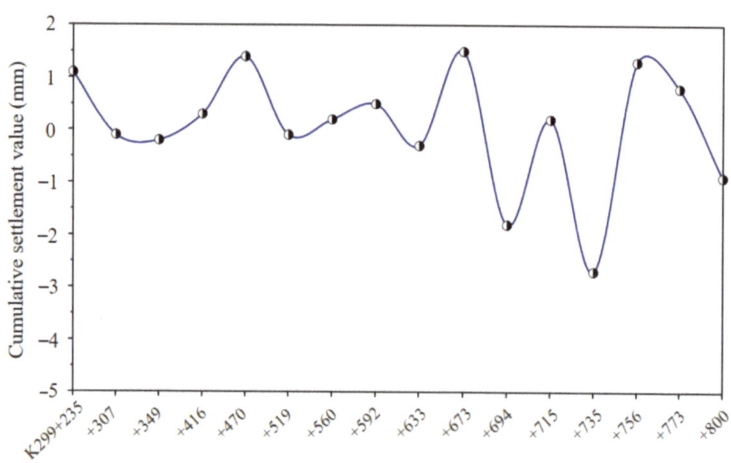

Figure 8. Monitoring data of accumulated settlement along sections K299 +235 to K299 +800.

To explain the consistent emergence of track geometry irregularity at the mileage point, K298 is difficult, while the deformation of the subgrade at this point is almost imperceptible. Therefore, we used the GPR technique to investigate this cause. The image of GPR showed that the reflected signals of the subgrade section were relatively weak from K299 +680 to +750, indicating disruption of the lower soil layer (Figure 9a). The field investigation showed that the slope toe of the subgrade was destroyed due to the construction of metro line 3, resulting in the tendency of the subgrade to incline to the outer side. The cause of subgrade defects in mileage point K235 was also investigated in detail using the GPR technique. The GPR image demonstrates distinct low reflection areas in contrast with the surrounding region at the bottom of the subgrade of this section (Figure 8). Since the subgrade is layered filling and compaction, thus the signal of each subgrade layer is represented as a layered texture on the GPR image. The upper part of the subgrade was well structured, but the lower part was degraded in sections K235 +540 to +680 (Figure 9b). The onsite investigation discovered that this location is close to the Yangtze River. An infiltrated surface and the outward hydraulic gradient after a rapid drop in water level would be formed during the dry season (November to April of next year). Consequently, soil particles were washed away from the subgrade due to seepage forces. This erosion phenomenon will damage soil structure (Figure 9b) and aggravate the consolidation of soil under the additional stress of high-speed trains. This was the cause of a significant increase in subgrade deformation of this section after 11 December 2012 (Figure 6).

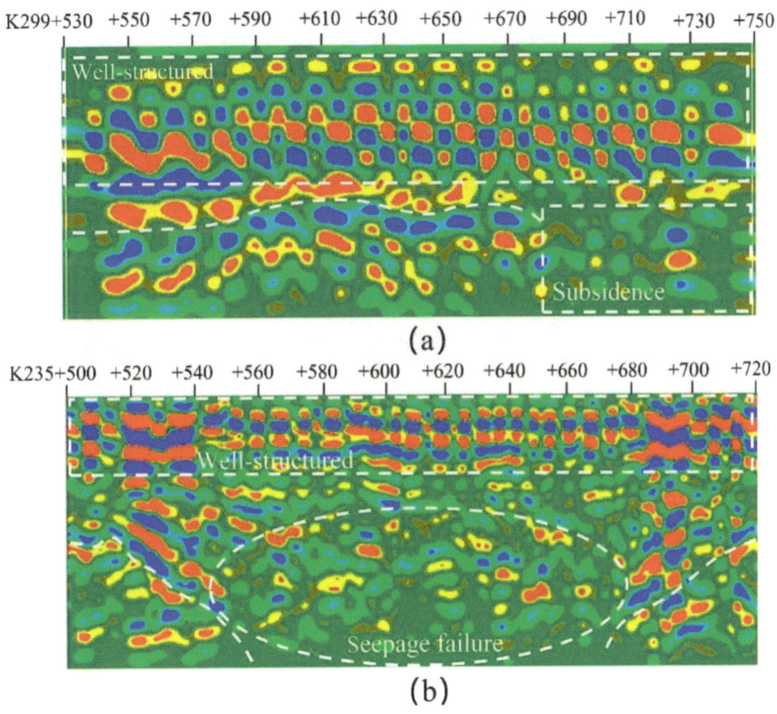

Figure 9. Ground-penetrating radar (GRP) scanning results for the sections of K235 and K299. (**a**) and (**b**) are GPR images of subgrade structures at K299 and K235, respectively, all GPR imges was obtained from Shanghai Railway Bureau.

In summary, it is concluded that the sections from K235 +540 to +680 and K299 +680 to +750 have been finally identified as subgrade defects using GRP. The subgrade structures at K235 and K299 are deteriorated, leading to varying degrees of deformation caused by groundwater fluctuations and human activities, respectively.

6. Discussions and Future Work

The subgrade is vulnerable to climatic and geological factors, human activities, and long-term train load. The deformation or defect is more likely to appear in the subgrade section. However, the defect of the subgrade is imperceptible because it is covered by a closed waterproof sealing structure. The detailed investigation and treatment of health status or defects of subgrade are the critical factors for ensuring the safe and smooth operation of the train. Thus, an efficient, economical, and promising method is expected to be developed to explore the potential defect in all railway mileage. To obtain detailed information on subgrade defects, it is necessary to investigate the factors that may affect the high-speed railway line. Therefore, three monitoring works should be carried out to the health status or defects of subgrade:

First, geologic hazards (e.g., land subsidence) will directly affect the stability of subgrade structures. The uneven settlement of the ground is vulnerable to the overexploitation of groundwater and the construction of metro and high-rise buildings. The subgrade near the river is more susceptible to water level fluctuation, just like the subgrade is being inundated by rising waters, causing decreases in strength and bearing capacity. The secondary consolidation settlement is caused by pore water pressure dissipation inside the subgrade after falling water level. As illustrated in this work, the D–InSAR technique can effectively capture regional ground deformation. Consequently, the K235 and K298~K300 sections severely affected by the land subsidence are roughly determined in this work.

Second, the state of railway infrastructure cannot be precisely measured using InSAR alone. Therefore, track geometry measurement is employed to locate the potential defect areas. Irregularities caused by minor deformations at the subgrade–bridge transition section and track deviation should be excluded by analyzing the time-series TQI data because those defects can be repaired by tamping devices or chemical grouting techniques. The combined results of the InSAR and TQI allow for the precise identification and reaffirmation of potential defect locations along the railway line. This study has highlighted that the mileage points at K235, K236, K237, and K299 are identified as potential defect mileage points. In addition, the detailed sites and defect causes should be investigated in the next step.

Third, deformation monitoring and detailed defect information for infrastructure are carried out by leveling and GPR technology, respectively. The excessive deformation can directly indicate that the subgrade structure has been severely damaged. The specific location of the defect can be determined by the high-accuracy measuring capability of the CPIII control network. The settlement monitoring of the infrastructure in mileage point K237 rules out the possibility of this position being affected by subgrade defects. Finally, K235, K236, and K299 are identified as subgrade defects. The GPR is used to survey the causes of defects and their development trends, and it possesses a function similar to "computed tomography (CT)" scanning in the medical field. GPR can identify shallow cavities, the depth of the water table, potential deformation inside the subgrade, and even the slip surface. Some of the contributions of GPR have been proved for the identification of defects at part of the K235 and K299 mileage points.

In conclusion, a novel "integrated" approach is presented to investigate railway subgrade defects in detail utilizing remote sensing, train measuring systems, and non-destructive techniques (Figure 10). The multisource, multiscale, and multitemporal information is collected by combining InSAR, TQI, precision leveling, and GPR techniques, encompassing the aerial, train-mounted, and ground-based platforms. A collaborative analysis method of multisource data is employed to find the target areas from large-scale ground subsidence and small-scale infrastructure status information. A case study of SNHR demonstrates that the proposed method can accurately identify the health status and locate defects in the subgrade. It can also quantitatively assess the development trends of defects and can provide suggestions for maintenance plans based on the service status of subgrade and threat degree of defects.

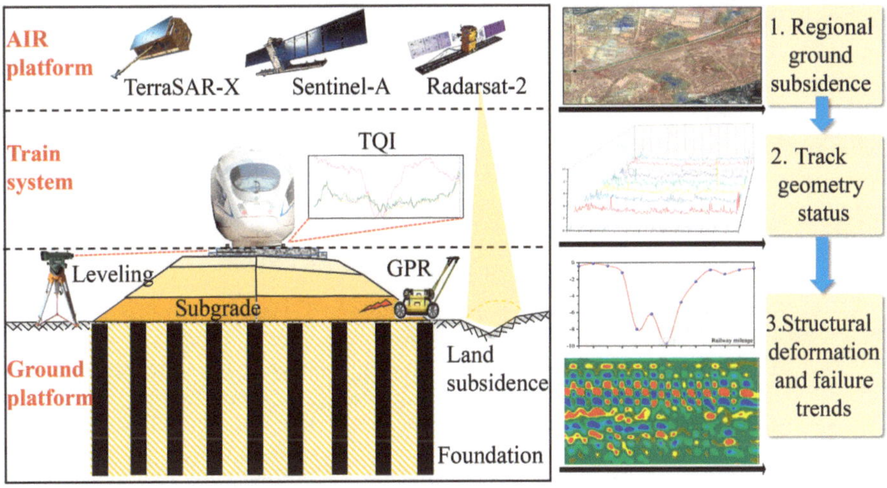

Figure 10. Integration of air, train system, and ground platform for identifying the potential subgrade defect.

In current practice, the monitoring of subgrade health status typically relies on one or two methods, including GPS, leveling, GPR, and InSAR, as well as the comprehensive inspection train [39–41]. For example, GPS can automatically capture long-term deformation information of the subgrade at specific locations, but it is very costly. However, the monitoring accuracy of these techniques depends on the transmission signal. In cases where there are obstructions from tall buildings, trees, or adverse weather conditions, the accuracy of deformation monitoring results may be compromised [42]. Traditional settlement monitoring methods can provide highly accurate deformation information, and GPR can assess the health status of the subgrade. However, these methods are quite time-consuming [43,44] and impractical for covering the whole subgrade. InSAR and comprehensive inspection trains boast their advantages in extensive coverage, yet the monitoring accuracy of both these methods tends to be relatively low [11,19].

In summary, depending solely on the sensitivity of the single monitoring method may result in blind spots or areas that are not being adequately monitored. Furthermore, a single monitoring method may only be able to detect one or a few of these types of degradation, missing the signs of others. In addition, some monitoring techniques are more susceptible to external factors such as weather conditions, traffic loads, or nearby construction activities. These factors collectively indicate that one or two monitoring methods alone cannot serve as a comprehensive solution for detecting defects covering the whole subgrade. Therefore, the "integrated" monitoring approach for subgrade defects in this work leverages the advantages of various monitoring techniques, allowing for cross-validation of the monitoring results. This approach enhances the accuracy and efficiency of subgrade defect detection.

However, the limitation of the study is that the land subsidence is determined using InSAR images for a relatively short period, ranging from August 2010 to November 2010, due to the unavailability of free InSAR data for the period from 2011 to 2014 [45]. This problem was solved after the Sentinel–1A satellite launched in April 2014. The Sentinel–1A images with high temporal resolution and a short revisiting period (12 d) can be adopted for frequent ground acquisitions in regional areas. Furthermore, the integration of multisensor SAR images will enable the high-precision and short-term monitoring of deformation along the railway with the increasing number of satellites and the availability of open data in the future.

7. Conclusions

The requirements for the safe operation and detection/monitoring of infrastructure deterioration are becoming higher due to the sustained growth of total railway mileage. However, adverse natural and environmental disasters jeopardize the sustainability of the railway network, which increases the probability of potential risks during high-speed train operations. Recent advancements in monitoring technology have made it possible to quickly survey and monitor subgrade health. This study provides a comprehensive overview of the primary methods employed for monitoring and detecting railway defects and deformation, and it also draws attention to the monitoring objects and "integrated" holistic monitoring technique for high-speed railway subgrade. The defect location, deformation trend, and the cause of deterioration are investigated in detail in SNHR. The procedure of this integrated method and the deficiency are discussed. The following conclusions are suggested:

First, the monitoring content of subgrade service status should contain the environmental and geological hazards that threaten the subgrade (e.g., land subsidence triggered by overexploitation of groundwater and engineering activities), track geometry status, deformation monitoring, and internal defect of the subgrade. An "integrated" holistic monitoring technique is carried out through the general investigation (locating the potential defects using InSAR and TQI), detailed surveys (long-term deformation monitoring by leveling), and exploration (defects diagnosis using GPR).

Second, the result of the D–InSAR method indicates that the K235 and K298~K300 sections of SNHR are severely affected by land subsidence. Time-series TQI data are combined to reconfirm the location of the potential defect area further. Subsequently, the local areas of the subgrade mileage points K235 and K299, as well as the bridge–subgrade transition mileage points K236 and K237, are estimated as the potential defect positions based on the frequency of TQI exceeding the threshold and settlement monitoring.

Third, the settlement monitoring data along the critical areas demonstrates that the settlement at K235 +587, K236 +050, and K236 +095 exceed the standard value of the deformation for high-speed railways. The maximum values of subgrade settlement are −46, −17.7, and −16.8 mm in these locations, respectively. Furthermore, the possibility of subgrade deterioration is excluded in the K237 mileage point (settlement not exceeding 15 mm). The differential settlement is identified as the main cause of track irregularities in the bridge–subgrade transition mileage point K237. Finally, the K235, K236, and K299 points are estimated as subgrade defects by cooperative analysis of INSAR, TQI, and settlement.

Fourth, the GPR technique is used to survey the health status at K235 and K299 mileage points. It is revealed that the lower part of the subgrade structure for K235 +540 to +680 deteriorated due to the influence of seepage failure caused by a rapid drop in water level and train dynamic load. Additionally, the lower soil layer of the subgrade is disrupted at K299 +680 to +750 due to the deformation of the toe triggered by engineering activity.

Fifth, this study has demonstrated that an "integrated" holistic monitoring technique can provide more comprehensive information on subgrade service status at different spatiotemporal scales. This technique can be more useful in the accurate identification of defects, development of maintenance plans, and effective implementation of fault recovery of railway infrastructure. Furthermore, this technique will continue to be improved with the increasing number of satellites and enhancement of Earth-observation capability.

Author Contributions: Conceptualization, Y.L. and Y.H.; software, Y.L.; validation, Y.H.; writing—original draft preparation, Y.L.; writing—review and editing, Y.L. and Y.H. supervision, L.X.; funding acquisition, L.X. All authors have read and agreed to the published version of the manuscript.

Funding: This work was funded by the National Natural Science Foundation of China (Grant Nos. U2268213, 42172322, and 42007419), National Key Research and Development Program of China (Grant No. 2018YFC1505403), Natural Sciences Funding Project of Hunan Province (Grant No. 2020JJ5981), Excellent Youth Fund Project of Hunan Provincial Education Department (21B0226), and the China Railway Group Limited Technology Research and Development Plan (2022—key program-01).

Data Availability Statement: The data are not publicly available due to confidentiality requirements.

Acknowledgments: We are grateful to ASF (Alaska Satellite Facility) and NASA (National Aeronautics and Space Administration) for providing the InSAR and DEM data.

Conflicts of Interest: The authors declare no conflicts of interest.

References

1. National Railway Administration of the People's Republic of China (NRAPR). Statistics of China's High-Speed Railway Operating Mileage. 2023. Available online: http://www.nra.gov.cn/wef/zxkn/202301/t20230104_339423.shtml (accessed on 4 January 2023).
2. National Railway Administration of the People's Republic of China (NRAPR), Announcement on Investigation of Major Railway Traffic Accidents. Available online: http://www.nra.gov.cn/zzjg/jgj/gzgl/gsgz (accessed on 30 April 2020).
3. Li, M.; Persson, I.; Spännar, J.; Berg, M. On the use of second-order derivatives of track irregularity for assessing vertical track geometry quality. *Veh. Syst. Dyn.* **2012**, *50*, 389–401. [CrossRef]
4. Xu, L.; Zhai, W. A novel model for determining the amplitude-wavelength limits of track irregularities accompanied by a reliability assessment in railway vehicle-track dynamics. *Mech. Syst. Signal Proc.* **2017**, *86*, 260–277. [CrossRef]
5. Sadeghi, J. Development of railway track geometry indexes based on statistical distribution of geometry data. *J. Transp. Eng.* **2010**, *136*, 693–700. [CrossRef]
6. Lasisi, A.; Attoh-Okine, N. An unsupervised learning framework for track quality index and safety. *J. Transp. Eng.* **2020**, *7*, 1–12. [CrossRef]
7. Lasisi, A.; Attoh-Okine, N. Principal components analysis and track quality index: A machine learning approach. *Transp. Res. Pt. C-Emerg. Technol.* **2018**, *91*, 230–248. [CrossRef]
8. Madejski, J.; Grabczyk, J. Continuous geometry measurement for diagnostics of tracks and switches. In Proceedings of the International Conference on Switches; Delft University of Technology: Delft, The Netherlands, 2002.
9. Liu, R.K.; Xu, P.; Sun, Z.Z.; Zou, C.; Sun, Q.X. Establishment of track quality index standard recommendations for Beijing metro. *Discret. Dyn. Nat. Soc.* **2015**, *2015*, 473830. [CrossRef]
10. Lu, C.; Cai, C. Overview on safety management and maintenance of high-speed railway in China. *Transp. Geotech.* **2020**, *25*, 100397. [CrossRef]
11. Čebašek, T.M.; Esen, A.F.; Woodward, P.K.; Laghrouche, O.; Connolly, D.P. Full scale laboratory testing of ballast and concrete slab tracks under phased cyclic loading. *Transp. Geotech.* **2018**, *17*, 33–40. [CrossRef]
12. Li, S.; Xu, W.; Li, Z. Review of the SBAS InSAR Time-series algorithms, applications, and challenges. *J. Geod. Geodyn.* **2022**, *13*, 114–126. [CrossRef]
13. Liu, C.; Li, N.; Wu, H.; Meng, X. Detection of high-speed railway subsidence and geometry irregularity using terrestrial laser scanning. *J. Surv. Eng.* **2014**, *140*, 04014009. [CrossRef]
14. Chen, H.; He, J.; Xue, Y.; Zhang, S. Experimental study on sinkhole collapse monitoring based on distributed Brillouin optical fiber sensor. *Optik* **2020**, *216*, 164825. [CrossRef]
15. Xing, X.; Chang, H.C.; Chen, L.; Zhang, J.; Yuan, Z.; Shi, Z. Radar interferometry time series to investigate deformation of soft clay subgrade settlement—A case study of Lungui Highway, China. *Remote Sens.* **2019**, *11*, 429. [CrossRef]
16. Chen, B.; Gong, H.; Chen, Y.; Lei, K.; Zhou, C.; Si, Y.; Li, X.; Pan, Y.; Gao, M. Investigating land subsidence and its causes along Beijing high-speed railway using multi-platform InSAR and a maximum entropy model. *Int. J. Appl. Earth Obs. Geoinf.* **2021**, *96*, 102284. [CrossRef]
17. Gutiérrez, F.; Galve, J.P.; Lucha, P.; Castañeda, B.; Bonachea, J.; Guerrero, J. Integrating geomorphological mapping, trenching, InSAR and GPR for the identification and characterization of sinkholes: A review and application in the mantled evaporite karst of the Ebro Valley (NE Spain). *Geomorphology* **2011**, *134*, 144–156. [CrossRef]
18. Wang, L.; Deng, K.; Zheng, M. Research on ground deformation monitoring method in mining areas using the probability integral model fusion D-InSAR, sub-band InSAR and offset-tracking. *Int. J. Appl. Earth Obs. Geoinf.* **2020**, *85*, 101981. [CrossRef]
19. Dheenathayalan, P.; Small, D.; Schubert, A. High-precision positioning of radar scatterers. *J. Geodesy.* **2016**, *90*, 403–422. [CrossRef]
20. Wang, R.; Yang, M.; Dong, J.; Liao, M. Investigating deformation along metro lines in coastal cities considering different structures with InSAR and SBM analyses. *Int. J. Appl. Earth Obs. Geoinf.* **2022**, *115*, 103099. [CrossRef]
21. Suo, Z.; Zhang, J.; Li, M.; Zhang, Q.; Fang, C. Improved InSAR phase noise filter in frequency domain. *IEEE Trans. Geosci. Remote Sens.* **2015**, *54*, 1185–1195. [CrossRef]

22. Jung, J.; Kim, D.; Lavalle, M.; Yun, S. Coherent change detection using InSAR temporal decorrelation model: A case study for volcanic ash detection. *IEEE Trans. Geosci. Remote Sens.* **2016**, *54*, 5765–5775. [CrossRef]
23. Bai, M.; Chen, Y.; Wang, C.; Li, X.; Tian, G. Radar spectral analysis and evaluation of the effect of grouting treatment in karst caves and soil caves. *Environ. Earth Sci.* **2018**, *77*, 795. [CrossRef]
24. Fontul, S.; Paixão, A.; Solla, M.; Pajewski, L. Railway track condition assessment at network level by frequency domain analysis of GPR data. *Remote Sens.* **2018**, *10*, 559. [CrossRef]
25. Dzurisin, D. A comprehensive approach to monitoring volcano deformation as a window on the eruption cycle. *Rev Geophys.* **2003**, *41*. [CrossRef]
26. Andronico, D.; Spinetti, C.; Cristaldi, A.; Buongiorno, M.F. Observations of Mt. Etna volcanic ash plumes in 2006: An integrated approach from ground-based and polar satellite NOAA–AVHRR monitoring system. *J. Volcanol. Geotherm. Res.* **2009**, *180*, 135–147. [CrossRef]
27. Allison, R.S.; Johnston, J.M.; Craig, G.; Sion, J. Airborne optical and thermal remote sensing for wildfire detection and monitoring. *Sensors* **2016**, *16*, 1310. [CrossRef] [PubMed]
28. Hua, L.; Shao, G. The progress of operational forest fire monitoring with infrared remote sensing. *J. For. Res.* **2017**, *28*, 215–229. [CrossRef]
29. Dai, K.; Li, Z.; Xu, Q.; Bürgmann, R.; Milledge, D.; Tomás, R.; Fan, X.; Zhao, C.; Liu, X. Entering the era of earth observation-based landslide warning systems: A novel and exciting framework. *IEEE Geosci. Remote Sens. Mag.* **2020**, *8*, 136–153. [CrossRef]
30. Lei, T.; Wang, J.; Li, X.; Wang, W.; Shao, C.; Liu, B. Flood Disaster Monitoring and Emergency Assessment Based on Multi-Source Remote Sensing Observations. *Water* **2022**, *14*, 2207. [CrossRef]
31. Hu, Q.; Ma, R.; Chen, F.; Jiang, A.; Wang, Y. Urban landscape monitoring based on high-resolution spaceborne TerraSAR-X data: A case study of Nanjing City, China. *Remote Sens. Lett.* **2018**, *9*, 447–456. [CrossRef]
32. Lin, P.; Chen, X.; Jiang, M.; Song, X.; Xu, M.; Huang, S. Mapping shear strength and compressibility of soft soils with artificial neural networks. *Eng. Geol.* **2022**, *300*, 106585. [CrossRef]
33. Xiao, Y.; Shi, C. On the Division of the Quaternary in the Nanjing region, Eastern China. *J. Nantong Univ. (Nat. Sci. Ed.)* **2008**, *7*, 60–65. (In Chinese)
34. Liu, L.; Cai, G.; Liu, S.; Chen, Y. Deformation characteristics and control for foundation pits in floodplain areas of Nanjing, China. *Bull. Eng. Geol. Environ.* **2021**, *80*, 5527–5538. [CrossRef]
35. Zebker, H.A.; Rosen, P.A.; Hensley, S. Atmospheric effects in interferometric synthetic aperture radar surface deformation and topographic maps. *J. Geophys. Res.-Solid Earth* **1997**, *102*, 7547–7563. [CrossRef]
36. Zhang, B.; Li, J.; Ren, H. Using phase unwrapping methods to apply D-InSAR in mining areas. *Can. J. Remote Sens.* **2019**, *45*, 225–233. [CrossRef]
37. Xu, X.J.; Lei, Y.; Yang, F. Railway Subgrade Defect Automatic Recognition Method Based on Improved Faster R-CNN. *Sci. Program.* **2018**, *2018*, 4832972. [CrossRef]
38. Zhao, Y.; Zhou, L.; Wang, C.; Li, J.; Qin, J.; Sheng, H.; Huang, L.; Li, X. Analysis of the spatial and temporal evolution of land subsidence in Wuhan, China from 2017 to 2021. *Remote Sens.* **2022**, *14*, 3142. [CrossRef]
39. Silvast, M.; Nurmikolu, A.; Wiljanen, B.; Levomäki, M. Identifying frost-susceptible areas on Finnish railways using the ground penetrating radar technique. *Proc. Inst. Mech. Eng. Part F J. Rail Rapid Transit* **2013**, *227*, 3–9. [CrossRef]
40. Jia, C.; Yang, X.; Wu, J.; Ding, P.; Bian, C. Monitoring analysis and numerical simulation of the land subsidence in linear engineering areas. *KSCE J. Civ. Eng.* **2021**, *25*, 2674–2689. [CrossRef]
41. Hou, Z.; Zhao, W.; Yang, Y. Identification of railway subgrade defects based on ground penetrating radar. *Sci. Rep.* **2023**, *13*, 6030. [CrossRef] [PubMed]
42. Schipperijn, J.; Kerr, J.; Duncan, S.; Madsen, T.; Klinker, C.D.; Troelsen, J. Dynamic accuracy of GPS receivers for use in health research: A novel method to assess GPS accuracy in real-world settings. *Front. Public Health* **2014**, *2*, 21. [CrossRef] [PubMed]
43. Xu, J.; Li, X.; Yang, J.; Gao, Y.; Wu, S. A remote subgrade settlement monitoring system based on optical method. *Urban Rail Transit* **2019**, *5*, 202–206. [CrossRef]
44. Elseicy, A.; Alonso-Díaz, A.; Solla, M.; Rasol, M.; Santos-Assunçao, S. Combined use of GPR and other NDTs for road pavement assessment: An overview. *Remote Sens.* **2022**, *14*, 4336. [CrossRef]
45. Li, G.; Yang, F.; Ke, Z.; Sun, X.F.; Liu, B.; Zhao, W.; Zhou, Y. Long-Term Monitoring of Local Track Irregularity and Its Influence for Simply Supported Girder Bridge of HSR. *Buildings* **2022**, *12*, 445. [CrossRef]

materials

Article

Effects of Polyethylene Glycol/Porous Silica Form-Stabilized Phase Change Materials on the Performance of Asphalt Binders

Hao Wang [1,2,*], Gui Pan [3], Lihong He [4] and Ling Zou [2,5]

[1] School of Civil Engineering, Chongqing Jiaotong University, Chongqing 400074, China
[2] CCCC First Highway Consultants Co., Ltd., Xi'an 710075, China
[3] Yuexiu (China) Transport Infrastructure Investment Ltd., Guangzhou 511458, China
[4] School of Materials Science and Engineering, Chongqing Jiaotong University, Chongqing 400074, China; sunnyhlh@126.com
[5] School of Transportation, Southeast University, Nanjing 211189, China
[*] Correspondence: 622160980007@mails.cqjtu.edu.cn

Abstract: The road performance and temperature-regulating properties of asphalt binders modified with novel polyethylene glycol (PEG)/porous silica (PS) form-stabilized phase-change materials (PEG/PS-fs-PCMs) were studied. PS and PEG were used as the supporting substance and PCMs. The results showed that PEG/PS-fs-PCMs could maintain a maximum weight percentage of 70% without leakage, at temperatures as high as 90 °C. The PEG/PS-fs-PCMs exhibited stable chemical structures, excellent thermal stability, high heat storage density, and suitable phase-change temperature. Based on conventional physical tests, the addition of PEG/PS-fs-PCMs can increase the viscosity and the degree of hardness of asphalt binders; thus, achieving an excellent comprehensive performance of the modified asphalt binder depends on determining the optimal dosage of PEG/PS-fs-PCMs. Additionally, incorporating PEG/PS-fs-PCM particles into the asphalt binder can enhance its ability to withstand permanent deformation at elevated temperatures, while PEG/PS-fs-PCMs mainly act as a filler, weakening the cohesive force of the asphalt molecules, and preventing the ductility of asphalt from expansion, according to DSR and BBR tests. Moreover, the use of PEG/PS-fs-PCMs can enhance the heat transfer properties of the asphalt binders, resulting in an improved temperature regulation performance. However, the accumulation of PEG/PS-fs-PCM particles on asphalt binders can negatively impact the storage stability of the modified asphalt binders, because of the difference in density between the two materials.

Keywords: asphalt binders; sol–gel method; PEG; road property; temperature-regulating performance

Citation: Wang, H.; Pan, G.; He, L.; Zou, L. Effects of Polyethylene Glycol/Porous Silica Form-Stabilized Phase Change Materials on the Performance of Asphalt Binders. *Materials* **2023**, *16*, 5293. https://doi.org/10.3390/ma16155293

Academic Editor: Jorge Carvalho Pais

Received: 29 June 2023
Revised: 18 July 2023
Accepted: 26 July 2023
Published: 27 July 2023

1. Introduction

Asphalt pavement is a popular choice for road construction, because of its affordability, reliable performance, recyclable properties, excellent wear resistance, and comfortable driving experience [1]. However, asphalt binders are a type of temperature-sensitive material, and their properties are vulnerable to extreme temperature variations, which shortens their lifespan, and increases the cost of road maintenance [2]. Moreover, the black surface of asphalt binders can easily absorb more solar radiation at high temperature, resulting in many pavement and environmental problems, such as asphalt aging, permanent deformation at high temperature, and the UHI effect [3,4]. Numerous methods have been employed by researchers to enhance the durability and functionality of asphalt pavements. The incorporation of modifiers into asphalt binders is widely recognized as an effective method to significantly improve the properties of the binders. Numerous bitumen modifiers, such as nanomaterials [5,6], polymers [7,8], and bio-oil [9] have been reported to date. Among the above modifiers, polymeric materials are widely regarded as the most commonly utilized substances for enhancing the various properties of asphalt binders against most types of distress. Due to their unique nanoscale dimensions, nanomaterials have consistently been shown to significantly enhance the properties of asphalt binders. However, although the aforementioned technologies have been proved to enhance the performance of asphalt pavements to some extent, the thermal damage to asphalt pavements caused by high temperatures occurs, which also exacerbates the urban heat island effect [10,11]. Moreover, these methods are passive, and may not even be representative, given the differences in material, traffic load, and climatic conditions, so they may not be applicable to different asphalt pavements [12,13]. Hence, it is urgent and essential to propose an effective approach to regulate the temperature of, and prevent thermal damage to, asphalt pavements.

Phase-change materials (PCMs) are known for their high efficiency in storing and releasing latent heat, which makes them ideal for maintaining a constant temperature. As a result, they have been used in applications in various fields within heat storage and conversion [14–16]. PCMs have been employed in extensive applications in various industries, such as in greenhouses, textiles, solar energy utilization, temperature-regulating fiber manufacturing, medicine, and food packaging [17–19]. This characteristic has attracted great interest from a large number of researchers regarding the application of PCMs in thermal regulation in asphalt pavements. Zhang et al. [20] prepared PAC and PACM samples with different PEG contents, and they found that incorporating PCMs in asphalt binders could improve the elastic property of the PAC, store heat energy, and regulate the temperature of asphalt pavements. According to the property of the phase-change thermoregulation agent, Si et al. [21] investigated and compared the thermoregulation of CPCMs at different temperatures, aiming to solve rutting distress in asphalt pavements. Ren et al. [22] synthesized different types of fs-PCMs using the sol–gel method, and the results showed that ACPCMs demonstrate a better stability at a high temperature, and can be applied to asphalt mixture. Zhang et al. [23] employed a vacuum impregnation technique to produce shape-stabilized EG/PEG PCMs, and the aim was to investigate the feasibility of enhancing properties, and controlling the temperature, through this method. Muhammad et al. [24] revealed that the use of suitable PCMs did not adversely affect the rheology of asphalt. Moreover, it was found to be an effective method for regulating temperature variations.

As a solid–liquid transformation PCM, PEG is widely applied in asphalt binders because of its excellent properties, such as its non-toxicity, competitive price, suitable working temperature, and high energy storage density [25–27]. However, the original PEG has its inherent disadvantage: liquid leakage in the molten state. Previous research has shown that the addition of PEG in asphalt binders results in a decrease in stiffness, which negatively impacts the high-temperature properties of the asphalt [23,28,29]. To prevent leakage of PEG in practical applications, a promising and effective method is to incorporate PEG into porous materials, to obtain form-stabilized PCMs [30]. In general, there are many types of supporting materials, such as comb-like polymers [31], silicon dioxide [32,33], and white carbon black [34]. In order for fs-PCMs to withstand the high mixing temperature of the asphalt mixture, which typically reaches 185 °C, the supporting materials must have a sufficient thermal stability. Porous silica (PS) is an inorganic porous material that has drawn considerable interest because of its various desirable properties, such as its flame retardancy, commendable thermal stability, and suitable thermal conductivity [35,36]. More importantly, PS contains a large number of porous network structures, and can be utilized as a supporting substance in preparing fs-PCMs, to avoid the leakage of the PEG at high temperatures. Therefore, it is crucial to conduct a thorough investigation into the impact of PEG/PS-fs-PCMs on binder-and temperature-regulating properties.

In this study, PEG and PS were selected as the working materials (PCMs) and support materials, respectively. The heat storage performance, chemical composition and structure, and thermal property of the PEG/PS-fs-PCMs were characterized. On this basis, PEG/PS-fs-PCM-modified asphalt binders were obtained using the melt-blending method, and the physical properties and temperature adjustment performance of the modified asphalt binder were investigated.

2. Materials and Methods

Polyethylene glycol (PEG), AR grade, was provided by Jiangsu Haian Petrochemical Plant, Nantong, China; the average molecular weight was 4000 (China). Sodium hydroxide was produced by Changsha Deyue Co., Ltd., Changsha, China. Silica sol was provided by Chongqing Jinglian New Materials Co., Ltd., Chongqing, China. The SK-70 asphalt (KOCH Bitumen Co., Ltd., Chongqing, China) was used for the base asphalt binders, and their performance-testing index is presented in Table 1.

Table 1. Technical testing index of the base asphalt.

Test Indices	Technical Requirements	Measured Values	Specification
Softening point (°C)	≥46	48.1	T0606-2011
Dynamic viscosity (60 °C Pa s)	≥180	223	T0620-2011
Ductility (10 °C cm)	≥25	37	T0605-2011
Penetration (25 °C mm)	60–80	66.3	T0604-2011

2.1. Preparation of PEG/PS-fs-PCMs

PEG/PS-fs-PCMs were obtained using the sol–gel method, in accordance with the following steps. Firstly, silica sol was mixed with PEG, and stirred vigorously at room temperature, until complete dissolution was achieved. A little sodium hydroxide was then gradually added as an accelerator to the PEG/silica sol mixture, to form the silica sol gel. The mixture was stirred for approximately 10–15 min, until the gelation process was complete. The composites, with the mass ratios of PEG to PS of 9:1, 8:2, 7:3, and 6:4, were tested for leaks, after being kept in an oven at 85 °C for several hours, to determine the maximum ratio of PEG to PS. The test results showed that as long as the impregnation ratio of PEG was equal to, or lower than, that of fs-PCM (70% by weight), the PEG did not leak out from the composite. The fabrication process for the composite material is shown in Figure 1.

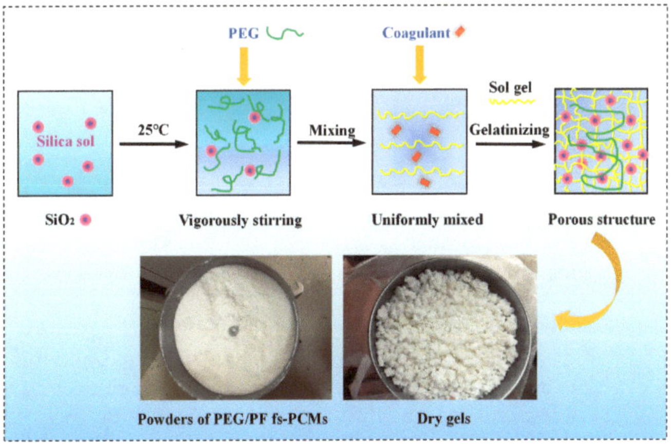

Figure 1. Schematic view of the preparation of PEG/PS-fs-PCMs.

2.2. Preparation of PEG/PT fs-PCM-Modified Asphalt Binders

This study utilized a high-speed shearing method to modify asphalt binders with PEG/PS-fs-PCMs. The process involved melting the base asphalt at 165 °C, and gradually adding the PEG/PS-fs-PCMs, while stirring at 3000 rpm for 30 min to ensure an even dispersion. Modified asphalt binders with varying PEG/PS-fs-PCM content (0, 5, 10, 15, and 20 weight%) were obtained through this process.

2.3. Characterization of Preparation of PEG/PS-fs-PCMs

2.3.1. SEM

The microstructures of the composites were observed using SEM (ZEISS Sigma, Oberkochen, Germany).

2.3.2. DSC

The enthalpy and temperature of the resulting PEG/PS-fs-PCMs were measured using DSC (DSC6000, PE Inc., Long Beach, CA, USA). The experiment was carried out under a nitrogen atmosphere, with a heating rate of 10 °C/min, and a temperature range of 20 °C to 90 °C. Each sample weighed approximately 5–8 mg.

2.3.3. FT-IR

Chemical structures were examined at room temperature using FT-IR (Nicolet iS5, Thermo Fisher, Waltham, MA, USA). All samples were prepared using the KBr plate technique.

2.3.4. TG

The thermal stabilities of the PEG/PS-fs-PCMs at 30–600 °C under a nitrogen atmosphere were evaluated using TGA (TGA 550, New Castle, DE, USA).

2.3.5. Leakage Test

To investigate the leakage resistance of the PCMs, PEG and PEG/PS-fs-PCMs were subjected to an hour-long exposure to 90 °C in an oven. The observation and recording of the process were carried out using a digital camera.

2.4. Experimental Methods of the Modified Asphalt Binders

2.4.1. Conventional Properties Test

The softening point, penetration (15 °C), and ductility (15 °C) of the PEG/PS-fs-PCM-modified asphalt binders were determined, according to the standard methods JTG E20 T0606 [37], JTG E20 T0604 [38], and JTG E20 T0605 [39].

2.4.2. Dynamic Shear Rheometer Test (DSR)

The rheological properties of the binders were investigated using a DSR (AASHTO T-315). This test was performed at a of 10 rad/s over a temperature range of 52–76 °C (6 °C/point).

2.4.3. Bending-Beam Rheometer Test (BBR)

The binders' ability to resist cracking at low temperatures was evaluated via the BBR test, in accordance with AASHTO T313. At the temperatures of −12 °C and −18 °C, the logarithmic creep rate (m-value) and creep stiffness (S) were measured.

2.4.4. Storage Stability Test

The samples were poured into a glass tube, and this tube was enclosed and placed in a refrigerator for 4 h. Then, the thermal conductivity of the two ends of the samples was calculated, and the thermal conductivity difference between them could be used as an alternative index to characterize the storage stability of the binders.

2.4.5. Temperature-Regulating Test

For the indoor irradiation experiments, a 400 W solar simulator (QVF133, PHILIPS, Hamburg, Germany) was used as the solar light source. A temperature recorder equipped with thermocouples was used to record the temperature–time curve of the modified asphalt binders, as shown in Figure 2.

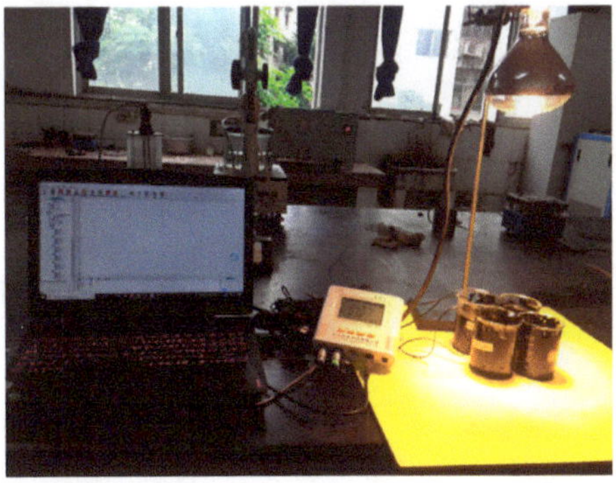

Figure 2. Experimental setup for evaluating the temperature-regulating properties.

3. Results

3.1. Properties of PEG/PS-fs-PCMs

3.1.1. Morphology Characterization

As shown in Figure 3, the surface of the PEG appears flat and smooth. As for the composite, it is evident that the PEG molecules are absorbed into the network structure of the PS. Additionally, the pore structure of the silica gel is predominantly occupied by these molecules, with only a few surface pores remaining unfilled. This demonstrates that the silica gel has a highly porous structure, making it an ideal supporting material for PEG. By confining PEG within its porous structure, the silica gel effectively prevents the solid–liquid leakage of PEG at high temperatures.

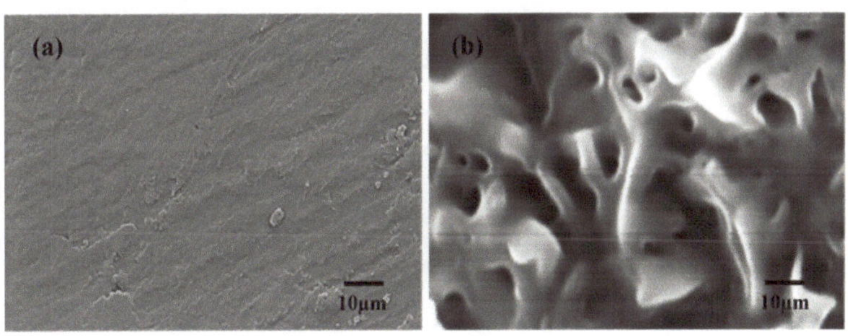

Figure 3. SEM images of the PEG (**a**) and PEG/PS-fs-PCMs (**b**).

3.1.2. Form-Stability Property

Form-stable behavior is a crucial index for fs-PCMs performance. The prepared PEG/PS-fs-PCMs were heated to 90 °C for 1 h, to evaluate their exudation stability. As shown in Figure 4, a clear liquid PEG leaked onto the filter paper after 1 h, implying that its form cannot be retained when the temperature exceeds its melting point. By contrast, PEG/PS-fs-PCMs can retain their original form without any leakage during heating. The results suggest that adding PS to PCMs can improve their form-stability properties, and make them more effective in preventing melt leakage at high temperatures. As a result, the PEG/PS-fs-PCMs prepared in this study are considered to be highly stable when exposed to elevated temperatures.

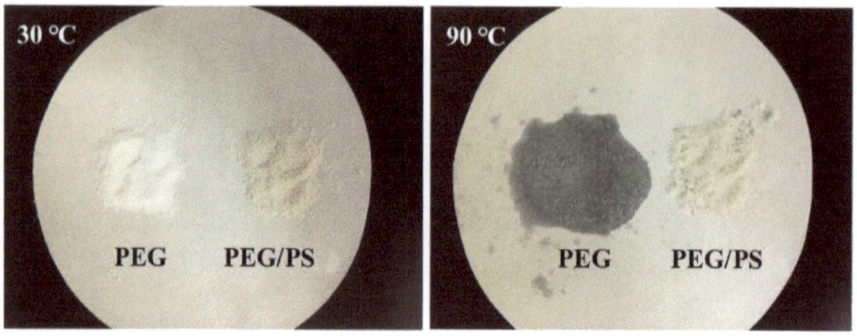

Figure 4. Images of the PEG and PEG/PS-fs-PCMs before and after heating.

3.1.3. Chemical Structure

Figure 5 shows the FT-IR absorption spectra of the samples. The peak at 3432 cm^{-1} in the spectrum of the PEG is caused by the stretching vibration in the OH group. The peak at 1643 cm^{-1} corresponds to the C=O elongation and the asymmetric CO elongation at 1113 cm^{-1}. The peaks at 1101, 794, and 468 cm^{-1} in the infrared absorption curve of the PS are typical Si-O-Si peaks. The asymmetric stretching peak of the -OH and the bending vibration of the molecular water are 3444 cm^{-1} and 1732 cm^{-1}, respectively, caused by the adsorption of part of the water or capillary pores on the surface of the porous SiO_2 particles. Obviously, the characteristic peaks of PEG/PS-fs-PCMs are almost the same as those of pure PEG. Furthermore, no new peaks are observed in the composite FT-IR spectrum, indicating that only a physical reaction occurred between the PEG and PS.

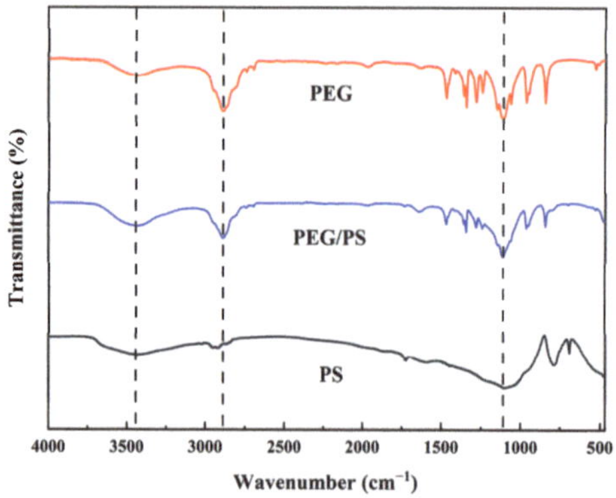

Figure 5. FT-IR spectra of the PEG, PS, and PEG/PS-fs-PCMs.

3.1.4. Thermal Property

The DSC curves are illustrated in Figure 6. According to the figure, the two samples exhibit similar phase-transition properties, which contain an exothermic peak and endothermic peak when the phase changes. The detailed results of the DSC test are shown in Table 2. As can be seen, the melting and solidification temperatures of the pure PEG were 61.2 °C and 34.9 °C, respectively, and those of the PEG/PS-fs-PCMs were 61.5 °C and 33.8 °C, respectively. Notably, the melting temperature of the PEG/PS-fs-PCMs is quite close to that of the PEG, but the solidification temperature decreases obviously after the addition of the PS, which is possibly caused by the volume restriction effect of the porous structure network of the PS [31].

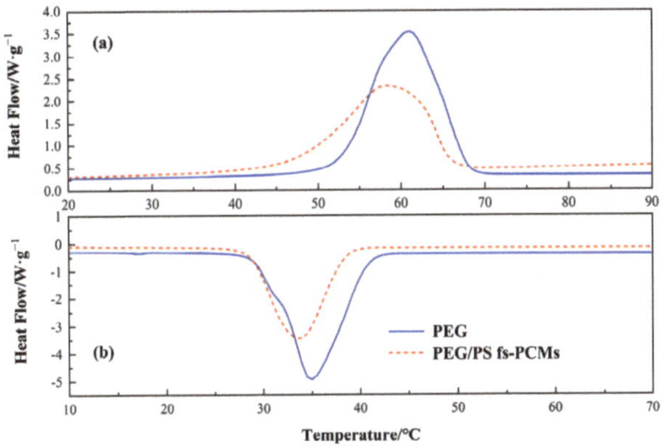

Figure 6. DSC curves of the PEG and PEG/PS-fs-PCMs during heating (**a**) and cooling (**b**) process.

Table 2. Data for the DSC curves of PEG and PEG/PS-fs-PCMs.

Samples	Tm (°C)	ΔHm (J/g)	Tc (°C)	ΔHc (J/g)
PEG	61.2	182.4	34.9	171.3
PEG/PS-fs-PCMs	61.5	121.6	33.8	114.7

Table 2 shows that the fusion and solidification latent heat for the pure PEG are calculated to be 182.4 J/g and 171.3 J/g, respectively. On the other hand, the PEG/PS-fs-PCMs have a fusion and solidification latent heat of 121.6 J/g and 114.7 J/g, respectively. Obviously, the latent heat of the PEG/PS-fs-PCMs slightly declines when compared to the PEG. The decrease in the transition enthalpy is primarily caused by a reduction in the mass percentage of the PEG after the incorporation of the PS. PEG is important for thermal energy storage during melting or solidification. Despite this reduction, the latent heat of the PEG/PS-fs-PCMs is higher compared to previous fs-PCMs studied by other researchers. Moreover, the phase-transition temperatures of the PEG/PS-fs-PCMs are similar to the deformation temperatures of asphalt pavements during summer. Therefore, the DSC test results demonstrate that PEG/PS-fs-PCMs have a desirable heat storage density, and favorable phase-transition temperatures, making them a practical option for use in hot mixed asphalt.

3.1.5. Thermal Stability of PEG/PS-fs-PCMs

To investigate the ability of PEG/PS-fs-PCMs to withstand high temperatures in asphalt mixtures, the thermal stability properties of the samples were analyzed using TGA. The results are presented in Figure 7, which shows the thermogram of the pristine PEG displaying an onset of weight loss at approximately 311.2 °C. Additionally, there is a significant decline observed from 388.4 to 407.8 °C, with a weight loss of 96.7%. This weight loss corresponds to the evaporation and decomposition of the PEG. Otherwise, the PEG/PS-fs-PCMs exhibit a relatively stable TG curve, and both the initial degradation temperature (Ton) and the temperature of the maximum decomposition rate (Tmax) are a little higher than those of the PEG. Specifically, the Ton and Tmax of the PEG/PS-fs-PCMs reach 314.5 °C and 410.9 °C, respectively, indicating that the dimensional pore structures of the PS have a positive effect on the thermal stability of the composites. Moreover, the amount of residual PEG/PS-fs-PCMs in the composite depends on the loading amount of PS, and the quantity of PEG impurities present. The results of the TG analysis indicate that the PEG/PS-fs-PCMs have a sufficient thermal stability, and can be utilized as a thermoregulating agent in asphalt at a mixing temperature of up to 185 °C.

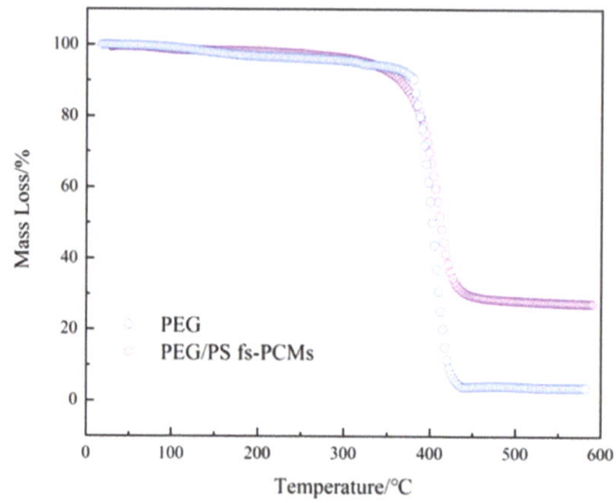

Figure 7. TGA curves of the PEG and PEG/PS-fs-PCMs.

3.2. *Performance of PEG/PS-fs-PCM-Modified Asphalt Binders*

3.2.1. Conventional Physical Performance

Generally, the permeability value is widely used to evaluate the degree of hardness of asphalt at intermediate temperatures. A low permeability value reflects that the asphalt is hard and stiff. Figure 8 shows the results of the permeability test. As can be seen, the permeability values of all the samples decreased almost linearly with the increase in PEG/PS-fs-PCMs content. The linear fitting is convincible, because R^2 is higher than 0.95. Specifically, the permeability value reduces by 6.1% compared with the control sample after 10% PEG/PS-fs-PCMs are added into virgin asphalt, and when the dosage of PEG/PS-fs-PCMs is increased to 20%, the permeability value is reduced by 10.3%. The obtained results reveal that the addition of the PEG/PS-fs-PCMs can improve the stiffness rutting resistance of asphalt binders. This may be explained by the fact that the oily phase can be absorbed by PEG/PS-fs-PCMs particles, thus increasing the asphaltene component of the asphalt, and restricting the flow of the asphalt, thereby enhancing its stiffness.

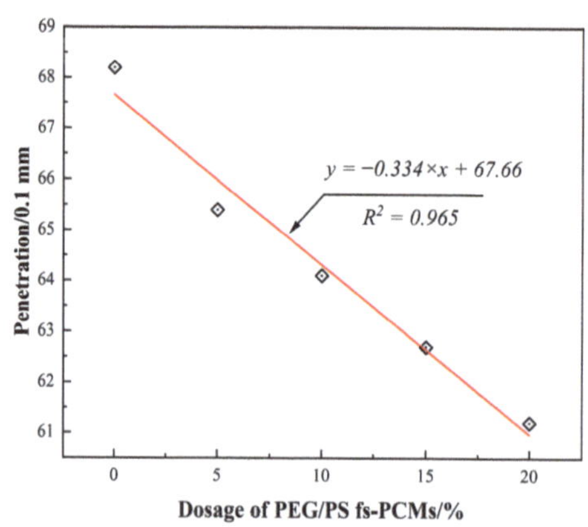

Figure 8. Correlation between the dosage of PEG/PS-fs-PCMs and the penetration of asphalt binders.

The softening point is a crucial parameter that indicates the stability of asphalt binders under high service temperatures. In general, a higher softening point indicates a better high-temperature stability in asphalt binders. According to Figure 9, the softening points linearly increase with the increase in the PEG/PS-fs-PCMs dosage, which is different from the permeability value. For example, the softening point of the modified asphalt with the modifier dosage of 5%, 10%, 15%, and 20% increases up to 5.6%, 8.2%, 12.1%, and 15.8%, respectively, when compared with that of the virgin asphalt, thus proving that the PEG/PS-fs-PCMs enhance the high-temperature property of asphalt. Hence, it is well supported that the viscosity and the degree of hardness of asphalt binders can be increased by adding PEG/PS-fs-PCMs. As a result, the asphalt binders become more compact and stable, thus eventually enhancing their high-temperature stability.

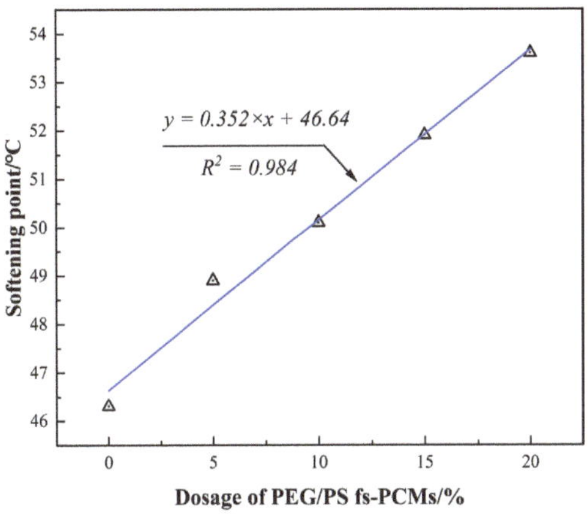

Figure 9. Correlation between the dosage of PEG/PS-fs-PCMs and softening point of asphalt binders.

The ductility is a crucial factor in evaluating the low-temperature performance. It is important to note that a higher ductility value indicates a poorer low-temperature performance. The test results for the ductility at 15 °C are shown in Figure 10. It is obvious that the ductility value is reduced after PEG/PS-fs-PCMs are added. Specifically, when the dosages of PEG/PS-fs-PCMs are 5%, 10%, 15%, and 20%, respectively, the asphalt binder ductility values decrease to 17.4%, 36.5%, 47.8%, and 60.6%, respectively, indicating that PEG/PS-fs-PCMs negatively affect the asphalt low-temperature performance, due to an increased hardness and stiffness. In conclusion, the results of conventional physical tests show that the optimal dosage of PEG/PS-fs-PCMs is the key factor in obtaining a modified asphalt binder with an excellent comprehensive performance.

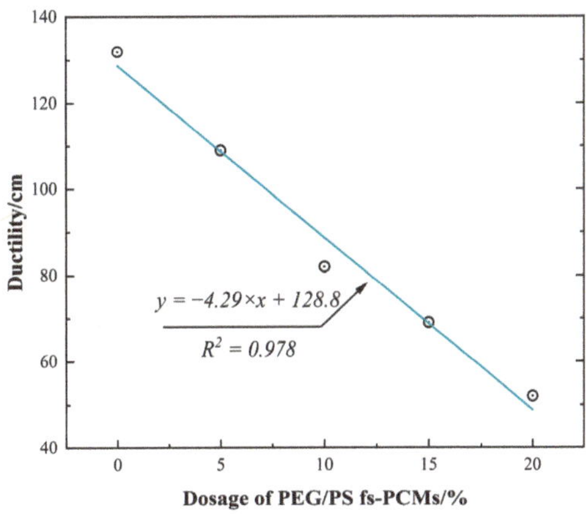

Figure 10. Correlation between the dosage of PEG/PS-fs-PCMs and the ductility of asphalt binders.

3.2.2. High-Temperature Performance

The dynamic shear rheometer (DSR) is a commonly employed technique to assess the ability of asphalt binders to withstand permanent deformation at high temperatures. The rutting resistance index (G*/sinδ) is an important parameter that is strongly correlated with the rutting behavior of asphalt binders. Higher values of G*/sinδ indicate greater resistance to permanent deformation. This study examines the impact of various dosages of PEG/PS-fs-PCMs on the rheological characteristics of asphalt binders.

As Figure 11 indicates, the rutting factor G*/sinδ of all the samples exhibits an apparent downward trend as temperatures increase, implying the typical temperature-dependency properties of asphalt. Moreover, the G*/sinδ obviously increases as the content of PEG/PS-fs-PCMs increases, indicating that the high-temperature anti-rutting performance can be greatly improved by adding PEG/PS-fs-PCMs. Specifically, compared to virgin asphalt, the increased amplitude of G*/sinδ in the PEG/PS-fs-PCM-modified asphalt with the contents of 10% and 20% is the greatest at 64 °C, at 52.3% and 125.7%, respectively. This can be explained as follows. On one hand, the PEG/PS-fs-PCMs particles with a greater surface area can adsorb light components in the asphalt, thus enhancing its anti-rutting property. On the other hand, some heat energy can be absorbed by the PEG/PS-fs-PCMs particles during the heating process; thus, the asphalt binder faces a lower actual test temperature. Moreover, the high-temperature properties of the asphalt binder are also enhanced due to the presence of silica, as demonstrated in other studies.

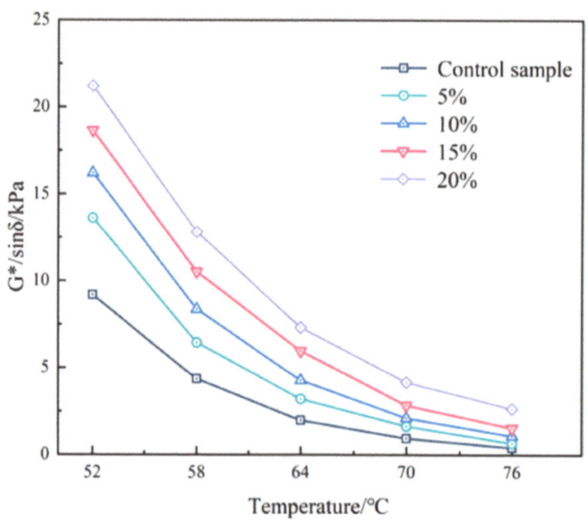

Figure 11. Rutting resistance of the PEG/PS-fs-PCM-modified asphalt binders.

Furthermore, the failure temperatures of the samples can be determined by calculating the linear fitting curves of log(G*/sinδ) at a G*/sinδ value of 1.0 KPa. Figure 12 and Table 3 reveal the failure temperature of the PEG/PS-fs-PCM-modified asphalt. The failure temperature of the control sample is about 68.1 °C. When the content of the loaded PEG/PS-fs-PCMs increases from 0% to 20%, however, the failure temperature of the modified asphalt increases from 68.1 °C to 82.2 °C, resulting in a 20.7% growth rate compared to the virgin asphalt. This suggests that the rutting resistance has improved.

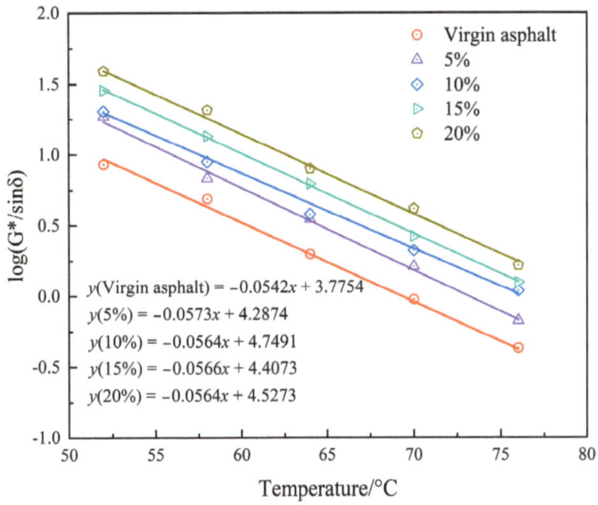

Figure 12. Correlation between the log(G*/sinδ) and temperatures of the modified asphalt binders.

Table 3. Failure temperature of PEG/PS-fs-PCM-modified asphalt binders.

Samples	Control Sample	5%	10%	15%	20%
R^2	0.975	0.964	0.951	0.972	0.968
Failure temperature	68.1 °C	73.2 °C	77.4 °C	78.1 °C	82.2 °C

3.2.3. Low-Temperature Performance

To examine the impact of PEG/PS-fs-PCMs on the low-temperature rheological characteristics of binders, we conducted a BBR test, and measured the creep rate (m) and stiffness modulus (S) as the evaluation criteria. Generally speaking, a lower creep stiffness indicates less tensile stress, and a higher creep rate indicates more stress relaxation. Therefore, the asphalt binders with a larger creep rate and/or smaller creep-stiffness modulus are believed to exhibit a better low-temperature performance.

As observed from Figure 13, the creep-rate value of all the PEG/PS-fs-PCM-modified asphalt binders decreases, while the creep stiffness modulus value increases, as the temperature decreases, indicating that the temperature dependency of the asphalt binders does not change after PEG/PS-fs-PCMs are added. However, the addition of PEG/PS-fs-PCMs to asphalt binders results in an increase in S values, and a slight decrease in the creep rate. This indicates that the rheological performance of the binders at low temperatures is reduced, making them more susceptible to cracking. This is possibly because the PEG/PS-fs-PCMs mainly act as filler, and weaken the cohesive forces of the asphalt molecules, thus preventing the ductility extension of the asphalt. According to the Superpave standard, the asphalt binder's S-value should be lower than 300 MPa, and the m-value should not exceed 0.3. It is impressive that all the samples meet the specified requirements for the m-values and S-values. Therefore, it is crucial to control the dosage of PEG/PS-fs-PCMs within 20%, to avoid further degradation of the low-temperature performance.

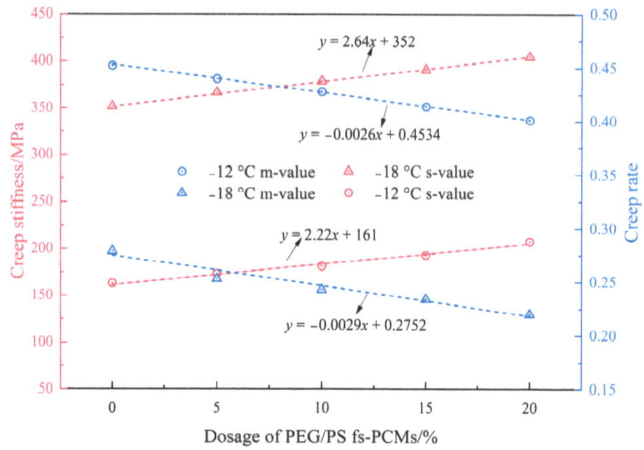

Figure 13. BBR test results for the asphalt samples.

3.2.4. Compatibility Stability Performance

It is universally known that modified asphalt binders are multiphase systems, and there is a difference in the density between the asphalt and the modifying agent. The phase separation and performance inhomogeneity of binders during high-temperature storing and pumping are significantly influenced by the compatibility stability. This can be observed in the results of the compatibility stability test, as shown in Figure 14.

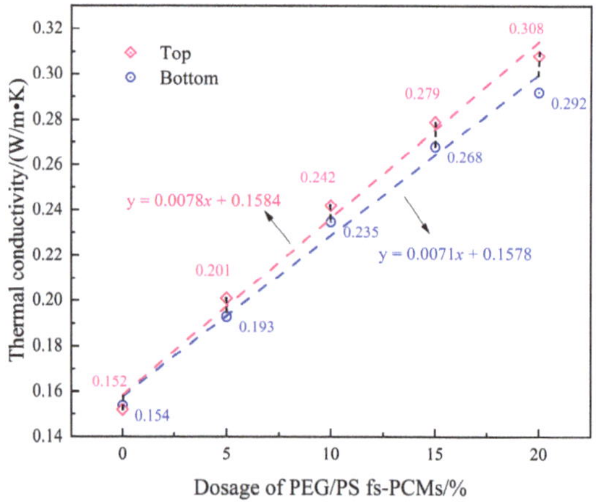

Figure 14. Storage stability of the PEG/PS-fs-PCM-modified asphalt binders.

It is observed that the thermal conductivity of the top and bottom parts of the virgin asphalt binder is relatively stable, with values of 0.154 and 0.152 W/m · K, respectively, indicating that the virgin asphalt is in a stable storage state. The addition of PEG/PS-fs-PCMs to modified asphalt binders results in an increase in thermal conductivity, indicating an improvement in the heat transfer performance. It is important to note that the difference in thermal conductivity between samples becomes more pronounced as the dosage of PEG/PS-fs-PCMs increases. For example, when the content of PEG/PS-fs-PCMs is 10% and 20%, the difference in the thermal conductivity of the modified asphalt binders increases by 0.007 and 0.016 W/m · K, respectively. This is probably because a small dosage of PEG/PS-fs-PCMs is aggregated on the top of the samples, due to the density difference. The results suggest that a high concentration of PEG/PS-fs-PCMs has an adverse effect on the storage stability of asphalt binders. Therefore, to achieve a modified asphalt with an exceptional performance, it is crucial to determine the optimal amount of PEG/PS-fs-PCM modifier, in conjunction with the ductility test and BBR test.

3.2.5. Temperature-Regulating Performance

To more intuitively characterize the impact of the prepared samples on the temperature-regulating performance of asphalt binders, an indoor irradiation test was conducted in the laboratory. Figure 15 depicts the temperature–time profiles of both the virgin asphalt and the PEG/PS-fs-PCM-modified asphalt binders during the irradiation process. The figure shows that the heating rate of the modified asphalt is initially similar to that of the virgin asphalt under light irradiation. As the irradiation time increases, and the temperature reaches the fusion point of the PEG/PS-fs-PCMs, an inflection point appears in the temperature variation curves, indicating a phase transition, and the absorption of thermal energy. Consequently, the temperature of the modified asphalt binders increases at a slower rate. Meanwhile, the thermal-regulating performance of the samples is related to the incorporated content of PEG/PS-fs-PCMs. As the dosage of PEG/PS-fs-PCMs increases, the thermal-regulating performance becomes better and better. The difference in the maximum peak temperature between the virgin asphalt and the 20% modified asphalt binders reaches 6.4 °C at the end of irradiation. Therefore, PEG/PS-fs-PCMs have the ability to absorb thermal energy, and control the temperature of the hot mix asphalt while it is being heated.

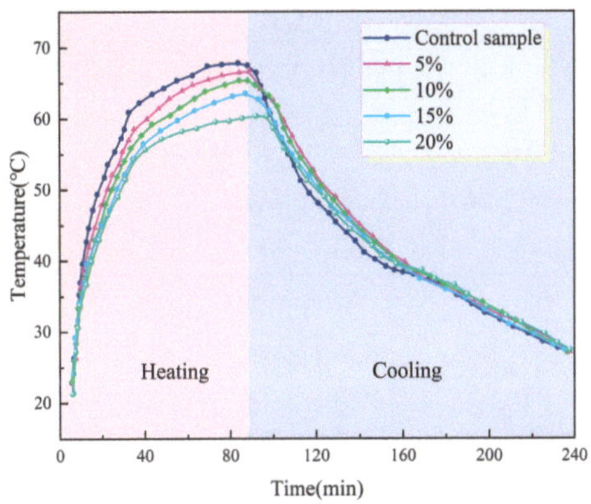

Figure 15. Temperature curves of the PEG/PS-fs-PCM-modified asphalt binders.

4. Conclusions

In this work, the experimental method used in the study contributes to a better knowledge of the effects of the prepared PEG/PS-fs-PCMs on the performance of the asphalt binder. From the experimental results and analyses, the following conclusions can be summarized.

(1) The prepared PEG/PS-fs-PCMs exhibit excellent thermal stability, commendable heat storage density, and appropriate phase-change temperatures for use as a thermoregulation modifier of an asphalt binder, as demonstrated via FT-IR, DSC, and TG experiments.

(2) Incorporating PEG/PS-fs-PCMs into the asphalt binder results in a decrease in the penetration and ductility, and an increase in the softening point. This may be attributed to the addition of PEG/PS-fs-PCMs weakening the cohesive force of the asphalt molecules, and limiting the ductility of the asphalt.

(3) The rutting resistance index ($G^*/\sin\delta$) of the asphalt binder improves with the increasing content of PEG/PS-fs-PCMs, and so does the failure temperature, suggesting that the addition of the fs-PCMs is beneficial in enhancing the rutting resistance of the asphalt binder at high temperatures.

(4) The creep rate of the asphalt binder decreases, while the creep stiffness increases, due to the incorporation of PEG/PS-fs-PCMs, resulting in worse low-temperature properties. Therefore, it is not recommended to further increase the dosage of fs-PCMs.
(5) Temperature regulation tests demonstrate that the addition of PEG/PS-fs-PCMs induces significant hysteresis in the asphalt binder temperature profiles, implying the outstanding temperature regulation capability of the fs-PCMs. The modified asphalt binder shows a maximum temperature difference of approximately 6.4 °C, compared to the virgin asphalt.
(6) Overall, considering the high-temperature performance and excellent thermoregulation efficiency, the appropriate dosage may be determined to be 10%. Furthermore, an aging test and workability analysis will be performed in the future, for a more comprehensive evaluation of the performance of the PEG/PS-fs-PCM-modified asphalt binder.

Author Contributions: Conceptualization, H.W. and L.H.; methodology, H.W.; validation, G.P., L.Z. and H.W.; formal analysis, H.W.; investigation, L.H.; resources, G.P.; data curation, H.W.; writing—original draft preparation, H.W.; writing—review and editing, L.H. and L.Z.; supervision, L.H.; funding acquisition, G.P. All authors have read and agreed to the published version of the manuscript.

Funding: This work was supported by the National Key Research and Development Program of China (No. 2018YFB1600100), the Science and Technology Research Program of Chongqing Municipal Education Commission (Grant No. KJKJZD-K202200701), and Innovative Talent Promotion Plan (2022KJXX-110).

Institutional Review Board Statement: Not applicable.

Informed Consent Statement: Not applicable.

Data Availability Statement: Data sharing not applicable.

Conflicts of Interest: The authors declare no conflict of interest.

References

1. Zhang, D.; Chen, M.; Wu, S.; Liu, Q.; Wan, J. Preparation of expanded graphite/polyethylene glycol composite phase change material for thermoregulation of asphalt binder. *Constr. Build. Mater.* **2018**, *169*, 513–521. [CrossRef]
2. Han, M.; Li, J.; Muhammad, Y.; Hou, D.; Zhang, F.; Yin, Y.; Duan, S. Effect of polystyrene grafted graphene nanoplatelets on the physical and chemical properties of asphalt binder. *Constr. Build. Mater.* **2018**, *174*, 108–119. [CrossRef]
3. Hu, J.; Wu, S.; Liu, Q.; García Hernández, M.I.; Wang, Z.; Nie, S.; Zhang, G. Effect of ultraviolet radiation in different wavebands on bitumen. *Constr. Build. Mater.* **2018**, *159*, 479–485. [CrossRef]
4. Siddig, E.A.A.; Feng, C.P.; Ming, L.Y. Effects of ethylene vinyl acetate and nanoclay additions on high-temperature performance of asphalt binders. *Constr. Build. Mater.* **2018**, *169*, 276–282. [CrossRef]
5. Sukhija, M.; Saboo, N.; Yadav, A.K.; Rath, C. Laboratory study on the suitability of nano-silica as a modifier for asphalt binders. *Constr. Build. Mater.* **2021**, *302*, 124406. [CrossRef]
6. Yadykova, A.Y.; Ilyin, S.O. Rheological and adhesive properties of nanocomposite bitumen binders based on hydrophilic or hydrophobic silica and modified with bio-oil. *Constr. Build. Mater.* **2022**, *342*, 127946. [CrossRef]
7. Ratajczak, M.; Wilmanski, A. Evaluation of Laboratory Methods of Determination of SBS Content in Polymer-Modified Bitumens. *Materials* **2020**, *13*, 5237. [CrossRef]
8. Su, M.; Si, C.; Zhang, Z.; Zhang, H. Molecular dynamics study on influence of Nano-ZnO/SBS on physical properties and molecular structure of asphalt binder. *Fuel* **2020**, *263*, 116777. [CrossRef]
9. Yadykova, A.Y.; Ilyin, S.O. Bitumen improvement with bio-oil and natural or organomodified montmorillonite: Structure, rheology, and adhesion of composite asphalt binders. *Constr. Build. Mater.* **2023**, *364*, 129919. [CrossRef]
10. Wang, X.; Fan, Z.; Li, L.; Wang, H.; Huang, M. Durability Evaluation Study for Crumb Rubber–Asphalt Pavement. *Appl. Sci.* **2019**, *9*, 3434. [CrossRef]
11. Xing, X.; Pei, J.; Shen, C.; Li, R.; Zhang, J.; Huang, J.; Hu, D. Performance and Reinforcement Mechanism of Modified Asphalt Binders with Nano-Particles, Whiskers, and Fibers. *Appl. Sci.* **2019**, *9*, 2995. [CrossRef]
12. Wei, K.; Ma, B.; Huang, X.; Xiao, Y.; Liu, H. Influence of NiTi alloy phase change heat-storage particles on thermophysical parameters, phase change heat-storage thermoregulation effect, and pavement performance of asphalt mixture. *Renew. Energy* **2019**, *141*, 431–443. [CrossRef]

13. Ma, B.; Chen, S.-S.; Wei, K.; Liu, F.-W.; Zhou, X.-Y. Analysis of thermoregulation indices on microencapsulated phase change materials for asphalt pavement. *Constr. Build. Mater.* **2019**, *208*, 402–412. [CrossRef]
14. Khan, Z.; Khan, Z.; Ghafoor, A. A review of performance enhancement of PCM based latent heat storage system within the context of materials, thermal stability and compatibility. *Energy Convers. Manag.* **2016**, *115*, 132–158. [CrossRef]
15. Ming, H.; Yan, L.; Yi, Z.; Min, X.; Hu, J. Preparation and Characterization of Microencapsulated Phase Change Materials with Binary Cores and Poly (allyl methacrylate) (PALMA) Shells Used for Thermo-regulated Fibers. *Thermochim. Acta* **2017**, *655*, 262–268.
16. Zhang, X.; Huang, Z.; Ma, B.; Wen, R.; Zhang, M.; Huang, Y.; Fang, M.; Liu, Y.; Wu, X. Polyethylene glycol/Cu/SiO_2 form stable composite phase change materials: Preparation, characterization, and thermal conductivity enhancement. *RSC Adv.* **2016**, *6*, 58740–58748. [CrossRef]
17. Tang, Y.; Jia, Y.; Alva, G.; Huang, X.; Fang, G. Synthesis, characterization and properties of palmitic acid/high density polyethylene/graphene nanoplatelets composites as form-stable phase change materials. *Sol. Energy Mater. Sol. Cells* **2016**, *155*, 421–429. [CrossRef]
18. Liu, X.; Rao, Z. Experimental study on the thermal performance of graphene and exfoliated graphite sheet for thermal energy storage phase change material. *Thermochim. Acta* **2017**, *647*, 15–21. [CrossRef]
19. Giropaloma, J.; Alshannaq, R.; Fernández, A.I.; Farid, M.M. Preparation and Characterization of Microencapsulated Phase Change Materials for Use in Building Applications. *Materials* **2015**, *9*, 11. [CrossRef]
20. Du, Y.; Liu, P.; Wang, J.; Wang, H.; Hu, S.; Tian, J.; Li, Y. Laboratory investigation of phase change effect of polyethylene glycolon on asphalt binder and mixture performance. *Constr. Build. Mater.* **2019**, *212*, 1–9. [CrossRef]
21. Wei, S.; Zhou, X.Y.; Ma, B.; Ning, L.; Ren, J.P.; Chang, Y.J. The mechanism of different thermoregulation types of composite shape-stabilized phase change materials used in asphalt pavement. *Constr. Build. Mater.* **2015**, *98*, 547–558.
22. Ren, J.; Ma, B.; Si, W.; Zhou, X.; Li, C. Preparation and analysis of composite phase change material used in asphalt mixture by sol–gel method. *Constr. Build. Mater.* **2014**, *71*, 53–62. [CrossRef]
23. Zhang, D.; Chen, M.; Wu, S.; Riara, M.; Wan, J.; Li, Y. Thermal and rheological performance of asphalt binders modified with expanded graphite/polyethylene glycol composite phase change material (EP-CPCM). *Constr. Build. Mater.* **2019**, *194*, 83–91. [CrossRef]
24. Kakar, M.R.; Refaa, Z.; Worlitschek, J.; Stamatiou, A.; Partl, M.N.; Bueno, M. Thermal and rheological characterization of bitumen modified with microencapsulated phase change materials. *Constr. Build. Mater.* **2019**, *215*, 171–179. [CrossRef]
25. Sarı, A.; Alkan, C.; Biçer, A. Synthesis and thermal properties of polystyrene-graft-PEG copolymers as new kinds of solid–solid phase change materials for thermal energy storage. *Mater. Chem. Phys.* **2012**, *133*, 87–94. [CrossRef]
26. Alkan, C.; Günther, E.; Hiebler, S.; Himpel, M. Complexing blends of polyacrylic acid-polyethylene glycol and poly(ethylene-co-acrylic acid)-polyethylene glycol as shape stabilized phase change materials. *Energy Convers. Manag.* **2012**, *64*, 364–370. [CrossRef]
27. He, L.; Li, J.; Zhou, C.; Zhu, H.; Cao, X.; Tang, B. Phase change characteristics of shape-stabilized PEG/SiO_2 composites using calcium chloride-assisted and temperature-assisted sol gel methods. *Solar Energy* **2014**, *103*, 448–455. [CrossRef]
28. Liu, Z.; Wei, K.; Wang, S.; Ma, B.; Wang, X.; Shi, W.; Xu, J. Effect of high-temperature-resistant epoxy resin/polyethylene glycol 2000 composite stereotyped phase change material particles on asphalt properties. *Constr. Build. Mater.* **2021**, *300*, 124007. [CrossRef]
29. Du, Y.; Huang, H.; Hu, X.; Liu, S.; Sheng, X.; Li, X.; Lu, X.; Qu, J. Melamine foam/polyethylene glycol composite phase change material synergistically modified by polydopamine/MXene with enhanced solar-to-thermal conversion. *Renew. Energy* **2021**, *171*, 1–10. [CrossRef]
30. He, L.; Wang, H.; Yang, F.; Zhu, H. Preparation and properties of polyethylene glycol/unsaturated polyester resin/graphene nanoplates composites as form-stable phase change materials. *Thermochim. Acta* **2018**, *665*, 43–52. [CrossRef]
31. Karaman, S.; Karaipekli, A.; Sarı, A.; Biçer, A. Polyethylene glycol (PEG)/diatomite composite as a novel form-stable phase change material for thermal energy storage. *Sol. Energy Mater. Sol. Cells* **2011**, *95*, 1647–1653. [CrossRef]
32. Fang, G.; Li, H.; Chen, Z.; Liu, X. Preparation and properties of palmitic acid/SiO_2 composites with flame retardant as thermal energy storage materials. *Sol. Energy Mater. Sol. Cells* **2011**, *95*, 1875–1881. [CrossRef]
33. Qian, T.T.; Li, J.H.; Ma, H.W.; Yang, J. The preparation of a green shape-stabilized composite phase change material of polyethylene glycol/SiO_2 with enhanced thermal performance based on oil shale ash via temperature-assisted sol-gel method. *Sol. Energy Mater. Sol. Cells* **2015**, *132*, 29–39. [CrossRef]
34. Zhang, X.; Huang, Z.; Ma, B.; Wen, R.; Min, X.; Huang, Y.; Yin, Z.; Liu, Y.; Fang, M.; Wu, X. Preparation and performance of novel form-stable composite phase change materials based on polyethylene glycol/White Carbon Black assisted by super-ultrasound-assisted. *Thermochim. Acta* **2016**, *638*, 35–43. [CrossRef]
35. Li, Y.; Li, J.; Deng, Y.; Guan, W.; Wang, X.; Qian, T. Preparation of paraffin/porous TiO_2 foams with enhanced thermal conductivity as PCM, by covering the TiO_2 surface with a carbon layer. *Appl. Energy* **2016**, *171*, 37–45. [CrossRef]
36. Sheng, N.; Nomura, T.; Zhu, C.; Habazaki, H.; Akiyama, T. Cotton-derived carbon sponge as support for form-stabilized composite phase change materials with enhanced thermal conductivity. *Sol. Energy Mater. Sol. Cells* **2019**, *192*, 8–15. [CrossRef]
37. *JTG E20 T0606-2011*; Softening Point of Bitumen (Ring-and-Ball Apparatus). Ministry of Transport of the People's Republic of China: Beijing, China, 2011.

38. *JTG E20 T0604-2011*; Penetration of Bituminous Materials. Ministry of Transport of the People's Republic of China: Beijing, China, 2011.
39. *JTG E20 T0605-2011*; Ductility of Asphalt Materials. Ministry of Transport of the People's Republic of China: Beijing, China, 2011.

sensors

Article

Field Measurements of Wind-Induced Responses of the Shanghai World Financial Center during Super Typhoon Lekima

Xu Wang [1,2], Guoliang Zhang [1,2], Yongguang Li [1,2], Hu Kong [3], Lang Liu [1,2,*] and Cheng Zhang [4]

1 State Key Laboratory of Mountain Bridge and Tunnel Engineering, Chongqing Jiaotong University, Chongqing 400074, China; xuwang@cqjtu.edu.cn (X.W.); liang_zgl@163.com (G.Z.); yongguangli@mails.cqjtu.edu.cn (Y.L.)
2 School of Civil Engineering, Chongqing Jiaotong University, Chongqing 400041, China
3 Department of Civil and Transportation Engineering, Hohai University, Nanjing 210098, China; 18271219660@163.com
4 Shenzhen Y. S. Mao Bridge Design Group Co., Ltd., Shenzhen 518110, China; y.s.mao@188.com
* Correspondence: janice_liu@cqjtu.edu.cn

Abstract: In this paper, the wind-induced responses of the Shanghai World Financial Center (SWFC) under Super Typhoon Lekima are measured using the health monitoring system. Based on the measurements, the characteristics of vibration, including probability density distribution of accelerations, power spectra, and mode shapes are studied. The curve method and the standard deviation method are used to analyze the relationship of the first- and second-order natural frequencies and damping ratios with amplitudes and the mean wind speed. The results show the following: (1) The structural wind-induced responses in the X and Y directions have high consistencies, and the vibration signals exhibit a peak state; moreover, response amplitudes and acceleration signals disperse when the floor height increases. (2) The first- and second-order natural frequencies in the X and Y directions decrease with the increasing amplitudes and are negatively correlated with mean wind speed; the maximum decrease in natural frequency is 5.794%. The first- and second-order damping ratios in the X and Y directions increase with the increasing amplitudes and are positively correlated with the mean wind speed; the maximum increase in damping ratio is 95.7%. (3) The curve method and the standard deviation method are similar in identifying dynamic characteristic parameters, but the discreteness of the natural frequencies obtained by the curve method is lesser. (4) Under excitations of various typhoons, the mode shapes of SWFC are basically the same, and the mode shapes in the X and Y directions increase with the height and have nonlinearity.

Keywords: super-high-rise building; dynamic characteristics; field measurement; natural frequency; damping ratio; Typhoon Lekima

Citation: Wang, X.; Zhang, G.; Li, Y.; Kong, H.; Liu, L.; Zhang, C. Field Measurements of Wind-Induced Responses of the Shanghai World Financial Center during Super Typhoon Lekima. *Sensors* **2023**, *23*, 6519. https://doi.org/10.3390/s23146519

Academic Editor: Jandro L. Abot

Received: 25 June 2023
Revised: 15 July 2023
Accepted: 16 July 2023
Published: 19 July 2023

1. Introduction

The rapid development of new materials, construction technologies, and design concepts has resulted in the remarkable rise of super-high-rise buildings [1]. Due to their light mass, highly reinforced material, and low damping, supertall buildings are wind-sensitive structures. Strong wind load becomes one of the dominant loads in structural design and has significant impacts on structural safety, occupant comfort, and construction costs [2]. Therefore, it is necessary to analyze wind-induced characteristics to provide beneficial references for wind-resistant design [3].

Normally, wind tunnel tests, numerical simulations, and field measurements are used to study the wind-induced responses of super-high-rise buildings. With the development of monitoring technology, field measurements have become the most accurate and reliable method for studying structural wind effects, since this method not only provides reliable information on wind effects in high-rise buildings [4] but also avoids the modelling and scaling errors implicit in numerical simulations and wind tunnel tests, such as unfavorable Reynolds number effects.

Over the past few decades, many scholars in China and abroad have carried out a series of empirical studies on the wind-induced responses of high-rise buildings. Ohkuma et al. [5], Denoon et al. [6], Kijewski-Correa et al. [7–13], and Yi et al. [14] conducted studies on the buildings of 68 m, 84 m, 245.7 m, 264 m, and 420 m in height, respectively. The main contents include wind-induced vibration response patterns, time-varying acceleration characteristics, and dynamic characteristic parameters, as well as the development of real-time monitoring systems and research into comfort assessment. Although the dynamic characteristics of high-rise buildings under wind action have been extensively studied by full-scale measurements, comprehensive empirical studies of wind effects on super-high-rise buildings of nearly 500 m height or even higher are still very limited, especially the dynamic characteristic parameters under strong typhoon action.

In general, the dynamic characteristics of high-rise buildings are related to their vibration amplitudes, which are usually excited by wind loads. Therefore, buildings and the wind environment can be seen as a time-varying system. Li et al. [15–28] used the random decrement technique (RDT), the Hilbert–Huang transform (HHT), and other methods to study the wind characteristics and wind-induced responses of various super-high-rise buildings in Hong Kong, Taipei, Shanghai, Guangzhou, and Shenzhen. These studies have different focuses, including the variation patterns of natural frequencies and damping ratios with amplitudes, comparisons of the measured results with wind tunnel tests, verification of wind tunnel tests, and finite element models. Zhang et al. [29] used a fast Bayesian approach to analyze the environmental vibration data of the Shanghai Center Tower at different stages of construction, identified the dynamic parameters of the structure, and compared them with the results of the finite element analysis. Fu et al. [30] compared the variations of the first-order natural frequency and damping ratio at low and high amplitudes for the Shanghai Center Tower, based on accurate wind data and the monitored data during the landfall of Typhoon Ampil. Currently, the discrete Fourier transform (DFT) method is commonly used to calculate the power spectral density (PSD) of discrete-time digital signals, but it has insufficient resolution for identifying lower frequency bands. For example, the variations in natural frequency caused by the nonlinear stiffness of structures are often at a level of 10^{-3} Hz or even less.

Although many scholars have conducted extensive research on wind vibration response measurements, the methods for extracting structural dynamic characteristics parameters for super-high-rise buildings in coastal regions have not been sufficiently discussed. Further research is required to develop suitable empirical models for wind-resistant design of structures. In this study, the wind-induced responses of the SWFC under Typhoon Lekima are explored. Firstly, the characteristics of the wind-induced responses are analyzed, including acceleration probability density, power spectrum, etc. Then, the curve method and the standard deviation method in the envelope random decrement technique (E-RDT) are applied. The amplitude dependence of natural frequencies and damping ratios and their variations with the mean wind speed are mainly studied and further compared to the findings of other scholars. Finally, the variations of overall mode shapes under four typhoons, i.e., Ampil, Rumbia, Jongdari, and Lekima, are analyzed.

2. Introduction to Field Measurements

2.1. *Super Typhoon Lekima*

The Japan Meteorological Agency named Super Typhoon Lekima at 15:00 on 4 August 2019, and it made landfall off the coast of Zhejiang at 01:45 on 10 August, with a maximum instantaneous wind speed of 52 m/s. Figure 1b depicts the moving path of Typhoon Lekima, which was characterized by a high intensity and long duration of landfall and finally caused severe disasters.

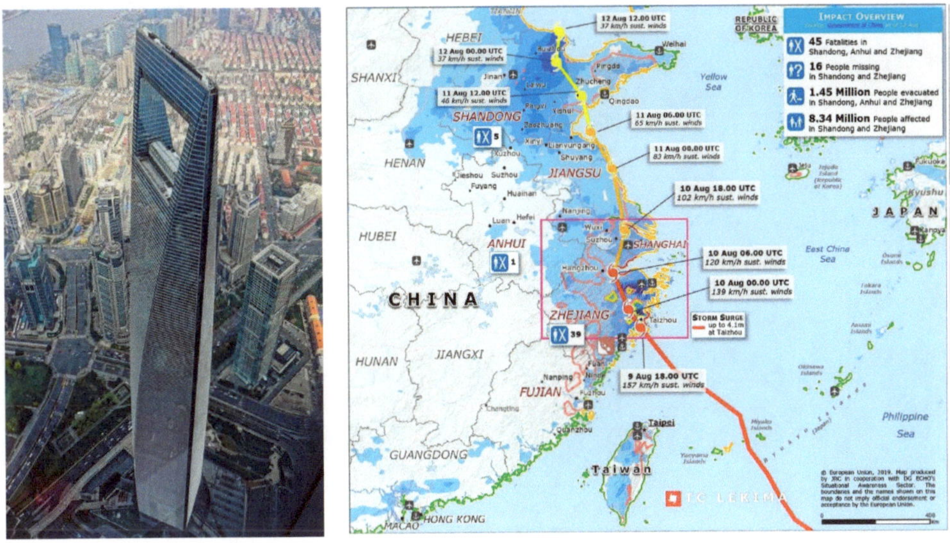

(a) (b)

Figure 1. Landfall of the typhoon and building profile image. (**a**) Building profile image (image © 2023 Google; image by authors). (**b**) Path of Super Typhoon Lekima (image © 2023 Google; map produced by JRC in cooperation with DG ECHO's Situational Awareness Sector).

2.2. The Shanghai World Financial Center and Monitoring System

The Shanghai World Financial Center (SWFC) is located in the Lujiazui Financial and Trade Zone in Pudong New District, Shanghai, between longitude 121°30′8.70″ east and latitude 31°14′9.63″ north. It is a D-class landform surrounded by many super-high-rise buildings. The structure has a total height of 492 m, with 101 stories aboveground and 3 stories underground, and a total floor area of approximately 350,000 square meters. The outline of the building is shown in Figure 1a.

The SWFC uses a triple resistance system—a megaframe structure—to carry the overturning moments caused by wind and earthquakes [31]. The structure consists of giant columns, giant diagonal braces, and perimeter band trusses to improve the stiffness and integrity of the structural system. In order to suppress the vibrations of the structure under strong wind and seismic loads, two active mass tuned dampers (ATMD) were installed on the 90th floor of the SWFC. The dampers are effective in reducing building sway [32], ensuring a safe and comfortable environment for office occupants.

To obtain the dynamic response characteristics of super-high-rise buildings under natural excitation, accelerometer sensors and anemometers were installed on the roof to form a health monitoring system. The measurement sites are installed as shown in Figure 2, where "A" is the three-dimensional accelerometer and "D" is the anemometer. The accelerometer is a TDA-33M accelerometer with a sampling frequency of 100 Hz, a sampling range of ±2 g, and a sensitivity of 1.25 V/g. As shown in Figure 3, the accelerometer takes the east–west direction of the SWFC as the X axis and the north–south direction as the Y axis. The anemometer uses 81,000 series of three-dimensional ultrasonic anemometers produced by R.M. Young, U.S.A. The working temperature is −50~+50 °C, the wind speed range is 0~40 m/s, and the measured accuracy of wind speed and wind direction are 0.01 m/s and 0.1°, respectively. The sampling frequency is 10 Hz, and the wind direction is defined as 0° in the positive north direction, and negative in the top view of the counterclockwise direction.

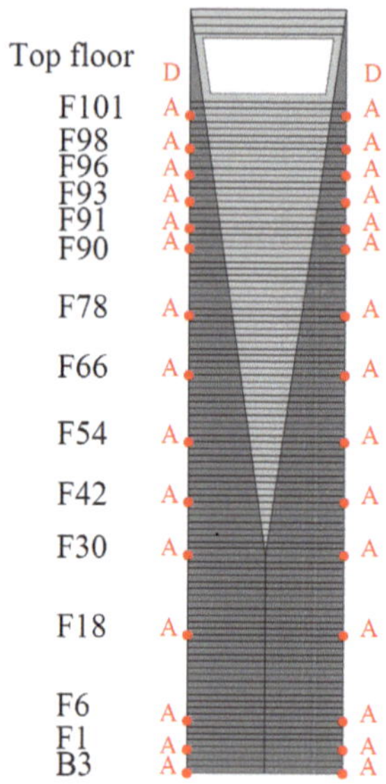

Figure 2. Measurement site layout of the SWFC.

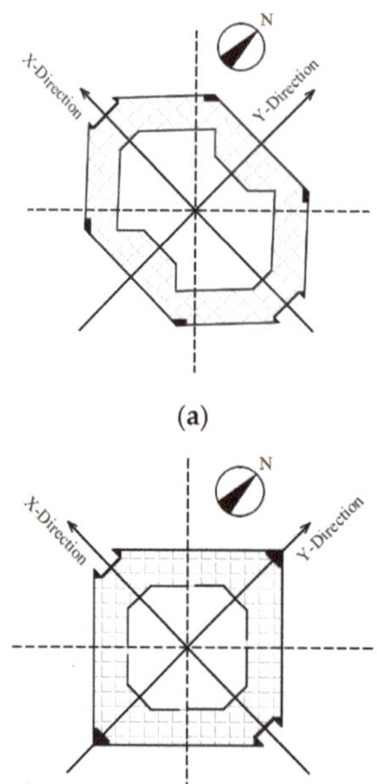

(**a**)

(**b**)

Figure 3. Floor plan of the SWFC. (**a**) Upper layout plan. (**b**) Lower layout plan.

3. Data Processing

Acceleration measurements at 72 h intervals from the top of the SWFC under Typhoon Lekima were used, recorded from 0:00 on 8 August to 0:00 on 11 August 2019, in the east–west (X) and north–south (Y) directions. The curve and standard deviation methods in the envelope random decrement technique (E-RDT) were used to identify structural dynamic parameters and conduct comparative analyses as follows [33]:

(1) The curve method.

① Make modal decomposition of the original acceleration responses to obtain the timescales of the first-order and second-order modal acceleration responses a_1 and a_2;

② Use the Hilbert–Huang transform (HHT) to find the envelope of the first- and second-order modal timescales:

$$A(t) = \sqrt{y(t)^2 + \left[\frac{1}{\pi}\int_{-\infty}^{+\infty} \frac{y(\tau)}{t-\tau} d\tau\right]} \tag{1}$$

where $A(t)$ is the envelope curve, $y(t)$ is the response timescale, and τ is the length of one response segment;

③ Set 20 intercept amplitudes by 0.5~1.5 times the standard deviations of a_1 and a_2;

④ Calculate the decay curve of free vibrations based on the E-RDT method;

⑤ Fit the decay curves of free vibrations by cubic spline curve fitting to obtain the first- and second-order natural frequencies and damping ratios.

(2) The standard deviation method.

① Make modal decomposition of the original acceleration responses to obtain the timescales of the first-order and second-order modal acceleration responses a_1 and a_2;

② Calculate the standard deviation S of a_1 and a_2, and set S as the interception threshold to find the zero point of the acceleration response;

③ Calculate the decay curves of free vibrations based on the E-RDT method;

④ Fit the decay curves of free vibrations by cubic spline curve fitting to obtain the first- and second-order natural frequencies and damping ratios.

4. Vibration Response Characteristics of the SWFC under Typhoon

4.1. Mean Wind Speed and Direction

Figure 4 depicts the 10 min mean wind speed and wind direction angle of measurements. As shown in the figure, the mean wind speed increased slowly with time from 0:00 on 8 August to 4:00 on 9 August 2019, and then increased at a fast rate. The mean wind speed reached its peak at 05:00 on August 10, when the corresponding mean wind speed was 39.1 m/s and the wind angle was 90°, indicating that the measurement sites were closest to the typhoon at this time. After that, the average wind speed then started to drop, with the wind angle fluctuating in the range of 107° to 200°, which was caused by the changing direction of the typhoon entering the Yellow Sea.

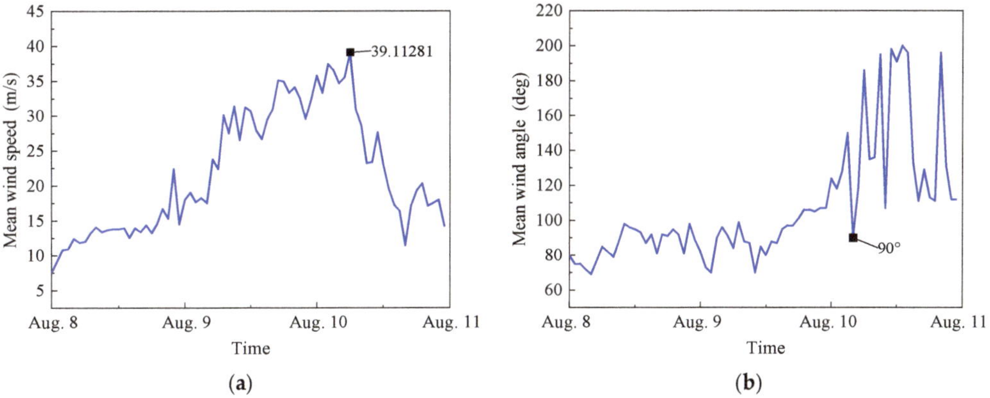

Figure 4. Mean wind speed and direction at the top of the SWFC: (**a**) 10 min mean wind speed, and (**b**) 10 min mean wind direction.

4.2. Wind-Induced Vibration Analysis

Wind-induced acceleration signals in the X and Y directions were also collected by the accelerometers, which are installed on 15 floors of the SWFC, at the same time. As seen in Figure 4, the peak of vibration acceleration appeared at 05:00 on August 10, so the 1 h acceleration signals near the maximum mean wind speed were chosen to plot the time history, as depicted in Figure 5a,b. As shown, the SWFC similarly vibrates in the east–west and north–south directions during the time period of interest. The maximum acceleration of 3.518 cm/s^2 occurs in the Y direction, and the one in the X direction is 2.602 cm/s^2. They are less than the specified value of 24.5 cm/s^2 recommended in design codes [34–36], which means the SWFC meets the provisional requirements under Typhoon Lekima.

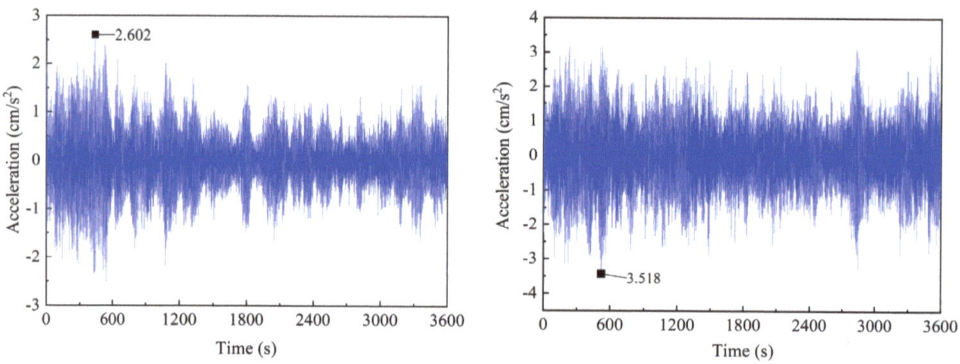

Dynamic responses in the X-direction

Dynamic responses in the Y-direction

(**a**)

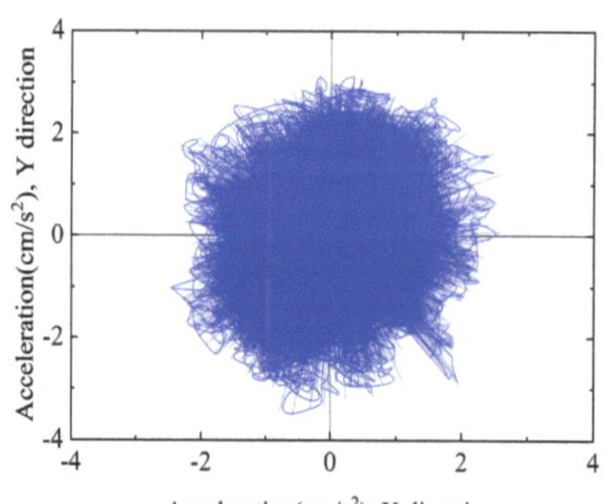

(**b**)

Figure 5. Acceleration response timescales and trajectories of the 101st floor of the SWFC: (**a**) measured acceleration response timescale at the 101st floor, and (**b**) acceleration trajectories on the 101st floor.

To further investigate the relationship between the acceleration amplitude and floor height, based on the measured results, the standard deviation and peak and mean values of the acceleration responses as the functions of the floor height are given in Figure 6. As shown, they follow similar trends with the increasing floor height in both X and Y directions. The maximum acceleration appeared on the 101st floor, demonstrating that the vibration increases with the structure height, and the dispersion of the acceleration signals becomes progressively greater.

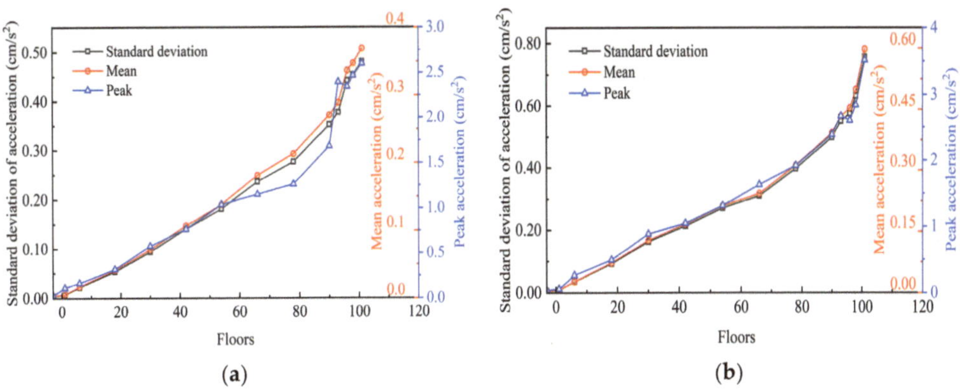

Figure 6. Mean accelerations, peak accelerations, and standard deviations of acceleration responses: (**a**) in the X−direction, and (**b**) in the Y−direction.

4.3. Probability Density Function of Acceleration

The probability density functions (PDFs) of the measured vibration signals are calculated and shown in Figure 7. The graphs show that the acceleration signals have kurtosis coefficients of K = 3.8565 in the X direction and K = 3.2209 in the Y direction. They are compared with the Gaussian distribution [37]: when the skewness coefficient S > 0, the PDF is right-biased; while the skewness coefficient S < 0, the PDF is left-biased; when the kurtosis coefficient K > 3, the PDF is highly peaked; while the kurtosis coefficient K < 3, the PDF is lowly peaked. As seen, the present PDFs are highly peaked [38–40], which means the probability of strong vibration occurring on high floors is large, and thus, more attention should be paid to structural design and health monitoring.

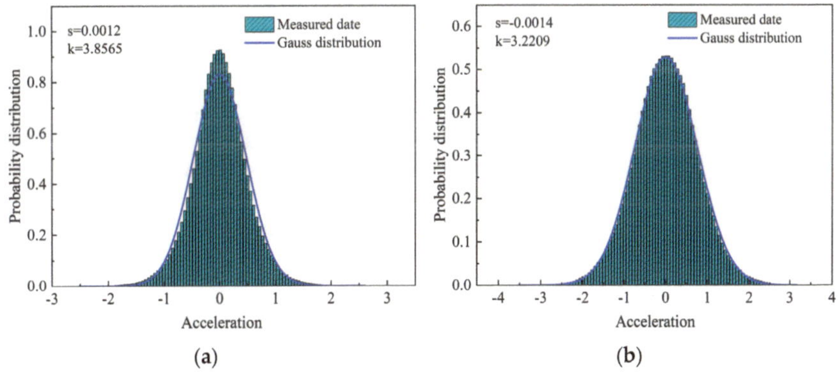

Figure 7. Acceleration probability density distributions: (**a**) in the X−direction, and (**b**) in the Y−direction.

4.4. Power Spectrum of Accelerations

The power spectral density of the time-domain acceleration signals is obtained using the mean periodogram method. This method involves segmenting the response signal data without overlapping and then averaging the power spectrum of individual segments [24].

As shown in Figure 8, the spectra are clear and distinct, with the peaks neatly arranged in a regular pattern. The building's wind-induced vibrations are highly consistent in both directions, and the first-order modes play a significant role. In the X direction, the first-order natural frequency is 0.143 Hz and the second-order natural frequency is 0.546 Hz; in the Y direction, the first- and second-order natural frequencies are 0.146 Hz and 0.516 Hz, respectively. The difference in the first-order natural frequency between the two directions is 0.003 Hz, and the difference in the second-order natural frequency is 0.03 Hz.

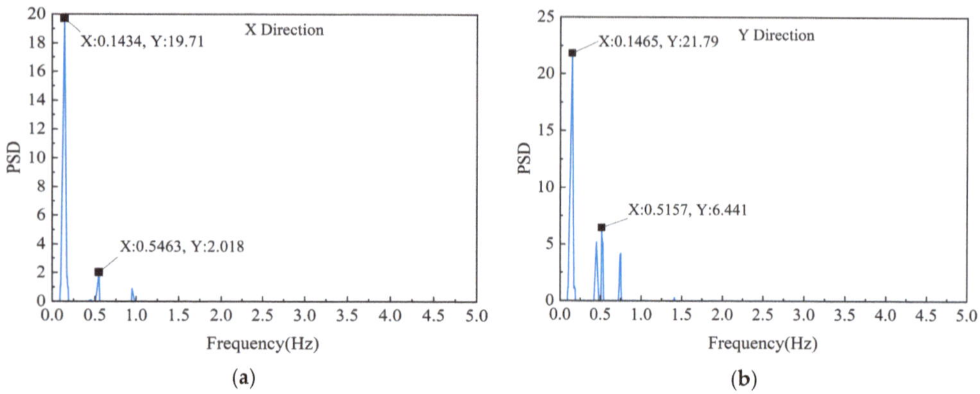

Figure 8. Power spectra of accelerations: (**a**) in the X−direction, and (**b**) in the Y−direction.

5. Parameter Identification Results

The natural frequencies, damping ratios, and mode shapes are the most basic parameters of structural dynamic properties, so accurate estimation of these three types of properties in structural design is of great importance [41]. In this section, the curve method and the standard deviation method are used to analyze the time history of accelerations of the 101st floor at the top of the SWFC in the X and Y directions.

5.1. Variation of Natural Frequencies and Damping Ratios with Amplitudes

The first method is the curve method. Using the same acceleration time history, set 20 intercepted amplitudes in the range of 0.5~1.5 times the standard deviation to obtain the natural frequencies and damping ratios of the SWFC. Figures 9 and 10 show the correlation curves of the first two natural frequencies and damping ratios with amplitudes, respectively.

As shown in Figure 9, the first- and second-order natural frequencies in both X and Y directions decrease as the amplitudes increase. The main reason is that nonlinear responses occur when structural amplitudes increase, and the steel joints slip, which makes the interactions between the structural and nonstructural elements greater. The natural frequencies identified by the curve method fluctuate considerably in the low-amplitude region, and the dependence becomes progressively more apparent as the amplitudes increase. The general trends are as follows: when the X-direction acceleration amplitudes increase from 0.04 cm/s^2 to 0.17 cm/s^2, the first-order natural frequency decreases by 0.0035 Hz, and the second-order natural frequency decreases by 0.0012 Hz; when the Y-direction acceleration amplitudes increase from 0.03 cm/s^2 to 0.18 cm/s^2, the first-order and second-order natural frequencies decrease by 0.0027 Hz and 0.0043 Hz, respectively. The maximum drop is 2.303% and the minimum drop is 0.261%, which are the first- and second-order natural frequencies in the X direction, respectively.

As shown in Figure 10, the first- and second-order damping ratios in the X and Y directions increase with the increasing amplitudes. The damping ratios identified by the curve method fluctuate considerably in the low-amplitude region, and their dependence increases with the increasing amplitudes. At the same time, in the high-amplitude region, the damping ratios tend to stabilize in different directions and even tend to increase in one direction. When the X-direction acceleration amplitudes increase from 0.04 cm/s^2 to 0.17 cm/s^2, the first-order damping ratio increases by 0.82%, and the second-order damping ratio increases by 0.40%. When the Y-direction acceleration amplitudes increase from 0.03 cm/s^2 to 0.18 cm/s^2, the first-order damping ratio increases by 0.98%, and the second-order damping ratio increases by 0.45%. The second-order damping ratio in the Y direction increases at most by 95.7%.

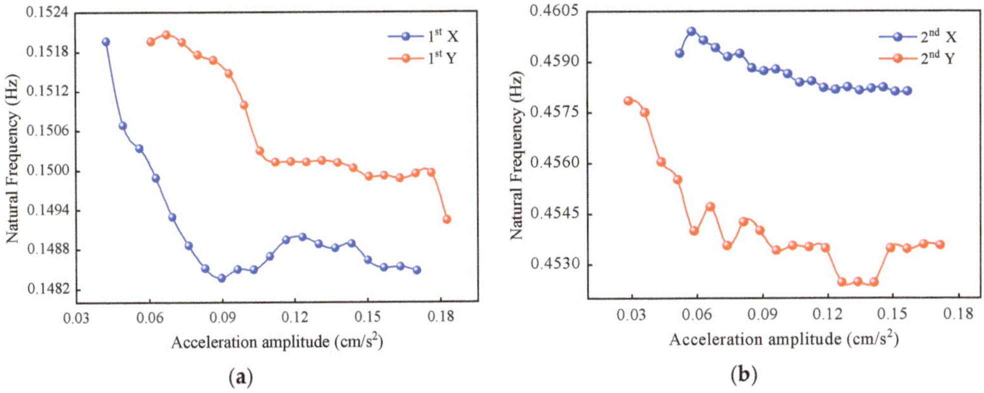

Figure 9. Natural frequencies determined by the curve method: (**a**) first-order natural frequencies, and (**b**) second-order natural frequencies.

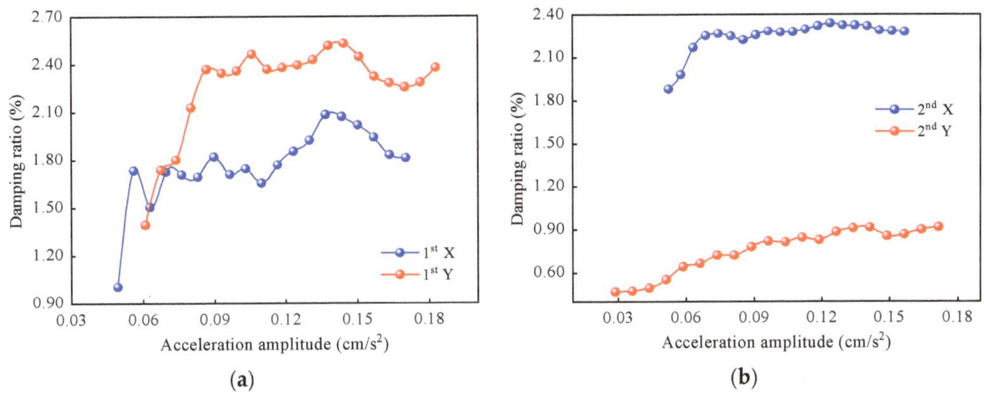

Figure 10. Damping ratios determined by the curve method: (**a**) first-order damping ratios, and (**b**) second-order damping ratios.

The second method is the standard deviation method. On the other hand, by setting the time interval as one hour and the standard deviation of the accelerations as amplitudes, structural frequencies and damping ratios are obtained. Figures 11 and 12 show the trends of the first- and second-order natural frequencies and damping ratios with amplitudes for the standard deviation method, respectively. Equations (1) and (2) are used to make a linear fitting, and the fitting parameters are shown in Table 1.

$$f = \alpha_0 + \beta_0 x \tag{2}$$

$$\xi = \alpha_1 + \beta_1 x \tag{3}$$

where f is the natural frequency, ξ is the damping ratio, α_0 and α_1 are the natural frequency and damping ratio when the amplitude x is 0, respectively, and β_0 and β_1 are the change rates of the natural frequency and damping ratio with amplitude x, respectively.

As shown in Figure 11, the first- and second-order natural frequencies in both directions decrease as the amplitudes increase. From the linear fitting, it can be seen that the first-order natural frequency is well fitted and the scatter points lie within the 95% prediction band. The natural frequencies identified by the standard deviation method are concentrated in the low-amplitude region, and the amplitude dependence diminishes as the amplitudes increase. The general trends are as follows: when the X-direction acceleration amplitudes increase from 0.03 cm/s^2 to 0.75 cm/s^2, the first-order natural frequency decreases by 0.007 Hz, and the second-order natural frequency decreases by 0.0212 Hz; when the Y-direction acceleration amplitudes increase from 0.04 cm/s^2 to 1.35 cm/s^2, the first-order and second-order natural frequencies decrease by 0.0089 Hz and 0.007 Hz, respectively. The maximum drop is 5.794% and the minimum drop is 1.527%, which are the first- and second-order natural frequencies in the Y direction, respectively.

Table 1. Fitting parameters of the curve method.

Fitting Parameters	First Order		Second Order	
	X Direction	Y Direction	X Direction	Y Direction
Natural frequencies	$\alpha_0 = 0.15123$, $\beta_0 = -0.00997$, $R^2 = 0.796$	$\alpha_0 = 0.15331$, $\beta_0 = -0.00506$, $R^2 = 0.836$	$\alpha_0 = 0.45693$, $\beta_0 = -0.00642$, $R^2 = 0.6$	$\alpha_0 = 0.4564$, $\beta_0 = -0.0039$, $R^2 = 0.724$
Damping ratios	$\alpha_1 = 0.00169$, $\beta_1 = 0.04027$, $R^2 = 0.69$	$\alpha_1 = 0.00235$, $\beta_1 = 0.02361$, $R^2 = 0.66$	$\alpha_1 = 0.01151$, $\beta_1 = 0.00645$, $R^2 = 0.33$	$\alpha_1 = 0.00544$, $\beta_1 = 0.00437$, $R^2 = 0.47$

(R^2, also known as the coefficient of determination of the equation, indicates the degree to which the variable X in the equation explains Y. R^2 values are between 0 and 1, and the closer to 1, the better X explains Y in the equation.)

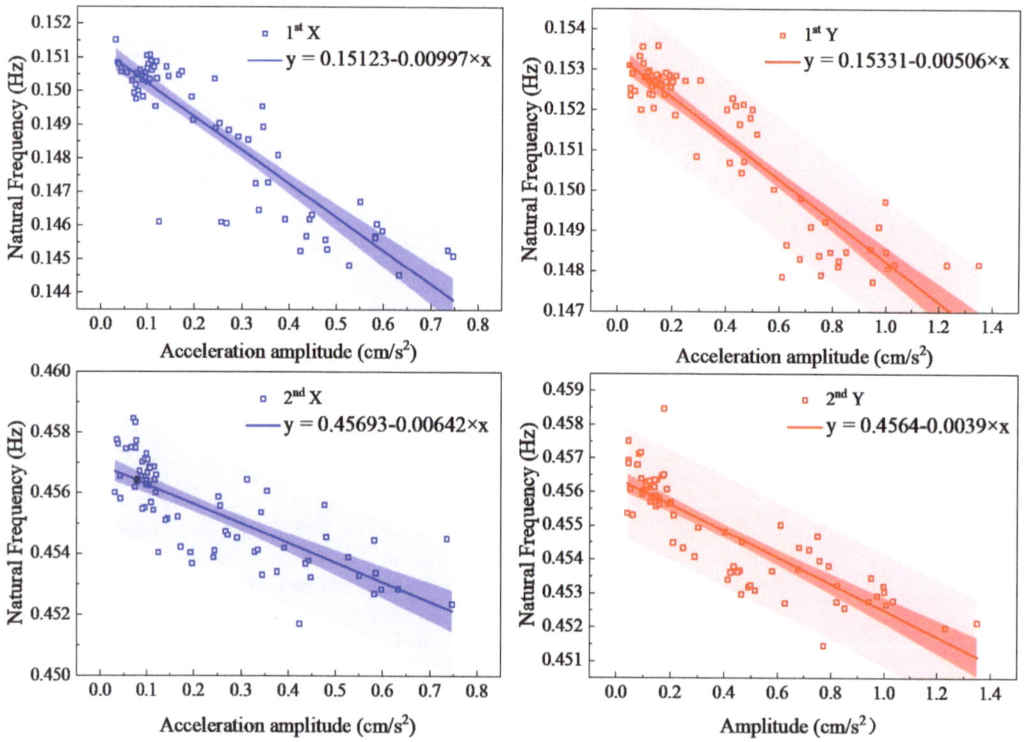

Figure 11. Natural frequencies determined by the standard deviation method.

As shown in Figure 12, the first- and second-order damping ratios in both directions increase with increasing amplitude. It can be seen that the identified results of second-order damping ratios have a large discretization and a wide range of prediction bands. The first-order damping ratio is less scattered, which is identical with the measurements of the Guangzhou Tower by the Hong Kong Polytechnic University [42]. The general trends are as follows: when the X-direction amplitudes increase from 0.03 cm/s^2 to 0.75 cm/s^2, the first-order damping ratio increases by 2.86%, and the second-order damping ratio increases by 0.46%; when the Y-direction amplitudes increase from 0.04 cm/s^2 to 1.35 cm/s^2, the first-order damping ratio increases by 3.25%, and the second-order damping ratio increases by 0.62%.

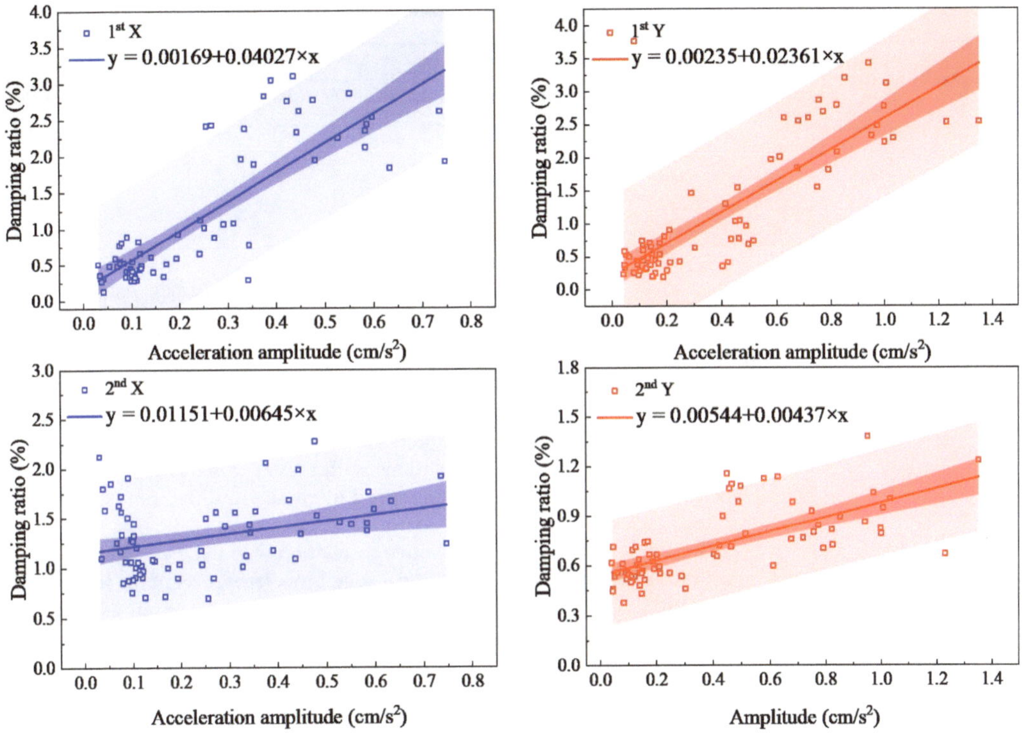

Figure 12. Damping ratios determined by the standard deviation method.

The natural frequencies and damping ratios identified by the curve method and the standard deviation method are compared with other results in the literature. The comparisons are shown in Tables 2 and 3. It was found that the two methods produced comparable results. Nonetheless, the dispersion of natural frequencies derived from the standard deviation method is large, the regularity is obscure, and the variation of the acceleration amplitudes is small. In contrast, the dispersion of the natural frequencies derived via the curve method is small, but the acceleration amplitude varies greatly. In addition, the first two orders of natural frequencies in the X and Y directions derived by the two methods are highly consistent with the calculations by other researchers. However, the first two orders of damping ratios are different. The first two orders of damping ratios derived from the curve method differ from the results of the three scholars by a maximum of 2.178 Hz, while the results derived from the standard deviation method fluctuate more.

Table 2. Comparison of the natural frequencies under the excitation of Typhoon Lekima.

Method or Study	First-Order Natural Frequencies (Hz)		Second-Order Natural Frequencies (Hz)	
	X Direction	Y Direction	X Direction	Y Direction
Standard deviation method	0.1512	0.1533	0.4569	0.4564
Curve method	0.1519	0.1521	0.4600	0.4579
Ya Jun Huang [43]	0.1513	0.1535	/	/
Yun Cheng He [44]	0.1510	0.1530	0.4660	0.4660
Zeng shun Chen [45]	0.1511	0.1526	/	/

Table 3. Comparison of the damping ratios under the excitation of Typhoon Lekima.

Method or Study	First-Order Damping Ratios (%)		Second-Order Damping Ratios (%)	
	X Direction	Y Direction	X Direction	Y Direction
Standard deviation method	0.13~3.104	0.19~3.77	0.70~2.28	0.38~1.38
Curve method	2.087	2.532	2.341	0.917
Ya Jun Huang [43]	3.349	3.778	/	/
Yun Cheng He [44]	2.310	0.354	2.090	0.753
Zeng shun Chen [45]	0.651	0.688	/	/

5.2. Variation Patterns of the Natural Frequencies and Damping Ratios with the Mean Wind Speed

On the basis of the measured results, the relationships of the natural frequencies and damping ratios with the mean wind speed are also found, as shown in Figures 13 and 14. Comparing the mean wind speed (Figure 4a) with Figures 13 and 14, the following summary can be obtained.

(1) The trends of the natural frequencies in the X and Y directions are basically the same, and both show negative correlations with the mean wind speed, i.e., the natural frequencies decrease as the mean wind speed increases, and the trend of the first-order natural frequencies is more pronounced.

(2) The damping ratios show a positive correlation with the mean wind speed, i.e., the damping ratios increase as the mean wind speed increases, except for the second-order damping ratio in the X direction, which may be due to the non-Gaussian nature of the original acceleration signals.

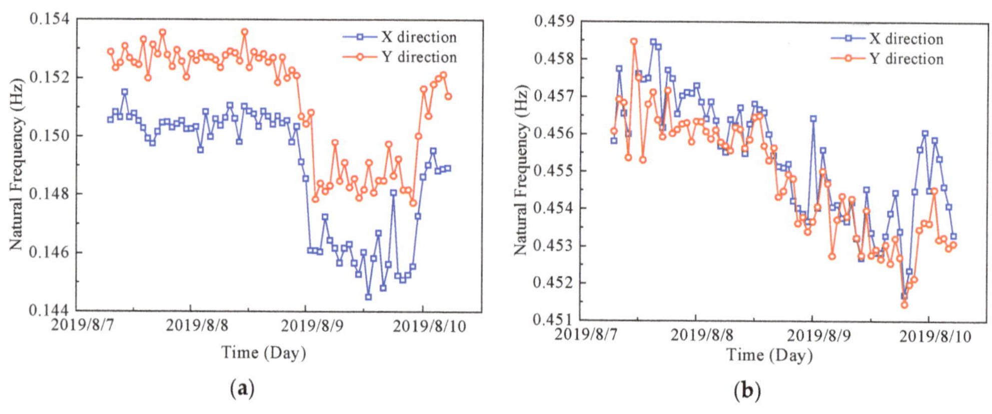

Figure 13. Variations of the natural frequencies with the mean wind speed: (**a**) first-order natural frequencies, and (**b**) second-order natural frequencies.

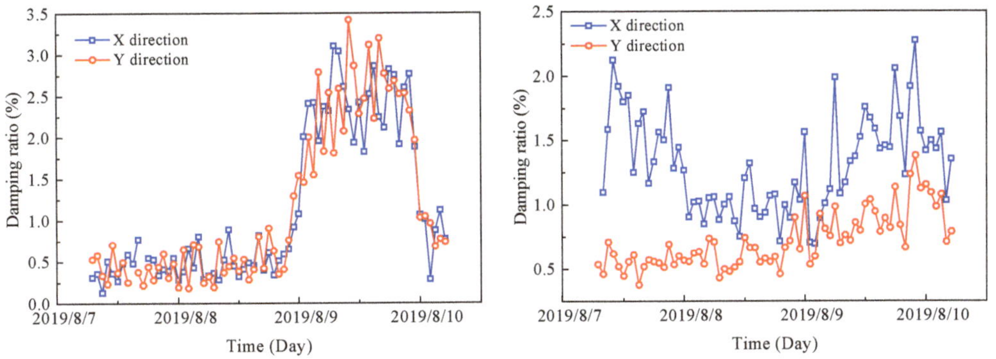

(**a**) (**b**)

Figure 14. Variations of the damping ratios with the mean wind speed: (**a**) first-order damping ratios, and (**b**) second-order damping ratios.

5.3. Integral Mode Shapes of the Structure

For integral mode shape identification, the frequency domain approach (FDA) is utilized, i.e., the ratio of the mutual spectrum to the self-spectrum of the acceleration signal is used as an approximation to determine the ratio of mode shapes [45]. The mode shapes on the 101st floor are used for normalization to obtain the first three orders of mode shapes. The results are also compared with those excited by the other three different typhoons, i.e., Ampil [30], Rumbia [30], and Jongdari [46], as shown in Figure 15.

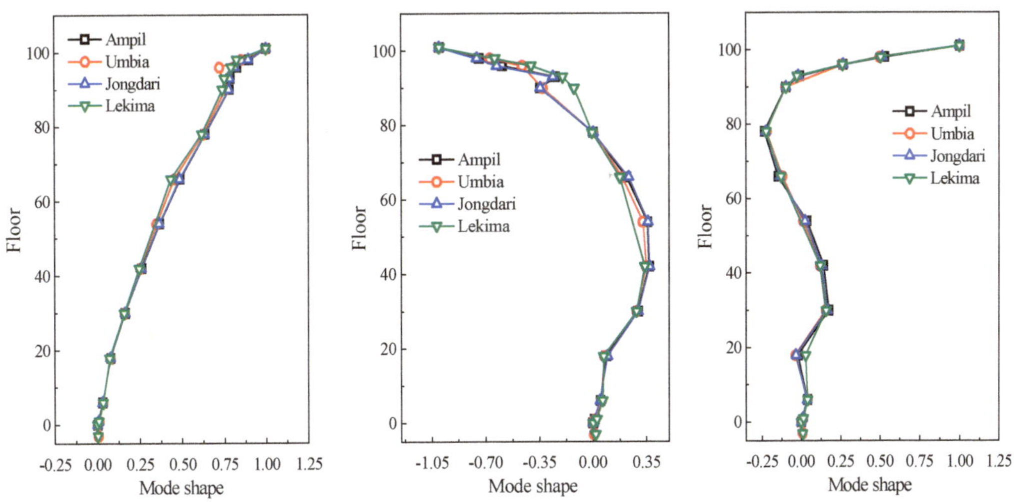

1st X−direction 2nd X−direction 3rd X−direction

Figure 15. *Cont.*

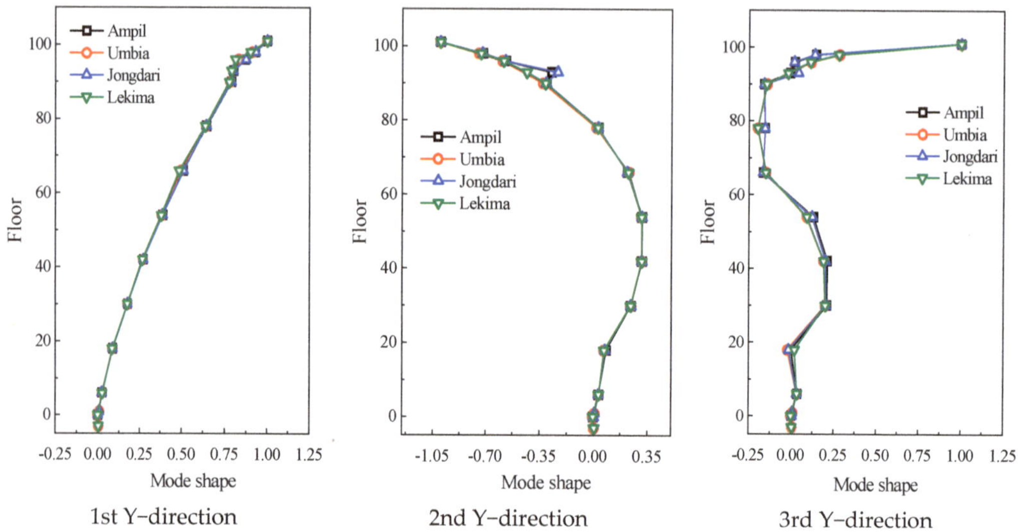

Figure 15. Comparisons of the mode shapes of the SWFC.

As seen, the mode shapes of the SWFC were essentially the same under Typhoons Lekima, Ampil, Rumbia, and Jongdari. The first-order mode shape has one zero point, the second-order has two, and the third-order has three. Both the X- and Y-directional mode shapes present an increasing nonlinear trend with the increasing floor level, especially the third-order mode shapes. The X-direction and Y-direction trends are virtually the same, indicating that the difference between the stiffness of the SWFC in the X direction and Y direction is not substantial, which complies with the findings of the natural frequency.

6. Conclusions

Based on the measured wind-induced vibration responses of the SWFC under Typhoon Lekima, this paper studies the amplitude correlations and variation patterns of the dynamic parameters, and the following main conclusions are obtained:

(1) The wind-induced vibration responses in the X and Y directions are highly consistent. The maximum acceleration response occurs in the Y direction, which is about 3.518 cm/s^2, and the maximum instantaneous acceleration in the X direction is 2.602 cm/s^2, both of which meet the design requirements.

(2) The curve method and the standard deviation method are effective and accurate in identifying dynamic characteristic parameters. The first-order natural frequency in the X direction is about 0.151 Hz, and the one in the Y direction is about 0.153 Hz; the second-order natural frequency in both X and Y directions is about 0.46 Hz. The first- and second-order damping ratios in the X and Y directions are less than 1%, which may be related to the dampers installed in the SWFC.

(3) The first- and second-order natural frequencies in both the X and Y directions decrease with the increasing amplitudes; the first- and second-order damping ratios in both the X and Y directions increase with the increasing amplitudes. In addition, it is found that the natural frequencies show negative correlations with the mean wind speed, while the damping ratios show positive correlations.

(4) The mode shapes of the SWFC are essentially the same under different typhoon excitations. The first- and second-order mode shapes in the X and Y directions grow with the increase in floor levels, displaying a nonlinear trend, and the third-order mode shapes have more obvious nonlinear trends.

Author Contributions: Conceptualization, X.W.; writing—original draft preparation, H.K.; writing—original draft preparation, G.Z.; supervision, Y.L.; writing—review and editing, L.L.; project administration, C.Z. All authors have read and agreed to the published version of the manuscript.

Funding: This research was funded by the Natural Science Foundation of Chongqing (Grant No.: CSTB2022NSCQ-MSX1655) and the State Key Laboratory of Structural Dynamics of Bridge Engineering and Key Laboratory of Bridge Structure Seismic Technology for Transportation Industry Open Fund (Grant No.: 202205). And The APC was funded by the Natural Science Foundation of Chongqing (Grant No.: CSTB2022NSCQ-MSX1655).

Institutional Review Board Statement: Not applicable.

Informed Consent Statement: Not applicable.

Data Availability Statement: All data generated or analyzed during this study are included in this article. All data included in this study are available upon request by contact with the corresponding author.

Conflicts of Interest: The authors declare no conflict of interest.

References

1. Su, J.; Xia, Y.; Weng, S. Review on field monitoring of high-rise structures. *Struct. Control Health Monit.* **2020**, *27*, e2629. [CrossRef]
2. Fujino, Y.; Siringoringo, D.M.; Ikeda, Y.; Nagayama, T.; Mizutani, T. Research and implementations of structural monitoring for bridges and buildings in Japan. *Engineering* **2019**, *5*, 1093–1119. [CrossRef]
3. Norén-Cosgriff, K.; Kaynia, A.M. Estimation of natural frequencies and damping using dynamic field data from an offshore wind turbine. *Mar. Struct.* **2021**, *76*, 102915. [CrossRef]
4. Hou, F.; Jafari, M. Investigation approaches to quantify wind-induced load and response of tall buildings: A review. *Sustain. Cities Soc.* **2020**, *62*, 102376. [CrossRef]
5. Ohkuma, T.; Marukawa, H.; Niihori, Y.; Kato, N. Full-scale measurement of wind pressures and response accelerations of a high-rise building. *J. Wind. Eng. Ind. Aerodyn.* **1991**, *38*, 185–196. [CrossRef]
6. Denoon, R.O.; Kwok, K.C. Full-scale measurements of wind-induced response of an 84 m high concrete control tower. *J. Wind. Eng. Ind. Aerodyn.* **1996**, *60*, 155–165. [CrossRef]
7. Kijewski-Correa, T.; Kilpatrick, J.; Kareem, A.; Kwon, D.K.; Bashor, R.; Kochly, M.; Baker, W.F. Validating wind-induced response of tall buildings: Synopsis of the Chicago full-scale monitoring program. *J. Struct. Eng.-ASCE* **2006**, *132*, 1509–1523. [CrossRef]
8. Kijewski-Correa, T.; Kochly, M. Monitoring the wind-induced response of tall buildings: GPS performance and the issue of multipath effects. *J. Wind. Eng. Ind. Aerodyn.* **2007**, *95*, 1176–1198. [CrossRef]
9. Kijewski-Correa, T.; Kwon, D.K.; Kareem, A.; Bentz, A.; Guo, Y.; Bobby, S.; Abdelrazaq, A. SmartSync: An integrated real-time structural health monitoring and structural identification system for tall buildings. *J. Struct. Eng.-ASCE* **2013**, *139*, 1675–1687. [CrossRef]
10. Kijewski-Correa, T.L. *Full-Scale Measurements and System Identification: A Time-Frequency Perspective*; University of Notre Dame Press: South Bend, IN, USA, 2003.
11. Kijewski-Correa, T.; Pirnia, J.D. Dynamic behavior of tall buildings under wind: Insights from full-scale monitoring. *Struct. Des. Tall Spec.* **2007**, *16*, 471–486. [CrossRef]
12. Pirnia, J.D.; Kijewski-Correa, T.; Abdelrazaq, A.; Chung, J.; Kareem, A. Full-scale validation of wind-induced response of tall buildings: Investigation of amplitude-dependent dynamic properties. In Proceedings of the New Horizons and Better Practices, Long Beach, CA, USA, 16–19 May 2007; pp. 1–10.
13. Kijewski-Correa, T.; Bentz, A. Wind-induced vibrations of buildings: Role of transient events. *Proc. Inst. Civ. Eng.-Struct. Build.* **2011**, *164*, 273–284. [CrossRef]
14. Yi, J.; Li, Q.S. Wind tunnel and full-scale study of wind effects on a super-tall building. *J. Fluid. Struct.* **2015**, *58*, 236–253. [CrossRef]
15. Li, Q.S.; Wu, J.R.; Fu, J.Y.; Li, Z.N.; Xiao, Y.Q. Wind effects on the world's tallest reinforced concrete building. *Proc. Inst. Civ. Eng.-Struct. Build.* **2010**, *163*, 97–110. [CrossRef]
16. Xu, Y.L.; Chen, S.W.; Zhang, R.C. Modal identification of Di Wang Building under typhoon York using the Hilbert–Huang transform method. *Struct. Des. Tall Spec.* **2003**, *12*, 21–47. [CrossRef]
17. Li, Q.S.; Xiao, Y.Q.; Wong, C.K.; Jeary, A.P. Field measurements of typhoon effects on a super tall building. *Eng. Struct.* **2004**, *26*, 233–244. [CrossRef]
18. Li, Q.S.; Wu, J.R. Correlation of dynamic characteristics of a super-tall building from full-scale measurements and numerical analysis with various finite element models. *Earthq. Eng. Struct. D* **2004**, *33*, 1311–1336. [CrossRef]
19. Li, Q.S.; Xiao, Y.Q.; Wong, C.K. Full-scale monitoring of typhoon effects on super tall buildings. *J. Fluid. Struct.* **2005**, *20*, 697–717. [CrossRef]
20. Li, Q.S.; Xiao, Y.Q.; Wong, C.K.; Jeary, A.P. Field measurements of wind effects on the tallest building in Hong Kong. *Struct. Des. Tall Spec.* **2002**, *12*, 67–82. [CrossRef]

21. Li, Q.S.; Yang, K.; Wong, C.K.; Jeary, A.P. The effect of amplitude-dependent damping on wind-induced vibrations of a super tall building. *J. Wind. Eng. Ind. Aerodyn.* **2003**, *91*, 1175–1198. [CrossRef]
22. Yi, J.; Zhang, J.W.; Li, Q.S. Dynamic characteristics and wind-induced responses of a super-tall building during typhoons. *J. Wind. Eng. Ind. Aerodyn.* **2013**, *121*, 116–130. [CrossRef]
23. Li, Q.S.; Zhi, L.H.; Yi, J.; To, A.; Xie, J. Monitoring of typhoon effects on a super-tall building in Hong Kong. *Struct. Control Health Monit.* **2014**, *21*, 926–949. [CrossRef]
24. Li, X.; Li, Q.S. Observations of typhoon effects on a high-rise building and verification of wind tunnel predictions. *J. Wind. Eng. Ind. Aerodyn.* **2019**, *184*, 174–184. [CrossRef]
25. Li, Q.S.; Fu, J.Y.; Xiao, Y.Q.; Li, Z.N.; Ni, Z.H.; Xie, Z.N.; Gu, M. Wind tunnel and full-scale study of wind effects on China's tallest building. *Eng. Struct.* **2006**, *28*, 1745–1758. [CrossRef]
26. Li, Q.S.; Xiao, Y.Q.; Fu, J.Y.; Li, Z.N. Full-scale measurements of wind effects on the Jin Mao building. *J. Wind. Eng. Ind. Aerodyn.* **2007**, *95*, 445–466. [CrossRef]
27. He, Y.C.; Li, Q. Dynamic responses of a 492-m-high tall building with active tuned mass damping system during a typhoon. *Struct. Control Health Monit.* **2014**, *21*, 705–720. [CrossRef]
28. Li, Q.S.; Zhi, L.H.; Tuan, A.Y.; Kao, C.S.; Su, S.C.; Wu, C.F. Dynamic behavior of Taipei 101 tower: Field measurement and numerical analysis. *J. Struct. Eng.-ASCE* **2011**, *137*, 143–155. [CrossRef]
29. Zhang, F.L.; Xiong, H.B.; Shi, W.X.; Ou, X. Structural health monitoring of Shanghai Tower during different stages using a Bayesian approach. *Struct. Control Health Monit.* **2016**, *23*, 1366–1384. [CrossRef]
30. Fu, G.; Quan, Y.; Gu, M.; Huang, Z.; Feng, C. Dynamic performance evaluation of a 492 m super high-rise building with active tuned mass dampers during four consecutive landfall typhoons within a month. *J. Build. Eng.* **2022**, *61*, 105259. [CrossRef]
31. Lu, X.; Li, P.; Guo, X.; Shi, W.; Liu, J. Vibration control using ATMD and site measurements on the Shanghai World Financial Center Tower. *Struct. Des. Tall Spec.* **2014**, *23*, 105–123. [CrossRef]
32. Zhou, K.; Li, Q.S.; Li, X. Dynamic behavior of supertall building with active control system during Super Typhoon Mangkhut. *J. Struct. Eng.-ASCE* **2020**, *146*, 23–31. [CrossRef]
33. Huang, N.E.; Shen, Z.; Long, S.R.; Wu, M.C.; Shih, H.H.; Zheng, Q.; Liu, H.H. The empirical mode decomposition and the Hilbert spectrum for nonlinear and non-stationary time series analysis. *Proc. R. Soc. London. Ser. A Math. Phys. Eng. Sci.* **1998**, *454*, 903–995. [CrossRef]
34. Xin, J.; Zhou, C.; Jiang, Y.; Tang, Q.; Yang, X.; Zhou, J. A signal recovery method for bridge monitoring system using TVFEMD and encoder-decoder aided LSTM. *Measurement* **2023**, *214*, 112797. [CrossRef]
35. Liu, S.; Jiang, Y.; Qiao, K.; Peng, L.; Liu, D. Record-based simulation of three-component long-period ground motions: Hybrid of surface wave separation and multivariate empirical mode decomposition. *Soil. Dyn. Earthq. Eng.* **2023**, *172*, 108037. [CrossRef]
36. Tang, Q.; Xin, J.; Jiang, Y.; Zhou, J.; Li, S.; Fu, L. Fast identification of random loads using the transmissibility of power spectral density and improved adaptive multiplicative regularization. *J. Sound. Vib.* **2022**, *534*, 117033. [CrossRef]
37. Wu, J.; Hu, N.; Dong, Y.; Zhang, Q. Monitoring dynamic characteristics of 600 m+ Shanghai Tower during two consecutive typhoons. *Struct. Control Health Monit.* **2021**, *28*, e2666. [CrossRef]
38. Zhang, L.; Hu, X.; Xie, Z.; Shi, B.; Zhang, L.; Wang, R. Field measurement study on time-varying characteristics of modal parameters of super high-rise buildings during super typhoon. *J. Wind. Eng. Ind. Aerodyn.* **2020**, *200*, 104139. [CrossRef]
39. Kumar, K.S.; Stathopoulos, T. Wind loads on low building roofs: A stochastic perspective. *J. Struct. Eng.-ASCE* **2000**, *126*, 944–956. [CrossRef]
40. Zhu, Z.; Lei, W.; Wang, Q.; Tiwari, N.; Hazra, B. Study on wind-induced vibration control of linked high-rise buildings by using TMDI. *J. Wind. Eng. Ind. Aerodyn.* **2020**, *205*, 104306. [CrossRef]
41. Hu, J.; Li, Z.; Zhao, Z. Full-Scale Measurements of Translational and Torsional Dynamics Characteristics of a High-Rise Building during Typhoon Sarika. *Materials* **2022**, *15*, 493. [CrossRef]
42. Xie, Z.; Liu, C.; Yu, X. Field measurements and wind tunnel experimental investigations of wind effects on Guangzhou West Tower. *Struct. Des. Tall Spec.* **2020**, *29*, e1774. [CrossRef]
43. Huang, Y.J.; Gu, M.; Huang, Z.F. Wind speed and acceleration measurements at the top of the Shanghai World Financial Center. *Tongji Daxue Xuebao/J. Tongji Univ.* **2017**, *45*, 821–826.
44. Chen, Z.S. Dynamic characteristics of a super high-rise building using a smart monitoring system. In Proceedings of the World Transport Convention (WTC 2008), Beijing, China, 19–20 June 2008.
45. Chen, J.; Chen, Y.; Peng, Y.; Zhu, S.; Beer, M.; Comerford, L. Stochastic harmonic function based wind field simulation and wind-induced reliability of super high-rise buildings. *Mech. Syst. Signal Process.* **2019**, *133*, 106264. [CrossRef]
46. Quan, Y.; Fu, G.Q.; Huang, Z.F.; Gu, M. Comparative analysis of the wind characteristics of three landfall typhoons based on stationary and nonstationary wind models. *Wind. Struct.* **2020**, *31*, 269–285. [CrossRef]

buildings

MDPI

Article

Nonlinear Finite Element Analysis and Fatigue Damage Assessment of Wind-Induced Vibration for the Tension Cable-Supported Power Transmission Structure

Jingyang Li [1], Bangjie Wang [1], Tao Wang [2,3,*] and Zhengliang Li [1,4]

[1] School of Civil Engineering, Chongqing University, Chongqing 400045, China; jingyanglisc@163.com (J.L.); bangjiewang@cqu.edu.cn (B.W.); lizhengl@hotmail.com (Z.L.)
[2] School of Transportation Science and Engineering, Harbin Institute of Technology, Harbin 150090, China
[3] Chongqing Research Institute of Harbin Institute of Technology, Harbin Institute of Technology, Chongqing 401151, China
[4] Chongqing Key Laboratory of Wind Engineering and Wind Resources Utilization, Chongqing 400045, China
* Correspondence: taowang@alu.cqu.edu.cn

Abstract: The tension cable-supported power transmission structure (TC-PTS) is a new type of power transmission structure suitable for mountainous terrain, and is sensitive to wind load. In this regard, a nonlinear finite element analysis model of wind-induced vibration is proposed for the TC-PTS, and the wind-induced vibration response of the structure is analyzed. Firstly, the tangent stiffness matrix of the three-dimensional truss element for the supporting suspension cable and transmission line, considering the geometric nonlinearity of structures, is derived through the relationship between the element elastic energy and its displacement. Subsequently, the element mass matrix and damping matrix of the supporting suspension cable and transmission line, as well as the element nodal load vector obtained from wind load equivalence, are given. Then, based on the nonlinear finite element theory, the nonlinear dynamic equation of wind-induced vibration is established for the TC-PTS and solved using the Newmark-β method combined with the Newton–Raphson iterative method. Furthermore, the rain-flow counting method and Miner's linear fatigue cumulative damage theory were used for wind-induced fatigue damage assessment. Finally, a two-span TC-PTS was selected as an example, and the wind-induced nonlinear vibration and fatigue damage assessment were analyzed through the proposed model. The results show that the proposed model has high computational accuracy and efficiency. The first three order vibration modes of the supporting-conductor part of the two-span TC-PTS were antisymmetric vertical bending, symmetric side bending, and antisymmetric side bending. With the increase in wind speed and wind direction angle, the maximum lateral displacement and tension of the supporting suspension cable and transmission line increased, and their degree of increase showed a nonlinear trend. In terms of the wind-induced fatigue analysis results of TC-PTS, the fatigue damage at the end of the supporting-conductor suspension cable was greater than the fatigue damage at its midpoint. Compared to the fatigue damage at the midpoint of the conductor, the fatigue damage at the end of the conductor was less affected by the wind direction angle, and both were more significantly affected by the wind speed.

Keywords: tension cable-supported power transmission structure; wind-induced vibration; nonlinear finite element; wind-induced fatigue damage

Citation: Li, J.; Wang, B.; Wang, T.; Li, Z. Nonlinear Finite Element Analysis and Fatigue Damage Assessment of Wind-Induced Vibration for the Tension Cable-Supported Power Transmission Structure. *Buildings* **2023**, *13*, 2924. https://doi.org/10.3390/buildings13122924

Academic Editor: Fabio Rizzo

Received: 22 September 2023
Revised: 5 November 2023
Accepted: 21 November 2023
Published: 23 November 2023

1. Introduction

As a part of electric energy infrastructure, a high-voltage transmission tower-line system is an important part of the transmission and distribution systems in the power grid [1]. With the boom in the development of power grid construction, the number of power transmission structures erected in mountainous terrains has been increasing [2]. However, the paths required for transmission line corridors in mountainous areas are becoming tighter, and the selection of tower locations is becoming increasingly difficult. For example, in the karst terrain of Guangxi, the Danxia terrain of Xinjiang, and other special and complex mountainous terrain in China, tower positions of the transmission tower-line system are often difficult to set up, owing to terrain restrictions, resulting in a wide range of line rerouting.

Therefore, in recent years, the electric power industry has begun the investigation of new power transmission structures that are well-suited for mountainous areas. A typical structure is the tension cable-supported power transmission structure (TC-PTS), which is a new form of overhead power transmission structure. The TC-PTS primarily consists of a conductor, ground wire, supporting-conductor suspension cable, supporting-ground wire suspension cable, tension-resistant insulator strings, and fixed brackets. The structural schematics are illustrated in Figure 1. Among them, the supporting suspension cable made of high-strength steel strands is the main load-bearing structure, and is usually fixed in mountain rock or rigid fixed brackets, with the transmission line through the connection of fittings and the supporting suspension cable forming a coupling system.

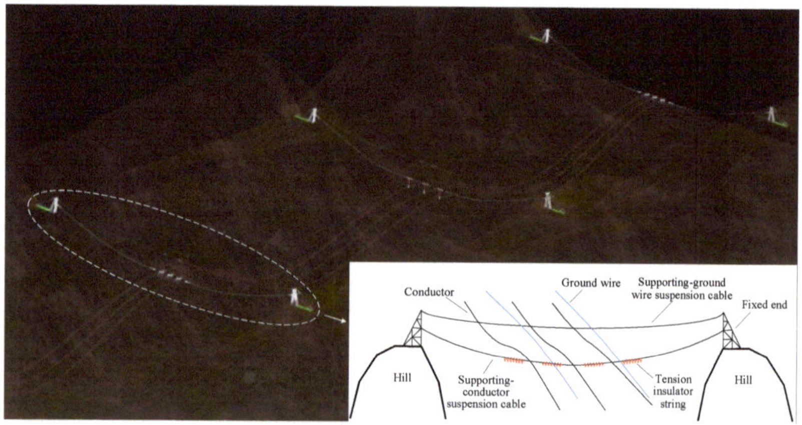

Figure 1. Schematic diagram of the tension cable-supported power transmission structure.

The supporting suspension cables and transmission lines in TC-PTS are large-span, highly flexible spatial suspension structures, and are typical wind-sensitive structures. In addition, the structures have mechanical characteristics such as large deformation and small strain, and the geometric nonlinear effect is significant. Compared with the transmission line and transmission tower-line systems, the transmission line of the TC-PTS forms a more flexible coupled system with supporting suspension cables, and its geometric nonlinear effect is more significant. Scholars have conducted a series of studies to address the wind-induced nonlinear vibration problem of transmission lines and transmission tower line systems. In early studies, transmission lines were usually considered to be suspension structures, and their static and dynamic characteristics were analyzed [3–5]. With the development of the finite element method and large deformation theory, the wind-induced vibration analysis of transmission lines and tower-line systems is generally performed using a nonlinear finite element method. The mechanical model of the transmission line and the tower-line system based on the truss element is convenient for performing various types of wind vibration and other static and dynamic analysis [6,7]. The mechanical model of the transmission line and tower-line system established using a cable element and beam element can consider the torsional and bending capacity of the transmission line and then study complex wind vibration problems such as galloping [8–13]. Scholars have examined the dynamic models of transmission lines and tower-line systems in depth, as well as the nonlinear vibration issues generated by wind, using rigorous theoretical investigations and numerical simulations. These studies have resulted in noteworthy study outcomes.

However, nonlinear dynamic modeling and wind-induced vibration analyses of TC-PTS have rarely been reported. In addition, TC-PTS is prone to wind-induced fatigue damage under the wind load, and it is necessary to evaluate the wind-induced fatigue of TC-PTS. The rain-flow method and the equivalent structural stress method [14,15] are often used for the wind-induced fatigue assessment. In order to evaluate the safety of TC-PTS, a framework for the wind-induced fatigue assessment needs to be proposed.

To this end, this study establishes a finite element model of TC-PTS based on 3D truss elements, then the nonlinear dynamic equations of the structure are formed, and the iterative solution process of the Newmark-β method combined with the Newton–Raphson method is provided. In addition, wind vibration fatigue assessment was performed by combining the rainfall counting method with the Miner linear fatigue cumulative damage theory. The remainder of this paper is organized as follows: Section 2 exhibits the simplified mechanical model of TC-PTS. The nonlinear finite element equation of wind-induced vibration and fatigue damage assessment method for TC-PTS is shown in Section 3. Subsequently, Section 4 provides an example and the corresponding discussion. Finally, some conclusions are summarized in Section 5.

2. Simplified Mechanical Model of TC-PTS

2.1. Structural Parameters and Boundary Conditions

The simplified model and the structural parameters of the TC-PTS are shown in Figure 2. l_s is the horizontal distance between the ends on both sides of the same suspension cable, l_c is the horizontal distance between two adjacent spans of the suspension cable, h_s is the height difference between the ends of both sides of the same suspension cable, h_c is the height difference between the positions of the same span of the conductor or ground wire in the adjacent hanging points of the suspension cable, and h_{wg} is the height difference between the ends of the supporting-ground wire suspension cable and the corresponding ends of the supporting-conductor suspension cable.

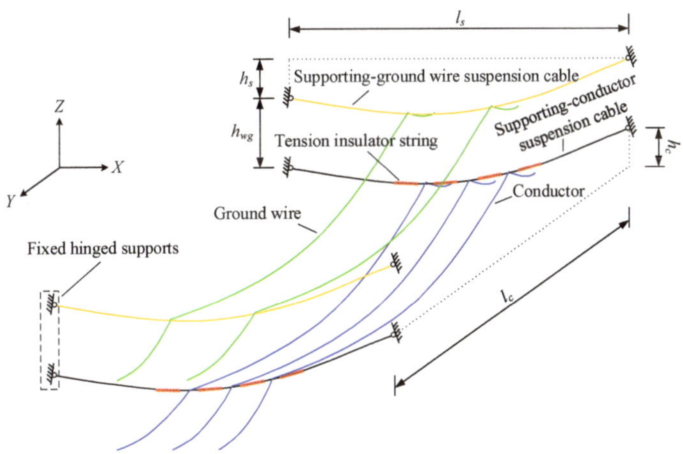

Figure 2. Simplified model and structural parameters of the tension cable-supported power transmission structure.

The connections between the ends of the supporting suspension cables, the fixed supports, and the connections between the transmission lines and corresponding suspension cables at the boundaries are hinged. Therefore, all the supports at the TC-PTS boundary can be considered as fixed hinged supports.

2.2. Initial Shape of TC-PTS

To establish a finite element model of the TC-PTS that considers the geometric nonlinear effect, it is necessary to first obtain the initial shape and initial internal force under gravity loading. For the supporting suspension cables and transmission lines of the TC-PTS, the initial shape can be simplified as a parabolic model, and the functional expression required to establish this parabolic model is

$$y = 4xf_m(1 - x/l)/l \tag{1}$$

where y is the vertical distance from the calculation point to the starting point, x is the horizontal distance from the calculation point to the starting point, l is the horizontal distance from the end point to the starting point, and f_m is the spanning mid-arc droop calculated according to Equation (2):

$$f_m = ql^2/(8\sigma_0 \cos\beta) \tag{2}$$

where q is the gravity load per element volume of the supporting suspension cable or transmission line, σ_0 is the horizontal stress of the suspension cable, β is the angle of elevation difference, and tanβ = h/l, where h is h_s or h_c and l is the corresponding l_s or l_c.

2.3. The Mechanical Characteristics of TC-PTS

In TC-PTS, the supporting suspension cables and transmission lines are mainly subjected to tension, and to a lesser extent, they will also be subjected to bending and torsional moments. It is reasonable to neglect the bending and torsional capacities in wind vibration response analysis without considering torsion [16]. Therefore, in the TC-PTS wind vibration analysis, the supporting suspension cables and transmission lines were primarily considered for the axial tensile capacity.

3. Nonlinear Finite Element Analytical Model of Wind-Induced Vibration and Fatigue Damage Assessment for TC-PTS

3.1. Element Stiffness Matrix for Supporting Suspension Cables and Transmission Lines

According to the mechanical characteristics of the TC-PTS, its supporting suspension and transmission line wind vibration are large deformation and small strain problems. Combined with the force characteristics of this power transmission structure, the finite element model shown in Figure 3 can be established using a 3D truss element by considering geometric nonlinear effects. Generally, the element stiffness matrix can be established using both the local and global coordinate systems. Compared with the stiffness matrix established in the local coordinate system [13,17,18], the element stiffness matrix in the global coordinate system [19–21] can be directly grouped into the system stiffness matrix. It can be directly grouped into a system stiffness matrix to effectively avoid the coordinate conversion process. For the three-dimensional truss element in the global coordinate system, the stiffness matrices are mostly derived using the principle of virtual work [19] and by deriving the truss ends [21]. However, the virtual work principle requires multiple integrals, and often results in implicit stiffness matrices that are difficult to apply directly, whereas the truss end strives for derivatives that require more complicated vector operations. Therefore, this study simplifies the operation through the relationship between the elastic energy of the 3D truss element and its deformation, and derives the explicit stiffness matrix of the 3D truss element under the global coordinate system.

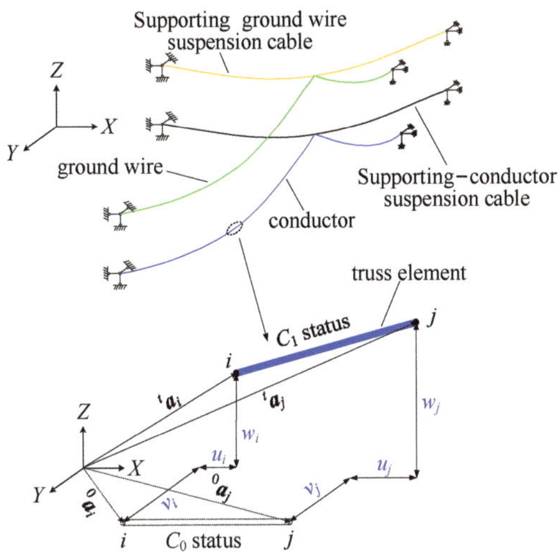

Figure 3. Simplified finite element model and node displacements.

In Figure 3, the physical quantities at the initial moment (C_0 state) are known quantities, and the node vectors at both ends of the element in the global coordinate system are

$$\begin{cases} {}^0\boldsymbol{a}_i = \{\ x_i \quad y_i \quad z_i\ \}^T \\ {}^0\boldsymbol{a}_j = \{\ x_j \quad y_j \quad z_j\ \}^T \end{cases} \tag{3}$$

The principle of vector subtraction is then used to obtain the element vectors of the truss element in the global coordinate system.

$${}^0\boldsymbol{a}_e = {}^0\boldsymbol{a}_j - {}^0\boldsymbol{a}_i = \{x_j - x_i \quad y_j - y_i \quad z_j - z_i\}^T \tag{4}$$

After time t, the truss element undergoes a displacement deformation movement to C_1 state, and the displacement vector of the nodes at both ends of the element under the global coordinate system is

$$\begin{cases} \boldsymbol{u}_i = \{\ u_i \quad v_i \quad w_i\ \}^T \\ \boldsymbol{u}_j = \{\ u_j \quad v_j \quad w_j\ \}^T \end{cases} \tag{5}$$

From Equations (3) and (5), the node vectors at both ends of the element in the global coordinate system are

$$\begin{cases} {}^t\boldsymbol{a}_i = {}^0\boldsymbol{a}_i + \boldsymbol{u}_i \\ {}^t\boldsymbol{a}_j = {}^0\boldsymbol{a}_j + \boldsymbol{u}_j \end{cases} \tag{6}$$

Therefore, the element vector of the truss element at time t is

$${}^t\boldsymbol{a}_e = {}^0\boldsymbol{a}_e + \boldsymbol{u}_j - \boldsymbol{u}_i \tag{7}$$

According to the node displacement vector, the element displacement vector can be expressed as

$$\boldsymbol{u}_e = \{u_i \quad v_i \quad w_i \quad u_j \quad v_j \quad w_j\}^T \tag{8}$$

Equation (7) can be further expressed as Equation (8) as

$${}^t\boldsymbol{a}_e = {}^0\boldsymbol{a}_e + [-\boldsymbol{I}_{3\times3} \quad \boldsymbol{I}_{3\times3}]\boldsymbol{u}_e \tag{9}$$

where $I_{3\times3}$ is the 3rd order element matrix.

At C_0 state, the truss element in the state has no strain and its element length is

$$^{0}l = \left|{}^{0}\boldsymbol{a}_e\right| = \sqrt{{}^{0}\boldsymbol{a}_e^{T}{}^{0}\boldsymbol{a}_e} \tag{10}$$

At C_1 state, the length of the truss element in the state is

$$^{t}l = \left|{}^{t}\boldsymbol{a}_e\right| = \sqrt{{}^{t}\boldsymbol{a}_e^{T}{}^{t}\boldsymbol{a}_e} \tag{11}$$

At this point, the strain in the truss element is

$$^{t}\varepsilon = ({}^{t}l - {}^{0}l)/{}^{0}l \tag{12}$$

The truss element elastic energy ^{t}U at time t can be calculated using Equation (13):

$$^{t}U = \int_V \int_0^{{}^{t}\varepsilon} \sigma(\varepsilon)d\varepsilon dV \tag{13}$$

where $\sigma(\varepsilon)$ is the stress of the truss element, and the relationship between stress and strain in the linear elastic range is $\sigma(\varepsilon) = E\varepsilon$, E is the Young's modulus, V is the element volume, and when the cross-section area of the truss element is a constant value of A, Equation (13) can be further expressed as

$$^{t}U = \frac{1}{2}EA{}^{0}l{}^{t}\varepsilon^2 \tag{14}$$

The element tangent stiffness matrix at time t in the global coordinate system can be calculated using the Hessian matrix, as follows:

$$^{t}\boldsymbol{K}_e = \frac{\partial^2 {}^{t}U}{\partial^{t}\boldsymbol{u}_e \partial^{t}\boldsymbol{u}_e^{T}} \tag{15}$$

The final result of the element tangent stiffness matrix at time t is:

$$^{t}\boldsymbol{K}_e = \frac{EA{}^{t}\varepsilon}{{}^{t}l}\begin{bmatrix} \boldsymbol{I}_{3\times3} & -\boldsymbol{I}_{3\times3} \\ -\boldsymbol{I}_{3\times3} & \boldsymbol{I}_{3\times3} \end{bmatrix} + \frac{EA}{{}^{t}l^3}\begin{bmatrix} {}^{t}\boldsymbol{a}_e{}^{t}\boldsymbol{a}_e^{T} & -{}^{t}\boldsymbol{a}_e{}^{t}\boldsymbol{a}_e^{T} \\ -{}^{t}\boldsymbol{a}_e{}^{t}\boldsymbol{a}_e^{T} & {}^{t}\boldsymbol{a}_e{}^{t}\boldsymbol{a}_e^{T} \end{bmatrix} \tag{16}$$

The element tangent stiffness matrix $^{t}\boldsymbol{K}_e$ shown in Equation (16) is derived by introducing only the linear elasticity assumption and not the small deformation assumption, which is an exact tangent stiffness matrix at any displacement. Therefore, it is applicable to the case of arbitrarily large deformations and small strains.

3.2. Element Mass Matrix and Damping Matrix

The TC-PTS finite element model mass matrix is constructed using a consistent mass matrix with an element-consistent mass matrix, as follows:

$$\boldsymbol{M}_e = \frac{{}^{0}l\rho A}{6}\begin{bmatrix} 2\boldsymbol{I}_{3\times3} & \boldsymbol{I}_{3\times3} \\ \boldsymbol{I}_{3\times3} & 2\boldsymbol{I}_{3\times3} \end{bmatrix} \tag{17}$$

where ^{0}l is the initial length of the element and ρ is the mass density of the element, respectively.

The element damping matrix is constructed using Rayleigh damping:

$$^{t}\boldsymbol{C}_e = \alpha_v \boldsymbol{M}_e + \beta_v {}^{t}\boldsymbol{K}_e \tag{18}$$

where α_v and β_v are the Rayleigh damping constants calculated using Equations (19) and (20):

$$\alpha_v = 2\xi_{k1}\omega_{k1}\omega_{k2}/(\omega_{k1} + \omega_{k2}) \tag{19}$$

$$\beta_v = 2(\xi_{k2}\omega_{k2} - \xi_{k1}\omega_{k1})/(\omega_{k1}^2 + \omega_{k2}^2) \quad (20)$$

where ω_{k1} and ω_{k2} are the intrinsic frequencies of the k1st and k2nd order modes, respectively; ξ_{k1} and ξ_{k2} are the damping ratios corresponding to the k1st and k2nd order modes, respectively. This can be considered as $\beta_v = 0$ [22]. In this case, ${}^t\boldsymbol{C}_e$ is not affected by the change in structural displacement, namely, $\boldsymbol{C}_e$.

3.3. Equivalent Nodal Load Vector

The primary loads to which the TC-PTS was subjected were gravity and wind. The static and dynamic displacements of this power transmission structure are based on the gravity load equilibrium position. Therefore, when calculating the structural equivalent nodal load vector, the gravity load is equivalent to the internal nodal load vector at the reference position. For the truss elements of the TC-PTS finite element model, the wind load per unit length was calculated using Equation (21):

$$F_D = \frac{1}{2}\rho_{air} V_z^2 C_D A_m \quad (21)$$

where $\rho_{\rm air}$ is the air density, V_z is the wind speed, C_D is the drag coefficient, and A_m is the windward projected area per unit length of the truss element, for supporting suspension cables A_m = $d\cos\theta$, for transmission lines A_m = $d\sin\theta$, d is the diameter of the truss element, and θ is the angle of the wind direction (shown in Figure 4).

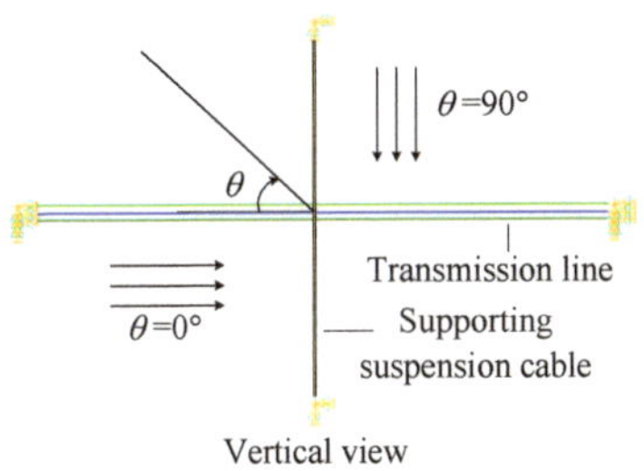

Figure 4. Schematic diagram of wind direction angle θ.

V_z can be calculated using Equation (22):

$$V_z = V_s(z) + V(t) \quad (22)$$

where $V_s(z)$ is the average wind speed and $V(t)$ is the pulsating wind speed.

The pulsating wind speed $V(t)$ is generated by the harmonic synthesis method [23]. The average wind speed $V_s(z)$ can be calculated from the exponential law according to Equation (23):

$$V_s(z) = (z/10)^{\alpha} V_{10} \quad (23)$$

where V_{10} is the reference average wind speed at the height of 10 m, z is the height, and α is the ground roughness coefficient. As TC-PTS is in the mountainous terrain, its ground roughness category can be considered as C. According to the "Code for Structural Loads of Buildings" [24], α is taken as 0.22.

The wind loads on the element are converted to element equivalent nodal loads, namely:

$${}^t\boldsymbol{F}_e = \frac{1}{2}{}^t l \begin{bmatrix} F_{ix} & F_{iy} & F_{iz} & F_{jx} & F_{jy} & F_{jz} \end{bmatrix}^T \quad (24)$$

where F_{ix} is the equivalent nodal load of the wind load in the X-direction at node i, and the rest are similar.

3.4. Establishment of the Nonlinear Dynamic Equation

The mass matrix, stiffness matrix, damping matrix, and equivalent nodal load vectors of each element of the TC-PTS were assembled, respectively, and the nonlinear dynamic equations of the structure were established based on the updated Lagrangian formulation at the reference position:

$$\boldsymbol{M}^{t+\Delta t}\ddot{\boldsymbol{u}} + \boldsymbol{C}^{t+\Delta t}\dot{\boldsymbol{u}} + {}^{t+\Delta t}\boldsymbol{Q} - {}^{0}\boldsymbol{Q} = {}^{t+\Delta t}\boldsymbol{F} \tag{25}$$

where ${}^{t+\Delta t}\ddot{\boldsymbol{u}}$ and ${}^{t+\Delta t}\dot{\boldsymbol{u}}$ are the acceleration and velocity vectors of the structure at time t+Δt, and $\boldsymbol{M}$ and $\boldsymbol{C}$ are the mass and damping matrix of the structure. ${}^{t+\Delta t}\boldsymbol{Q}$ and ${}^{0}\boldsymbol{Q}$ are the internal load vectors; that is, the internal nodal load vectors at time $t + \Delta t$ and the reference position, respectively. In addition, ${}^{t+\Delta t}\boldsymbol{F}$ is the equivalent nodal load vector of the structure at time $t + \Delta t$.

When Δt is sufficiently small, the displacement increment of the structure within a time step is small, and ${}^{t+\Delta t}\boldsymbol{Q}$ and ${}^{t}\boldsymbol{K}$ can be approximated using the following linear relationship:

$${}^{t+\Delta t}\boldsymbol{Q} = {}^{t}\boldsymbol{Q} + {}^{t}\boldsymbol{K}\boldsymbol{u} \tag{26}$$

where ${}^{t}\boldsymbol{Q}$ is the internal nodal load vector at time t, ${}^{t}\boldsymbol{K}$ is the stiffness matrix of the structure at time t, and $\boldsymbol{u}$ is the displacement increment within Δt; that is $\boldsymbol{u} = {}^{t+\Delta t}\boldsymbol{u} - {}^{t}\boldsymbol{u}$.

Combining Equations (25) and (26), the nonlinear dynamic equation in the form of displacement increments can be obtained:

$$\boldsymbol{M}^{t+\Delta t}\ddot{\boldsymbol{u}} + \boldsymbol{C}^{t+\Delta t}\dot{\boldsymbol{u}} + {}^{t}\boldsymbol{K}\boldsymbol{u} = {}^{t+\Delta t}\boldsymbol{F} - [{}^{t}\boldsymbol{Q} - {}^{0}\boldsymbol{Q}] \tag{27}$$

3.5. Solution of the Nonlinear Dynamical Equation

The Newmark-β method is used to solve the nonlinear finite element equations shown in Equation (27). Because Equation (26) is a linear approximation, the displacement increment of the solution produces a large error accumulation as the time step increases. Therefore, to improve the accuracy of the solution and avoid the accumulation of numerical instability within each time step, the Newton–Raphson iterative method was used in this study for balanced iterative calculation. The solution process is summarized as follows.

Step 1: Calculate the individual element characterization matrices ${}^{0}\boldsymbol{K}_e$, $\boldsymbol{M}_e$, and $\boldsymbol{C}_e$ based on the initial shape and strain of the element, and then assemble them to obtain the characterization matrices ${}^{0}\boldsymbol{K}$, $\boldsymbol{M}$, and $\boldsymbol{C}$ of the structure.

Step 2: Apply the initial equivalent nodal load ${}^{0}\boldsymbol{F}$, and then calculate ${}^{0}\ddot{\boldsymbol{u}}$ from the initial conditions ${}^{0}\boldsymbol{u}$ and ${}^{0}\dot{\boldsymbol{u}}$:

$${}^{0}\ddot{\boldsymbol{u}} = \boldsymbol{M}^{-1}[{}^{0}\boldsymbol{F} - \boldsymbol{C}^{0}\dot{\boldsymbol{u}} - {}^{0}\boldsymbol{K}^{0}\boldsymbol{u}]$$

Step 3: Select the time step Δt and calculate the integration constant [25] using the known physical quantities at time t. The effective stiffness matrix was calculated before performing the iterations of ${}^{t}\hat{\boldsymbol{K}}^{(0)}$:

$${}^{t}\hat{\boldsymbol{K}}^{(0)} = {}^{t}\boldsymbol{K} + c_0\boldsymbol{M} + c_1\boldsymbol{C}$$

Step 4: Perform Newton–Raphson iterations. Calculate the effective load vector ${}^{t+\Delta t}\hat{\boldsymbol{F}}$ at time $t + \Delta t$:

$${}^{t+\Delta t}\hat{\boldsymbol{F}} = {}^{t+\Delta t}\boldsymbol{F} + \boldsymbol{M}[c_0{}^{t}\boldsymbol{u} + c_2{}^{t}\dot{\boldsymbol{u}} + c_3{}^{t}\ddot{\boldsymbol{u}}] + \boldsymbol{C}[c_1{}^{t}\boldsymbol{u} + c_4{}^{t}\dot{\boldsymbol{u}} + c_5{}^{t}\ddot{\boldsymbol{u}}]$$

From this, calculate the unbalanced force vector for the first iteration ${}^{t+\Delta t}\hat{\boldsymbol{F}}^{(0)}$:

$${}^{t+\Delta t}\hat{\boldsymbol{F}}^{(0)} = {}^{t+\Delta t}\hat{\boldsymbol{F}} - [{}^{t}\boldsymbol{Q}^{(0)} - {}^{0}\boldsymbol{Q}] - [c_0\boldsymbol{M} + c_1\boldsymbol{C}][{}^{t+\Delta t}\boldsymbol{u}^{(0)} - {}^{t}\boldsymbol{u}]$$

where ${}^{t}\boldsymbol{Q}^{(0)}$ is the internal nodal load vector before the first iteration, and ${}^{t+\Delta t}\boldsymbol{u}^{(0)}$ is the displacement before the first iteration, namely ${}^{t}\boldsymbol{u}$.

Solve for the displacement increment $\boldsymbol{u}^{(1)}$ after the first iteration at time $t + \Delta t$:

$$\boldsymbol{u}^{(1)} = [{}^{t}\hat{\boldsymbol{K}}^{(0)}]^{-1}[{}^{t+\Delta t}\hat{\boldsymbol{F}}^{(0)}]$$

Step 5: Update ${}^{t}\boldsymbol{K}^{(1)}$ and ${}^{t}\boldsymbol{Q}^{(1)}$ and update the effective stiffness matrix ${}^{t}\hat{\boldsymbol{K}}^{(1)}$ and the displacement before iteration ${}^{t+\Delta t}\boldsymbol{u}^{(1)}$:

$${}^{t+\Delta t}\boldsymbol{u}^{(1)} = {}^{t+\Delta t}\boldsymbol{u}^{(0)} + \boldsymbol{u}^{(1)}$$

and the unbalanced force vector ${}^{t+\Delta t}\hat{\boldsymbol{F}}^{(1)}$:

$${}^{t+\Delta t}\hat{\boldsymbol{F}}^{(1)} = {}^{t+\Delta t}\hat{\boldsymbol{F}} - [{}^{t}\boldsymbol{Q}^{(1)} - {}^{0}\boldsymbol{Q}] - [c_0\boldsymbol{M} + c_1\boldsymbol{C}][{}^{t+\Delta t}\boldsymbol{u}^{(1)} - {}^{t}\boldsymbol{u}]$$

and calculate the displacement increment after the second iteration $\boldsymbol{u}^{(2)}$:

$$\boldsymbol{u}^{(2)} = [{}^{t}\hat{\boldsymbol{K}}^{(1)}]^{-1}[{}^{t+\Delta t}\hat{\boldsymbol{F}}^{(1)}]$$

Step 6: Determine whether the L2 norm of the unbalanced force vector ${}^{t+\Delta t}\hat{\boldsymbol{F}}^{(1)}$ satisfies the convergence condition; if not, loop Step 5 until the convergence condition is satisfied; if so, end the Newton–Raphson iteration to obtain the final nonlinear displacement ${}^{t+\Delta t}\boldsymbol{u}$ at time $t+\Delta t$, and then obtain the velocity ${}^{t+\Delta t}\dot{\boldsymbol{u}}$ and acceleration ${}^{t+\Delta t}\ddot{\boldsymbol{u}}$ at time $t + \Delta t$.

3.6. Wind-Induced Fatigue Damage Assessment of TC-PTS

Using the displacement time history and velocity time history obtained from the above TC-PTS wind vibration nonlinear finite element analysis model, the stress time history of each part of the TC-PTS could be derived. The rain-flow counting method was then used to process the stress time history and obtain the stress amplitude distribution statistically. Finally, the number of stress cycles in which fatigue damage occurs can be obtained using the *S–N* curve. The logarithmic form of the *S–N* curve is shown in Equation (28):

$$\lg N = C - m\lg S \tag{28}$$

where N is the number of stress cycles in which fatigue damage occurs, C and m are the material fatigue parameters, and S is the stress amplitude.

The fatigue parameters for support suspension materials [26] were obtained from $C = 13.84$ and $m = 3.5$. The fatigue parameters of the transmission line materials [27] are available for $N \le 2 \times 10^7$, $C = 13.27$, and $m = 5$, as well as for $N > 2 \times 10^7$, $C = 14.40$, and $m = 5.95$.

According to Miner's linear fatigue cumulative damage criterion, the wind vibration fatigue damage of each part of the TC-PTS is as follows:

$$D = \sum_{i=1}^{k} n_i/N_i \tag{29}$$

where D is the total fatigue damage value. When $D = 1$, fatigue damage occurs, k is the total number of stress amplitudes, n_i is the number of times the ith stress amplitude occurs, and N_i is the number of stress cycles corresponding to the ith stress amplitude where fatigue damage occurs.

In addition, wind vibration fatigue damage analysis of the supported suspension cables and transmission lines must consider the effect of the average stress. The Goodman model is often used to correct for the effect of the nonzero average stress S_m, namely:

$$S = S_0(1 - S_m/S_t) \tag{30}$$

where S is the actual stress amplitude, S_0 is the equivalent stress amplitude at zero average stress, S_m is the average stress, and S_t is the tensile strength of a material.

4. Example Analysis

4.1. Example Description

In this study, a two-span TC-PTS, as shown in Figure 5, was considered as the research object for wind vibration response analysis and fatigue damage assessment. The Kaimal spectrum was used as the wind speed power spectral density function, the wind speed time duration was taken as 600 s, the structural parameters of the two-span TC-PTS were l_s = 1000 m, l_c = 800 m, h_{wg} = 20 m, h_s = 0 m, and h_c = 0 m, respectively, and the endpoints of the supporting-conductor suspension cables were considered to be fixed at a height of 300 m. All the calculations required for the analysis in this study were performed on a computer with an Intel i5-9400F CPU (manufactured by Intel, Santa Clara, CA, USA) and 32 GB of RAM (manufactured by Corsair Memory, Fremont, CA, USA).

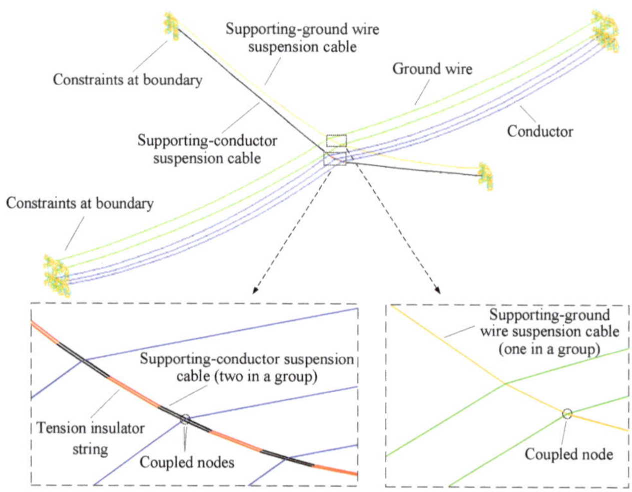

Figure 5. A two-span tension cable-supported power transmission structure.

Using the proposed model, the first five orders of natural frequency and the first three orders of the vibration mode of this power transmission structure can be calculated, as shown in Table 1 and Figure 6, respectively. The results of the frequency analysis in Table 1 show that the first five orders of intrinsic frequencies of the supporting-conductor part are 77.89%, 78.65%, 78.28%, 78.30%, and 78.30% of the corresponding values of the supporting-ground wire part. The lower-order natural frequency of the supporting-conductor part was lower than that of the supporting-ground wire part. As shown in Figure 6, for the supporting-conductor part, the first-, second-, and third-order vibration modes were antisymmetric vertical bending, symmetric lateral bending, and antisymmetric lateral bending, respectively.

Table 1. First five orders of natural frequency of TC-PTS.

Parts	Frequency Order	Proposed Model/(Hz)	ANSYS/(Hz)
The supporting-conductor part	first order	0.08063	0.08063
	second order	0.08174	0.08174
	third order	0.08186	0.08186
	fourth order	0.08188	0.08188
	fifth order	0.08188	0.08188
The supporting-ground wire part	first order	0.1035	0.1035
	second order	0.1039	0.1039
	third order	0.1046	0.1046
	fourth order	0.1046	0.1046
	fifth order	0.1046	0.1046

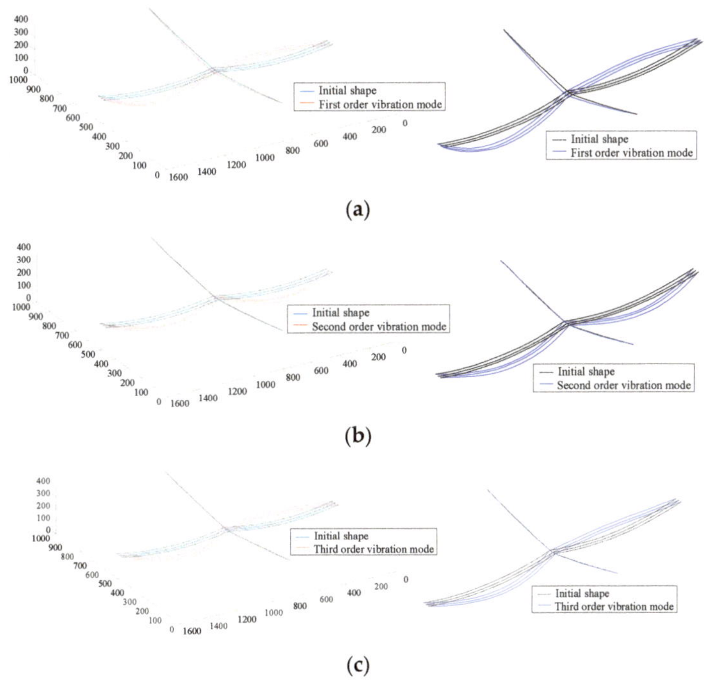

Figure 6. First three orders mode of vibration of the supporting-conductor part obtained by the proposed model (left) and ANSYS (right). (**a**) First order vibration mode; (**b**) second order vibration mode; (**c**) third order vibration mode.

To verify the correctness and calculation accuracy of the proposed model, ANSYS (ANSYS, Inc., Canonsburg, PA, USA) was used to perform a modal analysis, and the results of the two calculations were compared. In the ANSYS model, the supporting suspension and transmission lines were simulated using the Link180 element, and the structural parameters, boundary conditions, and other parameters were consistent with those of the model in this study. The ANSYS calculation results are also given in Table 1 and Figure 6, respectively, and it can be seen that the maximum absolute value of the relative error of the frequency results of the proposed model is 0.001%, which is more accurate, and the vibration modes are also consistent.

4.2. Wind-Induced Vibration Response of TC-PTS

4.2.1. Time History Analysis of TC-PTS Wind-Induced Vibration Response

The time history analysis of the wind-induced vibration response of TC-PTS was performed by considering the working conditions of the wind speed $V_{10} = 15\mathrm{m/s}$ and wind direction angle $\theta = 90^{\circ}$. The standard deviation (σ) of the lateral displacement time history results at each node position of a single span conductor in TC-PTS is shown in Figure 7, and it can be seen that the standard deviation of lateral displacement in the center of the conductor is the largest, which is 2.34 m, indicating that the location in the center of the conductor is more significantly affected by wind load, and the standard deviation values of lateral displacement in the remaining locations are roughly symmetrical along the location in the center of the conductor. The lateral displacement time histories at the center of the conductor are shown in Figure 8, and the average and maximum values of the lateral displacement time histories in Figure 8 are listed in Table 2. The results in Table 2 and Figure 8 show that the average value of lateral displacement in the conductor under wind load in this condition is 23.33 m, and its maximum value is 29.70 m. The lateral displacement is distributed from the minimum value of 16 m to the maximum value of 29.70 m, with a large variation.

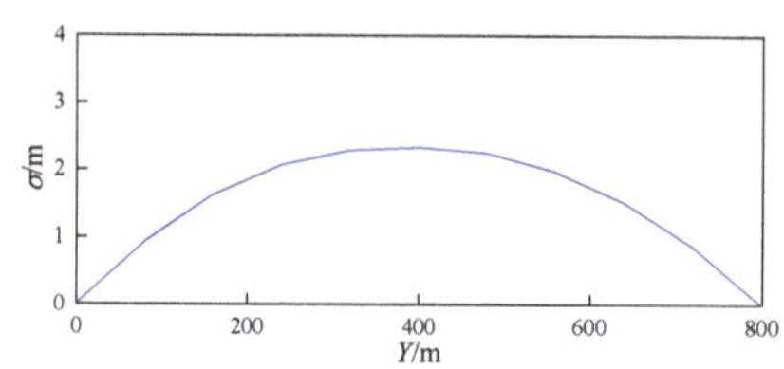

Figure 7. Mean square deviation of lateral displacement at each position of a single-span conductor.

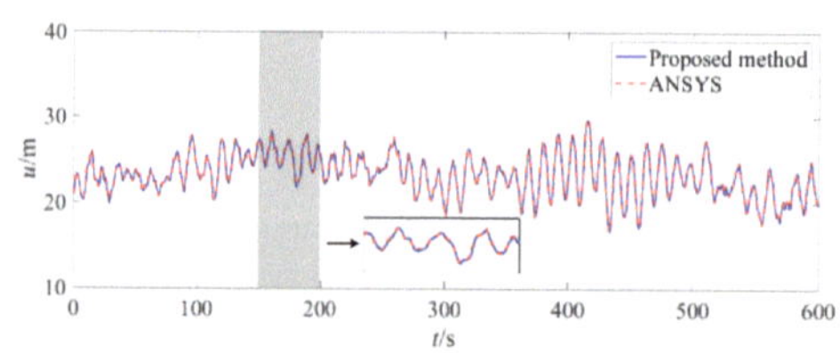

Figure 8. The lateral displacement time history of the conductor midpoint.

Table 2. Average and maximum values obtained by the proposed model and ANSYS.

Terms	Average Values	Maximum Values
Proposed model	23.33m	29.70 m
ANSYS	23.36m	29.95 m
Relative error/(%)	0.13	0.83

Similarly, the time history of the midpoint displacement of the conductor computed by ANSYS is plotted in Figure 8 and Table 2, whereas the computational efficiency of the proposed model is listed in Table 3. From Figure 8 and Table 2, it can be observed that the relative errors of the mean and maximum values of the lateral displacement time history obtained by the proposed model were 0.13% and 0.83%, respectively, and its accuracy was high. Simultaneously, the lateral displacement time history curves obtained by the proposed model had a consistent trend with the ANSYS results, which had a high degree of agreement. The results in Table 3 show that the computation time required by ANSYS to analyze the lateral displacement in the conductor was 35.88 min, whereas the computation time of the proposed model was 1.50 min, and its computational cost was only 4.19% of the former. Therefore, the proposed model can be applied to the wind-induced vibration nonlinear finite element analysis for TC-PTS more efficiently.

Table 3. Calculation efficiency obtained by the proposed model and ANSYS.

Terms	Computing Time/min	Calculation Efficiency/%
Proposed model	1.50	4.19
ANSYS	35.88	-

4.2.2. Effect of Different Wind Direction Angles on TC-PTS Wind-Induced Vibration Response

In this section, the influence of different wind direction angles (θ taken as 0°, 45°, 60°, and 90°) on the maximum lateral displacement of the transmission line and the maximum tension of the supporting suspension cables of TC-PTS at $V_{10} = 20\text{m/s}$ are investigated. The variation in the maximum lateral displacement of the transmission line and the maximum tension of the supporting suspension cable with the wind direction angle θ is shown in Figure 9. In Figure 9a, it can be seen that the maximum lateral displacement of the transmission line increases with the increase in θ, and initially it tends to be faster, and then slower, subsequently. In Figure 9b, the maximum tension of the supporting suspension cable increases with the increase in θ. The change in the incremental magnitude of the maximum lateral displacement of the transmission line was also owing to the stress-stiffening effect of the structure. Considering the conductor as an example, the maximum lateral displacements of the conductor were 0.0071, 21.15, 28.12, and 32.96 m for wind direction angles of 0°, 45°, 60°, and 90°, respectively, and the most unfavorable wind direction angle for the maximum lateral displacement of the transmission line was 90°. The maximum tension of the supporting suspension cables increased with the wind direction angle because the downwind projected area of the transmission line was larger than that of the supporting suspension cables. As the wind direction angle increased, the global wind load on the structure increased. Therefore, the wind direction angle had a more significant effect on both the lateral displacement of the transmission line and the tension of the supporting suspension cables, and the most unfavorable wind direction angle for the two-span TC-PTS was 90°.

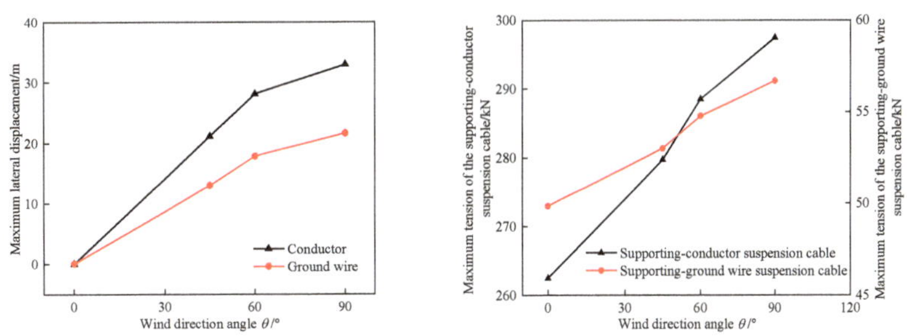

(a) (b)

Figure 9. The influence of wind direction angle θ. (**a**) Maximum lateral displacement of the transmission line; (**b**) maximum tension of the supporting suspension cable.

4.2.3. Effect of Different Wind Speeds on the Wind-Induced Vibration Response of TC-PTS

Based on the conclusions of the above analysis, this section examines the effect of the change in wind speed V_{10} on the maximum lateral displacement of the transmission line and the maximum tension of the supporting suspension cables under the most unfavorable wind direction angle of 90° working conditions. The variation in the maximum lateral displacement of the transmission line and the maximum tension of the supporting suspension cable with the wind speed V_{10} at $\theta = 90°$ is shown in Figure 10. In Figure 10a, it can be seen that the maximum lateral displacement of the transmission line increases with the increase in V_{10}, and the magnitude of the increase tends to be faster, initially, and becomes slow subsequently. This phenomenon is more obvious on the conductor. In Figure 10b, the maximum tension of the supporting suspension cables increases with the increase in V_{10} all the time. The change in the maximum lateral displacement increment of the transmission line is due to the stress stiffening effect of the structure. According to Equation (16), the element tangent stiffness increases with the increase in displacement, and the lateral stiffness of the structure also increases gradually with the increase in displacement. When the lateral stiffness is larger, the increase in lateral displacement of the transmission line will slow down.

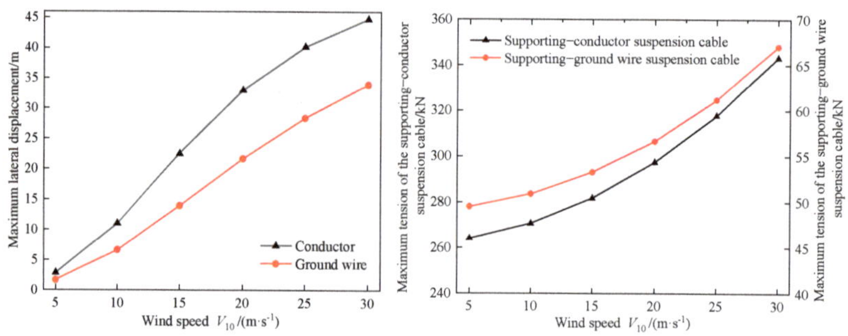

(a) (b)

Figure 10. The influence of wind speed V_{10}. (**a**) Maximum lateral displacement of the transmission line; (**b**) maximum tension of the supporting suspension cable.

The maximum lateral displacement of a transmission line refers to the greatest horizontal movement experienced by the line due to external forces or factors. In this study, the maximum lateral displacement of the conductor is taken as a case in point. The wind speeds considered are 5, 10, 15, 20, 25, and 30 m/s, while the corresponding maximum lateral displacements of the conductor are measured as 2.83, 11.03, 22.48, 32.96, 40.16, and 44.78 m, respectively. The wind speed is from 5 m/s, and with every increase of 5 m/s, the maximum lateral displacement of the conductor increases by 8.20, 11.45, 10.48, 7.21, and 4.62 m, respectively. Therefore, in the case of smaller wind speed, the increase in wind speed will be higher. The maximum lateral displacement of the conductor increases by 8.20, 11.45, 10.48, 7.21, and 4.62 m. Therefore, the increase in wind speed increases the maximum lateral displacement of the transmission line more significantly when the wind speed is low, whereas the maximum lateral displacement of the transmission line is less affected by the wind speed when the wind speed is high. The change in the maximum tension of the supporting suspension cables is due to the fact that the suspension tension is required to balance the equivalent nodal loads, and the increase in wind speed increases the suspension tension all the time.

For the maximum tension of supporting suspension cables, taking the maximum tension of supporting-conductor suspension cable as an example, the wind speed starts from 5 m/s, and for every increase of 5 m/s, the maximum tension of the supporting-conductor suspension cable increases by 6.59, 11.05, 15.62, 20.38, and 25.47 kN, respectively, and the increment of the maximum tension is approximately linearly related to the wind speed, and the maximum tension and the quadratic of the wind speed are approximately linearly correlated. When the wind speed is small, the maximum tension of the supporting suspension cable is less affected by wind speed, while when the wind speed is larger, the maximum tension of the supporting suspension cable is more significantly affected by wind speed.

Therefore, at lower wind speeds V_{10}, the lateral displacement of the transmission line was more affected by the wind speed than the tension of the supporting suspension cable, and at higher wind speeds V_{10}, the tension of the supporting suspension cable was more notably affected by wind speeds than the lateral displacement of the transmission line.

4.3. *Wind-Induced Fatigue Damage of TC-PTS*

4.3.1. Wind-Induced Fatigue Damage Assessment of TC-PTS

To comprehensively analyze the wind vibration fatigue damage condition of the two-span TC-PTS, the ends and midpoint of the supporting-conductor suspension cable and conductor were selected as fatigue analysis parts. The ends and midpoint of the supporting-conductor suspension cable and conductor are points A, B, C, and D, respectively. The wind-induced fatigue damage of each fatigue analysis part of the two-span TC-PTS was calculated by considering the working conditions of wind speed $V_{10} = 15\text{m/s}$ and wind direction angle $\theta = 90^{\circ}$ as examples. The stress time history of each fatigue analysis part is shown in Figure 11, and the corresponding mean stress, mean square deviation, and fatigue damage are listed in Table 4. It can be observed that, compared with point A, the stress mean value of point B is larger, but the mean square deviation is smaller, and its fatigue damage is smaller because the stress amplitude has a more significant effect on the fatigue damage value: the larger the mean square deviation, the more frequently the high stress amplitude occurs, and the more serious the corresponding fatigue damage. The mean stress value and mean square deviation of point C were greater than those of point D, and the fatigue damage value was also greater.

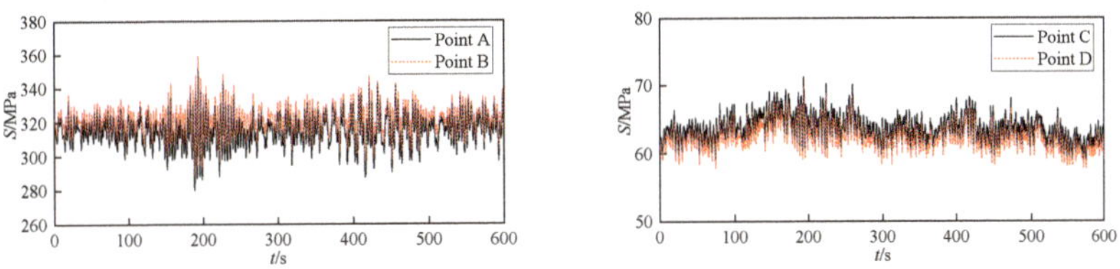

(a) (b)

Figure 11. Stress time history of each part. (**a**) The supporting-conductor suspension cable; (**b**) the conductor.

Table 4. Fatigue damage of each part.

Part	Mean Value of Stress/(MPa)	Mean Square Deviation of Stress/(MPa)	Fatigue Damage
Point A	315.04	9.12	9.70×10^{-11}
Point B	320.96	8.37	7.79×10^{-11}
Point C	63.70	1.87	1.41×10^{-12}
Point D	62.36	1.83	4.96×10^{-13}

4.3.2. Effect of Different Wind Direction Angles on Wind-Induced Fatigue Damage of TC-PTS

At $V_{10} = 20$ m/s, the influences of different wind direction angles θ on the fatigue damage of various parts of the TC-PTS are listed in Table 5. With a gradual increase in the wind direction angle, the wind load acting on the conductor is larger because of the coupling effect between the conductor and the supporting-conductor suspension cable, and the load borne by the supporting-conductor suspension cable also gradually increases. Therefore, the fatigue damage of each part increases. In the same working conditions, the fatigue damage of the supporting-conductor suspension cable end (point A) is larger than that of the midpoint (point B), while the fatigue damage of the conductor end (point C) is smaller than that of the midpoint (point D). It is easy to see that, among all the wind direction angles, the 90° wind direction angle has the most significant effect on the wind-induced fatigue damage of the TC-PTS.

Table 5. Fatigue damage of each part under different wind directions.

Case	Wind Direction/(°)			
	0	45	60	90
Point A	2.91×10^{-12}	4.43×10^{-11}	2.72×10^{-10}	1.80×10^{-9}
Point B	3.30×10^{-12}	1.29×10^{-11}	1.43×10^{-10}	8.87×10^{-10}
Point C	2.01×10^{-14}	1.17×10^{-13}	1.33×10^{-12}	6.13×10^{-11}
Point D	1.50×10^{-14}	5.25×10^{-13}	9.28×10^{-12}	4.11×10^{-10}

4.3.3. Effect of Different Wind Speeds on Wind-Induced Fatigue Damage of TC-PTS

Under the 90° wind direction angle condition, the effect of wind speed V_{10} change on the fatigue damage of each part of TC-PTS is shown in Table 6. As shown in the table, the wind-induced fatigue damage of each part increases with an increase in the wind speed V_{10}, and the fatigue damage at the end of the supporting-conductor suspension cable (point A) is larger than the fatigue damage of the midpoint (point B) under the same working conditions. With the gradual increase in wind speed V_{10}, the fatigue damage to all parts of the supporting-conductor suspension cable and conductor showed a tendency to increase faster and then slower. The fatigue damage of all parts of the supporting-conductor suspension cable increased gradually from 10^{-16} to 10^{-8}, whereas the fatigue damage of all parts of the conductor increased gradually from 10^{-25} to 10^{-7}. When the wind direction angle was 90°, the wind load per unit length of the conductor was greater than that of the supporting-conductor suspension cable, and the fatigue damage of each part of the conductor increased more than that of the supporting-conductor suspension cable.

Table 6. Fatigue damage of each part under different wind speeds.

Wind Speed/(m·s^{-1})	Point A	Point B	Point C	Point D
5	1.40×10^{-15}	1.31×10^{-16}	1.14×10^{-24}	2.64×10^{-25}
10	2.87×10^{-12}	1.59×10^{-13}	5.13×10^{-18}	1.47×10^{-16}
15	9.70×10^{-11}	7.79×10^{-11}	1.41×10^{-12}	4.96×10^{-13}
20	1.80×10^{-9}	8.87×10^{-10}	6.13×10^{-11}	4.11×10^{-10}
25	7.72×10^{-9}	3.84×10^{-9}	1.59×10^{-8}	8.39×10^{-8}
30	5.81×10^{-8}	1.70×10^{-8}	8.03×10^{-7}	3.32×10^{-7}

5. Conclusions

The TC-PTS is a new class of power transmission structures suitable for mountainous terrain. A nonlinear finite element model of the wind-induced vibration of the TC-PTS was proposed using the nonlinear finite element method. Wind-induced vibration response analysis and fatigue damage assessment were performed for a two-span TC-PTS, and the effects of wind speed and wind direction angle on the wind-induced vibration response and fatigue damage of the TC-PTS were discussed. The main conclusions are as follows:

(1) The proposed model calculates the natural frequency and displacement time history with high accuracy and efficiency, and the computational cost of obtaining the dynamic response is only 4.19% of that of ANSYS.

(2) For the mean wind load, at smaller wind speeds, the lateral displacement of the transmission line is more affected by wind speeds than the tension of the supporting suspension cable; at larger wind speeds, the tension of the supporting suspension cable is more notably affected by wind speeds than the lateral displacement of the transmission line.

(3) The fatigue damage value at the end of the supporting-conductor suspension cable was slightly greater than that at the midpoint. For the conductor, under the wind

direction angle cases, the midpoint fatigue damage value is greater than the end fatigue damage value, whereas under the wind speed cases, the end and midpoint fatigue damage values have no obvious comparison.

(4) The wind-induced vibration response and fatigue damage of TC-PTS were significantly affected by the wind direction angle, and the results of the arithmetic analysis showed that the most unfavorable wind direction angle for TC-PTS was 90°.

Although this paper proposes a framework for the wind-induced vibration response analysis and fatigue damage assessment of TC-PTS, there still remain aspects that can be improved further. For example, the design wind speed or extreme wind speed characteristics of the terrain where TC-PTS is located may differ from the characteristics of the flat ground wind field. The design wind speed or extreme wind speed of the terrain where TC-PTS is located can be obtained by filed measurement or other methods, and then the impact of its uncertainty on the wind-induced vibration response and fatigue damage of TC-PTS can be considered.

Author Contributions: Conceptualization, Z.L. and T.W.; methodology, T.W.; software, J.L. and B.W.; validation, Z.L., T.W., and J.L.; formal analysis, T.W.; investigation, T.W., B.W., and J.L.; resources, Z.L.; data curation, B.W.; writing—original draft preparation, J.L.; writing—review and editing, J.L. and B.W.; visualization, B.W.; supervision, T.W.; project administration, T.W.; funding acquisition, T.W. All authors have read and agreed to the published version of the manuscript.

Funding: The research presented in this paper was supported by the Special Support of Chongqing Postdoctoral Research Project (the funding number 2022CQBSHBT3009) and the Special Postdoctoral Support Project of Chongqing Research Institute of HIT (the funding number KY506023002).

Data Availability Statement: The data presented in this study are available on request from the corresponding author. The data are not publicly available due to the ongoing research work.

Conflicts of Interest: The authors declare no conflict of interest.

References

1. Bai, H.; Yi, T.; Li, H.; Ren, L. Multisensors on-site monitoring and characteristic analysis of UHV transmission tower. *Int. J. Distrib. Sens. Netw.* **2012**, *8*, 545148. [CrossRef]
2. Lou, W.; Wu, D.; Liu, M.; Zhang, L.; Bian, R. Properties of mountainous terrain wind field and their influence on wind-induced swing of transmission lines. *China Civ. Eng. J.* **2018**, *51*, 46–55+77.
3. Irvine, H. *Cable Structure*; The MIT Press: New York, NY, USA, 1981.
4. McClure, G.; Lapointe, M. Modeling the structural dynamic response of overhead transmission lines. *Comput. Struct.* **2003**, *81*, 825–834. [CrossRef]
5. Kempner, L.; Smith, S. Cross-rope transmission tower-line dynamic analysis. *J. Struct. Eng. ASCE* **1984**, *110*, 1321–1335. [CrossRef]
6. Tian, L.; Pan, H.; Ma, R.; Dong, X. Seismic failure analysis and safety assessment of an extremely long-span transmission tower-line. *Struct. Eng. Mech.* **2019**, *71*, 305–315.
7. Fu, X.; Li, H.; Tian, L.; Wang, J.; Cheng, H. Fragility analysis of a transmission line subjected to wind loading. *J. Perform. Constr. Facil. ASCE* **2019**, *33*, 04019044. [CrossRef]
8. Yan, Z.; Savory, E.; Li, Z.; Lin, W. Galloping of iced quad-conductors bundles based on curved beam theory. *J. Sound Vib.* **2014**, *333*, 1657–1670. [CrossRef]
9. Yan, Z.; Li, Z.; Savory, E.; Lin, W. Galloping of a single iced conductor based on curved-beam theory. *J. Wind Eng. Ind. Aerodyn.* **2013**, *123*, 77–87. [CrossRef]
10. Chunming, W.; Bin, M.; Tingting, S. Research on the wind-induced vibration coefficient of transmission tower-line system. *Phys. Procedia* **2012**, *24*, 149–154. [CrossRef]
11. Saffari, P.R.; Sirimontree, S.; Thongchom, C.; Thira, J.; Pouyan, R.S.; Suraparb, K.; Suphanut, K. Free and forced vibration of sandwich fgm porous variable thickness nanoplates integrated with magneto-electro-elastic layers via nonlocal strain gradient theory. *Eng. Sci.* **2023**, *24*, 918. [CrossRef]
12. Huang, H.; Xian, Y.; Zhang, W.; Guo, M.; Yang, K.; Xi, K. Analysis of wind-induced vibration of a spoke-wise cable-membrane structure. *J. Mar. Sci. Eng.* **2020**, *8*, 603. [CrossRef]
13. Liu, X.; Yan, B.; Zhang, H.; Zhou, S. Nonlinear numerical simulation method for galloping of iced conductor. *Appl. Math. Mech.* **2009**, *30*, 489–501. [CrossRef]
14. Rong, X.; Xie, A.; Zhao, P.; Yang, M.; Bu, J.; Wei, X. Fatigue performance of composite trough-girders with corrugated-steel-webs. *J. Constr. Steel Res.* **2023**, *207*, 107974. [CrossRef]

15. Wang, S. *Damage Detection and Evaluation of Stud Connector of Composite Girder Bridge Based on Acoustic Emission Technology*; Shijiazhuang Tiedao University: Shijiazhuang, China, 2023.
16. Yan, Z.; Huang, J.; Li, Z. Finite element model of bundle lines based on 6-dof node. *Eng. Mech.* **2012**, *29*, 325–332.
17. Zhu, Z.; Meguid, S. Analysis of three-dimensional locking-free curved beam element. *Int. J. Comput. Eng. Sci.* **2004**, *5*, 535–556. [CrossRef]
18. Zhu, Z.; Meguid, S. Vibration analysis of a new curved beam element. *J. Sound Vib.* **2008**, *309*, 86–95. [CrossRef]
19. Chen, Z.; Zeng, Q.; Yan, Q. A UL formulation for internal force analysis of spatial frame structures with large displacement. *China Civ. Eng. J.* **1992**, *25*, 34–43.
20. Shi, H.; Salim, H. Geometric nonlinear static and dynamic analysis of guyed towers using fully nonlinear element formulations. *Eng. Struct.* **2015**, *99*, 492–501. [CrossRef]
21. Liu, S. Accurate analysis method on the tangential stiffness matrix of pinned straight rod element. *Chin. J. Comput. Mech.* **2014**, *31*, 48–53.
22. Desai, Y.; Yu, P.; Popplewell, N.; Shah, A. Finite element modelling of transmission line galloping. *Comput. Struct.* **1995**, *57*, 407–420. [CrossRef]
23. Jakob, M. Wind field simulation. *Probabilistic Eng. Mech.* **1998**, *13*, 269–282.
24. *GB 50009-2012*; Load Code for the Design of Building Structures. China Architecture & Building Press: Beijing, China, 2012.
25. Wang, X. *Finite Element Method*; Tsinghua University Press: Beijing, China, 2003.
26. Yan, L.; Li, J.; He, X. Effect of central clamps on fatigue damage of suspenders of suspension bridge under joint action of fluctuating wind and vehicle loads. *J. Cent. South Univ.* **2023**, *54*, 1393–1406. [CrossRef]
27. Kong, D.; Li, L.; Long, X.; Ye, Z. Parametric investigation on fatigue life of transmission line subjected to Aeolian vibration. *J. Wuhan Univ. Technol.* **2010**, *32*, 53–57.

buildings

MDPI

Article

Seismic Performance and Calculation Method of Precast Reduced Beam Section Connection

Hao Qi [1], Shaowei Hu [2,*], Xiang Xue [1], Changxi Shan [1], Wenhao Li [1], Yajia Wang [1] and Peiyi Wang [3]

1 School of Civil Engineering, Chongqing University, Chongqing 400045, China; qihaocqu@outlook.com (H.Q.)
2 Yellow River Laboratory, Zhengzhou University, Zhengzhou 450001, China
3 China Construction Fifth Engineering Division Co., Ltd., Changsha 410004, China
* Correspondence: hushaowei@zzu.edu.cn

Abstract: To prevent brittle damage and improve the post-earthquake rapid repair capability of beam-column connections, a precast reduced beam section (PRBS) connection joint that can be rapidly repaired under earthquake action was proposed in this study. Four specimens, including a repaired specimen, were subjected to a quasi-static test to investigate the seismic performance and repair ability of the connection. Seismic performance indices such as the failure mode, hysteresis curve, skeleton curve, strain distribution, and ductility were obtained through observations and analyses. The results indicated that the novel connection exhibited superior load-bearing, energy dissipation, and rotation capacities, compared to the welded flange-bolted web and traditional bone-weakened connections. This novel connection effectively relocated the plastic hinge to alter the failure mode and prevent brittle damage. Additionally, rapid post-earthquake repair was achieved by replacing the dog-bone-style splice section, maintaining a high load-bearing capacity and seismic performance. Finite element (FE) models were established to analyze the mechanical behavior of the specimens, and a parametric analysis was conducted to study the influence of different parameters on the load-bearing capacity of the connection. Based on the experimental and FE analysis results, the possible yield and failure modes of the connection were analyzed, and a calculation method for the bearing capacity of the PRBS connection was proposed. A comparative result demonstrates that the proposed calculation method can accurately predict the load-carrying capacity of a connection.

Keywords: precast reduced beam section connection; earthquake resilient; dog-bone-style weakening; relocation of the plastic hinge; seismic rehabilitation

Citation: Qi, H.; Hu, S.; Xue, X.; Shan, C.; Li, W.; Wang, Y.; Wang, P. Seismic Performance and Calculation Method of Precast Reduced Beam Section Connection. *Buildings* **2023**, *13*, 2653. https://doi.org/10.3390/buildings13102653

Academic Editor: Annalisa Greco

Received: 9 September 2023
Revised: 13 October 2023
Accepted: 18 October 2023
Published: 21 October 2023

1. Introduction

The traditional beam-column connection mainly includes two forms: welded flange-bolted web and full-section welding connections, which are prone to brittle failure under earthquake action, leading to the failure and collapse of building structures [1–4]. The plastic hinge of the beam can be effectively transferred away from the connection surface of the beam and column through structural measures such as bone weakening of the flange, web opening, and web slotting [5–10]. The plastic hinge transfer method can significantly improve the energy dissipation capacity and ductility of a joint. However, it causes a large deformation at the plastic hinge after an earthquake, which leads to difficulties in realizing post-earthquake reinforcement and repair of the structure. Recently, assembled joints have been widely promoted because of their modular production and rapid construction [11–13]. Therefore, it is important to investigate the seismic performance of the assembled beam-column joints and their rapid repair capability after an earthquake.

Considerable research has been conducted on structural damage control and rapid repair performance of connections after an earthquake. Oh et al. [14] proposed a damper with a long and narrow hole for beam-column connections. The existence of the damper not only resulted in good seismic performance of the joints but also made the connection easy to repair after the earthquake. Reza et al. [15] proposed a rubber–steel core damper for beam-to-column connections, where the combination of core steel and rubber improved the post-earthquake repairability of the connection. Huang et al. [16] conducted rapid repairs of a frame structure that had been damaged by an earthquake, and their study showed that the reinforced structure still exhibited good seismic performance. Ma et al. [17] proposed an artificial energy-dissipation plastic hinge and conducted an experimental study, which showed that the proposed plastic hinge can effectively control the plastic damage location and concentrate the energy dissipation at the plastic hinge. Ye et al. [18] proposed a type of assembled connection with a recoverable function, and their results indicated that the damage and failure of the connections were concentrated in the replaceable energy-dissipation hinge. Zhang et al. [19,20] proposed a series of splicing connections made with a cantilever beam that were subjected to pseudo-static loading. The results indicated that the main components of the connection could not be destroyed by a reasonable design. Zhang et al. [21] proposed a high-strength steel framed-tube structure with replaceable link beams and performed low-cycle reciprocating loading tests on its substructures, showing that the damage under the action of an earthquake was concentrated in the energy-dissipation beam section, and the remaining main components were in the elastic phase. In recent years, Özkılıç et al. [22–30] proposed various innovative connections to address the challenging issue of post-earthquake structural repair. An extensive series of experiments, FE analysis, and theoretical investigations were developed to study the load-bearing mechanisms of these novel connections. The results indicated that these novel connections exhibited favorable seismic performance and repairability characteristics.

Most studies on post-earthquake repair methods are complex. Therefore, a novel prefabricated reduced beam section (PRBS) connection joint with a bone-weakened beam section is proposed based on the concept of controllable damage and rapid repair following an earthquake. As shown in Figure 1, the key components of the PRBS connection joint include precast concrete-filled square steel tubular (CFST) columns with a cantilever short beam, a dog-bone-style splice section, and an H-beam. To ensure the structural strength of the connection between the short beam and the CFST column, and to mitigate the risk of brittle failure, the application of a bolted web-welded flange connection, as recommended in the specification [31], was employed. Each component was manufactured in a prefabrication plant and efficiently assembled at the construction site using web-connecting plates and frictional high-strength bolts. During seismic events, a plastic hinge of the connection is formed in the dog-bone-weakened beam section, and the structure can be restored for reuse by replacing the dog-bone-style splice section.

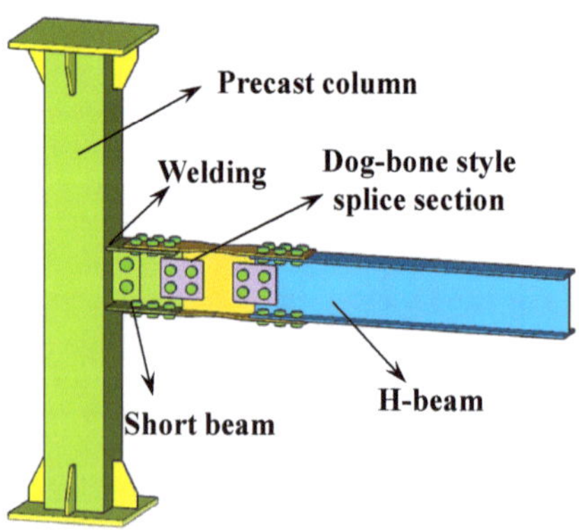

Figure 1. Proposed PRBS connection.

To investigate the seismic performance of the PRBS connection, quasi-static tests were conducted on welded flange-bolted web connection, a conventional bone-weakened, and a PRBS connection joints. A comparative analysis was performed on the failure modes, hysteresis curves, strain distributions, and energy consumption of different joints. Additionally, the repaired PRBS connection joint was subjected to a quasi-static test to explore the seismic performance of the PRBS connection after repair. Furthermore, FE models were established to analyze the mechanical behavior of the specimens, and a parametric analysis was conducted to study the influence of different parameters on the load-bearing capacity of the connection. Ultimately, a load-carrying capacity calculation method for the PRBS connection was proposed based on experimental and FE results. These findings contribute to the theoretical foundation for the widespread adoption of this novel connection.

2. Experimental Program

2.1. Test Specimens

Four composite structural frame joint specimens were designed and fabricated in this study: a bolted web-welded flange joint, a traditional bone-weakened joint, a PRBS connection joint, and a repaired PRBS connection joint, named BASE, RBS, PRBS, and PRBS-R, respectively. The geometric dimensions of the PRBS-R specimen are the same as those of the PRBS specimen. The BASE specimen was a bolted web-welded flange joint designed according to the Chinese code [31]. The precast column was a CFST column with a cross-sectional size and the height of the square steel tube was 200 mm × 200 mm × 10 mm (length, width, and thickness) and 1500 mm, respectively. The cross-sectional size and length of the H-beam were 200 mm × 120 mm × 8 mm × 6 mm (height, width, flange thickness, and web thickness) and 1500 mm, respectively. The RBS specimen was designed based on the BASE specimen, and the dog-bone-weakening dimensions of the flange were determined according to the specifications [31]. The circular weakening length and depth were 160 and 25 mm, respectively.

Additionally, the PRBS specimen applied a dog-bone-weakening configuration to the splice section. The column size of the PRBS was identical to that of the BASE specimen. In accordance with AISC 358 [32], for the purpose of mitigating the failure risk associated with beam-column welding and adhering to the "strong column weak beam" design principle, it is imperative that the probable maximum moment at the face of the column (M_f) remains below the plastic moment of the beam, determined by the expected yield stress (M_{pe}) as illustrated in Equation (1) and Figure 2.

$$M_f \leq \varnothing M_{pe} \tag{1}$$

where $\varnothing$ is the resistance factor for ductile limit states, the recommended value for this parameter is 1 in this study. M_f and M_{pe} can be calculated using Equations (2)–(4).

$$M_f = M_{pd} L / L_1 \tag{2}$$

$$M_{pd} = C_{pr} R_y W_u f_y \tag{3}$$

$$M_{pe} = R_y W_u^b f_y \tag{4}$$

where M_{pd} is the probable maximum moment at the bone-weakening section; C_{pr} is the factor accounting for peak connection strength, the recommended value for this parameter is 1.1; W_u is the plastic section modulus for the bone-weakening section; W_u^b is the plastic section modulus for the full section of the beam at the face of the column, and R_y is the ratio of expected yield stress to specified minimum yield stress.

In addition, the splice between the dog-bone-style splice section and beam was designed using the principle of equal strength [33]; that is, the high-strength bolts on each side of the flange must endure the tensile load-bearing capacity of the flange, and the high-strength bolts on each side of the web must endure the shear load-bearing capacity of the web. Based on the above design requirements, the cross-sectional size and length of the H-beam of the PRBS specimen were 200 mm × 120 mm × 8 mm × 6 mm (height, width, flange thickness, and web thickness) and 1030 mm, respectively. In addition, the dimensions of the flange and web plates of the dog-bone-style splice section were 600 mm × 120 mm × 8 mm (length, width, and thickness) and 70 mm × 200 mm × 6 mm (length, height, and thickness), respectively. The dimensions of the weakened area were identical to those of the RBS specimen. To ensure that the plastic hinge was concentrated in the splice section, the width of the short beam flange was increased to 160 mm. The cross-sectional size and length of the short beam were 200 mm × 160 mm × 8 mm × 6 mm (height, width, flange thickness, and web thickness) and 230 mm, respectively. The column, beam, and splice sections of the PRBS specimen satisfied the dimension limitations stipulated in the Chinese code [31]. Friction-type high-strength bolts of grade 10.9 s were used to connect various components of the joint; the bolt type at the web of the beam-column connection was M20 and the bolt type used in the other splicing parts was M16. The PRBS-R specimens were obtained by replacing the dog-bone-style splice sections of the PRBS specimens after the first test. There was no obvious plastic deformation in the flange and web bolt holes of the cantilever beam after the test of the PRBS specimen, which created construction conditions for replacing the dog-bone style splice section for the new fabricated connection. The geometric dimensions of the specimens are shown in Figure 3.

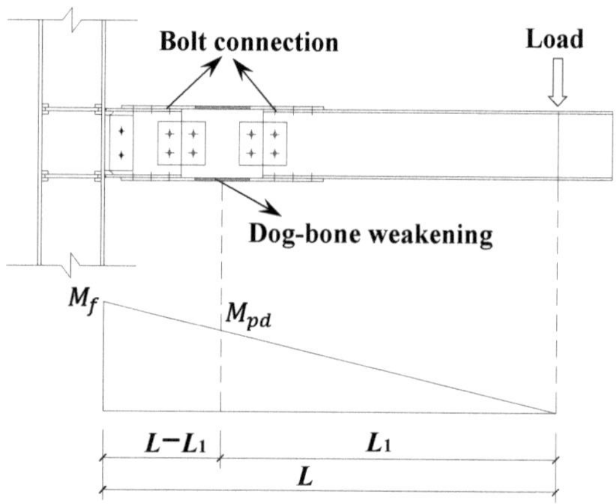

Figure 2. The gradient of the bending moment in PRBS connection.

2.2. Materials

A square steel tube was filled with C50 concrete. The compressive strength of the concrete cubes was tested according to the specifications stated in ref. [34] using 150 mm × 150 mm × 150 mm cube specimens. Three test cubes were examined, and the average cube compressive strength was 64.45 MPa.

Additionally, the steel grade of all the specimens was Q235 [31]. Tensile tests were performed on each type of steel plate according to the relevant material performance test specifications [35,36]. Three specimens were sampled and processed for each type of steel plate. The test results are presented in Table 1. In the table, the ultimate strain indicates the strain value corresponding to the maximum stress.

Table 1. Material properties of steel components.

Member	Coupon Location	Thickness t (mm)	Yield Strength f_y (MPa)	Ultimate Strength f_u (MPa)	Young's Modulus E_s (MPa)	Ultimate Strain ε_u
Column	Square steel tube	10	303.3	448.3	202,115.0	0.178
Beam	Flange	8	281.7	435.0	201,738.0	0.185
	Web	6	348.3	513.3	216,503.7	0.152

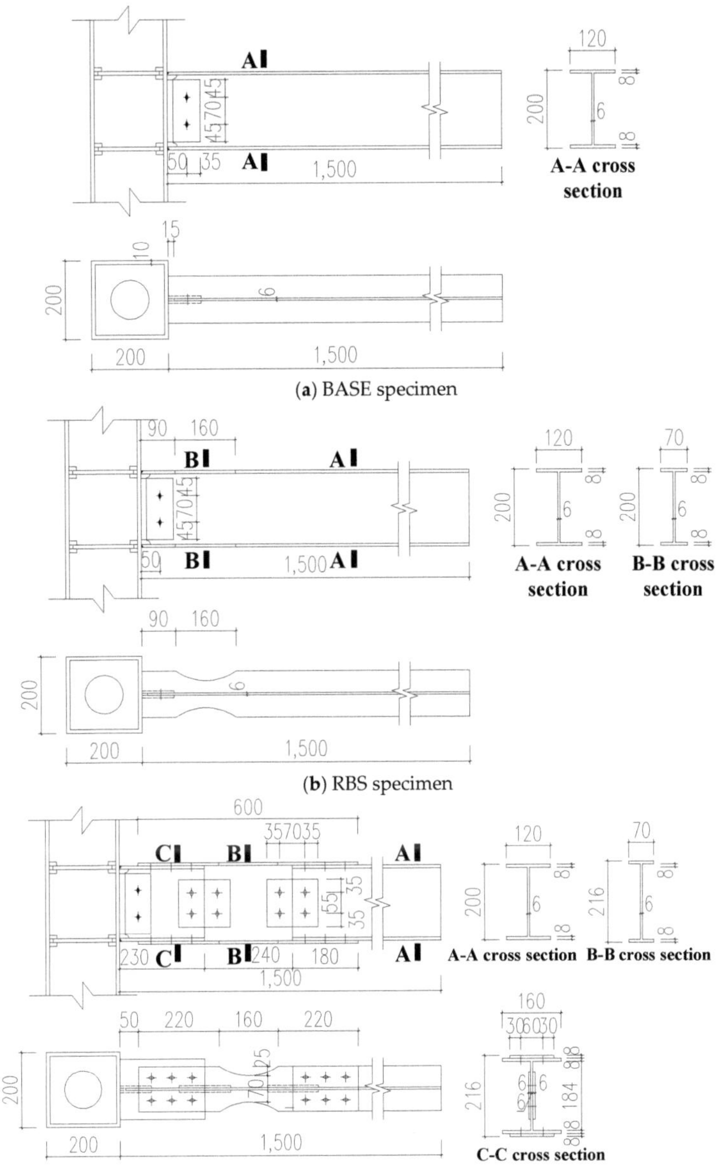

Figure 3. Dimensions of test specimens (Unit: mm).

2.3. *Loading Device and Measurement Scheme*

To accurately simulate the actual stress of the joint, the specimens were placed vertically and loaded onto the beam end. The loading device was composed of a self-balancing counterforce frame, electro-hydraulic servo actuator, lateral bracing, and horizontal bracing. The upper and lower ends of the column were connected to the actuator and the bottom beam of the counterforce frame, respectively, through the hinged device, and a 300 kN axial force was applied to the top of the column through the actuator considering the bearing capacity of the counterforce frame to simulate the actual upper load on the column. The beam end was connected to the actuator through a hinged device that applied cyclic loading to the beam end. The horizontal displacement of the column top was limited by the horizontal bracing to ensure that the boundary condition of the column top was hinged. Lateral bracing limited the out-of-plane displacement of the beam to prevent out-of-plane instability during testing. The loading device used in the test is illustrated in Figure 4.

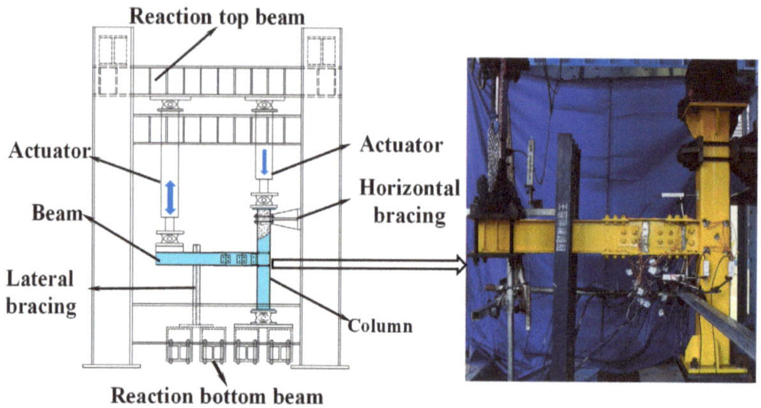

Figure 4. Loading device used in the test.

The test loading procedure adopted low-cycle reciprocating displacement-loading methods [33]. Each level of loading contained two cycles, and the inter-story drift angles of each level were 0.375%, 0.5%, 0.75%, 1%, 1.5%, 2%, 3%, 4%... In addition, the test was stopped when the load dropped to 85% of the peak load or the specimen was damaged. The specific inter-story drift angle and displacement amplitude are shown in Figure 5.

The displacement, load, and strain were measured in real time during the tests. The displacement of the load point in the beam was measured using the displacement sensor of the actuator, and the loads at the column top and beam end were measured using the force sensor of the force actuator. To accurately capture the stress distribution, time of the plastic hinge, plastic hinge distribution, and plastic development of the specimens during the experimental process, strain gauges were arranged at the corresponding positions of the square-steel tubular web, beam flange plate, and beam web plate. Because of the local damage to the repaired beam-column joint specimens, the residual strain caused by local plasticity affected the accuracy of the measurement results, and only the force and displacement data of the PRBS-R specimen were measured, not the strain data. The strain measurement scheme for the specimens is illustrated in Figure 6.

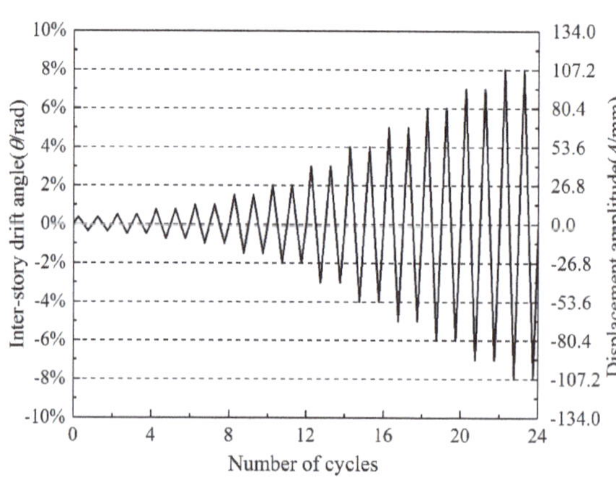

Figure 5. Loading procedure.

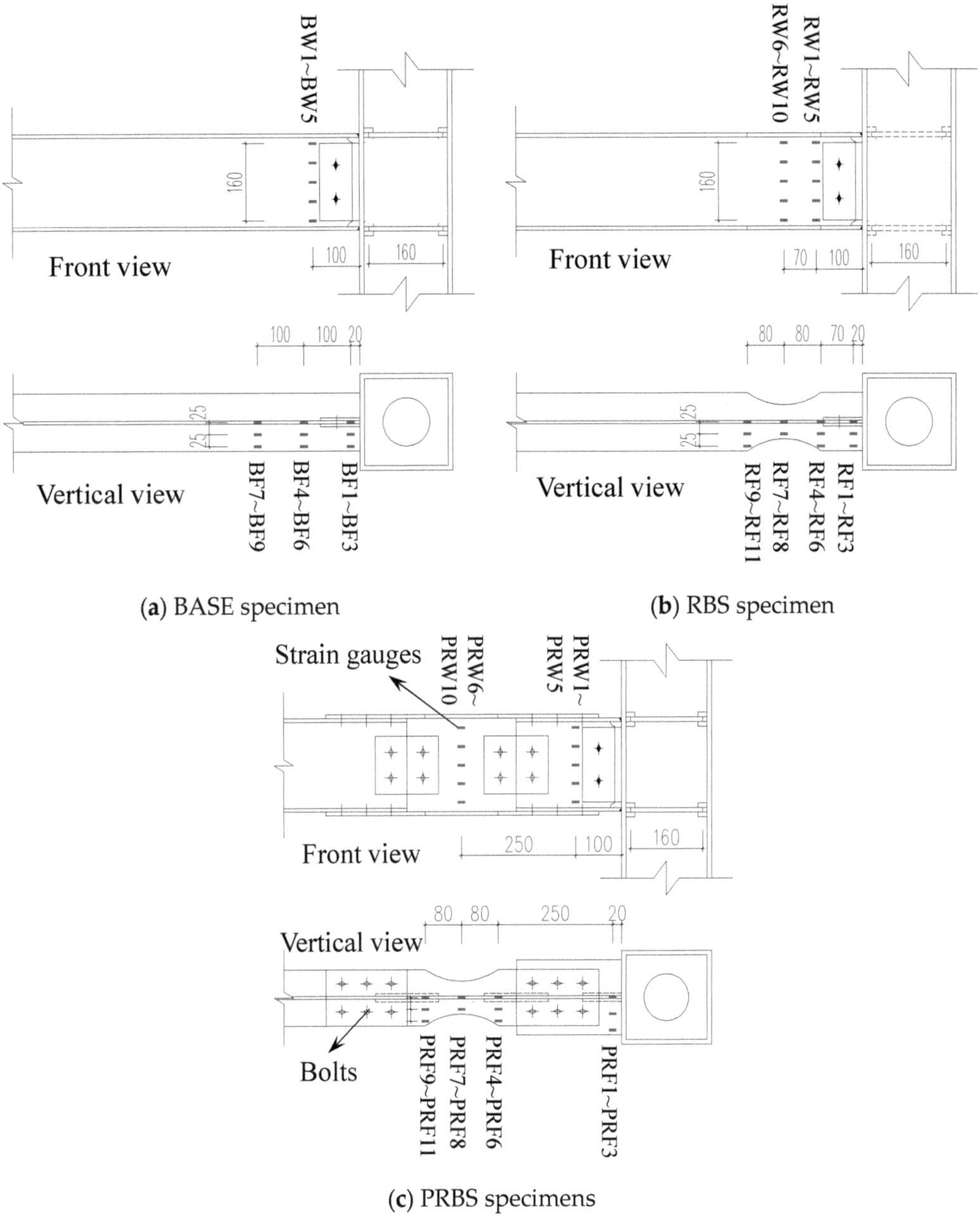

Figure 6. Strain measurement scheme.

3. Test Phenomena and Failure Modes

3.1. BASE Specimen

In the early stage of the test, the hysteresis curve of the BASE specimen was linear, and the specimen was in the elastic stage with no evident phenomena. By increasing the loading cycle at the 1st cycle of the 1.5% story drift ($\Delta = -20.1$ mm), the upper flange area near the column side of the steel beam began to show slight paint wrinkles due to compression, and the wrinkles flattened as the loading direction changed. At the first cycle of the 3%

story drift (Δ = −40.2 mm), local buckling occurred in the lower flange area of the steel beam near the column owing to compression, which was not fully recovered as the loading direction changed; this implies the generation of residual deformation. At the first cycle of the 4%-story drift (Δ = +53.6 mm), local buckling occurred in the roots of both upper and lower flanges, meanwhile, the slippage occurred in the high-strength bolt-connection area near the column side. As the displacement of the loading point reached the 2nd cycle of the 4% story drift (Δ = +53.6 mm), there was a non-penetrating crack in the middle of the upper flange near the side of the column, and the load-displacement curve showed that the bearing capacity decreased with the appearance of the crack. During the loading from the 2nd cycle of the 4% story drift (Δ = −53.6 mm) to the first cycle of the 5% story drift (Δ = +67 mm), the upper flange crack penetrated with a significant load drop when the load-displacement was −21 mm, and the test was declared over, as shown in Figure 7a.

3.2. RBS Specimen

Before loading to the 4% story drift, the test phenomenon of the RBS specimen was similar to that of the BASE specimen. When loaded to the 4% story drift (Δ = +53.6 mm), the paint wrinkles were more evident than in the previous stages. By increasing the loading cycle in the first cycle of the 5% story drift (Δ = +67 mm), local buckling occurred in the weakened upper flange because of compression, meanwhile, the slippage occurred in the web bolt-connection area. In the first cycle of the 6% story drift (Δ = −80.4 mm), buckling deformation of the upper flange of the beam and a slight torsion of the beam occurred. During loading to the first cycle of the 7% story drift (Δ = +93.8 mm), the weld heat-affected zone of the upper flange cracked when load displacement was +56 mm, and the test was stopped with a sharply decreasing load. The failure mode of the RBS specimen is shown in Figure 7b.

3.3. PRBS Specimen

The PRBS specimens were in an elastic state during the first three stages of the test. By increasing the loading cycle, in the second cycle of the 1% story drift (Δ = +13.4 mm), the load-displacement curve changed from the original linear trend to a fusiform envelope surface, which marked the beginning of the plastic stage of the specimen. At a 1.5–2% story drift, the envelope shape of the test load-displacement curve and the test phenomenon of the specimen indicated that the plasticity of the PRBS specimen was fully developed with no slippage at this stage of the bolt connection. At the first cycle of 3% story drift (Δ = +40.2 mm), the bolt slipping sounds occurred, the slippage of the bolt connection was observed in the flange-splicing area, and local buckling occurred on the flange of the dog-bone-style splice section near the side of the short beam. Simultaneously, the load-displacement curve began to show a serrated mutation and tended to be stable during subsequent loading, as shown in Figure 8. As the reciprocating displacement load was applied, slippage in the splicing section and local buckling occurred. At the first cycle of the 7% story drift (Δ = +93.8 mm), the net section of the flange bolt hole of the dog-bone-style splice section broke when load displacement was −93.8 mm, and the test was completed. The failure modes of the PRBS specimens are shown in Figure 7c.

3.4. PRBS-R Specimen

The phenomenon of the PRBS-R specimen in the test process showed no evident change from that of the PRBS specimen; the difference was that the slippage of the PRBS-R specimen in the high-strength bolt connection was more uniform than that of the PRBS specimen, resulting in a smoother load–displacement curve for PRBS-R specimen, as shown in Figure 8. The final failure mode of the PRBS-R specimen was the fracture of the net section of the upper flange bolt hole of the dog-bone-style splice section. The failure modes are shown in Figure 7d.

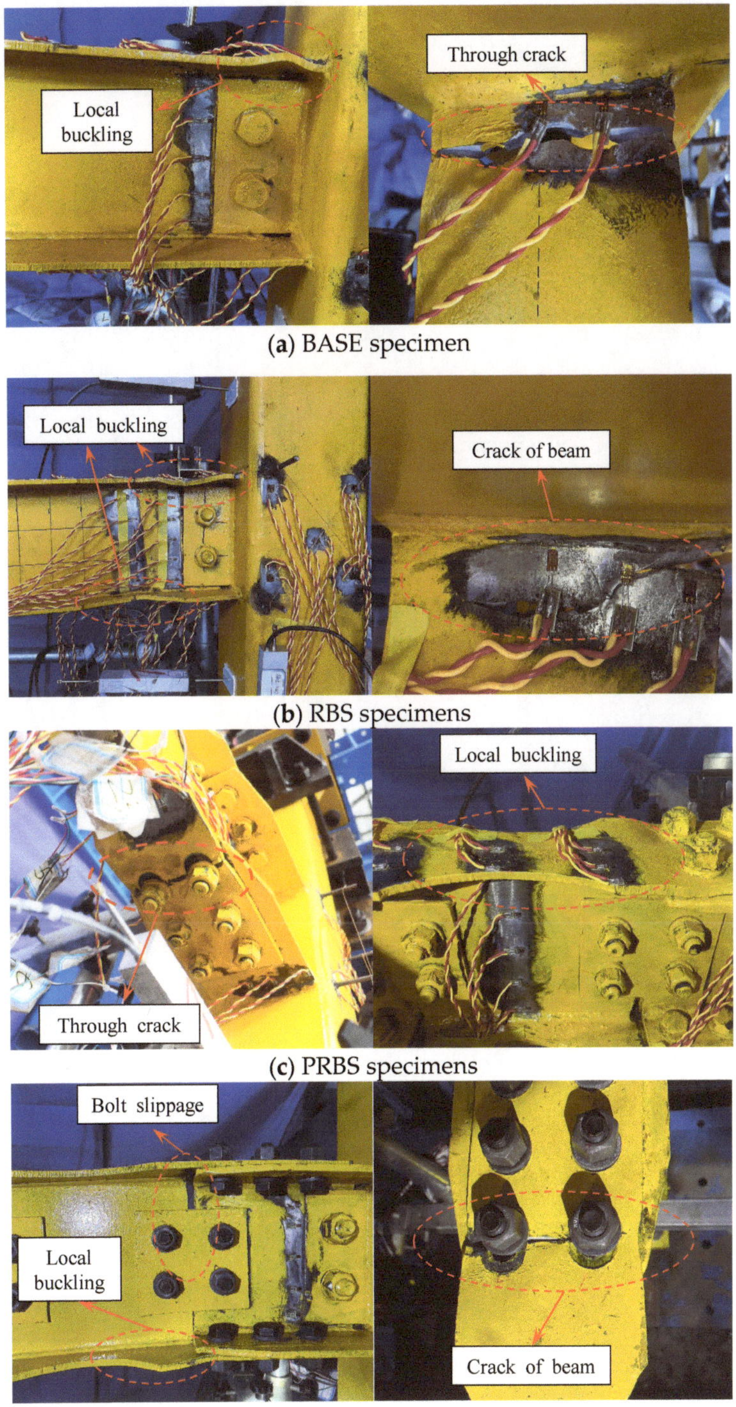

(**a**) BASE specimen

(**b**) RBS specimens

(**c**) PRBS specimens

(**d**) PRBS-R specimen

Figure 7. Failure mode of specimens.

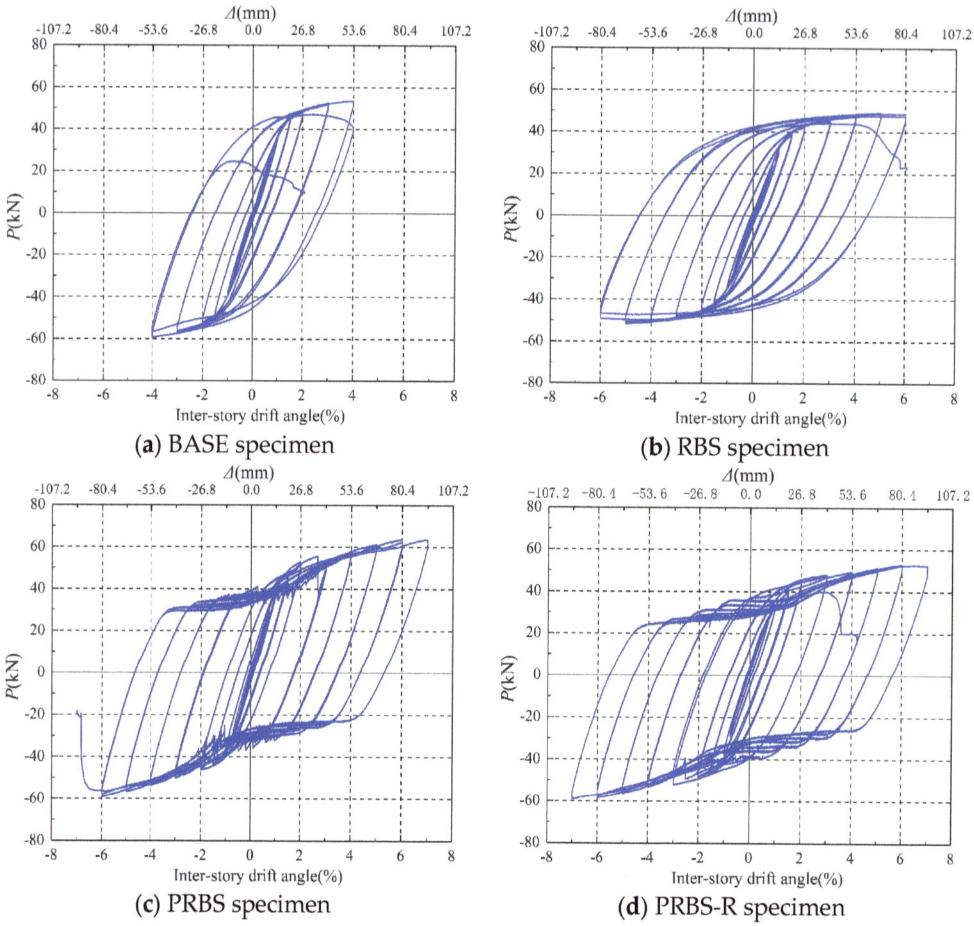

Figure 8. Hysteresis loops of specimens.

3.5. Failure Modes of Specimens

As for the BASE specimen, the flange of the beam near the surface of the column first yielded. With a gradual increase in loading displacement, the specimen completely lost its bearing capacity with crack penetration. The specimen exhibited clear characteristics of brittle failure. Unlike the BASE specimen, the RBS specimen initially yielded in the bone-weakening region, with plastic deformation primarily concentrated there. With the increase in loading displacement, the plastic region gradually extended toward the column until a crack appeared in the weld and its heat-affected zone, leading to specimen failure. It is worth noting that the plasticity of the bone-weakening region had already extensively developed before specimen failure. The primary reason for cracks appearing in the weld and its heat-affected zone, rather than the vulnerable area, might be attributed to construction quality issues in the weld.

As for the failure modes of PRBS and PRBS-R specimens, during the initial phase of load testing, the weakest area was concentrated within the bone-weakening region which entered the yielding stage first. As the applied load increased, the bolt friction connection near the short beam experienced failure, with a slippage of the bolt connection, which rendered it the weakest area. As the loading progresses, the flange bolt-hole net section develops cracks due to the material's tensile limit being exceeded, culminating in failure. Thus the final failure modes of the PRBS and PRBS-R specimens were the fracture of the flange bolt-hole net section of the dog-bone-style splice section and the plastic hinge occurring centrally in the dog-bone-style splice section. In addition, the test revealed that the energy dissipations of both the PRBS and PRBS specimens were mainly composed of the energy dissipations of the bolt slippage and plastic deformation of the dog-bone-style splice section.

4. Comparative Analysis of Results

4.1. Hysteresis Loop

The load-displacement hysteresis curves of the four specimens illustrated in Figure 8 indicate that the energy dissipation capacity of the remaining specimens was significantly better than that of the BASE specimens. In addition, for the PRBS and PRBS-R specimens, the plastic deformation of the dog-bone-style splice section and slippage of the bolt connection jointly dissipated energy, resulting in a higher rotational capacity and ductility of the novel connection joint.

A comparison of the hysteretic curves of the BASE and RBS specimens reveals their load-displacement curves were similar before the failure of the BASE specimen. However, because of the dog-bone-weakening structure of the flange, the bearing capacity of the RBS specimen was less than that of the BASE specimen, whereas the rotational capacity of the RBS specimen was better than that of the BASE specimen. The formation of a plastic hinge in the weakened dog-bone area could be observed in combination with the failure mode, which indicated that the purpose of plastic-hinge relocation of the RBS specimen was achieved. The plastic hinges formed in the weakened dog-bone area could prevent premature brittle fracture in the weld zone near the column under an earthquake load.

A comparison of the load-displacement curves of the RBS and PRBS specimens reveals that the hysteresis curve of the RBS specimen was fully spindle-shaped, and the energy dissipation was mainly through the plastic deformation of the plate. The load-displacement curve of the PRBS specimen showed a slender spindle shape in the early stage of loading and a full "Z" shape in the later stage of loading, indicating that the energy dissipation capacity in the later stage was better than that in the early stage. The main reason was that in the early stage of the test, the specimen energy dissipation was mainly achieved through the plastic deformation of the plate, but after the slippage of the bolt connection, the plastic deformation of the plate and the slippage of the bolt connection jointly dissipated energy. The load–displacement curve shows that the ductility of the PRBS specimen was better than that of the RBS specimen owing to the existence of the bolt connection slippage.

The shapes of the load-displacement curves of the PRBS and PRBS-R specimens were similar; however, the bearing capacity of the PRBS-R specimen was slightly lower than that of the PRBS specimen, mainly because the contact surface of the splicing area was worn with the application of a reciprocating load in the first test. The test results indicated that the repaired PRBS connection joint exhibited good seismic performance. Furthermore, the failure of the PRBS and PRBSR specimens was a net section fracture of the bolt connection. Both the plastic deformation of the plate and slippage of the bolt connection were fully developed before failure, implying that the damage mode of the PRBS and PRBSR specimens was ductile failure.

4.2. Skeleton Curve

The skeleton curves of the four specimens are shown in Figure 9. Before the failure of the BASE specimen, its skeleton curve was consistent with that of the PRBS. In addition, the RBS specimen showed good rotation ability compared with the BASE specimen, and owing to the dog-bone weakening of the flange, the bearing capacity of the RBS specimen was slightly lower than that of the BASE specimen.

Compared with the PRBS specimen, the skeleton curves of the PRBS and PRBS-R specimens exhibited the same trend, and both exhibited slip characteristics. However, owing to plastic damage in the first test, the bearing capacity of the PRBS-R specimen was lower than that of the PRBS specimen. Unlike the BASE and RBS specimens, after the bolt slip was complete, the bolt and hole wall began to contact and squeeze each other, causing the bearing capacities of the PRBS and PRBS-R specimens to continue increasing.

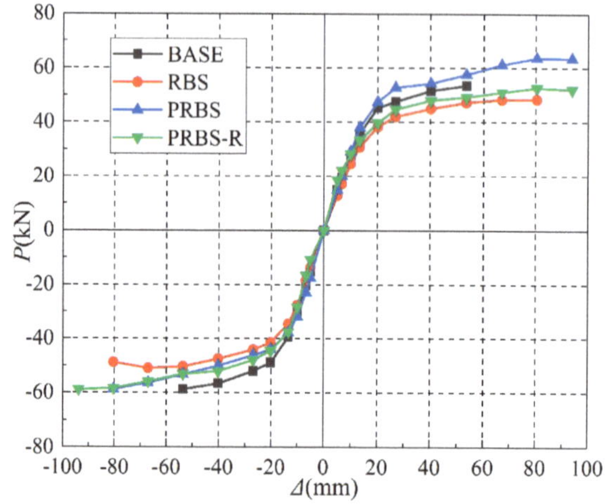

Figure 9. Skeleton curves of specimens.

4.3. Stiffness Degradation

During the test process, cyclic loading led to the accumulation of structural damage, and the structural stiffness decreased with an increase in the loading level. The secant stiffness of the first cycle of each loading stage was used to represent the structural stiffness at that stage [37]. The structural stiffness degradation curves of the specimens are presented in Figure 10.

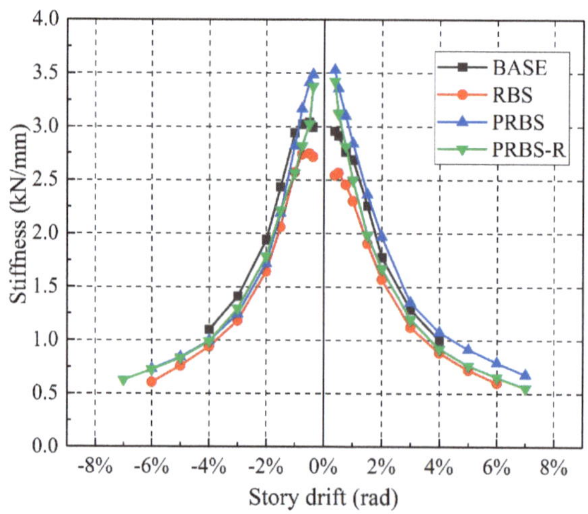

Figure 10. Stiffness degradation curves of specimens.

As shown in Figure 10, the stiffness of all the specimens consistently maintained a decreasing trend throughout the test. With an increase in the load stage, the specimens entered the hardening stage, and the stiffness degradation began to slow until the failure of the specimens. The main causes of stiffness degradation were the slippage of the bolt connection, buckling and plastic development of the beam, and accumulated damage to the specimens.

For the PRBS and PRBS-R specimens, the cantilever short beam was widened while designing the specimens to ensure that damage occurred in the dog-bone-style splice section. Therefore, the initial stiffness values of the PRBS and PRBS-R specimens were significantly higher than those of the BASE and RBS specimens. In addition, owing to the dog-bone weakening of the flange, the initial stiffness of the RBS specimen was lower than that of the BASE specimen.

4.4. Strain Distribution

During the test process, the strain of the panel zone did not exceed the steel yield strain; therefore, only the beam strain distribution is analyzed in this section.

Figures 11–13 show the strain distribution curves of the BASE, RBS, and PRBS specimens. The flange and web of the BASE specimens were in the elastic phase during the early stages of the test. With an increase in the load, the beam section near the column surface yielded first, implying that a plastic hinge was formed. The plastic deformation of the BASE specimen gradually accumulated until the beam section near the column surface cracked. For the RBS specimen, the first yielding of the flange occurred in the weakened region, indicating that the weakened beam flange could achieve plastic hinge relocation. However, the flange and web near the column surface yielded before the failure of the RBS specimen, indicating that damage did not occur only in the weakened region.

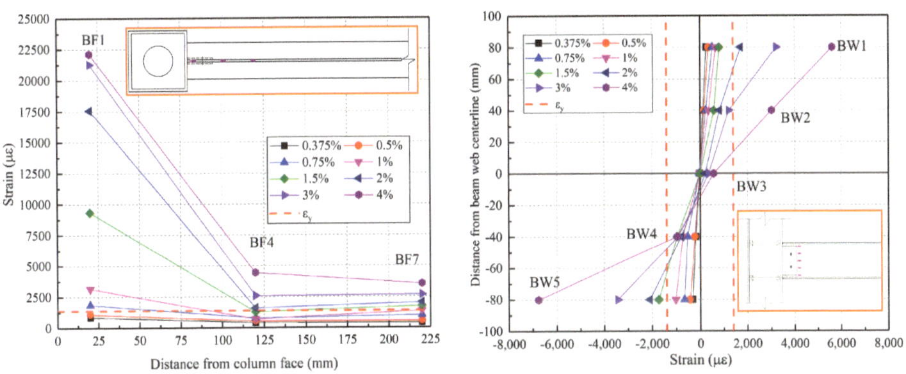

(**a**) Flange strain distribution (**b**) Web strain distribution

Figure 11. Strain distribution of BASE specimen.

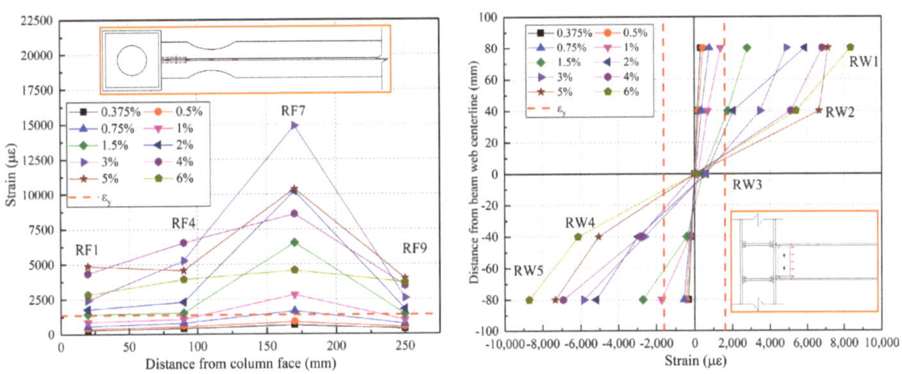

(**a**) Flange strain distribution (**b**) Web strain distribution

Figure 12. Strain distribution of RBS specimen.

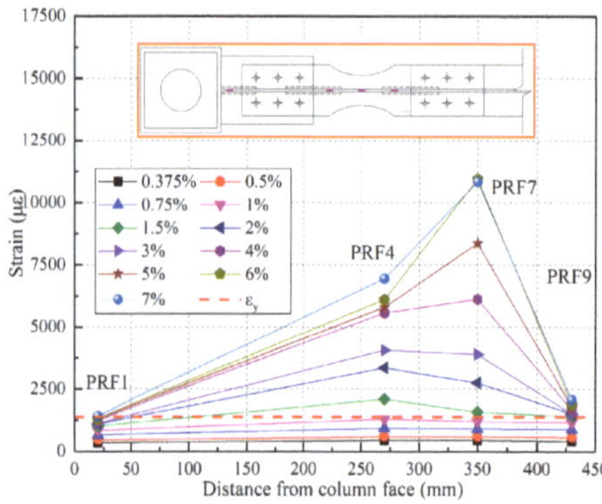

(**a**) Flange strain distribution

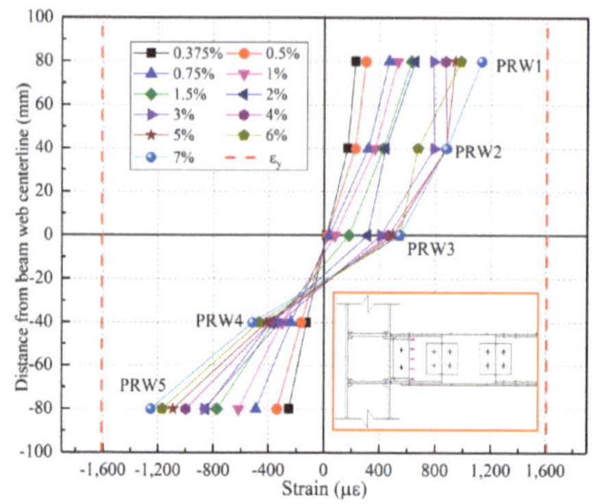

(**b**) Web strain distribution

Figure 13. Strain distribution of PRBS specimen.

In addition, for the PRBS specimen, the cantilever beam did not reach the yield strain and remained elastic during loading. In the middle and late stages of the test, the strain in the dog-bone-style splice section reached the yield strain, indicating the formation of a plastic hinge. In addition, after the test, the damage was primarily concentrated in the dog-bone-style splice section, and the cantilever beam and H-beam did not exhibit plastic damage during loading, which provided conditions for rapid repair after the connection was destroyed.

4.5. Seismic Performance Index

The skeleton curves of all the specimens in the test were processed and analyzed, and the seismic performance indices of each specimen were obtained, including the yield displacement (Δ_y), yield load (P_y), peak load (P_m), ultimate displacement (Δ_u), and displacement ductility (μ). To eliminate the influence of asymmetric loading on the calculation results, the average values of the performance index parameters were considered for the analysis. The calculation results are listed in Table 2. The yield loads and displacements of the specimens were calculated using the general yield-bending moment method [38].

Table 2. Seismic performance index of specimens.

Specimen	Δ_y/mm	P_y/kN	P_m/kN	Δ_u/mm	μ
BASE	23.32	48.44	56.12	53.60	2.30
RBS	24.39	42.00	49.76	80.40	3.30
PRBS	26.11	49.14	61.23	87.10	3.33
PRBS-R	27.91	45.66	55.82	93.80	3.53

As shown in Table 2, the peak load of the PRBS specimen was the largest, indicating that the PRBS connection had a higher bearing capacity. In addition, even though the peak load of the PRBS-R specimen decreased by approximately 7.1% after repair, it was still greater than that of the RBS specimen, indicating that the repaired PRBS-R specimen retained a significant bearing capacity. Except for the BASE specimen, the displacement ductility exceeded 3.0, indicating that the novel connection had good ductility [39].

4.6. Energy Dissipation Capacity

Figure 14 shows the total energy dissipations of the specimens. The diagram shows that the BASE specimen had the lowest energy dissipation, mainly because premature damage to the BASE specimen occurred to the extent that its beam plasticity was not fully developed. In addition, the total energy dissipation of the RBS specimen was higher than that of the BASE specimen, which was due to the local buckling in the bone-weakening area of the RBS specimen before failure, and the well-developed plasticity. Because of the local buckling and bolt slippage of the splicing section, the plasticity of the novel connection was fully developed, and the total energy dissipation was the highest. Compared with the BASE specimen, the total energy dissipations of RBS and PRBS specimens increased by 146.74% and 194.23%, respectively. Moreover, compared to the PRBS specimen, the energy dissipation of the RPRBS specimen was only reduced by 0.84%, which was caused by the initial imperfection of the repaired specimen.

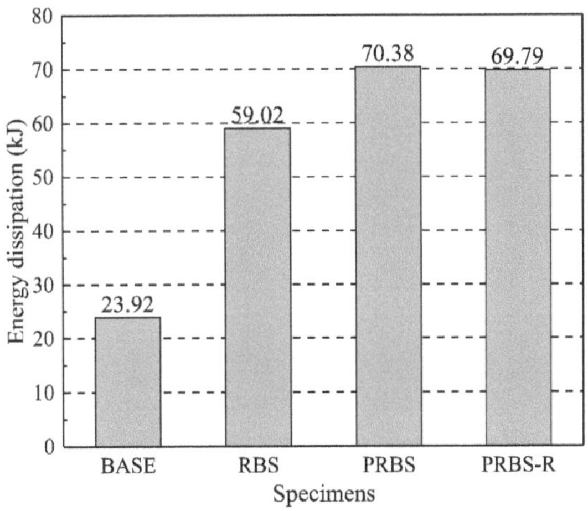

Figure 14. Energy dissipation of specimens.

5. Finite Element Analysis

5.1. Finite Element Model

In this study, the FE software ABAQUS 6.14 was used to model the BASE, RBS, and PRBS specimens. The model sizes were the same as those of the actual specimens. In the FE models, an eight-node solid element (C3D8R) was used to model the concrete, steel plate, and bolt [33,40]. To ensure that the force states of the models were consistent with those of the specimens, the model boundary conditions were consistent with those of the test, that is the hinge connection in the column top and bottom, lateral displacement limitation in the beam, displacement load applied to the beam end, axial load (300 kN) applied to the column top, and bolt load (100 kN for M16 bolts, 155 kN for M20 bolts) applied to the bolts. According to the literature [33], the mesh size of the beam was 15 mm after many trials and calculations, whereas that of the dog-bone-weakened area of the RBS and dog-bone-style splice section of the PRBS was 10 mm. The precast column was meshed to a size of 30 mm to improve computational efficiency. In addition, the multiple linear kinematic hardening mode was used to simulate the mechanical properties of steel. The strain-stress relationship of steel is shown in Figure 15a, where the values of f_y, f_u, E_s and ε_u were taken from Table 1. Furthermore, the concrete damage plasticity (CDP) mode was used to simulate the mechanical properties of concrete, as shown in Figure 15b. In the figure, f_{ck} is the standard value of axial compressive strength; f_{tk} is the standard value of axial tensile strength; ε_{ck} is the peak compressive strain of concrete and ε_{tk} is the peak tensile strain of concrete. The values of parameters and repeated unloading and reloading stress path of concrete were calculated by the method recommended in the literature [33].

The contact between concrete and square still tube, steel plate and steel plate, and bolt and steel plate were simulated using "surface-to-surface contact" provided by ABAQUS, which includes normal and tangential contact. The normal and tangential contact were simulated using "hard contact" and the Coulomb friction model, respectively. The friction coefficient between steel and concrete was taken as 0.25 [33], and the friction coefficient between steels was taken as 0.31, which was obtained from an anti-slip test. Nonlinear geometric deformations were considered in the FE model. The mesh and boundary conditions of the model are shown in Figure 16.

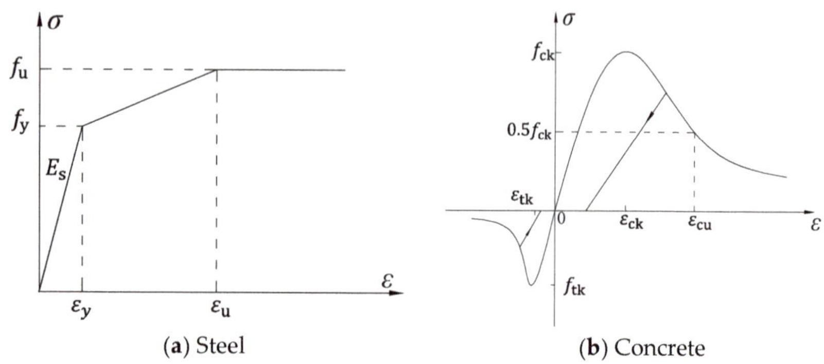

Figure 15. The strain-stress relationship of steel and concrete.

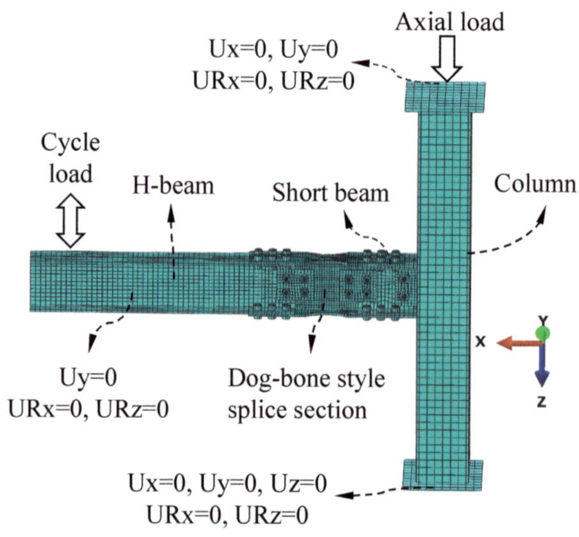

Figure 16. FE model of PRBS specimen.

5.2. Model Verification

A comparison of the deformation and plastic strain distribution between the FE model and the experimental test is shown in Figure 17. The figures unequivocally demonstrate that the FE models formulated in this study accurately reproduced the mechanical behavior and failure mode of the specimens. In the BASE model, plastic deformation and local buckling were initially manifested in the beam flange adjacent to the column. Under progressive loading, the plastic deformation escalated and extended outward, culminating in the formation of a plastic hinge at this location. Consequently, the beam flange near the column fractured, resulting in brittle failure.

In the RBS model, plastic deformation and local buckling initially occurred in the weakened flange. With the continued loading, the plastic region gradually extended towards the beam flange adjacent to the column, with the plastic hinge mainly concentrated in the weakened area. Ultimately, failure occurred in the weakened region of the flange, effectively achieving the objective of plastic hinge relocation and postponing structural failure.

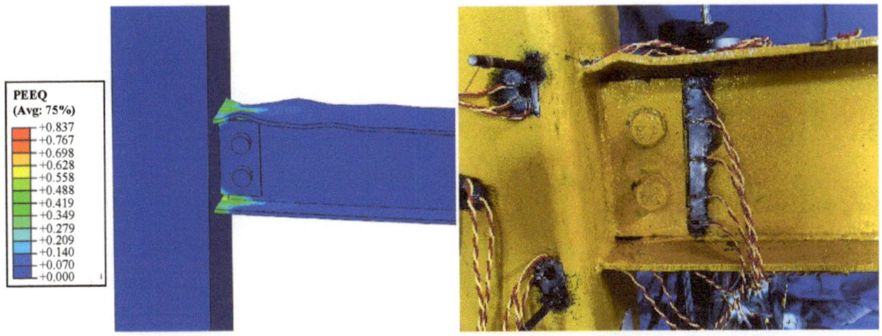

(**a**) BASE specimen

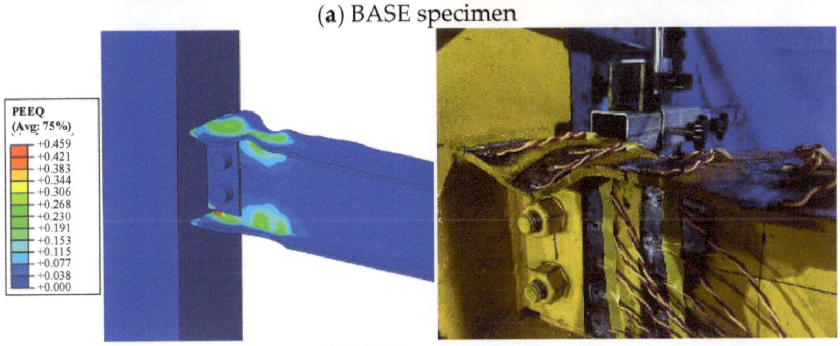

(**b**) RBS specimen

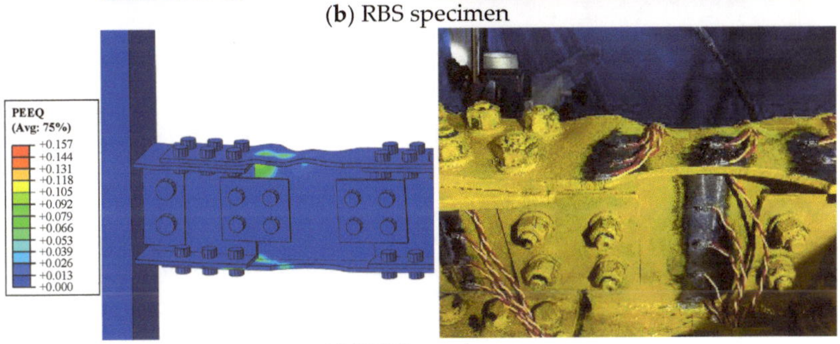

(**c**) PRBS specimen

Figure 17. Comparison of the failure modes of the test and FE.

In sharp contrast to the BASE and RBS models, the PRBS model initially exhibited plastic deformation and local buckling in the splice section. As the loading progressed, plastic deformation was predominantly concentrated in the splice section, which was consistent with the strain measurement data. Furthermore, the graph shows that the PRBS model, apart from the dog-bone-style splice section, remains in the elastic stage. Consequently, the swift restoration of the PRBS structure can be accomplished by replacing the dog-bone-style splice section. The PRBS connection effectively achieves the objectives of plastic hinge relocation, postponing structural failure, and post-earthquake rapid repair.

Hysteresis loops comparing the finite FE model with the test results are shown in Figure 18. The figures exhibit a notable agreement between the hysteresis loops of the FE models and test results, confirming the accurate prediction of the stiffness and bearing capacity of the specimens using the FE model. These comparison results indicate the suitability of the FE model established in this study for further investigation of this specific type of joint.

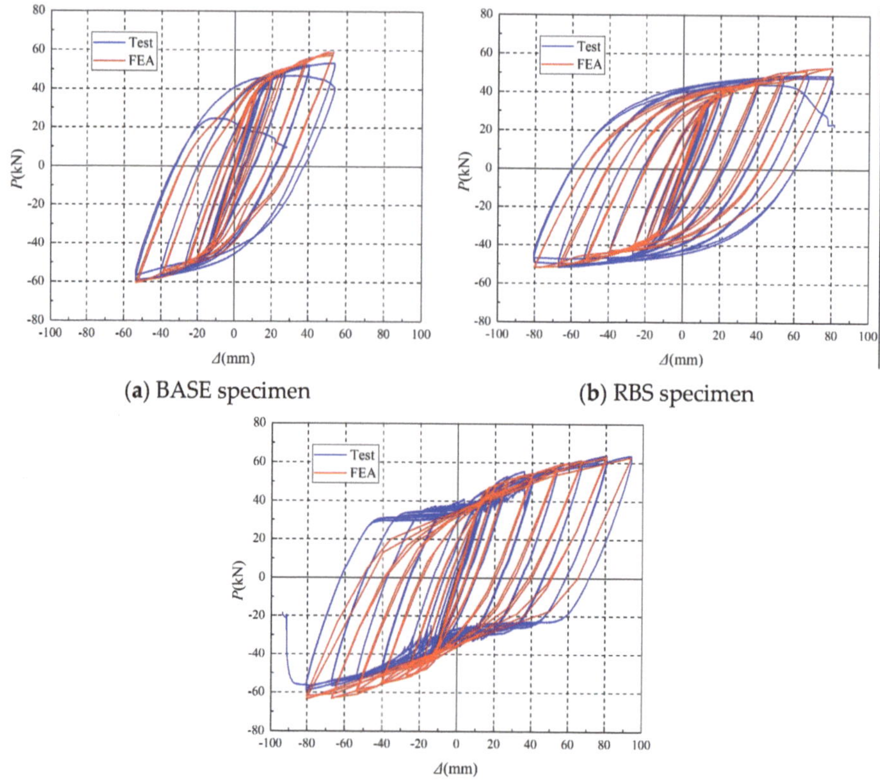

(**a**) BASE specimen (**b**) RBS specimen

(**c**) PRBS specimen

Figure 18. Comparison of the hysteresis loops of the test and FE.

5.3. Parametric Analysis

To investigate the influence of the dog-bone-style flange-weakening width, splice-section flange thickness, and flange bolt diameter on the load-bearing capacity of the PRBS connection, 15 finite element models were developed for parametric analysis using the ABAQUS software. All models were created using the methodology described in Section 5.1. To expedite computational efficiency, monotonic loading was applied to all the models. The model parameters are listed in Table 3 and are named as "P-D-t_f-d_b", where D represents the weakening width, t_f denotes the flange thickness, d_b denotes the flange bolt diameter, d_h denotes the bolt hole diameter and its value for various bolt types is dictated by the relevant code [31], as depicted in Figure 19. The skeleton curves and performance indices of the models are shown in Figure 20 and Table 4, respectively.

Figure 20a and Table 4 show that with a gradual increase in the weakening width, both the ultimate and yield loads of the models decrease progressively. As the weakening width increased from 0 to 35 mm, the ultimate load and yield load experienced a reduction of 9.09% and 20.46%, respectively. The decrease in yield load was mainly attributed to an increase in the weakening depth, which modified the model's yield mode. This alteration shifted the location of yield from the bolt splice connection to the bone-weakening section, resulting in an earlier initiation of the yield phase and a subsequent reduction in the model's yield load. Furthermore, as the increase in the weakening depth did not alter the model's ultimate failure mode, its impact on the model's ultimate load was relatively minor.

Figure 20b and Table 4 show that with a gradual increase in the flange thickness, both the ultimate and yield loads of the models progressively increase. Compared with the weakening width, the influence of the flange thickness on the load-bearing capacity of the model was more pronounced. As the flange thickness increased from 4 mm to 12 mm, the ultimate load experienced a significant increase of 114.33%. This substantial increase was mainly attributed to the enhancement of the overall flexural stiffness of the weakened section owing to the increased flange thickness. Consequently, the load-bearing capacity at the weakest interface improved. Under the effect of beam-end loading, the plastic deformation at the weakening section diminished with an increase in the flange thickness, leading to an increase in the load-bearing capacity of the model.

Figure 20c and Table 4 show that with an increase in the bolt diameter, both the ultimate and yield loads of the models first increased and then decreased. The maximum load-bearing capacity was achieved when the flange bolt diameter was 14 mm. The primary reason for this behavior was the change in the failure mode of the model with varying bolt diameters. When the flange bolt diameters were 12 and 14 mm, the failure mode of the model was compression failure of the flange bolt hole wall. Consequently, the load-bearing capacity increased with the bolt diameter. However, when the flange bolt diameter exceeded 14 mm, the failure mode of the model was tension damage to the splicing section flange at the bolt. This resulted in a decrease in load-bearing capacity with an increase in the bolt diameter.

Table 3. Characters of FE models.

Model	Weakening Width D/mm	Flange Thickness t_f/mm	Flange Bolt Diameter d_b/mm	Bolt Hole Diameter d_h/mm
P25-8-16	25	8	16	17.5
P0-8-16	0	8	16	17.5
P5-8-16	5	8	16	17.5
P10-8-16	10	8	16	17.5
P15-8-16	15	8	16	17.5
P20-8-16	20	8	16	17.5
P30-8-16	30	8	16	17.5
P35-8-16	35	8	16	17.5
P25-4-16	25	4	16	17.5
P25-6-16	25	6	16	17.5
P25-10-16	25	10	16	17.5
P25-12-16	25	12	16	17.5
P25-8-12	25	8	12	13.5
P25-8-14	25	8	14	15.5
P25-8-18	25	8	18	20.0
P25-8-20	25	8	20	22.0

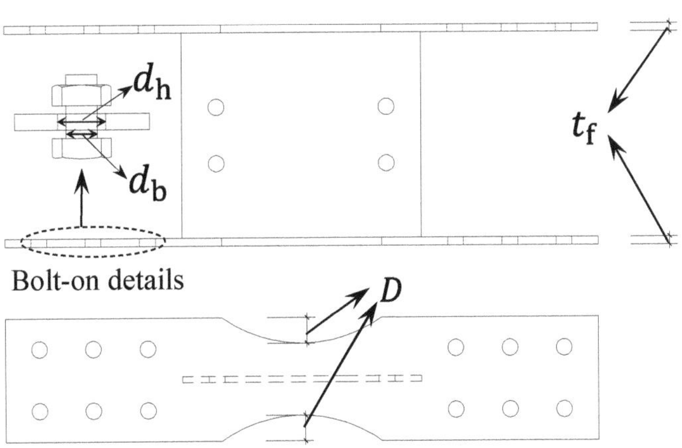

Figure 19. Research parameters of the splice section.

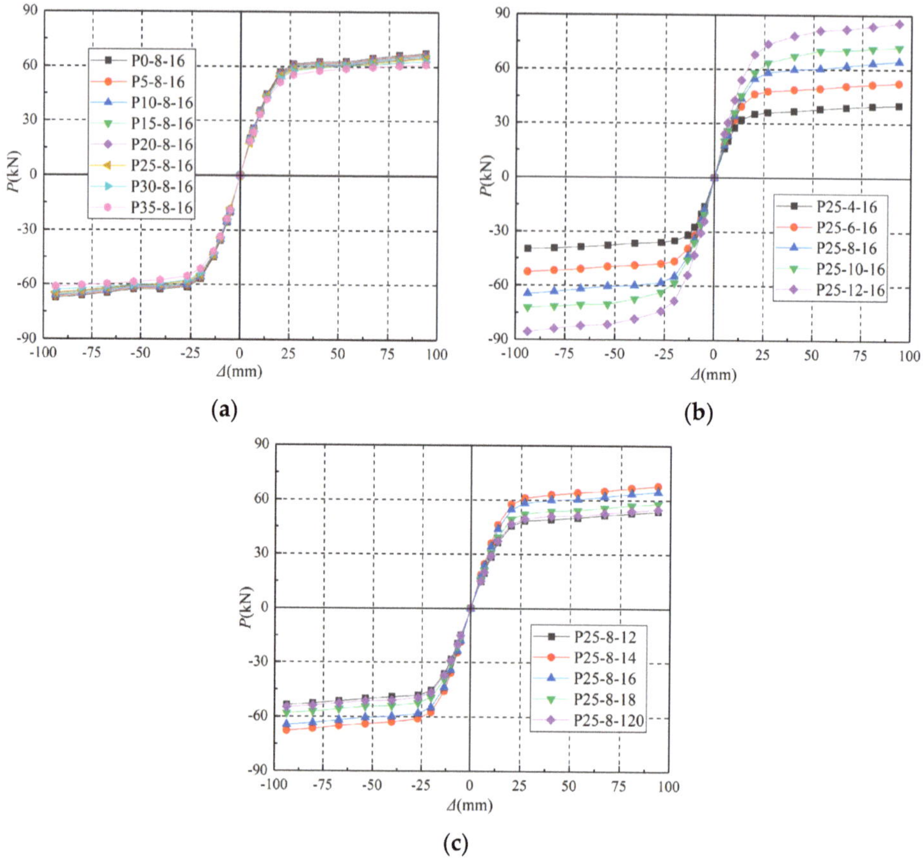

Figure 20. Skeleton curves of FE models. (**a**) Skeleton curves of different weakening widths. (**b**) Skeleton curves of different flange thicknesses. (**c**) Skeleton curves of different bolt diameters.

Table 4. Comparison between FE and theoretical analyses.

FE Model	FE Value		Calculated Value		Error	
	P_{ye}/kN	P_{ue}/kN	P_{yc}/kN	P_{uc}/kN	δy	δu
P25-8-16	51.06	64.53	47.22	64.08	−7.53%	−0.70%
P0-8-16	52.35	67.35	52.92	64.08	1.08%	−4.85%
P5-8-16	52.01	66.69	52.92	64.08	1.74%	−3.92%
P10-8-16	51.96	66.04	52.92	64.08	1.84%	−2.97%
P15-8-16	51.45	65.39	52.92	64.08	2.86%	−2.00%
P20-8-16	50.12	64.76	51.95	64.08	3.65%	−1.06%
P30-8-16	46.76	62.98	42.48	64.08	−9.15%	1.75%
P35-8-16	41.64	61.22	37.75	64.08	−9.35%	4.67%
P25-4-16	30.53	39.92	27.88	36.90	−8.69%	−7.56%
P25-6-16	40.96	52.56	38.69	50.36	−5.53%	−4.19%
P25-10-16	55.77	72.25	55.90	78.06	0.23%	8.04%
P25-12-16	59.12	85.56	64.74	92.30	9.52%	7.87%
P25-8-12	45.36	53.56	43.58	58.27	−3.93%	8.79%
P25-8-14	51.65	67.76	47.22	66.90	−8.58%	−1.27%
P25-8-18	45.35	58.08	47.22	61.06	4.12%	5.15%
P25-8-120	42.62	54.85	47.22	57.29	10.79%	4.46%

6. Bearing Capacity for the PRBS Connection

This section considers the PRBS specimen as the research object and analyzes the possible yielding and damage modes of the PRBS connection based on the test measurement data and the observed damage phenomena; thus, a simplified calculation method for the yielding bearing and ultimate bearing capacities of the PRBS connection is presented. In addition, the calculated results were compared with the experimental results. A schematic of the calculation is presented in Figure 21.

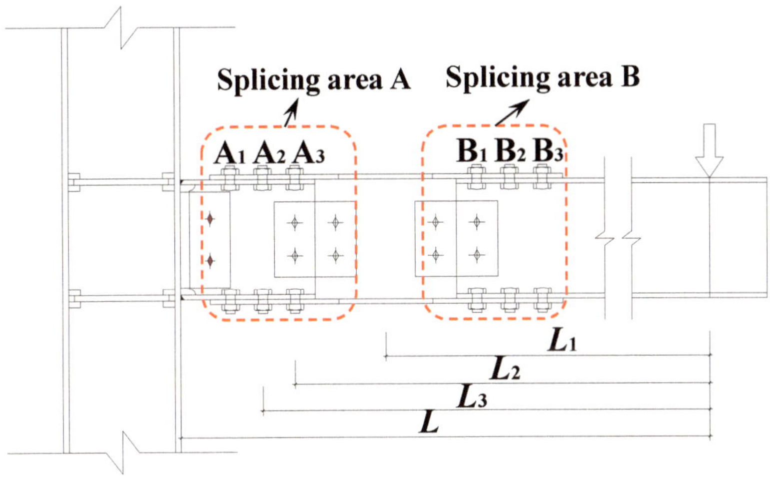

Figure 21. Calculation sketch for the bearing capacity. In the figure, A and B denote the splicing area, A_1–A_3, B_1–B_3 denote the flange bolt numbers, L and L_1–L_3 denote the distance between the loading point and the locations in the figure.Calculation of the yield-bearing capacity.

When the specimens were designed before the test, to avoid the risk of beam-column welding failure, the cantilevered short beam flange was widened to ensure that the probable maximum moment at the face of the column section (M_f) was less than the plastic moment of the beam based on the expected yield stress (M_{pe}) [32,33], such that the yielding and damage of the beam section at the face of the column occurred later than that of the weakened dog-bone section. According to the force mechanism, the plasticity of the PRBS connection joint may be caused by several scenarios: (1) yielding of the dog-bone-weakened section (P_{y1}), (2) yielding of the splice section flange at the A3 row of bolts (P_{y2}), (3) yielding of the flange bolt-hole wall by compression at splice area A (P_{y3}), and (4) yielding of the flange bolt screw by shear at splice area A (P_{y4}). The load at the time of possible yielding is calculated separately, and the minimum value is taken as the yield-bearing capacity (P_y). The formula is as follows:

$$P_{y1} = W_e f_y / L_1, \tag{5}$$

$$P_{y2} = [b_f t_f f_y (h - t_f) + n_1 \alpha_1 P \mu (h - 2t_f)/2] / L_2, \tag{6}$$

$$P_{y3} = \left[d t_f f_y (h - t_f) + \alpha_1 P \mu (h - 2t_f)\right] n / L_3, \tag{7}$$

$$P_{y4} = \left[A_{eff} f_v^b (h - t_f) + \alpha_1 P \mu (h - 2t_f)\right] n / L_3 \tag{8}$$

$$P_y = min\{P_{y1}, P_{y2}, P_{y3}, P_{y4}\} \tag{9}$$

where W_e is the elastic section modulus of the dog-bone-reduced section; f_y is the yield strength of the steel plate; b_f is the flange width of the splicing section in Area A; t_f is the flange thickness of the splicing section in Area A; h is the height of the dog-bone-style splice section; n_1 is the number of single-side flange bolts at the calculated section; α_1 is the reduction coefficient of the bolt friction, which is 0.7 [41]; P is the bolt preload; μ is the friction coefficient; d is the diameter of the screw hole; n is the number of bolts distributed on the upper or lower flange in Area A; A_{eff} is the effective cross-sectional area at the thread; f_y^b is the yield strength of bolt and f_v^b is the shear strength of the bolt, where $f_v^b = 0.58 f_y^b$ [42].

1. Calculation of the Ultimate Bearing Capacity

With an increase in the load, the failure of the PRBS connection joint may be caused by several scenarios: (1) plastic failure of the dog-bone-weakened section (P_{u1}), (2) tension damage of the splicing section flange at the A3 row bolt (P_{u2}), (3) compression failure of the flange-bolt hole wall in splicing area A (P_{u3}), and (4) shear failure of the flange bolt screw at splicing area A (P_{u4}). The load at the time of possible failure was calculated separately, and the minimum value was considered the ultimate bearing capacity (P_u). The formula is as follows:

$$P_{u1} = \frac{W_u f_u}{L_1} \tag{10}$$

$$e_f = I/I_f \tag{11}$$

$$P_{u2} = [(b_f - 2d)t_f f_u(h - t_f)]e_f/L_2 \tag{12}$$

$$P_{u3} = [dt_f f_u(h - t_f)]ne_f/L_3 \tag{13}$$

$$P_{u4} = \left[0.58 A_{eff} f_u^b (h - t_f)\right] ne_f/L_3 \tag{14}$$

$$P_u = min\{P_{u1}, P_{u2}, P_{u3}, P_{u4}\} \tag{15}$$

where W_u is the plastic section modulus of the dog-bone-reduced section, f_u is the ultimate strength of the steel plate, I is the moment of inertia of the calculated section, I_f is the moment of inertia of the calculated section flange, and f_u^b is the ultimate strength of the bolt.

The theoretical analysis results were compared with the experimental and FE analysis results, as shown in Tables 4 and 5 and Figure 22. Table 5 and Figure 22 show that the experimental and theoretical results for the PRBS specimen exhibit a satisfactory agreement, with an error of less than 5%. However, for the PRBS-R specimen, the yield-bearing capacity error is 3.40%, whereas the ultimate bearing capacity error is 14.8%. The primary reason for this discrepancy is that the PRBS-R specimen was a repair specimen with initial imperfections prior to testing, resulting in a noticeably lower ultimate load than the PRBS specimen. An analysis of Table 4 and Figure 22 reveals that the errors in yield-bearing and ultimate bearing capacities between the FE and theoretical analyses were predominantly below 10%. Consequently, these results suggest that the theoretical analysis aligns well with both the experimental and FE analyses.

Table 5. Comparison between experimental and theoretical analyses.

Specimen	Experimental Value		Calculated Value		Error	
	P_{yt}/kN	P_{ut}/kN	P_{yc}/kN	P_{uc}/kN	δy	δu
PRBS	49.14	61.23	47.22	64.08	−3.91%	4.65%
PRBS-R	45.66	55.82	47.22	64.08	3.40%	14.80%

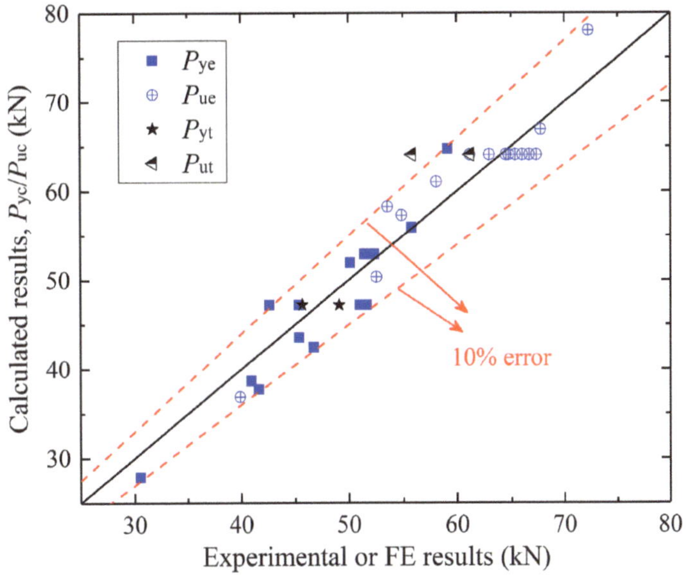

Figure 22. Comparison of theoretical analysis with experimental and FE analyses.

7. Conclusions

In this study, a precast reduced beam section connection joint with a dog-bone-style splice section was proposed, and quasi-static tests and FE analyses were conducted on welded flange-bolted web connection joints, traditional bone-weakened joints, PRBS connection joints, and repaired PRBS connection joints. The following conclusions are drawn from the comparative analysis:

(1) Compared with the BASE and RBS connection joints, the PRBS connection had a higher bearing capacity, ductility, and energy dissipation capacity. The energy dissipation of the PRBS connection was realized by bolt slippage and plastic deformation of the dog-bone-style splice section. Moreover, the novel connection could effectively relocate the plastic hinge and cause plastic damage concentrated in the dog-bone-style splice section.
(2) After the test, the main components of the PRBS connection were undamaged, resulting in a rapid repair of the joint. In addition, the repaired PRBS still exhibited stable mechanical properties, which could ensure the safe service of the overall structure.
(3) The FE model formulated in this study demonstrated excellent accuracy in predicting the mechanical behavior, failure mode, stiffness, and bearing capacity of the specimens. Furthermore, the results of the parametric analysis revealed that the flange thickness exerted the most significant influence on the load-bearing capacity of the model. The load-bearing capacity of the model was negatively correlated with the weakening width but positively correlated with the flange thickness. Moreover, as the flange bolt diameter increased, the load-bearing capacity of the model initially increased and then decreased.
(4) Based on the experimental data and observed failure phenomena, the possible yield and damage modes of the PRBS connection were analyzed, and a simplified calculation method for the yield and ultimate bearing capacities of the PRBS connection was proposed. The calculation results were in good agreement with the experimental results.

Author Contributions: Conceptualization, S.H.; Software, H.Q. and Y.W.; Investigation, H.Q.; Resources, C.S. and P.W.; Data curation, C.S.; Writing—original draft, H.Q.; Writing—review & editing, X.X., W.L. and Y.W.; Visualization, H.Q., P.W. and W.L.; Supervision, S.H.; Funding acquisition, S.H. All authors have read and agreed to the published version of the manuscript.

Funding: This research was funded by Chongqing Technology Innovation and Application Development Project, China (cstc2019jscx-gksbX0013), the Innovation Group Science Foundation of the Natural Science Foundation of Chongqing, China (cstc2020jcyj-cxttX0003), the National Natural Science Foundation of China (52130901) and the National Natural Science Foundation of China (51739008).

Data Availability Statement: Experimental data and numerical models in the study are available from the corresponding author upon reasonable request.

Conflicts of Interest: The authors declare no conflict of interest.

References

1. Tremblay, R.; Timler, P.; Bruneau, M.; Filiatrault, A. Performance of steel structures during the 1994 Northridge earthquake. *Can. J. Civ. Eng.* **1995**, *22*, 338–360. [CrossRef]
2. Xu, Q.; Chen, H.R.; Li, W.Y.; Zheng, S.S.; Zhang, X.H. Experimental investigation on seismic behavior of steel welded connections considering the influence of structural forms. *Eng. Fail. Anal.* **2022**, *139*, 106499. [CrossRef]
3. Fanaie, N.; Nadalipour, Z.; Sarkhosh, O.S.; Faegh, S.S. Elastic drift amplification factor in steel moment frames with double reduced beam section (DRBS) connections. *J. Build. Eng.* **2021**, *43*, 102563. [CrossRef]
4. Horton, T.A.; Hajirasouliha, I.; Davison, B.; Ozdemir, Z. More efficient design of reduced beam sections (RBS) for maximum seismic performance. *J. Constr. Steel Res.* **2021**, *183*, 106728. [CrossRef]
5. Mansouri, A.; Shakiba, M.R.; Fereshtehpour, E. Two novel corrugated web reduced beam section connections for steel moment frames. *J. Build. Eng.* **2021**, *43*, 103187. [CrossRef]
6. Deylami, A.; Tabar, A.M. Promotion of cyclic behavior of reduced beam section connection restraining beam web to local buckling. *Thin Walled Struct.* **2013**, *73*, 112–120. [CrossRef]
7. Chen, S.J.; Yeh, C.H.; Chu, J.M. Ductile steel beam-to-column connections for seismic resistance. *J. Struct. Eng.* **1996**, *122*, 1292–1299. [CrossRef]
8. Lee, C.H.; Jeon, S.W.; Kim, J.H.; Uang, C.M. Effects of panel zone strength and beam web connection method on seismic performance of reduced beam section steel moment connections. *J. Struct. Eng.* **2005**, *131*, 1854–1865. [CrossRef]
9. Paul, S.; Deb, S.K. Experimental study on a new V-cut RBS and CFT connections with bidirectional bolts under cyclic loadings. *J. Build. Eng.* **2022**, *46*, 103688. [CrossRef]
10. Tabar, A.M.; Alonso-Rodriguez, A.; Tsavdaridis, K.D. Building retrofit with reduced web (RWS) and beam (RBS) section limited-ductility connections. *J. Constr. Steel Res.* **2022**, *197*, 107459. [CrossRef]
11. Xu, Y.; Dai, Y.; Jia, Y.L. Study on the mechanical performances of aluminum assembled hub joints under axial loads. *J. Build. Eng.* **2022**, *50*, 104147. [CrossRef]
12. Zhang, R.J.; Zhang, Y.; Li, A.Q.; Yang, T.Y. Experimental study on a new type of precast beam-column joint. *J. Build. Eng.* **2022**, *51*, 104252. [CrossRef]
13. Li, S.; Xu, T.T.; Ding, S.H.; Li, X.L.; Guo, H.C.; Liu, Y.H. Seismic performance of a new assembled beam-column joint with cantilever beam splicing. *J. Constr. Steel Res.* **2022**, *196*, 107432. [CrossRef]
14. Oh, S.H.; Kim, Y.J.; Ryu, H.S. Seismic performance of steel structures with slit dampers. *Eng. Struct.* **2009**, *31*, 1997–2008. [CrossRef]
15. Soltanabadi, R.; Farhad, B. Experimental studies on a combined damper for repairable steel moment connections. *Int. J. Steel Struct.* **2018**, *18*, 211–224. [CrossRef]
16. Huang, L.; Sun, Q.; Deng, P.; Zhang, M.L.; Shan, R.; Peng, B. Study on seismic behavior of earthquake-damaged frame structures with rapid reversibility. *Build. Struct.* **2021**, *51*, 81–87. (In Chinese)
17. Ma, Z.H.; Zhang, J.G.; Liang, H.Z.; Li, L.Y. Numerical research on prefabricated frame joint based on artificial dissipative plastic hinge under low-reversed loading. *China Civ. Eng. J.* **2020**, *53* (Suppl. S2), 162–168. (In Chinese)
18. Ye, J.F.; Yan, G.Y.; Liu, R.Y.; Xue, P.R.; Wang, D. Hysteretic behavior of replaceable steel plate damper for prefabricated joint with earthquake resilience. *J. Build. Eng.* **2022**, *46*, 103714. [CrossRef]
19. Zhang, A.L.; Wang, Q.; Jiang, Z.Q.; Yang, X.F.; Zhang, H. Experimental study of earthquake-resilient prefabricated steel beam-column joints with different connection forms. *Eng. Struct.* **2019**, *187*, 299–313. [CrossRef]
20. Jiang, Z.Q.; Yang, X.F.; Dou, C.; Pi, Y.L.; Li, R. Design theory of earthquake-resilient prefabricated beam-column steel joint with double flange cover plates. *Eng. Struct.* **2020**, *209*, 110005. [CrossRef]
21. Zhang, H.; Su, M.Z.; Lian, M.; Cheng, Q.Q.; Guan, B.L.; Gong, H.X. Experimental and numerical study on the seismic behavior of high-strength steel framed-tube structures with end-plate-connected replaceable shear links. *Eng. Struct.* **2020**, *223*, 111172. [CrossRef]

22. Özkılıç, Y.O.; Bozkurt, M.B. Numerical validation on novel replaceable reduced beam section connections for moment-resisting frames. *Structures* **2023**, *50*, 63–79. [CrossRef]
23. Özkılıç, Y.O. A new replaceable fuse for moment resisting frames: Replaceable bolted reduced beam section connections. *Steel Compos. Struct.* **2020**, *35*, 353–370.
24. Özkılıç, Y.O.; Bozkurt, M.B.; Topkaya, C. Mid-spliced end-plated replaceable links for eccentrically braced frames. *Eng. Struct.* **2021**, *237*, 112225. [CrossRef]
25. Özkılıç, Y.O. Optimized stiffener detailing for shear links in eccentrically braced frames. *Steel Compos. Struct.* **2021**, *39*, 35–50.
26. Özkılıç, Y.O. Interaction of flange and web slenderness, overstrength factor and proposed stiffener arrangements for long links. *J. Constr. Steel Res.* **2022**, *190*, 107150. [CrossRef]
27. Özkılıç, Y.O.; Topkaya, C. Extended end-plate connections for replaceable shear links. *Eng. Struct.* **2021**, *240*, 112385. [CrossRef]
28. Özkılıç, Y.O. Cyclic and monotonic performance of stiffened extended end-plate connections with large-sized bolts and thin end-plates. *Bull Earthq. Eng.* **2022**, *20*, 7441–7475. [CrossRef]
29. Özkılıç, Y.O.; Un, E.M.; Topkaya, C. Frictional mid-spliced shear links for eccentrically braced frames. *Earthq. Eng. Struct. Dyn.* **2023**, 1–22. [CrossRef]
30. Özkılıç, Y.O. Cyclic and monotonic performance of unstiffened extended end-plate connections having thin end-plates and large-bolts. *Eng Struct.* **2023**, *281*, 115794. [CrossRef]
31. *GB/50017-2017*; Standard for Design of Steel Structures. National Standards of the People's Republic of China. China Architecture & Building Press: Beijing, China, 2017. (In Chinese)
32. *AISC/ANSI 358-16*; Prequalified Connections for Special and Intermediate Steel Moment Frames for Seismic Applications. American Institute of Steel Construction (AISC): Chicago, IL, USA, 2016.
33. Qi, H.; Hu, S.W.; Wang, P.Y.; Xue, X.; Shan, C.X.; Li, W.H.; Wang, K.; Li, J.H. Seismic performance of precast tubular web reduced beam section connection. *Structures* **2022**, *43*, 1259–1274. [CrossRef]
34. *GB/T 50081-2019*; Standard for Test Methods of Concrete Physical and Mechanical Properties. National Standards of the People's Republic of China; China Building Industry Press: Beijing, China, 2019. (In Chinese)
35. *GB/T 2975-2018*; Steel and Steel Products-Location and Preparation of Samples and Test Pieces for Mechanical Testing. National Standards of the People's Republic of China; Standards Press of China: Beijing, China, 2018. (In Chinese)
36. *GB/T 2281-2021*; Metallic Materials-Tensile Testing-Part 1: Method of Test at Room Temperature. National Standards of the People's Republic of China; Standards Press of China: Bejing, China, 2010. (In Chinese)
37. *JGJ/T 101-2015*; Specification for Seismic Test of Buildings. Industry Standards of the People's Republic of China; China Architecture & Building Press: Beijing, China, 2015. (In Chinese)
38. Wu, K.; Xu, C.; Cao, P.Z.; Lin, S.Q.; Chen, F. Experimental study on the flexural behavior of profile steel-steel fiber reinforced concrete composite beams. *China Civ. Eng. J.* **2019**, *52*, 41–52. (In Chinese)
39. Guo, Z.H.; Shi, X.D. *Reinforced Concrete Theory and Analyse*; Tsinghua University Press: Beijing, China, 2003. (In Chinese)
40. Wang, K.; Wang, Y.H.; Zhou, X.H.; Su, H.; Tan, J.K.; Kong, W.B.; Peng, T. Lateral behaviour and failure modes of buckling-restrained beam-only-connected steel plate shear walls. *Eng. Fail. Anal.* **2022**, *142*, 106833. [CrossRef]
41. Zhang, A.L.; Guo, Z.P.; Liu, X.C. Seismic performance of Z-type cantilever beam splices of column–tree connection. *J. Constr. Steel Res.* **2017**, *133*, 97–111. [CrossRef]
42. *JGJ 82-2011*; Technical Specification for High Strength Bolt Connections of Steel Structures. Industry Standards of the People's Republic of China; China Architecture & Building Press: Beijing, China, 2011. (In Chinese)

sensors

MDPI

Article

Graph Feature Refinement and Fusion in Transformer for Structural Damage Detection

Tianjie Hu [1,2], Kejian Ma [1,2] and Jianchun Xiao [1,2,*]

1 Research Center of Space Structures, Guizhou University, Guiyang 550025, China; gz_htj@hotmail.com (T.H.); makejian2002@163.com (K.M.)
2 Key Laboratory of Structural Engineering of Guizhou Province, Guiyang 550025, China
* Correspondence: jcxiao@gzu.edu.cn

Abstract: Structural damage detection is of significance for maintaining the structural health. Currently, data-driven deep learning approaches have emerged as a highly promising research field. However, little progress has been made in studying the relationship between the global and local information of structural response data. In this paper, we have presented an innovative Convolutional Enhancement and Graph Features Fusion in Transformer (CGsformer) network for structural damage detection. The proposed CGsformer network introduces an innovative approach for hierarchical learning from global to local information to extract acceleration response signal features for structural damage representation. The key advantage of this network is the integration of a graph convolutional network in the learning process, which enables the construction of a graph structure for global features. By incorporating node learning, the graph convolutional network filters out noise in the global features, thereby facilitating the extraction to more effective local features. In the verification based on the experimental data of four-story steel frame model experiment data and IASC-ASCE benchmark structure simulated data, the CGsformer network achieved damage identification accuracies of 92.44% and 96.71%, respectively. It surpassed the existing traditional damage detection methods based on deep learning. Notably, the model demonstrates good robustness under noisy conditions.

Keywords: structural damage detection; deep learning; CGsformer; graph convolutional network; global and local features; noise robustness

Citation: Hu, T.; Ma, K.; Xiao, J. Graph Feature Refinement and Fusion in Transformer for Structural Damage Detection. *Sensors* **2024**, *24*, 4415. https://doi.org/10.3390/s24134415

Academic Editors: Simon X. Yang, Jingzhou Xin, Hong Zhang and Yan Jiang

Received: 16 May 2024
Revised: 30 June 2024
Accepted: 4 July 2024
Published: 8 July 2024

1. Introduction

As the foundational support system of buildings, the structure directly affects the building safety. Throughout the entire service life of buildings, structural damage is inevitable due to the coupling of adverse factors such as environmental corrosion, material aging, fatigue damage, sudden disasters, and the long-term effects of abnormal loads [1–3]. With the extension of service time, the load-bearing capacity and resistance to natural disasters of the structure decrease, thus leading to unforeseeable safety challenges for buildings [4–6]. Therefore, the early diagnosis and localization of structural damage play a vital role in the timely maintenance of structures [7]. Structural damage detection [8–10], as the core technology of Structural Health Monitoring (SHM), is essential for grasping the structural working status and evaluating structural safety [11]. Consequently, structural damage detection methods have emerged as a prominent research focus in the field of SHM [12,13]. Structural damage detection is categorized into local-based [14,15] and global-based [16,17] methods. Local-based methods are mainly exploited to detect small regular structures or the local structure, which can be challenging to assess damage from large and complex structures. To overcome the limitations of local-based methods, many global-based structural detection methods have been developed. Among them, vibration-based structural damage detection methods have garnered widespread attention due to their flexibility, efficiency, and wide application. Traditional vibration-based structural damage detection methods use natural frequency, modal shape, and other modal measurements to detect structural damage. Although these traditional methods exhibit excellent performance in specific application scenarios, they still have obvious limitations when confronting with complex recognition tasks.

With the rapid advancement of Machine Learning (ML) technology, ML provides new ideas for vibration-based structural damage detection methods and has been extensively applied [18–21]. Compared to traditional vibration-based methods, ML-based structural damage detection methods are able to automatically learn and identify damage characteristics from large amounts of collected data. These methods not only have the advantages of high efficiency, accuracy, and adaptability, but they also do not need to rely on the subjective experience of experts. Zhang et al. [22] put forward a SHM method based on incremental Support Vector Regression (SVR) and substructure strategy, which realized the health monitoring of large and complex structures. Hua et al. [23] proposed a method to compress features using Principal Component Analysis (PCA), which improved the classification accuracy of their SVR and reduced the computational cost. Salkhordeh et al. [24] designed a decision tree classifier based on the Bayesian optimization algorithm to select effective features for structural damage classification. Wang et al. [25] proposed a concrete corrosion prediction model, which applied eXtreme Gradient Boosting (XG-Boost) based on the Bayesian optimization algorithm to extract corrosion-related features, and they used a Random Forest (RF) model for prediction.

ML-based structural damage methods have achieved promising performance, but their key drawback is their reliance on predefined hand-crafted features. These hand-crafted features often contain elements irrelevant to the task, thus resulting in reduced model performance. Deep Learning (DL), as the most representative technology of machine learning, has demonstrated powerful capabilities in various data analysis fields. Many DL-related methods have been proposed [26,27] for vibration-based structural damage identification. These methods typically take acceleration response signals as input and then extract features through a predefined network, thus achieving better classification performance. In 2017, Abdeljaber et al. [28] proposed a structural damage detection algorithm based on nonparametric vibration and a Convolutional Neural Network (CNN), which extracted sensitive features from the original acceleration response signal to achieve vibration-based damage detection and location. Zhang et al. [29] trained a one-dimensional (1D) CNN to detect local structural stiffness and mass changes. In addition to a 1D CNN, Khodabandehlou et al. [30] introduced the 2D CNN, and the method employed multiple acceleration response signals collected by multiple sensors as the input of the 2D CNN. Tang et al. [31] divided an original time series and used a 2D CNN to extract the time–frequency image of the time series to obtain local features for structural anomaly detection. Mantawy et al. [32] encoded a time series into images and trained a 2D CNN to extract local information to describe structural health and damage status. Although the CNN has shown a powerful ability to represent local information of input signals for structural damage detection tasks, it has difficulty describing the long-term dependence of time series. To capture the global information of the data, Lin et al. [33] introduced the Long Short-Term Memory (LSTM) network for damage detection, which used the acceleration response signal as input to learn the signals' long-term properties. Sony et al. [34] proposed an LSTM network to classify and locate structural damage by using structural acceleration response data. However, the local feature extraction capabilities of temporal modeling networks, the Recurrent Neural Network (RNN), and its variants are inefficient. To this end, subsequent work [35–37] has designed hybrid models, such as CNN and RNN (CNN-RNN), which combines the local feature extraction faculty of the CNN—with the ability to capture long-term dependencies—with the time series data of the RNN. Typically, these hybrid models integrate local and global features through parallel architectures [35,36] or learning strategies from local to global [38]. However, the former ignores the dependence between local and global features, and the latter leads to information loss and increases the complexity of the model. Therefore, how to effectively combine hybrid models to extract the local and global features of input data has become a challenging problem.

Although the aforementioned networks are able to extract local and global information from signals, they have difficulty in mining the intrinsic correlations of the data. Consequently, some researchers [39–42] have utilized irregular graphs to reflect the interconnections between structural damage data. Dang et al. [39] used the Graph Convolutional Network (GCN) to attain the intrinsic spatial coherence of sensor positions directly from structural vibration response data. Zhan et al. [41] exploited the GCN to construct graph structures in the wavelet domain for multisensor signals, and they trained the network with node classification as the goal for structural damage detection tasks. Wang et al. [40] proposed a waveguide-based dual GCN method for damage detection. The method employed Short-Time Fourier Transform (STFT) to obtain the spatial–temporal feature representation of the original waveguide signals. A local graph network was built using features obtained from samples, and then local graphs were formed by grouping nodes with similar types of damage, thus enabling damage detection on the structure. The successful applications of GCN-based methods show that the methods can effectively understand and represent the intrinsic correlations between original data and features.

There are still challenges in designing a deep learning method that accurately identifies structural damage. Firstly, both the CNN-based and RNN-based approaches have limitations in their abilities to model long-term dependencies and capture global temporal dependencies. Although RNNs are designed to deal with long-term dependencies, they may still encounter the problem of vanishing or exploding gradients when processing longer sequences. Secondly, some hybrid models usually use parallel networks, or they adopt the learning way from local to global, and there is a lack of investigation into the learning manner from global to local. Thirdly, in addition to the signal components related to the structural self-oscillation characteristics, the measured acceleration response signal inevitably contains noise. The direct input of such noisy data into deep learning models can affect the generalization ability and damage detection performance. Finally, research on structural damage methods based on graph convolution is limited.

According to the above analysis, we present a novel Convolutional Enhancement and Graph Features Fusion in Transformer Network (CGsformer) for detecting structural damage information. The proposed CGsformer network contains two contributions. On the one hand, the proposed network uses a hierarchical learning method from global to local to extract the characteristics of the acceleration response signal. Based on previous research work, global information can identify the periodic relationship of the acceleration vibration response signals, while local information can help to define and analyze the subtle differences in the acceleration response signals of short adjacent moments before and after damage occurs. Thus, a multihead self-attention was used to acquire the global information of the signals, and a convolution module was applied to represent the local information. On the other hand, the graph convolution module was embeded to enhance robustness against noise contamination. The graph convolution network constructs a graph structure for global features and filters out noise in global features through node learning, thus prompting the convolution module to extract more effective local features. On this basis, an extensive verification testing has been conducted by using the numerical model of the International Association for Structural Control (IASC)– American Society of Civil Engineers (ASCE) benchmark structure [43] and a four-story steel frame structure experiment. The study has compared the proposed CGsformer with deep learning-based models such as CNN, LSTM, and Transformer, especially in cases of limited datasets and noisy pollution scenarios.

2. Methodology

2.1. The Equation of Motion

Damage can cause detectable changes in the structural dynamic response. Therefore, it is usually possible to determine the structural health state by analyzing the structural response acceleration signals before and after damage. Without loss of generality, consider a linear multi-degree-of-freedom structure, whose motion equation is as follows [44]:

$$M\ddot{u}(t) + C\dot{u}(t) + Ku(t) = F(t) \tag{1}$$

where $\boldsymbol{M}$, $\boldsymbol{C}$, and $\boldsymbol{K}$ represent the structural mass, damping, and stiffness matrices with the dimension of $\boldsymbol{Z} \times \boldsymbol{Z}$, respectively. $u(t)$, $\dot{u}(t)$, and $\ddot{u}(t)$ denote the displacement, velocity, and acceleration with the dimension of $\boldsymbol{Z} \times \boldsymbol{1}$, respectively; The superscript is a derivative of time. In addition, $\boldsymbol{F}(t)$ represents the external loads with the dimension of $\boldsymbol{Z} \times \boldsymbol{1}$.

It is generally believed that the damage will not lead to the change in the mass of the structure, but it will cause a decrease in the stiffness. Hence, the changes in stiffness of the structure before and after damage can be defined as $\Delta\boldsymbol{K}$

$$\Delta K = K_u - K_d \tag{2}$$

This can be further expanded as the linear superposition of element stiffness matrices as follows:

$$\Delta K = \sum_{i=1}^{n} a_i K_u^i \tag{3}$$

where K_u^i defines the stiffness matrix of the ith element in the structure, and a_i is the damage coefficient, which varies from 0 to 1. For example, $a_i = 0.95$ means 5% stiffness lost in the ith element for stiffness.

The changes in stiffness can be reflected in the structural dynamic response. For example, damage can lead to a decrease in the structural vibration frequency and changes in the vibration modes. Therefore, collecting and analyzing structural responses can help us understand the health states of the structure. In practical applications, acceleration sensors, installed at different locations on the structure, are commonly used to capture these changes. The acceleration response refers to the measurement of the acceleration experienced by a structure under various loading conditions or external forces. This acceleration response data are usually collected by acceleration sensors.

This paper proposes a novel CGsformer network for structural damage detection. The proposed CGsformer network is not only able to extract global and local features in the acceleration response signal, but it also embeds the graph structure to better extract local features from global features. As far as we know, the proposed CGsformer unifies global features, local features, and graph structures for structural damage detection for the first time. At the same time, our experiments have proven that the proposed CGsformer has better classification performance and noise robustness.

2.2. Global and Local Feature Extraction

Research [35,36,45] has shown that the CNN-LSTM model has the ability to capture local and global features of signals, and its classification accuracy on structural damage detection tasks can surpass single CNN or RNN models. Inspired by previous studies, the multiheaded self-attention module and convolutional module were used to extract global and local features from structural damage detection signals.

Multihead self-attention module: This module comprises layer normalization, multihead self-attention, and dropout. Among them, multiheaded self-attention was proposed in [46,47]. To better understand multihead self-attention, we first describe self-attention. The self-attention mechanism aims to describe contextual information and capture the global characteristics of signals. The process of the self-attention mechanism is illustrated in Figure 1. Assume that matrix $\boldsymbol{X} \in \mathbb{R}^{M \times N}$ passes through linear matrices $\boldsymbol{W}_q \in \mathbb{R}^{M_q \times M}$,

$W_k \in \mathbb{R}^{M_k \times M}$, and $W_v \in \mathbb{R}^{M_v \times M}$ to obtain query matrix $Q_X \in \mathbb{R}^{M_q \times N}$, key matrix $K_X \in \mathbb{R}^{M_k \times N}$, and value matrix $V_X \in \mathbb{R}^{M_v \times N}$ as follows:

$$\begin{aligned} Q_X &= W_q X \\ K_X &= W_k X \\ V_X &= W_v X \end{aligned} \tag{4}$$

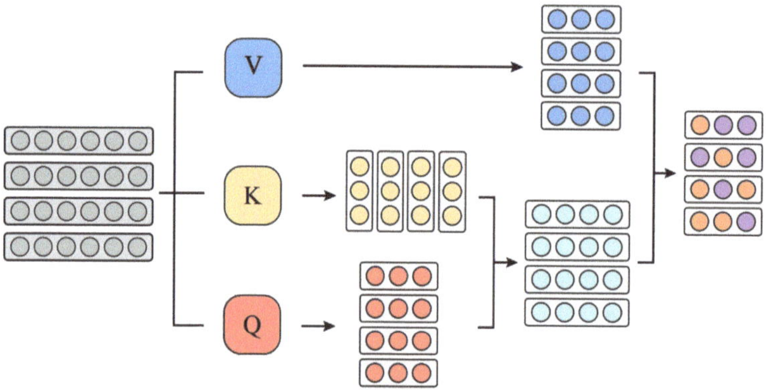

Figure 1. Self-attention mechanism.

Subsequently, the attention score matrix X_{score} is obtained by multiplying Q_X and K_X and by passing the softmax function. Finally, the attention coefficients matrix X_{coff} is multiplied by the V_X to acquire the attention coefficients. Similarly, the multiheaded self-attention mechanism uses multiple linear matrices W_q^i, W_k^i, and W_v^i to obtain the matrices Q^i, K^i and V^i, as shown in Figure 2. All attention coefficients are concatenated into multiheaded attention coefficients, which are obtained as follows:

$$\begin{aligned} M_{ac}(X) = Atten\left(W_q^1 X, W_k^1 X, W_v^1 X\right) \oplus \ldots\ldots \\ \oplus Atten\left(W_q^i X, W_k^i X, W_v^i X\right) \end{aligned} \tag{5}$$

where $\oplus$ represents the concatenate operation, and $Atten(\cdot)$ denotes the self-attention operations.

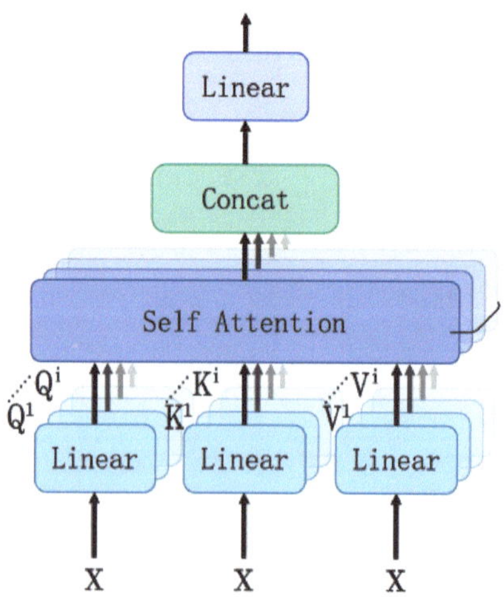

Figure 2. Multiheaded self-attention mechanism.

Convolution module: The convolution module includes a pointwise convolution operation with 1x1 convolution kernel, and it doubles the number of channels. Next, the GLU activation function is utilized to control the number of output features. Batch normalization is carried out to stabilize the internal distribution of the network after applying a depthwise convolution. Finally, a Swish activation function and pointwise convolution are employed to complete the processing of the whole module. The convolution module is depicted in Figure 3.

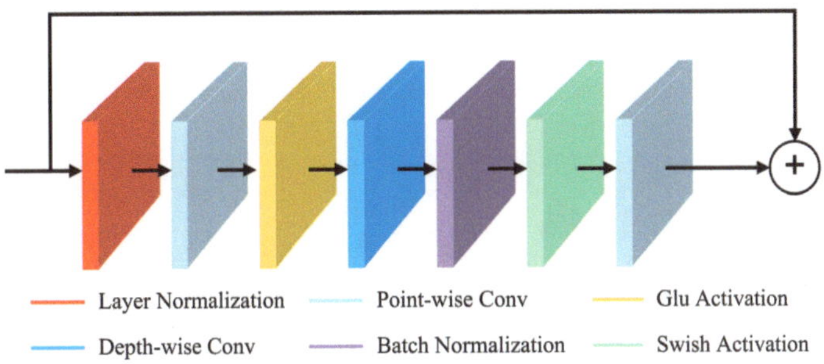

Figure 3. The convolution module.

2.3. *Graph Convolution Network*

As a data structure, a graph can effectively describe the association between two nodes, which is represented as follows [48],

$$G = \left(G_{nodes}, G_{edges}\right) \tag{6}$$

where G_{nodes} indicates the set of nodes, and G_{edges} represents the set of edges. The Graph Neural Network (GNN) is a deep learning model designed for processing graph-structured data, which are widely used in fields such as social network analysis, molecular structure modeling, and recommendation systems. The GCN is an important variant of GNNs, which is able to construct a graph structure from the input sequence or features and learn the adjacency relationships between nodes to enhance the understanding and representation of global information.

We assume that the input sequence data features are $\boldsymbol{X} \in \mathbb{R}^{n \times d}$. To construct a graph for $\boldsymbol{X}$, we first need to create nodes from the sequence and then define edges based on these created nodes. Each element in the sequence is considered as a node in the graph network. After node construction, the creation of edges is done through a self-attention mechanism on the nodes. When we have completed the construction of the nodes and edges, a graph is formed that contains rich and reliable connections between relevant nodes. Specifically, an input $\boldsymbol{X}$ can be represented by a GCN layer as follows:

$$X^{(l+1)} = M_{Lap} X^{(l)} W^{(l)} \tag{7}$$

where $X^{(l)}$ and $W^{(l)}$ are the input and learnable matrix for the lth layer, respectively, and M_{Lap} is the Laplacian matrix used to represent the topological structure of the graph, which is denoted as

$$M_{Lap} = D^{-\frac{1}{2}} \hat{A} D^{-\frac{1}{2}} \tag{8}$$

where $D = \text{diag}(d_1, d_2, \ldots, d_n)$ is the degree matrix used to describe the number of edges d_n corresponding to node n, and $\hat{A}$ is the adjacency matrix used to represent the relationships between nodes, which is defined as

$$\hat{A} = M_{re}\left(\frac{(W_d X)^T (W_n X)}{d}\right) \in \mathbb{R}^{n \times n} \tag{9}$$

where W_d and W_n denote the learnable matrices, d represents the feature dimension of the whole sequence, T denotes the transpose operation, and $M_{re}(\cdot)$ represents ReLU activation function, which is utilized as the activation function filters out negative links between nodes. Negative links imply that there is no necessary direct connection between these two nodes. Overall, the GCN enhances the correlation between local features and suppresses noise through graph structure, thus achieving better global modeling of features.

2.4. Extracting Robust Features via CGsformer

Traditional hybrid models usually adopt parallel architecture or learn from local to global methods. However, the former assumes that global information and local information are independent, while the latter will lead to the loss of some global information and increase the complexity of the model. To this end, we proposed the CGsformer network, which adopts a learning manner from global to local.

Figure 4 illustrates the proposed CGsformer network. The innovation and advantage of the proposed CGsformer network lies in the hierarchical learning from global to local to extract acceleration response signal features for representing structural damage. Meanwhile, embedding GCN into the learning process of global to local features aims to build a graph structure for global features. By learning node features of global information to filter out noise and retain important global features, the convolution module learns more effective local features.

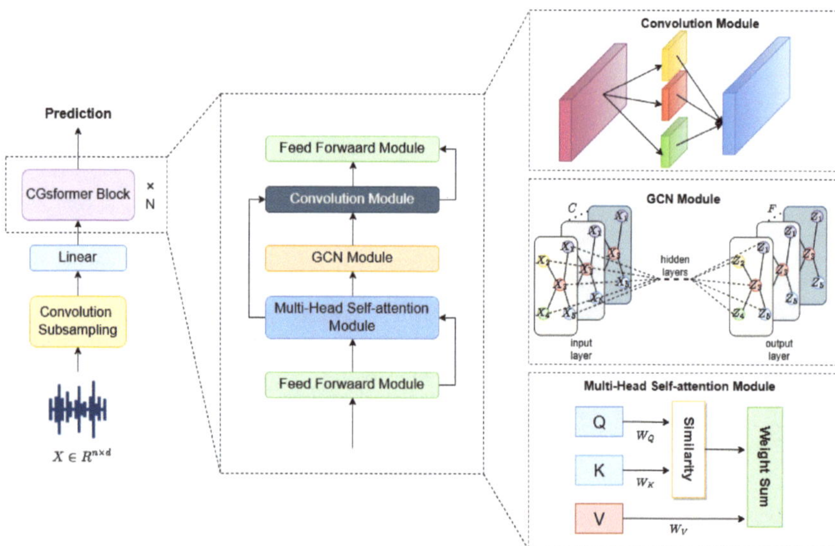

Figure 4. The overall structure of the CGsformer. The proposed CGsformer is mainly composed of four parts: feedforward module, self-attention mechanism module, convolution module, and graph network module.

The shallow features of the acceleration response signal are obtained through convolutional subsampling and linear layers, and they are input into the CGsformer block to extract deep global and local features. In the CGsformer block, the feedforward module, as shown in Figure 5, is first used to achieve independent mapping of each position in the sequence. The feedforward module consists of two linear projections and an intermediate nonlinear activation function, where the first linear layer extends the feature dimensionality of the data by four times, and the other linear layer projects it to the original model dimension. We normalize the network using the Swish activation function, dropout, and layer normalization operation in the feedforward module. The Swish function is used to nonlinearly transform the input signal, and dropout randomly discards neurons to prevent the network from overfitting. The layer normalization operation speeds up network training and convergence. Simultaneously, the entire module follows the prenormalized residual unit.

Then, the multiheaded self-attention module realizes the long-term dependency modeling of the structural acceleration response data so that the weights of each position can be dynamically adjusted according to the contextual information of different positions. The graph convolution module captures the connection relationship between different nodes to better understand the local and global dependencies in the structural acceleration response data from the perspective of graph structure. The convolution module completes the further capture of local features of the structural acceleration response data.

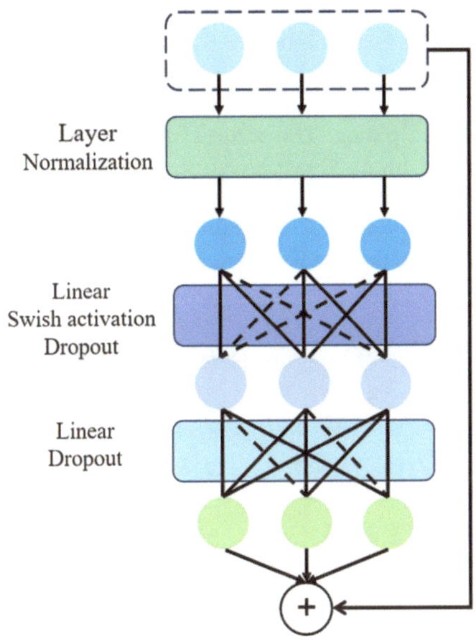

Figure 5. The feedforward module.

According to the above description, the process of the CGsformer model is unfolded as follows:

$$\begin{aligned} \widetilde{X} &= X + \frac{1}{2}M_{ff}(X) \\ \hat{X} &= \widetilde{X} + M_{ac}(\widetilde{X}) \\ \hat{X} &= M_{gcn}(\hat{X}) \\ \bar{X} &= \hat{X} + M_{conv}(\hat{X}) \\ Y &= M_{norm}(\bar{X} + \frac{1}{2}M_{ff}(\bar{X})) \end{aligned} \quad (10)$$

where $M_{ff}(\cdot)$ defines the feedforward module, $M_{ac}(\cdot)$ is the multiheaded self-attention module, $M_{gcn}(\cdot)$ is the graph convolution module, $M_{conv}(\cdot)$ is the convolution module, $M_{norm}(\cdot)$ is the layer normalization operation, and $Y \in \mathbb{R}^{n \times d}$ denotes the prediction result of the damage patterns after CGsformer output.

2.5. Prediction

Finally, we pass the features Y obtained from the CGsformer model through two Fully Connected (FC) layers to obtain the final damage category. The process is as follows:

$$M_{pre} = W_2(M_{re}(W_1 Y + b_1)) + b_2 \in \mathbb{R}^{d_{out}} \quad (11)$$

where W_1 and W_2 represent weights of the FC layers, b_1 and b_2 represent biases of the FC layers, and d_{out} represents the dimension of the output classes.

3. Verification by Simulation

In this section, we validated the damage detection performance of the CGsformer model through the phase I IASC-ASCE SHM numerical benchmark structure [43].

3.1. Test Setup and Data Preparation

IASC-ASCE SHM benchmark dataset: The numerical model of the IASC-ASCE SHM benchmark has been jointly established by the IASC and the ASCE [43], which provides a standardized and unified benchmarking platform for comparing and evaluating various structural health monitoring methods. As shown in Figure 6, the IASC-ASCE SHM benchmark structure is a $\frac{1}{3}$-scale steel frame model with four stories. The model has a story height of 0.9 m and a total height of 3.6 m. The plan size of the model is 2.5 m × 2.5 m, with each bay spanning 1.25 m. Each story consists of nine steel columns and eight diagonal braces. Each floor slab is composed of four uniformly distributed mass plates. The first floor has four 800 kg plates. The second and third floors each have four 600 kg plates. The fourth floor has three 400 kg plates and one 550 kg plate, which are arranged to give the structure an asymmetric mass distribution. The structure is excited by an external excitation acting in the diagonal direction at the top. For more detailed information on the IASC-ASCE SHM simulated benchmark structure, please consult reference [43].

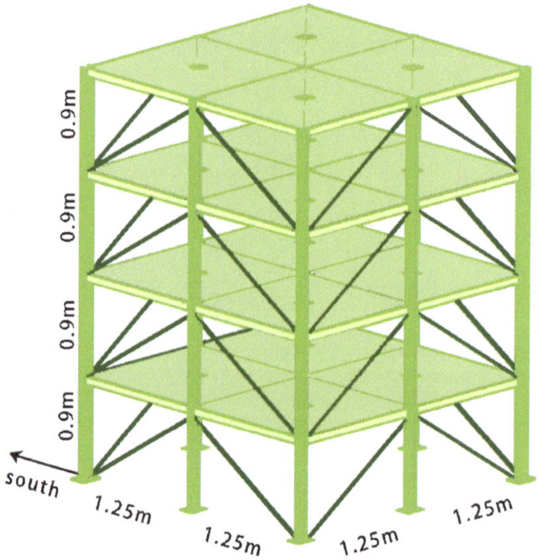

Figure 6. IASC-ASCE SHM Benchmark structure model.

The model defines a total of six damage patterns. Figure 7 illustrates three of these damage patterns: D.P.1, D.P.2, and D.P.4. This paper focuses on these damage patterns, and the original acceleration vibration response data were generated using the 12-degree-of-freedom finite element model in the MATLAB program from the IASC-ASCE SHM Research Group. The acceleration response data for these four damage modes (D.P.0, D.P.1, D.P.2, and D.P.4) were collected from four sensors on each floor, as shown in Table 1. The noise levels used were 0%, 20%, and 50%, respectively. The data were preprocessed, thus resulting in training (validation and testing) data for the classifiers. The dataset among the 24 contains a total of 6248 samples, with 4000 samples used for training, 1000 used samples for validation, and 1248 used samples for testing. The preprocessing procedure is described in detail in the next step.

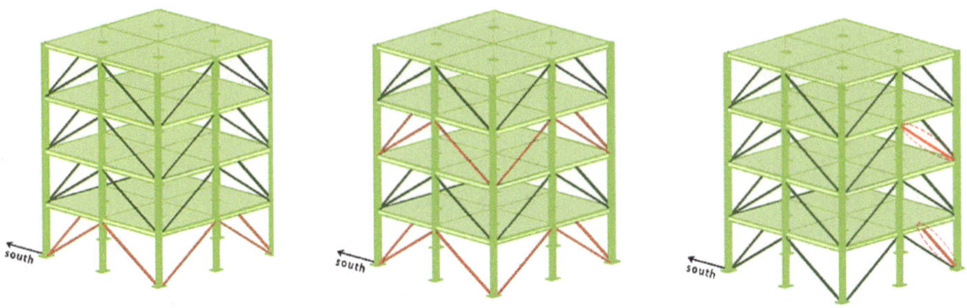

Figure 7. From left to right, they correspond to damage modes D.P.1, D.P.2, and D.P.4.

Table 1. Normal pattern and three cases of damage patterns as defined in the IASC-ASCE SHM simulated benchmark structure.

Damage Pattern	Pattern Description
D.P.0	No damage.
D.P.1	All braces on the first floor have no stiffness.
D.P.2	All braces on the first and third floors have no stiffness.
D.P.4	One brace of the first floor and the third floor has no stiffness.

Using Gaussian white noise to simulate environmental excitation is a common method. Gaussian white noise is a random process with a uniform power spectral density, which is characterized by equal and independent power components at all frequencies. The power spectral density of white noise is constant across all frequencies.

$$S(f) = \frac{N_0}{2} \tag{12}$$

where N_0 is the intensity of the noise power spectral density.

Data preprocessing: First, the response data collected on two acceleration sensors in the same direction on each floor were fused to obtain the fused translation data of each floor from the x and y directions [49] as follows:

$$\begin{cases} acc_{x,d} = 0.5 \times (acc_{1,d} + acc_{3,d}) \\ acc_{y,d} = 0.5 \times (acc_{2,d} + acc_{4,d}) \end{cases} \tag{13}$$

where $acc_{1,d}$, $acc_{2,d}$, $acc_{3,d}$, and $acc_{4,d}$ denote the acceleration time history response data collected by the four sensors in the IASC-ASCE SHM benchmark structural model for floor d, as depicted in Figure 8, respectively, and $acc_{x,d}$ and $acc_{y,d}$ denote the translational acceleration in the x and y directions of the d floor, all of which are 1D temporal data.

Next, the sampling points of the sensor were calculated. Assume that the sampling rate of the sensors is f_s, and the sampling time is t_s when collecting the original acceleration time history response data of the four sensors in floor d. Correspondingly, the number of sampling points S_j for one sensor in sampling time t_s is $S_j = f_s \times t_s$. The sampling points S_j are divided into m nonoverlapping data segments with fixed lengths of 128. The translational acceleration datasets for completing the segmentation process in floor d are represented as $acc'_{x,d}$ and $acc'_{y,d}$. Meanwhile, $acc'_{x,d}$ and $acc'_{y,d}$ are normalized and shuffled, and the processed data are defined as $acc''_{x,d}$, and $acc''_{y,d}$. According to the above method, the data from each sensor on every floor under other damage modes are processed, thus resulting in a dataset for each direction of every floor under each working condition. Assuming that the damage detection task contains a total of p damage patterns, the translational acceleration dataset D for all floors in the x and y directions can be expressed as

$$\begin{aligned} D = [&[acc''_{x,1,1}acc''_{y,1,1}\cdots acc''_{x,d,1}acc''_{y,d,1}\cdots acc''_{x,4,1}acc''_{y,4,1}], \\ &[acc''_{x,1,2}acc''_{y,1,2}\cdots acc''_{x,d,2}acc''_{y,d,2}\cdots acc''_{x,4,2}acc''_{y,4,2}], \\ &\cdots \\ &[acc''_{x,1,p}acc''_{y,1,p}\cdots acc''_{x,d,p}acc''_{y,d,p}\cdots acc''_{x,4,p}acc''_{y,4,p}]]. \end{aligned} \tag{14}$$

The translational acceleration dataset D is then divided into columns to obtain eight data subsets. Each subset represents the acceleration response data of a certain translational direction in a certain floor, and contain p damage patterns.

Therefore, the proposed CGsformer with first floor x direction acceleration time history response data will be introduced, i.e., $D_{x,1,p} = [acc''_{x,1,1}\ acc''_{x,1,2}\ \cdots\ acc''_{x,1,p}]$ containing p damage patterns. More generally, we define the $D_{x,1,p}$ as $X \in R^{n\times d}$.

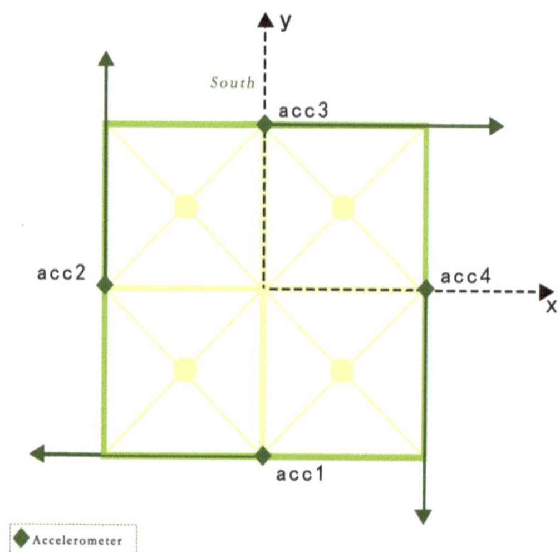

Figure 8. Distribution of IASC-ASCE SHM Benchmark model measurement points.

Implementation of details: The hyperparameters of the CGsformer are set as shown in Table 2. All hyperparameters were determined by performing ablation experiments on damaged structures in the first floor in the x direction. And, the same hyperparameter settings were used for all experiments. The entire model used the Adam optimizer for gradient descent, and the learning rate was 0.001. The loss was computed using the crossentropy function. Training and testing were performed on a machine with a Tesla A100 GPU using a batch size of 32.

Table 2. The hyperparameter settings adopted in the CGsformer.

Setting	Value
Encoder Layers	4
Encoder Dim	512
Attention Heads	2
Conv Kernel Size	19
Multihead Attention Dropout	0.4
CGsformer Dropout	0.1

3.2. Comparison with Other Models

To verify that the accuracy performance of CGsformer is better than other models, the performance of the proposed CGsformer structural damage detection method was compared with other methods on the IASC-ASCE SHM benchmark structural dataset with an initial layer x direction noise level of 0%, and the related acceleration response curves are illustrated in Figures 9 and 10. The comparative models include following:

- **CNN** [28]: In this experiment, a one-dimension (1D) convolution operation with two convolutional kernels of size five constructed the network.
- **LSTM** [33]: In this experiment, a bidirectional LSTM with two hidden layers and a dimension of 128 constructed the network.
- **CNN-LSTM** [37]: The spatial features were first extracted using a 1D CNN with a convolutional kernel size of 15, and then these features were input into a two-layer LSTM with a hidden layer dimension of 256 for temporal modeling.
- **Multihead CNN** [50]: Multihead CNN learns different-scale or different-type features by introducing multiple parallel convolutional branches. Each branch can focus on different spatial or frequency domain information, and their results are fused to more comprehensively describe structural damage information.
- **Transformer** [46]: In this experiment, four Transformer blocks were used with eight heads in the multiheaded attention mechanism, and the dimension was set to 512.
- **Conformer** [51]: Conformer combines the advantages of CNN and self-attention mechanisms, effectively handles long input sequences, and possesses strong modeling and contextual understanding capabilities. The experimental hyperparameter settings for Conformer were consistent with the CGsformer, as illustrated in Table 2.

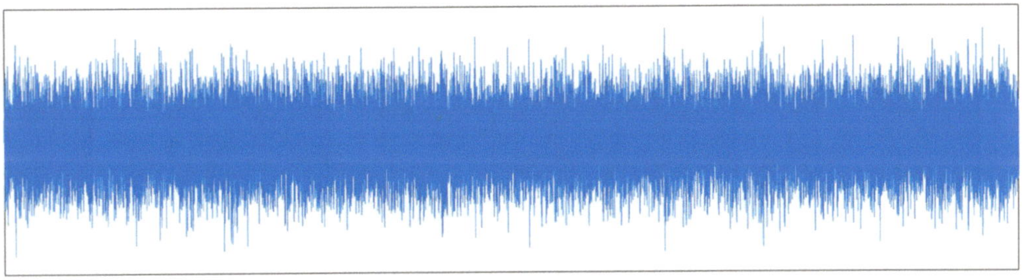

Figure 9. Acceleration time response curve of D.P.1 damage pattern, which was collected from sensor NO.1 (acc 1) on the first floor at 0 noise level.

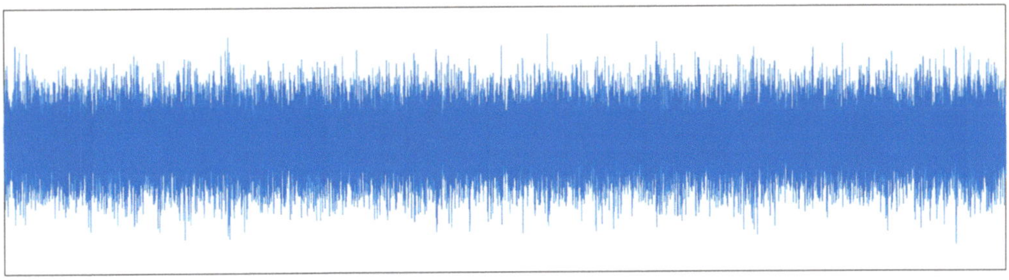

Figure 10. Acceleration time response curve of D.P.1 damage pattern, which was collected from sensor NO.3 (acc 3) on the first floor at 0 noise level.

To measure the classification performance of the different models, the accuracy performance ACC and F1 score (F_{score}) were adopted as indicators. The accuracy performance ACC and F1 score are defined as follows

$$ACC = \frac{X_{TP} + X_{TN}}{X_{TP} + X_{TN} + X_{FP} + X_{FN}} \tag{15}$$

$$F_{score} = \frac{2X_{pre}X_{rec}}{X_{pre} + X_{rec}} \tag{16}$$

where X_{TP}, X_{TN}, X_{FP}, and X_{FN} represent the true positive, true negative, false positive, and false negative values of data samples, respectively; X_{pre} and X_{rec} are the precision and recall rates, respectively, which are denoted as follows:

$$X_{pre} = \frac{X_{TP}}{X_{TP} + T_{FP}} \tag{17}$$

$$X_{rec} = \frac{X_{TP}}{X_{TP} + T_{FN}} \tag{18}$$

Table 3 presents the classification performance of the different models on the dataset. The proposed CGsformer achieved the best performance. The single CNN and LSTM models only considered the local information or global information of the input signals, thus resulting in relatively poor results. Compared to the CNN, the multihead CNN could learn features of different scales, and its accuracy was 3.45% higher than the CNN, thus reaching 92.39%. However, the multihead CNN still failed to capture the long-term information of the signals. The CNN-LSTM model took into account both local information and long-term dependencies to better capture the characteristics of the structural damage, with an accuracy of 92.55%. Compared to the CNN-LSTM model, the Transformer model captured more robust global dependencies through the self-attention mechanism and achieved a gain of 1.52%. The Conformer model employed the advantages of the CNN in capturing local features and the Transformer in capturing global features, with an accuracy of 95.27%. The proposed CGsformer model embedded graph structures into both global and local features. This not only effectively filtered out noise interference from global features, but it also helped the convolution module better understand global information, thereby extracting more effective local features. In Table 3, the identification accuracy of the proposed CGsformer is shown at 96.71%, thus achieving the best classification performance.

Table 3. Comparison results between CGsformer and other models.

Method	ACC	F_{score}
CNN	0.8910	0.8903
LSTM	0.8822	0.8826
CNN-LSTM	0.9255	0.9258
Multihead CNN	0.9239	0.9238
Transformer	0.9407	0.9406
Conformer	0.9527	0.9528
CGsformer	**0.9671**	**0.9672**

In order to test whether the proposed CGsformer model is significantly relevant to other models, the 95% confidence interval (CI) was utilized to test the statistical significance of the accuracy performance between the proposed model and the other models. In Table 4, we present the accuracy differences (ΔACC), 95% confidence intervals (CIs), and confidence intervals for the accuracy differences between the proposed model and other models (ΔCI), where $\Delta\text{ACC} = \text{ACC}_a - \text{ACC}_b$ represents the difference in accuracy between the proposed model ACC_a and other models ACC_b—the 95% CI—which is denoted as

$$CI = ACC \pm z \cdot SE, SE = \sqrt{\frac{ACC(1 - ACC)}{n_t}} \tag{19}$$

where z was set to 1.96 at the 95% CI, n_t is equivalent to the size of test samples, and ΔCI is formulated as follows:

$$\Delta\text{CI} = \Delta\text{ACC} \pm z \cdot \sqrt{SE_a^2 + SE_b^2} \tag{20}$$

Table 4. Statistical significance comparison of the proposed model with other models.

Model	ACC	ΔACC	CI	ΔCI
CNN	0.8910	0.0761	[0.8724, 0.9078]	[0.0562, 0.0960]
LSTM	0.8822	0.0849	[0.8630, 0.8996]	[0.0645, 0.1053]
CNN-LSTM	0.9255	0.0416	[0.9104, 0.9402]	[0.0240, 0.0592]
Multihead CNN	0.9239	0.0432	[0.9086, 0.9387]	[0.0255, 0.0609]
Transformer	0.9407	0.0264	[0.9261, 0.9532]	[0.0100, 0.0428]
Conformer	0.9527	0.0144	[0.9394, 0.9638]	[−0.0010, 0.0298]
CGsformer	0.9671	-	[0.9557, 0.9763]	-

It can be intuitively seen from Table 4 that the proposed CGsformer model had the most accurate predictive performance. Moreover, the CI of the proposed model was the narrowest, which indicates that the proposed model is more robust. Meanwhile, the proposed model had significant statistical significance compared to other models (except for Conformer). In conclusion, the proposed model is superior to the other models.

3.3. Ablation Study

In this subsection, the performance of the GCN module is compared at different positions in the model, and more conclusions are drawn. As shown in Figure 11, three combinations were set. The graph convolution module was placed before the multihead self-attention module, after the convolution module, and in between the two for the proposed CGsformer model. They are named Attention Before, Conversion After, and CGsformer, respectively.

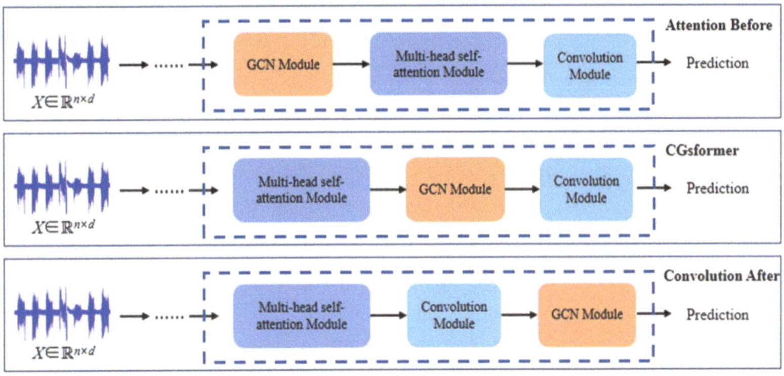

Figure 11. Illustration of ablation experiments with GCN placed at different positions in the CGsformer block.

The accuracy performance outcomes are presented in Table 5, and it can be found that the results of the three combinations were better than that of Conformer. Among the three combinations, the proposed CGsformer model achieved the best performance, followed by Attention Before and the relatively poor Convolution After. Based on these observations, the following conclusions can be drawn: (1) All three models outperformed the Conformer model: this means that graph structure learning can help the model select more discriminative features. (2) Placing the GCN module before the multihead self-attention module failed to capture the higher-order information of the current sequence features and only relied on shallow features for sequence representation learning, thus resulting in poorer performance. Placing the GCN module after the convolution module aggregated the final decision features, but this operation failed to fully utilize the semantic information represented by attention and loses effectiveness. (3) The proposed CGsformer model achieved the best performance, with an accuracy of 96.71%. Graph convolution learning of features after the self-attention module can further propagate global features to adjacent nodes, which helps to better understand and express node features by combining local adjacency information.

Table 5. The effectiveness of GCN at various positions in the CGsformer.

Method	ACC	F_{score}
Conformer	0.9527	0.9528
CGsformer	**0.9671**	**0.9672**
Attention Before	0.9631	0.9632
Convolution After	0.9623	0.9623

3.4. Comparative Analysis on the Four-Story Numerical Model with 24 Classifiers

As mentioned earlier, the acceleration response data from all measurement points of the IASC-ASCE SHM benchmark structure collected under four damage patterns were used to validate the proposed CGsformer-based damage detection method in this paper. The acceleration sensors at all measurement points on each floor had a sampling frequency of $f_s = 250$ Hz for all damage patterns, and the vibration response was measured over a duration of 800 s. Therefore, the number of sampling points S_j for the sensor within the sampling time was 200,000.

A total of 24 CGsformer classifiers (3 noise levels × 4 floors × 2 translational directions) were trained using the acceleration response data collected from all acceleration sensors of the IASC-ASCE SHM benchmark structure under three different noise levels (0%, 20%, and 50%). The collected data were tackled with the steps outlined in Section 3.1. Therefore, the dataset of the acceleration time history response data for each direction of every floor under each noise level contains 6248 samples (4 damage patterns × 1562 data segments). Among these, 4000 samples were used for training the classifiers, 1000 samples for validation, and 1248 samples for testing the trained classifiers.

The ACC results are reported in Table 6, and some observations can be made: (1) At different noise levels, increasing noise levels led to a decrease in accuracy. The highest average accuracy of 97.25% was achieved at a noise level of 0% (with the highest detection accuracy classifier appearing in the third floor, y direction, with an accuracy of 98.24%). At a noise level of 20%, the average accuracy reached 96.74% (with the highest detection accuracy classifier appearing in the fourth floor, y direction, with an accuracy of 98.32%). At a noise level of 50%, the average accuracy was 95.44% (with the highest detection accuracy classifier also appearing in the fourth floor, y direction, with an accuracy of 96.15%). The confusion matrix with the highest accuracy for the three noise levels is shown in Figure 12. As can be seen from Figure 12, misclassifications primarily occurred between D.P.0 and D.P.4 at all three noise levels, with the error rate increasing as the noise levels rose. This result is attributed to the training samples being insufficiently diverse and the model's limited sensitivity to the highly similar overlapping features in their acceleration response signals. Although increasing noise levels can lead to decreased accuracy, comparing the damage detection results between 0% and 50% noise levels shows that the average detection accuracy of the four-floor, eight-classifier model only decreased by 1.81%. This indicates that the proposed CGsformer-based damage detection model has strong noise resistance. (2) The CGsformer model exhibited stronger robustness as the noise increased. Taking the first floor, x direction, as an example, even with an additional 20% noise data, the detection accuracy reached 96.07%. When the noise levels were 0%, 20%, and 50%, the average accuracy values of the Conformer model were 96.18%, 95.25%, and 94.58%, respectively. Comparing the results of the CGsformer model in Table 6, the CGsformer model achieved better results, with an improvement of 1.07%, 1.49%, and 0.86% for the noise levels at 0%, 20%, and 50%, respectively. (3) Despite the suboptimal performance compared to low noise levels, the model still achieved results comparable to the multihead CNN model (92.39%) when dealing with a noise level of 50%. This indicates that the CGsformer model has further learned the correlation between local and global features, thus exhibiting stronger generalization even when half of the data is noisy.

Table 6. Experimental results of 24 CGsformer classifiers (3 noise levels × 4 floors × 2 translational directions) using the acceleration response data collected from all acceleration sensors of the IASC-ASCE SHM Benchmark structure under 0%, 20%, and 50% noise levels.

Direction/Floor	Noise 0%	Noise 20%	Noise 50%
First Floor, x direction	0.9671	0.9607	0.9279
First Floor, y direction	0.9688	0.9712	0.9375
Second Floor, x direction	0.9776	0.9704	0.9583
Second Floor, y direction	0.9816	0.9752	0.9391
Third Floor, x direction	0.9824	0.9631	0.9383
Third Floor, y direction	0.9671	0.9535	0.9191
Fourth Floor, x direction	0.9575	0.9607	0.9543
Fourth Floor, y direction	0.9776	0.9832	0.9615
Average	**0.9725**	**0.9674**	**0.9544**

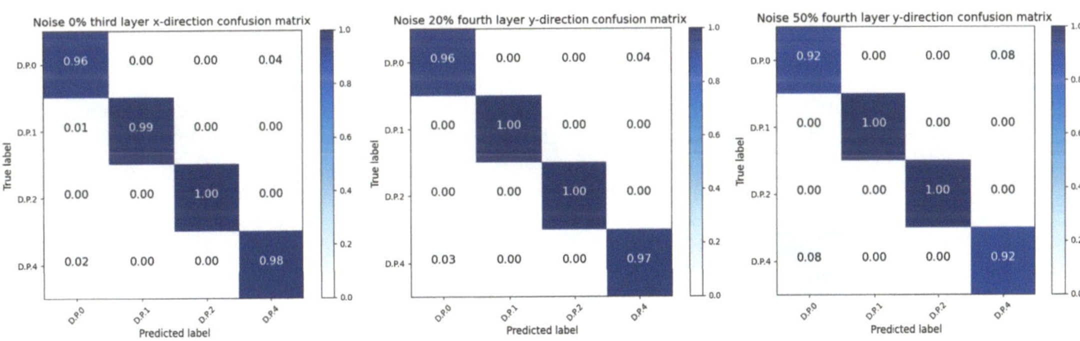

Figure 12. Diagram of the best-performing confusion matrix for the three noise levels.

4. Experimental Verfication

In this section, the performance of the proposed CGsformer model is further compared with other models in a four-story, single-span, steel frame structure to verify the effectiveness of the proposed CGsformer model in different structures through testing.

4.1. Experiment Description

As depicted in Figure 13, the experimental structure is a four-story, single-span, steel frame structure. The plan dimension of the structure is 260 mm × 320 mm, with a total height of 672 mm. All floor slabs have been constructed from 16 mm thick steel plates and are supported by four solid round steel columns. Each column has a height of 152 mm and a cross-section diameter of 16 mm. The structural members are made from grade Q235 steel, which has a nominal yield stress of 235 MPa.

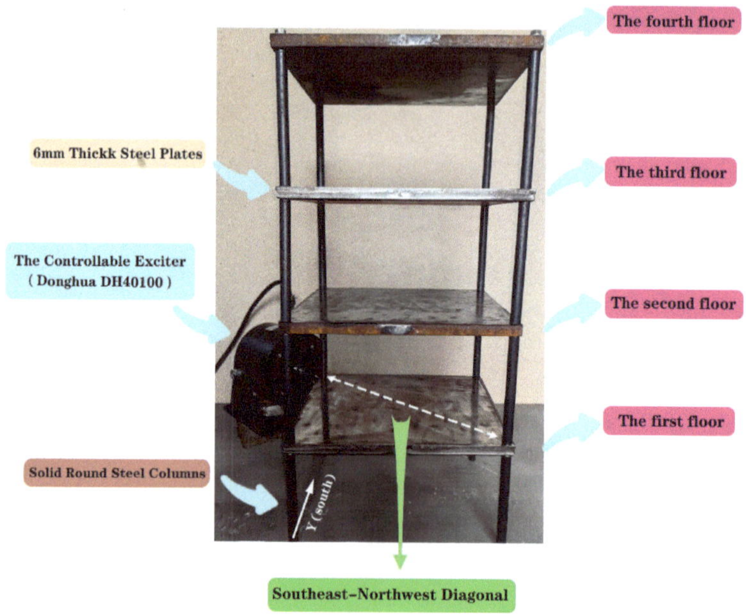

Figure 13. The four-story steel frame structure experimental model.

During the long-term service of the structure, corrosion of the structural metal surface can lead to a decrease in the net cross-sectional area of the components, thereby reducing the load-bearing capacity. It can lead to serious effects on the structural safety. Therefore, the focus of this study was to accurately identify changes in the net cross-sectional area of structural components. In Figure 14, the structure is considered to be in a healthy state, with a net cross-sectional diameter of 16 mm for all columns. The experiment replaced the 16mm original column at the southeast corner of one or more floors in the structural model with columns with a cross-sectional diameter of 14 mm or 12 mm. The purpose was to simulate different degrees of corrosion damage. Table 7 summarizes the six damage scenarios simulated in this paper.

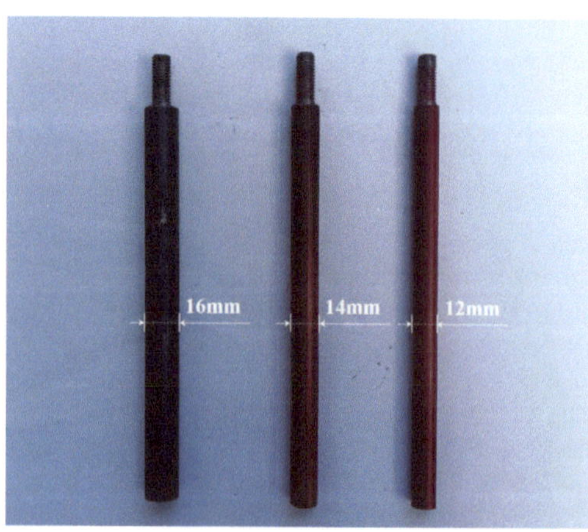

Figure 14. Three types of replacement columns.

Table 7. Damage patterns as defined in the experimental structure model.

Damage Case	Description
D.P.0	Without damage (The columns at the southeast corner of floors 1–4 all have a diameter of 16 mm)
D.P.1	Replaced the column on the first floor with a 14 mm diameter column.
D.P.2	Replaced the column on the second floor with a 14 mm diameter column.
D.P.3	Replaced the column on the third floor with a 14 mm diameter column.
D.P.4	Replaced the column on the forth floor with a 14 mm diameter column.
D.P.5	Replaced the columns on the first and second floors with 14 mm and 12 mm diameter columns, respectively

As shown in Figure 13, a controllable exciter (Donghua DH40100) with a stinger was used to apply a zero mean white Gaussian noise excitation to the first floor of the structure along the southeast–northwest diagonal. Four Donghua 1A401E acceleration sensors were installed at midspan positions on each of the four sides of every story slab to precisely measure the structural responses. A Donghua DH8303 data acquisition system was used to collect acceleration data. The acquisition rate was set at 250 Hz for all damage patterns, with each acquisition lasting 800 s and 200,000 data points being collected each time. Figures 15 and 16 present the y direction acceleration response signals for the first floor in the undamaged state.

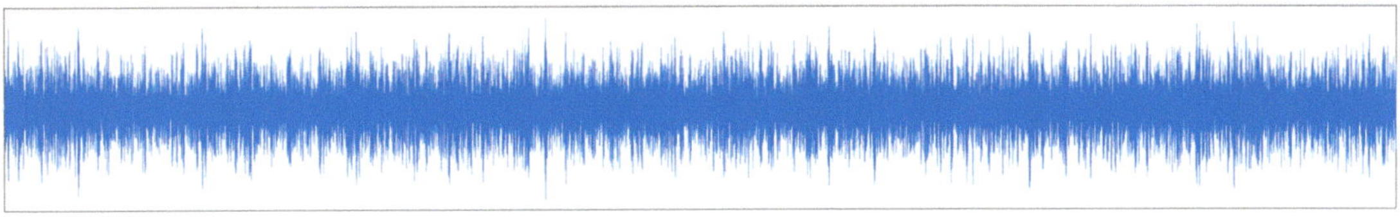

Figure 15. The acceleration response curves of the D.P.0 damage pattern in the south side on the first floor.

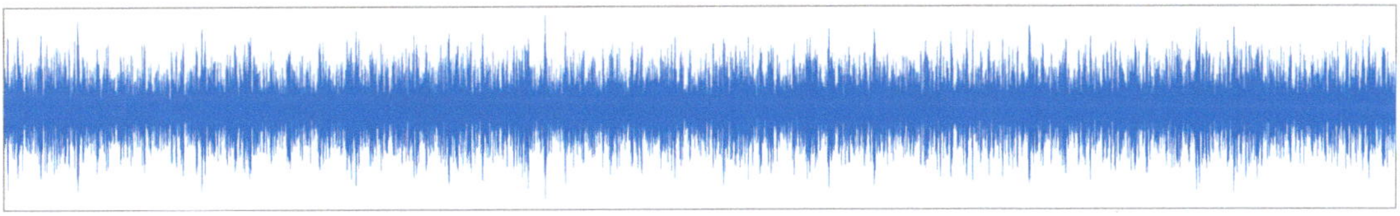

Figure 16. The acceleration response curves of the D.P.0 damage pattern in the north side on the first floor.

4.2. Comparative Analysis of Models in the Experimental Structure

To accurately assess the effectiveness and generalization capabilities of CGsformer, the acceleration response data for each direction on each floor were processed according to the data preprocessing method described in Section 3.1. Then, the dataset for each direction came out to 9372 samples. From these samples, 7500 samples were allocated for training the classifiers, and 1872 were set aside for testing after training. Finally, the preprocessed acceleration response time history data (4 floors × 2 translational directions) were used to train eight CGsformer classifiers.

In Table 8, the performance of the proposed CGsformer model is compared with the other four models, i.e., CNN, LSTM, Transformer, and Conformer. The CGsformer model demonstrated superior performance, thus achieving an average accuracy of 92.44%, which significantly outperformed the Conformer (91.71%), Transformer (87.60%), LSTM (83.92%), and CNN (78.43%) models. Noteworthy are the accuracies observed in the

y direction of the first floor and the x direction of the second floor, where CGsformer reached accuracies of 93.91% and 92.97%, respectively. These outcomes are attributed to the model's hierarchical learning approach. The CGsformer networks employs a multihead self-attention mechanism to capture the global information of signals, which helps to identify the periodic relationships of acceleration response signals. Meanwhile, it utilizes a convolution module to precisely capture the slight differences in acceleration response signals at short and adjacent moments. Furthermore, the graph convolution module embedded in the CGsformer enhances the model's robustness against noise pollution. It does this by constructing a graph structure for global features and filtering out noise in global features through node learning, thereby enabling the convolution module to extract more effective local features. These innovative aspects ensure the efficacy and reliability of CGsformer in structural damage identification, with good generalization performance across various types of damages.

Table 8. Identification accuracy results of the four stories and one-span steel frame structure from different models.

Floor/Direction	CNN	LSTM	Transformer	Conformer	CGsformer
First Floor, x direction	80.98	84.45	89.52	91.18	93.16
First Floor, y direction	77.78	82.10	87.01	93.37	93.91
Second Floor, x direction	76.01	86.59	86.43	91.93	92.97
Second Floor, y direction	82.05	84.93	87.23	93.48	93.56
Third Floor, x direction	81.20	83.60	84.61	88.67	90.03
Third Floor, y direction	76.33	86.37	85.95	91.72	92.25
Forth Floor, x direction	74.47	79.22	88.19	91.93	91.83
Forth Floor, y direction	78.63	84.13	91.82	91.40	92.20
Average	**78.43**	**83.92**	**87.60**	**91.71**	**92.44**

5. Conclusions

This paper presents an innovative deep learning model, called CGsformer, for detecting structural damages. The proposed CGsformer effectively extracts the global and local features of signals by employing a hierarchical learning approach from global to local. Additionally, the GCN is embedded after the multihead self-attention module for further propagating global features to adjacent nodes, which helps to better understand and express node features by incorporating local adjacency information. The proposed damage detection method based on the CGsformer was verified using simulation data from the IASC-ASCE benchmark structure and experimental data from a four-story, single-span, steel frame structure. Some valuable conclusions can be drawn from the validated results:

- The proposed damage detection model has demonstrated its feasibility in test setups with the IASC-ASCE simulated benchmark structure and a four-story, single-span, steel frame structure, thus achieving damage identification accuracies of 96.71% and 92.44%, respectively. These results not only validate the effectiveness of the CGsformer in identifying structural damage but also provide valuable insights for future research.
- The proposed CGsformer model exhibited high accuracy and robustness in limited datasets and noise-contaminated conditions. In the example of the IASC-ASCE benchmark structure, despite the noise level increasing from 0% to 50%, the detection accuracy only decreased by 1.81%. This means that the CGsformer can more effectively extract features from the acceleration response signal, thus showcasing strong noise resistance.

Although the proposed method achieved surprising performance through its learning manner from local to global, some limitations should also be noted. From an application perspective, the proposed method has not yet been tested in practical engineering. To this end, we plan to collaborate with industry partners to implement and validate our methods in real-world structural health monitoring scenarios. From a technical perspective, although the Transformer can achieve parallel computing compared to the RNN, it also increases the number of model parameters. Thus, we will consider compressing or pruning the model in the future work. Moreover, obtaining acceleration response data for structures with damage is a challenge. Thus, future research directions are intended to combine transfer learning with structural numerical simulation models for structure damage detection.

Author Contributions: Conceptualization, T.H., K.M. and J.X.; methodology, T.H., K.M. and J.X.; software, T.H. and J.X.; validation, T.H. and J.X.; formal analysis, T.H. and J.X.; investigation, T.H. and J.X.; resources, K.M. and J.X.; data curation, T.H. and J.X.; writing—original draft preparation, T.H. and J.X.; writing—review and editing, T.H., K.M. and J.X.; visualization, T.H. and J.X.; supervision, K.M. and J.X. All authors have read and agreed to the published version of the manuscript.

Funding: The research is supported by the National Natural Science Foundation of China (No. 50978064/Z091015) and the Natural Science Foundation of Guizhou Province of China (No. 2017[1036]).

Institutional Review Board Statement: Not applicable.

Informed Consent Statement: Not applicable.

Data Availability Statement: The code for generating the datasets used in the numerical verification is available on the datacenterhub website at https://www.dropbox.com/sh/zpkqy5w371mnzam/AAA-Omuvwx72tjv5NhnhnPuMa?e=1&dl=0 (accessed on 7 July 2024). The datasets used in the experimental verification are available upon request from the corresponding author. The datasets are not available to the public, as they are the preliminary results of an ongoing research project carried out in collaboration. Furthermore, this information will be used in future technological developments and will be subject to intellectual property protection.

Conflicts of Interest: The authors declare no conflicts of interest.

References

1. Fan, W.; Qiao, P. Vibration-based damage identification methods: A review and comparative study. *Struct. Health Monit.* **2011**, *10*, 83–111. [CrossRef]
2. Aabid, A.; Parveez, B.; Raheman, M.A.; Ibrahim, Y.E.; Anjum, A.; Hrairi, M.; Parveen, N.; Mohammed Zayan, J. A review of piezoelectric material-based structural control and health monitoring techniques for engineering structures: Challenges and opportunities. *Actuators* **2021**, *10*, 101. [CrossRef]
3. Xiao, F.; Hulsey, J.L.; Balasubramanian, R. Fiber optic health monitoring and temperature behavior of bridge in cold region. *Struct. Control Health Monit.* **2017**, *24*, e2020. [CrossRef]
4. Gkoumas, K.; dos Santos, F.; Pekar, F. Research in bridge maintenance, safety and management: An overview and outlook for europe. In *Bridge Maintenance, Safety, Management, Life-Cycle Sustainability and Innovations*; CRC Press: Boca Raton, FL, USA, 2021; pp. 1755–1761.
5. Malla, P.; Khedmatgozar Dolati, S.S.; Ortiz, J.D.; Mehrabi, A.B.; Nanni, A.; Ding, J. Damage detection in frp-reinforced concrete elements. *Materials* **2024**, *17*, 1171. [CrossRef] [PubMed]
6. Tang, Q.; Xin, J.; Jiang, Y.; Zhang, H.; Zhou, J. Dynamic Response Recovery of Damaged Structures Using Residual Learning Enhanced Fully Convolutional Network. *Int. J. Struct. Stab. Dyn.* **2024**, 2550008. [CrossRef]
7. Lucà, F.; Manzoni, S.; Cerutti, F.; Cigada, A. A damage detection approach for axially loaded beam-like structures based on gaussian mixture model. *Sensors* **2022**, *22*, 8336. [CrossRef] [PubMed]
8. Azimi, M.; Eslamlou, A.D.; Pekcan, G. Data-driven structural health monitoring and damage detection through deep learning: State-of-the-art review. *Sensors* **2020**, *20*, 2778. [CrossRef]
9. Dao, P.B.; Staszewski, W.J. Lamb wave based structural damage detection using stationarity tests. *Materials* **2021**, *14*, 6823. [CrossRef] [PubMed]
10. Xiao, F.; Sun, H.; Mao, Y.; Chen, G.S. Damage identification of large-scale space truss structures based on stiffness separation method. *Structures* **2023**, *53*, 109–118. [CrossRef]
11. Tong, K.; Zhang, H.; Zhao, R.; Zhou, J.; Ying, H. Investigation of smfl monitoring technique for evaluating the load-bearing capacity of rc bridges. *Eng. Struct.* **2023**, *293*, 116667. [CrossRef]
12. Angeletti, F.; Iannelli, P.; Gasbarri, P.; Panella, M.; Rosato, A. A study on structural health monitoring of a large space antenna via distributed sensors and deep learning. *Sensors* **2022**, *23*, 368. [CrossRef]
13. Altabey, W.A.; Wu, Z.; Noori, M.; Fathnejat, H. Structural health monitoring of composite pipelines utilizing fiber optic sensors and an ai-based algorithm—A comprehensive numerical study. *Sensors* **2023**, *23*, 3887. [CrossRef]
14. Kot, P.; Muradov, M.; Gkantou, M.; Kamaris, G.S.; Hashim, K.; Yeboah, D. Recent advancements in non-destructive testing techniques for structural health monitoring. *Appl. Sci.* **2021**, *11*, 2750. [CrossRef]

15. Hassani, S.; Dackermann, U. A systematic review of advanced sensor technologies for non-destructive testing and structural health monitoring. *Sensors* **2023**, *23*, 2204. [CrossRef]
16. Hou, R.; Xia, Y. Review on the new development of vibration-based damage identification for civil engineering structures: 2010–2019. *J. Sound Vib.* **2021**, *491*, 115741.
17. Xiao, F.; Hulsey, J.L.; Chen, G.S.; Xiang, Y. Optimal static strain sensor placement for truss bridges. *Int. J. Distrib. Sens. Netw.* **2017**, *13*, 5.
18. Avci, O.; Abdeljaber, O.; Kiranyaz, S.; Hussein, M.; Gabbouj, M.; Inman, D.J. A review of vibration-based damage detection in civil structures: From traditional methods to machine learning and deep learning applications. *Mech. Syst. Signal Process.* **2021**, *147*, 107077.
19. Nick, H.; Aziminejad, A.; Hosseini, M.H.; Laknejadi, K. Damage identification in steel girder bridges using modal strain energy-based damage index method and artificial neural network. *Eng. Fail. Anal.* **2021**, *119*, 105010.
20. Fallahian, M.; Ahmadi, E.; Khoshnoudian, F. A structural damage detection algorithm based on discrete wavelet transform and ensemble pattern recognition models. *J. Civ. Struct. Health Monit.* **2022**, *12*, 323–338.
21. Indhu, R.; Sundar, G.R.; Parveen, H.S. A review of machine learning algorithms for vibration-based shm and vision-based shm. In Proceedings of the 2022 Second International Conference on Artificial Intelligence and Smart Energy (ICAIS), Coimbatore, India, 23–25 February 2022; pp. 418–422.
22. Zhang, J.; Sato, T.; Iai, S. Support vector regression for on-line health monitoring of large-scale structures. *Struct. Saf.* **2006**, *28*, 392–406. [CrossRef]
23. Hua, X.; Ni, Y.; Ko, J.; Wong, K. Modeling of temperature–frequency correlation using combined principal component analysis and support vector regression technique. *J. Comput. Civ. Eng.* **2007**, *21*, 122–135.
24. Salkhordeh, M.; Mirtaheri, M.; Soroushian, S. A decision-tree-based algorithm for identifying the extent of structural damage in braced-frame buildings. *Struct. Control Health Monit.* **2021**, *28*, e2825.
25. Wang, Y.; Su, F.; Guo, Y.; Yang, H.; Ye, Z.; Wang, L. Predicting the microbiologically induced concrete corrosion in sewer based on xgboost algorithm. *Case Stud. Constr. Mater.* **2022**, *17*, e01649.
26. Lingxin, Z.; Junkai, S.; Baijie, Z. A review of the research and application of deep learning-based computer vision in structural damage detection. *Earthq. Eng. Eng. Vib.* **2022**, *21*, 1–21.
27. Eltouny, K.; Gomaa, M.; Liang, X. Unsupervised learning methods for data-driven vibration-based structural health monitoring: A review. *Sensors* **2023**, *23*, 3290. [CrossRef] [PubMed]
28. Abdeljaber, O.; Avci, O.; Kiranyaz, S.; Gabbouj, M.; Inman, D.J. Real-time vibration-based structural damage detection using one-dimensional convolutional neural networks. *J. Sound Vib.* **2017**, *388*, 154–170.
29. Zhang, Y.; Miyamori, Y.; Mikami, S.; Saito, T. Vibration-based structural state identification by a 1-dimensional convolutional neural network. *Comput.-Aided Civ. Infrastruct. Eng.* **2019**, *34*, 822–839.
30. Khodabandehlou, H.; Pekcan, G.; Fadali, M.S. Vibration-based structural condition assessment using convolution neural networks. *Struct. Control Health Monit.* **2019**, *26*, e2308.
31. Tang, Z.; Chen, Z.; Bao, Y.; Li, H. Convolutional neural network-based data anomaly detection method using multiple information for structural health monitoring. *Struct. Control Health Monit.* **2019**, *26*, e2296.
32. Mantawy, I.M.; Mantawy, M.O. Convolutional neural network based structural health monitoring for rocking bridge system by encoding time-series into images. *Struct. Control Health Monit.* **2022**, *29*, e2897.
33. Lin, Z.; Liu, Y.; Zhou, L. Damage detection in a benchmark structure using long short-term memory networks. In Proceedings of the 2019 Chinese Automation Congress (CAC), Hangzhou, China, 22–24 November 2019; pp. 2300–2305.
34. Sony, S.; Gamage, S.; Sadhu, A.; Samarabandu, J. Vibration-based multiclass damage detection and localization using long short-term memory networks. *Structures* **2022**, *35*, 436–451. [CrossRef]
35. Zou, J.; Yang, J.; Wang, G.; Tang, Y.; Yu, C. Bridge structural damage identification based on parallel cnn-gru. *IOP Conf. Ser. Earth Environ. Sci.* **2021**, *626*, 012017. [CrossRef]
36. Yang, J.; Yang, F.; Zhou, Y.; Wang, D.; Li, R.; Wang, G.; Chen, W. A data-driven structural damage detection framework based on parallel convolutional neural network and bidirectional gated recurrent unit. *Inf. Sci.* **2021**, *566*, 103–117. [CrossRef]
37. Bao, X.; Fan, T.; Shi, C.; Yang, G. Deep learning methods for damage detection of jacket-type offshore platforms. *Process Saf. Environ. Prot.* **2021**, *154*, 249–261. [CrossRef]
38. Fu, L.; Tang, Q.; Gao, P.; Xin, J.; Zhou, J. Damage identification of long-span bridges using the hybrid of convolutional neural network and long short-term memory network. *Algorithms* **2021**, *14*, 180. [CrossRef]
39. Dang, V.-H.; Vu, T.-C.; Nguyen, B.-D.; Nguyen, Q.-H.; Nguyen, T.-D. Structural damage detection framework based on graph convolutional network directly using vibration data. *Structures* **2022**, *38*, 40–51. [CrossRef]
40. Wang, S.; Luo, Z.; Shen, P.; Zhang, H.; Ni, Z. Graph-in-graph convolutional network for ultrasonic guided wave-based damage detection and localization. *IEEE Trans. Instrum. Meas.* **2022**, *71*, 2502011. [CrossRef]
41. Zhan, P.; Qin, X.; Zhang, Q.; Sun, Y. A novel structural damage detection method via multi-sensor spatial-temporal graph-based features and deep graph convolutional network. *IEEE Trans. Instrum. Meas.* **2023**, *72*, 2504814. [CrossRef]
42. Liang, Z.; Li, D.; Ren, W. Structural damage identification method based on recursive graph for automatic feature extraction. In Proceedings of the 29th National Conference on Structural Engineering (Volume II), Wuhan, China, 16–18 October 2020.

43. Johnson, E.A.; Lam, H.-F.; Katafygiotis, L.S.; Beck, J.L. Phase i iasc-asce structural health monitoring benchmark problem using simulated data. *J. Eng. Mech.* **2004**, *130*, 3–15. [CrossRef]
44. Hwang, H.; Kim, C. Damage detection in structures using a few frequency response measurements. *J. Sound Vib.* **2004**, *270*, 1–14. [CrossRef]
45. Yessoufou, F.; Zhu, J. Classification and regression-based convolutional neural network and long short-term memory configuration for bridge damage identification using long-term monitoring vibration data. *Struct. Health Monit.* **2023**, *22*, 14759217231161811. [CrossRef]
46. Vaswani, A.; Shazeer, N.; Parmar, N.; Uszkoreit, J.; Jones, L.; Gomez, A.N.; Kaiser, Ł.; Polosukhin, I. Attention is all you need. In Proceedings of the Annual Conference on Neural Information Processing Systems 2017, Long Beach, CA, USA, 4–9 December 2017.
47. Dai, Z.; Yang, Z.; Yang, Y.; Carbonell, J.; Le, Q.V.; Salakhutdinov, R. Transformer-xl: Attentive language models beyond a fixed-length context. *arXiv* **2019**, arXiv:1901.02860.
48. Zhang, Z.; Cui, P.; Zhu, W. Deep Learning on Graphs: A Survey. *IEEE Trans. Knowl. Data Eng.* **2020**, *34*, 249–270. [CrossRef]
49. Chen, F. Improvement of Kalman Filter and Kalman Estimator in the Application of Structural Damage Detection. Ph.D. Thesis, Xiamen University, Xiamen, China, 2014.
50. Junior, R.F.R.; dos Santos Areias, I.A.; Campos, M.M.; Teixeira, C.E.; da Silva, L.E.B.; Gomes, G.F. Fault detection and diagnosis in electric motors using 1d convolutional neural networks with multi-channel vibration signals. *Measurement* **2022**, *190*, 110759. [CrossRef]
51. Gulati, A.; Qin, J.; Chiu, C.-C.; Parmar, N.; Zhang, Y.; Yu, J.; Han, W.; Wang, S.; Zhang, Z.; Wu, Y.; et al. Conformer: Convolution-augmented transformer for speech recognition. *arXiv* **2020**, arXiv:2005.08100.

buildings

MDPI

Article

Analysis of Mechanical Properties during Construction Stages Reflecting the Construction Sequence for Long-Span Spatial Steel Structures

Gang Yao [1], Rui Li [1], Yang Yang [1,*], Xiaodong Cai [1], Yan Zhou [1], Canwei Zhou [2] and Ting Lei [2]

1 School of Civil Engineering, Chongqing University, Chongqing 400045, China
2 Chongqing Railway Group, Chongqing 401120, China
* Correspondence: 20121601009@cqu.edu.cn; Tel.: +86-18883876194

Abstract: When constructing long-span spatial steel structures, the unformed structure is often incomplete and unstable. The construction sequence significantly influences the mechanical state of the structure during the construction stages (CSs), affecting both the path and time effects. This study examined the mechanical properties of the construction process using an actual project as a case study, comparing two methods: one-step forming and stage-by-stage forming. Critical turning points of stress and displacement during the CSs were identified as the initial installation and unloading stages. Stress concentrations frequently occurred at temporary support points, and peak displacements often appeared at the outer overhanging bars of the structure. A well-planned construction sequence can effectively manage the structure's formation, boundaries, and loading to ensure construction safety and stability. The conclusions and analysis methods from this study provide valuable references for the design and construction of similar long-span spatial steel structures.

Keywords: long-span spatial steel structures; process control; construction sequence; finite element analysis; construction safety

Citation: Yao, G.; Li, R.; Yang, Y.; Cai, X.; Zhou, Y.; Zhou, C.; Lei, T. Analysis of Mechanical Properties during Construction Stages Reflecting the Construction Sequence for Long-Span Spatial Steel Structures. *Buildings* **2024**, *14*, 2389. https://doi.org/10.3390/buildings14082389

Academic Editor: Francisco López-Almansa

Received: 25 June 2024
Revised: 28 July 2024
Accepted: 31 July 2024
Published: 2 August 2024

1. Introduction

Structural state analysis during construction stages (CSs) is pivotal for ensuring the structural safety and efficiency of building projects [1–3]. By examining the mechanics and structural behavior during different CSs, engineers can identify potential instabilities and risks early [4]. The insights gained from such analyses assist in adjusting construction techniques and plans, optimizing resource allocation, and reducing costs, thereby enhancing the success and economic efficiency of building projects [5–8].

Long-span spatial steel structures often experience uneven stresses during the CS, as some members do not yet form a complete mechanical system [9]. The lack of adequate support after installing some structures can lead to additional loads on the installed parts. These uneven forces may cause deformation, stress concentration, or early failure, increasing safety risks and maintenance costs [10,11].

During the structure-forming process, the partially completed structure, supported by a temporary system, behaves as a time-varying structure [12]. The structure's behavior changes over time during the CSs, directly affecting its stress performance during construction and in service [13,14]. The construction sequence significantly influences the path and time effects during construction, making it critical to analyze the structural state of long-span spatial steel structures [15,16]. A well-planned construction sequence ensures the structure receives the necessary support and stability across all CSs, preventing damage or failure from premature or improper loading [17].

Assessing the reasonableness of the construction sequence involves evaluating the technical requirements and structural safety of each CS. Pre-construction simulations that accurately reflect the construction sequence and provide reliable, intermediate, and final results are important indicators for assessment [18–20].

With the rapid development of computer technology, tools such as building information modeling [21] and finite element analysis (FEA) [22] have been widely used in CS analysis. These advanced technologies allow for more accurate simulation and optimization of the construction process for long-span spatial structures [23–26].

Zheng et al. [27] simulated the jacking construction of a steel truss arch bridge using finite element software and analyzed the stresses and deflections at each jacking stage. Liu et al. [28] used finite element software to simulate and analyze the whole process of prestressing tension cable construction and found that the coupling effect would increase the calculated internal force of the structure and effectively reduce the influence of scaffolding dislodgement on the internal force of the structure. Qin et al. [29] proposed the concept of stress-free state variables to derive equations for controlling the mechanical equilibrium and geometry of structures constructed in stages, providing an effective method for construction control and calculation. Li et al. [30]. effectively assessed the local stress state of steel box girders during jacking construction by finite element analysis and revealed the effect of jacking asynchrony on structural forces.

Currently, the analysis of the structural state during the CS of long-span spatial structures has not garnered adequate attention. Specifically, the construction sequence plays a pivotal role in determining the mechanical behavior and stability of the structure as it is being built. To address this gap, our study aims to integrate theoretical analysis, numerical simulation, and real-world engineering examples. Through this comprehensive approach, we seek to delve deeply into the mechanical behavior of long-span spatial structures during their construction phase. Furthermore, we aim to elucidate the precise mechanism through which the construction sequence impacts the structural safety of such projects. By conducting this research, we hope to offer practical guidelines and recommendations for constructing similar long-span spatial structures. Our goal is to ensure not only the safety of these structures but also the efficiency of the construction process itself. This study promises to bridge the knowledge gap, providing valuable insights for engineers and construction professionals alike.

This paper is organized as follows. Section 2 provides the background to this study, including the theory of nonlinear time-varying analyses during the CS, commonly used steel construction techniques, and the engineering background. Section 3 presents the methodology of this study. The construction scheme is developed for an actual project to guide the specific construction process. Section 4 analyzes the construction sequence's effect on the construction process's mechanical properties. FEA software Midas Gen 2022 is used to analyze the mechanical properties of the whole construction process of a long-span spatial steel structure that explicitly reflects the construction sequence. The difference between one-step and stage-by-stage forming of the structure is compared. Section 5 presents the conclusions of this study.

2. Background

2.1. Nonlinear Time-Varying Analysis of CS

The construction of a long-span spatial steel structure is a complex and gradual process of the structural system. Throughout the construction process, the structure's geometry, stiffness, loading, and boundary conditions represent time-varying characteristics [31]. This makes for a certain degree of nonlinear relationship between the force state of the structure during CS and time [32,33]. Iterative or incremental methods generally solve nonlinear problems [34]. In particular, the incremental method involves dividing the load into several steps, where each load increment ΔP corresponds to a displacement increment $\Delta\delta$ in different loading steps, and the final displacement is obtained by accumulating the incremental steps. This method can obtain the analysis results of each stage in the loading process and is an effective method for solving geometric nonlinear problems in CS [35].

Based on incremental finite element theory, the birth and death unit method is a nonlinear analysis method of construction mechanics with more current applications [36]. The birth and death unit method achieves dynamic changes in the model by building a complete structural model at once and gradually activating or deactivating the corresponding units at each CS. The birth and death unit method ensures the continuity of the model. The structural analysis of each CS is based on the previous stage's internal force and deformation results, reflecting the stage-by-stage effects of the construction process. This study uses the node-corrected birth and death unit method to achieve the mechanical simulation of the construction process. The three-section cantilever structure shown in Figure 1 is used as an example to illustrate the basic principle of this analysis method.

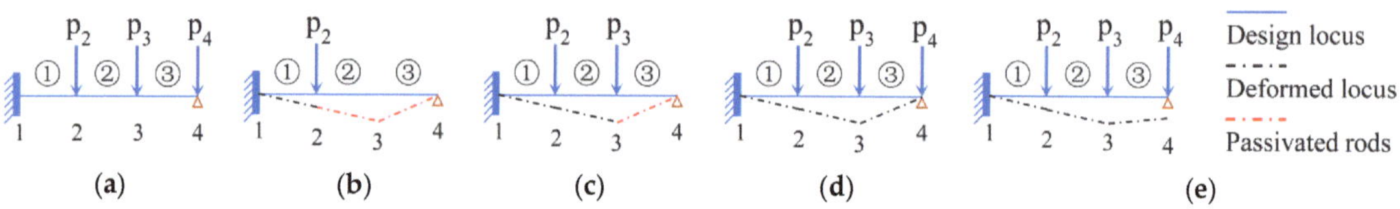

Figure 1. The analytical process of the node-corrected birth and death unit method. (**a**) Structural calculation model; (**b**) First construction stage. (**c**) Second construction stage. (**d**) Third construction stage. (**e**) Fourth construction stage.

In the first two CSs, the results of the activated unit calculations under the node-corrected birth and death unit method were identical to those of the birth and death unit method. However, the displacement of node 4 is constrained due to the temporary support. The drift of the dead unit is suppressed, controlling the installation pattern of the construction process. When analyzing the 3rd CS, due to the presence of temporary supports, the 3rd section of the beam is activated at this point and coupled with geometrical nonlinearities, forming the controlling equations of the structure at this stage in Equation (1).

$$\begin{bmatrix} K_{11,1} & K_{12,1} & 0 & 0 \\ K_{21,1} & K_{22,1}+K_{22,2} & K_{23,2} & 0 \\ 0 & K_{32,2} & K_{33,2}+K_{33,3} & K_{34,3} \\ 0 & 0 & K_{43,3} & K_{44,3} \end{bmatrix} \begin{bmatrix} u_1 \\ u_2 \\ u_3 \\ u_4 \end{bmatrix} = \begin{bmatrix} F_1 \\ F_2 \\ F_3 \\ F_4 \end{bmatrix} \tag{1}$$

where $K_{ij,m}$, u_i, and F_i are sub-matrices of the unit stiffness matrix, node displacements matrix, and load column matrices, respectively. $i = j = m$ has values in the range 1, 2, 3, 4.

The load P_4 is introduced into Equation (1) to act directly on the temporary support with boundary conditions $u_1 = 0$ and $u_4 = 0$. The controlling equation is now as in Equation (2).

$$\begin{bmatrix} 1 & 0 & 0 & 0 \\ 0 & K_{22,1}+K_{22,2} & K_{23,2} & 0 \\ 0 & K_{32,2} & K_{33,2}+K_{33,3} & 0 \\ 0 & 0 & 0 & 1 \end{bmatrix} \begin{bmatrix} u_1 \\ u_2 \\ u_3 \\ u_4 \end{bmatrix} = \begin{bmatrix} 0 \\ P_2 \\ P_3 \\ 0 \end{bmatrix} \tag{2}$$

The dead units are activated now, and the stiffness matrix is no longer singular. The displacement value under this CS can be obtained directly. Finally, the 4th CS is analyzed based on the first three CSs, and the control equation is shown in Equation (1). However, its boundary condition becomes $u_1 = 0$, and the corresponding control equation is shown in Equation (3).

$$\begin{bmatrix} 1 & 0 & 0 & 0 \\ 0 & K_{22,1}+K_{22,2} & K_{23,2} & 0 \\ 0 & K_{32,2} & K_{33,2}+K_{33,3} & K_{34.3} \\ 0 & 0 & K_{43,3} & K_{44,3} \end{bmatrix} \begin{bmatrix} u_1 \\ u_2 \\ u_3 \\ u_4 \end{bmatrix} = \begin{bmatrix} 0 \\ P_2 \\ P_3 \\ P_4 \end{bmatrix} \tag{3}$$

The displacement value at each node at completion can be obtained through Equation (3). After the displacements at each CS have been calculated, the corresponding structural strains and stresses can be derived from the geometric and physical equations.

2.2. Construction Methods

Long-span spatial steel structures adopt different construction methods for different structural forms and states [37]. Through scientific and reasonable selection of construction methods, construction efficiency, quality, and safety can be effectively improved [38,39]. At present, the most often used construction methods include the overhead bulk installation method [40], installation methods in strips or blocks [41], the aerial skidding method [42], the overall hoisting method [43], the overall lifting method [44], and the overall jacking method [45]. Relevant cases are shown in Figure 2.

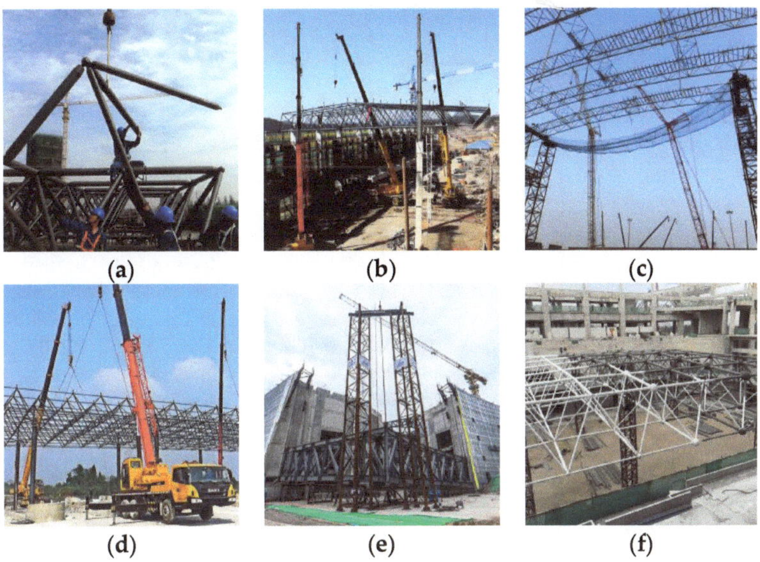

Figure 2. Common construction methods. (**a**) Overhead bulk installation method; (**b**) installation method in strips or blocks; (**c**) aerial skidding method; (**d**) overall hoisting method; (**e**) overall lifting method; (**f**) overall jacking method.

2.3. Engineering Background

Chengdu Tianfu International Airport in China contains terminals T1 and T2 in a "T" configuration. The building outline is symmetrical. T1 has a floor area of 387,400 m^2 and a land area of 126,000 m^2. It is a class I integrated terminal building. The building is 1290 m long from north to south, 520 m wide from east to west, and the highest point of the roof is 45 m. The roof grid elevation of T1 is hyperboloidal in shape, and it forms several spatial structures with significant differences in elevation through three skylight bands. The overall steel roof is a steel mesh frame with orthogonal and orthogonal four-angle cone-shaped welded-ball mesh frames. The height of the mesh frame is 4~6 m. The minimum height of the overhanging part is about 1.5 m. The grid size of the standard internode is about 5.2 m × 5.2 m. The specification of the rods is 180 mm × (10~500) mm × 40 mm, and the main specification of the welded balls at the node position is D300 × 14~D800 × 50, among which the balls with a diameter of more than 500 mm are the ones with reinforced ribs. The total steel consumption of the project is about 24,000 t, the number of mesh frame bars is about 160,000 and the projected area of the mesh frame is about 180,000 m^2. The actual view of the project is shown in Figure 3.

Figure 3. Actual view of the project.

T1 contains the central hall of the D area and three finger corridors: A, B, and C. Seismic joints separate them to form four independent structural units. The central hall of the D has a large project volume and is in the core dominant position in the construction process. Hence, this study focuses on the analysis of the central hall of the D. The plan dimensions of the central hall of the D are 522 m × (107~324) m, and the central column network dimensions are 9 m × 18 m, 12 m × 18 m, and 18 m × 18 m, with four floors above-ground, five floors of local commercial content, and a roof apex elevation of 45 m. The overview of the project is shown in Figure 4. This steel structure project has a large volume, wide construction scope, long timeframe, and, at the same time, high requirements for installation accuracy, which brings enormous challenges to on-site construction.

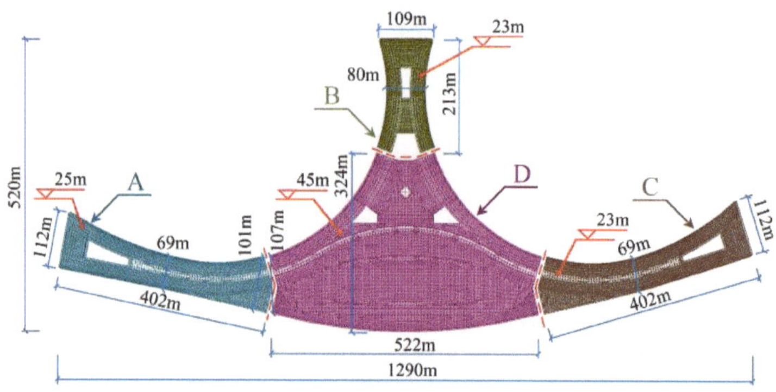

Figure 4. Overview of the project.

3. Methodology

The T1 terminal building adopts a long-span spatial steel structure system. In order to ensure the safety and stability of the roof's mesh frame of the T1 terminal building during the construction process, the mechanical performance analysis of the CSs of the long-span spatial steel mesh frame, which clearly reflects the construction sequence, was carried out. The framework of the analysis method is shown in Figure 5.

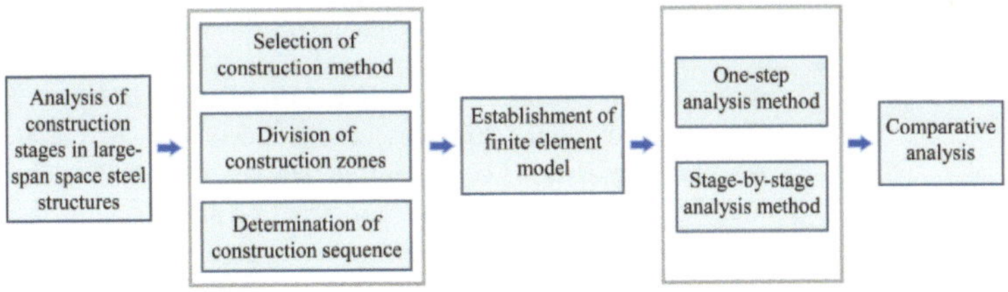

Figure 5. Analytical methodology framework.

The analysis of the CSs of the long-span spatial steel structure project begins with the corresponding construction design. Including the selection of the construction method, the division of the construction zones, and the determination of the construction sequence. The FEA model is established close to the actual project. The structural and mechanical properties and displacements obtained by the one-step and stage-by-stage forming methods are analyzed, respectively. Then, the results obtained by the two methods are compared, and finally, conclusions are drawn that are useful for guiding the construction.

3.1. Selection of Construction Methods

The long-span spatial steel structure in this project is constructed by combining floor or ground assembly and then the overall cumulative hoisting of blocks and local-zone block lifting. Each block hoisting unit is divided into several cumulative hoisting assembly zones according to the structural floor assembly elevation below the mesh frame projection and the mesh frame's elevation. The cumulative lifting of the zones during construction can effectively use the working surface of the concrete floor. The amount of overhead work is small, ensuring construction safety while accelerating the construction progress.

3.2. Division of Construction Zones

The spatial steel mesh frame is supported on the lower steel columns connected to the roof using column-top supports. The structure was partitioned based on the concrete construction sequence. As shown in Figure 6a, the hall's spatial steel grid was divided into 10 integral hoisting block units and 4 lifting block units.

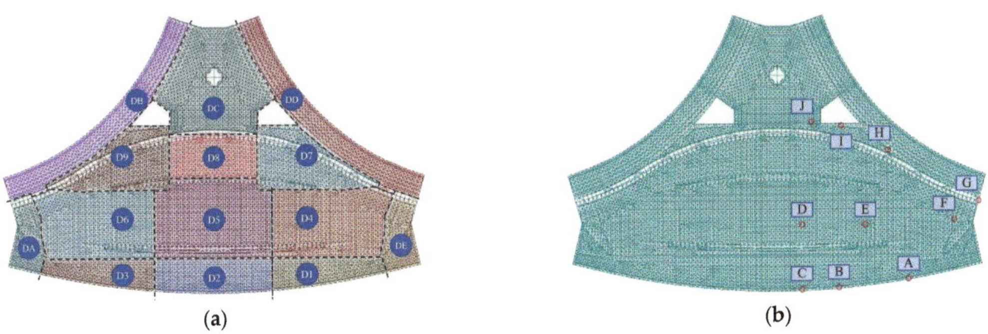

Figure 6. (**a**) Construction area division; (**b**) monitoring point arrangement.

Critical locations during construction were selected for continuous monitoring of stress and displacement changes. Due to the symmetry of the structure in the plan, one half of the structure was selected for monitoring. The nodes were selected as a reference based on the occurrence of maximum stress and displacement change values during each CS. The arrangement of monitoring points is shown in Figure 6b.

3.3. Determination of Construction Sequence

The project features a hyperbolic spatial mesh frame structure notable for its complexity, expansive spatial span, numerous rods, and stringent demands on installation accuracy. The scope of steel structure construction is extensive, and the timeframe is protracted. Additionally, there is considerable overlap with civil construction, electromechanical work, roofing, and other units. The construction site is confined, necessitating a piecemeal provision of workspaces. Therefore, the construction sequence is as follows: 2 sets of crawler cranes start construction from the left zone of the C zone and advance the construction in the direction of the A zone. The mesh frame hoisting installation process is interspersed with the local block zone rod embedded in it. Construction of the lifting block zone for part of the lifting operation. Temporary support partition unloading is interspersed in the process of mesh frame installation.

The specific CSs were as follows: hoisting sub-block D1 and installing it into the column-top support, hoisting sub-blocks D2 to D4, and carrying out the rod insertion between adjacent sub-blocks. After completing the installation of the D4 sub-block, lifting the DA and DE sub-block, and handling the connection between adjacent structures. Next, remove the lifters of sub-blocks D4 and D1 and turn them into supports. After that, hoist sub-blocks D5 to D9 in sequence, nesting the rods between neighboring structures after each hoist while dismantling the lifters. During this period, the lifting of the DB and DD sub-blocks was interspersed with removing the hoists on the D5 and D6 sub-blocks. Eventually, the DC sub-block was hoisted, and after removing its hoists and the remaining tire frames, all zones of the central-span support pylons were unloaded, completing the installation of the entire structure. Figure 7 shows the critical stages in the structure's formation process.

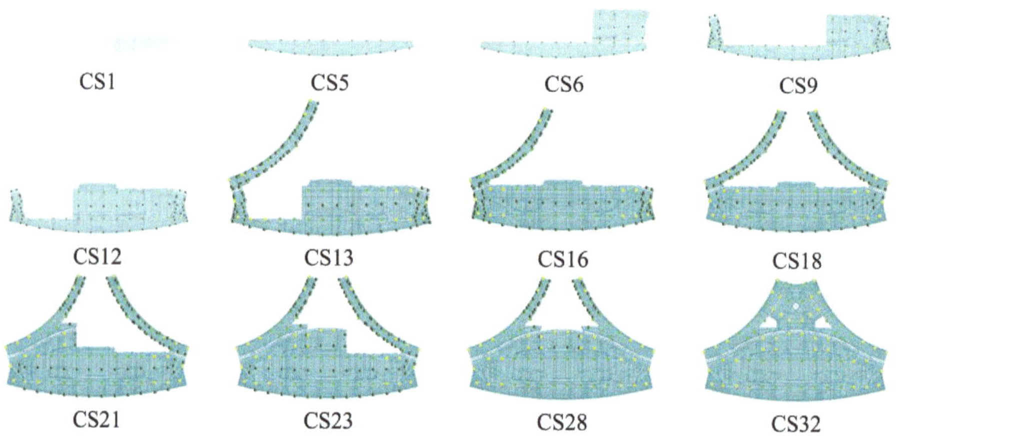

Figure 7. Key stages in the structure-forming process.

4. Experiments and Results

The construction of long-span spatial steel structures is affected by various factors, including the installation sequence, the transformation of the stress system, and the construction loads. The stiffness of the structural system is related to the stress–strain path, so the boundary time-varying process of its construction will directly affect the stress–strain path, which leads to different mechanical states of the overall structure. In this study, the finite element calculation software Midas Gen 2022 is used to perform CS simulation analysis to verify the reasonableness of the construction scheme. The structure is verified to meet the design and code requirements during and after forming. The pre-conditioning deformation is also set according to the displacement results to eliminate the adverse effects of construction deformation and to achieve the structural appearance requirements and functionality.

4.1. Establishment of the Model

The time-varying structure consists of the partially completed structure and the temporary support system, which constantly changes as construction proceeds. Moreover, each CS's specific conditions and results impact subsequent CSs. Therefore, establishing a correct model is essential to ensure the accuracy and reliability of the analysis results. The steps in the construction sequence analysis include the setting of material properties, which in turn leads to the creation of a complete model. Structural, boundary, and load groups are defined according to the construction scheme. The definition of the CS is then achieved by activation and inactivation of the structure, boundary, and load groups. The structural model was built in Midas Gen 2022 software, and the following is a detailed description of the computational model.

1. Material properties. According to the structural design program, the main materials are Q345 steel and 650-grade alloy steel [46–48]. The steel tie rods are made of 650-grade alloy steel, and the rest of the structural parts are made of Q345 steel;
2. Load value. The load conditions involved in the construction process analysis include the constant load and live load. The constant load includes the self-weight of the steel structure and accessory structures. Considering the bolted ball nodes in the grid structure, the self-weight coefficient is enlarged to 1.05. The live load includes the construction loads, such as temporary personnel and machinery weight during construction;
3. Power amplification factor. The structure is subjected to dynamic loads due to the action of lifting equipment during installation. Therefore, the dynamic load coefficient is set to 1.4;
4. Unit type. The grid structure is simulated by a beam unit, considering the bending moment. The truss unit simulates the steel tie rods without considering the bending moment;
5. Boundary conditions. Fixed supports simulate the connection between fixed supports at the top of columns and the mesh frame. Articulated bearing simulates the connection between a temporary support and the grid structure. The sliding bearing simulates the connection between the sliding bearing at the top of the column and the grid. The elasticity coefficients of the sliding bearing are set to 8 kN/mm, 6 kN/mm, and 4 kN/mm, respectively.

4.2. One-Step Analysis Method

The one-step analysis method can quickly and accurately predict and demonstrate before the structure is actually constructed, the stresses and displacement deformations of the structure once it is fully formed, and any potential structural safety hazards that may arise. The rapidity of the methodology helps the project team to evaluate the effects of different construction design options and select the best solution, thereby reducing construction design time.

4.2.1. Stress Analysis

The stress state of the structure after one-step forming is analyzed, and the stress cloud diagram and the stress values at each monitoring point are shown in Figure 8.

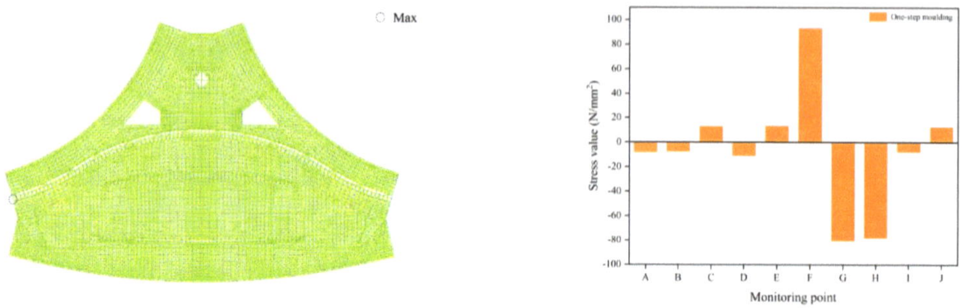

(a) (b)

Figure 8. (**a**) Stress cloud diagram after one-step forming; (**b**) stress values at each monitoring point.

As shown in Figure 8a, the overall stress distribution after the structure is molded is relatively uniform. The stresses showed a symmetrical distribution. The maximum stress value is located in the outermost suspension member in the DA zone, and its value is 113.2 N/mm^2. It is influenced by the self-weight of the suspension structure and the longer span. The presence of fixed supports led to the concentration of some highly stressed bars in the D6 and D4 zones, but the stress values were less than 100 N/mm^2, which satisfied the safety requirements. Figure 8b shows the stress values at key monitoring points. The maximum stress is at point F, with a value of +93.16 N/mm^2; the minimal stress is at point B, with a value of -7.37 N/mm^2. The higher stress monitoring points F, G, and H are all located in the structure's perimeter in the overhanging zone, which reflects the critical zone for monitoring the structure's stress after forming.

4.2.2. Displacement Analysis

The displacement state of the structure after one-step forming is analyzed, and the displacement cloud diagram and the displacement values at each monitoring point are shown in Figure 9.

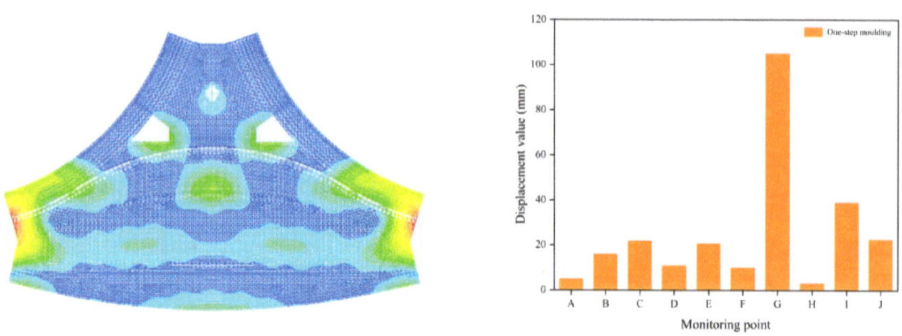

(a) (b)

Figure 9. (**a**) Displacement cloud diagram after one-step forming; (**b**) displacement values at each monitoring point.

As shown in Figure 9a, the larger displacements after forming the structure are concentrated in the overhang and mid-span regions of the structure. Similarly, the displacements are symmetrically distributed. The maximum displacement is located in the outermost bars of the overhangs in the DA and DE zones, with a value of 107.84 mm. Although the value meets the specification requirements, it is appropriate to set the deflection value of the constant load in the reverse arch according to the deflection value of the constant load in order to ensure the accuracy of the structure's formation. Figure 9b shows the displacement values at key monitoring points. The maximum displacement is at point G, with a value of 105.24 mm. The minimum displacement is at point H, with a value of 3.24 mm. The displacement value of monitoring point G, located in the outer part of the overhanging bars in the DE zone, is much higher than that of the rest of the monitoring points. This is because point H is affected by the support of the temporary support frame below and point G has the largest overhang distance and is affected by the effect of self-weight. Therefore, more attention should be paid to the structure's displacement and the deformation of the overhang and central-span zone.

4.3. *Stage-by-Stage Analysis Method*

The stage-by-stage forming method predicts the stress state and displacement deflections at each CS by modeling the step-by-step construction process of the structure. By modeling the loading conditions and structural interactions specific to each stage, this method provides a detailed understanding of the cumulative impact of successive construction activities on the overall structural integrity.

4.3.1. Stress Analysis

According to the division of CSs, the structure is divided into crucial CSs in the process of block installation, steel tie-rod tensioning, and overall unloading in the zones. Therefore, the stress cloud diagram results of the critical stages are selected to be presented, as shown in Figure 10.

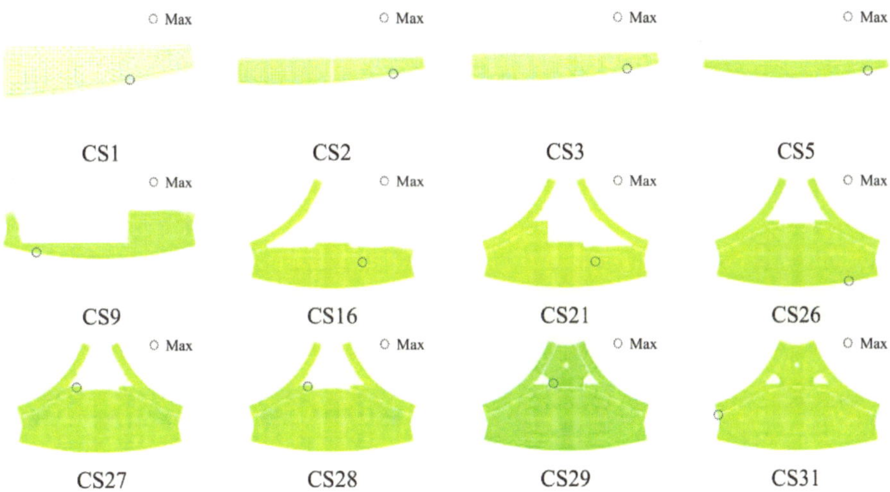

Figure 10. Stress cloud diagram in key construction stages.

The stress cloud diagrams visualize the stress distribution of the unshaped structure at various CSs. It can be seen that there is an evident stress concentration at the support points of the structure, while the structure has a more uniform distribution of stresses in other areas. From CS1 to CS8, the maximum stress value in the structure is located at the temporary support point in the D1 zone. At CS9, the temporary bracing of D1 zone was unloaded, the maximum stress value within the structure was transferred to the D3 zone. At CS16, when the temporary support frame was uninstalled at D3, the maximum stress value shifted to the D4 zone. Until CS26, when the temporary bracing frames of DD and the D7 zone were removed, the maximum stress value again appeared to be highest at the initial position of the D1 zone. After that, the construction operations consisted mainly of zone unloading, and the maximum stress value always appeared in the D9 zone. At CS31, the dismantling of the temporary bracing in subzones DB, DC, and DD caused the maximum stresses to occur in the outermost overhangs of subzone DA, which continued until the construction was formed. The forming and unloading of the structure was the leading cause of stress distribution throughout the construction process. It was necessary to ensure that the temporary bracing in the high-stress zones could reliably support the superstructure loads.

The maximum values of stresses occurring at each CS are recorded in Table 1. Their variations can be seen in Figure 11a.

Table 1. Maximum stresses in each construction stages.

Construction Stage	Tensile Stress (N/mm^2)	Compressive Stress (N/mm^2)	Construction Stage	Tensile Stress (N/mm^2)	Compressive Stress (N/mm^2)
CS1	+215.53	−158.65	CS17	+173.37	−181.97
CS2	+215.53	−158.65	CS18	+173.65	−182.15
CS3	+215.67	−158.78	CS19	+173.65	−182.15
CS4	+215.67	−158.78	CS20	+173.66	−182.15
CS5	+215.67	−158.78	CS21	+175.70	−182.18
CS6	+215.67	−158.78	CS22	+175.70	−182.18
CS7	+215.28	−158.50	CS23	+175.71	−182.21
CS8	+216.18	−159.27	CS24	+175.71	−179.74
CS9	+212.10	−176.13	CS25	+175.71	−179.27
CS10	+212.10	−176.13	CS26	+180.77	−177.82
CS11	+212.11	−179.11	CS27	+115.38	−154.87
CS12	+211.95	−181.95	CS28	+112.03	−154.07
CS13	+211.89	−181.95	CS29	+112.03	−166.54
CS14	+211.89	−181.95	CS30	+120.09	−186.28
CS15	+211.88	−181.95	CS31	+111.99	−101.48
CS16	+173.24	−181.97	CS32	+111.97	−101.88

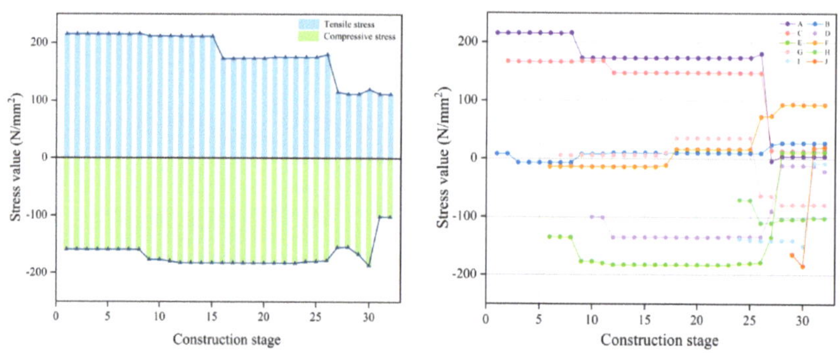

(**a**) (**b**)

Figure 11. (**a**) Variation of the maximum value of stress during the construction stages; (**b**) variation of the value of stress at the monitoring point.

The maximum tensile stress of 216.18 N/mm^2 and the maximum compressive stress of 186.28 N/mm^2 appear in CS8 and CS30, respectively. Both are less than the strength design value of the Q345 material, 295 N/mm^2. This shows that the strengths of the structural materials during construction meet requirements and have more safety reserves. Based on the trend in stress change, the data can be divided into the following four main stages.

The tensile and compressive stresses from CS1 to CS8 are relatively high and stable. They are around 215 N/mm^2 and −158 N/mm^2, respectively. This shows that the structure is subjected to high but infrequent changes in tensile and compressive stresses in this stage. The construction work in this stage mainly consists of installing individual sub-blocks and mending the linkage bars between neighboring sub-blocks. The stage is named the initial stage of stress.

There was a significant increase in compressive stresses from CS9 to CS15, with a total increase of 22.68 N/mm^2. In contrast, tensile stresses decreased slightly, with a total decrease of only 4.3 N/mm^2, due to the first removal of the lifters on the lifting sub-blocks and their conversion to supports during CS9. A case of redistribution of structural stresses was carried out during construction. This stage is named the stress adjustment stage.

The tensile stress decreased significantly from CS16 to CS25, from 211.88 N/mm^2 to 173.24 N/mm^2. The removal of the lifters on the D6 and D3 sub-blocks resulted in stress redistribution in the structure, which led to a decrease in the tensile stress. Moreover, the compressive stress remained stable, showing the continuity and consistency of the construction process. This stage is named the stress acceleration adjustment stage.

Both tensile and compressive stresses decreased significantly from CS26 to CS32. The total reduction in tensile stress is 68.8 N/mm^2, and the total reduction in compressive stress is 75.94 N/mm^2. The main construction task in this stage is the step-by-step demolition of the temporary structure, and the structural stresses are continuously redistributed. At this time, the construction has been nearly completed, and the structure has been relatively stable. This stage is named the stress stabilization stage.

Figure 11b shows the stress changes at the monitoring points throughout construction. From CS1 to CS25, the stress values at monitoring points A, C, D, E, and I are above 100 N/mm^2. From CS26 to CS32, the stress values at monitoring points D, E, H, and I are above 100 N/mm^2. This requires that attention be focused on the structural stresses in the vicinity of the locations of these monitoring points.

Combined with Figure 11a,b, the structural stress values after forming tend to be relatively stable and low. Stress maxima tend to occur during construction. These stress peaks may be due to stress concentrations resulting from changes in structure, loads, or boundary conditions during a particular CS. Therefore, to ensure a smooth construction process and the stability and safety of the structure for long-span spatial steel structures, the construction sequence must be fully considered and accurately modeled.

4.3.2. Displacement Analysis

Displacement changes are the focus of construction analyses. Monitoring displacements can help engineers assess the structure's performance during CS to ensure that it meets the design criteria and to identify potential safety issues in a timely manner. Therefore, the displacement cloud diagram results of the critical stages are selected for presentation, as shown in Figure 12.

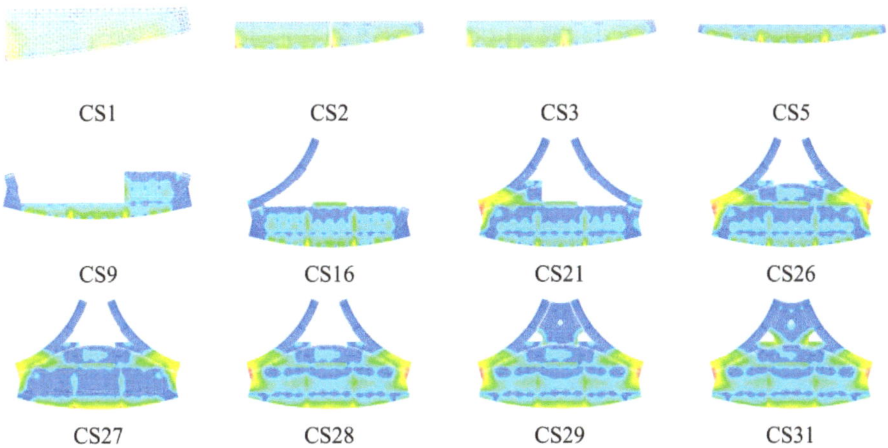

Figure 12. Displacement cloud diagram in key construction stages.

As shown in Figure 12, the higher displacement deformations during the construction of the structure occurred in the overhanging areas of the structure and the mid-span areas of the structural subdivisions. The remaining locations have more uniform displacement changes. From CS1 to CS20 and CS27, the overhang in the D1 and D3 zones adjacent to the D2 zone is where the displacement deformation is at its maximum. The other stages also showed large displacement deformations. From CS21 to CS26 and CS28 to CS32, the outermost overhangs of DA and DE showed the maximum displacement deformations. After CS26, the structure has been primarily formed. The construction activities after that mainly included unloading in the zones and replenishing rods in localized areas. At CS27, after unloading the temporary bracing frames of D1, D2, and D3, the displacement and deformation of the outermost overhanging bars in this area were concentrated. At CS28, after unloading the temporary bracing frames in the zones of D4, D5, and D6, the displacement and deformation in the overhanging zone of D1, D2, and D3 decreased, and the displacement and deformation in the central-span position of D4, D5, and D6 increased as a whole. The displacement distribution in the displacement cloud diagram was stable in the subsequent CS.

The maximum values of displacements occurring at each CS are recorded in Table 2. Their variations can be seen in Figure 13a.

Table 2. Maximum displacement in each construction stage.

Construction Stage	Displacement (mm)	Construction Stage	Displacement (mm)
CS1	70.86	CS17	68.72
CS2	71.23	CS18	68.74
CS3	74.43	CS19	68.74
CS4	74.43	CS20	68.72
CS5	74.83	CS21	89.18
CS6	74.83	CS22	89.18
CS7	74.83	CS23	89.18
CS8	74.83	CS24	89.19
CS9	74.84	CS25	89.19
CS10	74.84	CS26	89.17
CS11	74.75	CS27	110.76
CS12	71.37	CS28	106.47
CS13	71.38	CS29	106.47
CS14	71.38	CS30	106.47
CS15	71.07	CS31	106.48
CS16	68.71	CS32	106.52

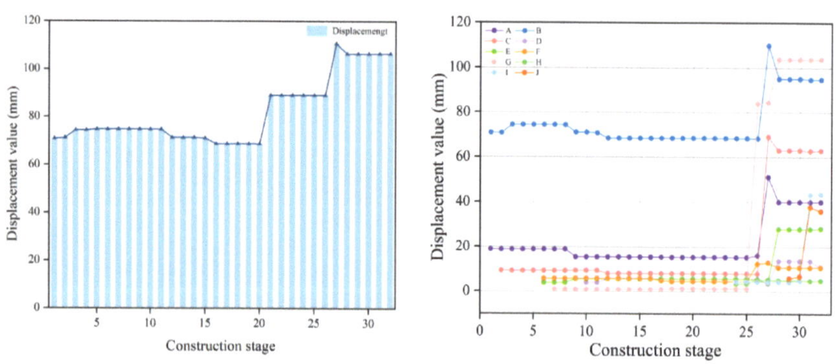

(a) (b)

Figure 13. (**a**) Variation of the maximum value of displacement during the construction stages; (**b**) variation of the value of displacement at the monitoring point.

The maximum displacement value of 110.76 mm occurred at CS27, where the deflection tolerance value is $L/250$, and the short span is 53,664 mm, so the displacement value of the frame in this area during construction meets the requirement. In order to ensure the accuracy of structure forming, it is appropriate to set up the reverse arch according to the deflection value of constant load here. The data can be divided into three main stages based on the trend in displacement changes.

The displacement values from CS1 to CS20 are relatively stable, with an extreme difference of 6.13 mm. The displacement maximum point is located at monitoring point B or the symmetry point of B. This is mainly due to the sizeable overhanging span and the self-weight of the structure's frame bars at that point. The stage is named the displacement stabilization stage.

From CS21 to CS26, the temporary support frame of the DB portion and the lifter of the D9 subsection were dismantled at CS21. This caused a displacement maximum of 89.18 mm at the symmetry point of point G in the DA zone. The stage is named the displacement fluctuation stage.

From CS27 to CS32, the temporary support frames of the D1, D2, and D3 overhanging areas were unloaded at CS27. Therefore, the maximum displacement value of the monitoring point rises to 110.76 mm, which is the maximum value during the construction process. With the continuous forming and unloading of the structure, the maximum displacement value was stabilized at about 106 mm and continued until the final forming of the structure. This stage is named the peak displacement stage.

Figure 13b shows the changes in displacement at the monitoring points throughout the construction process. The displacement values at monitoring points B, C, G, and their symmetry points are high. B, G, and their symmetry points can reach the maximum displacement value.

Combined with Figure 13a,b, unlike stresses, the displacement values of the structure are an accumulating process during construction. The structural displacement values after forming tend to be at a high level. This is determined by the stress characteristics of the structure and the cumulative effect during the construction process.

4.4. Construction Sequencing Impact Analysis

The one-step and stage-by-stage analysis methods have unique applications and advantages in engineering construction design. This study compares and analyses the simulation results of the two methods on the structure's stress state and displacement and deformation, using the actual project as a case study.

4.4.1. Stress Analysis

Stress is a physical quantity that describes the force applied to a unit area within a material, and it can reflect the internal mechanical behavior and state of a structure when subjected to external forces. The stress values after one-step forming and the maximum values in the stage-by-stage forming were recorded at the critical monitoring points. The results are shown in Figure 14.

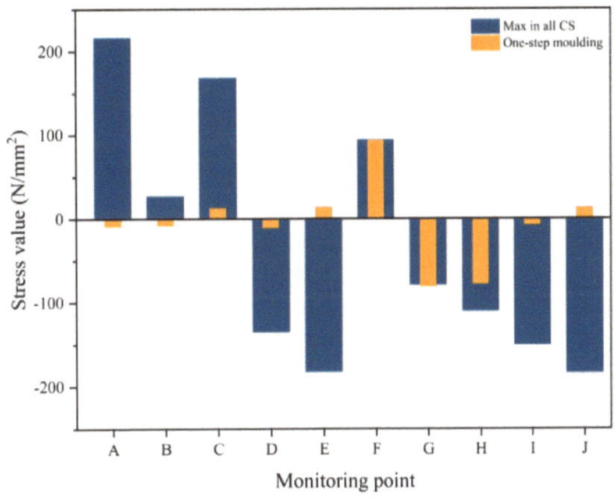

Figure 14. Comparison of maximum stress values at monitoring points obtained by the two methods.

As shown in Figure 14, the stresses in one-step and stage-by-stage forming are significantly different at most of the monitoring points. Often, a particular stage in the construction process will show higher absolute values of stress than in the forming condition. Only monitoring points F and G better match the stress values in stage-by-stage and one-step forming. In contrast, at the rest of the monitoring points, the difference between the stress values obtained by the two methods is too significant. In particular, the absolute stress value obtained by the two methods at monitoring point A is 208.21 N/mm^2, and the stress states is one of tension and the other of compression.

4.4.2. Displacement Analysis

Displacement is the change in shape or position of a structure after a force has been applied. By measuring and analyzing a structure's displacement response, the structure's overall stability and possible structural damage can be assessed. The values of displacement deflections after one-step forming and the maximum values of displacement deflections in stage-by-stage forming were recorded at the critical monitoring points. The results are shown in Figure 15.

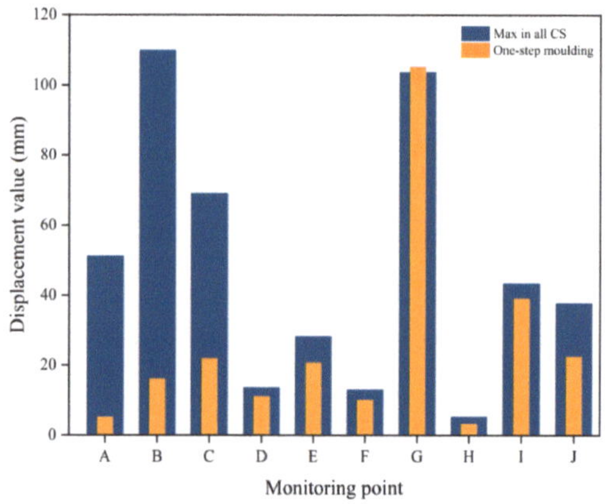

Figure 15. Comparison of maximum displacement values at monitoring points obtained by the two methods.

As shown in Figure 15, the values of displacement deformation obtained by the two methods also exhibit significant differences. Most of the maximum values of displacement occurring during the construction process are higher than the one-step forming displacement values. The phenomenon is more evident at monitoring points A, B, and C, where the difference between the displacement values obtained by the two methods reaches 93.75 mm at monitoring point B. The displacement value of the one-step forming method at monitoring point G is slightly higher than the maximum displacement value during construction by 1.44 mm, which may be due to the influence of the forming sequence on the structural displacement change. The displacement and deformation values obtained by the two methods are relatively consistent with those of the rest of the monitoring points. However, the displacement and deformation values obtained by the one-step forming method are generally low and cannot reflect the dangerous state of the structure.

Significant differences exist between one-step and stage-by-stage forming analysis methods in simulating stress states and displacement deformations while constructing long-span spatial steel structures. Although the one-step forming method can quickly give a prediction after forming, it ignores the dynamic factors during the construction process, such as structural forming and boundary condition changes. On the other hand, the stage-by-stage forming method can ultimately show the structural state after each critical CS, which helps to find and solve the problems in time to ensure construction safety. Therefore, when carrying out the construction of long-span spatial steel structures, the path and time effects caused by the construction sequence should be fully considered. The correct simulation method should be adopted to obtain reliable intermediate and final state simulation results, which can accurately predict the deformation and optimize the construction process.

5. Conclusions

This study uses the engineering background of the T1 terminal building of Chengdu Tianfu International Airport in China. It uses FEA software to analyze the mechanical properties of long-span spatial steel structures in the whole process of construction that explicitly reflects the construction sequence. Comparative analyses were conducted of the one-step and stage-by-stage forming methods for long-span spatial steel structures. The conclusions are as follows.

1. The initial installation and unloading process of the long-span spatial steel structure's subsection is the turning point where the stress state and displacement changes often occur. During the construction process, the stress concentration phenomenon mainly occurs in the position of the temporary support frame, and the displacement maximum value is in the position of the structure's outer overhang. Construction technicians should carry out targeted monitoring of the stress state and displacement deformation of the structure, which can quickly identify potential safety hazards.
2. In the process of structural forming, the stress changes through the initial stress stage, stress adjustment stage, stress accelerated adjustment stage, and stress stabilization stage, showing a tendency to develop from high stress to low stress. The maximum tensile stress decreases from +215.53 N/mm^2 to +111.97 N/mm^2, and the maximum compressive stress decreases from -158.65 N/mm^2 to -101.88 N/mm^2. While the overall displacement is an accumulative process, the displacement change passes through the stages of displacement stabilization, fluctuation, and peak displacement. Compared with the initial displacement value of 70.86 mm for CS1, the maximum displacement value of the structure after forming reached 106.52 mm. For areas of stress concentration, construction workers should focus on monitoring. At peak displacements, counter-arches should be set up based on the displacement values.
3. The one-step forming analysis method can quickly provide the structure's stress state and displacement and deformation after forming. The stage-by-stage forming analysis method provides a detailed picture of the structural stresses and displacement deformations after each critical CS. During the construction process, most structures' peak stresses and displacements occur at certain stages. Therefore, it is necessary for construction technicians to perform CS analyses that reflect the construction sequence.
4. The construction of a long-span spatial steel structure is a complex and continuous gradual process of the structural system. Due to the mutual coupling of structural geometric nonlinearity, time-varying physical properties, and boundary condition changes, different construction sequences will lead to different stress and deformation paths in the construction process. The construction path effect leads to different final mechanical states of the same structure under different construction paths.
5. Future research could explore different types of long-span structures, such as steel skywalks, steel truss bridges, and hybrid structural systems. Additionally, studying new construction techniques like modular construction, high-strength steel, or composite materials may further optimize construction efficiency and structural performance. By expanding the scope of research to include various structural forms and innovative construction methods, we hope to advance the safe and efficient construction of long-span spatial steel structures.

Author Contributions: Conceptualization, G.Y. and Y.Y.; methodology, G.Y. and Y.Y.; software, R.L. and Y.Z; validation, R.L.; formal analysis, R.L.; writing—original draft preparation, R.L., X.C. and Y.Z.; writing—review and editing, R.L. and Y.Y.; visualization, X.C. and C.Z.; supervision, T.L. and Y.Y.; funding acquisition, G.Y. and Y.Y; data curation, C.Z. and T.L. All authors have read and agreed to the published version of the manuscript.

Funding: This research was funded by Integration and Application of Smart Factory Technology for Prefabricated Buildings (CSTB2022TIAD-KPX0145) and the Chongqing Urban Rail Express Line Full Life Cycle CIM Technology Application Research and Demonstration Research Project (S20220413).

Data Availability Statement: The data presented in this study are available on request from the corresponding author.

Conflicts of Interest: Authors Canwei Zhou and Ting Lei were employed by the company Chongqing Railway Group. The remaining authors declare that the research was conducted in the absence of any commercial or financial relationships that could be construed as a potential conflict of interest.

References

1. Zucca, M.; Tattoni, S.; Di Castri, M.; Simoncelli, M. On the collapse of a post-tensioned reinforced concrete truss bridge during the construction phases. *Eng. Fail. Anal.* **2024**, *158*, 107999. [CrossRef]
2. Pei, H.; Jia, L.; Li, J.; Li, K.; Xia, R. A Study of the Mechanical Behavior of a Steel-Concrete Hybrid Beam Bridge during Construction. *Buildings* **2023**, *13*, 1781. [CrossRef]
3. Chen, H.; Yin, S.; Qiao, C.; He, J.; Hu, M. Construction Effects on the Mechanical States of a Truss Structure. *J. Perform. Constr. Facil.* **2022**, *36*, 1152. [CrossRef]
4. Ding, L.; Sun, Y.-J.; Zhang, W.-Z.; Bi, G.; Xu, H.-Z. Stress Monitoring of Segment Structure during the Construction of the Small-Diameter Shield Tunnel. *Sensors* **2023**, *23*, 8023. [CrossRef] [PubMed]
5. Yao, G.; Chen, Y.; Yang, Y.; Ma, X.; Men, W. Investigation on Buckling Performance of Prefabricated Light Steel Frame Materials under the Action of Random Defects during Construction. *Materials* **2023**, *16*, 5666. [CrossRef] [PubMed]
6. Fan, Y.; Xin, J.; Yang, L.; Zhou, J.; Luo, C.; Zhou, Y.; Zhang, H. Optimization Method for the Length of the Outsourcing Concrete Working Plane on the Main Arch Rib of a Rigid-Frame Arch Bridge Based on the NSGA-II Algorithm. *Structures* **2024**, *59*, 105767. [CrossRef]
7. Yao, H.; She, J.; Zhou, Y. Risk assessment of construction safety accidents based on association rule mining and Bayesian network. *J. Intell. Constr.* **2024**. [CrossRef]
8. Yao, G.; Wu, C.; Yang, Y. Scientometric Analysis for Mechanical Performance of Broken-Line Long-Span Steel Structure in Construction Considering Geometric Nonlinearity. *Symmetry* **2021**, *13*, 1229. [CrossRef]
9. Hu, J.; Chen, W.; Ren, S.; Zhang, S.; Qu, Y.; Yin, Y.; Yang, D. Building performance monitoring and analysis of a large-span aerogel-membrane airport terminal. *Eng. Struct.* **2020**, *219*, 110837. [CrossRef]
10. Deng, D.; Zou, J.; Cai, J.; Chen, Z.; Wang, H. Non-contact measurement method of displacement of spherical joints in long-span space steel structures. *Structures* **2024**, *61*, 106089. [CrossRef]
11. Guo, S.; Xu, D.; Shang, K.; Yang, S.; Wang, D.; Li, G.; Zheng, B.; Jiao, Y. Analysis of progressive collapse of a super-long span latticed steel arch structure. *J. Theor. Appl. Mech.* **2023**, *61*, 103–117. [CrossRef] [PubMed]
12. Fu, Y.; Dong, F.; Wang, J. Wind Resistance Performance Assessment of Long-Span Cable-Supported Bridges Based on Time-Varying Reliability Theory. *Sustainability* **2024**, *16*, 923. [CrossRef]
13. Fu, R.-C.; Wang, H.-N.; Jiang, M.-J. Exact analytical solution for deep tunnels in viscoelastic-plastic rock considering the actual loading path. *Appl. Math. Model.* **2024**, *128*, 370–391. [CrossRef]
14. He, Z.; Lu, Y.; Liu, F.; Pan, F.; He, S. An estimate on in-construction differential settlement of super high-rise frame core-tube buildings. *Struct. Des. Tall Spec. Build.* **2020**, *29*, e1737. [CrossRef]
15. Hu, Y.; Lei, H.; Zheng, G.; Shi, L.; Zhang, T.; Shen, Z.; Jia, R. Ground movement induced by triple stacked tunneling with different construction sequences. *J. Rock Mech. Geotech. Eng.* **2022**, *14*, 1433–1446. [CrossRef]
16. Ling, C.; Ruan, Y.; Wu, P.; Li, J.; Zhao, J.; Yuan, B. Influence of Different Excavation Sequence of Double-Side Heading Method on Supporting Structure. *Adv. Civ. Eng.* **2022**, *2022*, 2258594. [CrossRef]
17. Yang, S.-S.; Zhang, D.-W.; Wang, M.; Xu, J.-M.; Shen, C.; Zhang, C.-Z. Ground settlement caused by pipe-roof pre-construction method: Effect of the sequence of jacking pipe groups. *J. Cent. South Univ.* **2024**, *31*, 576–588. [CrossRef]
18. Liu, C.; Zhang, F.; Zhang, H.; Shi, Z.; Zhu, H. Optimization of assembly sequence of building components based on simulated annealing genetic algorithm. *Alex. Eng. J.* **2023**, *62*, 257–268. [CrossRef]
19. Soomro, M.A.; Mangi, N.; Xiong, H.; Kumar, M.; Mangnejo, D.A. Centrifuge and numerical modelling of stress transfer mechanisms and settlement of pile group due to twin stacked tunnelling with different construction sequences. *Comput. Geotech.* **2020**, *121*, 103449. [CrossRef]
20. Yao, G.; Sun, W.; Yang, Y. Analysis and Identification of Building Construction Accident Risk in China basing Exclusively Database. *IOP Conf. Ser. Earth Environ. Sci.* **2022**, *1101*, 072009. [CrossRef]
21. Waqar, A.; Othman, I.; Radu, D.; Ali, Z.; Almujibah, H.; Hadzima-Nyarko, M.; Khan, M.B. Modeling the Relation between Building Information Modeling and the Success of Construction Projects: A Structural-Equation-Modeling Approach. *Appl. Sci.* **2023**, *13*, 9018. [CrossRef]
22. Mudiyanselage, S.T.; Wijesundara, K.; Venkatesan, S.; De Silva, S.; Dissanayake, R.; Neluwala, P. Significance of construction sequence and the initial behaviour in concrete-faced rockfill dams. *Aust. J. Struct. Eng.* **2024**, *25*, 125–143. [CrossRef]

23. Jiao, Y.; Cao, P. Research on Optimization of Project Design Management Process Based on BIM. *Buildings* **2023**, *13*, 2139. [CrossRef]
24. Mayouf, M.; Jones, J.; Elghaish, F.; Emam, H.; Ekanayake, E.M.A.C.; Ashayeri, I. Revolutionising the 4D BIM Process to Support Scheduling Requirements in Modular Construction. *Sustainability* **2024**, *16*, 476. [CrossRef]
25. Yilmaz, G.; Akcamete, A.; Demirors, O. BIM-CAREM: Assessing the BIM capabilities of design, construction and facilities management processes in the construction industry. *Comput. Ind.* **2023**, *147*, 103861. [CrossRef]
26. Tang, Q.; Xin, J.; Jiang, Y.; Zhang, H.; Zhou, J. Dynamic Response Recovery of Damaged Structures Using Residual Learning Enhanced Fully Convolutional Network. *Int. J. Struct. Stab. Dyn.* **2024**, 2550008. [CrossRef]
27. Zheng, X.; Li, W. Finite element simulation analysis of steel truss arch bridge jacking construction. *Civ. Eng. J.-Staveb. Obz.* **2023**, *32*, 519–532. [CrossRef]
28. Liu, J.; Meng, F.; Chen, Y.; Chen, Z. The influence of coupling on the whole process of construction simulation of chord dome structure. *E3S Web Conf.* **2020**, *143*, 01046. [CrossRef]
29. Qin, S.; Wei, K.; Qin, J.; Yuan, R.; Xu, L.; Dan, Q. Stress-free-state based structural analysis and construction control theory for staged construction bridges. *Adv. Bridge Eng.* **2020**, *1*, 1. [CrossRef]
30. Li, Q.; Guo, H.; Guo, B. The Dual-Parameter Control of Synchronization in Steel Box Girder Incremental Launching Construction. *Appl. Sci.* **2023**, *13*, 12074. [CrossRef]
31. Xin, Y.; Li, J.; Hao, H.; Yang, N.; Li, C. Time-Varying System Identification of Precast Segmental Columns Subjected to Seismic Excitations. *J. Bridge Eng.* **2022**, *27*, 1–18. [CrossRef]
32. Sun, M.; Li, Q.; Li, Y. Investigation of time-varying natural frequencies of high-rise buildings under harsh excitations using a high-resolution combined scheme. *J. Build. Eng.* **2022**, *57*, 104859. [CrossRef]
33. Yang, L.; Wang, L.; Yu, B. Time-varying behavior and its coupling effects with environmental conditions and cementitious material types on surface chloride concentration of marine concrete. *Constr. Build. Mater.* **2021**, *303*, 124578. [CrossRef]
34. Huang, J.; Tan, P.; Zhang, Y.; Zhou, F. Endurance time analysis of seismic performances of long-span continuous rigid-frame bridges with corrugated steel webs. *Structures* **2022**, *43*, 990–1001. [CrossRef]
35. Yu, L.; Xu, W.; Zhang, D.-B.; Ma, X.-M.; Wu, Y.-H. Strain Incremental Adjustment Method of Cable Force of Composite Saddle Anchor Span of Single-Tower Single-Span Ground-Anchored Suspension Bridge. *Math. Probl. Eng.* **2021**, *2021*, 2956496. [CrossRef]
36. Gou, H.; Pu, Q.; Wang, J.; Chen, Z.; Qin, S. Spatial mechanical behaviors of long-span V-shape rigid frame composite arch bridges. *Struct. Eng. Mech.* **2013**, *47*, 59–73. [CrossRef]
37. Ruan, R.; Chen, Y.; Lin, W.; Wei, L.; Huang, J. The Construction Technology of Column Replacement Integral Accumulation Sliding at Uneven Elevation for Steel Structures. *Buildings* **2023**, *13*, 1958. [CrossRef]
38. Pei, B.; Chong, A.; Xia, H.; Kang, X. Design and key construction technology of steel-concrete-steel sandwich composite pylon for a large span cable-stayed bridge. *Sci. Rep.* **2023**, *13*, 6626. [CrossRef]
39. Wen, F.; Liang, X.; Chen, C.; Xu, L.; Feng, Q. Construction Control and Monitoring Platform of a Large-Segment Steel Box Girder with Hoisting Installation. *Appl. Sci.* **2023**, *13*, 9573. [CrossRef]
40. Wang, Y. In A closed coal yard grid construction simulation. In Proceedings of the 3rd International Conference on Civil, Architectural and Hydraulic Engineering (ICCAHE), Hangzhou, China, 30–31 July 2014; pp. 1672–1675. [CrossRef]
41. Yang, Q.; Yu, S.; Zhang, X.; Wang, Z.; Yan, J.; Chen, X. The Construction Technology of Roof Steel Structure in YanCheng NanYang Airport. *Adv. Civ. Eng.* **2018**, *2018*, 6386020. [CrossRef]
42. Lu, C.; Yang, Z.; Li, P.; Xu, Q.; Chen, L.; Zhang, C. Integral sliding of a 800 T steel roof truss for a cultural and art center building. *Case Stud. Constr. Mater.* **2022**, *17*, e01345. [CrossRef]
43. Yang, Y.; Du, H.; Yao, G.; Ma, X.; Men, W. Time-Varying Mechanical Analysis of Long-Span Spatial Steel Structures Integral Lifting in Construction Basing Building Information Model. *Sustainability* **2023**, *15*, 1256. [CrossRef]
44. Ruan, R.; Lai, M.; Jiang, C.; Wang, J.; Lin, Y. Integral Lifting of Steel Structure Corridor between Two Super High-Rise Buildings under Wind Load. *Buildings* **2023**, *13*, 2441. [CrossRef]
45. Shen, Y.; Lin, Z.; Wang, Z. Single-Side Accumulative Jacking Construction Method for Large-Span Arched Latticed Shells. *J. Constr. Eng. Manag.* **2022**, *148*, 06022001. [CrossRef]
46. Huang, Y.; Yang, J.; Feng, R.; Chen, H. Resistance of cold-formed sorbite stainless steel circular tubes under uniaxial compression. *Thin-Walled Struct.* **2022**, *179*, 109739. [CrossRef]
47. Akduman, S.; Karalar, M.; Mert, N.; Ozturk, H. Investigation of the Post-Fire Behavior of Different End-Plated Beam-Column Connections. *Buildings* **2024**, *14*, 1013. [CrossRef]
48. *GB 50017-2017*; Standard for Design of Steel Structures. Chinese Ministry of Housing and Urban-Rural Development, China Architecture & Building Press: Beijing, China, 2017.

atmosphere

MDPI

Article

The Wind Profile Characteristics of Super Typhoon Lekima Based on Field Measurement

Yanru Wang [1], Qianqian Qi [2], Shuqin Zheng [3,*], Bin Fu [1], Maoyu Zhang [1], Xu Wang [4], Chuanxiong Zhang [3] and Lei Zhou [4]

1 School of Civil Engineering, Taizhou University, Jiaojiang 318000, China; yanrupiaoyang@163.com (Y.W.); dorofubin@tzc.edu.cn (B.F.) ryanzmy0577@tzc.edu.cn (M.Z.)
2 School of Civil Engineering, Xinyang University, Xinyang 464000, China; 622200970065@mails.cqjtu.edu.cn
3 Key Laboratory of Intelligent Lifeline Protection and Emergency Technology for Resident Aty, Wenzhou University of Technology, Wenzhou 150080, China; zhangcx@wzu.edu.cn
4 State Key Laboratory of Mountain Bridge and Tunnel Engineering, Chongqing Jiaotong University, Chongqing 400074, China; xuwang@cqjtu.edu.cn (X.W.); lixuezl@tju.edu.cn (L.Z.)
* Correspondence: zhengshuqin@wzut.edu.cn

Abstract: Many cities in coastal areas are prone to typhoon disasters due to their location on the Pacific storm path, and the direct effect of catastrophic winds can lead to the destruction of low-rise buildings and severe damage to high-rise structures. The purpose of this study was to enhance the understanding of boundary layer wind profiles of strong typhoons in coastal areas and reduce property losses and casualties caused by wind disasters. Based on the field measurements of wind profile acoustic radar in coastal areas, the variation characteristics of the boundary layer wind profile during the passage of super typhoon Lekima were first studied in depth, and the evolution law of the typhoon boundary layer profile was summarized. Then, the effects of typhoon horizontal structure, topography, wind speed, and time distance on the characteristics of the typhoon profile were discussed, respectively. Finally, the evolution characteristics of wind profile parameters were obtained by fitting three wind profile theoretical models. Due to the strong variability of typhoon profile morphology, the theoretical model of wind profile is only applicable to the wind profile from the bottom to the low-level jet height of typhoons, while wind parameters are closely related to the spatial location of the typhoon wind field.

Keywords: strong typhoon; wind profile; evolution; wind parameters; field measurements

Citation: Wang, Y.; Qi, Q.; Zheng, S.; Fu, B.; Zhang, M.; Wang, X.; Zhang, C.; Zhou, L. The Wind Profile Characteristics of Super Typhoon Lekima Based on Field Measurement. *Atmosphere* **2024**, *15*, 558. https://doi.org/10.3390/atmos15050558

Academic Editor: Jimy Dudhia

Received: 8 March 2024
Revised: 21 April 2024
Accepted: 26 April 2024
Published: 30 April 2024

1. Introduction

The evolution of average wind speed within the atmospheric boundary layer with vertical altitude has always been a hot topic in structural wind resistance research. With the emergence of modern towering buildings, their structural shapes have become increasingly complex. Strong typhoons can seriously damage building structures, affecting their usability and durability. Studying the characteristics of typhoon wind fields, especially mastering the characteristics of typhoon profiles and their influencing factors and the evolution laws of related wind parameters, can further avoid the vibration and even damage of tall wind-sensitive structures under wind load excitation, thereby improving the safety and economy of high-rise building design [1].

Due to the high temperature of the ocean surface in tropical areas and the large amount of sea water evaporation, it is easy to form a low-pressure center dominated by the upwind. The low-pressure center gradually moves and strengthens with the rotation of the earth, thus driving more airflow around the deflection and forming a regular rotation of tropical cyclones. When the wind force near the bottom center of a tropical cyclone reaches 12 (32.7 m/s), it is considered a typhoon. The typhoon profile is related to topography, atmospheric humidity, surface roughness, atmospheric stability, Coriolis force, and other factors. Wind engineers have studied wind field characteristics through methods such as field measurements [2,3], numerical simulations [4,5], and wind tunnel tests [6,7].

As the most direct and effective way, field measurement is often used by scholars to study typhoon wind characteristics. Franklin [8] studied the wind profile characteristics of 17 hurricanes in the eye wall area and the outer circulation area by using GPS dropsonde. It was found that the hurricane eyewall area had higher wind speeds, and the maximum wind speed in the outer circulation area occurred at a height of 1 km. Powell et al. [9] found that the average wind speed in a tropical cyclone profile increases logarithmically with altitude, with the maximum value occurring at 500 m. Based on the measured data of gradient mast towers, Song et al. [10,11] found that the characteristics of typhoon wind profile were significantly different from those of good wind profile. Lin et al. [12] studied the wind speed profile and related parameters in typhoon-prone areas along the southeastern coast of China. They found that the wind profile index before the typhoon eye passed was higher than that after the typhoon eye passed, while the roughness length was the opposite. He et al. [13] studied the downwind profile characteristics of typhoons with different ground roughness based on a 356-m meteorological gradient tower. In fact, normal winds with high Beaufort scales also occur frequently. Liu [14] analyzed the average wind speed and wind direction profiles of 16 strong wind days based on a wind profiler and believed that the wind resistance design of super high-rise buildings should mainly consider Class I.

Real-time monitoring of meteorological satellites also plays a crucial role in typhoon research [15]. It can monitor and track the location and movement path of typhoons in real-time, collect key meteorological data, and provide valuable information for studying and improving typhoon prediction models [16–18]. With the development of computer technology, numerical simulation is widely used in wind research. For example, large eddy simulation (LES) is used to simulate turbulence [19]. Ren et al. [20] used CFD and LES to simulate the spatial wind field and turbulence of complex terrain and established a spatial wind field prediction model. The WRF model has been widely used in typhoon wind field simulation in recent years [21,22]. Zhang [23] used the WRF-UCM coupling model to simulate the urban wind field during the landfall of a strong typhoon. Due to the complexity and randomness of wind, the lack of observation data in East Asia and the low spatial and temporal resolution of data, the study of wind characteristics still has various obstacles and is difficult to master [24]. There are few studies on the characteristics and evolution of typhoon low-level jet [25]. More importantly, the existing research mainly focuses on the observation of 100-m height, which makes it difficult to reflect the real wind profile of the typhoon boundary layer, and there are few systematic studies on the characteristics of disaster-causing wind profile and its influencing factors [26].

In this study, the profile characteristics of super typhoon Lekima were studied in depth based on the measured data of ultra-efficient and long-distance wind profile acoustic radar in the coastal area of Taizhou, Zhejiang Province, China. The evolution law of the vertical distribution of wind speed in the atmospheric boundary layer under different influence stages of strong typhoons was systematically analyzed. The factors affecting the profile characteristics and the evolution law of related wind parameters were explored. It was intended to enhance the understanding of the profile characteristics of the boundary layer under strong typhoons in coastal areas, to provide a reference for typhoon boundary layer simulation, and to promote the wind resistance design of high-rise wind-sensitive buildings in coastal areas. This has important significance for the wind resistance design of civil engineering structures and disaster prevention and reduction.

2. Overview of Field Measurements

2.1. Measured by Typhoon Lekima

Super typhoon Lekima formed over the western North Pacific Ocean at 2:00 Beijing time on 4 August 2019, and moved northwestward. At 1:45 on the 10th, it made landfall along the coast of Cheng Nan Town, Wenling City, Zhejiang Province, China, with a maximum wind speed of 52 m/s near the center of the bottom layer and a minimum pressure of 930 hPa. Lekima is the third strongest typhoon to land in East China in nearly 70 years. Due to its long duration and wide range of disaster impacts, on 30 April 2021, the Typhoon Committee decided to remove the name of super typhoon Lekima and replace it with the new typhoon name "Co-may".

Figure 1a is the optimal track map of typhoon Lekima provided by the China Meteorological Administration. Figure 1b is the wind profile acoustic radar from the measured data source of this study, and Table 1 shows the detailed parameters of the instrument. Wind profile acoustic radar can continuously measure the distribution characteristics of atmospheric wind speed, wind direction, and disturbance in the vertical range of 30–950 m with ultra-high efficiency and long-distance continuous measurement, and can provide horizontal wind components updated every minute, including average wind speed per minute, standard deviation of average wind direction and air volume, temperature, signal-to-noise ratio, turbulence intensity in wind energy applications, stability in air quality applications, etc. It probes 47 height layers at 20 m intervals in a vertical range of 30–950 m.

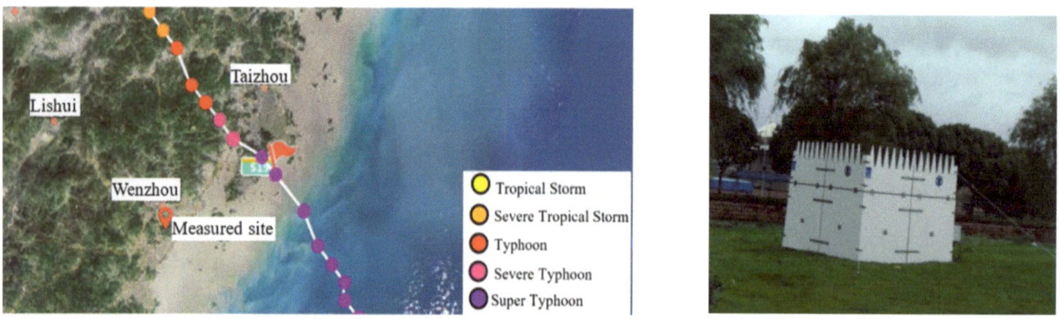

(**a**) (**b**)

Figure 1. Best track of typhoon and the Sodar Wind Profiler, including (**a**) Lekima (in CST, or UTC + 8). The tropical cyclone intensity scale from China Meteorological Administration: Tropical Depression (10.8–17.1 m/s); Tropical Storm (17.2–32.6 m/s); Typhoon (32.7–61.2 m/s); Super Typhoon (61.2 m/s); (**b**) The Sodar Wind Profiler.

Table 1. Specifications of the Sodar Wind Profiler.

Parameter	Value
Beamwidth	0°, ±22°, ±29°
Vertical resolution	10 m
Minimum measurement height	20 m
Maximum measurement height	1000 m
Sampling period	60 s
Horizontal wind speed accuracy	0.1~0.3 m/s
Vertical wind speed accuracy	0.03~0.1 m/s
Wind direction accuracy	1.5°
Horizontal wind speed range	0~50 m/s
Vertical wind speed range	−10~10 m/s

2.2. *Measurement Location and Sample Selection*

2.2.1. Measured Site

The measurement site is located in the northern campus of Wenzhou University, Ouhai District, Wenzhou City, Zhejiang Province, which belongs to the southeastern coastal area with many bays. The topography of the measurement point and its surrounding area is shown in Figure 2. About 2 km east of the measured point is Daluo Mountain, with the highest altitude of 707.4 m. The southeast side is about 17 km away from the coastline of the East China Sea, where a large number of typhoons land or pass through every year. The northeast side has the second largest river in Zhejiang, the Oujiang River; the northwest side is the open Sanyang wetland, and the west side is the urban area of Wenzhou, which belongs to the suburbs of the city. The distance between the typhoon landing site and the observation point is about 82 km. During the typhoon period, at 3:00 on 10 August 2019, the observation point is the closest distance from the typhoon center, about 70.8 km, and the intensity level of the typhoon is a strong typhoon at this time.

Figure 2. The measured site and surrounding terrain.

2.2.2. Sample Selection

The measured typhoon data started of the Lekima generation, from 14:00 on 4 August 2019, to 14:00 on 13 August 2019, in China Standard Time (CST = UTC + 8 h), totaling 216 h. The observation time interval was 1 min, and the average time interval was processed according to 10 min. Considering the time and space factors, a total of 1,192,320 wind speed and wind direction samples were collected. Some data obtained by remote sensing equipment have low quality and cannot be used for any analysis. To ensure the reliability and scientific validity of the analysis results, it is necessary to remove outliers and missing data. Due to the large sample size of the total observed data, invalid samples are scattered among valid samples, and the elimination of invalid samples in data processing does not affect the overall reliability and credibility of the data. In the subsequent calculation and analysis process of this paper, the invalid samples were excluded and only the valid samples were studied.

3. Measured Data Processing

The vertical distribution of wind speed in the atmospheric boundary layer is an important theoretical basis for the wind-resistant design of high-rise buildings in coastal areas. In this study, three commonly used wind profile theoretical models were used to analyze the global profile characteristics of super typhoon Lekima.

3.1. Theoretical Model of Wind Profile

3.1.1. Log-Law Model

Considering the flow characteristics of the atmospheric boundary layer and under neutral atmospheric stability conditions, the expression for the wind profile log-law model can be derived from the solution of the average wind speed differential equation [27]:

$$u(z) = \frac{u_*}{\kappa} \ln \frac{z}{z_0}$$

where, z_0 is the roughness length, u_* is the friction velocity, κ is the von-Kármán constant, generally 0.4, and $u(z)$ is the wind speed at the vertical height z.

According to the Monin–Obukhov similarity theory [28], when the atmosphere is in a stable or unstable state, the wind profile can be described in a more general form by adding a correction function:

$$u(z) = \frac{u_*}{\kappa}\left[\ln\frac{z}{z_0} + \psi\left(\frac{z}{L}\right)\right]$$

where, ψ is the Monin–Obukhov function and L is the Monin–Obukhov length. $\psi(z/L)$ has different values under different atmospheric stability conditions, and 0 is taken in the case of neutral stratification. Under neutral stratification, Equation (1) is a widely used wind speed profile expression, which is widely used in the field of micrometeorology because of its good theoretical basis and experimental support, and can better describe the wind profile in the near-surface range.

3.1.2. Power-Law Model

By analyzing wind speed data with different ground roughness in cities, open grasslands, and rural areas, Davenport [29] pointed out that the change of horizontal average wind speed with height satisfies the exponential function. Research has shown that the power law model can satisfy the wind profile characteristics in various terrains and has a simple expression [30,31]. This model is widely adopted in standards of the United States, China, and Japan [32–34]:

$$u(z) = u(z_1)\left(\frac{z}{z_1}\right)^{\alpha}$$

where, $u(z)$ and $u(z_1)$ are the wind speed at altitude z and reference height z_1, respectively, and α is the wind profile index, which is related to altitude, topography, horizontal wind speed, and atmospheric stratification, representing the change law of wind speed with the increase or decrease of altitude. A positive value indicates compliance with the power law model, while a negative value suggests that the change of wind speed with vertical height deviates from the power law profile.

3.1.3. Deaves-Harris Model

Deaves and Harris [35] proposed an improved wind profile model based on the logarithmic law model by introducing the atmospheric boundary layer height and fully considering the influence of ground roughness:

$$u(z) = \frac{u_*}{\kappa}\left[\ln\frac{z}{z_0} + 5.75\frac{z}{h} - 1.88\left(\frac{z}{h}\right)^2 - 1.33\left(\frac{z}{h}\right)^3 + 0.25\left(\frac{z}{h}\right)^4\right]$$

where, h is the height of the boundary layer, and the expression is as follows:

$$h = \frac{1}{\beta}\frac{u_*}{f}$$

where, β is an empirical parameter, generally taken as 6; f is the Coriolis force parameter, which is related to the latitude of the observation point, and is taken as 9.375×10^{-5}/s.

3.2. *Time Characteristics of Measured Average Wind Speed and Wind Direction*

Figure 3 shows the contours of the measured 10-min horizontal average wind speed and wind direction during the period from 8:00 on 9 August 2019, to 0:00 on 11 August 2019. The black dashed line in the figure indicates the time when the typhoon landed, which is 01:45 on 10 August 2019. Some white areas are caused by data exclusion. Figure 3a shows that the wind speed reached two peaks after the typhoon landed, forming a typical M shape structure. Moreover, the overall wind speed in the later stage of the measured typhoon is greater than that in the early stage. The maximum 10-min average wind speed in the later stage is 18 m/s, and the maximum instantaneous wind speed reaches 27.3 m/s. It is worth noting that there is an obvious low-level jet phenomenon within 150 m above the ground. From 10:40 to 11:00 on August 10, the 10-min average wind speed at the lower level exceeded 12 m/s.

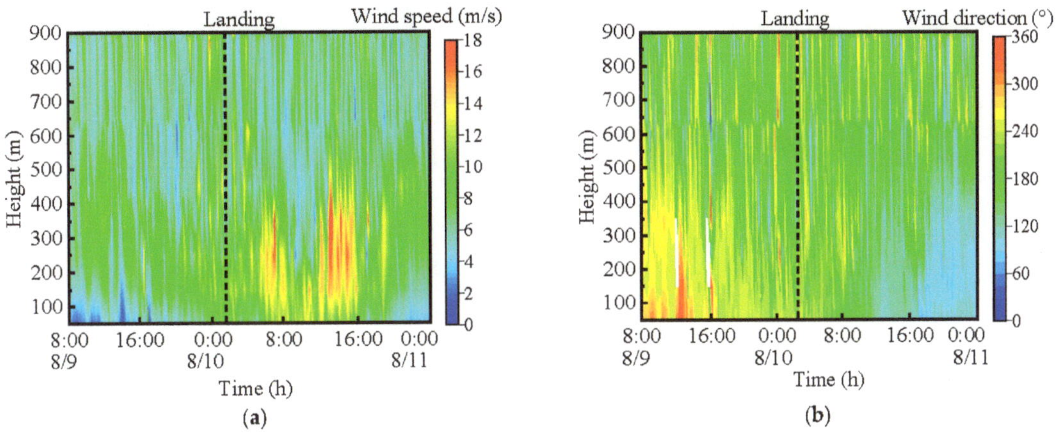

Figure 3. Measured 10-min horizontal wind speed and wind direction, including (**a**) Horizontal wind speed; (**b**) Horizontal wind direction; Time format: hh: mm/dd.

Figure 3b shows that during the passage of the typhoon, the average wind direction at the center of the measured points changed from southwest to southeast, and the average wind direction at 50 m changed by about 150°. The wind speed and direction changed significantly along the height direction. The measured height is below 550 m, and the wind speed increases significantly with the increase of height, with a significant change in wind direction; above 550 m, the changes in wind speed and direction with height and time are not significant. The wind speed gradually approaches consistency at different heights, and the change in wind direction angle is not significant, indicating that the gradient wind height has been reached.

4. Evolution Characteristics of Wind Profiles

4.1. Evolution of Typhoon Boundary Layer Wind Profile

The analysis of the evolution of the typhoon boundary layer wind profile from 8:00 on 9 August 2019, to 0:00 on 11 August 2019, based on the measured points within the influence range of the typhoon's 7th level wind circle, lasted a total of 40 h. Figures 4 and 5 show the evolution process of the measured boundary layer wind speed and wind direction profiles of typhoon Lekima, respectively. Each interval is divided based on 2 h, for a total of 20 intervals. Figure 4 also provides the extreme and peak values of the average wind speed span for each interval. The dotted line in the figure is the average wind speed profile in each interval, and the shaded part represents the peak of the horizontal average wind speed in this time interval. In Figure 5, the wind direction is 0° (360°) due north and 180° clockwise due south. The vertical height is 0 to 1000 m.

Intervals 6–8 (18:00 on the 9th to 0:00 on the 10th): As the distance between the typhoon center and the observation point decreases, the height of the maximum wind speed decreases from 300 m to 120 m, and the wind speed profile at the bottom shows anti-C shape; the horizontal average wind speed gradually increases, and the 10-min average wind speed below 500 m can reach 14 m/s. In Figure 5, the wind direction gradually rotated counterclockwise, the wind direction at the lower level was stable, and the wind direction above 750 m was relatively discrete.

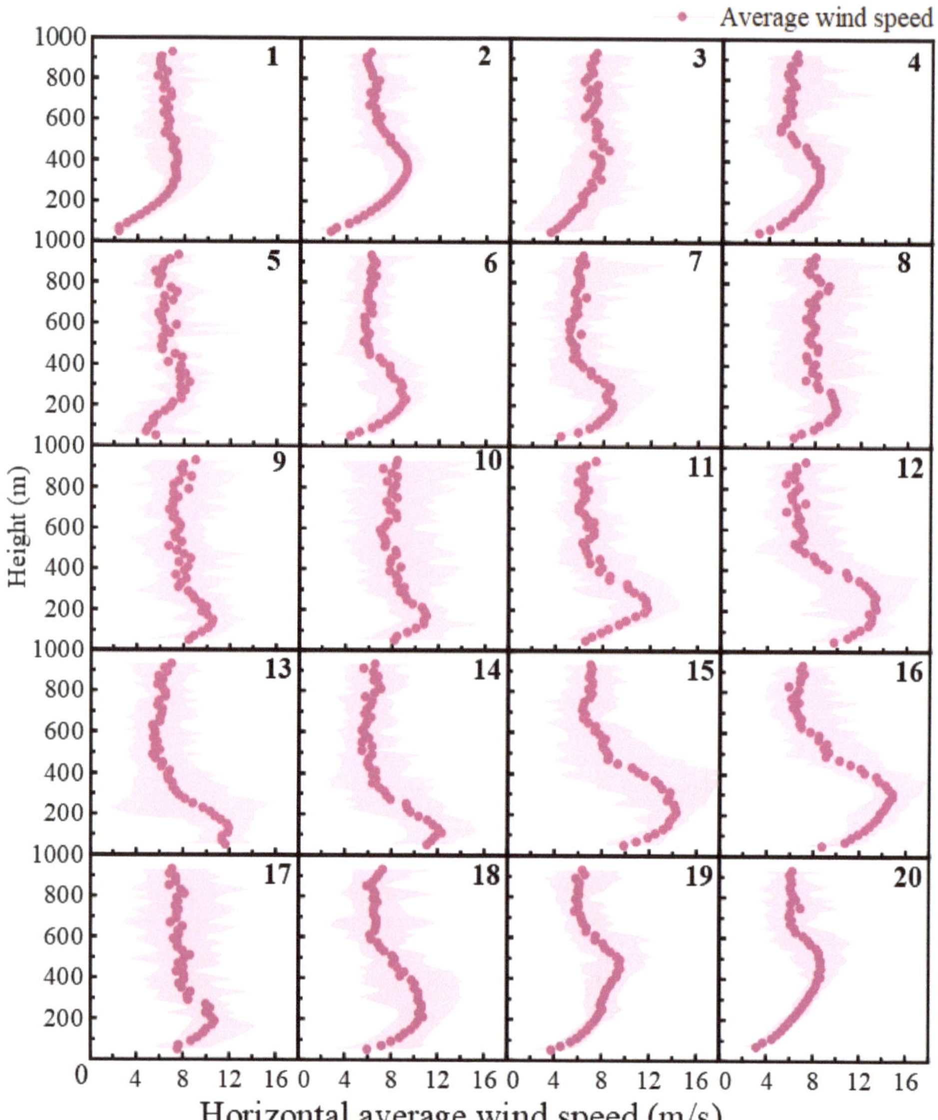

Figure 4. Evolution process of the boundary layer wind speed profile of typhoon Lekima.

Interval 9 (0:00 to 2:00 on the 10th): The typhoon makes landfall, and the measured point is affected by the 10-level wind circle.

Interval 10–16 (2:00 to 16:00 on the 10th): The distance between the measured point and the bottom center of the typhoon in interval 11 is the smallest, about 70.6 km. The wind speed above 800 m showed a trend of further increase during the interval between 11 and 12, but the second increase in wind speed did not surpass the lower-level extreme wind speed. In interval 14, the maximum horizontal average wind speed increases to about 14 m/s, appearing in the 120 m height layer, and the peak value of the 10-min average wind speed can reach 16 m/s. In Figure 5, when the typhoon center approaches and passes through the observation point, the wind direction changes significantly, and the change becomes more intense with the increase in height. The wind profile shapes in intervals 11, 12, 15, and 16 are S shape.

Intervals 1–5 (8:00 to 18:00 on the 9th): The measured point is far away from the typhoon path. Starting from Interval 2, it is affected by the 7-level wind circle. From Figures 4 and 5, it can be seen that the wind speed and direction profile begin to change.

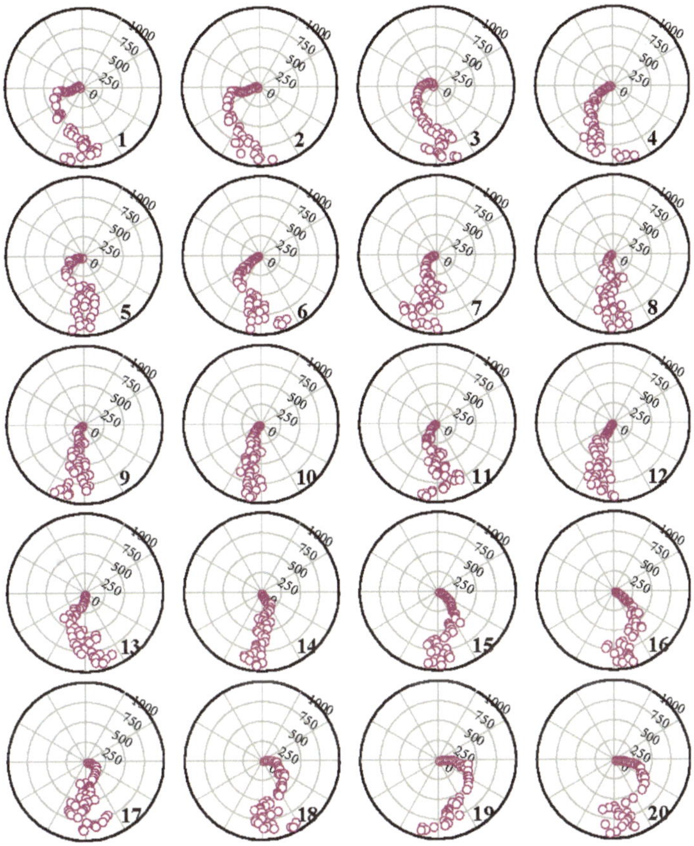

Figure 5. Evolution process of the boundary layer wind direction profile of typhoon Lekima.

During the process of a typhoon making landfall and moving inland, a significant low-level jet can be observed in the wind speed profile. The higher the average wind speed in the lower layer, the lower the height of the near-Earth jet layer. The height of the near-Earth jet layer refers to the altitude at which the low-level jet appears and is also the height of the low-level jet core (the apex of the "nose" profile). Kepert's research [36] indicates that the low-level jet is a common feature of the average wind profile in the boundary layer of tropical cyclones. Studies [9] have shown that the maximum average wind speed at the near-surface of Atlantic hurricanes is about 500–600 m vertical height, which is much higher than the maximum wind speed in this study. This may be related to the different ocean surfaces where the typhoon is generated. It is very unfavorable to the wind resistance of super high-rise structures in coastal areas.

Interval 17–20 (16:00 on the 10th to 0:00 on the 11th): The measured point gradually moves away from the typhoon and is still within the influence range of the 7-level wind circle. With the typhoon moving away and weakening, the wind speed at the bottom decreases, and the height of the maximum wind speed increases from 200 m to about 500 m.

In Figure 5, the wind direction changes completely, with a difference of about 180° from the initial wind direction.

In general, the characteristics of the typhoon boundary layer profile mainly have the following characteristics: (1) A low-level jet appears within the vertical height range of 120–500 m, with the maximum wind speed and a 180° change in wind direction. (2) The shape of the typhoon profile differs significantly from the theoretical profile model, depending on the radial position from the typhoon center to the outer edge of the typhoon.

4.2. Factors Affecting Wind Profiles

4.2.1. Influence of Typhoon Horizontal Structure on Wind Profile

Wind profile is the main factor affecting the wind load characteristics of high-rise buildings. To explore the influence of typhoon structure on the characteristics of wind profile, the wind profile of typhoons in different periods is discussed by time series. According to the evolution of the typhoon wind profile in Section 3.1, typhoons are divided into four periods, as shown in Table 2. They are the typhoon influence period, the typhoon landing period, the typhoon wind speed rising period, and the typhoon stable period.

Table 2. Different stages of typhoon.

Typhoon Stage Name	Corresponding Interval (NUM)	Wind Circle Range (Wind Scale)
Impact period	6–8, 17–18	7
Landing period	9	7–10
Wind speed rising period	10–12, 15–16	10
Stable period	13–14	7–10

Figure 6 shows the relationship between the maximum wind speed and the height of the maximum wind speed during the typhoon influence period and landing period with the distance (relative distance) between the measured ground and the typhoon center. According to the hour time path provided by the CMA, this is a total of 44. It can be seen that with the increase of relative distance, the maximum average wind speed gradually decreases, while the height of the maximum wind speed gradually increases, indicating that the greater the wind speed at the lower level, the lower the typhoon boundary layer and the more significant the low-level jet phenomenon.

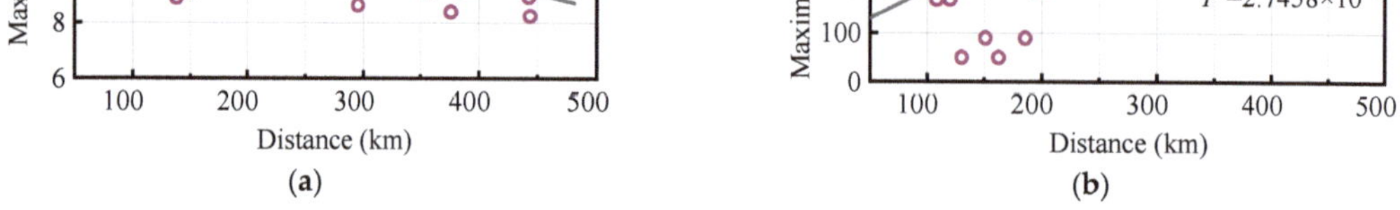

Figure 6. Changes of maximum wind speed and height of maximum wind speed with relative distance, including (**a**) maximum wind speed; (**b**) maximum wind speed height.

Figure 7 shows wind profiles during different periods of typhoons. The wind characteristics in different parts of a typhoon vary significantly. The typhoon influence period refers to the period when the typhoon approaches the coastline and moves away from the observation point, both within the influence range of the 7-level wind circle. The observation point is far away from the bottom center of the typhoon, and the bottom wind speed is the smallest during this period. During the typhoon landing period, the observation points begin to be affected by the 10-level wind circle, the maximum wind speed height decreases, and the wind direction changes. The bottom wind speed rises sharply during the typhoon wind speed rise. In the stable period, the bottom wind speed is the largest and the maximum wind speed height is the lowest.

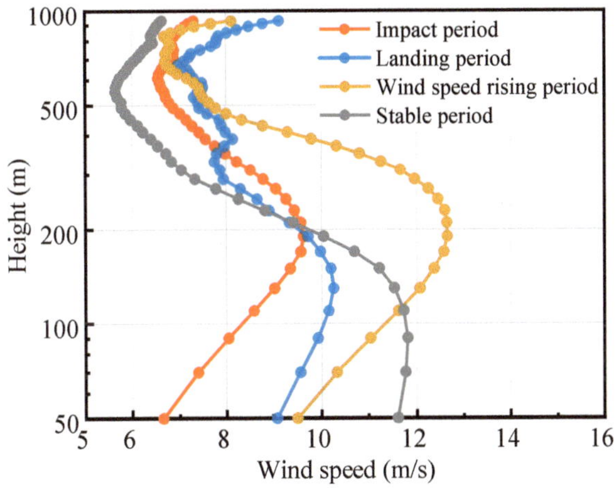

Figure 7. Wind profiles of different stages of typhoon.

Figure 8 compares the normalized average wind speed profiles with the empirical model for different stages of the typhoon. U_{50} represents the average wind speed at a vertical height of 50 meters. Considering the characteristics of the fitting model and combined with the characteristics of civil engineering structures, the measured data from the bottom to the height of the maximum wind speed or the wind speed with obvious inflection points in each period are fitted. The dynamic parameters such as wind profile index, roughness length, friction velocity, and boundary layer height are fitted, as shown in Table 3.

Table 3. Wind profile fitting parameters of typhoon at different stages.

Typhoon Period	Maximum Wind Speed Height (m)	Power-Law Model	Log-Law Model		D-H Model
		Index	Friction Velocity (m/s)	Roughness Length (m)	Boundary Layer Height (m)
Impact period	210	0.29	0.857	2.08	1524
Landing period	130	0.13	0.471	0.03	837
Wind speed rising period	210	0.23	0.917	0.739	1630
Stable period	110	0.04	0.182	0.01	324

Figure 8 shows that the three models can fit the wind profile from the ground to the height below the maximum wind speed well, but none of them can reproduce the anti-C shape wind profile and the phenomenon of secondary increase in wind speed. During the period of the typhoon's 7-level wind circle, the log-law and power-law fit the wind profile well at a height of 210 m. During the typhoon's landing period, the D-H model has the worst fitting effect. During the wind speed rising period, the three wind profile models can reflect the wind profile within 200 m near the ground. During the wind speed stable period, the normalized wind speed has an obvious inflection point at 110 m, and the wind profile index is the smallest at this time. The typhoon wind profile shows a low-level jet near the ground, and the wind profile model may fail at heights above 200 m. This shows that the wind velocity profile below the height of the low-level jet follows the exponential law, logarithmic law, and D-H model, but the fitting parameters are greatly affected by the spatial position inside the typhoon wind field.

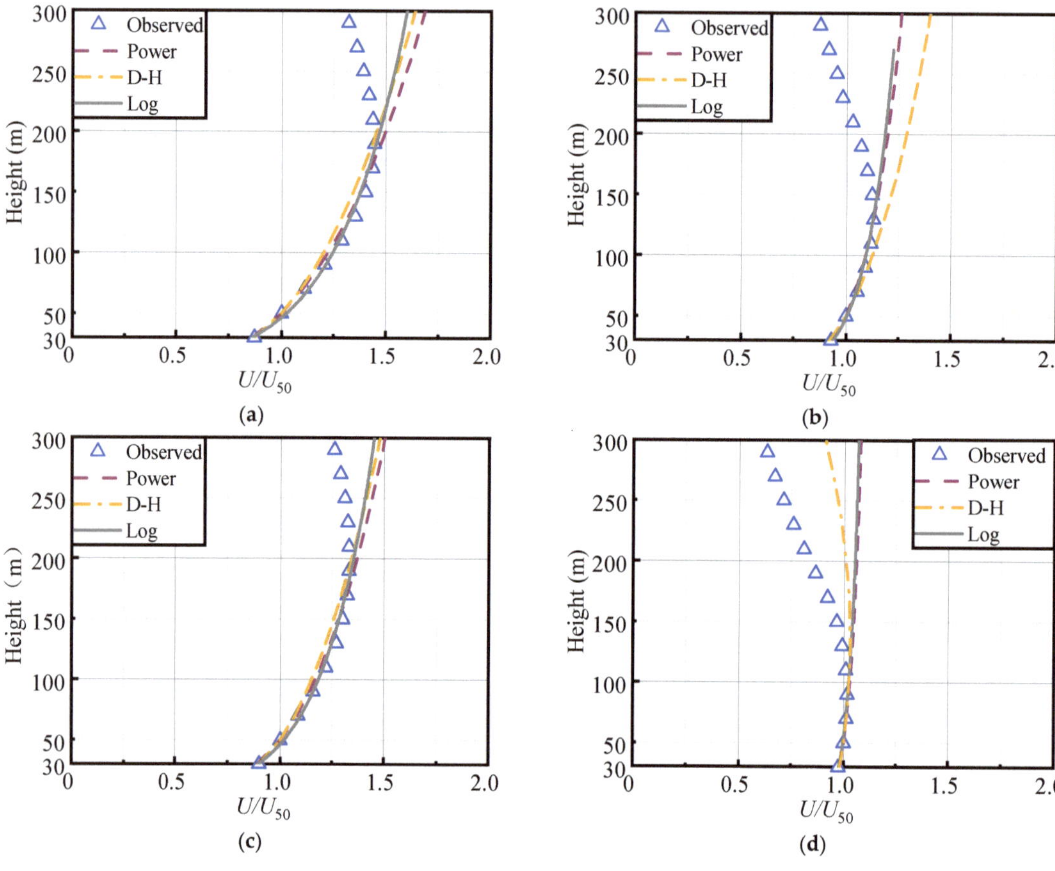

Figure 8. Comparison of wind profile models at different periods during typhoon Lekima, including (**a**) impact period; (**b**) landing period; (**c**) wind speed rising period; (**d**) stable period.

4.2.2. Influence of Incoming Wind Direction on Wind Profile

To study the influence of the average typhoon wind direction on the wind profile characteristics, the average wind speed at a height of 50 m (U_{50}) is used as the normalization to process the measured wind speed data at different heights with a basic time interval of 10 min. Considering the topographic effect near the observation point, the measured wind speed data at different heights are grouped based on the average wind direction at a height of 50 m. With a wind direction interval of $\theta = 45°$, the north direction is defined as 0° and divided into eight groups θ_i, i = 1, 2, 3, . . ., 8, which are 0°~45°, 45°~90°, 90°~135°, . . ., 315°~360°, respectively. To avoid the influence of the thermal effect, the wind data with wind speed less than 6 m/s are eliminated in the preprocessing of wind direction data. The average wind profile in each direction is shown in Figure 9. The missing wind direction intervals indicate that there is no data in that direction during the measurement period, or the data in that direction are removed because the wind speed at a height of 50 m is less than 6 m/s. The black squares in the figure represent the average wind speed records in each direction interval at each height level, the red dots represent all the measured data at each height level in that direction interval, and the black horizontal bars are error bars, indicating the standard deviation of that layer. *N* represents the number of sets of measured data in that direction interval that meet the conditions.

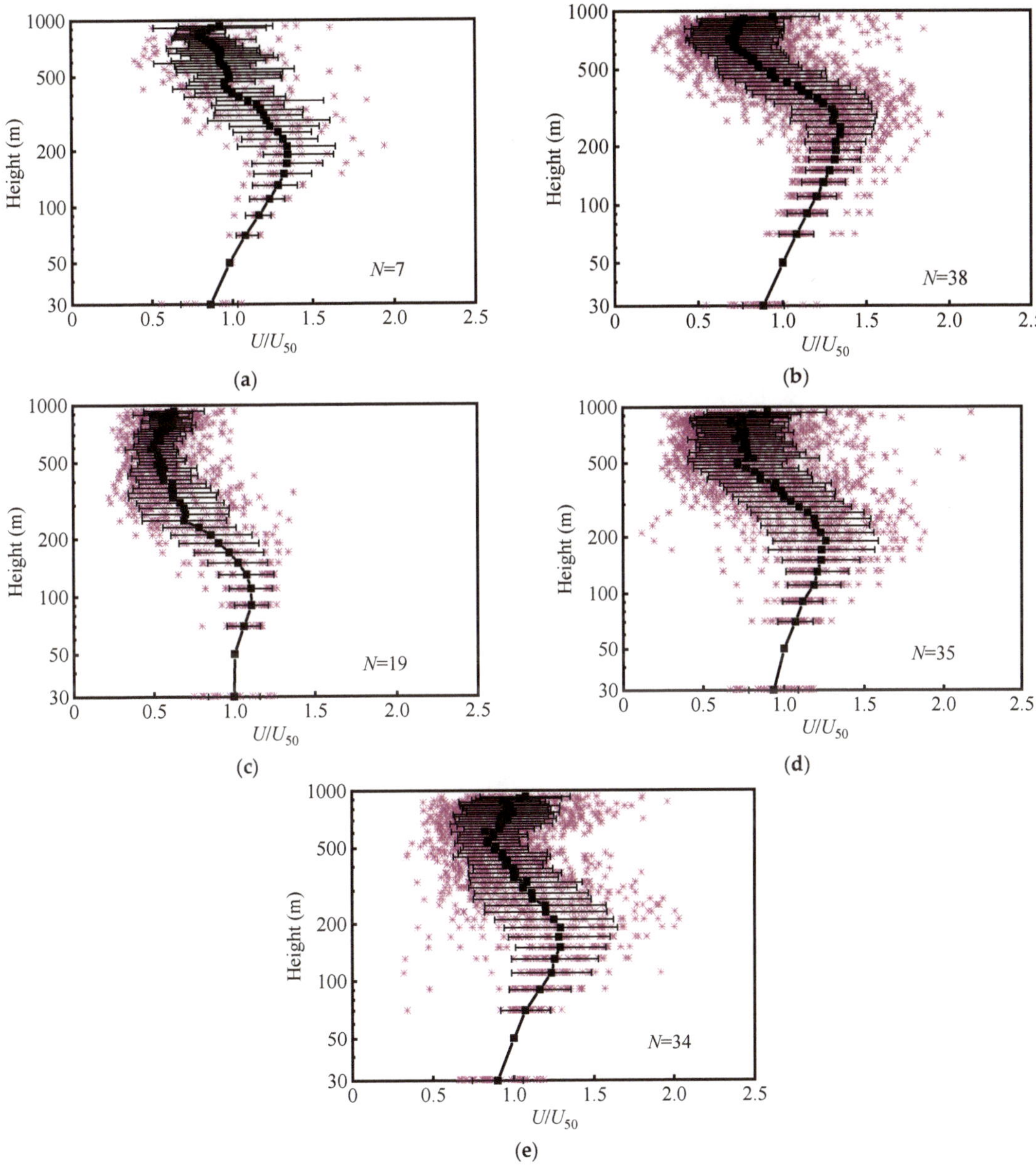

Figure 9. Wind profiles of typhoon Lekima in different average wind direction, including (**a**) $\theta2$; (**b**) $\theta3$; (**c**) $\theta4$; (**d**) $\theta5$; (**e**) $\theta6$.

It can be seen from the wind direction grouping that there are obvious differences in the average wind speed profiles in different wind direction intervals, which is more apparent in the 500 m and above height layers. In $\theta2$, the normalized wind speed at the ground increases to 1.4 at the 200 m height layer, and then gradually decreases between 200–900 m height layers. For $\theta 3$ and $\theta 4$, an obvious acceleration effect can be observed in the wind speed profile at a height of about 600 m; that is, the normalized wind speed in the upper layer tends to increase again, showing an S shape profile, but the normalized wind speed increment does not exceed the extreme value of the bottom wind speed. In addition, the wind profile of $\theta 4$ at about 100 m height obviously deviates from the exponential or logarithmic profile, which may be due to the fact that the upwind direction of the measuring point is hilly terrain. For $\theta 5$ and $\theta 6$, the normalized wind speed continues to increase from the surface below about 200 m height, showing an exponential or logarithmic law of increase, and the normalized average wind speed value can reach about 1.3. There is a significant deceleration effect in the wind profile between 200–500 m. The normalized wind speed in the upper layer remains stable, indicating that it is not affected by the terrain.

4.2.3. Influence of Wind Speed on Wind Profile

In order to explore the influence of different wind speeds on wind profile characteristics, the wind profile during the typhoon measurement is grouped and integrated according to the wind speed at a height of 50 m, and the wind speed interval is 2 m/s, as shown in Figure 10.

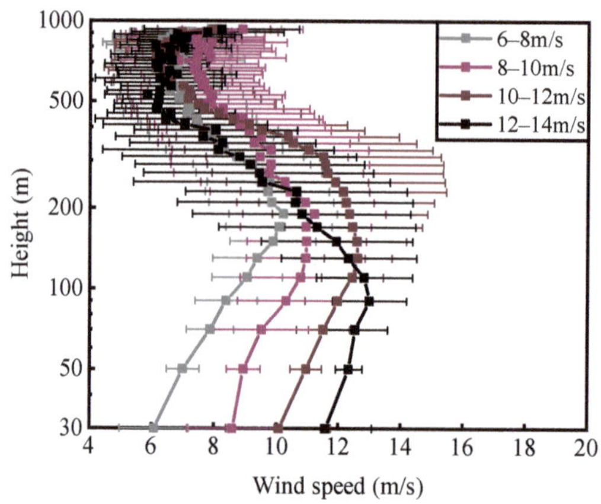

Figure 10. Wind profiles in different average wind speed intervals.

To avoid the influence of the thermal effect, data with wind speed of less than 6 m/s at a height of 50 m were also excluded. Based on the wind speed at the 50 m altitude level, the average wind speed profiles were divided into four groups: 6–8 m/s (46 profiles), 8–10 m/s (34 profiles), 10–12 m/s (37 profiles), and 12–14 m/s (16 profiles). The results show that the average wind velocity profile curve is different with different wind velocities, indicating that the average wind velocity profile curve is related to the wind velocity. The larger the wind speed range, the lower is the height of the maximum wind speed, the more drastic is the change of average wind speed with height, and the steeper is the wind profile curve. It is more obvious when the average wind speed is greater than 12 m/s. This is consistent with the conclusion of Powell [9] and Shu [37] in analyzing the typhoon wind profile. For a weather system with a larger average wind speed, the height of the maximum wind speed is lower.

4.2.4. Influence of Different Time Distance on Wind Profile

The average wind speed is calculated at intervals of 3 s, 10 min, and 1 h by engineering requirements or specification criteria. Different calculation results are obtained for the average wind speed at different intervals, leading to variations in the wind profile. Based on the measured data of typhoon Lekima, the wind velocity profiles of short time (10 min) and long time (60 min) are compared.

Figure 11 shows the time history of the horizontal wind speed at different altitudes and different time distances. Part of the white space in the figure is caused by data quality control. From the wind speed time-history curve, it can be seen that with the increase of the average time interval, the average wind speed at heights of 600–1000 m decreases significantly. The larger the time interval, the gentler is the three-dimensional wind speed surface, indicating that the changing trend of the average wind speed with time interval becomes more stable. In the figure, the 10-min average wind speed time-history curve basically fluctuates around the 60-min time-history curve. Compared to the 10-min average wind speed interval, the 60-min average wind speed interval is no longer able to reflect the development trend of wind speeds at different heights during the passage of typhoons. Using a 10-min average wind speed interval to represent the intensity of the average wind is considered more reasonable by our country's standards.

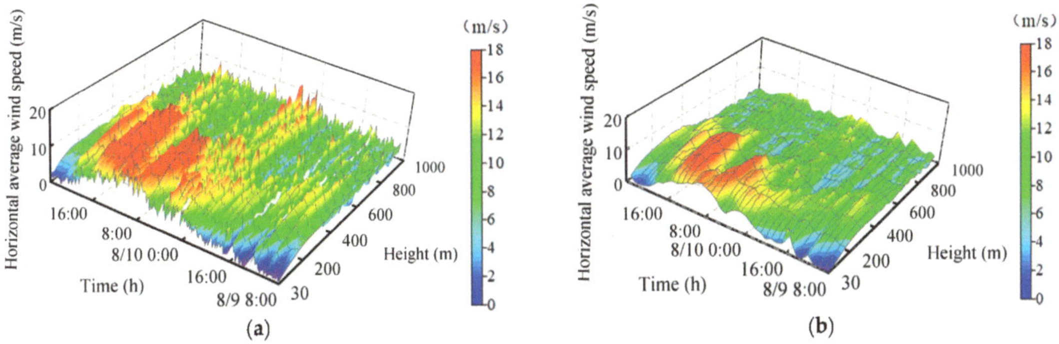

Figure 11. Horizontal wind speed time history at different time intervals, including (**a**) 10-min horizontal wind speed time history; (**b**) 60-min horizontal wind speed time history.

5. Evolution Characteristics of Wind Profile Parameters

5.1. Wind Profile Index

The vertical profile of the average wind speed within the boundary layer depends on a range of parameters, including upwind topography, surface roughness, atmospheric stability, and meantime distance. The wind profile index can reflect the variation of average wind speed with vertical height and is one of the important parameters for the wind resistance design of high-rise structures.

Figure 12 shows the change in typhoon wind profile index with 50 m average wind speed (0–5, 5–10, and 10–15 m/s) and vertical height (150 m and 250 m). In different height levels, the wind profile exponent tends to decrease as the average wind speed increases. Within the same wind speed group, the average wind profile exponent also decreases with increasing height. With the increase of U_{50}, the exponential distribution gradually changes from discrete to convergent. The wind profile exponent within the minimum wind speed range in the 150 m height level is 3.9 times that of the maximum wind speed range. In addition, the U_{50} is in the range of 10–15 m/s, and there is obviously a negative wind profile index, indicating that the wind profile distribution deviates from the power-law, log-law, and D-H model. There is a phenomenon of low-level jets.

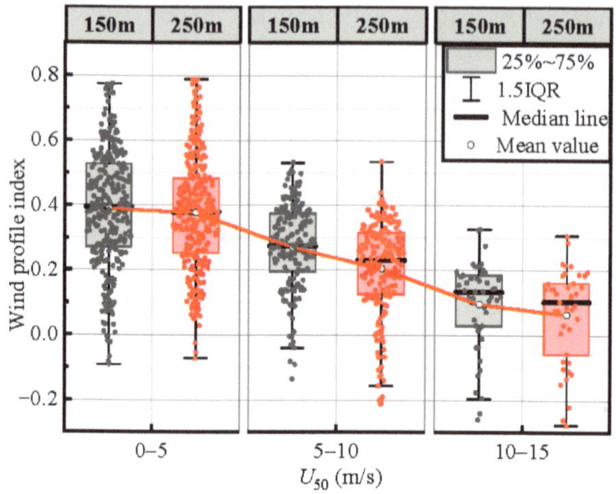

Figure 12. The change of wind profile index with average wind speed and height.

5.2. Roughness Length

The roughness length z_0 represents the roughness of the surface, and the greater the roughness length, the more obvious is the wind deceleration effect. Figure 13 shows the fitting comparison between the wind profile and the wind profile model of typhoon Lekima in different wind speed grades. The wind profile index, roughness length, and other parameters obtained by using the least square method [38] are listed in Table 4. R^2 represents the goodness of fit. The closer its value is to 1, the better the fitting degree. The fitting results show that the smaller the wind speed grade, the greater is the surface roughness length, which is consistent with the results of different impact stages of the typhoon. During the impact period of the typhoon's 7-level wind circle, the observation point is far away from the typhoon and is greatly affected by the terrain. When the average wind speed is greater than 12 m/s, the fitting effect of the wind profile model is the worst, which may be due to the small number of wind profile samples.

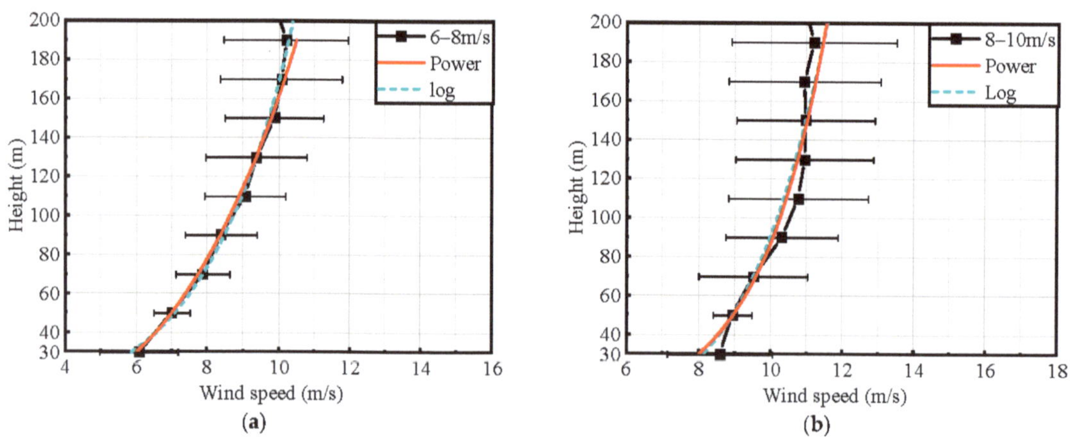

Figure 13. *Cont.*

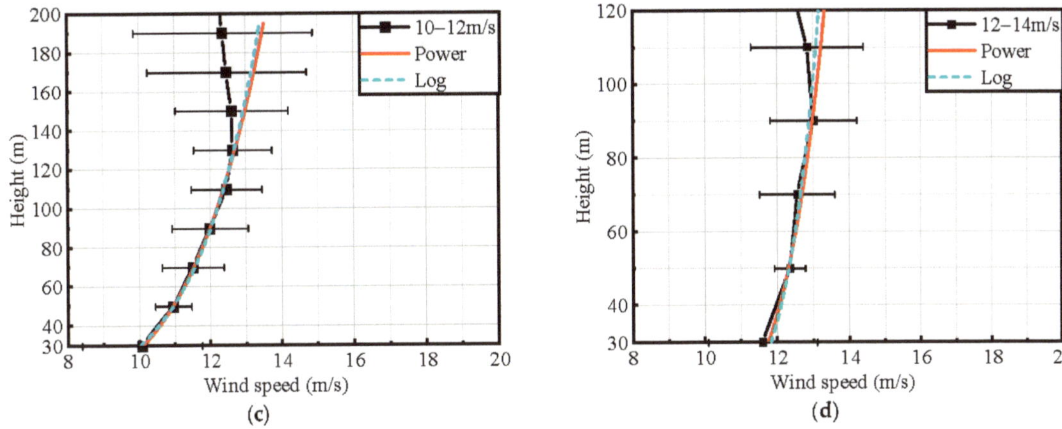

Figure 13. Comparison of wind profile and fitting model of typhoon with different wind speed classes, including (**a**) $6 < U_{50} < 8$; (**b**) $8 < U_{50} < 10$; (**c**) $10 < U_{50} < 12$; (**d**) $12 < U_{50} < 14$.

Table 4. Wind profile fitting parameters of typhoon with different wind speed levels.

Wind Speed (m/s)	Sample Number *N*	Fitting Range (m)	Power-Law Model		Log-Law Model		
			Index	R^2	u_* (m/s)	z_0 (m/s)	R^2
6–8	46	30–190	0.3	0.994	0.958	2.607	0.993
8–10	34	30–150	0.186	0.944	0.774	0.489	0.974
10–12	37	30–130	0.155	0.997	0.716	0.111	0.997
12–14	16	30–110	0.091	0.938	0.385	3.82×10^{-4}	0.894

5.3. Boundary Layer Height and Maximum Wind Speed Height

The height of the atmospheric boundary layer refers to the height where the effect of surface friction can be ignored. Different countries' specifications provide suggested values based on the underlying terrain. However, existing research indicates that h is also influenced by factors such as the different stages of a typhoon and the average wind speed.

Based on Equation (5), it is calculated that the boundary layer height in the typhoon wind speed stable period is 324 m, which is slightly less than the boundary layer height of Class B terrain in the specification (350 m). During the impact period, landfall period, and wind speed rising period of the typhoon, the boundary layer heights are 1524 m, 837 m, and 1630 m, respectively. Compared with the specification, they increased by 335%, 139%, and 365%, respectively. This may be due to the small value of the specification. Zhang [39] obtained an average boundary layer height of 1421 m based on typhoon "Maria," which is 306% larger than the specification. The boundary layer height calculated by Li [40] measuring a super typhoon far exceeds the standard value.

6. Conclusions

Based on the field measured data of typhoon Lekima captured by wind profile acoustic radar, this paper systematically studied the global profile characteristics of the boundary layer of a super typhoon and deepened the understanding of the variability characteristics of the typhoon wind field in coastal areas. The main conclusions can be summarized as follows:

(1) The measured wind speed and wind direction profiles of typhoon Lekima showed strong variability. A low-level jet phenomenon existed near the vertical height of 0.12–0.5 km, and the maximum wind speed height varied with the evolution of the typhoon wind field. It is far lower than the approximate 500–600 m maximum wind speed height of the hurricane studied by Powell [9], which may be related to the

different ocean surface of typhoon generation. Below 500 m, the wind direction deflects 180°; however, above the altitude level of 500 m, the wind direction changes more violently with the increase of height, with great dispersion, which is mainly affected by the structure of the typhoon wind field.

(2) The evolution process of the typhoon wind velocity profile can be divided into four stages: influence period, landing period, wind velocity rising period, and stable period. The log-law, power-law, and D-H models can fit well the wind profile in the height range of 50–200 m. The "anti-C" wind profile and the "S" wind profile with secondary increase of wind speed cannot be reproduced. The "S" shape profile of secondary increase wind speed is less than the extreme wind speed at the bottom of the typhoon. In the wind resistance design of high-rise structures in coastal areas, the influence of extreme wind speed at the bottom of typhoon should be fully considered and evaluated.

(3) The measured typhoon wind profile is affected by many factors. When the wind direction is hilly, the wind profile in the jet layer will deviate from the theoretical model. The higher the wind speed, the lower is the maximum wind height, and the steeper is the wind profile curve. The larger the time interval, the smoother is the 3D surface of the average wind speed. The conversion of wind speed time interval should be avoided as far as possible in structural wind resistance design.

(4) Wind parameters are greatly affected by the spatial position inside the typhoon wind field. The wind profile index decreases with the increase of average wind speed and vertical height. With the increase of average wind speed, the index distribution of wind profile tends to converge from discrete. The smaller the range of wind speeds, the larger is the length of surface roughness z_0. Compared to the standard, the boundary layer height during the typhoon impact period, landing period, and wind speed increase period increased by 335%, 139%, and 365%, respectively.

Author Contributions: Conceptualization, Y.W.; methodology, Q.Q.; software, B.F.; validation, X.W.; formal analysis, B.F. and C.Z.; investigation, S.Z. and B.F.; resources, X.W.; data curation, L.Z. and M.Z.; writing—original draft preparation, B.F., Q.Q. and Y.W.; visualization, Y.W. All authors have read and agreed to the published version of the manuscript.

Funding: This research was funded by Zhejiang Province Natural Science Foundation Project (LY19E080022), Natural Science Foundation of China (51508419, 51678455), Zhejiang Provincial Department of Education Project (The title of the Project: Wind Field Characteristic Monitoring and Wind Load Monitoring Research on Coastal Low-rise Buildings and the number: Y202147409), Chongqing Natural Science Foundation General Project (The title of the Project: Study on the roof vortex motion pattern of low rise buildings considering aerodynamic structure characteristics and its influence on wind pressure characteristics and the number: CSTB2022NSCQ-MSX1655) and the Science and Technology Research Program of Chongqing Municipal Education Commission (Grant No. KJQN202000713).

Institutional Review Board Statement: Not applicable.

Informed Consent Statement: Not applicable.

Data Availability Statement: The data presented in this study are available on request from the corresponding author.

Conflicts of Interest: The authors declare no conflicts of interest.

References

1. Fang, G.; Wei, M.; Zhao, L.; Xu, K.; Cao, S.; Ge, Y. Site-and building height-dependent design extreme wind speed vertical profile of tropical cyclone. *J. Build. Eng.* **2022**, *62*, 105322. [CrossRef]
2. Wang, Y.; Li, Y.; Qi, Q.; Zhang, C.; Wang, X.; Fan, G.; Fu, B. Experimental Study of the Fluctuating Wind Characteristics of Typhoon Jangmi Measured at the Top of a Building. *Sustainability* **2022**, *14*, 9266. [CrossRef]
3. Chen, T.; Fu, J.Y.; Chan, P.W.; He, Y.C.; Liu, A.M.; Zhou, W. Wind characteristics in typhoon boundary layer at coastal areas observed via a Lidar profiler. *J. Wind Eng. Ind. Aerodyn.* **2023**, *232*, 105253. [CrossRef]

4. Fang, G.; Zhao, L.; Cao, S.; Zhu, L.; Ge, Y. Estimation of tropical cyclone wind hazards in coastal regions of China. *Nat. Hazards Earth Syst. Sci.* **2020**, *20*, 1617–1637. [CrossRef]
5. Cheng, X.; Yan, B.; Zhou, X.; Yang, Q.; Huang, G.; Su, Y.; Yang, W.; Jiang, Y. Wind resource assessment at mountainous wind farm: Fusion of RANS and vertical multi-point on-site measured wind field data. *Appl. Energy* **2024**, *363*, 123116. [CrossRef]
6. Yao, Y.; Wu, H.; Li, X.; Zhao, B.; Chen, B.; Yi, T. A review of the transmission tower-line system performance under typhoon in wind tunnel test. *Wind Struct.* **2019**, *29*, 87–98.
7. Han, X.L.; Li, Q.S.; Zhou, K.; Li, M. Investigation of the Aerodynamic Forces on a 600-m-High Supertall Building by Field Measurements and Wind Tunnel Test. *J. Struct. Eng.* **2023**, *149*, 04023101. [CrossRef]
8. Franklin, J.L.; Black, M.L.; Valde, K. GPS dropwindsonde wind profiles in hurricanes and their operational implications. *Weather Forecast.* **2003**, *18*, 32–44. [CrossRef]
9. Powell, M.D.; Vickery, P.J.; Reinhold, T.A. Reduced drag coefficient for high wind speeds in tropical cyclones. *Nature* **2003**, *422*, 279–283. [CrossRef]
10. Song, L.; Chen, W.; Wang, B.; Zhi, S.; Liu, A. Characteristics of wind profiles in the landfalling typhoon boundary layer. *J. Wind Eng. Ind. Aerodyn.* **2016**, *149*, 77–88. [CrossRef]
11. Song, L.L.; Mao, H.Q.; Tang, H.Y.; Liu, A.J. Observation and Analysis of Guangdong Coastal Gales in the Near-surface Layer. *J. Trop. Meteorol.* **2004**, *20*, 731–736.
12. Lin, L.; Chen, K.; Xia, D.; Wang, H.; Hu, H.; He, F. Analysis on the wind characteristics under typhoon climate at the southeast coast of China. *J. Wind Eng. Ind. Aerodyn.* **2018**, *182*, 37–48. [CrossRef]
13. He, J.Y.; He, Y.C.; Li, Q.S.; Chan, P.W.; Zhang, L.; Yang, H.L.; Li, L. Observational study of wind characteristics, wind speed and turbulence profiles during Super Typhoon Mangkhut. *J. Wind Eng. Ind. Aerodyn.* **2020**, *206*, 104362. [CrossRef]
14. Liu, Z.; Zheng, C.; Wu, Y.; Song, Y. Investigation on characteristics of thousand-meter height wind profiles at non-tropical cyclone prone areas based on field measurement. *Build. Environ.* **2018**, *130*, 62–73. [CrossRef]
15. Zhuge, X.Y.; Guan, J.; Yu, F.; Wang, Y. A new satellite-based indicator for estimation of the western North Pacific tropical cyclone current intensity. *IEEE Trans. Geosci. Remote Sens.* **2015**, *53*, 5661–5676. [CrossRef]
16. Wang, Y.; Xia, B.; Chen, Y.; Chen, H.; Xie, J. Revisiting the Characteristics of Super Typhoon Saola (2023) Using GPM, Himawari-9 and FY-4B Satellite Data. *Atmosphere* **2024**, *15*, 290. [CrossRef]
17. Qi, W.; Yong, B.; Gourley, J.J. Monitoring the super typhoon lekima by GPM-based near-real-time satellite precipitation estimates. *J. Hydrol.* **2021**, *603*, 126968. [CrossRef]
18. He, Y.; Chen, T.; Tang, J.; Chan, P.; Fu, J. Thermodynamic and Kinematic Structures in the Rainband Region of Typhoon Lekima (2019) at Landfall. *Atmosphere* **2022**, *13*, 312. [CrossRef]
19. Lu, B.; Li, Q.S. Investigation of the effects of wind veering and low-level jet on wind loads of super high-rise buildings by large eddy simulations. *J. Wind Eng. Ind. Aerodyn.* **2022**, *227*, 105056. [CrossRef]
20. Ren, H.; Laima, S.; Chen, W.-L.; Zhang, B.; Guo, A.; Li, H. Numerical simulation and prediction of spatial wind field under complex terrain. *J. Wind Eng. Ind. Aerodyn.* **2018**, *180*, 49–65. [CrossRef]
21. Peng, S.; Liu, Y.; Li, R.; Wei, Y.; Chan, P.-W.; Li, S. Error Features in Predicting Typhoon Winds: A Case Study Comparing Simulated and Measured Data. *Atmosphere* **2022**, *13*, 158. [CrossRef]
22. Mukherjee, P.; Ramakrishnan, B. Investigation of unique Arabian Sea tropical cyclone with GPU-based WRF model: A case study of Shaheen. *J. Atmos. Sol.-Terr. Phys.* **2023**, *246*, 106052. [CrossRef]
23. Zhang, Y.; Cao, S.; Zhao, L.; Cao, J. A case application of WRF-UCM models to the simulation of urban wind speed profiles in a typhoon. *J. Wind Eng. Ind. Aerodyn.* **2022**, *220*, 104874. [CrossRef]
24. Li, X.; Li, Y.; Zhou, J.; Wang, Q.; Wang, X. Characteristics of Typhoon "Fung-Wong" Near Earth Pulsation. *Shock Vib.* **2021**, *2021*, 9972981. [CrossRef]
25. Liu, H.; He, M.; Wang, B.; Zhang, Q. Advances in low-level jet research and future prospects. *Acta Meteorologica Sinica* **2014**, *28*, 191–206. [CrossRef]
26. Tao, T.; Wang, H.; Zhao, K. Efficient simulation of fully non-stationary random wind field based on reduced 2D hermite interpolation. *Mech. Syst. Signal Process.* **2021**, *150*, 107265. [CrossRef]
27. Simiu, E.; Scanlan, R.H. *Effect of Wind on structure: An Introduction to Wind Engineering*; Tongji University Press: Shanghai, China, 1992.
28. Monin, A.S.; Obukhov, A.M. Basic laws of turbulent mixing in the ground layer of the atmosphere. *Tr. Geofiz. Inst. Akad Nauk* **1954**, *24*, 163–187. Available online: https://scholar.google.com/scholar_lookup?title=Basic%20laws%20of%20turbulent%20mixing%20in%20the%20surface%20layer%20of%20the%20atmosphere&journal=Contrib%20Geophys%20Inst%20Acad%20Sci%20USSR&volume=151&issue=163&pages=163-187&publication_year=1954&author=Monin,A&author=Obukhov,A (accessed on 7 March 2024).
29. Davenport, A.G. Rationale for determining design wind velocities. *J. Struct. Div.* **1960**, *126*, 39–68. [CrossRef]
30. Wang, X.; Huang, P.; Gu, M. Field Measurements About Integral Scales of Near-ground Turbulence During Typhoon 'Muifa'. *J. Tongji Univ. Nat. Sci.* **2012**, *40*, 1491–1497.
31. Wang, K.; Guo, Y.; Wang, X. Comparative Study of the Near-Surface Typhoon Wind Profile Fitting between Offshore and Onshore Areas. *Math. Probl. Eng.* **2021**, *2021*, 7422926. [CrossRef]
32. *G.B. 50009-2012*; Load Code for the Design of Building Structures. China Architecture and Building Press: Beijing, China, 2012.

33. ASCE. Minimum Design Loads for Buildings and Other Structures (ASCE/SEI 7-10). *Am. Soc. Civ. Eng.* **2014**, *559*, 996.
34. Architectural Institute of Japan. *AIJ2004 Recommendations for Loads on Buildings*; Architectural Institute of Japan: Tokyo, Japan, 2004.
35. Deaves, D.M.; Harris, R.I. *A Mathematical Model of the Structure of Strong Winds*; Construction Industry Research and Information Association: London, UK, 1978.
36. Kepert, J.; Wang, Y. The dynamics of boundary layer jets within the tropical cyclone core. Part II: Nonlinear Enhancement. *J. Atmos. Sci.* **2001**, *58*, 2485–2501.
37. Shu, Z.R.; Li, Q.S.; He, Y.C.; Chan, P.W. Vertical wind profiles for typhoon, monsoon and thunderstorm winds. *J. Wind Eng. Ind. Aerodyn.* **2017**, *168*, 190–199. [CrossRef]
38. Yao, J.W.; Hang, C.Y.; Zhuang, Z.; Liu, J.T.; Ban, Q.C. Study on Precise calculation method of urban surface roughness parameters in numerical simulation of urban ventilation. *Build. Sci.* **2020**, *36*, 99–106.
39. Chuanxiong, Z.; Yanru, W.; Zhangqi, H. Field measurement study on wind structure characteristics of specific topography under typhoon Maria. *J. Nat. Disasters* **2019**, *28*, 100–110.
40. Li, Q.S.; Zhi, L.; Hu, F. Field monitoring of boundary layer wind characteristics in urban area. *Wind Struct.* **2009**, *12*, 553–574. [CrossRef]

atmosphere

Article

CMIP6 Simulation-Based Daily Surface Air Temperature and Precipitation Projections over the Qinghai-Tibetan Plateau in the 21st Century

Kangming Wang [1], Xinyi Song [2], Fan Lu [1,*], Songbin Yu [1], Yuyan Zhou [1] and Jin Sun [1]

1 State Key Laboratory of Simulation and Regulation of Water Cycle in River Basin, China Institute of Water Resources and Hydropower Research, Beijing 100038, China; wangkangming@edu.iwhr.com (K.W.); yusongbin@edu.iwhr.com (S.Y.); sunjin@edu.iwhr.com (J.S.)
2 School of Hydraulic and Environmental Engineering, Changsha University of Science and Technology, Changsha 410114, China; songxy@csust.edu.cn
* Correspondence: lufan@iwhr.com

Abstract: The Qinghai-Tibetan Plateau (QTP), the source of many major Asian rivers, is sensitive to climate change, affecting billions of people's livelihoods across Asia. Here, we developed high-resolution projections of precipitation and daily maximum/minimum temperatures at 0.1° spatial resolution over the QTP. The projections are based on the output from seven global climate models (GCMs) from the Coupled Model Intercomparison Project Phase 6 (CMIP6) for historical (1979–2013) and projected (2015–2100) climates across four scenarios (SSP1-2.6, SSP2-4.5, SSP3-7.0, and SSP5-8.5). An updated nonstationary cumulative distribution function matching method (called CNCDFm) was used to remove model systemic bias. We verify the necessity of taking into account altitude in downscaling processes and the validity of nonstationary bias correction. Compared to the historical period, the climate in the QTP in the 21st century is warmer (1.2–5.1 °C, for maximum surface temperature) and wetter (3.9–26.8%) according to the corrected GCM projection. For precipitation, the Indus River (IDR), Tarim River (TMR), Inner of Qiangtang Basin (IQTB), Yarlung Zangbo (YLZBR), and Qaidam Basin (QDB) showed growth well above the global average across high radiative forcing scenarios, which could have a profound impact on the regional hydrological cycle. However, there is great uncertainty in precipitation prediction, which is demonstrated by a very low signal-to-noise ratio (SNR) and a large difference between Bayesian model averaging (BMA) and multi-model averages (MMAs). This bias-corrected dataset is available for climate change impact research in the QTP at the subregion scale.

Keywords: Qinghai-Tibetan Plateau; CMIP6; downscaling; nonstationary bias correction; uncertainty; reanalysis data

Citation: Wang, K.; Song, X.; Lu, F.; Yu, S.; Zhou, Y.; Sun, J. CMIP6 Simulation-Based Daily Surface Air Temperature and Precipitation Projections over the Qinghai-Tibetan Plateau in the 21st Century. *Atmosphere* **2024**, *15*, 434. https://doi.org/10.3390/atmos15040434

Academic Editor: Nicola Scafetta

Received: 12 February 2024
Revised: 22 March 2024
Accepted: 28 March 2024
Published: 30 March 2024

1. Introduction

The Qinghai-Tibetan Plateau (QTP), called the Asian Water Tower, is the source of many major Asian rivers, such as the Yellow, Yangtze, Mekong, Indus, Salween, and Brahmaputra [1]. Climate change in the QTP affects billions of people in Asia by changing regional water resources, agricultural production, energy security, and disaster risk [2–5]. That is why the QTP has attracted the attention of academics and policymakers. In previous studies, the QTP was found sensitive to climate change. Based on historical data and reanalysis data, previous research showed that the QTP area had a more significant increase in temperature than the surrounding areas [6]. However, it is difficult to estimate the spatial distribution of changes in meteorological elements. Because of complex topographic conditions, meteorological factors of the QTP have high spatial heterogeneity, especially in near-surface areas [7,8]. These distribution characteristics may exhibit different evolutionary trends under different climatic backgrounds [9]. Furthermore, low-resolution meteorological data lack representativeness and practical value on the plateau [10].

Numerous downscaling and multi-source data methods have been applied in large-scale regions to provide high-resolution data. In recent years, several high-resolution reanalysis datasets have been evaluated and applied in the Qinghai-Tibetan Plateau (QTP), with precipitation and temperature being the most crucial variables [11–15]. These studies not only demonstrate the possibility of generating high-resolution meteorological data in ungauged regions but also provide forcing data to drive a variety of terrestrial models, which is essential for climatic impact studies [16–18]. Nevertheless, due to differences in input data, numerical assimilation models, parameterization schemes, and the spatiotemporal resolutions of final products, these reanalysis datasets exhibit varying performances in different regions [19,20]. Therefore, the question of how to apply the reanalysis data remains worthy of discussion.

Global climate models (GCMs) are effective tools for simulating Earth's climate on continental scales and larger [21]. They are the most important tools and infrastructure for projecting future climatic conditions and assessing the climate change impact [22]. GCMs have been widely used in climate change research [23–25], showing growing potential when used as forcing data for land surface models and hydrological models. The Coupled Model Intercomparison Project Phase 6 (CMIP6) indicates an overall improvement in GCMs over the last decade [22,26]. Typically, the cold bias of CMIP5 on air temperature is significantly reduced in CMIP6 for the QTP [27].

However, major problems still exist and prevent researchers from using GCM results directly. At first, the models in CMIP6 still have considerable systematic biases in meteorological simulation of regional scale. Lun et al. [28] found that the multi-model ensembles (MME) of 23 GCMs in CMIP6 overestimated the QTP precipitation by about 84.1% in historical simulation, and the wet biases of the same models were even higher than historical observation. Secondly, the outputs of GCMs still feature a low resolution that is difficult to align with the input requirements of land and hydrological models, especially when compared to the more detailed scales needed for accurately simulating land and hydrological processes [29]. In general, the resolution of GCM outputs is between 0.5° and 2.5°, which is much larger than most land and hydrological processes. For instance, Su et al. [2] developed a hydrological model to simulate the impact of glacier melt in the QTP, and the model was set up on $0.083° \times 0.083°$ grids because glacier areas are from 100 km^2 to 10,000 km^2. Hence, downscaling GCMs has become a pre-processing step in many studies, but existing studies tend to use the linear interpolation method in downscaling [2,5,30–32], ignoring the influence of the underlying surface on the spatial distribution of climate variables, which may cause further systematic errors. Moreover, GCMs still exhibit large uncertainty that shows up in different models and produces widely different results when they simulate the same variable in the same forcing scenario. In the QTP, Zhang and Yang [27] found that GCM-simulated daily precipitation (including both liquid and solid phases, units: mm d^{-1}) values have large variation, with most values ranging from 1.4 mm d^{-1} to 3.5 mm d^{-1}. Such large variations make it difficult for users to decide which model is the best model or how to ensemble prediction based on several models. These problems become worse when looking at particular uses of GCM simulation output [22]. But on the other hand, although it is still challenging to obtain optimal projection over the QTP, smaller deviation forecasts (compared to the raw output from CMIP6) based on effective statistical methods are essential for regional-scale studies.

Therefore, this study attempts to produce a $0.1° \times 0.1°$ high-resolution projection over the entire QTP (HRP-QTP) based on reanalysis data and CMIP6 simulations, which includes the historical period (from 1979 to 2013) and future period (from 2015 to 2100), and can be used in watershed scale studies (variables information presented in Table 1). In the present study, combined reanalysis data were applied as historical data for ungauged regions. In order to deal with the complex terrain conditions of the QTP, we adopted a downscaling method different from the linear interpolation commonly used in previous studies to fit the effects of altitude on climate change. Furthermore, an updated bias correction method was used to remove systematic bias from GCM projections. The present study also produced ensemble data from several models. The projection is available for researchers who want to learn about climate change and its impact on the QTP. It also provides a robust and universal method to deal with CMIP6 data, which can be used in other regions confronting climate change risk.

Table 1. Definition of variables in the HRP-QTP.

Variables	Variable Name	Unit	Physical Meaning
Precipitation	Pr	mm d^{-1}	Daily precipitation, both liquid and solid
Maximum temperature	Tasmax	K	Daily surface (2 m) maximum temperature in one day
Minimum temperature	Tasmin	K	Daily surface (2 m) minimum temperature in one day

2. Materials and Methods

2.1. Definition and Division of Study Area

As for the scope of the Qinghai-Tibet Plateau, there are different definitions in different studies. We adopted the high-frequency use scope of the QTP (HF-QTP) as defined by Zhang et al. [33], which included several major river headwaters and can match most research requirements. The QTP ranges from 25°59′37″ N to 39°49′33″ N and from 73°29′56″ E to 104°40′20″ E, covering an area of 2542.30 × 10^3 km^2 [34]. Its altitudes range from 85 m to 8233 m, and approximately 90.9% belong to high altitude (>3000 m) areas. This range includes the world's highest mountains and massive deep valleys in terms of topography. The complex terrain of the QTP results in a complicated distribution of precipitation and surface temperature. A brief overview can be seen in Figure 1. These regions can be divided into ten subregions based on watershed [35,36], as follows: I. Tarim River (TMR); II. Qaidam Basin (QDB); III. Hexi Basin (HXB); IV. Yellow River (YR); V. Inner of Qiangtang Basin (IQTB); VI. Yangtze River (YZR); VII. Indus River (IDR); VIII. Salween River (SWR); IX. Mekong River (MKR); and X. Yarlung Zangbo (YLZBR). Among these, IQTB and QDB belong to the inland river basin, and some of the river channels are seasonal. We obtained the river distribution of QDB from National Basic Geographic Information [37]. Currently, there are only 131 long-observed meteorological stations in the QTP (Figure 1) maintained by the China Meteorological Administration (CMA). They are primarily situated in the eastern part of the QTP, with observations lacking in the western part of the plateau. The digital elevation model (measured from sea level, DEM, units: m) used in this section and Section 2.5 is provided by the National Tibetan Plateau Data Center "https://data.tpdc.ac.cn/ (24 March 2021)" [38].

In Figure 1, the lower Yarlung Zangbo valley (LYZV) is highlighted with red dots. Its boundary is defined by the watershed of lower Yarlung Zangbo and is the joint part of the Yarlung Zangbo River and Brahmaputra River. Unlike other regions in the QTP, LYZV belongs to the south of the Himalayas, is controlled by the Indian monsoon [39–43], and has greatly varied elevations (from 85 m to 6941 m). Because of water vapor transport by monsoons and the uplift effect of the valley, precipitation in LYZV is greater than in other regions [44]. Although LYZV is an ungauged region, we can estimate its annual precipitation ranging from 2500 mm to 3500 mm by using observations from similar neighborhoods in the south of the Himalayas and early investigation findings [43,45–47]. Thus, it is essential to provide reasonable estimations in LYZV. Special methods are required to deal with meteorological data in LYZV. We designed a method that combines two high-resolution grid datasets to reduce the underestimation of precipitation, which can be seen in Section 2.3.

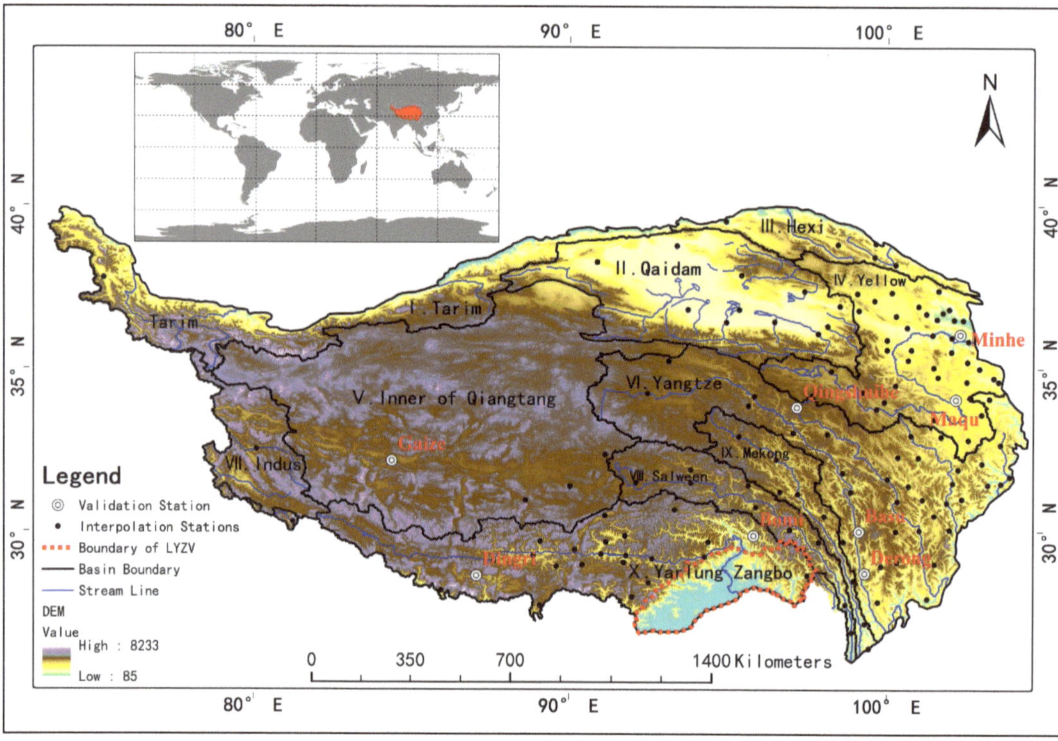

Figure 1. Location and geographical domains for HF-QTP. Black dots and white dots denote the meteorological stations from CMA, white dots are used as verification stations, and black dots are used as interpolation stations in Section 2.6. The QTP's major river basin boundaries are in black and the main streams are in blue. Due to the absence of surveying and mapping data, the rivers within the Qiangtang region are not depicted on the map. Names of the major river basin boundaries are written within each basin boundary. DEM is shown as the color scale in the background. The boundary of the low Yarlung Zangbo Valley (LYZV) is shown with a red dotted line.

2.2. Historical Meteorological Data

(1) Observation data from CMA.

As Figure 1 demonstrates, the Qinghai-Tibet Plateau is indeed a data-sparse region, with limited and unevenly distributed meteorological stations across the area. In the CMA published ground observation data (from "http://data.cma.cn (24 March 2021)"), there are about 147 observation meteorological stations with continuous long series on the Qinghai-Tibet Plateau, and the density of stations is 0.058 per 1000 km^2. Ma et al. [12] have demonstrated the impact of overly sparse sites on downscaling representativeness, so in previous studies, CMA site observations were generally used to validate reanalysis methods and results [13–15,19,20]. In the present study, CMA data were used to verify the availability of the downscaling method (Section 2.6). Among them, 8 stations are used as verification stations to represent climate characteristics at different altitudes (white dots in Figure 1) and the remaining 139 stations are interpolation stations (black dots in Figure 1). The precipitation and temperature data of interpolation stations from 1979 to 2013 were interpolated into the verification station grids (the grids are consistent with HRP-QTP) by different methods to compare the fitting effects of different interpolation methods on the verification stations. Further, a suitable interpolation method is selected for downscaling. Basic information on the verification stations is shown in Table 2.

Table 2. Basic information on verification stations.

Name	Altitude (m)	Latitude (°)	Longitude (°)	Year of Initial Observation
Qingshuihe	4415.4	33.80	97.13	1956
Basu	2589.2	30.00	99.10	1952
Derong	2422.9	28.72	99.28	1960
Gaize	4414.9	32.15	84.42	1973
Dingri	4300.0	28.63	87.08	1959
Bomi	2736.0	29.87	95.77	1955
Minhe	1813.9	36.33	102.83	1955
Maqu	3471.4	34.00	102.08	1967

(2) Reanalysis data.

It is crucial to select appropriate meteorological data for the historical period because multi-model ensemble prediction and bias correction are highly dependent on historical data. In other words, historical data determine whether the output reanalysis data are reasonable and available. We used two high-resolution reanalysis (0.1 × 0.1°) datasets from He et al. [11] (the China Meteorological Forcing Dataset, CMFD, from 1979 to 2013) and Ma et al. [12] (Gridded Precipitation for Quantile Mapping, GPQM, from 1980 to 2009) as historical data in our study. Data were downloaded from the National Tibetan Plateau Data Center "http://data.tpdc.ac.cn (24 March 2021)"; basic information about CMFD and GPQM is presented in Table 3.

Table 3. Basic information about CMFD and GPQM.

Name	Resolution (°)	Data Time Range	Coverage Area	Variables Used in Present Study
CMFD	0.1	1979–2018	Mainland of China	Precipitation rate and temperature
GPQM	0.1	1980–2010	A square area covers 70° to 99° east and 25° to 41° north	Precipitation

2.3. Combination of Reanalysis Data

CMFD provides a high-resolution meteorological forcing dataset across China and has been evaluated for its temperature performance in western China. It shows better performance in surface temperature results in the QTP's ungauged regions compared to other reanalysis data [11]; hence, the present study uses CMFD as the historical temperature source directly. However, CMFD precipitation is unverified in the western part of the QTP because of the lack of surface observational data, and it shows obvious bias in specified areas. What stands out is that it underestimates the precipitation in the LYZV compared with observations from ground stations in the Motuo and its immediate surrounding areas [48]. Hence, we adopt GPQM as a historical precipitation source in LYZV, which is validated with no obvious bias. GPQM considers observation stations both inside and surrounding the QTP [12]. It includes decades of gauge data from the south of the Himalayas, which can comprehensively reflect the precipitation characteristics of the QTP. To align the GPQM with the historical period of the HRP-QTP, we extend it using CMFD, and the extended method is shown in Figure 2.

Figure 2 employs three background colors to denote grid data from distinct sources: orange for GPQM data, purple for CMFD data, and green for the final 0.1° × 0.1° grid output of QTP's precipitation data. Initially, we extracted QTP's precipitation data, excluding the LYZV areas, from CMFD to serve as the historical precipitation source for other regions. Then, precipitation data for the LYZV were extracted from GPQM to the same grids as CMFD. Given that GPQM and CMFD share the same resolution, we applied GPQM data grid by grid to correct CMFD within the LYZV region. This correction allowed us to extend the GPQM data, ensuring that precipitation data from both sources matched the time dimension. We utilized an updated nonstationary CDF-matching method (CNCDFm), detailed in Section 2.5, for this correction. Finally, the extended GPQM precipitation data for LYZV were combined with the CMFD-extracted precipitation data for other regions, creating a comprehensive historical precipitation dataset for the entire QTP. This combined dataset will serve as the observed precipitation in subsequent steps.

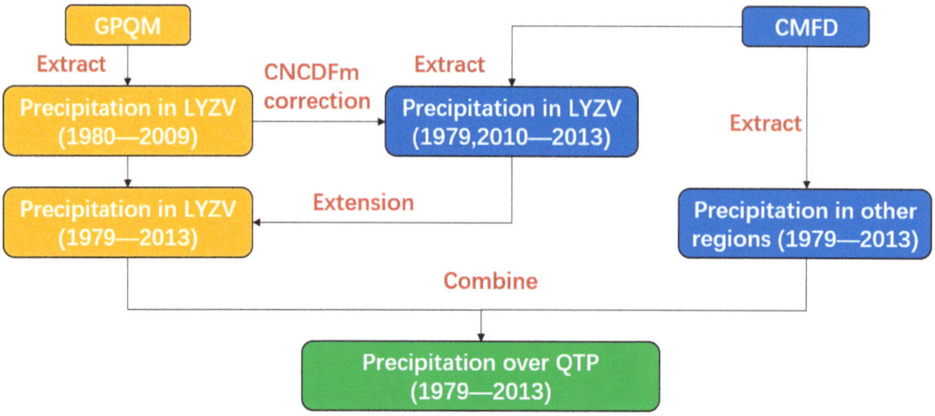

Figure 2. Flowchart for producing historical precipitation data over the QTP based on GPQM and CMFD.

2.4. Simulation and Prediction Data from GCMs

Seven CMIP6 GCMs were used to produce HRP-QTP (See Table 4). The GCM outputs were obtained from "https://esgf-node.llnl.gov/search/cmip6/ (24 March 2021)". These GCMs were selected because precipitation, maximum temperature, and minimum temperature data were available for the 7 GCMs across most future scenarios. Moreover, they represent different resolutions, from 1.12° to 2.5°. In an assessment by Lun et al. [28], the 7 GCMs overestimate the precipitation of the QTP by 0.8–1.2 mm d^{-1} in the historical period, compared with CMFD. We interpolated them to a resolution of 0.1° and used historical data to correct their systemic bias. The CMIP6's global simulation results of these climate models come from institutions worldwide, using different modeling methods. Compared with the representative concentration pathways (RCPs) of CMIP5, CMIP6 developed a new framework called shared socioeconomic pathways (SSPs) to quantitatively describe the relationship between climate change and socioeconomic development [26]. In this study, we selected SSP1-2.6, SSP2-4.5, SSP3-7.0, and SSP5-8.5 to represent different development and concentration pathways. These scenarios correspond to a sustainable development scenario, a medium challenge scenario for mitigation and adaptation, a high challenge scenario for mitigation and adaptation, and a high radiative forcing scenario with fossil fuels in the future period [49]. The four SSPs present varied global scenarios of radiative forcing, ranging from low to high. SSP1-2.6, SSP2-4.5, and SSP5-8.5 each align with the radiative forcing assumptions of RCP2.6, RCP4.5, and RCP8.5, respectively. In contrast, SSP3-7.0 serves as an alternative to the high radiative forcing scenario. Although it projects lower greenhouse gas concentrations than SSP5-8.5, it suggests that human society will face a greater risk of climate hazards [50]. We chose the first 3 SSPs to align our results with those of CMIP5 and SSP3-7.0 to represent a more likely high radiative forcing scenario.

Table 4. 7 GCMs used in this study.

Modeling Group	Model Name	Approximate Grid Spacing (Latitude × Longitude)
Beijing Climate Center (China)	BCC-CSM2-MR	1.125° × 1.125°
Bjerknes Centre for Climate Research, Norwegian Meteorological Institute (Norway)	NorESM2-MM	2.5° × 1.875°
Max Planck Institute for Meteorology (Germany)	MPI-ESM1-2-HR	0.9° × 0.9°
Institute for Numerical Mathematics, Russian Academy of Science (Russia)	INM-CM4-8	1.5° × 2°
Institute Pierre Simon Laplace (France)	IPSL-CM6A-LR	1.2676° × 2.5°
Meteorological Research Institute, Japan Meteorological Agency (Japan)	MRI-ESM2-0	1.121° × 1.125°
Geophysical Fluid Dynamics Laboratory (America)	GFDL-ESM4	1.25° × 1.00°

2.5. Methods of Processing CMIP6 Data

Methods used to process CMIP6 data are divided into three steps: downscaling, multi-model ensemble forecasting, and bias correction. At first, we used a statistical method to downscale 7 GCM outputs from different resolutions. Secondly, a prior-based method was used to generate an ensemble forecast from GCMs. Finally, all results from the downscaling and multi-model ensemble were corrected by a nonstationary method. Details of the three steps are as follows:

(1) Downscaling.

Most previous studies used linear interpolation methods to downscale GCM data and ignored the relationship between altitude and meteorological elements [2,4,25]. However, orographic studies demonstrated an approximately linear correlation between precipitation and temperature with elevation in high-altitude regions [51]. Several studies showed that it is necessary to consider altitude when generating high-resolution meteorological forcing [52,53]. Hence, we utilized ANUSPLIN (V4.36) to downscale the GCM data. ANUSPLIN is a software package that is based on the theory of thin-plate smoothing splines, and is an extension of linear regression, which considers distance, altitude, and other variables. Ma et al. [12] confirmed that ANUSPLIN is more accurate than inverse distance weighted and ordinary kriging methods in monthly precipitation interpolation over the QTP. In this study, ANUSPLIN was used to downscale daily precipitation and surface temperature, and altitude was considered a covariate in ANUSPLIN. More details about ANUSPLIN can be seen in the paper from M.F. Hutchinson and P.E. Gessler [54].

(2) Multi-model ensemble forecast.

Due to the uncertainty in GCMs, the results of different climate models are widely distributed, and it is necessary to provide certain projection results for specific studies [55]. In this research, Bayesian model averaging (BMA) was used to produce a multi-model ensemble forecast result based on 7 GCM outputs in the historical period and 4 SSPs. BMA is a statistical method that is based on the posterior probability density functions (pdfs) of model predictions [56]. BMA, which is different from the multi-model average (MMA), could be described as a weighted average, as follows:

$$M_{BMA} = \sum_{k=1}^{K} w_k M_k \quad (1)$$

where M_{BMA} is the output result of *BMA*, M_k is the output of model *k*, and w_k is the weight of model *k*, denoting the posterior probabilities of model *k* conditioned. If we assume that the actual climate change process, y, is a combination of k model outputs, w_k could be estimated based on fitting the data to the following relationship:

$$p(y|M_1, M_2, \ldots M_K) = \sum_{k=1}^{K} w_k p(y|M_k) \tag{2}$$

where the left side of Equation (2) denotes the pdf of the truth value in the historical period, and $p(y|M_k)$ denotes the conditional probability distribution of M_k; it represents the probability that y follows the pdf of the output from model k. When $p(y|M_k)$ is assumed to follow a normal distribution, it could be estimated as follows:

$$p(y|M_k) = \frac{1}{\sqrt{2\pi\sigma^2}} exp\left\{-\frac{[y(t) - M_k(t)]^2}{2\sigma^2}\right\} \tag{3}$$

with $y(t)$ and $M_k(t)$ representing the y and M_k values at time, t, respectively, and σ^2 is the variance of $y - M_k$. In this study, a Markov Chain Monte Carlo (MCMC) method proposed by Jasper et al. [57] was used to sample the weights of 7 GCM outputs. The weights were calculated based on historical GCM outputs in every 0.1° grid to maintain the spatial distribution characteristics of meteorological variables. The grid's daily temperature and precipitation data obtained in Section 2.3 were used as truth values. BMA, which differs from taking the mean value of several models, considers the model's performance during the training period, which could reduce the uncertainty of the ensemble result effectively [56–58]. MMA was also calculated in this study to compare with BMA.

(3) Bias correction of nonstationary sequence.

Because climate processes are complex and cannot be described accurately by existing methods, all GCMs still have considerable biases in their output results. Previous studies have adopted many statistical methods to eliminate the systematic errors of the model, such as linear and nonlinear corrections or cumulative distribution functions (CDFs) [31,59,60]. Moreover, it is important to retain nonstationary information in future scenarios and avoid the occurrence of abnormality during bias correction. In this research, we adopted a nonstationary bias correction technique proposed by Miao et al. [61], called updated nonstationary CDF-matching (CNCDFm), to correct downscaled GCM data and BMA results. CNCDFm model methods can be presented as follows:

$$\tilde{x}_{m\_p} = \begin{cases} I(x), & I(x) > 0 \\ g(x), & I(x) < 0 \\ 0, & x_{m\_p} = 0 \end{cases} \tag{4}$$

$$I(x) = x_{m\_p} + F_o^{-1}(F_{m\_p}(x_{m\_p})) - F_{m\_h}^{-1}(F_{m\_p}(x_{m\_p})) \tag{5}$$

$$g(x) = x_{m\_p} \times \frac{F_o^{-1}(F_{m\_p}(x_{m\_p}))}{F_{m\_h}^{-1}(F_{m\_p}(x_{m\_p}))} \tag{6}$$

where x denotes the meteorological variable from historical observations (o, denoted here as the reanalysis data obtained from Section 2.3) or models (m) during the historical period (h), current climate (c), or future projection (p) periods, $F(\cdot)$ is the CDF for a variable, and $F^{-1}(\cdot)$ is its inverse. CNCDFm is an updated version of the CDF technique. By combining the method by Li et al. [62] with the method from Wang and Chen [63], it effectively retained nonstationary information in future scenarios and avoided the occurrence of abnormality. For instance, this approach could avoid negative values for precipitation or unreasonably high values (when the denominator on the right-hand side of Equation (6) is very small). Miao et al. [61] provided validation within the CMIP5 to approve the method's efficiency.

Estimating the probability distribution of CDF is the key point of CNCDFm. We used the empirical cumulative distribution function (ECDF) to estimate the CDF in all periods for robust application. We used the 1979–2013 period as a training period to estimate $F_o(\cdot)$

and $F_{m\text{-}p}(\cdot)$ by the ECDFs of observed data and downscaled model output (obtained from Section 2.5 "Downscaling"). The wet day threshold was set at 0.1 mm/day to correct for the drizzle effect in precipitation data [31,64]. Model simulations in the historical period and the output of BMA were corrected in this step.

2.6. Effectiveness Evaluation of Methods

(1) Applicability test of downscaling method.

This study designed an experiment to compare the effectiveness of different interpolation methods. Using CMA historical observations, 8 stations (white dots in Figure 1) were used for validation due to their long and complete series of observations. Observation data from the remaining 139 stations (black dots in Figure 1) were interpolated to the validation stations using three downscaling methods, namely, ANUSPLIN, inverse distance weight (IDW) [65], and original Kriging (OK) [65]; the latter two methods do not take altitude into consideration, which is used to compare with ANUSPLIN.

Here, IDW is used to represent linear two-dimensional interpolation, defined as follows:

$$Z(x_0) = \frac{\sum_{i=1}^{n} \frac{Z_i}{d_i}}{\sum_{i=1}^{n} \frac{1}{d_i}} \tag{7}$$

where $Z(x_0)$ represents the interpolation result at verification station x_0, Z_i represents the observation from interpolation station i; n is equal to 139, and d_i is the planar Euclidean distance in kilometers between x_0 and station i.

OK is used to represent non-linear two-dimensional interpolation, which is equivalent to thin plate spline interpolation without considering covariates. In this paper, OK is also based on ANUSPLIN; for technical details, please refer to Hutchinson and Gessler [54].

The interpolation results were compared with ground observation data by the Wasserstein distance (WD) [66] and quantile-quantile (Q-Q) plot. WD, also named the earth mover's distance (EMD), is able to measure the distribution similarity between two samples very efficiently [67]; it could be defined as follows:

$$WD(P_1, P_2) = \inf_{\gamma \sim \prod(P_1,P_2)} E_{(x,y)\sim\gamma}[\|x - y\|] \tag{8}$$

where $\gamma \sim \prod(P_1, P_2)$ denotes all possible joint probability distributions between P_1 and P_2, inf indicates the greatest lower bound, and $E_{(x,y)\sim\gamma}[\|x - y\|]$ denotes the expectation of distance between samples x and y (which follow P_1 and P_2 distributions, respectively, representing the interpolated result and the observation at a verification station) when their joint distribution follows γ. In other words, WD indicates the minimum distance needed to move x into the same distribution as y. The smaller the WD, the closer the probability distribution between the interpolation results and the observations. The interpolation method with the minimum WD means that it most effectively captures the statistical distribution characteristics of the verification station.

Due to the inherently high randomness and uncertainty of weather systems, traditional Euclidean metrics, such as the mean absolute error (MAE) and root mean square error (RMSE), are limited to evaluating the effectiveness of interpolation methods [12]. This is because almost all interpolation methods struggle to capture the daily variations at unknown points [68]. In contrast, evaluating the fitting distribution characteristics through WD is more meaningful [69]. Long-term climate change studies may not necessarily require precise daily precipitation or temperature data for specific days, but accurate probability distribution characteristics within specific time periods are essential. This is crucial for research on extreme statistics and is important for both historical simulations and future predictions. Therefore, we adopt the WD as the evaluation metric.

(2) Statistical validation of bias correction.

To evaluate the absolute mean bias (AMB) of each model before and after correction, we verified the validity of the correction method. The evaluation was performed as follows:

First, we randomly selected 20 years from the historical period of 34 years; GCM outputs from this 20-year period were divided into two equal-length parts, x_{m_h} and x_{m_p}, as seen in Equations (5) and (6). Secondly, x_{m_p} was corrected by x_{m_h} and reanalysis data in the corresponding period; the pre-correction and post-correction AMBs against these observations are calculated as follows:

$$AMB = \frac{\left|\sum x_o - \sum x_{m-p}\right|}{n} \quad (9)$$

where x_o denotes the values extracted from the reanalysis data obtained in Section 2.3, and n denotes the sample size in the validation period.

The above steps were repeated 10 times in each 0.1° resolution grid to ensure the performance of the multi-model ensemble forecast. Bias correction was evaluated by the mean value of the results from these 10 iterations to guarantee the robustness of the evaluation [50], and the AMB was taken using absolute values to prevent the positive and negative from canceling out.

Quantile bias (QB) is used as a supplementary test to evaluate the ability of bias correction to extreme values. The QB of the specified percentage p can be defined as follows:

$$QB = F_{m-h}^{-1}(p) - F_o^{-1}(p) \quad (10)$$

where F_{m-h}^{-1} and F_o^{-1} are the same as in Equations (5) and (6), and F_{m-h}^{-1} is estimated by ECDF before and after correction.

(3) Evaluation of multi-model uncertainty.

Uncertainty is a key factor affecting the usability of model prediction results; the signal-to-noise ratio (SNR) was used to quantify the proportion of valid information amid uncertainty. SNR is defined as follows:

$$SNR = \frac{DS}{DN} \quad (11)$$

with

$$DS = |\overline{x} - x_h|$$

$$DN = \sqrt{\frac{1}{n+1}\sum_{i=1}^{n}(x_i - \overline{x})^2}$$

where, $\overline{x}$ means the prediction result of MMA, x_h denotes the historical simulation result, n denotes the number of climate models, and x_i denotes the prediction result of model i. It indicates that reliability is higher than uncertainty when SNR is greater than 1. Furthermore, because SNR is dimensionless, it can be used to compare different variables.

3. Results

3.1. *Performance of Downscaling Methods in the Historical Period*

Table 5 and Figure 3 display the effects of different interpolation methods. ANUSPLIN shows the best-fitting results at all eight validation stations for precipitation as well as the minimum temperatures. For maximum temperature, it also presents the best-fitting results at six validation stations. This suggests that considering altitude as a covariate can significantly reduce statistical distribution errors in downscaling. According to Table 5, ANUSPLIN achieves the smallest average WDs (0.49 mm d^{-1} for precipitation, 1.18 K for maximum temperature, and 1.39 K for minimum temperature, respectively) among the three downscaling methods for all variables. Only in the cases of maximum temperature at Dingri and Bomi and minimum temperature at Bomi and Maqu do IDW and OK methods perform slightly better than ANUSPLIN, respectively.

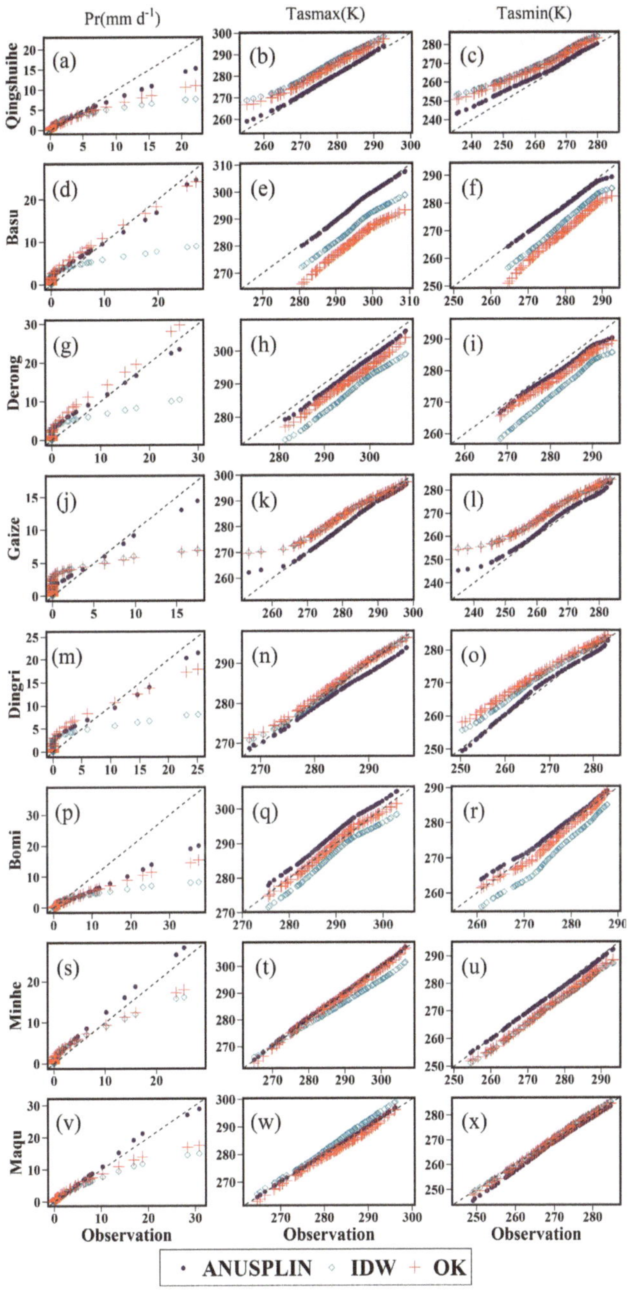

Figure 3. The Q-Q plot between downscaling results and observations from CMA in validation stations. The subfigures are numbered in sequence from (**a**–**x**) and labeled in the top left corner of each subfigure according to their order. Different colors and dot types denote different downscaling methods. Horizontal axes denote observation values and vertical axes denote downscaling results; they have the same units. The gray dotted line, "x = y", indicates the 1:1 line. Values closer to this line indicate a more accurate statistical distribution.

Table 5. WD for different methods at 8 independent validation stations. The bold terms indicate the best method for statistical distribution validation.

Name	Altitude (m)	Pr (mm d^{-1})			Tasmax (K)			Tasmin (K)		
		ANUSPLIN	IDW	OK	ANUSPLIN	IDW	OK	ANUSPLIN	IDW	OK
Qingshuihe	4415.4	**0.25**	0.82	0.57	**1.07**	7.69	5.97	**2.67**	8.68	7.27
Basu	2589.2	**0.55**	1.24	0.82	**0.41**	8.05	13.04	**0.84**	6.59	9.94
Derong	2422.9	**0.69**	1.34	0.98	**2.37**	8.28	4.72	**1.96**	8.25	3.71
Gaize	4414.9	**0.24**	1.05	1.08	**0.39**	3.66	3.51	**1.46**	5.95	5.94
Dingri	4300.0	**0.56**	1.15	0.72	1.8	**0.39**	0.87	**0.78**	3.21	4.59
Bomi	2736.0	**1.05**	1.5	1.14	2.86	3.04	**0.55**	1.23	4.83	**1.04**
Minhe	1813.9	**0.39**	0.63	0.63	**0.39**	3.11	0.52	**0.66**	4.74	4.2
Maqu	3471.4	**0.15**	0.59	0.51	**0.15**	1.74	0.99	1.5	0.78	**0.55**
Average	——	**0.49**	1.04	0.81	**1.18**	4.50	3.77	**1.39**	5.38	4.66

Figure 3 demonstrates that ANUSPLIN provides a better fit to the 'x = y' line at most stations. For precipitation, OK consistently overestimates light rain and underestimates heavy rainfall, as indicated by the red crosses in Figure 3. IDW exhibits the same issue at Gaize and Qingshuihe stations, represented by blue prisms in Figure 3a,j. At other stations, IDW tends to overall overestimate precipitation. For temperature, both IDW and OK consistently overestimate temperatures at high-altitude stations (Qingshuihe, Gaize, and Dingri, whose altitudes are higher than 4000 m) and underestimate at relatively low-altitude stations (Basu, Derong, and Bomi, whose altitudes are lower than 3000 m). This is likely because they do not account for the linear relationship between temperature and elevation. Overall, ANUSPLIN (indicated by purple dots in Figure 3) outperforms the other two methods in addressing the above issues, especially when fitting precipitation extremes, which is crucial for regional hydrological research.

Overall, the results in this section suggest that ANUSPLIN is available for the meteorological element interpolation of the QTP; that is, it is more reasonable than IDW and OK in the performance of precipitation extremes and temperatures. This advantage is likely attributed to its superior ability to model the nearly linear relationship between climate variables and altitude, taking elevation into account as a significant covariate. It also highlights that neglecting altitude's impact in downscaling efforts within the QTP can lead to significant statistical distribution biases, a finding that aligns with prior studies [8,12,70–72].

3.2. Multi-Model Ensemble and Bias Correction Performance

The AMBs before and after correction are shown in Figures 4–6, showing a substantial reduction in the AMBs after bias correction across all three variables from the seven GCMs and BMAs for the historical period (1979–2013). The outputs of these three variables from the seven GCMs all showed large biases before correction, as depicted in Figures 4, 5 and 6b–h. For precipitation, the AMBs of seven GCMs ranged from 0.87 to 1.73 mm d^{-1}, some of them were even higher than the mean precipitation over the QTPs (1.24 mm d^{-1} approximately); the high bias value ($\geq$3.00 mm d^{-1}) occurred in the southeast of the QTP, covering the southern parts of IDR, YLZBR, YZR, MKR, and SWR, before bias correction (Figure 4b–h). For temperature, seven GCM AMBs ranged from 1.35 to 6.84 K (maximum surface temperature, Figure 5b–h) and from 1.23 to 10.63 K (minimum surface temperature, Figure 6b–h). For maximum surface temperature, the high bias value ($\geq$8.50 K) occurred at the edge of QDB and most of TMR, IDR, and YLZBR; for minimum surface temperature, the high bias value ($\geq$8.50 K) occurred in the western part of the QTP (TMR, IDR, and western parts of IQTB and YLZBR) and small parts of QDB, before bias correction. Therefore, such uncorrected data cannot be used for climate research and decision-making at the regional scale [31]. After bias correction, the GCMs' absolute mean bias was reduced to an acceptable range. For precipitation, it was from 0.13 to 0.19 mm d^{-1} (about an 85% to 89% reduction Figure 4j–p). For temperature, they were from 0.40 to 0.64 K (maximum surface temperature, Figure 5j–p) and from 0.37 to 0.55 K (minimum surface temperature, Figure 6j–p). High values only occurred in smaller areas compared with uncorrected data. For precipitation, only NorESM2-MM exhibited high AMB ($\geq$3.50 mm d^{-1}) in a small part of LYZV (Figure 4k). The reduction in bias for mean annual precipitation, as well as the maximum and minimum temperatures, demonstrates the effectiveness of our bias correction based on CNCDFm.

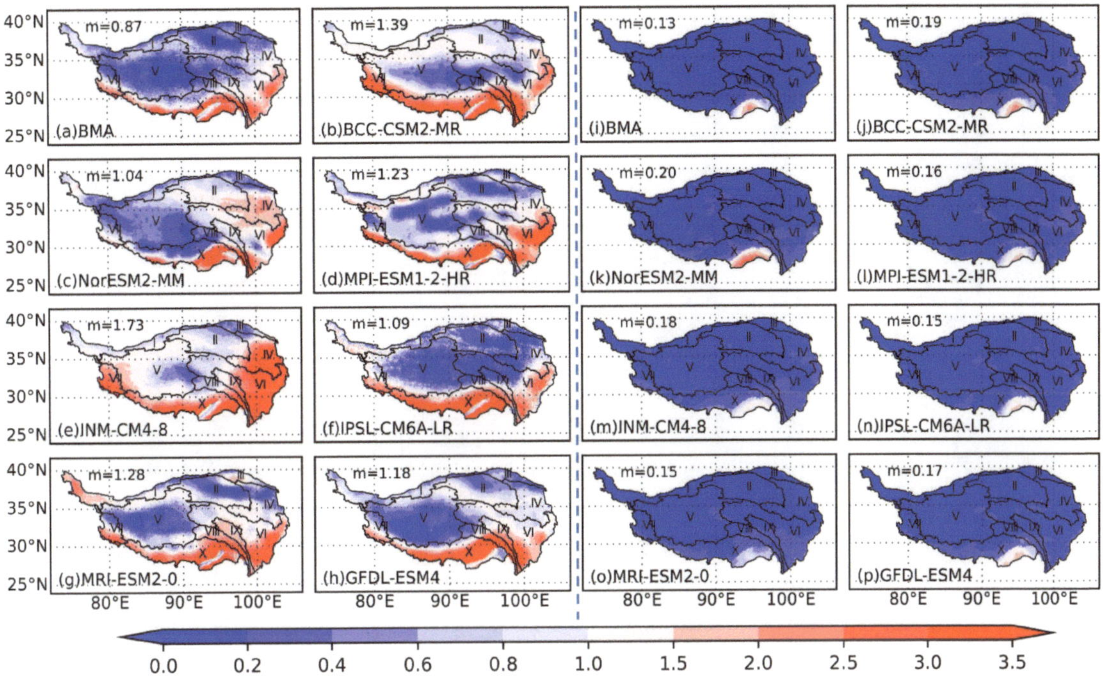

Figure 4. Absolute mean bias (AMB) for precipitation (mm d^{-1}) between individual models in CMIP6 and observational data over the QTP, AMB before correction on the left of the blue dotted line (**a**–**h**), AMB after correction on the right side (**i**–**p**). The mean value of AMB across the entire QTP is denoted in each subgraph as m; m is defined as the sum of all grid point values divided by the total number of grid points over the Tibetan Plateau, as shown in the following figure. The numbering (I–X) of each subregion (consistent with Figure 1) is also indicated in each subfigure.

To verify the correctness of the corrected data, we separately analyzed the seasonal cycles before and after bias correction (Figure 7) and the mean annual values after bias correction (Figures S1–S3) for three variables over the historical period. The data show a wet bias for precipitation and a relatively cold bias for temperature in the GCMs' seasonal cycles compared to observations from before bias correction (Figure 7a–c), typically due to systematic errors overlooking the monsoon effect and surface meteorological processes in GCMs [27,28,73]. Compared to the uncorrected data, the seasonal cycles of the corrected data more closely align with observations (Figure 7d–f), as CNCDFm has corrected biases for both high and low values [61]. The uncertainty in GCM outputs has been significantly reduced post-bias correction, as evidenced by the narrowed ranges (shaded areas in Figure 7d–f). Figures S1–S3 show the spatial distributions of mean annual values for the corrected data, indicating a strong agreement with observations. Precipitation data, in particular, show a consistent decline from the southeast to the northwest of the QTP (Figure S1c–i), aligning with prior studies [7,8,74]. Regarding temperature, all seven GCMs indicate lower values in the IQTB due to its high altitudes and latitudes, with higher temperatures primarily found in the southeastern part (Figures S2c–i and S3c–i), which is consistent with orographic findings [51]. Overall, our results indicate that the CNCDFm method has produced a reasonably corrected dataset based on CMIP6 GCMs, which is suitable for climate impact studies in the QTP.

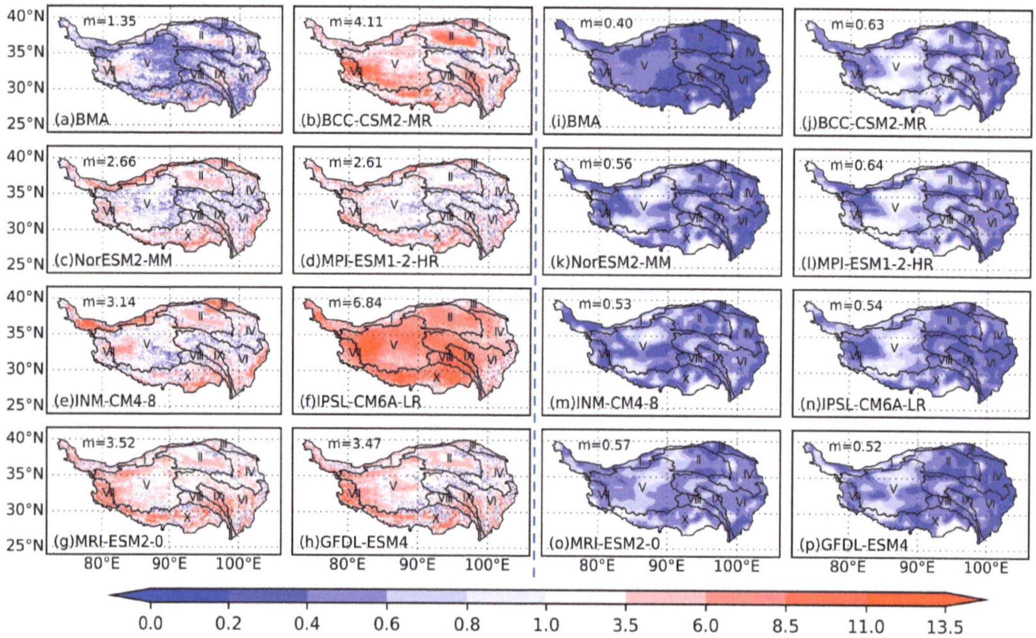

Figure 5. AMB is the same as in Figure 4 but for the maximum surface temperature (K).

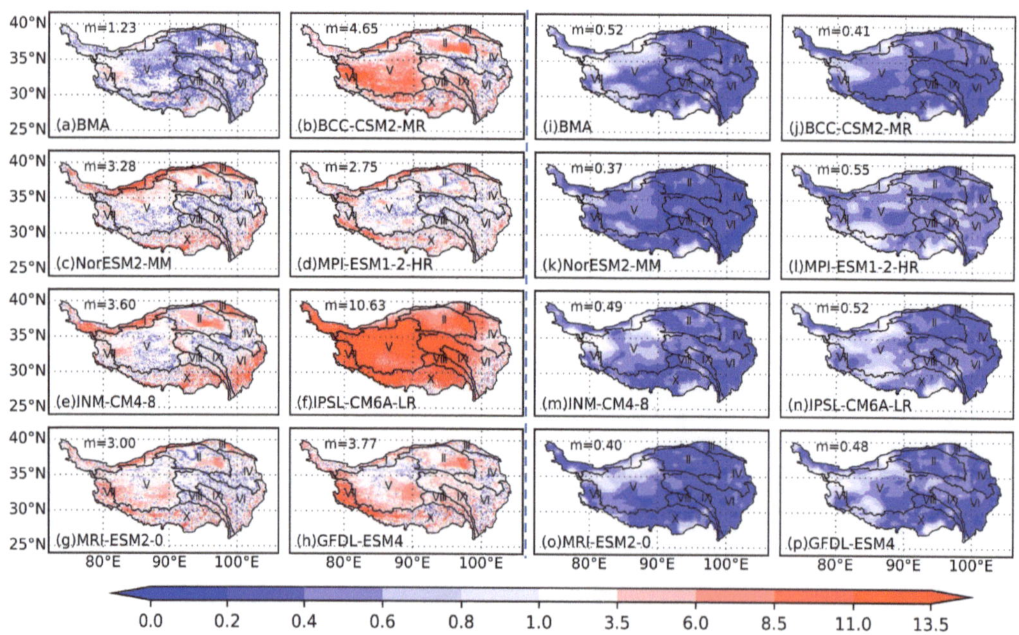

Figure 6. AMB is the same as in Figure 4 but for minimum surface temperature (K).

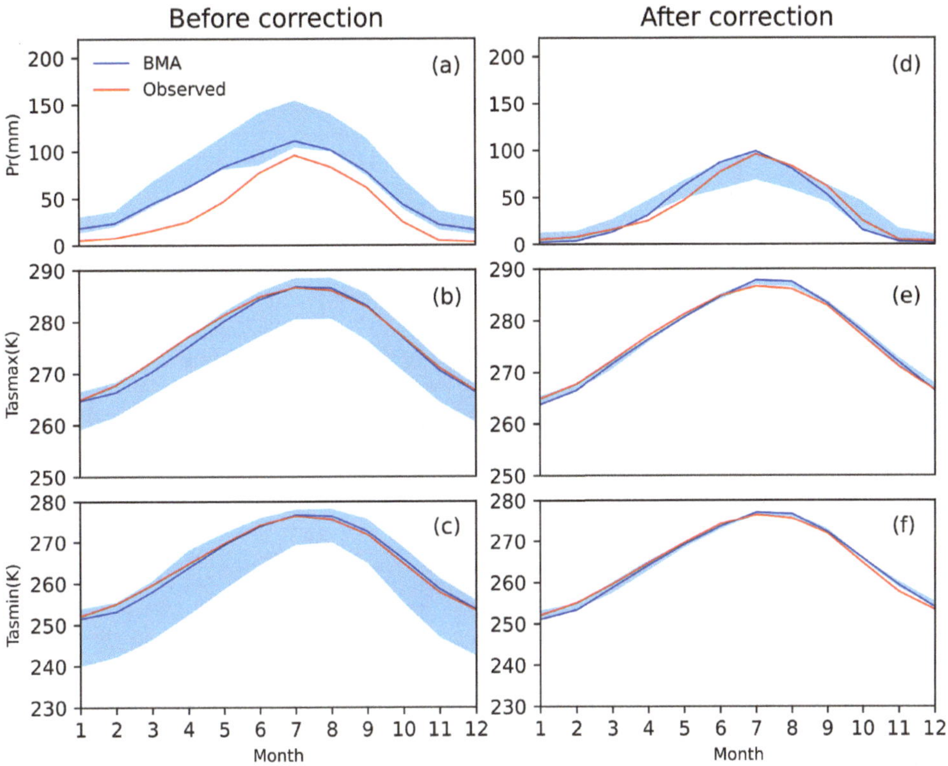

Figure 7. The seasonal cycle of precipitation, as well as maximum and minimum temperature, before ((**a–c**), on the **left** side) and after ((**d–f**), on the **right** side) bias correction. Comparison of the Bayesian model averaging (BMA, blue) mean seasonal cycle of bias-corrected (**a**,**d**) precipitation, (**b**,**e**) maximum temperature, and (**c**,**f**) minimum temperature against the observations for the 1979–2013 period (red). The shaded area represents the uncertainty (from minimum to maximum) of all 7 CMIP6 GCMs.

At the same time, we calculate the mean biases, seasonal cycles, and mean annual values for the BMA (as shown in Figures 4, 5 and 6a,i, the blue line in Figures 7 and S1–S3b). Before the bias correction, BMA exhibited the lowest absolute mean bias (AMB) compared to the outputs from the seven GCMs, attributed to the consideration of the models' posterior probability distributions (Figures 4, 5 and 6a). After bias correction, BMA showed the lowest AMB for precipitation (Figure 4i) and avoided the high-temperature AMB that occurred in every individual model (Figures 5i and 6i). In terms of seasonal cycles, the corrected BMA data align well overall with observations after bias correction (Figure 7d–f). Furthermore, the spatial distribution of the corrected BMA output is consistent with the observations (Figures S1–S3b). Overall, BMA provides a robust multi-model ensemble forecast result for precipitation, as well as maximum and minimum temperatures across the Tibetan Plateau.

Studying extreme events is a crucial aspect of climate change research. To evaluate the performance of bias correction in extremes, the QBs for uncorrected and corrected BMA 95th percentiles were compared against observations for the historical period (1979–2013) for three variables, as shown in Table 1 (Figure S4). The analysis revealed that biases in the 95th percentiles were successfully mitigated, with the corrected BMA data bias approaching 0 (Figure S4b,c,f) compared with the uncorrected. Thus, the corrected data also demonstrate improved performance for extreme values.

3.3. Future Climate Change Projection in the QTP

Projected changes in precipitation and maximum/minimum surface temperatures in the QTP for the 21st century, based on outputs from the corrected BMA and seven GCMs, span from 2025 to 2100. We estimated these changes for the near (2025–2049), medium (2050–2074), and far (2075–2100) future periods, using a historical baseline of 1985–2013, across all four scenarios (SSP1-2.6, SSP2-4.5, SSP3-7.0, and SSP5-8.5), as shown in Figures 8–10. Overall, the projections indicate consistent warming and wetting trends for the QTP in the 21st century, aligning with historical trends [6,75], except for SSP1-2.6, where the rate of increase in precipitation and temperature is projected to slow down significantly in the far future (Figures 8, 9 and 10c). Compared to the low radiative forcing pathway (SSP1-2.6), high radiative forcing pathways (SSP3-7.0 and SSP5-8.5) have considerably large increases in precipitation and temperature. For instance, under SSP5-8.5, mean precipitation is projected to rise by 26.8% in the far future (Figure 9i), with mean maximum and minimum surface temperature increases of 5.1 °C and 5.7 °C, respectively (Figures 9 and 10i). Excessive warming may accelerate the retreat of plateau glaciers, especially in the TMR and IQTB regions. Furthermore, it shows a consistent spatial pattern in the different SSPs. For instance, a higher precipitation increase was found in the northern parts of the IQTB and IDR, TMR, and QDB under SSP3-7.0 and SSP5-8.5 in the far future.

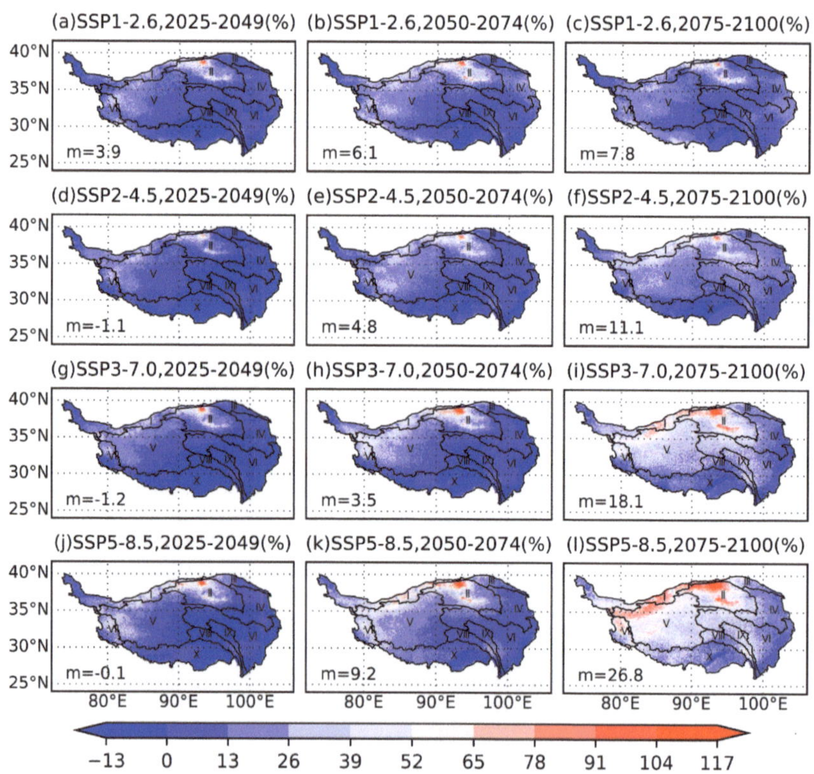

Figure 8. Future change of precipitation (%) against historical baseline (1985–2010) based on corrected BMA. (**a–c**) The projected change in mean annual precipitation for the SSP1-2.6 in the near (2025–2049), medium (2050–2074), and far (2075–2100) future; (**d–f**), (**g–i**) and (**j–l**), the same as (**a–c**) but for SSP2-4.5, SSP3-7.0, and SSP5-8.5, respectively. The mean change across the entire QTP (m) is shown in each panel as m. The numbering of each subregion (consistent with Figure 1) is also indicated in each subfigure. The numbering (I–X) of each subregion (consistent with Figure 1) is also indicated in each subfigure.

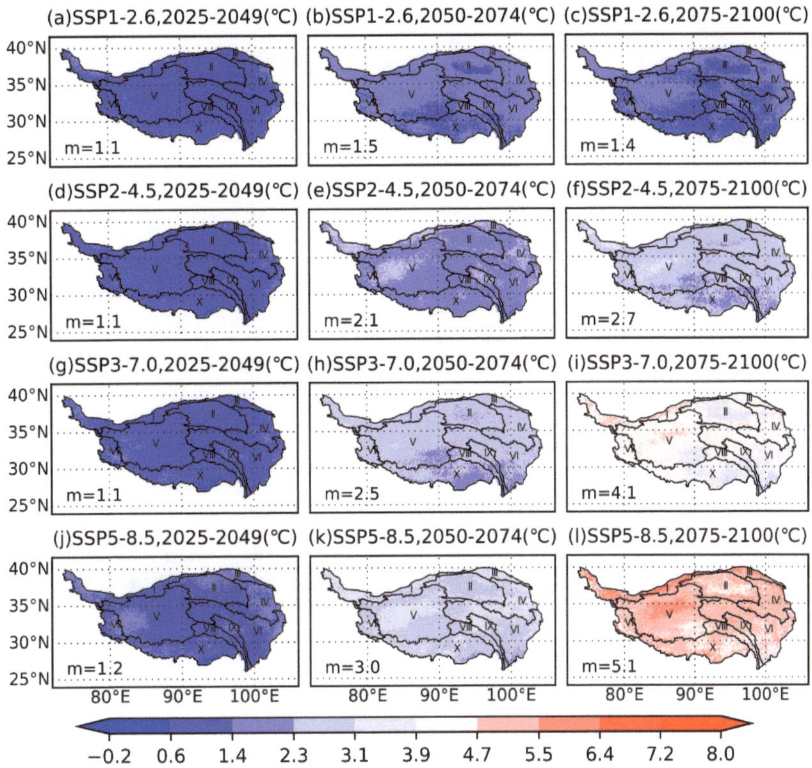

Figure 9. Future changes based on BMA; the same as in Figure 8 but for the maximum surface temperature.

Future precipitation and temperature were also estimated in 10 subregions based on corrected BMA (Tables S1–S3), as seen in the Supplementary Materials, Figures 8–10. The projections indicate increases in the mean annual precipitation and maximum and minimum temperatures across these subregions under middle and high radiative forcing pathways (SSP2-4.5, SSP3-7.0, and SSP5-8.5), with significantly higher increases across high radiative forcing scenarios compared to low radiative forcing ones, aligning with projections for South Asia [31]. Notably, under SSP3-7.0 and SSP5-8.5, all three variables are projected to continue to rise, which is consistent with the overall trend across the QTP. Under SSP1-2.6, the precipitation trend in HXB, IDR, IQTB, QDB, and TMR in the far future is lower than in the middle future, a trend likely attributed to the mitigating effects of reduced greenhouse gas concentrations on climate change. Overall, the bias-corrected dataset produced a reasonable projection for subregions in the QTP, and it is available for hydroclimatic research at the regional scale.

Figure 11 illustrates the time series for BMA (represented by solid lines) and MMA (represented by dotted lines), demonstrating a more continuous growth trend in temperature (both maximum and minimum) in SSP2-4.5, SSP3-7.0, and SSP5-8.5, as observed after applying a 30-year moving average. Conversely, a distinct turning point occurs mid-century in SSP1-2.6, where warming shifts to cooling across all subregions, highlighting the impact of carbon emission limitations on regional temperature trends. This warming trajectory parallels the global warming process projected by the IPCC's Sixth Assessment Report (AR6), which predicts a global surface temperature increase of 4 °C by the century's end under SSP5-8.5 compared to 1995–2014 [76]. Notably, the rate of warming in the QTP exceeds the global average, underscoring the region's sensitivity to climate change, as historically observed [6].

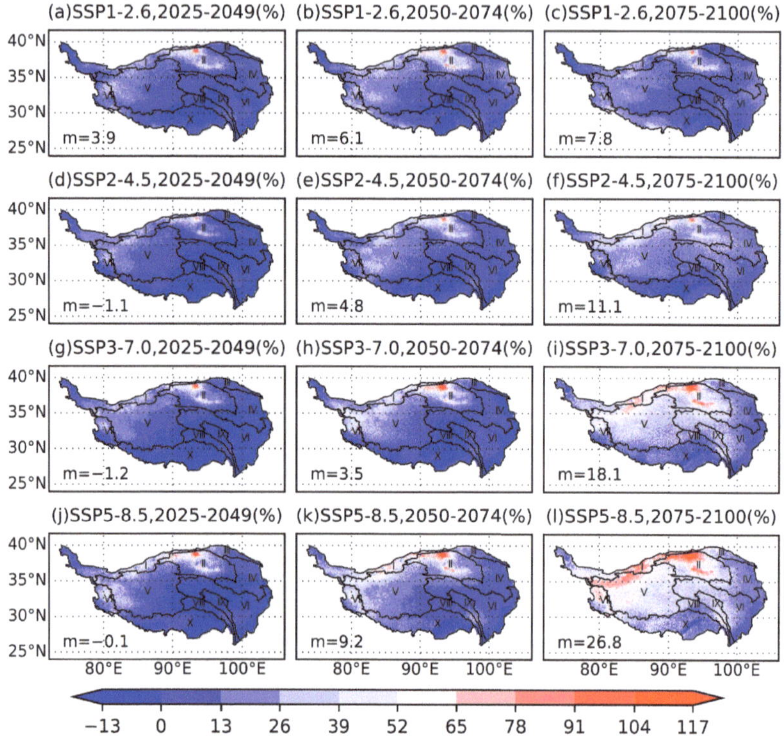

Figure 10. Future changes based on BMA; the same as in Figure 8 but for the minimum surface temperature.

Furthermore, noteworthy changes in the spatial distribution of precipitation and temperature across the Tibetan Plateau merit attention. Specifically, future projections indicate a significant increase in precipitation rates in IDR, QDB, TMR, and the northwest of IQTB (Figures 8 and 11(b1,c1,e1,g1), Table S1), which is not only far ahead of global future growth but also far ahead of the rest of the QTP. The future moistening trend in the northern region is consistent with historical trends. Zhang et al. estimated a historical precipitation growth rate of 1.5–3.0 mm yr^{-1} in this area [77]. The significant growth rate may be partly due to the historically low precipitation in this area (Table S1), and partly due to the enhanced atmospheric circulation caused by warming [78], which accelerates the local moisture recycling and increases the input of external moisture [77,79,80]. However, the drastic change in precipitation is bound to have a profound impact on the regional hydrological cycle. For temperature, there is no obvious spatial differentiation in the near future change (Figures 9 and 10a,d,g,j), while in the far future, spatial differentiation is prominent. For instance, IQTB and TMR have shown higher warming than other regions in SSP5-8.5, which may be due to the underlying positive feedback of the radiation absorption and will accelerate the ablation of regional glacial permafrost [81–84]. These spatial differences highlight the necessity of high-resolution projection for research in the QTP.

Table S4 provides statistics on the mean rate and change rate of the annual maximum five-day precipitation for future years based on BMA. Notably, IDR exhibits the highest change rate among the 10 sub-regions in the vast majority of scenarios and periods, with growth rates reaching 22.9 to 41.2% in the distant future. The significant increase in the annual maximum five-day precipitation, as an indicator of extreme precipitation, implies an elevated risk of hydrogeological disasters, such as floods, landslides, and debris flow [85–87]. Most areas in IDR are above 4000 m, given its higher elevation

(Figure 1). Geological disasters in this region will likely cause more significant damage downstream [88], highlighting the need for enhanced observation and monitoring in this area.

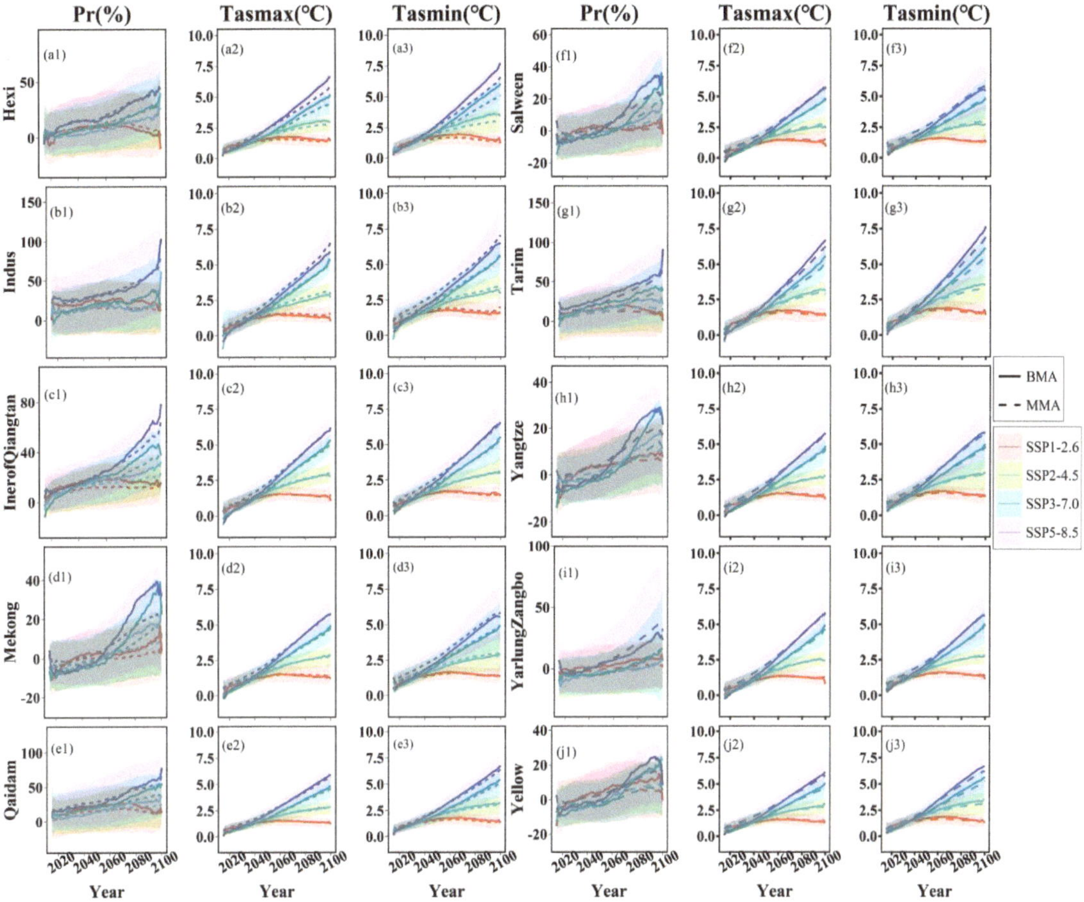

Figure 11. Future change time series of BMA and multi-model average (MMA) in precipitation, as well as maximum and minimum temperatures in different subregions (Numbered as (**a–j**), marked in the top left corner of each subfigure) of the QTP. Changes in the BMA and MMA for mean annual precipitation (%), maximum temperature (°C), and minimum temperature (°C) were estimated using a 30-year moving window against the historical reference period of 1979–2013. The shaded region shows uncertainty (estimated using one standard deviation) based on 7 GCMs. Different colors denote the 4 SSPs, and different types of lines denote BMA and MMA, respectively.

4. Discussion

In the present study, the effectiveness of downscaling and bias correction methods were evaluated. The projection results are presented at a basin scale and compared with previous research. Such results could verify the rationality of HRP-QTP, but some issues remain that deserve further discussion.

4.1. Availability of Reanalysis Precipitation

In numerous areas worldwide, the scarcity of observation data significantly hampers the execution of various environmental research initiatives, especially for precipitation [89]. He et al. [11] evaluated the accuracy of CMFD in simulating conditions in regions across western China that are poorly monitored, using additional data from measuring stations in IDR, IQTB, YZR, and SWR. The findings indicate that CMFD provides a more accurate representation of temperature in areas of the QTP that lack direct observational data, leading us to adopt CMFD temperature data as a proxy for historical observations. However, the dataset's precipitation measurements have not been similarly validated in the QTP.

Through the extension and correction of GPQM and CMFD data, we derived a combined dataset aimed at correcting the CMFD's underestimation of precipitation in the LYZV area, using additional station information from GPQM beyond the QTP. The historical annual mean precipitation across 12 subregions of the QTP (as shown in Figure 12) was estimated based on the data generated in Section 2.3. The results indicate that the QTP's historical annual mean precipitation is 456 mm/year, with a slight increase over historical periods, aligning closely with ground station measurements [75]. Spatial analysis reveals that the eastern and southern subregions, such as MKR, SWR, YZR, and YR, exhibit higher precipitation levels (602 mm/year, 653 mm/year, 598 mm/year, and 500 mm/year, respectively) compared to the central and northern subregions (HXB, IQTB, QTB, and TMR, with annual mean precipitation of 241 mm/year, 261 mm/year, 168 mm/year, and 145 mm/year, respectively), reflecting the patterns of water vapor transport in the QTP [8,47]. Notably, due to its high precipitation, at approximately 3120 mm/year, the YLZBR region records the highest precipitation among all subregions at 1108 mm/year.

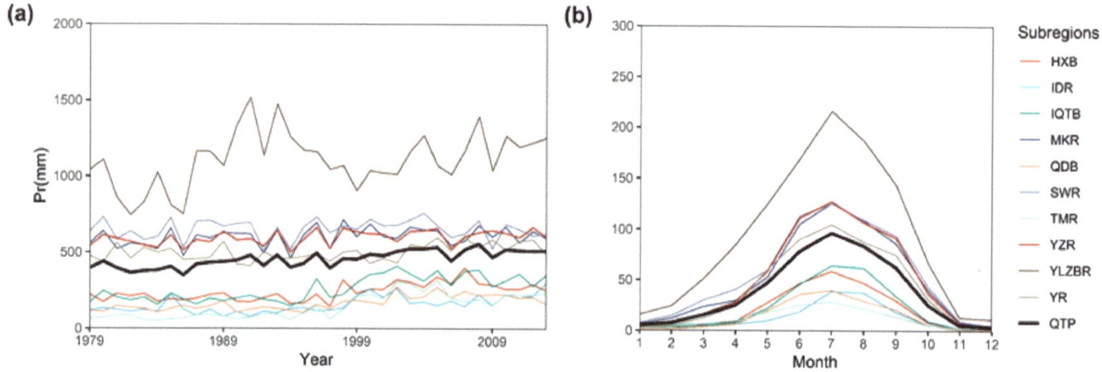

Figure 12. Annual mean from 1979 to 2013 (**a**) and seasonal cycle mean (**b**) of precipitation from combined data in different subregions of the QTP; the black lines represent the mean values of the QTP, and the colored lines denote multiple subregions.

Generally, gauge data are considered the most accurate precipitation data [90]. However, it is difficult to estimate and validate LYZV precipitation due to a lack of observations (Figure 1), and reanalysis based on multi-source data is the only method currently available to estimate precipitation in LYZV. Due to differences in methods and data sources, different reanalysis datasets have demonstrated a wide range of estimated results, ranging from 1000 mm/year (for instance, He et al. [11] estimated the LYZV annual mean precipitation at 1001 mm/year by combining upstream ground station data with satellite reanalysis data) to 4000 mm/year (the estimation by Ma et al. [12] is 3011 mm/year using interpolation of downstream gauge data; the estimation by Li et al. [8] is from 1000 to 4000 mm/year at different altitudes using grid data from the Global Precipitation Measurement, GPM). We tend to use the higher precipitation estimates for LYZV because they are more consistent with precipitation characteristics in the southern Himalayas, which are close to the same altitude regions in Bhutan [45] and corroborate early findings by Yang et al. [91]. Nonetheless, given the QTP's extensive area and complex topography, different reanalysis datasets offer distinct advantages across various regions [20]. Accurately estimating precipitation across the QTP remains an unresolved challenge that requires further observational evidence.

4.2. Availability of Downscaling Methods

In Section 3.1, we present a simple comparative experiment to demonstrate the effectiveness of ANUSPLIN, which takes elevation into account, in downscaling climatic elements in the QTP. Herein, IDW represents linear two-dimensional interpolation, and OK represents non-linear two-dimensional interpolation, both of which are compared with the non-linear three-dimensional interpolation of ANUSPLIN.

The interpolation result of IDW equals the weighted average of all known stations, with the weights being inversely proportional to the distance between the known points and the interpolation point. Therefore, once the locations of the sites are determined, the interpolation relationship is also established. OK and ANUSPLIN perform interpolation by constructing non-linear relationships to fit the climate surface of known stations [54]. The fitting relationship is time-varying (in this study, both use annual fitting). The difference lies in that ANUSPLIN considers covariates.

Compared to ANUSPLIN, the limitation of two-dimensional interpolations lies in their focus on spatial relationships between stations, leading to a reliance on information from nearby stations and overlooking the decisive role of elevation on plateau climates. Taking temperature as an example, a common rule is that temperature decreases with the increase in elevation [7]. The elevation of stations near the Basu station is higher than that of Basu (Figure 1), with the nearest five having an average elevation of 3082 m, significantly higher than Basu's 2589.2 m. This results in the two-dimensional interpolations significantly underestimating the temperature at Basu (Figure 3e,f). This bias is corrected in ANUSPLIN. Furthermore, the fitting performed by ANUSPLIN is time-varying, meaning that if the correlation between climate elements and elevation changes in future scenarios, the new correlation can also be captured [92]. This reduces the systematic bias caused by limitations in fitting capability during downscaling.

However, downscaling climate models continues to be an exceedingly complex issue. Identifying the optimal downscaling method and understanding how systematic errors are transmitted during the downscaling process urgently require further investigation. This study merely demonstrates the viability of applying ANUSPLIN in plateau areas.

4.3. Uncertainty and Reliability in Future Projections

Previous studies have indicated significant divergence in climate projections for the QTP from GCMs [27,28,30], leading to low SNRs. As shown in Figures 13–15, bias correction significantly increases the SNR for three variables: for precipitation, the SNR range increased from 0.23–0.40 to 0.43–0.90; for temperature, it rose from 0.37–1.12 to 3.67–5.52. This suggests that bias correction effectively eliminates systemic model biases, making future model projections more reliable. However, reliability varies greatly among different variables. For temperature, both maximum and minimum, the SNR exceeds 1 in all subregions (Figures 14 and 15) after correction, indicating a clear trend in the future temperature rise and reduced uncertainty. In contrast, precipitation shows an overall SNR below 1, with several regions, including LYZV, MKR, SWR, YZR, and YR (Figure 14b,d,f,h), exhibiting SNRs far below 1. This suggests that future changes in precipitation for these regions are not pronounced and are subject to considerable uncertainty, a conclusion supported by Figure 11(d1,f1,h1,i1,j1).

We compare our findings with those from Cui et al. [30], who projected precipitation for several major river basins using a broader set of GCMs, albeit at a lower spatial resolution. Their results, showing SNRs in Table 6, reveal that SNRs exceeding 1 only occur in the Indus, Mekong, Salween, and Yellow River basins under SSP2-4.5 when future temperatures increase by 3 °C. Contrasting these findings with ours (Figures 13–15) indicates that the SNRs for multi-model projections remain consistently low, suggesting that merely increasing the number of predictive models does not necessarily enhance prediction reliability. Mechanistically, precipitation predictions depend on factors such as wind, humidity, and aerosols [93–95]. The approximation and parameterization of these factors in GCMs might introduce more uncertainty into precipitation forecasts than temperature forecasts.

Furthermore, while current studies often prioritize MMA, we advocate for BMA based on our belief that historical simulations can inform and constrain future climate predictions. Unlike BMA, MMA does not incorporate prior information from historical periods, which is particularly relevant for temperature projections where the differences between BMA and MMA are minimal. However, for precipitation, especially in regions like IQTB, MKR, and YZR post-2050, BMA projections are significantly higher than those from MMA, as illustrated in Figure 11. Precipitation, subject to considerable model uncertainty, poses challenges. Ferguglia et al. [96] demonstrated that the emergent constraint approach—relying on a physically explainable link between historical and future data—does not robustly address systematic uncertainties, indicating the limitations of depending solely on multi-model precipitation outcomes. However, due to the lack of historical data and the unclear precipitation mechanism, statistical correction is still a widely used reanalysis method in existing research. We retained the correction results of each model in the final forecast dataset provided for further study.

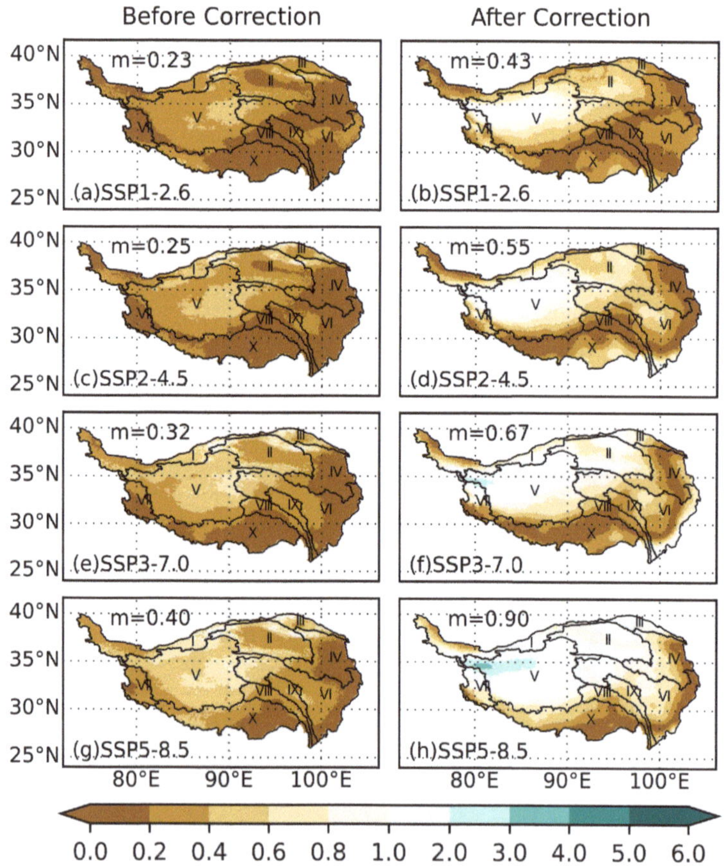

Figure 13. Signal-to-noise ratio (SNR) of MMA precipitation in different SSPs. The mean values of SNR across the entire QTP are denoted in each subgraph as m. The numbering (I–X) of each subregion (consistent with Figure 1) is also indicated in each subfigure.

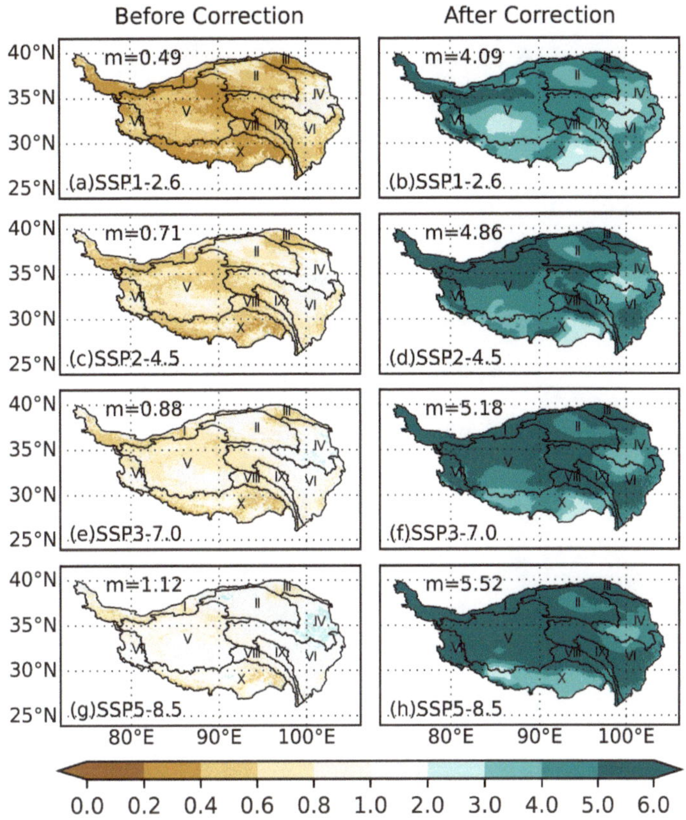

Figure 14. SNR, as in Figure 13 but for maximum temperature.

Table 6. SNRs for precipitation projection results from Cui et al. [30]. The bold terms indicate SNRs greater than 1; SNRs were calculated using Equation (11) ignoring models with no data, and the division of subregions is slightly different from our study.

Name	Scenarios	Temperature Rise Range		
		1.5 °C	2.0 °C	3.0 °C
Indus	SSP2-4.5	0.11	0.35	**1.09**
	SSP5-8.5	0.13	0.10	0.51
Mekong	SSP2-4.5	0.38	0.24	**1.14**
	SSP5-8.5	0.52	0.33	0.71
Salween	SSP2-4.5	0.28	0.16	**1.32**
	SSP5-8.5	0.61	0.43	0.80
Yangtze	SSP2-4.5	0.34	0.79	0.95
	SSP5-8.5	0.05	0.38	0.91
Yarlung Zangbo	SSP2-4.5	0.31	0.26	0.94
	SSP5-8.5	0.68	0.38	0.62
Yellow	SSP2-4.5	0.09	0.56	**1.15**
	SSP5-8.5	0.21	0.14	0.85

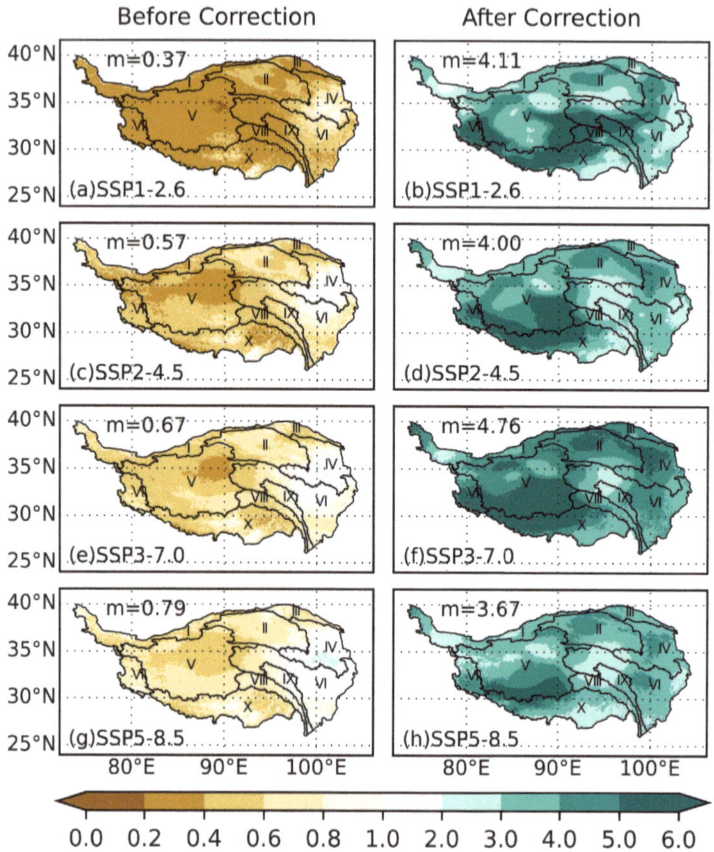

Figure 15. SNR, as in Figure 13 but for minimum temperature.

5. Conclusions

The QTP serves as Asia's most crucial water source, with climate change in the QTP poised to profoundly affect water security and geopolitical relations across the continent. The region's complex terrain necessitates high-resolution climate projections to underpin studies on regional climate change and guide adaptation strategies. To address this need, our research developed a high-resolution projection by utilizing reanalyzed meteorological datasets alongside outputs from seven GCMs. We employed a downscaling method that accounts for altitude variations and implemented an updated nonstationary bias correction technique, both of which were rigorously tested for validity. Key findings from our study include the following:

1. There is an approximately linear relationship between climatic factors and altitude in high-altitude areas. Therefore, the ANUSPLIN software, which incorporates DEM as a covariate, can more accurately fit the probability distributions of meteorological elements. Ignoring altitude's impact on the distribution of meteorological elements during downscaling leads to the overestimation or underestimation of temperatures and overlooks extreme precipitation events. Thus, incorporating elevation data as prior information in downscaling processes is crucial. Despite significant systematic biases in GCM simulations and predictions for the QTP climate, these can be substantially mitigated by employing an advanced nonstationary bias correction method, CNCDFm. For example, this method has been shown to minimize the deviation of

extreme values to nearly zero. Consequently, our research enhances the applicability and accuracy of GCM results for regional studies in the QTP.

2. Based on high-resolution projections, the QTP is expected to continue experiencing warming and wetting trends. The projected temperature change range remains consistent across the QTP subregions, mirroring the historical trend where the QTP temperature increase surpasses the global surface temperature rise. Under the Shared Socioeconomic Pathway SSP1-2.6, the region is anticipated to reach a warming inflection point by the mid-21st century. Furthermore, future precipitation projections indicate significant spatial variations, with growth rates in regions such as IDR, IQTB, QDB, and TMR far exceeding the global average. Such marked increases in temperature and precipitation are likely to fundamentally transform the regional hydrological cycle.
3. Two significant challenges hinder the accurate forecasting of future precipitation trends in the QTP. Firstly, the historical period lacks global high-resolution precipitation data that are convincing enough for detailed analysis, especially in ungauged regions such as the IQTB, YLZBR, TMR, and IDR. This scarcity complicates the analysis of historical precipitation changes and the calibration of GCMs. Secondly, there remains substantial uncertainty in GCM projections. While bias correction techniques can mitigate some systematic biases inherent in GCMs, the reliance on historical data to constrain future precipitation projections proves to be less effective than for temperature, rendering precipitation forecasts less reliable. This uncertainty also leads to significant discrepancies between the precipitation forecasts of BMA and MMA, both of which indicate an increasing trend across all basins. Further research is imperative to enhance the reliability of multi-model precipitation predictions.

Despite the limitations outlined above, this research provides a high-resolution future prediction for regional-scale studies of the QTP that is openly accessible for further testing, research, and decision-making purposes.

Supplementary Materials: The following supporting information can be downloaded at https://www.mdpi.com/article/10.3390/atmos15040434/s1, Figure S1: Mean annual precipitation (mm) for (a) observation and corrected data of (b) BMA and (c–i) individual GCMs for the historical period (1979–2013). The mean value of annual precipitation across the entire QTP is denoted in each subgraph as m; Figure S2: The mean annual value is the same as in Figure S1 but for the maximum surface temperature; Figure S3: The mean annual value is the same as in Figure S1 but for the minimum surface temperature; Figure S4: Quantile Bias (QB) of BMA in the 95th percentile of precipitation, as well as maximum and minimum temperatures, for the historical period (1979–2013); Table S1: Subregion-wise BMA-projected change in the mean annual precipitation (mm), the division of three futures is the same as in Figure 12, and the historical mean is from 1985 to 2010; Table S2: Subregion-wise BMA-projected change in the maximum surface temperature (°C); Table S3: Subregion-wise BMA-projected change in the minimum surface temperature (°C); Table S4: Subregion-wise BMA-projected change in the average annual maximum five-day precipitation (mm).

Author Contributions: Conceptualization, X.S. and K.W.; methodology, X.S. and K.W.; validation, K.W., S.Y., F.L. and J.S.; formal analysis, K.W.; data curation, K.W. and Y.Z.; writing—original draft preparation, K.W.; writing—review and editing, K.W., X.S., F.L. and Y.Z.; visualization, K.W. and S.Y.; supervision, F.L.; project administration, F.L.; funding acquisition, F.L. and Y.Z. All authors have read and agreed to the published version of the manuscript.

Funding: This research was funded by the Second Tibetan Plateau Scientific Expedition and Research Program (STEP, no. 2019QZKK0207-02), the National Natural Science Foundation of China (no. 51909275), IWHR Research & Development Support Program (no. WR110145B0052021), and Qinghai Central Government Guided Local Science and Technology Development Fund Project (no. 2022ZY020).

Institutional Review Board Statement: Not applicable.

Informed Consent Statement: Not applicable.

Data Availability Statement: All bias-corrected GCMs and BMA output for different scenarios were stored in Network Common Data Form (NetCDF) files. These files were named following a standard form as MODEL_scenario_variable.nc. For instance, bias-corrected precipitation from GFDL-ESM4 in SSP1-2.6 was named GFDL-ESM4_126_pr.nc, historical output was denoted 'hist' in the scenario, and BMA was denoted as a model name. The full dataset is available from the National Tibetan Plateau Data Center ("https://cstr.cn/18406.11.Atmos.tpdc.272514 (9 December 2023)"). The code used in this work is not published along with the dataset because a non-free software named ANUSPLIN is invoked by this code and cannot be redistributed without permission.

Acknowledgments: We acknowledge the data availability from the National Tibetan Plateau Data Center, Yang, K., He, J. ("https://cstr.cn/18406.11.AtmosphericPhysics.tpe.249369.file (24 March 2021)") and Ma, J., Li, H. ("https://cstr.cn/18406.11.Meteoro.tpdc.270993 (24 March 2021)"). Output from the CMIP6 models from "https://esgf-node.llnl.gov/projects/cmip6/ is greatly acknowledged (24 March 2021)".

Conflicts of Interest: The authors declare no conflicts of interest. The funders had no role in the design of the study; in the collection, analyses, or interpretation of data; in the writing of the manuscript; or in the decision to publish the results.

References

1. Tang, Q.H.; Lan, C.; Su, F.G.; Liu, X.C.; Sun, H.; Ding, J.; Wang, L.; Leng, G.Y.; Zhang, Y.Q.; Sang, Y.F.; et al. Streamflow change on the Qinghai-Tibet Plateau and its impacts. *Chin. Sci. Bull. Chin.* **2019**, *64*, 2807–2821. [CrossRef]
2. Su, F.; Zhang, L.; Ou, T.; Chen, D.; Yao, T.; Tong, K.; Qi, Y. Hydrological response to future climate changes for the major upstream river basins in the Tibetan Plateau. *Glob. Planet. Chang.* **2016**, *136*, 82–95. [CrossRef]
3. Zhou, Y.K.; Zhang, X.Y.; Yu, H.; Liu, Q.Q.; Xu, L.L. Land Use-Driven Changes in Ecosystem Service Values and Simulation of Future Scenarios: A Case Study of the Qinghai-Tibet Plateau. *Sustainability* **2021**, *13*, 4079. [CrossRef]
4. Yun, X.B.; Tang, Q.H.; Li, J.B.; Lu, H.; Zhang, L.; Chen, D.L. Can reservoir regulation mitigate future climate change induced hydrological extremes in the Lancang-Mekong River Basin? *Sci. Total Environ.* **2021**, *785*, 9. [CrossRef]
5. Ni, J.; Wu, T.H.; Zhu, X.F.; Wu, X.D.; Pang, Q.Q.; Zou, D.F.; Chen, J.; Li, R.; Hu, G.J.; Du, Y.Z.; et al. Risk assessment of potential thaw settlement hazard in the permafrost regions of Qinghai-Tibet Plateau. *Sci. Total Environ.* **2021**, *776*, 14. [CrossRef] [PubMed]
6. Duan, J.P.; Li, L.; Chen, L.; Zhang, H.X. Time-dependent warming amplification over the Tibetan Plateau during the past few decades. *Atmos. Sci. Lett.* **2020**, *21*, e998. [CrossRef]
7. You, Q.L.; Kang, S.C.; Pepin, N.; Flugel, W.A.; Yan, Y.P.; Behrawan, H.; Huang, J. Relationship between temperature trend magnitude, elevation and mean temperature in the Tibetan Plateau from homogenized surface stations and reanalysis data. *Glob. Planet. Chang.* **2010**, *71*, 124–133. [CrossRef]
8. Li, G.P.; Yu, Z.B.; Wang, W.G.; Ju, Q.; Chen, X. Analysis of the spatial Distribution of precipitation and topography with GPM data in the Tibetan Plateau. *Atmos. Res.* **2021**, *247*, 15. [CrossRef]
9. Yan, Y.P.; You, Q.L.; Wu, F.Y.; Pepin, N.; Kang, S.C. Surface mean temperature from the observational stations and multiple reanalyses over the Tibetan Plateau. *Clim. Dyn.* **2020**, *55*, 2405–2419. [CrossRef]
10. Lu, N. Scale effects of topographic ruggedness on precipitation over Qinghai-Tibet Plateau. *Atmos. Sci. Lett.* **2019**, *20*, e904. [CrossRef]
11. He, J.; Yang, K.; Tang, W.J.; Lu, H.; Qin, J.; Chen, Y.Y.; Li, X. The first high-resolution meteorological forcing dataset for land process studies over China. *Sci. Data* **2020**, *7*, 11. [CrossRef]
12. Ma, J.; Li, H.; Wang, J.; Hao, X.; Shao, D.; Lei, H. Reducing the statistical distribution error in gridded precipitation data for the Tibetan Plateau. *J. Hydrometeorol.* **2020**, *21*, 2641–2654. [CrossRef]
13. Wang, Y.W.; Wang, L.; Li, X.P.; Zhou, J.; Hu, Z.D. An integration of gauge, satellite, and reanalysis precipitation datasets for the largest river basin of the Tibetan Plateau. *Earth Syst. Sci. Data* **2020**, *12*, 1789–1803. [CrossRef]
14. Jiang, Y.Z.; Yang, K.; Qi, Y.C.; Zhou, X.; He, J.; Lu, H.; Li, X.; Chen, Y.Y.; Li, X.D.; Zhou, B.R.; et al. TPHiPr: A long-term (1979–2020) high-accuracy precipitation dataset (1/30 degrees, daily) for the Third Pole region based on high-resolution atmospheric modeling and dense observations. *Earth Syst. Sci. Data* **2023**, *15*, 621–638. [CrossRef]
15. Ding, L.R.; Zhou, J.; Zhang, X.D.; Liu, S.M.; Cao, R.Y. A long-term 0.01 degrees surface air temperature dataset of Tibetan Plateau. *Data Brief* **2018**, *20*, 748–752. [CrossRef]
16. Wu, W.B.; Ma, J.; Meadows, M.E.; Banzhaf, E.; Huang, T.Y.; Liu, Y.F.; Zhao, B. Spatio-temporal changes in urban green space in 107 Chinese cities (1990–2019): The role of economic drivers and policy. *Int. J. Appl. Earth Obs. Geoinf.* **2021**, *103*, 102525. [CrossRef]
17. Zhang, Z.Y.; Arnault, J.; Laux, P.; Ma, N.; Wei, J.H.; Shang, S.S.; Kunstmann, H. Convection-permitting fully coupled WRF-Hydro ensemble simulations in high mountain environment: Impact of boundary layer- and lateral flow parameterizations on land-atmosphere interactions. *Clim. Dyn.* **2022**, *59*, 1355–1376. [CrossRef]
18. Gao, M.L.; Li, Z.H.; Tan, Z.Y.; Liu, Q.; Shen, H.F. Simulating the Response of the Surface Urban Heat Environment to Land Use and Land Cover Changes: A Case Study of Wuhan, China. *Remote Sens.* **2021**, *13*, 4495. [CrossRef]

19. Huang, X.L.; Han, S.; Shi, C.X. Evaluation of Three Air Temperature Reanalysis Datasets in the Alpine Region of the Qinghai-Tibet Plateau. *Remote Sens.* **2022**, *14*, 4447. [CrossRef]
20. Bai, L.; Wen, Y.Q.; Shi, C.X.; Yang, Y.F.; Zhang, F.; Wu, J.; Gu, J.X.; Pan, Y.; Sun, S.; Meng, J.Y. Which Precipitation Product Works Best in the Qinghai-Tibet Plateau, Multi-Source Blended Data, Global/Regional Reanalysis Data, or Satellite Retrieved Precipitation Data? *Remote Sens.* **2020**, *12*, 683. [CrossRef]
21. Pitman, A.J.; Arneth, A.; Ganzeveld, L. Regionalizing global climate models. *Int. J. Climatol.* **2012**, *32*, 321–337. [CrossRef]
22. Touze-Peiffer, L.; Barberousse, A.; Le Treut, H. The Coupled Model Intercomparison Project: History, uses, and structural effects on climate research. *Wiley Interdiscip. Rev. Clim. Chang.* **2020**, *11*, e648. [CrossRef]
23. Immerzeel, W.W.; van Beek, L.P.H.; Bierkens, M.F.P. Climate Change Will Affect the Asian Water Towers. *Science* **2010**, *328*, 1382–1385. [CrossRef]
24. Vilasa, L.; Miralles, D.G.; de Jeu, R.A.M.; Dolman, A.J. Global soil moisture bimodality in satellite observations and climate models. *J. Geophys. Res. Atmos.* **2017**, *122*, 4299–4311. [CrossRef]
25. Liu, X.; Tang, Q.; Voisin, N.; Cui, H. Projected impacts of climate change on hydropower potential in China. *Hydrol. Earth Syst. Sci.* **2016**, *20*, 3343–3359. [CrossRef]
26. Eyring, V.; Bony, S.; Meehl, G.A.; Senior, C.A.; Stevens, B.; Stouffer, R.J.; Taylor, K.E. Overview of the Coupled Model Intercomparison Project Phase 6 (CMIP6) experimental design and organization. *Geosci. Model Dev.* **2016**, *9*, 1937–1958. [CrossRef]
27. Zhu, Y.Y.; Yang, S.N. Evaluation of CMIP6 for historical temperature and precipitation over the Tibetan Plateau and its comparison with CMIP5. *Adv. Clim. Chang. Res.* **2020**, *11*, 239–251. [CrossRef]
28. Lun, Y.R.; Liu, L.; Cheng, L.; Li, X.P.; Li, H.; Xu, Z.X. Assessment of GCMs simulation performance for precipitation and temperature from CMIP5 to CMIP6 over the Tibetan Plateau. *Int. J. Climatol.* **2021**, *41*, 3994–4018. [CrossRef]
29. Liu, J.F.; Lu, Y.Q. How Well Do CMIP6 Models Simulate the Greening of the Tibetan Plateau? *Remote Sens.* **2022**, *14*, 4633. [CrossRef]
30. Cui, T.; Li, Y.; Yang, L.; Nan, Y.; Li, K.; Tudaji, M.; Hu, H.; Long, D.; Shahid, M.; Mubeen, A.; et al. Non-monotonic changes in Asian Water Towers' streamflow at increasing warming levels. *Nat. Commun.* **2023**, *14*, 1176. [CrossRef] [PubMed]
31. Mishra, V.; Bhatia, U.; Tiwari, A.D. Bias-corrected climate projections for South Asia from Coupled Model Intercomparison Project-6. *Sci. Data* **2020**, *7*, 13. [CrossRef] [PubMed]
32. Thrasher, B.; Wang, W.L.; Michaelis, A.; Melton, F.; Lee, T.; Nemani, R. NASA Global Daily Downscaled Projections, CMIP6. *Sci. Data* **2022**, *9*, 262. [CrossRef]
33. Zhang, Y.; Ren, H.; Pan, X. *Integration Dataset of Tibet Plateau Boundary*; National Tibetan Plateau Data Center: Beijing, China, 2019. [CrossRef]
34. Zhang, Y.; Li, B.; Zheng, D. Datasets of the boundary and area of the Tibetan Plateau. *Acta Geogr. Sin.* **2014**, *69* (Suppl. 1), 65–68.
35. Zhang, G. Dataset of river basins map over the TP (2016). *Geophys. Res. Lett.* **2019**, *40*, 2125–2130. [CrossRef]
36. Ran, Y.; Wang, L.; Zeng, T.; Ge, C.; Li, H. *"One Belt, one Road" Boundary Map of Key Basins in Asia*; National Tibetan Plateau Data Center: Beijing, China, 2020. [CrossRef]
37. National, B. *1:250000 River Distribution Data Set of Qaidam River Basin (2000)*; National Tibetan Plateau Data Center: Beijing, China, 2020. Available online: https://data.tpdc.ac.cn/zh-hans/data/c77dce15-8d92-43e5-ace5-b5d781ff867c (accessed on 27 March 2024).
38. FAO. *SRTM DEM Data on the Tibetan Plateau (2012)*; National Tibetan Plateau Data Center: Beijing, China, 2019.
39. Xu, X.D.; Dong, L.L.; Zhao, Y.; Wang, Y.J. Effect of the Asian Water Tower over the Qinghai-Tibet Plateau and the characteristics of atmospheric water circulation. *Chin. Sci. Bull. Chin.* **2019**, *64*, 2830–2841. [CrossRef]
40. Li, C.Q.; Zuo, Q.J.; Xu, X.D.; Gao, S.T. Water vapor transport around the Tibetan Plateau and its effect on summer rainfall over the Yangtze River valley. *J. Meteorol. Res.* **2016**, *30*, 472–482. [CrossRef]
41. Chen, B.; Xu, X.D.; Yang, S.; Zhang, W. On the origin and destination of atmospheric moisture and air mass over the Tibetan Plateau. *Theor. Appl. Climatol.* **2012**, *110*, 423–435. [CrossRef]
42. Davis, M.E.; Thompson, L.G.; Yao, T.D.; Wang, N.L. Forcing of the Asian monsoon on the Tibetan Plateau: Evidence from high-resolution ice core and tropical coral records. *J. Geophys. Res. Atmos.* **2005**, *110*, 14. [CrossRef]
43. Karadan, M.M.; Raju, P.V.S.; Mishra, A. Simulations of Indian summer monsoon using RegCM: A comparison with ERA and GFDL analysis. *Theor. Appl. Climatol.* **2021**, *143*, 1381–1391. [CrossRef]
44. Guo, X.Y.; Wang, L.; Tian, L.D.; Zhou, J.; Wang, Y.W. Possible widths of Indian summer monsoon trajectories in Tibetan Plateau revealed by the direction of maximum summer precipitation decreases in recent decades. *Clim. Dyn.* **2022**, *60*, 2315–2330. [CrossRef]
45. Dorji, S.; Herath, S.; Mishra, B.K.; Chophel, U. Predicting summer monsoon of Bhutan based on SST and teleconnection indices. *Meteorol. Atmos. Phys.* **2019**, *131*, 541–551. [CrossRef]
46. Deka, R.L.; Mahanta, C.; Pathak, H.; Nath, K.K.; Das, S. Trends and fluctuations of rainfall regime in the Brahmaputra and Barak basins of Assam, India. *Theor. Appl. Climatol.* **2013**, *114*, 61–71. [CrossRef]
47. Li, X.Y.; Long, D. An improvement in accuracy and spatiotemporal continuity of the MODIS precipitable water vapor product based on a data fusion approach. *Remote Sens. Environ.* **2020**, *248*, 111966. [CrossRef]
48. Li, H.; Liu, L.; Koppa, A.; Shan, B.Y.; Liu, X.C.; Li, X.P.; Niu, Q.K.; Cheng, L.; Miralles, D. Vegetation greening concurs with increases in dry season water yield over the Upper Brahmaputra River basin. *J. Hydrol.* **2021**, *603*, 126981. [CrossRef]

49. Riahi, K.; van Vuuren, D.P.; Kriegler, E.; Edmonds, J.; O'Neill, B.C.; Fujimori, S.; Bauer, N.; Calvin, K.; Dellink, R.; Fricko, O.; et al. The Shared Socioeconomic Pathways and their energy, land use, and greenhouse gas emissions implications: An overview. *Glob. Environ. Change Hum. Policy Dimens.* **2017**, *42*, 153–168. [CrossRef]
50. O'Neill, B.C.; Kriegler, E.; Riahi, K.; Ebi, K.L.; Hallegatte, S.; Carter, T.R.; Mathur, R.; van Vuuren, D.P. A new scenario framework for climate change research: The concept of shared socioeconomic pathways. *Clim. Chang.* **2014**, *122*, 387–400. [CrossRef]
51. Korner, C. The use of 'altitude' in ecological research. *Trends Ecol. Evol.* **2007**, *22*, 569–574. [CrossRef]
52. Karger, D.N.; Schmatz, D.R.; Dettling, G.; Zimmermann, N.E. High-resolution monthly precipitation and temperature time series from 2006 to 2100. *Sci. Data* **2020**, *7*, 248. [CrossRef]
53. Karger, D.N.; Conrad, O.; Bohner, J.; Kawohl, T.; Kreft, H.; Soria-Auza, R.W.; Zimmermann, N.E.; Linder, H.P.; Kessler, M. Data Descriptor: Climatologies at high resolution for the earth's land surface areas. *Sci. Data* **2017**, *4*, 20. [CrossRef]
54. Hutchinson, M.F.; Gessler, P.E. Splines—more than just a smooth interpolator. *Geoderma* **1994**, *62*, 45–67. [CrossRef]
55. Gao, C.; Booij, M.J.; Xu, Y.P. Assessment of extreme flows and uncertainty under climate change: Disentangling the uncertainty contribution of representative concentration pathways, global climate models and internal climate variability. *Hydrol. Earth Syst. Sci.* **2020**, *24*, 3251–3269. [CrossRef]
56. Hoge, M.; Guthke, A.; Nowak, W. The hydrologist's guide to Bayesian model selection, averaging and combination. *J. Hydrol.* **2019**, *572*, 96–107. [CrossRef]
57. Vrugt, J.; Diks, C.; Clark, M. Ensemble Bayesian model averaging using Markov Chain Monte Carlo sampling. *Environ. Fluid Mech.* **2008**, *8*, 579–595. [CrossRef]
58. Radaideh, M.I.; Borowiec, K.; Kozlowski, T. Integrated framework for model assessment and advanced uncertainty quantification of nuclear computer codes under Bayesian statistics. *Reliab. Eng. Syst. Saf.* **2019**, *189*, 357–377. [CrossRef]
59. Watterson, I.G. Calculation of probability density functions for temperature and precipitation change under global warming. *J. Geophys. Res. Atmos.* **2008**, *113*, D12106. [CrossRef]
60. Ines, A.V.M.; Hansen, J.W. Bias correction of daily GCM rainfall for crop simulation studies. *Agric. For. Meteorol.* **2006**, *138*, 44–53. [CrossRef]
61. Miao, C.Y.; Su, L.; Sun, Q.H.; Duan, Q.Y. A nonstationary bias-correction technique to remove bias in GCM simulations. *J. Geophys. Res. Atmos.* **2016**, *121*, 5718–5735. [CrossRef]
62. Li, H.B.; Sheffield, J.; Wood, E.F. Bias correction of monthly precipitation and temperature fields from Intergovernmental Panel on Climate Change AR4 models using equidistant quantile matching. *J. Geophys. Res. Atmos.* **2010**, *115*, D10101. [CrossRef]
63. Wang, L.; Chen, W. A CMIP5 multimodel projection of future temperature, precipitation, and climatological drought in China. *Int. J. Climatol.* **2014**, *34*, 2059–2078. [CrossRef]
64. Gutowski, W.J.; Decker, S.G.; Donavon, R.A.; Pan, Z.T.; Arritt, R.W.; Takle, E.S. Temporal-spatial scales of observed and simulated precipitation in central U.S. climate. *J. Clim.* **2003**, *16*, 3841–3847. [CrossRef]
65. Burrough, P.A.; McDonnell, R.A.; Lloyd, C.D. *Principles of Geographical Information Systems*; Oxford University Press: Oxford, UK, 2015.
66. Piccoli, B.; Rossi, F. Generalized Wasserstein Distance and its Application to Transport Equations with Source. *Arch. Ration. Mech. Anal.* **2014**, *211*, 335–358. [CrossRef]
67. Ruschendorf, L. The Wasserstein Distance and approximation theorems. *Z. Wahrscheinlichkeitstheorie Verwandte Geb.* **1985**, *70*, 117–129. [CrossRef]
68. Cho, D.; Yoo, C.; Im, J.; Lee, Y.; Lee, J. Improvement of spatial interpolation accuracy of daily maximum air temperature in urban areas using a stacking ensemble technique. *Giscience Remote Sens.* **2020**, *57*, 633–649. [CrossRef]
69. Vissio, G.; Lembo, V.; Lucarini, V.; Ghil, M. Evaluating the Performance of Climate Models Based on Wasserstein Distance. *Geophys. Res. Lett.* **2020**, *47*, e2020GL089385. [CrossRef]
70. Zhao, H.; Huang, W.; Xie, T.; Wu, X.; Xie, Y.; Feng, S.; Chen, F. Optimization and evaluation of a monthly air temperature and precipitation gridded dataset with a 0.025° spatial resolution in China during 1951–2011. *Theor. Appl. Climatol.* **2019**, *138*, 491–507. [CrossRef]
71. Chen, C.F.; Hu, B.J.; Li, Y.Y. Easy-to-use spatial random-forest-based downscaling-calibration method for producing precipitation data with high resolution and high accuracy. *Hydrol. Earth Syst. Sci.* **2021**, *25*, 5667–5682. [CrossRef]
72. Jiang, Y.; Yang, K.; Shao, C.; Zhou, X.; Zhao, L.; Chen, Y.; Wu, H. A downscaling approach for constructing high-resolution precipitation dataset over the Tibetan Plateau from ERA5 reanalysis. *Atmos. Res.* **2021**, *256*, 105574. [CrossRef]
73. You, Q.L.; Cai, Z.Y.; Wu, F.Y.; Jiang, Z.H.; Pepin, N.; Shen, S.S.P. Temperature dataset of CMIP6 models over China: Evaluation, trend and uncertainty. *Clim. Dyn.* **2021**, *57*, 17–35. [CrossRef]
74. Yuan, X.; Yang, K.; Lu, H.; He, J.; Sun, J.; Wang, Y. Characterizing the features of precipitation for the Tibetan Plateau among four gridded datasets: Detection accuracy and spatio-temporal variabilities. *Atmos. Res.* **2021**, *264*, 9. [CrossRef]
75. Li, L.; Yang, S.; Wang, Z.Y.; Zhu, X.D.; Tang, H.Y. Evidence of Warming and Wetting Climate over the Qinghai-Tibet Plateau. *Arct. Antarct. Alp. Res.* **2010**, *42*, 449–457. [CrossRef]
76. IPCC. *Climate Change 2021: The Physical Science Basis. Contribution of Working Group I to the Sixth Assessment Report of the Intergovernmental Panel on Climate Change*; Cambridge University Press: Cambridge, UK; New York, NY, USA, 2021; *in press*.
77. Zhang, C.; Tang, Q.; Chen, D. Recent Changes in the Moisture Source of Precipitation over the Tibetan Plateau. *J. Clim.* **2017**, *30*, 1807–1819. [CrossRef]

78. Liu, L.; Zhang, W.; Lu, Q.; Wang, G. Variations in the Sensible Heating of Tibetan Plateau and Related Effects on Atmospheric Circulation Over South Asia. *Asia-Pac. J. Atmos. Sci.* **2021**, *57*, 499–510. [CrossRef]
79. Chen, H.; Zhu, L.P.; Hou, J.Z.; Steinman, B.A.; He, Y.; Brown, E.T. Westerlies effect in Holocene paleoclimate records from the central Qinghai-Tibet Plateau. *Palaeogeogr. Palaeoclimatol. Palaeoecol.* **2022**, *598*, 111036. [CrossRef]
80. Zhao, P.; Li, Y.L.; Jia, C.H.; Che, J.H. Atmospheric Meridional Circulation Between South Asia and Tibetan Plateau Caused by the Change of Planetary Boundary Layer Depth. *J. Geophys. Res. Atmos.* **2023**, *128*, e2023JD039506. [CrossRef]
81. Wu, T.H.; Zhao, L.; Li, R.; Wang, Q.X.; Xie, C.W.; Pang, Q.Q. Recent ground surface warming and its effects on permafrost on the central Qinghai-Tibet Plateau. *Int. J. Climatol.* **2013**, *33*, 920–930. [CrossRef]
82. Li, Y.J.; Zhang, C.J.; Li, Z.C.; Yang, L.W.; Jin, X.; Gao, X.Q. Analysis on the temporal and spatial characteristics of the shallow soil temperature of the Qinghai-Tibet Plateau. *Sci. Rep.* **2022**, *12*, 19746. [CrossRef]
83. Luo, D.L.; Jin, H.J.; He, R.X.; Wang, X.F.; Muskett, R.R.; Marchenko, S.S.; Romanovsky, V.E. Characteristics of Water-Heat Exchanges and Inconsistent Surface Temperature Changes at an Elevational Permafrost Site on the Qinghai-Tibet Plateau. *J. Geophys. Res. Atmos.* **2018**, *123*, 10404–10422. [CrossRef]
84. Xie, Z.H. Review of Qinghai-Tibet Plateau Impacting on Atmospheric Circulation. In Proceedings of the International Conference on Energy Development and Environmental Protection (EDEP), Beijing, China, 11–12 June 2016; pp. 355–359.
85. Maraun, D.; Knevels, R.; Mishra, A.N.; Truhetz, H.; Bevacqua, E.; Proske, H.; Zappa, G.; Brenning, A.; Petschko, H.; Schaffer, A.; et al. A severe landslide event in the Alpine foreland under possible future climate and land-use changes. *Commun. Earth Environ.* **2022**, *3*, 87. [CrossRef]
86. Ren, Z.; Ma, X.; Wang, K.; Li, Z. Effects of Extreme Precipitation on Runoff and Sediment Yield in the Middle Reaches of the Yellow River. *Atmosphere* **2023**, *14*, 1415. [CrossRef]
87. Touma, D.; Stevenson, S.; Swain, D.L.; Singh, D.; Kalashnikov, D.A.; Huang, X. Climate change increases risk of extreme rainfall following wildfire in the western United States. *Sci. Adv.* **2022**, *8*, eabm0320. [CrossRef]
88. Suhas, D.L.; Ramesh, N.; Kripa, R.M.; Boos, W.R. Influence of monsoon low pressure systems on South Asian disasters and implications for disaster prediction. *npj Clim. Atmos. Sci.* **2023**, *6*, 48. [CrossRef]
89. Araghi, A.; Adamowski, J.F. Assessment of 30 gridded precipitation datasets over different climates on a country scale. *Earth Sci. Inform.* **2024**, *17*, 1301–1313. [CrossRef]
90. Beck, H.E.; van Dijk, A.; Levizzani, V.; Schellekens, J.; Miralles, D.G.; Martens, B.; de Roo, A. MSWEP: 3-hourly 0.25 degrees global gridded precipitation (1979–2015) by merging gauge, satellite, and reanalysis data. *Hydrol. Earth Syst. Sci.* **2017**, *21*, 589–615. [CrossRef]
91. Yang, Y.; Gao, D.; Li, B. Preliminary study on water vapor channel in the lower Brahmaputra River Valley. *Sci. China* **1987**, *B*, 893–902.
92. Xu, X.M.; Zhang, X.Q.; Li, X. Evaluation of the Applicability of Three Methods for Climatic Spatial Interpolation in the Hengduan Mountains Region. *J. Hydrometeorol.* **2023**, *24*, 35–51. [CrossRef]
93. Wu, T.W.; Lu, Y.X.; Fang, Y.J.; Xin, X.G.; Li, L.; Li, W.P.; Jie, W.H.; Zhang, J.; Liu, Y.M.; Zhang, L.; et al. The Beijing Climate Center Climate System Model (BCC-CSM): The main progress from CMIP5 to CMIP6. *Geosci. Model Dev.* **2019**, *12*, 1573–1600. [CrossRef]
94. Muller, W.A.; Jungclaus, J.H.; Mauritsen, T.; Baehr, J.; Bittner, M.; Budich, R.; Bunzel, F.; Esch, M.; Ghosh, R.; Haak, H.; et al. A Higher-resolution Version of the Max Planck Institute Earth System Model (MPI-ESM1.2-HR). *J. Adv. Model. Earth Syst.* **2018**, *10*, 1383–1413. [CrossRef]
95. Boucher, O.; Servonnat, J.; Albright, A.L.; Aumont, O.; Balkanski, Y.; Bastrikov, V.; Bekki, S.; Bonnet, R.; Bony, S.; Bopp, L.; et al. Presentation and Evaluation of the IPSL-CM6A-LR Climate Model. *J. Adv. Model. Earth Syst.* **2020**, *12*, e2019MS002010. [CrossRef]
96. Ferguglia, O.; von Hardenberg, J.; Palazzi, E. Robustness of precipitation Emergent Constraints in CMIP6 models. *Clim. Dyn.* **2023**, *612*, 1439–1450. [CrossRef]

atmosphere

Article

An Improved Ensemble-Strategy-Assisted Wind Speed Prediction Method for Railway Strong Wind Warnings

Jian Liu [1], Xiaolei Cui [2], Cheng Cheng [1,*] and Yan Jiang [2,3,*]

[1] Chongqing Wukang Technology Co., Ltd., Chongqing 404100, China; liujian17@cmhk.com
[2] School of Civil Engineering, Chongqing Jiaotong University, Chongqing 400074, China; cqjtucuixl@mails.cqjtu.edu.cn
[3] College of Engineering and Technology, Southwest University, Chongqing 400715, China
* Correspondence: chengcheng@cmhk.com (C.C.); xnjtjiangyan@swu.edu.cn (Y.J.)

Abstract: Reliable short-term wind speed prediction is one of the core technologies in the strong wind warning system for railway applications, which is of great significance for ensuring the safety of high-speed train operations and ancillary railway facilities. To improve forecasting accuracy, decomposition-based methods have attracted extensive attention due to their superior ability to address complex data characteristics (e.g., nonstationarity and nonlinearity). Currently, there are two pre-processing schemes for decomposition-based methods, i.e., one-time decomposition and real-time decomposition. In order to apply them better, this paper first expounds the difference between them, based on a combination of DWT (discrete wavelet transform) and CKDE (conditional kernel density estimation). The results show that although the one-time decomposition-based method has an unexceptionable accuracy, it only can provide offline prediction and thus may not be practical. The real-time decomposition-based method possesses stronger practicability and is able to provide online prediction, but it has limited accuracy. Then, an improved ensemble strategy is developed by optimizing the selection of appropriate decomposed components to conduct the prediction on the basis of real-time decomposition. This improved ensemble strategy provides an effective guidance for this selective combination, including taking historical information into consideration in the data. Finally, numerical examples and practicality analysis using two groups of measured wind speed data demonstrate that the proposed method is effective in providing high-precision online wind speed prediction. For example, compared with CKDE, the average degrees of improvement achieved by the proposed method in terms of MAE, RMSE, and MRPE, are 16.25%, 17.66%, and 16.93, respectively, while those compared with the traditional real-time decomposition method are 17.11%, 18.54%, and 16.84, respectively.

Keywords: wind speed prediction; one time decomposition; real time decomposition; improved ensemble strategy; discrete wavelet transform; conditional kernel density estimation

Citation: Liu, J.; Cui, X.; Cheng, C.; Jiang, Y. An Improved Ensemble-Strategy-Assisted Wind Speed Prediction Method for Railway Strong Wind Warnings. *Atmosphere* **2023**, *14*, 1787. https://doi.org/10.3390/atmos14121787

Academic Editor: Massimiliano Burlando

Received: 13 November 2023
Revised: 23 November 2023
Accepted: 1 December 2023
Published: 5 December 2023

1. Introduction

In recent years, a large number of high-speed rail projects have been constructed around the world, and the corresponding scale of development is very rapid [1,2]. According to a statistical report by China Railway, China's railway service length had reached 155,000 km, including 42,000 km of high-speed railways, by the end of 2022. However, due to the light weight of high-speed train carriages and their high operating speeds, operational high-speed trains inevitably encounter large lift forces under the action of strong winds. Therefore, they are highly susceptible to derailment and overturning [3–5], which can lead to significant casualties and huge economic losses [6,7]. Meanwhile, safety and reliability of structures (e.g., bridges and transmission Lines) under the action of strong winds have also become an increasingly serious topic [8–15]. One of effective methods to addressing this predicament is the accurate prediction of strong winds (i.e., short-term wind speed prediction), which will enable the railway dispatching system to adjust dispatch plans in a timely manner [16]. Therefore, short-term wind speed prediction is one of the most important techniques in high-speed railway warning systems, which also plays a significant role in ensuring the operational safety of high-speed trains [17–19].

To perform short-term wind speed prediction, many attempts have been made to generate a plethora of forecasting methods [20,21]. Generally, these methods can be categorized into four different kinds, including physical models, statistical models, artificial intelligence- or machine learning-based models, and hybrid models [22]. Physical models mainly utilize meteorological data (such as temperature, humidity, air pressure, etc.) to carry out wind speed prediction. One representative model is the numerical weather prediction model (numerical weather prediction, NWP) [23]. However, this type of model requires a large amount of meteorological data and computational resources and is merely suitable for long-term wind speed prediction [24]. Statistical models (i.e., time series models) can directly analyze the linear relationship between historical wind speed records, from which the corresponding forecasting models are established. These models generally possess a simple and effective structure, and have been widely used in short-term wind speed prediction. They include the auto-regressive (AR) model, moving average (MA) model, auto-regressive moving average (ARMA) model, and difference auto-regressive integrated moving average (ARIMA) model [25,26]. However, these models usually perform well under the assumption of stationary time series and cannot address the prediction of wind speed data with high nonlinear features [27]. Fortunately, artificial intelligence- or machine learning-based models, such as Kalman filters [28], artificial neural networks (ANN) [29], fuzzy logic [30], support vector machines (SVM) [31], deep learning (DL) [32], etc., can effectively explain the nonlinear features embedded in the data. These models have been successfully applied in the field of wind speed prediction. However, the before-mentioned models mostly belong to the parametric model category and generally include complex parameter optimization processes [33,34]. Meanwhile, they easily fall into the problems of local optimization, overfitting, and low rate of convergence [35]. Recently, conditional kernel density estimation (CKDE) has been able to determine the probability density function (PDF) of the target variable and use it as a basis for addressing the time series prediction problem [36]. This method has data-driven attributes and can directly use the sample data to estimate the PDF without any parametric assumption. In addition to the simple single-modal density distribution problem, it can effectively deal with problems of multi-modal or skewed density distributions [22]. Therefore, the CKDE model can not only avoid errors and computational resources caused by parameter optimization processes, but also respond well to the non-linear and non-Gaussian characteristics hidden in wind speed time series [37]. Considering the advantages of CKDE, the integration of this method into wind speed prediction may be promising and challenging.

According to [38–43], wind speed time series usually present multiple characteristics (e.g., linearity, nonstationarity, nonlinearity, etc.). In this scenario, the above-mentioned single models may not provide satisfactory forecasting results. To pursue higher forecasting accuracy, the combination of different component methods (i.e., hybrid models) has attracted more and more attention. These hybrid methods can take advantage of the strength of each component model. There are four different hybrid models including decomposition-based models, parameter optimization-based models, weight allocation-based models, and error correction-based models [44]. Due to strong volatility and stationarity of wind speed time series, hybrid models based on signal pretreatment techniques are extensively employed. These pretreatment techniques main are empirical mode decomposition (EMD) [16], fast ensemble EMD (FEEMD) [45], discrete wavelet transform (DWT) [46], wavelet packet decomposition (WPD) [47], empirical wavelet transform (EWT) [48], variational mode decomposition (VMD) [49], and their developed versions. Generally, these types of hybrid models first apply signal pretreatment techniques to decompose the raw wind speed data into different stationary and regular subsequences. Then, one or several appropriate models are established for training each subsequence and performing the corresponding prediction.

Finally, the summation of all individual predictions in a linear manner can be regarded as the target prediction result [50]. For example, Liu et al. [51] used four different decomposition techniques (i.e., EMD, FEEMD, DWT, and WPD) and extreme learning machine (ELM) to implement the prediction of short-term wind speeds, and the forecasting result indicated that the accuracy of these decomposition-based models was superior to that of single models. In most relevant studies, the decomposition is only conducted once for the original data, and each decomposed subsequence is divided into training and testing sets (i.e., one-time decomposition). Although one-time decomposition-based models feature extremely high accuracy, the decomposition takes the future information into consideration and can only provide offline prediction [52]. Therefore, this method may not be practical. In contrast, the real-time decomposition-based method has a stronger practicability and is able to provide online prediction, but it forecasting accuracy is generally limited and sometimes even inferior to that of single models [50]. The possible reasons for this limited accuracy may be ascribed to the effect of the endpoint effect, mode mixing, improper or varying decomposition level number, and illusory components [24]. Meanwhile, the ensemble pattern (i.e., the direct summation of all individual predictions) may not be effective, while the selective ensemble of several individual predictions generally presents satisfactory results [53]. However, the difference between one-time decomposition-based methods and real-time decomposition-based methods is rarely reported, and the selective ensemble pattern in these methods sometimes depends on the user's experience, which generally lacks a theoretical basis.

From the above literature review, several main problems in the current methods of short-term wind speed prediction are summarized below: ① Physical models require a large quantity of meteorological data and computational resources, and are merely suitable for long-term wind speed prediction; ② statistical models only can perform well for the prediction of wind speed time series with strong linear and stationary characteristics; ③ artificial intelligence or machine learning models can capture linear and complex nonlinear characteristics in the wind speed data, but they cannot effectively address nonstationary characteristics in the data and their performance is vulnerable to model parameters; ④ although two kinds of decomposition-based methods have been widely used, the differences between then are rarely clarified, which sometimes brings about unsuitable applications. Meanwhile, forecasting accuracy is strongly related to the ensemble pattern of subsequence predictions, which generally lack a theoretical basis.

This paper first introduces the nonparametric CKDE model as the predictor and reviews the details of two traditional decomposition-based models and the differences between them on the basis of DWT. Then, the performance evaluation of these two methods is conducted based on measured wind speed data. Further, an improved ensemble-strategy-assisted real-time decomposition method is developed for short-term wind speed prediction, which can optimally select the decomposed components for ensemble predictions. Finally, numerical examples are performed to illustrate the performance of the proposed method. The forecasting results show that this improved method has higher prediction accuracy in comparison with traditional real-time decomposition-based models and the single CKDE model. For example, compared with CKDE, the average degrees of improvement realized by the proposed method in terms of MAE, RMSE, and MRPE are 16.25%, 17.66%, and 16.93, respectively, while those compared with the traditional real-time decomposition method are 17.11%, 18.54%, and 16.84, respectively.

2. Methodology

2.1. Discrete Wavelet Transform

As a convenient and efficient technique of multiscale signal processing, discrete wavelet transform (DWT) has been widely used in the analysis of nonlinear and non-stationary signals [54]. This technique can decompose the signal into specified quantity subsequences with different frequency bandwidths by employing a group of basis functions. These basis functions generally result from a series of translation and expansion operations based on the pre-specified mother wavelet function [54]. For the original wind speed time series $\{v_1, v_2, \ldots, v_n\}$ (denoted as $v[n]$), DWT can decompose this signal into a number of approximation and detail components, i.e., [55]

$$v[n] = D_1 + D_2 + \ldots + D_k + A_k \tag{1}$$

where n represents the number of original data points; k + 1 represents the specified decomposition level of DWT; D_1, D_2,..., D_k represent the detail components; and A_k stands for the approximation component.

With the augmentation of decomposition level numbers, the frequency on the corresponding subsequence gradually decreases, and thus the frequency overlap between two adjacent subsequences can be addressed to some extent [54]. Meanwhile, the number of decomposition levels can be set in advance. Therefore, this technique has the ability to address the time-frequency analysis of both single-channel data and multi-channel data [56]. In this technique, the critical problem is determining the mother wavelet function. According to [54], the Daubechies 10 wavelet has exhibited superiority in analyzing the wind speed time series, which is still used for the decomposition and reconstruction in this paper. The detailed illustration of DWT can be found in [55].

2.2. Conditional Kernel Density Estimation

After the decomposition by DWT is completed, the CKDE model is established for every decomposed subsequence. In this model, any assumption regarding the relationship between input and output variables can be avoided [57]. In addition to having the ability to describe the linear characteristics in the data, this model embodies an obvious advantage in addressing the data with multimodal or skewed probability distributions, and thus can explain the nonlinear or non-Gaussian characteristics in the wind speed time series [44]. The following gives the detailed modeling process of CKDE in wind speed prediction.

For any decomposed subsequence $\{x(t), t = 1, 2, 3, \ldots, n\}$, a series of input–output pairs with the L-step ahead forecasting task can be constructed, i.e.,

$$\begin{bmatrix} \boldsymbol{x}_1 \\ \vdots \\ \boldsymbol{x}_i \\ \vdots \\ \boldsymbol{x}_N \end{bmatrix} = \begin{bmatrix} x(1) & \cdots & x(m) \\ \vdots & \vdots & \vdots \\ x(i) & \cdots & x(i+m-1) \\ \vdots & \vdots & \vdots \\ x(N) & \cdots & x(n-L) \end{bmatrix}; \begin{bmatrix} y_1 \\ \vdots \\ y_i \\ \vdots \\ y_N \end{bmatrix} = \begin{bmatrix} x(m+L) \\ \vdots \\ x(i+L+m-1) \\ \vdots \\ x(n) \end{bmatrix} \tag{2}$$

where $\{\boldsymbol{x}_i, i = 1, 2, \ldots, N;\ N = n - m - L + 1\}$ denotes the m-dimensional input variable and $\{y_i\}$ stands for the one-dimensional output variable. On this basis, the joint probability density distribution of $(\boldsymbol{x}, y)$ can be expressed as:

$$\hat{f}(\boldsymbol{x}, y) = \frac{1}{N \cdot |\boldsymbol{H}_X| \cdot |H_Y|} \sum_{i=1}^{N} \left\{ K_m\left[\boldsymbol{H}_X^{-1}(\boldsymbol{x} - \boldsymbol{x}_i)\right] \cdot K\left[H_Y^{-1}(y - y_t)\right] \right\} \tag{3}$$

where $K_m[\cdot]$ is the m-dimensional kernel function and the Gaussian function is commonly used because of its simple structure and easy application; $\boldsymbol{H}_X$ and H_Y are the diagonal bandwidth matrixes for input and output variables, respectively [36]. These bandwidth matrixes are critical for ensuring modeling quality, and generally can be obtained via the normal reference criterion (NRC), i.e., [44]

$$\begin{aligned} &\boldsymbol{H}_X = diag\left[H_q\right]; \\ &H_Y = \sigma_Y \cdot (4/(m+2)N)^{1/(m+4)}; \\ &H_q = \sigma_q \cdot (4/(m+2)N)^{1/(m+4)},\ q = 1, 2, \cdots, m; \end{aligned} \tag{4}$$

in which σ_q and σ_Y correspond to sample standard deviations regarding the time series of $\{x(q), x(q+1), \ldots, x(q+n\text{-}m\text{-}L), q = 1, 2, \ldots, m\}$ and $\{x(m+L), x(m+L+1), \ldots, x(n)\}$, respectively.

Based on conditional probability theory, the density distribution of the output variable on the basis of the input variable can be defined as:

$$\hat{f}(y|\boldsymbol{x}) = \sum_{i=1}^{N}\left[\omega_i(\boldsymbol{x}) \cdot \frac{1}{|\boldsymbol{H}_Y|} \cdot K\left(\boldsymbol{H}_Y^{-1}(y - \boldsymbol{y}_i)\right)\right] \tag{5}$$

$$\omega_i(\boldsymbol{x}) = \frac{K_m\left(\boldsymbol{H}_X^{-1}(\boldsymbol{x} - \boldsymbol{x}_i)\right)}{\sum_{i=1}^{N} K_m\left(\boldsymbol{H}_X^{-1}(\boldsymbol{x} - \boldsymbol{x}_i)\right)} \tag{6}$$

and then the expectation of the output variable y can be given by

$$\overline{y} = \int y\hat{f}(y|\boldsymbol{x})dy = \sum_{i=1}^{N}\omega_i(\boldsymbol{x}) \cdot y_i \tag{7}$$

When the input variable $\boldsymbol{x}$ is updated to $\{x(n-m+1), x(n-m+2), \ldots, x(n)\}$, which can be marked as $\boldsymbol{x}_{n+L}$, the expectation provided by Equation (7) can be considered to be the L-step ahead forecasting value, i.e., [44]

$$\hat{x}(n+L) = \sum_{i=1}^{N}\omega_i(\boldsymbol{x}_{N+L}) \cdot \boldsymbol{y}_i \tag{8}$$

where $\hat{x}(n+L)$ is the L-step ahead forecasting value of $x(n+L)$. More information about this CKDE model can be found in [36].

2.3. Evaluation Metrics

For the sake of evaluating the prediction performance of the model intuitively, three commonly used evaluation indicators are employed, i.e., mean absolute error (MAE), root mean square error (RMSE), and mean relative percentage error (MRPE). Their calculation formulas are given as follows [20]:

$$\text{MAE} = \frac{1}{n_1}\sum_{i=n+1}^{n+n_1}|v_i - \hat{v}_i| \tag{9}$$

$$\text{RMSE} = \sqrt{\frac{1}{n_1}\sum_{i=n+1}^{n+n_1}(v_i - \hat{v}_i)^2} \tag{10}$$

$$\text{MRPE} = \frac{1}{n_1}\sum_{i=n+1}^{n+n_1}\left|\frac{v_i - \hat{v}_i}{v_i}\right| \times 100\% \tag{11}$$

where n_1 represents the number of the data used to evaluate the performance of the model, and for one-step ahead prediction, n_1 is equal to the number of the testing data (i.e., n_1 = 75). $\hat{v}_i$ denotes the predicted value of the measured wind speed v_i. In order to further exhibit the performance comparison, three indexes regarding the improved percentage between different forecasting methods in terms of the above indicators are defined as [16]:

$$\text{P}_{\text{MAE}} = \frac{\text{MAE}_1 - \text{MAE}_2}{\text{MAE}_1} \times 100\% \tag{12}$$

$$\text{P}_{\text{RMSE}} = \frac{\text{RMSE}_1 - \text{RMSE}_2}{\text{RMSE}_1} \times 100\% \tag{13}$$

$$P_{MRPE} = \frac{MRPE_1 - MRPE_2}{MRPE_1} \times 100\% \quad (14)$$

where MAE_1, $RMSE_1$, and $MRPE_1$ are evaluation indicators of the method under consideration, while MAE_2, $RMSE_2$, and $MRPE_2$ are those of the compared method.

3. Review of Traditional Decomposition-Based Prediction Models

There are two kinds of decomposition-based models in the field of wind speed prediction. This section takes the combination of DWT and CKDE to review the details of these two models and their differences. The specific illustration is presented below.

3.1. Steps of Traditional Prediction Methods

According to [52], there are two kinds of pretreatment manners in the traditional decomposition-based forecasting methods, i.e., the one-time decomposition-based forecasting method and the real-time decomposition based forecasting method. The main steps of the one-time decomposition-based forecasting method are shown in Figure 1a. In this method, DWT is employed only once to decompose all original wind speed data $\{v_1, v_2, \ldots, v_n\}$ into a number of subsequences with a fixed quantity, and then each decomposed subsequence is divided into the training set and the testing set for modeling and performance evaluation (i.e., the decomposition occurs before data set partitioning). The summation of all training sets is regarded as the training data, while that of all testing sets is considered to be the testing data. From Figure 1a, it can be seen that the one-time decomposition scheme simultaneously decomposes the known (training) data and the unknown (testing) data, i.e., the decomposition introduces an assumption that the expected data are available in advance. After the decomposition is completed, an appropriate forecasting model (i.e., CKDE) is established for training each decomposed subsequence and then performing the individual prediction. On this basis, the aggregation of all individual predictions, generally in a linear manner, is regarded as the final prediction. Updating all training sets using the corresponding testing set, the prediction is conducted based on the above procedures until all forecasting tasks are finished. Limited by the decomposition strategy, the one-time decomposition-based forecasting method can only provide offline prediction, and thus may not be directly applied in practice.

Differing from the above method, the real-time decomposition-based forecasting method can provide online prediction, and thus this method may have potential for railway strong wind warning systems. However, the performance of this method is unstable, and sometimes cannot generate satisfactory results because of the flaws of its component models, such as the end effect, mode aliasing, and redundancy in decomposition and the limited ability of the selected predictor. Figure 1b gives the details of the real-time decomposition-based forecasting method. In this method, all original data are divided into the training data and the testing data beforehand, and then the decomposition is implemented for the training data rather than all of the original data (i.e., the summation of the training data and the testing data), i.e., the decomposition occurs after data set partitioning. Therefore, the training data is equal to the training set and the testing data is equal to the testing set. Further, an appropriate predictor (i.e., CKDE) is selectively established for training and predicting each decomposed subsequence, and all individual predictions are summated to yield the final prediction. With the prediction being completed, the training data should be updated by inputting the testing data. Obviously, the decomposition is conducted in real-time and no future information is contained in the decomposition.

3.2. Experimental Data Illustration

In order to illustrate the performance of the traditional decomposition-based forecasting methods, wind speed data measured by an ultrasonic anemometer the along one of the high-speed rail lines in China are employed as the experimental data. According to [16], sampling with a one-minute time interval is enough for the wind speed time series in high-speed rail wind warning systems. On this basis, the measured wind speed data with continuous 5 h records in 2014 are used where the sampling time interval still adopts a one-minute interval, and thus a total of 300 wind speed data points are contained in the experimental data. Among them, the first 225 data points are used as the initial training data for model training, while the later 75 data points are used as the testing data for model performance evaluation. Figure 2 presents the details of the experimental data in a visual way and these data may have a weak nonstationary characteristic. Table 1 gives the statistical features of these data. From the table, it can be seen that these statistical features of all the original data and the training data are similar, and are slightly different from those of the testing data. Their differences may contribute to reflecting the performance of traditional decomposition-based forecasting methods.

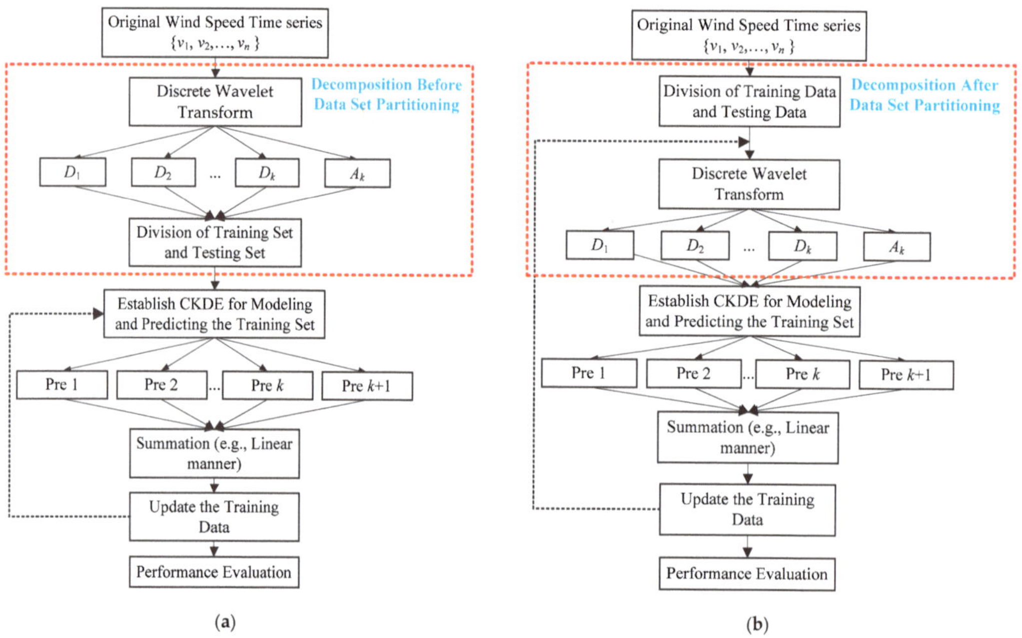

Figure 1. Flowchart of traditional decomposition-based forecasting methods. (**a**) One-time decomposition-based method. (**b**) Real-time decomposition-based method.

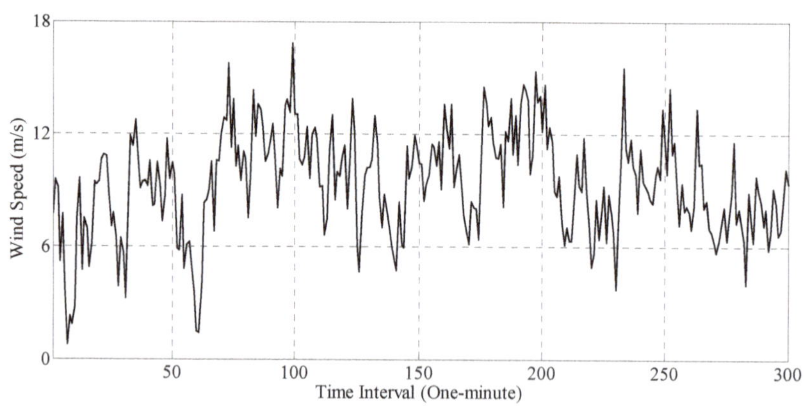

Figure 2. Original wind speed time series (continuous 5 h records).

Table 1. Statistical analysis of the experimental wind speed data.

Statistics	Mean (m/s)	Standard Deviation (m/s)	Maximum (m/s)	Minimum (m/s)	Skewness	Kurtosis
All Original Data	9.33	2.79	16.82	0.76	−0.22	3.10
Training Data	9.54	2.94	16.82	0.76	−0.48	3.09
Testing Data	8.69	2.17	15.54	3.74	0.63	3.89

3.3. Performance Comparison of Traditional Decomposition-Based Forecasting Methods

Based on the experimental wind speed data (see in Figure 2), the above two traditional decomposition-based forecasting methods (i.e., a hybrid of one-time DWT decomposition and CKDE, and a hybrid of real-time DWT decomposition and CKDE) are used to perform short-term wind speed predictions. Taking one-step ahead prediction as an example, the impact of the decomposition level number on forecasting accuracy is preliminarily analyzed, where the number of decomposition levels changes from 3 to 11. The results show that the optimal numbers of decomposition levels for the one-time decomposition-based forecasting method and real-time decomposition-based forecasting method were 8 and 9, respectively.

Figure 3 gives the decomposed results with two different decomposition schemes, where the number of decomposition levels is nine. In the figure, the solid line represents the decomposition in the one-time decomposition scheme for all original data, while the dashed line indicates the decomposition in the real-time scheme for the data in the testing set. Meanwhile, it can be observed that the decomposed results (i.e., the dashed lines in the right quarter) in the real-time decomposition scheme are significantly different from those in the one-time decomposition scheme, especially for the low-frequency items which exhibit stronger volatility than those of the one-time decomposition scheme. Figure 4 shows the forecasting results of these two decomposition-based forecasting methods, where both of them take their correspondingly optimal number of decomposition levels. Table 2 presents the corresponding error analysis results.

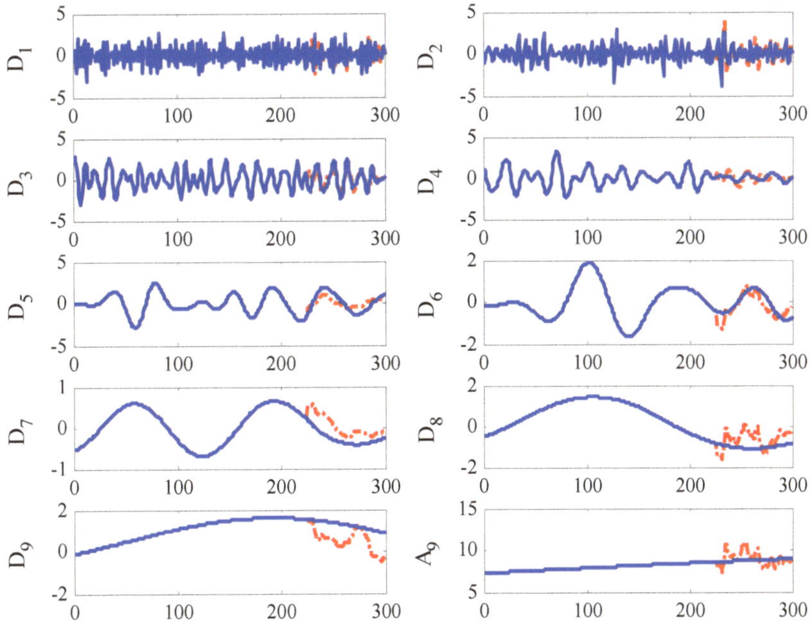

Figure 3. Subsequences from one-time decomposition (blue) and real-time decomposition (red).

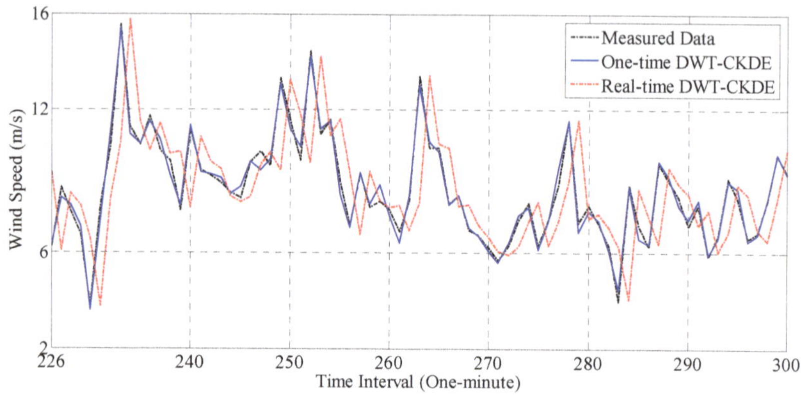

Figure 4. Forecasting results of the two traditional DWT-CKDE methods.

Table 2. Comparison of the two traditional DWT-CKDE methods.

Error Indicators	One-Time DWT-CKDE	Real-Time DWT-CKDE	Absolute Error
MAE	0.23	1.69	1.46
RMSE	0.30	2.15	1.85
MRPE/%	2.78	19.69	16.91

From Figure 4 and Table 2, it can be seen that the forecasting results of the one-time decomposition-based forecasting method are almost consistent with the trend of the actual wind speed, with the evaluation indexes of MAE, RMSE, and MRPE being 0.23, 0.30, and 2.78%, respectively. Compared with those indicators for the real-time decomposition-based forecasting method, the one-time decomposition-based forecasting method can realize an approximately 86% improvement in terms of these three indicators. On the surface, this method seems to have high potential. However, the decomposition in the one-time decomposition-based forecasting method takes the changing trend of unknown data into consideration, and thus fails to generate the actual required online prediction.

Fortunately, the real-time decomposition-based forecasting method can provide online prediction because the decomposition in this method takes the real-time pattern. However, its forecasting performance is not very stable, and sometimes even underperforms that of a single method. Table 3 summarizes the performance comparison between the real-time DWT-CKDE method and CKDE. It can be seen that the three error indicators of the former are 1.69, 2.15, and 19.69%, respectively, while those of the latter are 1.56, 2.01, and 18.28%, respectively, i.e., the decomposition brings out an obvious performance degradation. Therefore, in practical situations, the real-time decomposition-based forecasting method may not have an advantage. Although the decomposition scheme reduces the nonstationary and nonlinear characteristics hidden in the data, it may introduce other critical problems, such as strong volatility at the end of subsequences (see Figure 3), interferences of modal mixing and illusive components, which have some adverse effects on the performance of the involved predictor.

Table 3. Comparison of the real-time DWT-CKDE method and CKDE.

Error Indicator	CKDE	Real-Time DWT-CKDE	Absolute Error	P_{MAE}, P_{RMSE}, P_{MRPE}
MAE	1.56	1.69	−0.13	−7.69%
RMSE	2.01	2.15	−0.14	−6.51%
MRPE/%	18.28	19.69	−1.41	−7.16%

4. The Proposed Forecasting Method

This section first illustrates the details of the proposed method, then a numerical example based on the measured data in Figure 2 is conducted to exhibit the performance of this method. Finally, practicality analysis based on the other measured wind speed time series is carried out. The specific illustrations regarding these three parts are presented below.

4.1. Processes of the Proposed Method

Considering that not all of the decomposed subsequences are beneficial to the overall prediction accuracy of the real-time decomposition-based forecasting method, prediction based on the combination of certain or several subsequence predictions may provide a satisfactory result [20,53]. However, in most existing investigations, the summation of all subsequence predictions in a linear manner is the main ensemble strategy to generate the final prediction, and relevant studies aiming at the ensemble strategy are rarely reported. On this basis, this study explores the combination pattern (CP) of subsequence predictions and develops an improved ensemble-strategy-assisted DWT-CKDE method. This improved ensemble strategy can provide optimization selection for decomposed subsequences for prediction. The specific processes of this developed method are shown in Figure 5, and the corresponding main modeling steps are listed as follows:

(1) Divide the original wind speed time series $\{v_1, v_2, \ldots, v_n; n = 300\}$ into an initial training set $\{v_1, v_2, \ldots, v_{225}\}$ and a testing set $\{v_{226}, v_2, \ldots, v_{300}\}$;

(2) Perform the DWT decomposition on the initial training samples (the specified number of decomposition levels is set as 9, i.e., k = 8) to obtain the approximation component A_k and k detailed components $\{D_1,\ D_2,\ \ldots,\ D_k\}$. Then, the CKDE model is established for training each decomposed subsequence and implementing individual one-step ahead prediction. The predicted values of all subsequences are combined based on the following combination pattern (CP), i.e., $(\hat{D}_1 + \hat{D}_2 + \ldots + \hat{D}_k + \hat{A}_k)$, $(\hat{D}_2 + \hat{D}_3 + \ldots + \hat{D}_k + \hat{A}_k)$, ..., $(\hat{D}_k + \hat{A}_k)$, $(\hat{A}_k)$. On this basis, it is obvious that there $k + 1$ different CPs (i.e., ensemble strategies) are reserved for ensemble prediction in the next moment (i.e., the result of $\hat{v}_{226}$). In above CPs, $\hat{D}_j$, $(j = 1,\ 2,\ \ldots,\ k)$ denotes the one-step ahead forecasting value of the detailed component D_j and $\hat{A}_k$ stands for the one-step ahead forecasting value of the approximation component A_k;

(3) When the actual value of v_{226} is known, the ideal ensemble strategy (marked as the CP r_2) in Step (2) can be identified from $k + 1$ different CPs based on the minimum absolute deviation criterion, i.e., the prediction based on the ideal ensemble strategy has the smallest deviation with its corresponding actual value;

(4) Update the training set to $\{v_2, v_3, \ldots, v_{226}\}$ (i.e., the known value of v_{226}) and perform the corresponding DWT decomposition where one approximation component A_8 and eight detailed components $\{D_1,\ D_2,\ \ldots,\ D_8\}$ can be obtained. Then, the CKDE model is established to yield nine individual one-step ahead predictions and nine different CPs. With these predictions and the previous CP r_2, the one-step ahead prediction of the wind speed data point v_{227} (i.e., the result of $\hat{v}_{227}$) can be generated;

(5) When the actual value of v_{227} is known, the ideal ensemble strategy (marked as the CP r_3) in Step (3) can also be identified, and it has a similar deterministic process with that of CP r_2. This CP r_3 provides an ensemble strategy for the forecasting result $\hat{v}_{228}$;

(6) Similarly, update the training set to $\{v_i, v_{i+1}, \ldots, v_{i+224};\ i = 3,\ 4,\ \ldots,\ 76\}$ and repeat Step (4) and combine the ensemble strategy CP r_i to yield the forecasting result $\hat{v}_{i+225}$. When the actual value of v_{i+225} is known, the ensemble CP r_{i+1} can be obtained via a similar process with Step (5), which provides an ensemble strategy for the forecasting result $\hat{v}_{i+1+225}$, otherwise the prediction should be terminated. Note that the ideal combination pattern at the previous moment is regarded as the ensemble strategy at the current moment;

(7) As for the predicted value of $\hat{v}_{226}$ and CP r_1, they can be obtained by the following procedures: ① perform the DWT decomposition for the wind speed time series $\{v_1, v_2, \ldots, v_{224}\}$ and establish the CKDE model for training and predicting each

decomposed subsequence, thereby yielding nine different CPs; ② with the available of the value of v_{225}, the ideal ensemble strategy CP r_1 can be identified; ③ update the training set to $\{v_1, v_2, \ldots, v_{225}\}$ and repeat Step (4), then the predicted value of $\hat{v}_{226}$ can be generated;

(8) Perform the error analysis based on the performance evaluation indictors given in Section 3.1.

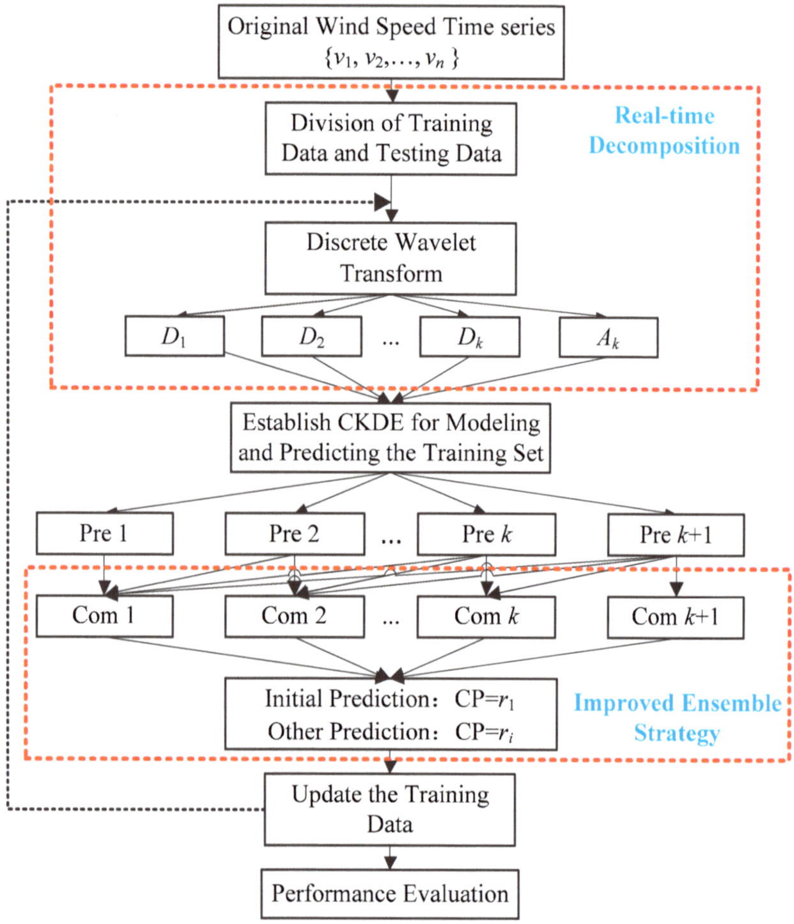

Figure 5. Flowchart of the proposed model.

From the above illustration, the proposed method is a hybrid of the real-time DWT decomposition, CKDE, and an improved ensemble strategy. Differing from traditional decomposition-based forecasting methods that generally require all decomposed subsequences involved in the prediction, this method has an ability to take a selective combination of individual predictions. The improved ensemble strategy provides effective guidance for this selective combination with consideration of the historical information in the data. Considering that the updating of training sets may lead to changes in data characteristics and the implied original information of each decomposed subsequence, the ensemble strategy may not be always the same. Therefore, it is necessary to update the combination pattern of the selected individual components in every real-time prediction. This time-varying ensemble strategy and the real-time decomposition scheme may be more beneficial to reflecting the actual complex wind environment.

4.2. Numerical Example

Based on the experimental data in Figure 2, the one-step ahead wind speed prediction is performed. Figure 6 takes the wind speed time series $\{v_2, v_3, \ldots, v_{226}\}$ as example to exhibit the decomposed results of DWT. In the figure, it is obvious that the decomposed subsequences are more stationary and regular than the original data. Therefore, the use of DWT to perform the pre-processing operation is effective in reducing the nonstationarity of the original data, which can lay a strong foundation for the application of statistical models, artificial or machine learning models for the prediction. For the sake of exhibiting the effectiveness of the selected combination, Table 4 presents the real-time prediction taking the ideal CP (i.e., each prediction adopts the optimal ensemble strategy which can make the prediction have the smallest deviation from its correspondingly actual value). Meanwhile, the influence of the number of decomposition levels on the forecasting accuracy is also examined. Table 5 gives the comparison of the prediction in the optimal ensemble strategy, real-time DWT-CKDE, and CKDE.

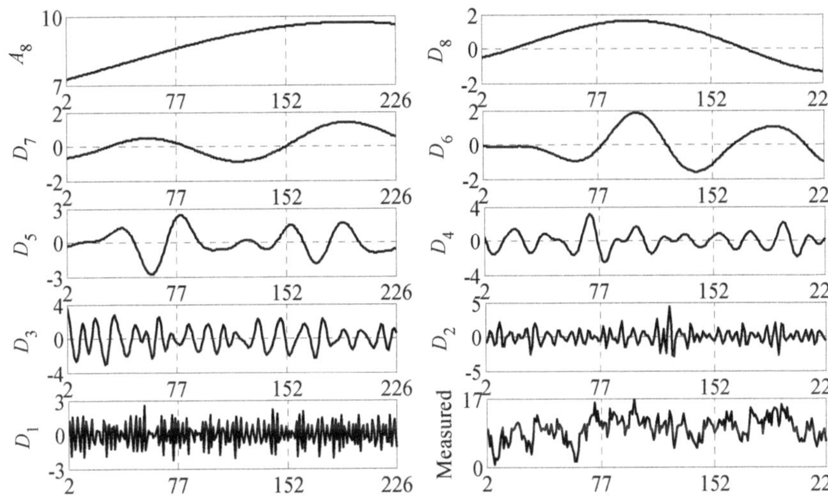

Figure 6. Decomposed subsequences of the wind speed time series $\{v_2, v_3, \ldots, v_{226}\}$.

Table 4. Forecasting results by the ideal CP considering the effect of decomposition level number.

Error Indicator	3	4	5	6	7	8	9	10	11
MAE	1.05	1.00	0.95	0.87	0.84	0.77	0.73	0.78	0.86
RMSE	1.52	1.47	1.41	1.35	1.32	1.23	1.21	1.24	1.37
MRPE/%	12.52	11.90	11.30	10.41	10.10	9.37	8.99	9.21	10.19

Table 5. Performance comparison of the ideal CP, CKDE, and real-time DWT-CKDE.

Error Indicator	Ideal Ensemble	CKDE			Real-Time DWT-CKDE		
		Result	Absolute Error	Improved Degree	Result	Absolute Error	Improved Degree
MAE	0.73	1.56	0.83	53.21%	1.69	0.95	56.21%
RMSE	1.21	2.01	0.80	39.80%	2.15	0.94	43.72%
MRPE/%	8.99	18.28	9.29	50.82%	19.69	10.70	54.34%

From Tables 4 and 5, it can be observed that the parameter of $k = 8$ (i.e., the number of decomposition levels is nine) corresponds to the minimum value of the evaluation indictors, and the prediction in the optimal ensemble strategy presents a higher accuracy than those of the real-time DWT-CKDE method and the CKDE method. For example, the values of three evaluation indicators in terms of the optimal ensemble strategy are 0.73 (MAE), 1.21 (RMSE), and 8.99% (MRPE), respectively. Meanwhile, the corresponding improved degrees of P_{MAE}, P_{RMSE}, and P_{MRPE} in comparison with the real-time DWT-CKDE method are 56.21%, 43.72%, and 54.34, respectively, while those compared with CKDE are 53.21%, 39.80%, and 50.82%, respectively. Obviously, these results demonstrate that the selective ensemble strategy (i.e., the selective combination of several individual predictions) is an effective way to improve the traditional real-time decomposition-based forecasting method. However, due to the lack of selection criteria, it is impossible to obtain an ideal CP in every time prediction.

The improved ensemble strategy provides effective guidance to determine the selection criterion. In this strategy, the ideal CP for the previous moment is regarded as the ensemble pattern for the current moment. Table 6 presents all the subsequence predictions $\hat{D}_j$, $(j = 1, 2, \ldots, 8)$ and $\hat{A}_8$ where the training set is updated to $\{v_2, v_3, \ldots, v_{226}\}$, i.e., the prediction of $\hat{v}_{227}$. In the table, it can be seen that the idea combination strategy in this moment is $\hat{D}_8 + \hat{A}_8$ (i.e., r_3) when the value of v_{227} is known, which can be used for the prediction of $\hat{v}_{228}$. To show the superiority of the proposed method, the real-time DWT-CKDE method and CKDE are used for comparison. Figure 7 intuitively shows the forecasting results of these three methods and the corresponding evaluation indicators are summarized in Table 7. From the figure, it can be seen that there is an apparent time-delaying effect in the prediction, and the proposed method is least affected. Meanwhile, the statistical results in Table 7 also demonstrate that the proposed method outperforms those compared methods. For example, the improved degrees realized by the proposed method in comparison with the real-time DWT-CKDE method regarding the indexes P_{MAE}, P_{RMSE}, and P_{MRPE} are 19.05%, 16.74%, and 16.66%, respectively, while those in comparison with CKDE are 12.82%, 10.95%, and 11.40%, respectively.

Table 6. Forecasting results of all decomposed subsequences and the ideal CP.

Subsequences	$\hat{D}_1$	$\hat{D}_2$	$\hat{D}_3$	$\hat{D}_4$	$\hat{D}_5$	$\hat{D}_6$	$\hat{D}_7$	$\hat{D}_8$	$\hat{A}_8$	**True Value** v_{227}
Prediction	−0.69	0.55	−0.19	−1.85	−0.81	0.53	1.18	−1.06	9.70	8.76 m/s
Ideal CP	$\hat{D}_8 + \hat{A}_8$ =8.64 m/s									

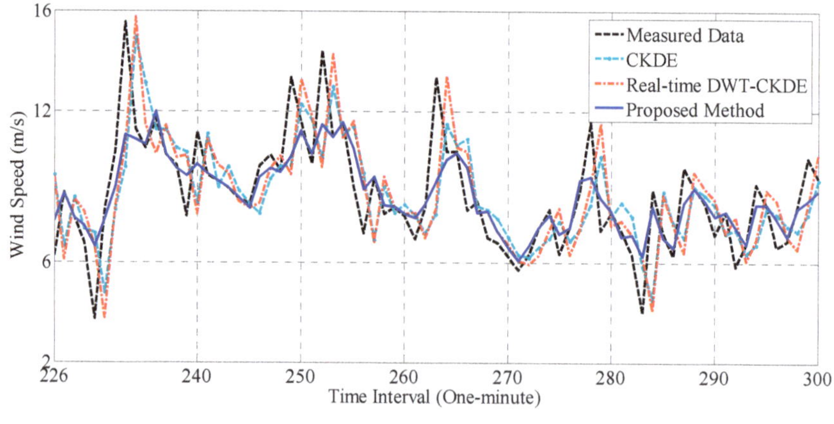

Figure 7. Forecasting results of the proposed model, real-time DWT-CKDE, and CKDE.

Table 7. Performance comparison of the proposed method, CKDE, and real-time DWT-CKDE.

Error Indicator	Proposed Method	CKDE			Real-Time DWT-CKDE		
		Result	Absolute Error	Improved Degree	Result	Absolute Error	Improved Degree
MAE	1.36	1.56	0.20	12.82%	1.68	0.32	19.05%
RMSE	1.79	2.01	0.22	10.95%	2.15	0.36	16.74%
MRPE/%	16.41	18.28	1.87	11.40%	19.69	3.28	16.66%

4.3. Practicality Analysis

In order to further express the practicability of the proposed method, an additional case study based on the measured wind speed date is performed. Figure 8 and Table 8 give some illustrations regarding the experimental data. From Table 8, it can be seen that the data characteristics in this experiment are significantly different from those in Figure 2 and Table 1. By comparison, these data may have stronger nonstationarity and non-Gaussianity. This difference is conducive to demonstrating the superiority of the proposed method in a more comprehensive way.

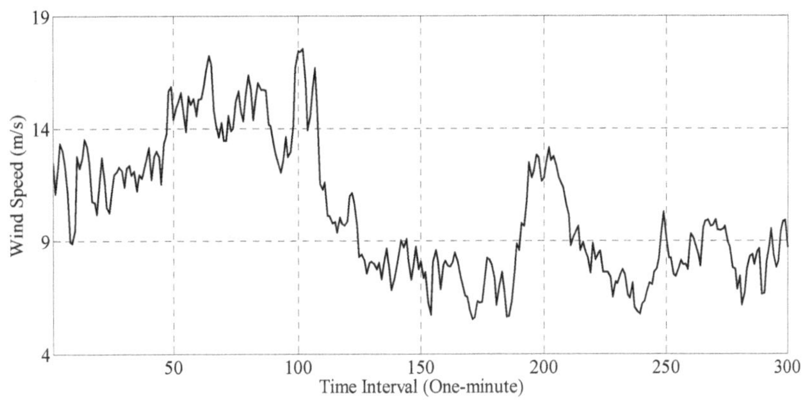

Figure 8. Original wind speed time series (additional case).

Table 8. Statistical analysis of the additional wind speed data.

Statistics	Mean (m/s)	Standard Deviation (m/s)	Maximum (m/s)	Minimum (m/s)	Skewness	Kurtosis
All Original Data	10.34	3.02	17.54	5.55	0.50	2.17
Training Data	11.10	3.08	17.54	5.55	0.12	1.94
Testing Data	8.09	1.16	10.28	5.77	0	2.10

Based on these wind speed data, the one-step ahead prediction is performed, on the basis of which the forecasting performance comparison between CKDE, real-time DWT-CKDE, and the proposed method is also conducted. Figure 9 intuitively presents the forecasting results of these methods, where all of them reflect the main trend of the actual wind speed. By comparison, the proposed method exhibits a higher consistence with the actual wind speed data. Table 9 quantificationally lists the forecasting performance comparison in terms of the evaluation indictors used in this paper. From the table, it can be seen that the proposed method has a higher forecasting accuracy than the other compared methods. For example, the indexes MAE, RMSE, and MRPE of the proposed method are 0.43, 0.54, and 5.38%, respectively, while those of CKDE are 0.54, 0.71, and 6.94%, respectively. Interestingly, in this experiment, the real-time DWT-CKDE method outperforms the single CKDE method, which is the opposite of the corresponding observation in Table 3. The reason for this observation may be ascribed to the advantages of the real-time DWT method (e.g., a good time frequency ability to address nonstationarity and nonlinearity), which contributes more to the forecasting accuracy than the adverse effect of the disadvantages of the real-time DWT on the accuracy (e.g., the adverse impacts of the end effect, modal aliasing, and illusory components).

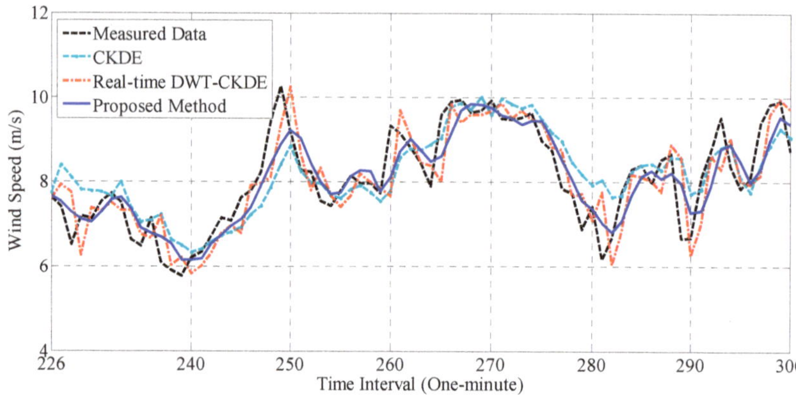

Figure 9. Forecasting results of the proposed model, real-time DWT-CKDE, and CKDE.

Table 9. Performance comparison of the proposed method, CKDE, and real-time DWT-CKDE.

Error Indicator	Proposed Method	CKDE			Real-Time DWT-CKDE		
		Result	Absolute Error	Improved Degree	Result	Absolute Error	Improved Degree
MAE	0.43	0.54	0.11	19.67%	0.51	0.08	15.17%
RMSE	0.54	0.71	0.17	24.36%	0.68	0.14	20.33%
MRPE/%	5.38	6.94	1.56	22.46%	6.48	1.10	17.01%

Computational complexity is another index by which to evaluate the practicality of the model. To this end, Table 10 gives a comparison of the computational costs of all of the involved models in each time prediction, where the time consumption is the average value of 10 operations. From the table, it can be found that the CKDE model generally has the lowest computational cost, while the proposed method possesses the largest time consumption. The reason for this observation is that the proposed method has the highest model complexity. Meanwhile, the difference in computation between one-time DWT-CKDE and real-time DWT-CKDE is very small, indicating that the DWT operation in the experiment takes very little time. Indeed, this time consumption will increase with increases in the amount of wind speed data. In addition, the computational burden of one-time DWT-CKDE is approximately equal to nine times the calculation time of CKDE, i.e., nine different CKDE models are, respectively, established for the corresponding subsequences in the one-time DWT-CKDE.

Table 10. Performance comparison of the proposed method, CKDE, and real-time DWT-CKDE.

Methods	One-Time DWT-CKDE	Real-Time DWT-CKDE	CKDE	Proposed
Time (s)	4.72	4.80	0.52	4.95

5. Conclusions

Short-term wind speed prediction is critical for the safe operation of high-speed railways and many methods have been put forward to improve forecasting accuracy. Among them, decomposition-based forecasting methods are widely applied due to their better data processing ability. On this basis, this paper first clarifies the procedures of traditional decomposition-based forecasting methods (i.e., the one-time decomposition-based method and real-time decomposition-based method) as well as their differences based on the commonly used discrete wavelet transform (DWT) and conditional kernel density Estimation (CKDE). To further enhance forecasting accuracy, the paper proposes an improved ensemble-strategy-assisted forecasting method, which is a hybrid of real-time DWT, CKDE, and an improved ensemble strategy. Finally, a numerical example and practicality analysis based on two groups of measured wind speed time series are employed to demonstrate the superiority of the proposed method. Some main observations and conclusions are summarized below.

(1) The one-time decomposition-based forecasting method has an extremely good forecasting performance. However, this method cannot provide online prediction because the one-time decomposition operation takes future data into consideration.
(2) The real-time decomposition-based forecasting method can provide online prediction, and thus may have the potential to be implemented in practice. However, the forecasting accuracy of this method is not stable and is sometimes unsatisfactory. Although decomposition methods reduce the nonstationary and nonlinear characteristics of data, they may also greatly increase the volatility of decomposed subsequences (especially for the end part of subsequences), thereby increasing the difficulty of prediction and ultimately leading to poor forecasting results.
(3) CKDE is still effective in the prediction of short-term wind speeds. This method can be regarded as the nonparametric model to some extent, and thus has a powerful applicability in addressing the time series problem. The numerical case in this paper shows that it even performs better than the decomposition-based forecasting method in some scenarios.
(4) The combination of several individual predictions (i.e., the prediction of each decomposed subsequence) may have a higher accuracy than the summation of all predictions in the short-term wind speed prediction. Along with this design concept, an improved ensemble strategy is developed which can perform selective combination prediction by analyzing information in the historical data. The experimental results indicate this strategy can clearly improve the forecasting accuracy of real-time decomposition-based methods and the single method. For example, compared with CKDE, the average degrees of improvement realized by the proposed method in terms of MAE, RMSE, and MRPE are 16.25%, 17.66%, and 16.93, respectively, while those in comparison with the traditional real-time DWT-CKDE method are 17.11%, 18.54%, and 16.84, respectively. Therefore, the proposed method may have great potential for railway strong wind warning systems.

Although the proposed method can provide satisfactory forecasting accuracy, it requires the largest computational burden because it possesses the highest model complexity, and its time consumption will increase with increases in the amount of wind speed data. Meanwhile, this method ignores the impact of the inherent limitations of DWT (e.g., the end effect, modal aliasing, illusory components, etc. [19]) on prediction accuracy. In addition, it fails to perform probabilistic prediction, which is critical for risk-orientated decision-making and safety scheduling in high-speed railways.

Author Contributions: Conceptualization, J.L., C.C. and Y.J.; data curation, C.C., J.L., X.C. and Y.J.; formal analysis, J.L., C.C., X.C. and Y.J.; funding acquisition, Y.J.; investigation, J.L., C.C. and X.C.; methodology, C.C. and J.L.; resources, J.L. and Y.J.; software, J.L. and C.C.; supervision, X.C. and C.C.; validation, J.L.; visualization, C.C. and Y.J.; writing—original draft, J.L., C.C., X.C. and Y.J.; writing—review and editing, C.C., J.L., X.C. and Y.J. All authors have read and agreed to the published version of the manuscript.

Funding: This research was funded by "China Postdoctoral Science Foundation (grant number: 2023M730431)", "Special Funding of Chongqing Postdoctoral Research Project (grant number: 2022CQBSHTB2053)", "the Science and Technology Research Program of Chongqing Municipal Education Commission (grant number: KJZD-M202300201)", "State Key Laboratory of Mountain Bridge and Tunnel Engineering (grant number: SKLBT-2202)", "Chongqing graduate education teaching reform research general project (grant number: YJG233030), "Wind Engineering Research

Center of Sichuan Key Laboratory (grant number: WEKLSC202303)" and "the Fundamental Research Funds for the Central Universities (grant number: SWU-KT22021)".

Institutional Review Board Statement: Not applicable.

Informed Consent Statement: Not applicable.

Data Availability Statement: The data presented in this study are available on request from the corresponding author.

Conflicts of Interest: The authors declare no conflict of interest. Authors Jian Liu and Cheng Cheng are employees of the company Chongqing Wukang Technology Co., LTD. The company had no roles in the design of the study; in the collection, analysis, and interpretation of data; in the writing of the manuscript or in the decision to publish the article. The paper reflects the views of the scientists and not the company.

References

1. Wang, M.; Wang, X.; Li, X.; Wang, P. Influence of spatial track alignment of long-span arch bridge on train operational stability. *Veh. Syst. Dyn.* **2023**, *61*, 3161–3180. [CrossRef]
2. Li, X.; He, H.; Wang, M.; Wang, P. Influence of long-span bridge deformation on driving quality of high-speed trains. *Int. J. Rail Transp.* **2023**, 1–19. [CrossRef]
3. Liu, D.; Wang, T.; Liang, X.; Meng, S.; Zhong, M.; Lu, Z. High-speed train overturning safety under varying wind speed conditions. *J. Wind. Eng. Ind. Aerodyn.* **2020**, *198*, 104111. [CrossRef]
4. Yan, N.; Chen, X.; Li, Y. Assessment of overturning risk of high-speed trains in strong crosswinds using spectral analysis approach. *J. Wind. Eng. Ind. Aerodyn.* **2018**, *174*, 103–118. [CrossRef]
5. Wang, M.; Chen, X.; Li, X.; Yan, N.; Wang, Y. A frequency domain analysis framework for assessing overturning risk of high-speed trains under crosswind. *J. Wind. Eng. Ind. Aerodyn.* **2020**, *202*, 104196. [CrossRef]
6. Chen, R.; Zeng, Q.; Huang, Y.; Xiang, J.; Wen, Y.; Guo, X.; Yin, C.; Dong, H.; Zhao, G. Analysis theory of random energy of train derailment inwind. *China Phys. Mech. Astron.* **2010**, *53*, 751–757. [CrossRef]
7. Cheli, F.; Ripamonti, F.; Rocchi, D.; Tomasini, G.I.S.E.L.L.A. Aerodynamic behaviour investigation of the new EMUV250 train to cross wind. *J. Wind. Eng. Ind. Aerodyn.* **2010**, *98*, 189–201. [CrossRef]
8. Duan, F.; Song, Y.; Gao, S.; Liu, Y.; Chu, W.; Lu, X.; Liu, Z. Study on Aerodynamic Instability and Galloping Response of Rail Overhead Contact Line Based on Wind Tunnel Tests. *IEEE Trans. Veh. Technol.* **2023**, *72*, 7211–7220. [CrossRef]
9. Fang, C.; Li, Y.; Wei, K.; Zhang, J.; Liang, C. Vehicle–bridge coupling dynamic response of sea-crossing railway bridge under correlated wind and wave conditions. *Adv. Struct. Eng.* **2019**, *22*, 893–906. [CrossRef]
10. Montenegro, P.A.; Barbosa, D.; Carvalho, H.; Calçada, R. Dynamic effects on a train-bridge system caused by stochastically generated turbulent wind fields. *Eng. Struct.* **2020**, *211*, 110430. [CrossRef]
11. Song, Y.; Zhang, M.; Øiseth, O.; Rønnquist, A. Wind deflection analysis of railway catenary under crosswind based on nonlinear finite element model and wind tunnel test. *Mech. Mach. Theory* **2022**, *168*, 104608. [CrossRef]
12. Song, Y.; Zhang, M.; Wang, H. A response spectrum analysis of wind deflection in railway overhead contact lines using pseudo-excitation method. *IEEE Trans. Veh. Technol.* **2021**, *70*, 1169–1178. [CrossRef]
13. Reis, E.D.; Carvalho, H.; de Oliveira, C.C.; de Vasconcelos, J.A.; Adriano, R.L.D.S.; Lopes, D.C. Geometric Nonlinear Analysis of Self-Supporting Structures for Overhead Transmission Lines. *Pract. Period. Struct. Des. Constr.* **2023**, *28*, 04023003. [CrossRef]
14. Aly, A.M.; Chapain, S. Large-scale open-jet testing for cladding design in high-rise buildings: Higher peak pressures compared to wind tunnels. *Pract. Period. Struct. Des. Constr.* **2022**, *27*, 04022016. [CrossRef]
15. Arvan, P.A.; Raju, R.D.; Arockiasamy, M. Offshore Wind Turbine Monopile Foundation Systems in Multilayered Soil Strata under Aerodynamic and Hydrodynamic Loads: State-of-the-Art Review. *Pract. Period. Struct. Des. Constr.* **2023**, *28*, 03123001. [CrossRef]
16. Liu, H.; Tian, H.Q.; Li, Y.F. An EMD-recursive ARIMA method to predict wind speed for railway strong wind warning system. *J. Wind. Eng. Ind. Aerodyn.* **2015**, *141*, 27–38. [CrossRef]
17. Gou, H.; Chen, X.; Bao, Y. A wind hazard warning system for safe and efficient operation of high-speed trains. *Autom. Constr.* **2021**, *132*, 103952. [CrossRef]
18. Liu, H.; Liu, C.; He, S.; Chen, J. Short-term strong wind risk prediction for high-speed railway. *IEEE Trans. Intell. Transp. Syst.* **2021**, *22*, 4243–4255. [CrossRef]
19. Hoppmann, U.; Koenig, S.; Tielkes, T.; Matschke, G. A short-term strong wind prediction model for railway application: Design and verification. *J. Wind. Eng. Ind. Aerodyn.* **2002**, *90*, 1127–1134. [CrossRef]
20. Jiang, Y.; Huang, G. Short-term wind speed prediction: Hybrid of ensemble empirical mode decomposition, feature selection and error correction. *Energy Convers. Manag.* **2017**, *144*, 340–350. [CrossRef]
21. Tao, T.; Shi, P.; Wang, H.; Ai, W. Short-term prediction of downburst winds: A double-step modification enhanced approach. *J. Wind. Eng. Ind. Aerodyn.* **2021**, *211*, 104561. [CrossRef]
22. Jiang, Y.; Liu, S.; Peng, L.; Zhao, N. A novel wind speed prediction method based on robust local mean decomposition, group method of data handling and conditional kernel density estimation. *Energy Convers. Manag.* **2019**, *200*, 112099. [CrossRef]

23. Landberg, L. Short-term prediction of the power production from wind farms. *J. Wind. Eng. Ind. Aerodyn.* **1999**, *80*, 207–220. [CrossRef]
24. Yu, C.; Li, Y.; Chen, Q.; Lai, X.; Zhao, L. Matrix-based wavelet transformation embedded in recurrent neural networks for wind speed prediction. *Appl. Energy* **2022**, *324*, 119692. [CrossRef]
25. Xin, J.; Jiang, Y.; Zhou, J.; Peng, L.; Liu, S.; Tang, Q. Bridge deformation prediction based on SHM data using improved VMD and conditional KDE. *Eng. Struct.* **2022**, *261*, 114285. [CrossRef]
26. Erdem, E.; Shi, J. ARMA based approaches for forecasting the tuple of wind speed and direction. *Appl. Energy* **2011**, *88*, 1405–1414. [CrossRef]
27. Jiang, Y.; Liu, S.; Zhao, N.; Xin, J.; Wu, B. Short-term wind speed prediction using time varying filter-based empirical mode decomposition and group method of data handling-based hybrid model. *Energy Convers. Manag.* **2020**, *220*, 113076. [CrossRef]
28. Bossanyi, E.A. Short-term wind prediction using Kalman filters. *Wind. Eng.* **1985**, *9*, 1–8.
29. Wang, D.; Yang, S.X. Intelligent feature extraction, data fusion and detection of concrete bridge cracks: Current development and challenges. *Intell. Robot.* **2022**, *2*, 391–406. [CrossRef]
30. Khasanzoda, N.; Zicmane, I.; Beryozkina, S.; Safaraliev, M.; Sultonov, S.; Kirgizov, A. Regression model for predicting the speed of wind flows for energy needs based on fuzzy logic. *Renew. Energy* **2022**, *191*, 723–731. [CrossRef]
31. Zhou, J.; Shi, J.; Li, G. Fine tuning support vector machines for short-term wind speed forecasting. *Energy Convers. Manag.* **2011**, *52*, 1990–1998. [CrossRef]
32. Wang, S.; Wang, J.; Lu, H.; Zhao, W. A novel combined model for wind speed prediction–Combination of linear model, shallow neural networks, and deep learning approaches. *Energy* **2021**, *234*, 121275. [CrossRef]
33. Lai, L.; Hajirasouliha, I.; Pilakoutas, K.; He, X.; Smyl, D. Applying the Continuous Hidden Markov Model to Structural State Estimation. *Pract. Period. Struct. Des. Constr.* **2023**, *28*, 04022073. [CrossRef]
34. Nayak, D.K.; Verma, G.; Dimri, A.; Kumar, R.; Kumar, V. Predicting the Twenty-eight day compressive strength of OPC-and PPC-prepared concrete through hybrid GA-XGB model. *Pract. Period. Struct. Des. Constr.* **2023**, *28*, 04023020. [CrossRef]
35. Sun, S.; Cao, Z.; Zhu, H.; Zhao, J. A survey of optimization methods from a machine learning perspective. *IEEE Trans. Cybern.* **2019**, *50*, 3668–3681. [CrossRef] [PubMed]
36. Hyndman, R.J.; Bashtannyk, D.M.; Grunwald, G.K. Estimating and visualizing conditional densities. *J. Comput. Graph. Stat.* **1996**, *5*, 315–336.
37. Han, Q.; Meng, F.; Hu, T.; Chu, F. Non-parametric hybrid models for wind speed forecasting. *Energy Convers. Manag.* **2017**, *148*, 554–568. [CrossRef]
38. Méndez-Gordillo, A.R.; Cadenas, E. Wind speed forecasting by the extraction of the multifractal patterns of time series through the multiplicative cascade technique. *Chaos Solitons Fractals* **2021**, *143*, 110592. [CrossRef]
39. Jiang, Y.; Zhao, N.; Peng, L.; Zhao, L.; Liu, M. Simulation of stationary wind field based on adaptive interpolation-enhanced scheme. *J. Wind. Eng. Ind. Aerodyn.* **2019**, *195*, 104001. [CrossRef]
40. Liu, M.; Peng, L.; Huang, G.; Yang, Q.; Jiang, Y. Simulation of stationary non-Gaussian multivariate wind pressures using moment-based piecewise Hermite polynomial model. *J. Wind. Eng. Ind. Aerodyn.* **2020**, *196*, 104041. [CrossRef]
41. Zhao, N.; Jiang, Y.; Peng, L.; Chen, X. Fast simulation of nonstationary wind velocity fields by proper orthogonal decomposition interpolation. *J. Wind. Eng. Ind. Aerodyn.* **2021**, *219*, 104798. [CrossRef]
42. Chen, X.; Jiang, Y.; Zhao, N.; Jia, S.; Wang, X.; Tan, S. Simulation of stationary and nonstationary wind velocity field along a long-span bridge using a numerical truncation method. *J. Wind. Eng. Ind. Aerodyn.* **2022**, *231*, 105206. [CrossRef]
43. Li, B.; Peng, L.; Jiang, Y.; Wu, F.; Hui, Y.; Luo, Y. Simulation of stationary non-Gaussian stochastic vector processes using an eigenvalue-based iterative translation approximation method. *Mech. Syst. Signal Process.* **2022**, *175*, 109128. [CrossRef]
44. Jiang, Y.; Huang, G.; Yang, Q.; Yan, Z.; Zhang, C. A novel probabilistic wind speed prediction approach using real time refined variational model decomposition and conditional kernel density estimation. *Energy Convers. Manag.* **2019**, *185*, 758–773. [CrossRef]
45. Sun, W.; Liu, M. Wind speed forecasting using FEEMD echo state networks with RELM in Hebei, China. *Energy Convers. Manag.* **2016**, *114*, 197–208. [CrossRef]
46. Li, Y.; Peng, T.; Zhang, C.; Sun, W.; Hua, L.; Ji, C.; Shahzad, N.M. Multi-step ahead wind speed forecasting approach coupling maximal overlap discrete wavelet transform, improved grey wolf optimization algorithm and long short-term memory. *Renew. Energy* **2022**, *196*, 1115–1126. [CrossRef]
47. Meng, A.; Ge, J.; Yin, H.; Chen, S. Wind speed forecasting based on wavelet packet decomposition and artificial neural networks trained by crisscross optimization algorithm. *Energy Convers. Manag.* **2016**, *114*, 75–88. [CrossRef]
48. Gilles, J. Empirical wavelet transform. *IEEE Trans. Signal Process.* **2013**, *61*, 3999–4010. [CrossRef]
49. Tang, Q.; Jiang, Y.; Xin, J.; Liao, G.; Zhou, J.; Yang, X. A novel method for the recovery of continuous missing data using multivariate variational mode decomposition and fully convolutional networks. *Measurement* **2023**, *220*, 113366. [CrossRef]
50. Jiang, Y.; Huang, G.; Peng, X.; Li, Y.; Yang, Q. A novel wind speed prediction method: Hybrid of correlation-aided DWT, LSSVM and GARCH. *J. Wind. Eng. Ind. Aerodyn.* **2018**, *174*, 28–38. [CrossRef]
51. Liu, H.; Tian, H.Q.; Li, Y.F. Four wind speed multi-step forecasting models using extreme learning machines and signal decomposing algorithms. *Energy Convers. Manag.* **2015**, *100*, 16–22. [CrossRef]
52. Wang, Y.; Wu, L. On practical challenges of decomposition-based hybrid forecasting algorithms for wind speed and solar irradiation. *Energy* **2016**, *112*, 208–220. [CrossRef]

53. Guo, Z.; Zhao, W.; Lu, H.; Wang, J. Multi-step forecasting for wind speed using a modified EMD-based artificial neural network model. *Renew. Energy* **2012**, *37*, 241–249. [CrossRef]
54. Su, Y.; Huang, G.; Xu, Y.L. Derivation of time-varying mean for non-stationary downburst winds. *J. Wind. Eng. Ind. Aerodyn.* **2015**, *141*, 39–48. [CrossRef]
55. Sundararajan, D. *Discrete Wavelet Transform: A Signal Processing Approach*; John Wiley & Sons: Hoboken, NJ, USA, 2016.
56. Huang, G.; Su, Y.; Kareem, A.; Liao, H. Time-frequency analysis of nonstationary process based on multivariate empirical mode decomposition. *J. Eng. Mech.* **2016**, *142*, 04015065. [CrossRef]
57. Liu, S.; Luo, Y.; Peng, L.; Jiang, Y.; Meng, E.; Li, B. Wind pressure field reconstruction based on unbiased conditional kernel density estimation. *J. Wind. Eng. Ind. Aerodyn.* **2022**, *223*, 104947. [CrossRef]

MDPI AG
Grosspeteranlage 5
4052 Basel
Switzerland
Tel.: +41 61 683 77 34

MDPI Books Editorial Office
E-mail: books@mdpi.com
www.mdpi.com/books

www.ingramcontent.com/pod-product-compliance
Lightning Source LLC
Chambersburg PA
CBHW040853230926
51850CB00014BA/3426